STANDARD HANDBOOK OF HEAVY CONSTRUCTION

Other McGraw-Hill Books of Related Interest

Brantley, Brantley · BUILDING MATERIALS TECHNOLOGY
Breyer · DESIGN OF WOOD STRUCTURES
Brockenbrough, Merritt · STRUCTURAL STEEL DESIGNER'S HANDBOOK
Brown · FOUNDATION BEHAVIOR AND REPAIR
Brown · PRACTICAL FOUNDATION ENGINEERING HANDBOOK
Civitello · CONSTRUCTIONS OPERATIONS MANUAL
Faherty, Williamson · WOOD ENGINEERING AND CONSTRUCTION HANDBOOK
Foster et al. · CONSTRUCTION ESTIMATES FROM TAKE-OFF TO BID
Gaylord, Gaylord · STRUCTURAL ENGINEERING HANDBOOK
Levy · PROJECT MANAGEMENT IN CONSTRUCTION
Merritt, Ricketts · BUILDING DESIGN AND CONSTRUCTION HANDBOOK
Merritt · STANDARD HANDBOOK FOR CIVIL ENGINEERS
Nardon · BRIDGE AND STRUCTURE ESTIMATING
Newman · DESIGN AND CONSTRUCTION OF WOOD-FRAMED BUILDINGS
Newman · STANDARD HANDBOOK OF STRUCTURAL DETAILS FOR BUILDING CONSTRUCTION
O'Brien · PRECONSTRUCTION ESTIMATING: BUDGET THROUGH BID
O'Brien · CPM IN CONSTRUCTION MANAGEMENT
Palmer, Coombs · CONSTRUCTION ACCOUNTING AND FINANCIAL MANAGEMENT
Parmley · FIELD ENGINEER'S MANUAL
Ritz · TOTAL CONSTRUCTION PROJECT MANAGEMENT
Tonias · BRIDGE ENGINEERING

STANDARD HANDBOOK OF HEAVY CONSTRUCTION

James J. O'Brien
Editor in Chief

John A. Havers

Frank W. Stubbs, Jr.

Third Edition

McGraw-Hill
New York San Francisco Washington, D.C. Auckland Bogotá
Caracas Lisbon London Madrid Mexico City Milan
Montreal New Delhi San Juan Singapore
Sydney Tokyo Toronto

Library of Congress Cataloging-in-Publication Data

Standard handbook of heavy construction / [edited by] James J.
  O'Brien, John A. Havers, Frank W. Stubbs, Jr.—3rd ed.
    p.   cm.
  Includes index.
  ISBN 0-07-047971-2 (hardcover)
  1. Engineering—Handbooks, manuals, etc. I. O'Brien, James
Jerome. II. Havers, John A. (John Alan).
III. Stubbs, Frank W. (Frank Whitworth).
TA151.S814   1996
624—dc20   95-45878
                                                                CIP

McGraw-Hill
   A Division of The McGraw·Hill Companies

Copyright © 1996, 1971, 1959 by The McGraw-Hill Companies, Inc. All rights reserved. Printed in the United States of America. Except as permitted under the United States Copyright Act of 1976, no part of this publication may be reproduced or distributed in any form or by any means, or stored in a data base or retrieval system, without the prior written permission of the publisher.

2 3 4 5 6 7 8 9 0   BKP/BKP   9 0 1 0 9 8 7 6

ISBN 0-07-047971-2

The sponsoring editor for this book was Larry Hager, the editing supervisor was Virginia Carroll, and the production supervisor was Suzanne W.B. Rapcavage. It was set in Times Roman by North Market Street Graphics.

Printed and bound by Quebecor/Book Press.

McGraw-Hill books are available at special quantity discounts to use as premiums and sales promotions, or for use in corporate training programs. For more information, please write to the Director of Special Sales, McGraw-Hill, 11 West 19th Street, New York, NY 10011. Or contact your local bookstore.

This book is printed on acid-free paper.

Information contained in this work has been obtained by The McGraw-Hill Companies, Inc. ("McGraw-Hill") from sources believed to be reliable. However, neither McGraw-Hill nor its authors guarantees the accuracy or completeness of any information published herein, and neither McGraw-Hill nor its authors shall be responsible for any errors, omissions, or damages arising out of use of this information. This work is published with the understanding that McGraw-Hill and its authors are supplying information but are not attempting to render engineering or other professional services. If such services are required, the assistance of an appropriate professional should be sought.

To *Fred C. Kreitzberg,* a *great** partner and *best friend.*

* Trust men and they will be true to you; treat them greatly and they will show themselves great.
RALPH WALDO EMERSON

CONTENTS

Contributors   xxvii
Preface   xxix

Part A Construction/Project Management

Chapter 1. Contractor's Organization   A1-3

Introduction / *A1-3*
Construction Project Delivery Systems / *A1-3*
   Introduction / *A1-3*
   Classic Owner-Designer-Constructor Organization / *A1-4*
   At-Risk Construction Management (CM) / *A1-5*
   No-Risk Construction Management / *A1-5*
   Design-Construct CM / *A1-6*
Contractor Management Organizations / *A1-7*
Recent Changes in Construction Relationships / *A1-10*
   Partnering / *A1-12*
   Private Construction Partnering / *A1-13*
   Management by Objectives, Management by Exception, Management
      by Chaos, Etc. / *A1-13*
   The Deming Management Method / *A1-15*
   The 14 Points and the Construction Industry / *A1-18*
Conclusion / *A1-20*
References / *A1-21*

Chapter 2. Heavy Construction Contracts   A2-1

Need for Good Contract Documents / *A2-1*
What Are Good Contract Documents? / *A2-1*
Use of Standard Forms / *A2-2*
EJCDC Construction-Related Contract Documents / *A2-2*
Bidding Procedures and Documents / *A2-3*
Bidding Documents / *A2-4*
The Agreement Between Owner and Contractor / *A2-6*
General Conditions / *A2-7*

Chapter 3. Construction Planning and Scheduling   A3-1

Creating a CPM Network / *A3-1*
   Introduction / *A3-1*
   The Baseline / *A3-3*

Sequencing / *A3-5*
Updating / *A3-11*
Using the Network / *A3-12*
   Preconstruction / *A3-12*
   Construction / *A3-13*
   Postconstruction / *A3-18*

Chapter 4. Construction Cost Control   A4-1

Introduction / *A4-1*
Definitions and Objectives / *A4-2*
   Definition of Cost Engineering / *A4-2*
   Objectives and Responsibilities / *A4-2*
   Principles of Effective Cost Control / *A4-3*
Classification of Accounts / *A4-4*
   General / *A4-4*
   Sample Classifications / *A4-5*
Materials and Equipment / *A4-5*
   General / *A4-5*
   Budget / *A4-5*
   Ordering of Materials / *A4-6*
   Coding Purchase Requisitions / *A4-6*
   The Cost Engineer and Purchasing Function / *A4-6*
   Purchase Order Register / *A4-8*
   Commitment Record / *A4-8*
   Cost Report—Materials and Equipment / *A4-10*
Subcontracts / *A4-12*
   Lump-Sum Price Contracts / *A4-12*
   Unit Price Contracts / *A4-13*
   Progress Payments / *A4-13*
   Reports / *A4-13*
Field Labor / *A4-16*
   General / *A4-16*
   Basic Accounting Data / *A4-16*
   Cost Control / *A4-16*
   Forms and Procedures / *A4-17*
   Basic Accounting Forms / *A4-18*
   Cost Control Forms / *A4-19*
Distributable Accounts / *A4-24*
   General / *A4-24*
   Field Office / *A4-26*
   Insurances—Injuries and Damages / *A4-26*
   Temporary Construction / *A4-28*
   Construction Equipment / *A4-29*
   Other Distributables / *A4-29*
Revised Estimate and Job-Cost Report / *A4-32*
   General / *A4-32*
   Revised Estimate Totals / *A4-34*
   Review of Job-Cost Report / *A4-34*
Summary / *A4-36*

Chapter 5. Project Management Information Systems   A5-1

Introduction / *A5-1*
Components of a Project Management Information System / *A5-2*
   Executive Information Systems (EIS) / *A5-3*

Functions of a PMIS / A5-3
Costs for a PMIS / A5-3
Benefits of a PMIS / A5-5
Establishing a Project Management System Network / A5-5
   Local Area Networks (LANs) / A5-6
   Wide Area Networks (WANs) / A5-8
Choosing a Project Management System / A5-9
   Choice of Project Management Software / A5-9
   Off the Shelf or Develop Your Own? / A5-10
   Report Writer / A5-11
Coding Structures / A5-11
   Work Breakdown Structure / A5-12
   WBS Dictionary / A5-12
   Project Commitment Structure (PCS) / A5-14
   Why a PCS and a WBS? / A5-16
Executive Information System (EIS) / A5-16
   What Is an EIS? / A5-16
   EIS Sample Screens and Functions / A5-17
   EIS Development and Implementation / A5-22
Project Management Information System Features / A5-23
   Budget and Cost Management Module / A5-24
   Estimating Module / A5-25
   Engineering Management Module / A5-26
   Planning and Scheduling Module / A5-26
   Integration of Cost and Scheduling Applications / A5-27
   Contract Administration Module / A5-28
   Document Control Module / A5-28

Chapter 6. Value Engineering for Construction Managers   A6-1

Introduction / A6-1
Definition of Value Work / A6-2
Benefits of Value Work / A6-2
Applications of VE by the Program/Construction Manager / A6-3
   Conceptualization of Design / A6-3
   During Design / A6-3
   VE of Facility Operations / A6-5
The Job Plan of Value Engineering / A6-5
Information Phase / A6-6
   Total Cost Management—Cost, Energy, and Life Cycle Cost Models / A6-7
   Cost Model / A6-7
   Energy Model / A6-8
   Life Cycle Cost Model / A6-11
   Function Analysis / A6-12
   FAST Diagramming / A6-15
Speculative/Creative Phase / A6-16
Evaluation/Analytical Phase / A6-16
Development/Recommendation Phase / A6-18
Report Phase / A6-18
Management of Value Programs / A6-23
VE Sequence and Typical Schedule / A6-23
Preworkshop Activity / A6-26
   Coordination Meeting / A6-26
VE Workshop / A6-26
Postworkshop Activity / A6-27
Typical Schedule / A6-27

Selecting a VE Team / *A6-28*
    VE Team Coordinator (VETC) / *A6-28*
    Conclusion and Summary / *A6-29*
Case Study: Conceptualizing Design—Bachelors' Enlisted Quarters (BEQ) / *A6-29*
    Design Concept, Scope, and Design Considerations / *A6-37*

Chapter 7. Quality Assurance/Quality Control   A7-1

Introduction / *A7-1*
Project Structure / *A7-3*
Predesign / *A7-3*
    Programming / *A7-4*
Design Stages / *A7-5*
    Schematic Development / *A7-5*
    Preliminary Design / *A7-5*
    Working Drawings / *A7-5*
    QA in Design / *A7-6*
Construction / *A7-7*
    Start-up / *A7-7*
QA Program Development / *A7-8*
QA Plan for Construction Manager / *A7-8*
Codes and Standards / *A7-8*
Licenses, Permits, Inspections / *A7-17*
Exculpatory Clauses / *A7-27*
Quality Control: Inspection Team / *A7-29*
    Inspectors / *A7-29*
    The Role of the Inspector / *A7-29*
    QC Field Testing / *A7-30*
References / *A7-31*

Chapter 8. Total Quality Management   A8-1

Introduction / *A8-1*
Mission Statement / *A8-1*
Objectives / *A8-2*
Foundation of the Program / *A8-2*
A Typical TQM Plan / *A8-3*
Specific Elements of a TQM Plan / *A8-3*
    Prebid Activity / *A8-3*
    Establishing Sound Contracts / *A8-4*
    Assembling the Project Team / *A8-5*
    Establishing the Project Plan / *A8-6*
    Managing Budget and Schedule / *A8-7*
    Project Phasing and Review / *A8-8*
    Project Control / *A8-9*
    Checking Procedures / *A8-10*
    Completed Project Evaluation / *A8-11*
Continuing Education in TQM / *A8-12*
Establishing Quality Teams / *A8-12*
    TQM Agenda Sheet / *A8-13*
References / *A8-13*

Chapter 9. Partnering — A9-1

Introduction / *A9-1*
What Is Partnering? / *A9-2*
   Partnering Fundamentals / *A9-3*
   Partnering Benefits / *A9-5*
   Key Elements of Partnering / *A9-6*
   Periodic Joint Evaluation / *A9-8*
   Partnering Implementation / *A9-8*
   Tangible Results / *A9-12*
Summary / *A9-12*
References / *A9-13*

Chapter 10. Safety — A10-1

Introduction / *A10-1*
Definitions / *A10-1*
Legal Sources of Liability / *A10-2*
Contractual Services That Affect a Construction Manager's Duty
   to Injured Workers / *A10-2*
Actions That Affect a Construction Manager's Duty to Injured Workers / *A10-8*
Liability Imposed by Statute / *A10-9*
State Safety Legislation / *A10-10*
OSHA / *A10-10*
   OSHA Regulations / *A10-11*
Risk Coverage, Immunity, and Indemnity / *A10-11*
Insurance / *A10-11*
Can Workers' Compensation Laws Immunize the Construction Manager
   from Liability to an Injured Worker? / *A10-12*
Is the Construction Manager Entitled to Indemnity for Safety Liability? / *A10-14*
Final Perspective / *A10-15*
References / *A10-15*

Chapter 11. Heavy Construction Claims — A11-1

Introduction / *A11-1*
Types of Heavy Construction Claims / *A11-2*
Preparation of Heavy Construction Claims / *A11-5*
Refutation of Heavy Construction Claims / *A11-6*
Proving and Pricing Heavy Construction Claims / *A11-6*
Roles of Lawyers, Consultants, and Experts / *A11-9*
Litigation in Heavy Construction / *A11-10*
Alternative Dispute Resolution in Heavy Construction / *A11-11*
Claims Avoidance in Heavy Construction / *A11-12*
Epilogue / *A11-14*

Part B Heavy Construction Equipment

Chapter 1. Engineering Fundamentals — B1-3

Introduction / *B1-3*
   Influence of Material on Equipment Selection / *B1-3*

Weight-Volume Relationships / *B1-4*
   Volume Measure / *B1-4*
   Load Factor / *B1-5*
   Payload / *B1-5*
   Shrinkage Factor / *B1-5*
   Soil Density / *B1-8*
   Soil Compaction / *B1-8*
   Material Investigations / *B1-9*
Equipment Characteristics / *B1-9*
   Power Units / *B1-9*
   Power Transmission and Controls / *B1-14*
   Power Ratings / *B1-15*
   Power Usable / *B1-17*
   Power Required / *B1-20*
Tires / *B1-24*
   Principles of Tire Engineering / *B1-24*

Chapter 2. Equipment Economics   B2-1

Introduction / *B2-1*
Acquisition / *B2-1*
   Forecasting Equipment Requirements / *B2-1*
   Selection of Equipment / *B2-2*
   Equipment Replacement / *B2-8*
   Financing / *B2-12*
Maintaining Availability / *B2-13*
   Reducing Downtime / *B2-13*
   Servicing / *B2-16*
   Repairs / *B2-17*
Ownership and Operating Costs / *B2-17*
   Value Accumulation / *B2-17*
   Loss of Value / *B2-19*
   Administrative and Overhead Cost / *B2-21*
Information Requirements for Financial Audit / *B2-22*
   Auditing Requirements / *B2-22*
Information for Operating Management / *B2-26*
   Estimating Costs / *B2-26*
   Equipment Records / *B2-27*
   Monitoring Costs / *B2-29*

Chapter 3. Tractors, Bulldozers, and Rippers   B3-1

Off-Highway Tractors / *B3-1*
   Classification / *B3-1*
   Equipment Characteristics / *B3-1*
Bulldozers / *B3-9*
   Blade Types / *B3-9*
   Tip, Tilt, and Angle / *B3-12*
   Applications / *B3-14*
   Production Estimating / *B3-16*
Rippers / *B3-18*
   Types / *B3-18*
   Operation / *B3-20*
   Production Estimating / *B3-20*
Reference / *B3-22*

Chapter 4. Scrapers   B4-1

Introduction / *B4-1*
Conventional Scrapers / *B4-1*
   Capacity Ratings / *B4-3*
   Equipment Specifications / *B4-5*
   Methods of Operation / *B4-7*
   Elevating Scrapers / *B4-11*
   Capacity Ratings / *B4-12*
   Equipment Specifications / *B4-13*
   Methods of Operation / *B4-13*
Multiple Scrapers / *B4-15*
Basic Cycle Considerations / *B4-16*
   The Cut / *B4-16*
   The Haul Road / *B4-17*
Equipment Selection / *B4-18*
   The Push-Loading Concept / *B4-18*
   Towed Scrapers versus Wheel Tractor-Scrapers / *B4-18*
   Number of Axles / *B4-19*
   Weight-to-Horsepower Ratio / *B4-19*
   Number of Engines / *B4-19*
   Size of Scraper / *B4-20*
   Power for Push-Loading / *B4-20*
   Track-Type versus Wheel-Type Pusher / *B4-21*
   Self-Propelled, Self-Loading Scrapers / *B4-22*
Productivity / *B4-24*
References / *B4-24*

Chapter 5. Excavators   B5-1

Introduction / *B5-1*
   Attachments / *B5-2*
   Telescopic Grading Hoes / *B5-8*
Single-Purpose Excavators / *B5-9*
   Mining Shovels / *B5-9*
   Walking Draglines / *B5-10*
   Stripping Shovels / *B5-22*
   Draglines and Clamshells / *B5-24*
References / *B5-30*

Chapter 6. Front-End Loaders   B6-1

Front-End Loaders / *B6-1*
Crawler-Type Loaders / *B6-1*
Wheel-Type Loaders / *B6-2*
Front-End Loader Ratings / *B6-6*
   Bucket Capacity, SAE Rating / *B6-6*
   Tipping Load, SAE Rating / *B6-10*
   Operating Load, SAE Rating / *B6-10*
Loader Buckets and Attachments / *B6-10*
Major Loader Applications / *B6-11*
   Loading / *B6-12*
   Hauling / *B6-12*
   Excavating / *B6-13*

Chapter 7. Hauling Units   B7-1

Introduction / *B7-1*
On-Highway Tractors / *B7-3*
   Classification / *B7-3*
   Equipment Characteristics / *B7-6*
   Production Estimating / *B7-6*
Types of Off-Highway Hauling Units / *B7-7*
   Rear-Dump Load-on-Back Truck / *B7-8*
   Bottom-Dump Tractor-Trailer / *B7-9*
   Side-Dump Load-on-Back Truck / *B7-11*
   Side-Dump Tractor-Trailer / *B7-12*
   Rear-Dump Tractor-Trailer (Rocker) / *B7-12*
Hauler Performance / *B7-13*
   Performance and Retarder Charts / *B7-13*
   Operating Conditions / *B7-17*
Estimating Hauler Production / *B7-19*
   Payload / *B7-20*
   Cycle Time / *B7-20*
   Loading Time / *B7-21*
   Travel Time / *B7-21*
   Turn-and-Dump Times / *B7-23*
   Spot Time / *B7-23*
   Delay Factors and Job Efficiency / *B7-24*

Chapter 8. Soil Surfacing Equipment   B8-1

Spreading Equipment / *B8-1*
   General / *B8-1*
   Equipment Types / *B8-1*
   Production / *B8-5*
Stabilization Equipment / *B8-6*
   General / *B8-6*
   Equipment Types / *B8-6*
   Equipment Selection / *B8-9*
Compaction Equipment / *B8-10*
   General / *B8-10*
   Equipment Types / *B8-10*
   Equipment Selection / *B8-16*

Chapter 9. Belt Conveyors   B9-1

Introduction / *B9-1*
Belt-Conveyor Systems and Subsystems / *B9-1*
   Plant Feed Systems / *B9-3*
   Storage-Reclaim Systems / *B9-4*
   Overland Haulage Systems / *B9-8*
   Ship and Barge Loaders / *B9-9*
   Portable Systems / *B9-9*
Belt-Conveyor Components / *B9-11*
   Belts / *B9-11*
   Pulleys and Idlers / *B9-13*
   Drives / *B9-15*
   Motors and Controls / *B9-15*

Transfer Chutes / *B9-16*
Supporting Structures / *B9-17*
Accessories / *B9-18*
Engineering Design / *B9-19*
Material Characteristics / *B9-19*
Capacity, Speed, and Sizing of Equipment / *B9-20*
Design Calculations / *B9-21*
Other Parameters / *B9-22*
Operation and Maintenance / *B9-25*
References / *B9-27*

Chapter 10. Cranes   B10-1

Introduction / *B10-1*
Mobile Mechanical Cranes / *B10-2*
Mobile Hydraulic Cranes / *B10-11*
  Crane Sizes / *B10-12*
  Capacity Enhancements / *B10-24*
  Shipping / *B10-25*
Tower Cranes / *B10-36*
Equipment Selection / *B10-43*

Chapter 11. Marine Equipment   B11-1

Introduction / *B11-1*
Barges and Scows / *B11-2*
  Selection / *B11-2*
  Inspection / *B11-3*
Tugs and Towboats / *B11-4*
  Selection / *B11-4*
  Inspection / *B11-5*
Dredging / *B11-5*
  Hydraulic Dredging / *B11-5*
  Ladder or Bucket Dredge / *B11-16*
  Dipper Dredge / *B11-17*
  Clamshell Dredge / *B11-17*
  Hopper Dredge / *B11-19*
  Dustpan Dredge / *B11-24*
  Drill Boat / *B11-25*
Floating Cranes, Derricks, and Pile Drivers / *B11-28*
  Cranes and Derricks / *B11-28*
  Pile Drivers / *B11-31*
Miscellaneous Equipment / *B11-31*
Auxiliary Diving Equipment / *B11-31*
  Transportation / *B11-31*
  Communications / *B11-32*
  Marine Risk Insurance / *B11-32*
  Hull Insurance / *B11-32*
  Protection and Indemnity Insurance / *B11-33*
  Worker's Compensation Insurance / *B11-33*
  Dredging and the Environment / *B11-34*

Chapter 12. Tunnel-Boring Machines   B12-1

Introduction / *B12-1*
TBM Selection / *B12-1*
   Geological Considerations / *B12-2*
   Borability / *B12-2*
   Stability / *B12-2*
   Other Geological Influences / *B12-2*
   Tunnel Design and Project Specifics / *B12-3*
Description of TBM Types / *B12-4*
   Soft-Ground Shields / *B12-4*
   Slurry Shields / *B12-5*
   EPB Shields / *B12-7*
   Excavator Shields / *B12-9*
   Hard-Rock TBMs / *B12-10*
   Hybrid TBMs / *B12-19*
Auxiliary Equipment / *B12-20*
   Logistics / *B12-20*
   Linings / *B12-24*

Part C   Heavy Construction Materials

Chapter 1. Conventional Concrete   C1-3

Section 1. Materials Selection and Mixture Proportioning / C1-4

Cement / *C1-4*
Aggregates / *C1-5*
   Water / *C1-6*
   Admixtures / *C1-7*
   Air-Entraining Admixtures / *C1-7*
   Accelerators / *C1-8*
   Retarders / *C1-9*
   Water-Reducing, Set-Controlling Admixtures / *C1-9*
   Plasticizers / *C1-9*
   Cementitious and Pozzolanic Materials / *C1-10*
   Waterproofers / *C1-10*
   Inhibitors of Alkali-Aggregate Reaction / *C1-10*
Proportioning Concrete Mixtures / *C1-11*
   Basic Physical Properties of Materials / *C1-12*
   Elements of the Trial Mix / *C1-13*
   Computation of Mix Proportions / *C1-18*
   Concrete-Mix Tests / *C1-25*
   Adjustments for Field Use / *C1-26*
Special Proportional Requirements / *C1-28*
   Six-In Aggregate Concrete / *C1-28*
   Lightweight Aggregate Concrete / *C1-29*

Section 2. Batching, Mixing, Placing, and Curing / C1-31

Material Handling and Storage / *C1-31*
   Material Characteristics / *C1-31*
   Plant Types / *C1-31*
   Material Handling / *C1-34*

Plant Storage / *C1-41*
Special Processing / *C1-43*
Batching / *C1-45*
   Aggregate / *C1-46*
   Cement / *C1-46*
   Water and Admixtures / *C1-47*
   Scales / *C1-52*
   Controls / *C1-53*
   Tolerances / *C1-56*
Mixing / *C1-59*
   Tilting Mixers / *C1-60*
   Nontilting Mixers / *C1-60*
   Paving Mixers / *C1-61*
   Truck or Transmit Mixers / *C1-62*
   Mixing Time / *C1-65*
Placing / *C1-66*
   Chutes / *C1-67*
   Buckets / *C1-67*
   Buggies / *C1-67*
   Pumps / *C1-68*
   Belt Conveyors / *C1-68*
   Vibrators / *C1-73*
   Screeding / *C1-82*
   Troweling / *C1-82*
Finishing and Curing / *C1-83*
   Finishing / *C1-83*
   Curing / *C1-84*

Section 3. Stationary Forms / *C1-86*

Planning for Formwork / *C1-86*
   Specification Requirements / *C1-86*
   Job Requirements for Formwork / *C1-87*
   Material Available for Formwork / *C1-88*
Form Materials / *C1-88*
   Lumber / *C1-88*
   Plywood / *C1-88*
   Metals / *C1-89*
   Plastics / *C1-89*
   Absorptive and Impervious Liners / *C1-90*
Formwork Requirements / *C1-90*
   Form Structure / *C1-90*
   Form Joints / *C1-92*
   Textures and Patterns / *C1-92*
   Form Ties / *C1-93*
   Form Coatings / *C1-95*
   Shores / *C1-95*
Job-Built Forms / *C1-96*
   Form Design / *C1-96*
   Form Hardware / *C1-97*
Prefabricated Forms / *C1-98*
Gang Forms / *C1-101*
Form Applications / *C1-101*
   Wall Forms / *C1-101*
   Column Forms / *C1-102*
   Beam Forms / *C1-102*
   Slab and Beam Forms / *C1-102*
Care of Forms / *C1-107*

Section 4. Slip Forms / C1-110

Design of Sliding Forms / C1-110
Jacking System / C1-113
Design of Working Deck and Bracing / C1-115
Reinforcing Steel / C1-117
Concrete Placing / C1-118
Control and Tolerance / C1-120
Finishing and Curing / C1-121
Connections for Beams and Slabs / C1-121

Section 5. Grouting / C1-122

Materials / C1-122
  Cement / C1-122
  Water / C1-122
  Bentonite and Other Clays / C1-123
  Admixtures / C1-123
  Special-Purpose Materials / C1-123
Proportioning / C1-123
  Fine Seams and Cracks / C1-124
  Cavity Filling / C1-124
  Filling Vertical or Steeply Sloping Spaces / C1-124
  Mixes for Precision Grouting / C1-125
  Post-Tensioned Cable Mixes / C1-125
  Grouting in Granular Materials / C1-125
Mixing and Pumping / C1-125
  Batching and Mixing Mortar / C1-126
  Transporting the Mortar / C1-129
Grouting Procedures / C1-130
  Hard-Rock Grouting / C1-130
  Grouting Cavities and Open Seams / C1-131
  Precision Grouting / C1-131
  Post-Tensioned Cable Grouting / C1-132
  Grouting in Granular Materials / C1-133
References / C1-133

Chapter 2. Reinforcing Steel   C2-1

Reinforcing Bars / C2-1
  Identification / C2-3
  Bar Mats / C2-3
  Coiled Stock / C2-5
Wire and Welded Wire Fabric / C2-5
Prestressing Reinforcement / C2-6
  Strand / C2-7
  Bars / C2-7
  Wire / C2-7
  Compacted Strand / C2-7
Coated Reinforcement / C2-8
  Coated Reinforcing Bars / C2-8
  Epoxy-Coated Wire and Fabric / C2-8
  Epoxy-Coated Strand / C2-8
Splices of Reinforcement / C2-9
  Lap Splices / C2-9
  Mechanical Connections / C2-9
  Welded Splices / C2-10
  Bundled Bars / C2-11
Detailing and Fabrication / C2-11

Handling and Placing Reinforcement / *C2-12*
    Handling and Storage / *C2-12*
    Placing / *C2-18*
    Tack Welding / *C2-18*
    Field Bending and Rebending / *C2-18*
    Bar Supports / *C2-18*
    Coated Reinforcement / *C2-19*
Metrication / *C2-19*
References / *C2-20*

Part D Heavy Construction Types

Chapter 1. Rock Excavation/Blasting   D1-3

Introduction / *D1-3*
Planning / *D1-3*
    Specifications and Local Laws / *D1-3*
    Geologic and Engineering Evaluation / *D1-4*
    Site and Equipment Factors / *D1-4*
    Operations Planning / *D1-4*
Drilling / *D1-5*
    Percussion Drills / *D1-5*
    Rotary Drills / *D1-6*
    Abrasive Drills / *D1-7*
Blast Design / *D1-8*
    Characterizing the Rock Mass / *D1-9*
    Rock Hardness and Density / *D1-10*
    Void and Incompetent Zones / *D1-10*
    Joint Spacing / *D1-10*
    Bedding Orientation and Location / *D1-10*
    Standard Blast Patterns / *D1-10*
    Criteria for Design / *D1-12*
Rules-of-Thumb Guidelines / *D1-13*
    Burden / *D1-13*
    Spacing / *D1-13*
    Subdrill / *D1-14*
    Bench Height / *D1-14*
    Collar Stemming Height / *D1-14*
Special Blasting Techniques / *D1-15*
    Buffer or Choke Blasting / *D1-15*
    Sinking Cuts / *D1-15*
    Overbreak Control / *D1-15*
    Controlled Blasting Techniques / *D1-16*
    Line Drilling / *D1-17*
    Presplitting / *D1-17*
    Smoothwall Blasting / *D1-18*
Explosives Selection / *D1-18*
    Properties of Explosives / *D1-19*
    Strength / *D1-19*
    Detonation Velocity / *D1-19*
    Density / *D1-19*
    Water Resistance / *D1-20*
    Fume Class / *D1-20*
    Detonation Pressure / *D1-20*

Borehole Pressure / *D1-20*
Sensitivity and Sensitiveness / *D1-20*
Explosive Products / *D1-21*
Primer Selection / *D1-21*
Initiation Systems / *D1-23*
Blasthole Loading / *D1-23*
General Loading Procedures / *D1-24*
Cartridge Explosives / *D1-24*
Bulk Explosives / *D1-24*
Priming / *D1-26*
Location / *D1-27*
Multiple Priming / *D1-28*
Side-Initiation / *D1-28*
Initiation Sequence / *D1-29*
Delay Time Interaction / *D1-29*
Environmental Effect of Blasting / *D1-30*
Flyrock / *D1-31*
Ground Vibrations / *D1-31*
Airblasts / *D1-32*
Bibliography / *D1-32*

Chapter 2. Paving   D2-1

Section 1. Bituminous Pavements / D2-2

Pavement Elements, Materials, and Types / *D2-2*
Pavement Elements / *D2-2*
Pavement Materials / *D2-3*
Pavement Types / *D2-6*
Construction of Surface Treatments, Bituminous Penetration Macadam,
    and Road-Mixed Pavements / *D2-7*
Tack Coat / *D2-8*
Prime Coat / *D2-8*
Fog Seal Coat / *D2-10*
Slurry Seal Coat / *D2-10*
Single-Pass Seal Coats / *D2-10*
Multiple-Pass Surface Treatments / *D2-11*
Plant-Mixed Seal Coat / *D2-13*
Penetration Macadam / *D2-13*
Road Mix / *D2-15*
Production of Hot-Mixed Bituminous Concrete / *D2-17*
Asphalt Plants / *D2-17*
Aggregate Feeding / *D2-17*
Aggregate Drying / *D2-21*
Dust Collection / *D2-25*
Screening / *D2-26*
Measuring and Mixing / *D2-30*
Plant Production / *D2-33*
Fines Feeding and Handling / *D2-34*
Plant Accessories / *D2-36*
Construction of Hot-Mixed, Hot-Laid Bituminous Concrete Pavements / *D2-40*
The Finisher / *D2-40*
Finisher Operation / *D2-41*
Automatic Screed Controls / *D2-44*
Finisher Applications / *D2-46*
Milling Machines / *D2-50*
Reclaimers/Stabilizer / *D2-53*

Compacting and Rolling / D2-54
Paving Tips / D2-59

Section 2. Portland Cement Concrete Pavements / D2-61

Preparation for Paving / D2-61
   Compaction Equipment / D2-61
   Automated Earthwork Equipment / D2-62
   Forms and Form Setting / D2-63
Placing Concrete and Steel / D2-63
   Slip-Form Paving / D2-63
   Paving Train / D2-65
   Concrete Delivery / D2-67
   Steel Reinforcement and Dowels / D2-69
   Machine Pavers / D2-70
Joints / D2-72
   Joint Forming / D2-73
   Joint Sealing / D2-73
Curing / D2-73
References / D2-74

Chapter 3. Piles and Pile Driving   D3-1

Introduction / D3-1
   History / D3-1
   Determining the Need for Piling / D3-1
   Types of Piles / D3-1
   Installation / D3-2
   Unique Relationship Among Engineer, Contractor, and Owner
      in Pile Foundations / D3-2
   Summary and Scope / D3-2
Pile Types / D3-2
   General / D3-2
   Timber Piles / D3-2
   Prestressed Concrete Piles / D3-7
   Concrete Cylinder Pile / D3-7
   Cast-in-Place Concrete Piles / D3-7
   Steel H Piles / D3-9
   Steel Pipe Piles / D3-9
   Monotube Piles / D3-10
Pile Stresses / D3-10
   Design Loads / D3-10
   Handling Stress / D3-10
   Driving Stress / D3-10
   Tension Stresses Due to Swelling Soils / D3-11
   Compression or Bending Stresses Due to Negative Skin Friction / D3-11
Pile Materials / D3-11
   Timber / D3-11
   Steel / D3-11
   Concrete / D3-12
   Grout / D3-22
Pile-Driving Equipment / D3-22
   Hammers / D3-23
   Equipment Placed Between Hammer and Pile / D3-25
   Leaders / D3-35
   Spotter / D3-35
   Preexcavation Equipment / D3-35

Pile Installation / *D3-35*
  General / *D3-35*
  Subsurface Conditions / *D3-36*
  Handling and Storage / *D3-36*
  Planning, Site Preparation, Pile Numbering, and Location / *D3-36*
  Hammer Selection, Wave Equation, and Pile Load Test / *D3-37*
  Preexcavation / *D3-37*
  Pile Driving / *D3-38*
  Augercast Piles / *D3-39*
  Pressure-Injected Footings (PIFs) / *D3-40*
Inspection / *D3-40*
  General / *D3-40*
  Equipment / *D3-40*
  Materials / *D3-41*
  Driven Pile Installation / *D3-43*
  Pressure-Injected Footing Installation / *D3-45*
References / *D3-46*

Chapter 4. Cofferdams and Caissons   D4-1

Introduction / *D4-1*
Cofferdams / *D4-3*
  General Considerations / *D4-3*
  Street Piling / *D4-9*
  Bracing Systems / *D4-20*
  Excavation / *D4-31*
  Pile Driving Within the Cofferdam / *D4-32*
  Bottom Seal / *D4-33*
  Dewatering / *D4-40*
  Cofferdam Difficulties / *D4-42*
  Slurry Wall Cofferdams / *D4-48*
  Cellular Cofferdams / *D4-49*
Box Caissons / *D4-50*
  General Considerations / *D4-50*
  Site and Foundation Preparation / *D4-51*
  Fabrication, Launching, and Deployment / *D4-53*
  Setting / *D4-56*
  Concreting / *D4-59*
  Examples / *D4-60*
Open Caissons / *D4-69*
  General Considerations / *D4-69*
  Cutting Edges / *D4-72*
  Setting / *D4-74*
  Sinking / *D4-75*
  Tipping and Sliding / *D4-79*
  Completing the Installation / *D4-81*
  Examples / *D4-83*
Pneumatic Caissons / *D4-89*

Chapter 5. Construction Dewatering   D5-1

Introduction / *D5-1*
Dewatering Methods / *D5-2*
  Surface Pumping / *D5-2*
  Well Method / *D5-3*
  Wellpoint Method / *D5-5*
  Ejector Method / *D5-10*
  Other Methods / *D5-11*

Pumps for Dewatering / D5-13
   Horizontal Centrifugal Pumps / D5-13
   Vertical Pumps / D5-13
   Submersible Pumps / D5-14
   Diaphragm Pumps / D5-14
   Air Lifts / D5-14
   Pump Selection / D5-15
Design of Dewatering Systems / D5-15
   Determination of Permeability / D5-17
   Well Formulas / D5-18
   Capacity of Wells / D5-20
Cost of Dewatering Operations / D5-21
References / D5-21

Part E Heavy Construction Projects

Chapter 1. Airports   E1-3

Section 1. Introduction / E1-4

History / E1-4
Case Studies Introduction / E1-5

Section 2. Case Study: John F. Kennedy International Airport, JFK Redevelopment Program / E1-6

Program Goal/Definition / E1-6
Airport History / E1-6
JFK Redevelopment Program Scope / E1-9
Airport Construction / E1-10
Operating Environment / E1-10
Passenger Considerations / E1-11
Restricted Construction / E1-11
Special Requirements / E1-12
AOA Restrictions / E1-13
Administrative Programs / E1-15
   Insurance / E1-15
   Tenant Construction or Alterations / E1-15
Building Construction / E1-15
   Air Traffic Control Tower Project Description / E1-15
   East Garage / E1-25
Civil Construction / E1-27
   Roadways / E1-27
   Utilities / E1-30
   Cogeneration / E1-32

Section 3. Case Study: Manchester Airport, New Terminal Project / E1-34

Chapter 2. Water Treatment Facilities   E2-1

Introduction / E2-1
   Wastewater Chronology / E2-1
   Clean Water Chronology / E2-2
   Construction Characteristics / E2-2
Case Study: Camden County, New Jersey / E2-2
   Delaware #1 WPTF / E2-3

Chapter 3. Highways   E3-1

Introduction / *E3-1*
Definition / *E3-1*
History / *E3-2*
Development of Highways in the United States / *E3-3*
Highway System / *E3-5*
   Federal-Aid Systems / *E3-6*
   State Systems / *E3-6*
   Local Road Systems / *E3-6*
   City Streets / *E3-6*
   Toll Roads / *E3-6*
Case Studies Introduction / *E3-7*
Case Study: Urban Construction, I-105 Glen Anderson Freeway / *E3-8*
   History / *E3-8*
   Description / *E3-9*
   I-105 Fact Sheet / *E3-13*
Case Study: Intermediate Terrain, Beaver Valley Expressway Segment
     of the Pennsylvania Turnpike / *E3-13*
   History / *E3-13*
   Economic Development / *E3-15*
   Schedule / *E3-16*
   Toll Collection / *E3-16*
   Land Acquisitions / *E3-16*
   Environmental Damage and Corrective Measures / *E3-16*
   Description of Work / *E3-17*
   Equipment / *E3-20*
   Geologic Features / *E3-20*
Case Study: Flat Terrrain, Superstition Freeway, Arizona / *E3-20*
Case Study: Swamp Reclamation, I-75 Alligator Alley / *E3-23*
Case Study: Rough Terrain, Glenwood Canyon / *E3-27*
   Pioneering in the Use of Materials and Methods / *E3-30*
   Innovations in Construction / *E3-32*
   Impact on Physical Environment, Unusual Aspects, and Aesthetic Values / *E3-32*
   Supporting Information / *E3-33*
Future Construction / *E3-35*
   Program / *E3-36*
   Surface Transportation Program / *E3-36*
   Congestion Mitigation and Air Quality Improvement Program / *E3-37*
   Bridge Replacement and Rehabilitation Program / *E3-37*
   Federal Lands / *E3-37*
   Special Programs / *E3-37*
Case Study: Urban Freeway Reconstruction, U.S. 59/Southwest Freeway HOV Lane Project,
     Houston, Texas / *E3-39*
   History / *E3-39*
   Project Description / *E3-41*
   Goals of the Project / *E3-41*
   Project Costs/Funding / *E3-42*
   Innovations in Construction and Management / *E3-43*
   Special Awards/Significant Notoriety / *E3-47*

Chapter 4. Pipelines   E4-1

Section 1. Underground Utility Construction / *E4-2*
Preliminary Work / *E4-2*
   Planning / *E4-2*
   Locating Crosslines / *E4-3*

Pavement Removal / *E4-3*
   Pavement Breaking / *E4-3*
   Pavement Removal / *E4-4*
   Sawing Pavement / *E4-4*
Excavating the Trench / *E4-5*
   Equipment and Methods / *E4-5*
   Rock Excavation / *E4-6*
   Dewatering / *E4-7*
   Shoring / *E4-8*
   Sand Excavation / *E4-9*
Installation of Pipe or Conduit / *E4-12*
   Water Mains / *E4-12*
   Sanitary and Storm Sewers / *E4-13*
   Gas Distribution / *E4-14*
   Electrical and Telephone / *E4-15*
Backfilling the Trench / *E4-17*
   Shading and Bedding / *E4-17*
   Equipment and Methods / *E4-18*
   Compaction / *E4-18*
Paving the Trench / *E4-20*

Section 2. Trenchless Pipelines / E4-21

Boring / *E4-21*
Micro Tunneling / *E4-23*
Pipe-Pushing Machines / *E4-24*

Section 3. Transmission Pipelines / E4-28

Planning Transmission Lines / *E4-28*
   Length / *E4-29*
   Location / *E4-29*
   Terrain / *E4-30*
   Weather / *E4-30*
   Labor Availability / *E4-30*
   Right-of-Way / *E4-30*
Construction Transmission Lines / *E4-30*
   Preliminary Work / *E4-31*
   Clearing / *E4-31*
   Trenching / *E4-31*
   Stringing / *E4-32*
   Bending / *E4-32*
   Welding / *E4-32*
   Cleaning / *E4-33*
   Coating and Wrapping / *E4-33*
   Lowering In / *E4-34*
   Backfilling / *E4-34*
   Cleanup / *E4-34*
   Testing Completed / *E4-34*
Reference / *E4-35*

Index follows Chapter E4

CONTRIBUTORS

Joseph Alcabes, P.E., C.C.E. (CHAP. A4)
P. Clay Baldwin   *O'Brien-Kreitzberg, Inc.* (CHAP. E1)
Gary S. Berman   *O'Brien-Kreitzberg, Inc.* (CHAP. A11)
Thomas A. Bryant II, P.E.   *O'Brien-Kreitzberg, Inc.* (CHAP. E3)
L. A. Caccese, C.E., P.E.   *U.S. Corps of Engineers (retired) and Dredging Consultant* (CHAP. B11)
Ross Caldwell   *O'Brien-Kreitzberg, Inc.* (CHAP. A3)
Richard D. Conner, Esq.   *Patton Boggs, L.L.P.* (CHAP. A10)
John A. Cravens   *Euclid, Inc.* (CHAP. B7)
Joseph Dixon   *O'Brien-Kreitzberg, Inc.* (CHAP. E1)
R. D. Evans   *Caterpillar Tractor Company* (CHAPS. B1, B7)
Richard Fennema   *O'Brien-Kreitzberg, Inc.* (CHAP. E1)
Ben C. Gerwick, Jr.   *Ben C. Gerwick, Inc.* (CHAP. D4)
Michael Giaramita, P.E.   *O'Brien-Kreitzberg, Inc.* (CHAP. E3)
Jerome Gold   *O'Brien-Kreitzberg, Inc.* (CHAP. E1)
F. H. (Bud) Griffis, Ph.D., P.E.   *Columbia University* (CHAP. A1)
David P. Gustafson   *Concrete Reinforcing Steel Institute* (CHAP. C2)
Charles Halboth, P.E.   *O'Brien-Kreitzberg, Inc.* (CHAP. A9)
Sverker Hartwig   *The Robbins Company* (CHAP. B12)
William J. Kennedy, Esq.   *Dechert, Price & Rhoads* (CHAP. A2)
Duncan MacLaren   *General Motors Corporation* (CHAP. B7)
Wesley F. Mikes   *O'Brien-Kreitzberg, Inc.* (CHAPS. A4, E2)
Joseph V. Mullin, Ph.D., P.E.   *Pennoni Associates* (CHAP. A8)
James J. O'Brien, P.E.   *O'Brien-Kreitzberg, Inc.* (CHAPS. A7, B1–B10, C1, D2, D5, E2, E4)
A. D. Pistilli, Ph.D., P.E.   *American Dredging Company* (CHAP. B11)
E. R. Santhin, M.E.   *American Dredging Company* (CHAP. B11)
E. V. Semonin   *Goodyear Tire & Rubber Company* (CHAP. B1)
Richard Smyth   *O'Brien-Kreitzberg, Inc.* (CHAP. E1)
Victor A. Sterner   *ICI Explosives USA Inc.* (CHAP. D1)
Peter K. Taylor   *Stone & Webster Engineering Corporation* (CHAP. D3)
Louis A. Tucciarone   *O'Brien-Kreitzberg, Inc.* (CHAP. A5)
Larry W. Zimmerman   *Lewis & Zimmerman Associates, Inc.* (CHAP. A6)

PREFACE

From the Preface to the second edition:

The *Handbook of Heavy Construction* was originally conceived by Frank W. Stubbs, Jr., as a comprehensive reference source for persons who are actively involved in heavy construction. The same objective has been adhered to during the preparation of the second edition. The contents of the book are directed primarily to those contractors who, in recognition of the increasing scope and complexity of their work, accordingly seek to extend their knowledge in allied areas of construction. The book will also be a useful reference source for equipment manufacturers, materials producers, owners' representatives, and students of the construction industry.

John A. Havers

The third edition continues the goals of editors Stubbs and Havers. It will also be a useful resource for engineers.

One of the tasks of the third edition is to preserve that part of the second edition still relevant to heavy construction today. Much was relevant. Most approaches, materials, and equipment were essentially the same. However, there have been changes and improvements over the 25 years, including:

Category	1971	1996
Engineering		
Computers	Mainframe service centers	PCs, LANs
	Custom software	Windows
Reports, letters	Typewriters	PC word processors
Drafting	Manual	CADD
Dimensions	English	Metric
Calculators	Slide rule	Electronic, PC
Tools and Equipment		
Tires	Tubes	Tubeless
Equipment	Cable controls	On-board computer
Tools	Manual	Powered
Materials		
Glue	Cements	Epoxies
Concrete	3000 lb/in^2	10,000 lb/in^2
Accelerators	Calcium chloride	Noncorrosive admixtures
Fill	Compacted	Reinforced earth
Structural steel	Rivets	H:S, bolts
Staff		
Workers	Male	Mixed genders

The third edition has been expanded from three major sections to five:

A. Construction/Project Management
   This section has more than doubled in size and scope. TQM, Partnering, QA/QC, Claims, Value Engineering, and PMIS chapters have been added.
B. Construction Equipment
   This section has been revised to reflect current equipment: tractors, bulldozers, scrapers, excavators, loaders, haulers, graders, conveyors, cranes, marine equipment, and TBMs.
C. Heavy Construction Materials
   Basic heavy construction materials are described: concrete, reinforcing steel, and structural steel.
D. Heavy Construction Types
   Types of heavy construction are described including: rock excavation, blasting, paving, piles/pile driving, cofferdams/caissons, and dewatering.
E. Heavy Construction Projects
   Case studies are presented for the following types of projects: Airports, water/wastewater plants, highways, and pipelines.

James J. O'Brien

STANDARD HANDBOOK OF HEAVY CONSTRUCTION

P·A·R·T A

CONSTRUCTION/ PROJECT MANAGEMENT

CHAPTER 1
CONTRACTOR'S ORGANIZATION

Professor F. H. (Bud) Griffis, Ph.D., P.E.
Department of Civil Engineering and Engineering Mechanics
Columbia University, New York City, New York

INTRODUCTION

As the construction industry plans to enter the 21st century, it must predict what the future will bring. Some key differences that will affect future operations may include

- Global markets for manufactured goods and for engineering and construction services
- New forms of project delivery systems such as build, own and transfer and build own and operate
- Continuous process improvement or total quality management
- Significant resource constraints, such as engineers, skilled craftspersons, and critical permanent equipment and materials
- Use of advanced technologies
- Increased off-site fabrication
- Increased concern for employee and public safety
- Decreased project size and greater emphasis on retrofit
- Waste minimization and environmental concerns

To remain competitive in markets that are increasingly global, owners of construction will place greater emphasis on productivity and quality of the constructed project. Construction contractors must organize to meet these new challenges.

CONSTRUCTION PROJECT DELIVERY SYSTEMS

Introduction

Civil and building construction have taken place over the centuries since humans started walking upright. Long before the age of the Roman Empire, the Egyptians built great structures in the desert. Construction in those days did not take place after bidding on a complete set of construction documents and being selected as the low bidder. Generally speaking, a master architect developed a concept. The architect then oversaw craft supervisors to accomplish the work using verbal explanations, sketches, and models. This type of construction organization is shown in Fig. A1-1. This method of construction management existed well into the 19th century and serves as a model for the European system today. However, in the mid- to late 19th century, construction documents started to become more formalized. Owners wanted more complete design before moving into construction. Drafting became a serious profession. In the United States, need for civil infrastructure increased dramatically. There was much work and not enough engineers and contractors. Early in the 20th century the engineering and contracting giants emerged as they are today. As more and more public monies were being spent on construction, more and more laws, rules, and regulations were put in place to protect the owners and the public. Public agencies moved to more open competition. This meant that, in general, construction documents had to be more complete before construction started. Private owners also moved to more competition. After all, it is axiomatic that the more competition, the lower the cost (initial contract award cost).

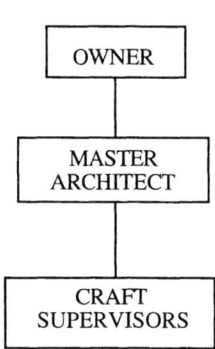

FIGURE A1-1 Ancient construction organization.

As the construction procurement process became more formalized, the designs became more formal. The construction documents for bidding, say for a medium-sized office building, consist of civil, architectural, structural, mechanical (HVAC and plumbing), electrical, and possibly specialty drawings plus a set of extensive specifications. This has, to a large extent, disconnected the designer from the construction (and the contractor from the design). There is no general obligation for the designer of the project to be involved at all in its construction, although design contracts usually include an option for the designer to perform some construction inspection.

To accommodate this type of construction process, several organizational structures that include the owner, designer, and constructor have evolved.

Classic Owner-Designer-Constructor Organization

Figure A1-2 shows what may be called a classic organization. In this type of organization, the designer submits the construction documents to the owner. The owner either bids or negotiates the documents. The general contractor then performs the construction using either their own forces or subcontracts. The designer may or may not provide inspection services. The contractor is usually responsible for quality control and the owner provides the quality assurance.

While most public owners continue to use the classic model as the preferred method of project delivery (there are many exceptions), private owners have moved to other methods of project delivery.

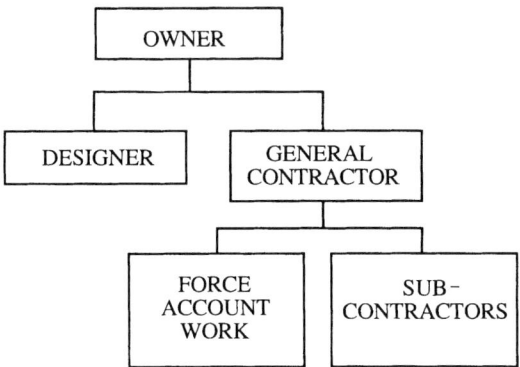

FIGURE A1-2 Classic owner-designer-constructor organization.

At-Risk Construction Management (CM)

Figure A1-3 shows a CM organization in which the CM shares the risk with the owner. Generally speaking, this type of CM has a background as a general contractor. The CM may guarantee the maximum price of a contract. The CM may hold the contracts for independent contractors or subcontractors. The preferred type of contract for this type of organization is a *guaranteed maximum price* (GMP) contract. A GMP is essentially a cost-reimbursable contract with a cap on the maximum expenditures (before contract modifications). This type of contract often lulls the owner into a false sense of security in that the GMP guarantees only what is shown on the original drawings. Any of the subsequent changes will add to the price of the project.

No-Risk Construction Management

A CM that is not at risk is one that assumes liability only to the extent of the fee and acts generally as a consultant to the owner. The construction is performed generally

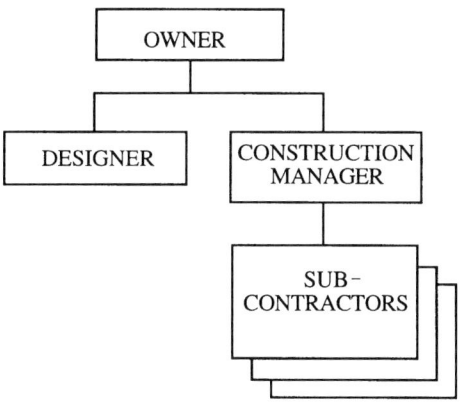

FIGURE A1-3 At-risk construction management organization.

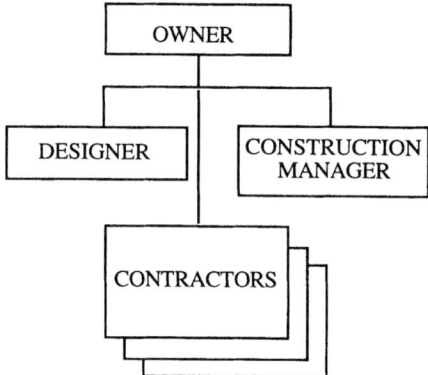

FIGURE A1-4 No-risk construction manager.

as a cost-reimbursable contract. The CM provides cost estimating, scheduling, contract administration, and quality assurance. The owner holds the contracts with the independent contractors or subcontractors. Figure A1-4 is an example of the organizational structure of a CM that is not at risk.

More and more, owners are looking for one stop services from the CM. Many private owners are moving towards having a CM perform the total project delivery process from planning to design to construction to start-up. Although there are many variations of the organization, the following is one such model.

Design-Construct CM

Figure A1-5 shows the structure of a design-construct CM structure. In this model, the CM may or may not be at risk. (The CM could even be a general contractor and be at total risk.) The owner expects the CM to retain the designer (the CM may even be the design firm), work with the designer in using construction technology to

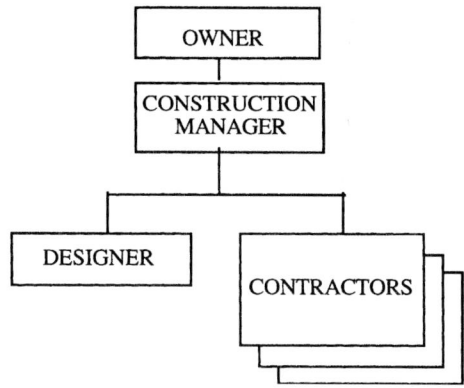

FIGURE A1-5 Design-construct CM organizational structure.

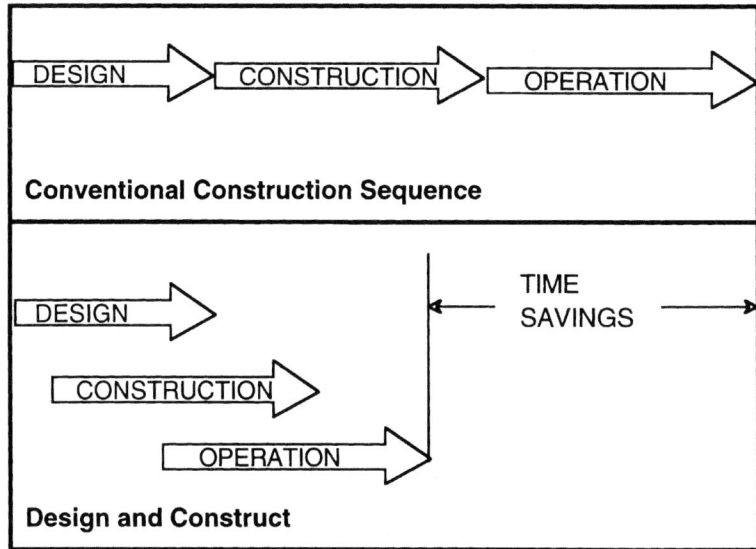

FIGURE A1-6 Contrast between conventional and design-and-construct sequences.

reduce the cost, and construct the project. This has the potential for saving considerable time for project delivery, as shown in Fig. A1-6.

CONTRACTOR MANAGEMENT ORGANIZATIONS*

There are many types of contractor organizations. These are classified in many ways: vertical, horizontal, regional, project, matrix, etc. The following section discusses some of the more popular organizational structures. Figures A1-7 through A1-10 outline the functions and the divisions for a large construction organization. The functions and organizational requirements are about the same regardless of whether the business is individually owned, a partnership type or company, or a closely held or publicly owned corporation. In certain instances, corporate officers, such as vice president or treasurer, are indicated by boxes in the charts. These titles are shown to indicate the degree of responsibility attached to the position and do not mean that a vice president necessarily heads the operation in question. Likewise, for example, the engineering department is headed by a chief engineer, but it is equally true that the chief engineer, in some cases, may also carry the title of vice president.

The organizational diagrams show a natural grouping or breakdown of functions such as may be found in any contractor's organization. Every organization will have variations in the groupings due to the process of natural growth, personalities, and local conditions, management requirements, and environments.

The functional arrangements of the organizational structures shown here are not unique. Different functions may be placed differently in different organizations.

---

* The bulk of this section is taken from "Section 1, Contractor's Organization" by Howard P. Maxton in *Handbook of Heavy Construction,* 2d ed., John A. Havers and Frank W. Stubbs, Jr. (eds.), McGraw-Hill, New York, 1971.

Estimating may be under the chief engineer or it may be under the cost engineer. It is not the purpose of this discussion to justify the location of the functions; it is merely to ensure that the functions are addressed.

The purpose of Fig. A1-7 is to present three primary divisions of functions:

1. Engineering and design
2. Field construction
3. Administration and control

Generally speaking, all construction management functions can be logically placed in one of these three groups. One of the real dangers in a contractor's organization is that of allowing the home office overhead to become too large. This will have a strong effect on the required markup and is a reason for some contractors' troubles.

Generally speaking, the actual field construction will be supervised by a director of operations or a similar construction executive. Often, a *project executive* will be assigned to a project to give the owner of the project entry into the contractor's upper management. The type of organizations shown in Fig. A1-8 will often be over-

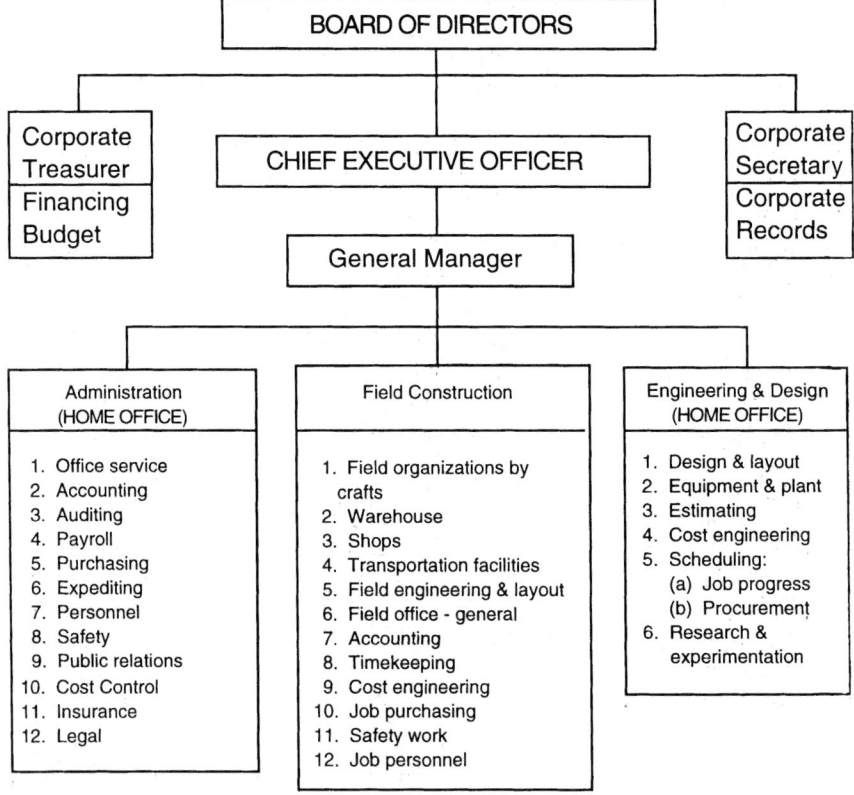

FIGURE A1-7 Functional contractor organization.

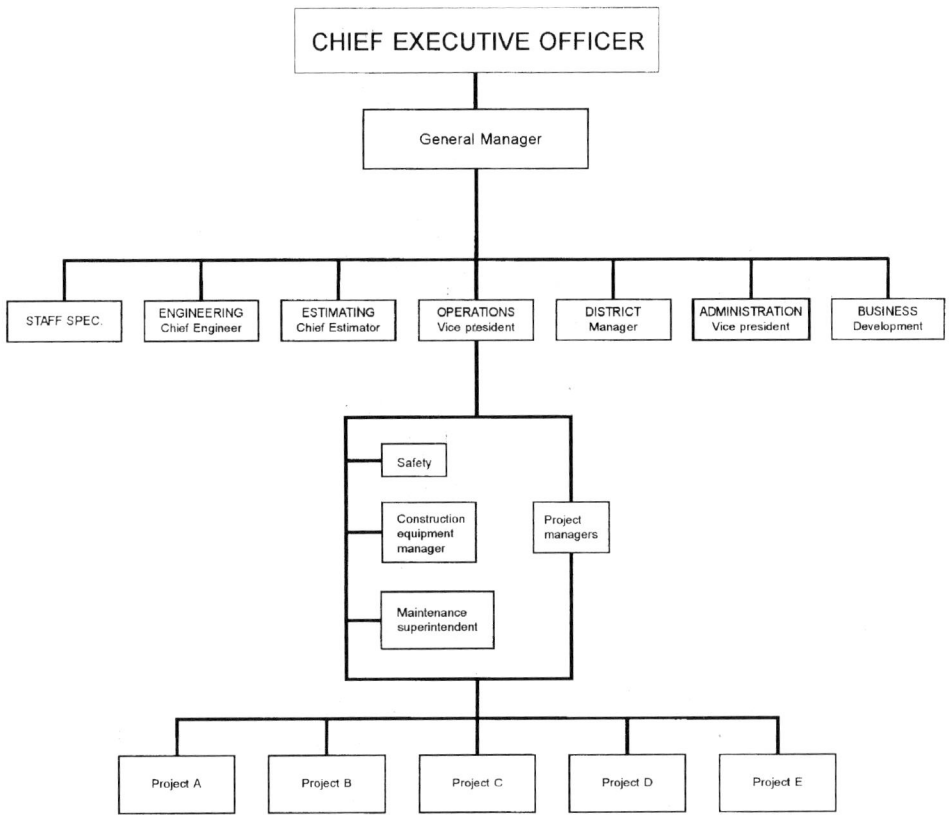

FIGURE A1-8 Home office organizational structure.

laid by a matrix-type organization with members of the other home office functional organizations assigned to a project manager for the accomplishment of project-related work.

The field operations of a contractor may be organized as shown in Fig. A1-9. The project manager or general superintendent may be assigned people from the home office to accomplish the functions necessary for the project. Engineering will provide for layout and shop drawing coordination, estimating assistance with major change order proposals, and progress reporting. The general superintendent will supervise the scheduling directly. On a large project, the superintendent will have craft superintendents reporting. These may or may not be from the contractor's own forces. Usually, these will be the supervisors of subcontractors assigned to the project.

A large contracting organization may have district offices overseeing the projects in a particular region. A typical district organization is shown in Fig. A1-10. It should be noticed that some of the functions from the home office are duplicated in the district office. This can often be a point of confusion to members of the project team. Policies and procedures should spell out completely the role of the functional elements in the district office so that there is no question as to where the ultimate responsibility for the functional accomplishments lies.

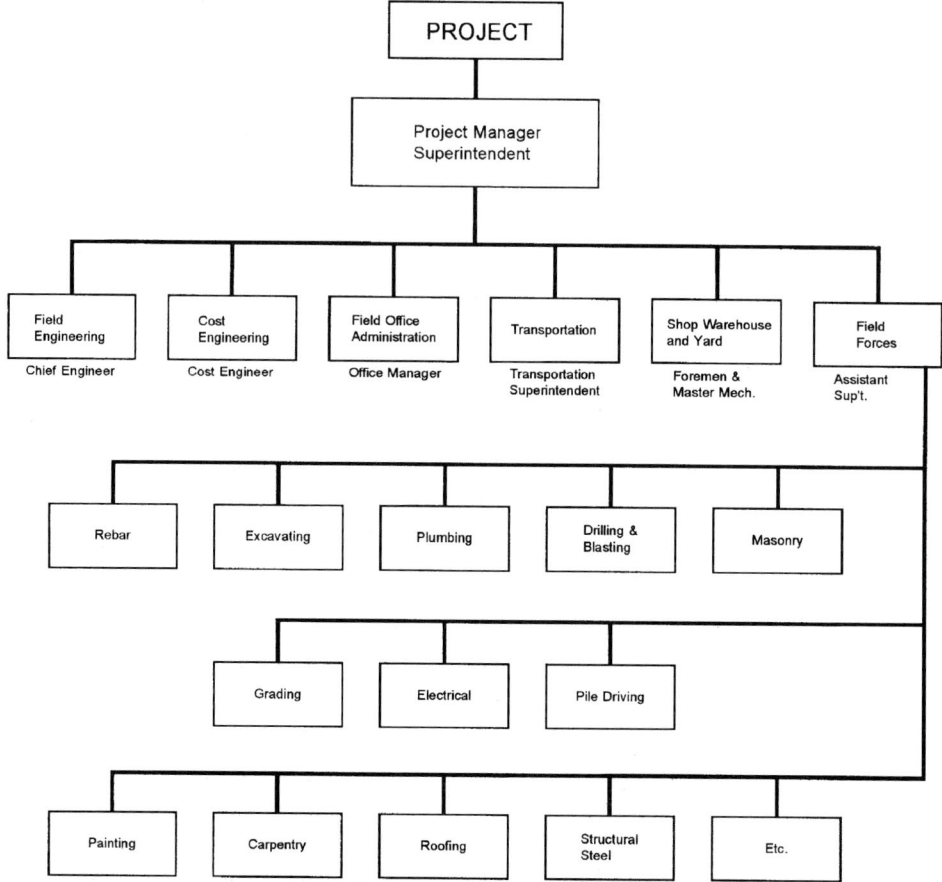

FIGURE A1-9 Contractor field organization.

RECENT CHANGES IN CONSTRUCTION RELATIONSHIPS

Total Quality Management (TQM) is a complete management philosophy that permeates every aspect of a company and places quality as a strategic issue. It is accomplished through an integrated effort between all levels of a company to increase customer satisfaction by continuously improving current performance. Developed in a manufacturing mass production setting, TQM is an effective, comprehensive management technique that has proven successful both overseas and in the United States, in services and in construction, notwithstanding the fact that construction is a one-time process. Every party in a production process is seen to have three roles: supplier, processor, and customer. The designer in a construction project is a customer of the owner, a processor of the design, and a supplier of plans and specifications to the contractor. The contractor is a customer for the designer's plans and

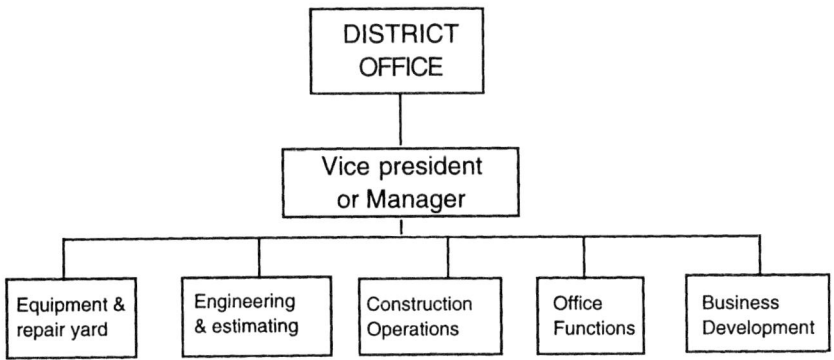

FIGURE A1-10 District office organization.

specifications, a processor of the construction, and a supplier of the completed structure to the owner.

Customer satisfaction is a key goal, whether customers are internal or external to the company. For engineering, the products are plans and specifications and the customers are the construction organization and the owner. For construction, the product is the completed facility and the customer is the final user of the facility. In design and construction organizations, internal customers receive products and information from other groups or individuals within their organization. Satisfying the needs of these internal customers is an essential part of the process of supplying the ultimate external customer with a quality product. Under TQM, management has two primary functions directed toward continual improvement of production process and subsequent increase in customer satisfaction. The first step is maintaining and incrementally improving current methods and procedures through process control. Then, efforts are turned toward achieving, through innovation, major technological advances in the production—i.e., engineering and construction processes.

The Construction Industry Institute feels that companies that do not implement TQM will not be competitive in the national and international market within the next 5 to 10 years.[1]

TQM is a great concept. However, with the same adversarial relationship brought on by the risk-shifting actions and contractual relationships in the current industry, there is no way to provide the incentives, training, or leadership to pull off TQM in a meaningful way in the public construction arena.

William Hayden, in his paper "Management's Fatal Flaw: TQM Obstacle,"[2] writes:

> In seven years I have seen a number of examples of the first flush of executive management-driven systems to deal with this quality "stuff." Examples of beginning TQM with "the right things at the wrong time" have included:
>
> 1. Update the quality manual, release it, and declare victory.
> 2. Project manager and communications training for the "troops."
> 3. Appoint a "corporate quality coordinator" with admonitions to try to be billable; make sure the coordinator believes improving quality is the job assigned.
> 4. Motivational speaker invited in to "get our people committed."
> 5. Try to get your hands on another firm's quality-assurance program "just to look at."

6. Unquestioning quality guru worship and compliance.
7. Forming work-improvement teams, with no training in the new basic and management analytical techniques.
8. Hiring an outside consultant to come in and install a proprietary "proven" TQM program.
9. President/CEO sends manager to seminar so the manager can come back and get "it" going.
10. Buying books and video tapes on quality and distributing them to top managers.
11. Sales-support groups see to it all proposals and brochures emphasize how important quality is within your firm, prior to actually doing anything substantive about it.
12. Using a senior semiretired technical manager to put the QA program together.
13. Rewriting Deming's[3] 14 points in AEC jargon and hanging it up on the wall.
14. Initial efforts begin in one corner of the firm to "try it."
15. Insist bottom-line results show up in six months or shut the effort down; repeat this every six months.

Each of these false starts to TQM has the potential, with some rethinking, to be a legitimate part of TQM. But a fatal flaw of management stands between good intentions and successful implementation . . . denial!

While not necessarily agreeing with Hayden's conclusions, there is no question that implementation of the TQM concept in the a-e-c industry faces some ferocious impediments.

Partnering

The U.S. Army Corps of Engineers, Portland District, has sought to resolve the problems associated with fixed-price, low-bid public contracting through the use of a concept called *partnering*. Partnering is a cooperative approach to contract management that they claim reduces cost, litigation, and stress among the owner, engineer, and contractor. They consider partnering as a formal management strategy. The Portland district claims that partnering offers a new paradigm for owner-contractor relationships. Under partnering, all parties agree from the beginning, in a formal structure, to focus on the creative cooperation and work to avoid adversarial confrontation. Working relationships are carefully and deliberately built, based on mutual respect, trust, and integrity. Partnering provides participants with a win-win orientation toward problem resolution and fosters synergistic teamwork. They found that the partnering management strategy was very successful. In two and one-half years, the Portland district used the partnering process on a variety of projects, with results that include 80 to 100 percent reductions in cost growth over the life of major contracts; time growth in schedules was virtually eliminated; paperwork was reduced by 67 percent; all project engineering goals were met or exceeded; projects were completed with no outstanding claims or litigation; safety records were significantly improved; and pleasure was put back in the process for all participants.

Partnering is a step in the right direction. It stresses that construction is performed by a team and attempts to develop that team. Unfortunately, the owner and the contractor are still bound by a low-bid, fixed-price contract managed by employees being judged in their performance evaluations on how much money they make for the contractor or how much money is saved by the employee of the owner. In addition, the internal objectives of many of the owner's staff may be considerably different. For example, in New York City, street reconstruction is supervised by the Bureau of Highways in the Department of Transportation. The DOT construction objective is to protect the taxpayer's interest and get the best and fastest construction for the least dollar cost. The DOT inspector's objective is to ensure that the project is constructed with the quality specified and that the contractor is maintaining a tidy work site. The DOT design representative has the objective of ensuring that the construction is in accordance with the plans and specification. The DOT postaudit analysts must ensure that the field personnel do not "give away the farm." The DOT traffic staff is concerned with ease of traffic movement and illegal parking. The Bureau of Water Supply and the Bureau of Sewers in the Department of Environmental Protection are responsible for usually half the cost of the project since water main and sewer reconstruction proceed as part of the project. The DEP inspectors are concerned with the quality of the water and sewer construction. In many cases, a project is partially funded with federal highway monies; thus, the New York State DOT and the Federal Highway Administration are also involved. They have an objective to ensure that the contract conformance is in accordance with their policies and procedures.

It is difficult to see how a partnership honeymoon can be a lasting relationship with this many partners.

Private Construction Partnering

Private construction partnering has been researched by the Construction Industry Institute and is being used by some and considered by other owners, contractors, and consultants. This concept anticipates that an owner will establish a long-term relationship with a contractor and designer. This relationship will be based on trust, with all parties participating in managing the total quality of the construction process and product. The Construction Industry Institute defines partnering as:

> *A long-term commitment between two or more organizations for the purpose of achieving specific business objectives by maximizing the effectiveness of each participant's resources. This requires changing traditional relationships to a shared culture without regard to organizational boundaries. The relationship is based upon trust, dedication to common goals, and an understanding of each other's individual expectations and values. Expected benefits include improved efficiency and cost effectiveness, increased opportunity for innovation, and the continuous improvement of quality products and services.*

This probably has a higher potential for success than any of the other concepts, but unless some laws and regulations are changed, this practice is not available to governmental agencies at any level.[4]

Management by Objectives, Management by Exception, Management by Chaos, Etc.

These phrases have been the product of business schools for half a century. During the first half of this century, construction was performed largely by a team, with the engineer and contractor proud to provide the owner with a quality product on time and on budget. For example, The Riverside Church in Manhattan was constructed in exactly that way. When the architect completed the design, John D. Rockefeller telephoned the president of the construction company and said, "John, we have the plans for the Church completed. Please get on with it." As the project progressed, periodic invoices were submitted to the owner and paid. A few weeks after the completion of the project, Joseph M. Markle, a New York construction attorney, visited the office of the company president. He showed Joe Markle two checks received from Mr. Rockefeller. The first was the final partial payment for the project. The second was a check for $50,000 with an inscription, "John, this is for a job well done!"[5]

After the Second World War, the production base in this country was in full steam. There was no Japanese or European competition and we had a voracious consumer society. People bought everything, regardless of quality. Construction boomed, and there was more work than contractors. It was a great time. Any management scheme worked, and it took talent not to be successful. Professors and management scholars could come up with any "new" management theory and say it was better. The country ran on automatic.

Enter the decades of the seventies, eighties, and nineties. World competition has America's industry outgunned. Figure A1-11 shows a recent *New York Times* comparison of national overall productivity among the United States, Federal Republic of Germany, and Japan. The construction industry productivity is even worse. What, if anything, can be done?

A case can be made that construction should be performed using *cost-plus* contracts. Recognizing that cost-plus contracts are very rarely managed competently and have justly earned the reputation for certain cost and schedule overruns, a properly managed cost-plus contract can be very successful. In fact, by using a team approach with global objectives, the adversarial relationships can be abolished, leadership from both the owner and the contractor can be brought to bear on the project, and productivity and quality can be optimized.[7] Unfortunately, this approach will never

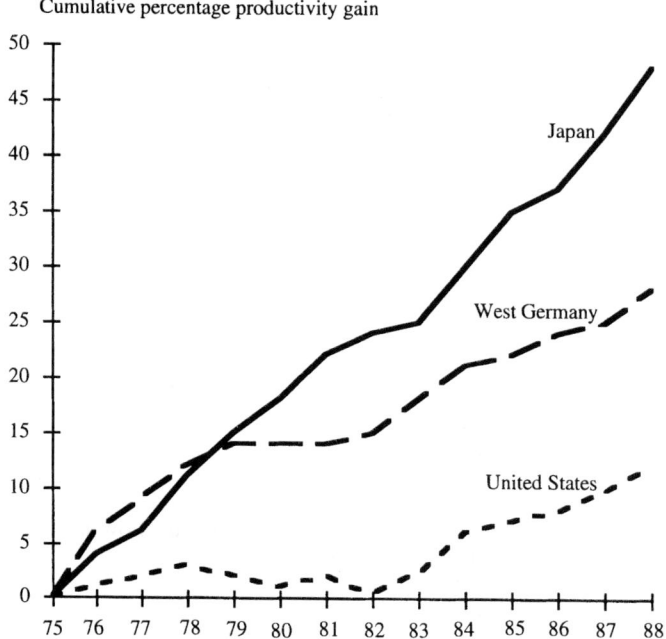

FIGURE A1-11 Productivity comparison among the United States, Federal Republic of Germany, and Japan.[6]

encroach on the firm-fixed price customs and traditions on which American procurement is largely based. Are there, then, other management concepts to explore?

The Deming Management Method

W. Edwards Deming, revered in Japan as the "Father of Quality Control," was the man who taught the Japanese how to produce goods of high quality at low cost. All of the more promising management approaches (Alternative Disputes Resolution or ADR, TQM, and partnering) are based on principles developed by Deming. They are the construction industry's attempt at implementing the Deming-developed management fundamentals. Deming has known for a quarter century that unless Americans changed their approach to productivity and quality, the economy would be destroyed. No one would listen.

Deming developed an entirely new concept of how to manage systems of machines and people. It is revolutionary, and it works. Deming's way is taught by few schools of management in the United States. Many things taught to managers in our schools and seminars about how to manage enterprises are contrary to Deming's way.

Deming's way has an interesting history. During World War II, U.S. statisticians, under the leadership of such men as Deming, Juran, and Shewart, pioneered new methods of control in the wartime industry. America's ability to produce large quantities of high-quality armaments using an unskilled labor force was the miracle that won the war. This miracle was due in no small measure to the methods introduced by these men. They were not just methods of statistical analysis. They represented the beginnings of an entirely new way to look at the operation of a factory.

But when the war ended, the mass markets of the United States were waiting to be filled, and skillful production management was not required. By 1950, many of the lessons of the war were discarded. New managers came to run new factories. They had little need, they believed, for methods to increase quality and productivity. They did not study them, and business schools did not teach them.

The U.S. market was a powerful deterrent to anyone who wanted to concentrate on high quality. Apparently insatiable markets swallowed goods of inferior quality. Americans accepted appliances of inadequate performance as though they represented the best that could possibly be done. Americans firmly believed that to increase the reliability of their appliances would require an increase in cost. And until Deming's way showed that this was false, they had no choice but to continue in their belief. By and large, they still do.

The years 1941 to 1945 taught Deming a profound lesson. He realized that a whole new philosophy of management was possible based on his experiences. And he saw that the basic idea could be put to work in any industry, even the service industries, and in ways not thought of during World War II.

A good manager in the construction industry understands scheduling, estimating, and quality control and assurance; is familiar with the latest techniques and software; can set up a management system within which employees, subcontractors, and suppliers can work; knows the construction business and how projects are constructed; directs the work through subordinates; gives clear assignments and sets standards of performance for employees to follow; sets goals and clear-cut objectives; appraises the performance of employees objectively, awarding the outstanding and educating, retraining, or dismissing the poor performer. Yet this individual is the same type of manager who provides the industry poor-quality projects behind schedule and overbudget.

A manager's job, according to Deming, is to provide a consistency and continuity of purpose for his or her organization and to always seek more efficient ways to meet its purpose. Unfortunately, corporate CEOs of today are concerned with about 40 quarters of earnings and 10 annual reports. The bottom line on the quarterly and annual report is all that matters. Students in management schools today look to join companies that pay the highest salaries, and there is no hesitation about leaving a company for more money. No loyalty is thought to be required to the organization. The management philosophy is that any manager can manage any organization, and that is wrong.

The Deming manager views the purpose as providing the highest-quality construction project at the least cost to the customer while staying in business and providing a system in which employees can perform their functions. The Deming manager is responsible for providing an ever-improving system, whereas employees have the responsibility for working within that system.

This manager knows that there is great variability within the system, especially with the construction industry's diversity. This manager also knows that the only means for improvement of and within that variability is with the workers themselves. There are countless ways for the system to go out of control, decreasing quality and increasing cost. Some of these occur randomly; some are caused by the system. For the manager and the workers to work together, they must regard the system the same way and speak a common language. For example, a Deming-method manager must learn some statistics and teach it to the workers. Thus, all employees keep their own statistics. Equipment operators keep track of their cycle times, truck drivers keep their own control charts, and first-line supervisors keep trend lines. All employees look for trends and correlations with other events—usually events beyond their control. Groups of employees meet from time to time to discuss trends and ways to improve the quality of the project and their group productivity. They provide data to the manager who attempts to improve the system based on worker input. Everyone in the system is involved in studying it and proposing how to improve it. Using this process, employees see that imposing work standards is a "dumb" idea in that it inhibits their ability to improve the system. "They will not need to manage by objectives because they will be engaged in consistently redefining their objectives themselves and recording the performance of the system."[8]

Workers may fear that speaking out might make them look bad to their supervisors and peers. They must be made to understand that it is axiomatic that approximately 80 percent of the problems are with the system, whereas only 20 percent are due to the workers. Thus, the workers should never fear speaking out about problems. The manager must cultivate this attitude with the workers.

There is a leadership principle taught by the U.S. Army and understood by Deming: Everyone wants his or her organization to perform well and works to improve its system. This is true of workers. They want to do their jobs and help improve the system.

The Deming manager offers the natural basis for team building and strives to prevent adversarial relations. The U.S. system of management, using almost any of the management theories, presumes that the relationship between the worker and the boss is adversarial. Thus, bosses have been trained to keep workers at arm's length so as not to lose their objectivity in judging their performance. A Deming manager and worker work closely together in a pleasant working atmosphere.

The Deming management philosophy is based on the idea (Deming's chain reaction) shown in Fig. A1-12. Improving quality reduces rework, thereby increasing productivity and reducing cost—the antithesis of the perception by most in the construction industry.

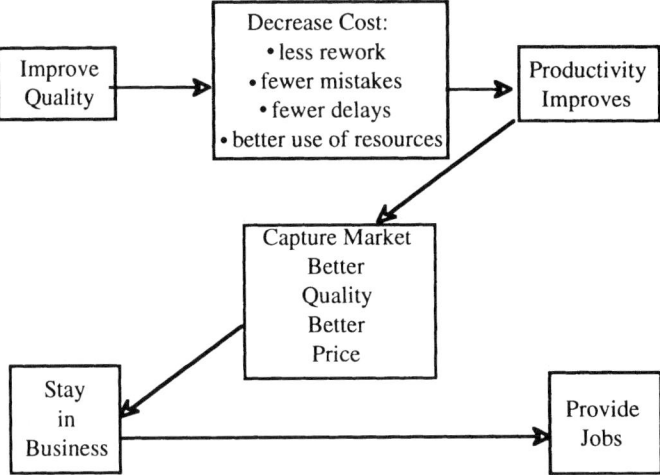

FIGURE A1-12 Deming's chain reaction.

The following is a condensation of the 14 points of the Deming management process[3]:

1. Create constancy of purpose toward improvement of product and service, with the aim to become competitive, stay in business, and provide jobs.
2. Adopt the new philosophy. We are in a new economic age. Western management must awaken to the challenge, learn their responsibilities, and take on leadership for change.
3. Cease dependence on inspection to achieve quality. Eliminate the need for inspection on a mass basis by building quality into the product in the first place.
4. End the practice of awarding business on the basis of price tag. Instead, minimize total cost. Move toward a single supplier for any one item, based on a long-term relationship of loyalty and trust.
5. Improve constantly and forever the system of production and service, to improve quality and productivity, and thus constantly decrease costs.
6. Institute training on the job.
7. Institute leadership. The aim of supervision should be to help people and machines and gadgets to do a better job. Supervision of management is in need of overhaul, as well as supervision of production workers.
8. Drive out fear, so that everyone may work effectively for the company.
9. Break down barriers between departments. People in research, design, sales, and production must work as a team, to foresee problems of production and in use that may be encountered with the product or service.
10. Eliminate slogans, exhortations, and targets for the workforce, asking for zero defects and new levels of productivity. Such exhortations only create adversarial relationships, as the bulk of the causes of low quality and low productivity belong to the system and thus lie beyond the power of the workforce.

11a. Eliminate work standards (quotas) on the factory floor. Substitute leadership.

11b. Eliminate management by objective. Eliminate management by numbers and numerical goals. Substitute leadership.

12a. Remove barriers that rob the hourly worker of the right to pride in workmanship. The responsibility of supervisors must be changed from sheer numbers to quality.

12b. Remove barriers that rob people in management and in engineering of their right to pride of workmanship. This means, inter alia, abolishment of the annual or merit rating and of management by objective.

13. Institute a vigorous program of education and self-improvement.

14. Put everybody in the company to work to accomplish the transformation. The transformation is everybody's job.

Can Deming's management philosophy be applied to the U.S. construction industry? The short answer might be that it cannot work in this industry. Or, as Brigadier General Jed Brown writes, "Perhaps in a post-industrial welfare state in decline, the industry has evolved to its necessary state."[9] If this is not the case, what stands in the way of the adoption of Deming's management fundamentals by the industry?

1. Government agencies at all levels will be hard-pressed to eliminate the firm-fixed-price, low-bid contract and it may not be in the public interest to do so. The Department of Transportation in the state of Florida has tried awarding contracts on the basis of a combination of low bid and efficient scheduling.

2. An even more serious roadblock is the way construction business is done. The worker's loyalty is less to the contractor or subcontractor than to the labor union. The worker is more likely to be responsive to the shop steward than to the craft superintendent.

3. A construction project is usually one of a kind. One does not throw away the first one completed and improve the process on the following projects. Deming's management methods were aimed at the manufacturing industries. Construction quality control is carried on throughout the project.

4. Finally, except in rare instances and certain industries, an owner and a specific construction company do not work together constantly. It is impossible to develop a single-party relationship with the bidding process required by law, custom, and regulation for governmental agencies.

There are other obstacles to the applications of Deming's philosophy in the construction industry. But for now, we will examine the 14 points one by one.

The 14 Points and the Construction Industry

Consider in turn the public owner, the private owner, the designer, and the contractor. Construction managers and other types of owner representatives are considered as owners or contractors, depending on the type of contract. For instance, a construction manager with a cost-plus contract may be considered as an owner, whereas a construction manager holding a guaranteed maximum price contract may be considered a contractor.

Point 1. It is certainly possible for a contractor's organization to create a constancy of purpose toward improvement in the quality of the constructed project, with the aim to become more competitive, stay in business, and provide jobs. The contractor's mindset must be changed from the conviction that improved quality means decreased productivity. It has been shown that up to 11 percent of a project cost is due to rework.[10] Dr. Deming asserts that quality improvement will bring productivity improvement. How is this achieved in a contractor's organization? *Leadership!* It requires monomaniacal drive to ensure that the necessary attitude will prevail throughout the organization. Public and private owner organizations, with intensive leadership from the top, can also create an atmosphere and attitude for constantly improving quality. Attitudes will have to change so as to regard the contractor and themselves as part of the construction team and not as adversaries.

Point 2. Adopting the philosophy of point 1, while seemingly the most simple point, may in fact be the most difficult. The key here for both contractors and public and private owners is leadership from the top and intensive training. The old habits are difficult to change. Integrity, trust, and cooperation are not traits of the industry. In partnering efforts, relatively extensive workshops are held in an attempt to foster trust and better working relationships. It is unfair for management to expect employees to change unless they are continuously trained in the new concepts. Point 2 is especially difficult for the contractors and subcontractors. Through training craft supervisors, educating the union leadership, and convincing the craftspersons that constant quality of the constructed project depends on the craftsperson's effort to continually improve the product. Since the trades know the work, it is they who should provide the most ideas for improvements.

Point 3. Controlled and discretionary inspections are a part of the industry and will not, in all probability, be eliminated from contract specifications. However, if the philosophy in points 1 and 2 are accomplished, inspections should become a formality. The constant striving for quality improvement will be what determines the quality of the product, not mass inspections.

Point 4. Public agencies are not going to end the practice of awarding business on the basis of the lowest responsible bidder. Nor will they move towards single suppliers. However, taking this as a given, can a contractor organization implement this point? The answer is yes. It makes infinite sense for a contractor to develop a long-term relationship with subcontractors and suppliers. The awards, while primarily considering price, should also consider quality, responsiveness of delivery, etc. A relationship based on loyalty can be developed, given the leadership from both organizations.

Point 5. Improving constantly and forever the system of production and service is the purpose of management in any type of organization. Most involved in the management of construction would probably say that they are doing this now, and they might be right. Regardless, there is no reason that a construction organization cannot apply point 5.

Point 6. It is absolutely essential that training be improved in the construction industry. There is a paucity of training in every organization associated with the construction industry. There must be more training, not only in the crafts, in construction management techniques, and in safety, but also in the application of these 14 points and personnel management and leadership.

Point 7. The supervision of management in the construction industry is in need of overhaul as well as supervision over craftspersons. The supervision must provide leadership stressing productivity through quality. The aim of supervision is to improve the system, removing the impediments that prevent construction workers from improving the quality of their work.

Point 8. This point requires intensive leadership and training in that it is extremely difficult to drive fear out of a worker's mind. Fear, in this sense, is not fear of injury but fear of peer and supervisor rejection. Workers and supervisors alike must understand that most problems are with the system, not with the workers. Leadership is necessary to ensure that the organization is a place where the workers can help improve the system by talking to management.

Point 9. Much improvement could be made if only point 9 were instituted in a contractor or owner organization. In owner agencies, different departments actually have different and sometimes opposing objectives. The same is true in contractor organizations. Construction should become a team effort once again.

Point 10. Slogans, exhortations, and targets are usually not abused in the construction industry.

Point 11. Construction supervisors do not often use leadership instead of quotas.

Point 12. Barriers that rob the worker of the right to pride in workmanship should be removed. In the U.S. construction industry, there are many barriers to pride in workmanship from peer pressure, union pressure, and pressure to meet a schedule. These can be removed with a change in attitude of both supervisor and craftspeople.

Point 13. Programs of education and self-improvement can and should be implemented.

Point 14. The leadership must communicate throughout the organization to ensure that everyone is on board for the transformation.

On a point-by-point basis, there is no reason that a contractor organization cannot follow Deming's 14 points to improve productivity through quality improvement. A key ingredient, though, requires Congressional action. Incentives for investment must be put back into the tax laws. With the implementation of the Tax Reform Act of 1986, the recession in the construction industry has made contractors bid jobs with no profit and little overhead. With economic conditions such as this, all energies are used just to try to stay in business. There is no energy left for change. The recession must end for contractors to exert the leadership necessary for change.

CONCLUSION

From the foregoing discussions, it appears that management philosophies in the construction industry have been focused, if not on the wrong problems, then certainly not on the complete problem. The management focus in construction in the recent past was on organizing, planning, directing, and controlling projects and people with emphasis on scheduling, estimating, and cost, schedule, and quality control. To improve the industry, the focus must move towards leading people to develop a system that will continually improve the constructed product. This means that the system must help the worker develop pride in workmanship and pride in productivity improvement through the elimination of rework.

The purpose of this chapter has been to start the construction industry thinking of how to change from its present adversarial nature to an industry that prides itself in the quality of its work and the economics of its productivity. Several approaches were discussed. ADR is helpful but does not go all the way. TQM is a step in the right direction, but incentive barriers keep the TQM approach from becoming universal, although it stems directly from the Deming management approach. Partnering, while pleasant when it works, does not get to the root of the problem. MBE (Management By Exception), MBO (Management By Objective), and other currently used management buzz words are not the answer to improving construction productivity. Although a suitable conclusion is difficult to reach, this chapter advocates that what is required is another revolution just like the one in Japan caused in part by W. Edwards Deming. This will be achieved through leadership and owners of construction companies demanding quality and paying for it as well as from contractors and subcontractors recognizing that quality improvement implies productivity improvement and from Congress revising the tax laws to encourage investment in building, research, and investment.

REFERENCES

1. Building Research Board, Inspection and Other Strategies for Assuring Quality in Government Construction, National Research Council, Washington, D.C., 1991.
2. W. M. Hayden, "Management's Fatal Flaw: TQM Obstacle," *J. Mgmt. Engrg.*, ASCE, vol. 8, no. 2, 1992, p. 122.
3. W. E. Deming, *Out of the Crisis*, MIT, CAES, Boston, 1982.
4. Construction Industry Institute, "In Search of Partnering Excellence," Special Publication 17-1, July 1991.
5. J. M. Markle, private conversation.
6. "Productivity Comparisons Between the U.S., the Federal Republic of Germany, and Japan," *New York Times*, Jan. 26, 1992.
7. F. H. Griffis and F. M. Butler, "A Case for Cost-Plus Contracting," *J. Const. Engrg. Mgmt.*, ASCE, vol. 114, no. 1, 1988, p. 83.
8. M. Tribus, "Deming's Way," Center for Advanced Engineering Study, MIT, Boston, 1989.
9. Gerald Brown, personal letter to the author, 1992.
10. K. Davis, W. B. Ledbetter, and J. L. Burati, "Measuring Design and Construction Quality Costs," *J. Const. Engrg. Mgmt.*, ASCE, vol. 115, no. 3, 1989, p. 385.

CHAPTER 2
HEAVY CONSTRUCTION CONTRACTS

William J. Kennedy, Esq.
Dechert, Price & Rhoads
Philadelphia, Pennsylvania

NEED FOR GOOD CONTRACT DOCUMENTS

A heavy construction project needs drawings and specifications that clearly and precisely call out the work to be accomplished. It is equally important that the powers, duties, and responsibilities of the principal participants in the project be set forth with the same degree of clarity and precision in an integrated set of project contract documents.

Contractors, owners, engineers, and construction managers all have their roles to play in bringing about the successful completion of a heavy construction project. In the best projects, they work together in a cooperative spirit to produce the mutually desired result. The concept of *partnering*, which is aimed at pushing the spirit of mutual cooperation from the top of the organization down to the workers in the field, is being tried on many projects. But the construction industry still has to operate in America's litigious society in which someone else "must" pay for any loss or damage you suffer. In administering or implementing a heavy construction project, it is all too easy to misstep onto someone else's turf, and adversarial attitudes still lurk beneath the surface. It is critical that the project players enter into contract documents that clearly and comprehensively allocate the players' respective duties and responsibilities.

WHAT ARE GOOD CONTRACT DOCUMENTS?

While it is difficult to define in the abstract what makes a contract document "good" for a heavy construction project, there are certain characteristics that generally acknowledged good contracts share: (1) carefully considered, (2) expressed clearly, (3) time-tested, (4) comprehensive, (5) fair, (6) balanced, and (7) applicable to the elements of a heavy construction project.

The contract represents the meeting of the minds of the parties about what project responsibilities are being accepted by each. In a heavy construction project, where there are a number of parties with different roles and interests coming from different disciplines, it is absolutely necessary that the contract not contain ambiguous or conflicting language or fail to address important aspects of the parties' working relationship. If the parties cannot clearly express their intentions, it is doubtful that they can carry them out without conflicts.

It is helpful if the language has been previously construed by the courts to establish its meaning. If not, then successful usage in previous projects is helpful. This should not be taken as an endorsement of thoughtlessly inputting language from a prior project. Certainly, if the language was tested and it worked on a prior project, it is a good sign. Obviously, if it was a source of confusion, conflict, or claims in a prior project, it needs to be changed or clarified for the next project.

It does little good to partner a project if the contract documents are tilted to favor one party to the prejudice of the other. Contracts that allocate risks and disclaim responsibilities without realistic regard for the roles, capabilities, expectations, and resources of the participants invariably push the unfairly burdened party to react in ways that produce misunderstandings, delays, and claims. In addition, there are many activities, procedures, and protocols that are common to virtually all heavy construction projects. The contract documents should recognize and account for these as well as any aspects that may be unique to the particular project.

USE OF STANDARD FORMS

Many owners and contractors have developed their own contract forms reflecting their own particular business philosophies and requirements. However, there are several advantages to using one of the national standard forms of construction contract either as printed or at least as a starting point to be customized to particular project demands. One advantage is that the standard contract language resulted from the cumulative lessons learned from large numbers and various types of projects. In many instances, the language will already have been court tested. Further, with a national standard contract document, the project participants know what to expect contractually and are able to plan and price their parts of the project accordingly.

EJCDC CONSTRUCTION-RELATED CONTRACT DOCUMENTS

The author believes that the Engineers Joint Contract Documents Committee's (EJCDC) series of construction-related contract documents embodies all the elements of a good heavy construction contract. EJCDC contract forms are generally recognized as articulating in its most complete form nationally accepted language describing the various functions, duties, and responsibilities of the participants in a heavy construction project.

EJCDC contract documents are developed and published by a special committee of the National Society of Professional Engineers, American Consulting Engineers Council, and American Society of Civil Engineers working in cooperation with delegates from the Association of General Contractors and with input from various owner and legal groups. The EJCDC documents are designed for use in engineered construction projects. Because the delegates who develop the contract language come from various parts of the country, from small-, medium-, and large-size companies or firms and from various professional disciplines or contracting specialties, the documents reflect a balanced view of nationally accepted practice in engineered construction. Where practices vary on a regional basis or by type of construction, the delegates attempt to incorporate the best practice, applying general tests of fairness and practicality in making their choices. Alternate approaches typically are set out in the guide to Supplementary Conditions or in commentary documents, thus providing the EJCDC series user with a degree of flexibility in using the forms.

EJCDC's Construction-Related Series of documents, its No. 1910-8 series, is updated on a seven-year cycle. The most recent edition was issued in 1990. Copies may be purchased by contacting any of the national headquarters of the national engineering societies previously referred to. The documents contemplate the traditional tripartite approach to heavy construction with responsibilities divided appropriately among the owner, the engineer, and the contractor. (See comments later in this chapter concerning multiprime projects, design-build projects, and contracts involving a construction manager.) Even if readers choose to use one of the other national organization's forms (for example AIA or AGC) or to develop their own contract forms, the author recommends obtaining a set of the EJCDC forms because they constitute a remarkably comprehensive checklist of what issues and items need to be addressed contractually in a heavy construction project. This chapter will refer to and rely heavily on the EJCDC approach to contract documents in discussing such issues.

BIDDING PROCEDURES AND DOCUMENTS

Publicly funded heavy construction projects generally are required by law to be competitively bid. Private owners have the option of requiring competitive bids or engaging in private negotiations with one or more contractors. In order to preserve a level playing field and encourage the best pricing, public bidding requirements are usually rigidly adhered to and governed by applicable laws and regulations. While the private project owner has much more flexibility in the way it deals with bidders and awards the contract, the private owner is well advised to adhere generally to public bidding principles and to weigh the potential advantages and consequences of doing so when it does deviate.

Although specific requirements may differ from jurisdiction to jurisdiction, most competitive bids will typically require the use of the bidding documents described subsequently. While the practice is not universally consistent on this point, the better view is that the bid documents are not part of the contract documents. This is because the contract documents supersede the bid documents after the award is made and the agreement is signed. It is the contract documents, not the bid documents, that govern the relationship between the owner and the contractor on the project. Nevertheless, what was disclosed in the bidding documents or said in the bid may be evidence of the parties' intentions and expectations. It is recommended that the contract documents specify, one way or the other, whether the bid form is part of the contract between the parties. It is also possible to incorporate specific parts of the bid into the contract. For example, it may be most convenient to attach a lengthy schedule of unit prices to the contract as an exhibit rather than retyping it at length.

BIDDING DOCUMENTS

The Advertisement or Invitation to Bid are brief descriptions of the project, the bidding schedules, and where more detailed information may be obtained. They are published or distributed in order to attract prospective bidders. They should contain at least enough information about the project to enable prospective bidders to decide whether to invest in obtaining copies of the bidding documents themselves.

The Instructions to Bidders are the rules of the game which the players (bidders) must follow in order to submit an acceptable bid. EJCDC publishes the Guide to the Preparation of Instructions to Bidders, No. 1910-12 (1990 ed.), which is not a form but a thoughtful and flexible series of guidelines and suggestions, adaptable to virtually any project, covering the procedures and requirements for bidding the typical heavy construction project, including the following.

Qualification of Bidders. Builders are generally required to supply information about their past job experience, manpower and equipment capabilities to do the job, financial capacity and data, and, sometimes, proposed principal subcontractors, suppliers, and equipment manufacturers. Owners generally reserve the right to obtain more detailed information on these matters from the apparently successful bidder. Because the information sought is often commercially sensitive, it is important that it be treated by the owner in strict confidence.

Examining Contract Documents and Site Prior to Bidding. Bidders are required to examine all portions of the Contract Documents, the physical conditions of the Site, and any legal requirements affecting the work prior to bidding. It is in everyone's interest that the contractor's bid take into account all of the site's problems as well as favorable conditions to the extent they can be reasonably ascertained in advance. It is particularly important that the requirements of the Instructions concerning investigation of subsurface conditions (both natural and human-made facilities) correspond with the language of the Contract Documents in allocating responsibility for "subsurface surprises." As will be explained later in this chapter, the EJCDC documents follow a modified federal construction contract approach in this regard.

Addenda. The procedure for issuing and acknowledging addenda that clarify, correct, or change the bid requirements prior to bid opening needs to be explained. In contrast with the AIA approach, EJCDC takes the position that in bid contracts no substitutes or equals should be considered during the period between bid opening and the effective date of the Agreement. Such changes are unfair to the bidders who were not selected as the apparent low bidder. Bids should be evaluated and the contract awarded on the basis of the materials and equipment described in the bidding documents.

Schedule. The bidding instructions should call attention to the chronology or schedule of dates of events including: (1) any bidders' conference, (2) bid opening, (3) notice of award, (4) notice to proceed, (5) effective date of the agreement, (6) time for substantial completion, and (7) time for final completion. EJCDC's Instructions include a suggested schedule of events from the day of bid opening through the day the contract time starts to run.

Bid Security. The requirements of bid security—cash, letter of credit, bid bond—to guarantee that the bidder will enter into a contract with the owner in accordance with its bid if the owner accepts that bid needs to be spelled out. EJCDC publishes two forms of Bid Bond (1910-28C and 28D). One calls for the forfeiture of a fixed penal amount under breach by the successful bidder. The other provides for the forfeiture of the penal amount or the damages suffered by owner by reason of the breach, whichever is less. Provision needs to be made in the instructions for the timing of returning bid security to the apparently unsuccessful bidders and the successful bidder.

Subcontractors and Suppliers. EJCDC Instructions include a helpful discussion of issues ranging from the requirement to designate those subcontractors and suppliers whom the owner retains the power to accept or reject to questions of whether or not a successful bidder will be held to its bid price if the owner requires a substitute subcontractor or supplier for the one on which the bid was based.

Bidding Procedures. The rules for the submission and award of bids should be explained. These include items such as:

1. Form and format of the bid
2. Whether telegrams or electronic facsimile bids will be accepted
3. How "mistakes" in bids will be dealt with
4. Whether bids will be opened publicly
5. What factors other than price will be evaluated
6. Whether the owner will conduct additional investigations of or engage in negotiations with the apparently successful bidder after bid opening but before award

The EJCDC Instructions discuss the ramifications and make suggestions with respect to these typical bid procedural matters.

The Bid Form is the actual bid itself. It should be standardized so that all bidders submit their bids in the same format covering the same subject matter in the same sequence. This puts all bidders on an equal footing and facilitates evaluation and comparison of bids. The bid language should contain an offer by the bidder to perform the work in accordance with the prescribed and specifically identified specifications and contract document terms. It is not necessary to print out the specifications and contractual provisions in the bid form at length. It is sufficient to reference them. Generally, the Bid Form will also include appropriate spaces for items such as: (1) price, including, where applicable, cost of the work, unit prices, alternatives or combinations, guaranteed maximum price, and basis for contractor's fee; (2) times for substantial completion and final completion; (3) contractor's representations concerning subjects such as prebid site investigation, subsurface investigation, disclosure and resolution of conflicts, ambiguities or omissions in the bidding documents, and noncollusion; and (4) bidders undertaking to keep the bid open and sign the agreement with the owner within prescribed time limits or forfeit the required bid security.

The EJCDC publishes a Suggested Bid Form and Commentary (EJCDC No. 1910-18, 1990 ed.), which may be conveniently adopted for either negotiated or public bids of heavy construction projects. The Commentary explains how to conform the Bid Form to deal with most situations likely to arise under varying legal, regulatory, and project-specific requirements.

THE AGREEMENT BETWEEN OWNER AND CONTRACTOR

The Agreement should contain the essential commercial terms, such as time and price, agreed on by the owner and contractor for the heavy construction project. This is in contrast to the terms that set forth the details of the duties and responsibilities of the parties who are carrying out the project, which are typically contained in the General Conditions and Supplementary Conditions. EJCDC publishes two forms of standard agreement between the owner and the contractor. The first is on the basis of a stipulated price (EJCDC No. 1910-8-A-1, 1990 ed.). The second form is similar but is on the basis of a cost-plus arrangement, including the option for a guaranteed maximum price (EJCDC No. 1910-8-A-2, 1990 ed.).

The effective date of the Agreement is important in the EJCDC system because the contract times start to run from that date, and the timing of subsequent project events are keyed to the start of the contract times.

The Agreement identifies the owner, the contractor, and the engineer as well as the project itself and provides that the contractor shall complete all the Work specified or indicated in the Contract Documents. The Agreement should also expressly identify each of the documents that are comprised by the Contract Documents, and copies of all such documents should be attached to the Agreement. In this way, all of the documents that constitute the entire agreement between the owner and contractor concerning the Work for the Project are collected in one project book. The EJCDC forms list a dozen different categories of contract documents typically issued in a heavy construction project.

As previously indicated, the Agreement should set forth the times for substantial completion and final completion of the Work. This may be expressed either by dates or by specifying the number of days in which completion shall be achieved. If liquidated damages are to be assessed for failure to complete the Work on time or for delay in achieving significant interim milestones on the way to completion, they should be specified in the Agreement. It is important that the amounts stipulated as liquidated damages bear a reasonable relation to the estimated damages the owner may actually sustain if the Work is late. Otherwise, there is a risk that the liquidated damages would be held to be an unenforceable penalty.

The EJCDC forms of Agreement provide several different ways of determining the contract price that may be applied separately or in combination for portions of the project. One typical method is the stipulated price commonly referred to as a lump-sum contract. The lump sum provides the owner with the greatest certainty as to the cost of the project and exposes the contractor to the greatest risk of loss. It is therefore necessary to define precisely and completely what is included in the lump-sum price, including items such as labor, materials, equipment, installation, overhead, profit, and cash allowances. Since the contractor is exposed to the risk of overruns, it is reasonable to assume that the contractor will attempt to cover the risk by including a contingency factor in the lump-sum price.

A second frequently used compensation method for all or part of a heavy construction project is unit price work. EJCDC's forms of agreements are set up to identify separate items of unit price work and an established unit price for each. The actual quantities and the classifications of unit price work actually performed by the contractor are determined by the engineer. The Agreement lists estimated quantities of each item for purposes of comparing bids and establishing the initial contract price. The EJCDC General Conditions provide that either the owner or contractor may make a claim for an adjustment in the contract price if the actual quantity of unit price work differs materially and significantly from the estimated quantity in the Agreement and there is no corresponding adjustment with respect to any other item of work.

The third variation of contract price in EJCDC's family of documents is the cost of the work plus a fee method. This is often coupled with a provision for a guaranteed maximum contract price to protect the owner against price overruns. Frequently, owners will also provide a bonus or sharing-of-savings provision as an incentive to bring the work in at as low a cost as is practicable. Cost of the Work is determined by ground rules set forth in the General Conditions which establish which costs are eligible to be included as the basis for a fee. The contractor's fee is determined either as a fixed fee or a specified percentage or percentages of various portions of the Cost of the Work. For projects where varying percentages are selected, there may be one percentage for work done directly by the contractor's forces, a higher percentage for specialized or risky work, a lower percentage for subcontracted work, and a still lower percentage applicable to the cost of materials and equipment supplied.

Changes in the work will be handled by change orders which should set forth any applicable increase or decrease in the amount of the contractor's fee and guaranteed maximum price.

Other financial matters covered in the Agreement include the procedures for progress payments, retainage, and final payment and whether and what interest is payable if payments are not made when due.

The last project-specific item typically covered in the Agreement is to repeat the contractor's representations as initially set forth in the bid documents previously described. Although these representations could be incorporated by reference to the bid, it is better practice to spell them out in the Agreement to have them all in one document for ease of reference and to avoid later arguments that some change in circumstances between the time of the bid and the signing of the Agreement made them no longer applicable.

While the Agreement for any heavy construction project will need to be tailored to particular project needs, the EJCDC forms provide a useful format and language to fit most projects.

GENERAL CONDITIONS

The General Conditions spell out the administrative and procedural aspects of the project as well as the roles, risks, and responsibilities assumed by the parties or assigned to the key participants, such as the engineer or construction manager. The EJCDC Standard General Conditions of the Construction Contract (EJCDC No. 1910-8, 1990 ed.) follows the traditional tripartite project delivery method: (1) the contractor is responsible for implementing the work; (2) the engineer who prepared the drawings and specifications is responsible for reviewing the work to ascertain that it is being implemented in accordance with the drawings and specifications; and (3) the owner is responsible for furnishing the site and paying for properly performed work in accordance with the terms of the agreement but is not responsible for the contractor's methods, sequences, or procedures of accomplishing the work. Within this framework there are a myriad of respective rights and responsibilities detailed by the General Conditions corresponding to various phases and events of the typical engineered heavy construction project. Key topics addressed by the EJCDC Standard General Conditions are discussed in the following.

Article 1—Definitions. EJCDC Standard General Conditions contain 45 separately defined terms. These definitions are critical because the meaning or sense of many of the provisions of this contract document often turns on the definition of a key word or term that has been used. However, the complete and exhaustive set of defined terms permits a laudable economy of expression in setting forth often complex concepts or relationships. Of all the defined terms, perhaps the most persuasively used yet difficult to define with comprehension and precision is the word *Work*, which when spelled with an initial capital is what the contractor is to provide. The EJCDC definition of Work is:

> The entire completed construction or the various separately identifiable parts thereof required to be furnished under the Contract Documents. Work includes and is the result of performing or furnishing labor and furnishing and incorporating materials and equipment into the construction, and performing or furnishing documents, all as required by the Contract Documents.

Thus Work focuses on the completed construction or completed parts of the construction. It is the result of the contractor's performing labor and installing materials and equipment into the construction. But it goes beyond the physical construction and includes also other services and documents the contractor is required to perform or furnish, for example, discharging liens or honoring warranties, or providing required bonds, insurance, or project equipment operating manuals. Work is differentiated from the *project,* which means the entire planned construction of which the contractor's Work may be only a part. Moreover, when *work* is used without an initial capital it indicates services, materials, or equipment to be performed or furnished by someone other than the contractor, such as a subcontractor, supplier, or another multiprime contractor.

Another frequently used defined term is *defective,* which, because it is an adjective, is the only defined term in the EJCDC General Conditions that does not appear with an initial capital. It is always printed in italics to indicate it too has a special and specific meaning. When it is used it is always used in conjunction with Work and refers to Work

> that is unsatisfactory, faulty or deficient in that it does not conform to the Contract Documents, or does not meet the requirements of any inspection, reference, standard, test or approval referred to in the Contract Documents, or has been damaged prior to Engineer's recommendation official payment (unless responsibility for the protection thereof has been assumed by Owner . . .).

The risk of *defective* Work is on the contractor unless it has been expressly shifted to the Owner.

Other definitions in the EJCDC system will be covered in the context of the particular topics discussed subsequently and are indicated by initial capitals.

Article 2—Preliminary Matters. This Article deals with a number of significant matters that should occur or be resolved before construction starts. These matters include: (1) the delivery of Bonds and insurance certificates; (2) the reporting of any conflicts, errors, ambiguities, or discrepancies discovered within the Contract Documents or between the Contract Documents and field conditions or applicable Laws or referenced standards or codes of a technical society (the Contractor is under a continuing duty to check and verify pertinent provisions and figures and field measurements, to report discrepancies and not proceed with the Work until discrepancies are resolved; (3) the scheduling of a preconstruction conference; and (4) the submission of schedules for the engineer to review and approve including a progress schedule indicating starting and completing times for the various stages of the Work, including any specified Milestones; a schedule of values for the various component parts of the Work; and a schedule of required Shop Drawings and Sample submittals including the times for submitting, reviewing, and processing each such required submittal. The Engineer will review the progress schedule to see that it provides an orderly progression of the Work to completion within the Contract Times. However, such approval will not impose responsibility on the Engineer for sequencing, scheduling, or progress of the Work. The Engineer will review the schedule of values to see that it breaks down the Work in sufficient detail to serve as the basis for progress payments during construction and that it includes quantities and prices of items of work, including an appropriate amount of overhead and profit for each item, that aggregates the Contract Price. The Engineer should not approve a progress payment until the schedule of values has been approved. The Engineer will review the schedule of Shop Drawings and Submittals to see that it includes only the complete list of submittals that the Engineer will require and that sufficient time is provided to review and process the required submittals. The Engineer should not review a submittal that is not listed in the approved schedule or an amendment thereof.

Other preliminary matters covered include administrative and procedural matters ranging from the number of copies of the Contract Documents that will be provided to provisions for when the Contract Times commence to run and when the Contract shall start to perform the Work. The defined term *Contract Times* is plural because it encompasses two distinct time periods: the number of days or dates to achieve Substantial Completion and the number of days or dates to complete the Work so it is ready for final payment.

Article 3—*Intent, Amending, and Reuse of Contract Documents.* This Article provides rules and assistance for interpreting and amending the Standard General Conditions. Some of the helpful rules of interpretation include the following:

> Any Work that may reasonably be inferred from the Contract Documents or from prevailing custom or trade usage as being required to produce the intended result will be furnished whether or not specifically called for.
>
> Words will be interpreted in accordance with their well-known technical or construction industry meaning.
>
> No provision of standards, specifications or order of technical societies or associations shall be effective to change the duties or responsibilities set forth in the Contract Documents.

There are six ways in which the Standard General Conditions may be modified:

1. Formal Written Amendment
2. Change Order
3. Work Change Directive (permits changed Work to proceed without a Change Order under the expectation that a Change Order will be subsequently issued
4. Field Order
5. Engineer's approval of a Shop Drawing or Sample
6. Engineer's written interpretation or clarification

Article 4—Subsurface and Physical Conditions. Subsurface surprises are rarely pleasant ones and frequently lead to claims. The EJCDC documents follow the federal approach of basically placing the risk of the unknown on the Owner. However, the EJCDC Standard General Conditions place a heavy emphasis on expressly identifying in advance what is known and on what information a Contractor may rely. The Contractor may rely on identified *technical data* contained in reports of subsurface tests to subsurface structures that were utilized by the Engineer in preparing the Contract Documents. Such technical data is identified in the Supplementary Conditions. The Contractor may not rely on: (1) the completeness of such reports and drawings for Contractor's purposes, (2) information and opinions not identified as technical data, or (3) Contractor's interpretations of the technical data. An equitable adjustment in the Contract price or Contract Times will be allowed to the extent caused by a subsurface surprise provided the subsurface condition (1) is of such a nature as to require a change in the Contract Documents or (2) differs materially from that indicated in the Contract Documents, or (3) is of an unusual nature and differs materially from conditions ordinarily encountered and generally recognized as inherent in work of the character provided for in the Contract Documents, (4) is of such a nature as to establish that the identified technical data is materially inaccurate. However, no equitable adjustment will be allowed if the Contractor knew about or reasonably should have discovered the subsurface surprise as the result of examinations or investigations required by the Bidding or Contract Documents. The Contractor will also be entitled to an equitable adjustment if Underground Facilities not indicated in the Contract Documents are encountered which the Contractor did not know about or could not reasonably have anticipated.

The Standard General Conditions also provide the procedure and responsibilities for dealing with Undisclosed Asbestos, PCBs, Petroleum, Hazardous Waste, or Radioactive Material. On discovery, the Contractor stops Work with respect to the hazardous conditions, the engineer initially assesses the problem, and the Owner is responsible for removing any such hazardous materials that are not identified in the Contract Documents as being within the scope of the Work.

Article 5—Bonds and Insurance. This Article sets forth the requirements for performance and payment bonds, Contractor's and Owner's liability insurance, builder's risk property insurance, and boiler and machinery insurance for the heavy construction project. The insurance market is volatile. Coverages available today may not be available tomorrow or new coverages may be underwritten that are not known today. Therefore, although the EJCDC General and Supplementary Conditions provide a comprehensive compilation of customarily provided coverage, it is best to rely on an insurance expert to tailor the insurance requirements of the project.

Sureties must be named in the current list of companies acceptable to post federal bonds. Further requirements for size or creditworthiness of the surety are covered in the Supplementary Conditions.

The Contractor's liability insurance is required to include workers' compensation, third-party bodily injury and property damage coverage, automobile liability, completed operations, contractual liability, and personal injury liability coverage. Specific limits, deductibles, aggregates, and other coverages required are as identified in the Supplementary Conditions. It is recommended that aggregate limits be on a per-project basis so that claims on the contractor's other projects do not diminish the amount of coverage available for the project in question. Contractor's liability insurance is required to remain in force through final payment and at all times thereafter when the Contractor is correcting defective Work. Completed operations coverage, and any coverage written on a claims-made basis, are required to remain in effect for at least two years after final payment.

The EJCDC General Conditions require that property insurance be maintained on the full replacement value of the Work. The coverage is to be written on a Builder's Risk *all-risk* or comparable form covering physical loss or damage to the Work, temporary buildings, false work, Work in transit, materials and equipment stored at the site or other agreed on location that has been included in an Application for Payment approved by the Engineer, and expenses incurred in the repair or replacement of insured property (including engineers' and architects' fees). The EJCDC form lists typical perils insured against and the Supplementary Conditions are used to expand the provided list.

The EJCDC form requires the Owner to supply the property insurance because many Owners have property insurance programs in place and feel that the transition from covering property under the Builder's Risk to covering it under their permanent property insurance is facilitated if they provide the property insurance during construction as part of their integrated program. In some instances, however, an Owner may not feel competent to provide this specialized insurance or the Contractor may be able to get a better price. In other cases, particularly where there is public funding or a grant for the project, it is advantageous to include the cost of the property insurance in the Contract Price. In these cases, the Contractor will supply the required property insurance. EJCDC's Supplementary Conditions contain alternative language to accommodate this scenario.

The philosophy of the EJCDC insurance requirements is to attempt to transfer and limit the risks of loss and liability to the insurance policies to the fullest extent possible and thereby eliminate, or at least substantially reduce, the need for litigation among the project's participants themselves or with their insurers. To this end, the Contractor is required to include the Owner, Engineer, Engineer's Consultants, and others identified in the Supplementary Conditions listed as additional insureds under the Contractor's liability policy. Similarly, the Owner will have the Contractor, Subcontractor, Engineer, Engineer's Consultants, and others identified in the Supplementary Conditions listed as additional insureds under the Owner's property insurance. The theory is if all the primary players are insured under one policy, there will be little incentive for the insurance company of one player to drag the others into a liability suit since the payment will all come out of the same pocket anyway. The concept is extended even further with respect to the property insurance policy provisions under which a series of express waivers of rights is required. The insurers are required to waive their subrogation rights against any insured or additional insured and the insureds and additional insureds all waive their rights against each other for any loss caused by a peril covered under the property insurance policy. In addition, the Owner waives all rights to recover against the other project participants for loss due to business interruption or other consequential loss caused by fire or other perils required to be insured against under the property insurance provisions.

Article 6—Contractor's Responsibilities. This Article contains an extremely comprehensive description or listing of the Contractor's responsibilities on the heavy construction project, including, in part, the following items.

Supervision and Superintendence. The Contractor supervises, directs, and inspects the Work. The Contractor is solely responsible for the means, methods, techniques, sequences, and procedures of construction. However, the Contractor is not responsible for an error or omission in design or in an expressly required specified construction means, method, technique, sequence, or procedure.

Labor, Material, and Equipment. The Contractor shall provide competent and qualified personnel to carry out the Work and will maintain good discipline and order at the site at all times. The Contractor shall also furnish all materials, equipment, tools, utilities, services, and other facilities and incidentals necessary for completion of the Work. All specified warranties expressly run to the benefit of the Owner.

Substitutes and "Or Equals." The procedures for submitting and obtaining Engineer's approval of a substitute or *or equal* item of material or equipment or a substitute for an expressly specified construction method or procedure is spelled out in detail.

Subcontractors. The Contractor is fully responsible for scheduling and coordinating the Work of subcontractors and suppliers and for all of their acts and omissions. It is also made clear that nothing in the Contract Documents creates a contractual relationship between the Owner or Engineer and any subcontractor or supplier for the benefit of the subcontractor or supplier. The Owner and Engineer have the right reasonably to object to any subcontractor or supplier. In the event the Owner revokes a previously accepted subcontractor, a Change Order will issue to account for the cost differential to Contractor of providing a substitute.

Safety. The EJCDC Standard General Conditions spell out in language that is clear, precise and comprehensive that the Contractor is responsible for the safety of all persons on the Work site or who may be affected by the Work and for all property at or adjacent to the site. In a climate where both OSHA and plaintiffs' attorneys seek to impose site safety responsibility on the Owner or Engineer, it is critical to have the respective roles and responsibilities of the Owner, Engineer, and Contractor in regard to site safety clearly spelled out. EJCDC places the responsibility for site safety on the Contractor because the Contractor controls the site, the workforce, the materials and equipment, and the way the Work is done during construction. It is the Contractor who is in the best position to anticipate and take precautions to safeguard against safety problems. While the Owner and Engineer may owe certain statutory or regulatory safety obligations to their own employees while they are on the site, the responsibility for safeguarding all others, the overall responsibility for site safety, and the contractual responsibility as between the Owner, Engineer, and Contractor for the safety of any person on the site or affected by the Work remains squarely on the Contractor's shoulders.

Safety Representative. The General Conditions require the Contractor to designate a qualified and experienced safety representative at the site whose duty is the prevention of accidents and the maintaining and supervising of safety precautions and programs. This language is flexible enough to accommodate the trend of some insurance companies to require a specially trained safety professional at or responsible for the site or proposed legislation that seeks to mandate that result.

Hazard Communication Programs. Recent legislation and regulations require employers to notify their employees of hazardous materials they may encounter or be exposed to in the workplace. This burden becomes extremely complicated in multi-employer workplaces such as a heavy construction project site where employees of the Owner, Contractor, Engineer, Engineer's Consultants, Subcontractors, Suppliers, Insurers, and governmental inspectors may all be on the site at one time or another interacting with each other. While each employer remains responsible under law for providing material data safety sheets or other hazard communication information to its employees or exchanging such data safety sheets and communications with other employers at the site, the Contractor is assigned the responsibility of coordinating the required exchange.

Shop Drawings and Submittal Procedures. Although the Engineer's approval of a Shop Drawing may effectively supplement or authorize a variation in the

Work, Shop Drawings are not treated as Contract Documents. Their principal purpose is to fill in the construction-related details of the design. They also provide a means for the Engineer to review the Contractor's interpretation of the design's requirements and the result the Contractor plans to achieve. EJCDC traditionally has been sensitive to the problem that unintended, construction-related liability might be imposed on the Engineer as the result of the review and approval of Shop Drawings. This concern is addressed in two ways. The General Conditions provide that the Engineer's review is "only to determine if the items covered by the submittals will, after installation and incorporation in the Work, conform to the information given in the Contract Documents and be compatible with the design concept of the completed Project as a functioning whole as indicated in the Contract Documents." In other words, the Engineer's review and approval of Shop Drawings is strictly end-result oriented. By way of contrast, the EJCDC documents also emphasize that the Contractor's prior review of Shop Drawings before submittal is equally important and includes such construction-related considerations as verifying (1) all field measurements, quantities, dimensions, performance criteria, installation requirements, and similar information, (2) all materials with respect to intended use, fabrication, shipping, handling, storage, assembly, and installation pertaining to performance of the Work, and (3) all information regarding means, methods, techniques, sequences, and procedures of construction and safety precautions incident thereto. The Contractor is also responsible for coordinating Shop Drawing and Sample submittals with other Shop Drawings and Samples and requirements of the Contract Documents.

Warranty, Indemnity. The Contractor is required to give a general warranty and guarantee that the Work will be in accordance with the Contract Documents and not defective (as defined). This general warranty is in addition to specific warranties and guaranties called for in the Specifications and to the Owner's other rights and remedies under the Contract Documents. However, it does not apply to misuse or to normal wear and tear.

The Contractor also is required to give a broad indemnification of the Owner and Engineer against bodily injury and property damage (other than to the Work itself) claims brought by employees of the Contractor or Subcontractors or Suppliers or by third parties where such claims are caused in whole or part by the negligence of the Contractor, Subcontractor, Supplier, or other person for whom the Contractor may be liable. This indemnification obligation does not extend to the Engineer's liability for professional errors or omissions.

The language of both the general warranty provision and the indemnity provision has been carefully drafted, after consultation with contractors' liability insurers, to track generally available insurance coverage. The Contractor is specifically required to carry contractual liability insurance to cover these undertakings. Moreover, it is expressly provided by the EJCDC documents that the Contractor's warranty and indemnification obligations and other representations will survive final payment and completion and termination of the Agreement.

Article 7—*Other Work.* In some projects, the Owner may perform work related to the project at the site using the Owner's own forces or by direct contracts with other contractors. If such other work was not noted in the Contract Documents, the Contractor may be entitled to make a claim for additional expenses incurred or additional time required in accommodating the other work. The Owner is responsible for coordinating the activities among the various prime contractors and the other work with the Contractor's Work. If someone other than the Owner is responsible for providing the coordination, the identity and extent of the coordinator's authority and responsibility and the specific matters covered are required to be itemized in the Supplementary Conditions. The Contractor's obligations when other work is being performed at the site include: (1) providing safe and reasonable access to other contractors or workers for purposes of storing materials and equipment and performing their work, and (2) doing all cutting, fitting, and patching required to make the applicable parts of the Contractor's Work come together properly with the other work performed by others.

Article 8—Owner's Responsibilities. The Owner's specific responsibilities with respect to various aspects of the project and the Work appear in the various sections of the General Conditions that specifically deal with those aspects. Article 8, in essence, provides a comprehensive index to where a particular Owner's responsibilities are located in the General Conditions. The Owner's responsibility, if any, to furnish the Contractor reasonable evidence that financial arrangements have been made to satisfy Owner's obligations under the Agreement and other Contract Documents is set forth in the Supplementary Conditions. Article 8 also contains a disclaimer which states that the Owner does not supervise and is not responsible for the Contractor's means, methods, techniques, sequences, or procedures of construction or safety program.

Article 9—Engineer's Status During Construction. The Engineer is the Owner's representative during the construction. All communications between the Owner and Contractor are intended to be through the Engineer. Because all of the Engineer's actions are for the benefit of the Owner only, and because the Engineer is not a party to the contract between the Owner and the Contractor, the Engineer's activities are not expressed as "responsibilities" per se but rather in terms of the Owner's and Contractor's agreement as to what role or status the Engineer will have during the Project. The Engineer's role includes visits to the site at intervals appropriate to the various stages of the construction: (1) to keep the Owner informed of the progress of the Work, (2) to try to guard the Owner against defective Work, and (3) to try to find out whether the completed Work will conform generally to the requirements of Contract Documents. The Engineer's job in this regard is subject to several expressed or implied limitations: (1) the Engineer is not on the site all the time and can't see everything that goes on; (2) even when there, the Engineer can't see everything that goes on; (3) the Engineer's efforts are focused on evaluating the part of the Work that has been completed to that point, not the part of the Work that is still in the process of being carried out; (4) the Engineer's visits and observations of Contractor's Work do not constitute supervision, direction, or control over the Contractor's means, methods, techniques, sequences or procedures of construction or safety precautions or programs; (5) it is assumed that the Owner is going to have a Resident Project Representative on site on a continuous basis to provide a greater degree of coverage, but that representative's responsibilities are subject to the same limitations as apply to the Engineer.

Articles 10, 11, and 12—Changes. Changes are an important feature of every heavy construction project. It is virtually impossible to anticipate every job condition and project requirement before the Work starts. EJCDC's contract devotes three articles to dealing with changes in the Work (Article 10), changes of Contract Price (Article 11), and changes of Contract Times (Article 12).

In addition to defining the general circumstances under which the Contractor will be entitled to a Change Order for a change in the Work, Article 10 provides for a Work Change Directive which documents the changed Work and permits it to proceed, subject to later agreement or other resolution of entitlement to a change in price or time (for example, an appeal from the Engineer's decision).

Article 11 sets up an order of priority for determining the value of any Work for which there is a claim for an adjustment of the Contract Price as follows: (1) specified unit prices, (2) if no unit price, then a mutually agreed lump sum, and (3) if (1) and (2) are not applicable, then on the basis of Cost of the Work as defined and a fee for the Contractor's overhead and profit as determined in Article 11.

The elements that do and do not constitute Cost of the Work are exhaustively delineated in Article 11. The listing is remarkably complete and is recommended as a reference whether or not the reader of this chapter chooses to use the EJCDC form.

CHAPTER 3
CONSTRUCTION PLANNING AND SCHEDULING

Ross Caldwell
O'Brien-Kreitzberg, Inc.
Pennsauken, New Jersey

CREATING A CPM NETWORK

Introduction

The two principal scheduling tools used in construction today are bar charts and CPM networks. For many smaller projects, only a bar chart is necessary. For heavy construction projects, however, a *critical path method* (CPM) network is a must.

There is no evidence that CPM networks were utilized during the construction of the Great Pyramids, the Taj Mahal, or the U.S. Capitol building. In spite of the lack of these important management tools, all of these historic projects were eventually completed. So why do we need a network today to successfully build a project? To help control costs. While it is unlikely that the pharaohs were too concerned about cost, it is likely that the total number of labor-hours expended on a project and worker productivity was not as important to them as it is to project owners today. A pharaoh had millions of devoted followers seeking a way to pay homage to their king. Unfortunately, the owner of the large project today does not command the same respect. If a pyramid was built today using the same means and methods as in ancient times, the project cost would be astronomical and the duration would span several decades. Of course, the use of modern equipment would reduce the cost and duration of the project significantly. However, the use of modern scheduling tools has also contributed to the reduction of overall project cost and time requirements. On any given project, both the contractor and owner benefit from the use of scheduling.

The contractor can benefit from properly scheduling its projects. Unfortunately, many contractors submit schedules only to satisfy a specification requirement. Most contractors that do put effort into scheduling realize some, if not all, of the following benefits:

- Control overhead by maintaining the project duration.
- Predict cash flows with accuracy.
- Reduce storage costs by coordinating deliveries with installations.

- Optimize crew size and equipment use.
- Justify time extensions.

An owner benefits from scheduling in two ways. An owner benefits indirectly from the contractor's savings just outlined. Over time, as contractors benefit from scheduling project after project, the bids on subsequent projects become more and more competitive. More directly an owner benefits from each of the following:

- Cash flow to various projects can be accurately predicted.
- Time extension requests can be evaluated.
- Owner-supplied equipment can be coordinated with contractor installation.
- Contractor's progress can be monitored by the owner and projections made for final completion and beneficial occupancy.
- The owner can coordinate move-in activities based on projections for beneficial occupancy.
- The owner can predict and take steps to avoid late finish with associated lost revenues.

The earliest example of what we now consider CPM scheduling was introduced in the 1930s in Poland by Karol Adamiecki in the form of the *harmonygraph*. The harmonygraph looked like a vertical bar chart with the addition of notations identifying interdependence between activities. What we now term "float" or "slack" (the free time associated with each activity) was also identified on the harmonygraph. Unfortunately for Mr. Adamiecki, the use of his harmonygraph was short-lived and not widespread.

CPM scheduling, as we know it today, was born at Du Pont in 1957. It began as an effort to better control construction projects with an eye towards the use of the computer, which was, at the time, a new tool for managers to explore. A team comprising Du Pont people and UNIVAC people designed a program, the principle of which has remained basically unchanged. Concurrently, the U.S. Navy created a system of its own to help in the development of the Polaris ballistic missile. The Navy's system, known as Performance Evaluation and Review Technique or PERT, is similar to the arrow diagramming techniques we use today, except in the assignment of activity durations. Each activity in a PERT network has three durations: the optimistic, the pessimistic, and the probable. PERT networks are generally best suited to research and development projects where durations cannot be definitively established. Through the 1960s, the use of networking increased as Du Pont continued to use CPM on construction projects and CPM/PERT systems were used to program a variety of weapons and aerospace projects.

Today, the specifications associated with most heavy construction projects require that some form of scheduling tool be utilized. One of the factors which has contributed most to the widespread use of scheduling has been the general acceptance of the microcomputer within the construction industry. With the availability of numerous user-friendly planning/scheduling software packages, even smaller, less sophisticated contractors have found network analysis useful regardless of specification requirements.

Anyone involved in the management of a construction project as a contractor, an architect, an owner's representative, or an owner should have some familiarity with scheduling. The study of CPM scheduling is divided between the creation of a schedule network and the use of the result. The creation of a network refers to the mechanics of schedule development, including the identification of individual work activities, the logical connection of the work activities, the calculation of schedule dates and float paths, and the ongoing maintenance of a project schedule using periodic updates. The use of the project schedule is dictated by the user's relationship to the project and by the stage of project development.

The Baseline

To better understand the development of a CPM schedule it is important to understand a number of principles.

Work Items/Activities. Work items are the building blocks of a schedule. The first step in the development of any schedule is to list the individual tasks which make up the total project. This list begins as one of relatively general work items such as site work, construct foundations, construct masonry walls, or install roofing systems. Such general activities are useful in developing the general sequence for the construction of a project. Each work item can be further broken into detailed activities in terms of scope, location, or subtasks. For example, "construct grade beam GB-7" may be a summary of the following

Excavate for grade beam GB-7
Form grade beam GB-7
Place rebar for grade beam GB-7
Place concrete for grade beam GB-7
Strip forms at grade beam GB-7
Backfill to top of grade beam GB-7

The level of detail is dictated by two things. First, many projects require that the schedule be used to generate cost requisitions. It is helpful if the work items in the schedule are scaled so that they are likely to be completed during a pay period. By doing this, arguments over progress on individual activities during the monthly requisition meeting are kept to a minimum. Similarly, a work breakdown structure or a contractor's billing system may govern level of detail.

The second element that determines the level of detail is the schedule logic. Additional detail is sometimes required to correctly show the relationship between two work activities. For example, generally the erection of structural steel is followed by an activity for plumbing and bolting. For a complex frame, a single erection activity and a single plumb-and-bolt activity would not be appropriate. Because it is typical for a plumb-and-bolt crew to follow an erection crew before the erection is complete, a more appropriate list of activities might be:

Erect structural steel—first bay
Plumb and bolt structural steel—first bay
Erect structural steel—second bay
Plumb and bolt structural steel—second bay
Erect structural steel—third bay
Plumb and bolt structural steel—third bay

There is no right answer. The list of activities must describe the work to be done without being so detailed that maintenance of the schedule is overwhelming. There are as many different ways to accomplish this as there are schedulers.

Durations. Once the list of activities has been established, each work item must be assigned an anticipated duration—the time required to perform that activity. Assigning a duration to a work item is not an exact science. The time required to perform an activity depends on:

- Available labor
- Worker productivity
- Anticipated weather
- Method of work
- Available equipment

A contractor that is committed to CPM scheduling has recorded the data from every job it has completed, calculating unit production rates and unit costs. It is then just a matter of dividing the quantity by the company unit rate to determine a duration. Contractors that have not analyzed previous jobs will "guesstimate" durations based on the general experience of the personnel assigned to the current project or on the assumptions made by the estimators for the bid.

Resource Loading. It is useful to those people using a schedule during the execution of a project to be aware of the assumptions made during the creation of the project schedule. Most scheduling software programs on the market today allow for the inclusion of resource information into the schedule. As previously noted, activity durations are assigned based on available manpower, worker productivity, available equipment, etc. The background information used by the scheduler can be included in the schedule as resource information. Resource information can be used merely to annotate the schedule, allowing the schedule users to compare actual circumstances with the assumptions made by the schedule developers. A more sophisticated use of resource information is to calculate schedule dates based on resource limitations.

For example, the work item Form panel 3—wall A may require the use of two carpenters, one laborer, one operator, and one crane. This would be a typical crew for the placement of a reusable wall form. If this were the only activity taking place on a given day, on a given project, it is likely that work on this activity would not be hampered by any labor or equipment shortages. If, however, this activity were taking place on a day during which carpenters were in heavy demand on the project, or the crane could not be used by this crew, this activity may not be accomplished. A resource-loaded schedule acts as a tool for the superintendent or project manager to make a decision as to which work activity the carpenters or the crane should be assigned to.

Resource leveling is an attempt by the schedule developer to head off labor and/or equipment usage problems such as these. To resource level a schedule, the scheduler first assigns required resources to all work activities. When the schedule dates and float values are calculated by the scheduling software, the labor and equipment usage of concurrent activities is checked and, if resource limits are exceeded, the software will reschedule the activity with the most float to a latter time frame. Resource loading is most often used, not as a specification requirement by an owner, but by a contractor looking to optimize its workforce. A contractor that is preparing to begin construction on a high-rise building may be trying to decide whether to use one or two tower cranes. By assigning tower crane usage to all appropriate work items, the contractor can calculate sample schedules utilizing one and two tower cranes. The contractor can then compare the resulting schedules with its contract requirements.

Cost Loading. One more class of information is often associated with work activities in the heavy construction project schedule. Most government contracts require that the contractor cost load the project schedule. By associating a dollar value with individual work activities, projectwide cost information can be calculated. Depending on the scheduling software being used, the cost information can be grouped, gathered, and/or summed any way necessary. For example, an owner might require that the contractor submit a cumulative cost curve showing the curve based on early schedule dates and on late schedule dates. The contractor may be interested in forecasting monthly billings. The owner may be concerned about monthly cash expenditures or, the contractor may need to calculate the costs of subcontractors' services. On many large-scale projects, the contractors' monthly requisition is developed from the project schedule. Even this application of cost information within the project schedule takes various forms. The development of a monthly requisition may be as simple as application of a percent complete to the dollar value associated with work items performed that pay period. Or, an owner may require an earned value report which compares actual costs for work performed against budgeted costs for work performed and calculates estimates to complete the remaining work.

Work Breakdown Structure. One of the most efficient methods of creating a work activity listing is through the use of a work breakdown structure. The creation of a work breakdown structure for a given project accomplishes many of the tasks involved in the development of a project schedule. A work breakdown structure is similar to an organization chart on which the project as a whole is represented by a single box centered at the top of the structure. The work breakdown structure may develop in any of a number of forms. If a contractor were to develop a work breakdown structure for the construction of a wastewater treatment plant, it would probably be most beneficial to the contractor to divide the overall project into work packages consistent with its subcontracts. Therefore, the second level of the contractor's work breakdown structure may consist of excavation, pile installation, concrete work, structural steel, process mechanical, HVAC, plumbing, and electrical. The next level of detail in the work breakdown structure might divide the work to be done by each subcontractor into the various facilities, such as preliminary treatment building, grit tanks, primary sedimentation tanks, aeration basins, secondary sedimentation tanks, sludge handling, and chlorination. It may be to the owner's benefit to first break down the project by facility and then by trade. The general form of a work breakdown structure is often outlined within contract specifications. In either case, as each branch of the work breakdown structure becomes more and more detailed, the identification of individual work activities becomes more and more apparent. For the creation of work activities, the work breakdown structure may be detailed to its sixth or seventh level on some branches, while the degree of detail of the third or fourth level of another branch may be sufficient to describe the work to be done. The result is a detailed project schedule which can be summarized, as needed, based on the work breakdown structure.

Sequencing

Logic. Scheduling logic is the most significant difference between bar charting and network analysis. Logic is the word used by schedulers to refer to the interrelationship between any two work items. It is specifically shown on network diagrams tying two activities together. It is the combination of scheduling logic with assigned durations that makes it possible to calculate early and late start dates, early and late finish dates, total float, and project completion.

Generally, on a given concrete pour, the placement of concrete follows the completion of formwork and the placement of reinforcing steel. This is an example of *absolute logic*. It is just how concrete work is done. A specific activity must physically precede another in order to accomplish the desired result. *Preferential logic* refers to relationships between activities which are imposed by the scheduler rather than the nature of the work. A scheduler's preferential logic may show the intent to work east to west. This is not to say that preferential logic is incorrect. There may be very good reasons for the scheduler's choice. Suppose a contractor wins a contract to build an addition on an existing facility. The contractor's decision to work from east to west may appear to be preferential. It is, however, necessary to build from the existing wall out. The distinction between absolute and preferential logic is made here because in the creation of a project schedule, it is only the preferential logic that can be changed to adjust the schedule. After all, it is fairly obvious that the application of paint cannot come before the construction of the wall to be painted. It is not always obvious, however, which wall should be painted first.

The addition of logic to the list of work items is what differentiates between a simple bar chart and a CPM network. And while absolute logic can sometimes be followed on a bar chart, it is the preferential logic that cannot be determined from the standard bar chart. Though the placement in time of bars on a bar chart could not be accomplished without someone's logic, it is impossible for anyone other than the scheduler to determine that logic. By definition, a network diagram shows both absolute and preferential logic.

Network Diagramming. The *arrow diagramming method* of network diagramming uses graphical symbols to represent work items. Arrows and circles are the elements used in arrow diagramming to build a schedule. The circles in Fig. A3-1 are called *nodes* or *events*. They represent points in time. The first circle is called the I node and the second circle is the J node. These designations date back to when this whole concept was being developed for the early computers. The arrow between the circles represents a work activity. Individual activities are joined together to create a network. The J node of one activity becomes the I node of another.

Precedence diagramming is another type of network analysis. It is calculated just like arrow diagramming. Figure A3-2 depicts the sample activity from Fig. A3-1 in its precedence form. Some precedence diagrams show the relationships between activities more clearly than arrow diagramming because each box contains all information for a given activity and the connecting lines clearly trace the relationships. This is different from arrow diagramming in which two activities share a node and restraints must be added to keep the logic clear. Another difference between precedence and arrow diagramming is that in precedence you can assign time durations

FIGURE A3-1 Activity-on-arrow.

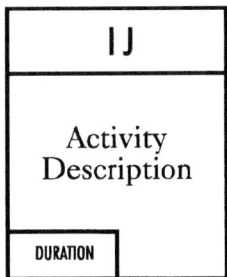

FIGURE A3-2 Activity in precedence method.

(called *leads* or *lags*) to the connecting lines to shift the start of an activity. There are also three different types of relationships available: start to start, finish to start, and finish to finish. Arrow diagramming uses only finish-to-start relationships.

Figure A3-3 shows a simple network containing 11 activities, A through K. This network, presented in ADM, not only presents a list of work items but also shows the logical relationships between them. The following statements describe the logic:

- Activity A precedes Activities B, D, I
- Activity H succeeds Activities G and K
- Activity K cannot begin until Activity J has been completed
- Activities C, E, and J must complete before Activity F can start
- Activity C is restrained by Activity B

The words *precede*, *succeed*, and *restrain* are used to describe the logical relationships between various activities. In the arrow diagramming method of scheduling, there is only one kind of relationship; one activity must end before another begins. It is the determination of this relationship that makes a CPM diagram more precise than a bar chart. In addition to the 11 activities identified by the letters A through K, there are three special activities called *restraints* used in this network example. Activity 40-80, Activity 70-80 and Activity 90-110 are all restraints. Restraints are not work items

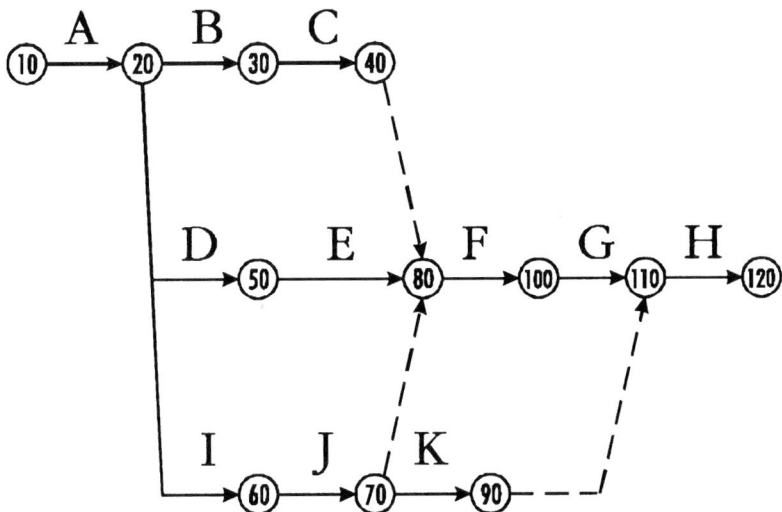

FIGURE A3-3 Arrow diagram or network.

but activities that have no duration and are only used to tie two other activities together to impose logic. For example, the restraint identified as 70-80 makes sure that Activity F (80-100) does not begin until Activity J (60-70) is complete. Figure A3-4 shows the precedence version of our sample network. Note that no extra restraint items are added. The 11 work activities are represented by 11 boxes with the relationships between individual work items depicted by connecting lines.

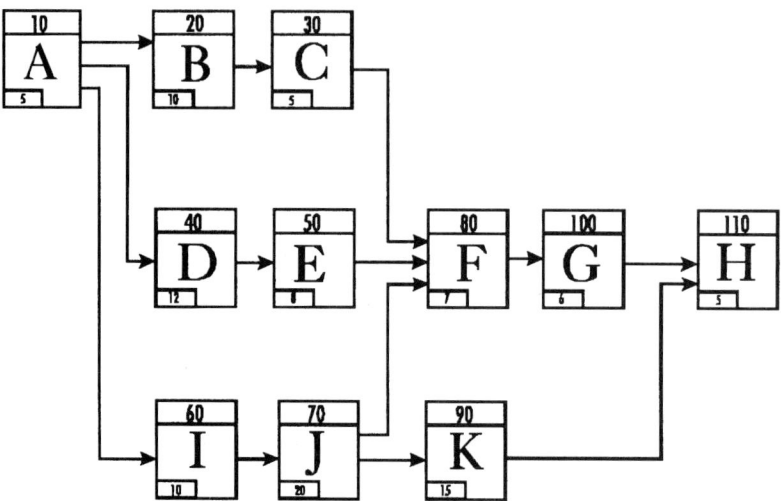

FIGURE A3-4 Precedence diagram format.

Fragnets.   A fragnet or fragment of a network can be as simple as the 11-activity example used here. While each individual work item represents the smallest building block to be used in the construction of a schedule, the fragnet can also be combined with other fragnets in the development of an overall project schedule. The fragnet in our example might represent the procurement and installation of one generator set. If a project calls for the installation of several generator sets, the fragnet could be duplicated the appropriate number of times, renumbered, and customized to describe each individual generator set. Finally, any logic ties between individual generator sets, such as crew restraints or the order of delivery, could be tied from fragnet to fragnet.

Fragnets are also used to help update a schedule during the construction period. A fragnet might be added to the update documentation to show how the actual sequence of events differed from those in the schedule. A fragnet might also be used to communicate revisions added to the baseline schedule for upcoming work. Fragnets should also be used to add change order work to the project schedule as part of the updating procedure.

Finally, fragnets are often used in claims presentations. The presenter generally makes use of summarized schedule activities with detailed fragnets highlighting trouble spots.

Calculations.   The planned start and finish dates for the activities in a network are calculated. By adding activity durations together, as directed by the logic in the network, the overall project duration and at what time each activity will take place can be

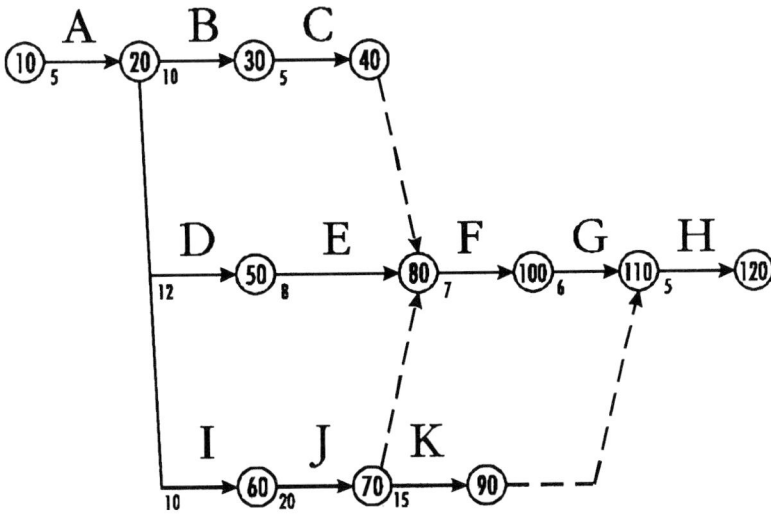

FIGURE A3-5 Arrow network (Fig. A3-3) with time durations.

determined. In Fig. A3-5, each activity in the fragnet has been given a duration. In this small schedule, the longest path through the network can be determined by inspection.

A-B-C-Restraint-F-G-H	totals 38
A-D-E-F-G-H	totals 43
A-I-J-Restraint-F-G-H	totals 53
A-I-J-K-Restraint-H	totals 55

The longest path is 55 days along and passes through the Activities A, I, J, K, and H by way of the Restraint 90-110. The longest path in the schedule is called the *critical path*. When calculating more complicated networks it is not always possible to find the longest path just by looking at the network. By adding all of the durations forwards and then backwards, we can find several pieces of information. Figure A3-6 shows the network with the forward pass and backward pass calculations done. The forward pass establishes the earliest point in time that each activity can start and end. The backward pass establishes the latest time that each activity can start and end. As an example, the path through A-D-E-F-G-H calculates as follows:

Forward pass

0 + 5: Activity A ends on Day 5
5 + 12: Activity D ends on Day 17
17 + 8: Activity E ends on Day 25
25 + 7: Activity F ends on Day 32

Notice that, in Fig. A3-6, Activity F finishes on day 42, not day 32 as calculated. When more than one path comes together, on the forward pass the higher start date is used. Therefore, because Activity J ends on day 35:

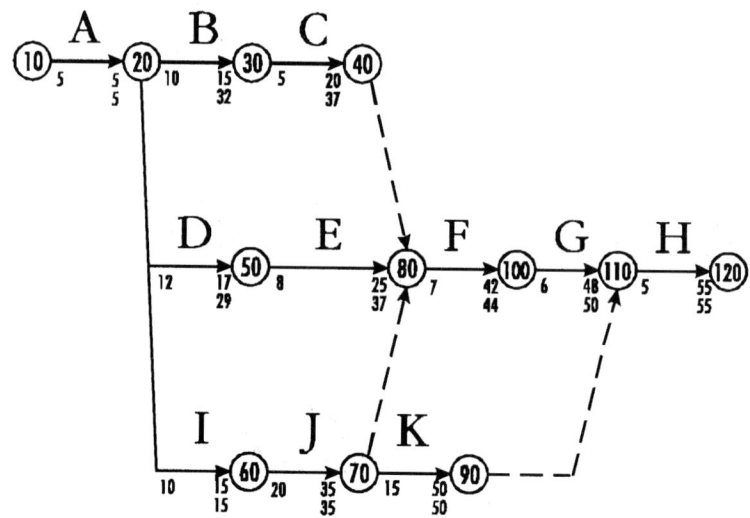

FIGURE A3-6 Arrow network with time calculations.

35 + 7: Activity F ends on Day 42

42 + 6: Activity G ends on Day 48

From Activity K:

50 + 5: Activity H ends on Day 55

Backward Pass

55 − 5: Activity H starts on Day 50

50 − 6: Activity G starts on Day 44

44 − 7: Activity F starts on Day 37

27 − 8: Activity E starts on Day 29

19 − 12: Activity D starts on Day 15

5 − 5: Activity A starts on Day 0

When both passes are complete there are some dates that do not match. Remember that the forward pass calculations result in the early dates and the backward pass calculations result in the late dates. The difference between the two is called *float*. Table A3-1 lists the calculations for the whole network.

The finish date of an activity is one day different from the start of the next because most calendars utilized today use a day that begins at 8:00 A.M. and finishes at 5:00 P.M.

The numbers in the total float column for each activity are calculated by subtracting the early start from the late start or the early finish from the late finish. The critical path is the string of activities with the least float, in this case zero. Notice that the critical path is the same path as the one we found to be the longest. Calculating the float for each activity and determining the critical path is a rather mechanical operation. That is, the scheduler and/or the scheduling software simply add all the

TABLE A3-1 Activity Time Calculations

I	J	Description	ES	EF	LS	LF	TF
10	20	A	1	5	1	5	0
20	30	B	6	15	23	32	17
30	40	C	16	20	33	37	17
20	50	D	6	17	18	29	12
50	80	E	18	25	30	37	12
80	90	F	36	42	38	44	2
90	110	G	43	48	45	50	2
110	120	H	51	55	51	55	0
20	60	I	6	15	6	15	0
60	70	J	16	35	16	35	0
70	100	K	36	50	36	50	0
40	80	Restraint	21	21	38	38	17
70	80	Restraint	36	36	38	38	2
100	110	Restraint	51	51	51	51	0

ES = early start; EF = early finish; LS = late start; LF = late finish; TF = total float.

durations in the forward pass and subtract the durations from the total in the backward pass, resulting in calculated float values. The real value of doing these relatively simple CPM calculations is in the determination of a range of dates for each work activity. Notice that the activities with zero float have early dates (start and finish) that are equal to their respective late dates. In other words, each activity on the critical path (with zero float) must take place between the start and finish dates identified by the calculation. There is no latitude in zero-float critical activities. However, in the activities with 17 days float, each respective activity can start and finish within a range of dates.

Updating

In a perfect world, the scheduler would complete the baseline schedule; the project would proceed in accordance with the baseline schedule and would complete as predicted by the baseline schedule. Perfection has proven to be an elusive thing in the construction industry. Late deliveries, changed conditions, change orders, union strikes, severe weather, and failed financing are just a sample of the conditions that change during a construction project. A project schedule is best used as a dynamic management tool. That is, as actual conditions surrounding a project change, so should the project schedule. Periodic changes in the schedule serve two functions. First, changes are made in order to sharpen predictions made based on information taken from the project schedule, such as project completion date and estimated cost to complete. Second, as changes are made periodically to the project schedule, an as-built record of the project is created. Periodic schedule updates formalize this process. An update period can be monthly on a large construction project. It can be quarterly during the design of a project. The update period may be daily for the maintenance work to be performed during an outage at a petroleum refinery. In general, the periodic update serves to define small portions of the overall project.

One function of the schedule update is to provide a cutoff point with which to measure progress to date. By applying percent completion, a status date, expected finishes, and remaining durations to the project schedule, a calculation of the schedule based on progress achieved during the update period can be made. It is also appropriate to incorporate approved change orders into the project schedule as part of the updating process. In so doing, the effect of a change order on the project can be determined.

General schedule maintenance may also be performed during the update of the project schedule. As mentioned previously, the actual sequence of events rarely follows the baseline schedule. Schedule maintenance refers to minor adjustments to logic, changes to durations, addition of activities, and the deletion of activities required to make the project schedule more accurately reflect actual conditions. It is important to keep the project schedule current so that predictions made based on the information derived from the project schedule are as accurate as possible. It should be noted that on many government contracts proposed changes to the schedule must be approved before being added officially.

Because the final as-built project schedule may differ somewhat from the baseline schedule, it is important that whoever is responsible for processing schedule updates be required to submit a narrative report along with the update. The narrative report serves to document all changes made to the project schedule in the course of updating.

USING THE NETWORK

Preconstruction

Each party involved in each phase of construction (preconstruction, construction, postconstruction) can benefit from the use of some form of scheduling tool. For the most part, during the preconstruction phase, only two parties are involved in a project: the owner and the designer. There may be additional entities involved in design-build type projects or if the construction manager is brought in early enough to assist with design packages, bid packages, or document preparation. However, during the preconstruction phase, two types of schedules are common.

On many large projects, a design schedule is required by the owner. The schedule for the design of a project is generally based on design milestones. Instead of standard work items as activities, the design schedule would be made up of activities representing predetermined design milestones such as conceptual design complete, process schematic complete, and rough calculation of design quantities. The milestones used by a designer or specified by an owner will vary greatly from project to project. It is often difficult to agree on the level of detail for design-type activities. A designer may develop an in-house schedule tracking individual sheets within a set of drawings. The other extreme would be the ill-defined milestones for 30, 60, and 90 percent design status.

The second type of schedule that is developed during the preconstruction phase of a project is what is sometimes called a *prebid schedule*. A prebid schedule is generally developed by the designer or may be developed by a construction manager, if brought on early enough in the project, and is included within the bid documents issued to potential bidders. A prebid schedule serves the owner by providing an early, summarized view of the project timetable. The prebid schedule serves the designer, because in order to create the prebid schedule, it is necessary for the designer to review staging, general design concepts, and potential trouble spots prior to the advertisement for bids. The prebid schedule serves the contractor by helping to visualize the designer's intent and expectations regarding staging and project duration.

Whether to use a prebid schedule should be carefully considered by an owner and/or designer. Because it is issued as part of the bid documents, any specific sequence of work depicted in a prebid schedule may be assumed to be a mandatory sequence by the potential bidders. This could lead to undesirably higher bids or possibly claims.

Generally, a prebid schedule should be used if a project requires very specific sequencing and/or staging as determined by the designer. Any relationship between work activities or work areas that are not specifically required by the contract should not be shown on a prebid schedule.

Construction

While schedules are sometimes used in the preconstruction phase of a project and are used extensively in postconstruction claims, the most beneficial use of a schedule is during the construction of a project. All parties involved in a project stand to gain from the proper development of a project schedule.

During construction, a comprehensive schedule serves the owner as a source of information. The information available to the owner takes the form of project status, percent complete, projected completion and milestone dates, cost information both current and projected, etc. CPM, the critical path method of scheduling, is an important tool for management by exception. By analyzing a properly maintained schedule, an owner need only consult a single sort of schedule information, the *total float sort,* to determine what part of a project is most important to maintaining the end date. Figure A3-7 is an excerpt taken from a sample total float–early start sort. As previously described, the total float for a given activity is the amount of time available between the earliest time that an activity can take place and the latest time that an activity must take place. The ideal schedule has one critical path with a total float equal to 0, which is one string of activities which must start on their respective start dates and end on their respective end dates in order to meet project completion. All other activities in the project would then have float values greater than 0. By using a total float–early start sort, an owner can quickly identify the list of activities which have the least leeway in the project and have them sorted chronologically. Provided the schedule is current, the total float–early start sort serves as a list of activities which must take place and lists when they must take place to maintain project completion.

The total float sort also acts as a gauge of current project status. By comparing a current report with a previous total float sort, it is possible to determine how the contractor is progressing. If the number representing the total float is a consistently smaller number, reaching 0 and then turning negative, it means that the contractor is unable to meet the dates presented in the schedule. If the float value grows, then the contractor is beating its schedule. If a whole new list of activities becomes the critical path, then it suggests that the contractor is progressing on activities that had been critical but is ignoring or experiencing difficulty with activities that were previously less critical.

A caution is that the critical path method highlights the one most important string of activities. It is often the case that an activity with 5 or even 10 days of float may slip and become the start of a new critical path. It is important not to just look at the *zero path* but to also review activities on what are called *subcritical paths.*

Having answered the owner's first question ("Where is the project today?") by using a total float sort, the next question an owner generally has is: "When will this project be finished?" A quick answer to this question is generally depicted on a sum-

Report: Schedule by TF
Layout: Sorted by TF/ES
Filter: All Activities

Highway Widening

C.M. Construction, Inc.

Page 1A of 2A

Act ID	Activity Description	Orig Dur	Rem Dur	%	Early Start	Early Finish	Late Start	Late Finish	Total Float
P1N2190	INSTALL SIGN SUPPORT STRUCTURE	8	8	0	02JUN99	11JUN99	02JUN99	11JUN99	0
P1N2195	ERECT SIGNS	3	3	0	14JUN99	16JUN99	14JUN99	16JUN99	0
P1N2100	PLACE CONST. BARRIER	12	12	0	02JUN99	17JUN99	02JUN99	17JUN99	0
P1N2115A	GRUB & STRIP TOPSOIL	8	8	0	18JUN99	29JUN99	18JUN99	29JUN99	0
P1N2105B	INSTALL EROSION CONTROL DEVICES	15	15	0	30JUN99	20JUL99	30JUN99	20JUL99	0
P1N2200	INSTALL TEMP. SHEETING	12	12	0	21JUL99	05AUG99	21JUL99	05AUG99	0
P1N2210	EXCAVATE RETAINING WALL	12	12	0	06AUG99	23AUG99	06AUG99	23AUG99	0
P2N2115B	REMOVE GUIDE RAIL	4	4	0	24AUG99	27AUG99	24AUG99	27AUG99	0
P2N3100	PLACE TEMP. CONST. BARRIER	6	6	0	30AUG99	06SEP99	30AUG99	06SEP99	0
P2N3225A	REMOVE TEMP. CONST. BARRIER	5	5	0	07SEP99	13SEP99	07SEP99	13SEP99	0
P2N2125B	INSTALL TEMP. ELECTRICS	15	15	0	30AUG99	17SEP99	30AUG99	17SEP99	0
P1N3110	REMOVE TEMP. PAVEMENT	9	9	0	07SEP99	17SEP99	07SEP99	17SEP99	0
P2N3120	REGRADE AREA	7	7	0	20SEP99	28SEP99	20SEP99	28SEP99	0
P1N2220	CONSTRUCT FOOTING	27	27	0	24AUG99	29SEP99	24AUG99	29SEP99	0
P2N2130	INSTALL ELECTRIC CONDUITS & STRUCTURES	19	19	0	20SEP99	14OCT99	20SEP99	14OCT99	0
P1N2230	CONSTRUCT RETAINING WALL	27	27	0	14SEP99	20OCT99	14SEP99	20OCT99	0
P1N2240	PLACE POUROUS FILL BEHIND WALL	9	9	0	21OCT99	02NOV99	21OCT99	02NOV99	0
P2N2140	INSTALL POWER & LIGHTING	15	15	0	15OCT99	04NOV99	15OCT99	04NOV99	0
P1N2250	BACKFILL RETAINING WALL	7	7	0	03NOV99	11NOV99	03NOV99	11NOV99	0
P2N2150	SUBGRADE PREPARATION	14	14	0	05NOV99	24NOV99	05NOV99	24NOV99	0
P2N2170	PLACE AGGREGATE & ASPHALT BASE COURSE	10	10	0	25NOV99	08DEC99	25NOV99	08DEC99	0
P1N2260	CONSTRUCT CONCRETE BARRIER AGAINST	20	20	0	12NOV99	09DEC99	12NOV99	09DEC99	0
P2N2180	PLACE ASPHALT STABILIZED COURSE	12	12	0	09DEC99	24DEC99	09DEC99	24DEC99	0
P1N2270	CONSTRUCT CONCRETE BARRIER AGAINST	24	24	0	03DEC99	05JAN00	03DEC99	05JAN00	0
P2N2185	PLACE 2" SURFACE COURSE	10	10	0	27DEC99	07JAN00	27DEC99	07JAN00	0
P2N2196	STRIPE ROADWAY	9	9	0	10JAN00	20JAN00	10JAN00	20JAN00	0
P2N2198	REMOVE TEMP BARRIER	4	4	0	21JAN00	26JAN00	21JAN00	26JAN00	0
P1N2300	CONSTRUCT CONCRETE BARRIER AGAINST	25	25	0	24DEC99 *	27JAN00	24DEC99	27JAN00	0
P1N2320	CONSTRUCT CONCRETE BARRIER AGAINST	24	24	0	14JAN00	16FEB00	14JAN00	16FEB00	0
P1N3200	CONSTRUCT CONCRETE BARRIER AGAINST	20	20	0	04FEB00 *	02MAR00	04FEB00	02MAR00	0
P1N3220	CONSTRUCT CONCRETE BARRIER AGAINST	6	6	0	03MAR00	10MAR00	03MAR00	10MAR00	0
P1N3225B	CONSTRUCT CONCRETE BARRIER AGAINST	6	6	0	13MAR00	20MAR00	13MAR00	20MAR00	0
P2G600	PROJECT COMPLETE	0	0	0		20MAR00		20MAR00	0
P2N2125A	EXCAVATE FOR ELECTRICAL	10	10	0	30AUG99	10SEP99	06SEP99	17SEP99	5d
P1G5200	CONSTRUCT CONCRETE BARRIER AGAINST	8	8	0	21OCT99	01NOV99	02NOV99	11NOV99	8d
P1NP100A	PROCUREMENT OF NOISE BARRIER POST &	10	10	0	02JUN99	15JUN99	10DEC99	23DEC99	137d
P1G098	BEGIN PROJECT	0	0	100	24SEP98 A		24SEP98 A		
P1G100	MOBILIZATION	4	0	100	24SEP98 A	29SEP98 A	24SEP98 A	29SEP98 A	
P1G105	CLEAR SITE	4	0	100	01OCT98 A	05OCT98 A	01OCT98 A	05OCT98 A	
P1N1140	PLACE TEMP MARKING TAPE	1	0	100	03NOV98 A	03NOV98 A	03NOV98 A	03NOV98 A	
P1G110	GENERAL CONDITIONS	13	0	100	08OCT98 A	04NOV98 A	08OCT98 A	04NOV98 A	
P1N1145	PLACE TEMP. CONST. BARRIER	1	0	100	04NOV98 A	25NOV98 A	04NOV98 A	25NOV98 A	
P1N1150	EXCAVATE	5	0	100	27NOV98 A	07DEC98 A	27NOV98 A	07DEC98 A	
P1N1170	PLACE AGGREGATE & ASPHALT BASE COURSE	0	0	100	14DEC98 A	22DEC98 A	14DEC98 A	22DEC98 A	
P1N1180	REMOVE TEMP. CONST. BARRIER	1	0	100	23DEC98 A	11JAN99 A	23DEC98 A	11JAN99 A	
P1N1202	CONSTRUCT TEMP. BENT #1 & #2	20	0	100	17DEC98 A	15JAN99 A	17DEC98 A	15JAN99 A	
P1N1185	PLACE 2" SURFACE COURSE	0	0	100	07JAN99 A	19JAN99 A	07JAN99 A	19JAN99 A	
P1N1190	DIVERT TRAFFIC ONTO TEMP. ROAD	1	0	100	20JAN99 A	28JAN99 A	20JAN99 A	28JAN99 A	
P1N1195	REMOVE TEMP. MARKING TAPE	1	0	100	01FEB99 A	09FEB99 A	01FEB99 A	09FEB99 A	
P1N1200	PLACE CONST. BARRIER & TEMP. STRIPING	4	0	100	17FEB99 A	17MAR99 A	17FEB99 A	17MAR99 A	
P1N1210	EXCAVATE RAMP & INSTALL 15" RCP PIPE WITH	10	0	100	18MAR99 A	30MAR99 A	18MAR99 A	30MAR99 A	
P1G200B	PROCUREMENT OF ESW/SL SIGNS	10	0	100	19APR99 A	29APR99 A	19APR99 A	29APR99 A	

FIGURE A3-7 Sample schedule report sorted by total float.

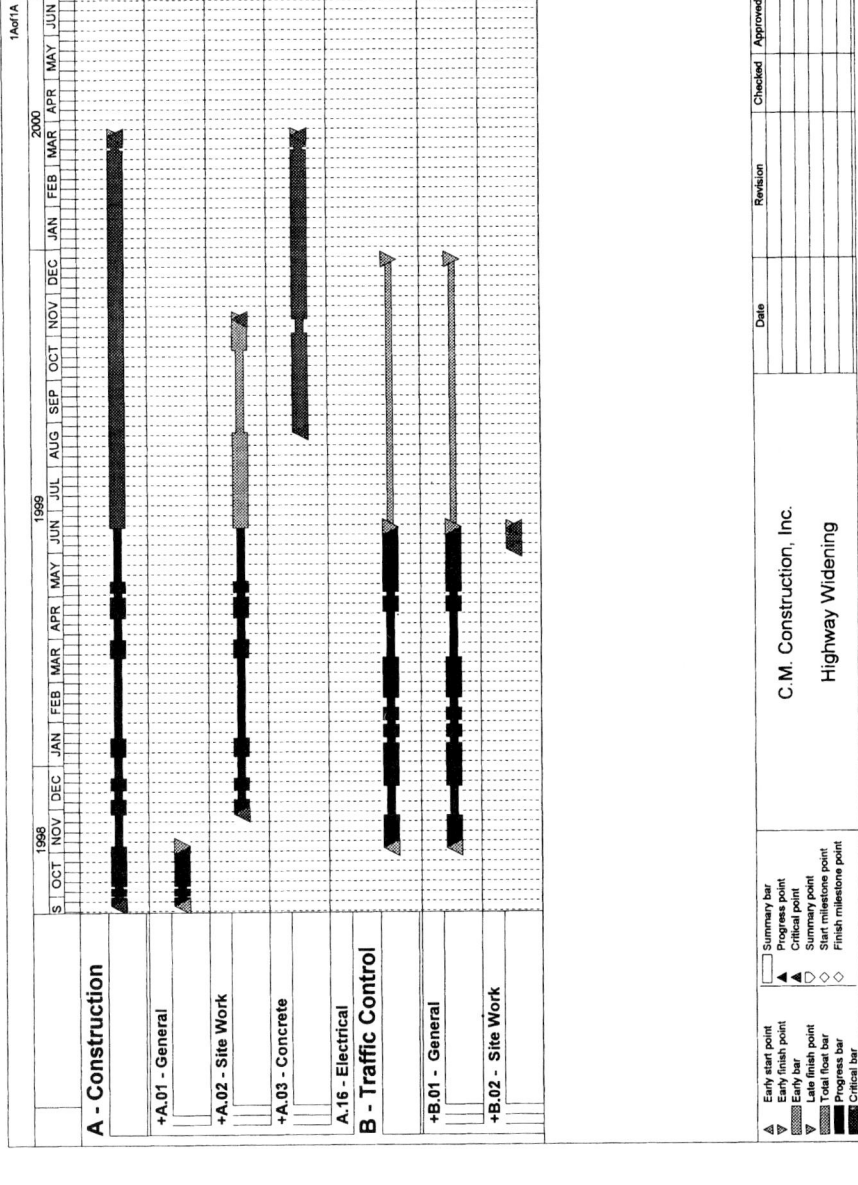

FIGURE A3-8 Sample summary bar graph reflecting work breakdown structure.

mary bar chart. Figure A3-8 shows an example of a summary bar chart. The detailed information found in a project schedule can generally be summarized in a number of different ways: by phase or stage, by work area, or by bid package or contract. In all cases, it provides an overview of projected dates for intermediate milestones and/or project completion.

If the summary bar chart shows that the milestone and/or completion dates have changed since previous reporting periods, the next question asked by an owner is: "What happened?" Most scheduling software allows for the comparison of the current schedule with a target schedule. The target might be the update from the previous reporting period or it may be the baseline schedule. While the reasons for a change in projected dates may not be apparent in a comparison schedule, this report can be used to determine which specific activities have changed enough to impact future dates. Once the individual activity has been identified, the source of delay or improvement can be investigated.

A schedule-based cost reporting system is often an important source of information for an owner. More and more, cost-loaded schedules are being used to develop partial payments. The use of a cost-loaded schedule has several benefits. By specifying that partial payments be developed from a cost-loaded schedule, the owner is guaranteeing that the schedule will be maintained by a contractor. While many contractors use CPM scheduling to their own benefit, many times a schedule is submitted to satisfy a contract specification. This often results in a questionable schedule. However, if schedule activities are correctly tied to contract pay items, and partial payments are generated from the updated schedule, great care is usually taken by all parties concerned to make sure that actual progress is correctly reported in the schedule. This leads to more accurate status estimates, as well as more accurate projections. By incorporating cost information into the project schedule, it is possible to make cost-related projections, as well as schedule-related projections. Projected values such as *estimate to complete, estimate at completion,* and *projected monthly cash flow* are all possible calculations.

A number of departments in the U.S. government now use cost-loaded schedules along with specific comparisons to determine contractor progress. The calculations and comparisons of *budgeted cost of work scheduled* (BCWS), *budgeted cost of work performed* (BCWP), and *actual cost of work performed* (ACWP) (see Fig. A3-9) are useful not only to the U.S. government, but to anyone interested in actual performance as it relates to the estimated cost and baseline schedule for any given project. The calculation of BCWS is simply the summation of costs associated with work activities in the baseline schedule. A BCWS may be determined for any single activity, group of activities (for example, concrete work), or for the entire project. It is simply the budgeted dollar amount assigned to each activity in the project schedule. The BCWP is calculated by applying actual progress or percent complete to the budgeted dollar amounts previously mentioned. The U.S. government and other owners are generally interested in the comparison between BCWS and BCWP, also referred to as *earned value,* because it tells the owner how a contractor is doing compared to the baseline schedule. In effect, it compares the anticipated cash flow as depicted in the baseline schedule and compares it with an actual cash flow. Earned value is schedule performance in terms of dollars. For example, the contractor may estimate approximately $2 million worth of concrete work on an airport project to reconstruct a runway. As the contractor progresses the work, percentage complete is tracked. Using scheduling software, the owner calculates earned value on the concrete item. The calculation is done by determining the quantity or dollar value of concrete, which should have been placed at the time the calculation is done, and compares that with the result of the reported percentage complete multiplied by the original budget values. Once the cal-

Report:
Layout: Cost/Schedule Variance
Filter: All Activities

Highway Widening

C.M. Construction, Inc.

Page 1A of 2A

Act ID	Activity Description	%	BCWP	ACWP	Cost Variance	BCWP	BCWS	Schedule Variance
STRT - Start-up								
P1G098	BEGIN PROJECT	100	0.00	0.00	0.00	0.00	0.00	0.00
P1G100	MOBILIZATION	100	100.00	100.00	0.00	100.00	100.00	0.00
P1G105	CLEAR SITE	100	3583.68	3583.68	0.00	3583.68	3583.68	0.00
P1G110	GENERAL CONDITIONS	100	19500.00	21375.00	-1875.00	19500.00	19500.00	0.00
		100	23183.68	25058.68	-1875.00	23183.68	23183.68	0.00
TRAF - Traffic control								
P1N1140	PLACE TEMP MARKING TAPE	100	288.32	288.32	0.00	288.32	288.32	0.00
P1N1145	PLACE TEMP. CONST. BARRIER	100	1790.52	1790.52	0.00	1790.52	1790.52	0.00
P1N1202	CONSTRUCT TEMP. BENT #1 & #2	100	22004.40	22004.40	0.00	22004.40	22004.40	0.00
P1N1180	REMOVE TEMP. CONST. BARRIER	100	1790.52	2238.15	-447.63	1790.52	1790.52	0.00
P1N1190	DIVERT TRAFFIC ONTO TEMP. ROAD	100	248.00	248.00	0.00	248.00	248.00	0.00
P1N1195	REMOVE TEMP. MARKING TAPE	100	288.32	0.00	288.32	288.32	288.32	0.00
P1N1200	PLACE CONST. BARRIER & TEMP. STRIPING	100	2588.12	2588.12	0.00	2588.12	2588.12	0.00
P1N1235	REMOVE TEMP. CONSTR. BARRIER	100	1666.52	1666.52	0.00	1666.52	1666.52	0.00
P1N1240	PLACE TEMP. STRIPING & DIVERT TRAFFIC	100	248.00	248.00	0.00	248.00	248.00	0.00
P1N2100	PLACE CONST. BARRIER	0	0.00	0.00	0.00	0.00	17799.84	-17799.84
P2N3100	PLACE TEMP. CONST. BARRIER	0	0.00	0.00	0.00	0.00	9999.12	-9999.12
P2N3225A	REMOVE TEMP. CONST. BARRIER	0	0.00	0.00	0.00	0.00	8332.60	-8332.60
P2N2196	STRIPE ROADWAY	0	0.00	0.00	0.00	0.00	0.00	0.00
P2N2198	REMOVE TEMP BARRIER	0	0.00	0.00	0.00	0.00	0.00	0.00
		46	30912.72	31072.03	-159.31	30912.72	67044.28	-36131.56
SIGN - Signage								
P1G200B	PROCUREMENT OF ESW/SL SIGNS	100	22000.00	22000.00	0.00	22000.00	22000.00	0.00
P1NP200C	PROCUREMENT OF SIGN STRUCTURES	100	20000.00	20000.00	0.00	20000.00	20000.00	0.00
P1N2190	INSTALL SIGN SUPPORT STRUCTURE	0	0.00	0.00	0.00	0.00	6895.36	-6895.36
P1NP100A	PROCUREMENT OF NOISE BARRIER POST &	0	4800.00	0.00	4800.00	4800.00	12000.00	-7200.00
P1N2195	ERECT SIGNS	0	0.00	0.00	0.00	0.00	5371.56	-5371.56
		49	46800.00	42000.00	4800.00	46800.00	66266.92	-19466.92
EART - Earthwork								
P1N1150	EXCAVATE	100	2846.60	2846.60	0.00	2846.60	2846.60	0.00
P1N1210	EXCAVATE RAMP & INSTALL 15" RCP PIPE	100	11149.20	11149.20	0.00	11149.20	11149.20	0.00
P1N2115A	GRUB & STRIP TOPSOIL	0	0.00	0.00	0.00	0.00	12533.12	-12533.12
P1N2105B	INSTALL EROSION CONTROL DEVICES	0	0.00	0.00	0.00	0.00	3760.32	-3760.32
P1N2200	INSTALL TEMP. SHEETING	0	0.00	0.00	0.00	0.00	9347.04	-9347.04
P1N2210	EXCAVATE RETAINING WALL	0	0.00	0.00	0.00	0.00	7542.24	-7542.24
P2N2115B	REMOVE GUIDE RAIL	0	0.00	0.00	0.00	0.00	3115.68	-3115.68
P2N2125A	EXCAVATE FOR ELECTRICAL	0	0.00	0.00	0.00	0.00	6141.20	-6141.20
P2N3110	REMOVE TEMP. PAVEMENT	0	0.00	0.00	0.00	0.00	6772.68	-6772.68
P2N3120	REGRADE AREA	0	0.00	0.00	0.00	0.00	6902.84	-6902.84
P1N2240	PLACE POUROUS FILL BEHIND WALL	0	0.00	0.00	0.00	0.00	11107.08	-11107.08
P1N2250	BACKFILL RETAINING WALL	0	0.00	0.00	0.00	0.00	7770.84	-7770.84
		14	13995.80	13995.80	0.00	13995.80	88988.84	-74993.04
ELEC - Electric and power elements								
P2N2125B	INSTALL TEMP. ELECTRICS	0	0.00	0.00	0.00	0.00	8884.80	-8884.80
P2N2130	INSTALL ELECTRIC CONDUITS & STRUCTURES	0	0.00	0.00	0.00	0.00	42243.08	-42243.08
P2N2140	INSTALL POWER & LIGHTING	0	0.00	0.00	0.00	0.00	16954.80	-16954.80
		0	0.00	0.00	0.00	0.00	68082.68	-68082.68
PAVE - Sub surface and paving work								
P1N1170	PLACE AGGREGATE & ASPHALT BASE	100	107477.52	107477.52	0.00	107477.52	107477.52	0.00
P1N1185	PLACE 2" SURFACE COURSE	100	23865.36	23865.36	0.00	23865.36	23865.36	0.00
P1N1220	PLACE AGGREGATE & ASPHALT BASE	100	75774.08	75774.08	0.00	75774.08	75774.08	0.00

FIGURE A3-9 Sample cost report identifying cost and schedule variances.

culation is done, the contractor's progress can be discussed, not in terms of days behind schedule, but rather in terms of cubic yards or dollars behind schedule.

It is generally the contractor's bookkeepers who are interested in the comparison between BCWS and ACWP. This is a direct comparison between what the bidders anticipated and what actually takes place in the field. As cost-loaded schedules see more and more use, the relationship between time and money gets closer.

Postconstruction

If a schedule has been well maintained through the course of a project, it benefits all concerned in the postconstruction time period as an as-built record of the project. On the positive side, contractors can use the information from the now complete project in future bids or in the development of future schedules. In the case of refinery turnarounds, owners may also have a need for the information contained in a project schedule for future work. Project schedules are also used in the postconstruction phase in support of or to refute delay claims.

In theory, the use of a CPM schedule in the evaluation of a delay claim is a simple idea. The project schedule is impacted with various delays, work stoppages, late material deliveries, rework, "acts of God," or any other event that disrupted the flow of the project. If the end date changes, the specific delay that caused the end date to change and the circumstances for that impact can be determined. In practice, however, it is much more complicated. In the construction industry, there are arguments such as whether a change should be incorporated into the schedule, who owns float, whether a baseline schedule should be used as the yardstick for the entire project, or whether the most recent update should be used for measuring the effect of a change on the end date. Each side of all of these arguments may have merit based on the specific circumstances. If a project schedule has been meticulously maintained with all previous changes incorporated into the schedule, then in evaluating a single change, the most recent schedule update should be used. If, however, the project schedule no longer resembles the project, it is necessary to rebuild the job on paper, compare the as-built with the as-planned, and identify all of the changes on the project which transformed the baseline into the actual. This approach is also necessary when both owner and contractor have caused delays on the project.

The issue of delay is always tied to dollars. A contractor may submit a delay claim either to avoid paying liquidated damages or in pursuit of delay damages. The most complicating issue in delay analysis is the determination of concurrent delay. To be able to collect on a delay claim, a contractor must be able to prove that the delay was by the owner. Even then, the contractor must be ready to defend against allegations that project delay was his (or her) own. Likewise, the owner must be "pure" in order to assess liquidated damages. When delays by both parties overlap, the overlap delay is referred to as concurrent delay. Concurrent delay is considered noncompensable. That is, if both the owner and contractor have delayed the project during the same time period, the appropriate amount of time is added on to the end of the contract without the payment of delay damages to the contractor. In that way, the contractor is not held responsible for liquidated damages, but neither is the owner held responsible for delay damages.

CHAPTER 4
CONSTRUCTION COST CONTROL

Wesley F. Mikes
O'Brien-Kreitzberg, Inc., Pennsauken, New Jersey

Joseph Alcabes, P.E., C.C.E.
West Hempstead, New York

INTRODUCTION

Everyone associated with a project will have an inherent interest in the cost for implementation and completion of construction. The interest may vary depending on each individual's association with the project. The owner will have an established budget to implement and provide a finished project to meet specific criteria and needs. The contractor, if working on a fixed-price contract, has a similar goal to complete the project within the bid price and to cover all costs of material and labor and ultimately make a profit. The more accurately cost records are maintained, the better the chances for the financial success of the project.

To assist in meeting these objectives, this chapter contains information for reporting and analyzing project costs. It describes an effective recording and reporting system that will give a measure of performance for the project's budget and will allow a reasonably accurate estimate of the cost to complete the project. Upon completion, accurately maintained records can then be utilized for future bidding, insurance, or tax purposes.

This chapter contains descriptions and samples of forms for recording and reporting project costs, along with notes describing the purpose and use of each. It is divided into eight sections which are convenient for describing the cost of the project:

1. Introduction
2. Definitions and objectives
3. Classification of accounts
4. Materials and equipment
5. Subcontracts

6. Field labor
7. Distributable accounts
8. Revised estimate and job cost report

These divisions allow the cost accounting system to be easily adapted and suited to various sizes and types of projects which may have different reporting needs. Reports may be prepared in the home office prior to mobilization and later in the field when job-site conditions allow. Each project must be carefully analyzed in advance and a cost reporting system to specifically meet the project's needs must be established. There is no need to overdo a cost reporting system which may later become a burden for site or office personnel.

Cost can be viewed from the position of the engineering contractor, who must deal with few self-managed subcontracts and a great deal of construction engineering design. In this type of an operation, a heavy burden is placed on the management skills of the project manager who has to coordinate the activities of the engineering and construction specialists who make up the project team. Costs may also be viewed from the vantage point of the construction contractor, who deals with many self-managed subcontracts and little construction engineering design. In this situation, architectural and engineering services are provided by others and the large engineering force is not required. Here, each subcontractor has to provide its own management skills, thus reducing the need for those skills on the part of the general contractor. The general contractor and the subcontractors can concentrate their efforts on the construction work rather than on the problems associated with engineering and design. From either of these positions, or from the third-party viewpoint of the owner, owner-operator, or construction manager, reporting and recording costs for cost control purposes are extremely important. The system provided in this chapter allows for this variation of viewpoints.

DEFINITIONS AND OBJECTIVES

Definition of Cost Engineering

Cost engineering is defined as the application of procedures to minimize cost in relation to the budget estimate prepared for any specific project. The budget estimate together with all documents defining the scope of work (and approved or finalized in a signed contract with a client) will be the control document for any specific project. In order to have effective cost control, any deviation from the scope of the budget estimate must be defined, evaluated, and presented to the client or to management for approval as a change in the scope of work.

Objectives and Responsibilities

The objective of the cost engineering function is to keep management advised of the current cost status of the project. This function can be accomplished by analyzing and controlling costs, maintaining an orderly system of records and reports, measuring the progress or work accomplished, and forecasting the total cost to complete the project in comparison with the budget or control estimate of the project.

The cost engineer is responsible for cost control through the preparation of cost engineering procedures or computerized programs, or revised estimates and project cost reports; classification of accounts; cost coding and distribution; cost control documents and reports; and evaluation and incorporation of scope changes and deviations into the revised budget. Such a system provides feedback of cost statistics for evaluation of the project, for use in the preparation of a final project cost analysis report, and for use by the estimating department on future projects. On smaller projects, this may be the responsibility of a single individual, whereas for larger projects, it can be the responsibility of an entire department.

Principles of Effective Cost Control

1. When the original budget estimate is established, all elements of the project should be adequately defined. A record should be made of specifically what the line item contains to avoid future confusion.

2. Every item in the budget estimate must be measured in relationship to the status of the completion of the line item and its expenditure. If the item did not appear in the budget estimate, there is strong indication that it was not in the initial scope of work. To have effective cost control, the cost trends must be established as soon as possible and measured against physical progress of the project so that remedial action by management can be taken in a timely fashion, if required.

3. A good definitive cost estimate and a realistic and tight master project schedule will provide excellent control documents from which the cost engineer can provide a measure of progress of the project for management's use.

4. When the initial engineering is managed effectively, both engineering and design are on schedule, and material deliveries are held on schedule for construction, management is achieving the end of reducing the final costs of the work.

5. Field cost engineering (after briefing by the home office as to the scope of work in the budget and instructions regarding the reports expected on a monthly basis or other predetermined intervals) is the keystone to forecasting cost performance of the project. The construction manager should be alerted immediately when unit costs and quantities are overrunning the budget estimate. Likewise, when unit costs are running substantially lower than anticipated. This can assist both the construction manager and the owner with respect to future change orders on the contract.

6. It is the responsibility of the project manager and the construction superintendent to specifically define in sufficient detail the scope of changes in work that departs from the contract documents so that an evaluation of the effects and cost of changes can be made for presentation to the client or management. It is also advantageous at this time to review the specific impact of a change order with respect to completion of the project. When properly evaluated, it may be to everyone's benefit when preparing a change-order cost estimate to include acceleration or buying-back time impact of the change.

7. For the cost engineer to provide management with an accurate, unbiased forecast of the anticipated project costs in comparison with the budget estimate, basic reports must be prepared and analyzed. These reports must be accurate, current, and measured against the progress of the project for them to work effectively.

Proper cost control necessitates getting a fix on the trend of the cost as soon as possible so that action can be taken if necessary. Cost reports that are running two or three months behind the actual project become historical documentation. The cost engineer's reports will sound a warning of any unfavorable cost trends to alert management to initiate corrective measures. Cost reports that are not current tend to reduce or eliminate the chances of timely reactions.

CLASSIFICATION OF ACCOUNTS

General

The backbone of the estimates and cost control system is the classification of accounts which is initially established for the project. As soon as a project has been authorized, a preliminary classification of accounts or an outline is prepared to cover the items of work included in the scope for the specific project. No one classification of accounts has been developed that can adequately cover every project. While various projects do have a number of similarities, they also have their own unique features. Hence, for each new project, it is essential that the project team prepare such a code of accounts from a master classification to cover the specifics of the project. There are various master systems which are utilized for different types of facilities.

Classification of accounts is the method of allocating actual expenditures and commitments for proper cost control. All estimates and material contract commitments are coded in accordance with the established classification breakdown. The classification code should be placed on all documents including purchase orders, receiving records, requisitions, and payments. It is the cost engineer's responsibility to maintain proper classification coding for all documents and to see that costs are properly allocated. The cost engineer must work closely with the field accountant, purchasing agents, craft supervisors, forepersons, and project engineers. This requires close personal contacts with forepersons and craft supervisors during each working day to be certain that craft labor costs are being properly classified. The results from the cost accounting system are only as reliable as the data which are entered. Every effort must be made to properly allocate both labor and materials. Preprinted time cards with descriptions of various accounts can substantially aid with proper allocations. This eliminates the need for craft labor to make decisions based on limited information in the construction field. The cost engineer and the timekeeper must also be alert and knowledgeable as to the actual work being performed in the field to eliminate obviously miscoded accounts.

The actual responsibility for preparation of cost ledgers in the field office rests with the field accountants. This frees the cost engineer to perform duties with respect to the cost analysis. Along with posting the cost ledgers, the field accountant with the assistance of the chief timekeeper will ascertain that the time cards are properly coded and prepare the necessary journal entries. The cost engineer will be provided with this information and other pertinent documents as required for the cost control function.

As noted, the classification of accounts is used throughout the project for identification purposes on all documents including requisitions, purchase orders, cost reports, and payments. The cost coding system also becomes the basis for historical retrieval of information for estimating purposes on future projects. Many times the cost engineer is faced with the realities of having to live with a coding system that is preestablished by the client or a regulatory agency. These could include the Federal

Energy Regulatory Commission, the Nuclear Regulatory Commission, or simply the client's in-house system. In order to maintain historical records, a construction manager or contractor may elect to maintain a dual system of coding for a specific project: one system for the client or the regulatory agency and a second system that is compatible with the company's own procedures. Through the use of on-site PCs, a cross-referencing system can readily be established.

Sample Classifications

As previously outlined, there are a number of established classification coding systems available for the cost engineer. Experienced cost engineers will most likely have their own preferred systems based on previous applications. If the experience is on a nonrelated type of project, it may be necessary for them to utilize a coding system which is more applicable to their current project assignment.

The most common coding system for typical building construction type projects is the Construction Specification Institute (CSI) *MasterFormat*™. This system is very compatible for computer application on site.

MATERIALS AND EQUIPMENT

General

Effective cost control for materials and equipment depends on the efforts of both home office and field office. Extremely close cooperation is required with both offices. If either party fails to maintain complete control, it will tend to dilute the efforts of the other party. Material costs must be controlled by constant review of both quantities and unit costs.

Quantities are controlled first during the design stage by the production of economic and cost-efficient designs which are a measure of the skill and experience of the design team. During the construction phase, costs are controlled by the economical use of materials (e.g., control of breakages, losses, theft, and wastage of materials, and in the judicious reuse of materials such as form lumber).

Initial cost savings are first controlled by the choice of the most economical types of material and equipment which are capable of performing the required task and then by the adherence to up-to-date and competitive purchasing practices (such as competitive bidding and bulk purchasing when practical).

Budget

When the budget estimate is prepared, it is most important that all material quantities be calculated carefully and unit or lump-sum costs be established on a subaccount basis. The budget estimate must include all costs associated with material deliveries such as taxes, freight, duty, and escalation where applicable. The worksheets and records that are prepared during the preparation of the budget estimate should be maintained for future comparison. These records will show the scope of work at the time the budget estimate was prepared. Deviations during and after construction can then be established and substantiated.

Ordering of Materials

As soon as a notice of award is received, the purchasing agent should meet with the project management and construction management teams to determine which items are to be purchased by the home office and which items will be obtained locally by the field office. Responsibilities thus are indicated with respect to each item on the control budget, which is prepared as soon as the budget estimate becomes available.

An equipment list showing all equipment ultimately required for the project and a materials list itemizing all required materials are prepared. As drawings are completed, bills of material should be prepared that list all items of equipment and materials covered by the drawings. To obtain materials or equipment, a requisition is then completed describing in appropriate detail the items required. It is extremely important that adequate descriptions, catalog numbers, finishes, etc., be provided for purchasing's use. Many times, individuals in the purchasing group will not fully understand the material or products which are being purchased. Incomplete information can result in incorrect equipment and poor quotations.

Coding Purchase Requisitions

The purchase requisition is implemented by engineering or construction for the purchasing department to procure the necessary materials and equipment for the project. The requisition is routed through the cost section for assignment of a proper cost code. Each item of the requisition must be individually coded if more than one item is contained therein. As soon as the code has been added, the requisition is then returned to the purchasing department for review and issuance to the proper vendor.

Each requisition must have the proper cost account code to allow for proper allocation into the system. Purchases made through the home office should be approved by the project engineer or manager and purchases made through the field office should be approved by the superintendent of construction. Each purchase order should show clearly the price of the item and other applicable costs including duty, freight, and taxes that may be applicable.

One definition should be made here which will apply throughout the chapter. *Commitment* is used to denote the sum of money that the company has obligated itself to pay. This includes the value of purchase orders placed and contracts signed and also payments for labor for which there may be no formal signed contract. Accurate records must be kept for unit price work which is a measure of work completed rather than a purchase order or invoice. Commitment should not be confused with *records of payments,* which are generally kept by the accounting group. The completion of a purchase order constitutes a commitment of funds. For proper control to be exercised at the appropriate time, when the project engineer or manager or the construction superintendent is asked to approve a purchase order, he or she must be provided with the proper data showing the budgeted allowance for the items involved. (When possible, this check should be carried out when the requisition is prepared.)

The Cost Engineer and Purchasing Function

The purchasing agent or purchasing department is generally responsible for all purchases required to support the project team. No purchases or commitments should be made without going through the proper channels. Unauthorized purchases can result in duplication of materials and equipment or unallocated costs. The cost engineer has a definitive role to play in the implementation of the purchasing functions. The engineer or construction superintendent initiates the purchase request, which is then prepared by the purchasing department or purchasing agent. The individual requesting the purchase order is responsible for properly describing the item to be purchased and including the cost account number and the budget evaluation for the items as indicated in the authorized budget. As the purchase order is routed through, the cost engineer must check that the proper cost code and budget have been entered onto the requisition. It is also the cost engineer's responsibility to alert the project manager when a requested item deviates from the control budget or does not exist in the control budget. Prior to the authorization for purchasing such items, a complete evaluation must be made to ascertain and verify the necessity for the specific item. This authorization becomes the responsibility of the project manager, who ultimately will be responsible for cost overruns. The budget allowance and delivery schedule of the item become two of the prime factors in evaluating and reviewing bids received by the purchasing department.

The purchasing department is responsible for the evaluation of *commercial aspects:* the estimated costs of transportation, escalation, payment terms, and other applicable conditions. The purchasing department should also participate with the project manager in the preparation of variance requests to use foreign material when the "Buy American Act" may be applicable to the contract. This requires an analysis of cost variations and delivery schedules. Upon receipt of bids, the purchasing agent should forward to the requesting engineer copies of the bids for review and recommendation. The engineer or construction superintendent is the individual who is responsible for the technical evaluation of the materials or equipment for specification compliance. The purchasing department should not issue purchase orders for technical equipment and materials without approval from either engineering or the construction superintendent. Compliance with these requirements is essential.

After a comparison of bids has been initiated by the purchasing department and reviewed by the engineers, and a recommendation of bidders to be selected has been made, the cost engineer will be issued a complete breakdown of bids received. This will allow the cost engineer to do a complete cost analysis with respect to the current budget allowed for the item. As soon as this analysis has been completed, authorization to issue the purchase order can proceed.

When quotations are received that are not in conformance with the current budget and delivery schedule, the cost engineer is responsible for notifying the project manager accordingly. Once given this information, the project manager can then determine what course of action may be required.

The cost engineer is the individual who can best perform this analysis because of association with the estimators and participation in the preparation of the original bid estimate. Every item of the budget represents a cost center that has a definite meaning and scope of work to be performed. The bids received through the purchasing group must be compared against the applicable cost center prior to issuing a purchase order. What the cost engineer must determine is what else must be added to this cost center that was included in the budget estimate. An example of this would be the procurement of a major piece of equipment. The cost center that was included in the budget may represent a total package and not just the major piece of equipment. The cost engineer will then have to incorporate the cost for piping, valves, etc., that will be required for systems operation. Many times a vendor will provide only the equipment and piping, and other utility connections remain to be completed. These may or may not be included as part of the specific cost center. In light of this, the actual bid prices received may be only a small part of the cost center; thus, the purchasing department is not in a position to fully evaluate the best prices. If the cost engineer is capable of providing this type of specific information, this will provide the means for control of purchased items prior to the issuance of a purchase order and not after the fact.

Purchase Order Register

Purpose. The purpose of the purchase order register is to record in numerical sequence of purchase order number each commitment of materials and equipment as it is made. The log is utilized for the preparation of the cost report.

Prepared By. The register should be prepared by the prime individual in charge of purchasing. If this is being done through a home office, the register should be prepared by them. If purchasing is being done both by the home office and the field office, a consolidated log should be maintained. When the field office is purchasing materials and equipment, the effort must be coordinated. With this in mind, it is an assignment that can readily be maintained by the cost engineer.

Frequency. The register should be continuously updated as purchase orders are issued.

Preparation. The purchase order register should be established immediately upon project authorization. Any and all material and acquisitions for the project should be included in the register.

As soon as site work begins, a single party, whether it be at the home office or the field office, should then have the responsibility of maintaining the coordinated log. From a convenience standpoint, whichever office is doing the primary purchasing should also maintain the register. If not being maintained by the cost engineer, this information should be provided to him or her on at least a weekly basis. A sample purchase order register is shown in Fig. A4-1.

Commitment Record

Purpose. The purpose of the commitment record is to record by subaccount all commitments for material and equipment. The commitment record is utilized for the preparation of the cost report and further delineates information which is similarly shown in the purchase order register.

Prepared By. This document should be maintained by the same individual that maintains the purchase order register. This is also a task which can be assigned to the cost engineer. This report also requires a coordinated effort between the home office and the field office when purchasing is a dual role.

Frequency. The commitment record should be maintained on a continuous basis as information is received.

Preparation. The project accountant should prepare the commitment record transparency sheet, filling in each appropriate subaccount immediately upon project authorization. Typical subaccounts are as follows:

CONSTRUCTION COST CONTROL A4-9

PURCHASE ORDER REGISTER

CLIENT: ABC CHEMICAL CO. JOB ORDER NO. SHEET OF

P.O. NO.	VENDOR AND DESCRIPTION	QUANTITY	UNIT PRICE	AMOUNT	OTHER EST. CHARGES INCL. OR TO BE ADDED TO P.O. FREIGHT / ESCALATION / OTHER	ESTIMATED P.O. TOTAL	ACCOUNT	REMARKS
EC-1	Delta Southern							
	Quench Tower T-1001	1	40,100	40,100	500 / 200 / –	40,800	A1.11	
	Amine Scrubber T-1002	1	43,500	43,500	600 / 200 / –	44,300	A2.11	
	Caustic Scrubber T-1003	1	45,000	45,000	600 / 200 / –	45,800	A3.11	
	Total – EC-1			128,600	1,700 / 600 / –	130,900		
EC-2	Fritz W. Glitsh & Sons Inc.							
	Ripple Trays – Quench Tower	5	1,200	6,000	Incl. / Firm / –	6,000	A1.12	
	Valve Trays – Amine Scrubber	34	250	8,500	Incl. / Firm / –	8,500	A2.12	
	Valve Trays – Caustic Scrubber	28	250	7,000	Incl. / Firm / –	7,000	A3.12	
	Total – EC-2			21,500		21,500		
EC-3	Industrial Steel Products Co. Inc.							
	Galvanized Steel – Pipe Racks	400 tons	340 ton	136,000	Incl. / Firm / –	136,000	D3.11	
	Galvanized Steel – Fractionation Structure	190 tons	360 ton	68,400	Incl. / Firm / –	68,400	D1.11	
	Total – EC-3			204,400		204,400		
EC-4	Samuel Moore & Co.							
	12 Tube Multiple Tube Instrument Cable	14,000 ft	2.26 ft	31,640	Incl. / Firm / –	31,640	K70.1	
	4 Tube Multiple Tube Instrument Cable	6,000 ft	895 ft	5,370	Incl. / Firm / –	5,370	K70.1	
	Total – EC-4			37,010		37,101		
EC-5	Chamman Division – Crane Co.							
	24" 125 Lb C.I. Tilting Disc Check Valves	3	2,600	7,800	Incl. / Firm / –	7,800	C2.71	
EC-6	Barton Instrument Corp.							
	Model 200 Flow Indicators	68	190	12,920	100 / Firm / –	13,020	K10.1	
EC-7	General Electric Company							
	2,300 V Substation	1		52,500	Incl. / Firm / –	52,500	E1.1	
	480 V Substation No. 1	1		14,800	Incl. / Firm / –	14,800	E1.1	
	480 V Substation No. 2	1		14,800	Incl. / Firm / –	14,800	E1.1	
	Total – EC-7			82,100		82,100		
EC-8	Byron Jackson Pumps Inc.							
	Condensate Pumps							
	P2001 Centrifugal Pump			1,500	Incl. / Firm	1,500	P1.11	
	P2002 Centrifugal Pump			1,800	Incl. / Firm	1,800	P2.11	
	P2003 Centrifugal Pump			1,500	Incl. / Firm	1,500	P3.11	
	P2004 Centrifugal Pump			1,800	Incl. / Firm	1,800	P4.11	
	Total – EC-8			6,600		6,600		

FIGURE A4-1 Purchase order register.

Estimated home office purchase—quantity

Estimated field office purchase—quantity

Budget—net amount

Budget, taxes, etc. (if applicable to the subaccount)

Budget—total

As soon as a purchase order or an amendment to a purchase order is received, it should be immediately recorded by the project accountant in the appropriate column. When authorized changes are approved which affect the subaccount, appropriate entries showing quantity, amount, etc., should be made in the line marked "authorized change number." The next line in each case may be used to show the current approved budget for the account (original budget amount plus the value of all authorized changes to date).

From the project authorization until the start of field operations, the commitment record (Fig. A4-2) should be maintained in the home office. Then, the maintenance

FIGURE A4-2 Commitment records.

Cost Report—Materials and Equipment

On a regular basis, a cost report of materials and equipment is prepared to show the quantity and amount of commitments to date, forecasts to complete, current estimated completion costs, and current approved budget allowances (see Fig. A4-3). Variations between the current estimated completion cost and the current approved budget allowances are shown in the last two columns.

The information for the preparation of this report is obtained from commitment records with appropriate references to the purchase order register, approved change orders, and pending engineering changes.

COST REPORT ______

MONTH ENDING ______ JO. NO. ______ PROJECT ______

REPORT NO. ______
SHEET ______ OF ______
DATE PREPARED ______
PREPARED BY ______

ACCT. NO.	DESCRIPTION	UNIT	COMMITTED						FORECAST TO COMPLETE			CURRENT EST TOTAL			APPROVED BUDGET (INCL. AUTHORIZED CHANGES)									OVERS +	UNDERS −
			1			2			3			4 = (2 + 3)			5			6			7 = (5 + 6)			8 = (4 − 7)	
			FOR PERIOD			TO DATE									ORIGINAL BUDGET			AUTHORIZED CHANGES			CURRENT BUDGET			DIFFERENCE BETWEEN CURRENT ESTIMATED TOTAL COSTS AND CURRENT BUDGET	
			QUANTITY	UNIT COST	AMOUNT	QUANTITY	UNIT COST	AMOUNT	QUANTITY	UNIT COST	AMOUNT	QUANTITY	UNIT COST	AMOUNT	QUANTITY	UNIT COST	AMOUNT	QUANTITY	UNIT COST	AMOUNT	QUANTITY	UNIT COST	AMOUNT		

FIGURE A4-3 Cost report form.

Purpose. This report will show by subaccount for material and equipment items, quantities and costs committed to date with forecasts to complete and to provide a comparison between committed estimated completion costs and current approved budget allowances.

Prepared By. This report requires a joint effort between the project accountant and the cost engineer. Ideally, this report is maintained in the field.

Frequency. The material and equipment report should be maintained on at least a monthly basis. At certain times throughout the project, when major groups of purchases are being made, the report should be updated more frequently. With the use of a PC at the job site, this report may be kept current on a weekly basis.

Preparation. As soon as the control budget has been completed, the cost engineer should prepare the master transparency sheets showing for each material and equipment subaccount:

- Account number
- Description
- Unit
- Original budget—quantity
- Original budget—amount

These reports should be maintained in a single location, which is most logically the site office. When purchasing is being done at the home office, the information must be coordinated on a regular basis between the office accountant and the site accountant. When the report is being prepared, the project accountant or field accountant uses the commitment records and fills in the quantity and amounts committed to date. From the change-order records, the cost engineer is then able to fill in the quantity and amounts of authorized changes.

After checking the pending engineering change record, making appropriate references to the purchase order register and commitment records, and, if necessary, talking to the project manager and other site personnel, the cost engineer then completes the quantity and amount of forecast to complete. With this information, the project accountant is then able to complete a report and return it to the cost engineer. The cost engineer must then review the complete report, explaining briefly any discrepancies, large overruns, or large underruns.

SUBCONTRACTS

Lump-Sum Price Contracts

Subcontracts issued on a lump-sum basis are treated as a commitment from the time the contract is executed. From a cost control standpoint, very little is achieved by the monthly or routine reporting of a contract broken into its component subaccounts. Unless the client insists on such a detailed regular report, the preferred method would be percentage complete of the subcontract agreement.

At the end of the project, each contract can then be broken down into appropriate subaccounts in the completion cost report. The subcontract agreements should contain a requirement that the subcontractor submit subaccount information at the completion of the subcontracted work.

Unit Price Contracts

With unit price contracts, the company's commitment at any time is usually the equivalent value of the work completed or in place by the subcontractor. When utilizing such contracts, it is necessary that a suitable procedure be established and followed whereby the completed work to date is measured or calculated at regular intervals and the appropriate details and backup information are forwarded to the project accountant for entry into the commitment records.

Progress Payments

For both lump-sum and unit price subcontracts, progress payments based on work in place are generally made on prestipulated intervals—most commonly, monthly. When such payments are submitted by a subcontractor, it is necessary for the project superintendent or the project manager to review construction progress and verify that the work has been satisfactorily completed. It is, therefore, necessary that a suitable procedure be followed for reporting progress.

Reports

Subcontractor Register
Purpose. The purpose of the subcontractor register is to record in numerical sequence by contract number each subcontract agreement entered into. The subcontract register is used for reference in preparation of the cost report for subcontracts.

Prepared By. This report should be prepared by the purchasing agent in the home office initially and may ultimately be maintained on-site.

Frequency. The register should be updated continuously each time a new subcontract agreement has been entered into.

Preparation. The register should be started when project authorization is granted. The form should be established prior to the issuance of any subcontract agreement, and updates should be recorded as soon as the first subcontract agreements are entered into. Ultimately, this report should be transferred to the site when site operations begin. Change orders will necessitate revisions or additions to the subcontract register based on the scope of work contained in the change orders. The field engineer who is handling change orders can generally maintain this register.

For a sample of the subcontract register form, see Fig. A4-1, which is the purchase order register; it serves the same purpose as the subcontractor register.

Commitment Records
Purpose. The purpose of commitment records is to record by account or subaccount the total value of lump-sum contracts, including contract changes, and the value of the work done on unit price contracts. This report is used in the preparation of the cost report for subcontracts.

Prepared By. These records should be initially prepared by the project accountant and maintained in the home office. When the field work begins, these records should be transferred to the site project accountant to maintain. (In some firms, the commitment record is maintained on-site by the cost engineer.)

Frequency. These records must be maintained on a continuous basis.

Preparation. As soon as the control budget is completed and available, the project accountant should prepare the commitment record transparency sheets, filling in the budget allowances for each subaccount as follows:

- Quantity
- Worker-hours (if required)
- Labor—amount
- Materials—amount
- Total—amount

Immediately on receipt of each contract or contract change, the individual maintaining the commitment records should enter the following:

Lump-sum price contracts
- Reference contract number
- Quantity (if available)
- Labor amount (if available)
- Material amount (if available)
- Total amount (if available)
- Unit price contracts
- Reference contract number
- Unit rate—in accordance with details of contract

When site operations commence, the project accountant should forward to the field accountant an up-to-date set of commitment records showing all home office commitments. From that time forward, the field accountant should maintain detailed records for all commitments made by both the home office and the field office.

Lump-Sum Price Contract. See preceding list under this heading. Similar details are entered each time a field order is issued or a copy of a contract change is received. It should be noted that a contract change can result from an engineering change originating in the home office or from a field order. The authorization of a field order constitutes a commitment, although the value of the commitment may be unknown at the time. Ultimately, a contract change is issued to formalize each field order or group of field orders; care should be taken not to duplicate the entry of any commitment.

Unit Price Contracts. See the preceding list under this heading. At a predetermined period monthly, the field accountant records the number of units completed and the value of the work in place. For a sample unit price commitment record, see Fig. A4-4.

Cost Report—Subcontracts

Purpose. The purpose of the cost report for subcontracts is to record subcontracted items, quantities, and costs committed to date with forecasts to complete and to provide a comparison between current estimated completion costs and current approved budget allowances.

FIGURE A4-4 Unit price commitment record.

Prepared By. These reports are prepared and maintained by the project accountant and the cost analyst before site work begins. Then the project accountant in cooperation with the cost engineer should maintain these records on-site, recording all subcontract activity for the project whether initiated in the home office or the field.

Preparation. As soon as the control budget is completed, the cost engineer should prepare the master spreadsheets showing for each subcontract account:

- Account number
- Description
- Unit (if applicable)
- Original budget—quantity
- Original budget—amount

Sets of spreadsheets are sent to the project accountant for home office–prepared reports and to the field accountant for site-prepared reports. When preparing these reports, the project accountant and the field accountant utilize the commitment records and subcontract register to obtain the quantity and amount committed to date.

From the change-order records, the cost engineer fills in the quantity and amount for authorized changes. After reviewing the pending engineering change records;

making appropriate references to subcontract register, commitment records, contract changes, and field orders; and, if necessary, speaking to the project manager and the project engineer or superintendent of construction, the cost engineer fills in the quantity and amount of forecast to complete.

The project accountant or field accountant finalizes the report and returns it to the cost engineer who uses the remarks column to briefly describe any discrepancies and differences from the previous reports. The cost report form (Fig. A4-3) can be utilized for this purpose.

FIELD LABOR

General

The field superintendent, in conjunction with the supervisory staff, is responsible for directing and controlling the efforts of the field forces to ensure that the work is performed in the most cost-efficient manner and within the project schedule. Inefficient labor performance is the most common cause of project overruns.

It is essential that an effective day-to-day reporting system be maintained to measure the performance of the field forces compared to either the budget estimate or productivity standards. The system described as follows utilizes the information on hours and cost available from the field accounting staff, supplements them by quantity reports, and analyzes them to produce a database which the superintendent of construction then uses to make key decisions.

Basic Accounting Data

1. The time-clock card provides the accounting team with the detailed hours worked by each employee and is used only in the preparation of the payroll journal.
2. The foreperson's time card provides detailed daily totals of craft hours, which are coded by the proper cost account classification. The foreperson's time cards are incorporated into the labor distribution schedule, which shows the daily hours expended by craft and identified by cost classification codes on a monthly basis.
3. The monthly analysis is used to record the weekly totals of worker-hours and cost by craft. At the conclusion of each month, the monthly average hourly rates per craft may be calculated and used on the labor distribution schedule showing the month's total labor costs by account and craft. At the end of the month, the monthly totals of worker-hours and labor costs per account are posted to the cost ledger sheets. These reports provide an accurate recording of craft hours and labor costs required for accounting purposes.

Cost Control

When combined with the accounting data, the methods described as follows for recording units of work performed provide the complete database for the analysis of productivity for craft labor control and reporting.

Foreperson's Quantity Report. The foreperson's records should contain detailed records of completed work coded to specific cost accounts as work progresses. The reports should be maintained minimally on a weekly basis; for critical work and for areas where significant progress is being made, they should be maintained on a daily basis.

Worker-Hour and Quantity Record. The worker-hour and quantity record is used to record the periodic and cumulative worker-hours expended and units of work completed for each cost accounting classification. The information for these records is obtained from the foreperson's time cards and the units of work from the foreperson's quantity report. This form also provides space for showing the estimated quantities and worker-hours in the approved budget. This provides a basis for review of performance and allows the user to immediately identify areas that may be overrunning the initial budget. There is also space provided for showing quantities and worker-hours for revised estimates in the event that the specific area is affected by a change order.

As work is started on a new account, the cost engineer should closely monitor the unit rates of worker-hours expended per unit quantity of work completed to ascertain progress in comparison with the budgeted rate. The cost engineer should observe the work in the field and discuss the productivity with the superintendent and the supervisory staff to assist them in forecasting the trend (e.g., will production rates increase or decrease as specific work progresses?). The cost engineer must alert the superintendent and supervisory staff at the earliest possible time so appropriate action may be implemented. The importance of this cannot be overemphasized: Prompt scrutiny of these records by the cost engineer may avoid costly overruns.

On at least a monthly basis, the field engineer should perform a physical progress survey in the field and check quantities in place utilizing contract drawings and bills of material. Discrepancies between these measurements and cumulative figures obtained from the foreperson's quantity reports must be adjusted in the cumulative column of the worker-hour and quantity record and in the "units completed to date" column of the next foreperson's quantity report.

Worker-Hour Report. The worker-hour report should be prepared weekly and will show in summary form worker-hours worked during the week for each account by craft. The report will also show by account the total worker-hours worked and the total labor cost for the week.

Cost Report—Field Labor. This report is prepared on a monthly basis and shows by account the quantities and unit labor costs for the period and to date with forecasts to complete in comparison with the budgeted amounts. Work completed and in place to date is posted from the worker-hour and quantity records and cost data from the cost ledger sheets. The forecast to complete is provided by the cost engineer after discussions with the project superintendent and the supervisory staff.

Forms and Procedures

Basic Accounting Forms. The following forms are standard record-keeping documents which are in general use in the industry: time card, payroll journal, monthly analysis, foreperson's time card, labor distribution schedule, and cost ledger.

Cost Control Forms. The preparation and use of the following forms are described in detail and are included as follows. It is essential that records of this nature be maintained if proper cost control is to be achieved on the project.

Foreperson's quantity report
Worker-hour and quantity report
Cost report—site labor

The following form is mentioned and described briefly to show how it can be utilized within the system if required: worker-hour report.

Basic Accounting Forms

Time Card

Purpose. The time card records when each individual employee arrives and leaves the project site. The data collected are used for the preparation of the payroll journal.

Prepared By. The time card is prepared by the project timekeeper.

Frequency. The time card is kept on a daily basis.

Preparation. Each card generally documents one person for each week. New cards are provided on a weekly basis. Standard time and overtime work are recorded separately.

Payroll Journal

Purpose. The journal is used daily to record the hours worked by the individual employee and weekly for calculation of payrolls. Each week, total worker-hours and labor costs by craft are summarized for preparation of the monthly cost analysis.

Prepared By. The payroll journal is prepared by the timekeeper.

Frequency. The payroll journal is kept on a weekly basis (worker-hours worked recorded daily).

Preparation. On a daily basis, the time cards from the previous day are reviewed and utilized to record days and hours in the proper columns of the payroll journal. Both standard time and overtime are recorded. At the conclusion of each week, the hours worked are totaled and entered into the "hours worked" column. The payroll is calculated, and a summary sheet showing the total worker-hours and cost for each craft is then prepared.

Monthly Analysis

Purpose. The purpose of the monthly analysis is to record during the month the weekly gross payroll costs by craft and to provide a method of determining the average wage rates for each craft during the reporting period.

Prepared By. The monthly analysis is prepared by the timekeeper.

Frequency. One set is used per month for recording data weekly.

Preparation. The total worker-hours and labor costs for each craft are transcribed weekly from the payroll journal summary sheets to the appropriate weekly portion of the monthly analysis sheet. At the conclusion of each month, the weekly worker-hours and labor costs are totaled and the cost divided by the total worker-hours to determine the average hourly labor cost per craft. These hourly rates will then be established and used on the appropriate labor distribution schedules to calculate the month's total cost for each subaccount.

Foreperson's Time Card

Purpose. The purpose of the foreperson's time card is to record by subaccount the worker-hours worked per employee and per crew. The data collected are used to calculate unit labor costs.

Prepared By. The foreperson's time card is prepared by the craft supervisor.

Frequency. It is prepared on a daily basis.

Preparation. Each day, the foreperson enters the subaccount number and a brief description of work being performed by the crew. The foreperson must record the hours worked per person and the total number of worker-hours for the crew for each subaccount. At the completion of the shift, the foreperson should sign the cards and turn them over to the craft supervisor.

Approved By. The foreperson's time card is approved by the craft supervisor.

Routing. The craft supervisor then turns the cards over to the timekeeper who cross-checks the hours shown with the time cards and records them on the payroll journal. The timekeeper is then responsible for recording the worker-hours worked onto the appropriate labor distribution schedules. The foreperson's time cards are then passed to the cost engineer who enters the worker-hour data on the appropriate worker-hour and quantity records.

Labor Distribution Schedule

Purpose. The purpose of the labor distribution schedule is to segregate into subaccounts labor hours worked by craft during the month so that monthly labor costs per subaccount may be obtained.

Prepared By. The labor distribution schedule is prepared by the timekeeper.

Frequency. One set maintained daily and aggregated weekly is used throughout the monthly period.

Preparation. A separate sheet is used monthly for each subaccount for the specific craft. Each day the hours are recorded on the foreperson's time cards and transferred to these sheets. Each week the worker-hours are totaled and balanced with those transferred from the payroll journal to the monthly analysis.

At the completion of each month, the worker-hours are totaled and multiplied by the average craft rate (obtained from the monthly analysis) to determine the total labor cost for the subaccount during the month.

Routing. The schedules are used by the field accountant for posting monthly labor costs to the appropriate cost classification at the completion of the month.

Cost Ledger

Purpose. The purpose of the cost ledger is to record by subaccount all costs and worker-hours expended. The data recorded are used to prepare the cost report for field labor.

Prepared By. The cost ledger is prepared by the field accountant.

Frequency. It is prepared on a monthly basis.

Preparation. At the completion of each month, the total worker-hours and total labor costs per subaccount are recorded from the labor distribution schedules and entered into the appropriate cost ledger sheets. The worker-hours expended and total labor costs to date are obtained by totaling the respective columns.

Cost Control Forms

Foreperson's Quantity Report

Purpose. The purpose of the foreperson's quantity report is to record each crew and, by subaccount, the units of work in place or completed on a daily basis and the foreperson's estimate of units to complete. The data recorded are used to prepare the worker-hour and quantity record.

Prepared By. This is prepared by the craft supervisor.
Frequency. It should be maintained on a daily basis.
Preparation. From the onset of construction, the cost engineer establishes the following items to be filled in on an appropriate number of forms for a monthly period to show each labor subaccount:

- Account number
- Craft
- Unit of quantity
- Description of work
- Original budget quantity (a)

When change orders are approved or revised estimates prepared, the cost engineer must then arrange for the following items to be filled in on future sheets:

- Budget including authorized changes quantity (b)
- Current estimate completion quantity (c)

At the start of each day, the craft supervisor must distribute appropriate report forms to the forepersons to cover the items which their crews will be working on that day. At the completion of the day's work and prior to returning the report to the supervisor, each foreperson fills in:

- Units completed today (e)
- Units completed to date (f)
- Foreperson's estimated units to complete (g)
- Total worker-hours worked today (h)

Routing. The craft supervisor transcribes figure (f) from today's card onto the following day's card against item (d), units previously reported. After signing the foreperson's quantity cards, the craft supervisor then forwards them to the cost engineer. A sample foreperson's quantity form is shown in Fig. A4-5.

Notes. This report is intended to be completed and submitted daily to correspond with the worker-hours reported by the foreperson's time cards. In cases such as erection of a specific piece of equipment taking several days, the foreperson's quantity report may be submitted instead on an event basis at the completion of the erection rather than daily. The craft supervisor should notify the cost engineer when reports are being accumulated to cover more than a one-day period.

Progress Measurement. On at least a monthly basis, it is necessary for the field engineer to physically check progress measurements in the field and compare them with quantities completed, using the drawings and bills of material. Any discrepancy between these check measurements and cumulative figures from the foreperson's quantity reports must be immediately adjusted by the cost engineer in the "cumulative" column of the worker-hour and quantity record in the "units completed to date" column of the next day's foreperson's quantity report.

Worker-Hour and Quantity Record

Purpose. The purpose of the worker-hour and quantity record is to record and accumulate by subaccount the worker-hours expended and the units of work com-

CONSTRUCTION COST CONTROL

```
                    FOREPERSON'S  QUANTITY  REPORT        DATE _____
J.O. NO _____ UNIT _____ CRAFT _____ ACCOUNT NO _____
DESCRIPTION OF WORK _____

(a) Original Budget Quantity
(b) Budget Incl. Auth. Changes No's ____ to _____ Quantity
(c) Current Estimated Completion Quantity

(d) Units Previously Reported
(e) Units Completed Today
(f) Units Completed to Date

(g) Foreperson's Estimated Units to Complete

(h) Total Worker-hours Worked Today | Signed | Foreperson | Supervisor

    Field Engineer's Progress Measurement
    Field Engineer's Remarks

                                       | Field Engineer | Date
```

FIGURE A4-5 Foreperson's quantity report.

pleted and in place so that a comparison with the budget allowance of worker-hours per unit is available. The data recorded are used in the preparation of the cost report for field labor.

Prepared By. This is prepared by the cost engineer.

Frequency. It is prepared on a daily basis.

Preparation. When construction begins, the cost engineer must arrange for sheets to be prepared for each labor subaccount showing:

- Account number
- Craft
- Unit
- Description of work
- Original budget—quantity
- Original budget—total worker-hours
- Original budget—worker-hours per unit

The cost engineer must on a daily basis obtain the previous day's foreperson's quantity report and foreperson's time cards and enter the day's quantities and worker-hours into the "actual for period" columns of the appropriate worker-hour and quantity record sheets (see Fig. A4-6). The cost engineer must also forward the cumulative total to date and calculate and enter the worker-hours per unit both for the day and for the cumulative totals. At the completion of the month, worker-hours recorded are checked against those on the labor distribution schedules and corrections or adjustments are made if necessary.

Progress Measurement. Periodically, but not more often than monthly, the field engineer should carry out a physical progress survey in the field and check quantities of work completed utilizing drawings and bills of material. The field engineer should also check the quantities to complete. Details of quantities are then established and forwarded to the cost engineer who enters these quantities in the "cumu-

WORKER-HOUR AND QUANTITY CONTROL ANALYSIS

PERIOD ENDING	SHEET OF	J O NO

ACCOUNT NUMBERS	DESCRIPTION	WORKER-HOURS DATA						BUDGET EST & AUTH CHANGES			ACTUAL			REVISED		
		BUDGET (1)	TO DATE (2)	PREVIOUS (3)	CURRENT (4)	% EXPEND (2) (5)	PHY COMP % (6)	QUANTITY	UNIT PRICE	AMOUNT	QUANTITY IN PLACE	UNIT PRICE	AMOUNT	QUANTITY	UNIT PRICE	AMOUNT

FIGURE A4-6 Worker-hour and quantity control analysis.

lative quantity" column and the "current established completion quantity" column on the next line of the appropriate worker-hour and quantity record. The cost engineer then rules off data previously recorded above this line and henceforth uses this corrected quantity to establish future quantities. If measured quantities differ, the cost engineer must reconcile the difference the cumulative obtained from the foreperson's quantity report and make the necessary corrections against the units previously reported on the next day's foreperson's time card.

Authorized Change Orders. Immediately upon receipt of an approved change order which affects a specific subaccount, the cost engineer must record the new budgeted quantities, worker-hours and worker-hours per unit in the "current budget" column and rule out the entry above.

Current Estimated Completion. Periodically, but not more often than monthly, the cost engineer analyzes all other data recorded to date and, after discussions with the superintendent, the supervisory staff, and the field engineer, records into the "current estimated completion" columns the best estimate of (total) quantities, total worker-hours, and worker-hours per unit for each subaccount. As this is recorded, the line above is then ruled out.

Notes. There are some labor items for which it is not practical to submit daily quantity reports. In such cases, in lieu of a daily entry, an entry for the period covered by the foreperson's quantity report is made and matched to the appropriate foreperson's time cards for the same period.

The worker-hour and quantity records give the earliest available information for comparing actual labor productivity with the productivity from the budget estimate. It is extremely important that in the early stages of each section of work, the cost engineer monitor closely the actual unit rate in comparison with the budgeted unit rate.

The cost engineer must observe the performance of the work in the field and discuss productivity with the project superintendent and the supervisory staff to assist in forecasting future trends and the project schedule. The cost engineer must also advise the superintendent of construction promptly of actual or suspected problem areas so that corrective action may be initiated.

Cost Report—Field Labor

Purpose. The purpose of the cost report for field labor is to record and show for each labor subaccount quantities, units, and total costs committed for the period, committed to date with forecasts to complete, and to provide comparisons between current estimated completion costs and current approved budget allowances.

Prepared By. This is prepared by the field accountant in conjunction with the cost engineer.

Frequency. It is prepared monthly, or more frequently if required.

Preparation. When the control budget is established, the cost analyst prepares master spreadsheets showing for each labor subaccount:

- Account number
- Description
- Unit
- Original budget—quantity
- Original budget—unit cost
- Original budget—amount

Sufficient sets of the spreadsheets are sent to the cost engineer to cover the number of reports required for the duration of the project.

For each report, the field accountant uses the cost ledger sheets and worker-hour and quantity records and fills in:

- Quantity for period (a)
- Quantity for date (b)
- Amount for the period (c)
- Amount to date (d)

The field accountant must also fill in the unit costs:

- Unit cost per period $\quad\dfrac{(c)}{(a)}$ (by calculation)
- Unit cost to date $\quad\dfrac{(d)}{(b)}$ (by calculation)

Utilizing the change-order records, the cost engineer must fill in the quantity, unit cost, and amount for all authorized changes.

After reviewing the pending engineering change records, making appropriate references to worker-hour and quantity records, and talking to the superintendent and the field engineer, the cost engineer then fills in the quantities, unit costs, and amounts for forecast to complete.

The field accountant then completes and returns the cost report to the cost engineer who uses the remarks column to explain discrepancies from the previous report. The cost report form (Fig. A4-3) can be used for this purpose.

Worker-Hour Report (Optional)

Purpose. The purpose of the worker-hour report is to show in summary form by subaccount the worker-hours per craft, total worker-hours, and total labor costs on a weekly basis.

Prepared By. It is prepared by the timekeeper.

Frequency. It is prepared weekly if required by the superintendent of construction.

Preparation. The total weekly worker-hours per craft per subaccount are transcribed from the labor distribution schedule, which may then be added to obtain the total worker-hours per subaccount. Total costs may be calculated by* multiplying the worker-hours per craft by the average rate per craft. For a sample worker-hour report form, see Fig. A4-7.

DISTRIBUTABLE ACCOUNTS

General

Distributable accounts often constitute a significant portion of the total construction budget, and it is important that they be properly recorded and controlled. In order

---

* For weekly reports throughout the monthly period, the average craft rates obtained from the previous month's monthly analysis should be used. For the last weekly report of the month, the average rates may be obtained from the current monthly analysis.

FIGURE A4-7 Worker-hour report.

for actual costs to be compared with budgeted allowances, they must be distributed and recorded in the same fashion that they were in the project budget.

When a project is partially or completely subcontracted, many of the items described are distributed over direct accounts and may never be shown as individual items. An example of this is the rental of construction materials and equipment. On a subcontracted project, the subcontractor will generally prorate the distributable costs into the cost of the work. To the subcontractor, equipment rental would be distributable; to the general contractor, it would be considered part of direct costs paid to the subcontractor. When a field organization exists, cost for distributable items such as the cost of the field office are incurred and must be periodically reported. If these costs have been included in the budgetary allocations as a line item, record keeping during construction will be simplified.

The distributable accounts are estimated and controlled under the following categories:

1. Field offices
2. Insurance—injuries and damages
3. Temporary construction
4. Construction equipment
5. Other distributables

Field Office

This account will be charged the expenditures of the field office staff which include, but should not be limited to, the following:

1. Construction supervision
2. Field office engineering staff
3. Accounting staff
4. Other staff
   a. Labor relations
   b. Purchasing
5. Quality control

These accounts will also be charged with the material expenses that are incurred for maintaining the field complex for the life of the project.

The field cost engineer should periodically (preferably on a monthly basis) update the personnel schedule and report summary, showing actual costs incurred as compared to what has been budgeted. Expenses are similarly collected and reported for comparison against the budgeted figures. For a sample nonmanual personnel schedule and report summary see Fig. A4-8.

Insurances—Injuries and Damages

This account will be charged for the expenditures for premiums of all classes of insurance protection for the project. In addition, costs for medical, hospital, and other services resulting from personal injuries not reimbursable by insurance carriers and payments to an injured person or to the owner or any other property which

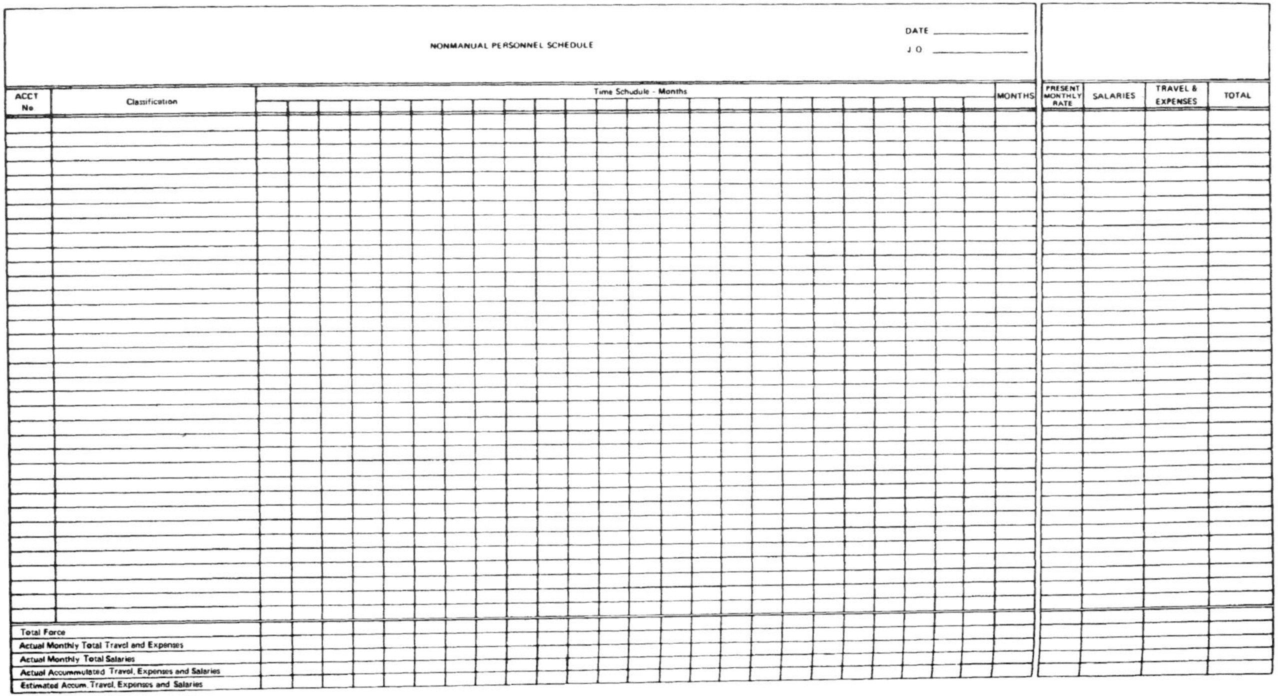

FIGURE A4-8 Nonmanual personnel schedule.

may be damaged will also be charged to this account. A partial listing of items included in this account are as follows:

    Worker's compensation insurance and employer's liability insurance
    Bodily injury and property damage insurance
    Property insurance
    State insurance funds
    Miscellaneous insurance (employee's personal insurance, fidelity and surety bonds, etc.)

Temporary Construction

This account is established to track the expenditures of all work of a temporary nature that is done solely for the purpose of construction, such as roads, storage facilities, job buildings, service systems and structures, supports and installation of equipment for construction plant, and construction equipment.

Temporary Roadways, Railroads, Parking Areas, Etc. Included in the account are all temporary transportation facilities necessary to get supervisory and craft personnel and material to and about the project. Associated items such as culverts for drainage should be included as part of the roadway work. Gin pole foundations and deadmen should be included in the account with crane mats.

Temporary Storage Facilities. Included in this account should be material storage yards and warehouses, including racks and other items for the storage of construction material. The account should also include ancillary costs associated with temporary warehouses such as plumbing, heating, and lighting installation, both within the building and in adjacent areas. Specific features of the building such as dimensions and type should be recorded in the cost ledger for future bidding purposes.

Temporary Buildings. Included in this account should be temporary buildings and facilities necessary for the performance of contract work. This should include, but not be limited to, the office buildings, shops, and locker and shower rooms. Costs associated with these facilities such as plumbing, heating, lighting, and air conditioning should be included. The account should also include costs for necessary relocation of existing buildings that will be used for the project. The building dimensions and type of construction should be recorded in the cost ledger for future reference.

Temporary Service Facilities. Included in this account are all costs for installing temporary services such as water lines, heating and gas lines, air lines, sewer and drains, culverts, and disposal equipment. Also included are light and power lines and equipment and oxygen, acetylene, and other piping manifolds. The costs for the installation start at the point of connection to the permanent service and the end at the point of connection to the temporary building or area. Also included is the cost of temporary services to permanent buildings or areas.

Miscellaneous. All expenditures for operation and supplies, including utility charges—e.g., water and electricity—required for the preceding temporary construction accounts are included in this account. Expenditures for items such as removing snow from roads and work areas, temporary closures, temporary heat, and weather protection should also be included in the miscellaneous account. For some projects, it may be desirable to break out the miscellaneous account into subaccounts to identify items such as temporary heat or winter protection.

Construction Equipment

1. This account contains types of equipment, tools, and supplies as follows:
   a. Automotive equipment
   b. Heavy construction equipment
   c. Light construction equipment
   d. Craft tools (salvageable)
   e. Expendable tools and supplies
2. Rental rates for all heavy and light construction equipment should be established for estimating and control purposes regardless of whether such equipment is owned, purchased for the client's account, or rented or leased from an independent company.
3. The heavy construction equipment (listed as item **b**) account will be charged with the expenditures for major construction equipment procured through an outside firm or purchased with the anticipation of resale value at the completion of the project and charged to the project on a rental basis. Charges for assembling and dismantling equipment should also be included in this account. Costs incurred for putting second-hand equipment into operating condition or into saleable condition at the completion of the project should be considered as a decrease in salvage value when establishing the applicable rental rates.

Operators and operation costs (fuel, oil, etc.) should be charged to the "and use" accounts. This includes operators of equipment rented on a fully maintained and operated basis.

For a sample of the construction equipment rental form and the construction equipment purchase form, see Figs. A4-9 and A4-10.

Other Distributables

Premium Pay. Costs should be included in the budget estimate if an extended work week is anticipated at the time that the estimate is prepared, covered by an authorized change if an extended work week is authorized in the change order, or accumulated from various daily premium overtime work.

Premium costs incurred for casual or incidental overtime labor should be charged to the appropriate account subdivisions. This will include the premium cost of day-to-day overtime other than scheduled for an extended work day or work shift in specific instances, such as concrete pouring or finishing.

The premium cost of overtime that is authorized over and above a standard work week or shift period should be charged to the appropriate account subdivisions.

The premium cost of scheduled overtime for one or more crafts over and above an authorized work week should be charged to the appropriate account subdivisions. Special scheduled overtime may be required and is advantageous when one or more crafts are working in confined areas or behind schedule.

RENTED CONSTRUCTION EQUIPMENT

PROJECT, CLIENT AND LOCATION	J. O. NO.
	DATE ISSUED

EQUIP. NO.	DESCRIPTION	P.O. REF.	RENTAL PERIOD			RENTAL RATE	ESTIMATED–ACTUAL			REMARKS
			START	FINISH	TIME		RENTAL AMOUNT	TRANSP. CHARGES	TOTAL	

FIGURE A4-9 Rented construction equipment.

PURCHASED CONSTRUCTION EQUIPMENT

PROJECT, CLIENT AND LOCATION	J. O. NO.
	DATE ISSUED

EQUIP. NO.	DESCRIPTION	P. O. REF.	NUMBER OF UNITS	TOTAL PURCHASE PRICE			ESTIMATED–ACTUAL			REMARKS
					NEW	USED	PERIOD OF USE ON JOB	SALVAGE CREDIT	COST TO JOB	

FIGURE A4-10 Purchased construction equipment.

Note. When overtime is scheduled for any period of time, there is a loss in productivity, and such costs must be taken into consideration when estimating the cost of work as well as the premium time portion. The project superintendent must participate in a decision to authorize overtime hours.

Employer's Labor Expenses and Benefits (Except Taxes and Insurances). Included in these accounts are transportation and subsistence allowances paid to workers, nonproductive time (reporting time, stand-by time, generated work day or work week, treatment time for job-incurred injuries, and idle time of construction equipment operators), recruiting expenses, and employer's contributions to union health and welfare and other benefit funds as set forth in working agreements. Contributions for Social Security, income tax withholding, state medical aid or group insurance and transportation (buses furnished by employer for employee transport) are included.

Survey Account. This account should include all expenditures for site layout, surveys, test pits, borings, test piles, river gauging, and the like that will be utilized for engineering studies and design.

Cleanup Account. The cleanup account covers all charges for final cleanup and final disposal of construction debris prior to turning the project over to the owner. This account should also include the day-to-day cleanup charges for work performed by an assigned crew for the duration of the project.

When cleanup work is performed by company forces, day-to-day cleanup associated with specific craft work should be chargeable to the respective "and use" accounts.

Subcontracted work that requires the contractor to perform all cleanup associated with his work will be allocated to the "and use" accounts.

General Condition Accounts. The accounts cover costs of guards and watchpersons, building permits, and licenses including inspection and other miscellaneous fees, operators of utility trucks and automobiles used for general work whose wages cannot be allocated to direct accounts, and wages of warehouse personnel.

Cost Report Form

1. Detailed cost reports similar to those prepared for material and labor must be maintained for each of the distributable accounts.
2. Each of the distributable accounts must be summarized for inclusion into the revised estimate and job-cost report.

REVISED ESTIMATE AND JOB-COST REPORT

General

Periodically, there should be a project oversight review performed in conjunction with a revised budget estimate and a job-cost report prepared. The purpose of this is to control and forecast the total cost of the project through completion. Based on specific projects, this may be warranted quarterly or semiannually. It is even more important that this evaluation be performed for projects that have experienced a

JOB TITLE

ABC Chemical Company

J.O. NO. 000
REPORT NO.   DATE

Process Unit

Account number	Description	Commitments Material	Commitments Labor	Estimate to complete Material	Estimate to complete Labor	Revised estimate Material	Revised estimate Labor	Total
	Process Equipment							
A	Towers	895,000	90,000	15,000	25,000	910,000	115,000	1,025,000
B	Boilers, steam superheaters	—	—	—	—	—	—	—
F	Process furnaces	1,050,000	53,000	160,000	62,000	1,210,000	115,000	1,325,000
G	General equipment	170,500	31,000	39,500	33,000	210,000	64,000	274,000
L	Reactors	75,000	4,000	5,000	1,000	80,000	5,000	85,000
M	Drums	306,000	12,000	12,000	6,000	318,000	18,000	336,000
Q	Storage tanks	204,000	8,000	8,000	4,000	212,000	12,000	224,000
P	Pumps and drivers	525,000	11,500	22,000	5,000	547,000	16,700	563,500
R	Compressors and drivers	1,225,000	26,500	53,000	17,000	1,278,000	43,500	1,321,500
S	Stacks	10,000	1,000	—	—	10,000	1,000	11,000
T	Heat exchangers	1,675,000	22,000	40,000	8,000	1,715,000	30,000	1,745,000
	Total—Process equipment	6,135,500	259,000	161,000	6,490,000	420,000	6,910,000	
	Process Materials							
C	Piping	1,718,500	806,000	1,081,500	1,094,000	2,800,000	1,900,000	4,700,000
D	Structures	255,500	100,500	14,500	39,500	270,000	140,000	410,000
E	Electrical	287,500	327,500	102,500	390,000	600,000	990,000	
H	Buildings	112,300	90,500	27,700	24,500	140,000	115,000	255,000
J	Civil	438,500	411,500	66,500	73,500	505,000	485,000	990,000
K	Instruments	377,500	25,500	172,500	64,500	550,000	90,000	640,000
N	Insulation and painting	485,000	636,000	95,000	94,000	580,000	730,000	1,310,000
	Total—Process materials	3,674,800	2,397,500	1,560,200	5,235,000	4,060,000	9,295,000	
	Distributable Accounts							
V	Insurance and taxes	190,000	205,000	130,000	235,000	320,000	440,000	760,000
O	Other distributable items	57,000	35,000	163,000	245,000	220,000	280,000	500,000
X	Temporary construction facilities	55,000	60,000	15,000	20,000	70,000	80,000	150,000
Y	Field office	30,500	190,500	29,500	189,500	60,000	380,000	440,000
Z	Construction tools and equipment	450,000	150,000	130,000	40,000	580,000	190,000	770,000
	Total—Distributable accounts	782,500	640,500	467,500	729,500	1,250,000	1,370,000	2,620,000
	Indirect Accounts							
U	Headquarters office	70,500	709,700	34,500	135,300	105,000	845,000	950,000
	Total—Project costs	10,663,300	4,006,700	2,416,700	2,688,300	13,080,000	6,695,000	19,775,000

FIGURE A4-11 Revised estimate and job-cost report.

A4-33

significant number of change orders or where work is being performed on a time-and-material basis. The overview should take into consideration any and all changed conditions that may have occurred during the life of the project. Obviously, the type of contract will dictate the requirements for the updated budget and its distribution. A typical sample of a revised estimate and job-cost report is shown in Fig. A4-11.

This report should be updated periodically and presented. It will be noted that the report is summarized into the following constituent parts:

- Account number and description
- Commitments
- Present material commitments
- Present labor commitments
- Estimate to complete
    Material estimate to complete
    Labor estimate to complete
- Revised estimate (material, labor, and total)
- Budget estimate (material, labor, and total)
- Differences—variations between revised estimate and budget estimate in terms of material, labor, and total

The preparation of the revised estimate should be consistent with the accounting system being used on a weekly and monthly basis for the project. All responsible groups involved in the project must submit this data to the cost engineer in the form of account code, present commitment, and projected estimate to complete.

The basis for reporting accurate cost projections is the standard working documents prepared by the various individuals on the project team. These include worker-hour charts; nonmanual personnel schedules; purchased and rented construction equipment forms; time, labor, and commitment records; purchase order register; materials committed records; and specially designed forms for the reporting and control of structural, mechanical, and electrical work items. These are all incorporated into the overall job-cost report and, when put to their ultimate use, provide a successful means of reporting and controlling project costs.

Revised Estimate Totals

The revised estimate totals are ascertained by adding the commitments to the estimate to complete. The revised estimate may be summarized as in Fig. A4-11. Items where a change is indicated from the budget estimate or the previous estimate should be verified, and if an increase or decrease is of significant value, it should be explained in a narrative form and submitted to the owner or client for review. It is essential that this report be provided at the earliest possible date to allow the maximum time to implement corrective measures.

Review of Job-Cost Report

The cost engineer has a significant, important assignment on a project and must place particular emphasis on checking the proper preparation of the following items:

1. Changes in scope—authorization and pending
2. Manual labor projections—rates and benefits used
3. Basis of calculation of insurance and taxes
4. Nonmanual projection
5. Projection of purchased and rented construction equipment including development of estimated salvage
6. Quantities, source, and use
7. Field order extras
8. Calculations of premium pay—manual and nonmanual (particularly if job is on an extended work week)
9. Check of estimated project schedules and completion date
10. Material and labor escalation
11. Contract terms regarding compensation and fee calculation

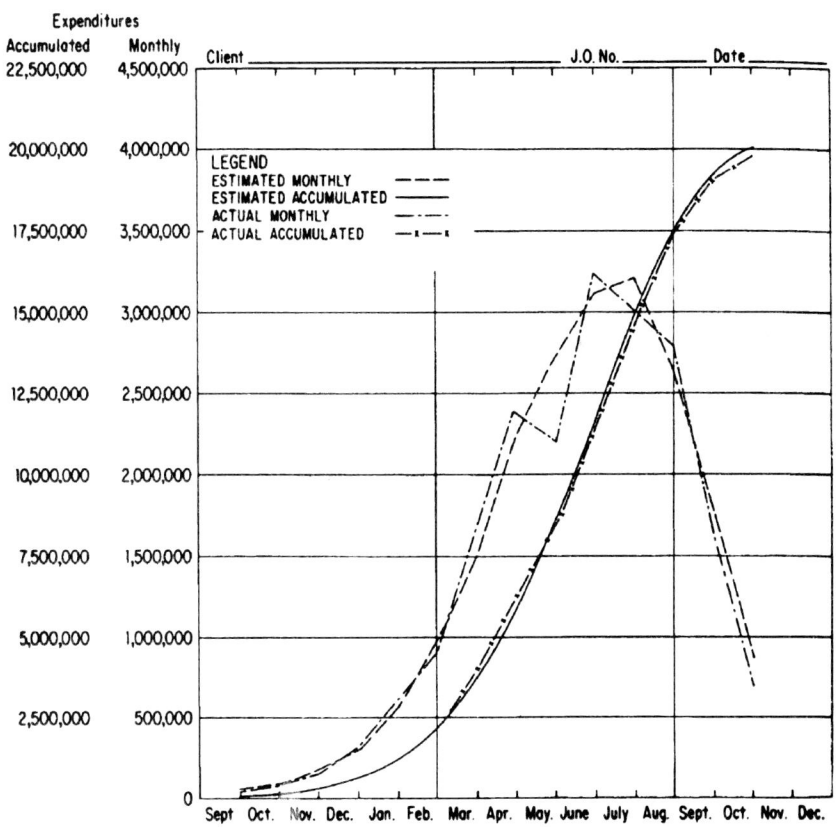

FIGURE A4-12 Typical estimated and actual expenditures chart: process unit-cost control.

The results of the cost engineer's efforts will be a complete and comprehensive report with up-to-date information. The report should be reviewed by the project management and supervision prior to issuance to the client.

SUMMARY

With today's projects becoming more and more complex and costly, a major project cannot succeed without proper controls. Maintaining proper records requires excellent documentation, accurate cost estimating, up-to-date cost controls, and up-to-date scheduling. Each of these is interrelated and ultimately will affect the success or failure of a project from a financial standpoint. Figure A4-12 is a typical estimated and actual expenditures chart graphically representing these interrelationships in a summary fashion. When utilizing proper cost controls, a chart of this nature can be maintained on a monthly basis and will work as an early warning system for management.

Figure A4-13 is a plot of cumulative costs for three types of cost: budget, actual value, and earned value. This reflects two types of problems: First, the earnings are half the budgeted showing schedule slip, and, second, actual cost is about 50 percent above earned, reflecting a significant out-of-pocket loss.

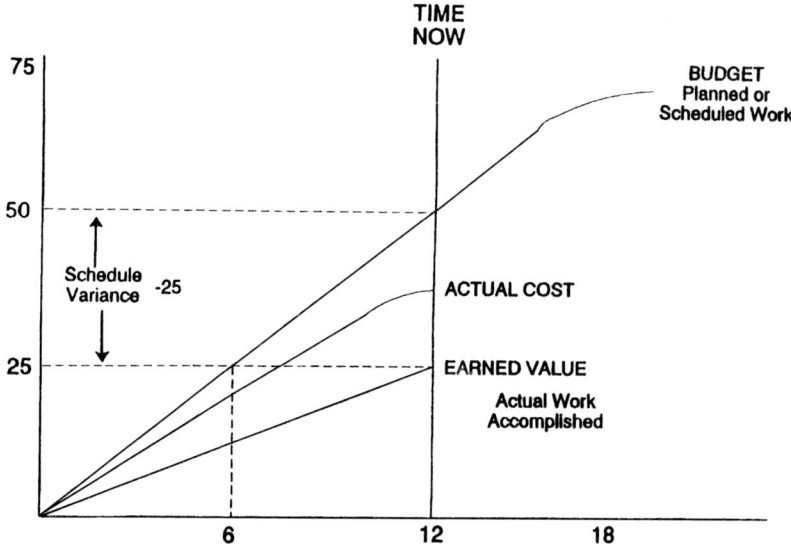

FIGURE A4-13 Cumulative actual cost and earned value plotted versus time.

CHAPTER 5
PROJECT MANAGEMENT INFORMATION SYSTEMS

Louis A. Tucciarone
O'Brien-Kreitzberg, Inc.
Toronto, Canada

INTRODUCTION

A project management information system (PMIS) is vital to the operation of any major construction project or business. These projects involve coordination of engineering, procurement, financing, and construction activities by many consultants, contractors, subcontractors, and third-party agencies, often including employees of the owner or project sponsor. These activities are usually undertaken at multiple sites in engineering company offices, manufacturing plants, and throughout the construction site which may be spread over several kilometers.

Fortunately, the revolution in information and communications technology is providing the engineering and construction industry with the powerful tools to manage and control these activities in a timely, cost-effective way. "Faster, smaller, cheaper" was the slogan of the Silicon Valley semiconductor industry during the last two decades. This provided the impetus for the proliferation of personal computer (PC) networks and client-server applications which have been replacing more expensive mainframe computer systems throughout the engineering and construction industry. As the cost of the computer hardware and software has plummeted, PC project management software has become easier to use. All members of the project team from the construction site to the executive suite can have access to a wealth of data at their desktops. However, this access may not by itself prove productive or worth the investment. Too often, "data rich and information poor" is a sad commentary expressed by many managers on the results of many project management system installations. The challenge for a well-conceived project management information system (PMIS) is to provide team members organized, reliable, and timely information accessible for proactive decision making.

COMPONENTS OF A PROJECT MANAGEMENT INFORMATION SYSTEM

The project management information system is the central nervous system of a major project or engineering or construction company. A well-designed and functioning system can improve the management productivity of each portion of the organization, including the engineering, finance, administration, procurement, and construction management functions.

The most common components of a PMIS for an engineering or construction organization include:

- Planning and scheduling (critical path method or CPM)
- Engineering control and performance measurement
- Job cost, project accounting, or cost engineering
- Cost estimating
- Contract administration
- Document control

Figure A5-1 indicates a typical configuration of these components. Since the latter 1980s, most construction projects are moving toward installation of these applications on PC local area and wide area networks. As indicated, these PMIS tools are often integrated on the desktop computer along with other office automation tools used by the project team members, including word processing, e-mail, spreadsheet, and graphics applications. Very often, the PMIS software is connected to corporate or owner financial systems, such as general ledger or accounts payable, which may

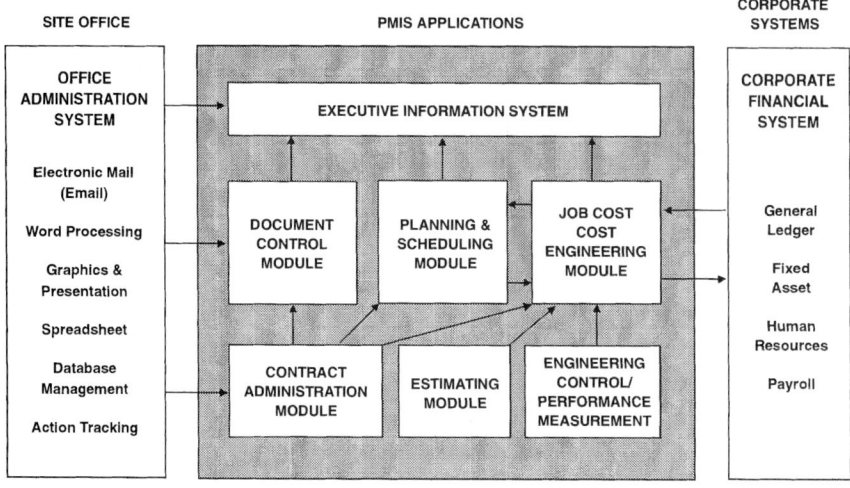

→ Denotes Data Links

FIGURE A5-1 Typical PMIS configuration.

reside in mainframe or minicomputer environments. These networks also now commonly include CADD (Computer Aided Design Development) equipment for use by engineering and construction management staff.

The application of these functions may vary somewhat among companies depending on whether they specialize in construction contracting, engineering, project or construction management, or provide a range of these services. A heavy construction project team will involve services from all these sectors. The organization responsible for project management, whether a contractor, consultant, or owner, will need to apply all of these modules in one form or another. One of the challenges for application to a major construction project is to integrate information from these applications, many of which reside in various "islands," i.e., satellite offices at the construction site, or in various engineering or construction company corporate offices.

Executive Information Systems (EIS)

Many companies and major construction owner organizations are using executive information system (EIS) applications to accomplish this integration and provide managers with consolidated, exception-based information in an easy-to-use form on their desktop computers. These EIS applications draw real-time data from the various databases of the project management modules previously noted and use powerful graphics and interactive tools to allow managers to focus on key performance or decision-making information.

Functions of a PMIS

Table A5-1 summarizes the key functions for each of the PMIS modules.

Costs for a PMIS

The use of these tools is becoming second-nature to most engineering and construction professionals; it is difficult to find many employees who can function without them. Nonetheless, a common question for project management organizations and companies is determining how much should be invested in project management systems and what are the key benefits from their implementation.

Ironically, advances in information technology have now made computer hardware and software a minor portion of the budget for project management systems. Most factories, offices or field headquarters are equipped with personal computers for day-to-day office productivity: payroll and timekeeping, word processing, e-mail, spreadsheet, graphics, or facsimile. The incremental cost is small for installing project management software applications in this environment.

A much greater investment is the staff required to use the systems for planning and scheduling, cost engineering, document control, or the other project management disciplines and the technical support staff for network operation and user and training. For major project engineering and construction management activities, the costs of project management and control staff are usually 80 to 90 percent of the total project management budget.

Most major construction projects involving annual cash outlays of more than $100 million U.S. can expect to budget a minimum of 2 percent of project cost for

TABLE A5-1 Project Management Information System Functions

Planning and scheduling	Engineering control/ performance measurement	Job-cost budgeting/cost engineering	Cost estimating	Contract administration	Document control
CPM network planning and scheduling	Worker-hour tracking by work package and task	Register project commitments, purchase order	Quantity take-off Digitized take-off	Change-order logs/ claims logs	Drawing control registers
Resource (equipment, materials, labor) loading and leveling	Performance measurement statistics:	Invoice and payment tracking	Pricing databases for materials and equipment and labor	Submittal tracking	Correspondence control registers
Procurement tracking	Budgeted cost work performed (BCWP)	Cash-flow forecasting	Estimate comparison reports	Daily inspection reports Progress payments	Intelligent searching and archiving retrieval
Project lookahead schedules and checklists	Budgeted cost work scheduled (BCWS)	Variance reports, budgeted, committed, estimated final cost	User-defined cost assemblies	Correspondence and transmittal tracking	Document status reporting
Recovery plans	Actual cost work performed (ACWP)	Authorizations for expenditure		Requests for information logs	Action item tracking
		Change-order details		Project photo documentation	Automated document routing
		Incurred cost/ progress measurement			Forms generation

project control staff functions (scheduling, cost estimating and engineering, and contract administration), including the amortized cost for the associated software.

Benefits of a PMIS

The potential paybacks can far exceed this investment and include:

- *Contract coordination and interfaces.* Critical path method (CPM) networks identify critical milestones and activities that will impact multiple contracts. This knowledge can be incorporated into contract packaging plans and contract schedule administration provisions to assist on-time completion of the facility and ensure revenue generation.
- *Scope control.* Potential changes to scope and their associated cost and schedule impacts can be highlighted for management at an early stage for effective decision making and authorization.
- *Claims avoidance and defense.* Rigorous documentation of as-built conditions and schedules, scope authorization, and cost details can be maintained. This discipline allows those administering contracts to avoid misunderstandings or resolve potential disputes, heading off adverse impacts on project schedule or cost. In those cases in which contractors file claims for damages, the system can provide the project sponsor with comprehensive records to defend against such claims. Such records are very expensive to compile after the job has been completed.
- *Cash-flow forecasting/cash management.* Accurate, current forecasts of cash flow can be provided to financing partners or funding agencies, based on a consistent approach by all projects or contractors. The project may also benefit from effective cash management of its project receivables and charges to third parties or subcontractors.
- *Performance measurement.* Quantitative estimates of work completed can be made to estimate the value of work performed prior to invoice payment whether for engineering, procurement, or construction contracts or subcontracts.
- *Configuration control.* Revisions to plans, specifications, and drawings can be tracked to ensure completeness and to ensure that documentation is communicated to the team members responsible. This can reduce errors and omissions in design or implementation.

ESTABLISHING A PROJECT MANAGEMENT SYSTEM NETWORK

An initial decision in establishing a project management information system is the choice of the computer platform, or hardware infrastructure.

Prior to the mid-1980s, most project management applications required the computing power and storage available only on mainframe or larger minicomputers. As the speed and storage capacity of personal computers increased exponentially, software developers have addressed this newly expanding market. Project management software is now plentiful for personal computers, and PC applications sacrifice neither function nor flexibility to their mainframe counterparts.

Cost is usually the key criterion for choosing a platform. Costs to consider include the initial acquisition and licensing costs for hardware, software, and telecommunications equipment as well as the ongoing technical support staff to assist users and modify software as required for the specific project application.

Some companies have significant investments already in place for mainframe or minicomputer information networks information system support staff, and in these cases using a mainframe project management application may be cost effective.

However, for the majority of new project setups or for companies upgrading or expanding their information system resources, personal computer networks provide the most cost-effective foundation. PC hardware and software is relatively inexpensive, technical support staff is abundant, and most companies install PCs as standard office equipment for many functions such as word processing, graphics and desktop publishing, spreadsheet, or e-mail. Adding project management software to these PCs is a relatively small incremental investment. PC networks can also be connected to company mainframe networks in a wide area network (WAN), so there is less risk of becoming isolated from other corporate information sources. (See Fig. A5-2.)

Local Area Networks (LANs)

The backbone of a personal computer project management system is a *local area network* (LAN). A LAN allows personal computer users on the cabled network to access common software applications and databases, share information storage resources, and communicate data. LANs have practical limitations on the distances among users; most LANs are located within single buildings or building complexes. *Wide area networks* (WANs) can connect LANs across greater distances, between the main office and various construction site offices or among satellite office complexes spanning one or more continents. WANs are also employed to connect LANs to minicomputer or mainframe networks which often exist among company facilities or among parent and subsidiary corporate facilities.

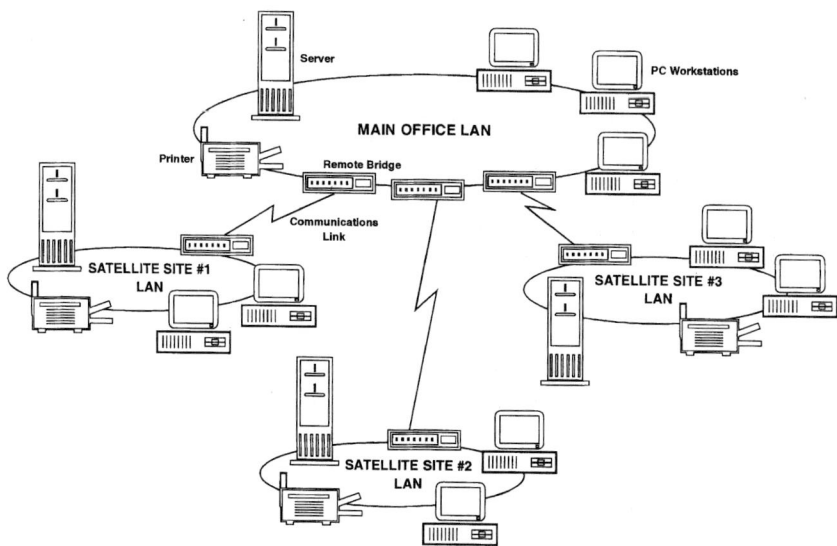

FIGURE A5-2 Project management system wide area network (WAN).

PROJECT MANAGEMENT INFORMATION SYSTEMS A5-7

The main components of a PC LAN are:

- File server equipment
- Personal computer workstations
- Peripheral equipment (printers, plotters, scanners, etc.)
- Cable plant

LANs vary in size and complexity according to the number of users, the number and type of applications that will be used, and the volume of data commonly being transmitted. The configuration and installation of a LAN should be done in consultation with a certified network engineer to ensure a reliable and cost-effective setup.

Table A5-2 gives a representative configuration and approximate pricing for a network of 60 PC workstations employing the project management applications described in this chapter and other common office automation tools. Such a setup would be a minimum configuration for an engineering or construction management office for a large civil construction project. Prices in the industry change rapidly; the costs quoted—$400,000 U.S.—are approximate only.

The file server is the brain of the network. The minimum recommended unit would be a high-speed (50 MHz or greater) personal computer with approximately 4 gigabytes (GB) of disk storage and 32 megabytes (MB) of memory. (A megabyte is 10^6 bytes; a gigabyte is 1000 MB or 10^9 bytes.) For networks that include CADD files or on-line document image storage and retrieval, additional disk capacity may be needed.

TABLE A5-2 Sample Hardware and Software Components: PC Network–PMIS Installation

Item	Number	Approximate initial cost (U.S.$)	
1. File server equipment			
1.1 Network file server	1	$25,000	
1.2 Communications server (bridge)	1	5,000	
1.3 Tape backup/uninterrupted power supply	1	5,000	
1.4 Network operating software	1	10,000	
		Subtotal file server	45,000
2. PC workstations			
2.1 PC hardware (including network cards)	60	150,000	
2.2 PC operating software	1	40,000	
		Subtotal PC workstations	190,000
3. Peripherals			
3.1 Printers	6	20,000	
3.2 Scanners	2	5,000	
3.3 Plotter	1	10,000	
		Subtotal peripherals	35,000
4. Cabling			10,000
5. Project management software (PMS)			100,000
		Total	$380,000
		Say	$400,000

Additional storage and memory can be added to the server incrementally as demands grow. An additional server can be added for redundancy and improved response as the number of users expands beyond 100.

A high-capacity tape drive should be acquired to allow secure backup of network information and to provide access to archived files that are not frequently used and can be stored off-line. Archival storage is becoming less important as the cost of on-line storage declines significantly. For example, a 4-GB storage upgrade costs less than $5,000 U.S. or less than one-tenth of an equivalent upgrade just five years ago.

Personal computer workstations are typically 486 processors with 8 to 16 MB of memory and 200 to 300 MB of hard disk storage. This configuration is now standard issue in the highly competitive PC market. The upgraded memory configuration is particularly important for adequate response time for new Windows-based applications and other graphics programs.

Peripheral equipment is shared by the network users and includes high-speed network printers (usually at least 20 pages per minute), plotters, facsimile servers, and scanners. More recently, CD-ROM or other optical storage media are being added for efficient storage of large amounts of data, including images. It is likely that future networks will feature multimedia peripherals as standard issue.

The cable plant is the medium which connects the workstations, server, and peripherals. There are several options in common use including token ring or Ethernet. These can support high-speed transmissions among the network stations between 10 to 16 Mb/s (a byte equals 8 bits). Network hubs complete the connections to and from the workstations and peripherals.

Wide Area Networks (WANs)

Many large projects require data communications among various satellite offices or between a main engineering or project management office and various remote construction sites.

The simplest, low-tech solution to sharing of information is via magnetic disks or other media which can be downloaded, transported, and uploaded at each site. This is common practice among stand-alone PC users and is usually sufficient for transferring data which does not change daily, such as weekly or monthly CPM progress updates, exchange of CADD files for as-built drawings, or weekly time reporting or invoice backup. However, this solution quickly becomes unwieldy if the volume of information that must be shared is large and the communications frequent, anywhere from daily to real-time.

If there are only a handful of computer stations in the satellite or site offices, a remote-control link to the main network may suffice. In this setup, a remote PC is used in tandem with a dedicated PC, called a *host,* which resides on the LAN. All processing occurs on the host machine: The remote station simply transmits keystrokes from the user and displays output on a monitor. However, this setup is not cost effective for several computers because each station must have a dedicated host. A larger satellite office warrants installation of its own local network which can be connected to other LANs on the project through a wide area network.

As shown in Fig. A5-2, LANs can be connected with remote bridges and transmission links. Remote bridges are usually dedicated PCs with specialized hardware.

Transmission speed between the bridges is usually the chokepoint for wide area communication. The network engineer should optimize the investment in these transmission links, balancing cost with performance required. Transmission links vary in capacity from 28 to 56 Kb/s (kilobit per second) for phone lines to higher-speed T1 lines operating at more than 1500 Kb/s. This T1 speed is still only one-tenth of what is common on the local network connection. Data compression alleviates this somewhat.

The communications logjams between offices may become less relevant as progress is made building the information superhighway to accommodate video and multimedia communication among faster processors. As fiber optic links become more prevalent and economical, data will pass at speeds more than 100 times the current capacity of T1 lines.

CHOOSING A PROJECT MANAGEMENT SYSTEM

To work effectively on a large program, the project management system must address several challenges:

- It must allow managers in various organizations in the central office and field offices to have access to the same information, in a timely manner.
- It must integrate and combine information from various sources and modules that are being used by the designers, contractors, and owner staff.
- It must put key performance and status information at the fingertips of decision makers, not just buried in detailed management reports which pile up unread on bookshelves or file cabinets in various offices.
- It must be able to expand incrementally and economically as the number of projects and office locations grows.

Choice of Project Management Software

Choosing project management software is becoming more and more akin to purchasing a home stereo/video entertainment system: There is a variety of off-the-shelf products based on industry-accepted standards available from various vendors. An organization can mix and match components or, in the case of systems, application modules, that best suit its budget and priorities. Just like changing or upgrading hi-fi components, organizations can add or replace individual modules as better products become available or project needs evolve.

The software vendors offer applications ranging in cost and functionality from industrial-strength applications to simpler starter applications. The industrial-strength systems can support projects requiring multiple coding structures, varying contract administration techniques, and detailed or complex networks and cost accounting requirements. The starter applications target organizations just beginning in project management or having less detailed project or program reporting requirements. These situations don't yet warrant more complex or expensive software tools.

For purposes of this handbook focusing on heavy construction projects, it is assumed that the more comprehensive, industrial-strength systems are necessary and justified. However, any systems procurement should review the requirements of the project team to ensure that all of the component modules are necessary or whether lower-cost applications could be substituted. There are many situations in which companies or projects have invested in elaborate application development or product procurement which far outpaced the staff capability or resources to use the tools effectively. For example, a project or company which primarily uses fixed-price or subcontracted engineering contracts would have little need to invest time in an engineering control or performance measurement application. Organizations which do not require contractors to prepare and update detailed or resource-loaded CPM networks may not need a higher-end scheduling package.

The major components of a project management information system include:

- Planning and scheduling (critical path method or CPM)
- Engineering control and performance measurement
- Job cost, project accounting, or cost engineering
- Cost estimating
- Contract administration
- Document control
- Executive information system

These components are shown in Fig. A5-1 for a typical major project setup. As indicated, the project management applications on a PC network are often linked to corporate financial information systems which may reside in mainframe or minicomputer environments and are integrated with other desktop office automation tools available to the project team, including word processing, e-mail, spreadsheet, graphics, or database applications.

Off the Shelf or Develop Your Own?

The array of commercially available project management software is expanding and evolving, so it is less common for organizations to develop their own applications. Most commercial applications that are procured can be customized by the company or project team if necessary to adapt to its procedures and forms.

When purchasing a commercial system it is important that the underlying programs use or link to an industry-standard (i.e., nonproprietary) relational database management software product, of which there are many on the market. This allows the team to customize the database structures and reports to meet project-specific requirements, and allows data to be transferred among modules or corporate applications which may be purchased from different vendors.

Many project management software vendors offer comprehensive management applications: Users can purchase one or more modules, sometimes with volume discounts for additional purchases or multiple user licenses. Other vendors specialize in certain applications or families of applications like scheduling or estimating applications. In addition to pricing discounts, the advantages of purchasing from one vendor are that the data structures and screen layouts are similar and the import/export routines for data transfer among applications are usually built in.

Purchasing components from separate vendors allows one to mix and match from the most suitable in each function and to upgrade as features change. The organization will need to spend some effort programming data transfer links among applications and users will not necessarily see a seamless interface among applications. As noted, if nonproprietary relational database software is used, these transfers are not difficult.

Report Writer

The purchase of a commercial report writer application is a low-cost but invaluable addition to any configuration. Most commercial project management packages provide many standard reports but invariably cannot meet the needs of all team members. The report writer allows the team to adapt the format and content of reports to the preferences of each level of the organization, presenting summarized or detailed information as necessary for the audience.

CODING STRUCTURES

The choice of hardware and software is not the only critical success factor for a PMIS installation. Just as important is the careful application of project coding structures to organize the data. For large or complex projects, these systems must store, access, and share millions of information records in the databases. The coding of this information is the key to ease and flexibility in meaningful reporting.

The design of the coding structures is one of the most important system start-up activities, but is often given inadequate attention in the rush to populate system modules with data and get the system up and running. Adding or modifying data coding retroactively is expensive, time consuming, and often impractical. The process of structure development should involve consultation with all parts of the project organization, from the accounting and purchasing department to the Board of Directors, to ensure that their reporting needs are understood. The coding can then be adapted to anticipate these future needs.

Project coding structures provide the backbone for the system. They allow integration or sharing of data among various applications or modules within the system, connections to information systems used by different contractors, and customized reporting from a central database that can be flexibly adapted for different levels of management or different parts of the organization.

There are numerous coding structures that can be used to classify information. During the 1970s the U.S. federal government fostered standards for project management that became embodied in the Cost and Schedule Control Systems Criteria (CSCSC) or what became commonly referred to as *C-Spec*. C-Spec popularized several coding structures that have become standards in most project management software systems including:

- *Work Breakdown Structure (WBS)*.   A hierarchical detail of project scope
- *Organization Breakdown Structure (OBS)*.   A hierarchical detail of staff responsibility for each project element

Each WBS item is linked to an OBS item so that responsibility for each element of project scope is clearly defined and project reporting can be sorted or grouped by scope or organizational area.

There are many other structures that are commonly employed to organize cost, schedule, and project management data including:

- Geographic location codes
- Financial asset codes

- Funding source codes (for projects with funding from various private or public sources)
- Contractor or subcontractor codes
- Cost element codes (or company charts of accounts)
- Project commitment structure or contract packaging structure

Most higher-end project management applications include various user-defined codes on its screens as required by the specific project or company. Use of relational database software also allows the user with some programming knowledge to add codes to the database and reports.

Work Breakdown Structure

The WBS organizes the scope of a program or project into hierarchical elements of sufficient and manageable detail to plan and control its execution. These elements are often represented in tree diagrams, as shown in Fig. A5-3.

The creation of the WBS is an important process involving all functional areas of the program team. This will ensure that all aspects of the program scope have been adequately identified, defined, and commonly understood by all team members. Each element of the WBS has assigned budgets, schedules, resources, and documentation reference. Thus, it becomes a unifying framework for project reporting and control.

By their nature, WBSs increase in levels of detail as programs or projects proceed through the planning, engineering, and construction phases. This parallels the need to plan and control in manageable units as the scope and tasks in each phase are further defined. However, it is important that the initial WBS be sufficiently comprehensive and, to a degree, generic to ensure that there are no significant scope omissions in preparation of baseline estimates, budgets, or schedules.

It is also not necessary that all branches or trees of the WBS be developed to the same level of detail concurrently. The level of definition for each branch can vary to the stage of the project implementation.

WBS Dictionary

The development of a Work Breakdown Structure Dictionary is also recommended to control the WBS and communicate its intent consistently to all members of the project team, which may include dozens of contractors and subcontractors. The WBS dictionary lists and defines the WBS elements in pictorial and tabular form (tree structures for each level), provides a reference for WBS coding, and includes descriptions of the scope of each element as well as important inclusions or exclusions. Figure A5-4 is a sample page from a WBS dictionary.

The WBS levels of detail for a large project usually develop incrementally as engineering proceeds and new procurement and construction packages are defined. There are also scope changes to a project which must be documented through changes to the WBS. Therefore, it is important that the program manager or owner maintain a master WBS dictionary and issue revisions to all team members in a controlled manner throughout the project life.

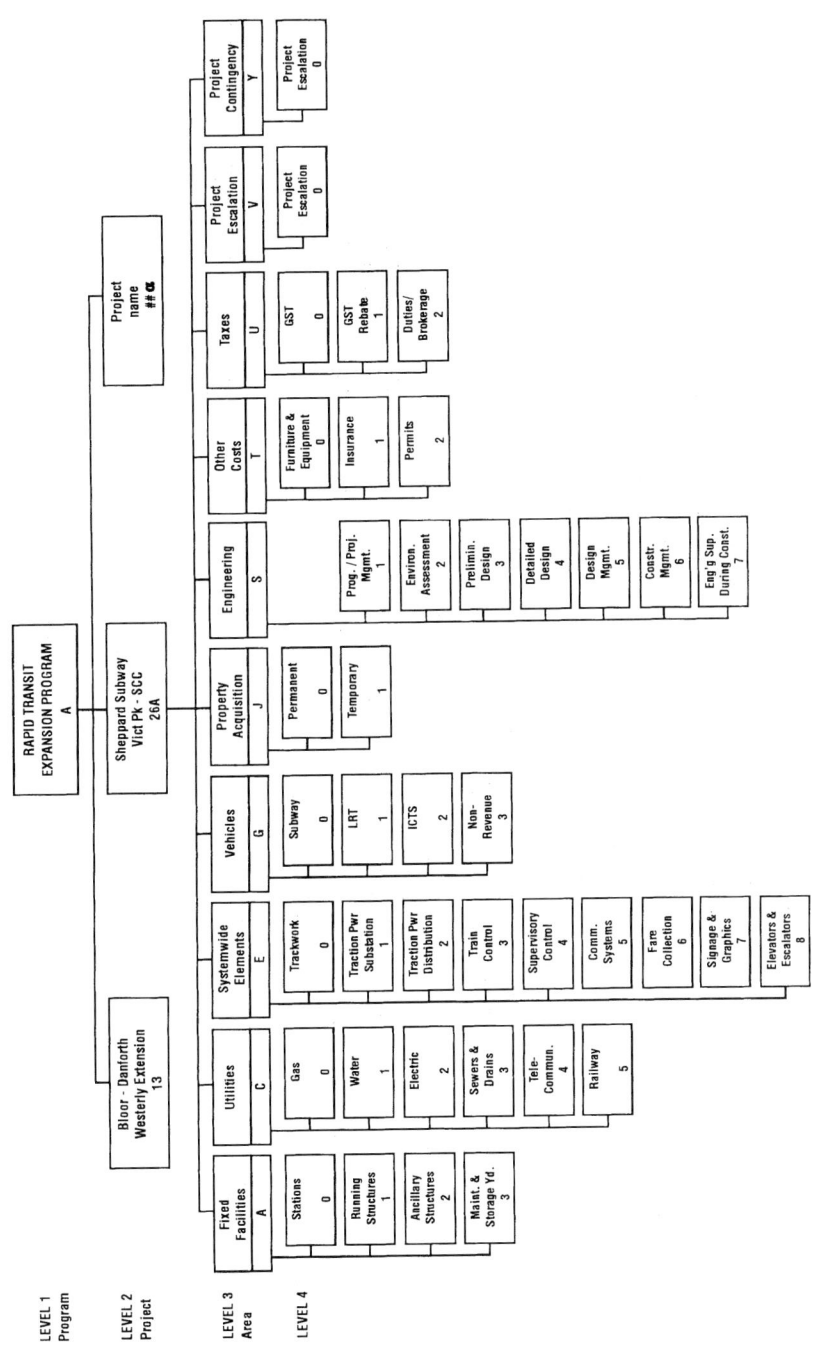

FIGURE A5-3 Sample work breakdown structure.

| RAPID TRANSIT EXPANSION PROGRAM | Rev. 93-E, Level 4 |
| Work Breakdown Structure | Date: July 1, 1993 |

DESCRIPTION: RUNNING STRUCTURES

SYMBOL/CODE: 1

Code Breakdown:

LEVEL	1 a	2 ##a	3 a	4 a	5 a	6 a	7 #####	CODE DESCRIPTION
Level 1: Program	A							Rapid Transit Expansion Program
Level 2: Project		##a						Specific Rapid Transit Expansion Project
Level 3: Element				A				Fixed Facilities
Level 4: Sub-Element				1				Running Structures
Level 5: Type								
Level 6: Location								
Level 7: Division								

Scope of Work:

The scope of the running structures WBS sub element encompasses material supply, construction, testing, and commissioning of underground, at-grade, or elevated structures upon which the trackwork or guideway will be placed. This includes demolition and removal of existing structure, cut and cover subway segments, tunnel segments, channel structures, open cut structures, at-grade segments, elevated guideway structures, 3 or 4 track structures, tapers, and Y connections, drainage, plumbing, ductbank installation, power supply and distribution for electrical other than traction power, lighting, fencing, restoration, landscaping and fencing along the right-of-way. Road work affected by the line structure such as access roads or road diversion will be identified with the running structure.

Notes:

NOT included in the scope of this element are operating systems required to run the vehicles (ie. trackwork, traction power supply and distribution, train signals, supervisory control, etc.) and building permits and construction insurance provisions.

FIGURE A5-4 Sample work breakdown structure.

Project Commitment Structure (PCS)

A commonly used coding structure which works in conjunction with the WBS is a *Project Commitment Structure* (PCS) or contract packages structure. This structure details purchase order, service agreements, and work packages in a hierarchical fashion.

Figure A5-5 is a sample project commitment structure. The lowest level is the payment item or bid item in a purchase order, work order, or contract. The highest levels can mirror the highest levels of the WBS.

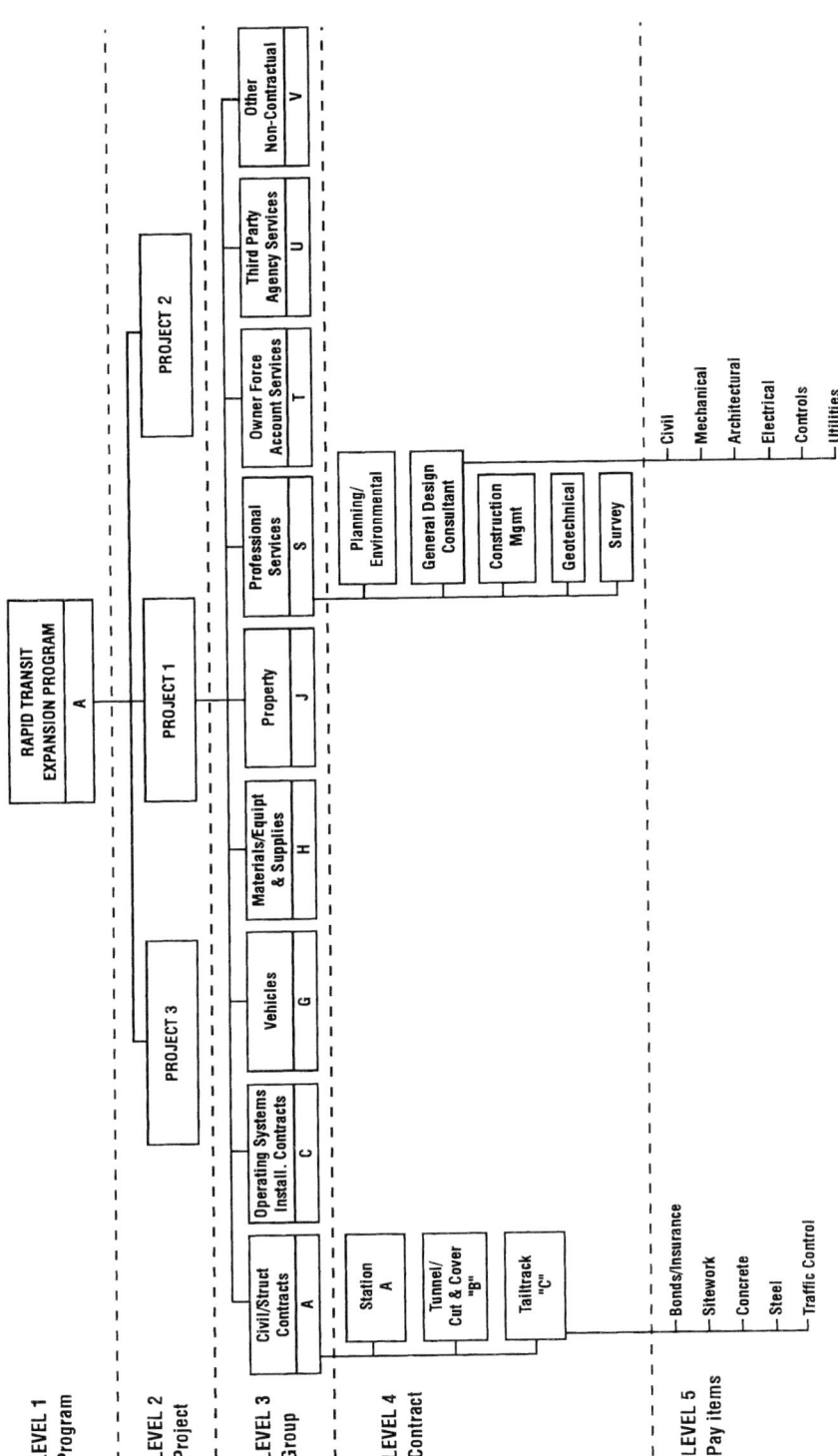

FIGURE A5-5 Sample project commitment structure.

Why a PCS and a WBS?

The WBS documents the scope of the undertaking; the PCS describes in detail the method by which the work will be executed. The PCS addresses how the engineering, procurement, and construction will be packaged for implementation. Usually, the planned work packages for procurement and construction change during the course of the design development. By providing the WBS and PCS as linked structures, the project management team may change the packaging assignments to reflect the most current implementation plan without continually redefining the baseline WBS.

The WBS and PCS can be linked in the management system at the lowest level of detail, i.e., each pay item in the commitment structure must relate to a unique WBS element. This allows information to be sorted and reported easily to different portions of the team, depending on the group's principal focus.

The commitment structure is the principal structure used for contract administration, cost and schedule data collection, forecasting, and reporting. The work breakdown structure is the principal structure used for estimating, budgeting and master scheduling, and establishing the project baseline. In combination, the two structures provide increased flexibility for management reporting. The dual structures allow the flexibility to administer contracts and work packages with great detail to control cost and schedule performance and still maintain status reporting against original project scope and budget objectives.

EXECUTIVE INFORMATION SYSTEM (EIS)

EISs have become more common in many industries during the last several years as corporations attempt to harness the information potential from many sources and systems throughout organizations to obtain a more coherent view of operations and performance. In many instances, these systems are spread over many geographic areas and different computer platforms as a result of the growth of decentralized LANs and minicomputer applications.

What Is an EIS?

An EIS is a computer program that provides on-line, direct access to key project information required by decision makers. There are several EIS software tools commercially available that allow users to customize applications for their specific organizations and purposes. Although applications vary among organizations and computer platforms, there are several common objectives:

1. Provide a single source of data that is consistent and auditable, combining information from disparate systems and sources.
2. Provide timely information to users.
3. Provide information in an easily accessible form for managers, who may not have the time or disposition to learn many computer programs.
4. Provide the capability for managers to identify trends or analyze key information on an exception basis from a large volume of data.

Most EIS systems for PCs are Windows applications. The Windows graphical user interface is easier to learn and operate and is quickly becoming a standard for home or office computer use. Thus, it is ideal for an executive system which should offer an uncomplicated, familiar interface for users.

EIS applications can be written in commercially available, nonproprietary database software, making it economical and flexible for customization or enhancement.

The EIS application draws information from the underlying project control software. Thus, it can be added as an enhancement to a variety of control software. It can also be customized to suit each company's installation without affecting the base control systems.

Each EIS application is usually customized to the specific project and management team, but there are several common features:

1. It integrates information in an EIS database from underlying project management applications and various other database and graphic files. Information is uploaded from these other databases at specified frequencies (daily, weekly, or monthly) depending on the application.

2. It allows read-only access from a filtered database: The EIS database is consistent with the underlying applications, but managers are not able to access those applications or databases directly. This ensures security of information.

3. It emphasizes easy-to-use interactive screens with powerful graphics and management-by-exception capability. The EIS allows managers to peruse summary information with a variety of graphic reports to highlight trends or comparisons. Managers can also zero in on details behind the summary comparisons. The system also includes exception query capability so that managers can select certain variance thresholds and identify only those contracts or activities that meet these threshold criteria.

4. It provides a management communication tool: All managers with a PC have access to the EIS. Information that was formerly exchanged verbally or by paper media can be shared on-line. Among the typical information suitable for electronic access are program procedure manuals, organization charts, responsibility matrices, daily news clippings of interest to the program, and all status or progress reports. The paper savings can be substantial and the access time reduced significantly. The application also is connected to the e-mail so that managers can pass comments or questions about the information they are reviewing to other managers or subordinates.

All information screens can also be annotated and printed so managers can share hard copies in presentations and meetings. Because it is a Windows application, users can also cut and paste among EIS and other applications and take advantage of other Windows tools.

EIS Sample Screens and Functions

The following exhibits extracted with permission from the EIS for the Toronto Transit Commission Rapid Transit Expansion Program (RTEP) demonstrate some of the functions and versatility of these management tools in use on a multibillion-dollar engineering and heavy construction program.

Ease of Use. The Executive Information System is designed to be used directly by all members of the project management team without the assistance of intermediaries. The application utilizes the Microsoft Windows graphical user interface and all functions are accessed by pressing large easy-to-find buttons and by the pull-down menus located at the top of every screen. Figure A5-6 shows the opening screen which allows managers to choose a project or area of information.

All functions are immediately available from the pull-down menus that are located at the top of each screen. Common functions such as accessing help and printing may also be accessed by pressing on-screen buttons. Menus may be tailored to suit each project's requirements by adding and deleting options as required.

Graphics Presentation. The EIS makes extensive use of graphics to summarize and highlight information. The use of graphics often brings out important information that might lie buried in a standard report. However, unlike static cost report graphs, the EIS allows the user to select what is being graphed as well as how it should look. The user can choose a variety of graphing types and options to customize the presentation to suit individual needs. (See Figs. A5-7 and A5-8.)

Structured Information. The EIS employs the same structures used throughout the Program Management Information System and may include a Work Breakdown Structure, a Commitment Structure, a Funding Structure, or an Organization Structure. A manager can quickly zero in on the areas of interest and choose what level of reporting (summary or detail) is desired.

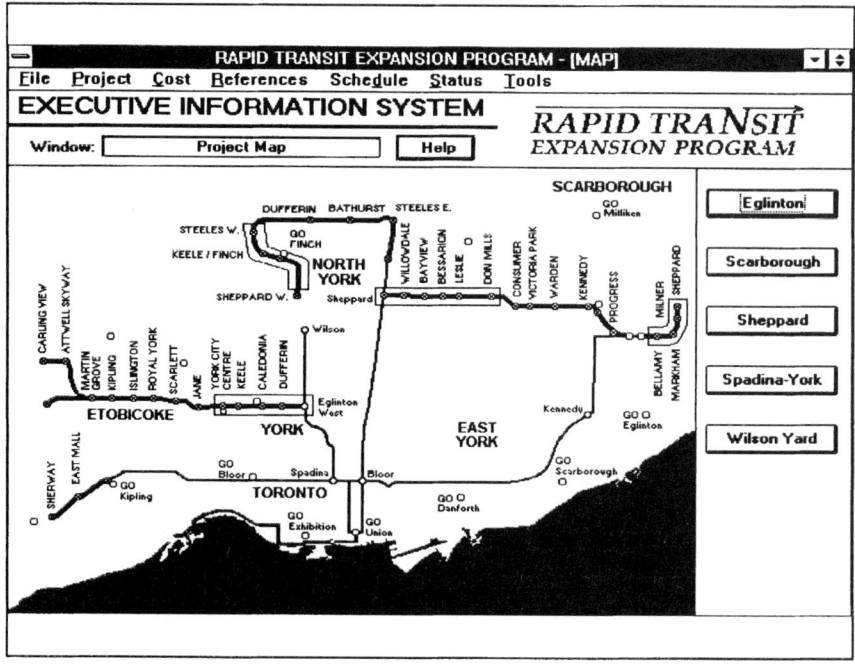

FIGURE A5-6 Executive Information System: opening screen. *(Toronto Transit.)*

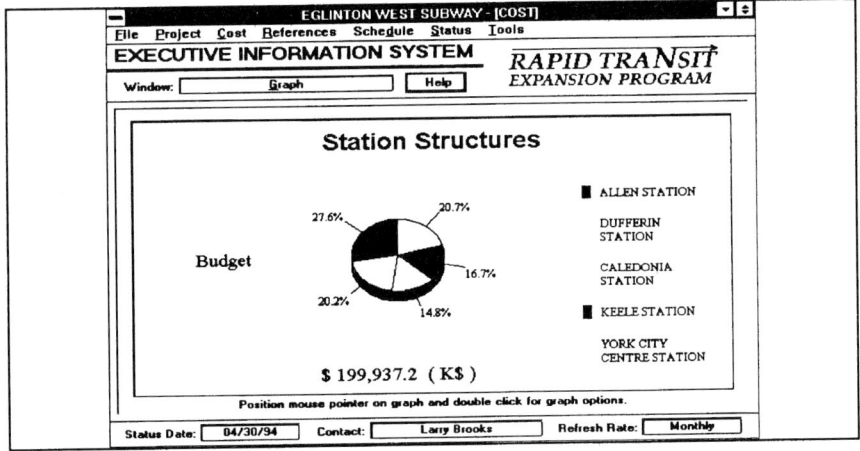

FIGURE A5-7 Executive Information System: graphical information.

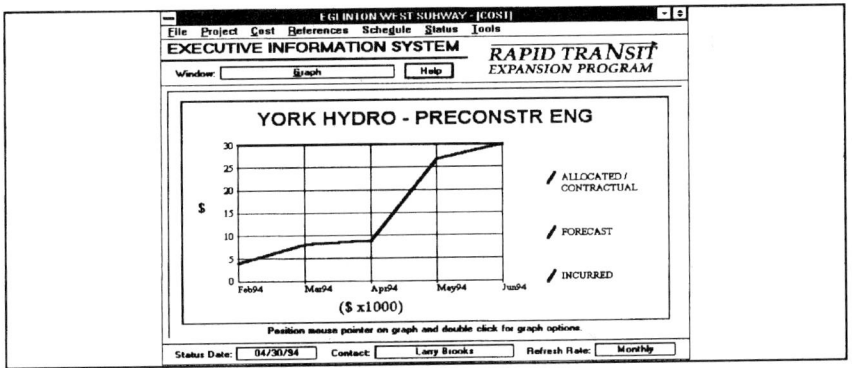

FIGURE A5-8 Executive Information System: cost trends.

The manager chooses from a selection screen a summary for an entire project or work breakdown category or the specifics of a contract. Once a structure selection has been made, the user can view a graph or the data itself, or look for exception cases within the selected structure. (See Figs. A5-9 and A5-10.)

Management by Exception. The EIS provides the capability to select variance thresholds and query the system for those items that fall within the selected threshold. This provides an easy method to concentrate attention on key indicators and manage by exception. This is particularly critical on large, complex projects with multiple contracts or work packages. Managers need a tool to focus on key factors and avoid data overload.

Figures A5-11 and A5-12 show the cost exception screen where a user can pose questions such as:

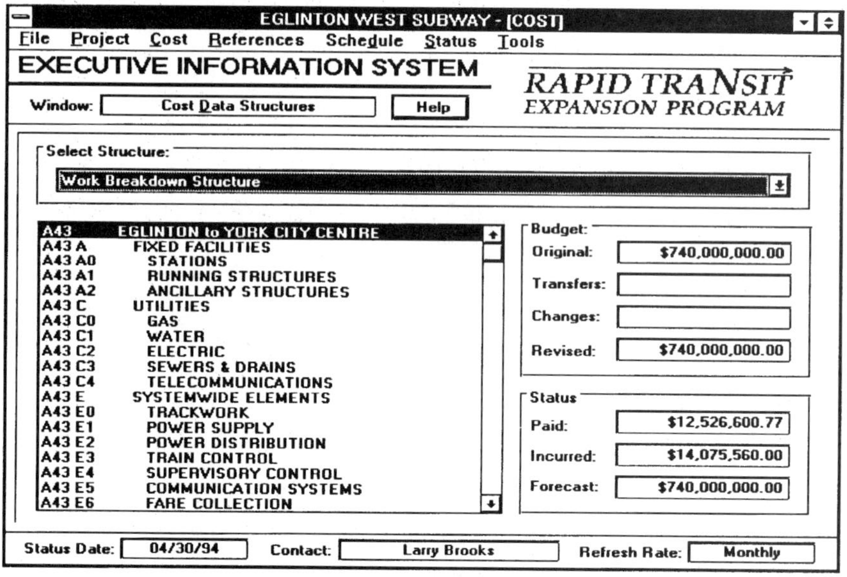

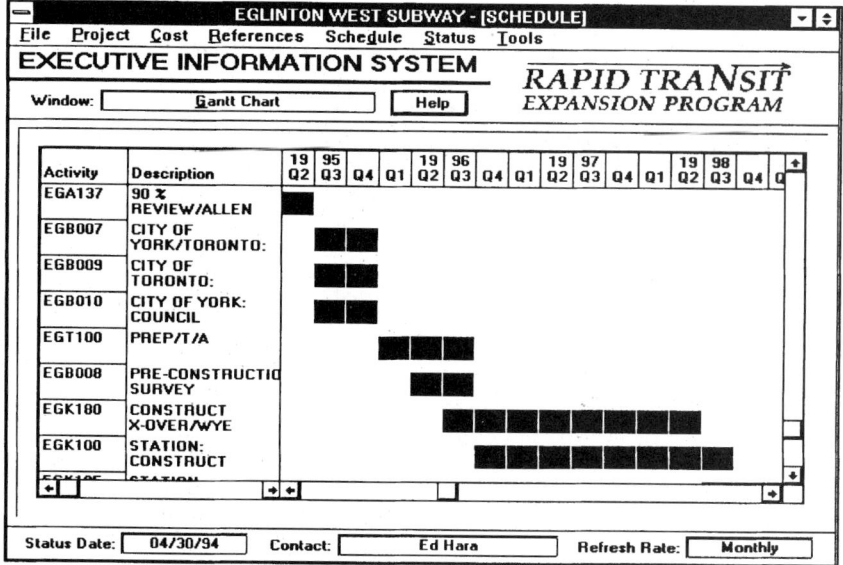

FIGURE A5-9 Executive Information System: cost data by WBS.

FIGURE A5-10 Executive Information System: schedule Gantt chart.

PROJECT MANAGEMENT INFORMATION SYSTEMS

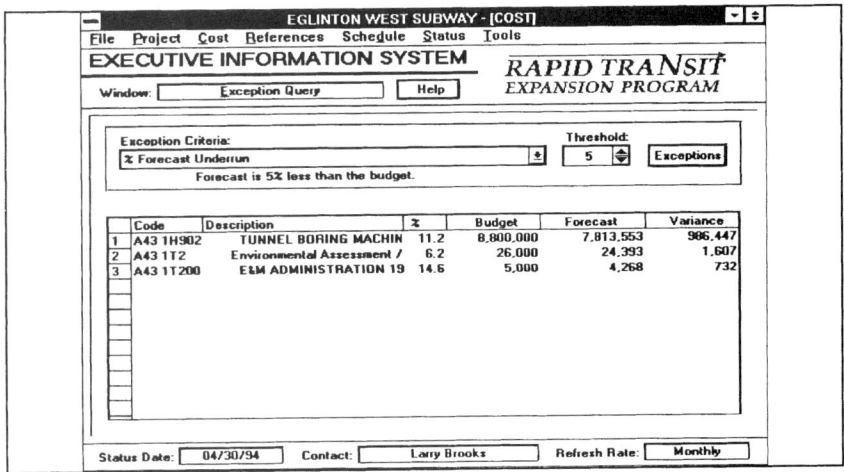

FIGURE A5-11 Executive Information System: Management by Exception—work packages with more than 5 percent overrun forecast.

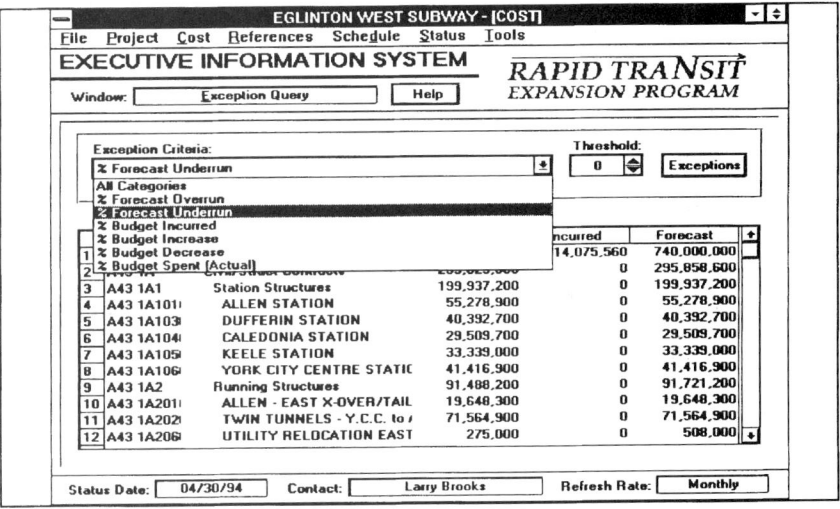

FIGURE A5-12 Executive Information System: Management by Exception—work packages with any underruns forecast.

- Which budget items have over 95 percent of the budget committed?
- Which items are forecast to exceed budget by more than 5 percent?
- Which items have an incurred cost greater than 90 percent of the contract amount?
- Which items have change orders greater than 10 percent of the base budget?
- Which items have a budget greater than $100,000 and a forecast overrun of more than 5 percent?

The schedule exception screen allows a manager to pose similar types of questions including:

- Which activities are within 10 days of being on the critical path?
- Which activities have started more than 15 days behind schedule?
- What activities will be in progress during the next 45 days?
- What activities are more than 95 percent complete or what activities are complete?

Additional exception conditions are easily added to tailor the Executive Information System to the needs of each project team. Other commonly used exception queries are the status of submittals, change orders, or contract approvals from the contract administration module, and the status of aged receivables or payables from the project accounting module.

Communications. The EIS facilitates project team communications by providing an up-to-date on-line source of commonly referenced information, such as Organization Charts, a Glossary of Terms, Program Procedures, and Current Status Reports.

Other reference material may be scanned into the system on a daily basis. Managers can view this information or print hard copy as required. Examples of scanned information include Program Advertisements and Relevant News Clippings.

Timely Information. All screens note the status date of the information and the team member responsible for update. This allows a user to direct questions or queries directly to the individual responsible either via phone, e-mail, or memo.

The update frequency varies by the type of information. Some data is updated daily, such as news clippings, payment information, or document logs; some information weekly or biweekly, such as CPM schedule updates; other information on a monthly basis, such as cost forecasts or incurred costs. The frequency is determined by the needs and procedures of the project management team. The program can accommodate any update cycle.

In actual operation, the RTEP managers place equal importance on data reliability and accuracy, even at the expense of some time lag. For example, comprehensive monthly cost control updates are routinely performed 10 days after month-end and then reflected in the EIS. This is preferred to an approach of various contracts within a project being updated on different cycles.

EIS Development and Implementation

EIS applications are suitable for any agency involved in significant engineering or construction management. The benefits compound in situations with multiple projects or large single projects in excess of $100 million capital cost with numerous contract packages.

Most large projects or programs already invest in PC local area networks for other office information systems. Similarly, project management software for scheduling and cost control and other database tools are common to most projects.

EIS software can be added to these applications at a minimal incremental cost. The EIS software licenses are less than $5000 for a 100-user network. The initial application setup time can be as little as two to four person-weeks for an experienced database analyst.

Comprehensive program management information systems are essential for successful engineering and construction management. However, too often agencies that have invested in hardware and software systems are disappointed with the results.

Information often remains remote, inconsistent, or difficult to interpret. Adding an EIS to a functioning project management system can give managers timely and accessible information that can be tailored to meet the managers' needs. Making it accessible and timely also promotes its use for communication and decision making by the management team. The value-added from an EIS is immense for a project organization already investing in computerized information systems.

PROJECT MANAGEMENT INFORMATION SYSTEM FEATURES

This section provides a summary of the principal functions and features of project management software. Much of the information is extracted, with permission, from the O'Brien-Kreitzberg documentation of its Project Management/Control System (PM/CS). The O'Brien-Kreitzberg Project Management/Control System (OK-PM/CS) is an integrated management tool consisting of PC-based project management programs, system policies, and procedures. OK-PM/CS was developed by OK construction and project management personnel with hands-on design and construction management experience. The benefits of such a system to control project time and cost have been proven on a wide range of assignments, both large and small.

The system is modular with various off-the-shelf programs integrated through the use of coding structures and industry-standard relational database software. An EIS application summarizes the information as a management-by-exception tool for program staff.

The features of the principal modules are as follows.

Budget and Cost Management: This module includes capabilities for budgeting, cost collection, forecasting, scope and change control, cash flow projections, project accounting, and cost reporting. In some settings these functions are labeled cost engineering or job costs.

Estimating: This module includes capabilities for estimators to develop efficient conceptual and detailed quantity take-offs and construction estimates using commercially available, industry databases or customized information developed by the team from bids or vendor pricing.

Engineering Control and Performance Monitoring: This module includes capabilities for managing engineering hours against work packages and calculating various performance measurement or earned value statistics in accordance with C-Spec standards.

Planning and Scheduling: This module includes CPM scheduling, resource management, and what-if? analysis. The graphic reports provide bar charts, time-scaled logic diagrams, and resource curves, as well as various comparisons to target schedules.

Contract Administration: This module includes capabilities for construction field administration, including management of submittals, correspondence, change orders, payments, and daily reports. The Contract Administration module works in conjunction with CPM schedules to assist in dispute/claim avoidance and resolution. This module also allows collection of construction cost information from the field and automatic transfer of these data in the Budget and Cost Management module. This provides a mechanism for early identification of cost trends and/or potential problems.

Document Control: This module includes indexing and reporting capabilities for management of all correspondence and submittals related to engineering or construction activity. The system's indexing and reporting capabilities are complemented by conventional storage technology such as microfiche or the addition of state-of-the-art optical storage and imaging technology to allow efficient storage and retrieval of large volumes of records. These high-tech imaging systems are fast becoming the systems of choice for large organizations or major capital improvement programs. For ease of implementation, the Document Control module allows an incremental approach to introducing such new technology.

All modules generate action-oriented information in hard-copy reports for all levels of project management from the executive field staff. Reports increase in detail as they are targeted for progressively lower levels of management in the organization chart. OK-PM/CS reports are focused on areas of concern and calls for action, rather than being simply historic records of achievement or failure. The various standard reports can be easily adjusted by project controls staff to fit each organization's needs by using commercially available report-writer software. This eliminates the need to engage experienced programmers to customize reports. Managers can see reports in the format and level of detail most useful to them.

Budget and Cost Management Module

The Budget and Cost Management module specifically addresses:

- Budgeting and scope change control
- Capital cost planning and contract management
- Package scheduling and cash flows
- Project cost accounting
- Monthly cost reporting

This module provides for entry and maintenance of the Work Breakdown Structure. It also provides a separate structure, the Project Commitment Structure, which tracks contracts and purchase orders, at either pre-award or post-award status, along with their planned or actual pay items. The PCS provides the vehicle for supporting cost planning, contract administration, and cost accounting. Pertinent working-level reports and audit trails, as well as summary cost and accounting reports, can be produced using the WBS or the PCS.

The Budget and Cost Management module monitors cost in accordance with the project cost-coding directives. It develops a budget baseline and provides a basis for monitoring costs throughout the duration of the program. Cost elements collected, categorized, and tracked include:

- Authorizations for expenditures and commitments by vendors or contractors
- Updated estimates of costs to complete or future commitments as work package scopes are added or modified
- Contingencies based on set percentages of contract values or as established through specific risk assessments for work packages
- Cost escalation factors, such as construction and manufacturing indices
- Incurred costs corresponding to work in place

- Actual costs based on invoices received or amounts paid
- Change orders for work packages either pending or approved

The module allows the following functions to be performed:

- Provides variance reports comparing estimated final costs to amounts budgeted or authorized for expenditure at each level of detail for the project. Reports can be suited and selected according to the WBS/OBS and PCS structures.
- Provides a means to identify the project contingency remaining and allows assignment to individual contracts or purchase orders.
- Tracks all invoices, funding claims, holdback, and payment amounts for each contract and purchase order. Allows reconciliation to corporate general ledgers.
- Identifies cost trends resulting from design or field changes and incorporates them into cost-to-complete forecasts for each package. Additionally, change-order submittals and contractor claims are reviewed for substance and subsequently evaluated for accuracy and potential impact.
- Tracks change orders to the contracts or purchase orders and maintains registers of the change-order details to assist in contract administration and claim substantiation.
- Generates projected cash flow curves using information defined at the level of the project and at the level of the commitment packages. This involves integration with the schedule progress updates from the CPM module for each contract or work package. These cash flows will be used for long-term fiscal budget planning and or project financing analysis. A variety of standard forecasting profiles are available and the system allows the user to create a specific profile for each package.

Estimating Module

The Estimating module assists with the development of program cost estimates through the definition of basic data structures and the quantification of required resources. By using a computerized tool conceptual estimates can be generated quickly from prior estimating history and detailed estimates can be revised efficiently to reflect changes in assumptions.

The Estimating module offers the following features:

- Standard component resources (labor, materials, equipment, services, and subcontracts) are defined and quantified within the project framework.
- Users may enter quantities manually by item and allow the system to sort and summarize by WBS or other coding structure. The estimator may also use a digitizing tablet to calculate quantities from drawings.
- The user can choose among commercially available pricing databases for various trade items or develop its own calculations of composite unit costs for various items. These are often referred to as cost assemblies. The cost per unit for each of the standard assemblies is determined.
- The range of foreign currencies that may apply and the conversion factors relative to the base currency of the estimates are defined.
- Escalation rate projections for labor and materials are established.
- The estimate is analyzed by use of summary and detailed reports and sorted by WBS, resource/discipline codes, and user codes.

- The estimates can be transferred to the budget or forecast fields of the cost engineering module by linking the bid items and WBS codes.
- Estimates can be revised quickly and compared to prior estimates based on changes to labor or equipment unit pricing.

Engineering Management Module

The Engineering Management module supports earned-value and performance analysis, as well as the identification and processing of engineering change notices.

The application defines the components and activities for all work packages. Actual workhours incurred can be imported or input from various corporate time collection systems at the various discipline and classification levels, and can include owner staff involved in engineering and administration.

Earned-value and performance analysis is a tool that is useful in monitoring engineering services. Performance analysis relies on the existence of a baseline plan against which performance is assessed. Performance against a baseline plan is measured using earned-value, which is essentially a computation of the value of work satisfactorily completed. The principal advantage of the earned-value approach is its correlation of work package, organizational responsibility, WBS, budget, and schedule. Standard reports provide comparisons of budget, incurred and forecast costs, as well as earned-value and performance analysis. Status and details of the project change notices are recorded to reconcile the impact of changes in scope to the schedule and budget for engineering work packages.

Planning and Scheduling Module

The Planning and Schedule module is used in planning, controlling, and managing the program and its component projects.

Planning the project means determining what needs to be done—defining specific activities and work tasks, coordinating these activities, preparing work schedules, and assigning and allocating resources to competing activities.

Controlling the project means measuring performance, suggesting corrective action when needed, evaluating options, and devising workarounds if plans cannot be achieved. For major projects, planning and scheduling involves several hierarchical levels:

- Master program schedules
- Project schedules
- Detail contract or work package schedules

A variety of computer programs for schedule control are available on the market. The advanced features of the industrial-strength scheduling software include:

- Critical path method scheduling
- Precedence and arrow diagramming methods
- Resource leveling
- Performance evaluation
- Intelligent alphanumeric activity identifiers

- Activity coding (24 custom codes)
- Date and duration constraints
- Summary and detail network links
- Milestones and flags
- Standard as well as customized reports

The systems also include graphic capabilities that enable the user to create a variety of detailed or summary charts, diagrams, profiles, and curves. Graphics features include:

- Gantt schedule charts (bar charts)
- Time-scaled logic diagrams
- Pure logic plots
- On-screen drawing tools to create lines, boxes, symbols, and logos, and to add text
- Current and target schedule comparison
- Highlighting of progress and critical activities

Procurement Management. Procurement is an integral part of any major construction program and includes materials, equipment, and professional services. Procurement management involves planning, controlling, and managing the various procurement activities. The procurement activities will be managed by use of special scheduling reports which are issued to monitor key procurement activities for each procurement package, including:

- Preparation of material and equipment specifications
- Preparation of requests for proposals (RFPs) for professional services
- Internal review of specifications and RFPs
- Tender request issuance
- Tender receipt
- Tender analysis
- Development of recommendations
- Receipt and approval of vendor data or shop drawings
- Equipment fabrication or delivery milestones

Integration of Cost and Scheduling Applications

Data elements in the Cost Control and Schedule subsystems are related through the program WBS and PCS structures. Schedule activities contain the WBS and PCS structures in the activity code fields. Although a one-to-one relationship between scheduling activities and WBS codes may not exist or be practical at all levels, all the activities will be coded to reflect and collect their costs in the proper cost account.

The integration of these activity codes allows CPM schedule progress information to update the cash flow forecast for each contract package in the cost module. It also provides integrated cost/schedule performance evaluation information for each package.

Contract Administration Module

The Contract Administration component is designed for monitoring all documentation related to contract administration, whether for design consultants or construction contractors, including submittals, correspondence, payment requests, and change orders.

This module helps avoid a common cause of project delays and costly claim disputes: the failure of the project team to address promptly the numerous requests for clarifications, approvals or responses to changed field conditions that arise during the course of major construction jobs. In those cases where disputes cannot be avoided, the module provides comprehensive, organized documentation to resolve disputes. Such documentation can be very difficult and expensive to piece together after the fact, or after claims are filed. A field-level configuration might typically include the following modules:

- Claims log
- Document log (general contract correspondence)
- Field instruction log
- Inspector's daily diary
- Meeting minutes
- Noncompliance notice log
- Photograph log
- Progress payment logs
- Punchlist log
- Requests for information log
- Safety concern notice log
- Submittal log

The use of field-level document management software helps to verify that each construction contract is properly documented for effective daily action. It facilitates the communication of vital information from the field to other team members.

The module also provides standard forms for transmittals and reports to improve the quality and comprehensiveness of record keeping.

Document Control Module

Project personnel spend countless hours generating and filing forms, reports, letters, memos, and drawings. They often spend as many hours looking for those documents, looking for related documents, and searching for information concerning the status of those documents.

Computerized document control systems allow all project correspondence to be profiled, indexed, and stored in an electronic library for ease of access and retrieval.

A document can be a specification, drawing, transmittal, photo, or other correspondence. The system uses profiles which describe the document and include information such as the document number, author, date created, latest revision, archive date, location, retention classification, keywords, and comments. In addition to the standard fields, profiles may also include user-defined alphanumeric, numeric, and date fields. Query-by-example allows users to search for documents by filling out a blank profile with any information they happen to know: author, approximate date, subject, keyword, file code, or even text within the document itself. The system responds with a "hit list" of documents that match the query. The user can view the document profile, view the document itself, or launch the application that created it. (See Figs. A5-13 and A5-14.)

There are several other powerful features of document control software:

Intelligent Full Text Searching: To assist in research, the system contains a sophisticated intelligent search capability that allows users to search for concepts and content within documents. For example, a search for "behind schedule" would recognize "late" as the same concept. (See Figs. A5-15 and A5-16.)

Network File Management: The Document Control module works with the network operating system to relieve users from such details as file naming and directory creation and organization. Security is implemented at the document, profile, and index levels. When users perform a search, they see only those documents to which they have security clearance. Once system parameters are set, the Document Control module manages daily operations with minimal maintenance from the system manager.

Storage and Retrieval: Both character-based text and images can be attached to a profile and managed with the Document Control module. A common approach is to start with profiles and network management and then add the scanning of incoming documents at a later date. System users can retrieve and edit documents in whatever format is available, such as a word-processing document, a spreadsheet, or a scanned image. When documents are edited, the original application is launched with the file loaded, ready to edit. The system checks out a document when a user edits it, essentially locking it so that no one else can revise it

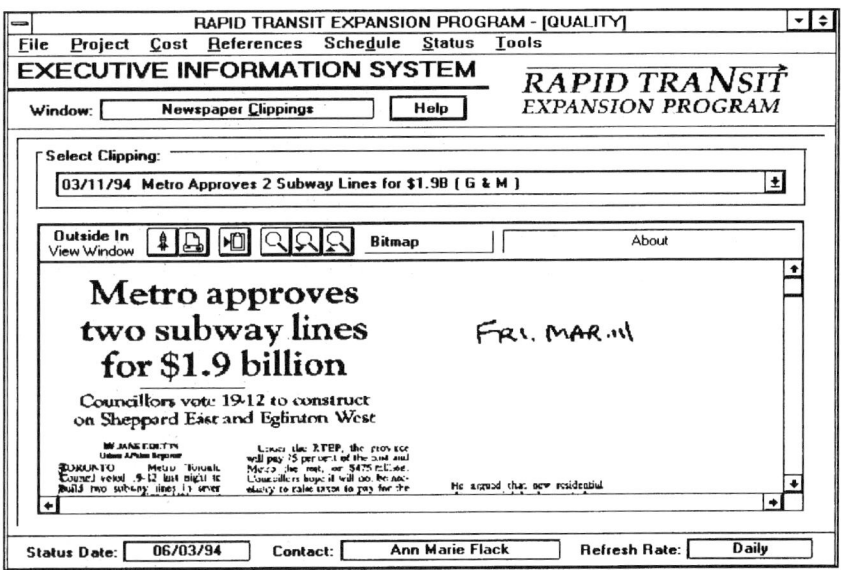

FIGURE A5-13 Recall of newspaper articles.

A5-30 STANDARD HANDBOOK OF HEAVY CONSTRUCTION

```
┌──────────────────── RAPID TRANSIT EXPANSION PROGRAM - [STATUS] ────────────────────┐
│ File  Project  Cost  References  Schedule  Status  Tools                            │
│ EXECUTIVE INFORMATION SYSTEM          RAPID TRANSIT                                 │
│ Window: [    Project Map    ]  [ Help ]   EXPANSION PROGRAM                         │
│  ┌─Select Progress Report::──────────────────────────────────────────────────────┐  │
│  │ 04/30/94 April '94 Progress Report                                         ▼ │  │
│  └──────────────────────────────────────────────────────────────────────────────┘  │
│  ┌──────────────────────────────────────────────────────────────────────────────┐  │
│  │ Outside In  [♪][🗎][▶□][🔍][🔍][🔍]  Document  │  WordPerfect 5.1/5.2         │  │
│  │ View Window                                                                   │  │
│  │                                                                               │  │
│  │   EGLINTON SUBWAY                                                             │  │
│  │                                                                               │  │
│  │      The request for tender proposals for the East Tail Track and main Sewer  │  │
│  │      relocation are in the final stages of completion. The relevant tender    │  │
│  │      packages are expected to be issued mid-May.                              │  │
│  │                                                                               │  │
│  │      The Allen Station Area is now in the preliminary design stage which is   │  │
│  │      scheduled to be completed at the end of July '94.                        │  │
│  └──────────────────────────────────────────────────────────────────────────────┘  │
│                                                                                     │
│  Status Date: [ 04/30/94 ]  Contact: [ Lou Tucciarone ]  Refresh Rate: [ Monthly ] │
└─────────────────────────────────────────────────────────────────────────────────────┘
```

FIGURE A5-14 Recall of newspaper article.

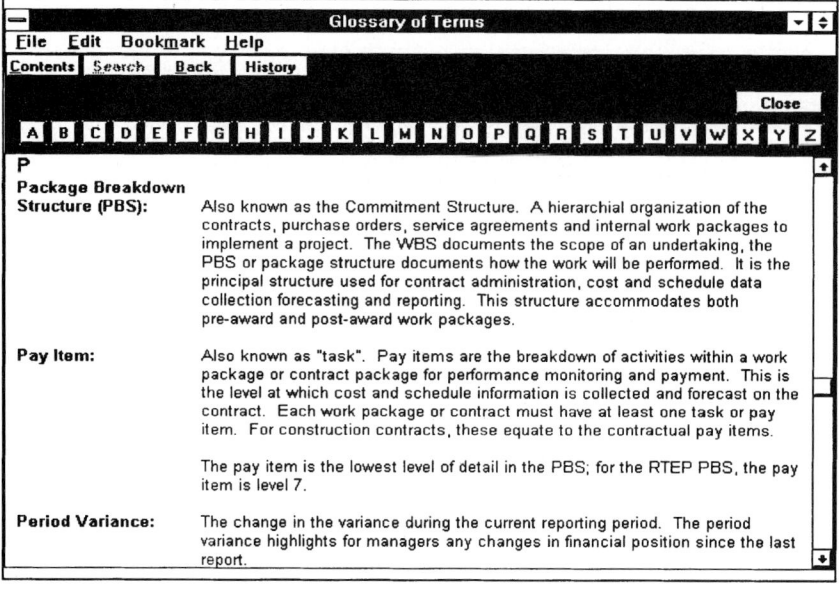

FIGURE A5-15 Glossary of terms.

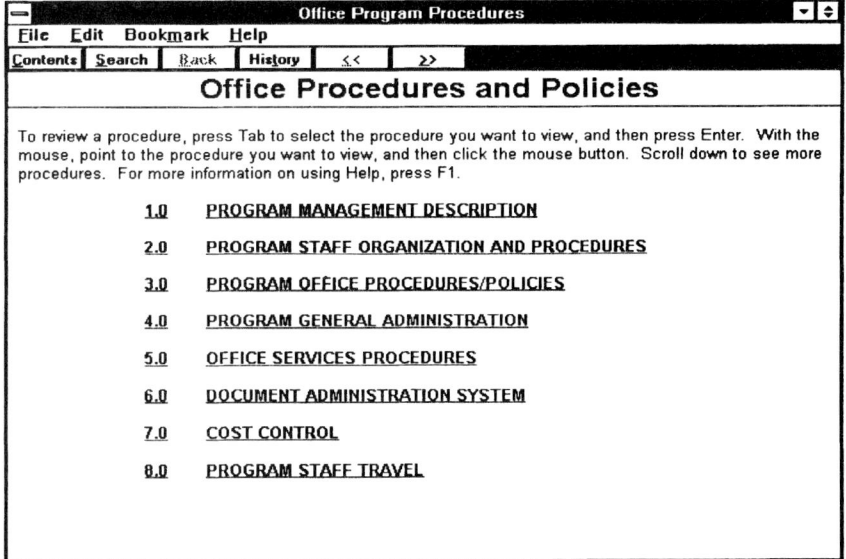

FIGURE A5-16 Menu of office procedures and policies.

until it is returned. Of course, others may still view it at any time. In fact, the system supports the ability to check out documents for home use and will automatically synchronize documents created in remote office locations or on portable computers, or those documents created on workstations while the network is unavailable.

Reporting: Action lists, ticklers, and a wide variety of system management reports can all be produced. The report writer allows existing reports to be customized and new reports to be added as required.

Customization: The profile, menus, text strings, help messages, screen defaults, and documents previewers can all be modified to the needs of the project team.

CHAPTER 6
VALUE ENGINEERING FOR CONSTRUCTION MANAGERS

Larry W. Zimmerman, P.E., FSAVE
Executive Vice President
Lewis & Zimmerman Associates, Inc., Rockville, Maryland

INTRODUCTION

The value engineering (VE) methodology is a systematic approach of identifying a project's functional requirements, features, and quality ratings and optimizing those design, construction, and operational criteria to achieve an improved project. The VE methodology is applied through the use of the VE *job plan* by a multidisciplined team. The job plan follows the format of the inventive or creative process, with each step in the process a building block.

Properly followed, VE will lead the construction manager to a full evaluation adding value to the owner's facilities. VE includes an evaluation of the needs for the facility, the layout, and its effect on the performance of the owner's mission or purpose; the level of quality and durability of the facility; the facility life cycle cost; staffing needs as affected by the facility layout; and the operating systems to enhance the business or organization being served. In short, defining, analyzing, and improving the customer needs are a big part of the VE results.

Having read this far, we know that there is a systematic approach that is important to VE. This systematic approach, the job plan, is the road map to success in VE. It is the key to the organization's success in applying VE. As the construction manager for a program or a project, VE is one of the best approaches available to evaluate the entire program, obtain consensus on the program features, and obtain the owner's input into the requirements and value objectives that they see for their project. VE has also been used to assess the project delivery system for a program.

DEFINITION OF VALUE WORK

Value engineering is a management methodology using a systematic approach (the VE job plan) to develop (conceptualize) or analyze and seek out the best functional balance between cost, performance, and reliability. For the P/CM it allows a means of increasing value for the client.

$$\text{Value engineering} = \textit{number 1 value-added service}$$

This section will outline the merits and the methodology of value work. More complete descriptions of value work may be obtained through the Society of American Value Engineers.

BENEFITS OF VALUE WORK

Facility owners and operators are the first to recognize the significance of increasing value. Whether in obtaining additional facilities, output from a process plant, increased life of the facility, better quality, or a host of other factors, the increase is vital to corporate profitability or to cost of services provided. The planning and design of facilities has the effect of setting the value parameters at a cost to the owner.

As program/construction managers, our services are also focused on added value to the owner. Value Engineering is the methodology to add value to a project.

$$\text{Value index} = \frac{\text{function} + \text{performance} + \text{quality}}{\text{cost}}$$

The value index gives a simple formula to added value. The numerator can include many things that an owner cites as important to the company's value index and the various value objectives each are at a cost. As P/CMs, we influence each aspect of the value index through the design and cost of construction and operations.

Benefits to the owner are accrued when value is increased and their required functions are performed at minimum cost. Other benefits to the project delivery are important to note:

- Focus on project criteria early in the design process
- Rapid assessment of concept and planning realistic cost estimates
- *Communications and team building*
- Increased value (more functions at less cost)
- Improved plans and specifications (lower change orders)
- Improved corporate profitability
- Improved internal rate of return
- Projects enabled to complete within the budget

Financial results from VE average over 5 percent savings in construction cost. Returns in savings generated from VE are $15 to $25 per each dollar spent on VE work. Some companies' results double these amounts. The costs of VE services are 0.4 percent for large, complex projects. Smaller jobs may have VE costs up to 0.8 percent of construction costs.

With the P/CM's interest in increased value and cost management, VE is a vital aspect of the project delivery system.

APPLICATIONS OF VE BY THE PROGRAM/CONSTRUCTION MANAGER

Value engineering can and should serve as the P/CM's methodology to advance and direct the project forward. It is one of the most important tools to improve the value (value-added) of the project.

Be careful—many P/CMs are prone to think of VE as cutting costs, often by sacrificing the long-term benefits of the design (life cycle costs). Many P/CMs view VE as the estimator's job in finding cheaper materials and products. Think again. VE is accomplished in an entirely different setting as a team of multidisciplined individuals using the VE methodology (in a workshop setting) to achieve results. The team concept and VE methodology are vital to success in VE.

Now that it is clear that VE is much more than a cost-cutting exercise, let's look at where VE can be applied to assist the P/CM in full program management. VE applications have expanded greatly in the last decade from analyzing designs midway through the project, to conceptualizing the program using VE as the methodology to build design programs and build consensus and ownership of the design. Applications for VE use are outlined as follows.

Conceptualization of Design

VE has advanced to where the workshop format, cost, and life cycle cost modeling are being used to develop the design concepts with the architect-engineer (A/E), owner, and P/CM. In a matter of weeks, the project team, acting in partnership, defines the owner's objectives and arrives at solutions that meet project needs. Recent experiences have shown excellent results and benefits:

- Consensus for design
- Months off the design schedule
- Budgets that are much more realistic
- Early evaluation of construction sequencing, contracts, and method of construction
- Early user input
- Less redesign as communications are vastly improved
- Teamwork

An example project used to conceptualize a design is included at the end of this chapter.

During Design

VE has traditionally found its niche during design development. Its use is well defined in this venue and accounts for savings from 5 to 15 percent of construction and life cycle costs while improving total value of the program.

Design decisions have a tremendous impact on the functionality and costs of a facility. The highest return on the VE effort can be expected when VE is performed to develop the concept or early in the design process before major design decisions have been completely incorporated. This principle is illustrated schematically in Fig. A6-1.

When two VE studies are performed during design, the first VE workshop should be held at the 20 to 30 percent stage of design completion and the second at 65 to 75 percent of design completion. If only one VE study is determined to be sufficient, the workshop should be performed at the 20 to 30 percent stage of design completion.

Typical study areas for a 20 to 30 percent VE workshop are the overall facility layout, concept layout for buildings, staffing profile, architecture, materials of construction, interior layouts, selection of unit processes, foundation designs, electrical concepts, and process control concepts. For the 65 to 75 percent VE workshop, typical study areas include piping layouts; structural, mechanical, HVAC, and electrical design drawings and specifications; and architectural details. The second study emphasizes constructability issues and details of the design.

When the VE studies are factored into the overall program master schedule from the start of the project, they can be accomplished concurrently with the design and not delay its completion.

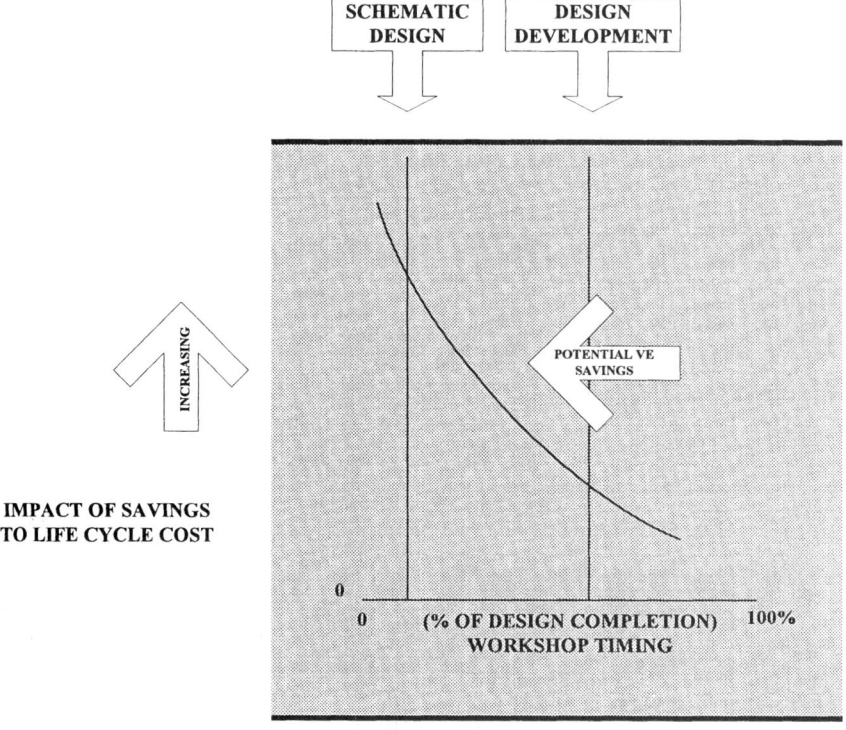

FIGURE A6-1 Potential impact of savings to develop the concept versus workshop timing.

VE of Facility Operations

VE is making headway into the facility operations as owners recognize the follow-on costs and their effects on bottomline profitability. Examples included the following:

- Toll card and fare systems for highways, transit systems, and arena management.
- Health care facility systems operations for record keeping, material supply, food services, security, and staffing.
- Disaster recovery to provide needed services in reduced time frames. The World Trade Center and Seattle Kingdome are examples.
- Industrial processes for manufacturing affected by profitability.

Studies in these areas are often performed on operating facilities and to improve systems and process applications for new planned activities.

THE JOB PLAN OF VALUE ENGINEERING

The VE job plan consists of the following five distinct phases:

1. Information phase
2. Speculative/creative phase
3. Evaluation/analytical phase
4. Development/recommendation phase
5. Report phase.

These phases are designed to, first, thoroughly define the task, program, or problem to be solved. They assist in finding solutions to meet the owner's value objectives and provide a creative atmosphere for development. For construction, these phases define the customer's needs, their program, and the functions that need to be performed. This phase of the VE job plan is vital for P/CMs as it leads to clear and concise project criteria, design definition, and cost projection. Secondly, the creativity and judgment phases create an open framework of communications and consensus building between the project's designers, users, owner, and operating staff. Progression through VE analysis gives a big lift to the P/CM as VE strives to achieve similar objectives. Development and reporting phases provide a framework for the completion and the development of ideas, concepts, designs, and alternatives to enhance the mission and goals of the owners (those needing the facility) of the facilities.

The VE job plan will assist the construction manager, in a number of ways:

- It is an organized approach that quickly focuses the VE team on the project requirements, functions required by the owner, their criteria and value objectives, the cost sensitivity of the project, and the cost of the facilities related to the functions and quality received. It identifies high cost to worth areas, and sets the stage for identifying selected alternatives to minimize costs while maximizing quality. VE teams that do not follow a formal job plan tend to be cost-cutting reviews rather than true value engineering studies.

- It encourages the VE team to take a more thoughtful and creative approach, i.e., to look beyond the use of common or standard approaches.

- It emphasizes total cost of the ownership of a project. If the facility is financed through a bond, it helps to confirm the cash flow and eventual user rates for financial data. For construction managers, the use of the job plan may be foreign to their preconceived idea of value engineering. The job plan coupled with the function evaluation of the projects is the basis for VE. This is vital to the understanding and use of VE, because most CMs think that VE is done in the estimating department by pulling out costs from buying out the job for less cost. In most cases, this leads to cheapening of the project, or a perception of cheapening the project, from those involved.

INFORMATION PHASE

During the information phase, the VE team solicits owner and designer background on the owner's objectives, value criteria, and technical and cost factors to develop an overall understanding of the project's functions and requirements. Most of the data, including the cost and energy models, will have been reviewed by the VE team members prior to an actual VE session. The information phase occurs before and during the gathering of VE teams.

An oral presentation by the owner and designer—a statement of owner needs and criteria—provides the VE team with an understanding and appreciation of the factors that define the project and that have influenced the project's design. These findings serve to open the lines of communication between the VE team members, the owner, and the designer. It allows the designer to expose the VE team to the difficulties encountered during the design of the project, and the owner to clearly state needs and goals.

Traditionally done by an oral presentation, the findings include a description of the rationale, evolution, constraints, alternatives, and percentage completion for the major design components. The quality and organization of the data presented by the owner and design professional are important since these factors directly impact the usefulness of the VE recommendations. For the P/CM, it includes early cost definitions to the project parameters before excessive cost is entrenched.

Following the initial VE work, the owner, and sometimes the designer, participate in the VE workshop to answer questions which may arise and to assist in project improvement. Frequently, the VE team solicits owner/designer comments on the creative ideas before proceeding with the evaluation/analytical phase of the workshop.

Discipline must be exercised by the VE team during this phase to ensure that sufficient time is taken to collect and verify data before beginning the development of alternative ideas. The development of inappropriate VE recommendations can often be avoided by the careful evaluation of project background information.

It is important for the VE team to appreciate the designer's effort in the development of the project's concepts, its drawings, and specifications. The team should understand the designer's rationale for the project's development, including the assumptions used to establish the design criteria and select the materials and equipment. The VE team should identify and review the alternatives considered by the design professional. Cost, energy, and life cycle cost models are developed if VE is used to develop design, and to analyze costs if VE is used during design or in the analysis of an operating system.

Total Cost Management—Cost, Energy, and Life Cycle Cost Models

In VE studies, the cost, energy, and life cycle data are organized in a manner to facilitate rapid analysis and identification of high-cost systems or components. This is accomplished by assembling the cost and energy data in the form of models. The P/CM or VETC typically prepares the cost, energy, and life cycle models with the assistance of a cost estimator.

The P/CM will recognize the importance of accurate, identifiable costs. The availability of accurate and comprehensive cost data is an essential element in the success of project VE studies. This point cannot be overemphasized. The VE team uses cost data as its primary tool for evaluating alternative ideas. The quality of the team's evaluations and recommendations will be only as good as the cost data. Inadequate or inconsistent cost data will result in inaccurate results.

The cost data should be prepared in a detailed and organized manner to serve as the basis for evaluating VE recommendations. Particular attention should be devoted to establishing operational, maintenance, and energy costs. The replacement frequency and costs of major subsystems or components with a service life less than the planning period should be established prior to the study. All cost data should be developed on the basis of market prices prevailing at the time of the VE study.

Cost Model

A cost model is a P/CM road map of cost for managing the project. It also serves as a VE study tool. There are two general types of cost models commonly used for VE studies. One type is a cost matrix, which presents estimated costs by subsystem, functional area, or construction trade. The cost matrix provides a one-page comparative display of each major cost element. The other type of cost model is a functional cost model, which presents both estimated and target construction costs distributed by subsystem or functional area. The target cost is determined during the VE workshop since it represents the VE team's estimate of the least cost to perform the function of each subsystem or functional area.

Figure A6-2 is an example of a cost matrix-type model. It is a one-dimensional matrix which presents the wastewater treatment works costs distributed by major cost category.

Figure A6-3 is an example of a functional cost model. This type of model breaks down the total cost for the facilities in terms of major functional area, such as process stream, operating units, site, buildings, and support. This process is continued to successively lower levels.

In a functional model, costs can be represented on either a dollar or parametric basis. The value of a parametric format, such as dollars per million gallons of flow, would be for ready comparison to historical cost data.

Cost models are used for quick identification of high-cost subsystems or functional areas, since these areas frequently offer the greatest potential for cost savings. Pareto, an Italian economist, formulated a law of economic distribution which lends credence to this approach. Pareto's law states that 80 percent of the cost will normally occur in 20 percent of the constituent items. Because this law of economic distribution holds true for construction projects, the cost models aid in the identification of the relatively few elements or components which constitute the bulk of the cost.

Energy Model

The significant cost of energy continues to have a substantial impact on facility operating costs. Energy optimization must be one of the goals of a VE study. Energy models present displays of energy consumption by subsystem or functional area. The models typically express energy in units of kWh per year or kWh per use area. An

PROJECT XYZ WWTP (65% Design)

ITEM COST MATRIX: CSI DIVISION NO. VS. UNIT PROCESS

MATRIX COST MODEL No. ______

TEAM NO. 1 **SHEET** 1 **OF** 1

No. CSI	Description	Cost Category: GENERAL & SITE	EXIST.* STRUCTS.	CONTROL BUILDING **	BLOWER BLDG.	AERATION TANKS	FINAL CLARIFIER	DISINFECTION	GRAVITY THICKENER	Total Cost
1	General Requirements	135,210	43,120	133,290	135,310	57,695	62,245	25,610	55,500	647,980
2	Sitework	985,010	-0-	13,455	11,900	64,170	113,860	47,165	15,410	1,250,970
3	Concrete	-0-	-0-	151,885	188,990	392,680	724,065	108,650	135,150	1,701,420
5	Miscellaneous Metals	-0-	-0-	79,865	47,160	43,870	25,500	12,760	25,375	234,530
6-9	Building Work	-0-	6,540	421,240	123,730	10,900	3,270	3,270	119,900	688,850
11	Mechanical Equipment	-0-	176,220	388,140	244,510	148,890	148,860	-0-	231,315	1,337,935
12	Furnishings	-0-	-0-	73,860	-0-	-0-	-0-	-0-	-0-	73,860
13	Instrumentation	-0-	20,575	38,400	27,125	-0-	-0-	-0-	22,700	108,800
15	Mechanical Components	456,550	125,660	248,960	132,740	39,975	-0-	25,875	85,750	1,115,510
16	Electrical	38,150	9,810	731,170	476,550	16,350	5,450	5,450	27,250	1,310,180
	SUBTOTAL (rounded)	1,615,000	382,000	2,280,000	1,388,000	775,000	1,083,000	229,000	718,000	8,470,000
	Contingency	81,000	19,000	114,000	69,000	39,000	54,000	12,000	36,000	424,000
	*Includes modifications to existing headworks, pump room, and primary sludge pump room.									
	**Includes belt filter presses									
	Total Cost	1,696,000	401,000	2,394,000	1,457,000	814,000	1,137,000	241,000	754,000	8,894,000
	Percent of Cost	19.1	4.5	26.9	16.4	9.1	12.8	2.7	8.5	100

FIGURE A6-2 Example matrix cost model.

energy model is normally based on average conditions with separate notations for peak demands. As in the functional cost model, target energy consumption estimates are assigned to each area. The target estimates represent the least possible energy consumption for each subsystem or functional area based on historical energy data and new concepts.

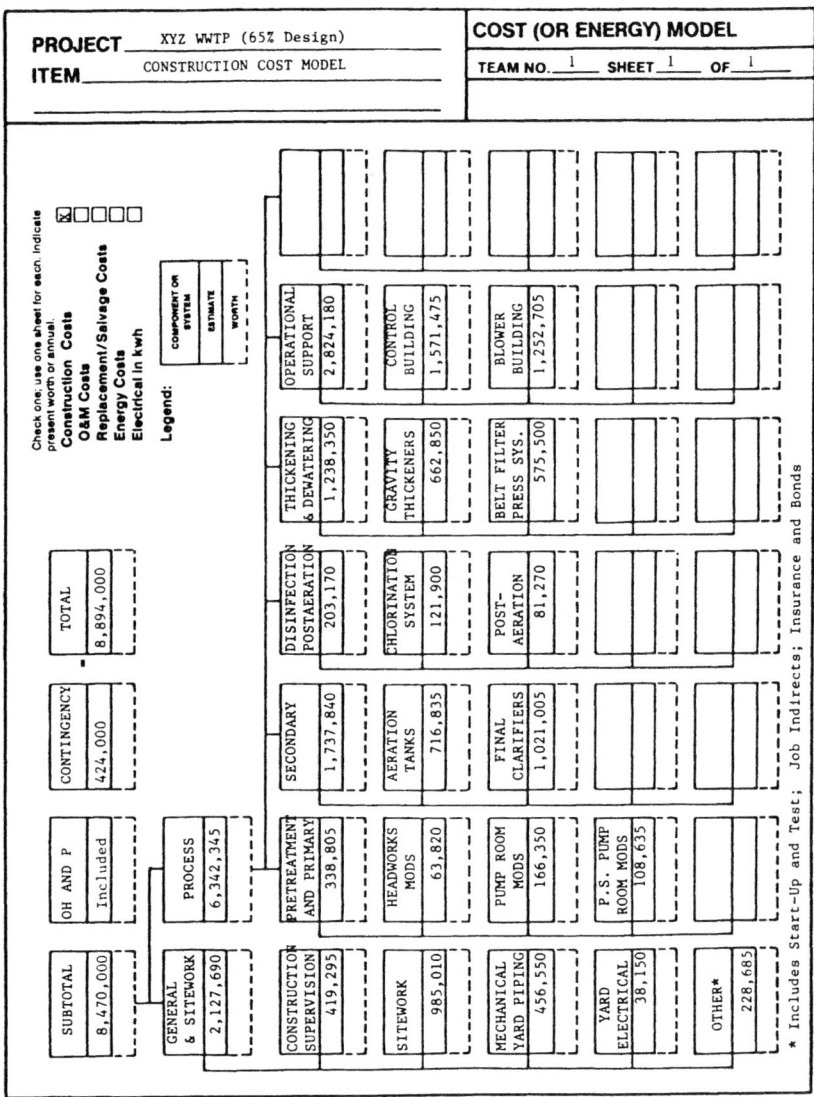

FIGURE A6-3 Example cost model.

Figure A6-4 is an example of an energy model for a wastewater treatment works. The energy model lists all major energy-consuming items such as motors, lighting, heating/cooling equipment, and emergency generators. The motor horsepower, electrical demand, fossil fuel consumption, or other appropriate energy parameters are converted into common energy units of equivalent kWh/yr before they are transferred to the energy model.

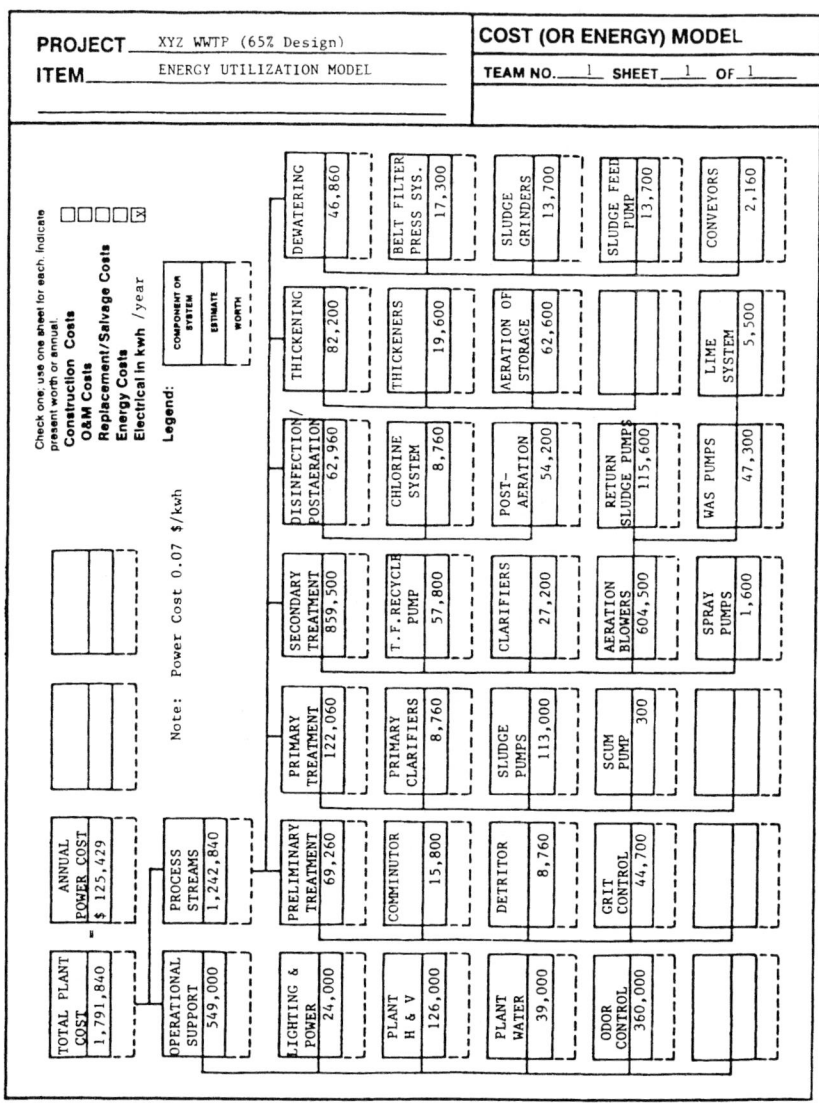

FIGURE A6-4 Example energy model.

Life Cycle Cost Model

Since the cost and energy models do not predict the total costs of owning and operating a facility, a life cycle cost (LCC) model is used. The LCC model provides a complete cost picture and serves as a baseline for the VE team's determinations of the cost impacts of VE recommendations. (See Fig. A6-5.)

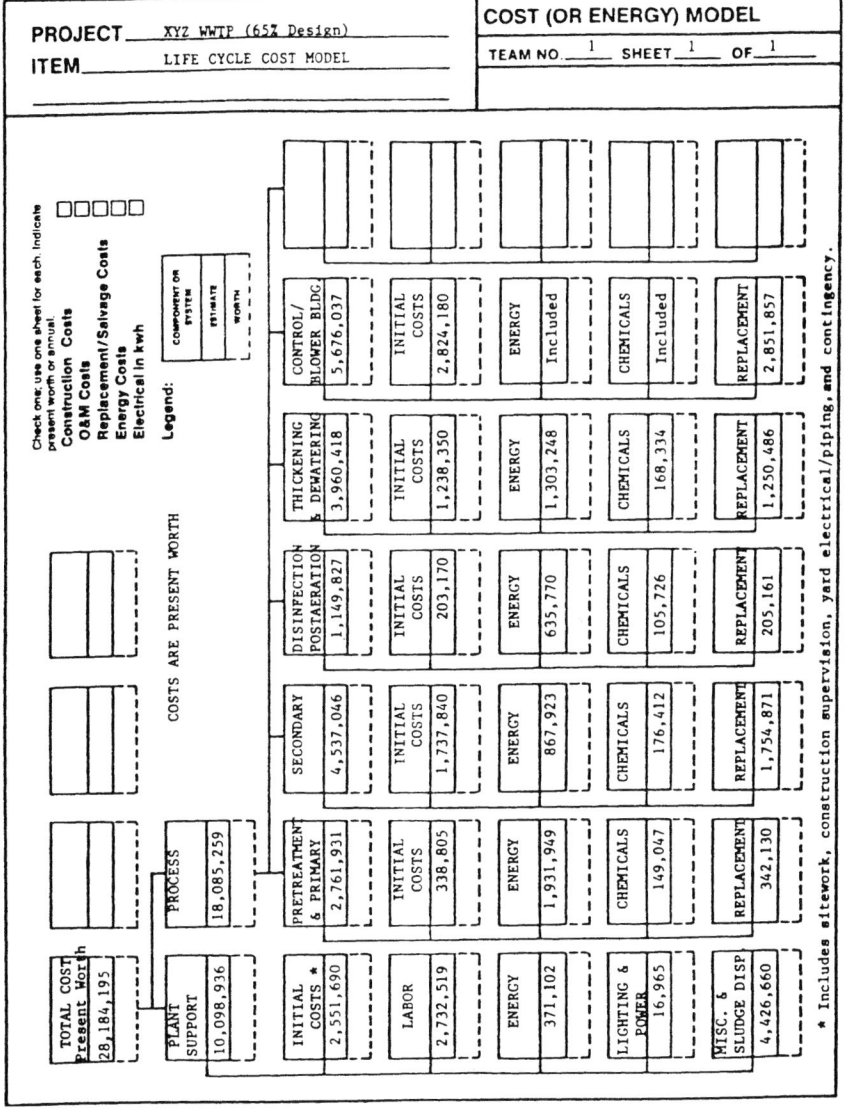

FIGURE A6-5 Example life cycle cost model.

Since the total cost of owning an asset consists of initial costs and all future costs, the latter must be discounted to present value (present worth) before they can be combined with initial costs to obtain the total life cycle costs. In other words, the time value of the future costs must be taken into account. For example, a $100,000 replacement cost 10 years in the future would have a present worth of $38,554 at a 10 percent discount rate.

The data from the cost and energy models are used in the development of the life cycle cost model. However, these models contain only a portion of the data needed to establish the total life cycle costs. The model must also include the additional costs of operation, staffing, maintenance, and equipment replacement.

Operation, maintenance, and replacement costs are typically the most difficult of all cost categories to estimate because of limited reference materials and historical data. Different operational philosophies cause operation and maintenance levels to vary greatly from facility to facility. Frequently, facilities with high levels of operation and maintenance have lower equipment replacement costs.

The most effective method for estimating operation and maintenance costs involves an examination of the historical data from existing facilities. Facilities such as hospitals, corrections facilities, and process plants are examples where staffing or other costs far outweigh the initial construction costs. Process plants may be more sensitive to time than to costs, as product life may warrant time to market as the key project criterion.

Function Analysis

Function analysis is the cornerstone of value engineering since it separates VE from other cost reduction techniques. The function analysis approach is used in value engineering to arrive at the basic purpose of systems facilities and subsystems. It aids the VE team in determining the required primary functions and peripheral or support functions at the least cost. While function analysis sounds like a very simple technique, it is probably one of the most misunderstood tools in value engineering. A VE team must be careful not to gloss over the function analysis and simply start listing creative ideas to reduce costs. Cost reduction studies simply list creative ideas for reducing costs, while value engineering focuses on a function analysis of the entire project. As the P/CM, be aware that you must perform function analysis or risk results that are cheap and nonfunctional. The function analysis approach provides the VE team with the mechanism for becoming deeply involved in the facility design and identifying costs which can be reduced or eliminated without affecting the performance or reliability of the facility.

Functions are identified by a two-word verb/noun description. The verb is an action verb and the noun is a measurable noun. As an example, the function of an electric cable is to *conduct current*. *Conduct* is the action verb and *current* is the measurable noun. Other examples are to *support load, convey flow,* and *house inmates*.

The basic function of an item is the specific task or work it must perform. Secondary functions are those functions that may be needed but are not actually required to perform the specific task or work. Required secondary functions are absolutely necessary to accomplish the specific task or work, although they do not exactly perform the basic function. As an example, the basic function of an aerator is to supply air; however, it also mixes the wastewater. In this case, mixing is a required secondary function for the aerator mechanism.

The following is a list of questions which are helpful in determining the functions of an item:

1. What is its purpose?
2. What does it do?
3. What is the cost?
4. What is it worth?
5. What alternative would accomplish the same function?
6. What would that alternative cost?

In function analysis, it is important to identify functional areas sequentially since the functions vary according to the selected area. For example, the function of the total facility would be established before functions are established for the unit processes or functional areas.

The most difficult part of the function analysis is establishing an estimation of the worth of each subsystem or component for comparison with its estimated cost. Since worth is an indication of the value of performing a specific function, extreme accuracy in estimating the worth is not critical. Worth is merely used as a mechanism to identify areas of high potential savings. Subsystems performing secondary functions have no worth because they are not directly related to the basic function. As an example, an access road to a hospital facility does not provide the primary function of curing illness or reducing suffering, even though the road may provide a required secondary function for the facility. Thus, the road is an area to examine for potential savings without affecting the basic function of the facility.

Value engineering looks for alternatives to the original design that might effectively increase the value and/or reduce the cost of the project. Alternatives may be developed by asking the basic question, "What else will perform the essential function, and what will it cost?" The alternatives for performing a function identified in determining worth become part of the creative idea listing. Thus, the creative phase of value engineering usually begins during the function analysis. When creative ideas are identified in the information phase, the VE team should simply record the ideas for later use in the speculative/creative phase.

Figure A6-6 (worksheet) can be used by the VE team to accomplish a function analysis. To complete this worksheet, the VE team would follow these sequential steps:

1. Identify the study area.
2. Identify the basic verb/noun function of the study area.
3. List the component parts of the study area.
4. List the verb/noun function of each component and subcomponent.
5. Identify whether each function is basic, secondary, or a required secondary function.
6. Identify the estimated construction cost of each function.
7. Speculate on the worth or the least cost to accomplish the function.

Figure A6-6 also illustrates the use of function analysis for a wastewater treatment facility. As shown, the function of the treatment facility, the unit processes, and other subsystems are identified in a verb/noun description with the type of function, i.e., basic, secondary, or required secondary. Similarly, the estimated cost and worth are indicated for each of the components.

As part of the function analysis, the VE team makes a comparison of the cost-to-worth ratios for the total facility and its subsystems. These cost-to-worth ratios are obtained by dividing the estimated cost of the system or subsystems by the total worth for the basic functions of the system or subsystem. High cost-to-worth ratios suggest areas of large potential cost savings and identify systems or subsystems that would be selected for further study by the VE team. Similarly, low cost-to-worth ratios indicate areas where further study efforts would not be justified due to diminished potential for cost savings. Cost-to-worth ratios greater than 2 usually indicate areas with the potential for substantial cost savings.

Function Analysis

PROJECT XYZ WWTP (65% Design) **FUNCTION ANALYSIS** No. 1
ITEM Wastewater Treatment Facility **TEAM NO.** 1 **SHEET** 1 **OF** 1
BASIC FUNCTION REMOVE POLLUTANTS

Sub Component Description	Function Verb	Function Noun	Kind[1]	Initial Cost[2]	Worth[3]	Cost/Worth	Comments
Control Building	House	Support Equipment	RS	1,571,000	1,200,000	1.6	Simplify construction
Blower Building	House	Equipment	RS	1,253,000	900,000	1.4	Reduce building size
Aeration Tanks	Convert	BOD & Ammonia	B	717,000	600,000	1.2	Modify to four sections
Final Clarifiers	Concentrate	Sludge	B	1,021,000	900,000	1.1	Consolidate; simplify construction & Reinforcing
Chlorine Tanks	Disinfect	Effluent	B	203,000	170,000	1.2	Simplify construction; change gates
Gravity Thickeners	Concentrate	Solids	RS	663,000	600,000	1.1	Increase loading; modify holding tank
Headworks	Remove	Grit & Screening	B	64,000	50,000	1.3	Reduce no. pumps; modify controls
Existing Pump Room	House	Equipment	RS	166,000	160,000	1.05	
Existing Pump Chamber	House	Equipment	RS	109,000	90,000	1.2	
Site & General	Support	Construction	RS	1,480,000	1,250,000	1.2	Consolidate; reconfigure
Other (including contingency)				1,647,000	1,580,650		
TOTAL				8,894,000	7,500,650	1.2	Basic function C/W only

[1] B = Basic Function S = Secondary Function
[2] Original Cost Estimate
[3] Worth - Least Cost to Accomplish Function.

FIGURE A6-6 Example function analysis worksheet.

To refine the identification of study areas offering potential for cost savings, each facility can be divided into subsystems which, in turn, can be divided into components. The cost-to-worth ratio of each component would be determined in the same manner described for the facility or its subsystems.

FAST Diagramming

FAST is an acronym for *function analysis system technique*. It is a tool that graphically shows the logical relationship of the functions of an item, subsystem, or facility. The FAST diagram is a block diagram based on answers to the questions of "Why?" and "How?" for the item under study.

A FAST diagram is most appropriately used on complex systems as a road map for clear delineation of the basic and secondary functions of a particular system.

FAST diagramming may be used in place of or to augment the function analysis portion of the information phase. FAST is used on complex projects to define the problem or facility needs and requirements. When unsure of the project goals, FAST is an excellent tool. Figure A6-7 is a FAST diagram for public works shop activities. Notice that the higher-order functions occur on the left and details (functional needs) occur on the right. More advanced forms of FAST identify responsibility matrices, obstacles to implementation, and quantities of input/output by assigning responsibility to each function.

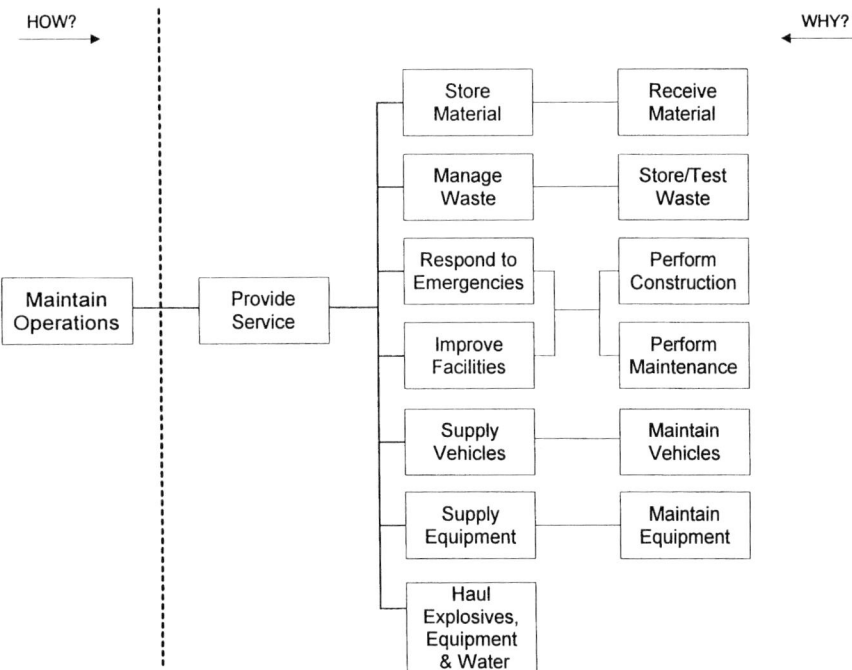

FIGURE A6-7 FAST diagram for public works shop activities.

SPECULATIVE/CREATIVE PHASE

The speculative/creative phase is another group interaction process which the VE team uses to identify alternative ideas for accomplishing the function of systems or subsystems associated with specific study areas. This phase involves an open discussion without any restrictions on the imagination or inventive thinking of individual team members. All analysis, evaluation, or judgment of the ideas generated is delayed until the evaluation/analytical phase. All ideas should be immediately recorded to avoid forgetting them as the discussion continues. The ideas should be listed by system, subsystem, and component to facilitate effective organization of the study. Many owners state that creativity is often missing in our design solutions. Here we aim to capture creativity which increases value.

The desired objective of the speculative/creative phase is to generate a completely free interplay of ideas among team members to create an extensive list of alternative ideas for later evaluation. The key to successful results is the deferral of any critical judgments or comments which might inhibit any of the team members.

Since a value engineering team is composed of a variety of personalities, some individuals will readily supply many ideas while others will have to be encouraged to express their ideas. The active participation of all team members is encouraged in the creative development of ideas.

To overcome this reluctance to venture outside familiar areas and risk the embarrassment of proposing an idea that might be subject to ridicule, the *yes/if* technique is often used. When a team member expresses an idea, another team member would respond with the statement, "*Yes* that idea might work, *if* we take the idea and improve it as follows," rather than condemning the idea. In this manner, team members build on the idea of a fellow team member to improve and refine the idea.

The following points should be considered during the speculative/creative phase.

1. When team members believe that improvements can be made to the project, they will work to achieve it.

2. There is always room for improvement in a project. Most designers will have many ideas for improving their project after observing its construction.

3. The word "impossible" should be eliminated from the team's thinking. The synergistic effect of free flow of information generated by a multidisciplined team can create extraordinary results.

4. Look for associations of ideas. Often a function can be performed by a technique currently applied to another area or industry.

5. Record all ideas as they are identified rather than risk forgetting them.

Speculative/creative ideas generated by the VE team can be listed on a worksheet as shown in Fig. A6-8. This worksheet is used for both listing and evaluating ideas.

EVALUATION/ANALYTICAL PHASE

During the evaluation/analytical phase, the ideas developed in the speculative/creative phase are examined. The VE team evaluates the feasibility of each idea by

VALUE ENGINEERING FOR CONSTRUCTION MANAGERS   A6-17

PROJECT XYZ WWTP (25% design)   **CREATIVE/EVALUATION** No.____
ITEM _____   TEAM NO. 1   SHEET 4   OF 5
                                     ABC, Inc.

CREATIVE

NO.	IDEA	ADVANTAGES	DISADVANTAGES	RATINGS*
PC-1	Reduce number of tanks	Reduce cost		8
PC-2	Review train concept	Simplify operation		8
PC-3	Check soluble BOD of pharmaceutical wastes	Review with PC-1		
PC-4	Use concrete planks vs. checker plates for tank covers in selected areas	Reduce cost		8
SC	SECONDARY CLARIFIERS (SC)			NA
SC-1	Remove weirs on inlet end and check effluent weir length			DS
SC-2	Revise influent channel	Improve function		9
SC-3	Remove walkways and handrail every other tank	Reduce cost	Some loss of convenience	DS
SC-4	Add metering for RAS for clarifiers 9–12	Improve function		7
SC-5	Use concrete planks vs. checker plates in selected areas	Reduce cost		8
EBB	EXISTING BLOWER BUILDING (EBB)			
EBB-1	Use centrifugal blowers for carbonaceous aeration	Reduce cost Improve maintenance		DS
	NITRIFICATION AERATION SYSTEM (NAS)			

*10 = MOST FAVORABLE   1 = LEAST FAVORABLE

FIGURE A6-8 Example speculative/creative worksheet.

identifying its advantages and disadvantages. The ideas are then rated on a scale of 1 to 10 or by matching them to the owner's objectives.

Even though detailed cost estimates for ideas are not developed until later in the study, the VE team would use its experience to estimate rough cost savings for ideas to aid in the evaluation process. It is important to note that some ideas with high cost savings potential may not benefit the facility's design because they may reduce the efficiency or quality.

The evaluation of creative ideas can be done on the right-hand side of the worksheet in Fig. A6-8.

DEVELOPMENT/RECOMMENDATION PHASE

In the development/recommendation phase, the best ideas from the evaluation/analytical phase are developed into workable VE recommendations. The VE team researches and develops preliminary designs and life cycle cost comparisons for originating concepts or for the original designs and the proposed alternative ideas.

During this phase, the technical expertise of each team member becomes very important. A multidisciplined team provides the resources essential for the development of sound VE recommendations. Frequently, VE team members must consult outside experts, vendors, and reference sources to obtain additional evaluation information before developing the VE recommendations.

The development of an idea into a recommendation should be thorough and address the key issues of accepting the idea. Comparison of the new concept to others is important. Sketches of the original design and each alternative idea are critical for communicating ideas. Life cycle cost analysis assesses the full costs. A realistic comparison of each alternative idea will lead to acceptance or rejection. And, finally, the path to implementation is important. The P/CM's job will be to help sort through the best enhancements and alternatives with the project team.

Developmental information for alternative ideas and for presentation of VE recommendations is shown in Figs. A6-9 through A6-12. Supporting documentation for a VE recommendation such as design calculations, life cycle cost estimates, and sketches should be attached.

It is important that the VE team be able to convey the concept of each VE recommendation in a clear and concise manner to avoid its rejection due to a lack of understanding by the owner or designer. In preparing VE recommendations, each team member should strive to view the recommendation from the designer's perspective for reliability, cost effectiveness, and implementation.

Frequently, a number of ideas are identified by the VE team which have little impact in terms of cost savings. However, these ideas may be worthwhile in terms of operation, maintenance, or design improvements.

REPORT PHASE

The report phase consists of both oral and written communication of the results from the VE study. The VE recommendations are presented by the VE team orally. The oral presentation should be a relaxed and informal meeting to review the team results. The presentation provides an opportunity for the project team, P/CM, owner, and designer to discuss the VE recommendations.

VALUE ENGINEERING FOR CONSTRUCTION MANAGERS   A6-19

The VE team leader usually prepares a written VE report which summarizes the results of the entire VE study. This report is used by the P/CM, owner, and designer in their review and evaluation. The report should be prepared and submitted to the owner/designer shortly following the VE effort.

PROJECT XYZ WWTP (25% Design)	VE RECOMMENDATION No. JC-5
ITEM REVIEW SIZE AND NUMBER OF PIPES TO EXCESS FLOW BASIN	TEAM NO. 1 SHEET 1 OF 1
	ABC, Inc.

ORIGINAL: (Attach sketch where applicable)

Flow from the junction chamber to the retention basin is via 2-48 inch pipelines. The maximum headloss at 90 MGD is some 7 feet, meaning the excess flow basin must be placed deep into the ground.

(See attached sketch JC-5-1)

PROPOSED: (Attach sketch where applicable)

Change the pipeline sizes to 1-36" diameter and 1-60" diameter to reduce headloss from 7 feet at maximum flow conditions to about 3.5 feet. The result is that the tanks may be raised some 3.5 feet. Scouring velocities are maintained by using a smaller pipe for low flow conditions and a larger pipe for higher flows. (See attached sketch JC-5-1)

DISCUSSION:

LIFE CYCLE COST SUMMARY	PRESENT WORTH COSTS		
	INITIAL COST	O & M COSTS	TOTAL
ORIGINAL	362,600	-0-	362,600
PROPOSED	170,000	-0-	170,000
SAVINGS	192,600	-0-	192,600

FIGURE A6-9 VE recommendation.

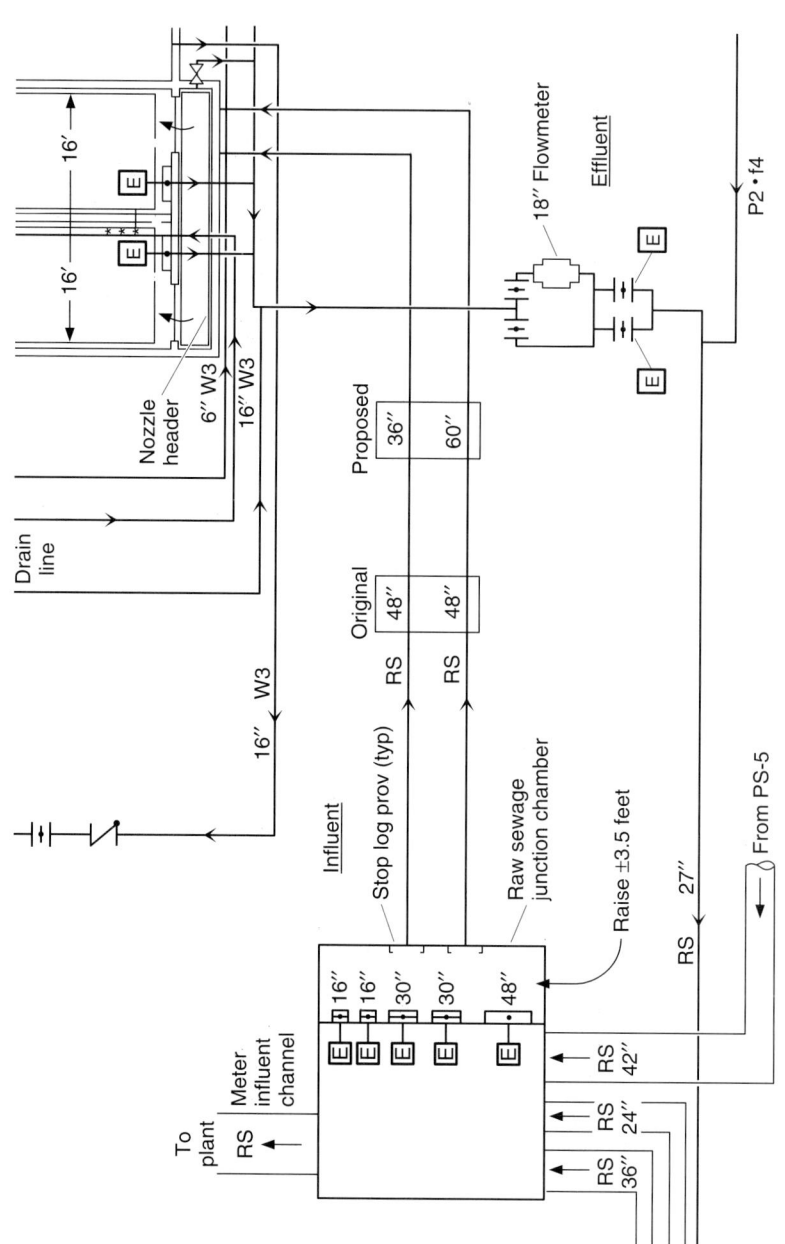

FIGURE A6-10 Sketch to support Fig. A6-9.

CALCULATION SHEET

Subject: __XYZ WWTP (25% DESIGN)__ Date: __4/13/84__ No. __JC-5__

Calculations By __ABC, Inc.__

<u>FLOW CONDITIONS ASSUMED FOR CONVEYANCE INTO EXCESS FLOW BASIN</u>: Low flow 4 MGD
High flow 90 MGD

ELEVATION AT JUNCTION BOX 679.10
LENGTH OF PIPE TO EXCESS FLOW BASIN 1000 Lf

PRESENT ELEVATIONS (BACKED-UP FROM BASIN)

	J.B.	BASIN	HL	
90 MGD	679.10	672.72	6.38	← MAX LOSS
50 MGD	674.93	672.49	2.44	
4 MGD	672.67	672.10	0.57	

ASSUME A MAXIMUM OF 2.5' HEADLOSS

$$\therefore 2.5'/1000 = \frac{0.25'}{100} = 66''\phi \text{ at } 90 \text{ MGD}$$

ASSUME A SMALLER PIPE FOR FLOWS UP TO 20 MGD. USE 36"φ.

$Q_{BIG\ PIPE\ (60''\phi)}$ = 90 - 20 MGD = 70 MGD   Use 60"φ.

REDUCED HEADLOSS ALLOWS RAISING OF TANK BOTTOM BY <u>3.5 feet</u>.

EXCAVATION ON SIDE SLOPES 10,950 cy

$$\frac{(750)(434)(3.5)}{27} = 42,200$$

53,150 cy
× $4
$ 212,600

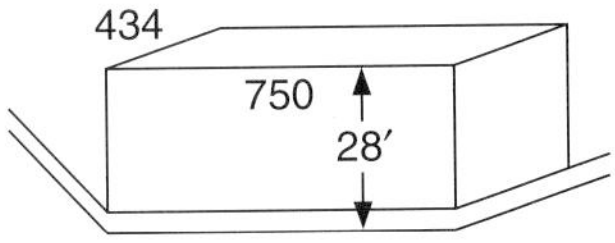

(diagram: box 434 × 750, 28')

FIGURE A6-11 Calculations to support Fig. A6-9.

CALCULATION SHEET

Subject: XYZ WWTP (25% Design)
Date: 4/13/84
Calculations By: ABC, Inc.
No.: JC-5
Page 4 **of** 4

PIPELINE 1   FOOTAGE (PCCP PIPING)

ORIGINAL 2 at 48" ⌀   2000 $75 $150,000
          1 at 36" ⌀   1000   50   50,000 ⎫
          1 at 60" ⌀   1000   120   120,000 ⎬ 170,000
                                                 ⎭

ORIGINAL

  EXCAVATION   212,600   NOTE: SOME DEWATERING
  2 at 48"   150,000   SAVINGS MAY ALSO
                -------   BE POSSIBLE.
                $362,600

PROPOSED

  1 - 36" ⌀ ⎫ $170,000
  1 - 60" ⌀ ⎭

              Δ = 192,600

FIGURE A6-12 Cost calculations for Fig. A6-9.

The most important element in the VE report is the VE recommendations. The information generated during the VE workshop can be further developed for accuracy and completeness before it is included in the VE Report.

MANAGEMENT OF VALUE PROGRAMS

Increasing the value of your construction program and the operation of your facility involves the cooperative participation of the project owner, its operations staff, the P/CM, the project designer, and the value coordinator. It is important to note that the value goal of all three parties is identical, i.e., to ensure that the final design represents the most efficient combination of cost, performance, and reliability. During the design of a facility, the owner, the P/CM, the project designer, and the value coordinator function as a team.

Value success depends heavily on the management and organization of VE work and the attitude and cooperative spirit of the participants. Diverse viewpoints and perspectives provide an excellent opportunity to enhance value and reliability.

VE SEQUENCE AND TYPICAL SCHEDULE

Value efforts are applied at various stages of a program. As outlined previously, VE is used to develop the concepts for the program, during the design, and also to act as a postoccupancy evaluation measure. Included as Figs. A6-13 and A6-14 are task flow diagrams for use in VE for conceptualizing the design and for VE work where design is in progress. Each follows the VE job plan with variations using function analysis for its basis.

As illustrated by Figure A6-13, VE to conceptualize the design is performed in a continuous workshop setting with a core team with involvement from owner groups, A/E, P/CM, users, VE coordinators, specialists, and stakeholders in the venture. Early definition of the owner's challenges and needs for the facilities as relating to business needs are key to early project definition.

Functional needs surveys assess all the known functions that the owner needs to perform with the new facility. It includes function analysis and/or FAST diagramming to set about the owner/customer needs. Various options are considered and early conceptual cost budgets prepared as the first steps in cost control and management. The owner's value objectives play heavily into the cost budgets and provide the owner a clear picture of cost and value decisions. Conceptual reports of findings include the program, conceptual plans, estimates, schedules, and a sign-off from all stakeholders in the concept plan.

The study takes two to six weeks depending on the level and complexity of design. Often, breaks occur for a few days as individual specialty work is performed.

Value efforts performed during design can conveniently be divided into these sequential periods of activity: (1) preworkshop activity, (2) VE workshop, and (3) postworkshop (report) activity. Figure A6-14 is a task flow diagram which outlines the effort which occurs during these periods of activity.

For most facility design projects, two VE studies are held at different stages of design completion to obtain maximum benefits. In these instances, the preworkshop activities, the workshop, and the postworkshop activities will be performed twice.

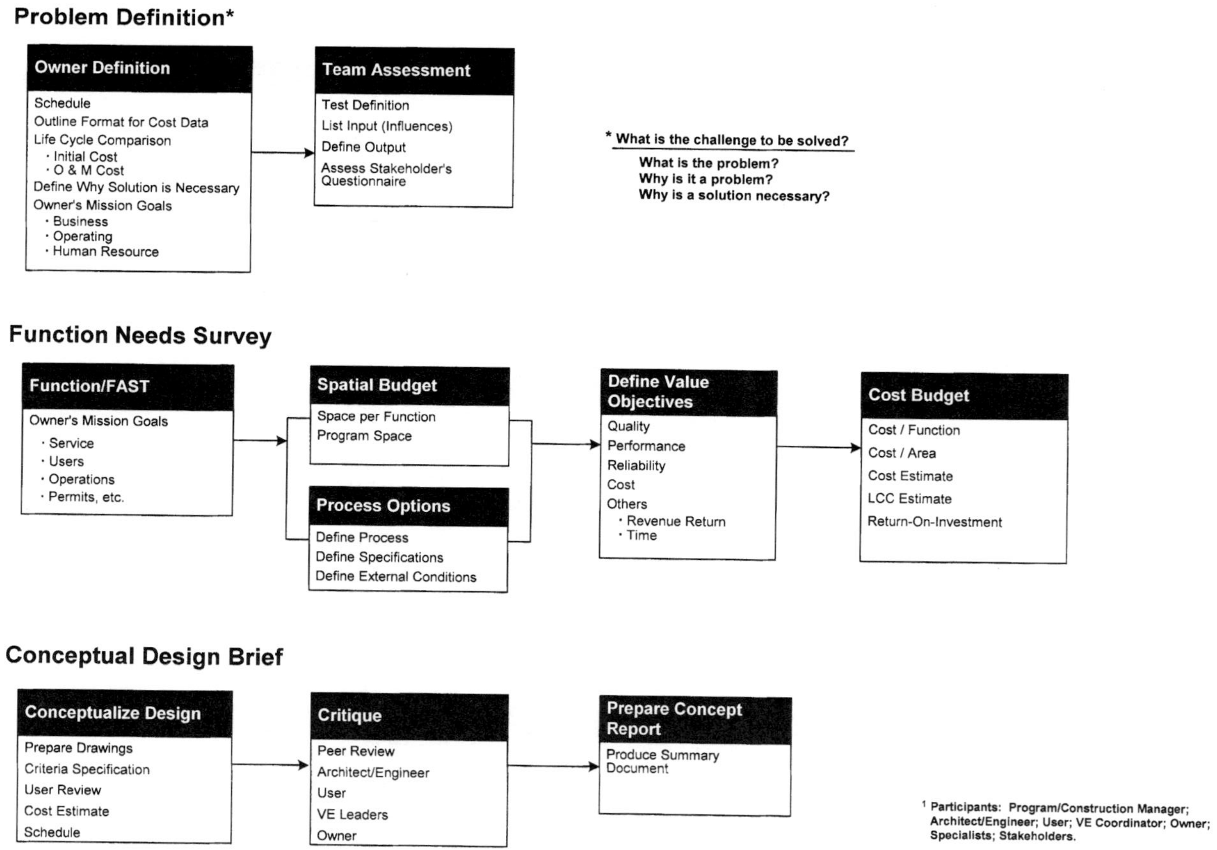

Problem Definition*

Owner Definition
Schedule
Outline Format for Cost Data
Life Cycle Comparison
· Initial Cost
· O & M Cost
Define Why Solution is Necessary
Owner's Mission Goals
· Business
· Operating
· Human Resource

Team Assessment
Test Definition
List Input (Influences)
Define Output
Assess Stakeholder's Questionnaire

* What is the challenge to be solved?

What is the problem?
Why is it a problem?
Why is a solution necessary?

Function Needs Survey

Function/FAST
Owner's Mission Goals
· Service
· Users
· Operations
· Permits, etc.

Spatial Budget
Space per Function
Program Space

Process Options
Define Process
Define Specifications
Define External Conditions

Define Value Objectives
Quality
Performance
Reliability
Cost
Others
· Revenue Return
· Time

Cost Budget
Cost / Function
Cost / Area
Cost Estimate
LCC Estimate
Return-On-Investment

Conceptual Design Brief

Conceptualize Design
Prepare Drawings
Criteria Specification
User Review
Cost Estimate
Schedule

Critique
Peer Review
Architect/Engineer
User
VE Leaders
Owner

Prepare Concept Report
Produce Summary Document

[1] Participants: Program/Construction Manager; Architect/Engineer; User; VE Coordinator; Owner; Specialists; Stakeholders.

FIGURE A6-13 Value engineering study task flow diagram (conceptualize design).

A6-24

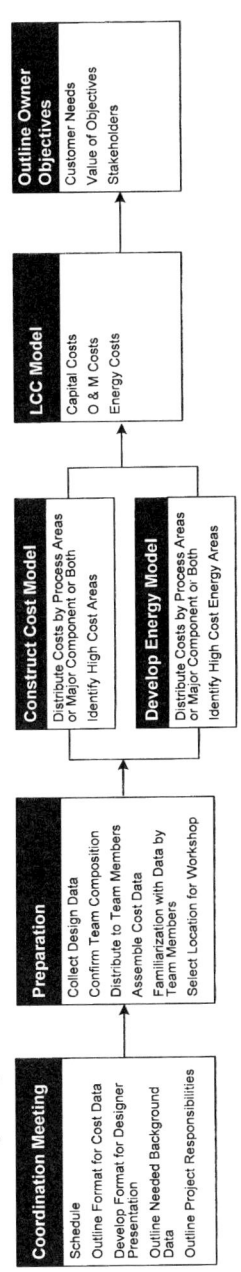
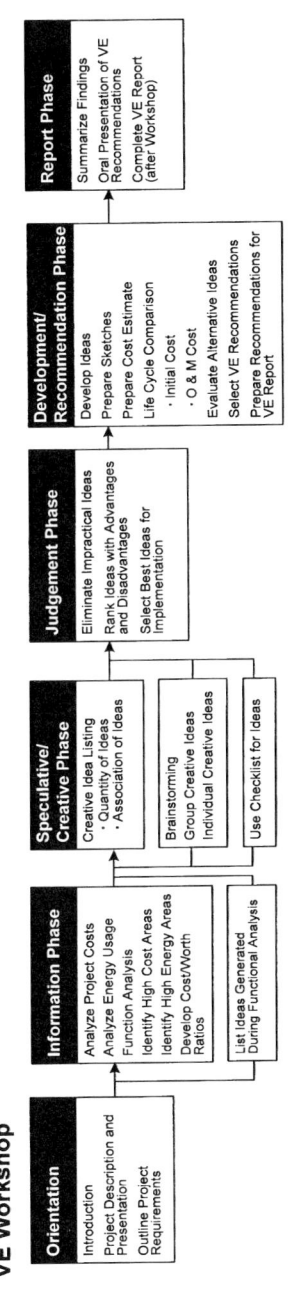
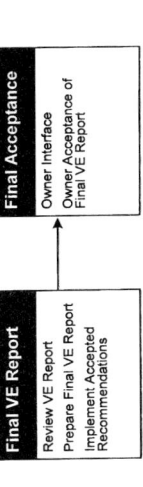

FIGURE A6-14 Value engineering study task flow diagram (design in progress).

PREWORKSHOP ACTIVITY

The VE team coordinator (VETC) uses this period to become familiar with the project, obtain and review the technical and cost data, complete logistical arrangements for the VE workshop, coordinate timing for the VE workshops, complete the selection of VE team members, and establish a comfortable working relationship with the owner and the designer. Team members become familiar with the design and the P/CM provides the updated project cost estimate.

The success of the overall VE study depends heavily on the organization and management of the preworkshop activity. During the three to six weeks of preparation for the VE workshop, the following activities should be accomplished in the following general sequence:

1. A coordination meeting between the owner, project designer, and value engineering team coordinator (VETC)
2. Accumulation of the project's technical and cost data
3. Confirmation of the composition of each VE team and logistical arrangements for the VE workshop
4. Preparation of cost, energy, and life cycle models
5. Distribution of the technical and cost data to VE team members

Coordination Meeting

A meeting is held with the owner, designer, and VETC at the beginning of the preworkshop activity to promote a common level of understanding about the objectives of the VE workshop, establish a productive working environment, confirm the schedule of events, and establish the responsibilities for completing the VE workshop preparations. Items discussed during the meeting would include the availability and format of technical and cost data, conduct of the VE workshop, processing of the VE recommendations, plus the date, location, and other logistical arrangements.

The format of the designer/owner's presentation of project information to the VE team at the start of the VE workshop should be established during the coordination meeting.

The effectiveness of the VE workshop is ultimately dependent on the technical and cost data available for the VE workshop. The designer and owner should supply the project data to the VETC well in advance of the VE workshop to allow sufficient time for review and development of the VE study models.

VE WORKSHOP

The VE workshop typically varies in duration, depending on size and complexity of the project. It culminates in the oral presentation of the VE team's recommendations.

The VE methodology (job plan) used by the VE team during the VE workshop has five distinct phases:

1. *Information phase:* During this phase, the VE team gains as much information as possible about the project design, background, constraints, and projected costs.

Estimates of cost, energy, staffing, and other factors are revealed. The team performs a function analysis of systems and subsystems to identify high cost areas.
2. *Speculative/creative phase:* The VE team uses a group interaction process to identify alternative ideas for accomplishing the function of a system or subsystem.
3. *Evaluation/analytical phase:* The ideas generated during the speculative/creative phase are screened and evaluated by the team. The ideas showing the greatest potential for cost savings and project improvement are selected for further study.
4. *Development/recommendation phase:* The VE team researches the selected ideas and prepares descriptions, sketches, and life cycle cost estimates to support the VE recommendations.
5. *Report phase:* VE recommendations are presented to the owner and designer during an oral presentation at the conclusion of the VE workshop. Shortly after completion of the workshop, the written VE report is prepared by the VETC.

POSTWORKSHOP ACTIVITY

Following the VE workshop, the owner and designer thoroughly review the VE report and decide whether to accept or reject each of the VE team recommendations. A final VE report documents the acceptance or rejection of each recommendation.

The postworkshop VE activity involves a thorough review and evaluation of each VE recommendation presented in the VE report and the preparation of the final VE report.

The owner, PCM, and designer evaluate each VE recommendation on the basis of technical, operational, and life cycle cost savings considerations. (Normally, redesign and implementation costs for the recommendations are not considered since these costs are usually insignificant when compared to the potential cost savings.)

The owner and designer consult with the VETC to clarify any questionable items which arise during their review of the VE recommendations. An in-depth evaluation of each VE recommendation provides the best basis for reaching a sound decision to accept or reject a recommendation.

The acceptance of a VE recommendation requires no justification. Such action requires only a statement of acceptance. When certain elements of a VE recommendation are acceptable and other elements of the recommendation are unacceptable, a justification should be provided for only the rejected portion of the recommendation. Occasionally, the designer may modify a VE recommendation before incorporating it into the design. These modifications would be fully described in the final VE report.

Accepted VE recommendations should be incorporated into the design as soon as possible by the designer.

TYPICAL SCHEDULE

Typical time periods for accomplishing VE are:

Preworkshop activity (each)	3–6 weeks
VE workshop (each)	1–2 weeks
Prepare and issue VE report (each)	1–3 weeks

SELECTING A VE TEAM

Team members should be selected based on the key factors of design emphasis, cost (high cost areas), risk, and influence in the design. User input is also vital to success. Additional considerations for the selection of a VE team include:

- When particular disciplines do not represent major cost areas or the design in a particular discipline is not sufficiently completed to warrant an in-depth study, consideration should be given to the use of part-time VE team members. For example, an HVAC or electrical engineer may be needed for only two or three days during a VE workshop conducted at 20 to 30 percent of design completion.
- Although electrical work represents a relatively small percentage of a facility's construction cost, electrical (energy) consumption will be a major operational cost. Accordingly, an electrical engineer is normally included on the VE team to aid in the identification of operational cost savings.
- Since operation and maintenance considerations and costs are a vital part of a VE study, one member of the VE team should have experience in the operation of process facilities.
- The VE workshop conducted at the 65 to 75 percent stage of design completion should have one or more VE team members with substantial construction experience. This experience stimulates VE recommendations related to the project's constructibility.

VE Team Coordinator (VETC)

The VETC plays a key role in the success of a VE study. This individual is responsible for managing all aspects of the VE study including management of the team members during the workshop. Therefore, the VETC must have extensive experience with VE. Experience in construction also helps. A typical level of experience for a VETC would include a *certified value specialist* (CVS)* whose qualifications would include a major portion of VE experience in the field of construction.

Additional attributes for the VETC include:

- Level of effort
- Strong leadership, management, and communication capabilities
- Knowledge of the abilities and work attitudes of the team members.

The level of effort required for a VE study is normally a function of the complexity of the facility's design. Frequently, a simple increase in the number of team members may be adequate to achieve sufficient disciplines and experience to maximize the potential for identifying a complete cross section of cost-saving ideas. For facility designs of average complexity, one VE team per workshop with five to eight members is generally sufficient. As the design complexity and construction cost increase, more than one VE team per workshop is needed to focus additional attention on particular subsystems. Therefore, the number of VE teams and team members will vary with the study areas and complexities of the project.

---

* CVS certification is administered by the Society of American Value Engineers (SAVE) as a national standard recognizing competence in the field of value engineering.

Conclusion and Summary

As owner and P/CM, your project program, cost, and schedule will be greatly influenced by the ability to, first, clearly define your project and, second, control the cost and schedule of the project as it progresses. Value engineering can be your vital tool to combine quality, criteria, cost, and function issues in your project control. As an owner, you are looking to obtain all your value objections at the lowest possible cost. Combining the talents of the P/CM in measuring costs with value experience will greatly benefit your program. The value methodologies are a natural supplement to the P/CM's talents and services.

CASE STUDY: CONCEPTUALIZING DESIGN— BACHELORS' ENLISTED QUARTERS (BEQ)

The worksheets and schemes in Figs. A6-15 through A6-22 represent a small sampling of a conceptualized program for the design of BEQs in Hawaii. This program is for a 132-room facility. The basic elements include one portion of each of the function analysis, creative listing, judgment analysis, recommendation of the concept, BEQ prospectus, and the general portion of the Design Concept, Scope, and Design Considerations (Criteria). These examples are a small portion of a larger report.

FIGURE A6-15   Perspective view of BEQ.

PROJECT:	P-141 Bachelor Enlisted Quarters	INFORMATION PHASE FUNCTION ANALYSIS	LEWIS & ZIMMERMAN ASSOCIATES, INC.
LOCATION:	Subase Pearl Harbor		
DATE:	July 20-30, 1993	ITEM: New BEQ Building	
PAGE	1 OF 1		

ITEM		FUNCTION		
#	DESCRIPTION	VERB	NOUN	TYPE
	Modernization of BEQ's	House	Personnel	B
		Improve	Morale	HO
		Promote	Safety	R
		Improve	Environment	B
		Provide	Security	R
		Allow	Parking	R
		Enable	Cooking	R
		Allow	Gathering	R
		Clean	Clothes	R
		Provide	Amenities	R
		Enhance	Aesthetics	A
		Improve	Privacy	R
		Remove	Materials (Hazardous)	R
		Improve	Finishes	A
		Landscape	Site	A
		Increase	Comfort	A
		Remove	Trash	R
		Secure	Rooms	R
		Improve	Surveillance	R

FIGURE A6-16 Function analysis BEQ.

PROJECT: P-141 Bachelor Enlisted Quarters	**INFORMATION PHASE FUNCTION ANALYSIS**	LEWIS & ZIMMERMAN ASSOCIATES, INC.
LOCATION: Subase Pearl Harbor		
DATE: July 20-30, 1993	ITEM:	
PAGE 1 OF 1	Civil - Site	

ITEM		FUNCTION		
#	DESCRIPTION	VERB	NOUN	TYPE
	Site Utilities	Access	Sewer	R
		Access	Water	R
		Transfer	Hot Water	R
		Upgrade	Transformer	R
	Fire Protection	Promote	Safety	R
	Site Improvements	Traverse	Site	R
		Access	Site	R
		Egress	Site	R
		Park	Vehicles	R
	Landscape	Enhance	Aesthetics	A
		Construct	Recreation	R
		Screen	Equipment	R
	Site Preparation	Remove	UST's	R
		Demolish	Buildings	R
		Remediate	Site	R

FIGURE A6-17 Function analysis.

Lewis & Zimmerman Associates, Inc.   **Creative Idea List**

PROJECT: P-141 - Bachelor Enlisted Quarters   LOCATION: Subase Pearl Harbor

CREATIVE IDEA LISTING (By Discipline)

Discipline: Architectural - Room Layout

Idea No.	Creative Idea	Advantages	Disadvantages	Q	$	NV	P	Total Rating
1	Typical rectangular room modified to QOL criteria	Minimizes exterior circulation space, familiar layout	Room gross area, occupants' "territory" difficult to define, privacy, poor natural ventilation	0	-1	-1	-1	7
2	Typical square room modified to QOL criteria	Privacy, natural ventilation, familiar layout, good occupant "territory" definition	Room gross area, excessive O/S circulation space, building footprint size	+2	-1	+1	+1	8
3	Trapezoidal shaped rooms	Privacy, natural ventilation, good "territory" definition, reduces room gross area, minimizes O/S circulation space	"Different", creates different shaped building	+2	+1	+1	+1	9
4	Double loaded corridor, semi-private room	Maximizes privacy	Requires A/C, natural ventilation difficult, excessive O/S circulation space, gross room area, cost	+2	-2	-2	+2	3
5	Delete washer/dryer in room provisions	Reduces interior area requirements	No in room washer/dryer provisions	-1	+1	0	0	7
6	Optimize trapezoidal concept with 20 SF closet	Reduces unneeded space	More difficult to arrange furniture	0	+1	0	-1	8

LIST ALL IDEAS BEFORE RANKING!   Significant Improvement +2, +1, 0, -1, -2 Significant Degradation

Q = Quality, $ = Cost
NV = Natural Ventilation, P = Privacy
DS - Design Suggestion

FIGURE A6-18 Creative idea list.

VALUE ENGINEERING RECOMMENDATION	Lewis & Zimmerman Associates, Inc.		
PROJECT: P-141 - Bachelor Enlisted Quarters Subase Pearl Harbor	**DATE:** July 20-30	**RECOMMENDATION NO.** RL-3	
ITEM: Trapezoidal Shaped Rooms		**SHEET NO.** 1 of 4	

ORIGINAL DESIGN: (Attach sketch where appropriate)

The original design concept did not meet the new Quality of Life criteria. The typical rectangular and square rooms were modified to meet the QOL criteria to serve as a baseline for the analysis.

PROPOSED CHANGE: (Attach sketch where appropriate)

The recommended design consists of a trapezoidal shaped room, with the narrow end as the entrance and common toilet/shower area. The wide end of the room contains the living/sleeping area.

ADVANTAGES:

- Reduces room gross area
- Minimizes outside circulation space
- Good privacy potential within the room
- Meets natural ventilation requirements
- Meets all QOL criteria within the rooms

DISADVANTAGES:

- Different from the standard room layout
- Creates a different shaped building.

DISCUSSION / JUSTIFICATION:

Analysis of the traditional square and rectangular rooms showed that it was not possible to achieve the total number of living modules (132) and sleeping rooms (264) specified in the DD 1391 and required by the Subase. The trapezoidal shape allows space in the room where it is most needed and significantly reduces inside and outside circulation space requirements. Approximately 88 SF per module can be saved with this configuration. The cost per SF difference between these rooms will be less than the $135-$150 cost/SF of the entire module, as the major changes are in floors, ceilings and related finishes. It is estimated that $57.00/SF is an appropriate cost for the area that will change between these room layouts. Based on this, over $660,000 can be saved with this design.

COST SUMMARY	INITIAL COST	ANNUAL O&M COST	LIFE-CYCLE COST
ORIGINAL DESIGN	$	$	$
PROPOSED CHANGE	$	$	$
SAVINGS			$ 660,000

FIGURE A6-19 VE recommendation (page 1 of 4).

A6-34 STANDARD HANDBOOK OF HEAVY CONSTRUCTION

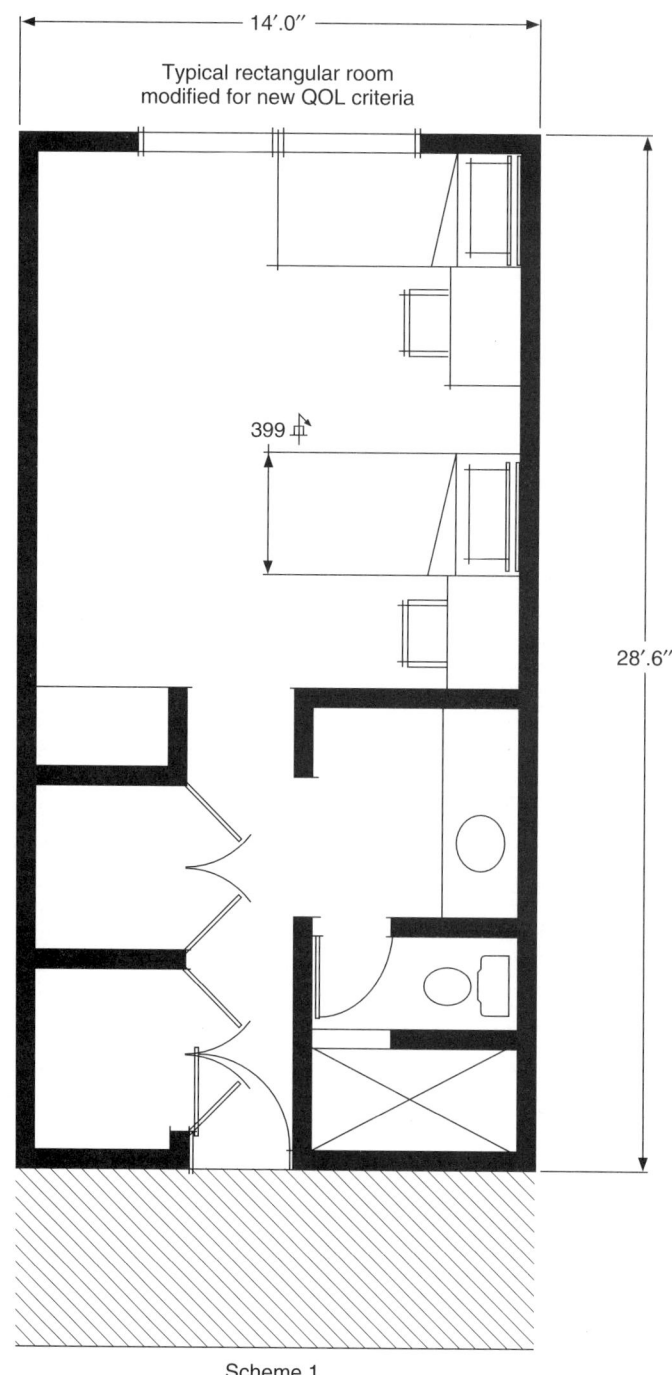

FIGURE A6-20 VE recommendation (page 2 of 4).

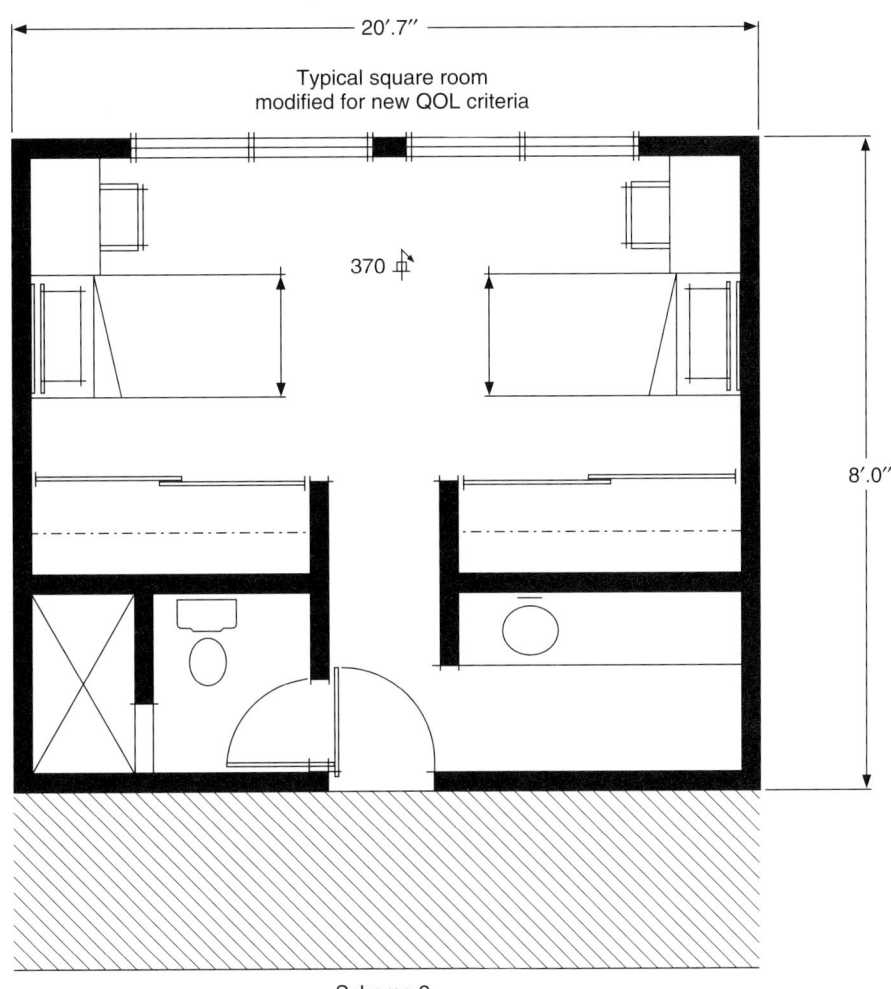

FIGURE A6-21 VE recommendation (page 3 of 4).

A6-36 STANDARD HANDBOOK OF HEAVY CONSTRUCTION

Proposed design
trapezoidal shaped room

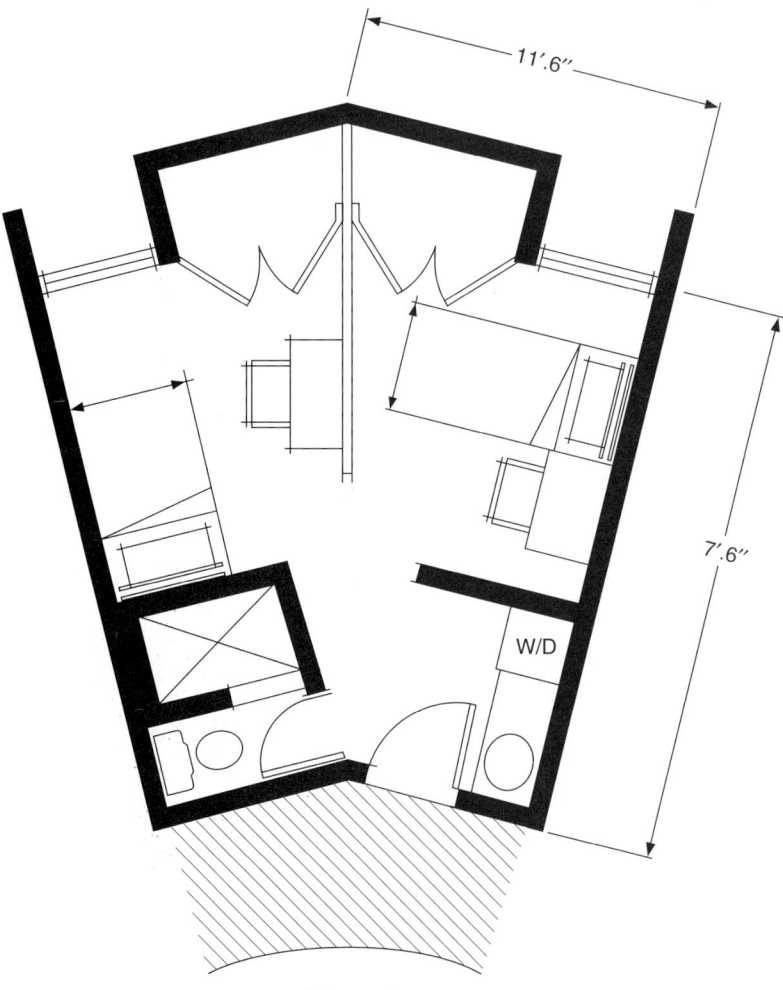

Scheme 3

FIGURE A6-22 VE recommendation (page 4 of 4).

Design Concept, Scope, and Design Considerations

Existing Conditions. There is a housing shortfall at the Subase and the personnel are forced to accept whatever housing is available on base. As a result, rooms have become overcrowded, exceeding established minimum allowable living area per person. This problem is compounded with the new Quality of Life criteria established for BEQs.

General Design Concept

1. Most of the living modules will be designed to house two E1 to E4 personnel per room. Each of these rooms offers semiprivate living conditions for the residents. These quarters could alternatively house one E5 or E6. Thirty-two atypical rooms will house only E5 or E6 personnel.
2. Each room will have its own bathroom. No more than two residents will share a bathroom.
3. A lobby, lounge, offices, bulk storage for residents, linen/supplies storage, and mechanical/electrical rooms will be located on the first floor. The lobby will be open to take advantage of the local climate. Other common elements, including a small lounge and washer/dryer facilities, are located on each floor.
4. Male and female handicapped restrooms will be provided on the first floor to accommodate H/C visitors.
5. Kitchenettes will not be provided in living modules. However, space for small refrigerators and microwave ovens will be provided. A kitchenette will be provided in conjunction with the lounge on each floor.
6. The lounge/reading room, kitchenette, and a washer with two dryers will be situated on each floor across from the elevators.
7. Landscaping of the site will be designed to enhance the atmosphere of the BEQ. Large trees on the site will be retained.
8. All living modules, lounge/reading rooms, kitchenettes, and washer/dryer areas will be naturally ventilated. Offices will be air conditioned.
9. The building will be equipped with a sprinkler system for fire protection.
10. Three high-speed elevators will serve the building.
11. Fifty-four parking stalls (26 standard, 26 compact, and 2 handicap stalls) are located along the north and east boundaries of the site. In addition, a loading area is located at the southeast corner of the site.

Architectural

1. General
    a. The building has been designed as a 17-story structure with 16 rooms at each of the typical floors.
    b. The first floor will accommodate eight rooms along with common areas, offices, storage areas, and mechanical/electrical spaces.
    c. The building has been designed and sited to optimize passive cooling through the use of natural ventilation. The final siting will be determined after wind tunnel tests to ensure that optimal natural ventilation exists.

2. Exterior architectural (BEA)
    a. Architectural style: The building derives its expression by its response to climate—recessed fenestration for shading and to enhance ventilation. Architectural treatments will be compatible with other high-rise BEQ buildings within the submarine base.
    b. Landscaping: All mature specimen trees will be retained to the extent possible. New landscaping will be in-filled to enhance the shaded character of the existing trees. Mechanical and electrical equipment and service areas will be heavily screened by planting.
    c. Graphics/signs: All graphic detail will comply with users' requirements. A new BEQ sign will be provided at the new entrance.
3. Design area tabulation
    a. Approximate gross floor area:
       First floor   6,150 ft^2
       Typical floors (16 floors)   6,150 ft^2 each floor
       Total   104,534 ft^2 (includes mechanical equipment areas)
    b. Design minimum square foot per occupant is 90 ft^2 for E1 to E4 and 180 ft^2 for E5 to E6
    c. 32 one-room living units for one person (E5 or E6)
    d. 116 two-room living modules for two people (E1 to E4)
    e. A total of 264 living units
    f. A maximum capacity of 496 residents based on applicable BEQ QOL criteria
4. Building design
    a. The exterior walls will be painted concrete. Interior walls will be insulated to reduce noise levels where necessary.
    b. Floor finishes will include carpet, vinyl composition tile, and ceramic tile. Wall finishes will include painting and ceramic tile wainscot. Ceilings will be painted gypsum board and concrete.
    c. Aluminum frame jalousie windows will be used to provide natural ventilation throughout the building.
    d. Exterior doors will be hollow metal with jalousie inset, except storefront will be provided for the public entrance into the BEQ. All interior doors will be painted solid-core wood doors with metal frame.
5. Energy conservation
    a. Solar tinted glass and architectural shading devices will be used to reduce heat gain.
    b. Window openings will be maximized to enhance natural ventilation.
    c. Ceiling fans will be used to enhance comfort levels when there are no trade winds.
    d. Water conservation plumbing fixtures will be used.
    e. To reduce solar heat gain, light-colored exterior surfaces and roofing will be selected.
    f. High-efficiency lighting fixtures and localized switches will be used to turn off artificial lighting in unoccupied areas.

6. Fire protection
   a. The buildings are residential occupancy with low hazard classification.
   b. A fire alarm system, smoke detectors, heat detectors, emergency lighting, portable fire extinguishers, and sprinkler system will be provided.
7. Living modules
   a. Living units will be accessible through single loaded external corridors.
   b. The living area exceeds the QOL guidelines of 90 ft^2 net per E1 to E4 and 180 ft^2 net for E5 to E6.
   c. Each living unit will have a private bath/shower. The lavatory/counter and under-counter refrigerator can be used simultaneously while the water closet, bath/shower is occupied.
   d. Individual 20 ft^2 gross closets with swing doors and locks are provided.
8. Laundry
   a. Each floor (2 through 16) will have one washer and two dryers next to the lounge, which is centrally located on the floor.
   b. The floor will be sloped to a floor drain.
   c. The floor surface finish shall be nonslip.
   d. A clothes folding table and hanging racks shall be provided.
   e. A janitor's sink will be provided within the laundry room with a floor sink.
9. Lobby and BEQ entrance
   a. The small lobby shall be located at the building entrance.
   b. The main lounge and kitchenette will be located off the lobby.

CHAPTER 7
QUALITY ASSURANCE/QUALITY CONTROL

James J. O'Brien, P.E.
O'Brien-Kreitzberg, Inc., Pennsauken, New Jersey

INTRODUCTION

It is generally agreed in the construction industry that the three control parameters for construction projects are quality, time, and cost. Until the advent of better project management tools in the late fifties and early sixties, quality was the element most controlled in the field. Since 1960, the focus has shifted to time and cost controls, with an emphasis on computerized scheduling (CPM). Undoubtedly, some of the increased emphasis on schedule and cost has detracted from the control of quality.

Ironically, in the nuclear industry, where cost and time have been runaway in the later plants, quality control has been developed to a science. Quality control procedures in the nuclear industry are the most demanding in the construction industry.

In *Civil Engineering Magazine,* the author wrote about quality control[1]:

> Quality in construction is too important to be left to chance. Engineers must educate owners to insist upon a quality control plan, comprehensive inspection and a competent testing program.
>
> A look at history gives some insight into the problem. Until the end of the 19th century, building projects were delivered by master builders, who combined architecture and building. Civil projects were usually undertaken, as they had been since Roman times, by military engineers. Aqueducts, highways, railroads, and even our interstate highway system, were influenced by military requirements.
>
> As architects separated themselves from hands-on delivery of buildings, and engineers became civil engineers first and military engineers second, methods changed. Through the first half of this century, engineers and architects were in total control during the design phase. During construction, they carried out a role described as "supervision," ensuring that the owner received his money's worth in terms of quality. Change orders were controlled rigidly, as they might roll back as "errors and omissions." Control of time was limited and often considered a function of luck.

In the 1950s and 1960s, owners became increasingly concerned with cost and schedule, areas where the design professionals were not providing good control. The emphasis continued to be on quality and control of exposure to liability.

As the role of the master builder narrowed to that of designer, construction was taken over by the general contractor who had to be concerned with the realities of maintaining a viable business. The general contractor had to spread out management, seek new jobs, and manage effectively to maintain profits.

In the public sector and, to a large degree, in the private sector, the sealed competitive bid became popular. This gave the owner the advantage of competitive pricing. It also forced the general contractor to look for every advantage during construction to control cost and to maintain a profitable stance.

As mechanical and electrical systems became more complex, the general contractor turned responsibility for such work over to subcontractors, including quality control of their work. In other specialty areas, the subcontractors are actually craftspersons.

When the general contractor was essentially in complete control of a project, quality control was an inherent duty. Today, with even topnotch contractors brokering out almost the entire project, who is in charge of quality control? Through contract, subcontract, and sub-subcontract, the general contractor has delegated responsibility for quality. At best, the situation is confusing. At worst, it becomes a subject of litigation. At the extreme, it becomes the basis for building failures.

Inspection is a vital part of quality control. It is one of the most substantial costs in project management. The traditional clerk-of-the-works approach provides just what is paid for—a clerk. Those owners willing to pay for additional field coverage by the design professional usually get an effective quality control service. It costs about 2 to 4 percent of construction.

There are three basic approaches to providing construction management services: AE-type services, professional CM services, and contractor-type construction management. In the first two, adequate field inspection is usually proposed as part of the CM package. If properly funded, more architectural and engineering firms provide the necessary quality control through inspection services.

The contractor CM, on the other hand, is not oriented to inspection, relying on the contractors employed to provide their own quality control. The contractor CM depends on the business relationships—if problems are discovered, business leverage is applied to ensure that they are corrected. Otherwise, serious matters can go unresolved.

One way in which more attention will be given to quality control is development of a project quality control plan. Presently, testing and inspection requirements are scattered throughout the contract specifications. To develop a firm plan, the testing and inspection requirements can be combined into a new division of the specs. This would emphasize quality control and provide an organized location in which all quality control requirements are identified to the bidders.

As part of the quality control plan, the manner in which the construction manager (or project manager) will apply quality control procedures should be described to the bidders. This will permit them to assign appropriate costs to the testing procedures.

The ASCE book *Quality in the Constructed Project*[2] compares quality assurance and quality control:

- *Quality assurance* (QA) comprises all those planned and systematic actions necessary to provide confidence that items are designed and constructed in accordance with applicable standards and as specified by contract.

- *Quality control* (QC) comprises the examination of services provided and work done, together with management and documentation necessary to demonstrate that these services and work meet contractual and regulatory requirements. QA/QC programs previously formulated by the design professional and the constructor are modified to meet the specific requirements of the project as defined by the owner/design professional agreement and the owner/contractor contract and by applicable regulatory requirements.

PROJECT STRUCTURE

Quality control (QC) of construction occurs only during the construction phase, which (while it is the most visible portion of the project) takes only about one-half, or in some cases one-third, of the time consumed from original concept and approval of the project through move-in and utilization. Although QC duties are confined to the construction period, it is helpful to the quality assurance (QA) program to understand the roles and attitudes of the various characters within the project cast. By the time the job reaches the field, the designer, owner, and various government reviewing bodies—such as building code specialists—have developed certain definite opinions, and the project itself may be popular within the community or highly unpopular. This history and set of attitudes are carried forward into the construction phase, and will affect the QC role.

There are four major categories of project progress:

1. *Predesign activities,* with the owner having primary responsibility for progress. Quality assurance policies and procedures should be in place.
2. *Design phase,* with the architect, engineer, A-E, or in-house staff having primary responsibility for progress and implementation of QA.
3. *Construction,* with the contractor or the in-house construction force responsible for progress, but the owner (and/or the owner's agents) responsible for QC.
4. *Furnish and move-in,* (also termed *start-up*) with the contractor having primary responsibility and the owner applying QC in accord with the QA procedures.

PREDESIGN

The predesign portion of a project is a poorly defined, intangible, and time-consuming phase, during which the owner, its technical staff, or consultants should be very busy with a number of important roles. Typically, these duties are accomplished by omission or default rather than rigorously managed. The seeds of many project problems are planted in this field of neglect. The antidote is a comprehensive quality assurance (QA) plan.

Most projects seem to appear from nowhere, characterized by an evolutionary style of aggregate thinking from many sources that gathers pressure, both political and personal, until the project has been articulated. Programs that develop in this manner include schools, hospitals, public buildings, industrial plants, and highways—almost any project that can be identified. Key project characteristics are power structure and consensus. Actually, a project should go through the following stages: goals, resource evaluation, decision making, budgeting, and funding.

Needs and goals (including the QA plan) must be preestablished, preferably as policy. Projects that can be utilized to meet the goals of the organization are reviewed against available resources. Usually this occurs as an intuitive relationship—making selections and establishing priorities based on intuition and personal preferences—rather than through any quantitative means.

The decision to go ahead requires the identification of specific projects and the development of preliminary cost estimates. These are usually accomplished in-house on the basis of square-footage costs, or they may be prepared by a consultant.

After the project has been placed in the budget and funding has been made available to meet the budget, the project—still under the direction of the owner—enters the predesign phase, but implementation has begun.

Programming

The last predesign activity should be the development of a specific program defining the intent of the owner in regard to the functional uses of the project. Although this philosophy is important, it often is lost during the project's design phase. It is clearly the owner's responsibility to make the project requirements known and to interpret them in terms of cost impact prior to the selection of the designer. Programming is a unique talent, usually requiring a combination of the owner's knowledge and consulting expertise.

A functional programming effort should reconfirm the budgetary estimate and establish what material will be included in the project. The functional program incorporates policy, and it should be approved by the appropriate owner authority.

Projects typically do not have formal program documents, perhaps deliberately. The result is uncertainty at the beginning of the design phase. Since the designer does not get compensated for uncertainty, the only defense is to proceed slowly at the early portion of the project, developing program-type statements that can be affirmed or revised by the client. Unfortunately, clients typically are tremendously inclined to change their minds almost constantly. From the designer's viewpoint, this is not only time consuming, but highly expensive. Once the design firm (or in-house design team) has been selected, it should play a significant role in developing the scope of services to be performed. An owner, particularly an inexperienced owner, threatens quality when he or she attempts to get more than bargained for. This can be avoided by:

- Identifying the owner's project objectives, and developing the scope to meet those objectives
- Identifying the design tasks required to accomplish the scope identified
- Establishing a budget based on worker-hours per task
- Identifying cost required to achieve a quality design within the project parameters and scope

It is important that the design professional identify for the owner those certain tasks which are necessary to assure quality, and to insist that these be included in the project's design phase costs. While it is a truism that additional dollars spent to assure quality during the design will reduce the cost of quality control measures during and after construction, the design professional should make this clear to the owner early rather than late.

Owners, particularly inexperienced ones, expect quality in the completed project, but are unaware of the effort required to deliver quality both in the design phase and the following construction. The design professional must be aware of the probability of loss of quality in accelerated projects, fast-track informal procedures, rejection of recommendations by professionals, budget cutting at midstream, and other management actions on the part of the owner. All too often, the design professional hopes for the best and accepts these impositions on the design team's ability to perform and incorporate quality. It is the design professional's responsibility to point out the probable implications of such actions. The design professional must point out to the owner the owner's responsibilities in meeting the owner's own expectations of quality.

Funding limitations can exert a negative influence on quality in the project. If emphasis is on providing program at cost, then any limitation of cost tends to squeeze quality-related activities.

Another dimension should be pointed out to the owner: Design fees are approximately 1 percent of the ultimate life cycle cost of a project, yet design is the most important influence on the life cycle costs. Saving money at the expense of design can have a 100-to-1 negative impact on the ultimate life cycle costs of the project. If these realities can be brought out to an owner, the owner can cooperate with the design professional to agree on the role of quality in the design phase.

The design professional, however, should recognize the reality that quality at any cost is not necessarily a good bargain. There must be an appropriate balance between minimal levels of quality and higher levels of quality which the owner may decide to fund. Conversely, the design professional is professionally constrained from delivering a design which has less than the minimum quality.

DESIGN STAGES

Project design is a relatively complex series of activities that become increasingly detailed as the project proceeds through the various design phases, which are outlined in the following paragraphs.

Schematic Development

This is also called the *sketch phase*. Concept plans are developed by the architect, and basic engineering systems analysis is made. Design criteria are specified, and schematic drawings are made. A perspective is prepared. Various systems are defined in terms of performance and, where appropriate, by single line diagram or sketch. The QA plan should include procedures for review and evaluation for this and following stages.

Preliminary Design

This is also called the *design development*. After approval of the sketch or schematic phase, the drawings are refined to a degree sufficient to permit the development of dimensioned space layouts, heating and ventilating systems main feeders, and electrical main feeders, as well as definite development of the structural framework. Requirements for utilities are also definitely developed, and specific equipment requirements are determined. The budgetary type cost estimate is revised, and a more firm estimate made. Within QA parameters, value engineering (VE) and constructibility reviews are best applied at this stage.

Working Drawings

These are also referred to as contract drawings and specifications. This phase comprises about two-thirds of the work, but fewer of the decisions. The design as defined in the prior stages is developed in complete detail, including dimensions, so that it can be specifically priced by prospective contractors. The contract documents include both drawings and specifications. In this stage, A-E QA procedures should guide.

As the project proceeds, each change becomes more difficult to implement. Each revision requires many more reviews and changes of related items. The range of changes that can be accepted gracefully narrows in a funnel effect, with the most changes being more acceptable early in the project and increasingly costly as the project proceeds. These factors can be quantified, but vary with each project.

The design phase often proceeds essentially unscheduled and uncoordinated. This disorder is partially understandable because specific interconnections between the phases, such as the structural, mechanical, and electrical, are difficult to express. Conversely, the design phase must be closely coordinated and interconnected at all stages and steps. Scheduling this at the daily liaison level of detail would require a gigantic scheduling network. The usual compromise is to schedule concurrent activities, with an implied understanding that continual liaison is carried out.

During the design there is constant interplay between the designer and the owner. The owner reviews the project at the major points, and also is available on a daily basis for information. Quite often, the owner is furnishing or specifying special equipment that requires attention. The owner and/or the architect is involved in various agency reviews. The design phase offers a great potential for time gains; many activities actually can occur concurrently, and the scheduling problem is usually one of resource allocation of the design staff, rather than a problem in sequential planning. Clear identification of the design alternate can be beneficial to the designer as well as the owner in anticipating review requirements, and in placing responsibility on the owner to expedite decisions. (It is typical for the owner to delay decisions, assuming that the designer can work around problems. Conversely, the designer exercises patience, waiting until external pressures force a decision from the owner.)

With so many people concerned with the project, all the time required is suspect, with the exception of the building time. Described variously as red tape, inefficiency, delays, and in similar disparaging terms, the review and planning stages have their definite roles. Very often, these parts are overplayed, as individuals tend to see their own activities as the most important, and are thus willing to take more than their share of project time. Also, in the early planning stages, individuals do not consider that planning time will affect the delivery date of the facility.

QA in Design

The ASCE *Quality in the Constructed Project*[3] comments on this subject:

> The design professional is responsible for formulating and implementing the QA/QC program for the design phase to meet requirements under the agreement for professional services. The owner may wish to review and approve the program and may require documentation.

Procedures. The design professional may have a QA/QC program already in place to cover normal activities. This base program is usually adapted and expanded as necessary to meet any special requirements imposed by the owner and required by the project.

The quality management procedures employed by the design professional are used to improve thought processes, clarify communications among team members, and translate the concepts and mental images of the project in the designer's mind to physical structures and systems to be built by the constructor. These means include:

- Staffing the design effort with experienced personnel; implementing the owner's statement of project requirements for design and seeking clarification or amplification as necessary; confirming field conditions as they influence design; establishing and maintaining open lines of communication among design team members and with the owner
- Preparing, reviewing, and coordinating designs, drawings, and specifications among design team members.
- Scheduling special reviews and progress reporting as appropriate for internal control and as required by the QA/QC program
- Forming review activities to include advisers on construction, operation, and maintenance, and design specialists not included in the day-to-day design activities
- Arranging for peer reviews as specified in the professional services agreement

Confirmation to the owner that the design professional meets his or her QA/QC responsibilities is provided through submittal of the progress reports, reports of project reviews, and appropriate documentation at the end of certain phases of the project in compliance with the design professional's quality assurance program.

CONSTRUCTION

In this stage, quality control implements the QA plan.

Some owners have a highly developed construction management team as part of their internal structure. Typical would be major school districts, cities, industrial firms, competitive builders, states, and federal agencies. Organizations that need a highly developed project management team but usually do not have the ongoing staff include universities, hospitals, airports, and medium-to-small industrial firms.

During the course of the project, it is almost inevitable that certain areas will be revised or omissions will be noted. Change orders will develop and can have a substantial impact on the relationship between the contractor and the owner. Another important but often neglected area is the coordination of shop drawing QC reviews between the owner and the contractors. Delays in returning shop drawings can affect the completion time; conversely, contractors will often submit imperfect shop drawings as a ploy to screen problems and delays.

Start-up

Very often, the completion of a project appears to occur on turnover of the facility from the contractor to the owner. The turnover usually involves a punch list prepared by the QC team, however. This list may be trivial, or it could entail years of work. The nature of the relationship between the owner and the contractor by the conclusion of the project may directly relate to the length and difficulty of the punch list.

The owner often fails to appreciate fully its own responsibilities once the contractor has phased over the building. For one thing, previously selected and trained personnel must be provided to operate the equipment in the building. The owner must staff and prepare personnel far enough in advance to effect a smooth phaseover.

QA PROGRAM DEVELOPMENT

In the nuclear industry, with its great cost and time overruns (in the 1980s plants), quality control has been developed to the most exacting standards. An important precedent for the development of a quality assurance program is found in the ANSI/ASME NQA-1 standard, Quality Assurance Program Requirements for Nuclear Facilities, which encompasses the following elements:

- Qualification of inspection, examination, and testing personnel
- Requirements for the collection and maintenance of QA records
- QA terms and definitions
- Auditing of quality assurance programs
- Quality assurance for control procurement

The front end of NQA-1-1983 is included as Figure A7-1.

QA PLAN FOR CONSTRUCTION MANAGER

Figure A7-2 is the generic OKA Quality Assurance Program Plan for construction management projects to be implemented by O'Brien-Kreitzberg, Inc. This plan must be customized for each assignment. In most cases, this customization deletes the scope section because the owner does not desire that level of professional services or is providing that scope in some other fashion.

CODES AND STANDARDS

The quality assurance plan of the owner may specify specific standards to be followed in the projects. Even if it does not, construction specifications often incorporate other standards that further define the quality requirements for the materials or methods to be used in the construction process.

The American Society for Testing and Materials promulgated ASTM standards used in building construction. Similarly, The American National Standards Institute promulgates construction-oriented standards. The standards of these two organizations represent the majority of those referenced in typical construction specifications. The field inspection team will need a library of appropriate standards in order to fulfill its task.

The QA program may require adherence to a specific building code, and building codes may also incorporate standards. For instance, the building code for the City of New York is 472 pages long, but has as an appendix with 371 pages of building code reference standards. Many of these, in turn, reference ASTM and ANSI standards.

When construction is accomplished within a jurisdiction, the building code of that jurisdiction, such as the City of New York, usually applies. There are certain excep-

AN AMERICAN NATIONAL STANDARD

Quality Assurance Program Requirements for Nuclear Facilities

ANSI/ASME NQA-1-1983 EDITION

SPONSORED AND PUBLISHED BY

THE AMERICAN SOCIETY OF MECHANICAL ENGINEERS

United Engineering Center • 345 East 47th Street • New York, N.Y. 10017

FOREWORD

(This Foreword is not a part of American National Standard for Quality Assurance Program Requirements for Nuclear Facilities, ANSI/ASME NQA-1-1983 Edition.)

Early in 1975, the American National Standards Institute (ANSI) assigned overall responsibility for coordination among technical societies, development, and maintenance of nuclear power quality assurance program standards to the American Society of Mechanical Engineers (ASME). The ASME Committee on Nuclear Quality Assurance was constituted on October 3, 1975, and began operating under the ASME Procedures for Nuclear Projects, which received accreditation on January 15, 1976. The ASME Committee on Nuclear Quality Assurance currently operates under the ASME Codes and Standards Development Committee Procedures, which received accreditation on June 21, 1979. This Committee prepared ANSI/ASME NQA-1, Quality Assurance Program Requirements for Nuclear Power Plants, which was first issued in 1979 as an American National Standard. Three Addenda

FIGURE A7-1 *Used with the permission of the ASME.*

have subsequently been published, NQA-1a-1981, NQA-1b-1981, and NQA-1c-1982. The 1983 Edition consists of the 1979 Edition, with all approved revisions through the NQA-1c-1982 Addenda.

Parallel with the development of ANSI/ASME NQA-1, American National Standards N46-2 Subcommittee, under the sponsorship of the American Institute of Chemical Engineers, prepared Revision I to ANSI N46.2, Quality Assurance Program Requirements for Post Reactor Nuclear Fuel Cycle Facilities, which was issued as an American National Standard in 1978.

The N46-2 Subcommittee included the following personnel during the preparation of ANSI N46.2:

J. A. Perry, *Chairman*	L. L. Gordon
W. F. Kelly, *Secretary*	R. A. Jacobus
W. E. Baker	L. W. Keith
R. C. Baldwin	J. A. Maneatis
F. D. Carpenter	D. L. Martin
K. G. Carroll	R. E. L. Stanford
K. E. Goad	

This Standard is an integration of ANSI/ASME NQA-1 and ANSI N46.2, Revision I. It contains an Introduction, Basic Requirements, and Supplements. In addition, nonmandatory guidance is provided in the Appendices, which do not set forth requirements. The 18-criteria structure of Appendix B of 10 CFR Part 50 has been retained. The Introduction, Basic Requirements, and Supplements together are intended to meet and clarify the criteria of Appendix B of 10 CFR Part 50, dated January 20, 1975.

This document is based upon the contents of ANSI/ASME N45.2-1977, Quality Assurance Program Requirements for Nuclear Facilities; ANSI N46.2, Revision I, Quality Assurance Program Requirements for Post Reactor Nuclear Fuel Cycle Facilities; and the following seven daughter Standards of ANSI/ASME N45.2.

N45.2.6	Qualification of Inspection, Examination, and Testing Personnel for Nuclear Power Plants
N45.2.9	Requirements for the Collection, Storage, and Maintenance of Quality Assurance Records for Nuclear Power Plants
N45.2.10	Quality Assurance Terms and Definitions
N45.2.11	Quality Assurance Requirements for the Design of Nuclear Power Plants
N45.2.12	Requirements for Auditing of Quality Assurance Programs for Nuclear Power Plants
N45.2.13	Quality Assurance Requirements for Control of Procurement of Items and Services for Nuclear Power Plants
N45.2.23	Qualification of Quality Assurance Program Audit Personnel for Nuclear Power Plants

This document does not incorporate the requirements of the following Standards, which provide related quality assurance requirements and guidance.

N45.2.1	Cleaning of Fluid Systems and Associated Components During Construction Phase of Nuclear Power Plants

FIGURE A7-1 *Used with the permission of the ASME. (Continued)*

N45.2.2	Packaging, Shipping, Receiving, Storage, and Handling of Items for Nuclear Power Plants (During the Construction Phase)	
N45.2.3	Housekeeping During the Construction Phase of Nuclear Power Plants	
N45.2.4	Installation, Inspection, and Testing Requirements for Instrumentation and Electric Equipment During the Construction of Nuclear Power Generating Stations	
N45.2.5	Supplementary Quality Assurance Requirements for Installation, Inspection, and Testing of Structural Concrete, Structural Steels, Soils, and Foundations During the Construction Phase of Nuclear Power Plants	
N45.2.8	Supplementary Quality Assurance Requirements for Installation, Inspection, and Testing of Mechanical Equipment and Systems for the Construction Phase of Nuclear Power Plants	
N45.2.15	Requirements for the Control of Hoisting, Rigging, and Transporting of Items at Nuclear Power Plant Sites	
N45.2.20	Supplementary Quality Assurance Requirements for Subsurface Investigations Prior to the Construction Phase of Nuclear Power Plants	

This document sets forth requirements and nonmandatory guidance for the establishment and execution of quality assurance programs during the design, construction, and operation, including decommissioning of nuclear facilities. The arrangement of the basic and supplementary requirements permits judicious application of the entire document or only portions of the document. The extent to which this document should be applied, either wholly or in part, will depend upon the nature and scope of the work to be performed and the relative importance of the items or services being produced. The extent of application is to be determined by the organization imposing this document. For example, it may only involve the Basic Requirements; Basic Requirements in combination with selected Supplements; Basic Requirements in combination with Supplements with appropriate changes; or the entire document. This document is written to allow application to any structure, system, or component that is essential to satisfactory performance of the facility.

Requests for interpretation or suggestions for improvement of this Standard should be addressed to the Secretary of the ASME Committee on Nuclear Quality Assurance, American Society of Mechanical Engineers, 345 East 47th Street, New York, New York 10017.

CONTENTS

Foreword		iii
Standards Committee Roster		v
I	**INTRODUCTION**	1
1	Purpose	1
2	Applicability	1
3	Responsibility	1
4	Definitions	1

FIGURE A7-1 *Used with the permission of the ASME. (Continued)*

II	**BASIC REQUIREMENTS**	2
1	Organization	2
2	Quality Assurance Program	2
3	Design Control	2
4	Procurement Document Control	2
5	Instructions, Procedures, and Drawings	2
6	Document Control	2
7	Control of Purchased Items and Services	3
8	Identification and Control of Items	3
9	Control of Processes	3
10	Inspection	3
11	Test Control	3
12	Control of Measuring and Test Equipment	3
13	Handling, Storage, and Shipping	3
14	Inspection, Test, and Operating Status	3
15	Control of Nonconforming Items	3
16	Corrective Action	3
17	Quality Assurance Records	4
18	Audits	4
		4
III	**SUPPLEMENTS**	5
S-1	Terms and Definitions	5
1S-1	Supplementary Requirements for Organization	7
2S-1	Supplementary Requirements for the Qualification of Inspection and Test Personnel	8
2S-2	Supplementary Requirements for the Qualification of Nondestructive Examination Personnel	10
2S-3	Supplementary Requirements for the Qualification of Quality Assurance Program Audit Personnel	11
3S-1	Supplementary Requirements for Design Control	13
4S-1	Supplementary Requirements for Procurement Document Control	16
6S-1	Supplementary Requirements for Document Control	18
7S-1	Supplementary Requirements for Control of Purchased Items and Services	19
8S-1	Supplementary Requirements for Identification and Control of Items	23
9S-1	Supplementary Requirements for Control of Processes	24
10S-1	Supplementary Requirements for Inspection	25
11S-1	Supplementary Requirements for Test Control	27
12S-1	Supplementary Requirements for Control of Measuring and Test Equipment	28
13S-1	Supplementary Requirements for Handling, Storage, and Shipping	29
15S-1	Supplementary Requirements for the Control of Nonconforming Items	30
17S-1	Supplementary Requirements for Quality Assurance Records	31
18S-1	Supplementary Requirements for Audits	34
IV	**APPENDICES**	37
1A-1	Nonmandatory Guidance on Organization	37
2A-1	Nonmandatory Guidance on the Qualifications of Inspection and Test Personnel	38

FIGURE A7-1 *Used with the permission of the ASME. (Continued)*

2A-2	Nonmandatory Guidance on Quality Assurance Programs	40
2A-3	Nonmandatory Guidance on the Education and Experience of Lead Auditors	41
3A-1	Nonmandatory Guidance on Design Control	43
4A-1	Nonmandatory Guidance on Procurement Document Control	48
7A-1	Nonmandatory Guidance for Control of Purchased Items and Services	53
17A-1	Nonmandatory Guidance on Quality Assurance Records	55
18A-1	Nonmandatory Guidance on Audits	58

Figures

2A-3.1	Sample Form for Record of Lead Auditor Qualification	42
3A-1.1	Division of Responsibilities	45
3A-1.2	Division of Responsibilities	46
3A-1.3	Drawing Issue Checklist	47
4A-1.1	Logic Chart for Determining Appropriate Quality Requirements	52

I INTRODUCTION

1 PURPOSE

This Standard sets forth requirements for the establishment and execution of quality assurance programs during the design, construction, operation, and decommissioning of nuclear facilities. Nonmandatory guidance is provided in the Appendices.

2 APPLICABILITY

The requirements of this Standard apply to activities that affect the quality of structures, systems, and components of nuclear facilities. Nuclear facilities include facilities for power generation, spent fuel storage, waste storage, fuel reprocessing, and plutonium processing and fuel fabrication. These activities include the performing functions of attaining quality objectives and the functions of assuring that an appropriate quality assurance program is established, and verifying that activities affecting quality have been correctly performed. Activities affecting quality include designing, purchasing, fabricating, handling, shipping, storing, cleaning, erecting, installing, inspecting, testing, operating, maintaining, repairing, refueling, and modifying. The application of this Standard, or portions thereof, shall be specified in written contracts, policies, procedures, or instructions.

3 RESPONSIBILITY

The organization invoking this Standard shall be responsible for specifying which Basic Requirements and Supplements, or portions thereof, apply, and appropriately relating them to specific items and services. The organization upon which this Standard, or portions thereof, is invoked shall be responsible for complying with the specified requirements.

4 DEFINITIONS

Terms used in this Standard that require unique definition are included in Supplement S-1, Terms and Definitions.

FIGURE A7-1 *Used with the permission of the ASME. (Continued)*

II BASIC REQUIREMENTS

1 ORGANIZATION

The organizational structure, functional responsibilities, levels of authority, and lines of communication for activities affecting quality shall be documented. Persons or organizations responsible for assuring that an appropriate quality assurance program is established and verifying that activities affecting quality have been correctly performed shall have sufficient authority, access to work areas, and organizational freedom to (1) identify quality problems; (2) initiate, recommend, or provide solutions to quality problems through designated channels; (3) verify implementation of solutions; and (4) assure that further processing, delivery, installation, or use is controlled until proper disposition of a nonconformance, deficiency, or unsatisfactory condition has occurred. Such persons or organizations shall have direct access to responsible management at a level where appropriate action can be effected. Such persons or organizations shall report to a management level such that required authority and organizational freedom are provided, including sufficient independence from cost and schedule considerations.

2 QUALITY ASSURANCE PROGRAM

A documented quality assurance program shall be planned, implemented, and maintained in accordance with this Standard, or portions thereof. The program shall identify the activities and items to which it applies. The establishment of the program shall include consideration of the technical aspects of the activities affecting quality. The program shall provide control over activities affecting quality to an extent consistent with their importance. The program shall be established at the earliest time consistent with the schedule for accomplishing the activities.

The program shall provide for the planning and accomplishment of activities affecting quality under suitably controlled conditions. Controlled conditions include the use of appropriate equipment, suitable environmental conditions for accomplishing the activity, and assurance that prerequisites for the given activity have been satisfied. The program shall provide for any special controls, processes, test equipment, tools, and skills to attain the required quality and for verification of quality.

The program shall provide for indoctrination and training, as necessary, of personnel performing activities affecting quality to assure that suitable proficiency is achieved and maintained.

Management of those organizations implementing the quality assurance program, or portions thereof, shall regularly assess the adequacy of that part of the program for which they are responsible and shall assure its effective implementation.

3 DESIGN CONTROL

The design shall be defined, controlled, and verified. Applicable design inputs shall be appropriately specified on a timely basis and correctly translated into design documents. Design interfaces shall be identified and controlled. Design adequacy shall be verified by persons other than those who designed the item. Design changes, including field changes, shall be governed by control measures commensurate with those applied to the original design.

FIGURE A7-1 *Used with the permission of the ASME. (Continued)*

4 PROCUREMENT DOCUMENT CONTROL

Applicable design bases and other requirements necessary to assure adequate quality shall be included or referenced in documents for procurement of items and services. To the extent necessary, procurement documents shall require Suppliers to have a quality assurance program consistent with the applicable requirements of this Standard.

5 INSTRUCTIONS, PROCEDURES, AND DRAWINGS

Activities affecting quality shall be prescribed by and performed in accordance with documented instructions, procedures, or drawings of a type appropriate to the circumstances. These documents shall include or reference appropriate quantitative or qualitative acceptance criteria for determining that prescribed activities have been satisfactorily accomplished.

6 DOCUMENT CONTROL

The preparation, issue, and change of documents that specify quality requirements or prescribe activities affecting quality shall be controlled to assure that correct documents are being employed. Such documents, including changes thereto, shall be reviewed for adequacy and approved for release by authorized personnel.

7 CONTROL OF PURCHASED ITEMS AND SERVICES

The procurement of items and services shall be controlled to assure conformance with specified requirements. Such control shall provide for the following as appropriate: source evaluation and selection, evaluation of objective evidence of quality furnished by the Supplier, source inspection, audit, and examination of items or services upon delivery or completion.

8 IDENTIFICATION AND CONTROL OF ITEMS

Controls shall be established to assure that only correct and accepted items are used or installed. Identification shall be maintained either on the items or in documents traceable to the items.

9 CONTROL OF PROCESSES

Processes affecting quality of items or services shall be controlled. Special processes that control or verify quality, such as those used in welding, heat treating, and nondestructive examination, shall be performed by qualified personnel using qualified procedures in accordance with specified requirements.

10 INSPECTION

Inspections required to verify conformance of an item or activity to specified requirements shall be planned and executed. Characteristics to be inspected and inspection methods to be employed shall be specified. Inspection results shall be documented. Inspection for acceptance shall be performed by persons other than those who performed or directly supervised the work being inspected.

FIGURE A7-1 *Used with the permission of the ASME. (Continued)*

11 TEST CONTROL

Tests required to verify conformance of an item to specified requirements and to demonstrate that items will perform satisfactorily in service shall be planned and executed. Characteristics to be tested and test methods to be employed shall be specified. Test results shall be documented and their conformance with acceptance criteria shall be evaluated.

12 CONTROL OF MEASURING AND TEST EQUIPMENT

Tools, gages, instruments, and other measuring and test equipment used for activities affecting quality shall be controlled and at specified periods calibrated and adjusted to maintain accuracy within necessary limits.

13 HANDLING, STORAGE, AND SHIPPING

Handling, storage, cleaning, packaging, shipping, and preservation of items shall be controlled to prevent damage or loss and to minimize deterioration.

14 INSPECTION, TEST, AND OPERATING STATUS

The status of inspection and test activities shall be identified either on the items or in documents traceable to the items where it is necessary to assure that required inspections and tests are performed and to assure that items which have not passed the required inspections and tests are not inadvertently installed, used, or operated. Status shall be maintained through indicators, such as physical location and tags, markings, shop travelers, stamps, inspection records, or other suitable means. The authority for application and removal of tags, markings, labels, and stamps shall be specified. Status indicators shall also provide for indicating the operating status of systems and components of the nuclear facility, such as by tagging valves and switches, to prevent inadvertent operation.

15 CONTROL OF NONCONFORMING ITEMS

Items that do not conform to specified requirements shall be controlled to prevent inadvertent installation or use. Controls shall provide for identification, documentation, evaluation, segregation when practical, and disposition of nonconforming items, and for notification to affected organizations.

16 CORRECTIVE ACTION

Conditions adverse to quality shall be identified promptly and corrected as soon as practical. In the case of a significant condition adverse to quality, the cause of the condition shall be determined and corrective action taken to preclude recurrence. The identification, cause, and corrective action for significant conditions adverse to quality shall be documented and reported to appropriate levels of management; follow-up action shall be taken to verify implementation of this corrective action.

17 QUALITY ASSURANCE RECORDS

Records that furnish documentary evidence of quality shall be specified, prepared, and maintained. Records shall be legible, identifiable, and retrievable. Records shall be protected against damage, deterioration, or loss. Requirements and responsibilities for record transmittal, distribution, retention, maintenance, and disposition shall be established and documented.

FIGURE A7-1 *Used with the permission of the ASME. (Continued)*

18 AUDITS

Planned and scheduled audits shall be performed to verify compliance with all aspects of the quality assurance program and to determine its effectiveness. These audits shall be performed in accordance with written procedures or checklists by personnel who do not have direct responsibility for performing the activities being audited. Audit results shall be documented and reported to and reviewed by responsible management. Follow-up action shall be taken where indicated.

FIGURE A7-1 *Used with the permission of the ASME. (Continued)*

tions. For instance, state and federal agencies are not subject to the sovereignty of a city, and therefore are not subject to its building codes. Often as a courtesy, and also a practicality, the state or federal agency chooses to abide by the building code, and requires its designers and inspectors to do so.

A major responsibility under a building code lies with the design professionals. It is their obligation to design the facility in accordance with applicable building codes. This responsibility usually culminates in the submittal of the completed design to the licensing body authorized by the building codes for review and approval. The inspection team, in turn, starts with an approved set of documents that conform to the appropriate building code. Even in the case of a design-builder, state law usually requires that the design documents be prepared by licensed architect-engineers and appropriate submittals made on behalf of the design-builder by these professionals. Accordingly, the field inspection team should be able to rely on the approved documents as being in conformance with the appropriate building code.

LICENSES, PERMITS, INSPECTIONS

As appropriate, the QA procedure should identify the requirements for licenses. For instance, Title B under Licenses describes eight different types of licenses required in the City of New York. It is the responsibility of the QC inspection team to ensure that individuals requiring licenses have, in hand, a current license before they work on a project. In certain cases, the license relates to supervision, as in the case of Master Plumber, Master Rigger, Special Rigger, Sign Hanger, or Concrete Testing Laboratory. Other licenses related directly to the craftspersons involved, including: Welder (structural), High Pressure Boiler Operator, Hoisting Machine Operator, and Oil Burning Equipment Installer.

As noted previously, it is the responsibility of the designer and/or owner to have plans reviewed and approved. The New York City Code states that plans will be stamped or endorsed "Approved," and the, "One set of such approved plans shall be retained in the department office of the Borough in which the building premises or equipment is located; and after the issuance of a work permit, a second set of such approved plans shall be retained at the place where the building premises or equipment is located and it shall be open at all times to inspection by the Commissioner and his authorized representatives until final inspection of the work is completed." Accordingly, it is the responsibility of the inspection team to make sure that such a set of approved documents is available. Further, even though the plans have been approved, typically it is still the responsibility of contractors to acquire permits for either all of or stages of the work.

O'BRIEN-KREITZBERG & ASSOC., INC.

POLICY STATEMENT

This Quality Assurance Program Plan establishes the O'Brien-Kreitzberg & Associates, Inc. (OKA) Quality Assurance Program, which is intended to control the quality aspects of construction management and construction inspection services provided by OKA. The specific applicability of the requirements of this Program Plan shall be established in contract and project documents.

OKA personnel shall effectively implement this Program Plan, as a primary commitment, to assure the quality of construction.

The Principal-in-Charge of each project shall regularly review the status and adequacy of the project quality program. Personnel responsible for project quality activities shall have free access to the Principal-in-Charge to identify quality problems.

James J. O'Brien, P.E.   Fred C. Kreitzberg, P.E.

OKA QUALITY ASSURANCE PROGRAM PLAN

TABLE OF CONTENTS

POLICY STATEMENT
TABLE OF CONTENTS

Section	Title	Page No.
1.0	**QUALITY PROGRAM SCOPE**	1
1.1	Construction Management and Inspection Services	1
1.2	Quality Assurance Program Implementation	2
1.3	Definitions	3
2.0	**APPLICABLE STANDARDS**	4
3.0	**QUALITY PROGRAM MANAGEMENT**	5
3.1	Organization	5
3.2	Management Review and Audit	6

FIGURE A7-2 OKA Quality Assurance Program Plan.

TABLE OF CONTENTS (cont'd)

Section	Title	Page No.
3.3	Construction and Inspection Planning	7
3.4	Inspection Procedures	8
3.5	Inspection Records	9
3.6	Corrective Action	10
4.0	**CONSTRUCTION DOCUMENTS AND STANDARDS**	12
4.1	Construction Document Control	12
4.2	Design Changes	13
4.3	As-Built Records	14
4.4	Inspection and Test Equipment Calibration	15
5.0	**CONTROL OF PURCHASES**	17
5.1	Supplier Evaluation	17
5.2	Control of Sub-Tier Suppliers	18
5.3	Source Inspection	20
6.0	**CONSTRUCTION CONTROL AND INSPECTION**	21
6.1	Material and Inventory Controls	21
6.2	Construction Process Controls	22
6.3	Inspection Control	23

1.0 QUALITY PROGRAM SCOPE

1.1 Construction Management and Inspection Services

1.1.1 This Quality Assurance Program Plan establishes the Quality Assurance Program for construction management and inspection services provided by O'Brien-Kreitzberg & Associates, Inc. (OKA).

1.1.2 The OKA commitment to excellence in construction management includes vigorous controls for cost and schedule, which, although compatible with and supported by the quality requirements in this Program Plan, are established and implemented in separate corporate and project documents.

1.2 Quality Assurance Program Implementation

1.2.1 The specific application of the requirements of this Program Plan for a project shall be defined by contract documents, which shall establish:

- Project organization and assignment of responsibilities.
- Definition of external interfacing quality organizations and functions.
- Project-unique requirements which supplement the Program Plan for a specific project.

1.2.2 The Quality Assurance Program is implemented in corporate and project procedures, specifications and instructions.

1.2.3 Corporate procedures are approved by the Principal-in-Charge. Project procedures and instructions are approved by the Project Manager.

FIGURE A7-2 OKA Quality Assurance Program Plan. (*Continued*)

1.2.4 The Project Manager shall provide for preparation, listing, and distribution of applicable project control documents to project personnel.

1.2.5 The Project Manager shall provide for the indoctrination and training of project personnel in the requirements and application of the project control documents.

1.3 **Definitions**

The word "shall" is used to denote a requirement. The word "should" is used to denote a recommendation. The word "may" is used to denote permission: neither a requirement nor a recommendation.

2.0 APPLICABLE STANDARDS

2.1 The OKA Quality Assurance Program Plan is responsive to the requirements of the following standards, as they apply to the construction management and construction inspection services offered by OKA:

- ANSI/ASME N45.2-1977, "Quality Assurance Program Requirements for Nuclear Facilities."
- Military Specification MIL-Q-9858A of 12/14/63, "Quality Program Requirements."
- Military Specification MIL-I-45208A of 12/16/63, "Inspection System Requirements."
- RDT F2-2/1973, "Quality Assurance Program Requirements."

3.0 QUALITY PROGRAM MANAGEMENT

3.1 **Organization [3.1]**

3.1.1 A typical project organization for construction management and inspection is shown below:

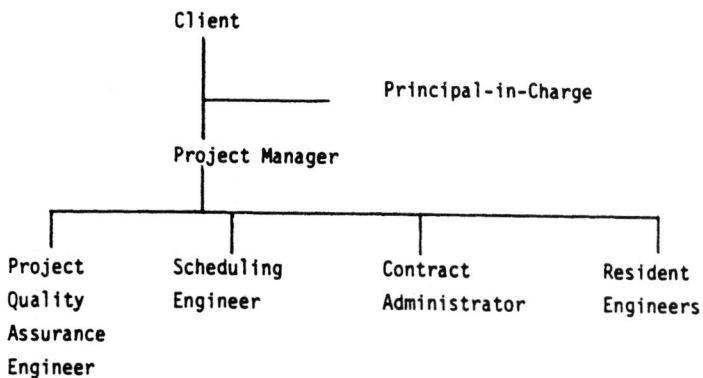

3.1.2 For each project, specific authority and lines of communication are established to report, control and resolve problems that could affect the quality of the work

FIGURE A7-2 OKA Quality Assurance Program Plan. (*Continued*)

effort. The project organization and a description of assigned responsibilities shall be documented and maintained current in the Project Procedures Manual.

3.2 **Management Review and Audit**

3.2.1 The Principal-in-Charge shall regularly review the adequacy of the project quality program, and implement necessary program additions or changes. This review may be accomplished by:
- Review of project reports.
- Discussions with project and client personnel.
- Formal audit by an independent audit team.

3.2.2 Formal audits shall include an evaluation of quality assurance practices, procedures, and instructions; the effectiveness of implementation; and conformance with policy directives. In performing this evaluation, the audits should include evaluation of work areas, activities, processes, and items; and review of documents and records.

3.2.3 The Principal-in-Charge shall define the scope of the audit and select an audit team experienced in auditing techniques, competent in the technical areas to be evaluated, and independent of project responsibilities. The audit shall be planned and accomplished in accordance with written procedures or checklists, with the results reported to the Principal-in-Charge, who is responsible for initiating and evaluating corrective action.

3.3 **Construction and Inspection Planning**

3.3.1 The Project Manager shall provide for the review of construction drawings and specifications with respect to:
- constructability, including, but not limited to, conformance to generally accepted construction practices, enforceability of specifications and avoidance of disputes.
- planning of inspection and testing requirements, methodology and documentation.

3.3.2 Inspection planning shall include:
- Construction/inspection sequencing and identification of mandatory hold points.
- Inspection procedure, personnel qualification and equipment requirements.
- Scheduling inspector training and certification.
- Identifying inspection and testing to be subcontracted.
- Scheduling inspections and tests.
- Identifying individuals authorized to request and to approve inspections and tests.

3.4 **Inspection Procedures**

3.4.1 The Project Quality Assurance Engineer shall prepare inspection procedures which establish:
- the characteristics to be inspected.

FIGURE A7-2 OKA Quality Assurance Program Plan. (*Continued*)

- the inspection methods.
- the acceptance and rejection criteria.
- the methods for recording inspection results.
- special preparation, cleaning, or measuring devices.

3.5 Inspection Records

3.5.1   Inspection and testing records shall be identified, collected and indexed to assure retrievability.

3.5.2   The records shall include the results of reviews, inspections, tests, audits, monitoring of work performance, materials analyses, and inspection logs. The records shall also include, as appropriate, closely related data such as qualifications of personnel, procedures, and equipment. Inspection and test records shall, as a minimum, identify the date of inspection or test, the inspector or data recorder, the type of observation, the results, the acceptability, and the action taken in connection with any deficiencies noted. Required records shall be legible, identifiable, and retrievable.

3.5.3   These records shall be reviewed to assure that the records are legible and complete.

3.5.4   The records which have been identified and collected shall be suitably protected against fire, theft, and damage.

3.5.5   The Project Quality Assurance Engineer shall identify, collect, review, index, maintain and arrange for the transfer of inspection records.

3.5.6   Inspection records shall be transferred to the client in accordance with project contract documents. Records may be transferred at various points in the project and at the end of the project. The Project Manager shall obtain the specific consent of the client prior to the destruction of any inspection records.

3.6 Corrective Action

3.6.1   Conditions Adverse to Quality

Conditions adverse to quality may be identified by a number of techniques:
- Audits of OKA by clients.
- Internal OKA audits.
- Audits of sub-tier suppliers by OKA
- Project reports.
- Principal-in-Charge review of projects.
- Discrepancy Reports.

Each of the above techniques has a mechanism to effect the correction of the condition adverse to quality: the audit technique has the audit report and response to the audit report mechanism; the project report and the Principal-in-Charge reviews result in management action; and the discrepancy report has the disposition mechanism.

FIGURE A7-2   OKA Quality Assurance Program Plan. (*Continued*)

QUALITY ASSURANCE/QUALITY CONTROL A7-23

3.6.2　Significant Conditions Adverse to Quality

Significant conditions adverse to quality are those which extend beyond a single condition or item. A significant condition adverse to quality must be generic in nature to a large number of items or be a deficiency in the quality program.

Each condition adverse to quality shall be analyzed to determine if it represents a significant condition adverse to quality, as defined above. This analysis shall be performed by the individual making the disposition of the condition adverse to quality.

The Principal-in-Charge shall perform an analysis to determine if there are any broad programmatic problem areas or if any negative trends are detectable. This analysis shall be performed at least annually, as part of the management review.

4.0　CONSTRUCTION DOCUMENTS AND STANDARDS

4.1　Construction Documents and Standards

4.1.1　The Project Manager shall establish procedures to control design documents used for construction inspection activities, including drawings, specifications, procedures and instructions, and changes to these documents, to preclude the use of unapproved or out-dated documents. These procedures shall control:

- Defining the issuing authority for various documents and changes.
- Establishing and updating distribution lists.
- Verifying the use of current documents for inspection activities.
- Removing obsolete drawings from use.

4.1.2　The inspection program (discussed in Section 6.3) shall include monitoring design document controls established by construction contractors, to verify that current documents and document changes are issued and used to control construction activities.

4.2　Design Changes

4.2.1　The Project Manager shall establish procedures to control engineering change request, approval, and issue for construction, in conformance with project contract requirements. These procedures shall include monitoring implementation of approved engineering changes.

4.2.2　The inspection program (discussed in Section 6.3) shall include monitoring field changes accomplished by construction contractors, for compliance with applicable construction specification requirements. The verification of as-built drawings (discussed in Section 4.3) shall include these field changes.

4.3　As-Built Records

4.3.1　The Project Quality Assurance Engineer shall provide for the collection and verification of as-built records as required by the construction specifications and contract documents.

4.3.2　Final as-built drawings shall include all approved engineering and field changes.

FIGURE A7-2　OKA Quality Assurance Program Plan. (*Continued*)

4.3.3 In addition to as-built drawings, as-built records include specifications, procedures and instructions used in control of configurations or in construction, inspection records (discussed in Section 3.5), and material certifications and test data.

4.3.4 The Project Quality Assurance Engineer shall provide for the indexing and transfer of as-built records to the client upon completion of the project, as required by the construction specifications.

4.4 Inspection and Test Equipment Calibration

4.4.1 The Project Quality Assurance Engineer shall provide for the control, calibration and adjustment of inspection and test equipment, to assure that tools, gages, instruments, and other measuring and testing devices used for inspection and testing are properly controlled, calibrated, and adjusted at specific periods to maintain accuracy within necessary limits. These requirements are not intended to imply a need for special calibration and control measures on rulers, tape measures, levels, and such other devices, if normal commercial practices provide adequate accuracy.

4.4.2 The calibration of inspection and testing equipment shall be accomplished in accordance with written procedures, which shall include the following requirements:
- Identification of equipment and traceability to calibration data.
- Calibration methods, frequency, maintenance, and control.
- Labeling and marking of equipment to indicate due date for next calibration.
- Provisions for determining the validity of previous measurements when equipment is determined to be out of calibration.
- Use of calibration standards with an uncertainty (error) of less than one-fourth the tolerance of equipment being calibrated, within the state-of-the-art.
- Traceability of reference and transfer standards to nationally recognized standards. When national standards do not exist, the basis for calibration shall be documented.

4.4.3 Calibration intervals shall be based on required accuracy, use of equipment, stabiltiy characteristics, or other factors affecting the measurement. Calibration may be performed on site or by qualified laboratories utilizing competent personnel. Equipment which is found to be frequently out of adjustment shall be repaired or replaced.

4.4.4 Special calibration shall be performed when the accuracy of the equipment is suspect. When inspection or test equipment is found to be out of calibration, an evaluation shall be made and documented of the validity of previous inspection or test results and of the acceptability of items previously inspected or tested.

4.4.5 These requirements for inspection and test equipment calibration shall be imposed on sub-tier suppliers, including inspection and testing services.

4.4.6 Installed instrumentation used in acceptance testing shall be calibrated in accordance with project acceptance testing documents.

FIGURE A7-2 OKA Quality Assurance Program Plan. (*Continued*)

5.0 CONTROL OF PURCHASES

5.1 Supplier Evaluation

5.1.1 The Project Manager shall provide for the evaluation of the capability of a supplier to provide an item or service in accordance with engineering and quality requirements. This evaluation shall be based on one or more of the following:

- Supplier's capability to comply with the elements of the quality standard applicable to the type of material, equipment, or service being procured.
- Past records and performance for similar procurements to ascertain the capability of supplying a manufactured product (or services) under an acceptable quality assurance system.
- Surveys of supplier's facilities and quality assurance program to determine his capability to supply a product which satisfies the design, manufacturing, and quality requirements. This survey should include, as appropriate, facilities, production capabilities, personnel capabilities, process and inspection capabilities, and organization, in addition to the supplier's quality assurance program.

5.2 Control of Sub-Tier Suppliers

5.2.1 The Project Quality Assurance Engineer shall review purchase orders for items and services supporting construction inspection activities, to verify:

- complete and correct statement of the technical and quality requirements, including reference to appropriate standards and specifications.
- identification of records to be prepared, maintained, submitted or made available for review, such as drawings, specifications, procedures, procurement documents, inspection and test records, personnel and procedure qualifications, and material, chemical, and physical test results. Record retention and disposition requirements shall be provided.
- provisions for extending applicable requirements of procurement documents to lower tier subcontractors and suppliers, including purchaser's access to facilities and records, if appropriate.

5.2.2 The Project Quality Assurance Engineer shall assess the effectiveness of the control of quality by suppliers supporting construction inspection activities. This assessment shall be accomplished at intervals consistent with the complexity of the item or service, with the quantity of material furnished, and with the duration of the service furnished, based on one or more of the following:

- Direct source inspection (discussed in Section 5.3)
- Reviews of objective evidence of quality furnished by the supplier, such as inspection and test records, personnel and procedure qualifications, material physical and chemical test results, and supplier licensing and certification.
- Periodic audits of the supplier quality program and procedures.
- Comparison or re-test of supplier products or services, by independent testing facilities, and/or against known standards.

FIGURE A7-2 OKA Quality Assurance Program Plan. (*Continued*)

5.3	**Source Inspection**
5.3.1	When specified in project contract documents, the Project Manager shall establish procedures for source inspection at supplier's facilities. The source inspection activities may include:

- Reviewing material acceptability, including associated expendable and consumable materials necessary for the functional performance of structures, systems, and components.
- Witnessing in-process inspections, tests, and non-destructive examinations.
- Reviewing the qualification of procedures, equipment, and personnel.
- Verifying that fabrication or construction procedures and processes have been approved and are properly applied.
- Verifying the implementation of the quality assurance/quality control systems.
- Reviewing document packages for compliance to procurement document requirements, including qualifications, process records, inspection and test records.
- Reviewing certificates of compliance for adequacy.
- Verifying that nonconformances have been properly controlled.

5.3.2	The results of source inspection activities shall be documented and shall include copies of certifications, chemical and physical analyses, inspection reports, test results, personnel and process qualification results, code stampings, and nondestructive test reports as required by the applicable specification.

6.0 CONSTRUCTION CONTROL AND INSPECTION

6.1	**Material and Inventory Controls**
6.1.1	The inspection program (discussed in Section 6.3) shall include monitoring material controls utilized by construction contractors, to verify that these controls are implemented in compliance with construction specification requirements. These material controls include:

- Receiving and receiving inspection.
- Material identification.
- Storage and preservation.
- Handling and rigging.
- Identification, segregation and disposition of non-conforming material.

6.1.2	The Project Manager may establish an inventory control system for equipment and material received at the construction site as required by the project contract documents.
6.2	**Construction Process Controls**
6.2.1	The inspection program (discussed in Section 6.3) shall include monitoring construction process controls utilized by construction contractors, to verify that these

FIGURE A7-2 OKA Quality Assurance Program Plan. (*Continued*)

controls are implemented in compliance with applicable construction specification requirements. Construction process controls include:

- Clear and complete instructions for performing work functions, appropriate to the complexity and importance of the activities involved.
- Workmanship standards and criteria.
- Control of special processes, such as welding, heat treating, and application of coatings, including qualification of personnel, procedures and equipment.
- Documentation of correct construction sequence and material identification, such as concrete lift release cards, wire pull and termination cards, and mechanical alignment data.

6.3 **Inspection Control**

6.3.1 The Project Quality Assurance Engineer shall implement construction inspection activities in accordance with project contract and specification requirements. These inspections shall be accomplished using approved inspection and test procedures (as discussed in Section 3.4). Inspection and test results shall be documented, and reviewed by the Project Quality Assurance Engineer.

6.3.2 The Project Quality Assurance Engineer shall document the training, qualification and certification of inspection personnel, in accordance with project contract and specification requirements.

6.3.3 The Project Quality Assurance Engineer shall maintain a system for identifying the inspection status of equipment, systems, and structures subject to construction inspections. Inspection status shall be indicated by stamps, marks, tags or labels attached to the item, or on documents such as drawings, construction travellers, or inspection records traceable to the item.

6.3.4 The inspection procedures shall include monitoring the following construction contractor activities for compliance with project contract and specification requirements:

- Design document control (discussed in Section 4.1).
- Field changes (discussed in Section 4.2).
- Material controls (discussed in Section 6.1).
- Construction process controls (discussed in Section 6.2).

FIGURE A7-2 OKA Quality Assurance Program Plan. (*Continued*)

EXCULPATORY CLAUSES

Exculpatory clauses in contracts seek to avoid responsibility for unanticipated situations, with a view toward shifting all risk to the contractor. One type of exculpatory clause the designer often inserts invokes existing codes and regulations as part of the specification in the hope that should any of these be violated, the exculpatory clause will require compliance. In effect, the designer hopes that if the specification has fallibilities they will be countered or covered by a standard regulation.

In a classic case decided at the turn of the century by the New York State Court of Appeals (*MacKnight Flintic Stone Co. v. the Mayor*, 160 N.Y.72), the contractor was required to provide a watertight boiler room for a courthouse and prison, but to do so "in a manner and under the conditions prescribed and set forth in the . . . specifications. . . ."

The specifications described in detail the manner in which the waterproofing was to be accomplished. The contractor faithfully followed the specifications, and serious leaks developed. The specifications required a five-year guarantee for waterproofing, but the contractor defended on the basis that he had followed the specifications, and it was the specifications themselves that did not result in a waterproof space. In finding for the contractor, the court said:

> This is not the case of an independent workman left to adopt his own methods, but of one bound hand and foot to the plan of the defendant. The plaintiff had no right to alter the specifications. . . . If the plan and specifications were defective, it was not the fault of the plaintiff, but of the defendant, for it caused them to be made, and it alone had the power to alter them.

The following are some typical exculpatory clauses:

- "This survey is furnished as information to bidders. Bidders shall verify survey information, and owner assumes no responsibility for accuracy of the survey."
- "Drilling logs shown on the plans are for information only, and the contractor is warned that reliance on this information shall be at his own risk, and neither the owner nor the engineer shall be liable for errors."
- "Contractor is responsible for interpretation of subsurface soil conditions as described in the boring data. Further, all soil and boring data is for information only and shall not be considered part of the contract document."
- "Contractor shall visit the site and satisfy himself as to actual site conditions, and shall verify all existing dimensions. Owner assumes no responsibility for variations in existing documents."
- "Whenever provisions of any section of the plans or specifications conflict with any union, trade association or agency regulations contractor shall make necessary reconciliation of such conflict without recourse to the owner."
- "It is the intent of these plans and specifications to produce first-class quality work. Should a conflict occur which would preclude the placing of first-class quality work, the contractor shall request in interpretation of the manner in which the work shall be placed, and shall perform in accordance with such interpretation."
- "Should a conflict occur between drawings and specifications, the contractor is deemed to have estimated the more expensive way of doing the work and shall so provide."
- "It is the intent of this contract to provide for complete installation of all portions of the work. All items, materials and equipment are to be furnished and installed complete and ready for operation or use. Contractor will be deemed to have based his bid upon an operating installation."
- "Where this contractor's work requires supports, connections, or installation of any group of items furnished by other contractors, the omission of any item from the drawing shall not relieve this contractor from the responsibility for installing, connecting or supporting such item at no increase in cost."
- "Wherever additional materials or work are required to complete the work in accordance with obvious intent of the drawings, contractor shall provide these

materials or work at no additional cost to the owner, even if not shown or specified specifically."

One purpose for using a complete and detailed specification is to describe a required item in such a manner that the contractor can accurately estimate its cost so the owner pays no more or no less than the owner should. The exculpatory clause represents an attempt to avoid the impact of errors or omissions in the specification by providing an umbrella of very broad and intangible requirements.

QUALITY CONTROL: INSPECTION TEAM

Inspectors

There are biblical references to inspectors, and early monumental projects such as the pyramids had working scribes as well as overseers. In fact, the term *clerk of the works* is probably a carryover from the early material counters.

The role implied by the title clerk of the works is as archaic as the term itself. No longer (if in fact it was ever true) is the inspector able to stand by and count materials coming into the job. Perhaps a hundred years ago when labor was cheap and materials were scarce, there were widespread instances of shortchanging through the use of shoddy materials. In today's construction industry, it is more usual to find that good material is used, but it is installed improperly, with connections not made or controls not operable. This is where quality control is important—making certain that the labor and materials bought and paid for are properly applied.

In most cases, mistakes benefit no one; everyone is the loser. The craftsperson becomes less productive, the contractors acquire tarnished reputations, and the owners (at some point) must dig deeper into their own pockets to make up for problems created by installation mistakes.

The Role of the Inspector

Historically, government agencies tend to view the role of inspection in the traditional, ultranarrow sense. Nevertheless, this role is a base for the things that an inspector can and should do. The following description is from the manual, *Construction Inspection Procedures,* prepared by the General State Authority of Pennsylvania, discussing the role of the inspector.

> The primary function of the field personnel of the construction division is inspection, and the persons assigned to this task are designated as inspectors. There are three classifications of field inspectors: general, mechanical, and electrical. . . . The Inspector must be able to look upon and view critically the particular phase of the construction project to which he is assigned. This requires some degree of experience in the construction field. In addition to experience, the Inspector must also have the ability to evaluate and analyze what he is inspecting. Therefore, a most important and necessary requirement is that the Inspector be able to fully read, comprehend, and interpret the contract plans and specifications. It is also very important that the Inspector have the ability to maintain records that will fully reflect the inspections performed.
>
> The Inspector must closely follow the progression of each stage of construction. He must be alert to existing conditions and be able to foresee future problems. When the Inspector notices through his daily inspections that certain phases of the work are not being done in accordance with the plans and specifications, or when other problems

occur, he is to immediately report these errors, violations, or problems to "management" for further action.

In effect, the Inspector is not authorized to revoke, alter, substitute, enlarge, relax, or release any requirements of any specifications, plans, drawings or any other architectural addenda. In addition, the Inspector must not approve or accept any segment of the work which is contrary to the drawings and specifications. At no time is the individual Inspector allowed to stop the construction work or interfere with the contractor's employees.

It must be recognized that the title "Inspector" creates a barrier between him and the contractors on the job. How effective he can be in his role depends mainly on how he handles himself in this relationship. He must display knowledge, experience, integrity, ability and the use of good judgement.

The role described is a narrow one of quality control. It suggests that the inspector cannot give up any of the prerogatives of the owner, and, at the same time, cannot delay the work. This situation is mitigated in many cases by the assignment of resident engineers with concomitant duties to inspect. Even in these cases, where the inspector represents management, work may not be disrupted except in accordance with the contract. In a cost-plus contract, the owner often has prerogatives to stop the work—if any such stoppage will be paid for directly by the owner. In fixed-price and negotiated contracts, the owner's representatives must be much more careful, or they will be liable for a claim of interference.

The Corps of Engineers offers the following suggestions in regard to the inspector's role:

> An Inspector should at all time be thoroughly familiar with all the provisions of the contract which he is administrating. This includes familiarity with the plans and specifications including all revisions, changes, and amendments. In addition, the Inspector must be thoroughly familiar with pertinent Corps of Engineers, individual district, and supervisor's administration policies.

Inspectors have different responsibilities and authorities, dependent on the organizational setup under which they are working and their own capabilities. Inspectors should know their part in the organization, and should be aware of the importance of high-quality construction. They should understand their own level of technical knowledge and accept their responsibilities without overstepping their authority.

In order to do this, inspectors must be aware of the extent of their authority. To that end, inspectors always have the authority to require work to be accomplished in accordance with the contract plans and specifications. Procedures and policies on stopping work for safety violations or construction deficiencies should be reviewed with the appropriate supervisors before being employed.

QC Field Testing

The quality control (QC) program usually includes field testing. Traditionally, test requirements are described in the specification wherever (and whenever) the design engineer (or architect) chooses. It is recommended that the quality assurance (QA) procedures call for an index of all test requirements (or the requirements themselves) to be listed in Division 1 of the specifications (01400 Quality Control). If this has not been done during the design phase, the QC group must set up an inspection plan.

Step one in the inspection plan is specification review, writing up each test requirement, using a form to record individual tests. The individual requirements should then be summarized by area/floor, etc.

The contract may require the contractor to conduct the QC tests with observation by the inspection team, or the inspection team may take samples and conduct tests.

To accomplish job site testing, appropriate reference documents must be available because the specifications usually refer to testing procedure by reference to standards such as the ASTM, ACI, ASME, ANSI, or others. The inspection team should review specifications and be certain that referenced standards are available and have been reviewed before testing is called for. It is often appropriate to conduct a trial or rehearsal test before actual job materials are mixed.

Often, testing may be done under the auspices of the materials manufacturer, although the obvious vested interest involved must be considered. Architect-engineers often recommend the use of special testing laboratories or groups as an adjunct to the inspection team. This practice can be both effective and economical, as it limits the amount of specialized testing equipment that must be purchased and does not require the inspection team to develop special skills that are used only to a limited extent on a project. Conversely, where large amounts of material are used on the job, it may be appropriate for the inspection team to learn the skills for specific tests. For instance, on large excavation or backfill projects, certain special moisture content tests are run so frequently that it is appropriate for the field team to conduct them.

In cases where materials have been mixed off the job site, such as ready-mix concrete, it may be appropriate to have a special test team assigned to the assembly or batching area. Typically, certified laboratory services are retained to perform the tests at the batch plant. This is particularly appropriate for premixed concrete, asphalt, gravel, and aggregate.

Off-site testing also is used for the inspection of materials at the point of manufacture. Usually, this is done during quality control testings or as required (for example, water pressure testing). The test is usually conducted using manufacturer's equipment and personnel under the supervision of the inspection team.

The test requirements for construction sites are many and varied. The specifications will provide explicit directions to the inspection team for the type of test to be run, equipment to be used, and conditions of the test.

REFERENCES

1. James J. O'Brien, "Quality: A Commentary," *Civil Engineering Magazine,* Feb. 1986, pp. 48–49, used with permission.
2. ASCE, *Manual for Quality in the Constructed Project, A Guideline for Owners, Designers and Constructors,* vol. 1, 1990, p. 85, used with permission.
3. *Ibid.,* p. 84–85, used with permission.

CHAPTER 8
TOTAL QUALITY MANAGEMENT

Joseph V. Mullin, Ph.D., P.E.
Senior Vice President, Pennoni Associates Incorporated
Philadelphia, Pennsylvania

INTRODUCTION

Early in the history of construction, the concept of quality was established as a commitment of both owner and constructor. Government agencies and trade organizations established codes and standards of quality which have evolved as technology changed and new materials were developed. Local and national code requirements have become the foundation of the industry, but these alone do not assure quality in the finished project. In recent years, the work of Demming[1] and Juran[2] have refined the concepts, and they are now global in impact. Essentially, the concept of *Total Quality Management* (TQM) defines a mission to continuously strive for zero defects in our daily performance. Often this is expressed in a *mission statement* written for a firm and expressing their goals relating to continuous quality improvements.

MISSION STATEMENT

A typical mission statement follows:

> The employees of ACME Constructors will consistently provide the highest level of service by successfully fulfilling the needs and wishes of our clients. We recognize the need to provide service in an accurate and timely fashion and our aim is to construct to the highest standards of quality.
>
> We will use Quality Vision to continuously provide a work environment which will heighten employee satisfaction among our internal customers, our fellow workers. We will strive to anticipate and exceed the expectations of our clients.

This mission statement should be established with input from the entire organization. It is not a management directive but a commitment of the entire workforce.

Many TQM seminars stress the formulation of the mission statement as the first step in developing a team approach to quality. It is often the focus of a brainstorming session where employees at various levels express their views on the proper direction of the firm and how to reflect it in the mission statement.

OBJECTIVES

The overall TQM objective is to meet client requirements, needs, and expectations the first time and every time we serve them. There are often specific goals incorporated into this objective. Here are some examples:

1. Improve service to clients.
2. Lower our costs to provide that service.
3. Increase productivity.
4. Enhance client loyalty.
5. Generate repeat business.
6. Improve employee morale.
7. Enhance profitability and growth.
8. Encourage professional growth of staff.

There are many other objectives which could be included but the list should be kept to a reasonable number to emphasize importance.

FOUNDATION OF THE PROGRAM

TQM postulates that good judgment and common sense, in conjunction with adherence to professional standards of practice and contract requirements, will ensure quality results at all times. Policies and procedures are based on sound business judgment and accepted practice in the construction industry. The following are major elements of TQM:

1. The establishment of an organizational structure, functional responsibilities, level of authority, and lines of communication for the management, direction, and execution of the Total Quality Program.
2. The establishment of measures to assure and document that all applicable regulatory requirements, as well as pertinent technical criteria, are addressed in the planning and scope of work activities.
3. The establishment of documented instructions and procedures that prescribe requirements for the performance of field measurements and operations, as well as the processing and interpretation of data.
4. The establishment of measures to improve quality standards and supporting technical procedures, including changes thereto, and to assure that those involved in work activities use current approved procedures for performing those activities.
5. The involvement of employees in the creative process of developing new and better ways to serve our clients.

6. The establishment of a system for review of all work for technical correctness and adequacy. This plan includes a quality assurance review of the contract documents, project plan, and schedule prior to bidding.

A TYPICAL TQM PLAN

There are 11 specific areas defined in this typical plan, from prebid activities to completed project evaluation. The last element is continuing education, wherein upgrading of policies and procedures is ongoing and everyone is trained to do a better job. The Quality Assurance Plan sets forth recommended procedures for establishing sound contracts, assembling project teams, preparing the project plan, and establishing necessary controls. This plan provides the opportunity for employee involvement in the creative process of continuous improvement. Employees and management learn from one another and the process of TQM perpetuates itself. In short, the commitment to quality is a dynamic process of continuous improvement—it permeates all that we do.

SPECIFIC ELEMENTS OF A TQM PLAN

The following sections of a typical plan address specific elements, but all require the creative input and the commitment of management and staff. The ideas that lead to better performance can be set down on paper but it is the dynamic of putting the plan in action that generates the results.

Prebid Activity

The construction business depends on the ability to identify and develop prospective clients before there is a need for construction services. There is also the ongoing need to retain existing clients. In most cases, selection by new clients is either informal or through a standard bid process. It is wise to exercise care in selecting who to serve. This section deals with the process prior to submitting a bid or entering into an agreement. It recognizes that avoiding potential problems is sometimes more important than attracting new business.

Client Selection

1. Establish the integrity and financial capacity of unknown clients.
2. Avoid clients lacking in professional ethics or having a propensity for litigation as a standard business practice.
3. Maintain a list of undesirable clients based on prior experience.
4. Avoid opportunities where risk outweighs potential awards.

Analysis of Project Content

1. Meet with the prospective client and visit the project site to determine the scope of the project and compare to the resources and plan put forth by the client to assure viability.

2. Determine source of funding and assess both sufficiency and timing.
3. Establish the budget and schedule to include all design liaison elements, regulatory approvals, and possible contingencies for inclusion in the bid.
4. Evaluate current workload and your ability to perform including:
   a. Licensing and special certification needed
   b. Knowledge/experience required
   c. Required subconsultants
   d. Delivery standards anticipated

Project Scheduling

1. Clearly define the work effort needed in each construction activity and consult with experts in each area when establishing scope and schedule.
2. Estimate both worker-hours and delivery items as a check on cost estimates. Make sure estimate is consistent with resources.
3. Set priorities and identify dependent tasks before establishing schedule.
4. Refine and adjust schedule to optimize sequencing before final choice is made and work plan is hardened.

This exercise before bidding a project is essential to be successful in pursuing new work. Each new undertaking is an adventure and presents some degree of risk. The goal here is to avoid unreasonable risk and to pursue only those projects on which the organization can provide the level of quality expected and consistently provided to existing clients. Knowing the strengths and limitations of the organization is a key element in performing quality work. There is always some degree of risk in any worthwhile undertaking. The goal is to manage that risk intelligently and perform successfully.

Establishing Sound Contracts

Contractual agreements are the basis for most work and protect the rights of both owner and contractor. Contracts are intended to facilitate doing business rather than serve as an obstacle to agreement. This section of the Total Quality Management plan deals with contractual agreements and their role in assuring successful results for both owner and contractor.

Well-Written Contracts

1. The contract must clearly define what services are to be performed, the responsibilities of all parties, and compensation.
2. Standard terms and conditions have been established at most firms and should be used in all contracts.
3. Standard contract forms issued by the Engineers' Joint Council are available and have been adopted by many firms because of their universal acceptance on large projects. They may not be usable in all circumstances, but they are a valuable basic reference.

Nonstandard Contract Forms

1. Where prospective clients have devised their own contract form, it is essential that these contracts be reviewed and compared to the Engineers' Joint Council form to identify potential risks and areas requiring modification.
2. Consistency of the owner/contractor agreement and the general terms and conditions is absolutely essential.

Contract Alerts

1. Nonstandard forms of agreement should be reviewed by counsel prior to signing.
2. Be especially alert to terms which set unreasonable or subjective standards. The law demands reasonable care, skill, and diligence, unless a higher standard of performance is voluntarily followed.
3. Responsibility for work by others over which the contractor has no control is a common problem area. Avoid contract terms where coordinating the work of another independent contractor or other third parties to the agreement over whom the contractor has no contractual control is expected.
4. In general, "time is of the essence" clauses should be avoided where the process cannot be controlled from beginning to end.

The overall goal is to enter into fair and reasonable contracts which protect the rights of clients and contractors without interfering with the delivery of quality services. It is important to diligently pursue sound agreement without alienating clients. This requires sensitivity and skill in preparing the agreement and fairness in establishing terms and conditions. There is a need for confidence and trust in every working relationship. Protecting against every eventuality with contract wording is difficult at best. A joint commitment to fairness is the principal ingredient of a quality contract.

Assembling the Project Team

When appropriately implemented, the team approach to project management offers both continuity and motivation to produce quality work. The team leader must clearly define the overall project requirements and keep all team members informed of day-to-day progress to meet objectives. This section deals with the elements of team building.

Management and Team Selection

1. Each construction activity required for the project must be represented on the project team with participants having responsibility for that activity.
2. The project manager is responsible for selecting appropriately skilled members of the project team and establishing the working relationship at the outset.
3. Both experience and special training are essential to quality performance. A mix of both is advantageous because the interaction strengthens the overall capabilities of the team.

Team Composition

1. For a team to operate effectively, it needs a strong leader who listens well and uses all team members effectively. The project manager must be a capable and experienced professional with the authority to deal with the client on all major issues. This person should be involved in developing the contract with the client and be fully aware of the overall goals of the project.
2. The project manager has primary responsibility for selecting the team. For large projects, the team must be sufficiently broad to include more than one person from each construction activity plus appropriate support people. Each specialty then becomes a subunit of the overall construction team and should have an identified leader.
3. The project manager deals with all outside consultants, monitors progress, and maintains budgetary and schedule control.
4. The project manager receives all information that comes into the office relevant to the project, conducts all necessary meetings, and disseminates information to all team members and outside consultants. He/she must also be copied on all correspondence and should assemble and send all submittals for approval or payment.

Team Member Responsibilities

1. Team members provide the expertise in their respective disciplines and support one another in areas of joint responsibility. All team members are responsible for their assigned tasks as established by the project manager. They must establish specific standards to assure quality and provide guidance to other team members and the project manager.
2. All team members are responsible for the quality of each segment of the work completed and adherence to all design and regulatory constraints.
3. Team members must be alert to alternative designs and their constructability to assure the best quality at reasonable cost.
4. Team members are responsible for adhering to the approved schedule and must meet the deadlines assigned.
5. Team members are expected to use the appropriate theory and standards of good construction practice to accomplish their assigned tasks. They must seek out those with specialized knowledge to assure quality in the constructed project.

Building the appropriate team and motivating everyone on the team to put forth their best efforts are key elements of TQM. Making certain that every team member is well informed and able to contribute creatively is most important. Anticipating problems and avoiding their consequences is essential to teamwork.

Establishing the Project Plan

A written project plan should be developed upon signing the contract with the client. This allows adjustment to the scope of the work as proposed and negotiated to avoid misunderstandings. The work plan thus incorporates any changes in approach or scope, terms or conditions, which may affect deliverables and billing procedures. The project plan will become the blueprint for how the project will proceed.

Defining the Project Plan. The following steps will help to define the project plan:

1. With the project manager, the project team must define, in the plan, the client aims which define the function of the current project and anticipated future expansion. This plan should be established by the project team at the first project meeting as a guide for further work.
2. The project plan must clearly define project limitations for each element of the project activity involved.
3. The project plan must define critical design requirements relating to space, access, applicable codes, permits required, potential problem areas, and alternate strategies for resolution.
4. Elements such as existing site plans, zoning information, and geotechnical and other site data should be obtained at the outset if this information is critical to project scope and can be incorporated into the plan.
5. All constraints, such as code restrictions, timing restrictions, site access restrictions, and long-lead delivery items which must be selected early, should be clearly identified in the project plan.

Administration of the Project Plan

1. The project manager should develop a distribution list for the plan and include client, team members, and consultants.
2. The project plan must be reviewed and updated regularly.
3. Responsibility for adherence to the project plan by all subcontractors is essential, with each responsible for its own activities and plan elements.
4. A thorough check of the written project plan should be undertaken at the completion of each phase, along with approval by the client.

The concept of a project plan is new to many contractors, but it has been used increasingly on large and complex projects. It is a valuable tool to provide creative inputs from subcontractors and all project team members. Perhaps its greatest value is to coordinate interfaces where the finished part of one construction activity must be built on by another team member.

Managing Budget and Schedule

The project plan defines the procedure for accomplishing the project by establishing specific responsibilities and constraints on each construction activity within the overall effort. Budgeting worker-hours and setting timetables to accomplish each task is the next element to assure a quality project.

Overall Budget

1. The total cost must be apportioned to each activity and subcontractor in terms of the design elements. A worker-hour budget is more reliable than a cost estimate in controlling costs as long as hourly rates are accurately determined and do not change during the project.
2. Appropriate overhead cost and profit must be added to the job cost to arrive at total bid price. Only by controlling both direct cost and indirect costs (e.g., purchases, travel, expenses) can the project manager assure overall budget compliance.

3. Unscheduled overtime or unanticipated expense as well as project delays must be minimized if the budget is to be successfully met.
4. Changes in scope of the project plan which are directed by the client and agreed to by all parties must be documented. Appropriate adjustments to the total cost and budget must be made immediately and documented to avoid subsequent billing problems and misunderstandings.

Time Schedule. Clients usually are eager to complete a project and realize its benefits as soon as possible. But a realistic rather than an optimistic schedule should be established. Sufficient time for regulatory approvals and anticipated changes must be allocated to avoid disappointment. Wherever possible fast-tracking may be employed to improve schedules, but the added coordination required to do so must be factored into the construction costs.

1. Owner requirements in timing must be clearly understood at the outset. Where financing is governed by meeting certain deadlines, the schedule becomes especially important, and all concerned must understand this at the outset.
2. Reliable scheduling requires that other in-house commitments be accurately predicted and factored into the schedule being prepared. Priorities must be established and adhered to if valid schedules are to be developed. Each construction activity or subcontractor must maintain an updated worker-hour load estimate to accomplish this goal. Being overextended is one of the principal reasons for failure to perform quality work.

Administering the Schedule and Budget

1. At the kickoff meeting for the project, the team will be given the final schedule and budget for the project along with the project plan.
2. Each discipline must confirm the final budget and schedule and commit to them as work begins.
3. The project manager and team leaders will monitor the budget, schedule, and project plan weekly to assure acceptable performance in each construction activity.
4. It is essential that project control eliminate any short-cutting of necessary checking time, resulting in undetected errors. Wherever possible, cross-checking for consistency will be a part of quality management.
5. Each activity interfacing or building on the work of another must be allocated time to check and/or verify the work preceding theirs to avoid errors. This is especially necessary in dealing with subcontractors outside the firm.
6. Small projects must be handled carefully with special attention to details since frequent meetings are less likely to occur.

Project Phasing and Review

Large projects should be divided into phases with sufficient review at the end of each phase to assure adherence to the project plan and satisfying client goals and technical requirements. Phases proposed here are general in nature and will vary from one project to another. Billing agreements may be keyed to phasing or independent of it, depending on circumstances.

Typical Project Phases

1. Site preparation and excavation
2. Foundation work
3. Structural work
4. Utilities and building systems
5. Walls, floors, and enclosures
6. Interior and exterior finishes
7. Grading, paving, and landscaping

Each phase has some impact on those that follow. Changes, alterations, or substitutions of materials must be coordinated with the design team, and alternative analysis must sometimes be employed. The generous use of photographs, video imaging, and first-hand site observations are key elements to success. Visual data not only enhances the quality of the evolving project plan but serves to document each phase of the work.

As each phase is completed, a review of actual-versus-budgeted costs allows adjustment to subsequent phases and highlights areas where changes in approach must be made. This can result in restructuring the project team, changing coordination procedures, or liaison with the client and/or the design team as the project plan is updated. In essence, learning from that which has been completed is essential to continuous quality improvement.

Project Control

As projects expand in scope and size, the project team may include many people and a variety of construction phases. Orderly procedures must be established and followed to avoid confusion and assure good communication as the project progresses. The goal is to accomplish all tasks correctly and on schedule. At the same time, there must be allowance for creative solutions and interactions which improve the quality of the constructed project.

This is an interactive process which must use resources effectively everyday within the project plan.

1. The project manager must establish milestones in each phase with the team leaders and obtain commitments from those who must accomplish those milestones on the schedule. The project plan should show time lines and graphic display of progress for use at all meetings.
2. The project manager should establish a flowchart for work furnished by each activity to others and determine critical path events at the outset. This includes chronological order of all dependent activities and exchanges of information to those other disciplines. The format and accuracy of information to be exchanged between related phases must be clearly documented.
3. Project team members must be aware of the work flow to meet the requirements of accuracy, completeness, and timely delivery to others as the project unfolds. Every person working on a project should be aware of the overall scope and schedule as well as their unique contributions.
4. Regular meetings are essential to good coordination, especially when changes occur. Meetings should be carefully planned and include a prepared agenda with

handouts detailing all changes which must be understood and acted on. Resolution of problems and creative interaction are key elements.
5. The project manager and team leaders must track and document the exchange of information at each milestone. Where clients must be involved in exchange, it is wise to plan project review meetings to gain concurrence or set new goals.
6. Project review meetings should be held weekly or at specific stages of completion, whichever is more cost effective. These meetings must have a prepared format and be brief but also encourage exchange of ideas to enhance quality and improve schedule.
7. Major projects should be reviewed by senior management at key points of phase completion to assure that the proper resources are being allocated. Both schedule and budget should be addressed, as well as technical challenges and staffing.

Checking Procedures

The pressures of schedule and complexity of projects today require technical solutions within ever-diminishing time frames. One of the areas which is always vulnerable to cost cutting and time saving is adequate checking time. It is essential that the schedule reflect the time to check completed work and to verify independently the validity of information exchanged between construction activities.

Checking Contract Drawings

1. All critical dimensions and calculations must be checked for completeness before beginning construction. Overall dimensional consistency must be verified at this point as well.
2. Lead engineers in each activity must be responsible for checking all material schedules, design constraints, and typical details within that discipline.
3. Checking by each team member as work is performed should include dimensions, quantities, and locations. Consistency in format of documents is important.

Checking Contract Specifications

1. Specifications should begin being reviewed early in the bid process with an accumulating notebook of critical items in the project.
2. Clarification of any items that are inconsistent or unclear is of the highest priority.
3. Manufacturer's specifications should never be accepted without thorough review and understanding of the details. Use of the same terms, symbols, and logic as the design drawings is essential to avoid confusion.
4. Lead engineers in each activity must review and understand the technical specifications for that discipline to check accuracy and consistency with other activities.
5. The project manager has overall responsibility for checking the accuracy and completeness of technical specifications. Careful evaluation of all substitutions and consistency of wording and format are essential.

Checking Shop Drawings

1. The lead engineer in each construction activity must check shop drawings with design assistance as needed.

2. The contractor must complete a check of the shop drawings before they are submitted to the designer. Approval by the contractor must be a requirement for submission of shop drawings to designers to expedite the process.

Checking Budget and Schedule

1. Periodic review of the budget and schedule should be accomplished at regular intervals during the project to assure compliance. Frequency depends on the project size and duration, but weekly review is suggested on most projects of longer duration.
2. Outside consultants must be included in all reviews of budget and schedule. The same standards of control and quality performance must be established with suppliers and subconsultants as is expected from the contractor's own construction team members.

Checking is one of the best measures to pick up errors and assure quality in the constructed project. Frequent visits to the job site and spot checks on progress are valuable to this process. Critical information and details should be checked independently by more than one person.

Completed Project Evaluation

When the project is constructed, it is important to provide a smooth transition from construction into owner operation, with good planning as the key element. This requires attention to detail and coordination of the contractor completion phase with the owner/operator moving on-site. Several steps must be undertaken to make this transition painless.

Evaluation by the Owner

1. Punch list items must be corrected and all necessary adjustments made to start up the operating systems. These services may be carried on by others and negotiated separately, but coordination by the construction representative is the key to quality performance at this stage of the project. Punch list items should address:
   a. Guarantees
   b. Certificates of inspection
   c. Schedules of maintenance
   d. Operating instructions
   e. Maintenance stock
   f. Record drawings
   g. Bonds
   h. Access control-security
2. Start-up services may include operator training, development of operations and maintenance manuals, and testing of systems to assure proper efficiencies. The extent of start-up services depends on the complexity of the project, but a quality project requires critical attention to detail and good planning of this phase.
3. A formal walk-through should be scheduled with the owner by the contractor to assure that aesthetic and functional goals are satisfied at the final project evaluation stage.

Evaluation by the Design Team

1. The project manager and design team are expected to analyze the performance of the construction team as to:
   a. Quality of work
   b. Meeting schedule requirements
   c. Meeting design requirements

   A final design review meeting is effective in identifying problems and making corrections.

2. Contractors should assist designers in providing as-built data to be recorded for future reference as part of the project. Information provided by systems suppliers should be provided in duplicate so that the owner and design team may both retain copies for future reference.

Client Follow-up.   A true test of any project is the test of time. A six-month follow-up with the owner can be a valuable experience to assess systems performance. It also provides the opportunity for continuing contact with valued clients.

CONTINUING EDUCATION IN TQM

Total Quality Management requires a dynamic approach to improving performance on a day-to-day basis. It requires continuous attention at all levels of the organization. Procedures and techniques must be continuously tested and upgraded as change occurs. Every employee is part of the process and must be empowered to work for better quality. Training is the basis for accomplishing that objective.

1. Policy and procedures are in place to support the Total Quality Management approach. Those policies and procedures must be made available to all employees and constantly updated as improvements are developed and tested.
2. A quality committee should be put in place and meet monthly to act on quality issues. Regular lunch-time forums addressing specific quality issues should be planned by the committee.
3. Regular seminars are suggested for all key employees with topics focused on improving business performance.
4. The quality committee should serve as the focus for coordination of training in quality improvement and concepts related to improving quality, but managers should also be encouraged to create an atmosphere which empowers all employees to suggest, work toward, and help implement quality improvements concepts in their day-to-day duties.

ESTABLISHING QUALITY TEAMS

When an area of concern is identified, a quality team can be formed to address the problem and develop recommendations to solve it. The typical quality team comprises six to eight people, at least three of whom are familiar with the problem at hand. Other team members should have some knowledge of the operation, even if they are unfamiliar with the specific area of concern. It is good to involve people who are involved in similar activities, even if they are unrelated. The goal is to bring creative ideas, objective thinking, and teamwork to the solution. The team elects their facilitator and record keeper to keep the meetings on target and on time. Everyone is encouraged to contribute their ideas, concerns, and honest opinions to solve the problem. The facilitator makes sure all are heard and keeps the group moving toward solutions rather than conflicting positions. The record keeper keeps the notes on points of agreement and keeps time to prevent the meeting from turning into an open-ended gripe session. Discussion can be wide open without fear of it being recorded. Only areas of agreement and proposed actions by the team are recorded.

Quality teams meet as frequently as they feel necessary to reach their objective. Often additional information must be gathered between meetings to better understand the problem. Team members are then assigned specific tasks in gathering information and performing analysis as the processing unwinds.

A typical quality team meets 6 to 10 times over three months, depending on the complexity of the problem at hand. A typical meeting agenda follows.

TQM Agenda Sheet

Implementing Recommendations. When a quality team completes its work and reports its recommendations to management, this should be done by the team in presentation. The managers responsible for the activity being addressed should be made aware of how the recommendations were developed and have the opportunity to apprise the quality team of related constraints if they have been overlooked. The manager then has the option of implementing the actions immediately or asking the team to address the added constraints. It should be emphasized that the ultimate responsibility of the quality team is to make intelligent recommendations, not to implement them. Implementation is the responsibility of management. Conversely, failure to implement well-conceived and positive recommendations will undermine TQM, since team members then feel that their work is ignored. Truly, TQM is everyone's responsibility.

REFERENCES

1. W. Edwards Demming, *Out of Crisis,* MIT-CAES, 1985.
2. J. M. Juran, *Juran on Leadership for Quality,* Free Press, MacMillan, New York, 1989.

CHAPTER 9
PARTNERING

Charles Halboth, P.E.
O'Brien-Kreitzberg, Inc.
Pennsauken, New Jersey

INTRODUCTION

Construction contracting has been frequently compared to marriage. Two parties with potentially different backgrounds, values, and goals agree to live and work together for mutual benefit. Sometimes it results in a good situation, where both parties benefit and prosper both individually and collectively. All too often, individual problems and differences grow into prolonged conflicts, where one or both parties suffer and lose as a result of the association.

Modern construction contracting can bring dozens of major participants together over the life of a project. Each party brings its own set of philosophies, priorities, and values. As each party pursues its individual goals, miscommunication, specific problems, and mutual distrust can form the basis for the development of an adversarial relationship that is ultimately detrimental to all involved. This type of relationship is reflected in construction delays, poor-quality workmanship, cost overruns, inability to resolve disputed issues, and, ultimately, costly and time-consuming litigation.

Over the past several decades, there have been a lot of bad marriages in the construction industry. On many projects, the effort spent developing support for a position or defending a claim seems to exceed the effort spent in planning and supervising the work. There are many reasons that at least partially explain this extremely contentious and litigious environment:

- *Fragmentation.* On a typical project, an owner relies on a wide variety of consultants and subconsultants to provide engineering and support work that allow the project to be constructed. Similarly, the construction prime contractor(s) rely on many subcontractors and suppliers for the materials and expertise that they cannot provide. Other important participants typically include those with financial or regulatory interests. This multitude of participants virtually guarantees a wide variety of miscommunication and varying interpretations of technical and commercial issues.

- *High risk.* Both owners and contractors face large risks in today's construction market. Both attempt to shift as much risk as possible to the other. The threat of litigation has resulted in the development of onerous and restrictive contract documents. Environmental concerns, project safety, third-party liability, governmen-

tal regulation, and similar issues can dramatically affect the outcome of a project for some or all of the participants.
- *Economic pressure.* Market uncertainties, high rates of business failures, and minimal profit margins have placed contractors, suppliers, and consultants under severe economic pressure. Tight budgets and economic pressures can adversely influence the quality of an owner's, contractor's, or consultant's work effort.
- *Tendency towards dispute.* A typical construction project attempts to do something never done before with a team that has never worked together. The members of each team commonly mistrust each other and possess preconceived notions regarding capability and integrity and little, if any, understanding of the motives and goals of their counterparts.

Partnering is an attempt to move away from the litigious nature of the recent past. It is premised on the philosophy that mutual cooperation will reduce the risks inherent in the construction process for all parties. It is a process that promotes teamwork and trust and focuses on resolving problems and developing attitudes and processes that will increase the likelihood of a successful "marriage."

WHAT IS PARTNERING?

Partnering is a cooperative approach to contract management that can reduce costs, litigation, and stress, as well as improve project quality and time of delivery. The objective of partnering is to encourage the participants in a construction project to abandon their traditional adversarial approach and substitute a cooperative, team-based management approach.

Partnering has been defined as:

> *a long term commitment between two or more organizations for the purpose of achieving specific business objectives by maximizing the effectiveness of each participant's resources. This requires changing traditional relationships to a shared culture without regard to organizational boundaries. The relationship is based on trust, dedication to common goals, and an understanding of each other's individual expectations and values. Expected benefits include improved efficiency and cost effectiveness, increased opportunity for innovation, and the continuous improvement of quality products and services.*[1]

Partnering is not a new concept. Many companies in the construction industry have long-term relationships with owners, subcontractors, suppliers, consultants, and other businesses that contain many elements of partnering. These relationships continue because they are found to be an efficient way to execute projects and are beneficial to both parties. The experiences and benefits of such relationships have led to the gradual evolution of what is now called partnering.[1]

Private-sector partnering generally involves a long-term commitment between an owner and a construction firm to meld together to achieve complementary objectives on a project or a series of projects. This long-term relationship allows both parties to work more effectively and efficiently. Through improved cooperation and sharing of resources, both parties win on individual projects through reduced costs, improved schedule performance, and better quality. Overall, both parties can achieve long-term benefits from work-load stability and reduced overhead and administrative costs.[1]

Public agencies are typically required to use a competitive, low-bid contracting process for construction projects. Consequently, it is not possible to establish the long-term, multiproject relationship that commonly characterizes the private sector. In the public sector, partnering usually begins after an individual contract is awarded and focuses on creating an atmosphere that is conducive to enhancing communication and minimizing disputes. The ultimate goal of public-sector partnering is to eliminate the traditional "us-them" mentality, replacing it with an emphasis on teamwork and cooperative problem solving.

Partnering offers many opportunities for the improvement of quality and cost-effectiveness, while promoting innovation and teamwork. It requires commitment, the right situation, and the right chemistry between organizations and individuals for it to succeed. Partnering is an attitude. It is a way of doing business that recognizes that owners, contractors, and consultants have common goals which can be achieved through mutual cooperation and communication.

Partnering Fundamentals

The effective implementation of partnering requires basic changes in the way owners, contractors, and engineers approach projects. These changes include the following.

Development of a Team Approach. Partnering, as discussed here, is not a legal relationship. The particular design, construction, or consultant contract defines the legal relationships, obligations, and responsibilities, and allocates risk among the parties, but it is silent on one of the most important components of a successful project: how the major participants interact. The partnering process attempts to establish working relationships among the parties (stakeholders) through a mutually developed, formal strategy of commitment and communication.[2]

The need for the establishment of a game plan for interaction among the major project stakeholders is underscored by the large number of participants on even a modest construction project. As Figure A9-1 illustrates, many of the most important working relationships exist in the complete absence of a contractual relationship.

Partnering involves the development of a cooperative management team composed of key players from each organization involved in the project. The team identifies all the individual goals that constitute a win and refines them into a group of common goals and benefits that are to be achieved through project execution. The team then develops specific processes and procedures to achieve these goals. The critical point is that contractors, owners, and engineers are able to jointly develop and mutually agree to a unified set of goals that will constitute a successful project.

How these many team members interact and work together can often be the difference between a successful and unsuccessful project. Partnering can provide a management plan that cuts across organizational and contractual boundaries. It addresses issues of concern for the project and the individuals involved and provides strategies and guidelines for the total project management effort.

Promote the Win-Win Attitude. This emphasis on common goals does not eliminate conflict altogether, but serves to focus effort on problem solving rather than posturing and litigation. It helps eliminate the "us-them" mentality and promotes forming a "we" attitude on the part of a project team. It reinforces the notion that a good job is good for everyone, while a bad job is good for no one.[3]

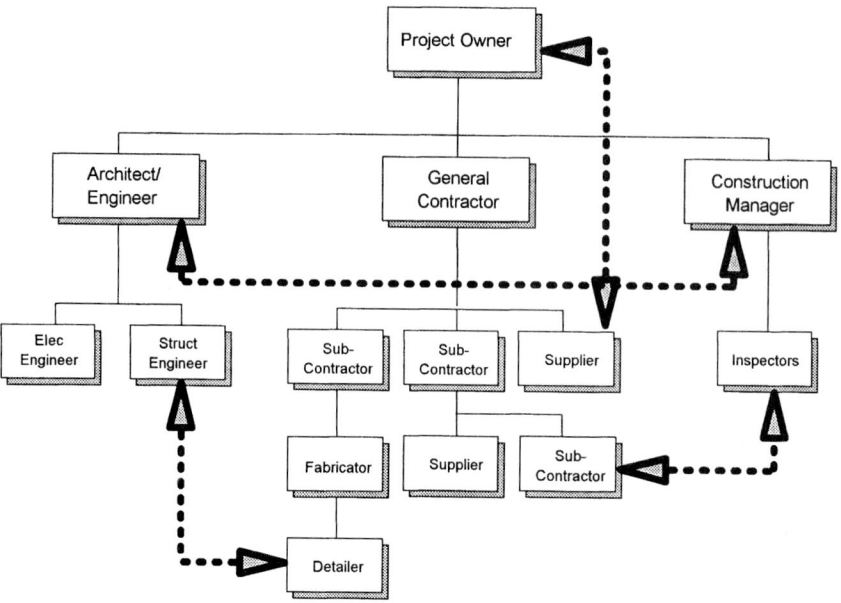

FIGURE A9-1 Typical project organization/key relationships.

Preemptive Dispute Resolution. The best way to resolve any dispute is to prevent it. Partnering seeks to reduce the time and effort spent building a case and divert that effort into constructive resolution of problems. Partnering recognizes that difficulties and disputes will occur on virtually all construction projects, and that successful projects are distinguished by how their problems are actively managed to a successful resolution for all involved. Partnering promotes the following general approach to long-term dispute resolution and prevention:

- Gain a better understanding of all relevant points of view.
- Do not to allow individual disputes to poison the entire relationship. If individual problems cannot be resolved, their effect should be minimized and not allowed to impact other elements of the project.
- Promote a managed resolution to contentious issues.
- Reduce counterproductive posturing and disputes.

Early implementation of partnering recognizes that the investment of a modest amount of time and effort in preventive measures early in the project can result in large dividends and savings throughout the life of the venture.

Proactive Management Agenda. Every project is unique. Each project team contains a unique blend of strengths and weaknesses. Through partnering, a unified *management agenda* for a project is formulated that serves to:

- Establish lines of communication.
- Develop specific, project-sensitive procedures.

- Clarify roles and responsibilities.
- Identify and troubleshoot critical project elements.
- Promote effective decision making.
- Develop creative solutions to problems.

Develop Project-Specific Processes. Partnering is not a set of abstract or generic, textbook solutions to all the problems of the industry. It is the opportunity for all parties involved to positively confront the difficult and contentious elements of their specific project and develop definite processes and solutions that will maximize the chances for overall project success.

Partnering Benefits[2]

Some of the benefits that can accrue to the different participants in a partnering program are as follows.

Benefits to the Project Owner

- Reduced exposure to litigation through open communication and issue-resolution strategies
- Lower risk of cost overruns and delays because of better time and cost control
- Better-quality product because energies are focused on the ultimate goal and not misdirected adversarial concerns
- Expedited performance of contract through efficient execution and administration
- Reduced administrative costs via the elimination of case building and other costs associated with traditional dispute resolution (legal and expert fees, staff time, etc.)
- Increased opportunity for innovation

Benefits to the Contractor

- Reduced exposure to litigation
- Increased productivity through elimination of posturing and case building
- Expedited decision making
- Lower risk of cost overruns and schedule delays
- Increased opportunity for a financially successful project

Benefits to A/E and Consultants

- Reduced exposure to litigation
- Minimized exposure to liability for document deficiencies through early identification of problems and cooperative, prompt resolution of problems
- Enhanced role in decision-making process as an active team member, providing interpretation of design intent and solutions to problems
- Reduced administrative costs through the elimination of defensive case building and avoidance of claim administration and defense costs
- Increased opportunity for financially successful project

Benefits to Subcontractors and Suppliers

- Reduced exposure to litigation
- Equity involvement in project increases opportunity for innovation and implementation of recommendations
- Potential for improved cash flow due to fewer disputes and withheld payments
- Improved decision making, avoiding problems and saving time and money
- Enhanced role in decision-making process as active team member
- Increased opportunity for financially successful project

The considerable similarity in these individual goals forms the logical basis for a true partnership. The basic fact that we all want much the same thing out of a project seems to have eluded most of us for the last several decades. The partnering process begins to make sense when individual team members understand that their individual goals are compatible with the goals of other team members. It becomes a persuasive force when team members realize that the achievement of their individual goals is actually enhanced when project goals (and the goals of other team members) are achieved.

The benefits of partnering can extend beyond the traditional owner/contractor relationship. It has been shown to be very useful when employed by multiple governmental agencies where there is an overlap of responsibility and authority. Partnering has been applied within an owner organization, where different departments must develop a high level of cooperation in order to complete a project. Owners and engineers have successfully utilized partnering to better establish roles and responsibilities and eliminate turf wars between staff and outside consultants. The same principles and processes are applicable to all these relationships. (See Fig. A9-2.)

Key Elements of Partnering

The three key elements in a partnering relationship are *trust, common goals,* and *long-term commitment.* As these three elements are realized, other subsidiary benefits will accrue and the benefits to all members of the project team will be maximized.

Traditional	Partnering
Adversarial Relationships	Cooperative Teamwork
Suspicion	Trust & Cooperation
Assigning Fault	Resolving Problems
Disputes & Litigation	Successful Project Completion
Misunderstood Requirements	Open Communication

FIGURE A9-2 Traditional versus partnering relationships.

The concepts of growth and continual improvement are basic to the partnering relationship. As such, it provides an ideal atmosphere for the introduction of Total Quality Management. Even in the public sector, the element of long-term commitment may provide substantial competitive advantages in obtaining future work for both contractors and consultants. Trust can promote a greater use of cost-plus contracting for certain elements of work. The sharing of common goals can result in a positive work environment that promotes innovation and high productivity.

Partnering generates a complex set of relationships that require a great deal of attention and maintenance. For those willing to dedicate the effort, there are substantial benefits. The key elements of partnering include the following.

Long-Term Commitment. The commitment to make partnering work must originate with top management. Project staff must see that it is OK to trust and cooperate with all members of the team. The benefits of partnering may not be achieved quickly. It may be difficult for some individuals or organizations to make the adjustment. A long-term commitment to the concept generates the atmosphere needed to honestly address issues and obstacles, promote an atmosphere of constant improvement, and allow team members to build on successes.

Trust. Teamwork is not possible where there is cynicism about others' motives. Trust permits the combining of the resources and knowledge of each partner for the maximum benefit of the project. The development of personal relationships and straightforward communication fosters an atmosphere of trust.

Mutual trust must be established to a much greater degree than is common in traditional contracting relationships. The first step in this process is to trust in the terms of the partnering agreement.

Common Goals. Through the partnering process, project participants identify individual goals which are common to all members of the project team. These jointly developed, mutually agreed upon common goals may include completing the project ahead of schedule, expediting technical review turnaround, containing costs, no lost-time injuries, reducing paperwork, or any other goals that are specific to the nature of the project.

The project goals should be achievable, quantifiable, and clearly understood and embraced by all members of the team. The project goals are the cement that holds the team together and the benchmark against which ultimate success will be measured.

Synergism. Freed from the constraints of adversarial relationships and the need to constantly justify and defend one's actions, the project team is able to concentrate on completing the project. An open exchange and consideration of ideas can result in creative solutions, increased productivity, and dramatic successes. With success comes a sense of accomplishment and pride that develops among the project team promoting even greater effort and a can-do attitude that is critical to project success.

Open Communication. Open and objective communication concerning progress, problems, and disputes that will undoubtedly arise will promote resolution and strengthen the partnership.

Continuing Improvement. A dedication to continuing improvement involves two key elements:

- All parties to the partnering agreement must realize that it is an evolutionary process. All must work toward continuous improvement if the program is to succeed. Lessons learned should be documented and shared.
- All team members can contribute to improved project performance and are urged to make suggestions pertaining to any technical or administrative issues.

Periodic Joint Evaluation

Evaluation of total project performance is vital. Traditional project evaluation is a one-dimensional top-down perspective:

- The owner evaluates contractor schedule performance.
- The contractor evaluates labor productivity.

In a partnering relationship, evaluation of project goals is a cooperative effort, performed by all project stakeholders jointly. The performance of all parties (owner, contractor, engineer, etc.) are jointly reviewed. In this environment, if certain goals are not being achieved, the ultimate reasons for the problem can be determined and corrective action initiated as necessary. The joint evaluation is necessary to ensure that the process is being implemented and that all stakeholders are carrying their fair share.

The evaluation should be a formal process, agreed on early in the project. The areas of evaluation, quantification of progress and relative value of each area must all be agreed to by the team. Goals that are easy to quantify are preferable, but most evaluations will be a combination of hard and soft goals. A brief written summary of the evaluation that includes a discussion of accomplishments over the current period and suggestions for improvement over the upcoming period should be included.

Partnering Implementation

Partnering requires considerable time, effort, and commitment at all stages of a project. While the following section describes a typical partnering program (see Fig. A9-3), it should be remembered that, at the core, partnering is an attitude that can be expressed and reinforced in any number of ways that are suitable to the specific participants and project conditions. Implementation typically progresses through the following stages.

Initiation. Early implementation is considered very important for a successful partnering program. Typically, there is an introductory period, where the concept of partnering is introduced to key project participants. A prepartnering meeting is commonly held with the executive management of major project participants. At this time, the need for partnering is discussed and specific implementation details can be planned. At least one public-sector owner includes a statement about partnering in their bid documents and initiates the process as soon as a successful bidder is selected.

Several key facets of partnering must be emphasized during this phase.

- *Partnering is voluntary.* The expenses of initiating and maintaining the process should be equitably shared among all the participants.

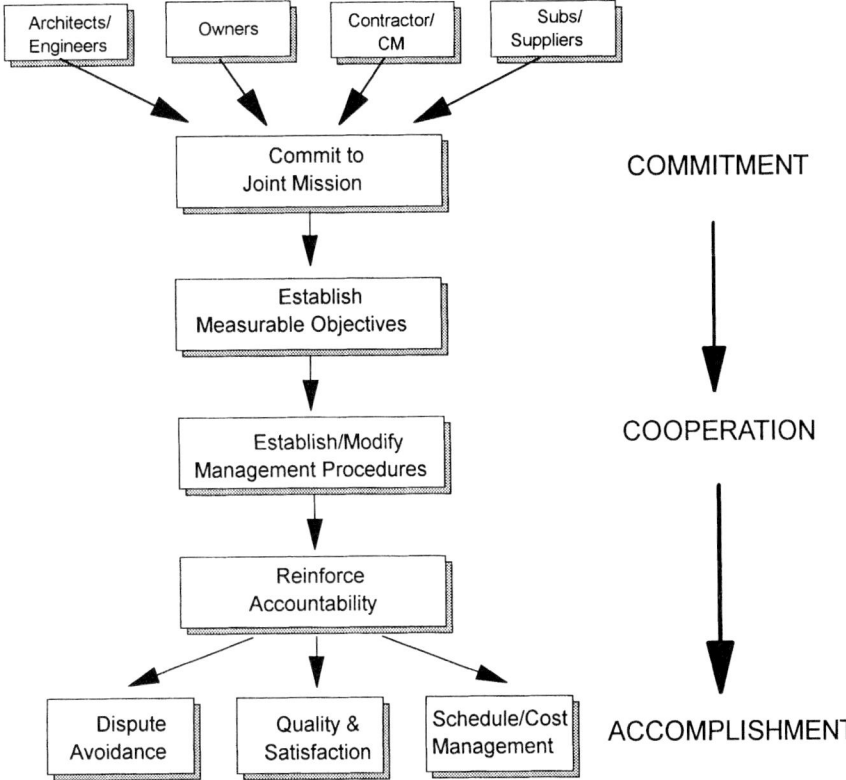

FIGURE A9-3 The partnering process.

- *Obtain commitment and active participation from top management of all participants.* This active participation of top management sets the tone and signals that it is OK to become involved in a process that for many will require a major change in attitude and approach.
- *Select workshop site and participants.* Identification of key organizations and individuals who are project stakeholders is undertaken. Team should be balanced so that neither contractor, owner, nor any other major interest feels outnumbered or not an equal in the process. The total size of the group must also be kept in mind. In general, the larger the group, the less efficient the process becomes. The workshop should be held at a neutral site away from the job site and the offices of any of the participants.
- *Select a facilitator.* The use of an outside facilitator is a decision to be made by the participants. The facilitator is someone who has no vested interest in the decisions made at the workshop, who will manage the workshop process and foster and promote the decision-making process while remaining neutral on the subjects being deliberated. This approach should be considered carefully. Factors to consider include the previous experiences of the group members with partnering, the size of the group, availability, and cost. It is important that all parties see the facilitator as

a neutral party in the process. A good facilitator is someone skilled in team building and group dynamics, with practical experience in the partnering process.

The Partnering Workshop. Partnering is typically initiated via organized workshops and follow-up meetings attended by key participants in the project. The initial workshop is usually held at a neutral location, away from the project site or the offices of any of the participants. Sessions can run from two to five days. Time should be allotted for social activities to permit the individuals to develop personal relationships and a better understanding of each other.

The workshop typically consists of large and small group discussions, personality awareness, and team-building exercises through which the common goals and objectives of the participants are identified. The workshop is the mechanism through which partnering is introduced to the majority of project participants. The ultimate goal of the workshop is to help eliminate the "us-them" type of mentality and develop project teamwork and a "we" attitude.

Typical products of a partnering workshop include:

Partnering Charter. The partnering charter is one of the key products of the workshop. The charter refines and records the common goals and objectives developed during the workshop. It should contain project-specific, quantitative (if possible) performance goals for the project. The charter not only is a symbol of the stakeholders' commitment to partnering but can also be used as the baseline against which implementation and success of the project can be measured. This charter is completely within the limits of the applicable contract(s) and governing procurement regulations.

The official signing of the charter, after the personal interaction necessary for its development, is an important formalization and commitment to the process. A typical partnering charter is shown in Fig. A9-4.

Problem Solving. At some point in the workshop, a list of project-specific problems or potential problems is developed. Issues that can be addressed range from

---

Our mission is to ensure that all parties involved achieve a successful project. We define a successful project as one which is completed on time, within budget, of highest possible quality, accident free, and satisfies both personal and professional goals. As professionals, we commit to cooperative relationships based on respect, trust, and honesty in order to achieve this mission.

Our project goals are as follows:

1. Improve time of performance on issue resolution.
2. Achieve critical schedule milestones
3. Reduce overall project duration by 60 days.
4. Streamline the management process by defining clear lines of project responsibility and reduce unnecessary paperwork.
5. Make a profitable and claim free project.
6. Maintain cost growth (change orders) to 5% or less of original contract value.
7. Foster an environment for quality and safe construction.
8. Strive for good public perception and enhance professional reputations.

FIGURE A9-4 Typical partnering charter.

general procedures such as change-order processing to specific matters such as checkout and acceptance of elevators. In smaller problem-solving sessions, teams composed of representatives from all stakeholders discuss individual problems and the manner in which they would like to see them handled. When each team completes its study, it presents its findings and recommendations to the entire workshop.

These sessions can result in any of the following:

- Revisions to management procedures
- Better understanding of technically complex issues
- Clarification of intent or expectation
- Dedication of additional resources to specific tasks
- Commitment to follow-up and further study

Partnering emphasizes decision making at the lowest possible level but recognizes that this cannot always occur. A vital component in problem resolution is an escalation process, where problems that cannot be solved at one level are concurrently elevated to the next-higher level of management within all organizations involved. This process has several advantages:

- Maintains the focus on the issues
- Helps to contain problems and minimize their impact on other portions of the project
- Provides a managed alternative to a deadlock situation
- Eliminates personality conflicts that may be an obstacle to resolution

Most important, the escalation process emphasizes that when disagreements or problems are encountered *no action is not an option.*

Follow-up.   One of the most difficult aspects of a partnering program is the follow-up phase. During this time, the team must evaluate its performance, identify shortcomings and problems, and initiate any corrective action that may be required. It requires additional work and is never a popular task. It also serves to reinforce the commitment of the team members and publicize the positive results that have been achieved.

Monthly meetings to discuss the status of the partnering effort are a common follow-up methodology. Similarly, partnering can become a regular agenda item on any and all regularly held progress meetings. Evaluation questionnaires should be sent to the project stakeholders at least one week before the meeting and returned in time to be tabulated and the results presented at the meeting.

Top management involvement is equally important during this phase. Many team members will require the reinforcement provided by top management involvement to maintain commitment to the process.

Celebration.   Partnering tends to humanize a project. Personal relationships are formed and a sense of project identity and unity is developed. A natural by-product of this process is the desire of the team members to socialize and celebrate significant accomplishments or achievements. This is one of the best methods to reinforce the concepts of partnering and foster its acceptance among all team members.

The up-front commitments and cost associated with a partnering agreement suggest that it is most appropriate for large, complex, high-risk, or otherwise sensitive projects. However, it must be remembered that partnering is primarily an attitude, not a sophisticated or mysterious process. The concept can be applied on a low-cost basis. In its most basic form, it is a personal commitment between project personnel to maintain open communication and work for the overall benefit of the project.

Tangible Results

The formal concept of partnering emerged in the mid-1980s. There have been few studies documenting the results of projects where partnering has been employed versus the results where it has not been utilized. One study, performed in the summer of 1992 by the Army Corps of Engineers,[4] compared 19 projects where partnering was employed against 28 projects where no form of partnering agreement was utilized. Measurable results that were compared include cost change, change-order cost, claims cost, value engineering savings, and duration change. These results are summarized in Table A9–1.

The results of this study indicate that partnered projects tend to perform better than nonpartnered projects in the areas of overall cost containment, schedule performance, change-order cost, and claims cost. The study also identified certain intangibles that were considered beneficial:

- Reduced administrative paperwork
- More enjoyable work environment
- Reduced communications barriers
- Less adversarial relationships

The Army Corps of Engineers study quantifies and verifies what has been the overall experience of the industry: Partnering can be a very effective means of improving overall project performance.

SUMMARY

Partnering is not a quick fix for all the ills of public-sector construction contracting. Partnering does not affect any contract terms or government procurement regulations. Partnering seeks to change attitudes and promote a managed method of

TABLE A9-1 Partnering/Nonpartnering Comparison

Comparison variable	Partnering (16 projects)	No partnering (28 projects)
Mean cost change, %	2.72	8.75
Mean schedule change, %	9.07	15.53
Mean change orders, %	3.89	7.74
Mean claims cost, %	0.67	5.01
Mean value eng savings, %	0.73	0.05
Mean contract award price	$10,368,643	$11,448,745

resolving problems and achieving successful project completion. It is not new and it is not mysterious. The process requires commitment, trust, and a willingness to experiment with a different working relationship among owners, engineers, and contractors.

The ultimate benefits of partnering include:

- Expedited decision making with project-specific issue and problem resolution strategies
- Minimizing project misunderstandings, tension, and administrative case building through open and frank communication
- Improved chances of financial success and schedule adherence for all involved

REFERENCES

1. "In Search of Partnering Excellence," Special Publication 17-1, Partnering Task Force, Construction Industry Institute, Austin, Tex., July 1991.
2. "Partnering: A Concept for Success," The Associated General Contractors of America, Washington D.C., Sept. 1991.
3. David P. Johnson, "Partnering, Who Cares?" *Construction Business Review,* September/October 1992, pp. 46–48.
4. David C. Weston, and G. Edward Gibson, "Partnering—Project Performance in U.S. Army Corps of Engineers," *Journal of Management in Engineering,* ASCE, Oct. 1993.

CHAPTER 10
SAFETY

Richard D. Conner, Esq.
Patton Boggs, L.L.P., Greensboro, North Carolina

From ages 1 through 44, injury is the leading cause of death in the nation. The cost to our nation exceeds $100 billion annually. Every day of the year an average of 32 workers die on the job, and 5500 suffer a disabling injury.

INTRODUCTION

Injury and death are tragic facts of everyday life in the construction industry. Thorough steps must be taken to ensure the safety of workers. This chapter explores (1) the legal sources of liability for safety and provides a legal perspective for understanding the benefits and risks of taking a proactive role in safety matters during construction and (2) risk management options and procedures.

DEFINITIONS

The term *construction manager* as used in this chapter refers to both an independent professional construction manager acting as the owner's agent that does not hold the construction subcontracts and is not responsible for construction means and methods, as well as a general contractor holding subcontracts that is responsible for construction means and methods and is managing the construction phase.

The term *proactive approach* as used in this chapter refers to both the professional construction manager and the general contractor as follows.

1. For the professional construction manager, it is those activities through which the construction manager meets the safety liability risk head on with a proactive approach, positively acting to provide such services as:

    preparing the project safety and health program; training contractors and their workers about safety matters

    monitoring and inspecting the contractor's compliance with safe working procedures

    maintaining safety records for the project

notifying, or whistle blowing, with respect to unsafe working conditions

advising the owner regarding disciplinary action necessary to correct unsafe conditions

2. For the general contractor, it is those activities through which the general contractor manages its obligation to provide and implement a safety program during the construction phase.

LEGAL SOURCES OF LIABILITY

The liability for personal injury of an employee is primarily based in negligence, and, increasingly, is imposed by statutes such as the Occupational Safety and Health Act. It is critically important to recognize that the most carefully drafted contract may not immunize the construction manager from liability to the injured worker, for the simple fact that the worker has not personally signed off on the exculpatory contract.

Most personal injury actions against a construction manager will involve traditional negligence analysis of whether the construction manager owed a duty of care to the worker, whether the construction manager breached its duty, and whether the worker was proximately harmed by the breach.

The fact that most actions are based on negligence does not mean that the contract between the construction manager and the owner is unimportant. To the contrary, the contract with the owner is usually the best evidence of the duties that the parties intend the construction manager to undertake with respect to safety. As the reported legal cases show, the most common sources of evidence on this issue are (1) the construction manager's contract, (2) the construction manager's actions (or inactions), and (3) obligations imposed by statute. By understanding these sources of liability and actively managing the drafting of contracts and performance in the field, the construction manager may exercise some control over its liability for safety.

CONTRACTUAL SERVICES THAT AFFECT A CONSTRUCTION MANAGER'S DUTY TO INJURED WORKERS

In *Wenzel v. Boyle Galvanizing Co.,*[1] the Eleventh Circuit, applying Florida law, determined that a construction manager who had contracted with an owner for professional construction management services only was liable to a contractor's injured construction workers. The workers were injured in a fall from scaffolding. The particular location where the workers were operating was extremely constricted and would not permit a safety net, although safety nets were used at other locations on the site. The workers also were not using safety belts.

According to the contract, the construction manager was required to provide, implement, and administer a site safety and health program consisting of, but not necessarily limited to, the following.

- Develop a project safety manual that will establish contract safety guidelines and requirements.
- Review project contractors' safety programs for compliance with the project safety manual.

- Provide daily surveillance of contractor work areas for compliance with the safety program.
- Develop and invoke procedures for advising contractors of safety violations and deficiencies.
- Develop and invoke procedures for initiating corrective action by the owner and back-charging the contractor, if the contractor does not comply with safety violation directives.

To fulfill these responsibilities, the construction manager had the authority to issue notices of safety violations, insist that a particular employee be removed from the job, and stop the work of a particular contractor.

In *Wenzel,* the court held that a construction manager owes a duty of care for safety to on-site workers if its contract mandates that it provide, implement, and administer a site safety and health program. The duty obligates the construction manager to perform with the diligence and skill of a reasonable construction manager under similar circumstances. Further, the court held that the traditional exculpatory clauses in professional construction management contracts imposing responsibility on the construction contractor for "means and methods" does not protect the professional construction managers from liability for an injury to a third-party worker arising from a safety violation.

The court determined that the construction manager had acted negligently because the construction manager had contract responsibilities for safety, had superior knowledge of safety measures and knew of the danger of an unprotected platform and untied safety belts. The court determined that the construction manager should have exercised its contractual rights to insist that safety nets be installed at every work site.

Cases such as *Wenzel* illustrate that the determination of whether a duty of care exists with respect to safety usually begins with a reference to the contract documents. The greater the contractual responsibility assumed by a professional construction manager for project safety, the greater the probability the construction manager will be deemed to have a duty of care to ensure the safety of workers.

It is possible to interpret cases like *Wenzel* as imposing safety liability on the construction manager because the construction manager held the affirmative power and authority to stop work, and was culpable for not exercising this power to avoid a known risk. The affirmative power to stop work has long been a source of liability for administering architects. Fear of this risk may well be the principal reason why modern editions of AIA Documents B141 (the owner/architect agreement) and A201 (the general conditions) have deleted the architect's power and authority to stop work.

For example, in addressing the application of OSHA construction standards to work performed by a design professional, the Occupational Safety and Health Review Commission in *Secretary of Labor v. Simpson, Gumpertz & Heger, Inc.,*[2] held that the design firm was not subject to the construction standards because it did not exercise substantial supervision over the construction work. The Commission declined to create an exception to the longstanding principles governing construction work sites by holding that design or engineering firms could be held responsible simply because their activities were related to the overall construction project, without regard to the extent to which they actually performed or supervised the construction work.

In a similar vein, design professionals have often been found not to have been "in charge" of the construction work and, therefore, not liable under state safe workplace statutes. For example, a design professional hired to "observe the construction for the purpose of determining [that] what is [on site] is generally in agreement with the documents" does not so involve the design professional with the construction work that it could be considered in charge of the work.[3]

It is more prudent, however, to recognize that the contractual power to stop work is one of several factors considered by the courts. The courts also note that the construction manager's superior skill, position, and construction experience gives the construction manager the professional ability to foresee harm. The broad range of administrative and management services typically exercised by the professional construction manager, including control of the scheduling and review and monitoring or implementation of safety programs, logically suggests that the professional construction manager is inescapably involved with safety, is chargeable with knowledge of job-site danger, and has powers to exercise to avoid danger, even if lacking the express authority to stop the work.

This is especially the case where the professional construction manager takes a proactive approach to safety. In other words, for purposes of assessing risks of safety liability, there may be little distinction between the express authority to stop work and the responsibility and authority to advise the owner, forcefully if necessary, to order a work stoppage. A breach of either duty by the professional construction manager can impose liability under the reasoning of current cases.

The prudent professional construction manager must recognize that courts are simply disposed to find a duty and breach where a professional construction manager was continuously on site, with knowledge of an unsafe condition and the skills to correct it, yet the accident was not avoided. Courts are not interested in contractual language limiting the construction manager's responsibility. Courts are interested in contractual language exhibiting the construction manager's participation in safety issues and affording the construction manager opportunities to cause the correction of unsafe conditions. In sum, the construction manager's contract exposes the professional construction manager to liability when it requires services related to safety, but may not immunize the construction manager from liability when it contains exculpatory language.

The construction manager has been found subject to OSHA Construction Standards despite an absence of the construction manager's performance of actual construction work. Typical is the *Kulka*[4] case in which the Occupational Safety and Health Review Commission held that the construction manager substantially supervised actual construction so as to subject it to the OSHA Construction Standards. The construction manager's contract with the owner required the construction manager to "provide recommendations and information to the Owner and the Architect regarding the assignment of responsibilities for safety precautions and programs [and verify] that the requirements and assignment of responsibilities are included the proposed Contract Documents." The construction manager also reviewed the safety programs of each of the trade contractors. In addition, the owner depended on the construction manager to maintain safe working conditions at the site. In similar cases, the Commission has held that, although construction managers perform no actual construction tasks, they have considerable authority over the performance of the work and the safety measures implemented by the contractors.[5] The construction manager's contractual administrative and supervisory functions are often inextricably intertwined with the actual physical labor performed on the job site.

Although individually tailored agreements may have a variety of provisions addressing safety, the most common standard form contracts treat safety issues in similar ways. In general, the standard form contracts provide for the construction manager to advise and/or to coordinate and report on safety issues, but make the contractor primarily responsible for maintaining a safe site. The standard form contracts do not articulate the extended scope of services provided by some construction managers engaging in a proactive approach. The standard form contracts do not address additional compensation for safety related services.

For example, AIA Documents B801/CMa (1992 edition), the Owner/Construction Manager Agreement, specifically provides in paragraph 2.3.12 that the construction manager is to review contractor safety programs and to coordinate these programs for the project. This provision, unfortunately, conflicts with Article 10 and paragraph 4.6.6 of the AIA General Conditions, Document A201/CMa (1992 edition). The General Conditions provide that the contractor must initiate, maintain, and supervise all safety programs, and that the construction manager will not be responsible for construction means, methods, techniques, or safety precautions and programs.

It is difficult to reconcile the AIA disclaimer of responsibility for safety and the AIA requirement that the construction manager is responsible to coordinate the safety programs for the project. The requirement to coordinate is sufficient to raise a risk of exposure to liability under cases like *Wenzel,* and suggests proactive steps to coordinate. Yet the AIA Documents give no definition or limit to the services the construction manager is to perform to coordinate the safety programs. Safety services are compensated as a part of basic services under the AIA Documents.

The Associated General Contractor Documents, AGC Doc. No. 8/500 ¶2.2.8.1 (1980 edition), requires the construction manager to review contractor safety programs and to make appropriate recommendations. Like the AIA Documents, the AGC Documents make it clear that the construction manager's duties do not relieve the contractors of final responsibility for safety. Unlike the AIA Documents, however, the AGC Documents give the construction manager the power to take necessary steps to ensure contractor compliance with the safety program. The AGC Documents do not define the services necessary to satisfy the construction manager's responsibilities, and do not provide for additional compensation for safety-related services.

The Construction Management Association of America's contract documents also address project safety concerns. Like other standard owner/construction manager agreements, CMAA Document No. A-1 (1993 edition) gives the construction manager authority to review the contractor's safety program, but does not make the construction manager responsible for implementation of the plan. Paragraph 3.5.1.12 of the Agreement provides:

> The CM shall require each Contractor that will perform Work at the site to prepare and submit to the CM for general review a safety program, as required by the Contract Documents. The CM shall review each safety program to determine that the programs of the various Contractors performing Work at the site, as submitted, provide for coordination among the Contractors of their respective programs. The CM shall not be responsible for any Contractors' implementation of or compliance with its safety programs, or for initiating, maintaining, monitoring or supervising the implementation of such programs or the procedures and precautions association therewith, or for the coordination of any of the above with the other Contractors performing the Work at the site. The CM shall not be responsible for the adequacy or completeness of any Contractor's safety programs, procedures or precautions.

The basic intent of this provision is reiterated in the CMAA general conditions, Document No. A-3 (1993 edition). Paragraph 2.1.6 of Document A-3 expressly states that neither the owner nor the construction manager have responsibility for the contractor's means, methods, and techniques and "safety precautions and programs." Paragraph 5.1.1 reiterates the basic intent that the contractor shall be solely responsible "for initiating, maintaining, monitoring and supervising all safety programs, precautions and procedures in connection with the Work and for coordinating its programs, precautions and procedures with those of the other Contractors performing the Work at the site."

The safety obligations for the construction manager having general contractor responsibility during the construction phase are specified in contracts to require the general contractor construction manager to plan, *control,* and conduct the physical activities necessary to provide a safe, healthy work site environment. Primary responsibility for construction safety is assigned to the general contractor construction manager by standard provisions contained in standard specifications spelling out a formal construction safety policy. The specification requirements differ by jurisdiction. The following provisions provide a few examples:

- *The United States Army Corps of Engineers, Safety and Health Requirements for all Corps of Engineers activities and operations.* The requirements are contained in the Corps Manual implementing safety and health standards and requirements contained in the Code of Federal Regulations, Executive Orders, Department of Defense, and Federal Acquisition Regulations. These standards are applicable to construction contract work under the provisions of the Federal Acquisition Regulations. Section 01.A.02 provides that the employer shall be responsible for initiating and maintaining a safety and health program which complies with U.S. Army Corps of Engineers Safety and Health Requirements.

- *Alaska Department of Transportation and Public Facilities Standard Specifications for Highway Construction (1988) § 107-1.06:* "The contractor shall provide and maintain in a neat, sanitary condition such accommodations for the use of employees as necessary to comply with the requirements of the State and local Board of Health, or of other bodies of tribunals having jurisdiction. Attention is directed to Federal, State, and local laws, rules, and regulations concerning construction safety and health standards. The Contractor shall not require any workers to work in surroundings or under conditions that are unsanitary, hazardous, or dangerous to health or safety."

- *State of California, Department of Transportation, Standard Specifications (1992) § 7-1.01:* "The Contractor shall keep himself fully informed of all existing and future State and Federal laws and county and municipal ordinances and regulations which in any manner affect those engaged or employed in the work, or the materials used in the work, or which in any way affect the conduct of the work, and of all such orders and decrees of bodies or tribunals having jurisdiction or authority over the same. He shall at all times observe and comply with . . . such laws . . ."; *§ 7-1.01E:* attention is directed to the provisions of § 6705 of the Labor Code concerning trench excavation safety plans; *§ 7-1:06:* "The contractor shall conform to all applicable occupational safety and health standards, rules, regulations and orders established by the State of California"; *§ 7-1.09:* "It is the contractor's responsibility to provide for the safety of traffic and the public during construction"; *§ 7-1.10:* "When explosives are used, the Contractor shall exercise the utmost care not to endanger life or property."

- *Department of Highways, Division of Highways State of Colorado, Standard Specifications for Road and Bridge Construction (1991) § 107.06:* "The Contractor shall observe all rules and regulations of Federal, State and local health officials. The Contractor shall not require any worker to work in surroundings or under conditions which are unsanitary, hazardous or dangerous to health or safety." Additional provisions are found in §§ 107.07, 107.08, 104.04, 104.06, 105.14, and 105.15.

- *Florida Department of Transportation, Standard Specifications for Road and Bridge Construction (1991) § 7-1.1:* "The Contractor shall familiarize himself and comply with all Federal, State, county and city laws, by-laws, ordinances and regulations which control the action or operation of those engaged or employed in the work or which affect materials used. Particular attention is called to the safety regulations promulgated by the State of Florida Department of Labor and Employment Security. . . ."

- *Hawaii Department of Transportation Standard Specifications for Road and Bridge Construction (1985) § 107.12:* "The Contractor shall provide and maintain in a neat, sanitary condition such accommodations for the use of his employees as may be necessary to comply with the requirements of the State and local Boards of Health, or other bodies or tribunals having jurisdiction. Attention is directed to Federal, State and local laws, rules and regulations concerning construction safety and health standards. The Contractor shall not require any worker to work in surroundings or under conditions which are unsanitary, hazardous or dangerous to his health or safety."

- *New York State, Department of Transportation, Standard Specifications Engineering Instruction Code EI-92-017 (4/13/92):* the NYDOT points out that the contractor is responsible for carrying out health and safety activities and that those requirements are spelled out in the Standard Specifications, and in some cases in the contract documents (such as in special notes and design details). Relevant sections include:

Sec. 107-01	Laws, Permits and Licenses
Sec. 107-05	Safety and Health Requirements
Sec. 619-1.02m	Basic Maintenance and Protection of Traffic—Project Site Patrol
Sec. 102-14	Itemized Proposal
Sec. 102-17, Art. 1	Work to Be Done
Sec. 105-02	Orders to Foreman
Sec. 105-08	Cooperation by the Contractor

Additional details are provided in EB 86-11, "Construction Division Policy Statement on Safety and Summary of Responsibilities."

The engineering instruction points out that contract compliance provisions are discussed in detail. The primary provisions of the Standard Specifications include:

Sec. 102-17, Art. 8	No Estimate on Contractor's Non-Compliance
Sec. 102-17, Art. 11	Right to Suspend Work and Cancel Contract
Sec. 103-03	Right to Suspend Work and Cancel Contract
Sec. 105-01	Stopping Work
Sec. 102-17, Art. 13	Delays, Inefficiencies and Interference
Sec. 619-5	Maintenance and Protection Traffic—Basis of Payment
Sec. 105-08	Cooperation by the Contractor
Sec. 108-03	Failure to Complete Work on Time
Sec. 108-04	Extension of Time

The policies that are established by the various jurisdictions are implemented in the construction contracts. The standard specifications are incorporated into the contract as a part of the contract documents. Also, each contract contains specific clauses dealing with safety and health requirements. For example, New York state in its Standard Form of Agreement requires the contractor to "perform all work in the contract in a workmanlike manner with due regard to the safety and health of the employees and of the public. The Contractor shall comply with Title 29, Code of Federal Regulations, Part 1926, Safety and Health Regulations for Construction (OSHA) regarding the safety and protection of persons employed in construction and demolition work." The typical construction contract also provides requirements for insurance including workman's compensation insurance and liability and property damage insurance.

In addition to incorporation of standard specifications and standard contract clauses, the typical construction contract will require the contractor to provide a written project safety and health plan documenting the contractor's policy relative to safety and identifying and addressing specific health and safety concerns to be encountered on the project. Such contractual requirements typically direct the contractor to include in its plan:

- Identification of project and company safety officers
- Hazardous materials communication plan
- Employee safety training programs
- Company safety policies
- Procedures to address project health and safety concerns
- Procedures for compelling worker compliance with health and safety requirements
- Recordkeeping

Additionally, contract provisions in connection with the safety and health plan require the contractor to ensure that each subcontractor complies with safety requirements.

In sum, it would appear that standard form agreements sufficiently involve the professional construction manager in safety matters so as to raise an exposure to liability under present case law. The standard form agreements require the professional construction manager to engage in a scope of positive action with respect to safety. Many construction managers deem it prudent to address the reality of their positive involvement with safety, and to develop a detailed scope for safety related services. These proactive services should be clearly articulated in the professional construction manager's contract. Typically, contract provisions in the general contractor construction manager's contract require the construction manager to plan, control, and ensure compliance with safety requirements. The element of control and implementation of safety obligations increases the potential of the liability of the general contractor construction manager.

ACTIONS THAT AFFECT A CONSTRUCTION MANAGER'S DUTY TO INJURED WORKERS

In determining whether the construction manager owes a duty to an injured worker, courts look not only at the terms of the construction manager/owner agreement, but also at the action (or inaction) of the construction manager in performing services.

Regardless of the services required by the construction manager's contract, if acts associated with safety are performed, the construction manager may be held to owe a duty to perform those acts with due care, or face liability to the foreseeable injured worker. Further, where a construction manager fails to act when presented with an obvious safety hazard and has the power to correct, the courts may hold the construction manager liable for failing to exercise power over the project to correct the hazard.

The chances are high that a court will find that the construction manager assumed a duty of care to an injured worker where the construction manager asserted a high level of control over the work site or gave instructions or orders involving safety matters. A proactive approach will increase the construction manager's exposure to liability.

As shown in *Hammond v. Bechtel*,[6] the court's focus on the acts of the construction manager usually accompanies an examination of contract language requiring safety-related services. For example, in *Plan-Tec v. Wiggins*[7] the construction manager had the contractual authority to direct that defective equipment not be used, and to suspend or terminate the job if a contractor did not comply with the construction manager/contractor contract. The construction manager could stop workers to "inspect" equipment. Safety problems were reported to the construction manager.

In addition to these contractual responsibilities, the construction manager had appointed a safety director, initiated safety meetings, and ordered certain safety precautions. The court held these actions to be sufficient evidence that the construction manager had assumed a duty for the overall safety aspects of the project. The court went on to state that once performance of safety measures is attempted, performance must be completed in the manner of a reasonably prudent person, or liability can be incurred.

Similarly, in *Phillips v. United Engineers & Constructors, Inc.*,[8] the court held that there was sufficient evidence to present to a jury the question of whether a construction manager had assumed a duty for the overall safety aspects of the project. The construction manager's "safety coordinator" held safety meetings biweekly for contractors' superintendents, conducted tours of the job site, during which safety violations or unsafe practices were noted, and notified the violators advising them to remedy the problems.

Conversely, in *Everette v. Alyeska Pipeline Service Co.*[9] an injured worker could not recover from a construction manager who did not retain the power to control safety procedures or revise job specifications. Here, the construction manager had only the right to direct the sequence of the contractors' performance, inspect the contractors' progress, or receive reports from the contractors.

LIABILITY IMPOSED BY STATUTE

There is a growing general awareness that construction is one of our most dangerous industries. The U.S. Department of Labor and National Safety Council statistics indicate that construction employees sustain 250,000 to 300,000 lost-time injuries and 3000 work-related fatalities per year. This accounts for 12 percent of all occupational injuries and illnesses and 20 percent of all work-related fatalities. The cost to the construction industry of these accidents and fatalities exceeds $10 billion per year.

State and federal legislatures have responded to this danger by enacting safe work and construction safety legislation, and by beefing up the enforcement of such legislation both in terms of the size of fines and the class of persons cited for violation. In general, state and federal safety legislation falls into two categories:

1. Safe workplace acts, which impose a general duty on the project owner and contractor to provide their own employees with a safe work site[10]
2. Specific duty legislation such as "scaffolding" or structural work acts, which impose on owners and contractors the duty to implement and maintain specific safety precautions for all workers engaged in work covered by the act[11]

Safety legislation presents the risk of fines or even imprisonment. Also, violation of a safety statute may impose strict civil liability.[12] At a minimum, violations of a statutory duty may be admissible as strong evidence of negligence in a civil action by an injured worker.[13]

One of the most difficult issues concerning safety legislation is the determination of who is a party intended by the legislation to be responsible for safety, and thus subject to fines or the adverse evidence of a safety violation. Most safety legislation imposes duties on "owner" and/or "contractors." It is critically important, however, for the construction manager to review the statutory definition of these terms to determine if the construction manager falls within the definition. This is particularly true of specific duty legislation, which commonly defines the responsible party in terms of control of the workplace and opportunity to comply with and enforce the statutory requirements. Either the construction manager's contract or the construction manager's actions may exhibit sufficient control and opportunity to impose liability, even where the construction manager is not the "contractor" as that term is intended by the contracting parties or understood in the industry.

STATE SAFETY LEGISLATION

Courts are showing a willingness to find that the construction manager falls within the definition of a responsible party. In *Carollo v. Tishman Construction & Research Co.*,[14] the court found a construction manager liable for the personal injuries of a subcontractor's employee pursuant to state safety law. The safety law defined *contractor* as "one who coordinates and/or supervises the project for an owner, assuming the on-the-job responsibilities of the owner as its alter-ego."

The construction manager was found to be a contractor for safety law purposes because of the duties it had contracted to perform. The construction manager was responsible for design consultation, monitoring of project costs, scheduling of design and construction phases, and reviewing the design of the project. The court held that the construction manager was to a great extent responsible for carrying out the entire job and that the construction manager therefore could not escape responsibility.

In *Nowak v. Smith & Mahoney*,[15] the court found a construction manager liable to an injured electrical worker under a state law requiring the owner or general contractor to provide a safe construction site. The court found that the construction manager was not a "contractor," but was nevertheless liable because the duty arising under the statute was based on contractual or other actual authority to control the activities that brought about the injury. The construction manager had the contractual authority to enforce safety standards and to choose responsible contractors.

OSHA

Although many states have work-site safety legislation, the best known, and most generally applicable legislation is the federal Occupational Safety and Health Act, first enacted in 1970. This Act imposes safety and health standards on industry, including the construction industry, through the promulgation and enforcement of regulations by the Occupational Safety and Health Administration (OSHA). Each state has the option of enacting its own job safety and health plan, to be operated jointly with OSHA, provided the state plan is at least as stringent as national OSHA standards. The construction manager must check to see if the state in which the project is located has enacted more stringent OSHA regulations.*

OSHA Regulations

OSHA regulates safety on the construction work site through regulations requiring:

- Establishment of safety and health programs for the project
- Implementation, monitoring, and enforcement of the programs
- Reporting and communication between employers and workers, and between the employer and OSHA
- Imposition of fines and/or imprisonment for violations of the preceding requirements

The applicability of OSHA regulations to the general contractor construction manager is not questioned. The considerable authority over the performance of the work and over the safety measures implemented by subcontractors allows for the application of OSHA regulations. The general contractor construction manager performs administrative and supervisory functions that are inextricably intertwined with the actual physical labor used to erect the improvement and is so directly and vitally related to the construction being performed that it is engaged in "construction work" within the meaning to OSHA Construction Standards.

RISK COVERAGE, IMMUNITY, AND INDEMNITY

Although the construction manager's risk of liability to the injured worker may be high, protection is available. Professional errors and omissions insurance is available to the professional construction manager to cover claims and the cost of defending claims. Under certain facts, statutory immunity may be available under state workers' compensations laws. Contractual indemnity, while not avoiding the worker's lawsuit, may provide reimbursement for sums paid the worker.

INSURANCE

Given the risk of loss posed by safety-related claims, construction managers will purchase insurance that provides coverage for job safety responsibilities. Special care should be taken to ensure that the insurance obtained will cover the risks presented by construction management work and, in particular, safety liabilities.

---

* States, territories, and commonwealths that have enacted their own plans include Alaska, Arizona, California, Connecticut, Hawaii, Indiana, Iowa, Kentucky, Maryland, Michigan, Minnesota, Nevada, New Mexico, New York, North Carolina, Oregon, Puerto Rico, South Carolina, Tennessee, Utah, Vermont, Virginia, Virgin Islands, Washington, and Wyoming.

As it grows and becomes more established as a discipline, construction management is increasingly recognized as a professional service. The courts view the liability of a construction manager for safety as based in negligence. A construction manager will also likely be held to a professional standard of care in the discharge of its duties and undertakings. Broadly speaking, this means that in order to recover on a legal claim against a construction manager, a party will have to demonstrate that the construction manager negligently violated the standard of care required of a reasonable construction manager acting under similar circumstances. The liability of a construction manager is, then, analogous to that of other professionals such as architects, engineers, doctors, and accountants.

The insurance industry recognizes this fact by endeavoring to cover construction management activities under a professional liability policy. Frequently, construction management, to the extent it is insured for, is listed among the professional services provided for under an architects' and engineers' policy. Under such an arrangement, professional liability coverage for construction management is contingent on the professional construction manager qualifying for architects' and engineers' insurance. The professional construction manager must therefore possess a professional license and be otherwise eligible for the underlying A/E policy.

Various architects' and engineers' underwriters take different approaches to construction management as an activity covered under an issued policy. Some policies may not provide coverage for professional construction manager activities. While other common wordings do extend coverage for construction management, insurers often issue guidelines or conditions that may affect the work done or the approach to management practiced by the insured.

Policy exclusions and special endorsements can also impact construction management activities. The provisions in the professional liability policy of insurance should be carefully consulted by construction managers and their counsel to ensure that coverage exists for the safety-related activities that will be undertaken on a project.

A non-architect/engineer construction management firm may experience difficulty obtaining insurance coverage for professional construction management activities. Lacking a professional license, a firm such as a contracting company will be unable to obtain the standard architects' and engineers' policy to cover construction management work. Professional liability coverage is available from a limited number of sources on a special risk basis for construction management firms lacking professional licenses. Additionally, certain underwriters at Lloyds of London may soon offer a policy designed for the non-architect/engineer construction management firm. This development may mean that professional liability coverage will become more readily available to the non-architect/engineer firms engaged in construction management, and will assist those companies in dealing with safety responsibilities and liability.

It is critically important for the construction manager to recognize that typical comprehensive general liability insurance does not provide coverage for professional activities such as construction management.[16]

CAN WORKERS' COMPENSATION LAWS IMMUNIZE THE CONSTRUCTION MANAGER FROM LIABILITY TO AN INJURED WORKER?

Generally, workers' compensation acts seek to balance the interests between a worker and an employer, and to avoid costly negligence litigation. The acts accomplish this balance by holding the employer strictly liable to the worker for job-site injuries, but limiting the worker to statutorily scheduled compensation for injuries. The employer who purchases workers' compensation insurance is liable only to provide insurance proceeds to pay statutory compensation, and is immune from a civil action in damages.

It appears that a professional construction manager may not be successful in claiming immunity as the "statutory employer" of an injured worker, simply because the construction manager is outside of the contractual chain of employment between the project owner and the injured worker.[17]

The immunity afforded the "statutory employer," however, generally is shared by all "statutory employees," who are deemed "statutory coemployees" of the injured worker. The law of the governing jurisdiction must, of course, be reviewed for modifications to this general scheme.* If the construction manager enjoys the status of statutory coemployee of the injured worker, the construction manager may be immune from a negligence action by the worker.

"Employer" under workers' compensation acts is not limited, generally, to the company that actually hires and pays wages to the injured worker. Rather, the statutory employer, for worker's compensation purposes, is the party whose trade, business, or occupation is that in which the worker is engaged.[18] Accordingly, for example, where an owner is a land developer whose trade and business is construction, the developer may be the statutory employer of a contractor's worker, liable to the worker to assure payment of workers' compensation benefits. The developer, and subordinates, including the construction manager providing services for the project, may be immune from the worker's suit.[19]

On the other hand, a textile corporation's principal trade and business is not construction, despite the corporation's continuous involvement with the construction and renovation of textile mills and plants. The corporation, generally, will not be deemed the statutory employer of construction personnel. The corporation, and the construction manager under contract with the corporation, will not enjoy statutory immunity.

A different analysis often is applied to public agencies, such as water, sewer, and transportation authorities created under statutory mandate to engage in construction. In this case, even though the agency may have no direct employees actively engaged in construction, the courts nonetheless find that the statutory mandate to construct compels a finding that the agency is in the trade or business of construction, and is the statutory employer. Under these facts, an agent construction manager under contract to such an authority may enjoy statutory immunity from liability to an injured construction worker.[20]

Where a construction manager is engaging in a private project for an owner whose principal trade, business, or occupation is construction, or a public project for

---

* Many states have statutory or case law exceptions to workers' compensation immunity. In general, the exceptions take two forms: (1) The imposition of civil liability upon the employer for an "intentional" or "willful" tort resulting in injury to the employee, and (2) various disparate treatment of the immunity afforded the fellow employee as compared to the immunity afforded the employer. For example, in *Woodson v. Rowland*, 407 S.E. 2d 222 (N.C. 1991), the court discussed the various exceptions under North Carolina law. The worker's direct employer is entitled to immunity except for intentional torts, defined to be intentional misconduct accompanied by knowledge that it is substantially certain to cause serious injury or death. The worker's coemployee is entitled to immunity, except for acts of willful, wanton, and reckless negligence. See also *Mandolidis v. Elkins Indus., Inc.*, 246 S.E. 2d 907 (W. Va. 1978); *Bozley v. Tortorich*, 397 So. 2d 475 (La. 1981); *Beauchamp v. Dow Chem. Co.*, 398 N.W. 2d 882 (Mich. 1986); *Jones v. VIP Dev. Co.*, 472 N.E. 2d 1046 (Ohio 1984); *VerBouwens v. Hamm Wood Prods.*, 334 N.W. 2d 874 (S.D. 1983). Many state legislatures have reacted to cases creating exceptions to the workers' compensation immunity, generally to limit the scope of the exception. See, e.g., Mich. Comp. Laws § 418.131 (Supp. 1990); Ohio Rev. Code Ann. § 4121.80 (1990); W. Va. Code § 23-4-2 (1983).

an agency statutorily mandated to construct, certain steps can be taken to maximize the potential that all parties involved in the project will share the workers' compensation immunity. For example, the construction manager may recommend that the statutory employer owner acquire a wrap-up workers' compensation policy insuring the construction manager, the architect, the general contractor, and all subs. Although the owner's purchase of a wrap-up policy may not be necessary to afford immunity,[21] it can be strong evidence that the owner is the statutory employer, and may be the most cost-effective and efficient method of providing workers' compensation coverage for the project.

In *Washington Metropolitan Area Transit Authority v. Johnson*,[22] the WMATA purchased a wrap-up policy for the Washington Metro project. WMATA was deemed the general contractor for the project because of its statutory mandate to construct the Washington Metro. By acquiring the insurance coverage, WMATA, and all parties under contract with WMATA for construction of the project, enjoyed statutory immunity from the civil claims of injured workers. (Note that Bechtel Associates, the safety engineer and "agent" for WMATA, was immune from liability under a section of the compact that created WMATA, providing that WMATA was exclusively liable for the torts of its agents.)[23]

The owner's purchase of a wrap-up policy, however, will not operate to "buy" immunity where the owner is not a statutory employer. For example, in the *Wenzel* case, discussed earlier in this chapter, injured workers on the Curtis H. Stanton Energy Center, owned by the Orlando Utilities Commission (OUC), sued the construction manager and safety consultant for job-site injuries. Not only had the OUC purchased a wrap-up workers' compensation policy, but also the Florida Workers' Compensation Act expressly granted immunity to the statutory employer's safety agent.

The court "commended" the OUC for purchasing the insurance, and the construction manager for assisting in safety matters. The court, however, denied the construction manager immunity because the OUC was not the statutory employer, and had no statutory duty to obtain the insurance. In short, OUC had volunteered in obtaining the insurance, and could not "buy" immunity that was not provided by statute.

IS THE CONSTRUCTION MANAGER ENTITLED TO INDEMNITY FOR SAFETY LIABILITY?

Indemnity is a party's right to compel reimbursement when it discharges an obligation for which another would otherwise be liable. Indemnification differs from contribution in that indemnity is the shifting of the entire loss from one party to another party which should bear the loss. Contribution is the distribution of the loss among the negligent parties by requiring each to pay a share of the injured party's loss.

Indemnity may be implied as a matter of law, or be express as part of a contract. In implied indemnity, a party who is only passively negligent is reimbursed by the actively negligent party. In the contractual indemnity, reimbursement is governed by the terms of the contract, provided the contractual terms do not run afoul of state law limiting the scope of indemnity to which the parties can agree. For example, a number of states have *anti-indemnity* statutes that control the right of contracting parties to shift the risk of loss among themselves.[24]

Most standard form contract documents contain express indemnity clauses. As provided in the form contracts, indemnity is both a source of construction manager liability and a source of financial protection from the claims of injured workers where safety was the primary responsibility of a party other than the construction manager.

For example, the Construction Management Association of America's Document No. A-1 (1993 edition), Article 9.5, provides that the construction manager, owner, and design professional will indemnify the others from any claims for bodily injury for which the indemnitor is liable as a result of its negligent acts or omissions. Further, this Article provides that the owner will indemnify the construction manager from claims caused by the contractors or the design professionals. Also, the owner will cause each contractor to indemnify the construction manager from claims arising from the contractor's wrongful acts and omissions. The contractor's indemnity agreement is found in CMAA Document No. A-3 (1993 edition), Article 4.15.

The CMAA documents provide a system of mutual indemnity among the owner, construction manager, and design professional, and unilateral indemnity by the contractors in favor of the owner, construction manager, and design professional. This system attempts to facilitate the shifting of risk to the most actively negligent party among the owner, construction manager, and design professional, and to pass that risk to the contractor if the injury arose from a wrongful act or omission of the contractor.

In view of the fact that the CMAA documents impose primary responsibility for safety on the contractor (see, e.g., CMAA Document No. A-3 [1993 edition], Article 5), the CMAA documents do a good job of shifting the risk of personal injury to the contractors. Thus, although the contract cannot avoid a personal injury action by an injured worker against the construction manager, the contract does provide a contractual vehicle for recoupment from the contractor.

FINAL PERSPECTIVE

Simply put, the combination of (1) the construction manager's contractual responsibilities for safety, (2) the construction manager's actual conduct in safety matters, (3) increased safety legislation, and (4) the courts' propensity to find construction managers liable exposes the construction manager to risks of liability to injured workers on most jobs. The construction manager can either face this challenge, or ignore the risk. No exculpatory clause will assure immunity from liability. Given this choice, the reasonable approach of many construction managers is to recognize and take hold of the risks through a proactive approach, placing the construction manager in control of the circumstances creating risks of worker injury.

The proactive approach places the construction manager squarely in the path of liability. If conducted under a clearly defined and maintained scope of services, for adequate compensation, and if insured and covered by the contractor's indemnity, the proactive approach gives the construction manager a manageable risk of liability.

Portions of this Chapter are taken from an article previously published in the Construction Law Adviser, a monthly newsletter offering practical advice for lawyers and construction professionals, and are reprinted with permission of the copyright holder, Clark Boardman Callaghan.

REFERENCES

1. 920 F.2d 778 (11th Cir. 1991).
2. 1992 WL 224829 (OSHRC Docket #89-1300, August 28, 1992), *aff'd on other grounds sub nom. Reich v. Simpson, Gumpertz & Heger, Inc.,* 3 F.3d 1 (1993).

3. See, e.g., *Wartenberg v. Dubin, Dubin & Moutoussamy,* 630 N.E.2d 1171 (Ill.App. 1994).
4. *Kulka Constr. Management Corp.,* 1992 WL 224824 (OSHRC Docket #88-1167, August 28, 1992).
5. See, e.g., *Bechtel Power Corp.,* 4 BNA OSHC 1005, 1975–76 CCH OSHD ¶20,503 (No. 5064, 1976); *Bertrand Goldberg Assocs.,* 4 BNA OSHC 1587, 1976–77 CCH OSHD ¶20,995 (No. 1165, 1976).
6. 606 P.2d 1269 (Alaska 1980).
7. 443 N.E.2d 1212 (Ind.App. 1983).
8. 500 N.E.2d 1265 (Ind.App. 1986); see also *Ivanov v. Process Design Assocs.,* 1993 WL 405299 (Ill.App. 1993).
9. 614 P.2d 1341 (Alaska 1980).
10. See, e.g., 29 U.S.C. § 654(a)(1); Cal. Lab. Code §§ 6400, 6401; Ill. Rev. Stat. Ch.48, ¶137.3; Md. Ann. Code art. 89, §32(a); Mont. Code Ann. §50-71-201; N.C.Gen.Stat. §95-129; N.Y. Lab. Law §200; Or. Rev. Stat. §654.010; Va. Code Ann. §40.1-51.1(a); Vt. Stat. Ann. tit.21, §223; Wash.Rev. Code §49.17.060.
11. See, e.g., 29 U.S.C. § 654(a)(2); Ill. Rev. Stat. Ch. 48 ¶ 60–69; La. Rev. Stat. Ann. § 40:1672; Mo. Rev. Stat. §§ 292.090, 292.480; Mont. Rev. Code Ann. §§ 50-77-100 to 103; Neb. Rev. Stat. § 48-425; N.Y. Lab. Law § 240.
12. See, e.g., *Binge v. J.J. Borders Constr. Co.,* 419 N.E. 2d 1237 (Ill. App. 1981); *Vegich v. McDougal Hartmann Co.,* 419 N.E. 2d 918 (Ill. 1981).
13. See Martel & Glewwe, "Construction Personal Injury Claims," *Construction Briefings,* Sept. 1984, at 7; *Valdez v. Cillessen & Son,* 734 P.2d 1258 (N.M. 1987).
14. 440 N.Y.S. 2d 437 (Sup. 1981).
15. 494 N.Y.S. 2d 449 (A.D. 1985).
16. See *Harbor Ins. Co. v. Omni Constr., Inc.,* 912 F. 2d 1520 (D.C. Cir. 1990) (excess liability policy insuring property damage and personal injury at the site excluded coverage for claim arising out of incidental professional services by contractor). See also C. Foster, et al., *Construction and Design Law* § 5.10, "Construction Management and Insurance" (1991).
17. See, e.g., *O'Boyle v. J.C.A. Corp.,* 538 A. 2d 915 (Pa. Super. 1988); *Cox v. Turner Constr. Co.,* 540 A. 2d 944 (Pa. Super. 1988); *Hope v. State,* 525 So. 2d 353 (La. App. 1988); *Brady v. Ralph Parsons Co.,* 520 A. 2d 717 (Md. 1987).
18. C. Foster, et al., *Construction and Design Law* § 36.2a.2a.3, "Statutory Employer Under Workers' Compensation Statutes" (1991).
19. See *Evans v. Hook,* 387 S.E. 2d 777 (Va. 1990) (developer deemed statutory employer and project architect afforded immunity as fellow employee).
20. See *Garcia v. Pittsylvania County Serv. Auth.,* 845 F. 2d 465 (4th Cir. 1988) (County Service Authority deemed statutory employer and project engineer afforded immunity as fellow employee).
21. See *Garcia v. Pittsylvania County Serv. Auth.,* 845 F. 2d 465, 468 (4th Cir. 1988).
22. 467 U.S. 925, 104 S. Ct. 2827, 81 L. Ed. 2d 768 (1984).
23. *Johnson v. Bechtel Assocs.,* 717 F. 2d 574 (D.C. Cir. 1983).
24. See, e.g., Alaska Stat. § 45.45.900 (1986); Ariz. Rev. Stat. § 34-226 (1981); Cal. Civ. Code §§ 2782, 2782.5 (1980); Conn. Gen. Stat. § 52-572k (Supp. 1990); Fla. Stat. Ann. § 725.06 (1972); Ga. Code Ann. § 13-8-2(b) (1989); Idaho Code § 29-114 (1971); Ind. Code Ann. § 26-2-5-1 (1971); Ill. Rev. Stat. ch. 29, § 61 (Supp. 1990); La. Rev. Stat. §§ 9:2780 (1981); Md. Code Ann., Cts. & Jud. Proc. § 29-114 (1971); N.C. Gen.Stat. § 22B-1 (1986); Ohio Rev. Code Ann. § 2305.31 (1975); R.I. Gen. Laws §6-34-1 (1976); Tenn. Code Ann. § 62-6-123 (1976); Utah Code Ann. §13-8-1 (1969); W. Va. Code § 55-8-14 (1975); Wyo. Stat. Ann. § 30-1-131 (1977).

CHAPTER 11
HEAVY CONSTRUCTION CLAIMS

Gary S. Berman
O'Brien-Kreitzberg, Inc.
New York City, New York

INTRODUCTION

Heavy construction projects involve the placing and/or removing of large quantities of natural and human-made materials. Whether the project is an airport, highway, treatment plant, or dam, the work to be performed is often of an indeterminate nature and, as such, is subject to question and interpretation. These questions and interpretations lead to disputes and changes which may escalate into claims and lawsuits.

Disputes and claims on heavy construction projects are not unlike those associated with commercial, institutional, and residential construction. What differs are the number of companies involved, their frequency, the size of the disputes and claims, and their impact on the project.

Disputes and claims in heavy construction are characteristic of projects that involve huge quantities of materials, extend over long periods of time, and require much of the work to be performed outdoors or under extreme environmental constraints. Differing site conditions, delays, material quantities, weather, labor, impossibility of performance, and constructive changes are just some of the problems which may lead to disputes and claims on heavy construction projects. These projects can also be disrupted by disputes and claims of maladministration, implied warranty, superior knowledge, defective design, and termination.

The expression *dispute* within the context of this chapter is defined as an issue or problem in which two or more involved parties disagree on the following points:

- That an issue or problem exists
- Which parties caused the issue or problem
- The solution of the issue or problem
- The impact on the project caused by the issue or problem
- When the issue or problem actually began to be an issue or problem
- All of the above

A *claim* is the result of an unresolved dispute that has been formally prepared and submitted in a format specified by the contract documents or industry standards. Depending on provisions in the contract documents, claims can escalate into lawsuits or can set into motion an alternative dispute resolution technique. Since the genesis of a dispute, claim, or lawsuit can evolve from a single issue or problem, this chapter addresses all issues, disputes, and problems as *claims*.

The chapter will also explore and define the technical aspects (not legal) of the various types of claims; the preparation and refutation of claims; what it takes to prove and price claims; and the use of outside consultants, alternative dispute resolution, litigation, and claims avoidance.

TYPES OF HEAVY CONSTRUCTION CLAIMS

Certain heavy construction projects are more susceptible to particular claims than others and, given the innovative project management and claims avoidance techniques presently used in heavy construction, some projects avoid formal claims. However, experience has proved that, despite all efforts to prevent claims, they will continue to occur.

The work in heavy construction projects, as in most construction projects, is segregated into four elements: time, cost, quality, and scope. Time is the duration, usually in days, in which the project is specified to be completed; cost is the price the owner or user has agreed to pay for the work; quality is the desired level of performance the owner or user expects; and scope is the actual description and breadth of the work to be completed. All claims have their roots in the quadrangle of issues comprising time, cost, quality, or scope. In fact, the issues of a claim could consist of any one, or all, of these elements, and rarely would one element not impact on the others.

Projects involving the rehabilitation of an existing structure or those in which the work extends below the surface strata are subject to claims of differing site condition. What distinguishes rehabilitation work is the representation in the contract documents of the as-built or existing condition which is to be changed. When the work involves excavation or deep foundations—common in pipeline, tunnels, and environmental remediation projects—the physical conditions of the material to be removed or treated are what differ at the site. For example, differing site conditions associated with the rehabilitation of a bridge, track work, or highway could mean conditions that not only vary from those represented in the contract documents but from those normally encountered on a project of that type.

Probably the most common type of claim in heavy construction, usually referred to as the *delay claim*, involves time and is the result of an event which delayed the agreed completion date of the project. The claim associated with time is usually a request for time extension due to a change in scope, or to compensate a contractor for unusual or unexpected conditions caused by, for example, weather, a strike, or a defective design. A change in scope could result from the owner or user modifying the type or quantity of work as described in the contract documents (sometimes referred to as a constructive change or extra work claim), or a differing site condition. Whatever the cause, something has changed which has affected the contractor's ability to complete the work in accordance with the time provisions of the contract documents.

Time of completion is also affected when a contractor is delayed due to events beyond its control. These include the unexpected events described earlier, plus maladministration on the part of the owner, user, or its agents, and delays in the delivery of owner-furnished equipment (often occurring on treatment plant, airport, and rail projects). The owner or its agents have the responsibility of administering and managing contracts on the project in a manner which does not negatively impact the performance of a contractor.

It is difficult to conceive of a claim today that does not include a time element. The reason it is important for a contractor to be granted time extensions is that contract documents in heavy construction projects almost always include penalties or liquidated damages should the contractor not complete the work in the specified time. The granting of a time extension changes the time of performance, either the duration or the completion date, thus allowing the project to be finished "on time," even though the project is not finished in accordance with the time provisions originally specified.

A related claim which can result from a delay on the project is termed *acceleration*. Primarily before bidding, but more specifically after notice of award, a contractor plans its work in accordance with the contract documents, in the time allowed by the contract, and in a manner to earn desired profit. When an unexpected event causes the contractor to abandon its original plan and complete the work in a shorter duration, or when the contractor is asked to complete a larger scope of work in the originally specified duration, the contractor is to said to have accelerated its performance. If the acceleration is not caused by the contractor, the contractor is entitled to be compensated for its increased costs due to this acceleration.

Under most contracts in heavy construction, in order for a contractor to be compensated for its increased costs due to acceleration, the increased pace of the work must be directed, in writing, by the owner. Otherwise, the owner has the choice of having the project completed at a different time by the issuance of a time extension.

Some delay claims seek only time extensions and no additional compensation. One example of this excusable delay claim is commonly referred to as *force majeure*. These delays are caused by unforeseeable causes beyond the control of the owner and contractor. Typical causes of force majeure are strikes, acts of government, fires, floods, extreme weather, health restrictions, embargoes, and unpredictable delays which affect lower-tier suppliers and subcontractors. Other examples of noncompensable, excusable delays are associated with contract termination and work suspensions, but in most cases the contract defines how these delays will be resolved.

A common cause of delay is loss of productivity. Again, based on the contractor having a plan for accomplishing the work, loss of productivity occurs when more labor hours or equipment than were planned are required to complete contract work. The causes of this increase in labor are usually claimed to be the result of events caused by the owner, its agents, or another party, and not due to inefficiencies by the contractor's workforce. The proof of loss of productivity is more an art than a science. Since the late 1950s, organizations in both the private and public sectors have been attempting to derive formulas which closely model the performance of the trades in the hope of identifying the causes of loss of productivity. Unfortunately, there is no one accepted method of proving a productivity loss.

Disruption is also a common cause of delay, although an actual delay may not occur even though a disruption has. The owner, through a directive, an act, or by not acting, can precipitate a disruption by causing the contractor to work out of its planned sequence or in a sequence which is not logical or conventional within the industry. This work performed out of sequence may not necessarily cause a delay and may not cost the contractor money. But, generally, the increased costs and an extension of time are recoverable. It is important to note that a disruption usually causes a loss of productivity and increased equipment costs.

A claim on the project which includes all the elements of time, cost, quality, and scope is defective design. Defective design is usually the result of the services performed, and work products produced, by design professionals associated with the project. In heavy construction—for example, in the upgrading or construction of an airport—literally scores of design professionals may be involved in the design of the project. Conversely, the design of a highway, pipeline, or bridge may involve one designer or only a few.

Defective design is usually characterized as an error or omission in the contract documents which include, but are not limited to, drawings, general or special provisions, and technical specifications. In heavy construction, most errors and omissions are found in the drawings and technical specifications. Examples of errors could include the specification of incorrect materials, dimensional or quantity errors, ambiguous provisions, and designs that cannot be constructed. Omissions could simply consist of not indicating the design in sufficient detail to show intent, or missing specification sections.

When a design is not constructible or practical, a contractor may have the right to submit an impossibility-of-performance claim. Although not encountered frequently, these claims have merit and can have a substantial impact on the project. Designs that include elements which cannot be constructed or are impractical to build should be discovered during the bidding phase, when the contractor is estimating the job. But this is not always the case, and in court the concept of "impossible" or "impractical" is a complex legal argument.

The difficulty associated with performance claims is convincing the owner and designer that the project cannot be built, or that it is impractical to build as designed, despite the contractor having entered into the contract. One factor affecting these claims is that if the contractor succeeds in building a part of the disputed work but not the entire project per the plans and specifications, one could argue that the scope of work is not impossible or impractical.

Another claim common to all project types arises over the interpretation of the contract documents. Contract documents are sometimes afflicted with conflicting provisions, ambiguities, or provisions that do not include enough detail. These characteristics could be viewed as being part of a defective design, but this is not necessarily the case. Provisions requiring interpretation can affect the project in terms of time, cost, quality, and scope—most often cost and scope.

If a contractor, for example, based its bid on substituting a piece of equipment that has been specified as "or equal," but the owner or its agents disagree that the equipment is truly "equal," a dispute may arise which delays the delivery and installation of the equipment. If the contractor is directed to buy the specified equipment, or equipment which the owner decides meets requirements, the contractor may have to purchase equipment at a greater cost than estimated in the (or its) bid. This type of dispute can have a negative rippling effect throughout the project if not brought under control quickly. Disputes like this are often ignored until too late.

Another type of claim becoming more commonplace today in heavy construction is the total-cost claim. Simply stated, a contractor submits a claim for additional compensation, calculated by subtracting the difference between what the project actually cost and the bid (or planned) cost. The contractor must meet three conditions if it is to prevail using the total-cost method of proving its claim: First, it must show that there is no other way to calculate its claim; second, it must show that its bid or planned costs were reasonable; and, third, it must prove that the causes of these increased costs are due to factors beyond the control of the contractor. A contractor might submit a modified total-cost claim, which accounts for increased costs caused by the contractor by subtracting these costs from the contractor's total costs prior to subtracting the bid or planned costs.

Other claims made less frequently include overinspection, superior knowledge, breach of contract, interference, and claims for extended overhead. Contractors, lawyers, and construction consultants have applied numerous names to claims and have created new forms of entitlement by combining two or more types of claims.

Claims in heavy construction take many forms and often occur as multiple claims, even though the root cause may be a single event, problem, or issue. A claim, whether verbal or written, should ultimately be presented in a certain format if it is to be successfully prosecuted or refuted.

PREPARATION OF HEAVY CONSTRUCTION CLAIMS

All legitimate claims can be divided into two distinct parts: entitlement and damages. Throughout the heavy construction industry different words have been used to describe these two parts, but the results are the same.

Entitlement is the portion of the claim that addresses why the party has the right to make a claim and seek damages as a result of the event. Damages are the resultant cost and time impacts the event has caused. Entitlement and damages are not mutually exclusive; you cannot be awarded damages unless you can prove entitlement. Similarly, you do not have to prove entitlement, or, for that matter, submit a claim, if you do not have damages.

Entitlement, or its basis of recovery, should be established in concert with the contract documents. The contract documents often define the circumstances for which a contractor is and is not entitled to be compensated. Contractors need to have owners agree on entitlement prior to producing a costly and time-consuming calculation of damages.

The composition and style of a claim can take many forms. Some contractors, lawyers, and consultants use standard forms, but it is generally a good rule to format the claim around the best way of proving entitlement and damages.

At a minimum, all claims should include certain components. There should always be a description of the event that caused the claim and what effect it had on the project, i.e., the cause and effect. There can be multiple causes and multiple effects, and there can be effects which occur as a result of multiple impacts, which are commonly referred to as *ripple effects*. Sometimes the issues are so complex that a simple explanation of the effect, or impact, is not sufficient. Therefore, an analysis must be performed that documents the impact. Today, analysis is usually computer-generated and uses software to document the cost and time considerations of the impact. Heavy construction claims are usually complex and require supplementing the cause-and-effect narrative and analysis with project documentation.

Lastly, some claims require a certification. On federal projects subject to the Federal Acquisition Regulations (FAR), claims must usually include a formal certification in a precise format described in the FAR. The certification is normally a narrative that describes the authenticity and accuracy of the claim, which is then certified by the signature of an authorized representative of the contractor, usually an officer of the firm or its attorney.

A general rule is that the more detailed a legitimate claim is, the faster it will be approved and the contractor will receive, with minimum hassle, what it is seeking: compensation in cost and time. But understanding how the other party will analyze the claim is an excellent first step in deciding how to prepare the claim.

REFUTATION OF HEAVY CONSTRUCTION CLAIMS

For every claim prepared and submitted, there is always someone seeking the reason to deny entitlement or to reduce or deny the associated damages. The refutation of claims is another reason to hire lawyers, consultants, and experts. Heavy projects are so complex today that the claims they produce are sometimes too complicated for anyone but a professional to refute. Also, the party refuting the claims often doesn't have the time, objectivity, or resources to devote to understanding and resolving the issues.

Claims can arrive in the form of a single letter or in twenty volumes, each a three-ring binder filled with hundreds of sheets of documentation and analysis. The refutation of complex claims is a meticulous process requiring not only knowledge of the construction industry, but of litigation, ADR, and negotiation processes. Negotiation is usually the second step in refuting a claim. It can be the most effective and least time-consuming process for resolving a claim if the consultant hired does a thorough and objective analysis of the issues. When the analysis is complete, the consultant will be able to inform its client of the strengths and weaknesses of the respective positions and will be able to estimate the chance of success in proceeding with a certain strategy. If the issues are legal, or if legal and technical issues are intertwined, a lawyer and consultant should work together on the refutation strategy.

If the claimant is unrelenting in its pursuit of the claim, regardless of an independent and objective analysis which refutes the issues, then litigation may ensue.

PROVING AND PRICING HEAVY CONSTRUCTION CLAIMS

Proving entitlement and calculating damages is the challenge to submitting a successful claim. As in criminal law, where the defendant is presumed innocent until proven guilty, parties rarely have equal status in contract law. In fact, many construction contracts drafted by owners attempt to shift many of the risks and liabilities to the contractors and their subcontractors. However, on the day the contract begins, two presumptions exist: First, a basic premise is that contract interpretation is often construed against the drafter of the contract. Second, the burden of proof lies with the party bringing forth the claim. It is this second presumption that proves most troublesome for heavy contractors.

Proof is important in demonstrating entitlement and documenting damages. For example, a contractor of a major rail system is instructed to take additional steps to complete the project on time, despite project delays. The owner states that if the contractor can prove that the delays were not caused by the contractor, the reasonable costs of acceleration will be borne by the owner. In this case, assuming that the contractor is entitled to recover these costs pursuant to the contract, the contractor must prove (1) that the delay, which caused the owner to issue a directive to accelerate, was caused by a party other than the contractor; (2) that acceleration did occur; and, (3) that the increased costs of the acceleration were reasonable. This all assumes that the project was completed on time.

It is not always clear that a project, or part of a project, was completed on time. The contract should clearly define the level of work or a milestone that distinguishes completion. Whether in reference to the construction of a transit system, tunnel, highway, or wastewater treatment plant, the reaching of a milestone can be elusive. But in this example, let's assume that the project was completed on time.

The contractor now must establish that acceleration actually took place. This can be proven by demonstrating an increase in labor, supervision, and equipment beyond those levels indicated in the bid or preconstruction planning. However, the contractor must establish that this increase in resources occurred as a result of mitigating the delays and is not the result of compensating for the contractor's inefficiencies. The increase in labor could be demonstrated by time slips, a daily accounting of labor by trade and equipment on-site (acknowledged daily by a representative of the owner), or by illustrating in tabular form or graphically an increase in resources at the point the acceleration began. Presuming that the contractor has proven an increase in resources not due to its own inefficiencies, it must prove the associated reasonable damages.

The contractor can take a number of approaches to documenting these costs, but the approach must be established and agreed to by the parties prior to the acceleration beginning. Although burdensome for the contractor, the best way is for the contractor to establish new accounting codes for the labor and equipment associated with the acceleration. At the end of the job, the new accounting code can be reconciled, the appropriate burdens added, and, thus, the cost to the owner calculated. The owner must monitor and verify that these additional resources are being accounted for correctly on a daily basis, ensuring that resources used in nonaccelerated contract work aren't being tallied in this new accounting code. This method in public contracts is commonly referred to as *force account work*.

Another method of documenting the acceleration costs is to use a modified total-cost approach, but the difficulty in meeting the criteria for using this method is burdensome. Another method is to shift the responsibility for determining these costs to the owner. In this approach, unit costs for labor and equipment, and the associated burdens, are agreed on prior to beginning the acceleration. The owner, through its own forces in the field or its agents, keeps the field records on labor and equipment. This information is agreed to daily by the contractor and this data subsequently becomes the basis for the damages calculation.

Although this example is not truly representative of the complexity and level of detail required to prove heavy construction claims, it does illustrate the various levels of proof required to successfully prosecute a claim. One of the important axioms provided by this example is that it is not enough to actually prove that the event which led to the claim occurred and resulted in an increase in cost; the contractor has the additional burden of proving that it was not some other event or circumstance which caused the increase in cost.

In any heavy construction claim, certain construction costs are recoverable and some are nonrecoverable. Although the contract is the determining factor, the contractor must consider the following list of costs when calculating its recoverable costs:

- Increased wage rates
- Increased material prices
- Storage costs
- Idle equipment
- Rental and lease costs and fees
- Administrative expenses
- Financing charges
- Interest
- Seasonal influences

- Profit
- Supervisory costs
- Loss of productivity and efficiency
- Field overhead
- Engineering costs
- Insurance
- Bonds
- Home office costs

Other costs incurred by the contractor that are generally thought of as nonrecoverable include:

- Claim preparation costs
- Attorney fees
- Expert fees
- Consequential damages

In heavy construction, whether pricing your bid, preparing a change order, or submitting a claim, three factors are always considered: labor, materials, and equipment. These three components combined constitute direct costs. The work originally specified in the contract documents to be performed by the contractor is designated as contract work. Simply, the contractor is, or should be, allowed to recover the increased labor, material, and equipment costs, referred to as *direct costs*, for all work not designated as contract work. This could be attributed to new work and work that must be redone (with the presumption that the contractor was not the cause of the rework). This is not true if the increase in labor, materials, and equipment is the responsibility of the contractor, or its partners, subcontractors, or suppliers.

For example, a new toxin is unearthed at a hazardous waste remediation site and construction is suspended and the project delayed five weeks. The contractor will incur increased recoverable costs associated with idle equipment and crews, and an increase in bond, rental, financing, and insurance costs due to the extended period of performance. In the same scenario, the work is disrupted, but the scope of work is not changed by the discovery of this new toxin and no delay occurs. Then the contractor could recover costs for loss of efficiency and productivity, increased supervision, seasonal costs, and material price increases.

All other project costs not attributed directly to labor, materials, or equipment are referred to as *indirect costs*. Most indirect costs are recoverable, but, again, the contract will define the parameters. Examples of indirect costs include bonds, insurance, field overhead, profit, administrative expenses, and home-office overhead, the latter being one of the most difficult to collect. These costs, exclusive of profit, are referred to as the "cost of doing business." As such, contractors are entitled to claim for these costs.

There are other types of recoverable costs defined in the contract. If a contractor is terminated for convenience, it is entitled to recover certain costs. These usually amount to being paid for the work which precedes the effective termination date—both direct and indirect costs. The contract may also allow for certain demobilization costs.

Because the total construction value of wastewater treatment plants, pipelines, and airports is high, so is the expected profit of a contractor. In terminations-for-convenience, contractors often seek to recover lost profits as a result of not completing the intended contract work. Unless there is a provision for such in the contract, lost profits, home-office overhead costs, and other fees are not recoverable.

In the case of a contractor submitting a claim, it is likely that the owner may have a counterclaim for its damages resulting from the actions of the contractor. Referred to in the public sector as *liquidated damages* and in the private sector as *penalties*, these costs are real and often used to offset construction claims. In many cases, when these costs are taken into account, the contractor may, in fact, owe money to the owner.

On an airport, transit system, or dam project it is difficult, if not impossible, to account for increases in costs for the owner due to a project delay. Therefore, stipulated in the contract, the parties agree on liquidated damages, which are typically in the form of dollars per day of delay (assuming the delay is the fault of the contractor). LDs, as they are often called, can amount to as little as $100 per day or as much as $100,000 per day, although some projects may warrant more. The amount varies, based on the contemplated costs of the owner during the period between the originally planned completion date and the actual completion date. These costs could include labor, materials, and equipment used by the owner or its agents.

It is not hard to see how a $250,000, 90-day delay claim on the part of a contractor is quickly offset by LDs of $3000 per day ($270,000). LDs are often highest when the public client is expecting a revenue stream as a result of the completed project, as in an airport or transit system.

In the private sector, an owner's project is also often tied to a revenue stream, and, therefore, certain profits will not be realized if the project is not commissioned on time. These contracts often have penalty provisions that can be based on numerous factors, but do not necessarily have to reflect the aggregate of real costs due to the delay. In other words, the penalty may be punitive rather than a reimbursement for actual lost profits or the increased costs of the private owner. Whatever the reasons or amount, these penalties can be used to offset claims.

Contractors, subcontractors, suppliers, and even owners do not have the time or resources to prosecute and refute construction claims. The hiring of competent consultants such as lawyers, engineers, construction experts, and claims consultants is an important step in prevailing, no matter what side of the table you are sitting on.

ROLES OF LAWYERS, CONSULTANTS, AND EXPERTS

Sometimes, the successful settlement of a construction claim requires supplementing the contractor's staff with additional resources to help prepare, present, and negotiate the issues. Since most, if not all, claims have their foundation in the contract between the disputing parties, and because the contract is a legal instrument, a party should consult an attorney during the course of a construction project. In fact, it is suggested that lawyers be consulted prior to the contractor or owner signing the contract.

Lawyers are best at developing the strategies for settling the legal issues in claims. Today's airports, treatment plants, tunnels, and environmental projects are fraught with so many liabilities that an experienced construction lawyer is required to untangle the web. Also, the construction industry as a whole is presently so litigious that the advice of counsel is a must. The rules of thumb are simple: Let a construction lawyer review your contract prior to signing and, if necessary, let him or her help you negotiate points that need to be resolved; prior to submitting a claim, obtain the advice of a lawyer on entitlement; and, last, do not accept legal opinions from anyone who is not a lawyer. However, in the complex construction industry of today, lawyers cannot, and should not, deal with the technical issues of the dispute.

There are literally thousands of firms and individuals portraying themselves as construction and engineering consultants and technical experts. A legitimate construction consultant should be brought in to review the contract prior to signing, if the parties are unfamiliar with the technical provisions of the contract documents. For the contractor, this ensures that its bid and plan are based on technical details it understands, rather than a request for information or clarification being submitted after construction has begun. This could prove costly to the contractor.

There is no conflict between lawyers and consultants. A lawyer handles the legal aspects of the dispute and is recognized as the so-called *program manager* of the dispute; the consultant handles the technical aspects of the dispute. The selection of both is crucial to having your claim approved or winning your case in trial. Both lawyers and consultants should be recognized as knowledgeable in the engineering and construction industries. Do not hire an individual or firm that does not have a specialty in heavy construction. As stated earlier, there is no shortage of lawyers and consultants who understand heavy construction. Sometimes the lawyer likes to choose the consultant, and sometimes the contractor or owner seeks the advice of the consultant to choose the lawyer. Both methods are acceptable and should be encouraged, since the lawyer and consultant must be comfortable with each other.

Last, the process for selecting a lawyer or consultant should be based on qualifications, with price an important, but secondary, consideration. First, request information from a firm in the form of a statement of qualifications; second, request a technical and budget proposal; and, last, interview the proposed project team. It is important to check references on the firm and on the members of the project team. Remember, what you are really hiring is not a firm, but the individuals who are proposing to work with the parties.

LITIGATION IN HEAVY CONSTRUCTION

The ultimate adjudication of a construction claim is at trial. But before arguments can be made before a judge, jury, or board, the litigation process begins and lays the framework for the trial. Litigation is usually a long, arduous process, whose procedures are prescribed by the courts and the legal community. An experienced trial lawyer with knowledge of the construction industry is essential to winning at trial. Equally important is the construction expert, for it is his or her expert testimony that may be the key factor in persuading the adjudicators.

The litigation process can be divided into a series of phases, which include discovery, pretrial, trial, and posttrial. Other variations of these four phases may exist or one of these phases can be further subdivided. Discovery is usually the longest, most expensive, and most time-consuming phase for the participants.

In the early 1990s, parties began to do everything possible to avoid litigation. New techniques for resolving disputes have become popular and attempts made to avoid expensive and time-consuming litigation. These alternative dispute resolution techniques are discussed in greater detail in the next section.

Although litigation is feared by the parties, a trial is sometimes the best and fairest forum for resolving the dispute. What is most attractive about trial is the fact that the decision is binding and the outcome, if other than a jury trial, may be reasonably predicted based on related, precedent-setting cases which are described in detail in case law.

Even the fact that the litigation process has begun doesn't preclude settling the case prior to the actual trial. At any time during the process, settlement discussions or negotiation can take place. However, these discussions are not usually begun until each side comprehends the strengths and weaknesses of the other's position. Sometimes the threat of a protracted litigation is the strategy behind a settlement.

ALTERNATIVE DISPUTE RESOLUTION IN HEAVY CONSTRUCTION

The litigiousness of the heavy construction industry exploded in the late 1980s and early 1990s. The value of new construction was declining, competition shaved profits, and owners were resigned to the fact that their projects would be plagued with claims. Contractors were spending their profits preparing and resolving claims, their time was spent negotiating rather than constructing or taking care of business, and owners were spending large sums of money refuting claims. These events led to the birth of *alternative dispute resolution,* as we know it today.

Alternative dispute resolution, commonly referred to as ADR, is not a new concept, but rather the rebirth of existing techniques. Also, innovative amalgamation of various ADR techniques has given rise to new methods for resolving disputes.

The most common ADR techniques used in heavy construction are arbitration, mediation, neutral advisors, and dispute review boards. Other ADR techniques in the construction industry include minitrials, use of special masters, advisory arbitration, and private judges. Naturally, good old negotiation has always been an alternative to fighting battles in the courts.

Arbitration has been around and used by the construction industry for more than half a century. Usually binding, the process is often administered by a third party which has no interest in the outcome. Not unlike trial, the parties are allowed to present their cases before a panel of one or three people who ultimately decide on the issues of entitlement and damages. What is different from trial is that the arbitrators are usually members of the heavy construction industry and understand what is being discussed. There is no need to educate the jury, so to speak.

Arbitrations were once thought of as a forum which resolved the issues in a short time and at a reasonable cost. Unfortunately, today, because of the binding nature of the decision and the courts upholding of these decisions, arbitrations have become almost like trials, although in arbitrations the arbitrators seem to have the right to make up the rules as they go along.

Mediation, on the other hand, is a forum in which the parties make all the decisions and agree on the resolution of the dispute. The mediator acts as a facilitator, attempting to make sure each party understands the dispute and the other's position, and seeks to develop innovative ideas to reach resolution. What is unique to mediation is that the mediator, at his or her discretion, can caucus with the parties privately and confidentially to discuss the ramifications of maintaining certain positions and discuss the likelihood of settlement based on certain consolations. The mediator would then take only the ideas authorized by the party it just caucused with to the other in an attempt to negotiate the issues. Once the issues have been resolved, the mediator will bring the parties together to be sure all agree on how the dispute is to be resolved and put these ideas in writing for the parties to sign. However, any agreements are nonbinding, except for good faith, and normally the discussions held during mediation may not be brought up during trial if the mediation fails to reach an accord.

The concept of the neutral advisor is increasing in popularity in the heavy construction industry. Rather than each side hiring its own set of construction experts and consultants, each side agrees to jointly hire and pay for one set. This consultant then objectively and independently reviews the issues, analyzes the facts, and recommends a solution. The findings of the consultant usually enable the parties to better understand each other's positions, thus setting the stage for settlement discussions.

Often, neutral advisors are engaged at the beginning of the project before a dispute has arisen. This affords the neutral the opportunity to visit the site and know the history of the job, thus facilitating an analysis of the facts.

The tunneling industry gave birth to the last popular ADR technique, known throughout the heavy construction industry as the DRB, or *dispute review board*. Because of the difficulty in anticipating the composition of the various subsurface strata in deep foundations and tunnels, differing-site-condition claims on these projects were commonplace. To help resolve these claims, a board was set up to adjudicate them.

The board consisted of three or more members, but usually an odd number. In the case of a three-member board, one member was selected by the owner, one member by the contractor, and the third member chosen by the two. As in an arbitration, the claims were presented before the board and a decision made. Depending on the DRB contract provisions, the decision of the board was most often binding, although that did not preclude a complaint being filed in the courts if a party was dissatisfied with the outcome.

Like project neutrals, the boards are often hired before construction begins. They visit the project periodically and are called in when a certain type of dispute is submitted. The type of dispute is described in an agreement between the two parties.

The use of minitrials, special masters, advisory arbitration, and private judges are just combinations of the various forms of ADR, with the outcomes sometimes binding and sometimes not.

CLAIMS AVOIDANCE IN HEAVY CONSTRUCTION

It is thought by many in the construction industry that claims cannot be avoided. But what is a fact is that all parties can take certain steps to minimize their frequency and impact on the project. Also, certain steps can be taken to mitigate the effect of claims once they are submitted.

In the early 1990s a revolution to prevent claims began to diminish the litigiousness of the construction industry. Business devices such as partnering, total quality management, and team-building became popular techniques for preventing claims. Their effectiveness spanned the entire length of the spectrum, from producing claimless projects to having no effect on the project whatsoever. Academics and self-proclaimed gurus of these techniques are still exploring variations of these methods in search of an answer to preventing claims. But there will be no answer. The project parties must enter into the contract in good faith and contractors must accurately represent their capabilities and experience in projects they are bidding. Other than these suppositions, a few procedures, when followed, will prevent or minimize the likelihood of construction claims and their impact on heavy projects.

Claims avoidance is most effective during the design phase of the project cycle. In fact, it is during this early stage of the project that disputes and issues not only cost the least to resolve, but save exorbitant unnecessary expense. There are four services that can be performed during the design phase to avoid claims during construction, aside from hiring competent consultants and contractors.

1. Design oversight by an agent of the owner, such as a construction manager, brings construction expertise and insight into the design process. Often, design professionals are not experienced in actual field construction, other than what they have seen before or read in the technical books and articles. By infusing real construction influence into the design there is a likelihood that design problems, either errors or omissions, will not occur.
2. *Construction document review* is another claims avoidance technique. Here, experienced construction people will review the entire contents of the construction documents for the following:

    Conflicts and inconsistencies between provisions

    Errors or omissions

    Ambiguous language

    Provisions that can lead to disputes

    Provisions that conflict with local, state, or federal regulations

    Provisions that are inconsistent with industry standards

    Ill-defined or nonspecific language (e.g., words like "reasonable" or "about")

    Needed provisions that could reduce change orders, disputes, and claims

    Proper utilization of ADR clauses

    Contractual conflict regarding contractor interfaces

    Whether the roles of the parties have been properly defined

    Provisions that are contested by case law

3. The constructibility review may be the single best tool for preventing claims. The assumption in using this technique is that there has been no construction oversight during the design phase. A constructibility review assesses the actual construction of the project by evaluating the manner in which the project will be constructed. The review would identify nonconstructible or impractical elements of the work, problems of interfacing between the trades and other contractors, interferences which can impact construction, the phasing necessary to complete the work on schedule in the time allotted, and construction conflicts in the various documents of the contract.
4. The review or development of administrative procedures during construction is another vital ingredient of claims avoidance. These procedures could cover everything from how to submit a request for information to defining responsibility for approving the contents of shop drawings.

During the next phase in the life cycle of the project, the bidding phase, a technique for claims avoidance, more applicable to owners, is to turn the bid analysis over to an independent and objective third party. Assuming the owner is not restricted to the lowest bid, but rather to the lowest "responsible" bidder, the owner has the latitude to select the contractor that will provide the best value at the lowest risk. Here, the third party, such as a construction manager, will qualify and rank the contractors based on qualifications, references, applicable experience, safety record, financial stability, bonding status, price, and their plan to complete the project on time, within budget, and at the quality levels desired by the owner.

During the construction phase, although late in the game to prevent claims, it's not too late to mitigate their effects on the project and prevent their escalation. A technique, applicable to the owner and contractor, is to engage a third party to participate in a threshold review analysis, referred to as a TRA. A TRA is implemented when predetermined milestones occur. The first milestone is achieved when a certain number of change orders have been submitted, independent of their cost. The second milestone occurs when change orders of a certain value are submitted, or the aggregate sum of change orders reaches a predetermined threshold.

For example, during a bridge rehabilitation project, a contractor has submitted 30 change orders, when the owner did not expect any. An independent party would review the overall scope of these change orders to decide whether or not these submissions are symptomatic of a much larger problem. The review would take place only after the 30th change order was submitted, assuming 30 was the threshold identified before construction began.

In the same example, no matter how many change orders have been submitted, a change order for $75,000 has been submitted. Assuming the threshold was established for all change orders in excess of ½ percent of the original contract value, a review of the change order would take place independently to confirm entitlement and its value and to again see if this change order be could symptomatic of a larger, unidentified problem.

The TRA gives all parties confidence that a larger problem doesn't exist that would, if allowed to remain unchecked, cause significant impact on the project.

Another claims avoidance technique used during the construction phase is to institute either random audit or continual oversight. The random audit offers a snapshot of the status of the project at any time during the project, at any frequency of occurrence. Continual oversight is being on site at the time work is progressing, even during multiple shifts and on weekends. These two methods of monitoring provide another tier of insurance that potential disputes, critical issues, and change orders are being actively dealt with and that any trends or patterns which may lead to bigger problems are being identified.

The last claims avoidance method is to have claims review by the experts, including lawyers and claims consultants. Their knowledge, judgment, and experience may lead to early settlements or even to the partial or total rescinding of the claim in total.

It's important to realize that flux is inevitable on the construction site and change precipitates disputes and claims. If parties deal with each other openly, accurately, and fairly, an equitable settlement of the issues can be obtained early and with little impact to the project.

EPILOGUE

Whether the project is a new airport construction, the building of a transcontinental pipeline, or the placing of millions of yards of concrete for a dam, claims may be justified. The initial hurdle to recovering an equitable adjustment is proving entitlement and the associated damages. Lawyers, claims consultants, and other experts play a necessary and critical role in the prosecution and refutation of construction claims in heavy construction. ADR includes a menu of techniques which, for the right project, and under certain conditions, will lead to settlement when implemented by experienced people. Litigation is a costly, time-consuming, and stressful way to resolve disputes, but sometimes it is the necessary and preferred method. There are many ways to prevent and minimize claims, but none is more powerful than understanding your contract before execution and dealing in good faith.

P · A · R · T   B

HEAVY CONSTRUCTION EQUIPMENT

CHAPTER 1
ENGINEERING FUNDAMENTALS*

R. D. Evans
Caterpillar Tractor Company, Peoria, Illinois
(Material, Equipment and Job Characteristics)

E. V. Semonin
Goodyear Tire & Rubber Company, Akron, Ohio
(Tires for Construction Equipment)

James J. O'Brien, P.E.
O'Brien-Kreitzberg, Inc., Pennsauken, New Jersey

INTRODUCTION

There are basic engineering fundamentals that affect the selection of equipment for handling bulk materials. Most important and significant of these are: (1) material characteristics, (2) equipment characteristics, and (3) job characteristics. There are also important interactions between these characteristics. Tires, although actually an equipment characteristic, will be discussed as a separate topic in this section.

Influence of Material on Equipment Selection

The construction person reserves the term *rock* for material which must be removed by blasting or some equivalent method. In construction work, bedrock may be encountered in surface outcrops or in excavations which extend below the depth of

---

* Some of the material in this chapter is from R. D. Evans, "Material Characteristics," "Equipment Characteristics," and "Job Characteristics" and E. V. Semonin, "Tires for Construction Equipment," in Stubbs and Havers (eds.), *Standard Handbook of Heavy Construction*, 2d ed., McGraw-Hill, New York, 1971.

soil cover. The influence of bedrock on equipment selection is related to rock type, bed thickness, and extent. Generally speaking, the user of quarry rock must drill and shoot, load out with some type of loader, and haul directly to a fill or to a crusher for further processing. Those procedures require special skills, i.e., proper spacing of blastholes for fragmentation, compatible explosives, and correct bench height, to mention a few. Heavy-duty tractors equipped with rippers are also used to rip or fracture many rocks to a degree where they can be dozed to a final position, loaded into trucks, and hauled, or loaded and hauled by conventional scraper units.

Common excavation consists predominantly of soil or earth. Its removal will generally not require the use of explosives, although tractors equipped with rippers can be used to loosen tight, consolidated earth and thus ease its handling by shovels, scrapers, dozers, loaders, or backhoes.

An *unclassified* excavation may consist of a combination of rock and soil and thus present the excavation problems of both material types. A careful analysis should then be made in order to select the proper equipment.

WEIGHT-VOLUME RELATIONSHIPS

Material weights must be considered in all phases of construction. For example, weight influences the choice of horsepower for acceptable performance of transporting equipment and may establish the basis for payment for hauling. Material weight per unit of volume is also a criterion in determining the acceptability of compacted fills.

Contractors are interested in the volumes occupied by rock and soil materials during the processes of excavation, transportation, and placement. These volumes are influenced by methods of excavation. For example, a shovel excavates an earth bank and loads out a truck, which in turn transports the load to a fill where the material is compacted: bank yard—loose yard (shovel)—loose yard (truck)—compacted yard.

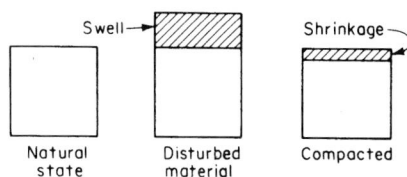

FIGURE B1-1 Volume changes in soils.

Volume Measure

Soil volume is defined according to its state in the earth-moving process. For simplicity, the following discussion will be confined to English system units, although their metric system equivalents may be directly substituted. The three commonly used measures of soil volumes are illustrated in Fig. B1-1 and are described as follows:

Bank cubic yard (bcy)	1 yd^3 of material as it lies in the natural state
Loose cubic yard (lcy)	1 yd^3 of material after it has been disturbed by loading, with consequent swelling relative to its natural state
Compacted cubic yard (ccy)	1 yd^3 of material after it has been compacted

Compaction usually results in shrinkage relative to both its natural and loose states. Assume that 1 bcy of material weighs 3000 lb and that, by virtue of its material characteristics, it swells to 1.3 yd^3 when loaded. The resulting 1.3 lcy still weighs

3000 lb, although the soil now has 30 percent swell relative to its bank condition. On the other hand, if either 1 bcy or 1.3 lcy is compacted, its volume may be reduced to 0.8 ccy. Its weight still remains unchanged at 3000 lb.

Generally, earth-moving jobs are calculated on the basis of bank cubic yards. Thus, in order to estimate production, the relationships between bank measure, loose measure, and compacted measure must be known. Typical values for a range of materials are listed in Table B1-1. Values are shown in pounds and kilograms for bank and loose states only, as compacted yards would vary widely depending on compaction effort.

Load Factor

The ratio between bank measure and loose measure is called *load factor* (LF):

$$LF = \frac{\text{bank cubic yard (bcy)}}{\text{loose cubic yard (lcy)}}$$

If the percent of swell of the material is known, the load factor may also be obtained by using the relationship:

$$LF = \frac{100\%}{100\% + \% \text{ swell}}$$

Load factors and percent of swell for various materials are listed in Table B1-2.

Payload

Maximum payload, measured in terms of weight, is the weight limitation on the net load which can be carried by a unit of equipment. The actual payload carried by a unit of known tare weight can readily be obtained by weighing. It is also possible to express payload in terms of bank yards, loose yards (in the unit), or compacted yards. The capacity of an equipment unit is then stated in terms of the volume of loose material, either struck or heaped, which it can contain. *Struck capacity* is the volume level-full, while *heaped capacity* is the volume of the unit with load heaped above the top of the body. For scrapers, the slope of the heaped portion of the load is often taken as 1:1. For trucks and trailers, it is variously taken as 1:1, 2:2, or 3:1.

To estimate the machine payload in bank cubic yards, the actual payload volume in loose cubic yards is multiplied by the load factor:

$$\text{Load (bcy)} = 1 \text{ cy} \times LF$$

Manufacturers' specification sheets supply load and volume ratings for their hauling equipment. In handling the lighter materials, heaped capacity may be reached before the permissible payload in pounds is attained; in such cases, sideboards might be added to increase the capacity.

Shrinkage Factor

The ratio between compacted measure and bank measure is called *shrinkage factor* (SF):

$$SF = \frac{\text{compacted cubic yards (ccy)}}{\text{bank cubic yards (bcy)}}$$

TABLE B1-1 Approximate Weight of Materials*

Material	lb/lcy	kg/loose m³	lb/bcy	kg/bank m³
Bauxite	2400	(1425)	3200	(1900)
Caliche	2100	(1250)	3820	(2265)
Cinders	960	(575)		
Carnotite, uranium ore	2770	(1630)	3680	(2185)
Clay:				
Dry excavated	1840	(1090)	2560	(1520)
Wet excavated	3080	(1825)	4280	(2540)
Dry lumps	1820	(1070)	2530	(1500)
Wet lumps	2700	(1600)	3750	(2225)
Natural bed	2130	(1265)	2960	(1755)
Clay and gravel:				
Dry	1940	(1150)	2290	(1360)
Wet	2220	(1255)	2620	(1555)
Coal:				
Anthracite, raw	2000	(1190)	2700	(1600)
washed	1850	(1050)	2500	(1485)
Bituminous, raw	1600	(950)	2160	(1280)
washed	1400	(830)	1890	(1120)
Decomposed rock:				
75% R-25% E[†]	3300	(1955)	4720	(2800)
50% R-50% E	2900	(1715)	3860	(2290)
25% R-75% E	2660	(1585)	3320	(1970)
Earth:				
Loam, dry excavated	2100	(1250)	2620	(1555)
Moist excavated	2430	(1440)	3040	(1805)
Wet excavated	2700	(1600)	3380	(2005)
Dense, packed	3100	(1840)	3880	(2300)
Flowing mud	2920	(1730)	3650	(2165)
Packed, dry	2560	(1520)	3200	(1900)
Granite:				
Broken	2780	(1650)		
Solid	—	—	4720	(2805)
Gravel:				
Loose, dry	2560	(1520)	2870	(1700)
Pit run	3240	(1920)	3630	(2155)
Dry, ¼–2 in	2840	(1680)	3180	(1685)
Wet, ¼–2 in	3380	(2005)	3790	(2250)
Sand and clay:				
Loose	2700	(1600)	3380	(2005)
Compacted	4050	(2405)		
Gypsum:				
Broken	3050	(1810)	5340	(3170)
Crushed	2700	(1600)	4720	(2800)
Solid	—	—	4720	(2800)
Hematite, iron ore	4150	(2465)	4900	(2905)
Limestone:				
Broken	2620	(1550)		
Solid	—	—	4400	(2605)
Magnetite, iron ore	4680	(2785)	5520	(3275)
Peat:				
Dry	680	(400)	1140	(675)
Wet	1350	(800)	2250	(1335)

TABLE B1-1 Approximate Weight of Materials* (*Continued*)

Material	lb/lcy	kg/loose m³	lb/bcy	kg/bank m³
Pyrite, iron ore	4340	(2580)	5120	(3040)
Sandstone:				
Broken	2550	(1505)	4260	(2525)
Solid	3920	(2305)	6550	(3885)
Sand:				
Dry, loose	2400	(1440)	2690	(1595)
Slightly damp	2850	(1680)	3190	(1895)
Wet	3120	(1860)	3490	(2070)
Wet, packed	3120	(1860)	3490	(2070)
Sand and gravel:				
Dry	2920	(1730)	3240	(1920)
Wet	3380	(2005)	3750	(2225)
Slag:				
Broken	2970	(1760)	4960	(2945)
Solid	3670	(2105)	6130	(3635)
Snow:				
Dry	220	(130)		
Wet	860	(515)		
Stone, crushed	2700	(1600)	4510	(2675)
Taconite	4050–5400	(2405–3200)	7090–9450	(4208–5600)
Topsoil	1620	(960)	2320	(1375)
Traprock:				
Broken	2950	(1745)	4420	(2620)
Solid	4870	(2880)	7300	(4320)

* Varies with moisture content, grain size, degree of compaction, etc. Tests must be made to determine exact material characteristics.
† R, rock; E, earth.

TABLE B1-2 Correlation Between Swell Percentage and Load Factor

Swell (%)	Load factor	Swell (%)	Load factor
5	0.952	80	0.556
10	0.909	85	0.541
15	0.870	90	0.526
20	0.833	95	0.513
25	0.800	100	0.500
30	0.769	5.3	0.95
35	0.741	11.1	0.90
40	0.714	17.6	0.85
45	0.690	25.0	0.80
50	0.667	33.3	0.75
55	0.645	42.9	0.70
60	0.625	53.8	0.65
65	0.606	66.7	0.60
70	0.588	81.8	0.55
75	0.571	100.0	0.50

Shrinkage factor is either estimated or obtained from job plans or specifications which show the conversion from compacted measure to bank measure.

Example. Construct a 10,000-yd³ bridge approach of dry earth with a shrinkage factor of 0.75. Haul unit is rated 14 yd³ struck and 20 yd³ heaped.

How many bank yards are needed? How many heaped loads are required, assuming 25 percent swell? (LF of 0.80 from Table B1-2.)

$$\text{bcy} = \frac{\text{ccy}}{\text{SF}} = \frac{10{,}000}{0.75} = 13{,}333$$

$$\text{Load (bcy)} = 1 \text{ cy} \times \text{LF} = 20 \times 0.80 = 16.0$$

$$\text{Number of loads required} = \frac{13{,}333}{16} = 833$$

Soil Density

This is defined as the weight of a unit volume of the soil. *Dry density* refers to the weight of dry solids in this unit volume, and *wet density* is the weight of the dry solids plus the weight of water in a unit volume.

Soil density is dependent on the relative volumes of the solid particles and the void spaces. It is also dependent on the specific gravity of the solids and, in the case of wet density, on the amount of water in the soil. The relative volumes of solid particles and void spaces are dependent on the type, nature, and extent of compactive effort. For a given set of conditions, the maximum wet density increases with increasing water content until the voids are almost entirely filled.

The specific gravity of the soil solids is the true specific gravity of the soil; it is equal to the weighted average of the specific gravities of all the mineral particles. Although the specific gravity of minerals varies widely, soils normally contain a preponderance of quartz and quartzlike minerals. This narrows the usual range of true specific gravity of soils to values between about 2.55 and 2.75. Within limited geographical areas, the actual range of variation may be much less.

Soil Compaction

Most earth structures require compaction to ensure a strong, uniform foundation. In the case of large earth fills, the material is usually spread in controlled layers and rolled or compacted by specialized equipment. Cuts also must be compacted; this is accomplished by first excavating below grade and then building up and compacting them. In either case, a foundation or subgrade is provided so that other structures such as buildings, pavement, or airstrips can be constructed.

The nature of soils is such that, as moisture is added to a dry sample, the density to which the sample can be compacted by a given compactive effort increases until an optimum moisture is reached. The density then decreases as more moisture is added, and the sample eventually turns into a plastic mud. Some soils are far more critical with respect to moisture content than others; that is, certain soils will obtain their maximum densities over a fairly wide range of moisture content, whereas others lose stability rapidly as moisture is increased beyond the optimum point.

ENGINEERING FUNDAMENTALS

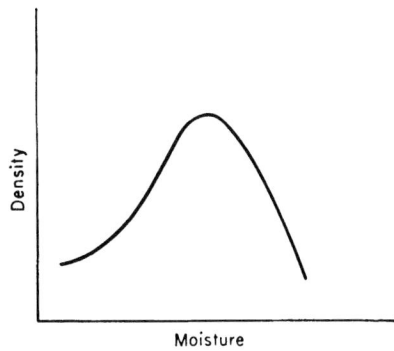

FIGURE B1-2 Typical moisture-density curve.

To predict the amount of moisture that should be in the soil at the time of compaction, a series of laboratory standard compaction tests is run with progressively increasing quantities of moisture. The results are plotted on a graph, with the density recorded along the vertical axis and the percentage of moisture recorded along the horizontal axis. A smooth curve can then be drawn through the plotted points, and the optimum moisture point with its corresponding density is read off this curve as it peaks (Fig. B1-2). This optimum moisture content is valid for the type and amount of compactive effort employed in the test. Under other conditions, such as those experienced in field compaction, a different optimum moisture content might be applicable.

Material Investigations

The values listed in Table B1-2 are consensus figures and are representative of commonly occurring soils and rocks. There are many instances where careful analysis of readily available data will permit more accurate estimating and help to minimize bidding uncertainties. For example, careful study of soil profiles on the plans will yield information as to specific soil types, locations, and extent. In some cases, moisture content and densities of soils in place may appear on the plans or may be obtained from the contracting agency. In addition, agricultural soil maps, air photos, geologic maps, and well logs are available and useful.

The contractor may also find it advisable to perform supplementary soil studies. Soil moisture and density measurements at shallow depths are readily and easily carried out and are often of major value. Careful examination of all available information, a minimum of exploration, and some testing may yield a surprising amount of information at a low cost.

EQUIPMENT CHARACTERISTICS

Power Units

Gasoline and diesel engines have been used for a considerable period of time as basic power sources for heavy construction equipment. Gas turbine engines have been more recently introduced as power sources and are used in conjunction with electric generators in much the same way as diesel electric units. Electric power is generated by the primary power source and is used to power electric motors. These motors then convert the electricity back into mechanical power. Both gas turbine and diesel electric generator units up to the 1000-hp class are now used in large off-highway trucks.

Gas and Diesel. There are many points in common between diesel- and gasoline-fueled engines. If this discussion is confined to reciprocating-piston four-stroke-cycle engines of the type used today, the principal components of both engines are quite similar. Such differences as exist are largely in the method of introducing fuel and in the process by which it is ignited. Nevertheless, these differences are important in the use and selection of engines.

As early as gasoline engines were developed and improved, high compression ratios were tried to obtain increased power and economy. It was soon learned that higher pressures were limited by preignition. This is a form of self-ignition which takes place when heat of compression, combined with heat passed to the air-fuel mixture by hot engine parts, becomes great enough to cause premature firing. Preignition limited early engines to a compression ratio of approximately 4:1; these engines were heavy, expensive, and uneconomical. Only about 7 percent of the fuel's potential heat energy was transformed into work—about the same as for a steam engine.

Improved gasolines and additives, with better ignition characteristics and impeded self-ignition, now permit the use of higher compression pressures. The upper limit on gasoline-engine compression ratio is presently about 12:1, although many engines operate in the range of 7:1 to 9:1 in order to burn cheaper lower-octane gasoline.

Largely because of the inherent limitations of the gasoline engine, the diesel engine was developed. Dr. Diesel applied the principles, outlined by earlier theoreticians, of a cylinder in which air is compressed and thereby heated. When atomized fuels are subsequently introduced into this cylinder, the resulting combustion is more complete and efficient as a result of the higher compression pressures. Compression ratios in diesel engines range up to 22:1, but 16:1 is a commonly used value.

At this point it might be well to consider the fuels which are used in gasoline and diesel engines. Both engines are adaptable to a wide range of fuels. Gas engines basically use gasoline but can be adapted with fair success to use natural gas, liquefied petroleum gases, manufactured gas, sewage gas, methanol, and light distillates. Diesel engines have a broad appetite for all kinds of diesel fuels, fuel oils, and crude, and they can be converted readily, with an ignition system, to burn natural gas. The use of the term *octane* in connection with gasoline and the term *cetane* in connection with diesel can be explained as follows.

- *Octane rating* is a measure of gasoline antiknock quality. A gasoline with a high octane rating can be used in a higher-compression engine, and generally it is a better fuel. Originally, gasolines were 50 octane, but now automobiles use 70 to 95 octane and aviation fuels exceed 100.

- *Cetane rating* is a measure or grading of diesel fuels according to ignition lag and length of time between injection and ignition. The quicker-burning fuel has the higher cetane number. In diesel fuels the range is from 30 to 60.

Most gasoline engines employ a carburetor to mix air and gasoline in the proper amounts. Approximately 15 lb of air are required for the combustion of 1 lb of gasoline; expressed on a volumetric basis, this is roughly the equivalent of 9000 ft^3 of air for each cubic foot of gasoline. These proportions are influenced by engine speeds, engine loads, ambient air pressure, temperature, and humidity. It is a difficult task to maintain optimum air-to-fuel ratios over the usual wide range of operating conditions. The system of mechanically metering and injecting fuel, as employed by a diesel, is generally more precise than a carbureted system.

Ignition is accomplished in gas engines by means of a spark plug which conveys a heat impulse to the air-fuel mixture. The spark plug is backed up by a battery to supply electrical energy, a coil to raise the voltage, a breaker to furnish high voltage, a condenser to provide a quick, clean electrical break, a distributor to time the spark impulses, a wire to complete the electrical path from the condenser to frame, and a ground to act as a return path for circuits.

The system is quite complex in comparison to the diesel engine system wherein a single, cam-operated plunger pumps fuel to an injector. The injector then sprays the fuel into the cylinder, as timed from a camshaft. There are various types of diesel systems—direct injection, precombustion chambers, separate or combined pumps and nozzles—but all are simple compared to the gas system. A precise air-fuel ratio is not required in diesel engines and, in fact, there is always excess air which is furnished by naturally aspirated, supercharged, turbocharged, and/or aftercooled systems. The excess air which is available in diesel engines, particularly turbocharged and aftercooled versions, accounts for lower combustion and exhaust gas temperatures and is an advantage in satisfying cooling requirements.

Temperatures of combustion vary widely. Naturally aspirated and turbocharged diesels will experience about 4300 to 4700°F, while aftercooled diesels may be in the neighborhood of 4000 to 4100°F. A typical gasoline engine will have peak temperatures of 5000 to 5200°F, at least 10 percent higher than naturally aspirated or turbocharged engines and a full 25 percent higher than aftercooled diesels. Exhaust gas temperatures are also higher in gasoline engines.

Peak cylinder pressures of a gasoline truck engine will run about 750 to 900 lb/in^2, somewhat higher than for the naturally aspirated, precombustion-chamber diesel. Turbocharged prechamber diesels have peak cylinder pressures of about 1000 to 1700 lb/in^2, depending on the degree of turbocharging. Naturally aspirated direct-injection diesels run to about 1700 lb/in^2, and when turbocharged to about 300 lb/in^2 higher. Diesel engines are built more heavily than similarly powered gasoline engines and can compress the charge more highly while keeping internal temperatures lower. Since they work at higher pressures, diesels extract a greater percentage of the available fuel energy than do gasoline engines.

Diesel fuel has more heat energy to begin with than does gasoline, and diesel engines use a higher percentage of the available fuel energy. Fuel consumption of a gasoline engine at full load will approximate 0.6 lb per bhp-hr while a comparable medium-duty diesel will average less than 0.4 lb per bhp-hr. At idling speeds the difference is even more marked, with fuel consumption rates of gasoline engines running three times those of diesels. Diesel fuel also costs less than does gasoline, although the differential per gallon may vary widely according to the locality.

Although it might seem from this discussion that there are no areas in which the heavy-duty gasoline engine excels, such is not the case. Gasoline engines, being lighter in weight, are favored where minimum weight is an overriding criterion. The light weight of their moving parts permits them to accelerate quickly, although this advantage is largely offset by the greater torque capability of the diesel. Since the gasoline engine works at lower pressures, it can be built to wider tolerances. Coupled to its light weight, this results in a low purchase price.

Gasoline engines, because of their wider tolerances, will run on fuel and oil too dirty for a modern diesel. Since less mass is involved, working at more liberal tolerances and at much lower cranking pressures, gasoline engines will start more easily at low temperatures than will diesels. However, this situation is exactly reversed if high humidity has caused severe water condensation on the electrical ignition system of the gasoline engine.

The ignition and carburetion problems experienced by gasoline engines, especially as they become older, are a major source of downtime which is not shared by diesels. Although service for gasoline engines is widely available, not every automobile dealer is able to adequately service the heavy-duty gasoline engine. On the other hand, as diesels have become more and more common, excellent service facilities can readily be located.

The diesel is designed to give many more trouble-free miles of service than its gasoline-engine counterpart. Top overhauls, especially, are more frequent on gasoline engines since the high combustion and exhaust temperatures lead to warped or pitted valves and valves seats, top ring wear, and piston cracking. This high heat load is further compounded by the fact that the lightly built (so-called "heavy-duty") gasoline engine operates at a BMEP* of 100 to 125 lb/in^2 at full load, and from 140 to 160 lb/in^2 at maximum torque. This compares with a range of 65 to 95 lb/in^2 at full load and 90 to 120 lb/in^2 at maximum torque for the medium-duty diesel.

Because of the volatility of the fuel, gasoline-powered vehicles are less safe in a collision. Gasoline stored either in the vehicle or in storage tanks is prone to fire and explosion hazards, and this is reflected in insurance rates. Many contractors are finding that diesel power gives them greater security.

Generally speaking, the heavy-duty modern diesel engine has virtually supplanted gasoline engines for most classes of construction equipment. Its former image as a huge, slow-speed engine no longer fits. Medium-duty direct-injector diesel engines are now competing with heavy-duty gasoline engines. Comparing a medium-duty diesel (up to 250 hp) and a popular heavy-duty gasoline engine, we find that the latter weighs only 400 lb less.

A final advantage of diesels—one sure to become increasingly important in a society growing ever more alert to air pollution—is the lower emission of carbon monoxide. Not only in city traffic, but anywhere there is a high engine density or low air circulation, the much lower carbon monoxide value in exhaust from diesels will constitute a major advantage.

In summary, the diesel offers:

1. Less need for servicing
2. Longer life
3. Lower fuel consumption
4. Lower-priced fuel
5. Lower fire hazard
6. Low CO emission

Gas Turbines. History records many uses of turbinelike machines to harness the energy of moving fluids and thus perform work. An obvious example is the windmill, which is powered by moving air. A modern turbine, simply stated, is a rotary engine actuated by reaction of a current of fluid. Air (fluid) is compressed, heated, and expanded over a bladed turbine assembly to produce power. The basic turbine cycle consist of four phases: compression, heat addition, expansion, and exhaust.

One type works as follows: Atmospheric air is drawn into the turbine and compressed by a rotating axial-centrifugal flow compressor. The compressor increases density and temperature of intake air. After compression, air is directed to combustion chambers where burning of fuel produces heat. Hot gases—directed by stator vanes—expand rapidly over turbine assemblies, applying force to blades and causing rotation. Gas-producer turbine wheel drives inlet air compressors; power turbine wheel drives gear reduction system. Finally, gases are ducted back to atmosphere.

---

* BMEP, or *brake mean effective pressure*, is a product of the mean effective pressure by mechanical efficiency, and is expressed in pounds per square inch.

Advantages of gas turbines are:

1. Lighter and smaller than heavy-duty reciprocating engines of comparable hp, both diesel and gas
2. No radiator or jacket water systems
3. Fewer moving parts—no pistons, rings, valves, rocker arms, crankshafts, etc.
4. Split shaft units, with one connected to compressor and the other an output section, are linked together only by gas flow. As load is applied to the output shaft and power turbine, causing lug, the gas-producer turbine wheel maintains full speed and power. At output-shaft stall the torque multiplies 2.2 times and thus provides a torque-converter performance.

Disadvantages of gas turbines are:

1. High fuel consumption
2. High initial price
3. Components are subjected to high temperatures and stresses

Table B1-3 compares a gas turbine with a diesel unit.

TABLE B1-3 Comparative Data for Gas Turbine and Diesel Engine

	Turbine	Diesel*
Hp, continuous	300	300
Weight, lb	450	6500
Weight/hp ratio	1.5	21.7
Length, in	40	95
Height, in	27	57
Width, in	25	40

* Weight includes torque converter and radiator.

Load Factor. The horsepower output of an engine is set by its manufacturer to ensure good service life provided the engine is used for the purpose for which it was designed. The engine will consume a given quantity of fuel when developing this rated horsepower (full load) and a lesser quantity at partial load. The load factor, expressed either as a decimal or as a percentage, reflects the ratio between actual and full load.

For example, an engine which is used on a pumping application which requires its full horsepower continuously will have a load factor of 1.0, or 100 percent. If the same engine is used on applications such as dozing or excavating, where it seldom works continuously at full load, its load factor will be less than 100 percent and its fuel consumption may be appreciably reduced.

A unit which is set to consume 10 gal/h at full load will consume only 8 gal/h at 80 percent load factor. Actual load factors depend on type of service, and they range from virtually zero up to 1.0. A tractor will have a high load factor, perhaps up to 0.9, when ripping or pulling plows at full throttle. A medium load factor of 0.75 is typical when dozing, pulling scrapers, or push loading, while a load factor of 0.5 might represent traveling with light loads or idling for long intervals.

Power Transmission and Controls

Transmissions. Some form of transmission of power—mechanical, hydrodynamic, hydrostatic, or electric—is used in practically every type of modern construction equipment. The possible variations are manifold as well as sophisticated.

The Society of Automotive Engineers developed the following terminology to describe transmissions:

Transmission is a device for transmitting power at a multiplicity of speed and torque ratios.

Mechanical transmission is one in which the speed and torque ratios are obtained by gears or other mechanical elements.

Countershaft transmission is a mechanical transmission in which gears are mounted on parallel shafts.

Sliding gear countershaft transmission is a countershaft transmission in which ratios are obtained by axial relocation of gears to mesh with other gears.

Constant mesh countershaft transmission is a countershaft transmission in which gears are constantly in mesh and ratios are obtained by engagement of positive or friction clutches.

Synchronized countershaft transmission is a constant mesh countershaft transmission in which means are provided for synchronizing the speeds of engaging elements.

Planetary transmission is a mechanical transmission using planetary gears.

Hydrodynamic transmission is a transmission using a hydrodynamic drive unit (fluid coupling or torque converter).

Hydrostatic transmission is a transmission using fluid placed under pressure by a positive displacement pump to drive a positive placement motor.

Automatic transmission is a transmission in which normal ratio changes are effected automatically without manual assist.

Semiautomatic transmission is a transmission in which some of the functions of normal ratio changes are effected automatically.

Power-shifted transmission is a transmission in which manually selected ratio changes are accomplished through application of power.*

Power-assisted shift transmission is a transmission in which the effort of manual ratio change is assisted by the application of power.*

Manually shifted transmission is a transmission in which all ratio changes are effected manually.

Interrupted drive transmission is a transmission in which ratio changes cannot be effected without interruption of drive.

Continuous drive transmission is a transmission in which ratio changes are effected without interruption of drive.

Split torque drive comprises two or more parallel paths of torque transmission, one of which is mechanical and one or more hydrodynamic.

---

* Normally hydraulic, pneumatic, or electric.

Electric-driven units are not included in the definition, but the power-shifted and power-assisted types use electric power in addition to hydraulic and pneumatic linkages to accomplish their functions.

Running Gear. Mobile construction equipment can generally be equipped with either tracks or wheels as its running gear. Each type has its advantages and disadvantages, and the choice for a particular job should be the one which obtains the desired production at the least cost. The basic factors which influence this choice will now be discussed.

When speed and mobility are important considerations, rubber-tired running gear has its biggest advantages over crawlers. Equipment that spends much of it workday shuttling between jobs (such as a utility machine) or on a job that does not call for a lot of traction (a dozer spreading fill, for example) probably should be on rubber. Speed and mobility become relatively less important when the unit is given a permanent job in a limited area.

Where traction requirements are high, the crawler running gear can offer significant advantages over rubber tires. Consider, for example, a tractor whose effective pulling force is limited by the tractive reaction which can be mobilized. In most materials of interest to contractors, tracks can develop more traction than rubber tires. As a consequence, a crawler-tractor will develop a greater effective drawbar pull than a wheel-type tractor of similar weight and horsepower. Thus, traction is the crawler's ally for jobs where high drawbar pull requirements exert more influence on the cost per yard than does speed.

The tractive limitation on pull power is the weight of the machine itself. No mobile unit of existing design can exert more pounds pull than the weight on its driving wheels or tracks. For instance, if a conventional automobile has only 40 percent of its total weight on the driving wheels, then the maximum pulling force which it can exert is limited to 40 percent of its total weight. Remember, this is a maximum figure. Underfoot conditions can make it considerably less, as anyone who has ever had an automobile stuck in a snowbank will verify.

In steep terrain, where high speeds are not obtainable, maximum machine gradeability may be important. This gradeability is also related to developable tractive reaction; hence, steep grades will tend to favor the use of crawler running gear.

Rolling resistance refers to the resistance encountered when a wheeled vehicle traverses a surface; particular importance is attached to the situation where this surface is a soil of low bearing capacity. Tires penetrate a soft surface, leading to increased rolling resistance. Tracks "build their own road," with rolling resistance essentially constant. The superior *flotation* of tracks will favor their use in unstable soils.

Power Ratings

Horsepower is defined as *an expression of rate of work output,* and all equipment manufacturers can supply specification sheets which show the rated power of their units. While these ratings may be entirely correct, they do not necessarily indicate the *working horsepower* in the sense that this term is interpreted by the contractor. Various factors can influence a horsepower rating.

Basically, there are two types of vehicle horsepower ratings: one based on *performance* and the other on *capacity.* Performance horsepower ratings report the actual work output of the engine as applied to the vehicle. Capacity horsepower ratings are generally described as *maximum* and tell the potential of the basic engine. Both are valid ratings, but they give the engine output under entirely different conditions and thus are not comparable. The four basic factors involved in determining engine horsepower are sea-level correction, rating speed or r/min, accessory load, and load duration.

A sea-level correction allows for the theoretical effect on engine rating caused by differences in ambient temperatures and barometric pressures. The air density where an engine is rated varies according to the location at which the test is performed. By correcting ratings to sea-level standard conditions, a basis for comparison is provided. The ratings, however, show only potential outputs under the specified conditions.

Rating speed, or r/min, indicates the engine speed at which the horsepower rating is produced. It is not necessarily the same r/min at which the engine is set for use in a particular vehicle. While the engine might have the same capacity rating in all uses, its actual output and r/min will be set to correspond with the power needed for a particular unit.

Accessory load defines how the engine is equipped for a given rating. Accessories (fan, generator, and air compressor) use up power, and a higher horsepower rating can be obtained by excluding them in rating calculations. The fan is the largest power consumer, in some cases taking as much as 30 hp from an engine.

Load duration specifies how long the engine must hold the load to establish the horsepower rating. Extreme output can be taken from an engine only for a short time. A maximum rating generally states that the load is held for a period of 5 min. A flywheel rating is based on unlimited operation of a vehicle when used for the purpose for which it was designed. Throughout the industry, six general rating descriptions are used:

1. *Maximum:* the bare engine, sea-level and temperature correction, extra r/min (beyond what will be applied in the vehicle), and 5 min of load
2. *Maximum:* bare engine, sea-level and temperature correction, vehicle r/min, and 5-min load
3. *Maximum:* bare engine, sea-level and temperature correction, vehicle r/min, and engine's rated load
4. *Flywheel:* only fan excluded, sea-level and temperature correction, vehicle r/min, and vehicle load (comparable to service conditions)
5. *Flywheel:* with accessories, sea-level and temperature correction, vehicle r/min, and vehicle load
6. *Flywheel:* with accessories, factor conditions, vehicle r/min, and vehicle load

The problem that arises now is trying to compare unlike horsepowers when choosing machinery for various jobs. To illustrate the effect of these rating definitions, let us take as an example an engine set at 100 hp and define this horsepower as *flywheel with accessories* (description No. 6). Correcting this particular engine to sea-level standard conditions—the no. 5 flywheel definition—could then bring the horsepower rating to 104. Moving to flywheel definition no. 4 and removing only the fan could raise the horsepower rating to 114.

The next alternative, no. 3, is a capacity-type rating. The bare engine is corrected to sea level at the engine manufacturer's rated load, which could result in a horsepower rating of 120. The no. 2 maximum rating is still based on the vehicle operating speed, but the short-duration load could result in a possible 135-hp rating. The ultimate of capacity ratings, no. 1, might reach as high as 150. Thus, the same engine with a flywheel-horsepower output of 100 could, by these various definitions, have a maximum horsepower as high as 150 for rating purposes.

As another aspect of horsepower ratings, the intended use of a vehicle and the conditions under which it will normally work can result in different flywheel-horsepower settings for essentially the same engine—with equal service life anticipated. The same engine, but with different r/min settings, is used in an industrial wheel tractor rated at 360 flywheel horsepower and a rear-dump truck rated at 400 flywheel horsepower.

Though set at a lower horsepower, the wheel-tractor engine will work harder than the truck engine because of job conditions. The wheel tractor uses power while loading, travels on rough haul roads with high rolling resistance, uses power when dumping on the fill, and usually returns on a road with a high rolling resistance. The machine also weighs over 10 tons more than the truck, both empty and loaded. The off-highway truck does not require power for loading, usually travels on well-maintained haul roads with low rolling resistance, needs only a small amount of power while dumping, and returns on a road with relatively low rolling resistance.

Maximum horsepower ratings reflect capacity, or capability, and flywheel ratings indicate the actual output performance. Flywheel-horsepower output of an engine is set by the manufacturer to give a vehicle a good service life. Thus, no limitations are made on the operation of the vehicle as long as it is used for the purposes for which it was designed and built. It becomes the manufacturer's responsibility to set the engine output to match intended usage and still give good service life. Contractors who know the differences in horsepower ratings are able to buy or rent construction machinery with a better understanding of what it can do on the job.

As used in construction equipment, rated flywheel horsepower must be further corrected for transmission and other losses in order to predict available power for useful work. Transmission losses can be 10 to 20 percent, with minimum losses in manual transmission and final reduction assemblies and with maximum losses in hydraulic transmissions. Electric generator units which power electric traction motors provide highly efficient transmission systems with losses of 10 percent or less.

Power Usable

In the preceding discussion of power ratings, it was indicated that power is energy in action—a force performing work at a given rate. The power setting for the engine will establish the theoretical *power available* for a given machine. It is then necessary to establish whether this power available is, in fact, power usable. The machine is adequately powered for its intended application if *power usable* is at least equal to the power requirement.

The power potential in all classes of equipment is limited by many factors. Among some of the important factors are altitude, temperature, wear, and inadequate maintenance. In addition, traction is a limiting factor which applies to transportation equipment and will be discussed later.

Altitude and Temperature. Internal combustion engines use oxygen from the air in the combustion process. At low altitudes, the pressure of the atmosphere is relatively great and a larger volume of air flows into the cylinders of internal combustion engines on the suction stroke. Due to the fact that the atmosphere is also compressed, the oxygen content per unit of volume of air is increased, so power output varies with density or weight of air surrounding the engine. Likewise, temperature affects power as the atmosphere expands with heat and reduces the amount of oxygen content per unit of volume of air.

From these statements, it can be concluded that oxygen content of intake air on a volumetric basis varies directly with an increase in atmospheric pressure (low altitude) and inversely with a rise in atmospheric temperature. Relative humidity also has an effect, as the water in moisture-laden atmosphere takes up volume that could otherwise be occupied by oxygen.

Since these factors affect the performance of internal combustion engines, a common basis of comparison is provided by the Society of American Engineers Test Code. A correction factor converts observed data to specified standard conditions, as indicated in the following:

1. Barometric pressure—29.38 in Hg
2. Temperature—85°F
3. Vapor pressure—0.38 in Hg

Reduction in rated horsepower thus occurs at the higher altitudes and temperatures and must be recognized in order to estimate actual performance. Approximate corrections can be made by the following rules of thumb for naturally aspirated engines:

Altitude For gasoline and four-stroke-cycle diesel engines, deduct 3 percent from sea-level rating for each 1000 ft of altitude above sea level.

For two-stroke-cycle diesel engines, deduct 1 percent for each 1000 ft of altitude above sea level.

Temperature Deduct 1 percent of rated power at 85°F for each 10°F temperature rise and add 1 percent for each 10°F temperature drop.

The use of turbochargers and aftercoolers reduces loss of power at higher altitudes and temperatures, but the number of variables precludes any convenient formula for predicting the net change in power. It is necessary to consult engine manufacturers for derating data based on specific operation conditions.

The wide variety of engines available today permits a sophisticated approach to the selection of the best engine for a particular application. For example, one manufacturer uses a naturally aspirated engine which has flywheel horsepower ratings ranging from 65 at 1680 r/min to 105 at 2200 r/min. The net horsepower at the flywheel of the vehicle engine is measured when operating under SAE standard ambient temperature and barometric conditions—85°F (29°C) and 29.38 in (746 mm) of mercury. Vehicle engine equipment includes fan, air cleaner, water pump, lubricating-oil pump, fuel pump, and generator. The engine will maintain full horsepower up to 2500 ft (760 m) altitude. The same basis engine if turbocharged can be rated at 115 flywheel horsepower and maintain this horsepower up to 10,000 ft without derating.

Wear and Inadequate Maintenance. These are present, in varying degrees, in most equipment units. They can further reduce power usable below the rated value of power available. For example, ring wear can result in blow-by and a loss of power which is recognized by the tractor operator as a reduction in gradeability. A fouled fuel system can prevent the flow of sufficient fuel for combustion, with a consequent loss of power. Dirty filters, clogged oil coolers, and air leaks in suction lines are additional causes for loss of power. These problems should be recognized and corrected as soon as possible. A well-planned preventive maintenance and inspection program is required to maximize equipment performance.

Traction. The ability of tracks or wheels to grip the ground is termed *traction*. It is dependent on the weight on the tracks or driving wheels and the type of surface on which the vehicle operates. Tracks will pull better on firm earth than on concrete, and rubber tires will pull better on concrete than on firm earth. Traction conditions, therefore, may limit the usable drawbar pull or rimpull.*

A specific type of underfoot condition can be expressed in terms of a *coefficient of traction* (see Table B1-4). This coefficient is a function of the surfaces in contact. It is determined by experiment and is then applied to the total weight on the driving wheels or tracks. For instance, if we say that the coefficient of traction for rubber tires on ice is 0.12, this means that wheel slippage can be expected if the required pounds pull exceeds 12 percent of the machine weight on the driving wheels. This represents only one limiting condition, since the actual power available in the form of potential rimpull may be less than the rimpull at which slippage occurs.

TABLE B1-4 Coefficients of Traction

	Rubber tires	Tracks
Concrete	0.90	0.45
Clay loam, dry	0.55	0.90
Clay loam, wet	0.45	0.70
Rutted clay loam	0.40	0.70
Dry sand	0.20	0.30
Wet sand	0.40	0.50
Quarry pit	0.65	0.55
Gravel road (loose surface)	0.36	0.50
Packed snow	0.20	0.25
Ice	0.12	0.12
Firm earth	0.55	0.90
Loose earth	0.45	0.60
Coal, stockpiled	0.45	0.60

The usable pounds pull at which slippage is imminent can be determined as follows:

$$\text{Traction limitation on usable pounds pull} = \text{coefficient of traction} \times \text{weight on drivers or tracks}$$

Example. What is the traction limitation on the maximum usable drawbar pounds pull of a track-type tractor pulling a compactor in loose earth? Weight of tractor: 32,600 lb.

Usable pounds pull = total weight of tractor × coefficient of traction

= 32,600 lb × 0.60

= 19,560 lb pull

---

* *Rimpull* is the pulling force, measured at the ground contact point, which the engine delivers to the tires.

Power Required

This depends on the characteristics of the unit, its functions, and the job conditions. Power is required for each equipment function, but these functions may not be performed concurrently. A mobile unit, for example, must overcome rolling resistance and grade resistance in order to achieve mobility. An elevating scraper, in addition to its function of mobility, also requires power to excavate and load material. A loader requires power for mobility and for loading, but these functions are not performed concurrently.

Rolling Resistance. This can be defined as the resistance to movement of mobile equipment at constant speed over level ground (Fig. B1-3). The following major factors determine the rolling resistance for rubber-tired units:

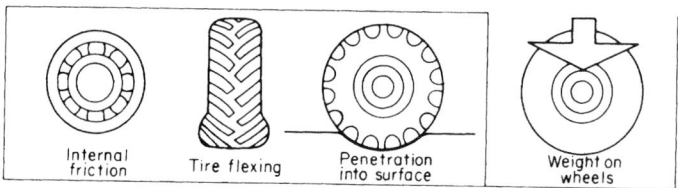

FIGURE B1-3 Factors affecting rolling resistance.

1. Ground penetration
2. Flexing of tire sidewalls
3. Speed
4. Wheel bearing friction
5. Total weight on tires

Factors other than ground penetration can be approximated, for all surface conditions, by a single value of 40 lb for each ton of gross vehicle weight (GVW) which is carried on the tires. It is also possible to relate the tire penetration on a given ground surface, approximately at least, as a function of this same GVW. By combining factors 2, 3, and 4 with a specified ground condition or surface (factor 1), the rolling resistance per unit of GVW can be stated as a single number, or rolling-resistance factor. It can also be expressed as a percentage of the GVW.

Table B1-5 lists typical rolling-resistance factors. Since these are functions of tire penetration, they can vary considerably for different segments of the same haul road. At a given point on the haul road, the rolling resistance may also change each time a unit passes. Rolling resistance is a very important factor and will frequently determine the maximum speed at which a given piece of equipment can operate.

To calculate the force or rimpull necessary to overcome the rolling resistance of rubber-tired hauling units, the entire weight of the unit, including both tractor and the loaded wagon or scraper, must be considered. For a tractor-scraper combination of this type, loaded with a gross weight of 56 tons and with a rolling resistance on the haul road of 100 lb/ton, the power required to overcome rolling resistance when moving at constant speed over level ground would be

TABLE B1-5 Typical Rolling-Resistance Factors for Rubber-Tired Equipment

Description of haul road	lb/ton	Percent of GVW
Hard, smooth, stabilized, without penetration under load	40	2
Firm, smooth, flexing slightly under load	65	3.25
Rutted dirt, flexing considerably under load	100	5
Rutted dirt, no stabilization, somewhat soft under load	150	7.5
Soft, rutted mud or sand, deep penetration under load	200–400	10–20

$$56 \text{ tons} \times 100 \text{ lb/ton} = 5600 \text{ lb rimpull}$$

or

$$112{,}000 \text{ lb} \times 0.05 = 5600 \text{ lb rimpull}$$

To estimate the force or *drawbar pull** required to overcome the rolling resistance offered by a track-type tractor towing a wheeled scraper, it is necessary to consider only the gross weight of the scraper unit. The rolling resistance of a track-type tractor is essentially a constant value, independent of ground penetration, since the unit lays its own travel surface as it advances. The values of drawbar pull which are supplied by tractor manufacturers have already been corrected for rolling resistance.

The following example illustrates the drawbar pull which is required to overcome the rolling resistance of a track-type tractor which pulls a wheeled scraper. Assume a tractor weight of 20 tons, with scraper and load weighing 44 tons, for a total weight of 65 tons. Assume also that the combination will be operating on a level haul road with a rolling resistance of 150 lb/ton. Since the rolling resistance of the track-type tractor has already been recognized in establishing its drawbar pull rating, it is only necessary to calculate the rolling resistance of the loaded scraper. The power required to move the unit at constant speed over level ground would then be:

$$44 \text{ tons} \times 150 \text{ lb/ton} = 6600 \text{ drawbar pounds pull}$$

Grade Resistance. Grade resistance (GR) is the force of gravity which must be overcome when going uphill. It is a function of the total weight of any vehicle, regardless of whether it is a track- or wheel-type machine.

In earth-moving work, grades are most frequently measured in *percent* slope, which is the ratio between vertical rise (or fall) and the horizontal distance in which the rise (or fall) occurs. For instance, a vertical rise of 5 ft in 100 ft horizontal distance would be a 5 percent grade.

When the grade is uphill or *adverse,* the effect is a demand for more power. Grade resistance then is a hindering force. If the grade is downhill or favorable, the effect is a helping force, tending to produce additional pounds pull to propel the vehicle. This downhill effect is commonly called *grade assistance* (GA). Regardless of whether the terrain is uphill, downhill, or level, *rolling resistance* (RR) is always present and must be considered.

---

* *Drawbar pull* is the pulling force, measured at the drawbar of a track-type mobile unit, which the engine delivers to the tracks.

When traveling uphill, a vehicle must overcome rolling resistance *plus* grade resistance.

When traveling over level terrain, a vehicle must overcome only rolling resistance.

When traveling downhill, a vehicle must overcome rolling resistance *less* grade assistance.

Both grade resistance and grade assistance are estimated in the same way. A rule of thumb states that each 1 percent of grade produces a hindering or helping force of 20 lb/ton of vehicle weight. This is in addition to rolling resistance. The formula can be expressed as:

GR (or GA) = (total equipment weight + load) × 20 lb/ton × units of % grade

Example. For a loaded track-type tractor-scraper combination weighing 64 tons (tractor, 20 tons; scraper and load, 44 tons) operating on a haul road with an adverse grade of 10 percent, the required drawbar pull to overcome grade resistance would be:

$$64 \times 20 \times 10 = 12{,}800 \text{ lb}$$

The maximum gradient (gradeability) for a mobile equipment unit may be fixed either by available power (drawbar or rimpull) or by tractive reaction which can be developed before slippage occurs.

$$\text{Maximum gradeability (\%)} = \frac{\text{useable drawbar or rimpull (lb)}}{\text{GVW (tons)} \times 20 \text{ lb/ton}}$$

$$= \frac{\text{useable drawbar or rimpull (lb)} \times 100}{\text{GVW (lb)}}$$

$$\text{GR drawbar (rimpull) hp} = \frac{\text{GVW (tons)} \times 20 \text{ lb/tons} \times \% \text{ grade} \times \text{mi/h}}{375}$$

Other Requirements. Mobile construction equipment may have either a singular function or multiple functions. The "pure" hauler is an example of a single-function unit, and a power shovel, a self-loading scraper, and a front-end shovel are examples of multiple-function units. Power must be adequate for all functions.

In the case of a "pure" hauler, the function is transportation of material, and the power requirement in terms of rimpull can be calculated by the formulas in the preceding paragraphs. In other cases, power requirements are complex and we must rely on observed results and experience.

Controlling Requirements. It is evident that equipment functions require power that is compatible with application and job conditions. The controlling power requirement for acceptable performance can be a combination of rolling resistance and grade resistance, or it can be within the category of complex requirements mentioned previously.

Neglecting the force required to overcome wind resistance and to provide acceleration, the total resistance to the motion of a mobile equipment unit can be taken as the sum of rolling resistance and grade resistance. The rolling resistance of a track-mounted self-propelled unit is considered to be constant, as was explained earlier, and has already been accounted for in the drawbar pull rating which is supplied by its manufacturer. The rolling resistance of a wheel-mounted mobile unit is a function of the particular ground surface and the weight which is carried on rubber tires.

If the entire weight of the mobile unit is carried on tires, as for a towed scraper or a wheel tractor, its rolling resistance and its grade resistance can be combined into a single figure for ease in calculation. This combined resistance can be expressed as required pounds pull per ton of GVW on the assumption that 1 percent adverse grade is equivalent to a 20 lb/ton rolling resistance. Alternatively, it can be expressed as an equivalent adverse grade by increasing the actual gradient by 1 percent for each 20 lb/ton of rolling resistance.

Graphically, these power considerations can be shown as follows: Figure B1-4 is a typical rimpull-speed graph (rimpull or pounds available on the vertical axis and speed on the horizontal). The usable power area is limited by the applicable traction factor, which is superimposed at the top of the graph.

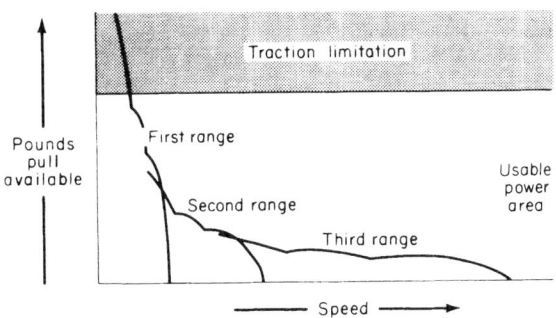

FIGURE B1-4 Traction limitation on usable power.

Figure B1-5 shows the same chart, but with the power required (total resistance) now superimposed at its bottom. The operating range thus falls between this *power required* and the *power usable,* which was shown to be limited by traction.

In the case of the power shovel that moves to a new location only when material is out of operating range, rolling and grade resistances are of little importance.

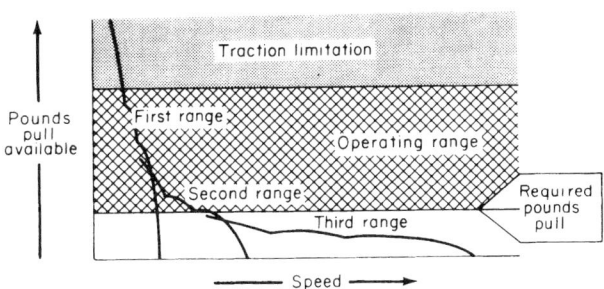

FIGURE B1-5 Operating range of usable power.

Engine horsepower becomes significant in order to provide power for digging and crowding into a bank and for swinging to load a truck or otherwise dispose of the material. Basic assumptions must be made by the designer regarding line pulls, line speeds, and swing speeds in order to produce a unit with acceptable digging power and rates. Clutches, brakes, cable sizes, shafts, gears, and machine stability must be adequate and contribute to a total design. A working balance of these elements depends on years of design and operating experience.

TIRES

Principles of Tire Engineering

A brief summary of some basic engineering concepts of the pneumatic tire should prove helpful in understanding the practical aspects of tire performance.

Fundamentally, a tire is a nonrigid torus. Its main structural members are high-tensile cords of textile fibers or steel cable. These cords are wrapped around wire bead foundations which fasten the entire structure to the rim wheel under compressive forces.

The uninflated tire is not capable of carrying any load beyond that which can be supported by the inherent stiffness of its envelope. In this state, the cord and bead structure are in a comparative relaxed condition. When the tire is inflated but unloaded, its cord structural members are put in tension. The inflation pressure exerts a force acting radially outward, and the tire envelope must contain this force. The cords, now under equal tension, exert a pulling force on the bead wires. The bead wires are in turn put in tension. This enables the tire structure to transmit bending and compression loads, which it could not do when uninflated.

When the tire is loaded, the compressive load stresses appear mostly as a relief of the tension stresses within the deflected contact area of the tire. The vertical components of tension in the tire walls above the ground are relaxed. Excluding the small increment of load supported by the tire structure itself, the total load on the tire exactly equals the reduction in tension within the walls. These reactions are shown in Fig. B1-6.

As indicated in Fig. B1-6, the loaded tire deflects at the ground-tire interface, resulting in a contact zone. The area of this ground contact and the pressures which are created at the ground-tire interface are of critical importance to tire life and performance. The area of the ground contact equals the applied load divided by the sum of the tire pressure and the pressure component exerted by the tire stiffness. This relationship can be expressed as

$$A = \frac{W}{P + P_1}$$

where A = gross contact area of tread-ground contact, in^2
W = applied load, lb
P = tire inflation pressure, lb/in^2
P_1 = tire stiffness component, lb/in^2

The relevance of tire load, pressure, and contact area to overall tire function is readily apparent. The load on a tire determines the percent of deflection at a particular pressure. These factors, with appropriate allowance for the tire stiffness compo-

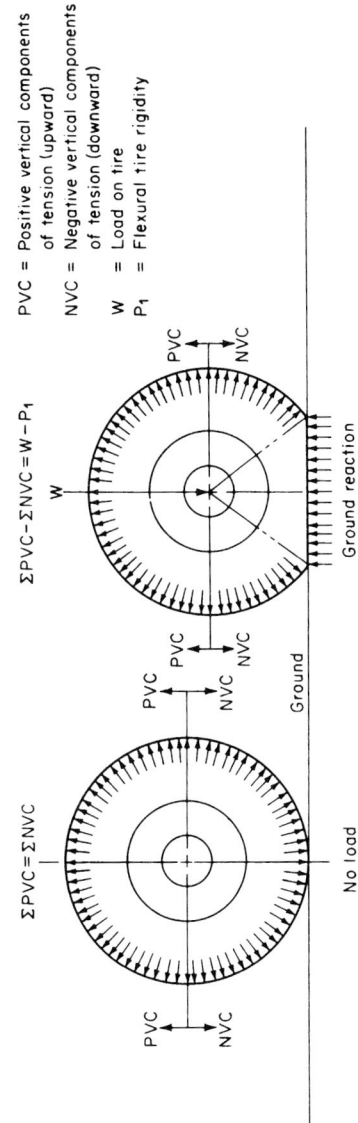

FIGURE B1-6 Tire load reactions. The sum of the positive vertical components (above the axle) minus the sum of the negative vertical components (below the axle) is equal to the tire load minus the increment of flexural tire rigidity.

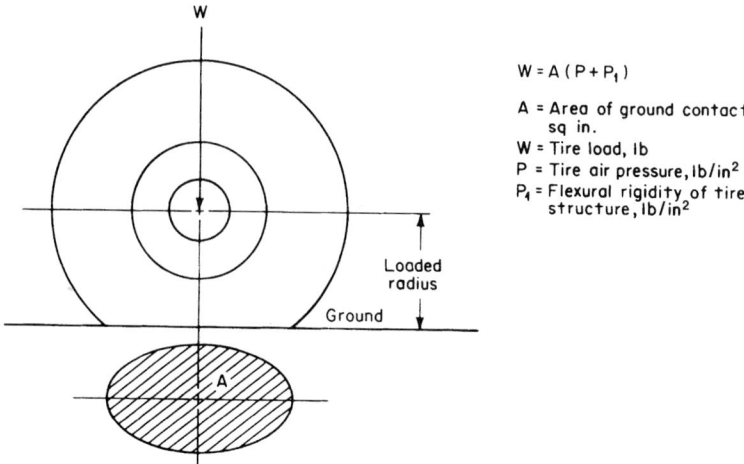

FIGURE B1-7 Load, pressure and ground contact.

nent, determine the contact area. Obviously, holding inflation pressure constant, the higher the load on a tire, the greater the amount of deflection. A greater amount of deflection increases the contact area and at the same time increases the work-load cycle of the tire (Fig. B1-7). This is of crucial importance because the hysteresis varies with the rates and magnitudes of the stress reversals which occur when a moving tire deflects under load. The greater the energy loss, the higher the temperature increase. Stress relaxation, heat generation, and other factors are manifestations of basic molecular processes which occur when a rubber structure adapts to various stresses.

Briefly, a small tire can carry a greater load by an increase in pressure. However, increasing inflation pressure beyond the recommended amount increases the carcass stresses. It also increases the average contact pressure on the tread and the side thrust on the rim flange. The accepted industry concept is, instead, to increase tire size when loads are increased. This results in a larger contact area while maintaining a moderate tire inflation pressure, and both tread and carcass stresses are kept within acceptable limits.

CHAPTER 2
EQUIPMENT ECONOMICS*

James J. O'Brien, P.E.
O'Brien-Kreitzberg, Inc., Pennsauken, New Jersey

INTRODUCTION

As equipment grows in size and becomes more complex, contractor-management must improve its ability to evaluate and utilize the new hardware. To survive financially, highway and heavy contractors must be knowledgeable in acquiring, utilizing, maintaining, and trading equipment. On most highway contracts today, the cost of equipment allocated against the job will exceed the cost of job labor. On some heavy contracts that are either high-volume, high-production jobs or predominantly rock excavation jobs, the cost of equipment may be double or triple the cost of labor.

This section covers the general concepts of evaluating equipment needs and is limited to the fundamentals of equipment economics. Other sections are devoted to specific construction requirements and include the economics of alternate methods of performing the work.

ACQUISITION

Forecasting Equipment Requirements

A strategy for building a specific construction capability requires good communications between those responsible for financial planning, bidding, and equipment use and maintenance and those who are responsible for construction management. Market surveys are needed to identify the potential work for the years ahead. Planned projects are evaluated as to soil conditions and equipment requirements. Forecasts based on the probability of obtaining projects are translated into future equipment needs by type and size of equipment. For example, a smaller highway contractor may conclude to obtain work for only two categories of scrapers, i.e., a 300-hp elevating scraper and a spread of the 400-hp wheel tractor-scraper size. A larger contractor

---

* Some of the material in this chapter is from E. A. Cox, "Equipment Economics," in Stubbs and Havers (eds.), *Standard Handbook of Heavy Construction,* 2d ed., McGraw-Hill, New York, 1971.

may conclude to justify a spread of high-production scrapers such as the 950-hp size in addition to the other two spreads. The number of units in each type of spread will depend more on the contracts already obtained and projects being advertised in the very near future.

When the contractor must acquire additional equipment to accomplish work under contract, the decision to purchase, lease with option to purchase, or rent without purchase option depends on the probability of obtaining sufficient work for the equipment to amortize the cost. The decision also depends on whether the equipment is normally available on a lease or rental basis when needed. Standard equipment is usually readily available on a rental basis, but it is often purchased due to the high probability of obtaining work for it. Special-purpose equipment such as a ringer for a Manitowoc crane often entails greater shipping expense when obtained on a rental basis, but this may be justified by job needs.

Based on market forecast, a contractor may elect to phase out a type of work such as pipeline construction by allowing the spread to become noncompetitive. When market conditions change, the contractor may choose to reenter the field with a major investment in new equipment. Under these conditions, the capital investment decision is much like a proposed investment in any other property. An analysis is required of the anticipated cash flows and earnings at varying levels of use. Occasionally, a piece of equipment will pay for itself on work already under contract and there is no question of profitability. Alternatives should always be evaluated.

In order for the larger contractor to keep track of equipment commitments and maintain high utilization, some system of displaying the planned use of equipment is required. For the smaller contractor, a schedule board may be used for a Gantt chart form of presentation. For the larger contractor, such a display may be automated through the use of a computer, as illustrated in Fig. B2-1. Such a system is of value only if accurate, up-to-date information can be obtained as to the location and condition of the equipment and all known requirements for its uses, including projected overhaul times. Marketing, in the heavy-construction sense of the word, is the matching of plant and personnel requirements of prospective jobs to the available resources of the contractor.

Selection of Equipment

Most new projects provide some opportunity to consider the selection of new equipment. The first consideration in any endeavor is getting the job done, and the selection of construction equipment should be based primarily on production considerations. In order to meet competition, however, and achieve productivity gains, cost reduction is essential. The contractor must select that equipment which may be used most effectively over its life, taking into account its effect on the utilization of the other equipment available.

Production estimates should always be based on the entire equipment system, with provision for move-in and move-out as well as for steady-state production rates. Allowances may then be made for requirements such as additional storage bins and for loss of production during start-up and shutdown. One of two approaches may be used. First, if a fixed time for accomplishing the operation has been specified, the contractor must size equipment to meet the time requirements. Second, if there is no stipulated time requirement, the contractor may make the most effective use of existing equipment and personnel. The first condition is the most prevalent, and the usual approach is to compute the average production rate required and the peak production rate required. The net working time is estimated by allowing for normal delays

MAJOR EQUIPMENT PROJECTION
AS OF 1 JUL

DESCRIPTION	EQUIP NR	JOB NR	STATUS	REMARKS
CMI AUTO GRADE SP30 66	329501	1562	RENT	
GR3301				
A C DD MOTOR GRADER 69	330104	1509	AVAIL	
GR3305				
CAT 112 GRADE 99HP D 65	330501	1557	AVAIL	
J D 570 MOTOR GRADER 68	330502	0024	RENT	
A C DD MOTOR GRADER 69	330503	1545	RENT	
GR3306				
NO. 12 MOTOR GRADER		1584		01
CAT 12 GRADE 115HP D 60	330629	1530	RENT	
CAT 12 GRADE 115HP D 63	330639	1545	RENT	
CAT 12 GRADE 115HP D 65	330640	1576	STORE	
CAT 12 GRADE 115HP D 65	330641	0024	RENT	
CAT 12 GRADE 115HP D 66	330642	0006	AVAIL	
CAT 12 GRADE 115HP D 66	330643	1545	RENT	
CAT 12 GRADE 115HP D 66	330644	1566	RENT	
CAT 12 GRADE 115HP D 66	330645	0024	RENT	
CAT 12 GRADE 115HP D 68	330646L	1561	RENT	
GR3309				
NO. 14 MOTOR GRADER		1584		01
CAT 14 GRADE 150HP D 63	330908	1450	RENT	
CAT 14 GRADE 150HP D 63	330911	0006	AVAIL	
CAT 14 GRADE 150HP D 64	330913	1559	RENT	
CAT 14 GRADE 150HP D 65	330915	1530	RENT	
CAT 14 GRADE 150HP D 66	330916	1529	RENT	
CAT 14 GRADER 150 HP 66	330917	1530	STORE	
CAT 14 GRADER 150 HP 66	330918	1487	RENT	
CAT 14 GRADE 150HP D 67	330919	1555	OTHER	
CAT 14 GRADE 150 HP 68	330920L	1561	RENT	
CAT 14 GRADE 150 HP 69	330921L	1566	RENT	
CAT 14 GRADE 150HP D 69	330922L	1562	RENT	
GR3310				
NO. 16 MOTOR GRADER		1584		01
CAT 16 GRADE 225HP D 65	331001	1581	1530	
CAT 16 GRADE 225HP D 66	331002L	1559	RENT	
CAT 16 GRADE 225HP D 67	331003L	1566	RENT	
CAT 16 GRADE 225HP D 67	331004L	1553	AVAIL	
CAT 16 GRADE 225HP D 68	331005L	1576	RENT	
CAT 16 GRADE 225HP D 68	331006L	1566	RENT	
CAT 16 GRADE 225HP D 68	331007L	1562	RENT	

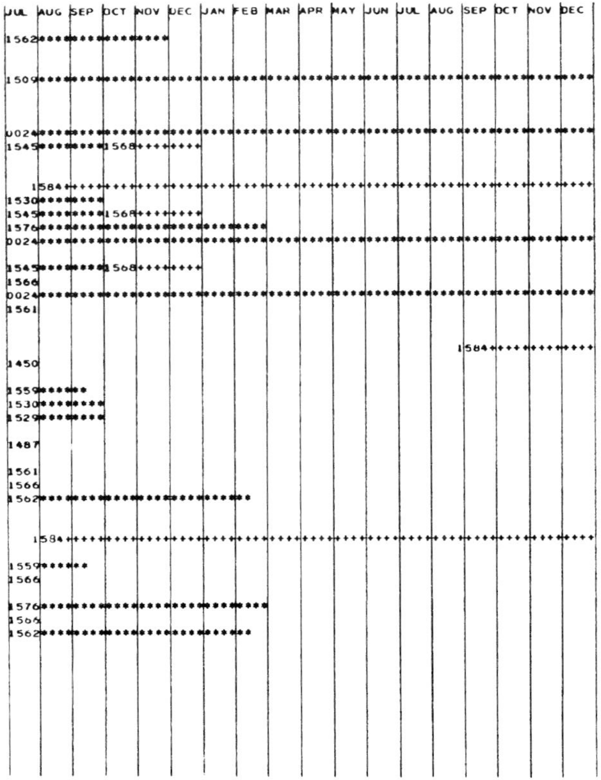

FIGURE B2-1 Projected equipment backlog of work. *(H. B. Zachry Company.)*

such as move-in time and normal weather loss, plus a reserve for contingent losses due to abnormal weather, labor strikes, and supplier delays. The quantity to be produced divided by the net working time in hours will give the effective production rate required. For example, 60 days may be allowed for excavating a channel containing 350,000 bank yards. Based on a 40-h week, there is a total of 343 h available, and the average production rate required is 1,020 bcy/h. Estimating a loss of 15 percent or 51 h for weather and another 5 percent or 17 h for contingent delays, the net working time becomes 275 h and the effective production rate required becomes 1273 bcy/h. Allowing for 83 percent working efficiency, the peak production rate becomes 1273 divided by 0.83 or 1534 bcy/h. The allowance for working efficiency provides time for personal delays, repositioning, minor servicing, unusual material characteristics, operator fatigue, and cleanup. In earth-moving estimates, this job efficiency is often recognized by basing production forecasts on a 50-min hour rather than on a 60-min hour. In a new or unique operation such as erecting a new type of structural component, there also may be a significant decrease in initial performance rates while the crew is learning the process. In such cases, it may be necessary to reduce the peak production rate not only by an efficiency factor but also by a learning factor in order to arrive at the required average production rate.

Fluctuation in production rates is absorbed in many ways. Aggregates are stockpiled to ensure an uninterrupted supply for the batch plant. On a pipeline, sufficient open ditch is maintained to ensure that the stringing and bending crews do not catch up with the ditching machine. Enough roadway is fine-graded before starting to place base material to ensure that the operation will be able to continue at its maximum production rate. Prior to starting a slab crew, a sufficient number of beams will be put in place to enable the crew to keep working without subsequently waiting for the placement of beams. Unless sufficient time and work are allowed between crews to minimize interference, efficiency may be lowered because equipment breakdowns and similar delays will affect a greater portion of the project. When this condition exists, equipment must be sized for a higher peak rate in order to obtain the required average rate. Contractors are becoming more cognizant of the need to be able to estimate the variations in the production rate as well as the average production rate. Where job conditions vary significantly, the production rates will vary accordingly.

Most equipment selection is based on production rates which are taken from textbooks, manufacturers' handbooks, or contractors' records. Manufacturers and their dealers are also performing equipment selection studies for contractors. These studies are usually made on the larger projects where contractors have a greater opportunity to amortize a large percentage of the cost of new equipment. Table B2-1 contains the results of a typical study. Computer programs for computing cycle times are available from equipment manufacturers if contractors prefer to use them in their own studies. The production figures are computed from machine payload, machine weights, rimpull curves, rotating-mass constants, average fixed times, rolling resistances, grade resistances, length-of-haul segments, and tire loading in tons per tire. Other algorithms include engine specifications, converter specifications, final drive ratio, engine brake horsepower and r/min, and other data for computing performance. Production may be based on either a 50- or 60-min hour (but no allowance is made for downtime); operator efficiency; haul-road conditions other than length, grade, and rolling resistance; and other job efficiency factors. The studies generally try to take into account the more significant job conditions, and earth-moving studies are often based on a seismic analysis of subsurface conditions.

These computer simulation techniques are being used by the larger contractors as well as by equipment manufacturers. As an example, Table B2-2 analyzes the

TABLE B2-1 Typical Equipment Production Study

			Caterpillar 666		Caterpillar 660		Caterpillar 650		Caterpillar 633	
Haul	Quantity, bcy	Distance one way, ft	Cycle, min	Production, bcy/h	Cycle, min	Production, bcy/h	Cycle, min	Production, bcy/h	Cycle, min	Production, bcy/h
A	1,897,965	5,700	6.26	409	8.36	306	8.01	260		
B*	210,885	8,200	8.11	316	11.15	230	10.81	192		
C	4,218	700	2.25	1139	2.67	958	2.32	897	2.70	534
D	4,702,735	5,967	6.47	396	8.66	296	8.31	250		
E	4,246,818	5,533	6.14	417	8.18	313	7.82	266		
F*	1,202,044	8,200	8.11	316	11.15	230	10.81	192		
G	421,770	6,700	7.00	366	9.48	270	9.13	228		
H	632,655	2,165	3.56	720	4.41	580	4.03	516	4.34	332
I*	3,078,921	13,050	11.68	219	16.54	155	16.26	128		
J	1,370,572	1,000	3.00	855	3.95	648	3.67	566	3.75	384
K	2,319,735	1,000	3.00	855	3.95	648	3.67	566	3.75	384

Note: Due to high ambient temperatures and long haul distances, the ton-mile-per-hour rating is exceeded when using the 666 scraper on the 8,200- and 13,050-ft hauls. Alternatives include reducing the average speed or payload of the machine. Tires are available for the other machines that are within the tmph.

TABLE B2-2 Typical Earthmoving Estimate Using Computer Simulation

Haul stations	Dist(ft) grade (%)	Quantity, yd³ fin area, yd²	No.	Class	Daily rate	Hourly rate	h/day	Min/h	Prod/unit	Yd³/h	Unit cost	Total cost	Total hours
CUT 333+94.84 − 336+0.0	1869.6	29213.1 HAUL	3	HU64570	77.70	59.82	10.0	50	243	731	0.333	14202	39.9
FILL 315+63.47 − 316+71.95	−4.9	13124 LOAD	2	TA73500	53.22	41.52	10.0	50	365	731	0.153		
CUT 322+36.31 − 324+0.0	642.2	6040.9 HAUL	3	HU64570	77.70	59.82	10.0	50	300	901	0.270	2382	6.7
FILL 316+71.95 − 316+98.26	−4.1	7147 LOAD	2	TA73500	53.22	41.52	10.0	50	450	901	0.126		
CUT 318+0.0 − 319+0.0	111.4	2144.9 HAUL	2	HU64570	77.70	59.82	10.0	50	339	679	0.240	867	3.2
FILL 316+98.26 − 318+0.0	−4.2	8544 LOAD	2	TA73500	53.22	41.52	10.0	50	339	679	0.165		
CUT 321+0.0 − 322+36.31	245.4	1947.0 HAUL	3	HU64570	77.70	59.82	10.0	50	327	983	0.246	705	2.0
FILL 319+0.0 − 321+0.0	−3.5	14650 LOAD	2	TA73500	53.22	41.52	10.0	50	491	983	0.114		
CUT 330+0.0 − 333+94.84	604.2	46155.9 HAUL	3	HU64570	77.70	59.82	10.0	50	297	893	0.273	18379	51.7
FILL 324+0.0 − 330+0.0	−7.4	25508 LOAD	2	TA73500	53.22	41.52	10.0	50	446	893	0.126		
CUT 371+29.92 − 373+61.08	3292.7	247816.9 HAUL	5	HU64570	77.70	59.82	10.0	50	186	931	0.435	137730	265.9
FILL 336+0.0 − 343+3.33	−7.0	44341 LOAD	2	TA73500	53.22	41.52	10.0	50	465	931	0.120		
CUT 349+0.0 − 353+0.0	455.0	58771.9 HAUL	3	HU64570	77.70	59.82	10.0	50	306	919	0.264	22750	63.9
FILL 343+3.33 − 349+0.0	−9.4	31573 LOAD	2	TA73500	53.22	41.52	10.0	50	459	919	0.123		
CUT 365+0.0 − 371+29.92	1138.7	639819.9 HAUL	3	HU64570	77.70	59.82	10.0	50	244	733	0.333	310374	872.5
FILL 353+0.0 − 365+0.0	−13.0	87848 LOAD	2	TA73500	53.22	41.52	10.0	50	366	733	0.153		

Source: H. B. Zachry Company.

earth-moving requirements for a highway project. Because of the cost of this type of analysis, a mathematical model is more likely to be employed where a given job situation will exist for a relatively long time. The major deterrent to its use is the cost of obtaining accurate input information as to elemental cycle times and the variation in cycle times.

Regardless of the method of estimating production rates for a particular project, the prudent contractor falls back on previous experience to verify answers. For this reason, most estimators have benchmark production rates for equipment which provides a cross-check on the job under consideration. These benchmark production rates have been derived from prior time studies or other measurements of production and are generally applicable to well-defined job conditions. Often, a new project can be classified as harder than one of the benchmark jobs but easier than another. Since obsolescence is shortening the economic life of much of the heavy equipment, it is important that good benchmark production rates be obtained as early as possible in the life of a unit.

After determining that an operation may be performed at a satisfactory production rate by two or more types of equipment, the cost of the alternatives must be evaluated. In comparing the return on investment from two or more alternatives, the additional factors of inflation, taxes, resale value, and the time value of money must be considered, together with dealer service and the equipment's reliability, adaptability, obsolescence risk, standardization potential, and other factors. The most important factors to be analyzed are the probability of future work at a profitable price and the accuracy of forecast maintenance cost.

A study of the used equipment being auctioned off or a walk through a large contractor's equipment yard, will invariably show evidence of misjudgments as to the utilization of equipment. Part of the equipment will reflect the high risk of obsolescence. Other equipment will reflect the unstable market conditions of the construction industry. The continued trend toward larger equipment has increased the pressure on contractors to keep it busy. Utilization is the most important economic factor for high-cost, low-maintenance equipment such as cableways and fixed cranes.

Maintenance cost of some equipment may more than double in severe operating conditions. As an example, the *Caterpillar Performance Handbook* shows a repair factor of 0.04 for a wheel loader operating in a free-flowing, low-density material. When loading shot rock, the repair factor jumps to 0.09, an increase of 125 percent. The economics of whether to use a shovel or a wheel loader for rock excavation depend primarily on the projected maintenance costs under these service conditions.

The analysis of the design characteristics of each manufacturer's equipment to determine its suitability to job conditions is an important aspect of equipment selection. Design criteria vary among manufacturers, and component options provide a wide range of cost and ruggedness. A balance among reliability, investment cost, and operating cost should be selected, since a management policy of selecting the lowest-priced equipment can often lead to higher costs. Consideration of design reliability should include criteria such as:

1. Reputation of the vendor
2. Warranty or guarantee
3. Service organization of the vendor
4. Use of reputable standard components
5. Adequacy of power train or drive mechanisms
6. Structural design at connecting points

Another consideration in the selection of equipment is the benefits to be derived by standardization. Cost savings due to family standardization may range from 5 to 15 percent of the purchase price of the equipment. The advantages of standardization come from flexibility in scheduling, reduced labor cost for maintenance and repairs, lower spare-parts-inventory cost, and greater purchasing power associated with a larger volume from a single supplier. The disadvantages of standardization include:

1. *Loss of operational efficiency*—due to timing of new models, small gains in efficiency often will be sacrificed to stay with one make. Also, the manufacturer may fall behind competitors in technological improvements.
2. *Loss of bargaining power*—the loss of competitive activity may lead to the manufacturer to give lower trade-in allowances or permit service to deteriorate.

In summary, a list of factors to be considered in comparing equipment alternatives should include:

1. Applicability of the equipment to the planned operation and project conditions
2. Reliability of the equipment and predicted maintenance requirements
3. Service support by the vendor and manufacturer
4. Flexibility and utilization potential of the equipment
5. Investment costs
6. Fuel consumption and operating requirements
7. Operator acceptability and training requirements
8. Safety features
9. Supervisor acceptability
10. Standardization considerations

Equipment Replacement

Success in attaining the optimum investment policy is fundamental to the economic growth of a contractor and, in an industry as competitive as heavy construction, the consequences of an unwise equipment investment or replacement policy can easily jeopardize the financial position of the company. Replacement of equipment is an investment and must be strictly considered as such.

When considering replacement or other equipment investment alternatives, it is important to remember that the past may be used only as a guide to the future. The price paid for equipment and the depreciation taken to date are of interest only as they pertain to taxes on capital gains. Investment decisions must be based on present circumstances and the best estimate of future conditions. Many variables will affect the return on investment. Some of these may be defined mathematically, while others, such as employee morale and safety, cannot as yet be accurately defined mathematically and must be judged by construction management.

A distinction between replacement of equipment due to its physical condition and displacement of the equipment due to obsolescence should be noted. Each new project, each new equipment model, and even each improvement to existing equipment models raises the issue of an investment opportunity by displacement. Contractors generally consider equipment replacement when downtime becomes excessive or when the time for a major overhaul approaches. Some owners review their equipment condition when awarded a new job and make replacement decisions at that time. Other contractors may review their equipment at year end and, based on tax position and available capital, make their replacement decisions then. However it may be solved, the problem of replacing equipment is a frequently recurring one.

Several general rules for replacement have been offered. One is to replace when the anticipated operating and overhaul costs plus the decrease in salvage value during the next period of use are the same as or greater than the operating and fixed-charge costs for a new piece of equipment. A similar rule is "as long as the average cost is greater than the marginal cost of extending the life of equipment by one additional year, do not replace—as soon as the marginal cost of one additional years' service exceeds the average cost, the asset should be replaced." The Caterpillar Tractor Company states in simpler terms that "whether cumulative cost per hour becomes progressively higher or lower with added machine hours is the key to the replacement decisions."

In practice, each time a piece of equipment is repaired, the cumulative cost per service hour is increased. The question becomes one of whether the repaired equipment can then earn enough to provide economic justification for the repair. The decision may be based on either previously established guidelines or a study of the specific piece. Guidelines may be used as a matter of policy; for example, previous replacement studies may have consistently shown that the type of equipment in question should be replaced immediately prior to the second overhaul. By using such a policy, the losses are restricted to the dispersion of individual cases from the average. Records will generally show that the magnitudes of the losses warrant individual consideration. For example, the condition of tractors of equal age may vary considerably due to varying degrees of job severity and maintenance expertise.

Mathematical models to describe the replacement situation have been in use since the early 1920s. They are constantly being improved by increasing the number of parameters considered in the estimates, and some are very comprehensive. Equipment replacement models will generally contain the following parameters:

1. *The value of money.* Interest rates are used for determining the cost of maintaining an investment in equipment and also for adjusting cost to a common time value. When comparing interest rate with rate of return after taxes, the relationship *rate of return equals interest rate times one minus the tax rate* may be used. The interest rate or rate of return after taxes should always be greater than the present inflation rate. If the inflation rate is considered separately in the study, however, it should not be included in the interest rate also.

2. *Inflation.* The inflation rate used should be applicable to the cost adjusted. There has been a significant difference between increased prices for new equipment and inflation for mechanics' wages.

3. *Taxes.* The method used should convert costs to a common tax basis before comparing them.

4. *Salvage values.* If a trade-in offer is more than the cash value of a piece of equipment, the difference is actually a reduction in the price of new equipment.

5. *Utilization.* In comparing alternatives, the cost should be based on the amount of work that will be accomplished by the alternative rather than on expenditures.

6. *Standby uses.* Alternative uses of present equipment, such as standby uses, should be weighed against keeping the old equipment on the job or replacing it with new.

Figure B2-2 shows the expenditures incurred to date for a specific D-8 tractor. It also shows the projected costs for a new D-8 tractor, based on the current purchase price and the average operating cost experience, after adjustment for the item value of money. An overhaul for the old D-8 is estimated at $57,000. An approach to making the decision whether to overhaul or replace is as follows:

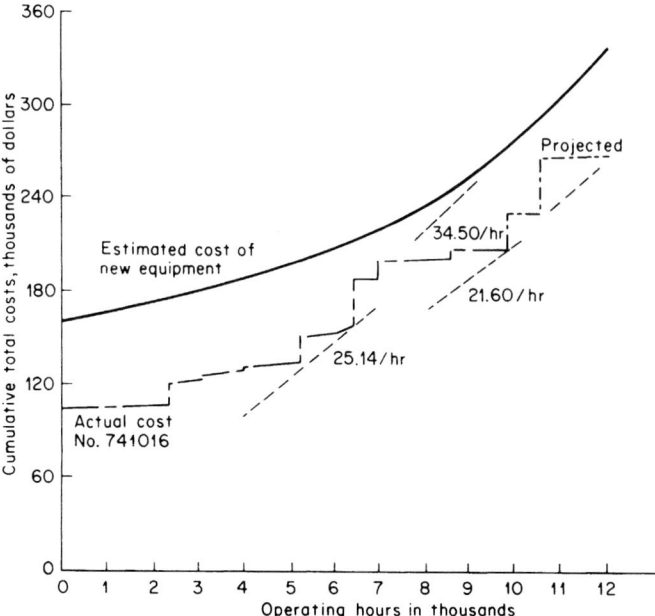

FIGURE B2-2 Comparison of actual expenditures for a D-8 tractor with projected expenditures for a replacement unit.

Old Tractor with $57,000 Overhaul
　Operating cost per hour next 3,000 hours:
　　　Repairs　　　　　　　　　　$6.00
　　　Downtime loss*　　　　　　　3.00
　　　Productivity loss†　　　　　　2.25
　　　Total　　　　　　　　　　　$11.25
　　　Amortization of overhaul:

Hours worked following overhaul	Total overhaul cost, $/h		Trade-in gain on overhaul, $/h		Overhaul amortization, $/h		Operating cost, $/h		Relative cost, $/h
1000	57.00	−	24.00	=	33.00	+	11.25	=	44.25
1500	38.01	−	12.00	=	26.01	+	11.25	=	37.26
2000	28.50	−	6.00	=	22.50	+	11.25	=	33.75
2500	22.80	−	3.00	=	19.80	+	11.25	=	31.05
3000	18.99	−	1.50	=	17.49	+	11.25	=	28.74
3500	16.29	−	0.75	=	15.54	+	11.25	=	26.79
4000	14.25			=	14.25	+	11.25	=	25.50
4500	12.66			=	12.66	+	11.25	=	23.91

*Downtime costs to the contractor are those incurred by idle workers, equipment, and other resources while waiting for a piece of equipment which is being repaired.
†Productivity loss is the difference in performance between the old tractor and a new one of the same type and is caused by wear, clogged fuel injectors, and similar factors affecting engine horsepower of the old tractor as well as improvements to the new tractor.

The alternatives under consideration are to increase any remaining unamortized investment in the old tractor by a $57,000 overhaul or to invest in a new tractor. For either alternative, the average hourly costs can be projected over periods of use ranging up to the expected life of the investment. Figure B2-2 indicates that the projected cost curve for the new tractor has a minimum slope at approximately 10,000 h of use and an hourly cost of $21.60. Since the projected hourly costs for continued use of the old tractor provide for amortization of overhaul costs only, with no allowance for any residual value of the tractor prior to the overhaul, any trade-in allowance for the old tractor can be treated as an effective reduction in the acquisition cost of the new tractor. The project minimum cost of $34.50/h for the new tractor will be correspondingly lowered by approximately $1.50 for each $15,000 of trade-in allowance.

New tractor (life assumed to be 10,000 working hours)

With trade-in of, $	Relative cost, $/h	With trade-in of, $	Relative cost, $/h
0	34.50	45,000	30.00
15,000	33.00	60,000	28.50
30,000	31.50	75,000	27.00

If a trade-in of $45,000 is offered, the new tractor will cost about $30/h of use and the old tractor must work approximately 2800 h after overhaul to recover the cost of the repair at the $30/h rate. If the trade-in amount were increased to $75,000, the new tractor would cost about $27/h, and the old tractor would need to work about 3400 h at $27/h to recover the overhaul cost.

Figure B2-3 shows a break-even chart for the overhaul or trade decision. As an example, if a trade-in of $60,000 were offered, the overhaul would have to be amortized over a 3000-h period in order to be justified.

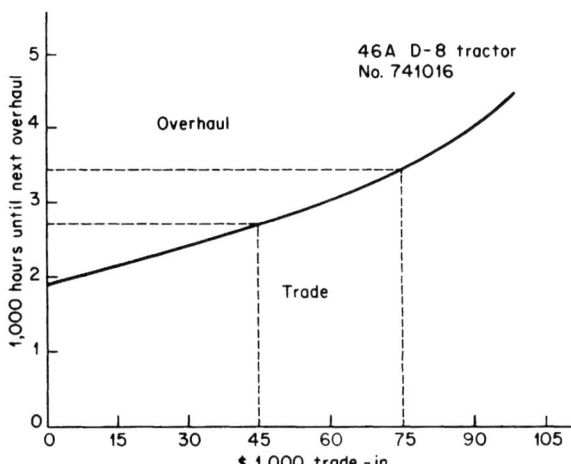

FIGURE B2-3 Break-even chart for overhaul or trade.

The method just described is a general one for construction equipment and may not be applicable to other types of equipment or to equipment operated under extreme conditions. The most important aspect of replacement is a good analysis of the difference in productive capability. It is also important to make a good appraisal of the value left in a piece of equipment; there is sometimes a tendency to forget that equipment expenditures do not become costs until they are consumed at some later date. For example, an expenditure for a new generator adds value that becomes a part of the cost of using the equipment in the future. The new generator may have substantially more value to the contractor as a new piece of hardware with many hours of life remaining than as a piece of hardware to be immediately traded in or sold. The overall value of many replaced parts in terms of additional useful life is often underestimated.

Although it is possible to allocate some downtime costs to the piece of equipment causing it, this would be very impractical with most accounting systems presently in use. Also, the average downtime costs for a D-8 tractor could have no relationship to the future work planned for the unit. A policy of downgrading the use as a tractor gets older is frequently followed.

All successful replacement systems have the common requisite that a contractor must know equipment condition and cost to be able to intelligently make the best trades for it.

Financing

If the investment in equipment represents old, unreliable, noncompetitive equipment, the contractor may find it impossible either to bid jobs profitably with existing equipment or to borrow funds to purchase new equipment. The company with modern, high-production equipment may still require protection against the unforeseen loss of contracts or other contingencies that are inherent in the construction industry. Additional financial leverage may also be desired to provide the necessary cash flow for a profitable expansion. When financing of equipment becomes necessary, there are four common alternatives. These are dealer financing, leasing, bank financing, and financing through a commercial finance company.

Most dealers have convenient, uncomplicated financing plans available at reasonable interest rates. They are often in a better position to require a smaller down payment. Financing through the dealer does not give as much freedom to shop the market as do other plans, however.

Leasing offers several attractive advantages and has become a very popular form of financing. There may be a tax advantage by expensing monthly rentals as they are incurred. Whether this will be allowed by the tax authority depends on the nature of the lease and, specifically, the provisions for capturing title to the equipment after expiration of the lease. Personal property taxes may favor leasing. Manufacturer's warranty may also be passed along to the contractor by the leaser. Equipment from several manufacturers may be combined under one lease and, with a firm commitment from a lessor in hand, a contractor may shop as on a cash basis. Usually no down payment is involved, so leasing may appeal to companies short on cash since it will enable them to present stronger balance sheets.

Commercial banks often limit the amount they will lend to an individual contractor, and it is possible that equipment financing will restrict the amount available for unsecured loans.

One of the large commercial finance companies may finance a contractor's equipment as a chattel mortgage loan, or it may purchase the dealer's installment sales contract. Under a chattel mortgage loan the title passes to the purchaser and debt collateral often includes greater assets than the specific piece. The interest as well as the default risk is generally higher than for a title retention plan, and generally the latter is preferred.

The extent to which favorable cash flow can be maintained and taxes are prudently managed compounds the significance of having acceptable debt alternatives available. Financing is certainly one of the most important aspects of equipment economics and should be accorded careful investigation based on competent advice.

MAINTAINING AVAILABILITY

The importance of adequate equipment maintenance continually increases. The trend toward larger, more sophisticated machines keeps the cost of construction equipment constantly increasing and, along with the greater production rates, there is a greater loss whenever a machine cannot be producing. Improved reliability has been engineered into machinery, and better materials and more sophisticated controls are now being used. However, even the best equipment can be misused by the operator who tries to maintain production under adverse job conditions after a bad night off the job. Good operating practices are overlooked, breakdowns occur, and valuable time and resources are wasted.

Reducing Downtime

It is a project manager's prime responsibility to make a profit on the project while exercising reasonable prudence in the care and utilization of resources. To get the most out of the equipment dollar, it is essential that construction management appreciate the relative costs of maintenance and of downtime. An overemphasis on production costs may influence contractors to push their equipment to obtain maximum production performance. For example, sideboarding scrapers and trucks is still common practice.

The additional equipment costs resulting from adverse job conditions are often underestimated or ignored. If a load on a bearing is increased to twice the design load, its expected life is decreased to one-tenth its design life. The life of a bearing varies inversely with the load raised to the $3\frac{1}{3}$ exponential power. Reduction in gear life under excess loads is even greater than bearing-life reduction. The effect of speed on bearing life is almost a straight-line relationship. If speed is increased by 20 percent, then bearing life is reduced by 20 percent. Abuse is sometimes justifiable when an unusual project or circumstance develops, but construction management needs to be cognizant of the consequent price that is paid.

Figure B2-4 shows a concept that has been receiving increasing attention recently. As construction processes become larger, more automated, and more highly integrated into well-balanced systems, the adverse effect of downtime is more pronounced. The use of larger assemblies or standby units reduces both downtime and the skill level required of field mechanics. At one extreme, no spare parts may be carried on the project. At the other extreme, for example, a standby central-mix plant could be kept at the job site. Somewhere between these extremes is the most economical level of standby inventory.

Most construction projects lend themselves to segmentation, where an interruption on an early operation will not affect a later operation unless the interruption is of an

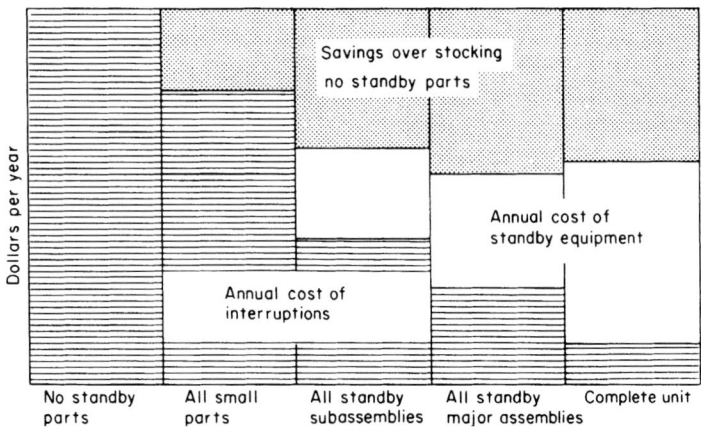

FIGURE B2-4 Concept of optimum inventory component level.

unusually long duration. For example, sufficient aggregate can be stockpiled so that a crusher breakdown can be repaired before the central-mix plant runs out of aggregate. Shutdown of the central-mix plant, however, stops the hauling units and paving train as well. As the continuous-process plant becomes more common in construction, analysis of the economics of standby parts or components may be desirable.

Consider a Euclid belt loader pulled by two D-8 tractors servicing a balanced hauling spread of 10 Euclid wagons. An analysis of maintaining standby loader components may be approached as follows:

Assume:

Cost of standby loader	$122,250		
Less salvage	$24,500		
Annual cost at 8 years		$12,500	
Debt service (40%)(10%)(56%)(122,500)		2,744	
Maintenance		7,000	
Total annual standby cost		$21,994	
Spread downtime cost per hour	$525/h		

2—D-8 (loader)
10—LDT34 bottom dump
2—No. 16 motor grader
1—Compactor (sheeps foot)
2—D-8 dozer (cut and fill)
1—Water wagon
Miscellaneous

	Without standby	With standby
Engine		
100-h delivery—6 h installation		
Major repair 1.5 times/yr		
Annual downtime	159 h	9 h

Belt
  30-h delivery—16 h installation
  10 times/yr
  Annual downtime   460 h   160 h
Other
  Frequency—75 times/yr
  Average duration—2 h
  Annual downtime   150 h

Alternatives:

Ci = Cost of interruption caused by component failure

Ci = Annual frequency × avg h duration × downtime cost/h

1. Complete standby loader
   Annual cost of standby unit* $21,994
   $Ci - (86.5)(0.3)(525)$ 13,624
   Total downtime costs $35,618

2. Both belt and engine standby
   Cost of standby belt $1,925
   Cost of standby engine   5,513
   Ci belt = (10)(16)(525)   4,000
   Ci engine = (1.5)(6)(525)   4,725
   Ci other = (75)(2)(525)   78,750
                                                    $174,913

3. Standby belt only
   Cost of standby belt $ 1,925
   Ci belt = (10)(16)(525)   84,000
   Ci engine = (1.5)(106)(525)   83,475
   Ci other = (75)(2)(150)(525)   78,750
                                                    $248,150

4. Standby engine only
   Cost of standby engine $ 5,513
   Ci belt = (10)(46)(525)   241,500
   Ci engine = (1.5)(6)(525)   4,725
   Ci other = (75)(2)(525)   78,750
                                                    $330,488

5. No standby
   Standby cost $ 0
   Ci belt   241,500
   Ci engine   83,475
   Ci other   78,750
                                                    $403,725

---

* It is assumed that 0.3 h will be required to shift to the standby loader in the event of a breakdown of the primary loader.

A major project involving precast components may warrant consideration of the length of time a piece of equipment such as a gantry crane could be down for repairs. It may be possible and desirable to build an inventory of the precast components which require the use of the crane so that construction can be continued by using up the inventory when the crane must be taken out of service for repairs.

To be complete, an evaluation of an acceptable level of downtime must take into account the economics of field repairs versus home shop repairs versus dealer repairs. It must consider the level of parts a dealer may consign to a job as opposed to a purchase inventory. For the more complex assemblies, the specialized training of the dealer must be weighed against the training of the contractor's mechanic.

Current practice for establishing inventories is to select the percentage of original equipment cost that will be maintained and then develop a list of spare parts to equal this amount. On domestic work where a fast part supply is available, the average inventory level is around 3 percent of original cost. Most major manufacturers are able to locate parts if available anywhere within the United States and have them on the job site within 24 h. On overseas work, the inventory level will vary more, depending on the urgency in completing the project. Public works projects will require inventory levels in the 6 to 10 percent range. A defense installation may require 10 to 18 percent or higher.

Inventories for individual equipment models may be maintained at any desired level up to and including a complete standby unit. The factors to be considered include (1) the amount of construction that would be stopped by the failure of the machine; (2) the time required to obtain the necessary part at the job site; (3) the probability of failure considering hours of use since new or last overhaul as well as severity of use; (4) the complexity of the unit or assembly; and (5) the disposition of excess spare parts after project completion.

Servicing

The costs and benefits of preventive maintenance programs are difficult to evaluate, and practice varies widely. An inadequate lubrication program can cause quick deterioration of wearing parts and loss of performance through contamination. Breakdown repair cost and downtime can quickly become excessive. A blown gasket permitting glycol antifreeze to contaminate the oil can rapidly necessitate an engine overhaul.

A good preventive maintenance program prolongs equipment life, reduces cost of downtime, and minimizes lubrication cost. It can mean the difference between 85 and 95 percent availability. This 10 percent increase in availability may represent an effective reduction in equipment costs of about 3 percent. Standardization of lubricants can reduce purchase cost and inventory, and oil drain intervals may be extended through analysis of oil samples. Testing may be accomplished by differential infrared analysis, membrane filtration analysis, spectrographic analysis, spot tests, and other laboratory procedures. Many of the oil distributors can provide a testing service at moderate cost.

Table B2-3 shows a computer-produced report from a system employing a spectrographic analysis. The permissible amount of contaminant in the oil depends on the type of engine and the oil used. Heavy-duty S-1 oil, recommended for some equipment, has a high boron content as opposed to S-3 oil, recommended for other types of equipment. The effective use of a sampling program requires frequent consultation with equipment dealers to diagnose the possible causes of high contaminant levels.

Repairs

With the best of care, equipment breakdowns will still occur. The contractor must then seek the most economical level of maintenance capability to meet the needs. The economics of an optimum balance among number of mechanics, parts inventory, operator training, movement of equipment for repairs, and dealer arrangements are very complex and intangible. Linear programming techniques can be used to plan and schedule maintenance works to best utilize the facilities available and thus minimize repair costs. For example, a contractor with more than one shop will frequently be able to do particular maintenance jobs more economically in one shop than in another. A dealer may also do certain repairs that the contractor cannot perform.

The establishment of repair priorities and the effective scheduling of repair jobs will reduce repair costs and improve the utilization of equipment. The success of such efforts will depend on knowledge of the construction schedule and the condition of working equipment.

OWNERSHIP AND OPERATING COSTS

The cost of using a given piece of equipment may vary widely due to the influence of the many relevant variables. These variables include local weather conditions, timing of the purchase, obsolescence due to the introduction of new equipment, and other difficult-to-predict factors. Recommended procedures for estimating costs vary from simple straightforward forms that do not attempt to relate cost to the users' specific applications to very sophisticated models whose requirements of input data are well beyond the ability of the average contractor to provide. The following paragraphs will present an overview of equipment economics, and subsequent paragraphs will identify accounting information needs and outline procedures for estimating and monitoring equipment costs. From the viewpoint of establishing the right price for a given project, the contractor is interested in the equipment investment which must be committed to a project and in the equipment value that will be consumed in order to complete the project. Also of concern are the administrative costs of owning and managing the equipment.

Value Accumulation

By the time a contractor has moved a piece of equipment onto a project, there has already been a substantial investment. This includes the unamortized cost of the equipment and any major modifications to it, plus the miscellaneous costs of ownership. If it is not a new piece of equipment, the investment may include a prior overhaul plus preventive and field maintenance costs.

The acquisition cost of a piece of equipment is normally regarded as the machine price including attachments, the sales tax if any, and the freight charges to the contractor's yard or initial project. In the event of a trade-in, the equipment acquisition cost may be regarded as the purchase price adjusted by the difference between the trade-in allowance and the expected price the contractor could receive after selling expenses on the open market.

Miscellaneous costs of ownership include insurance, ad valorem or personal property taxes, and the cost of special service equipment. Interest is generally included as a miscellaneous ownership cost but cannot be properly considered without relating it to the overall capitalization policy of the company. If a contractor

TABLE B2-3 Laboratory Report Showing Oil Contaminants

Date processed:
Job no:
Job location:

Unit number	Sys code	Equipment description	Sample date	Total eng. h/mi	Visc. 210°	Flash point
453308	A1B	CAT 944 LOAD 2YD 63	04-08	2284	062	440
453308	A1B	CAT 944 LOAD 2YD 63	04-26	2401	069	440
453308	A1B	CAT 944 LOAD 2YD 63	05-24	2528	000	067
453308	A1B	CAT 944 LOAD 2YD 63	06-14	2586	067	420
453311	A1B	CAT 944 LOAD 2YD 64	09-08	5811	073	443
453311	A1B	CAT 944 LOAD 2YD 64	09-21	5885	072	465
453311	A1B	CAT 944 LOAD 2YD 64	10-05	5951	000	000
453311	A1B	CAT 944 LOAD 2YD 64	11-09	6097	065	435
453311	A1B	CAT 944 LOAD 2YD 64	11-23	6146	071	450
453311	A1B	CAT 944 LOAD 2YD 64	12-14	6204	068	445
453311	A1B	CAT 944 LOAD 2YD 64	01-04	6228	064	430
453311	A1B	CAT 944 LOAD 2YD 64	02-01	6267	065	430
453311	A1B	CAT 944 LOAD 2YD 64	04-26	6415	067	440
453401	A1B	EUC L20 LOAD 2½Y 63	02-12	1485	053	350
453402	A1B	EUC L20 LOAD 2½Y 67	02-06	3489	066	440
453403	A1B	CAT 950 LOAD 2½Y 68	09-28	1075	000	000
453403	A1B	CAT 950 LOAD 2½Y 68	03-29	1821	071	470
453403	A1B	CAT 950 LOAD 2½Y 68	04-26	1924	067	445
453403	A1B	CAT 950 LOAD 2½Y 68	05-24	1987	064	395
453403	A1B	CAT 950 LOAD 2½Y 68	06-14	2063	065	420
453505	A1B	CAT 966 LOAD 2¾Y 65	11-09	2002	072	460
453505	A1B	CAT 966 LOAD 2¾Y 65	12-14	2034	066	440
453505	A1B	CAT 966 LOAD 2¾Y 65	01-18	2084	063	000
453505	A1B	CAT 966 LOAD 2¾Y 65	01-31	2093	063	435
453505	A1B	CAT 966 LOAD 2¾Y 65	05-24	6654	066	435
453505	A1B	CAT 966 LOAD 2¾Y 65	06-14	6735	066	415
		LOAD 2¾Y 65	11-23	2018	073	470
		LOAD 2¾Y 65	11-09	1707	071	460

Source: H. B. Zachry Company.

elects to finance half of the equipment to invest capital in a new nonconstruction venture, the cost of interest would not be an equipment cost. On the other hand, if financing becomes necessary in order to obtain larger or more modern equipment, the financing cost would rightfully be allocated to the equipment.

Major modification costs should be regarded in the same manner as investment in new equipment. Equipment modification may serve a need that is not satisfied by available standard equipment, or it may be a method of realizing greater value than could be obtained by trading in old equipment or selling it outright.

A major overhaul restores equipment value that may be realized either through additional use or through a higher sale price. In either event, the overhaul cost represents an additional investment in equipment value to be realized at some future date.

Preventive and field maintenance also restore equipment value which is subsequently consumed through equipment use. A broken hydraulic line, fouled spark plug, or even contaminated oil will make the equipment less valuable, whether to the project or to a different user. Replacing the parts or the contaminated oil increases the value of the equipment to the project or to a subsequent purchaser. In order to

EQUIPMENT ECONOMICS

SAE grp.		Spectrographic analysis							
Field	Lab	Boron	Lead	Iron	Alum	Copper	Chrome	Tin	Silicon
000	000	011	007	006	004	000	001	015	004
000	000	003	018	049	012	000	014	021	005
000	425	002	013	034	007	000	006	015	004
000	030	002	010	015	005	000	004	015	004
030	030	001	007	017	007	000	005	023	004
030	030	003	007	017	006	001	003	018	004
030	000	002	007	021	008	001	009	027	004
030	000	002	007	022	006	000	008	023	004
030	040	003	007	017	004	002	009	023	004
030	000	002	007	014	004	001	005	029	004
030	000	001	008	017	004	002	004	015	004
030	000	003	007	031	007	001	007	022	004
030	000	002	008	017	006	001	000	013	004
030	000	087	016	118	024	005	005	033	028
030	000	095	007	070	013	000	000	010	018
030	000	005	015	077	016	006	011	028	005
000	000	005	008	038	013	000	007	022	004
000	000	003	007	038	008	000	005	018	004
000	030	003	007	022	005	000	000	015	004
000	030	003	007	032	007	000	002	024	004
030	000	091	011	046	016	000	016	014	016
030	000	012	007	014	004	001	002	025	004
030	000	007	008	030	009	000	005	021	010
030	000	006	009	008	004	000	000	014	004
000	030	004	009	012	003	000	000	012	004
000	030	003	008	011	005	001	000	015	004
030	040	084	007	017	006	000	001	008	004
030	000	068	007	026	010	000	001	018	005

simplify accounting and also to better control expenditures, field maintenance costs are often carried as project costs in a contractor's accounts. It is difficult to define precisely where a field repair ends and an overhaul beings, and it is more desirable to think of all repairs as adding or restoring value. Since fuel and lube are immediately consumed, it is appropriate to treat them as project costs. Move-in costs are also immediately consumed and fall within this latter category.

Generally, any equipment expenditure that will not be used up on the first project should be regarded as an additional investment. In return, the contractor expects to receive a number of standard hours of usage during which the equipment must be earning a profit.

Loss of Value

As equipment becomes more sophisticated, with an increased number of moving parts and more complex systems for control, the consumption of ownership and maintenance value becomes more difficult to predict. The factors contributing most to the loss of value of a particular piece of equipment to a contractor include (1) use of equipment causing wear on moving parts, (2) operational obsolescence, (3) time-dependent physical deterioration, (4) risk cost of ownership, (5) utility cost of ownership, and (6) miscellaneous ownership costs.

The ranges of ownership and operating costs for different types of equipment under varying job conditions are shown in Table B2-4. Figure B2-5 shows the influence of job conditions and utilization on D-8 tractors. Under unusual job conditions, the variance may be greater. As the estimated repair cost approaches the upper value of the ranges given, management should actively seek other methods of doing the job. Equipment repair costs under "severe" and "very severe" working conditions are difficult to estimate because the type of breakdown becomes less predictable. Frame failures, as an example, occur in a random fashion. The rule of thumb that a 100 percent overload on a bearing results in a 90 percent reduction in bearing life should be remembered.

TABLE B2-4 Equipment Repair Costs as Percentages of Initial Cost Less Tires

Equipment	Easy	Medium	Severe
Track-type tractors	70	90	130
Tractor-drawn scrapers	30	40	60
Pipe layers	20	30	40
Wheel-tractor scrapers	20	90	130
Wheel-tractor bottom dumps	40	50	70
Off-highway trucks	60	80	110
Wheel-type tractors	40	60	90
Track-type tractors	70	90	130
Wheel loaders	40	60	90
Motor graders	30	50	70

Source: Caterpillar Performance Handbook.

A survey of used equipment in large contractors' yards will inevitably disclose pieces of equipment in usable condition that cannot be economically used or sold for more than scrap value.

Physical deterioration takes place even when equipment is not working. Tires deteriorate from age, exposed metals rust, seals dry out and leak, seat coverings and upholstery become brittle and develop cracks, and theft occurs. These costs continue during the times when equipment is not being utilized, and appropriate allowances must be made for them.

The use of equipment by contractors normally carries a higher risk than does its use in other industries. Dam and channel construction exposes equipment to flooding. Work around steep grades increases the likelihood of runaways and complete wreckage of equipment. Booms and power lines are a disturbing mixture. The costs that may result from these factors must be borne as accident losses or as insurance costs.

Unlike other industries, construction does not involve a fixed plant. When work at one project runs out, the equipment loses some of its value as an income-producing asset. Additional moving cost must be incurred to relocate it to another

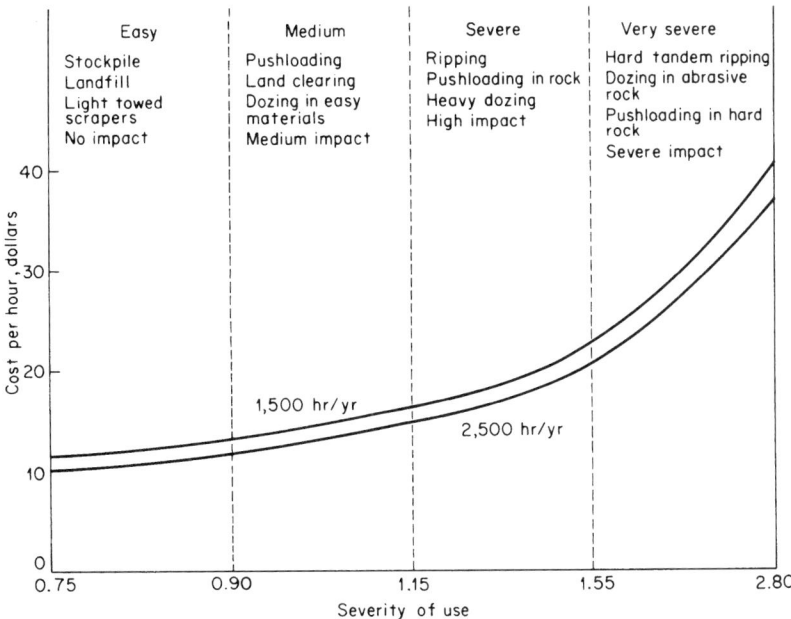

FIGURE B2-5 Hourly cost of D-8 tractors under varying job conditions.

income-producing location. It may also lose some value in its present configuration, and expenditures may be required to change or readjust attachments for the next available project. Since the magnitude of these costs depends on the nature of the next project, they are normally estimated as job costs and expenditures are charged directly to the project. In evaluating equipment cost and technological obsolescence, the cost of teardowns, moves, and setups may be significant, however, and should be considered in any replacement study.

Administrative and Overhead Cost

Certain administrative and overhead costs must be borne by the contractor who uses heavy equipment. These costs, while not adding to or taking away from the value of the equipment, are generally proportional to the size of the equipment pool. They include accounting expenditures, storage facilities for idle equipment, nonallocable taxes and insurance, and maintenance facility cost. These costs should be applied as a burden, along with the hourly ownership and operating cost.

Shop cost poses a particular problem, and care should be taken not to distort cost records by the manner in which overhead is handled. A common practice is for contractors to estimate and charge field shop cost directly to the project, while home-office shop costs are applied to the equipment through an applied burden charge. Under this policy, however, costs are distorted to the extent that major overhauls are done in the field shop. If field shop costs are to be handled as job overhead, it is preferable that central shop cost be treated as a general overhead item and budgeted accordingly.

The preferred way of allocating both central shop and field shop overhead to the equipment is by applying a burden to all repair labor and part charges. Shop overhead cost will generally be about 20 percent of labor and part charges. Normally, a central shop will be more efficient than a field shop.

Some taxes and insurance are difficult to charge directly to an equipment unit. A blanket liability insurance policy on vehicles is an example. For the average contractor, the broad-coverage insurance and taxes will amount to about 5 percent of total equipment cost. Insurance on specific pieces of equipment should either be charged to the equipment as a miscellaneous ownership cost or charged to the project, whichever is more applicable.

The cost of warehousing equipment parts and supplies can best be allocated as a handling charge on the warehouse issues. A small percentage will normally cover warehouse cost.

INFORMATION REQUIREMENTS FOR FINANCIAL AUDIT

Actions by the federal government to regulate spending for planned improvements through tax legislation are felt immediately by the construction industry, and a thorough understanding of the economic effect is essential to profitable operation. Adequate equipment records are needed not only for tax requirements but also for establishing a financial capability. Public contracts and most private projects require performance bonds that may be obtained only by demonstrating a financial position satisfactory to a surety. Good equipment records go beyond these minimum requirements so as to aid in the analysis of tax decisions and to provide a measure of contractor's performance in equipment management.

Auditing Requirements

Some contractors apply their depreciation and maintenance charges directly to the project to which the equipment is assigned. The common practice in the larger companies is to use a standard rate approach and allocate the cost of equipment to the individual contract and often to an item within the contract. A system combining a calendar time charge and a use (hours worked) charge has been shown to have substantial advantages over other systems. The most obvious advantage of separate time and use charges is that estimates of utilization are not as sensitive to error as they are in a single-charge system. From a practical standpoint, usage cannot be accurately recorded on small equipment and, where measurement of usage is impractical, a time charge only must be used. The calendar time charge to the project starts when the equipment is assigned to the job and continues until it is released for transfer to the next project. In the H. B. Zachry Company, all equipment operating costs are normally allocated as use charges. Ownership costs are allocated in equal proportions on a time charge and on a use charge. This method further reduces the effects of inaccurate estimates of utilization. By allocating the ownership charges 50 percent on a time basis and 50 percent on a use basis, the problem of second-shift work is handled in an acceptable manner. In a few of the larger companies, different rates may be used under different conditions, and age is often a basis for using a reduced rate.

Where a usage charge is applied, the equipment costing system is usually similar to a payroll system with labor distribution. Usage is reported on time cards, the appropriate rates are applied, and an equipment usage register and equipment distribution journal are prepared for each project.

Two uncommon features of the Zachry system include a double allocation of the time charge and the use of an adjustment factor for severe conditions. Figure B2-6 shows 50 percent of the ownership cost allocated to a project overhead account. The charges are then reallocated to direct cost items. Table B2-5, taken from an equip-

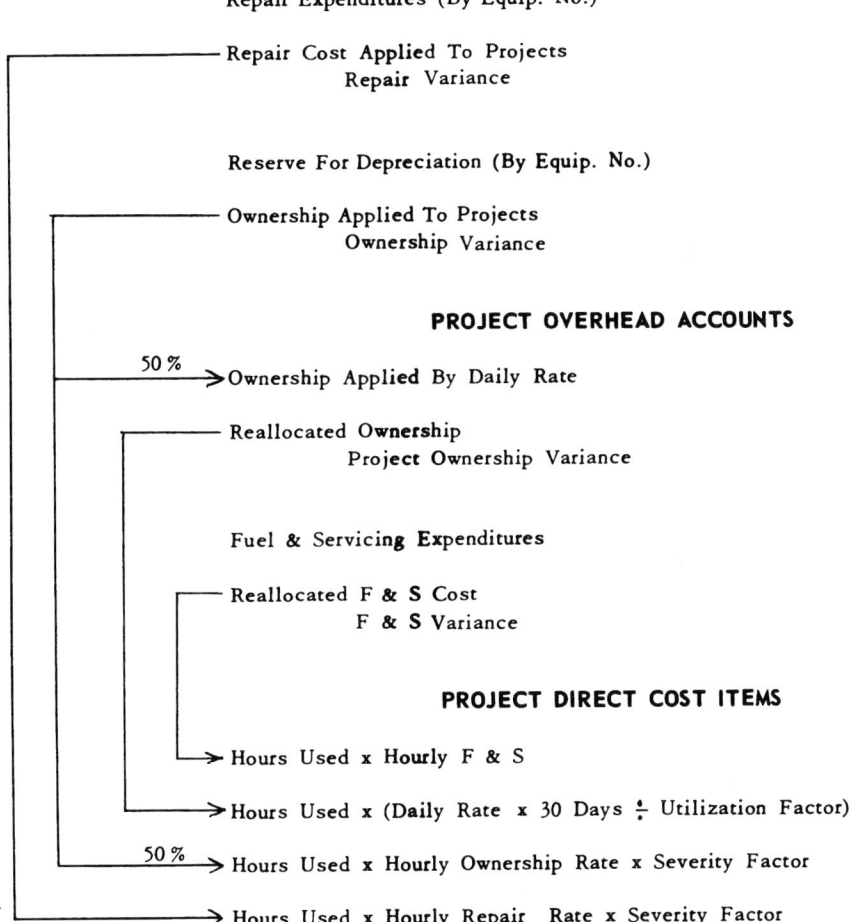

FIGURE B2-6 A method of allocating equipment costs to project items. Utilization factor is the estimated hours used per month on item. Severity factor is the estimated ratio of item equipment severity to average company severity. *(H. B. Zachry Company.)*

TABLE B2-5 Equipment Distribution Journal

Period ending 04/27
G L or job 1450 DE CORDOVA DAM

G L or job	Distribution code	Subdistribution code	Equipment number	D/H	Rental days	Prod hours	Util factor	Sever factor	Daily	Hourly	Hourly alloc	Rental amount
400	831	831	74017	H		58	130	1.00	2.72	2.33	.63	172
						58*						172
400	841	841	900049	H		6	130	1.00	.28	.42	.06	2
400	841	841	901408	H		16	130	1.00	.90	.76	.21	14
						22*						16
400	1025	1025	151298	H		5	130	1.00	.05	.66	.01	3
400	1025	1025	151304	H		9	130	1.00	.05	.66	.01	6
400	1025	1025	151308	H		7	130	1.00	.05	.66	.01	5
400	1025	1025	151309	H		10	130	1.00	.05	.66	.01	7
400	1025	1025	270206	H		10	130	1.00	.56	.20	.13	3
400	1025	1025	270512	H		5	130	1.00	1.23	1.15	.28	6
400	1025	1025	270608	H		9	130	1.00	1.64	1.73	.38	18
400	1025	1025	600401	H		5	130	1.00	6.87	4.95	1.59	32
400	1025	1025	645287	H		5	130	1.00	10.62	4.35	2.45	34
400	1025	1025	645806L	H		5	130	1.00	26.19	14.70	6.04	104
400	1025	1025	647001	H		10	130	1.00	29.92	22.17	6.90	291
400	1025	1025	647002	H		50	130	1.00	29.92	22.17	6.90	1454
400	1025	1025	732705	H		10	130	1.00	7.69	4.48	1.77	63
400	1025	1025	741019	H		5	130	1.00	15.16	9.74	3.50	67
400	1025	1025	741609	H		9	130	1.00	22.80	13.45	5.26	168
400	1025	1025	741611	H		7	130	1.00	22.80	13.45	5.26	132
400	1025	1025	767408	H		5	130	1.00	15.80	12.46	3.65	81
400	1025	1025	770806L	H		5	130	1.00	19.83	14.35	4.58	95
400	1025	1025	886503	H		2	130	1.00	11.00	12.72	2.54	30

				173*						
400	1140	1143	151298	H	45	130	1.00	.05	.66	.01	2599
400	1140	1143	151304	H	41	130	1.00	.05	.66	.01	29
400	1140	1143	151308	H	43	130	1.00	.05	.66	.01	27
400	1140	1143	254733	H	30	130	1.00	1.56	2.17	.36	28
400	1140	1143	255002	H	40	130	1.00	1.11	1.81	.26	76
400	1140	1143	270512	H	45	130	1.00	1.23	1.15	.28	83
400	1140	1143	270608	H	41	130	1.00	1.64	1.73	.38	65
400	1140	1143	515505	H	2	130	1.00	3.87	3.35	.89	86
400	1140	1143	549402	H	5	130	1.00	1.60	1.14	.37	9
400	1140	1143	591509	H	33	130	1.00				8
400	1140	1143	600401	H	4	130	1.00	6.87	4.95	1.59	26
400	1140	1143	645287	H	45	130	1.00	10.62	4.35	2.45	306
400	1140	1143	645805L	H	45	130	1.00	26.19	14.70	6.04	934
400	1140	1143	645806L	H	45	130	1.00	26.19	14.70	6.04	934
400	1140	1143	741019	H	45	130	1.00	15.16	9.74	3.50	597
400	1140	1143	741609	H	41	130	1.00	22.80	13.45	5.26	768
400	1140	1143	741611	H	43	130	1.00	22.80	13.45	5.26	804
400	1140	1143	767408	H	45	130	1.00	15.80	12.46	3.65	726
400	1140	1143	770603	H	33	130	1.00	16.72	12.20	3.86	529
400	1140	1143	770702	H	27	130	1.00	19.83	14.35	4.58	512
400	1140	1143	770805L	H	45	130	1.00	19.83	14.35	4.58	852
400	1140	1143	770806L	H	45	130	1.00	19.83	14.35	4.58	852

Source: H. B. Zachry Company.

ment distribution journal, shows the use of a severity factor and a utilization factor. Equipment costs are allocated by daily and hourly rates to distribution codes within a project. The daily rate is converted to an hourly rate by the use of a utilization factor. This factor is estimated for each item and identified by a distribution code at the time the project is bid. The severity factor reflects an estimated deviation from the average condition on which the standard rates are based. The cost difference between the estimated utilization and the utilization actually reported is charged to a project overhead account. Therefore, the overhead account receives a credit if the equipment is used more than the estimated amount and a debit if the utilization is below the estimated. This system allocates all the cost to the items where the cost was estimated and makes provision for some operations to be worked on a double shift while others are worked single shift. The use of the two factors has resulted in a system in which estimated costs and allocated costs are very comparable.

INFORMATION FOR OPERATING MANAGEMENT

Taylor recognized that a good system was as valuable an asset as a piece of equipment. Certainly, in the management of equipment, facts rather than intuition should be the basis for decisions. The information system should provide the communications necessary to attain the management goals of achieving the optimum levels of (1) the most suitable equipment; (2) in the best condition economically feasible; (3) where needed, when needed; (4) with adequate support to provide acceptable service; and (5) at lowest cost consistent with company objectives. Management from the top is essential for attaining these goals and requires proper purchasing decisions, strong maintenance policies, and the establishment of clear standards for performance. Equipment cost estimates must be made with reasonable accuracy. Reliable communications must be maintained between field mechanics and top management. A systematic analysis of meaningful records is essential, and information must be communicated to those responsible for equipment management.

Estimating Costs

Effective management is essential in contracting, but even the most efficient contractor must obtain work at the right price to achieve long-term success. Since equipment expenditures are consumed over a long period of time and under unforeseeable conditions, cost will not be exactly as anticipated. In order to facilitate intelligent bidding and to ensure the maintenance of a profitable average, however, cost must be forecast with reasonable accuracy. How close the estimate needs to be depends on the facts and circumstance of each individual project. Where job conditions are very uncertain, national averages may be the only pertinent data available. The Associated Equipment Distributors compile annually a reference book of nationally averaged rental rates. Equipment manufacturers' handbooks normally show average cost as well as adjustment factors for tailoring a rate to the predicted job conditions. These costs and adjustments are continually being refined through additional testing and measuring procedures. Manufacturers are particularly progressive in determining component life under differing conditions.

It is more desirable to base a particular estimate on specific costs than to depend on national averages. Wage rates, training opportunities, and many other circumstances influence operating cost. The use of national averages may lead to the loss of projects that could have been constructed profitably or to the awarding of contracts at an insufficient price to meet equipment cost. Accuracy will be improved if careful consideration of life is given to major components of the large units. For example, it may be desirable to consider a diesel engine and pump set as two separate pieces of equipment with separate lives and cost.

Figure B2-7 shows a typical format for estimating the average hourly rate for owning and operating a piece of equipment. For a given model of equipment, the contractor may make only one estimate based on his average or standard job conditions and equipment use. The estimate may then be adjusted for individual project conditions.

Equipment cost estimates are used most extensively for bidding, project planning, allocation of cost to projects, and budgetary control of maintenance. The format shown in Fig. B2-7 is generally satisfactory for these purposes. It should be used with the recognition that factors such as inflation must be considered. The delivered price, line item 1, should include freight as well as attachments. Overall company debt policy should be considered when using item 7 to ensure that interest, insurance, and taxes are not duplicated anywhere in the estimating or accounting system. If overhead cost of a central shop and administrative cost are to be included within a standard rate used by estimating and accounting, they may be added under item 11—Special Items. It is often more convenient to exclude the operator's hourly wage from equipment cost and estimate it as a separate figure.

As company policies change to adapt to different project mixes, the cost of maintaining equipment will change. If a larger percentage of repair work is returned to the dealers, equipment overhead ratios may change. Equipment cost is constantly changing, and the only way to keep accurate estimates of it is by collecting reliable equipment records.

Equipment Records

To be meaningful, records must be more than listings of expenditures and more than just averages. The records must be pertinent to the day-to-day decisions and must be reliable enough to instill confidence in their use. They must be where they are needed at the time they are needed. They must be acquired at reasonable cost and convenience. The information system must be simple and flexible enough to survive in the construction-project environment. All these features are necessary, but it is also of primary importance that the information system be wanted and used by those who need the facts.

The diversity of the locations at which contractors build complicates the record-keeping task. Transferring records between job sites as the equipment is transferred is sometimes awkward. Documents that should be transferred with each machine include its maintenance jacket, preventive maintenance records, maintenance instructions, lubrication charts, parts catalog, tire records, and any operating instructions. Certain parts that are carried in inventory should be transferred with the equipment, and inventory records should be adjusted accordingly.

To facilitate the analysis of equipment records, more contractors are turning to centralized files and data processing systems. The interest in utilizing computers has stimulated more communication between groups using the hardware, and considerable progress has been made in the past decade toward reconciling the conflicting information needs of the different functions within contractors' organizations. Part of the information is often retained in machine-readable form on a combination of magnetic tape and disk files. Table B2-6 shows typical contents and use of files

Machine or Component		D7E PS

DEPRECIATION VALUE

1. Delivered Price (including Attachments) — 136,629.00
2. Less Tire Replacement Costs:
   - Front _____
   - Drive _____
   - Rear _____
3. Delivered Price Less Tires
4. Less Resale Value or Trade-in
5. NET VALUE FOR DEPRECIATION — 136,629.00

OWNERSHIP COSTS

6. Depreciation: $\dfrac{\text{Net Depreciation Value}}{\text{Life in Hours}} = \dfrac{136{,}629}{10{,}000}$ — 13.66
7. Interest, Insurance, Taxes:
   Annual Rates: Int. 9% Ins. 2% Taxes 2% = 13%
   Estimated Annual Use in Hours 2,000

Yrs. Life	3	4	5	6	7	8	9	10
% Invest.	66.7	62.5	60.0	58.3	57.1	56.3	55.6	55.0

   Hourly Cost = $\dfrac{\text{Price} \times \%\text{ Invest} \times \text{Tot. Ann. Rate}}{\text{Annual Use in Hours}}$ — 5.34
8. TOTAL HOURLY OWNERSHIP COSTS — 19.00

REPAIR COSTS (Allocated by Standard Rate)

9. Tires: $\dfrac{\text{Replacement Cost}}{\text{Estimated Life in Hours}}$
10. Repairs: $\dfrac{\% \times \text{Del. Price - Tires}}{\text{Life in Hours}}$ 90% — 12.30
11. Special Items: _____
12. TOTAL HOURLY REPAIR COST — 12.30

SERVICE COSTS (Direct Job Charge)

Unit Price x Consumption

13. Fuel: 1.00 x 6.7 — 6.70
14. Lubricants, Filters, Grease:

    Unit Price x Consumption
    - Engine 3.30 x .04 — .13
    - Transmission 4.00 x .03 — .12
    - Final Drives 3.30 x .02 — .07
    - Hydraulics 3.30 x .03 — .10
    - Grease .60 x .05 — .03
    - Filters 4.00 x .16 — .64
15. Service Labor 19.10 x .07 — 1.34
16. TOTAL SERVICE COST — 2.43

17. Operator's Hourly Wage — 24.75
18. TOTAL COST PER HOUR — 65.18

FIGURE B2-7 Equipment cost estimate form.

retained by equipment management. Additional information retained may include trends of cost, utilization, and various ratios to be used as follow-up on management policies.

It is impossible at present to diagnose the amount of wear on all equipment parts, and most wear must be judged from the life of the part and knowledge of job conditions. Therefore, it is essential to retain records of part replacements for review when considering future overhauls. Minimum information retained pertaining to a repair should include the equipment number, date, service meter reading, and a list of parts installed. One successful system is to file invoices as they are paid, for easy reference by equipment number. This centralized equipment folder then provides a history of all charges made to that equipment number and includes a copy of the work order that accompanies an invoice. The file, therefore, contains an itemized list of parts installed on the machine. The file needs to be easily accessible to the equipment manager who makes replacement decisions.

Monitoring Costs

Equipment cost estimates should continually be monitored and maintained to provide a measure of performance for project managers and to permit budgetary control over equipment cost. The level at which budgetary control of maintenance is

TABLE B2-6 Equipment Information Retained by a Large Contractor

File	Contains	Use
Equipment category	Utilization estimates Cost estimates Shipping data	Estimating new work Allocating cost to projects (work breakdown) Controlling maintenance expenditures Selection of categories for maintenance and scheduling system
Equipment by individual piece	Depreciation data Location Availability status Mechanical condition Usage Last maintenance check date Financing or leasing information	Depreciation schedules Allocation of cost to projects Control of utilization Control of diagnostic checks Project equipment investment
Cost history (active equipment)	Cost history Utilization history	Maintenance cost control Establishment of cost estimates Comparison of equipment types
Usage transactions	Usage history	Control of equipment policies Analysis of severity cost
Diagnostic file	Oil contaminant amounts	Indicates equipment malfunction or excessive wear
Cost history (inactive equipment)	Cost history Utilization history	Special studies such as resale value analysis

Source: H. B. Zachry Company.

attempted varies significantly among companies. Most companies record their cost to the individual piece of equipment by type of expenditure. A typical breakdown of the types of expenditure used is (1) labor, (2) parts, (3) tires and tubes, (4) lubrication and preventive maintenance, (5) accident repairs, and (6) special modifications. Along with the maintenance expenditures, depreciation and other ownership costs are normally charged to the individual equipment numbers. In increasing numbers, construction companies are extending their budgetary control to a more detailed level by equipment system. A typical breakdown of systems will include (1) engine, (2) electrical and ignition, (3) fuel, (4) cooling, (5) undercarriage, (6) steering, (7) power train, (8) hydraulic, (9) brake system, (10) body, and (11) accessory equipment. The optimum level depends on the variability of the equipment's use. With more uniform job conditions, greater detail can be effectively utilized. Cost may be summarized and compared by type of equipment, model, and size (i.e., Caterpillar 660 scrapers), or, at a still higher level, by functional type of equipment (i.e., hauling units).

Cost for a given period has little meaning when taken by itself. At a minimum, an inception-to-date cost and estimate should be compared, and it is very desirable to maintain cost in such a manner that trends are apparent. A unit cost per hour provides the best control for most situations. Production quantities may be appropriate units of cost for some applications and for very stable job conditions, but are unsatisfactory for most contracting work.

A maintenance cost system is necessary to protect a contractor's high capital investment in construction equipment. It permits the control of good service procedures and the prevention of excessive operating costs due to overmaintenance or inadequate maintenance. It should also enable the contractor to price work more intelligently. The fulfillment of these objectives should govern efforts to develop effective information systems.

As long as construction projects are of shorter duration than the life of equipment, the control of equipment costs will be a major challenge.

CHAPTER 3
TRACTORS, BULLDOZERS, AND RIPPERS*

James J. O'Brien, P.E.
O'Brien-Kreitzberg, Inc., Pennsauken, New Jersey

OFF-HIGHWAY TRACTORS

Classification

Off-highway tractors or prime movers fall into two general classes: crawlers or track-type (Fig. B3-1), and rubber-tired or wheel-type (Fig. B3-2). The latter class will be viewed as consisting of two-axle units with either 2- or 4-wheel drive. This excludes the single-axle tractor shown in Fig. B3-2a which requires another unit such as a single-axle scraper or trailer for support and stabilization, i.e., a two-axle motor scraper.

Figures B3-3 and B3-4 illustrate off-highway units. These are self-contained prime movers which provide tractive power for drawbar work. Such work includes bulldozing and push-loading scrapers as well as towing scrapers, sheepsfoot and rubber-tired rollers, dirt and rock wagons, plows and disks, rippers, logging arches, and skid pans.

These prime movers also function as mobile mountings for various attachments and accessories. Prominent among these attachments are blades (straight, angle, tilt, push cups, brush, etc.), rippers, pipelayers, towing and logging winches, backfillers, snowplows and wings, and push blocks.

Equipment Characteristics

Prime Movers. Specification sheets for prime movers provide the best means to evaluate equipment characteristics. They indicate the type and horsepower rating of the engine, the type of transmission, final drive, axles, steering, brakes, suspension, tires or tracks, weight, and other pertinent information. Equally important are the drawbar-speed and rimpull-speed curves for track-type and wheel-type tractors,

---

* Some of the material in this chapter is from R. D. Evans, "Tractors, Bulldozers, and Rippers," in Stubbs and Havers (eds.), *Standard Handbook of Heavy Construction,* 2d ed., McGraw-Hill, New York, 1971.

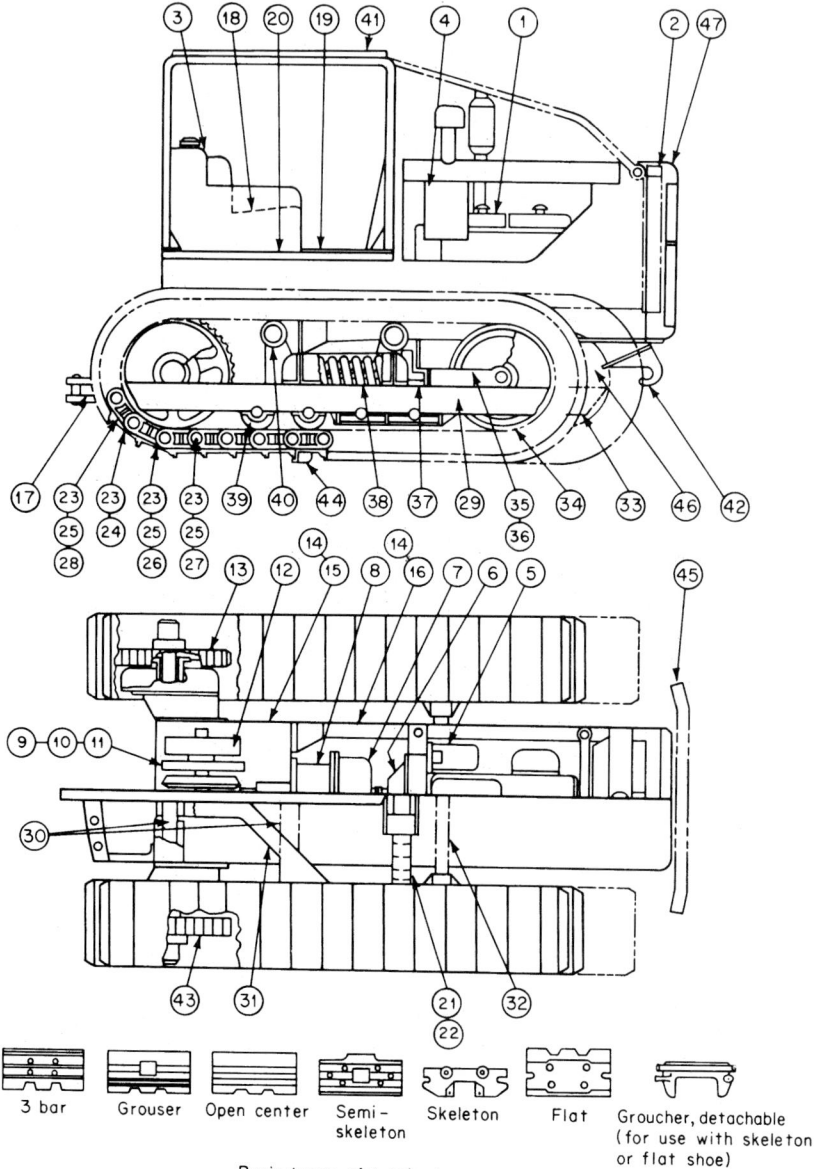

FIGURE B3-1 Nomenclature—track-type tractors. *(SAE Handbook.)*

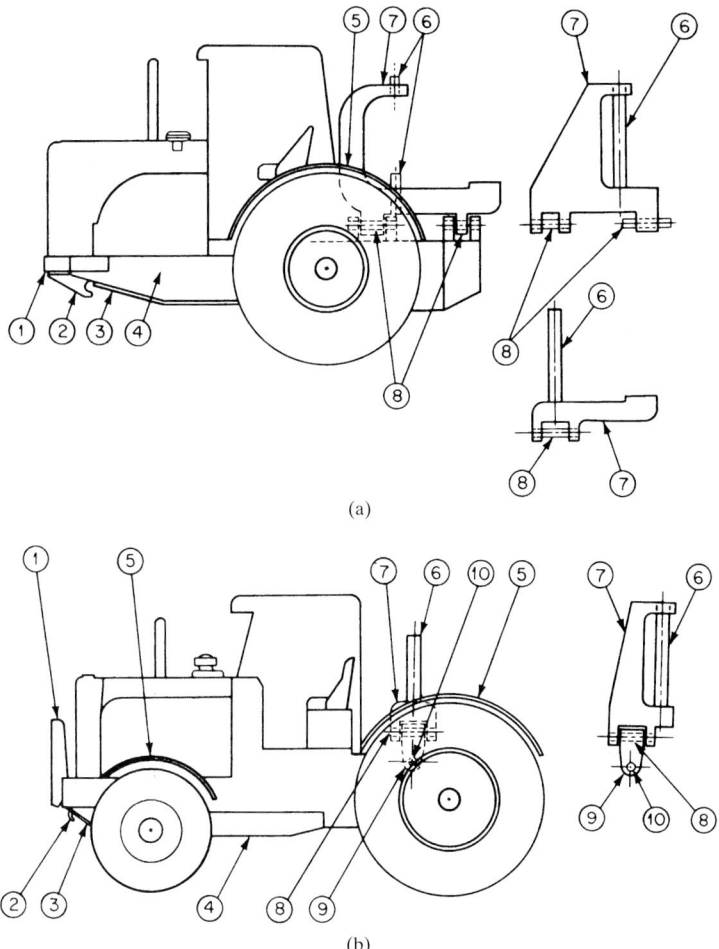

FIGURE B3-2 Nomenclature—wheel-type tractors: (*a*) Single axle; (*b*) two axle. (*SAE Handbook.*)

respectively. Figure B3-5 shows a track-type tractor curve and Fig. B3-6 shows a wheel-type tractor curve.

Running Gear. Prime movers for modern off-highway construction equipment are usually available with either tracks or wheels as their running gear. Track-type tractors and power shovels (excavators) have been used since the turn of the 20th century. Wheel-type (rubber-tired) tractor-scrapers were introduced in the late 1930s and were followed by rubber-tired drawbar units, dozers, and loaders. All the latter types have counterpart units mounted on tracks. For any job, a choice of running gear type must be made in order to properly utilize the prime movers and obtain optimum production at the least cost.

FIGURE B3-3 Model 2755 general-purpose 87-hp wheel-type tractor. *(John Deere Co.)*

FIGURE B3-4 Model G50 G 90-hp track-type tractor. *(John Deere Co.)*

TRACTORS, BULLDOZERS, AND RIPPERS

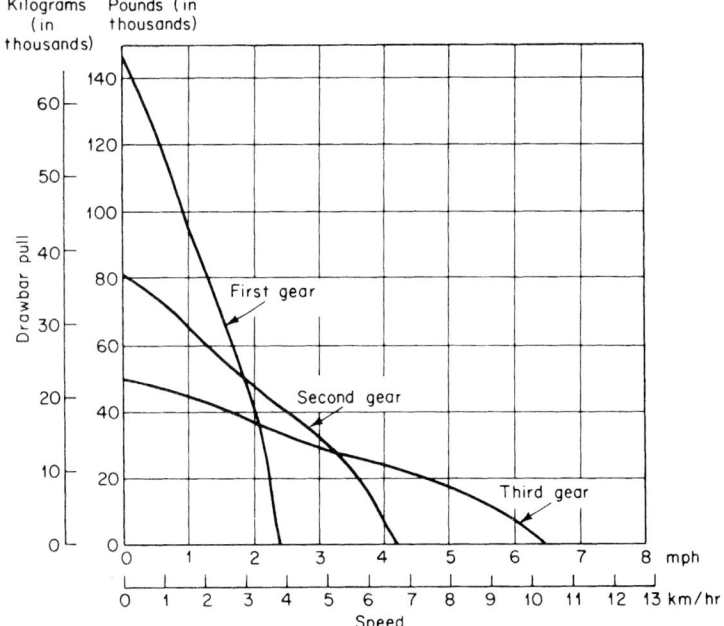

FIGURE B3-5 Track-type drawbar—speed curve.

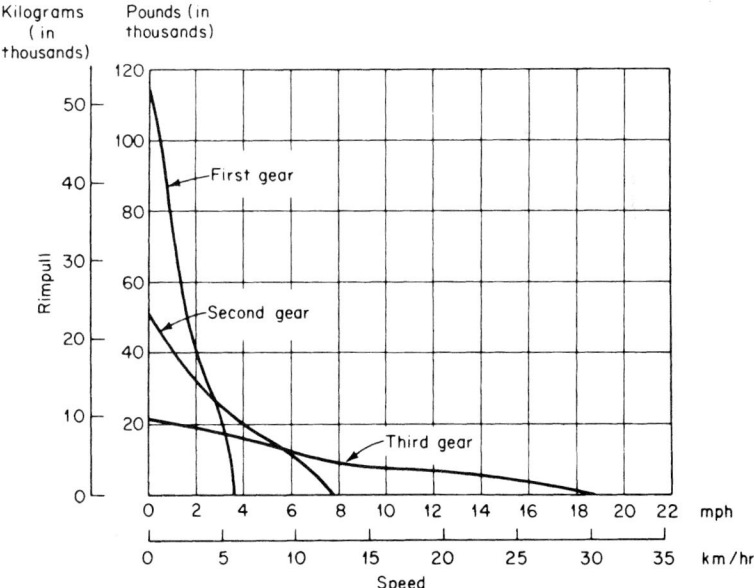

FIGURE B3-6 Wheel-type rimpull—speed curve.

When speed and mobility of the prime mover are of major importance, wheels are the obvious choice for the running gear. Speed and mobility become relatively less important when traction requirements are high. In these circumstances, crawler running gear can prove greatly superior to rubber tires, since it can normally develop more traction. The crawler tractor can then normally exert more usable drawbar pull than can a wheel-type tractor of similar weight and horsepower. This situation favors the use of crawler running gear on jobs where high drawbar-pull requirements exert more influence than speed on the cost per yard.

The coefficient of traction (maximum drawbar pull or rimpull at imminent slippage divided by gross vehicle weight) is generally considered to have a maximum value of 0.6 for rubber-tired units operating on all surfaces excluding concrete. For track-type tractors, this maximum figure is about 0.9. The curves in Fig. B3-7 match the rimpull of a typical dozer-equipped wheel-type tractor with the drawbar pull of a comparable track-type machine. Detail A in the graph examines more closely the usually working range for both types. The shaded areas represent the ranges of developable traction for soil types varying from dry clay loam to loose earth—0.6 to 0.9 traction coefficients for tracks and 0.4 to 0.6 coefficients for wheels.

The crawler-mounted tractor of Fig. B3-7, for example, weighs 82,000 lb equipped with dozer and a control unit. At 0.9 coefficient of traction, it can develop 73,800 lb of pull before its tracks start to slip. The rubber-tired unit, working weight 91,000 lb,

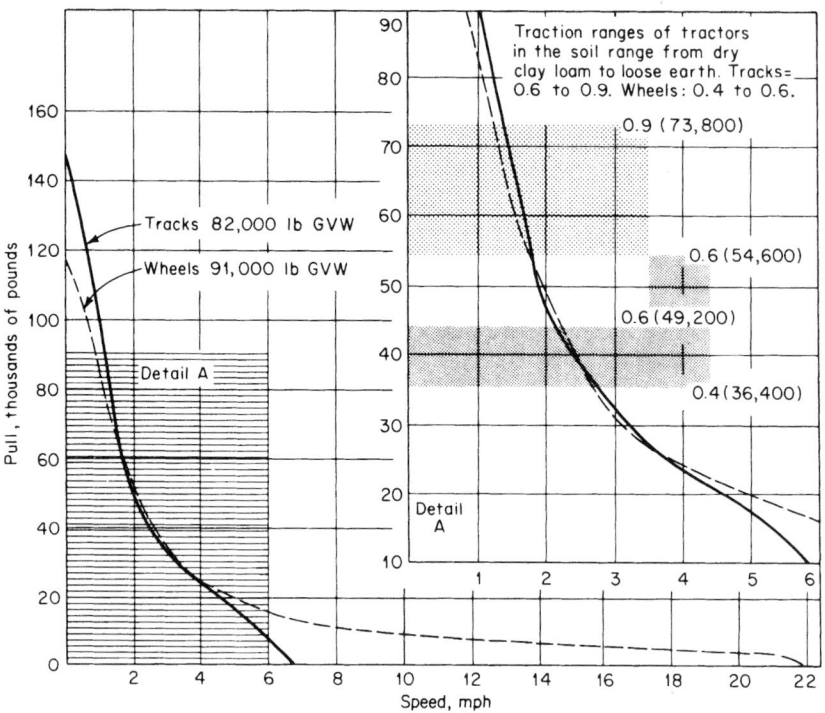

FIGURE B3-7 Comparison of wheel-type tractor rimpull with the drawbar pull of track-type (power shift) tractors.

can produce 54,600 lb of pull at its maximum coefficient (0.6). The working speeds for the two machines, when operating under these conditions, will not be substantially different.

As a machine weight increases, available pounds of pull go up. Mounting an 11,000-lb ripper on the crawler might increase its drawbar pull by 9900 lb.

To look at the weight-drawbar-pull relationship another way, the weight of machine required to develop a specified usable drawbar pull is directly related to the applicable coefficient of traction. Accordingly, a rubber-tired machine must usually be heavier than a crawler tractor if it is to develop an equivalent amount of drawbar pull. Of course, the heavier machine will also have to have more horsepower to keep its power-to-weight ratio within reasonable limits. Operating in a material which corresponds to the maximum coefficients of traction for each type of running gear, a wheeled machine needs half again as much weight as a crawler (0.9/0.6 = 1.5) to deliver the same drawbar pull.

On wet ground, the coefficient of traction for rubber falls off more rapidly than for tracks. As a result, the wheel tractor must be even more than 1½ times the weight of a track-type tractor for comparable drawbar pull during rainy weather. At the other end of the tractive scale—in loose sand, where traction coefficients for both types are about 0.3—a 91,000-lb wheel unit with proper flotation might develop a slightly greater drawbar pull than an 82,000-lb crawler.

Maximum drawbar pull, however, does not tell the full story of the work capabilities of two types of machines. Track-type tractors perform best at speeds up to about 3 mi/h, while wheel tractors are designed to work best between 2 and 5 mi/h. It can be seen in Fig. B3-7 that the two curves almost match between 2 and 4 mi/h. Applications such as fill spreading and high-speed production dozing show off the wheel tractor's rimpull-speed relationship to best advantage.

Wheeled machines, unlike their track-type counterparts, have to expend some of their rimpull in overcoming rolling resistance. Since rolling resistance varies so much, no attempt has been made to allow for it in curves. An actual rimpull curve for the wheel tractor at work could be somewhere below that shown in Fig. B3-7.

Since a wheel tractor needs more weight and power than a crawler of similar drawbar-pull capabilities, it will generally be more expensive. The higher price, when reflected as an hourly depreciation rate, contributes to a higher hourly owning and operating cost and thus to a higher price per pound of usable drawbar pull. In some applications, the faster operating speed of the wheel machines will more than make up the difference, but contractors will want to observe units working side by side before concluding that wheels will give a lower cost just because they are faster.

One cost item deserves special attention—that of tires and tracks. When wheel tractors first became available, relatively high track costs were a telling argument in favor of wheels. The later years have seen major advances in both metallurgy and design which, in most cases, have extended track wear more than track prices have increased.

On the other hand, tires for construction equipment have also been improved. Further, they can usually be recapped one, two, and sometimes three times before the carcass must be discarded. All in all, the relative costs of tracks and tires have little effect in choosing between wheel- and track-type tractors. In fact, the most influential element in this decision cannot be any *one* contributing item but has to be the final figure—cost per yard.

The relative advantages and disadvantages of wheel and track running gears can be summarized as follows.

	Wheel advantages	Track advantages
Working conditions	Firm clay, concrete Abrasive soils but no sharp edges Level or downhill work Poor in wet weather	Variety of soils Sharp edges not so destructive Any terrain Good in all weather
Effect on ground	Good compaction; variable with counterweight and ballast	Good flotation; low ground pressure with different shoe sizes and types
Application	Long distances Loose soils Fast return speeds, 8–12 mi/h Moderate blade loads and long, thin cuts when dozing Highly mobile	Short distances Tight soils Slow return speeds, 4–7 mi/h Large blade loads and short, heavy cuts when dozing Restricted mobility

Power Transmission and Controls. Track- or wheel-type prime movers are equipped with hydraulic electrical or electronic controls.

Most track- and wheel-type tractors are offered with a choice of direct-drive or power-shift transmission. If we consider two tractors with identical engine flywheel horsepower, any difference in performance characteristics will be due to the direct-drive transmission having a little more drawbar pounds pull in a narrow speed range compared with the power-shift transmission having slightly less drawbar pounds pull in a wider range. In a dozing or drawbar application with a variable load, the direct-drive transmission will provide more drawbar pounds pull with more frequent gear shifting compared with the faster and less frequent shifting of the power-shift transmission which generates slightly less drawbar pounds pull. It is now generally accepted that the power-shift tractor will give superior production in an application which has variable load as compared with the superior production of the direct-drive transmission in an application with a constant load.

In 1993, John Deere announced the availability of an optional torque converter transmission for its G-series crawler bulldozers. Machines equipped with the torque converter are designated as the 450G-TC, 550G-TC, and 650G-TC with 70, 80, and 90 hp respectively.

Purchasers can choose between a direct drive or torque converter version of John Deere's full powershift transmission. The transmission provides four speeds forward and four speeds in reverse where most other machines in this class provide only three. The result is three good working speeds and a fast transport speed as compared to one working speed, a creeping speed, and a high transport speed typically found in a three-speed transmission.

The optional torque converter provides a simple means of both increasing the pushing power at slower track speeds and making the machine easier to control. The torque converter enhances operation in tight confines next to buildings, walls, or curves where slow speed operation and precise control are required. It allows the operator to ease up to objects by simply working the brake and decelerator as opposed to feathering a clutch. Finish grade work requires less operator effort and skill.

The optional direct-drive transmission is retained for applications where heavy drawbar pull and brute power at higher speeds are desired, such as fire plow and log-skidding application, or where long, heavy pushes are the norm.

Table B3-1 is a partial listing of Caterpillar track-type tractors with some characteristics.

TABLE B3-1 Partial Listing of Caterpillar Track-Type Tractors[1]

Model	Horsepower (flywheel)	Weight, lb	Low ground pressure (LGP) weight, lb
D3C Series II	70	15,435	17,170
D4C Series II	80	16,661	17,427
D5C	90	18,650	19,800
D4H Series II	95	24,035	
D4H XL (Series III)	105	25,750	27,080
D5E	105	25,755	
D5E Series II	120	29,200	
D5H XL	130	30,830	35,400
D6D	140	32,210	38,300
D6E	155	32,987	
D6H Series II	165	39,075	
D6H XL Series II	175	41,811	
D6H XR Series II	175	41,192	43,590 (170 hp)
D7G	200	45,560	
D7H Series II	215	54,401	
D7H XR Series II	215	54,939	59,169 (230 hp)
D8N	285	81,222	81,025
D9N	370	94,196	
D10N	520	131,817	
D11N	770	217,847	

BULLDOZERS

The bulldozer is widely used and is probably the most versatile of any of the earthmoving tools.

A bulldozer is a frame-mounted unit with a blade, curved in its vertical section, extending in front of the tractor. Blades range in width from some 6½ ft to around 24 ft and in height from about 5 ft down to approximately 2½ ft.

Blade Types

There are four basic blade types: straight, angle, U, and cushion. The track-type tractor can have any of the four basic blades mounted on it, whereas the wheel-type tractor is generally restricted to either the S or C blade or a lightweight U blade for the handling of light materials such as coal and wood chips.

Straight Blades. The S blade attaches by push arms and braces to the tractor frame. It can be adjusted or tilted so that one corner will be approximately 2 ft below ground level on the large blades and 1 ft below ground level on the smaller units. Tip (forward or backward) adjustment can be made to a maximum of 10°.

The S blade is primarily a heavy-duty blade used in excavation and pioneering work. It is also suitable for drifting material and for trenching and backfilling. Since it can be tilted, it makes an excellent penetrating tool in hard excavation and can be successfully used to dislodge large boulders. By changing the pitch (tip), light materials can be drifted or hard soils may be dug, depending on conditions.

Figure B3-8 is a Model 750B dozer with a straight blade. Figure B3-9 shows the dimensions including tilt and digging depth of the Model 750B dozer.

FIGURE B3-8 Model 750B LGP with straight blade (LGP stands for low ground pressure). *(John Deere Co.)*

Angling Blades. An A blade is generally wider and heavier than the straight blade for the same tractor, but it is lighter in cross section. It mounts on a C-shaped frame, each end of which is attached to the tractor frame. As its name implies, the blade can be angled to the right or left of a normal S-blade position, up to a maximum angle of 25°. Angling blades can also be tilted in the same manner as the straight blade. However, due to the mounting on the C frame, the blade cannot be tipped. For comparable machines, the angling blade will be some 1 to 2 ft wider than the straight blade. A blades are most effectively employed in side casting, as in the construction of a sidehill road or in backfilling a trench.

Figure B3-10 shows the dimension symbols of a Model 850B dozer. Table B3-2 shows the dimensions for three types of blades: straight, semi-U, and angle.

U Blades. Most manufacturers build a U blade for larger machines. This blade has the same cross-sectional strength as the straight blade and can be tilted and tipped in the same manner. The U blade is wider than the straight blade and its outside edges are canted forward 25° to minimize spillage in loose material.

U blades perform much the same duty as the S blade. Their use is more suited to the noncohesive soils or rocks, whereas the S blade is primarily a cohesive-soil tool. The U blade is a good match on a ripping tractor and can also excel in stockpiling coal, drifting material into a grizzly, pioneering, and clearing.

TRACTORS, BULLDOZERS, AND RIPPERS B3-11

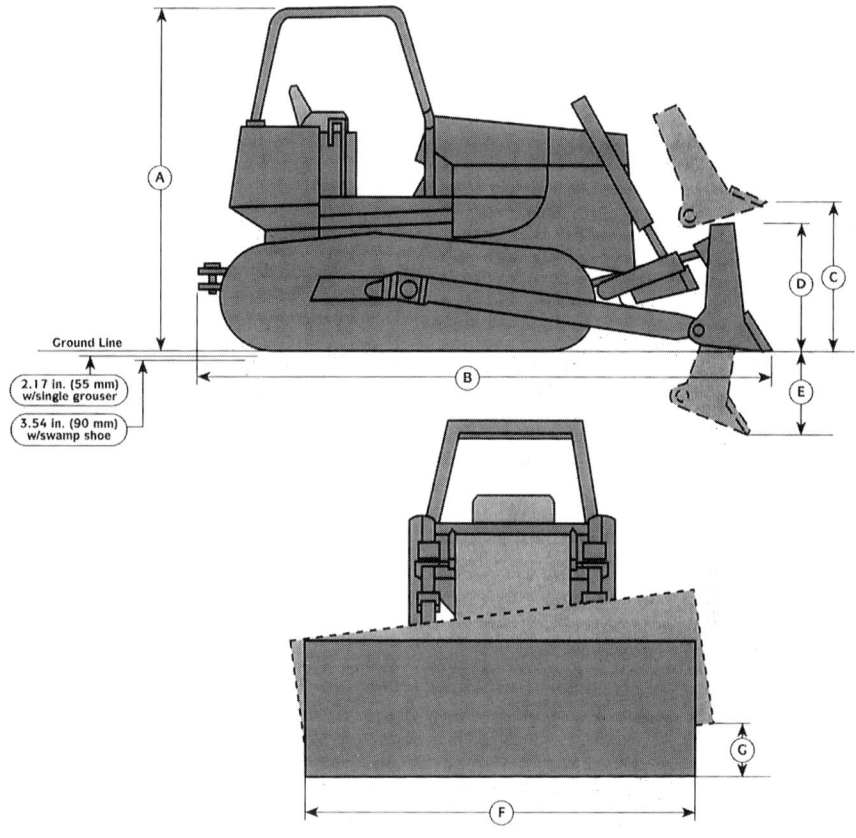

Key:
A Overall height .... 122.6 in. (3114 mm)
B Overall length .... 198 in. (5024 mm)
C Blade lift height .... 41.5 in. (1054 mm)
D Blade height .... 38 in. (965 mm)
E Digging depth .... 20.2 in. (513 mm)
F Blade width .... 132 in. (3350 mm)
G Blade tilt .... 15 in. (381 mm)
Blade capacity .... 3.23 yd.³ (2.47 m³)

FIGURE B3-9 Dimensions including blade tilt and digging depth of the Model 750B dozer with straight blade. *(John Deere Co.)*

Cushion Blades. The C blade is mounted to the inside of the tractor main frame. It is narrower than the S blade and is used mainly to push-load scrapers. Because the blade is compact, the tractor can maneuver quickly in the cut area to contact scrapers. The reduced blade length also minimizes the chances of cutting tires during the push cycle. It is used instead of a push cup because it can also clean up in the cut and thus help to maintain high productivity. The blade is a utility one and not a production tool from the bulldozing aspect.

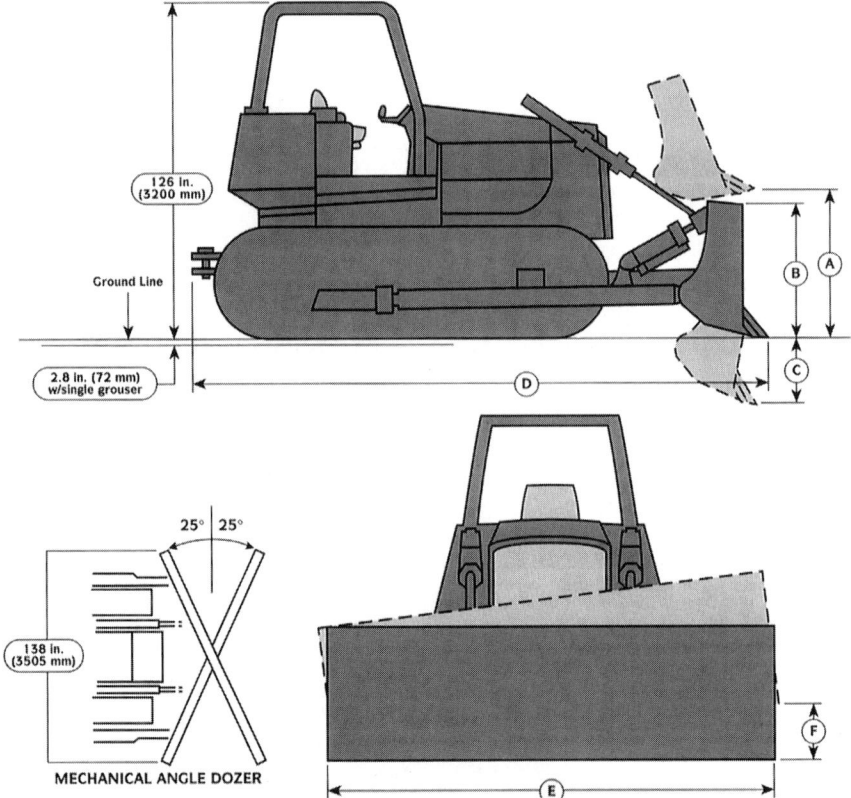

FIGURE B3-10 Dimension symbols for a Model 850B dozer. Use with Table B3-2. *(John Deere Co.)*

The blade mounting is usually through a series of rubber and steel disks onto the push arms. Those disks absorb the shock when the blade contacts a scraper at relative speeds up to 3 mi/h. The C blade cannot be tilted, tipped, or angled as can the S and U blades.

Tip, Tilt, and Angle

To change the tilt position of S or U blades, the tilting braces are rotated in opposite directions. The end result is that one corner of the blade will be considerably lower than the opposite corner. Some manufacturers have replaced one of the braces by a hydraulic cylinder (called a *tilt* cylinder) which performs the tilting operation. Some tilt controls are separate from the dozer controls (raise, lower) and are hand or foot operated. Others are incorporated with the dozer control lever so that all movements can be made with one lever. In either case, the dozing operation is not interrupted.

To change the tip position of an S or U blade, the tilting braces are rotated in the same direction. They expand to move the angle of tip of the blade forward and contract to move the angle of tip of the blade backward. This adjustment can be manual,

TABLE B3-2 Dimensions of 850B Dozer for Three Blades: Straight, Semi-U, and Angle (Reference Fig. B3-10)

Dozer specifications

Blade	Blade capacity per SAE J1265 yd³	m³	A blade lift height in	mm	B blade height in	mm	C digging depth in	mm	D overall length (tractor with blade) in	mm	E overall width* (tractor with blade) in	mm	F maximum tilt in	mm	Weight lb	kg	Total operating weight (tractor with blade) lb	kg
Straight	3.89	2.97	42.0	1067	44.5	1130	20.6	523	214	5436	127	3226	15.0	381	4600	2086	39,315	17,833
Semi-U	5.40	4.13	43.0	1092	46.5	1181	20.2	513	222	5630	138	3500	15.0	381	5020	2277	39,700	17,995
Angle	3.77	2.88	38.8	985	40.0	1016	18.7	475	214	5436	152	3860	13.23	336	5035	2284	39,715	18,015

* Includes cupped end bit.
Source: John Deere Co.

B3-13

hydraulic, or even electric. In the hydraulic adjustment, both braces would be replaced by hydraulic cylinders.

To change the angle of an A blade in relation to the tractor center line, the blade is pivoted about its central connection to the C frame. The blade braces are then pinned into the proper holes on the C or blade-support frame.

Applications

Land Clearing. Generally speaking, a bulldozer is the first piece of equipment on the job site. One of its first tasks will be that of land clearing. Brush and small trees are removed by lowering the blade a few inches into the ground to strike and cut the roots. This is usually done in the lower speed ranges, and backing up may occasionally be necessary to clear the blade so that it can always cut cleanly. Medium trees and brush 4 to 10 in in diameter usually require more than one pass. About 1 to 3 min is required for this operation, using tractors in excess of 100 drawbar horsepower.

For large trees 12 to 30 in in diameter, more care and more time are required. First, contact should be made with the blade high and centered for maximum leverage. This contact should be made gently, pushing the tree at half throttle while inspecting for dead limbs which may break and fall. A bulldozer should not charge into trees at full throttle. The direction of fall, usually the direction of lean, is determined. Then a cut is made on the opposite side to a depth sufficient to cut some of the larger roots. The roots on both adjacent sides are cut in a similar manner. After building an earth ramp on the initial side of the tree to obtain greater leverage, the tree is pushed. As the tree starts to fall, the tractor should be reversed quickly to avoid incurring damage as the roots are pulled out of the ground. When stumps are removed, the holes should be filled and the area leveled to prevent water from accumulating. The time consumed for this operation will be 5 to 20 min per tree.

Stripping. Stripping consists of the removal of topsoil that is not usable as fill material or as stable subgrade. Stripping operations should be planned so as to minimize haul distance. In dozer work, the maximum haul distance is approximately 300 ft, and for maximum economy the material should be moved only once. In rough or steep terrain where hillsides and steep ravines must be stripped, hauls are usually longer. For these conditions, stripping should be done so that the topsoil is dozed downgrade to scrapers working at the bottom of the grade. The scrapers then perform the necessary hauling. All stripping operations should be planned so that other earthmoving units can begin excavation as soon as possible.

Pioneering and Sidehill Cuts. Sidehill cuts should always be started or pioneered from the top and then worked downward. Working downhill gives the advantage of gravity. It may be necessary to reach the starting point by climbing up a more gradual slope on the opposite side of the hill. In exceptionally steep or rocky terrain, a block hung on a tree or anchor point may be used to pull the pioneering tractor back up the hill, either by another tractor or by a winch on the pioneering tractor. Very steep areas can be worked by hanging on the line until a bench is well started.

On average terrain, cuts are started by working straight down the hill, making short passes to bench out an area large enough so that the tractor can eventually turn and work parallel to the road. Pioneering cuts should slope into the uphill side for maximum earthmoving efficiency. Short, swinging passes should be made as the dozer works downhill to drift the material over the side of the cut. Pioneering cuts should be wide enough so that the scrapers that follow have room to maneuver.

Backfilling. Dozers are the best equipment for backfilling because material may be pushed directly ahead of the machine over embankments into ditches or directly against a structure. Angling dozers are excellent for backfilling ditches because they can drift material into the trench while maintaining forward motion. With straight dozers, best results are obtained by approaching at an angle, ending up each pass by swinging in toward the trench, culvert, or structure. When backfilling culverts, crawler tractors should not cross the structure unless there are at least 12 in of solid material on top.

Dozing Rocks and Frozen Ground. Rocks are generally removed by using one corner of the blade, as the full power of the tractor can then be applied to a short section of the blade. Best penetration can be obtained by tilting the corner of the blade. Large single rocks can be removed by using the blade to lift the rock, simultaneously applying power to the tracks. When a large rock is embedded in a group of smaller ones, the smaller ones should be removed before starting to move the large one.

Formations such as shale and sandstone are found in the inclined position. The operator should dig under the outcrop rather than attempt to penetrate from the opposite side. The formation can then be crushed by driving the tractor over it, following which it can easily be removed.

When working in frozen ground, the blade must be tilted for penetration. Once the area is penetrated, the blade will work against the exposed edge and, by pushing and lifting, the ground will be broken.

Ditching. Although the motor grader is better suited for ditching work, a bulldozer can be used effectively for rapid construction of rough ditches. The rough ditch is constructed with a straight blade by working at right angles to the length of the ditch.

Spreading. When trucks, scrapers, or wagons are used for hauling, bulldozers are ideal spreading tools at the fill. The blade should be kept in the straight position so that the material is drifted directly under the cutting edge. Depth of spread is usually set in the job specifications to obtain the desired compaction. Tractor-dozers are also excellent tools for pulling compaction equipment on the fill.

Downhill and Slot Dozing. Whenever possible, dozing should be done downhill for greatest production. In downhill work, it is not necessary to travel down with each load. Several loads can be piled up at the brink of the hill and pushed to the bottom with one pass.

Slot dozing uses spillage from the first few passes to build a windrow on each side of the dozer's path. This forms a trench, preventing spillage on subsequent passes. Cut sections, where possible, should be alternated as "slots" between narrow uncut sections. These uncut sections can then be removed in normal dozing. With favorable grades and soil conditions, the increase in production may amount to as much as 50 percent.

Blade-to-Blade Dozing. This gives increased output for haul distances of 50 to 300 ft. At less than 50 ft, the extra yardage obtained is offset by the extra time required to maneuver the second dozer into position. Two or more dozers may be used effectively in blade-to-blade dozing. However, there are limitations to this method. Any delay which affects one machine will simultaneously affect the other machines, with a consequent decrease in production.

Bulldozing Tips

1. Do not back up farther than necessary, and do not push earth greater distances than required.
2. Know where the next pass is going to be made and where material will be placed. Always plan the operation thoroughly.
3. Alternate the tilt and angling adjustments periodically to balance wear on the blade and steering clutches.
4. When dropping down a steep hill or over the side of a fill, use the blade for a brake.
5. When traveling, carry the blade low or about 14 in above the ground; this practice will protect the underside of the tractor.

Production Estimating

Many bulldozer applications such as ditching, grubbing, stumping, pioneering, and stripping do not lend themselves to an exact production analysis in terms of volume of material moved, but in most cases an approximation can be made of the total machine hours required to complete the job. The work can then be bid as a lump sum, or in some cases on a per-acre or lineal-foot basis, respectively, for clearing and ditching. In contrast, production dozing on short hauls or feeding belt loaders can generally be analyzed to give reliable production figures. Table B3-3 indicates typical bulldozer production under these conditions. In general, the economical hauling distance for a dozer ranges from 25 to 300 ft.

TABLE B3-3 Estimated Track-type Tractor-Bulldozer Production
(loose cubic yards per 60-minute hour with power-shift (PS) tractor and S blade)

Tractor flywheel hp	One-way haul distance, ft					
	50	100	150	200	300	400
275–450	1070	557	379	352	268	216
200–275	940	482	328	305	232	188
125–200	685	355	240	225	171	138
75–125	500	259	176	164	125	101
50–75	295	153	104	97	74	60

For A blade use 75 percent of values for S blade.
For U blade use 125 percent of values for S blade.
For DD (direct drive) use 80 percent of values for power shift (PS).
The table is based on the following conditions:
0 grade { Dozing speed 160 fpm
Return speed 220 fpm up to 150 ft
Return speed 400 fpm beyond 150 ft
No slot or trench.
Material is in the bank state (no stockpile).
Each of the following will increase production as indicated:
  Doze on stockpile, add 15 to 20 percent.
  Doze downhill on 10 to 20 percent grade, add 20 to 40 percent.
  Use of slot dozing technique, add 15 to 20 percent.

More detailed estimates of bulldozer production can be obtained by analyzing the working cycle. Consideration must then be given to material type, blade capacity, travel and return distances and speed, and working efficiency. Blade capacities for various sizes of track-type tractors and for different blade types are listed in Table B3-4.

TABLE B3-4 Blade Capacities for Track-type Bulldozers (loose cubic yards measure)

Tractor flywheel hp	A blade	S blade	U blade
275–450	7.4	10.5	13.5
200–275	6.5	9.1	11.8
125–200	4.7	6.7	8.7
75–125	3.4	4.9	7.2
50–75	2.0	2.9	NA

Production in loose cubic yards (lcy) per 60-min hour is figured in the following manner:

$$\text{Yd/h} = \frac{60 \text{ min} \times \text{dozer capacity}}{\text{fixed time} + \text{haul time} + \text{return time}}$$

where fixed time = 0.05 min when only forward and reverse lever is required to shift (PS)
= 0.10 min when both forward and reverse lever and gear selector are required to shift (DD)

Haul time and return time are computed from the travel distance and tractor speeds.

Dozing will generally be done at slow speeds from 100 to 170 fpm forward and 220 to 400 fpm in reverse.

Example. Consider a 200 to 275 flywheel hp tractor, straight blade, capacity 9.1 lcy, 50-ft travel distance, haul and return in second gear.

$$\text{Fixed time} = 0.05 \text{ min}$$

$$\text{Travel time} = \frac{50 \text{ ft}}{160 \text{ fpm}} = 0.31 \text{ min}$$

$$\text{Return time} = \frac{50 \text{ ft}}{225 \text{ fpm}} = 0.22 \text{ min}$$

$$\text{Production} = \frac{60 \text{ min} \times 9.1 \text{ lcy}}{0.05 \text{ min} + 0.31 \text{ min} + 0.22 \text{ min}} = 940 \text{ lcy/h}$$

To estimate the expected production, load factor and job efficiency should be included. The production loose cubic yards can be converted to bank cubic yards (bcy) by multiplying lcy by a load factor whose value depends on the type of material. Assuming a 0.80 load factor and a 50-min-hour efficiency, the expected production would be

$$940 \text{ yd/h} \times 0.80 \text{ load factor} \times 50/60 = 626 \text{ bcy/h}$$

The next step is to compute the unit cost of dozing the material in question. To accomplish this, it is necessary to obtain an estimate of the hourly owning and operating cost for the unit. This cost can be obtained from the user's records or estimated from data such as that supplied in Table B3-4. Assuming an owning and operating cost of $200/h for the bulldozer in the example, we have

$$\text{Unit production cost} = \frac{\$200}{626 \text{ bcy/h}} = \$0.32/\text{bcy}$$

RIPPERS

The tractor-mounted ripper is used in construction, mining, and quarry operations. Due to the greater weight and higher horsepower of track-type tractors as well as to advances in ripper design, this combination can rip material which would have been blasted a few years ago. Ripping production and costs vary from material to material, and all rock formations cannot be ripped. However, if a rock can be ripped with a minimum production of 150 to 200 yd^3/h, the costs are generally lower than for drilling and blasting. In some cases, a cost reduction of 30 to 70 percent has been recorded. In addition, ripped material can be scraper-loaded to effect greater savings.

Types

Various types of rippers are illustrated in Fig. B3-11.

Towed Units. These are virtually obsolete, having been replaced by the integral types: heavy-duty hinge and parallelogram. The towed units are cumbersome, hard to maneuver, and do not use tractor weight effectively compared to the integral units. The towed unit was usually cable operated while the integral ripper is hydraulically operated, and in conjunction with the tractor on which it is mounted, provides a more compact and efficient unit. The tractor itself is usually equipped with crawlers, thus avoiding the excessive damage to tires that might be expected in rocky conditions. Figure B3-11*a* illustrates a towed ripper.

Miscellaneous Units. One type of ripper, generally referred to as a *scarifier,* is mounted on motor graders and loaders as shown in Fig. B3-11*b*. It can also be mounted on a small crawler tractor.

Backrippers are mounted on dozer blades, as shown in Fig. B3-11*c*. As the dozer proceeds forward, the back rippers pivot in the mounting brackets and drag over the ground. Then, as the tractor backs up to its original dozing point, the rippers engage in the ground. They are used in less severe conditions than the heavy-duty units. The shanks and points can be removed or pinned in the brackets above the blade base when not in use.

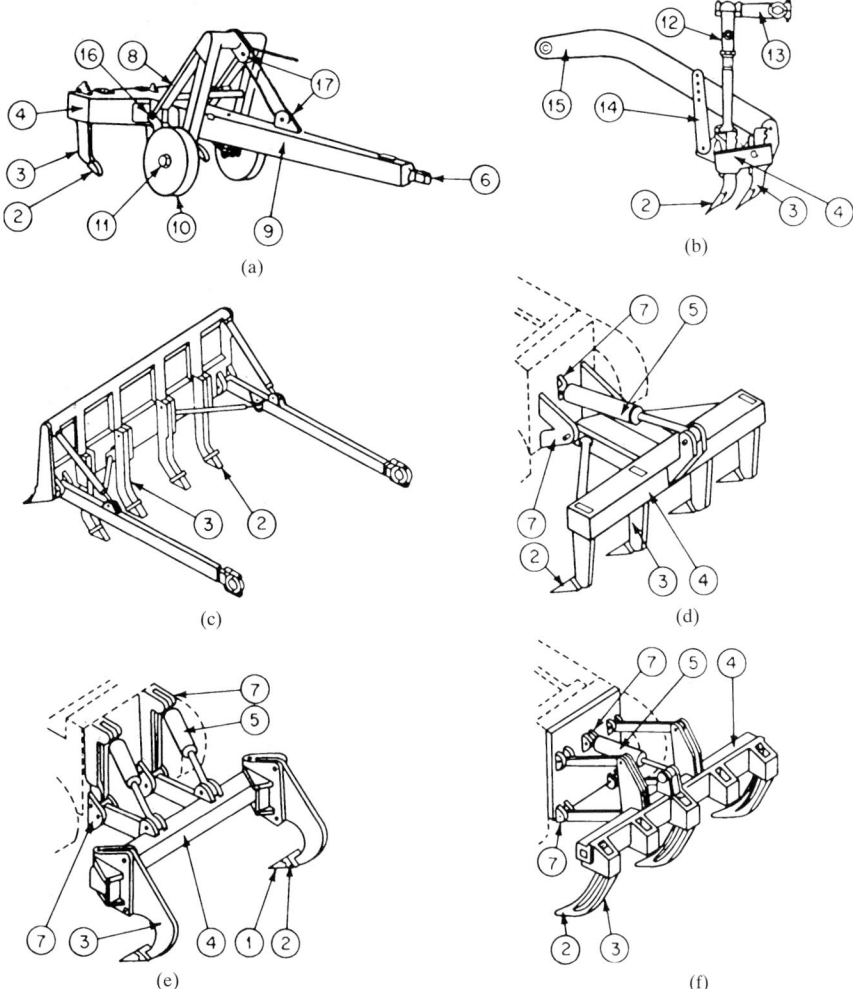

FIGURE B3-11 Types of rippers: (*a*) Towed ripper, hinge type; (*b*) ripper, motor-grader mounted; (*c*) back ripper, dozer mounted; (*d*) hinge-type ripper, tractor mounted; (*e*) hinge-type ripper, swivel shanks, tractor mounted; (*f*) parallelogram-type ripper, tractor mounted. *(SAE Handbook.)*

Hinge and Parallelogram Units. The two basic types in general use today are the hinge and parallelogram rippers shown in Fig. B3-11*e* and *f.* The beam of the hinged ripper pivots on link arms to raise or lower the ripper shank. This causes the angle between the ripper point and the ground to vary as the point enters the ground, which causes a penetration problem. The parallelogram ripper linkage allows the angle between the ripper point and the ground to remain constant as the point enters the ground, which improves penetration capability. Shanks are fixed or swivel and will penetrate up to 50 in.

Figure B3-12 shows a three-tooth ripper in action behind a Model 450G dozer.

FIGURE B3-12 Three-tooth ripper in use with Model 450G dozer. *(John Deere Co.)*

Operation

Generally, slow speeds—1 to 1.5 mi/h—should be used for most ripping operations, as point and track wear increases very rapidly with only small increase in speed. In most cases, a ripping job is begun with only one shank. If the material is easily penetrated and fractures into small pieces, a second shank can then be used. In tough conditions, however, two shanks cause greater track slippage and impose severe off-center loads on both tractor and ripper. It is normally desirable to rip as deep as possible, but if stratification is encountered it may be best to remove the material in its natural layers.

Tandem ripping may prove profitable under some conditions. Adding a second tractor to push the first tractor-ripper combination will extend the useful service range of the ripper into harder materials. Many times only a small part of the total rock on a job is too tough for a single tractor and ripper to handle, and in these cases tandem ripping can prove more economical than bringing in a drill crew. In very hard rock it is not uncommon for tandem ripping to increase production three or four times over that obtained with a single tractor, even though the cost may double.

Some rock that is too difficult to rip can be preblasted with a light charge of explosives and then ripped successfully.

Production Estimating

With refraction seismograph information, it is possible to investigate the rippability of various rock formations. The seismograph measures the velocities of seismic waves through subsurface materials. These velocities can be compared to data obtained from previous tests, wherein rippability and seismic velocity have been correlated for typical materials and for selected machines. An example of such correla-

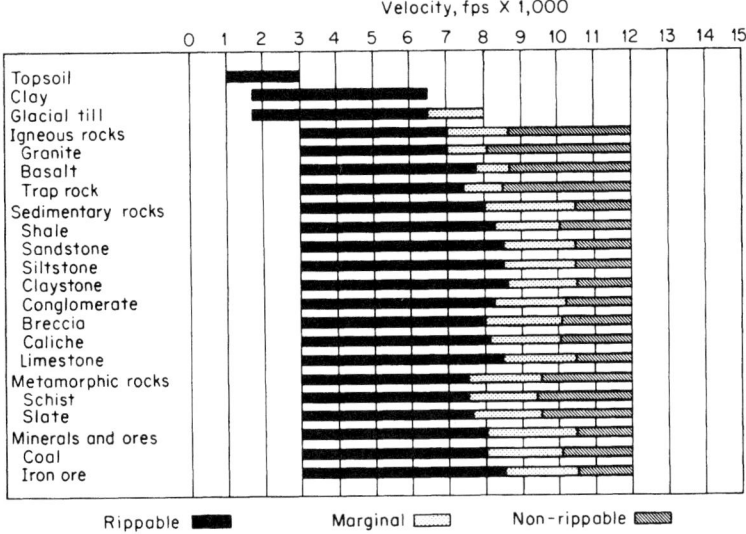

FIGURE B3-13 Ripper performance as related to seismic wave velocities (385-hp track-type tractor with parallelogram single-shank ripper).

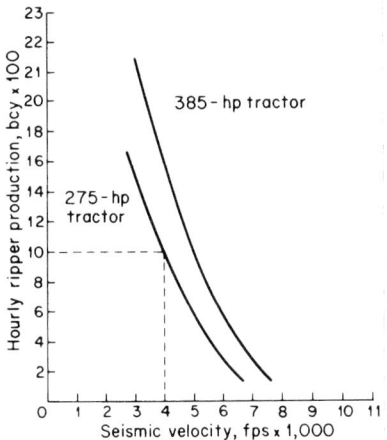

FIGURE B3-14 Ripper production. Considerations: power shift tractors; single-shank rippers; production figures based on 100 percent working efficiency (60-min hour). Chart is for ripping in all classes of material under ideal conditions. For igneous rocks with seismic velocities of 6000 fps or higher, the production figures should be reduced by 25 percent. If thick laminations, vertical laminations, or other adverse factors exist, the indicated production figures should be derated appropriately.

tion for a 385-hp track-type tractor with a parallelogram single-shank ripper is shown in Fig. B3-13. For tractors in the 200- to 275-hp class, the rippability ranges would be reduced by 15 to 25 percent.

Ripping production can also be estimated from seismic velocities. Figure B3-14 shows production for 385- and 275-hp crawler-tractors with parallelogram single-shank rippers, operating as single units. For example, this figure indicates that the single 275-hp tractor can rip approximately 1000 bcy/h in a rock whose seismic velocity is 4000 fps. Figure B3-13 provides data for preliminary selection of ripping units over a wide range of seismic conditions. The final selection of equipment should be based on job production requirements.

Ripping ownership and operating costs will vary from 25 to 100 percent above bulldozing costs for comparable tractor sizes. Assuming the owning and operating cost for the 275-hp tractor (including operator's wages) to be

$200/h for bulldozing and 50 percent greater for ripping, then for the preceding example the unit ripping cost would be $300/1000 = $0.30/bcy. Since the production rate is based on a 60-min hour, an approximate adjustment should be made to obtain the expected production cost.

REFERENCE

1. *Caterpillar Performance Handbook,* Caterpillar, Inc., 1993.

CHAPTER 4
SCRAPERS*

James J. O'Brien, P.E.
O'Brien-Kreitzberg, Inc., Pennsauken, New Jersey

INTRODUCTION

The scrapers used in modern earth-moving and heavy construction include towed wheel scrapers, Fig. B4-1; wheel tractor-scrapers, Fig. B4-2; and elevating scrapers, Figs. B4-3 and B4-4. All scrapers are either powered or towed by a prime mover which can be (1) a track-type tractor, (2) a two-axle wheel-type tractor, (3) a single-axle wheel-type tractor, or (4) a twin-engine unit. This last, illustrated in Fig. B4-5, is identical to types (2) and (3) except that an additional engine is mounted on the scraper to drive the scraper axle.

CONVENTIONAL SCRAPERS

Towed wheel scrapers and wheel tractor-scrapers (both two-axle and three-axle) are frequently referred to as *conventional scrapers*. Figure B4-6 shows the components and standard nomenclature for both towed and wheel tractor-scrapers. The towed unit has a truck (component 10) and gooseneck (component 2) arrangement which is suitable for towing, whereas the wheel tractor-scraper has a gooseneck only (component 2) which attaches to the tractor portion to make an integrated, self-propelled unit. The principal components of conventional scrapers are the apron, cutting edge, bowl, ejector, push frame and plate, draft frame, hitch, and running gear.

The primary functions of the conventional scraper are to load (generally with assistance from a pusher), haul, spread, and return to a loading point. These basic operations constitute the materials-handling working cycle. The conventional scraper can also be used for light clearing, rough or fine grading, bank sloping, and ditching.

---

* Some of the material in this chapter is from R. D. Evans, "Scrapers," in Stubbs and Havers (eds.), *Standard Handbook of Heavy Construction*, 2d. ed., McGraw-Hill, New York, 1971.

B4-2 STANDARD HANDBOOK OF HEAVY CONSTRUCTION

FIGURE B4-1 Caterpillar D8H and 463E towed scraper (18/26).

FIGURE B4-2 Caterpillar 631 wheel tractor-scraper (21/30).

FIGURE B4-3 John Deere 762B elevating scraper. *(John Deere Co.)*

SCRAPERS

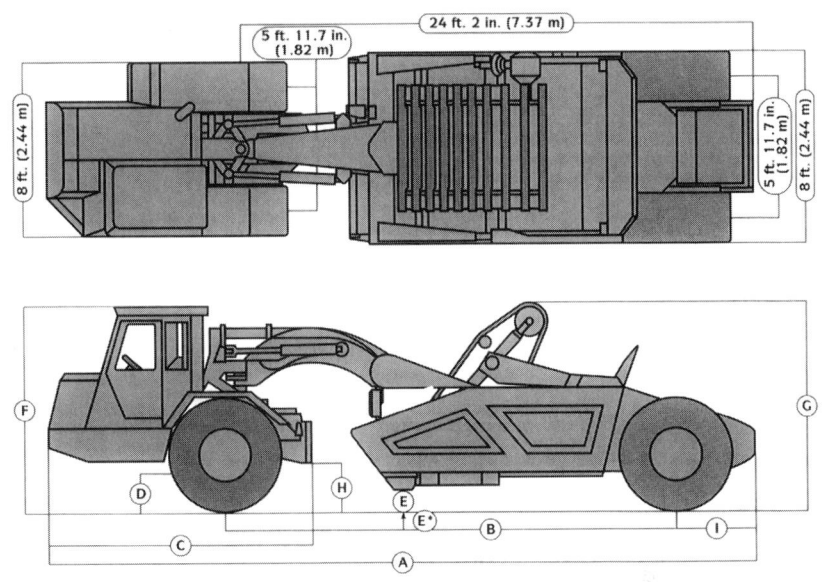

	BOWL AT GROUND LEVEL	BOWL UP	BOWL LEVEL
A	33 ft. 7 in. (10.24 m)	33 ft. (10.06 m)	33 ft. 3 in. (10.14 m)
B	21 ft. 1 in. (6.42 m)	20 ft. 4.5 in. (6.21 m)	20 ft. 6 in. (6.25 m)
C	13 ft. (3.97 m)	13 ft. 1.5 in. (4.00 m)	13 ft. 0.5 in. (3.98 m)
D (axle clearance)	18.5 in. (470 mm)	18.5 in. (470 mm)	18.5 in. (470 mm)
E	—	18.7 in. (474 mm) w/o teeth 16.4 in. (417 mm) w/teeth	11.6 in. (295 mm) w/o teeth 8.9 in. (226 mm) w/teeth
F	10 ft. 2.5 in. (3.11 m)	9 ft. 10 in. (2.99 m)	9 ft.11.5 in. (3.04 m)
G	8 ft. 5.5 in. (2.58 m)	9 ft. 3 in. (2.82 m)	8 ft. 9.5 in. (2.68 m)
H	17.3 in. (440 mm)	22.4 in. (570 mm)	19.2 in. (488 mm)
I	4 ft. 2.5 in. (1.28 m)	4 ft. 5 in. (1.35 m)	4 ft. 5 in. (1.35 m)

E*: cut below ground level: 7.1 in. (180 mm) w/o teeth
10.3 in. (262 mm) w/teeth

FIGURE B4-4 John Deere 762B elevating scraper, dimensions. *(John Deere Co.)*

Capacity Ratings

Conversion of yardage ratings for conventional scrapers into capacities in terms of pay yards is required for each specific project. Scraper yardage ratings are generally determined on the basis of SAE J741a[1] as follows:

FIGURE B4-5 Caterpillar 627 twin-engine scraper.

1. Capacities shall be rated on the basis of cubic yards. Standard ratings shall include both struck and heaped capacities (see Fig. B4-7).
2. Struck capacity of a scraper shall be the actual volume enclosed by the bowl and apron, struck off by a straight line passed along the top edge of the side plates or adjacent load-carrying mechanism or extensions thereof. Struck capacity shall be given to nearest 0.1 yd^3.
3. When the top of the front apron, in the closed position, is below the top edge of the side plates, the capacity shall be limited, either by a plane from the top edge of the apron to the forward corners of the side plates, or by a plane at a slope of 1:1 extending from the top edge of the front apron to the plane formed by the top edge of the side sheets, whichever gives the smaller capacity.
4. To determine the 1:1 slope for the limiting plane, the scraper shall be set in its normal carrying position with the apron closed.
5. Top extensions of the ejector above the side plates shall not be included in the determinations of struck capacity.
6. The volume occupied by apron arms, sheave frames, or other internal projections shall be disregarded in calculating the struck capacity.
7. Heaped capacity of a scraper shall be the sum of the struck capacity and the volume enclosed by the four planes at a 1:1 slope extending upward and inward from the top of the solid portion of the front apron, from the top of the solid portion of the ejector or rear plate, and from the top edges of the side plates. Small barred or screened openings in the apron may be ignored in determine the solid top line.
8. The scraper shall be set in the same carrying positions as used for determining struck capacity.
9. For scrapers of less than 12 yd^3 struck capacity, the heaped capacity shall be given to the nearest ½ yd^3; for scrapers of 12 yd^3 struck capacity and larger, the heaped capacity shall be given to the nearest 1.0 yd^3.
10. If the top edge of the side plate (or extension thereof), front apron, or ejector is not a straight line, a mean line through its configuration shall be used to establish the base line of the plane enclosing the heaped capacity.
11. The possible interference of overhead structures, such as sheave guides and cables, with the heaped capacity shall be ignored.

SCRAPERS

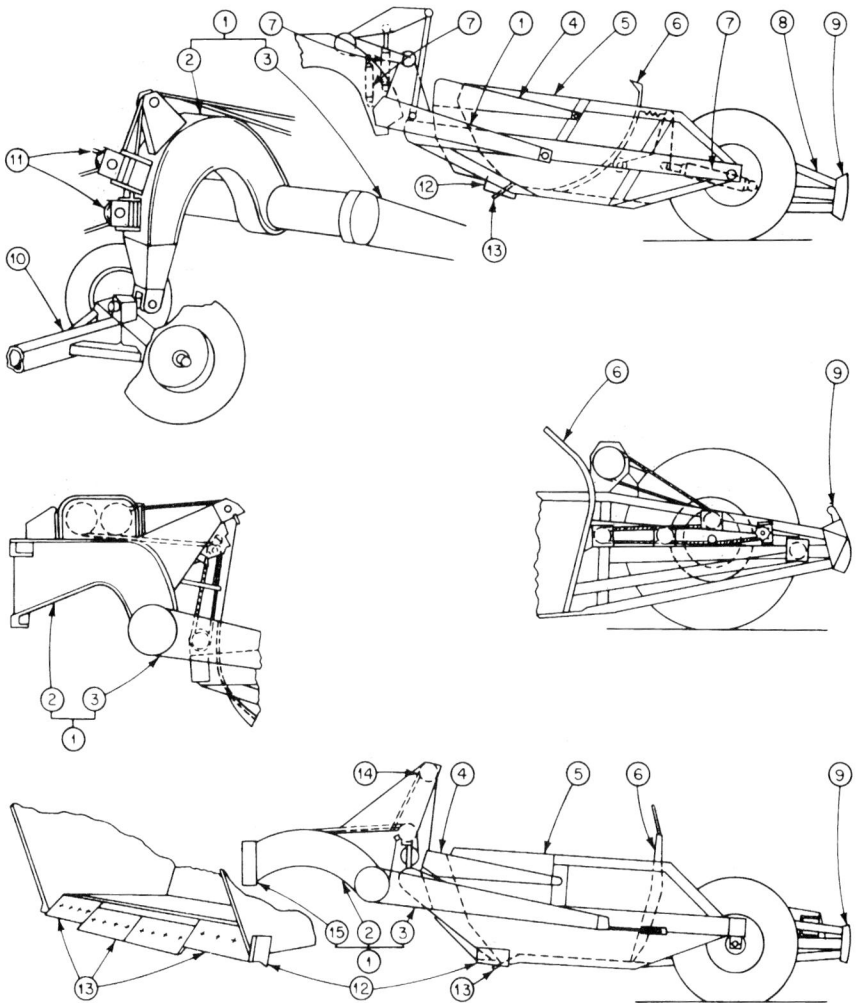

FIGURE B4-6 Components and nomenclature for conventional scrapers. (1) Frame, draft, (2) gooseneck, (3) yoke, (4) apron, (5) bowl, (6) ejector, (7) cylinder, (8) frame, push, (9) plate, push, (10) truck, front, (11) fairlead, (12) bit, side, (13) cutting edge, (14) sheave, (15) housing, kingpin.

12. Both the struck and heaped capacities as defined are, with minor reservations, definitely measurable quantities providing a standard method of comparing the volume of scrapers.

Equipment Specifications

Specification details for representative scrapers of the conventional type are summarized[2] in Tables B4-1 to B4-6.

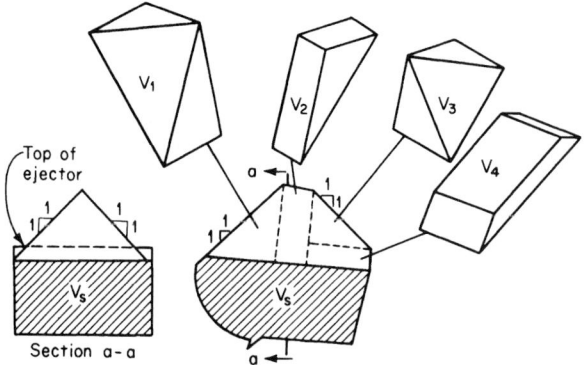

FIGURE B4-7 Typical scraper volume computation. Volume struck = V_s (express to nearest 0.1 yd³). Volume heaped = $V_H = V_s + (V_1 + V_2 + V_3 + V_4)$. If less than 12 yd³ struck, express to nearest 0.5 yd³. If 12 yd³ struck or more, express to nearest whole cubic yard.

TABLE B4-1 Specifications for Caterpillar Wheel Tractor-Scrapers, Standard Type[2]

Model	Horsepower (flywheel)	Weight lb-empty	Rated load, lb	Capacity-CY Struck	Capacity-CY Heaped
621 E	330	67,195	48,000	14	20
631 E Series II	450	96,880	75,000	21	31
651 E	550	134,370	104,000	32	44

TABLE B4-2 Specifications for Caterpillar Wheel Tractor-Scrapers, Tandem-Powered, Push-Pull Type[2]

Model	Tractor horsepower (flywheel)	Scraper horsepower	Weight lb-empty	Rated load, lb	Capacity-CY Struck	Capacity-CY Heaped
627 E	330	225	77,510	48,000	14	20
637 E Series II	450	250	112,090	75,000	21	31
657 E	550	400	151,810	104,000	32	44

TABLE B4-3 Specifications for Caterpillar Wheel Tractor-Scrapers, Elevating Type[2]

Model	Horsepower (flywheel)	Weight lb-empty	Rated load, lb	Capacity-CY heaped
613 C	175	33,650	26,400	11
615 C Series II	265	56,450	40,800	17
623 E	365	74,300	55,000	23
633 E	475	112,000	82,000	34

TABLE B4-4 Specifications for John Deere Wheel Tractor-Scrapers, Elevating Type

Model	Horsepower (flywheel)	Weight lb-empty	Rated load, lb	Capacity-CY heaped
762 B	180	35,600	27,600	11.5
862 B	265	49,050	40,000	16

Source: John Deere Co.

TABLE B4-5 Specifications for Caterpillar Wheel Tractor-Scrapers, Standard Auger Type[2]

Model	Horsepower (flywheel)	Weight lb-empty	Rated load, lb	Capacity-CY heaped
621 E	365	79,750	48,000	21
631 E Series II	490	101,370	75,000	31
651 E	594	146,770	104,000	44

TABLE B4-6 Specifications for Caterpillar Wheel Tractor-Scrapers, Tandem-Powered Auger Type[2]

Model	Tractor horsepower (flywheel)	Scraper horsepower	Weight lb-empty	Rated load, lb	Capacity-CY heaped
627 E	330	225	89,035	48,000	21
637 E Series II	450	250	120,235	75,000	31
657 E	550	400	167,270	104,000	44

Methods of Operation

The materials-handling working cycle of the conventional scraper will be described in terms of an earthwork application.

Loading. Figure B4-8a shows a loading attitude for the conventional scraper. The operator approaches the cut with the ejector to the rear and the apron raised approximately 15 in (Fig. B4-8b). The bowl is then lowered to the desired depth of cut (Fig. B4-8c) and the tractor transmission shifted to L load range to keep engine speed high. For work other than finish grading operations, available power and length of cut are factors to consider in determining the optimum depth of cut.

Positive apron control permits the operator to choose the correct opening when loading (Fig. B4-8d). Increasing the opening will prevent dirt from piling up in the front of the lower apron lip, and decreasing the opening will keep dirt from rolling out of the bowl. In loose material, the bowl should be raised and lowered rapidly; this can be repeated as necessary to pump material into the bowl for maximum loading.

When the bowl is full, the apron is closed and the bowl is then raised (Fig. B4-8e). In loose, dry material, the bowl should be raised only slightly, then the apron closed partially, and the bowl again raised.

Traveling Loaded. Figure B4-9a illustrates the traveling attitude for the loaded scraper. If the scraper controls are of the hydraulic type, as will usually be the case, a

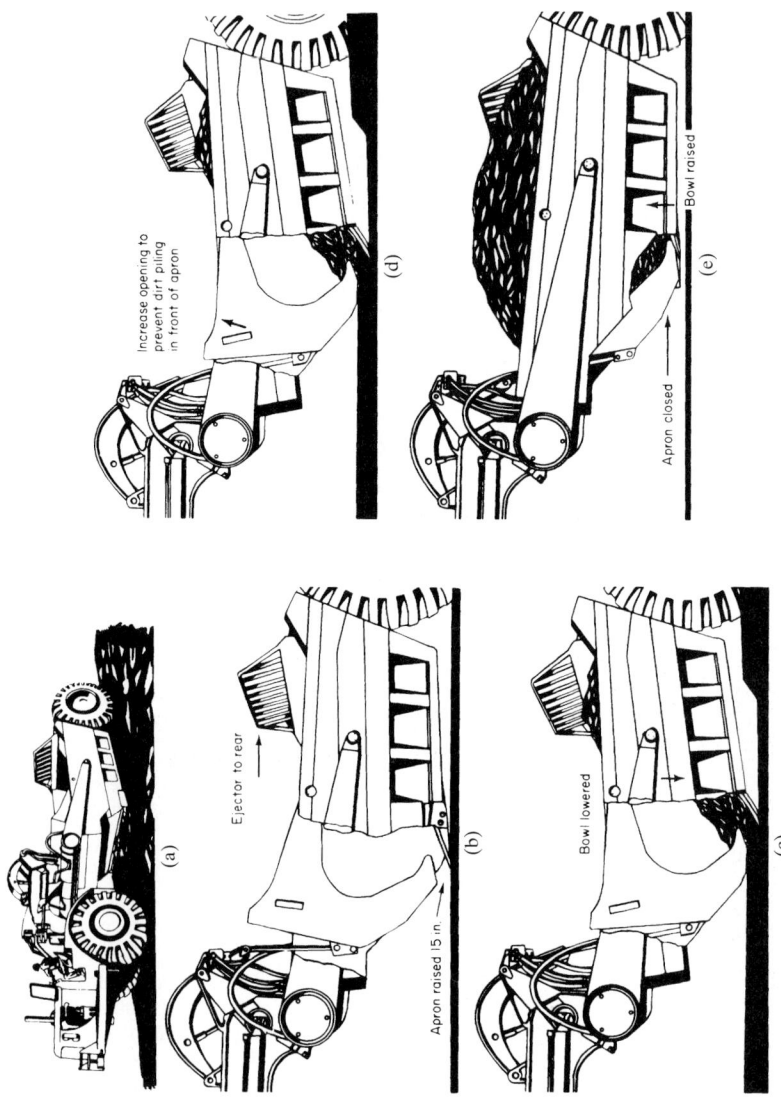

FIGURE B4-8 Loading a conventional scraper: (*a*) Loading attitude; (*b*) approaching the cut; (*c*) making the cut; (*d*) apron opening during loading; (*e*) finishing the cut.

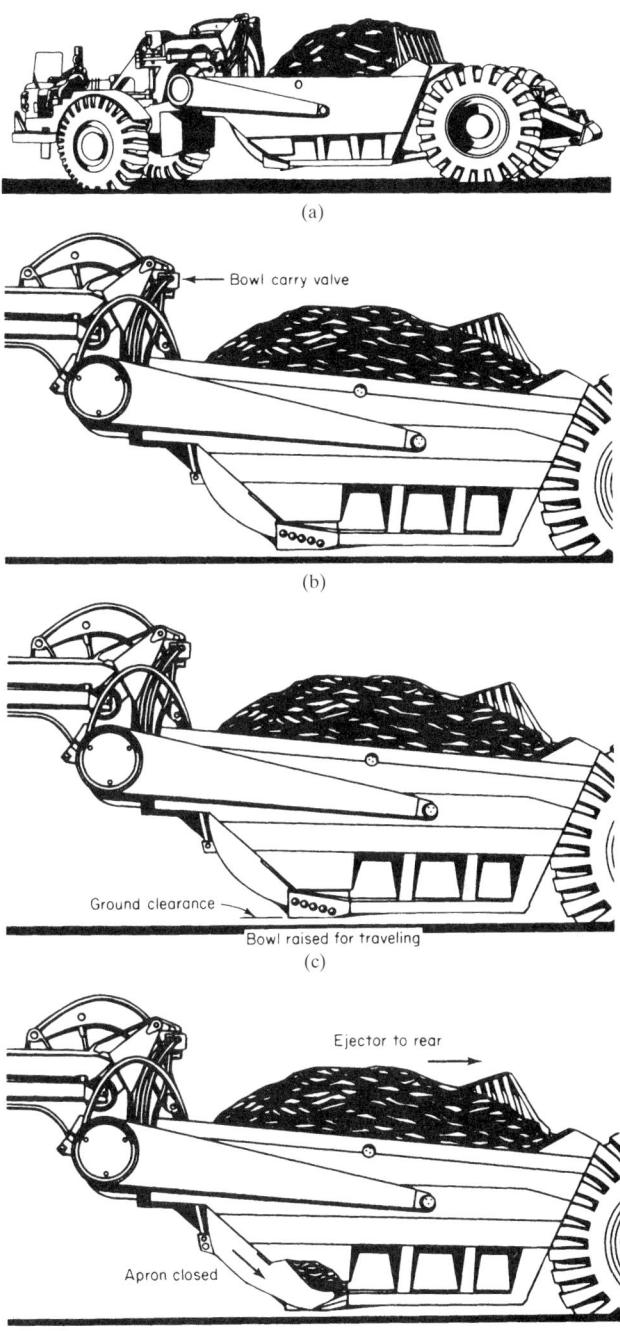

FIGURE B4-9 Traveling loaded with a conventional scraper: (*a*) Traveling attitude; (*b*) bowl-carry check valve; (*c*) bowl position; (*d*) apron and ejector position.

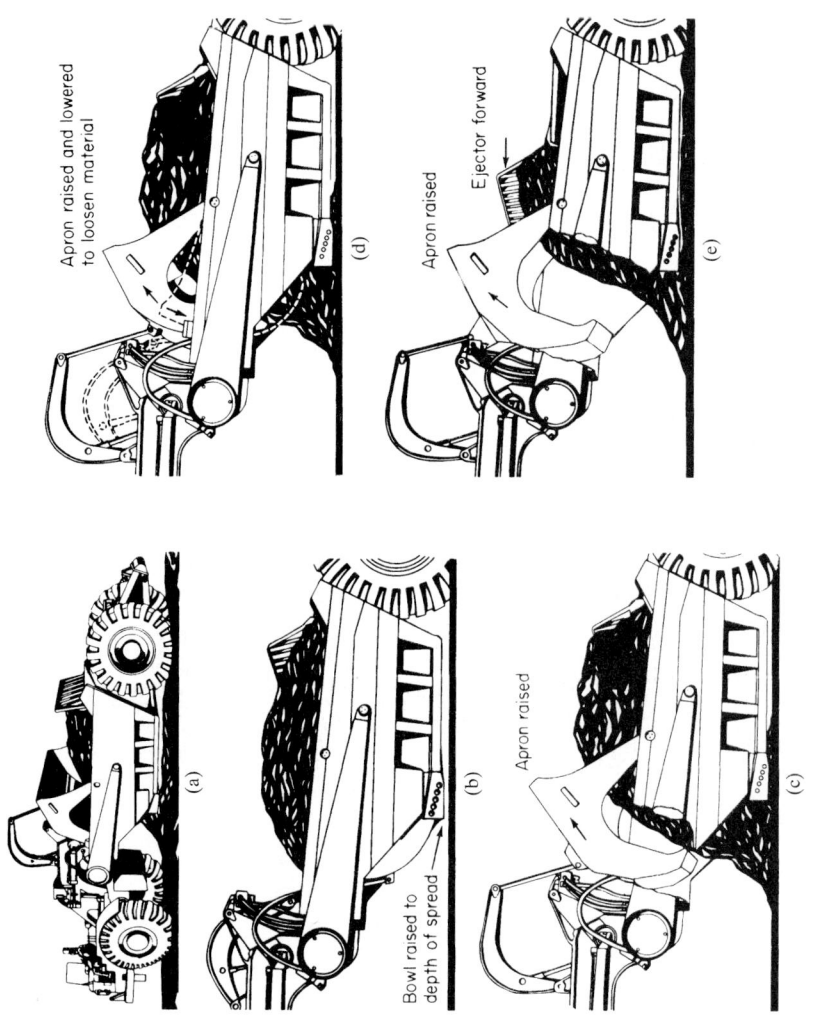

FIGURE B4-10 Unloading a conventional scraper: (*a*) Unloading attitude; (*b*) bowl position; (*c*) apron position (loose material); (*d*) apron position (sticky material); (*e*) ejector movement.

B4-10

bowl-carry check valve will operate automatically in the bowl-lift hydraulic circuit (Fig. B4-9*b*). This valve locks the bowl cylinders hydraulically so that the bowl cannot drop in case a large hose should break.

The bowl should be raised to provide sufficient clearance for high-speed travel (Fig. B4-9*c*). The apron should be fully closed to prevent loss of the material, and the ejector should remain in the rear position (Fig. B4-9*d*).

Unloading. The unloading scraper attitude is illustrated in Fig. B4-10*a*. The bowl should be positioned to spread the material to the desired depth (Fig. B4-10*b*). A partial opening of the apron during the initial unloading process will assist in maintaining an even spread of loose material (Fig. B4-10*c*). If the material is wet and sticky, the apron can be raised and lowered repeatedly until the material behind it is loosened and drops out of the bowl (Fig. B4-10*d*). When the material has fallen from the back of the apron, the ejector is moved forward to push the remaining material out of the bowl at a uniform rate (Fig. B4-10*e*). It is recommended that the apron be raised completely before moving the ejector forward in any type of material.

In some hydraulically controlled scrapers, an automatic two-speed ejector-control system moves the ejector forward slowly when considerable force is required to push the material out of the bowl. As this required force is reduced, the control system automatically increases the speed of ejector movement.

Returning to Cut. The traveling attitude of the unloaded scraper while returning to the cut is illustrated in Fig. B4-11.

FIGURE B4-11   Returning to cut.

Elevating Scrapers

In contrast to the conventional scraper, which usually depends on a pusher tractor for loading, the elevating scraper is considered to be self-loading. It is similar to a conventional scraper except that its apron has been replaced by an elevator or *ladder*. This elevator is made up of heavy bicycle-type chain with flights or paddles bolted to it. The chain is rotated by a power source which is independent of the forward travel of the scraper. It generally has variable speed capability and can be reversed. A conventional scraper depends on pusher power to force material over the cutting edge and then backward into the bowl. The elevating scraper must only force the dirt over the cutting edge, since the ladder will carry it from there into the bowl.

Self-loading elevating scrapers were first considered as utility tools—grading between concrete forms, hauling backfill to structures, and the like. Their production capabilities have now been recognized, and they compete with the other scraper classes for large embankment jobs. They load well in most materials, with the following exceptions: shot rock, ripped rock, boulders, "bony" material that is too large to pass between the cutting edge and elevator flights, and cohesive material with high moisture content that tends to ball up and stick to flights. The cost of a pusher and its operator is eliminated for the elevating scraper because of its self-loading capability. However, due to the dead weight of its loading mechanism, it has a reduced efficiency in the traveling portion of its working cycle.

Capacity Ratings

Yardage ratings for self-loading scrapers are generally determined on the basis of SAE J957, as follows:[1]

1. Capacity shall be rated on the basis of cubic yards.
2. The standard rating shall be the total volume of the heaped capacity.
3. The elevating scraper bowl shall be set at its normal carrying position when calculating its standard rated capacity.
4. To determine the standard rating, it will be necessary first to determine the struck capacity.
5. The struck capacity of an elevating scraper shall be the actual volume enclosed by the bowl and elevator, struck off by a straight line passed along the top edge of the bowl sides or adjacent load-carrying mechanism, or extensions thereof.
6. If the top edge of the bowl sides, or extensions thereof, is not a straight line, a mean line through its configuration and parallel to the ground shall be used.
7. The forward edge of the bowl shall be the plane, or planes, formed by the upward straight path, or paths, of the inner edge of the flights on the elevator adjacent to the load.
8. The elevator shall be located in the manufacturer's recommended or normal operating position.
9. The boundary of the lower forward bowl will be a line normal to the cutting edge and passing through the centerline of the elevator bottom idlers, or a line from the cutting edge to the nearest point of the elevator flight path.
10. The heaped capacity shall be the sum of the struck capacity and the volume enclosed by four planes as follows:
    a. Planes extended upward and inward at a 1:1 slope from the top edges of the bolt sides.
    b. A continuation of the uppermost plane formed by the upward path of the inner edge of the flights on the elevator.
    c. A plane extending from the top of the solid portion of the ejector drawn tangent to the path of the outer edge of the flights at the top of the elevator and stopping at the intersection with the plane established by item **b**.
11. The volume occupied by flights, levers, or other internal projections shall be disregarded in calculating the standard rated capacity.
12. For scrapers of less than 12 yd^3, the standard rated capacity shall be given to the nearest 0.5 yd^3; for scrapers of 12 yd^3 and larger, the standard rated capacity shall be given to the nearest whole cubic yard.

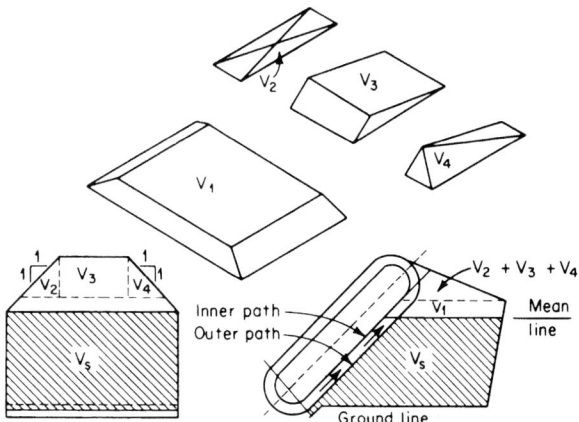

FIGURE B4-12 Elevating scraper volume computation. Volume struck = V_s. Standard rating volume heaped = $V_H = V_s + (V_1 + V_2 + V_3 + V_4)$.

13. The heaped capacity as defined, with minor variations, is a definitely measurable quantity which provides a standard method of comparing the volume of elevating scrapers. (See Fig. B4-12.)

Equipment Specifications

Specification details for representative elevating scrapers[2] are summarized in Tables B4-3 and B4-4.

Methods of Operation

The working cycle of the elevating scraper is quite similar to that of the conventional scraper.

Loading. In making the cut, the bowl is first lowered to a depth that will permit the elevator and tractor to operate at a high, constant engine speed. Shallow cuts, as illustrated in Fig. B4-13a, are a necessity. Deep cuts will tend to force material into the bowl, and this will slow or stall elevator flights. The elevator flights sweep the materials into the bowl, and any attempt to force material into the bowl will actually increase the time required to load.

The elevator has four speeds forward and one reverse. Materials such as sand, silt, and topsoil, which are easy to penetrate, are loaded at high speeds. This keeps the cutting edge swept clean and prevents the material from being pushed along in front of the cutting edge. If the operator repeatedly lifts and lowers the bowl during the loading operations, the advantage of high-speed elevator operation will be lost. Low speeds are used in loading tough materials, such as hard-packed clay, gumbo, etc. Low elevator speed will lessen elevator bounce and permit the elevator flights to sweep material into the bowl.

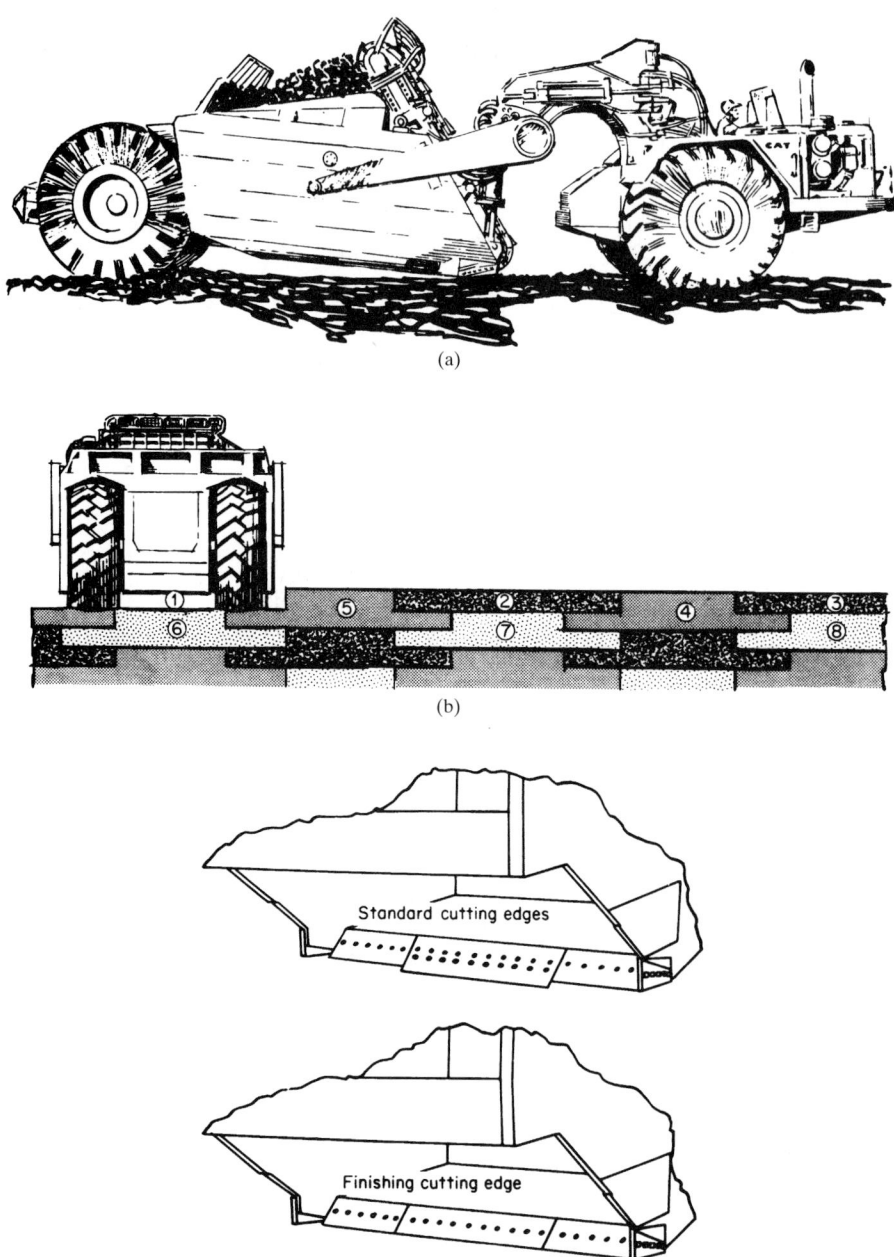

FIGURE B4-13 Loading an elevating scraper: (*a*) Shallow cuts to assure elevator efficiency; (*b*) straddle loading; (*c*) cutting edges.

FIGURE B4-14 Travel attitude with proper bowl clearance.

If possible, ridges 5 to 6 ft wide should be left between successive loading passes. Then, on the following passes, the operator can straddle load to pick up the ridges at a depth below previous cuts. This will leave other ridges, thus permitting the cycle to be repeated (Fig. B4-13*b*).

In addition to a straight cutting edge, an offset cutting edge or stinger is available to assist in the loading cycle (Fig. B4-13*c*). The center section of this cutting edge is extended as a stinger for normal loading. For hard loading, protruding ripping teeth can be bolted to the cutting edge. For doing finish work, loading windrowed material, or doing cleanup work, a finishing center section should be used. When the bowl is fully loaded, the elevator should be stopped. If this is not done, material will be swept from the face of the load and spilled over the sides of the bowl.

Traveling Loaded. When the scraper is loaded, the elevator should be stopped and the bowl raised until it clears the ground by 1 to 2 in. Travel with the bowl in this position should be continued until any loose mound of material is smoothed out. This will leave a smooth cut in the loading area. The bowl should be raised sufficiently for clearance when traveling (Fig. B4-14).

Unloading. The bowl should be lowered to permit the desired depth of spread. Condition of fill and depth of spread will change the speed and power requirements. Throughout the unloading cycle, the tractor should be operated at full engine speed and without allowing the engine to lug.

With the machine in motion, the ejector floor is opened. The material in the bowl will then begin to unload itself, and the leading edge of the ejector floor will strike off the unloaded material in a smooth, even layer.

Returning to Cut. With unloading completed, the bowl can be raised to the desired height for the haul-road conditions. The ejector floor is closed, and the scraper returns to the cut.

MULTIPLE SCRAPERS

The first scrapers in this category emerged as single-bowl units with power on two axles. In most cases, they were push-loaded, except in extremely easy loading materials and/or when favorable grades were present to provide additional power (gravity).

Configurations for the multiple scrapers vary depending on the manufacturers. Some multiple units are permanently coupled and require only one operator station, while others have disconnect features which permit independent use of the individual units as desired. In the latter case, individual operator's controls are overridden by a central control station when operating as a multiple unit.

Loading and spreading are generally done in sequence—first bowl, second bowl, etc. The all-wheel drive units, like the elevating-type units, eliminate pusher and extra operator cost as well as wait time in the cut. However, they have increased maintenance repair and tire costs in comparison with conventional scrapers. They load well in all stable materials but may be limited in their traction capabilities when loading in poor underfoot conditions. All wheel-drive units—single, twin, triple—present a complex problem as far as productivity estimating is concerned. It is advisable to consult their manufacturers for production and cost information.

A push-pull concept (Fig. B4-15) provides a self-loading feature which is comparable to that of multiple units. Here two units assist each other only during the loading cycle, while on all other portions of the work cycle they operate as individual haul units. The first unit to arrive in the cut becomes the front unit and is loaded first. This takes advantage of the partial load the machine will get while positioning, as most operators self-load as much as possible. In addition, loading the front unit first permits a shorter cut length and results in a smoother cut. The rear unit can eliminate a possible gap by overlapping the first cut slightly, as illustrated in Fig. B4-16.

BASIC CYCLE CONSIDERATIONS

The Cut

When loading in the cut, material enters the scraper bowl rapidly at first and slows down as the loading progresses. Figure B4-17 is a typical load-growth curve which relates the time that a scraper spends in the cut to the load obtained. Note that the load finally reaches an area of the curve where further loading consumes excessive

FIGURE B4-15 Caterpillar 627 push-pull scrapers.

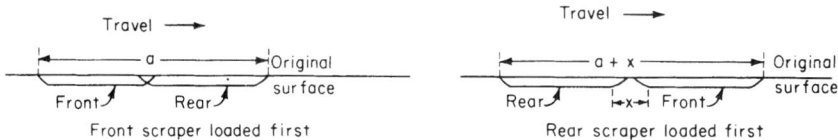

FIGURE B4-16 Push-pull loading technique.

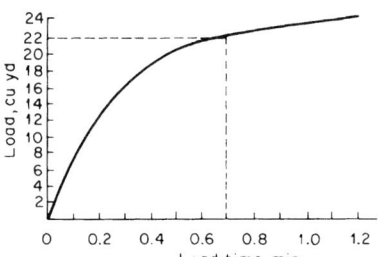

FIGURE B4-17 Load growth curve.

time, and the scraper should then pull out of the cut. Table B4-7 is an example of 400-hp wheel tractor-scraper, working on an approximate 3000-ft haul, which is loaded first with 22 and next with 24 yd^3 of material. Notice that the load time in the cut must be doubled to pick up the last 2 yd^3 of the 24-yd^3 load, resulting in a production loss of 240 yd^3/h.

Adequate pushers are necessary for efficient operation in the cut. Table B4-8 shows the production capability of four 400-hp wheel tractor-scrapers with one pusher. Although the cost per hour for the equipment spread is increased by adding another pusher, increased production lowers the cost per cubic yard. The lowest cost per yard occurs when full advantage is taken of the maximum potential of the hauling units.

Tandem pushing may pay off with large scrapers of 28 yd^3 and greater capacity when working in difficult loading material such as loose sand and tough clay. Poor traction in the cut, long hauls, and rock cuts are conditions which also warrant tandem pushing. Cushioning devices and power-shift transmissions, introduced in recent years, have made tandem pushing increasingly practical and popular.

The Haul Road

A large part of a haul unit's time is spent on the haul road. A poorly maintained haul road causes extreme bounce, which results in load loss, machine and operator fatigue,

TABLE B4-7 Controlling Load Time

Payload, bcy	22.0	24.0
Load time, min	0.75	1.5
Total cycle time, min	5.0	6.0
Trips/60-min h	12.0	10.0
Scraper production, bcy/60-min h	264.0	240.0
Pusher cycle time, min	1.5	2.25
Scrapers loaded/60-min h	40.0	26.7
Max pusher production, bcy/60-min h	880.0	640.0
Lost production, bcy/60-min h	—	240.0

TABLE B4-8 Adequate Pushers

4 scrapers, 21/30		1 pusher, 385 hp	
Payload	22 bcy	Payload	22 bcy
Load time	0.75 min	Load time	0.75 min
Scraper cycle time	5 min	Pusher cycle time	1.5 min
Unit production	264 bcy/h	Loads per hour	40
Potential spread production	1056 bcy/h	Maximum pusher production	880 bcy/h
Potential lost production			176 bcy/h

and excessive downtime. The slower speed at which the haul unit must be operated lowers production and raises cost per yard.

EQUIPMENT SELECTION

The Push-Loading Concept

More construction dirt moving is done with push-loaded fleets of scrapers than with all other systems combined. The economics of the push-loading system are such that its profitable application spans the greatest breadth of the actual conditions encountered in the field. The key to a pushed-scraper spread's economy is the fact that both pusher and scraper share in the work of obtaining the load. The resulting loading costs are lower than for almost any form of top-loading system and for other self-loading systems—elevating or multiple units—providing the pusher cost can be prorated over three to four scraper units.

A top-loading hauling unit will generally have a lower tare weight than its scraper counterpart, and thus it can carry a higher payload. However, the savings which are realized in the haul cost must be sufficient to offset the lower loading cost of the scraper spread before the "pure" haul system becomes economical. The heavier weight of an elevating scraper penalizes it on the haul portion of the cycle, and this weight penalty must be overcome through the elimination of the pusher before the economical performance of the elevating scraper can equal or exceed that of the scraper-pusher combination. The interrelated load-and-haul cost of the scraper-pusher system has proved to be an economical choice for a wide range of haul distances, underfoot conditions, grades, and materials.

In average, common earth materials, a single 385-hp pusher will typically load out 35 to 40 loads of 22.5 bcy/h. That is a pit production of 800 to 900 bcy/h. While hauling costs with the scraper will be slightly higher than with a pure hauler, the cycle must become quite long before this cost will offset the difference in the loading costs. Additionally, the scraper offers an unmatched ability to spread its load in an even lift on the fill. Frequently, this can result in substantially lower fill-processing costs than can be obtained with other types of haulers.

With the advent of special-application scrapers, which are designed expressly for rock handling, the push-loaded scraper has one of the widest material appetites of any system. It has been used economically to handle materials ranging from blow sand to gumbo to shot granite. This versatility can be a key factor in picking a system where a variety of materials is to be encountered.

Since push-loaded scrapers occupy such a dominant position in earth-moving systems, the science of matching particular types of scraper equipment to the job has become an important one. There are proper applications for both towed scrapers and wheel tractor-scrapers, two-axle and three-axle rigs, various size scrapers behind the same basic power unit, and for different size units of the same basic type.

Towed Scrapers Versus Wheel Tractor-Scrapers

The advantageous economics of the crawler-drawn scraper (towed) as compared to the wheel tractor-scraper stem are evident from its:

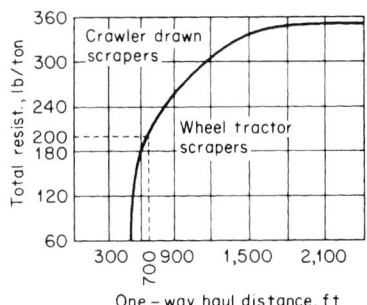

FIGURE B4-18 Application zones—crawler versus wheel scrapers.

Ability to minimize the adverse effects of high rolling resistance by laying down its own road

Superior tractive effort in mud

Ability to handle severe grades

Ability to economically self-load some materials if the tractor is matched with a proper size of scraper

Typically lower undercarriage costs in severe rock work when compared to tire costs on a wheel tractor-scraper

The limiting factor of the system is speed—normally a maximum of 4 to 5 mi/h. Thus, this method must be classed as a short-haul system—the maximum economical haul distance increases as total resistance increases.

Figure B4-18 roughly defines the economic areas, comparing crawler-drawn units with wheel tractor-scrapers of equivalent capacity. As the total resistance increases, the economical haul distance for crawler-drawn scrapers is extended, so that over a 200 lb/ton haul road the crawler speed would be economical out to approximately 700 ft. This relationship will vary somewhat, depending on the specific units considered.

Number of Axles

Which is needed—traction and maneuverability, or speed? Two-axle scrapers can climb grades that other machines cannot and, with hydraulic steering, they can duckwalk through mud. However, the better riding qualities of three-axle units permit higher top speeds (40 versus 30 mi/h) which can pay off on long, flat runs where climbing and mudding abilities are secondary.

Weight-to-Horsepower Ratio

Where scrapers of different sizes are available with the same engine power, other choices are possible. The larger load can mean more production per horsepower hour if underfoot conditions and grades do not greatly hamper cycle times, but rough going or adverse grades may demand lower weight-to-horsepower ratios.

Number of Engines

Scrapers with engine power on two axles (twin power) can climb adverse grades of 50 percent or more, fully loaded. They can also outperform other units in extremely soft, muddy footing. However, they require an investment up to 25 to 27 percent larger than their single-drive axle counterparts. Thus, the question is: When will they produce enough additional yardage to justify the extra cost?

Many still consider the tandem-powered scraper as a specialized tool—for opening up jobs, extreme adverse grades, short hauls, etc. However, these units are being used in more and more "average-conditions" work.

Size of Scraper

In general, unit costs of production are lower with large equipment than with smaller machines. Machine cost rises at a slower rate than machine capacity, and operator wages will generally remain constant regardless of machine size. However, if a large machine is to pay its way, its productivity must be fully utilized in a balanced spread. Large units also require more pushing power, which adds costs, and job layouts must provide adequate maneuvering room.

Power for Push-Loading

Pusher selection can be the most important single element of a properly applied scraper spread, since the entire spread's output is dependent on it. Two general conclusions which can be drawn regarding a properly equipped cut are:

1. There must be a close balance between the hourly load-count capabilities of the pusher equipment and the total loads which can be hauled per hour.
2. Excess capacity of either loading or hauling equipment cannot be fully utilized and detracts from the system's potential unit-cost economy.

The conclusions from the data are:

1. A 3:1 ratio is the closest practical balance—it delivers lowest unit cost.
2. Either 2:1 or 4:1 ratios result in a significant cost penalty.
3. An improvement could be made in 4:1 ratio economy by lowering load time and carrying smaller load. This still results in some penalty, however.

Balance between loading and hauling equipment is essential for economic application for any interdependent load and haul system. Normal matching of push power to scraper size is as follows:

1. 14/20—270 hp usually; 180 hp in good material; 385 hp in very hard, severe conditions
2. 21/30—385 hp usually; 270 hp in good material; tandem 2-385 hp in very hard, severe conditions
3. 28/38—tandem 2-385 hp usually; 385 hp in good material
4. 32/44 and up—tandem 2-385 hp
5. Tandem 32/44—occasionally can use 385 hp, but usually tandem 2-385 hp
6. 40/54 and up—never less than tandem 2-385 hp; sometimes triple pushers

A dual 385-hp unit can also be ideal pusher equipment where tandem 2-385-hp capacity is called for, as typified by the results of a field loading study of 32/44 (500-hp) units.

Users go to duals when they feel they have reached maximum pit or cut efficiency with tandems. They are looking for that extra 5 to 10 loads per hour which can be gained with duals under the proper set of conditions—primarily with a good pit and good supervision. (Jobs have been observed where the loading rate exceeded 90 loads per hour.) The duals have the capability of:

Faster load times
   Full power at the instant of contact
   No wait time for positioning second tractor
   Greater pushing effort
   Faster ground speed and faster loading rate
Quicker boost
Faster return
Less approach and maneuver time

All these factors contribute to a faster cycle time and thus more loads per hour. The economy in using duals results from more loads per hour, not from eliminating one operator.

Track-Type Versus Wheel-Type Pusher

The big advantage of wheel-type running gear is mobility—speeds to 20+ mi/h. In general, wheel pushers are justifiable when this mobility can be used, such as where cuts are thin and the spread moves from cut to cut several times daily.

Basically, the picture is this: a 385-hp track-type unit equipped for pushing weights around 85,000 lb. With a coefficient of traction of 0.9 (normally applied to tracks), the machine is capable of a tractive effort of some 76,000 lb. For a wheel tractor to provide a comparable pushing effort, it would need to weigh over 125,000 lb—assuming a coefficient of traction of 0.6, which is a good figure for rubber in most conditions. A 400-hp wheel-type unit with full counterweights weighs about 7000 lb more than the track unit. The net result is a 15,000- to 20,000-lb differential in drawbar pounds pull—favoring the tracks. For underfoot conditions which are less favorable for the wheel unit, the tractive advantage of the track-type unit could be considerably greater.

Wheels push longer—have a longer cycle time—and will load out fewer loads per hour than will tracks. In addition, there is a slightly larger payload in the scraper pushed by tracks. The average hourly cost also favors the track-type machines.

Conclusions

1. Wheels will seldom equal track production for comparable units.
2. Hourly cost of investment is generally higher for wheel pushers.
3. Loading cost per cubic yard favors tracks.

There can be exceptions to these generalizations. For example, rubber-tired dozers may be justified in extreme cases of abrasive sand on the basis of tire costs versus track costs. However, the case must be truly an extreme one to offset the basic economics of the track-type unit.

Using a wheel and two single-track units in tandem offers an economic possibility. The mobility of the wheel unit permits it to push one track unit and then quickly pick up another in a shuttle-type operation. In this manner, three pushers (two track and one wheel) can handle a spread instead of the customary four pushers.

Self-Propelled, Self-Loading Scrapers

The self-loading scraper concept has become an earth-moving system of growing importance as contractors continue to strive for increased efficiencies by minimizing the mismatch and bunching effect inherent in push-loaded scraper fleets. Self-loading to date has been achieved with two basic methods: elevating design or multiple-bowl, all-wheel-drive units.

Elevating Scraper. This unit, in itself, is a balanced spread and is capable of productivity approaching that of a push-loaded scraper on short hauls. Being a more complex machine, it weighs some 8500 to 11,000 lb more than its conventional counterpart and represents a higher investment to attain essentially the same payload capacity. Yet the elevating scraper has proved very economical for utility work—cleanup and finishing behind production spreads—and is a natural choice for small jobs where productivity of a single unit is all that is needed. One of its most significant advantages is balanced productive capacity that can be bought in increments, compared to a minimum balanced-spread investment for two comparable conventional scrapers and a pusher. The concept also can be an economical production earth-moving system under the right conditions.

In short, the basic considerations in selecting elevating scrapers are:

No pusher required

Not vulnerable to mismatch

Minimum balanced spread investment

No rock appetite

Elevator is dead weight when hauling

When examining the economics of elevating scrapers, it is simply a case of the elevator working for you ± 1.00 min, during loading only. For the remainder of the cycle, the additional 10,000 lb of elevator weight works against you. Thus, on short hauls in good conditions, if the elevator works for you 1.20 min and against you only 2.00 min, it may look attractive. On hauls of greater length, on extreme grades, or with high rolling resistance, when you find it working for you for 1.20 min and against you 4 to 5 min, it is probably less economical.

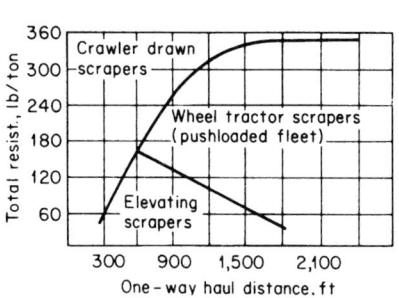

FIGURE B4-19 Economical application zones.

Figure B4-19 supplies a graphic explanation of the approximate areas of economical application of each type of machine. Note that along the break-even limit for elevating scrapers, the haul distance gets shorter as the resistance increases—the elevator works against you a greater portion of the time. Any mismatch between regular scraper and pusher would move the practical intersection somewhat to the right (in favor of the elevator). This relationship identifies the elevator as an economical short-haul system.

SCRAPERS

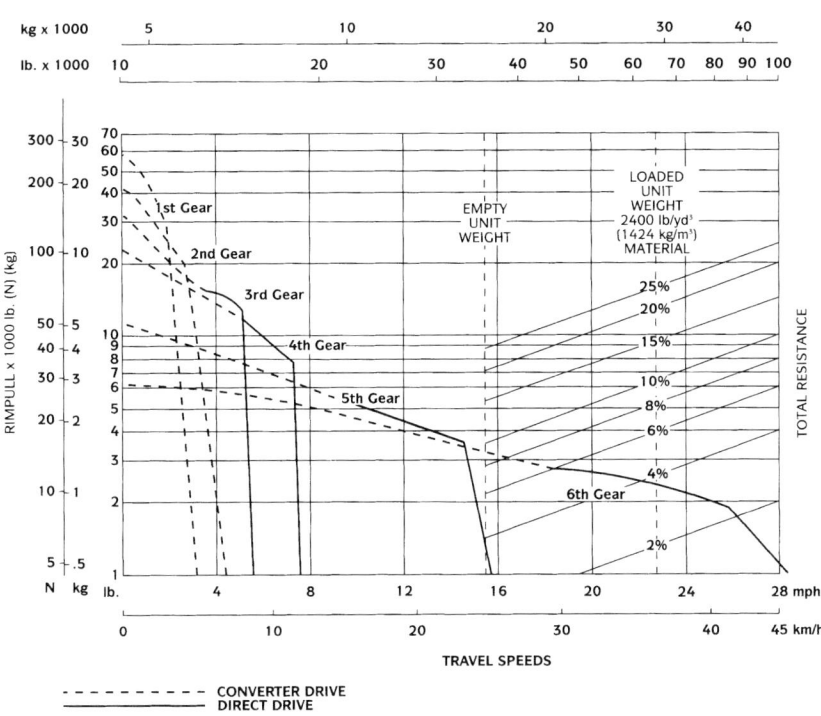

GRADEABILITY/SPEED/RIMPULL
To calculate gradeability performance: Read down from vehicle weight to the percent of total resistance. (Total resistance equals actual percent of grade plus one percent for each 20 lb./ton of rolling resistance.) From this weight-resistance point, read horizontally across to the curve with the highest obtainable speed range, then down to maximum speed. Usable rimpull is affected by tire size, traction available, and weight on drive wheels.

*Ambient temperature, length of haul, and weight of material moved could affect the tire ton-mph capacity. To prevent premature tire failure under adverse conditions, consult the tire manufacturer.

FIGURE B4-20 Grade ability: rimpull versus travel speeds. *(John Deere Co.)*

Another way of looking at the economics of elevating scrapers is in terms of the optimum ratio of scrapers to pushers. Some feel that elevating scrapers are more economical until this ratio approaches 4:1. At this point, the cost of the pusher is spread over four scrapers and is more than offset by the shorter load time and the lower haul cost of the conventional scraper, thus resulting in a favorable economic situation for the push-loaded scraper fleet.

The operational range of the elevating scraper may be extended in actual practice. Many contractors do not find it convenient, for one reason or another, to add a scraper every time the haul lengthens out and unbalances the scraper-to-pusher ratio. Instead, they will frequently work with basic spreads of 3:1, 4:1, etc., and the economic range of elevating scrapers is thus extended past that which might be calculated on paper.

TABLE B4-9 Haul Capacities, Towed Scrapers (Rome Industries), All-Hydraulic, Self-Loaded, Caterpillar Tractors[2]

		Self loaded			
Scraper	Tractor	Struck capacity, yd^3	Load time, min	Hourly haul 400 ft, yd^3	Production haul 800 ft, yd^3
R 56 H	D-5	9	1.5	121	88
	D-6	9	1.2	146	106
R 67 H	D-6	12	1.2	178	124
	D-7	12	1.0	233	156
	D-7	14	1.0	252	178
	D-8	14	0.8	277	190
R 89 H	D-7	18	1.0	284	194
	D-8	18	0.8	323	220

TABLE B4-10 Haul Capacities, Towed Scrapers (Rome Industries), All-Hydraulic, Push-Loaded, Caterpillar Tractors[2]

		Push-loaded			
Scraper	Tractor	Struck capacity, yd^3	Load time, min	Hourly haul 400 ft, yd^3	Production haul 800 ft, yd^3
R 56 H	D-5	9	1.0	140	96
	D-6	9	1.0	164	117
R 67 H	D-6	12	0.8	195	134
	D-7	12	0.6	268	182
	D-7	14	0.6	286	195
	D-8	14	0.5	308	205
R 89 H	D-7	18	0.6	318	209
	D-8	18	0.5	370	338

PRODUCTIVITY

Figure B4-20 shows rimpull versus travel speeds for a John Deere Elevating Scraper. Table B4-9 shows production for towed scrapers, self-loaded at hauls of 400 and 800 ft, respectively. Table B4-10 shows the same thing for push-loaded towed scrapers.

REFERENCES

1. *Society of Automotive Engineers (SAE) Handbook,* Standards and Recommended Practices.
2. *Caterpillar Performance Handbook,* Edition 24.

CHAPTER 5
EXCAVATORS*

James J. O'Brien, P.E.
O'Brien-Kreitzberg, Inc., Pennsauken, New Jersey

INTRODUCTION

The excavators covered in this section are primarily the full-revolving, transportable digging machines. These include power shovels, backhoes, draglines, and clamshells.

The past two decades have seen an enormous evolution in the areas of shovels and backhoes. First, hydraulic-activated excavators have replaced cable-activated machines in predominance. Second, backhoes (hydraulic-activated) have replaced the power shovel as the predominant tracked excavator. Further, current shovel configuration is very similar to a backhoe configuration (such as the Caterpillar 235D Front Shovel).[1] Third, the transition from electric to microprocessor controls has optimized the capabilities of the hydraulically activated machines.

The various digging actions of these excavators are powered by hydraulic cylinders and motors and are controlled by hydraulic valves or variable-stroke pumps. Hydraulic excavators have the advantage of a maneuverable bucket with a wrist action. In addition, when a hydraulic excavator is engaged in heavy digging, one motion can stall without stopping other actions or killing the engine. This feature permits the operator to maintain a stalling force momentarily while manipulating the other controls.

The hydraulic-activated backhoe has the additional advantage that its digging forces do not tend to lift the bucket out of a deep trench (as the cable pull on a cable power hoe did).

Backhoe dippers should be accurately spotted and controlled during the digging stroke. Backhoe booms and dippers experience high shock and fatigue loads in normal digging use. The backhoe can be abused by an improper chopping action or by side loading when the dipper is dropped into a trench while still swinging at high speed. Figure B5-1 is the self-propelled, tire-type, John Deere Co. Backhoe Model 495D. Figure B5-2 is the similar, larger Model 595D. Figure B5-3 compares the specifications for the 495D and 595D.

---

* Some of the material in this chapter is from E. O. Martinson, "Excavators," in Stubbs and Havers (eds.), *Standard Handbook of Heavy Construction*, 2d ed., McGraw-Hill, New York, 1971.

FIGURE B5-1 Self-propelled, tire-type backhoe, Model 495D. *(John Deere Co.)*

Table B5-1 has specifications for seven types of backhoe buckets and a selection chart. The selection chart is based on resistance (lb/cy) of the material to be handled.

Figure B5-4 is a Model 790E-LC crawler mounted excavator with long reach. Figure B5-5 shows the digging depth and reach of the 790E-LC.

Figure B5-6 is a large, track-mounted Model 992D-LC at work. This unit has a maximum digging depth of 30 ft.

Table B5-2 lists 13 track and 4 wheeled Caterpillar backhoe excavators. Because some models have more than one configuration, the table lists 26 configurations.

Attachments

A wide range of attachments is available to be used with backhoe excavators. Figure B5-7*a* shows a LaBounty concrete cracker in action. Figure B5-7*b* shows three types of jaws for the Universal Processor™ (UP). Table B5-3 is a table of specifications for the UP, ranging from a 6.7 to 44 in.

Figure B5-8*a* shows a LaBounty Material Breaker™ (MB) in action. Figure B5-8*b* is a detailed sketch of the MB. Table B5-4 has the specification for MB 130 with a jaw opening of 77 in.

Figure B5-9*a* shows a LaBounty On-Grade Processor (OGP) in action; fig. B5-9*b* is a detailed sketch of the OGP. Table B5-5 is the specification for the OGP40. LaBounty makes many other attachments such as the grapple in Fig. B5-10.

EXCAVATORS

(a)

FEATURES

John Deere turbocharged diesel engine
95 SAE net hp (71 kW) in digging mode
100 SAE net hp (75 kW) in travel mode

36,575 lb. (16 590 kg) maximum operating weight

20 ft. 8 in. (6.30 m) maximum digging depth

30 ft. 8 in. (9.35 m) maximum reach at ground level

18.6 mph (30 km/h) travel speed—rubber-tired mobility

High-efficiency, variable-flow hydraulic system with fuel-saving, mode control features

Automatic engine idling system

Large cab for improved operator comfort and visibility. Includes all controls for transporting and excavating

Two-lever, low-effort, all hydraulic pilot control of boom, arm, bucket, and 360-degree continuous swing

Complete instrumentation/warning system continuously monitors vital machine functions

Hydrostatic drive with Hi-Lo ranges provides excellent on- and off-road versatility

Vandal protection with lockable service doors

(b)

FIGURE B5-2 Self-propelled, tire-type backhoe, Model 595D. *(John Deere Co.)*

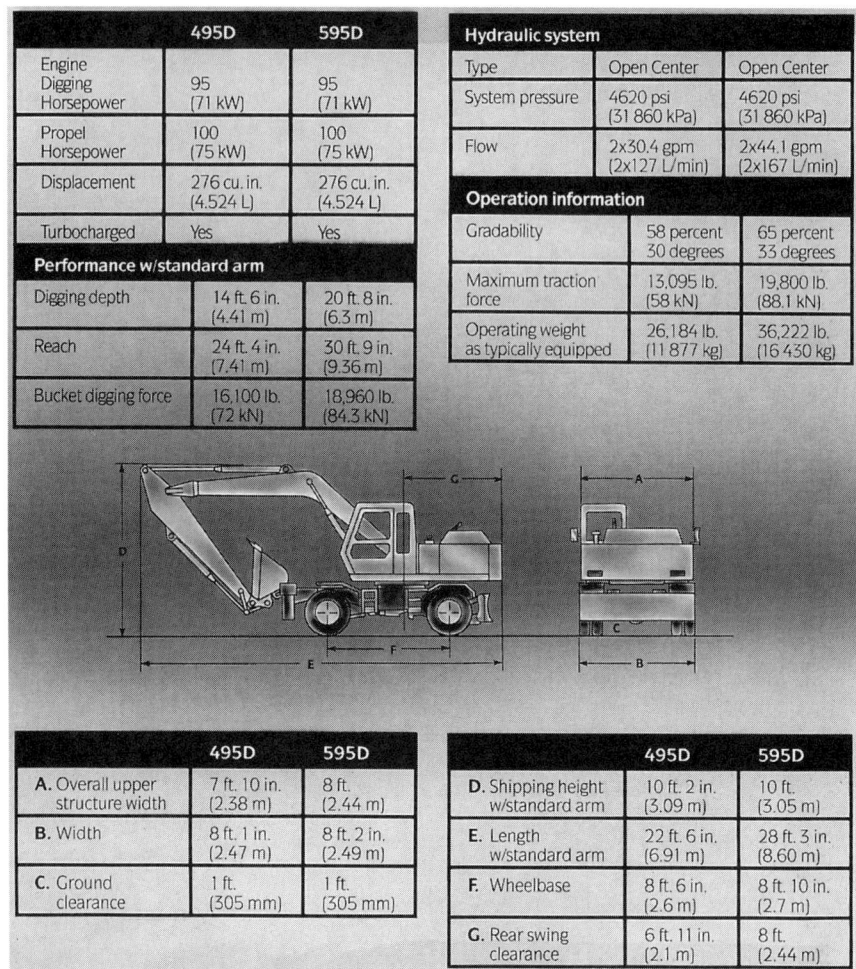

FIGURE B5-3 Specifications comparison for Models 495D and 595D. *(John Deere Co.)*

An ultimate attachment is the John Deere 9310G backhoe (Fig. B5-11). This attaches to the rear end of their 450G, 550G, 650G series dozers and their 455G and 555G series crawler loaders. The unit can be quickly removed.

The Backhoe Loader (see Fig. B5-12) is a backhoe with a loader attached. The combination allows the backhoe to work without front loader support. It also lets one operator do the work of two.

Figure B5-13 shows the operating characteristics for the 710D backhoe. Table B5-6 lists the 710D loader specifications. The John Deere Backhoe Loaders have an extendable *dipperstick* that extends the backhoe reach by 5 ft (Fig. B5-14).

Table B5-7 is a listing of five John Deere backhoe loaders. Table B5-8 is a similar list of six Caterpillar units.

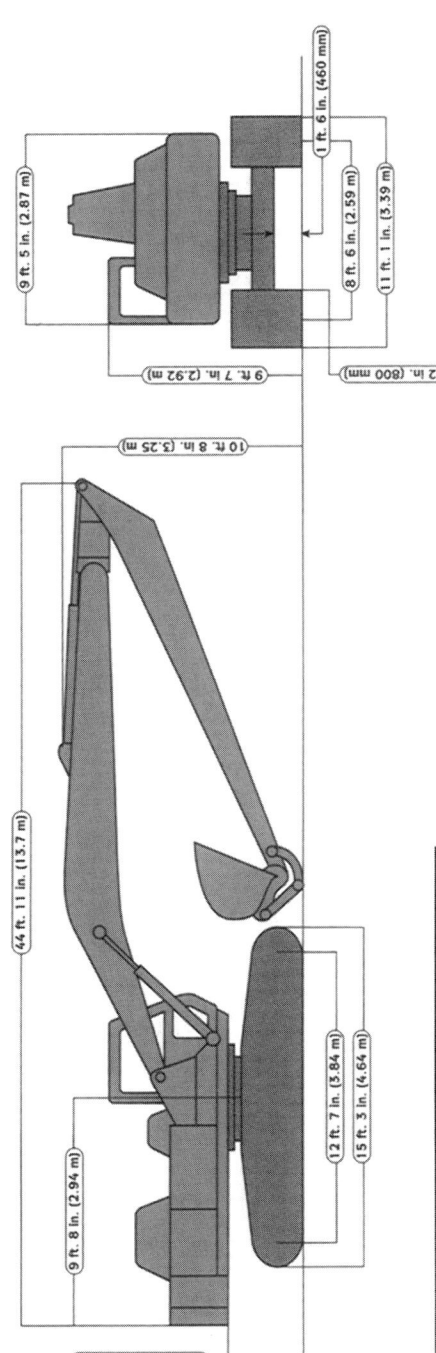

OPERATING INFORMATION

Arm length .... 25 ft. 5 in. (7.86 m)
Arm force with 60 in. (1525 mm)
  bucket .... 11,900 lb. (52.9 kN)
Bucket tangential force with 60 in.
  (1525 mm) bucket .... 16,100 lb. (72 kN)
Bucket cutting edge tip radius .... 37 in. (940 mm)
Lifting capacity over front or rear @
  ground level 20 ft. (6.1 m) reach ... 11,622 lb. (5272 kg)
A Max. reach .... 59 ft. 8 in. (18.19 m)
A Max. reach @ ground level .... 59 ft. 5 in. (18.11 m)
B Max. digging depth .... 45 ft. 4 in. (13.83 m)
C Max. cutting height .... 51 ft. 9 in. (15.78 m)
D Max. dumping height .... 44 ft. 5 in. (13.53 m)

FIGURE B5-4 Model 790E-LC with long front specifications. *(John Deere Co.)*

TABLE B5-1 Backhoe Bucket Specifications and Selection Chart

Buckets

A full line of buckets is offered to meet a wide variety of applications. All capacities are SAE heaped* ratings. The buckets have an adjustable bushing feature for side clearance, with the exception of the ditching bucket. Tooth selection includes either the John Deere Fanggs® Standard, Tiger, Twin Tiger, Abrasion panel, or Flare, or the ESCO (Vertalok) Standard, Tiger, Twin Tiger, or Flare tooth. Replaceable cutting edges are available through John Deere parts. Optional side cutters add 6 in (150 mm) to bucket widths.

Type bucket	Bucket width in	Bucket width mm	Bucket capacity* yd³	Bucket capacity* m³	Weight lb	Weight kg	Bucket dig force lb	Bucket dig force kN	Arm dig force 8 ft 9 in (2.7 m) lb	Arm dig force 8 ft 9 in (2.7 m) kN	Arm dig force 10 ft 6 in (3.2 m) lb	Arm dig force 10 ft 6 in (3.2 m) kN	Arm dig force 13 ft 1 in (4.0 m) lb	Arm dig force 13 ft 1 in (4.0 m) kN	Bucket tip radius in	Bucket tip radius mm	No. teeth
General-purpose plate lip	30	760	1.20	0.92	1770	803	41,620	185.1	36,970	164.4	30,645	136.3	26,340	117.2	62.5	1588	4
	36	915	1.48	1.13	1872	849	41,620	185.1	36,970	164.4	30,645	136.3	26,340	117.2	62.5	1588	4
	42	1065	1.75	1.34	1998	906	41,620	185.1	36,970	164.4	30,645	136.3	26,340	117.2	62.5	1588	5
	48	1220	2.03	1.55	2115	959	41,620	185.1	36,970	164.4	30,645	136.3	26,340	117.2	62.5	1588	6
	54	1370	2.30	1.76	2215	1005	41,620	185.1	36,970	164.4	30,645	136.3	26,340	117.2	62.5	1588	7
	60	1525	2.59	1.98	2338	1060	41,620	185.1	36,970	164.4	30,645	136.3	26,340	117.2	62.5	1588	7
General-purpose high capacity	30	760	1.26	0.96	2420	1097	37,430	166.5	35,490	157.9	29,550	131.4	25,525	113.5	69.5	1765	4
	36	915	1.56	1.19	2550	1156	37,430	166.5	35,490	157.9	29,550	131.4	25,525	113.5	69.5	1765	4
	42	1065	1.85	1.41	2710	1229	37,430	166.5	35,490	157.9	29,550	131.4	25,525	113.5	69.5	1765	5
	48	1220	2.15	1.64	2815	1277	37,430	166.5	35,490	157.9	29,550	131.4	25,525	113.5	69.5	1765	6
	54	1370	2.45	1.87	2982	1352	37,430	166.5	35,490	157.9	29,550	131.4	25,525	113.5	69.5	1765	6
	60	1525	2.74	2.10	3089	1401	37,430	166.5	35,490	157.9	29,550	131.4	25,525	113.5	69.5	1765	7
Heavy-duty plate lip	36	915	1.48	1.13	2138	970	41,620	185.1	36,970	164.4	30,645	136.3	26,340	117.2	62.5	1588	4
	42	1065	1.75	1.34	2210	1002	41,620	185.1	36,970	164.4	30,645	136.3	26,340	117.2	62.5	1588	5
	48	1220	2.03	1.55	2324	1054	41,620	185.1	36,970	164.4	30,645	136.3	26,340	117.2	62.5	1588	6
	54	1370	2.30	1.76	2557	1160	41,620	185.1	36,970	164.4	30,645	136.3	26,340	117.2	62.5	1588	6
Heavy-duty high capacity	30	760	1.26	0.96	2516	1141	37,430	166.5	35,490	157.9	29,550	131.4	25,525	113.5	69.5	1765	4
	36	915	1.56	1.19	2781	1261	37,430	166.5	35,490	157.9	29,550	131.4	25,525	113.5	69.5	1765	5
	42	1065	1.85	1.41	3120	1415	37,430	166.5	35,490	157.9	29,550	131.4	25,525	113.5	69.5	1765	5
	48	1220	2.15	1.64	3318	1505	37,430	166.5	35,490	157.9	29,550	131.4	25,525	113.5	69.5	1765	6
	54	1370	2.45	1.87	3562	1615	37,430	166.5	35,490	157.9	29,550	131.4	25,525	113.5	69.5	1765	6
Severe-duty cast lip	42	1065	1.75	1.34	2774	1258	40,020	178.0	36,425	162.0	30,245	134.5	26,045	115.8	65.0	1651	5
	48	1220	2.03	1.55	2815	1277	40,020	178.0	36,425	162.0	30,245	134.5	26,045	115.8	65.0	1651	5
Severe-duty plate lip	30	760	1.26	0.96	2850	1292	35,150	156.3	34,595	153.9	28,885	128.5	25,030	111.3	74.0	1880	3
	36	915	1.56	1.19	3024	1371	35,150	156.3	34,595	153.9	28,885	128.5	25,030	111.3	74.0	1880	4
	42	1065	1.85	1.41	3343	1516	35,150	156.3	34,595	153.9	28,885	128.5	25,030	111.3	74.0	1880	4
	48	1220	2.15	1.64	3522	1597	35,150	156.3	34,595	153.9	28,885	128.5	25,030	111.3	74.0	1880	5
Ditching	72	1830	1.66	1.27	2531	1148	51,005	226.9	39,695	176.6	32,635	145.2	27,795	123.6	51.0	1295	0

Bucket selection chart
Recommended bucket size*

			General-purpose		Heavy-duty	
lb/yd³	kg/m³	Material (loose weight)	yd³	m³	yd³	m³
700	420	Wood chips	9.0	6.9	—	—
750	440	Peat, dry	8.0	6.1	—	—
950	560	Cinders	5.5	4.2	—	—
1170	690	Peat, wet	5.0	3.8	—	—
1600	950	Topsoil	4.0	3.0	—	—
1780	1050	Coal	3.5	2.7	3.25	2.5
2100	1250	Caliche	1.75–2.50	1.3–1.9	1.50–2.50	1.1–1.9
2100	1250	Earth, loam	2.75	2.1	2.50	1.9
2250	1330	Shale	2.75	2.1	2.50	1.9
2400	1420	Sand, dry	2.75	2.1	2.50	1.9
2500	1480	Clay, dry	2.00–2.50	1.5–1.9	1.75–2.25	1.3–1.7
2550	1510	Earth, dry	2.00–2.50	1.5–1.9	1.75–2.25	1.3–1.7
2600	1540	Limestone, broken or crushed	1.63–2.25	1.2–1.7	1.50–2.00	1.1–1.5
2700	1600	Earth, wet	2.00–2.50	1.5–1.9	1.75–2.25	1.3–1.7
2800	1660	Clay, wet	2.00–2.50	1.5–1.9	1.75–2.25	1.3–1.7
2800	1660	Rock, granite, blasted and broken	1.63–2.75	1.2–2.1	1.50–2.50	1.1–1.9
2850	1690	Sand, moist	2.25	1.7	2.10	1.6
2900	1720	Sand and gravel, dry	2.25	1.7	2.15	1.6
3100	1840	Sand, wet	2.15	1.6	2.00	1.5
3400	2020	Sand and gravel, wet	2.00	1.5	1.85	1.4

* Contact your John Deere dealer for optimum bucket and attachment selections. These recommendations are for general conditions and average use. Larger buckets may be possible when using light buckets, for flat and level operations, less compacted materials, and volume loading applications such as mass excavation applications in ideal conditions. Smaller buckets are recommended for adverse conditions such as off-level applications and uneven surfaces. Bucket capacity indicated is SAE heaped.

Source: John Deere Co.

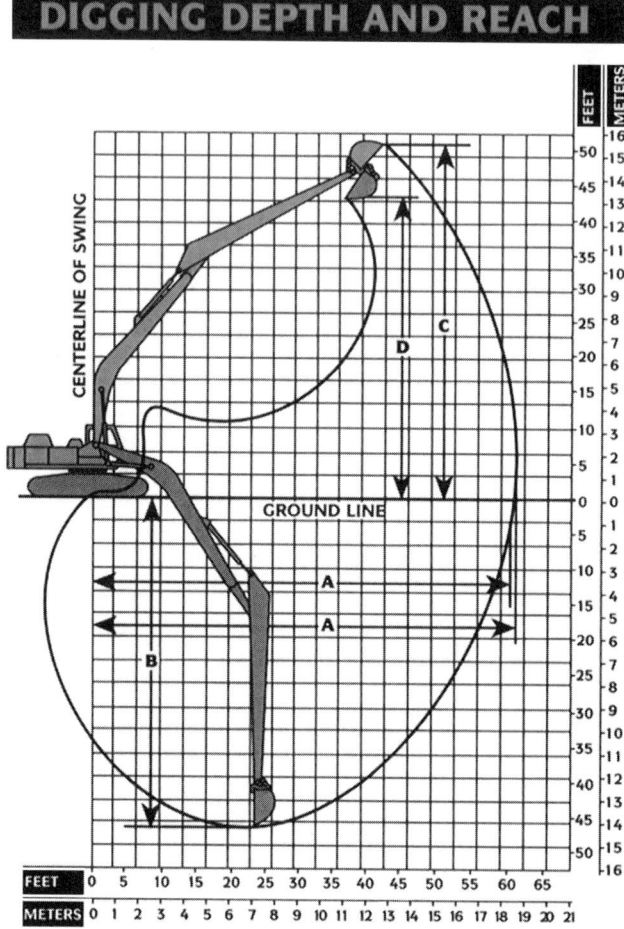

FIGURE B5-5 Digging depth and reach, Model 790E-LC. *(John Deere Co.)*

Telescopic Grading Hoes

The telescopic grading hoe shown in Fig. B5-15 is a specialized hydraulic-actuated excavator. It can grade slopes and flat areas more readily than other attachments, since it has a straight-line action even when the digging line is not parallel to the boom axis. Various buckets, blades, and hooks are also available for this machine.

The bucket of the hydraulic grading hoe has both a wrist action and a rotation right and left about the longitudinal digging action. These actions make it useful in trimming as well as in excavating. Its carrier can be equipped either with crawler-type or wheel-type running gear. Truck carriers will sometimes include quick-set outriggers for added lifting capacity.

FIGURE B5-6 Model 992D-LC at work. *(John Deere Co.)*

Figure B5-16 is a side view of a truck-mounted Gradall hoe. The hoe has a reach of over 30 ft. The boom can telescope, raise, lower, dig, swing, and tilt 220°. Options are available which increase reach to 45 ft and 360° boom tilt. Figure B5-17 is the truck-mounted Gradall at work.

Figure B5-18 is a side view of a track-mounted Gradall. Figure B5-19 is the Gradall at work.

SINGLE-PURPOSE EXCAVATORS

Mining Shovels

Next in size above the commercial-size shovel is the two-crawler electric mining shovel of the type shown in Fig. B5-20. Shovels in this class are generally similar to those in the smaller size range, but they are not convertible to other attachments. They are built in sizes from 5 to 25 yd^3, and all except the smallest of them are electrically driven. Some of these excavators have a main ac motor which drives the main hoist through eddy-current clutches and the other motions by dc Ward-Leonard control, while other units use dc motors on all motions. Mining shovels are used for direct loading into hauling units and occasionally into the receiving hopper of a conveyor belt.

TABLE B5-2 Partial Listing, Caterpillar Backhoe Excavators[1]

Model	Type	HP	Weight, lb	Bucket range* (CY heaped)
E170B	Track	54	15,200	0.2–0.4
311	Track	79	24,270	0.4–0.8
312	Track	84	26,360	0.4–0.8
E140	Track	89	30,800	0.4–1.0
211BLC	Track	105	36,600	0.3–1.1
213BLC	Track	110	41,030	0.6–1.3
206BFT	Wheel	105	29,760	0.4–1.0
212BFT	Wheel	110	32,500	0.4–1.0
214B	Wheel	110	40,470	0.6–1.3
214BFT	Wheel	135	40,470	0.6–1.3
224B	Wheel	135	47,170	0.5–1.6
320	Track	128	42,150	0.9–1.8
320L	Track	128	44,910	0.9–1.8
320N	Track	128	43,560	0.9–1.8
E240C	Track	148	50,700	0.8–1.9
EL240C	Track	148	52,000	0.8–1.9
325	Track	168	56,270	1.1–2.1
325L	Track	168	59,560	1.1–2.1
330	Track	222	70,830	0.9–2.8
330L	Track	222	73,877	0.9–2.8
235D	Track	250	103,780	1.0–3.5
235LC	Track	250	108,620	1.0–3.5
350	Track	286	104,960	1.7–3.4
350L	Track	286	106,970	1.2–2.9
375	Track	428	166,380	2.0–5.8
375L	Track	428	171,280	2.0–5.8

* Rounded to 0.1 CY

Walking Draglines

These draglines are supported on circular bases or tubs which bear directly on the ground, as shown in Fig. B5-21. A pair of long shoes is suspended from the excavator sides, and these shoes are lowered to the ground when the machine is to be moved. The entire machine and its base are then partially lifted and dragged backward away from the boom direction. The walking shoes have a lift and travel of several feet per stroke and can move the machine at 15 to 20 fpm or 0.2 mi/h. The ground-contact pressure is around 10 to 18 lb/in^2 under the base and approximately twice this under the shoes.

Walking draglines are made with drag buckets in sizes from 12 to 220 yd^3 and with booms from 170 to 300 ft long. All units except the smallest are electrically driven. Power distribution is generally at 4160 V on the smaller machines and 7200 V on the larger ones. The peak load can be four times the average load or perhaps two times the installed horsepower of the excavator, and the power supply and distribution system must be sized accordingly.

Walking draglines are generally employed in coal mine stripping and mineral mining. They can dig one-half to two-thirds of the boom length below the machine base, can lift the spoil above base level about two-fifths of the boom length, and have a reach of approximately the boom length from their centers. On this basis, the ver-

FIGURE B5-7 (*a*) LaBounty UP4 Universal Processor with concrete-cracking jaws. (*b*) Three types of jaws for the Universal Processor. (*LaBounty division of Stanley Hydraulic Tools.*)

TABLE B5-3 Specifications, LaBounty Universal Processor

Model	Jaw configuration	(1) Approximate excavator weight (2d member) (lb)	(M. tons)	(1) Approximate excavator weight (3d member) (lbs)	(M. tons)	(2) Approximate attachment weight (lbs)	(kg)	Jaw opening (in)	(mm)	Jaw depth (in)	(mm)	(3) Reach (ft)	(m)
UP4-SII	Shear	Miniexcavator, skid steer loaders, and loader backhoes		Consult the factory		525	238	6.7	170	6.7	170	3'-9"	1.1
	Concrete cracking					525	238	13.5	340	6.6	170	3'-9"	1.1
UP20-SII	Shear	20,000	9	42,000	20	4,100	1,860	20	508	18	470	6'-6"	2.0
	Concrete pulverizer	20,000	9	42,000	20	4,100	1,860	26	670	19	480	6'-5"	2.0
	Concrete cracking	20,000	9	42,000	20	4,000	1,810	42	1,070	23	600	6'-3"	1.9
UP40-S II	Shear	38,000	18	70,000	30	6,500	2,900	24	610	21	540	9'-0"	2.7
	Concrete pulverizer	38,000	18	70,000	30	6,700	2,950	32	810	21	540	9'-0"	2.7
	Concrete cracking	38,000	18	70,000	30	6,500	2,900	49	1,260	27	700	9'-0"	2.7
UP50-S II	Shear	50,000	22	90,000	40	9,475	4,170	27	700	24	620	12'-6"	3.8
	Concrete pulverizer	50,000	22	90,000	40	9,475	4,170	36	910	23	590	12'-6"	3.8
	Concrete cracking	50,000	22	90,000	40	9,475	4,170	56	1,430	31	800	12'-6"	3.8
UP70-S II	Shear	70,000	30	120,000	55	11,500	5,080	36	910	30	760	14'-0"	4.3
	Concrete pulverizer	70,000	30	120,000	55	11,900	5,216	48	1,220	32	810	14'-0"	4.3
	Concrete cracking	70,000	30	120,000	55	11,400	5,035	70	1,800	39	1,000	14'-0"	4.3
UP90-S II	Shear	90,000	40	150,000	70	16,600	7,530	42	1,070	31	800	13'-0"	4.0
	Concrete pulverizer	90,000	40	150,000	70	16,600	7,530	62	1,570	35	890	13'-0"	4.0
	Concrete cracking	90,000	40	150,000	70	16,600	7,530	72	1,800	44	1,110	13'-0"	4.0

Source: LaBounty, a division of Stanley Hydraulic Tools.

(a)

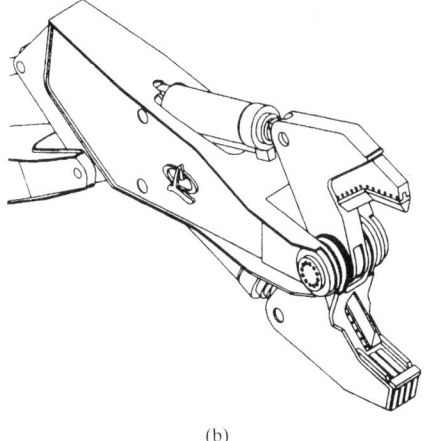

(b)

FIGURE B5-8 (*a*) and (*b*) LaBounty MB Material Breaker. *(LaBounty, a division of Stanley Hydraulic Tools.)*

TABLE B5-4 Specifications, LaBounty Material Breaker

	(1) Approximate excavator weight (2d member)		(2) Approximate attachment weight		Jaw opening		(3) Throat to tip		Reach	
Model	(lb)	(M. tons)	(lb)	(kg)	(in)	(mm)	(in)	(mm)	(in)	(m)
MB 130	130,000	59	24,000	10,900	77	1,960	44	1,117	16'-2"	4.9

Source: LaBounty, a division of Stanley Hydraulic Tools.

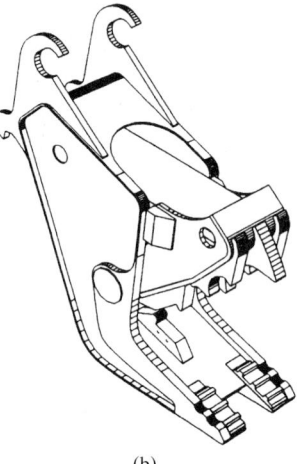

(a)   (b)

FIGURE B5-9 (*a*) and (*b*) On-grade process. *(LaBounty, a division of Stanley Hydraulic Tools.)*

EXCAVATORS   B5-15

FIGURE B5-10 Grapple picking up scrap. *(LaBounty division of Stanley Hydraulic Tools.)*

TABLE B5-5 Specifications, LaBounty On-Grade Processor (OGP)

Model	(1) Approximate excavator weight (3d member) (lb)	(M tons)	(2) Approximate attachment weight (lb)	(kg)	(A) Overall length (in)	(mm)	Jaw opening (in)	(mm)	Throat to tip (in)	(mm)	(3) Reach (in)	(m)
OGP 40	40,000–60,000	18–27	4,200	1,909	75	1,905	39	991	41	1,041	6'-0"	1.83

Source: LaBounty, a division of Stanley Hydraulic Tools.

B5-16 STANDARD HANDBOOK OF HEAVY CONSTRUCTION

FIGURE B5-11 Backhoe mounted on the rear of a 550G crawler dozer. *(John Deere Co.)*

FIGURE B5-12 Model 710D Backhoe Loader. *(John Deere Co.)*

EXCAVATORS

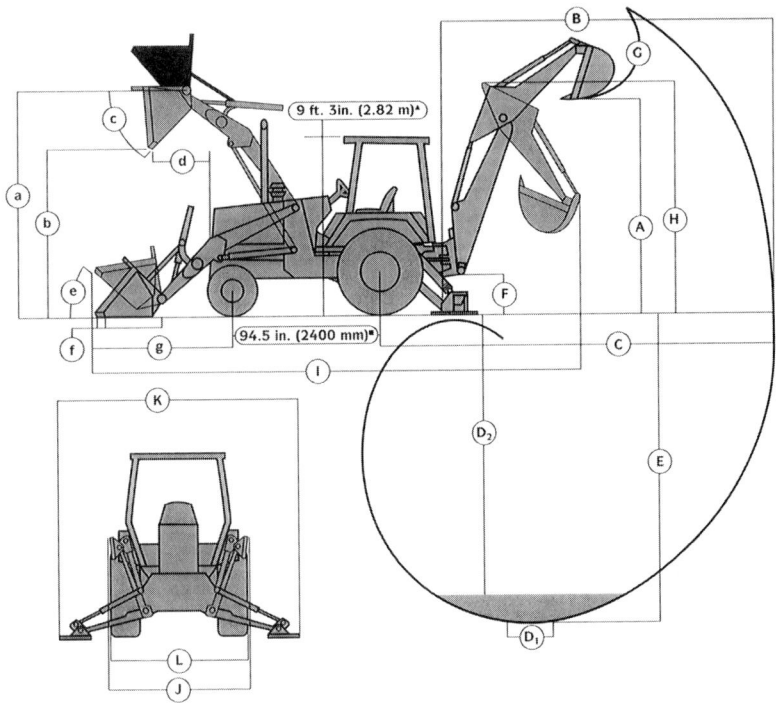

▲ ROPS and Cab
■ MFWD 96.5 in. (2450 mm)

Key:	Backhoe*	Extendable Dipperstick Retracted	Extended
A. Loading height, truck loading position	13 ft. 3 in. (4.03 m)	13 ft. 1 in. (3.98 m)	15 ft. 3 in. (4.65 m)
B. Reach from center of swing mast	22 ft. 8 in. (6.90 m)	22 ft. 8 in. (6.90 m)	27 ft. 3 in. (8.31 m)
C. Reach from center of rear axle	26 ft. 9 in. (8.15 m)	26 ft. 9 in. (8.15 m)	31 ft. 4 in. (9.56 m)
D. Digging depth (SAE):			
(1) 2-ft. (610 mm) flat bottom	18 ft. 2 in. (5.53 m)	18 ft. 1 in. (5.51 m)	23 ft. 1 in. (7.03 m)
(2) 8-ft. (2440 mm) flat bottom	17 ft. 3 in. (5.26 m)	17 ft. 2 in. (5.24 m)	22 ft. 4 in. (6.82 m)
E. Maximum digging depth	18 ft. 2 in. (5.55 m)	18 ft. 1 in. (5.52 m)	23 ft. 1 in. (7.04 m)
F. Ground clearance, minimum	14 in. (356 mm)	14 in. (356 mm)	14 in. (356 mm)
G. Bucket rotation	162 deg./152 deg.	162 deg./152 deg.	162 deg./152 deg.
H. Transport height	13 ft. 10 in. (4.21 m)	13 ft. 11 in. (4.31 m)	13 ft. 11 in. (4.31 m)
I. Overall length, transport	26 ft. 7.9 in. (8.13 m)	26 ft. 7.9 in. (8.13 m)	26 ft. 7.9 in. (8.13 m)
J. Stabilizer width – transport	8 ft. (2.44 m)	8 ft. (2.44 m)	8 ft. (2.44 m)
K. Stabilizer spread – operating	11 ft. 5 in. (3.48 m)	11 ft. 5 in. (3.48 m)	11 ft. 5 in. (3.48 m)
L. Overall width (less loader bucket)	88 in. (2235 mm)	88 in. (2235 mm)	88 in. (2235 mm)

*See backhoe performance

BACKHOE PERFORMANCE

Digging force, bucket cylinder (power dig position)	12,940 lb. (57.5 kN)	12,940 lb. (57.5 kN)	12,940 lb. (57.5 kN)
Digging force, crowd cylinder	9600 lb. (42.7 kN)	9600 lb. (42.7 kN)	6700 lb. (29.8 kN)
Swing arc	180 deg.	180 deg.	180 deg.
Operator control	Two levers	Two levers	Two levers
Bucket positions	-2 to -14 deg. rollback	-2 to -14 deg. rollback	-2 to -14 deg. rollback
Stabilizer angle rearward	13 deg.	13 deg.	13 deg.
Lifting capacity, maximum, boom at 65 deg.	9130 lb. (4141 kg)	8890 lb. (4032 kg)	6030 lb. (2735 kg)
Leveling angle	13 deg.	13 deg.	13 deg.

*Backhoe specifications are with 24-in. x 11.1 cu. ft. (610 mm x 0.31 m³) bucket.

FIGURE B5-13 Specifications for 710D backhoe. *(John Deere Co.)*

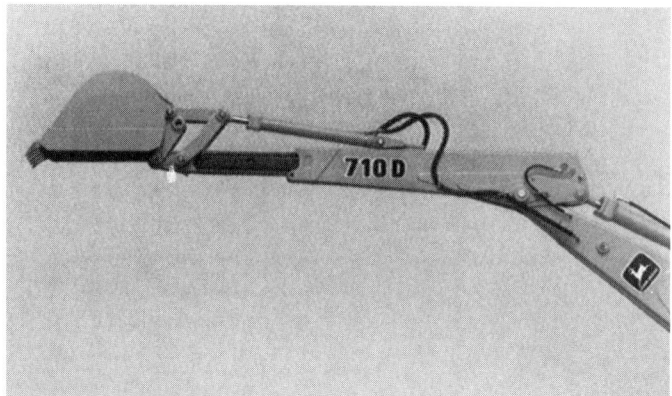

FIGURE B5-14 Dipperstick extension on 710D backhoe. *(John Deere Co.)*

TABLE B5-6 Loader Characteristics for 710D Backhoe Loader

	w/1.5 yd^3 (1.15 m^3) bucket	w/1.7 yd^3 (1.3 m^3) bucket and MFWD	w/1.375 yd 96 in multipurpose bucket and MFWD
Operator control	single lever	single lever	
Breakout force	14,050 lb (62.5 kN)	13,100 lb (58.4 kN)	13,500 lb (60.1 kN)
Lifting capacity, full height	7,733 lb (3503 kg)	7,257 lb (3287 kg)	7,295 lb (3305 kg)
Height to bucket hinge pin, max	11 ft 8 in (3.57 m)	11 ft 10 in (3.61 m)	11 ft 10 in (3.61 m)
Dump clearance, bucket at 40°	9 ft 6 in (2.88 m)	9 ft 5 in (2.87 m)	9 ft 5 in (2.87 m)
Bucket dump angle, max.	40°	40°	40°
Reach at full height, bucket at 40°	33.5 in (0.85 m)	36.8 in (0.93 m)	32.9 in (0.84 m)
Rollback at ground level	40°	40°	40°
Digging depth below ground, bucket level	5.8 in (147 mm)	4 in (101 mm)	6 in (152 mm)
Distance from front axle centerline to bucket cutting edge	80.5 in (2044 mm)	82.6 in (2098 mm)	81.2 in (2063 mm)
Raising time to full height	4.3 s	4.3 s	4.3 s
Bucket dump time	1.3 s	1.3 s	1.3 s
Bucket lowering time	3.2 s	3.2 s	3.2 s

Source: John Deere Co.

TABLE B5-7 Backhoe Loaders

Model	HP, STD	HP, turbo	Weight (base), lb
300D	60	—	12,200
310D	70	75	13,600
410D	75	85	15,600
510D	95	—	16,000
710D	115	—	23,000

Source: John Deere Co.

EXCAVATORS

TABLE B5-8 Backhoe Loaders

Model	Type	HP, gross	Weight
416B	Center pivot	77	13,700
426B	Center pivot	82	14,970
436B	Center pivot	87	15,086
446B	Center pivot	103	19,603
428B	Side shift	77	17,480
438B	Side shift	87	17,600

Source: Caterpillar Co.

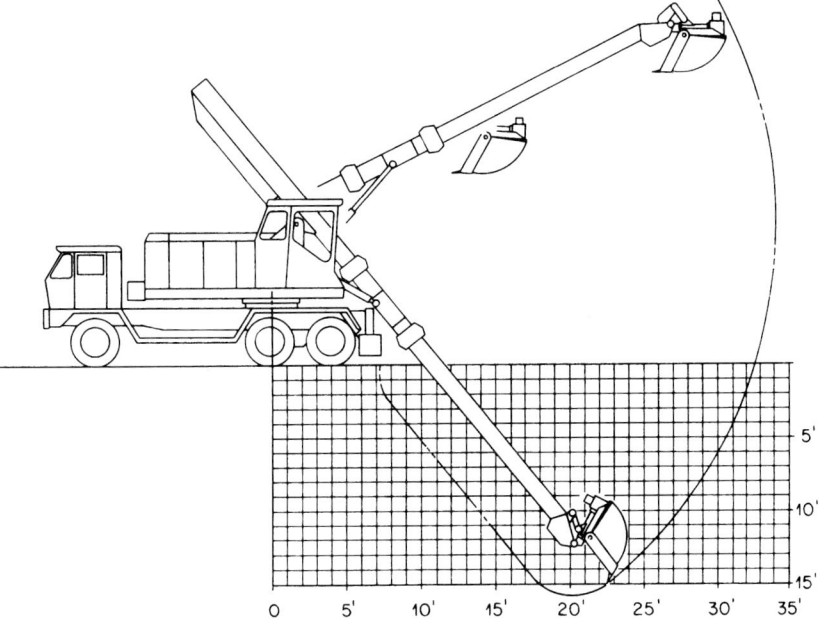

FIGURE B5-15 Hydraulic telescopic boom grader. The 12- to 15-ft straight-line action and long reach are useful in trimming of slopes and ditches, slab lifting and loading, and general excavating.

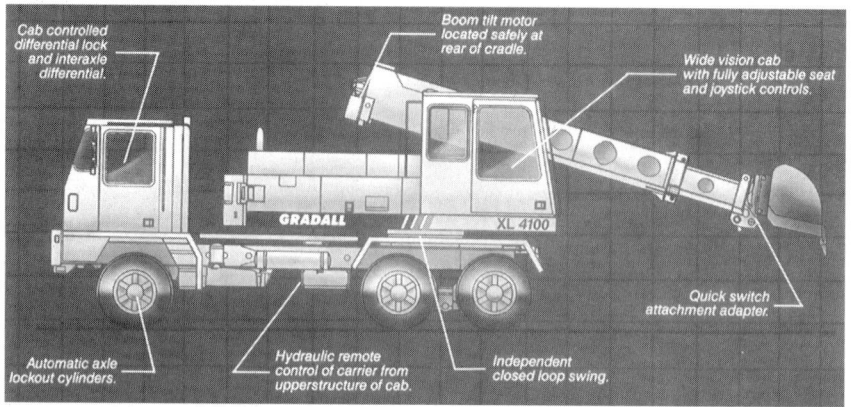

FIGURE B5-16 Gradall® hoe. *(Gradall.)*

FIGURE B5-17 Gradall at work. *(Gradall.)*

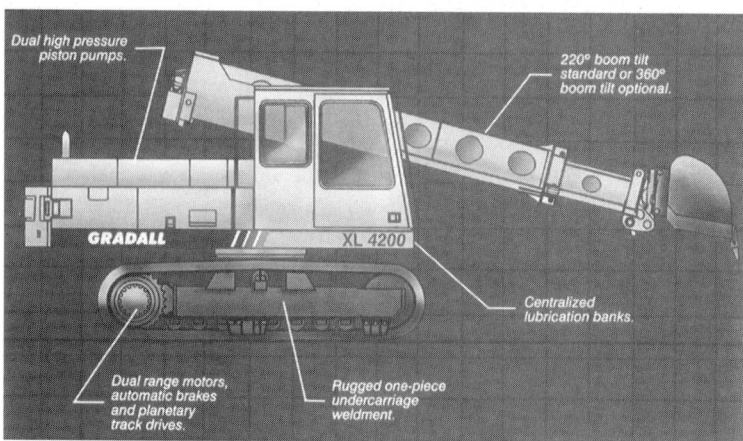

FIGURE B5-18 Gradall, side view. *(Gradall.)*

B5-20

FIGURE B5-19 Gradall at work. *(Gradall.)*

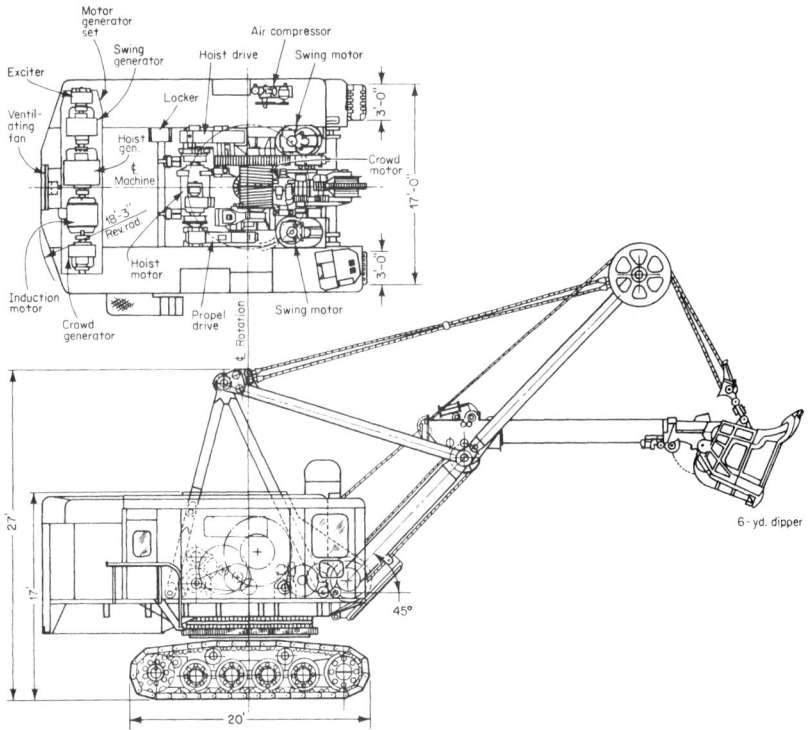

FIGURE B5-20 Electric mining shovel, two-crawler machines in sizes from 4- to 25-yd³ dipper capacity. Power supply by trailing electric cable.

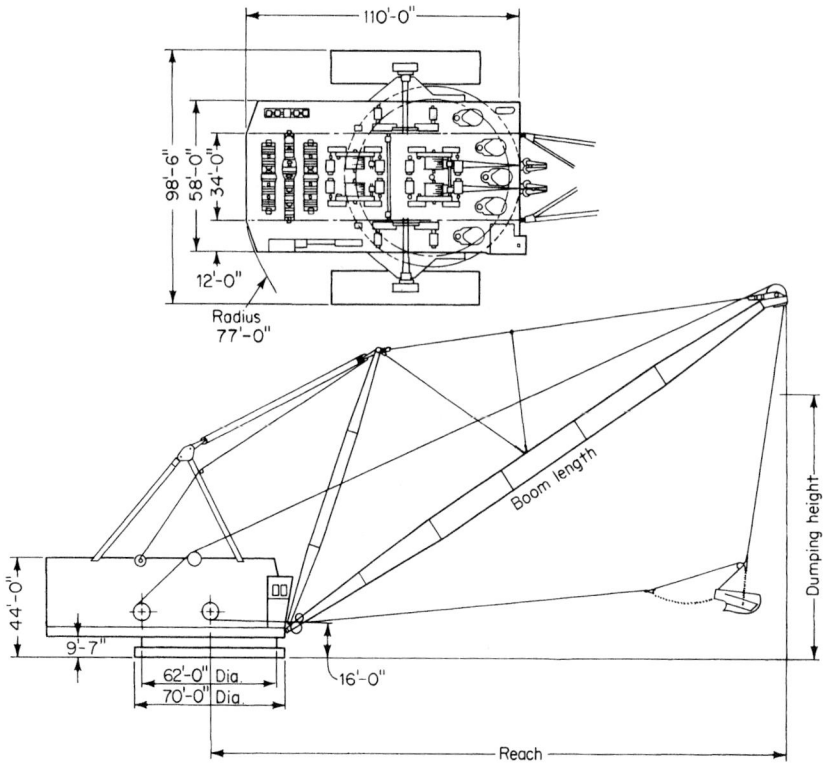

FIGURE B5-21 Electric walking dragline, 8- to 200-yd³ bucket capacity, 200- to 300-ft booms. 70-yd³ 6,000,000-lb weight shown.

tical distance from the bottom of the deepest pit to the peak of the spoil bank could be about the same as the boom length.

The cycle time for the walking dragline is on the order of 60 s, with hoist speeds of 400 to 700 fpm and drag speeds of 300 to 500 fpm. The larger sizes with longer booms have the higher speeds, and thus they require a disproportionately large increase in power in comparisons with the smaller sizes.

Stripping Shovels

Like walking draglines, these excavators are used in permanent (10 to 40 years) operating situations in stripping coal overburden. The stripping shovel shown in Fig. B5-22 is mounted on a carrier with four sets of crawlers, one set at each corner of its base. Stripping shovels are presently built in sizes from 40- to 140-yd³ dipper capacity. They cannot lift spoil as high as walking dragline, and their dumping height is limited to seven-tenths of the boom length. Both classes of machines have similar weight, cost, and power for a given size of bucket. The shovel has a faster cycle and the ability to dig harder material, while the dragline has a greater reach and lifting height.

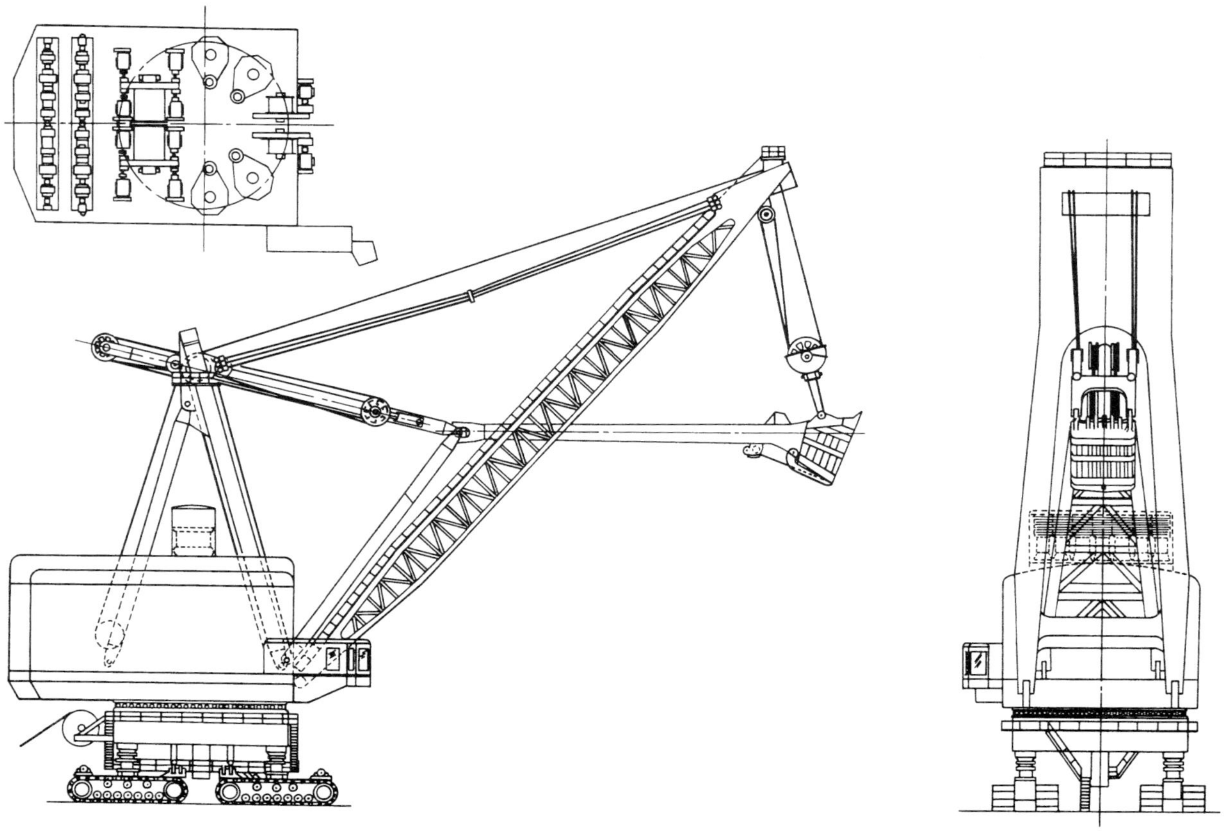

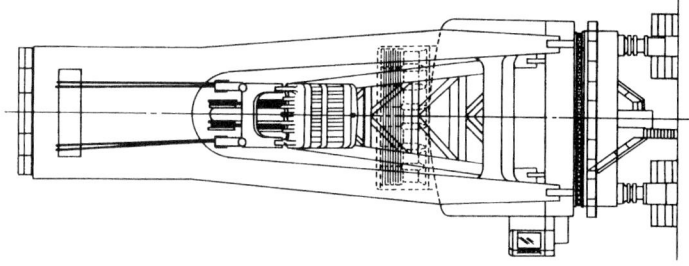

FIGURE B5-22 Electric stripping shovel, mounted on four hydraulically leveled crawler sets. Digs harder material than other types. Built in 40- to 140-yd^3 dipper sizes.

The stripping shovel crawlers have much less bearing area than the dragline supports, with contact pressures on the order of 60 lb/in^2. These shovels can travel at 0.15 to 0.2 mi/h.

Draglines and Clamshells

The dragline attachment and the clamshell attachment to a crane are illustrated in Fig. B5-23. Both of those attachments require the length of crane boom which is frequently used for mobile cranes in the medium-lift class. However, an excavator with the dragline or clamshell attachment will require additional counterweights and longer and wider crawlers than will shovels and backhoes.

Clamshells can dig vertically at considerable depths and, when necessary, through small passages. They also have the ability to lift their loads to appreciable heights above grade. Light rehandling buckets with two- to four-part reeving are used for filling overhead bins with loose material from pits, stockpiles, or railway gondola cars. General-purpose buckets, with four to six parts of line reeved in the

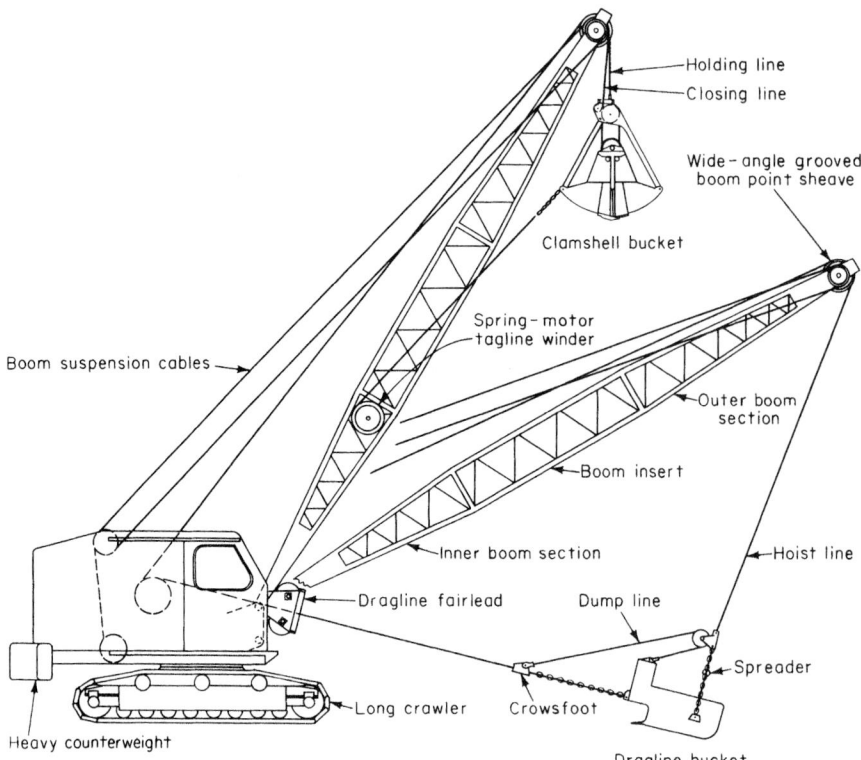

FIGURE B5-23 Dragline and clamshell attachment arrangement diagram with nomenclature shown.

closing line, are used in excavating shafts or trenches where clearances are restricted by shoring and cross bracing. Still heavier buckets may be needed for underwater digging.

Clamshells are usually reeved to have the same line speeds for closing and for holding, so that both lines will stay tight during hoisting. Should there be any difference in these line speeds, the closing line should be at the higher speed. The clutch should then be set loose enough so that it will not raise the loaded bucket on the closing line alone. A bucket should not be dropped too rapidly, because a sudden stop at the bottom of the excavation can damage its hinge stops. The bucket should also be fully open when lowered, to prevent bending the lip. (See Fig. B5-24.)

A dragline is capable of digging at a distance away from its mounting which is about equal to its boom length and to a depth which is equal to one-half of its boom length. Where necessary to clear high banks or spoil piles, the bucket can be hoisted to elevations of one-half the boom length less the dump height indicated in Table B5-9. Although the dragline bucket cannot be spotted as accurately and positively as can shovels and hoes, a good operator can use it to load hauling units. The dragline

FIGURE B5-24 Barge-mounted clamshell. *(Manitowoc.)*

TABLE B5-9 Vertical Dumping Heights Required by Clamshells and Draglines

Bucket size, yd^3	Clamshell, vertical height	Dragline dumping height below boom point
½	9 ft	11 ft 9 in
¾	10 ft	12 ft 9 in
1	10 ft 6 in	13 ft 9 in
1½	11 ft	16 ft
2	11 ft 6 in	18 ft
2½	12 ft	19 ft
3	12 ft	21 ft
4	13 ft	22 ft

is especially useful where the area to be excavated is large, too soft for other machines, or under water (Fig. B5-25). Fig. B5-26 is a dragline range diagram.

Clamshell-bucket rated capacity is equal to the actual struck or plate-line inside volume in cubic yards, measured from a surface across the top edge of the side plates and back plate. The dragline-bucket rated capacity is 90 percent of the inside struck volume with the front edge of the lip (not teeth) being the line of the vertical front measuring plane. The 10 percent deduction is to allow for angle of repose of material at the front. Another way of calculating this rated dragline-bucket yardage is to divide the *vertical front* volume in cubic feet by 30.

Long booms for lift cranes can be safely designed with a relatively low factor of safety of 1.6, based on a critical column buckling load which acts in conjunction with a 2 percent side load. The crane will operate only a limited number of cycles at peak

FIGURE B5-25 Manitowoc 4600 dragline. *(Manitowoc.)*

38" DRUMS WITH 1⅝" DRAG ROPE

A	Boom Length	100'			120'			140'		
B	Boom Angle	30°	35°	40°	30°	35°	40°	30°	35°	40°
C	CL of Rot. to Ctr. Dump	95'	90'	85'	112'	106'	100'	129'	123'	115'
D	Depth Cut (Approx.)	87'	82'	75'	113'	104'	96'	103'	93'	83'
E	Digging Reach (Approx.)	108'	105'	102'	128'	125'	122'	148'	146'	141'
F	Dumping Height	38'	45'	52'	48'	57'	65'	58'	68'	78'
H	Ht. from Grade to CL of Boom Pt.	58'	65'	72'	68'	77'	85'	78'	88'	98'
J	Casting Distance	13'	15'	17'	16'	19'	22'	19'	23'	26'

The working ranges in the table above are based on the following conditions:
1. Dimension 'F' is based on 'G' dimension of 20'.
2. Dimension 'D' is based on a 16' floor level bottom to fill the bucket and casting distance 'J' equal to approximately 1/3 the dumping height 'F'. The two dimensions are also based on maximum drum capacity with one layer of rope. Front drum (hoist) capacity —184' of 1⅜" diameter rope —Drum No. 48142. Rear drum (drag) capacity —153' of 1⅝" diameter rope —Drum No. 48143. The depth of cut, casting distance and digging reach may vary considerably depending on digging conditions, design of bucket and operator's skill. Maximum digging depths are attainable under ideal conditions and cannot be guaranteed.
3. New machines equipped with wire rope to dig to a depth of 60', 70' and 80' for boom lengths of 100', 120' and 140' respectively.

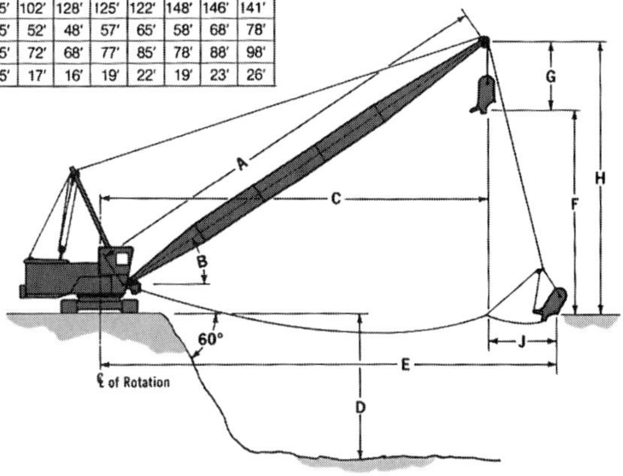

FIGURE B5-26 Dragline range diagram. *(Manitowoc.)*

load, and fatigue is not a significant design consideration. Draglines, in contrast, will operate a large number of cycles at their full lifting loads and with heavy side loads. It follows that draglines require shorter and heavier booms than the lightweight, tubular-alloy, heat-treated booms which are suitable for the high-lift cranes. The single boom-point sheave of the dragline is designed with a wide-angle groove to accept the wide side deflections due to accelerating the swing or holding the bucket against a side slope.

Hoisting is performed at about 180 fpm. The drag speed is slightly reduced from this, usually to 160 fpm, by means of a smaller drum. The hoist action is partially opposed by the drag-cable pull, which is slowly released by the drag brake unit when dumping position is reached. Operating experience indicates that the weight of the loaded drag bucket should be limited to between 35 and 50 percent of the rated maximum engine line pull. Clamshell loaded weight can be somewhat greater, ranging between 45 and 60 percent of this maximum power pull.

Drag inhaul cables are usually one size larger than the hoist and are of the Seale pattern. This has the larger wires on the outside of the cable, even though the smaller ratio of drum-to-cable diameters might suggest small-wire strand cable construction. The operator should prevent spoil from piling up in front of the machine, since otherwise the drag cable may abrade unnecessarily against rocks. Drag buckets must be gently lowered into the ground as the digging pull starts. Draglines, with their long booms and heavy counterweights, have high swing inertias which can cause large heat inputs to swing clutches. This can be a problem if the operator tries to accelerate or stop the swing motion too quickly.

Practical boom lengths and operating ranges for various sizes of clamshells and draglines are indicated in Fig. B5-27. The hoisting capacities of several sizes of excavators are shown in Fig. B5-28. Dragline and clamshell bucket sizes are shown in the rectangles on this chart; the position of each rectangle indicates the practical radius at

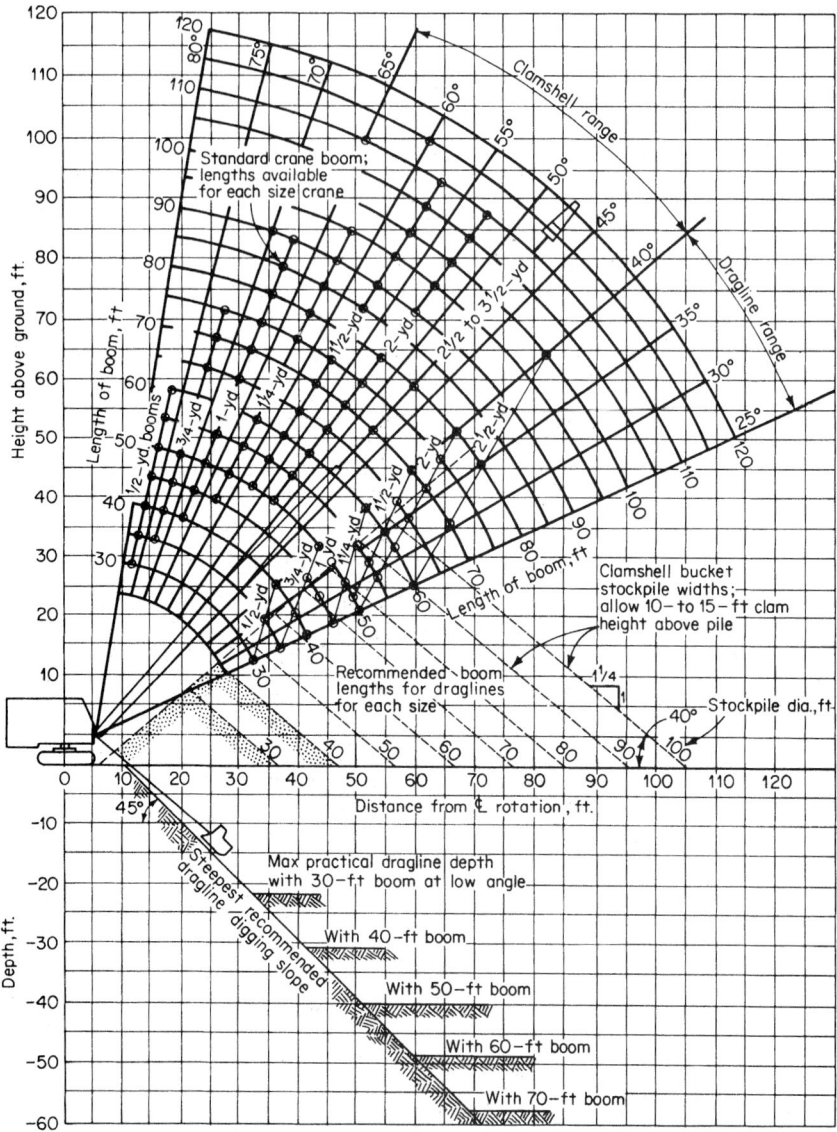

FIGURE B5-27 Dragline and clamshell working ranges. Commonly used boom lengths shown for each class of work for graphically determining required boom length to clear stockpiles or hoppers and the working radius or reach.

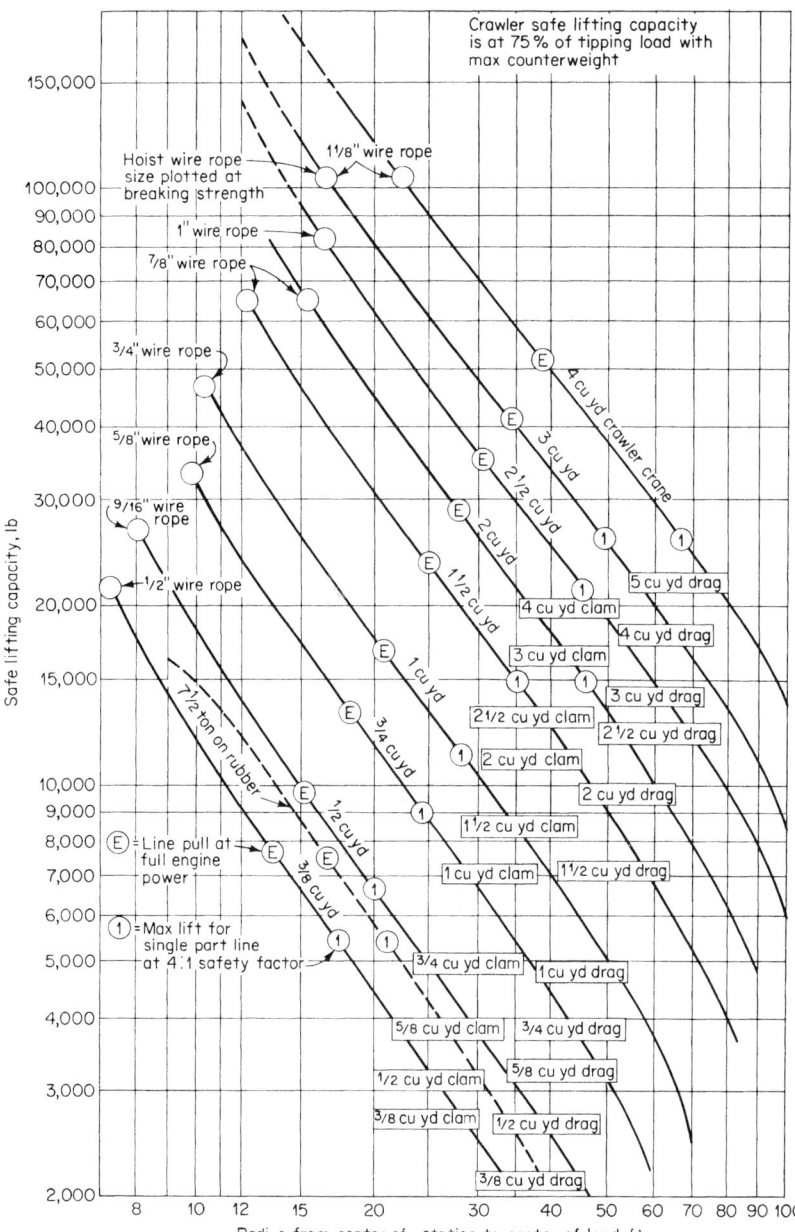

FIGURE B5-28 Clamshell and dragline selection for safe load against tipping for nine sizes of crawler cranes. Loaded-bucket weights are plotted to show size of machine required. The maximum operating radius is the intersection of the bucket rectangle with safe-load capacity of one or two sizes for each bucket size.

which the excavator size which it represents might be used. The chart uses an encircled "E" to indicate the usual maximum line pull for each excavator. The encircled "1" on each line size indicates the continuous working line pull which corresponds to one-quarter of the breaking strength of the cable size shown. It should be noted that the line loading for the largest clam in each excavator size comes near this point.

REFERENCE

1. *Caterpillar Performance Handbook,* Edition 24, pp. 4-125–4-136.

CHAPTER 6
FRONT-END LOADERS*

James J. O'Brien, P.E.
O'Brien-Kreitzberg, Inc., Pennsauken, New Jersey

FRONT-END LOADERS

Front-end loaders are versatile, self-propelled machines mounted either on crawler or wheel-type running gear. They are regularly equipped with a front-mounted, general-purpose bucket with which they dig, scoop, lift, carry, and dump into hauling units, bins, hoppers, conveyors, and stockpiles. They also transport, spread, and compact fill material. In addition, a variety of buckets, grapples, blades, and other front- and rear-mounted attachments enables them to do such work as doze, scrape, clam, grub, forklift, rip, ditch, trench, and winch.

CRAWLER-TYPE LOADERS

Except for their integrally mounted equipment, crawler-type loaders are closely related to crawler tractors. On most makes of loaders the track frames are rigidly held to the chassis instead of oscillating. Loader tracks generally extend farther forward for better bucket lifting and carrying stability. Track gauge width between track centers is sometimes greater than on tractors with equal power. Overlapping flat, low-profile, triple-grouser track shoes provide for easier turning and permit some movement over pavement.

The fabricated steel loader frame and the tractor front frame are often integral. Double-acting hydraulic cylinders on each side of the frame actuate the loader lift and bucket mechanism. The hydraulic system includes the hydraulic pump on a live drive from the engine, a pressure-relief valve, an oil reservoir, filters, and loader lift and bucket controls. Additional controls are available to actuate a rear-mounted ripper and/or the clam of a multipurpose bucket.

A power-shift transmission with torque-converter drive makes bucket crowding and lifting easy without killing the engine. It also permits effortless forward-reverse shifts and speed changes. The transmission includes a power takeoff drive for rear-mounted equipment, such as a winch.

---

* Some of the material in this chapter is from E. A. Braker, "Front-End Loaders," in Stubbs and Havers, *Standard Handbook of Heavy Construction*, 2d ed., McGraw-Hill, New York, 1971.

Crawler-type loaders have the advantages of good flotation and traction on soft or uneven terrain. They operate over sharp objects which would be destructive to rubber tires, and they have compact design for close-quarter maneuverability. However, on-highway transport equipment, such as a truck and tilting trailer, is needed to move them from job to job over public roads and pavements.

Figure B6-1 is a Model 655B tracked loader. Figure B6-2 provides specifications for the 655B. Figure B6-3 is the Model 755B in action. Table B6-1 is a listing of Caterpillar and John Deere crawler front loaders.

Figure B6-4 shows a Model 755B counter-rotating. With the 755B's dual-path hydrostatic design, the operator can counter-rotate the tracks for space-saving spot turns. This boost in productivity is especially useful on small job sites and in close quarters. There is less need to operate the crawler in reverse, and therefore more opportunity for extended life of the undercarriage.

WHEEL-TYPE LOADERS

In 1948, Frank G. Hough began production of the original 4-wheel-drive integral front-end loader. This was equipped with rear-mounted engine, up-front operator, and a 1½-yd^3 bucket. Numerous manufacturers have since entered the market with 4-wheel-drive loaders of up to 15-yd^3 regular rated capacity.

Diesel engines power most of the loaders. Some loaders are available with optional makes of diesels, and gasoline engines are also available for the smaller-size loaders. The drive is through a torque converter, power-shift transmission, front- and rear-axle differentials, and planetary gear reductions in all wheels. Differentials are generally of the limited-slip, torque-transferring type in either front, rear, or both axles. The steel main frames of the loaders may be either rigid or articulated.

FIGURE B6-1 Model 655B front loader crowding and lifting load. *(John Deere Co.)*

FRONT-END LOADERS

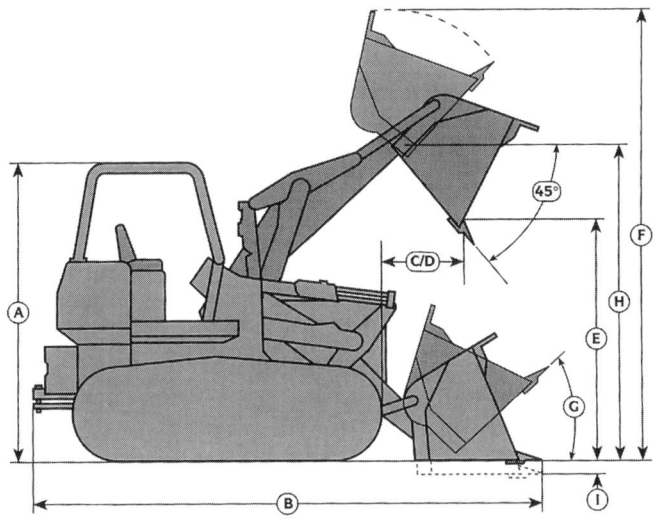

```
Key:                                                     General Purpose
Bucket Type                                 Standard        Wide-Track        Multipurpose
Capacity, heaped, SAE .................................2 cu. yd. (1.5 m³)   2 cu. yd. (1.5 m³)   2 cu. yd. (1.5 m³)
A. Overall height ........................................121.5 in. (3086 mm)  121.5 in. (3086 mm)  121.5 in. (3086 mm)
B. Overall length .........................................222.5 in. (5652 mm)  218.7 in. (5555 mm)  223 in. (5664 mm)
C. Reach at maximum height (45° discharge) ........48 in. (1219 mm)    45.3 in. (1151 mm)    47.4 in. (1204 mm)
D. Reach at 84 in. (2134 mm) clearance (45° discharge) ....66.5 in. (1689 mm)  63.8 in. (1621 mm)   65 in. (1651 mm)
E. Dump clearance, maximum height (45° discharge) .........110.5 in. (2807 mm)  113.2 in. (2875 mm)  111.6 in. (2834 mm)
F. Maximum operating height ............................194 in. (4928 mm)    191.2 in. (4857 mm)   190 in. (4826 mm)
G. Rollback angle
   Concrete level ...................................................38 degrees       36 degrees        36 degrees
   Carry position ..................................................45 degrees       45 degrees        45 degrees
H. Height to hinge pin ...............................................142 in. (3607 mm)   142 in. (3607 mm)    142 in. (3607 mm)
I. Digging depth ....................................................5.5 in. (140 mm)    5.5 in. (140 mm)     5.5 in. (140 mm)
```

FIGURE B6-2 Specifications for Model 655B. *(John Deere Co.)*

TABLE B6-1 Crawler-Type Front Loaders, Caterpillar and John Deere

Model	Horsepower	Weight, lb	Capacity-range CY (heaped)
Caterpillar			
913C Series II	70	17,742	1.1
935C Series II	80	19,311	1.3
953B	120	31,887	2.0–2.4
963	150	40,490	2.3–2.9
973	210	54,899	3.3–4.2
John Deere			
455G	70	18,745	1.25
555G	90	21,058	1.5
655B	120	32,400	2.0
755B	140	36,150	2.25

FIGURE B6-3 Crawler-type loader Model 755B. *(John Deere Co.)*

Most loaders have an automatic boom lift kickout which is adjustable for various dumping heights up to the maximum clearance in inches from the ground to the bucket cutting edge at a 45° dumping angle as shown in the comparative specifications. The corresponding reach at the maximum dumping height is measured from the tires or frontmost part of the tractor. An automatic bucket positioner that can be adjusted to the desired bucket digging or scooping angle may be either optional or regular equipment.

Some loaders have two foot pedals for applying their 4-wheel power brakes. One of the brake pedals disengages the transmission, allowing the operator to divert full engine power to the live-driven hydraulic pump or pumps for load prying and lifting.

Wheel-type loaders have much higher speeds than the crawler type and operate best on firm surfaces. They can be run on pavement and moved from job to job under their own power. A wider variety of sizes is available to match job requirements. Their maintenance cost is lower where wet, sandy soils would be highly abrasive to crawler tracks and undercarriages. However, wheel loaders normally have less traction than crawler-type loaders of equal weight, and this limits the usable power which they can develop for heavy digging and grade climbing. Their load-handling capacity is also reduced on boggy ground.

Figure B6-5 is a front view of John Deere Model 744E at work. Figure B6-6 is the specifications for the 744E. Figure B6-7 is the bucket selection guide for the 744E. Table B6-2 is loader operating information for the 744E loader.

FIGURE B6-4 Model 755B counter-rotating. *(John Deere Co.)*

FIGURE B6-5 Model 744E at work. *(John Deere Co.)*

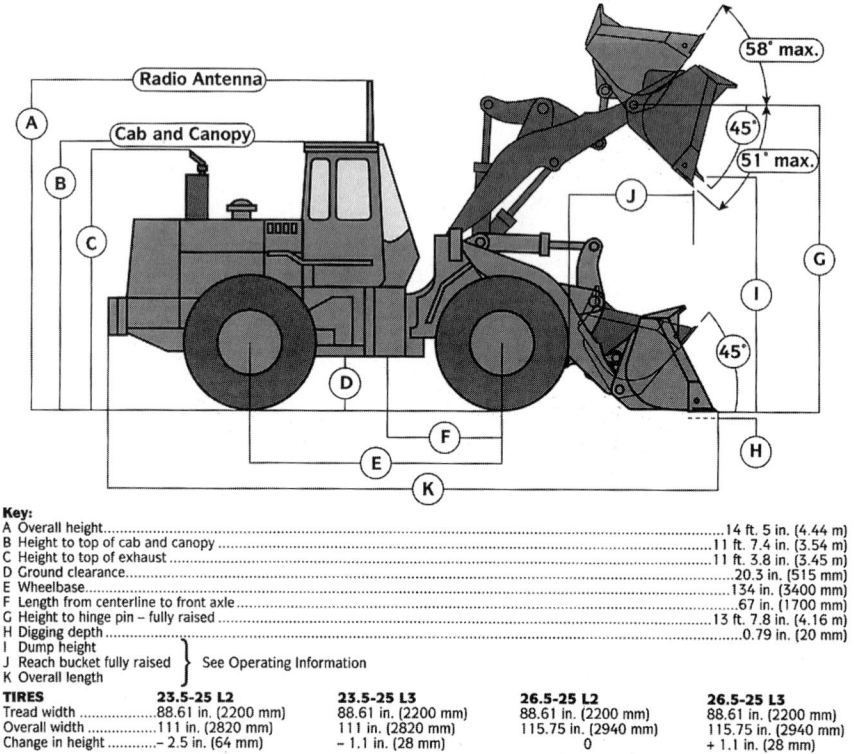

Key:
A Overall height...14 ft. 5 in. (4.44 m)
B Height to top of cab and canopy...11 ft. 7.4 in. (3.54 m)
C Height to top of exhaust...11 ft. 3.8 in. (3.45 m)
D Ground clearance...20.3 in. (515 mm)
E Wheelbase...134 in. (3400 mm)
F Length from centerline to front axle...67 in. (1700 mm)
G Height to hinge pin – fully raised..13 ft. 7.8 in. (4.16 m)
H Digging depth...0.79 in. (20 mm)
I Dump height
J Reach bucket fully raised } See Operating Information
K Overall length

TIRES	23.5-25 L2	23.5-25 L3	26.5-25 L2	26.5-25 L3
Tread width	88.61 in. (2200 mm)	88.61 in. (2200 mm)	88.61 in. (2200 mm)	88.61 in. (2200 mm)
Overall width	111 in. (2820 mm)	111 in. (2820 mm)	115.75 in. (2940 mm)	115.75 in. (2940 mm)
Change in height	– 2.5 in. (64 mm)	– 1.1 in. (28 mm)	0	+ 1.1 in. (28 mm)

FIGURE B6-6 Specifications/dimensions for Model 744E. *(John Deere Co.)*

Figure B6-8 shows a 624G rear view. Table B6-3 is a listing of Caterpillar wheeled loaders. Table B6-4 is a listing of John Deere wheeled loaders.

FRONT-END LOADER RATINGS

The Society of Automotive Engineers, Inc., has established and published industry SAE standards ratings for bucket volume capacity and tipping load weight as well as a recommended operating load rating for front-end loaders. Selection of the proper bucket capacity for the material being handled involves these ratings, which may be briefly described as follows.

Bucket Capacity, SAE Rating

SAE rating (nominal heaped) is described in SAE Standard J742b. The volume in cubic yards (or cubic meters) is based on physical dimensions of the bucket only, without consideration of the bucket action on any specific machine. It applies to a bucket with opening oriented in an upright, level position and filled with material heaped at a 2:1 angle of repose.

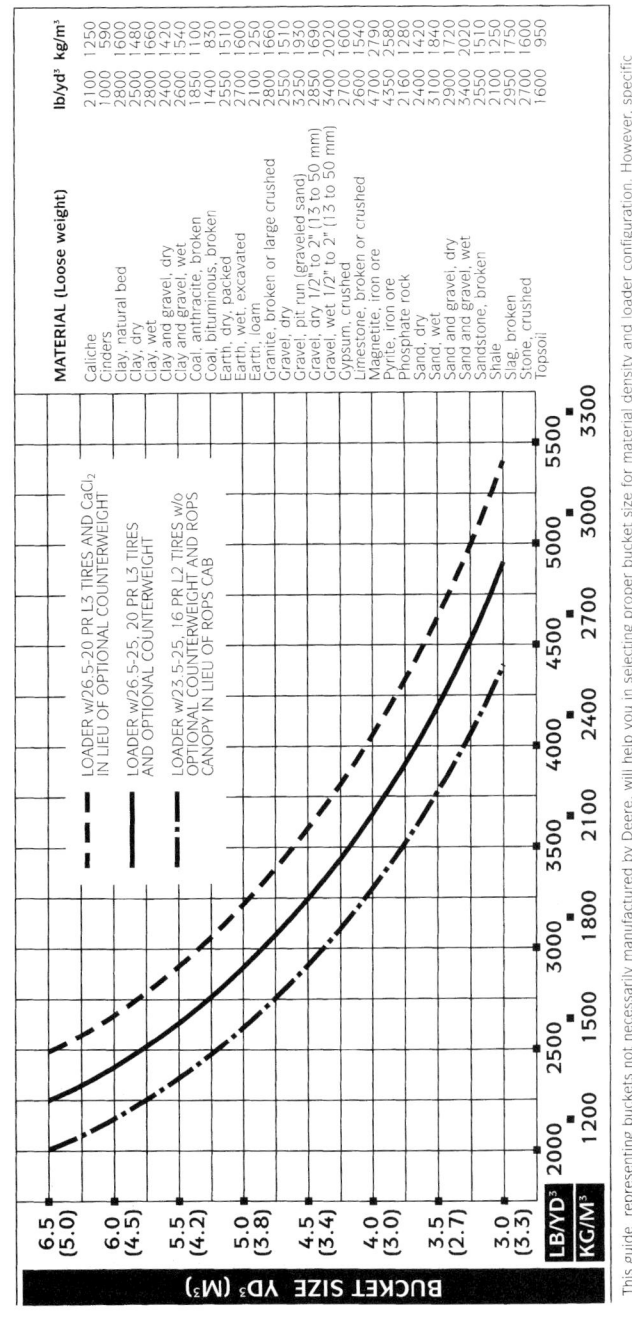

FIGURE B6-7 Bucket selection guide for the 744E. *(John Deere Co.)*

TABLE B6-2 Operating Information for the 744E Loader

Operating Information	Bucket type/size	Stockpiling w/BOC	Stockpiling w/teeth	Excavating w/BOC	Excavating w/teeth
Capacity, heaped SAE	yd³	5.0	4.75	4.25	4.0
	m³	3.8	3.7	3.3	3.1
Capacity, struck SAE	yd³	4.3	4.0	3.6	3.35
	m³	3.3	3.1	2.7	2.6
Bucket width	in	119.7	119.7	119.7	119.7
	m	3.04	3.04	3.04	3.04
Breakout force, SAE J732C	lb	42,463	38,078	45,885	40,770
	kN	189	169	204	181
Tipping load, straight	lb	33,755	34,130	34,136	34,507
	kg	15 311	15 481	15 484	15 652
Tipping load, 35° turn	lb	29,705	30,035	30,040	30,367
	kg	13 474	13 624	13 626	13 774
Tipping load, full turn, SAE 40° turn	lb	28,506	28,823	28,828	29,141
	kg	12 930	13 074	13 076	13 218
Reach, 45° dump, 7 ft (2.13 m) clearance	in	73.47	78.03	68.3	76.03
	mm	1866	1982	1735	1931
Reach, 45° dump, full height	in	50.08	54.65	46.46	54.17
	m	1272	1388	1180	1376
Dump clearance, 45°, full height	in	117.64	113.15	120.2	115.73
	mm	2988	2874	3053	2939
Overall length	ft-in	26-4	26-10	26-0	26-6
	m	8.02	8.18	7.92	8.08
Loader clearance circle, bucket in carry position	ft-in	44-3.5	44-8.2	44-1.1	44-5.9
	m	13.5	13.62	13.44	13.56
Operating weight	lb	46,310	46,017	45,931	45,638
	kg	21 006	20 873	20 834	20 701

NOTE: BOC = bolt-on cutting edge.
Loader operating information is based on machine with all standard equipment, 26.5-25, 16 PR L2 tires, optional counterweight, ROPS cab, 175-lb (79 kg) operator and full fuel tank. Information is affected by tire size, ballast, and attachments.

FIGURE B6-8 Model 624G at work; rear view. *(John Deere Co.)*

For bucket capacities of from ¾ to 3 yd³, the SAE standard rating interval is ⅛ yd³; for buckets over 3 yd³, the interval is ¼ yd³. If the calculated heaped capacity falls below a given rating interval by more than 2 percent, the next lower interval is deemed to be the SAE rating. Struck capacity is shown decimally to three significant figures when given in addition to the SAE rating.

TABLE B6-3 Caterpillar Wheeled Loaders

Model	Horsepower	Weight, lb	Capacity-range CY (heaped)
910F	80	15,982	1.3–1.7
918F	98	19,900	2.0–2.3
928F	120	23,874	2.6–2.8
930T	105	21,362	2.3
936F	140	27,220	2.8–3.3
950F Series II	170	36,316	3.8–4.0
966C	170	35,093	3.5–4.0
960F	200	36,733	4.0–4.3
966F Series II	220	40,096	4.8–5.0
970F	250	49,775	5.0–5.3
980F	275	60,803	5.5–7.0
988F	400	97,727	7.8–8.0
990	610	161,994	11.0–11.2
992D	690	193,108	14.0
994	1250	385,000	13–26

Source: Caterpillar Company.

TABLE B6-4 John Deere Wheeled Loaders

Model	Horsepower	Weight, lb	Capacity-range CY (heaped)
244E	55	11,746	1.0
344E	75	17,276	1.25–2.75
444E	95	21,064	1.5–3.25
544G	115	23,655	1.75–3.5
544GH	125	24,449	2.0–3.5
624G	145	28,103	2.25–4.25
624GH	150	28,246	2.25–4.25
644G	170	35,357	3.25–5.25
644GH	185	38,245	3.25–5.25
744E	230–250	47,150	3.0–6.5

Source: John Deere Co.

Tipping Load, SAE Rating

This is defined in SAE Standard J732c. It is based on the loader being at operating weight and stationary on hard, level ground. Operating weight with specified bucket is that of a fully serviced loader, including a full fuel tank; counterweight in amount specified, if any; and a 175-lb operator. The bucket is rolled fully back with the center of rated-load gravity at the maximum forward position in the raising cycle. Tipping load is the minimum weight in pounds or kilograms that causes the front rollers of crawler-type loaders to clear the tracks or the rear wheel-type loaders to clear the ground. Articulated-steering loaders should be rated in full turn position.

Operating Load, SAE Rating

To conform with SAE recommended practice, J818a, the SAE-rated operating load should not exceed 50 percent of the SAE-rated tipping load for wheel-type front-end loaders or 35 percent for the crawler type. The higher figure is permissible for wheel-type loaders because they normally operate on harder surfaces and smoother terrain than the crawler type.

The SAE-rated capacity in cubic yards of the bucket attached to the loader should not exceed the loader's SAE operating-load rating divided by the weight per cubic yard of scooped material. Because of material swell, this weight may be less than that in-bank but more than the loose weight, depending on the digging, crowding, and lifting action of the bucket. For efficient, long-life performance of the loader, it is important to know the loader application and the unit weight of the material to be handled in order to select the right type and capacity bucket.

LOADER BUCKETS AND ATTACHMENTS

General-purpose buckets are available for each model loader in a range or SAE-rated yardage capacities approved for handling specified unit-weight materials without exceeding the SAE operating-load rating of the loader. However, unless otherwise ordered, a standard general-purpose bucket is usually supplied for material weighing at least 3000 lb/yd^3 of SAE-rated capacity. General-purpose buckets are of one-piece all-welded steel construction with a straight cutting edge to which digging teeth can be added if required.

An optional multipurpose or four-in-one bucket can be used as a dozer, scraper, and clamshell. It consists of two major segments. The rear segment is essentially a dozer blade with a straight cutting edge. The front segment is a hydraulically operated clam with a straight cutting edge at the front, and it is usually equipped with digging teeth. Operation of the clam requires a third hydraulic control valve. The multipurpose bucket is much heavier than the standard-capacity general-purpose bucket, and a loader equipped with it will generally need more counterweight for the same operating load rating.

A one-piece bucket of heavy-duty construction with a protruding V-shaped cutting edge is also available. This is known as a *spade-nosed rock bucket* and is used for prying out and scooping up shot rock. Other special buckets include a quarry type with stub teeth to load fragmented shot rock and a skeleton rock bucket to sift out dirt and undesirable small pieces of rock when loading mine-run stock into hoppers or trucks. A side-dump bucket can be used for handling loose, stockpiled material and eliminates turning of the loader to dump into a hauling unit.

Straight or angling dozer blades, a grubber blade for land clearing, logging and lumber grapples, and other front-end attachments for special applications are available in place of buckets on many loaders.

Figure B6-9 shows some attachments which can be mounted on front loaders. Figure B6-10 shows attachments in action.

A rear-mounted ripper-scarifier or a towing winch not only provides counterweight but increases the utility of a crawler-type loader. Quick-attachable, rear-mounted backhoes are available for the smaller wheel- and crawler-type loaders for ditching and trenching.

Canopies, cabs, cab pressurizers, heaters and air conditioners, rollover protection devices, safety belts, and backup warning lights are among the attachments available for operator comfort and safety. Instrument panel covers, fuel and oil caps, and hood side doors, all of which lock into place, are good investment features to protect against vandalism. Tires other than standard are available to suit operating conditions.

MAJOR LOADER APPLICATIONS

Front-end loaders are regularly used for loading, hauling, excavating, clearing, and cleanup. Ability to carry, compact, and spread cover material for sanitary landfill makes them excellent units not only for air- and water-pollution control but also for land making.

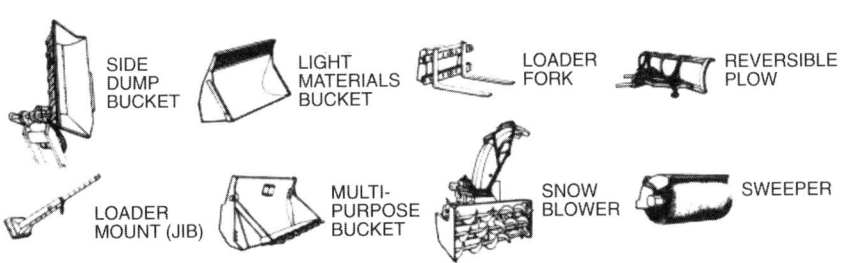

FIGURE B6-9 Attachments for front loaders. *(John Deere Co.)*

Loading

This is a major application in which operating conditions are often favorable for wheel-type loaders. It consists of scooping, lifting, turning, and dumping materials such as sand, gravel, and crushed or shot rock from stockpile, bank, or construction area into hauling units. The larger, articulated wheel-type loaders, because of their mobility and high productivity, are particularly satisfactory for loading shot rock into off-highway hauling units.

Time for each pass depends on kind of material; size and type of loader; bucket filing, lifting, and dumping conditions; and the proficiency of the loader and hauler operators. Table B6-5 gives estimated times in minutes under average conditions to load the material, including shot rock, into hauling units of various truck/heaped capacities with power-shift loaders of 1¼ to 15 yd^3 SAE bucket capacity. Table B6-5 allows a minimum of 0.5 min per pass for stockpile or free-running bank loading with rigid-frame wheel loaders.

Hauling

Rubber-tired loaders are excellent for moving loose materials over short distances to dump into hauling units, hoppers, conveyors, bins, or where needed on construction sites. Production in loose cubic yards at 100 percent working efficiency in approximately the SAE-rated capacity of the bucket in cubic yards multiplied by 60 min and divided by the cycle time in minutes.

Forks handle pipe, pallets, and a variety of other materials.

Snow blower has its own 60 horsepower engine. Clears a path 7 feet wide.

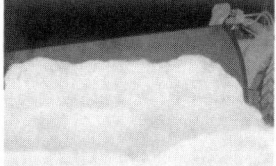
Snow plows have manual or hydraulic angle. Have automatic trip feature.

High-capacity, light-duty buckets speed handling of lighter weight materials.

Adjustable jib boom adds height and reach when placing loads.

Sweeper picks up debris in sweeping process. Has water sprinkler for dust control.

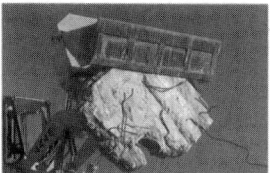
Multipurpose bucket acts as a normal bucket, a dozer, scraper or clam.

Side dump bucket speeds work in confined quarters, lays bedding quickly without waste.

Engine puller rips engine and transmission from car in seconds (444E and larger).

FIGURE B6-10 Front loader attachments in action. *(John Deere Co.)*

TABLE B6-5 Estimated Times (min) to Load Hauling Units from Stockpiles or Free-Running Banks with Rigid-Frame Wheel Loaders

Loader standard capacity, yd³	Hauling-unit nominal struck/heaped capacity, yd³, and tonnage rating							
	⅘, yd³, 6 tons	⁶⁄₇.₅ yd³, 9 tons	⁸⁄₁₀ yd³, 12 tons	¹⁰⁄₁₂ yd³, 15 tons	¹²⁄₁₅ yd³, 18 tons	¹⁴⁄₁₇ yd³, 21 tons	¹⁶⁄₂₀ yd³, 24 tons
1¼	2.0	3.0	4.0				
1½	1.5	2.5	3.3	4.0			
1¾	—	—	3.0	3.5	4.5		
2	—	2.0	2.5	3.0	4.0		
2½	—	1.5	2.0	2.5	3.0	3.5	
3	—	—	—	2.0	2.5	3.0	3.5
4	—	—	—	1.5	2.0	2.0	2.5

Cycle time consists of two elements: (1) fixed time for loading, lifting, and dumping the bucket, four changes of direction, and two turns; and (2) travel time for load carrying and empty returning. High reverse speeds enable faster time cycles when turns are not over 90°, all made in the bucket-loading area.

A fixed time allowance of 0.30 min is generally adequate. However, if 180° turns must be made at each end of the haul, an additional 0.06 to 0.10 min may be required. Travel time depends on the loader's average forward and reverse speeds over the distance and terrain involved. Allowance must be made for acceleration and braking as well as for the maximum speed at which full bucket loads can be safely carried without spilling.

Job conditions normally result in a 48- to 50-min productive hour, corresponding to working efficiencies of 80 to 83 percent. To convert loose yards to in-bank yards, multiply by the material swell factor.

Excavating

Crawler and heavy-duty wheel-type loaders are excellent for many excavations jobs. Figure B6-11 shows a typical basement digging operation arranged so that the loader can be driven forward out of the excavation on a moderately inclined ramp.

FIGURE B6-11 Basement excavation.

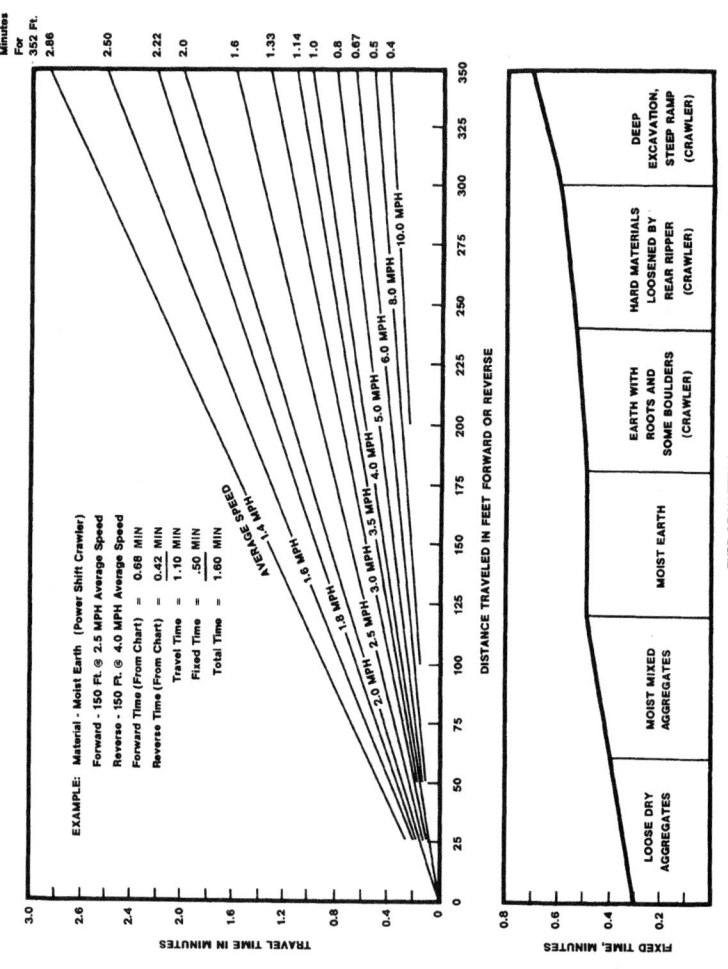

FIGURE B6-12 Loader cycle times. (Tabulated values are for power shift. For manual shift, add 0.05 min for each gear change.)

TABLE B6-6 Estimated Times (minutes) to Load Loose Materials into Hauling Units by Front-End Loaders

Loader standard capacity, yd³	$8/10$ yd³, 12 tons	$10/12$ yd³, 15 tons	$12/15$ yd³, 18 tons	$15/18$ yd³, 22 tons	$20/24$ yd³, 30 tons	$24/30$ yd³, 36 tons	$30/36$ yd³, 45 tons	$33/40$ yd³, 50 tons	$40/50$ yd³, 60 tons
Stockpile:									
2	2.1	2.5	3.3	3.8					
2½	1.7	2.1	2.5	3.0					
3	—	1.7	2.1	2.5	4.2				
3½	—	—	1.8	2.2	3.4	4.2			
4	—	1.3	1.6	2.0	3.0	3.8	3.8		
4½	—	—	—	1.7	2.4	2.9	3.4		
Shot rock:									
5	—	—	1.9	2.5	3.1	3.7	4.5		
6	—	—	—	1.9	2.5	3.1	3.8	4.5	
8	—	—	—	—	2.0	2.6	3.2	3.4	
9	—	—	—	1.3	2.0	2.6	2.7	3.4	3.9
10	—	—	—	1.3	2.0	2.0	2.7	2.8	3.5
12	—	—	—	—	1.4	2.0	2.1	2.7	2.8
15	—	—	—	—	—	1.4	—	2.1	

TABLE B6-7 Estimated Loading Capabilities of a Front-End Loader with 3½ yd³ SAE-Rated Bucket

(3000 lb/yd³ material)

One-way distance, ft	Production per 60-min hour					
	Average speed, mi/h	Cycle time, min	Work cycles	Yd³ (3000 lb/yd³ material)	Tons
25	2.9	0.496	121	424	636
50	4.3	0.564	106	371	556
75	5.1	0.635	94	329	493
100	5.8	0.691	87	304	456
150	7.0	0.786	76	266	399
200	8.0	0.868	69	241	361
300	8.8	1.07	56	196	294
400	9.4	1.27	47	164	246

Loaders have advantages over dozers of comparable size because they can be used to lift excess excavated material and dump it into trucks or over the edge of the excavation into stockpiles.

A rear-mounted ripper or scarifier can be used on a crawler loader to loosen such materials as compacted earth, hardpan, shale, slate, and decomposed rock that do not readily yield to direct bucket loading. The ripper can also be used to tear up old brick, asphalt, or broken concrete pavement. One operator can then both dig and load the material for removal.

The amount of material that can be excavated per hour depends on the kind of material and the distance it is to be moved. Figure B6-12 gives travel time in minutes for distances to 350 ft and the fixed time allowance for filling the bucket, dumping, four power-shift changes of direction, and a full load cycle of hydraulics. Although higher peak speeds can be realized on firm, level ground, it is seldom that average travel speeds for crawler-type loaders to exceed 2 to 2.5 mi/h forward and 4 mi/h reverse when used for excavation. Confined work areas, slippage, and steep ramps tend to reduce actual travel speeds.

Table B6-6 has estimated times (minutes) to load loose materials into hauling units by front-end loaders. Table B6-7 has estimated loading capabilities of a front-end loader with a 3½ yd³ SAE-rated bucket.

CHAPTER 7
HAULING UNITS

John A. Cravens
Vice President, Sales, Euclid, Inc., Cleveland, Ohio

Duncan MacLaren
District Manager, Southwest Region, Terex Division, General Motors Corporation, Dallas, Texas

R. D. Evans
Civil Engineer, International Sales, Caterpillar Tractor Co., Peoria, Illinois

James J. O'Brien, P.E.
O'Brien-Kreitzberg, Inc., Pennsauken, New Jersey

INTRODUCTION

While the contents of this section are applicable to both types of vehicles, it is important to recognize that the terms *on-highway* and *off-highway* refer not only to whether a hauling unit may or may not legally be operated on public highways but also to the construction of the unit itself.

From a legal standpoint, the distinction between on-highway and off-highway hauling units can be made as follows: On-highway haulers are vehicles whose design and manufacture are intended to permit their use on the public road, and as such they are subject to motor vehicle registration. In general, they will comply with gross vehicle weights, axle loadings, and width limitations imposed by the states. (See Fig. B7-1.) Off-highway haulers are vehicles whose design and manufacture are intended to preclude their use on public roads; thus, they do not comply with on-highway registration. Typical characteristics are widths in excess of 8 ft 6 in and/or axle loads at rated payload in excess of 24,000 lb.

In addition to legal distinctions, there are specific design features which characterize the off-highway hauling units:

1. Body construction is specifically designed to absorb high-impact loading. The floor of the body may be constructed of high-tensile steel (100,000 lb/in^2) or aluminum and will generally be fitted with steel wear plates to reduce abrasion damage.
2. The power train will consist of a diesel engine—often turbocharged—coupled to a torque converter and power-shift or automatic transmission. The transmission will be of the planetary or countershaft design and will often be mounted remotely from the engine. A means is provided for the final gear reduction at the drive wheels. This is usually a planetary-type axle or an internal driving gear and ring gear in the wheel hub or final drive case.
3. The ratio of pounds of gross vehicle weight to horsepower (flywheel or net) is generally lower than for an on-highway hauling unit. Top speed is slower than for a comparable on-highway hauling unit (about 40 to 45 mi/h), but gradeability under adverse conditions of grade and rolling resistance is from 25 to 40 percent better.
4. In most cases, a rigid-type frame is used to hold all driving components in alignment. Box sections, I-beams, or ship channel frame rails—generally separated by cylindrical torque tubes—are used. On large hauling units, the frame may be built in two sections and joined by means of two or more vertical pins so as to articulate for better maneuverability.
5. Wheel and drive configuration may be 4 by 2, 4 by 4, 6 by 2, or 6 by 4. Tires designed for earth moving, mining, and logging service are standard equipment.

FIGURE B7-1 Typical on-highway rear-dump hauler. *(White Truck Division, White Motor Co.)*

In addition, for the off-highway hauler, extra attention is given to oversize air cleaners and filters on all engine openings, vents in the drive-train gear cases, shockproof mounting of all instruments, dustproofing of electrical systems, heavier cabs, cushioned mounting for radiators, and heavier mounting of fenders and fuel tanks. It follows that the initial purchase price of a specially designed off-highway hauler is considerably higher than that of a highway-type unit which has been modified for off-highway service. (See Fig B7-2.)

In construction or mining, it is important for a prospective buyer to analyze the differences between on-highway and off-highway hauling units. It can be prohibitively expensive to misapply a strictly highway-type unit to a rugged off-highway load-and-haul situation. Careful analysis of the job to be done is necessary to make the proper equipment choice.

ON-HIGHWAY TRACTORS

Classification

On-highway units are generally classed either as straight (or unit) trucks or as truck-tractors. The former class constitutes the integral mounting for conventional rear-dump bodies and mixers, while the latter class is used in conjunction with low beds (Fig. B7-3), bottom dumps (Fig B7-4), rear dumps (Fig. B7-5), and side-dump bodies. Another system of truck classification, particularly favored in military terminology, is by wheels and drivers and is given in that order. For example, a 6×4 (Fig. B7-6) has six wheels and four drivers, and a 6×6 (Fig. B7-7) has six wheels and six drivers.

Nominal tonnage rating, which should not be interpreted as an actual load rating, is another method of classifying trucks. This method is used in some states for licensing purposes. Tonnage rating for payload capacity is in general use in Europe, in addition to gross vehicle weight (GVW) and gross combination weight (GCW). In the United States, the most widely used method of truck classification is based on GVW and GCW.

All the states have rules and regulations which establish vehicle size and weight limits. The maximum permissible on-highway width is 8 ft in most states, with a few

FIGURE B7-2 Typical off-highway hauler, 35-ton rated payload. *(Euclid, Inc., subsidiary of White Motor Co.)*

FIGURE B7-3 Low-bed truck-tractor.

FIGURE B7-4 Bottom-dump truck-tractor.

FIGURE B7-5 Rear-dump truck-tractor.

permitting 8.5 ft. Length, height, tandem-axle spacing, and speeds are other limiting factors which vary from state to state.

Regulations for on-highway trucks are manifold. Since all classes of highways are involved—interstate, primary, secondary, rural, city, etc.—multiple levels of government (federal, state, county, township, municipal) may exert their jurisdictional authorities. Various highway types must also be considered—reinforced concrete,

FIGURE B7-6 Kenworth 6 × 4 unit.

FIGURE B7-7 Oshkosh 6 × 6 unit.

asphaltic, gravel, dirt, etc.—each of which may have different load-bearing characteristics.

In view of those factors, it is not surprising that nationwide conformity in truck regulations simply does not exist. This lack primarily affects the operation of freighters and cartage trucks, although all trucks which operate on a public highway are subject to applicable legal controls. Our concern, however, is limited to *semis* and tandem units, or *doubles,* which may be used to haul earth, base material, or aggregates. These trucks are primarily on-highway trucks, although most of them have been designed and manufactured to be used in off-highway applications as well.

Equipment Characteristics

Four basic types of trucks are used for bulk material handling in on- and off-highway service. Table B7-1 identifies these types and indicates the usual ranges of gross vehicle ratings and payloads. The basic comparison has been limited to these two items because of the wide selection of engines, transmissions, front and rear axles, and tires. By way of example, Table B7-2 lists manufacturers' specifications within the "usual" and "high" ranges for typical dump trucks in the three-axle, single-unit class.

It is frequently necessary for a contractor to select dump trucks from within the "usual" 30,000- to 45,000-lb range of GVW to comply with highway load restrictions. The same three-axle truck can be equipped with a larger engine, more capacity in its rear axle, tires with a higher load rating, and a body with a greater capacity. Thus, in those states where a 70,000 GVW is permissible by special permit, the actual payload can be increased up to 24 tons. This larger payload then permits the single unit to compete on an economic basis with semis and doubles.

Production Estimating

The performance of on-highway units is geared to a repetitive basic work cycle. The time required to perform each work cycle can be analyzed in terms of its fixed elements and its variable elements. It is frequently convenient to view as fixed elements

TABLE B7-1 Truck GVW (GCW) Ratings and Payload Capacities

Basic type of truck	Manufacturers' gross vehicle rating, lb		Payload capacity, tons	
	Usual	High	Usual	High
Two-axle	23,000	30,000	7.5	10
Three-axle	44,000	60,000	15.5	21
Four-axle	68,000	74,000	25	25
Tractor-trailer	72,000	80,000	22	26

TABLE B7-2 Specification Ranges for Typical Three-Axle, Single-Unit Dump Trucks

	Usual range	High range
Engine (diesel) hp	225–275	225–275
Transmission	10 speed	10 speed
Rear bogie, lb	32,000	38,000
Speed, mi/h	55–60	55–60
Tires	10.00–20	11.00–24
Body—yd^3, struck	12–14	16–18
Payload—tons	12–16	20–24
GVW, lb	30,000–45,000	46,000–70,000

the times required for loading, acceleration, gear shifts, braking, turns, and dumping. Variable time elements then consist of travel times on the haul and the return. Wait times resulting from loading delays and traffic delays can either be identified separately or treated as variable time elements. An efficiency factor can also be introduced to compensate for minor and routine delays which are not specifically identified in the basic time cycle. The production rate in cubic yards or tons per hour can then be expressed by the equation:

$$\text{Production, yd}^3 \text{ (tons)/h} = \frac{60 \text{ (min)}}{\text{fixed + variable time (min)}} \times \text{payload (yd or tons)} \times \text{percentage efficiency}$$

Example
Single-Unit Dump Truck

Payload capacity, tons	15
Fixed time, min	10
Variable time, min	5
Efficiency, present	80
Tons/h = 60/15 × 15 × 0.80 =	48

TYPES OF OFF-HIGHWAY HAULING UNITS

There are several distinct types of off-highway hauling units, each of which has specific design features which enable it to satisfy a particular set of job requirements. Major types of off-highway units include:

1. The rear-dump truck carrying a body on the truck chassis.
2. Bottom-dump tractor-trailer unit.
3. A side-dump truck carrying the body on the truck chassis. This unit may be dumped either by its own integral hoist mechanism or by an independent hoist or *skyhook*. Independent hoists are used in places where a fixed dump point is required, such as a hopper or crusher, and one dumping mechanism can thus handle many trucks.
4. Side-dump tractor-trailer unit.
5. Rear-dump tractor-trailer unit.

To select the right size and type of hauling unit, the estimator or equipment engineer must know thoroughly the nature of the job which has to be done and also be acquainted with the particular operating characteristics and the advantages and limitations of the various machines which could conceivably do the job. A tentative selection of the type of hauler can then be made, and estimates can be prepared for several suitable types and sizes of machines. Following this, the estimated hauling costs per cubic yard or ton of material can be compared. In selecting a hauling unit, these general criteria are used:

1. The versatility of the hauling unit; can it be efficiently used in other applications?
2. The compatibility between the unit being considered and currently used hauling or loading equipment.
3. Possible restrictions on maneuvering space at the loading or dumping areas or on side and overhead clearances.
4. Possible vehicle weight or width restrictions on haul roads, including bridges.
5. The effect of extreme grades, either favorable or unfavorable, particularly on the loaded haul.

Rear-Dump Load-on-Back Truck

The rear-dump truck with body mounted on the truck chassis is the most popular type of rubber-tired hauling vehicle. It can haul free-flowing material such as earth, sand, and gravel and more bulky material such as blasted rock, ore, shale, and coal.

The normal excavating type consists of a body shell with heavy rib reinforcements and an inner liner or wear plate which is usually welded to the body shell (Fig. B7-8). This double bottom can take the loading impact of rock dropped from large loading shovels. Special *low-profile* bodies have been developed to cater to the lower load-height capability of the wheel loader, which is increasingly popular for loading off-highway haulers (Fig. B7-9).

When the rock is not well blasted, it is frequently necessary for the operator to carefully balance a huge piece of blasted rock on the teeth before carefully pacing it in the truck body. Even at best, the loading impacts are extremely severe in this type of work. For continuous heavy-duty rock service in quarries and in rock excavations, a heavy-duty rock option body is available. This has front and side liners in addition to floor wear plates. Liner plates in larger trucks will run from ¼- to 1-in-thick alloy-steel plate.

In the colder climates, design precautions can be taken to prevent material from freezing to the rear-dump body. Exhaust gases from the engine are carried through ducts in the front slope sheet, in the bottom of the body, and on a portion of the side walls. In most cases, this is adequate to prevent material from freezing in the body. In extremely severe conditions, it may be necessary to clean out the forward bulkhead, the corners, and along the junction of the sides and bottom of the body, despite the

FIGURE B7-8 Fifty-ton off-highway hauler with standard body. *(Euclid, Inc., subsidiary of White Motor Co.)*

HAULING UNITS

FIGURE B7-9 Fifty-ton off-highway hauler with low-profile body. *(Euclid, Inc., subsidiary of White Motor Co.)*

application of heat. Under these circumstances, the conventional body can become so clogged with frozen material that it requires cleaning after every second or third trip.

The heated body can also help considerably when handling wet, adhesive clay soils; bauxite; and other sticky materials. Heat which flows between the liner plate and the outer body shell creates a dried layer of material next to the body shell and facilitates the dumping of wet clay-type soils.

Use rear dumps when:

1. The material to be hauled is large rock, ore, shale, etc., with individual particle sizes over 12 in in diameter, or when the material involves both free-flowing and bulky components.
2. The hauling unit must dump into restricted hoppers or over the edge of a waste bank or fill.
3. The hauling unit is subject to severe loading impact while under a shovel, backhoe, dragline, or hopper.
4. Maximum maneuverability—such as a rapid spotting—is required at the loading or dumping areas.
5. Maximum gradeability is required.

Bottom-Dump Tractor-Trailer

The bottom-dump tractor-trailer consists of a diesel-powered prime mover or tractor with large earth-mover type single or dual tires on each drive wheel and a semi-trailer with drop bottom or clamshell-type doors (see Fig. B7-10). The trailer hopper is generally wider and deeper at the front to put more weight on the drive axle. It is also wider at the top than at the bottom so that the side walls of the trailer are sloped inward. This shape of hopper is suitable for use only with relatively free-flowing materials, since large pieces of rock or shale might jam in the smaller bottom opening when the vehicle is dumping. The tractor may be of the 4-wheel type, which is particularly well suited to long high-speed hauls; or it may be an overhung-engine type of prime mover without a steerable front axle. This latter type is particularly adapted to working in soft soil conditions or around structures where very good maneuverability is required.

The bottom dump has the following characteristics:

1. Adaptable to free-flowing materials such as earth, sand, gravel, crushed stone, and the like.
2. When fitted with large single tires having air pressures of 30 to 50 lb/in^2, it offers better flotation than the rear-dump unit. It can thus haul through soft soils or sand better than the rear dump.
3. Adaptable to long, level, high-speed hauls, particularly when fitted with large-diameter dual tires. The machine is generally geared for higher travel speed than a rear dump, resulting in reduced gradeability and performance characteristics. The rear dump is built for pulling up steep grades; the bottom dump is built for longer, level hauls.
4. Spreads its load in windrows by dumping while in motion. It can also dump in drive-over hoppers, in which case it stops during the dumping operation.
5. Ability to pull on steep grades is restricted because of unfavorable power-weight ratio compared to rear dumper and less weight on drive wheels, which limit traction in soft soils or wet weather. In general, long, adverse grades should not exceed 5 percent for best application.
6. Low cost per ton of carrying capacity, and mechanical simplicity of operation and maintenance.

When equipped with large, single, flotation-type tires, and, to a lesser degree, when fitted with large-diameter duals, the bottom dump offers a significant advantage over dual-tired rear-dump equipment when the bearing capacity of the soil is low. In general, the lower the air pressure in the tire, the better the flotation. While there are certain exceptions to this statement, the inflation pressure of a tire can be used as a rough index of its flotation ability. Generally, the dual-tire rear-dump trucks have tire inflation pressures of 60 to 80 lb/in^2, whereas the single-tired bottom-dump equipment uses tires inflated to 30 to 50 lb/in^2.

Use bottom dumps when:

1. The material to be hauled is relatively free-flowing.

FIGURE B7-10 One-hundred-ton bottom dump being loaded by belt loader. *(Euclid, Inc., subsidiary of White Motor Co.)*

2. The hauling unit has unrestricted dumping into a drive-over hopper or the load is to be spread in windrows.

3. The haul is relatively level, thus permitting high-speed travel.

4. Long grades on the loaded haul do not exceed 3 to 5 percent. This is recommended as a general rule for maximum unit efficiency, but it is not the measure of the maximum gradeability of a bottom dump.

Side-Dump Load-on-Back Truck

Side-dump trucks are widely used in quarry-rock hauling and stripping operations. Many of the cement plants and large stone quarries which formerly used rail equipment had crushers and hoppers which were adapted to side-dump railroad cars. It was a logical evolution that, in the changeover from rail to truck, the side-dump vehicle should be considered (see Fig. B7-11). Because of the necessity of dumping over the drive tires and the design requirement for stability, the predominant type of chassis used with side-dump bodies is the tandem-axle or 6-wheel variety, especially where an integral hoist is required. The 4-wheel type of truck has been used in quarry operations where independent hoists are used at the crusher or hopper. Side-dump bodies are used in quarry stripping to dump the overburden "on the fly," whereas the rear-dump truck must stop, reverse, and dump its load over a waste bank.

For example, an engineer selecting equipment for a stripping operation might have a choice of a tandem-axle vehicle with side-dump body carrying 20 tons payload or a 4-wheel vehicle with single-drive axle mounting a rear-dump body carrying 20 tons payload. The merits of side dumping versus end dumping and the differences in the two hauling vehicles in traction, complexity of maintenance, performance characteristics, tire capacities, dumping angle of the body, and other pertinent features would be analyzed. These factors are often more important than the basic question of side-dump or rear-dump method of getting rid of the load.

Load-on-back side dumps are custom built to suit specific applications, so stock models are not offered.

FIGURE B7-11 Twenty-ton Euclid chassis with side-dump body.

Side-Dump Tractor-Trailer

The side-dump tractor-trailer is of heavy-duty construction and is able to handle both free-flowing material and rock. The dump angle is approximately 55°, which is adequate for all but the most adhesive materials.

There are two kinds of side-dump trailers, both of which are of the *tub* type. One has fixed, flared sides and the other has a downfolding gate on one side so that there is no obstruction to the material when dumping. The first type of body is more often seen in quarry installations where there is a firm roadbed at the dumping point and the side of the body can extend down into the hopper area when dumping. The downfolding-side trailer is commonly used in stripping operations, where the side of the body chutes the load away from the trailer tires (see Fig. B7-12). The unit moves forward while dumping. In some designs, the body can dump either to the left or the right, depending on operating requirements.

Use side dumps when: Bottom-dump criteria apply but where dumping is restricted to one direction (e.g., over a bank, where other hauling units cannot maneuver).

Rear-Dump Tractor-Trailer (Rocker)

The overhung-engine type of prime mover which pulls a rear-dump semitrailer has some unusual and interesting operating characteristics (see Fig. B7-13). This type of machine has excellent maneuverability because the overhung-engine tractor can turn 90° either to the right or to the left without moving forward. In some designs, the wheel base can be shortened for turning by raising the body to the dump position. This moves the trailer wheels forward and closer to the prime-mover drive wheels, shortening the wheelbase and making it possible to obtain an extremely short turning diameter. This is of particular importance in tunnel work or pioneering on narrow mountain roads. Both the prime mover and the trailer are equipped with large single tires with rock treads. This gives the hauling unit better flotation characteristics than for a dual-tired rear dump.

The tractor-trailer rear dump does not have the overall gradeability or traction of a rear-dump truck. It generally has a greater gross weight per horsepower than the load-on-back truck. On steep grades, the tractor-trailer unit has a weight transfer off

FIGURE B7-12 Euclid tractor with 50-ton side-dump trailer.

FIGURE B7-13 Terex tractor with rear-dump trailer.

the drive axle and back to the trailing axle. This limits its traction on wet or slippery haul roads.

The prime mover for the tractor-trailer rear dump is also designed to haul a rubber-tired scraper. It is thus possible to use the prime mover with a rear-dump trailer for rock hauling; then, when *scraper dirty* is encountered, the rock hauler can be parked, its wheels and tires removed, and a scraper bowl installed. The prime mover can then continue to operate, hauling a scraper rather than the rear-dump trailer. The interchangeability of the prime mover is of particular interest in regions where there is both rock and dirt to be hauled yet insufficient volume to keep a specialized rock hauler or earth-moving scraper consistently at work.

Use rocker-type dumpers when:

1. Distance from loading area to dumping area does not exceed 500 ft.
2. The maximum grade on the loaded hauls does not exceed 10 percent. This is not a measure of the maximum gradeability of rocker-type dumpers, but it is a practical limit for efficient operation.

HAULER PERFORMANCE

Performance and Retarder Charts

Table B7-3 is a listing of Caterpillar construction and mining trucks. Table B7-4 is a listing of Caterpillar articulated trucks. The performance potential of gradeability of a hauling unit can be expressed by means of performance and retarder charts. Once the variables of rolling resistance and grade resistance have been identified for a given haul road, these charts specify the maximum attainable speed for a given hauling unit when operating in each transmission range or gear. Performance retarder charts for a 35-ton hauling unit are shown in Figs. B7-14 and B7-15. Each chart is actually two separate graphs placed side by side. The right-hand graph relates the power provided by the unit's power train, expressed as rimpull, to the vehicle speed in miles per hour. The left-hand graph relates the two major factors which dictate the required rimpull at any specified vehicle speed; these factors are vehicle weight and total resistance (grade resistance plus rolling resistance) to forward motion. The procedure for reading the chart is as follows:

1. **Determine the vehicle specifications.** Select the appropriate chart showing the correct engine, transmission, gear ratio, and tire size.
2. **Establish the vehicle weight both empty** (net vehicle weight—NVW) and loaded (gross vehicle weight—GVW). If the actual NVW and GVW are different from the *rated* NVW and GVW, draw the correct lines on the graph.
3. **Estimate the total resistance** (rolling resistance plus grade resistance). Rolling resistance is always a positive number; grade resistance is positive for a vehicle traveling uphill and negative for a vehicle traveling downhill. Remember that rolling resistance and grade resistance are measures of the rimpull required to keep a vehicle in motion under particular haul-road conditions. The hauler must develop rimpull equal to this total resistance in order to maintain a constant speed. Total resistance is expressed, in these charts, as a percent of the vehicle's weight.
4. **When the total resistance as a percent of vehicle weight has been established,** locate it on the vertical scale at the far left side of the chart. Read down the

TABLE B7-3 Caterpillar Construction and Mining Trucks

Model	Horsepower (flywheel)	Top speed, mi/h	Weight empty, lb	Gross weight	Capacity, tons	Capacity cy Struck	Capacity cy Heaped
769C	450	47	68,750	149,000	40	22.9	30.9
771C quarry truck	450	25	74,560	163,100	44	23.4	35.5
773B	650	38	86,869	204,000	58	34	44.8
775B quarry truck	650	28	93,325	234,300	65	37.2	51.4
77C	870	37	132,422	324,000	95	47.6	67.1
785B	1290	35	212,458	550,000	150	74	102
789B	1705	34	268,837	700,000	195	196	137
793B	2057	34	323,709	830,000	240	126	169

Source: The Caterpillar Co.

TABLE B7-4 Caterpillar Articulated Trucks

Model	Horsepower (flywheel)	Top speed, mi/h	Weight empty, lb	Gross weight	Capacity, tons	Capacity cy Struck	Capacity cy Heaped
Two axle							
D20D	180	28.6	33,070	72,754	20	9.7	12.7
D25D	260	30	43,428	93,428	25	13	18
D30D	285	32	48,278	108,278	30	16.5	22.5
D40D	385	34	61,800	141,800	40	22.1	29.3
Three axle							
D250D	214	26.5	38,150	88,350	25	13.1	17
D300D	285	30.2	45,600	105,600	30	15.7	21.6
D350D	285	30	54,221	124,221	35	20.9	26.8
D400D	385	34.4	61,800	141,800	40	21	28.6

Source: The Caterpillar Co.

HAULING UNITS

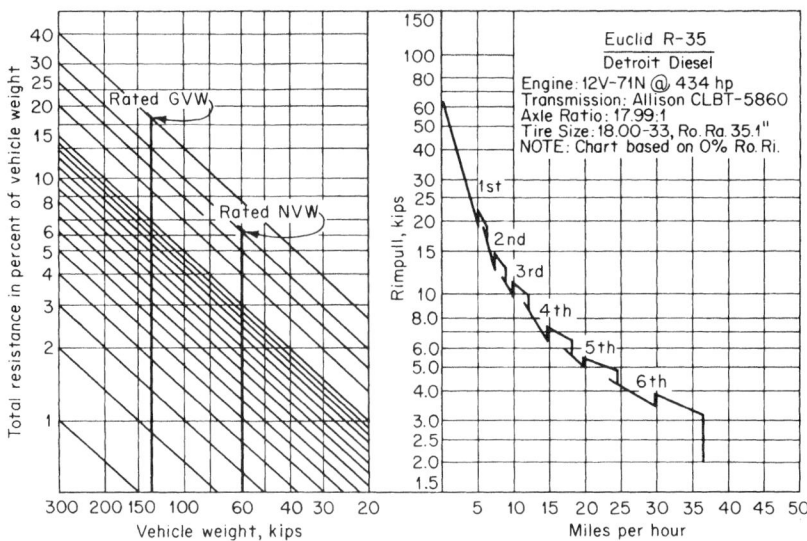

FIGURE B7-14 Hauler performance chart.

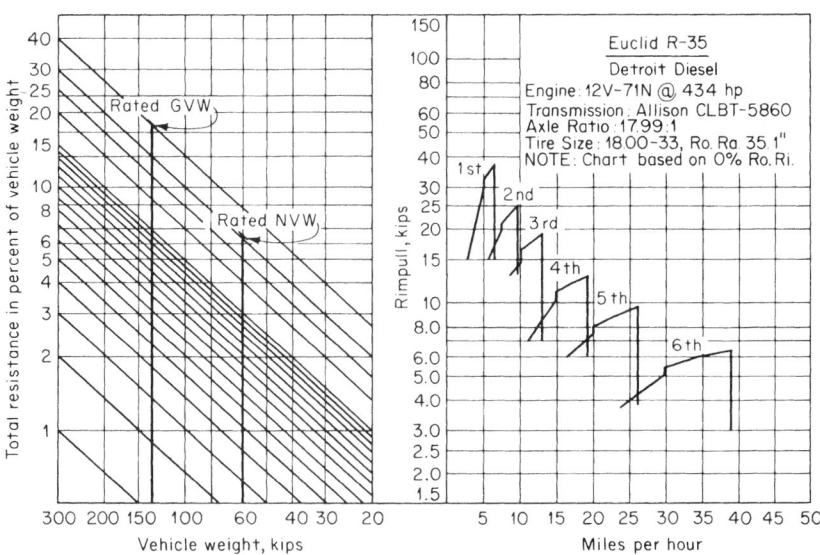

FIGURE B7-15 Hauler retarder chart.

appropriate slanted line. Stop at the vertical GVW line if the hauler is loaded, or at the vertical NVW line of the hauler is empty.

5. Extend the point of intersection horizontally to the performance curve. The intersection of this horizontal line and the performance curve gives the vehicle's speed.

6. When the horizontal *required rimpull* line intersects both the converter and lockup (i.e., *direct drive*) rimpull curve in a given gear range, two interpretations could be made. For example, as illustrated in Fig. B7-16, the speed selection could be either 14.3 mi/h in the converter or 16.7 mi/h in the lock-up. As a rule of thumb, select the lock-up speed range when accelerating and select the converter speed range when slowing down.

Retarder charts are read in a similar manner to a performance chart, although the values shown in the total resistance column are actual negative numbers. Thus, for example, a negative 13 percent grade with a 3 percent rolling resistance is read as 10 percent total negative resistance. The resulting horizontal line (from step 5) may cross two or more retarder rimpull curves. The retarding speed range falls between the first and last points of the intersection which the horizontal line makes with the curve.

Any of the speeds which fall within the maximum and minimum required rimpull values may be used by selecting the appropriate transmission range, by varying the amount of oil in the retarder, or by accelerating the engine above the speed at which it is driven by the other members of the power train.

In the event that detailed specifications are not available when a preliminary performance check is being made, vehicle speed in miles per hour can be determined when the total payload and net weight of the vehicle as well as the horsepower of the engine are known. The following formula is used:

$$\text{Vehicle speed in mi/h} = \frac{\text{net horsepower at drive wheels} \times 375}{\text{GVW in tons} \times [40 + (20 \times \text{percent of grade})]}$$

The net horsepower at the drive wheels equals the maximum engine horsepower at governed speed times 0.81. This assumes a 10 percent loss in efficiency due to the power requirements of driving accessories such as fan, air compressor, generator or alternator, and hydraulic pumps. Another 10 percent is deducted for friction losses in the hauler chassis.

Each percent of grade is approximately equivalent to 20 lb of rimpull or tractive effort per ton of gross weight. The preceding formula includes a figure of 40 lb/ton for rolling resistance, which is representative for hauling equipment which is operated over well-maintained haul roads. The hauler is thus working against the equivalent of 2 percent grade because of road resistance even when hauling on the level.

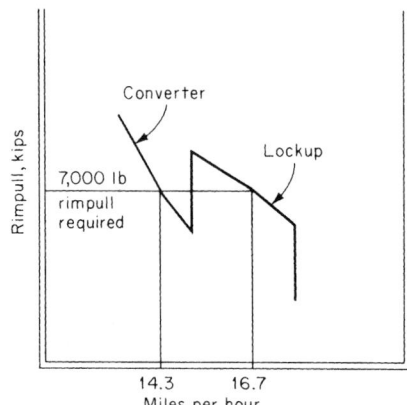

FIGURE B7-16 Converter speed versus lock-up speed.

While this figure is somewhat empirical and will vary according to the haul roads and basic vehicle design, it is typical for both on-highway and off-highway service under well-maintained conditions.

Applying this formula to a 35-ton hauler with a gross weight of 65.25 tons and 434-hp engine rating, the results for a 4 percent grade are as follows:

$$\text{Speed} + \frac{(434 \times 0.81) \times 375}{65.25 \times [40 + (20 \times 4)]} = 17 \text{ mi/h}$$

Now change the rolling resistance from 40 to 80 lb/ton. This would be a typical value for a soft or poorly maintained dirt surface:

$$\text{Speed} + \frac{(434 \times 0.81) \times 375}{65.25 \times [80 + (20 \times 4)]} = 12.5 \text{ mi/h}$$

Poor haul roads can drastically reduce the hauling speed. It is common practice in construction and mining projects to use motor graders for haul-road maintenance where high production is required. Graded and well-drained haul roads, permitting two-way traffic and providing a wide radius and good visibility on curves, are important to lowering hauling costs.

Operating Conditions

In the tables for estimating hauler production and costs, the data are frequently grouped under the headings *favorable, average,* and *unfavorable.* These terms assist in further identifying the hauling job.

Favorable Conditions:
1. Material being excavated and hauled:
   Topsoil
   Loam/clay mixture
   Compacted coal
   "Tight" earth (no rock)
2. Loading area (cut or borrow):
   Unrestricted in length or width
   Dry and smooth (or under constant maintenance)
3. Rolling resistance:
   Under 4 percent
4. Supervision:
   Constant at both the loading and dumping areas
5. Weather:
   Little rainfall or other weather delays
   Ambient temperatures of 40 to 60° F

Average Conditions:
1. Material being excavated and hauled:
   Clay with some moisture
   Soft or well-ripped shale

Loose sand with some binder
Mixtures of different earths
Sand/gravel mixture
2. Loading area (cut or borrow):
Some restrictions in length or width
Dry, with some loose material
3. Rolling resistance:
4 to 8 percent
4. Supervision:
Intermittent at both the loading and dumping areas
5. Weather:
Moderate rainfall or other weather delays
Ambient temperatures of 60 to 100° F

Unfavorable Conditions:
1. Material being excavated and hauled:
Heavy (dense) or wet clay
Loose sand with no binder
Coarse gravel (no fines)
Caliche or unripped shale
2. Loading area (cut or borrow):
Tight restrictions in length or width
Wet, slippery, and/or soft (not maintained)
Units load uphill or on a side slope
3. Rolling resistance:
Above 8 percent
4. Supervision:
None (e.g., no spotters) at either the loading or dumping area
5. Weather:
Excessive rainfall or other weather delays
Ambient temperatures in excess of 100 or below 0° F

Seldom is an entire job—considering individual segments of the haul cycle and equipment costs—given a blanket description of *favorable, average,* or *unfavorable.* Weather may be a consideration only if the hauling job must be completed with a fixed time period.

Example. A contractor is required to move 400,000 bcy of rock in 6 calendar months, or 180 days.

6 calendar months × 24 working days per month = 144 working days

The estimator determines that 312.5 bcy/h can be moved.

312.5 bcy/h = 3,125 bcy/day

400,000 bcy ÷ 3,125 bcy/day = 128 working days

The estimator has provided approximately 11 percent margin between the *days available* and the *days required*. This margin could provide for weather delays and other unforeseen events. It may also encompass equipment breakdown—poor machine availability. By proper job supervision, driver education, and other improvements, the actual hourly hauling rate might well be increased beyond the estimated value.

Machine availability is the percentage of the scheduled working hours during which a unit's mechanical condition is such that the unit is able to haul. Preventive maintenance, regular inspection, and servicing of hauling units are very important in maintaining high availability. Proper driver training to prevent misuse of hauling equipment is also important.

Table B7-5 lists average availabilities for hauling units as functions of hauling unit type and anticipated life. These values may be considered as adequate for estimating purposes, assuming good preventive-maintenance practices, "average" operating conditions, and operation for 50 percent or less of any 24-hour period.

One method of allowing for maintenance and repair time is to include spares or standby units in the hauling fleet. Spare units are an inexpensive form of production insurance. A standby unit does not incur any operating costs and should, therefore, be charged to the job at its ownership cost only. The hauling cost per ton or yard will be increased only slightly by having spare units on standby. On the other hand, the lack of spare units can result in costly production losses. To calculate the required total fleet, operating haulers plus spares, the following simple formula may be used:

$$\text{Required fleet} = \frac{\text{number of operating haulers} \times 100}{\text{percent machine availabillity}}$$

Assume that an estimate calls for the continuous use of 10 trucks and that their type, age, and condition indicate 85 percent machine availability. A fleet of 11.8 or 12 trucks will then be required to satisfy the production requirements. Two spares would be provided to ensure that 10 haulers are actually in operation at all times.

ESTIMATING HAULER PRODUCTION

Hauler production under optimum conditions can be estimated by means of the formula:

TABLE B7-5 Availabilities of Hauling Units*

	\multicolumn{5}{c}{Years of anticipated life}				
	1	2	3	4	5	6
Types of unit	\multicolumn{5}{c}{Average availability, percent}				
Rear dumps	90	89	88	86	84	83
Bottom dumps, side dumps	92	90	89	88	86	86
Rocker-type dumpers	88	86	83	80	78	78

* Factors such as severe operating conditions, multishift operations, and unsatisfactory preventive maintenance practices can reduce machine availability by as much as 20 percent below the levels considered "average." Similarly, favorable conditions may increase first-year values by as much as 5 percent and can show an increasing advantage over average conditions and subsequent years.

Hauler production per 60-min hour (tons or yard^3) =

$$\frac{60 \text{ min}}{\text{time to complete one cycle (min)}} \times \text{hauler's payload per cycle (tons or yard}^3\text{)}$$

Payload

The term *rated payload* refers to the maximum recommended load for a hauling unit as established by the manufacturer. Either the volumetric capacity or the rated payload will limit the allowable load, depending on the characteristics of the material which is to be hauled. To determine a hauler's payload per trip, the bank (in place) and loose (excavated) weights per cubic yard of the material should first be established. A check can then be made to determine whether the rated payload is exceeded when the hauling unit is filled to its heaped capacity with loose material. Whichever situation governs, heaped capacity or rated payload, will establish the maximum payload. The payload in pounds of a hauling unit should not be exceeded, since such overloading could adversely affect the life and performance of the unit.

Example. The maximum capacity of a typical 35-ton rear-dump unit is as follows:

Struck capacity (SAE)   23.3 yd^3

Heaped capacity (2:1 SAE)   29.0 yd^3

Rated payload 70,000 lb

This specification indicates that the hauler can be loaded (with the material heaped at 2:1 on the body) with 29 yd^3 of material providing that the 29 yd^3 weigh no more than 70,000 lb.

Assume that the unit is used to haul material with a unit weight of 90 pcf in its bank condition and a unit weight of 77.7 pcf in a loose state. The rated payload of 70,000 lb for the hauling unit will not be exceeded until the payload amounts to 33.3 yd^3. However, this hauling unit is not equipped with any devices (e.g., top extensions) for increasing its volumetric capacity beyond its rated capacity. Therefore, the maximum payload which can be carried without excessive spillage is estimated at 29.0 yd^3, loose measure. The hauler's payload per cycle or trip can also be expressed as 60,900 lb, or as 30.45 tons.

Cycle Time

In preparing an estimate of hauler production, each cycle is considered to be made up of the following parts:

Time to complete one cycle = loading time + haul time + dump-and-turn time + return time + spot-and-delay time

Example. A 35-ton rear dump, being loaded with shale by a 9 yd^3 wheel loader, might have the following cycle time:

Load: 3 passes a 0.90 min per pass	2.70 min
Haul: 2700 ft at 18.0 mi/h avg	1.70 min
Dump and turn	1.30 min
Return: 2700 ft at 21.0 mi/h avg	1.46 min
Spot and delay	0.25 min
Total time to complete one cycle	7.41 min

Loading Time

A hauling unit may be loaded by a power shovel, backhoe, dragline, clam, belt loader, front-end loader (either rubber-tired or track-type), or by a gravity hopper. Loading timer is simply the hauler's calculated payload divided by the loading rate (e.g., pounds per minute).

Front-End Loaders. In loading a hauling unit with a rubber-tired or track-type front-end loader, maximum efficiency can be obtained by placing the hauling unit as close to the loader as possible, so as to reduce travel distance between the two. In matching a hauling unit with a front-end loader, it is important to consider whether the loader's specified dump height (the distance from the tip of the bucket to the ground when the bucket is tilted forward to its maximum extent) is compatible with the hauler's loading height. Such consideration ensures that the top of the loader bucket will clear the top of the hauler body. It is also important to check the front-end loader's specified reach (the horizontal distance from the tip of the bucket to the front of the tires when the loader bucket is in the maximum dump-height position). The reach should be such that the load will be properly distributed over the width of the hauler body.

For estimating purposes, assuming minimum haul distances (e.g., between material pile and hauler) and average digging conditions, typical cycle times for rubber-tired front-end loaders are supplied in Table B7-6.

Travel Time

Whether the hauler is loaded and proceeding to the dump area or is empty and returning to the loading site, travel time is determined as follows:

$$\text{Travel time (min)} = \frac{\text{length of road section (ft)}}{\text{average speed over road section (mi/h)} \times 88}$$

The factor 88 is 5280/60 and is used to convert feet and miles per hour into minutes.

The determination of a hauler's maximum speed over a given section of road where both rolling resistance and grade resistance are known has already been dis-

TABLE B7-6 Typical Cycle Times for Front-End Loaders When Loading Hauling Units

Rated bucket capacity, lcy	Loader cycle time, min/pass
Less than 4	0.40–0.60
4–10	0.50–1.00
10–20	0.80–1.50

cussed under Hauler Performance. However, the hauler will not operate at its maximum speed over the entire length of a road section. The theoretical maximum speed must therefore be reduced to a practical average to compensate for such items as lower speeds during the acceleration or deceleration of the vehicle. This reduction can be accomplished by introducing a speed factor whereby

$$\text{Average speed (mi/h)} = \text{maximum speed (mi/h)} \times \text{speed factor}$$

A speed factor of 1.00 means that the vehicle will continuously maintain its theoretical maximum speed along the road section, while a speed factor less than or greater than 1.00 indicates that the average vehicle speed will be slower or faster, respectively, than this theoretical maximum speed. Several variables affect the speed factor: the weight-to-horsepower ratio for the hauling unit, the length of the road section, and the initial or end speed of the hauler. The weight-to-power ratio is determined as follows:

If the hauler is loaded:

$$\text{Weight-to-power ratio} = \frac{\text{gross vehicle weight}}{\text{flywheel or net horsepower}}$$

If the hauler is empty:

$$\text{Weight-to-power ratio} = \frac{\text{net vehicle weight}}{\text{flywheel or net horsepower}}$$

A hauler with a higher weight-to-power ratio will have, relatively, less power available to the drive wheels to overcome the resistance of the grade and the haul-road surface than will a unit with a lower weight-to-power ratio. Consequently, the higher a hauling unit's weight-to-power ratio, the longer it will take to build up to its maximum speed and the lower will be its average speed over a given haul-road section. While acceleration and momentum affect only the travel speeds at the beginning and end of a given road section, the length of the road section also is important. The longer the section, the more time will be spent at maximum speed and the nearer to 1.00 will be the speed factor. A hauler which enters or leaves a road section at or near the maximum speed will have a speed factor closer to 1.00 than a unit which starts from stop or must slow down at the end of the road section.

The effect of a vehicles momentum is evident when a hauler enters a road section which has a different grade than the previous section. If the vehicle enters a grade traveling at a faster speed than its theoretical maximum speed on the grade, then its average speed on that grade will be greater than the maximum speed (as determined from the performance charts) and the speed factor will be greater than 1.00. Likewise, if the hauler enters a grade traveling at a slower speed than the maximum attainable speed on that grade, then the average speed on the new grade will be less than the maximum speed and the speed factor will be less than 1.00. These effects are due to the momentum that the hauler accumulates on the previous road section.

Speed factors for varying road lengths and conditions are supplied in Tables B7-7 and B7-8. When the length of a road section and the average travel speed are known, the travel time can be obtained directly from Table B7-9.

HAULING UNITS

TABLE B7-7 Factors for Conversion of Maximum Speed to Average Speed—Unit Starting from Stop

Haul-road length, ft	Downgrade* 1	2	3	Level* 1	2	3	Upgrade* 1	2	3
0–250	0.45	0.40	0.35	0.38	0.34	0.30	0.30	0.25	0.20
251–500	0.54	0.48	0.42	0.44	0.40	0.36	0.36	0.31	0.26
501–750	0.63	0.56	0.50	0.50	0.46	0.42	0.42	0.37	0.32
751–1000	0.72	0.64	0.56	0.56	0.52	0.48	0.47	0.44	0.38
1001–1250	0.80	0.72	0.62	0.62	0.58	0.54	0.55	0.50	0.44
1251–1500	0.87	0.80	0.70	0.68	0.64	0.60	0.60	0.55	0.50
1501–1750	0.94	0.85	0.75	0.75	0.70	0.65	0.65	0.60	0.55
1751–2000	1.00	0.90	0.80	0.85	0.80	0.70	0.75	0.70	0.60

* Weight-to-power ratios: 1 = 300 lb/hp and under; 2 = 301–400 lb/hp; 3 = 401 lb/hp and over.

TABLE B7-8 Factors for Conversion of Maximum Speed to Average Speed—Unit in Motion When Entering Road Section

Haul-road length, ft	Downgrade* 1	2	3	Level* 1	2	3	Upgrade* 1	2	3
0–250	0.70	0.62	0.54	0.65	0.58	0.50	0.60	0.54	0.46
251–500	0.80	0.70	0.60	0.75	0.66	0.56	0.68	0.62	0.52
501–750	0.90	0.78	0.66	0.85	0.74	0.62	0.75	0.70	0.58
751–1000	1.00	0.86	0.72	0.95	0.82	0.68	0.85	0.78	0.64
1001–1250	1.10	0.94	0.78	1.05	0.90	0.74	0.95	0.86	0.70
1251–1500	1.20	1.00	0.84	1.15	0.98	0.80	1.00	0.95	0.76
1501–1750	1.30	1.10	0.90	1.25	1.05	0.85	1.05	1.00	0.80
1751–2000	1.40	1.20	0.96	1.35	1.15	0.90	1.10	1.05	0.84

* Weight-to-power ratios: 1 = 300 lb/hp and under; 2 = 301–400 lb/hp; 3 = 401 lb/hp and over.

Turn-and-Dump Times

The time required to dump the payload and to maneuver the hauling unit into position for the return route is dependent on the destination of the material, the type and maneuverability of the hauling unit, the type and condition of the material, and the road maintenance in the dumping area. The prediction of conditions at the dumping area is based on the estimator's personal experience and familiarity with either the hauling unit or the job. Representative times are listed in Table B7-10.

Spot Time

The time for the hauling unit to be spotted at the loading unit in a position ready for loading is dependent on the type of approach (straight ahead or back-in), the absence or presence of a spotter to guide the driver, and the time interval between hauling units. Typical spot times are supplied in Table B7-11.

TABLE B7-9 Conversion of Travel Distance (ft) and Average Travel Speed (mi/h) to Travel Time (min)

Speed, mi/h	Travel distance, ft									
	100	200	300	400	500	600	700	800	900	1000
3	0.379	0.757	1.136	1.515	1.893	2.272	2.652	3.030	3.409	3.788
4	0.284	0.568	0.853	1.136	1.420	1.705	1.989	2.273	2.557	2.841
5	0.227	0.454	0.681	0.908	1.136	1.363	1.591	1.818	2.045	2.273
6	0.189	0.378	0.568	0.757	0.946	1.136	1.325	1.515	1.705	1.894
7	0.162	0.324	0.487	0.649	0.811	0.974	1.136	1.299	1.461	1.623
8	0.142	0.284	0.426	0.568	0.710	0.852	0.994	1.136	1.278	1.420
9	0.126	0.252	0.378	0.505	0.631	0.757	0.883	1.010	1.136	1.263
10	0.113	0.227	0.341	0.454	0.568	0.681	0.795	0.909	1.023	1.136
12.5	0.091	0.182	0.273	0.363	0.454	0.545	0.636	0.727	0.818	0.909
15	0.075	0.151	0.227	0.303	0.378	0.454	0.530	0.605	0.681	0.757
17.5	0.065	0.129	0.194	0.259	0.324	0.389	0.454	0.519	0.584	0.649
20	0.057	0.113	0.170	0.227	0.284	0.341	0.397	0.454	0.511	0.568
22.5	0.050	0.101	0.151	0.202	0.252	0.303	0.353	0.404	0.454	0.505
25	0.045	0.090	0.136	0.181	0.227	0.272	0.317	0.363	0.408	0.454
27.5	0.041	0.082	0.124	0.165	0.206	0.248	0.289	0.330	0.371	0.412
30	0.038	0.076	0.113	0.151	0.189	0.227	0.265	0.303	0.341	0.379
32.5	0.035	0.070	0.104	0.139	0.174	0.209	0.244	0.279	0.314	0.349
35	0.032	0.065	0.097	0.129	0.162	0.194	0.227	0.259	0.291	0.324

TABLE B7-10 Typical Turn-and-Dump Times (min)

Operating conditions	Rear dumps	Bottom dumps	Side dumps
Favorable	1.0	0.4	0.7
Average	1.3	0.7	1.0
Unfavorable	1.5–2.0	1.0–1.5	1.5–2.0

TABLE B7-11 Typical Spot Times (min)

Operating conditions	Rear dumps	Bottom dumps	Side dumps
Favorable	0.15	0.15	0.15
Average	0.30	0.50	0.50
Unfavorable	0.80	1.00	1.00

Delay Factors and Job Efficiency

There are frequently found to be hazards or obstructions in the haul road for which time allowances should be made when computing the cycle time. In addition, conditions resulting in "'continuous" delays may exist, which again increases cycle time. Typical items which initiate intermittent or continuous job delays, thereby reducing cycle efficiency, are identified in Table B7-12.

Obviously, an estimate which is too optimistic in predicting the hourly production ability of each hauling unit will prove to be unrealistic. It will be found that the specified production cannot be maintained and that the number of haulers assigned to

TABLE B7-12 Typical Delay Factors

Intermittent factors	Continuous factors
(Estimate delay time for each item)	(Estimate delay time over the entire haul and return route)
One-way haul roads	Extremely high, or variable, rolling resistance
Delay at passing points	
Multiple curves or switchbacks	Wet or slippery haul roads
Blind corners	Unskilled operators
Bridges	Long, downgrade hauls
Underpasses	Inexperienced management and/or supervision
Railroad crossings	
Cross traffic	

the job is insufficient. An estimate must be based on a realistic evaluation of the sustained or average hauling production over a long period of time.

The *effective* working hour is a simple device which makes for the unavoidable delays which are encountered on all operations. The 50-min hour, which assumes that delays will average 10 min per clock hour, is frequently used as a basis for estimating. This is equivalent to an 83 percent overall job efficiency. The hauling cycle time, computed as explained earlier in this section and expressed in minutes, is then divided into 50 min to determine the average number of trips which will be completed each clock hour. An assumption of a 50-min hour is probably adequate for average conditions. For those jobs considered to be *favorable* from the standpoint of possible delays, a 55-min hour may be used, which is equivalent to a 92 percent overall efficiency. Conversely, projects which are subject to unusual delay and efficiency problems should be estimated on the basis of a 45- or even a 40-min working hour, equivalent to overall efficiencies of 75 or 67 percent.

CHAPTER 8
SOIL SURFACING EQUIPMENT*

James J. O'Brien, P.E.
O'Brien-Kreitzberg, Inc., Pennsauken, New Jersey

SPREADING EQUIPMENT

General

The art of soil spreading can be divided into two basic types: spreading the earth fill in embankment work and spreading highly graded granular materials on base course and shoulder work. Although there may be some resemblance between the two operations, they are actually separate and distinct. Each has its own procedures and problems.

On a typical embankment operation, earth fill is brought in from the borrow area in bulk haul units such as scrapers or rear- or bottom-dump trucks. In most cases this haul dirt must be spread to a uniform and controlled lift thickness. In this way it becomes possible to achieve a high uniform density during the compaction operation.

In base course or shoulder spreading, the granular soil is brought in, usually in rear-dump units, and placed on the finish subgrade. Base courses are generally placed in one or more lifts, each of which is compacted to the specified requirements. After compacting, final exact trimming removes all the excess material. Careful control is required in spreading each lift of base course, both to ensure compaction uniformity and to minimize the quantity of material which must subsequently be removed during the trimming operation.

Equipment Types

The category of soil-spreading equipment encompasses many varieties of construction equipment. The more commonly used types for each of the two areas of soil spreading are as follows:

---

*Some of the material in this chapter is from E. J. Haker and C. F. Riddle, "Soil Surfacing Equipment," in Stubbs and Havers, *Standard Handbook of Heavy Construction,* 2d ed., McGraw-Hill, New York, 1971.

Earth fill spreading:
  Bulldozers
  Scrapers (pan and elevating)
  Motor graders
  Large units
Base course and shoulder spreading:
  Motor graders
  Tractor-mounted box spreaders
  Automatic trimmer type of equipment

Bulldozers. Crawler tractors with rugged, front-mounted blades have for years been the workhorses in earth-fill spreading. They can also provide secondary services on the fill as push help for unloading haul vehicles. Bulldozers are manufactured in a wide range of sizes from 5000 to 218,000 lb gross weight and from 42 to 770 hp. (See Chap. B3.)

Scrapers. Although basically haul units, scrapers can be generally considered to have a spreading function on the fill. Scrapers are manufactured in both the self-propelled and towed versions. The self-propelled version is available in two styles, the pan type and the elevating type. (See Part B, Chaps. 3 and 4.)

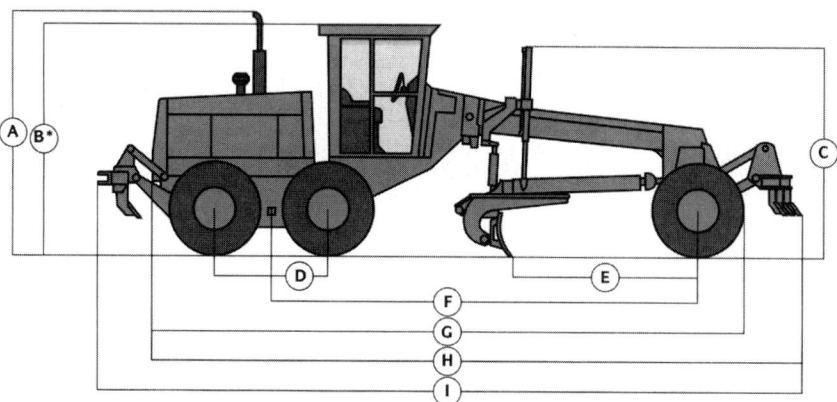

Key:
A Height to top of exhaust ....10 ft. 10 in. (3.30 m)
B Height to top of cab ....10 ft. 1.5 in. (3.09 m)
C Height to top of blade lift cylinders ....9 ft. 8 in. (2.95 m)
D Tandem axle spacing ....5 ft. 0.7 in. (1.54 m)
E Bladebase ....8 ft. 9 in. (2.67 m)
F Wheelbase ....19 ft. 7 in. (5.97 m)
G Overall length ....27 ft. 11 in. (8.51 m)
H Overall length with scarifier ....30 ft. 3 in. (9.22 m)
I Overall length with scarifier and ripper ....32 ft. 7 in. (9.93 m)
*Add 8.3 in. (210 mm) for full-height cab
 Add 1.0 in. (25.5 mm) for cab with air conditioning
 Add 0 in. (0 mm) for low profile canopy with ROPS

FIGURE B8-1 Dimensions, John Deere 772B-H motor grader. *(John Deere Co.)*

SOIL SURFACING EQUIPMENT   B8-3

FIGURE B8-2 Front scarifier and rear ripper. *(John Deere Co.)*

FIGURE B8-3 Spreading by motor grader. *(John Deere Co.)*

Motor Graders.   The most versatile tool in the category of spreading equipment is unquestionably the motor grader illustrated in Fig. B8-1. This unit is capable of effectively spreading most types of soils in addition to performing many other duties. In addition to the blade, a scarifier or dozer blade can be mounted on the front. A scarifier or ripper can be mounted on the back (Fig. B8-2). Originally developed some one hundred-odd years ago, it has consistently maintained its popularity and improved its performance level through the years. In addition to spreading (Fig. B8-3), the motor grader can handle the following operations effectively:

1. Fine finish grading
2. Ditching of all types (Fig. B8-4)
3. Haul-road maintenance (earth moving and logging) (Fig. B8-5)
4. Ripping and scarifying
5. Bank cutting
6. Snow removal (Fig. B8-6)

FIGURE B8-4 Fine grading next to gutter. *(John Deere Co.)*

FIGURE B8-5 Haul road maintenance. *(John Deere Co.)*

The John Deere B series motor graders (Table B8-1) has eight forward and four reverse transmission speeds. These are specifically designed for motor grader applications to make it possible to match ground speed to the task—even when r/min is high to maintain the hydraulic action. The direct-drive power shift transmission puts the power right in the wheels. Five transmission speeds below 9 mi/h (14.5 km/h) can move through heavy loads at a constant rate. In addition, heavy-duty inboard-mounted planetary gears evenly distribute shock loads to the axles. And there is a standard, operator-controlled "on-the-go" differential lock/unlock to get maximum traction and tighter turns when needed.

SOIL SURFACING EQUIPMENT

FIGURE B8-6 Snow removal using optional front blade. *(John Deere Co.)*

TABLE B8-1 John Deere Motor Graders

Model	Horsepower	Gross weight, lb	Max speed, mi/h
570B	90	20,225	22
670B	135	29,030	24
672B	135	30,230	24
770B	155	31,250	25
770B-H	155–185	31,300	25
772B-H	155–185	32,500	25

Source: John Deere Co.

The two B series dual horsepower graders (770B-H and 772B-H) are designed to provide high levels of performance at maximum operational efficiency. Power levels increase automatically, when shifted into the fourth-gear-and-above range. There is 19 percent more power to meet higher speed applications. That means 155 net hp (116 kW) in gears 1 to 3, 185 net hp (138 kW) in gears 4 to 8.

The B series has automatic all-wheel drive (patented). The 15-position control dial lets the operator adjust the speed ratio between front and rear wheels over a 7 percent range.

Table B8-2 lists characteristics of seven Caterpillar Company motor graders.

Production

Several procedures have been advanced for estimating the production from a piece of spreading equipment. Among these are formulas that pertain to the motor grader. One of the best accepted of these formulas is the following:*

---

* Caterpillar Tractor Company.

TABLE B8-2 Caterpillar Motor Graders

Model	Horsepower	Gross weight, lb	Max speed, mi/h	Blade, ft
120G	125	28,350	25	12
130G	135	28,770	24	12
12G	135	29,860	24	12
140G	150	31,090	26	12
140G AWD	180	32,880	26	12
14G	200	45,610	27	14
16G	275	60,150	27	16

Source: Caterpillar Co.

$$\text{Time to complete a job (h)} = \frac{\text{no. of passes} \times \text{distance in miles}}{\text{avg. speed in mi/h} \times \text{efficiency factor}}$$

The effects of variations in operator performance and in the type of soil can usually be compensated for in the efficiency factor. This factor has to be estimated on the basis of good judgment, but it should generally fall into the range from 0.70 to 0.90. Production estimating for other types of spreading equipment should be handled similarly, using the preceding formula as a general guide.

STABILIZATION EQUIPMENT

General

Soil stabilization, perhaps one of the least familiar aspects of highway and airfield construction, is considered to be one of the more important steps. Stabilization, as the name implies, is intended to make the soil structure more stable or less subject to change from outside influences. Mechanical stabilization can be accomplished by equipment which mixes the soil, using one or more rotors, to produce a more homogeneous material. In its simplest form, stabilization can consist merely of blending the concentrations of sand and aggregate which are naturally present in a soil so as to obtain a more homogeneous state. In the more practical usage, however, stabilization involves the mixing of an additive with the soil before blending the combination into a uniform mixture. This additive can be a complementary type of soil or it can be a manufactured agent such as portland cement, asphalt, or lime. Table B8-3 lists the more commonly used stabilization agents.

Stabilized soils are used in the base, subbase, and embankment layers. In order to carry traffic, they must be covered with either an asphaltic or concrete surface course to prevent abrading. The minimum surface treatment should be an asphalt-chip seal coat.

Equipment Types

The equipment which is used to stabilize soils is of two basic types: the in-place mobile mixer and the central-mix plant. The in-place mixer can be further divided into three separate categories: the multiple-pass units, the single-pass units, and the road-mix units.

TABLE B8-3 Common Stabilizing Agents

Agent or additive	Form used	Suitable soils	Reaction with soil	Construction area
Asphalt	Cutback liquid or aqueous emulsion	Sandy and granular types	Physical binder	Base course
Lime	Fine powder or aqueous slurry	Plastic (plasticity index higher than 10)	Chemical reaction to reduce plasticity, also a physical binder	Subbase and subgrade
Portland cement	Fine powder	Granular (plasticity index lower than 10)	Physical cementation	Base course
Salts (calcium chloride, sodium chloride)	Granules	Sands and gravels	Moisture retention	Base course

Multiple-Pass Stabilizer. This is the most popular type of stabilization equipment in use today. It consists basically of a carrier whose large engine drives an attached rotor. This rotor is rotating within an enclosure called a *hood* or *mixing chamber*. (See Fig. B8-7.)

The carrier consists of a frame mounted on four rubber-tired wheels, with a single engine providing power for both the rotor drive and propulsion. The rotor drive is mechanical through a two- or three-speed transmission. Hydrostatic power transmission is preferred for the propulsion drive, since it provides an infinite range of working speeds as well as dynamic braking control.

The rotor consists of a series of connected disks. These hold the cutting teeth or tines at their outer periphery and generally rotate in a forward down direction at speeds of from 100 to 250 r/min. The cutting teeth or tines are mostly of the quick-change variety to minimize the downtime when replacing a worn set. The rotor, which can be raised or lowered, is covered on all sides except the front with a steel-plate enclosure called the hood or mixing chamber. The rear surface is slanted and has a hinged lower portion called a *tailboard*, which regulates the release of the mixed soil. (See Fig. B8-8.)

A fluid system is available on all units to facilitate the introduction of a liquid stabilizing additive such as water, asphalt, or lime slurry. This fluid system basically consists of a pump driven by a separate engine, a meter, a spray bar mounted on the hood, hoses connecting these components, and a means of regulating the flow of the additive.

Central-Plant Stabilization. Soil stabilization is also accomplished in a central mixing plant which usually is located at the source of the soil or aggregate to be used. The central mix method, popular in some areas of the country, has been largely confined to portland-cement stabilization and, to a smaller degree, asphalt stabilization. It has been estimated that as much as 40 percent of the soil cement used in the United States is being mixed in a central plant.

Two basic types of central-mix plants are in use today: the batch type and the continuous-flow type. The continuous-flow plant appears to be the more popular,

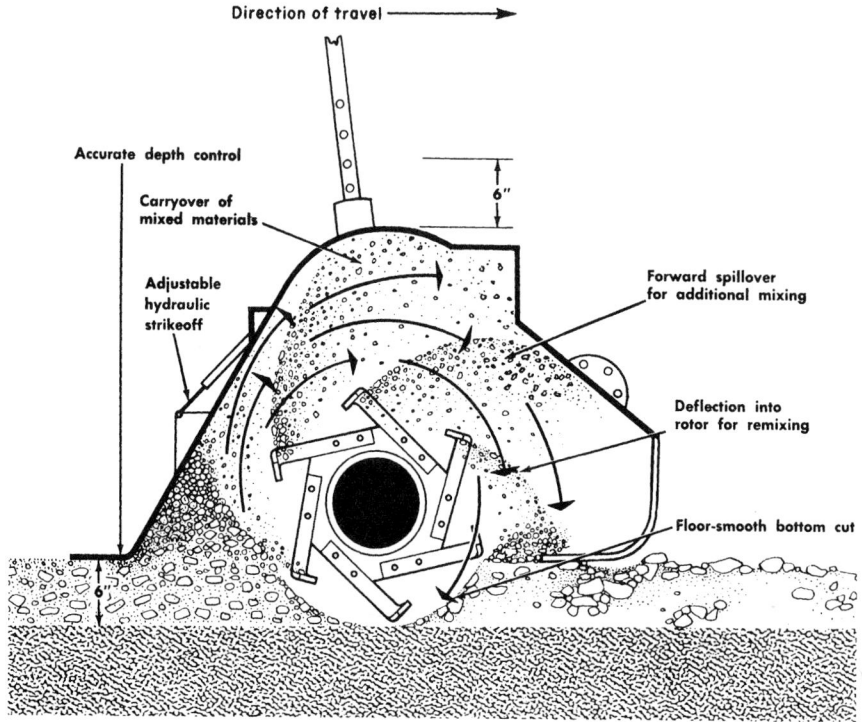

FIGURE B8-7 Rotor and hood mixing action (multiple-pass stabilizer).

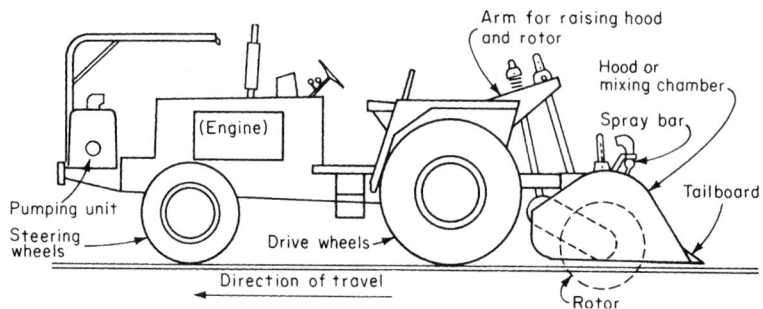

FIGURE B8-8 Typical multiple-pass stabilizer.

due to its higher production capabilities. Production of an average continuous-flow plant is about 400 to 600 tph, with the largest producing about 1000 tph.

The typical central mix plant includes a conveyor with a loading hopper, a cement storage tank with an accurate feed mechanism, and a pug mill with a water supply. A front-end loader or clam bucket is needed to keep the hopper loaded with the soil or aggregate, to ensure even distribution on the conveyor. The cement feed is timed to the conveyor speed. Cement is added to the material on the conveyor and the water is added in the pug mill. Figure B8-9 illustrates a typical central-mix plant.

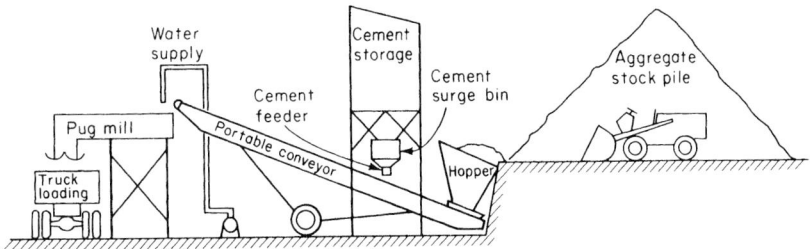

FIGURE B8-9 Typical central-mix plant (soil cement).

In the continuous-flow plant, the pug mill has a holding hopper to facilitate more continuous production of the plant.

Haul times from the central-mix plant to the job site can be critical, especially with soil cement, and 30 to 45 min is considered to be the limit. An open dump type of truck with a canvas cover is generally used as a haul vehicle. After the stabilized soil has arrived at the job site, it is quickly spread and compacted.

Equipment Selection

Several factors are involved in the selection of the best type of equipment for the job:

1. Type of job
2. Size of job
3. Contractor economics
4. Specifications of governing agency

General guidelines when considering these factors are as follows.

Type of Job. Individual considerations here include type of soil, type of additive, equipment function, depth of cut or mixing, and soil location. These considerations are discussed here, together with a general analysis of each.

   Type of Soil: Soil can vary from tough-cutting, high-plasticity-index (PI) clay to easily mixed sandy soil. Clay generally requires a multiple-pass stabilizer, preferably a larger, higher-horsepower unit. Easily mixed sandy soils can be handled by most types of equipment, including the road mix and central plant.

   Type of Additive:
      Portland cement, asphalts: Most types of equipment. Central-plant and road-mix units perform well.

      Lime: Generally a multiple-pass stabilizer, preferably the larger sizes.

      Salts: In-place stabilizers generally used.

   Equipment Function: This refers to whether a pulverizing or mixing operation is involved.

      Pulverizing: General a multiple-pass unit.

      Mixing: Most types of equipment. Central-plant and road-mix units perform well.

   Depth of Cut: This generally determines size of unit, as the larger units tend to cut deeper.

Soil Location: Soil location determines whether in-place mixing or a central plant will be used. If soil must be hauled in, some consideration should be given to central-plant mixing.

Size of Job. Although the size of the job and production expected per hour should provide some control of the type of equipment selected, other factors far outweigh this one. Production can generally be increased by adding to the fleet.

Contractor Economics. This important factor generally refers to the capital investment that a contractor should have in stabilization equipment. Initial cost and expected production must be weighed here, and a contractor's previous experience is perhaps the strongest influence. For a contractor first entering the stabilizing field, the selection of a standard size of multiple-pass stabilizer is generally the safest way to start. Experience gained with this relatively inexpensive unit can better help to further evaluate the purchase of the larger equipment in the future.

COMPACTION EQUIPMENT

General

Compaction might be considered the most important phase in the construction of any soil structure. If the compaction is inadequate this is generally reflected in the failure of any subsequent construction.

Compaction can be defined as the process of densifying or increasing the unit weight of a soil mass through the application of static or dynamic forces, with the resulting expulsion of air and, in some cases, moisture. The effects of compaction on the soil are twofold. First, it reduces future settlement by rearranging the particles to form a more compact soil mass. Second, it builds strength into the soil structure.

Static forces are considered to be those which are produced by a roller or compactor which accomplishes a soil densification primarily by its heavy weight. Dynamic forces are those which utilize a combination of weight and energy to produce a vibratory or tamping effect on the soil. Compactors utilize both static and dynamic forces to achieve the required soil density.

Equipment Types

Today's compaction equipment is represented by many large, highly responsive and versatile self-propelled units. These can produce a high degree of compaction with only a minimum of equipment breakdown problems. Compactors and rollers can be divided into four major classes: rubber tired, segmented pad or tamping foot, vibratory, and steel wheel. Each of these classes can be further subdivided into two or more type or size classifications. As an example, rubber-tired rollers are being manufactured in three distinct size categories, with tire diameter being the primary identification.

Rubber-Tired Compactor. This is frequently referred to as a *pneumatic* compactor, and has been called the *universal* compactor. The latter name may exaggerate the facts, but it suggests the broad field of usage for rubber-tired compactors. (See Fig. B8-10.) Rubber-tired compactors have achieved excellent results in almost

FIGURE B8-10 Medium-size (21-ton) rubber-tired compactor. *(Bros Division—American Hoist.)*

every type of compaction. They first appeared in the early 1930s as towed wagons with a rock-filled box, and they have now graduated to the level where they are being manufactured in distinctly different tire sizes.

The compaction tools are the rubber tires, and these are specially designed to provide a specific ground-contact pressure (GCP) on the soil. The magnitude of GCP is controlled by both wheel loading and tire inflation pressure. A national standardization of GCP values was achieved early in 1967, and each currently manufactured model of rubber-tired compactor will carry a decal which identifies its maximum allowable wheel load and the ground-contact pressures which can be attained with the various combinations of wheel loadings and tire inflation pressures. Figure B8-11 shows a typical decal covering the 7.50-×-15 tires.

Rubber-tired compactors have seen considerable usage in rolling asphaltic concrete surfaces. Here they perform the function of an intermediate roller, immediately following the breakdown steel-wheel roller. In most instances, the rubber-tired roller is followed by finish-steel-wheel rolling. When used as an intermediate roller, the rubber-tired roller contributes to compaction uniformity and to the orientation of aggregate particles in their most stable positions.

Rubber-tired compactors are designed to provide a uniform coverage and equal wheel loading even on rougher terrain. Uniform coverage is possible because the rear wheels are staggered to roll the spaces left by the front wheels. Equal wheel loading on uneven terrain is provided by means of their articulated axles.

The feature of changing tire inflation pressures on the go provides a very direct means of varying the ground-contact pressures; the higher the tire pressure, the higher the ground-contact pressures, assuming the ballasted weight has remained constant.

Two theories have been advanced to attempt to explain soil behavior beneath a loaded contact surface such as a rubber tire. These theories are referred to as the *bulb concept* and the *cone concept*. (See Fig. B8-12).

CERTIFIED MAXIMUM GROUND CONTACT PRESSURES
ISSUED BY
BITUMINOUS EQUIPMENT MANUFACTURERS BUREAU
UNDER THE SPONSORSHIP OF
CONSTRUCTION INDUSTRY CIMA MANUFACTURERS ASSOCIATION
135 S. LaSalle St., Chicago, Illinois 60603
FOR 7.50 × 15 SMOOTH TREAD COMPACTOR TIRES

TIRE PLY		4 PLY	6 PLY			10 PLY					12 PLY						14 PLY						
TIRE PRESSURE		35	35	50	60	35	50	60	70	90	35	50	60	70	90	110	35	50	60	70	90	110	120	130
WHEEL LOAD		GROUND CONTACT PRESSURES and CONTACT AREAS																						
1000	GCP	37	37	44	49	38	44	47	51	57	38	44	47	51	60	65	46	50	54	56	61	68	71	74
	CA	27	27	23	20	26	23	21	20	18	26	23	21	20	17	15	22	20	19	18	16	15	14	14
2000	GCP	43	43	50	55	46	52	56	60	67	46	53	56	60	69	75	54	59	62	65	72	78	82	86
	CA	47	47	40	36	43	38	36	33	30	43	38	36	33	29	27	37	34	32	31	28	26	24	23
2500	GCP	45	45	52	58	49	56	59	64	71	50	57	60	65	74	78	57	63	66	70	76	83	87	90
	CA	56	56	48	43	51	45	42	39	35	50	44	42	38	34	32	44	40	38	36	33	30	29	28
3000	GCP	47	47	55	61	53	60	65	67	75	53	60	64	69	77	83	60	66	70	73	80	87	91	94
	CA	64	64	55	49	57	50	46	45	40	57	50	47	43	39	36	50	45	43	41	38	34	33	32
3500	GCP			57	63		62	67	71	80		64	67	71	81	86		68	73	76	83	90	94	98
	CA			61	56		56	52	49	44		55	52	49	43	41		51	48	46	42	39	37	36
4000	GCP				65			68	73	82			70	75	84	89			75	79	86	94	98	101
	CA				62			59	55	49			57	53	48	45			53	51	47	43	41	40

GCP—Ground Contact Pressure
CA—Ground Contact Area

PERFORMANCE FIGURES HAVE BEEN APPROVED, SUBJECT TO TIRE MANUFACTURERS NORMAL TOLERANCE BY

MAXIMUM ALLOWABLE WHEEL LOAD THIS ROLLER 3500

GOODYEAR TIRE & RUBBER CO.
GOODRICH TIRE & RUBBER CO.
FIRESTONE TIRE & RUBBER CO.
U.S. RUBBER TIRE CO.
GENERAL TIRE CO.

FIGURE B8-11 Typical decal showing ground-contact pressures for rubber-tired compactors.

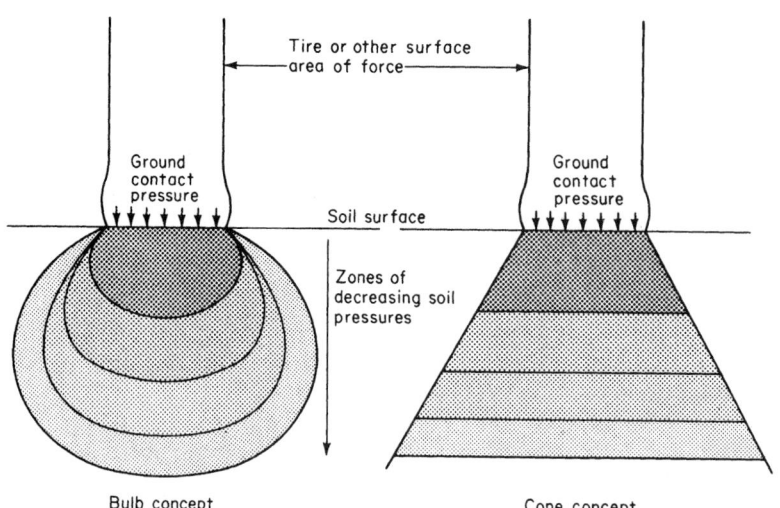

FIGURE B8-12 Theories of soil compaction.

Segmented-Pad or Tamping-Foot Roller. This self-propelled roller features the tamping pad or the tamping type of foot. The inherent dynamic beat developed by these tamping pads or feet can in many cases produce a high degree of compaction. (See Fig. B8-13.)

The need for these high-production earth-fill compactors has been generated by the increasing capacity of the hauling units. Factors such as increased haul yardage, higher speeds, and faster loading and unloading cycles have paced increased volumes in fill hauling.

FIGURE B8-13 Segmented-pad compactor (32,000 lb). *(Rex Chainbelt Inc.)*

The modern segmented-pad or tamping-foot compactor satisfied the need for faster compacting through its higher speeds, greater maneuverability, and increased pulling power. Its use reduces grade congestion, especially on the more crowded fill areas. Some units feature full-width compaction in one pass. A front-mounted blade can be attached to increase the versatility of the compactor.

The bulb and cone concepts as discussed under the rubber-tired compactors also apply to the segmented-pad or tamping-foot compactor. However, the pressures exerted by the pads or feet are of much greater magnitude than the pressures exerted by the rubber tires.

Walking out is the term normally associated with this type of compactor. It refers to a condition of deep penetration of the pads or feet on the first pass, with less penetration on each subsequent pass. The wheels in effect ride higher with each pass, hence, the term "walking out."

The tow-type sheepsfoot is increasingly being replaced by the large, self-propelled units. However, despite the fact that its popularity has decreased, the towed unit still fills a compaction need. With its slower speed and lower cost, it can provide economical compaction in the smaller types of fill areas.

Vibratory Compactors. These units combine static weight with a generated, cyclic type of force. The dynamic type of compaction is relatively new and really came into its own during World War II. It was introduced to satisfy the need for higher densities in granular types of soils.

B8-14 STANDARD HANDBOOK OF HEAVY CONSTRUCTION

FIGURE B8-14 Self-propelled vibratory compactor—roll type. *(Raygo Inc.)*

Two basic types of vibratory compactors are also available: the roll type and the plate type. The roll type, being the more popular, is manufactured in numerous styles and sizes. The plate type is generally available only in a self-propelled version. The roll types are available with either a smooth or a foot type of roll (Fig. B8-14).

Many units now offer a system which will indicate the vibrating amplitude of the roll, and this information is a guide to the vibration frequency at which the roller and the soil are in resonance. There is some evidence to indicate that vibration at this resonant point will produce faster compaction. The frequency can be varied in most vibratory compactors. Many vibratory rollers are also available in a suitable version to compact deep-lift asphalt bases, complete with spray system and smooth tires.

The generally accepted rating for vibratory compactors is *total applied force*. This is expressed in pounds or tons and is the numerical sum of the dynamic force plus the static weight of the vibrating surface on the soil.

A brief discussion of the basic terminology used in discussing vibratory compactors follows (see Fig. B8-15).

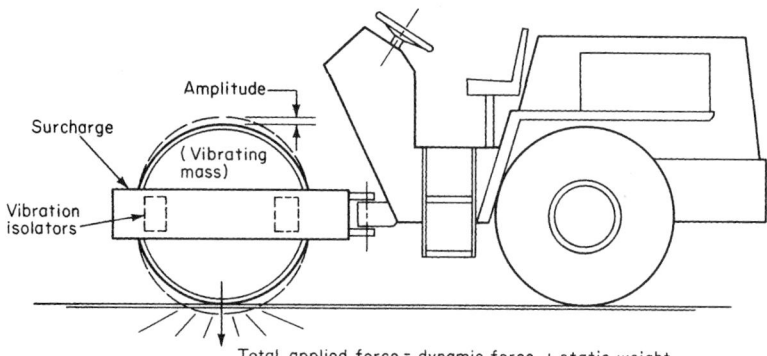

FIGURE B8-15 Vibratory compaction terminology.

1. *Vibrating mass:* The total entity that vibrates as a unit.
2. *Surcharge:* The nonvibrating mass that surrounds and is affixed to the vibrating mass through the isolators. As a general rule of thumb, the greater the surcharge weight, the greater the depth of compaction.
3. *Vibrating isolator:* The connection between the vibrating mass and the surcharge or nonvibrating members; generally of an elastic rubber type of compound.
4. *Frequency:* The number of vibrations in a unit of time (usually 1 min). Frequency generally relates to production; other factors being equal, the greater the frequency, the greater the production.
5. *Dynamic force:* The force generated, usually by centrifugal means, and generally described in pounds or tons.
6. *Amplitude:* The vertical movement of the vibrating mass from its position at rest to a maximum in one direction. Total vertical movement is actually double amplitude or displacement. Amplitude is not controlled by the vibratory compactor alone. It is the result of a vibrating system composed of the compactor's vibrating mass and the soil being compacted. Soil conditions such as type, moisture content, and degree of compaction affect amplitude considerably.

Steel-Wheel Rollers. As their name implies, these are rollers in addition to being compactors. They are designed primarily for pavement course rolling, and their rolls are machined to provide a smooth, concentric surface.

There are two basic styles of steel-wheel rollers: the tandem with two rollers of equal width, and the three-wheel with one wide guide roll and two narrow drive rolls. There is also a unique version of the tandem, called the *three-axle tandem*, which is similar to the tandem in appearance except that it has two guide rolls on a walking-beam type of arrangement. The three-axle tandem is generally the heaviest of the steel-wheel rollers in total ballasted tonnage. The walking beams allow the forces exerted by the guide rolls to be varied.

Features in steel-wheel rollers include hydrostatic drive, hydrostatic steering complete with steering wheels, a pressure-type fog-spray system for wetting the rolls, and self-contained portability wheels on the 4- to 6-ton tandems to facilitate towing without the use of a trailer. Hydrostatic drive provides infinitely variable speed control together with dynamic braking action. (See Fig. B8-16).

FIGURE B8-16 Tandem steel-wheel roller. *(Rex Chainbelt Inc.)*

Equipment Selection

Soil, the relatively thin crust covering the surface of our planet, actually provides the reason for compaction. Naturally occurring soil does not normally possess the strength required to support heavy loads, but this strength can frequently be built into the native soils through compaction. As vehicles become larger and heavier, soil strength requirements also become greater. In order to move these vehicles, it becomes necessary to construct highways and airfields with adequately strong foundations. The strength required in these foundations is built by means of compaction.

Soils suitable for highway or airfield construction are actually of two basic types: the granular or nonplastic soils exemplified by sands and gravels, and the fine-grained or plastic soils such as the clays. Naturally occurring soils are generally mixtures of these two types, resulting in an almost endless variety of soils. Each of these must be analyzed for the best methods of compaction. To simplify the identification and classification of these numerous soil mixtures, various agencies concerned with highway and airfield construction have developed soil classification systems. These systems are in general agreement and classify the granular-type soils as best and the plastic-type soils as poorest for construction.

The type of soil determines to a large extent the type of compactor that should be selected for efficient compaction. Other factors such as machine availability and soil moisture are also important, but the type of soil appears to be the biggest factor. Figure B8-17 shows a rule-of-thumb relationship between type of compactor to be used and type of soil. Each of the four types of compactors also has its own individual characteristics, including the compactor rating method applied to each type of compactor. (Table B8-4.)

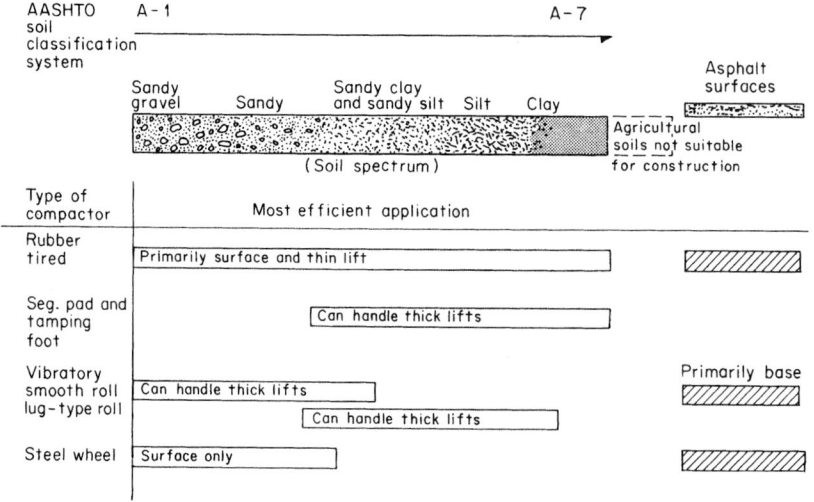

FIGURE B8-17 Compactor suitability.

TABLE B8-4 Compactor Rating Methods

Type of compactor	Method of rating	Terms used
Rubber-tired	Tire pressure on soil surface per unit area	Ground contact pressure (GCP) expressed in lb/in^2
Segmented pad and tamping foot	Pad or foot pressure on soil surface per unit area	
Vibratory	Total force applied to soil	Total dynamic force generated plus static weight on roll expressed in lb
	Frequency	Expressed in cycles or vibrations/per minute
	Dynamic force	Force generated expressed in lb
Steel wheel	Roll pressure on pavement surface per unit of roll width	Expressed in lb/in (lineal) of roll width (pli)

Another significant factor which provided stimulus for the new automatic trimmers is the steadily decreasing number of highly skilled motor-grader operators. Each year there is a greater shortage of operators who can precisely control the motor grader blade and thus achieve the necessary high degree of trimming accuracy.

CHAPTER 9
BELT CONVEYORS*

James J. O'Brien, P.E.
O'Brien-Kreitzberg, Inc., Pennsauken, New Jersey

INTRODUCTION

The principal advantages of a belt-conveyor system over other means of overland haulage are its low maintenance, ability to cross adverse terrain, low labor requirement, high reliability, and excellent safety record. When the total tonnage to be carried is sufficient to justify the higher initial investment, a belt-conveyor system nearly always results in the lowest overall per-ton-mile cost.

As early as the 1920s, long belt-conveyor systems were successfully used for earth movement. Five million cubic yards of dirt were moved for more than a mile from Denny Hill, through the business district in Seattle, onto barges. This was followed over a period of several decades by a series of large earth and dam projects.

Modern advances in technology have greatly enhanced the capabilities of belt-conveyor systems. The use of belt with a steel-cable core has made possible single-flight center distances of over 15,000 ft. High speeds (1200 fpm) and wide belts (120 in) have capacities to over 20,000 tons per hour (tph). Sophisticated controls and devices such as closed-circuit TV have drastically reduced operating costs and, at the same time, have increased systems reliability. For the operation of the systems, electronic systems provide efficient, indispensable tools. Developments in related systems and components, such as portable crushers, screens, etc., have also been instrumental in making the belt-conveyor system more versatile and adaptable to heavy construction.

BELT-CONVEYOR SYSTEMS AND SUBSYSTEMS

To attain optimum utilization, a belt conveyor must be viewed as an integral part of an overall system. Only after the objective and function of an entire project are clearly analyzed and defined can the most suitable and economical belt conveyor or belt-conveyor system be selected. The adequacy of a belt-conveyor system, including

---

* Some of the material in this chapter is from A. T. Yu, "Belt Conveyors," in Stubbs and Havers, *Standard Handbook of Heavy Construction,* 2d ed., McGraw-Hill, New York, 1971.

its subsystems and components, must be weighed in terms of the overall system which it serves and in terms of today's as well as tomorrow's needs. For example, a 36-in belt system may require considerably less investment than an equivalent 42-in network and will probably operate at a lower per-ton cost. Yet the equivalent 42-in system may be the more desirable choice if its use could result in substantially faster completion of the project and/or savings in the number of required shifts in the quarry, leading to lower overall project and unit costs.

A typical belt-conveyor system used for an earth dam project is shown in Fig. B9-1. Here, many million yards of rock and earth are simultaneously required to top out a large earth dam and to be barged out for a causeway. Various grades of material from a nearby borrow pit are excavated, stockpiled, and then moved to the construction site via an overland belt-conveyor system. Reclaim conveyors at the site then route the respective materials to truck bins or to a barge loader.

In the system illustrated, while the proper selection and sizing of each of the individual components is important, the overriding consideration is still the most economical yard of earth or rock excavated and hauled to the destination. A costly, wide belt may not be necessary if the bulk of the material to be hauled is alluvial, and a crusher can easily handle the occasional large boulders. If space is a problem at the dam site, a larger stockpile near the excavating pit may be the solution. If the terrain is rugged and steep downhill grades are involved, an overland belt system not only will minimize investment but may even generate power for running other equipment.

This system may be divided into several subsystems according to the specific functions each performs; they are described as follows.

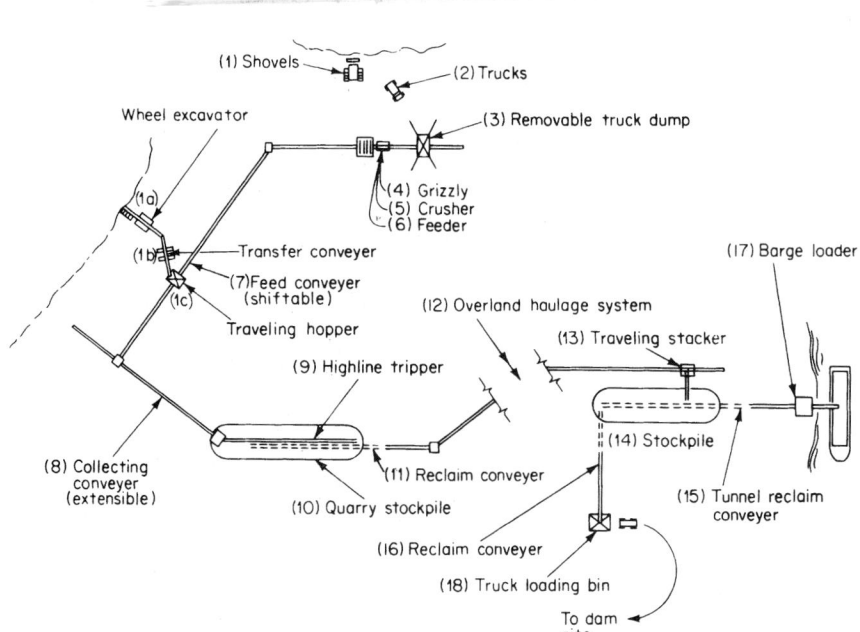

FIGURE B9-1 A typical belt-conveyor system in a construction project.

Plant Feed Systems

Since a belt conveyor is a continuous bulk-handling device, the system will attain its highest productivity and efficiency when it can be loaded uniformly at its maximum design rate. Thus the key to maximizing the utilization of a belt-conveyor system is its feed. A good plant system must be capable of adapting to the working cycles of equipment and converting intermittent random feed into steady, uniform flow.

In the earth-moving project example, alluvial or relatively easy digging material is dug by a bucket wheel excavator which is itself a continuous loader. Through a crawler-mounted portable transfer conveyor and a traveling receiving hopper, material is fed onto the feed conveyor nearly parallel to the face of the borrow pit. This conveyor should be of shiftable design to enable it to move sideways toward the pit with the advance of excavation.

For the rock zones, where large shot rock must be loaded by shovels, a different feed system is required. Here trucks are unloaded into portable or semiportable truck dumps, over a grizzly, prior to being crushed to sizes that are economical for belt-conveyor handling. Surge hoppers are provided when needed, and mechanical feeders receive the combined product to deposit it onto the long feed conveyor. The truck dump may be designed to accommodate rear- and bottom-, or side-dump trucks with greater than 100-ton capacity (see Fig. B9-2).

FIGURE B9-2 Removable truck dump in bolted construction is readily dismantled and reerected as excavation progresses.

Table B9-1 summarizes the major advantages and disadvantages of the most widely used types of feeders. Several of these feeders are illustrated in Fig. B9-3. On occasion, if the reclaim conveyor is relatively short, it can also serve as a feeder. It then receives material directly from the hopper bottom by gravity. This is possible only if the muck contains enough fines to cushion the impact of the large lumps. Even then, regulating the surge may be a problem.

Storage-Reclaim Systems

Since fluctuation in input and output of a system or a plant is unavoidable, a storage-reclaim system provides the flexibility and reserve needed for a continuous, smooth operation. It is insurance against breakdowns, strikes, weather, maintenance, and other unexpected contingencies.

How much insurance to provide depends on the requirements of the overall system. One must consider the normal rate of plant input and all the factors which might affect it plus the normally required rate of reclaim and the possible consequences if this is not maintained. The characteristics and types of material to be stored, the necessity for blending or shelter, and the most suitable equipment to use are additional factors. The cost of the storage itself should be included in this all-

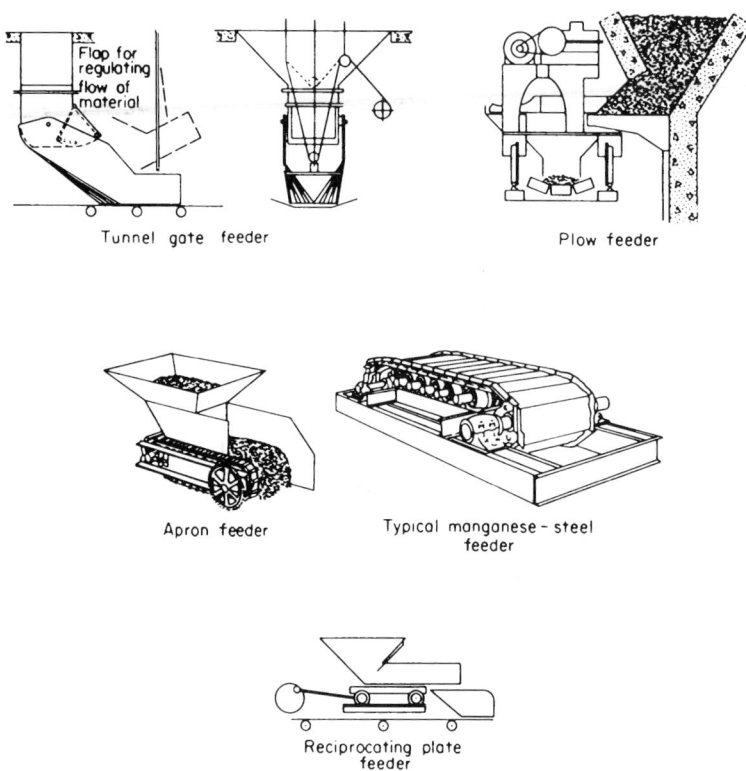

FIGURE B9-3 Some commonly used feeders.

TABLE B9-1 Feeders

Type	Major advantage	Disadvantage
Vibrating feeder (mechanical, magnetic, or pneumatically actuated)	Versatility for a variety of materials	Inclination adjustment may be difficult
Reciprocating plate feeder	Simplicity in operation and maintenance	Not readily adjustable
Apron feeder	Low cost, relatively rugged construction	Cleanup, maintenance
Manganese feeder	Rugged construction for heavy-duty service	Cleanup, maintenance, high cost
Belt feeder	Economical, simple, and clean operation	Not sufficiently rugged for hard lumps
Plow feeder	Positive, continuous flow	Relatively limited capacity—tunnel expense
Gate feeder	Extremely simple and economical operation	Difficult to regulate feed, poor wear resistance
Chain feeder	Simplicity	Regulation difficulty
Dozer trap-belt loaders	Mobility, simple operation	Difficulty in handling boulders

encompassing evaluation. Here one must consider the available space, terrain, soil and foundation requirements, water table, and other engineering factors.

When the quantity of material to be stored is relatively small, storage bins are often used. The major advantages of bin storage are cleanliness of operation and the ease of material reclaim. Simple gates under the bins are operated mechanically, pneumatically, or hydraulically and can feed trucks or rail cars directly and quickly. Prefabricated storage bins of up to 5000-yd^3 capacity are commercially available. For larger quantities, yard storage would normally be required.

The capacities of outdoor stockpiles can vary from several thousand tons to several million tons. Figure B9-4 gives the storage capacities of materials with different angles of repose. When a tunnel-gravity reclaim system is used, only the part of the stockpile reclaimable by gravity is considered *live storage* (see Fig. B9-5). The dead storage is normally regarded as a safety reserve to be used for emergencies.

For outdoor yard storage, if neither degradation nor dust is a problem, high-line trippers and shuttle conveyors may be used to form a long, rectangular stockpile. When dust or weather becomes a problem, the stockpile may be housed. Traveling stackers or stacker towers can build stockpiles of various shapes on the ground. A double-wing, track-mounted stacker (see Fig. B9-6) forms two longitudinal stockpiles, one on each side of the track. A track-mounted *radial* stacker builds a kidney-shaped pile. A rubber-tire mounted stacker may be over 100 ft in length. The crawler-mounted stacker, equipped with its own propelling drives, usually has the greatest flexibility. Throwers, or slingers, are occasionally installed at the end of the stackers to increase their reach. If dust is a problem, extendible telescopic chutes

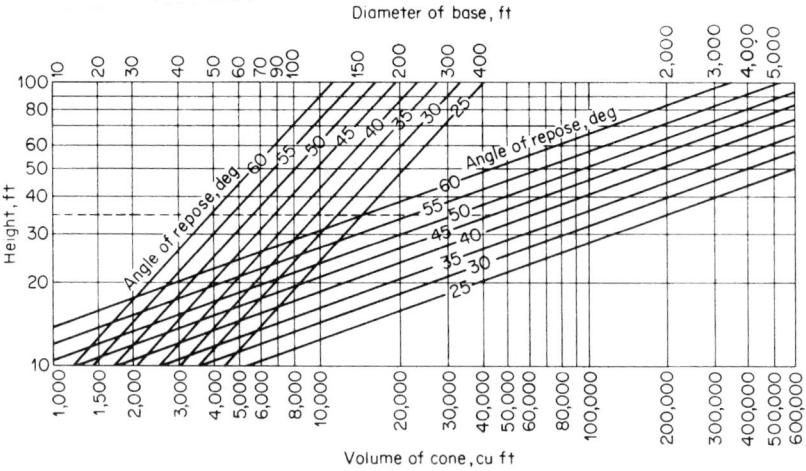

FIGURE B9-4 Capacities of conical stockpiles. Note: To find diameter of base and volume of a cone, enter chart at the left with height of cone. Follow across until reaching the first set of lines "Angle of repose" of the material to be stored; directly above, at the top of the chart, read the diameter of the base. Then resume reading along the height of the cone until intersecting with second set of lines "Angle of repose"; directly below, at the bottom of the chart, read volume (cubic feet) of the given cone. The heavy line example indicates that a cone 35 ft high, built up of material with a 45° angle of repose, will have a base diameter of 90 ft and a volume of 45,000 ft.^{2} The operations may be reversed to find height of a cone to give any required volume. Frustums of cones may also be calculated by finding the volume of the full cone and of the upper section, and subtracting the latter from the former.

BELT CONVEYORS B9-7

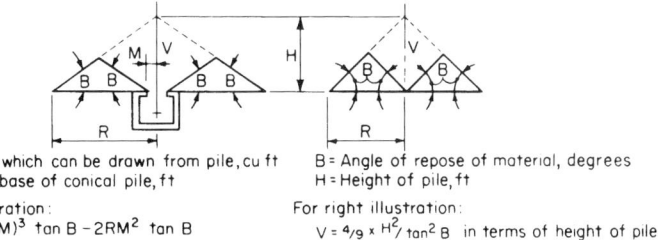

V = Volume which can be drawn from pile, cu ft
R = Radius, base of conical pile, ft

For left illustration:
$V = \frac{4}{9}(R+M)^3 \tan B - 2RM^2 \tan B$

B = Angle of repose of material, degrees
H = Height of pile, ft

For right illustration:
$V = \frac{4}{9} \times H^2 / \tan^2 B$ in terms of height of pile

For right illustration where M = 0 (formula sufficiently accurate for approximate calculations left hand illustration also) $V = \frac{4}{9} R^3 \tan B$

FIGURE B9-5 Method of determining the amount of material which can be withdrawn by a tunnel conveyor from a conical pile.

may be installed at the heads of the stackers, or the stackers may be fitted with luffing booms to decrease drop.

The most commonly used method for reclaiming yard stockpiles has been gravity feed to reclaim conveyors housed in a tunnel. The tunnel runs under the stockpile and is built of reinforced concrete, timber, or corrugated steel culvert. Openings in its roof are provided with feed gates or mechanical feeders to withdraw the material and feed it at a uniform rate onto the tunnel conveyor.

FIGURE B9-6 Typical double-wing rail-mounted stacker.

The bucket-wheel-type reclaimer, either rail- or crawler-mounted, is popular. When the controlled blending of material is required, a bridge-mounted bucket wheel which is a combination reclaiming-blending machine is also available. The wheel-mounted buckets dig into the storage pile and discharge, as they reach the highest or topmost position, onto an adjacent conveyor belt. The capacities of these machines vary from 150 to 4000 tph.

Overland Haulage Systems

Because of the large quantities of materials involved in heavy construction, the cost of hauling each yard of earth or rock can materially affect the profit of an entire project. Hence, all factors must be carefully weighed before the most economical haulage system is finally chosen. A belt-conveyor system is compared with other commonly used haulage systems in Table B9-2.

In addition to the systems identified in this table, there are other means of long-distance haulage for special applications. For example, the overhead cableway has been used very successfully for concrete dams and in mining operations. Cableways are particularly advantageous when the construction is in a deep gorge where employment of other equipment is not practical.

The cable-belt system utilizes two independent steel cables, one on each side of the rubber belt, to support the belt and to carry tension. Since the belt is no longer required to transmit drive force, its cost tends to be lower than that of equivalent

TABLE B9-2 A Belt-Conveyor System Compared with Other Commonly Used Haulage Systems

Haulage system	Advantages	Disadvantages
Railroad	Rugged equipment—resistance to abuse Same installation may be used to haul equipment, supplies, and personnel No need for power transmission nor large power plant if diesel equipment is used	High construction costs, especially in rugged terrain Inability to negotiate steep grades, consequently requiring much longer run Maintenance High accident rate in difficult terrain
Trucks	Low investment cost Highest flexibility Can carry equipment, supplies, and personnel	High maintenance—road and equipment Require a team of operators and shop personnel High operating cost High accident rate
Belt conveyors	Low operating cost Comparatively lower investment cost for high tonnage and rugged terrain Requires least number of operators Reliability and safety Noiseless—quieter operation	Normally cannot be used for hauling supplies, equipment, or personnel Rubber belt is vulnerable to damage unless all design precautions are exercised and operating restrictions are properly policed

belting for conventional conveyors. The speed of this system is comparatively lower, and cable wear has been a problem in some installations.

Table B9-3 compares the costs of alternative haulage systems with that for a 6.5-mi, six-flight conveyor system with a 36-in-wide belt. This system rises 700 ft and then plunges 1700 ft prior to arriving at the terminal. This rugged terrain can be traversed with only 6.5 mi of belt-conveyor system as compared to 14 mi of rail line or 10.8 mi of truck road.

TABLE B9-3 Relative Cost Comparison—Various Haulage Systems

	Conveyor	Railway	Cableway	Trucks
Length required (miles)	6.5	14.0	6.5	10.8
Relative cost per ton-mile	1.00	0.58	2.29	1.30
Relative cost per ton	1.00	1.26	2.29	2.16
Relative capital cost	1.00	1.30	0.81	0.97

Examples of long belt-conveyor haulage systems used for heavy construction and other industries are shown in Table B9-4. Other well-known examples are Grand Coulee Dam (48 in, 5000 ft, 1930), Shasta Dam (36 in, 9.6 mi, 1940), Bull Shoals Dam (30 in, 7 mi, 1948), Great Salt Lake Crossing (54 in, 2.5 mi, 1955), Trinity Dam (42 in, 3 mi, 1958), George Washington Bridge approach (42 in, 2 mi, 1961), Oroville Dam (54-, 72-, 96-in, 5-mi-plant and land, 1964).

Ship and Barge Loaders

When excavated material must be routed over wide stretches of water, ship or barge loaders equipped with belt conveyors serve as a link between water and land.

The simplest loader is a fixed-tower type. The tower supports an overhanging boom-conveyor which reaches out to trim the ship or the barge transversely. Lengthwise trimming is accomplished by moving the vessel with its own gear, a towboat, or winches on shore.

A traveling loading tower affords the added flexibility of not requiring movement of the vessel for trimming. The tower itself is fed by a trailing tripper, and both move along the pier together.

The capacity of a modern ship- or barge-loading system may vary from several hundred tons per hour to over 10,000 tph.[1] Before loader size, capacity, and type can actually be determined, consideration must be given to the size of vessels and the frequency of their arrivals, the loading cycle time, and climatic and water conditions at the pier (wind, tide, current, surf, waves, and ice). A ship- or barge-loading system is, in essence, a large batch-handling subsystem. A storage-reclaim system becomes a key to converting it into a smooth, continuous bulk-handling operation.

Portable Systems

Portable conveyor systems are usually lighter, lower in cost, and easily fitted to a job's needs. Since virtually no erection costs are involved, they may readily be moved to the most desirable locations to handle assignments. In practice, the com-

TABLE B9-4 Examples of Long Belt-Conveyor Systems

Width, in	System length, ft	Max single-flight length, ft	Speed, fpm	Capacity, tph	Material	Location	Year	Note
47	12,140	12,140	492	1,760	Excavated earth	Kobe, Japan	1964	Move mountain to sea through culvert and tunnel
66	15,000	15,000	1,100	12,000	Excavated earth	Portage Mtn. Dam, Vancouver, Can.	1965	Overland
42	54,048	10,591	492	1,200	Stone	Nagato, Japan	1964	Overland/tunnel
36/42	35,756	6,450	600	800/1,200	Coal	Greene City, Pa.	1965	Overland/underground
36	50,300	16,400	590	2,240	Iron ore	Marcona, Peru	1966	Overland
42	33,377	10,950	500	—	Bauxite	Jamaica	1967	Overland
42	47,520	23,760	1,200	3,600	Bauxite	Jamaica	1985	Overland

ponents and subsystems are unitized in a system or plant and broken down into logical module units such as feeders, screens, crushers, and stackers. Depending on how they are supported, the portable systems may be classified in five groups:

1. Rubber-tire running gear
2. Wheels running on rails
3. Crawler-type running gear
4. Skid mounted
5. Semiportable

The systems which travel on rubber tires are the lightest and are readily towed around the job site by dozers or tractors. A heavy unit can be loaded onto a truck-trailer for long-distance highway movement. The tires may be left on when the unit arrives at the new site, or they may be temporarily removed and the plant jacked up and supported on timber or concrete blocks to provide a firm foundation. Power is furnished by trailing cables, portable diesel generators, or directly connected gasoline or diesel engines. Because of its relatively low cost, the system with rubber-tire running gear is by far the most widely used. Among other applications, it has been used in ship and barge loading.

Machines which travel on rails are generally used in more permanent installations where a stable, ballasted rail bed can be prepared. The movement of these machines is confined to the rail line limits. Typical machines in this category are hoppers, stackers (Fig. B9-6), shuttle conveyors, trippers, blending-reclaim machines, feeder, ship and barge loaders, etc.

Machines which travel on crawler tracks have the greatest flexibility by virtue of their ability to move around in rough terrain. They are usually large, sturdy machines, especially those designed for excavation. The wheel-on-boom bucket excavators are equipped with their own boom conveyors to receive a discharge excavated material onto a connecting conveyor network. A transfer conveyor which moves on crawlers adds to the system flexibility.

A crawler-type machine is usually backed up by a shiftable feed-collecting conveyor. This shiftable conveyor has unitized deck stringer sections mounted on rail ties. Specially designed lugs enable the conveyor to snake into new positions without causing damage to belt, machinery, or structures. Shifting is accomplished by specially designed rigs mounted on dozers. A 60-in-wide, 1000-ft conveyor can be moved laterally up to 6 ft and realigned in a matter of hours.

When frequent movement is not required, semiportable plants are used. These plants are built with simple bolted construction, readily erected on light foundations. On completion of a project or a phase of it, they are easily dismantled and moved to another location.

Belt-Conveyor Components

Belts

In a belt conveyor, the most important and usually most costly component is the rubber belt. An oversimplified description of the construction of a conveyor belt is a carcass imbedded in rubber. The carcass carries tension and absorbs impact. It is made up of one or many plies or layers of fabric (cotton or synthetic), bonded together by a friction and/or skim coat of rubber. When the carcass is of single-layer construction, it could be fabric yards or steel-wire standard cables, the latter being the newest type of carcass with the highest tension rating. The rubber compound may be natural, synthetic, or a blend. *Breakers* with a cord or leno (an open mesh fabric) construction are sometimes placed between the cover and the carcass to increase the adhesion as well as the impact and puncture resistance of the belt.

The conveyor-belt industry has rated various types of belt rubber in three grades according to their cut, tear, and abrasion resistance.

Cover grade	Cut and tear resistance	Abrasion resistance	Application
1	Excellent	Excellent	Large-sized rock, sharp cutting materials, extremely rugged service
2	Good	Excellent	Heavy-duty service—sized materials with limited cutting action, primarily abrasion
3	Low	Good	General light-duty service—small, sized material

Allowable working tension varies according to the type of carcass employed. It may be as low as 27 lb per in per ply for a low-cost cotton fabric carcass and as high as 240 lb per in per ply for a high-tensile synthetic construction (see Table B9-5). A belt with a steel wire-rope core can carry up to 4000 lb per in of belt width, using ⅜-in stranded cables.

In selecting a conveyor belt, in addition to the quality of the rubber, one must also consider the tension requirements, troughability over idlers, longitudinal flexibility as a function of the pulley diameter, number of fabric plies, impact resistance, and cover thicknesses in order to find the most appropriate and economical choice.

Conveyor belts are usually shipped in rolls, to be made endless at the job site. Vulcanized splices are considered superior to mechanical splices in view of their higher strength and longer service life as well as the resultant cleanliness. On the other hand, when high initial costs are not justified and time does not allow for extensive vulcanizing procedures, mechanical fasteners such as Flexco metal links have been

TABLE B9-5 Tension Ratings of Conveyor Belts with Fabric Carcass *(lb/ply/in)*

Fabric identification, RMA	Normal mechanical fastener splice	Normal vulcanized splice
35	27	35
43	33	43
50	40	50
60	45	60
70	55	70
90	—	90
120	—	120
155	—	155
195	—	195
240	—	240

used successfully in many installations. A simple mechanical splice may be made in a matter of minutes, compared to many hours for making and curing a vulcanized splice. Mechanical fasteners have been lifesavers for quick repairs of cuts and rips in conveyor belting in an emergency. A compromise between the two methods of splicing is the so-called *cold-patch*. This method requires the same stripping and gluing procedure as used in the vulcanized splices, but without the application of heat after the splices are made. This greatly reduces the time required for the completion of a splice and still has many of the advantages inherent in a vulcanized splice.

Pulleys and Idlers

Pulleys are usually located at the terminals of a belt conveyor, where they support the belt and transmit driving power. Between the terminals, the conveyor belt is supported by troughing idlers on its loaded upper half and by return idlers on its normally empty lower half.

A conveyor pulley is usually of welded steel construction. Its continuous rim is supported by steel end disks which are fitted with compression hubs on shafting. Some pulleys are crowned at the center to help the training of the conveyor belt. Lagging of rubber or other material is used to increase the coefficient of friction between the belt and the drive pulley. It also helps to reduce wear on the pulley face and eliminate material buildup. Wing-type pulleys of slatted construction also help to eliminate material buildup on the pulley face. Pulley outfits with shop-assembled shafting and bearings are usually shipped as complete assemblies for ease of field erection.

Troughing idlers are commonly of the three-roll type, with the center roll horizontal and the two side rolls inclined to form a trough (Fig. B9-7a). The standard troughing angle for side rolls is 20°, but advances in belting technology are responsible for a trend toward *deep trough* idlers. These idlers permit greater belt capacity without increasing belt width. Their troughing angles may vary from 25 to 35 or even 45°. Transition idlers (Fig. B9-7b) with gradually varying troughing angles are required to properly support the belt as it approaches the flat terminal pulleys.

The idler rolls are usually made up of steel tubing, with welded end disks which house antifriction bearings. Two systems of lubrication are available. The so-called *sealed-for-life* idlers use factory-lubricated and sealed rolls and can be relubricated only by disassembling the seals in shops. The regreaseable idlers may be lubricated on site through grease fittings. Some of these idlers have internal or external piping connecting all the bearings to one grease fitting for one-shot lubrication.

For lighter duties, catenary troughing idlers (Fig. B9-7c) may be substituted for the three-roll type. Individual rubber rolls, or a continuous spiral assembly, are molded on a flexible catenary member (usually a stranded cable). The assembly rotates on fixed bearings at the ends of the catenary member. The rolls may also rotate on individual bearings supported by a fixed, flexible catenary member. The spiral type may be made up of a continuous steel coil spring, which also helps to absorb impact.

The spacing of troughing idlers is dictated by the permissible belt sag between the idlers and thus is a function of belt tension and the load carried on the belt. Impact idlers (Fig. B9-7d) at loading points are spaced as close as physical dimensions permit, still allowing room for cleanup. The spacing on the return idlers (Fig. B9-7e and f) is usually a function of the weight of the belt only and varies anywhere from 8 to 12 ft. For the economical design of long conveyors in which the tension varies substantially from one terminal to another, the spacing between troughing idlers

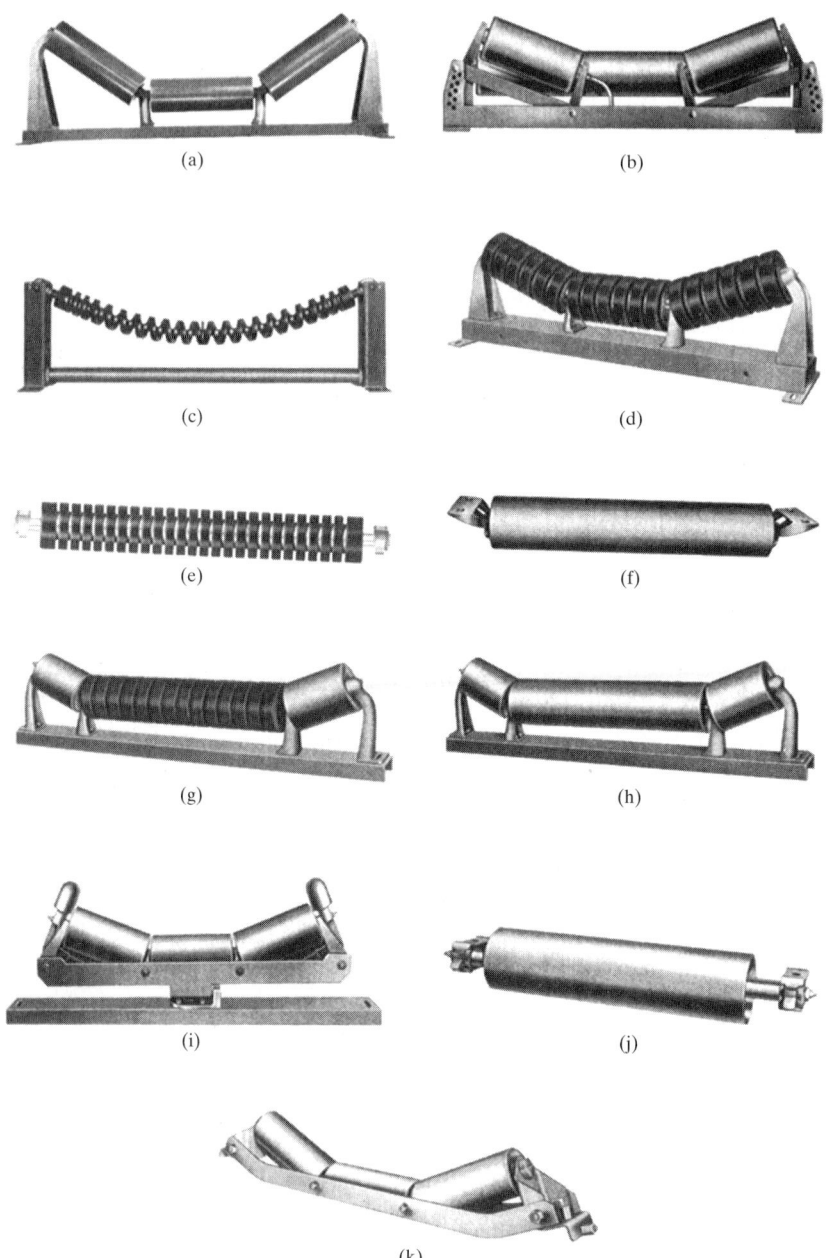

FIGURE B9-7 Troughing idlers: (*a*) standard troughing idler; (*b*) transition troughing idler; (*c*) spiral catenary idler; (*d*) rubber-disk impact idler; (*e*) rubber-disk return idler; (*f*) flat-roll return idler; (*g*) picking idler (impact); (*h*) picking idler; (*i*) troughing-training idler; (*j*) return-training idler; (*k*) wire-rope stringer idler.

may be graduated from around 3 ft at the low-tension sector to over 10 ft where high belt tension can support the load between the idlers through catenary action.

To cushion impact on the belt at feed points, the rubber-disk impact idlers illustrated in Fig. B9-7g are used. To facilitate hand picking, the length of the troughing side rolls may be reduced and the length of the center flat roll increased (Fig. B9-7h). This enables the belt to assume a relatively flat yet troughed configuration. Rubber-disk return idlers and helical or spiral return idlers are often used for very sticky material. These idlers have a self-cleaning feature to minimize excessive buildup on the return belt or on the return idlers or pulleys. Sections of rubber sleeves vulcanized on flat return idlers may achieve the same results.

Various types of training idlers are available for carrying (Fig. B9-7i) as well as return belt runs (Fig. B9-7j). These idlers induce auxiliary side forces on the belt if it begins to run out of line. A sophisticated installation may even employ a hydraulically actuated training idler.

Drives

Nearly all belt conveyors are driven by electric motors. Speed reduction may be achieved by chain and sprockets, V belts, or gearing. The most common device consists of reduction gears which are housed in a box as an entirely shop-assembled unit. The shaft-mounted reducer, with ratings up to 1200 hp, has the advantage of ease of replacement. The parallel-shaft base-mounted units have the greatest load capacity. Right-angle helical and worm-gear reducers provide added flexibility in machinery arrangement and layouts. These speed reducers may be directly coupled to drive pulley shafts or used in conjunction with chain sprockets and other speed reduction devices to achieve further reduction.

Flexible couplings between drive and driven shafts allow for minor misalignments. To prevent a loaded, inclined belt conveyor from running backward, a backstop can be used. This allows free motion of the drive pulley in the forward direction but automatically prevents rotation of the drive pulley in reverse. The most common types of backstops are ratchet and pawl, differential band brake, and overrunning clutch; sophistication and costs run in the same order.

To prevent a loaded, declined conveyor from running out of control, a brake is required. This may be in the form of a disk brake on the motor or a shoe brake on a brake-type coupling. Eddy-current brakes exert a retarding force through a magnetic field without direct mechanical connection.

Drives are usually located at the discharge ends of inclined conveyors. For stackers and booms, where weight at the discharge end is critical, the drive is normally placed internally near the feed end. The drive for a declined conveyor is usually located at the tail end, where the conveyor is fed.

Motors and Controls

The most commonly used motors for conveyor applications are the squirrel-cage and wound-rotor types. On rare occasions, dc motors are also used. A motor for a conveyor-belt drive must have sufficient torque to overcome the starting friction of the system. At the same time, it should not impose excessive stress and strain on the belting or on its associated mechanical and structural components.

The wound-rotor motor, with external resistance connected through a controller to the secondary winding, provides stepped-torque control between maximum and minimum limits. The number of steps is increased as the torque limits are moved closer together.

Because of its relatively simple and rugged construction, with low initial cost and maintenance, the squirrel-cage motor is by far the most popular in conveyor drive application. There are a number of standard designs of the squirrel-cage motor, and NEMA (National Electrical Manufacturers Association) classifies them into five categories according to speed-torque characteristics. The most commonly used is the NEMA-B design. When higher starting torque is required, the NEMA-C design can provide a minimum of 200 percent of the full-load torque at start.

The variable-speed motor provides added flexibility in conveyor drives. Speed control may be infinitely variable, or the motor may be of a definitive multispeed type. Torque control is also obtained with oil- and magnetic-type couplings.

Since conveyor applications frequently involve a dusty atmosphere, most of the motors used are the totally enclosed, fan-cooled type. Occasionally, open drip-proof motors are used to minimize initial cost.

To achieve the highest degree of automation, nearly all the modern conveyor systems are sequentially interlocked. It is generally so arranged that a conveyor cannot start until the conveyor which it feeds has reached its rated speed. To avoid pileups at the junctions during normal stopping and to clear the belt of material, a reverse sequential-stop system with appropriate time delays assures that no conveyor will stop until after its feed conveyor has stopped and it has completely cleared of material.

Transfer Chutes

Transfer chutes are provided at belt conveyor junctions to confine, guide, and feed material. These are usually custom-designed devices for a specific type of material and a specific application. A properly designed chute minimizes the impact on the belt it feeds. A properly designed *stone box* can achieve this without eliminating the beneficial velocity component in the direction of the belt travel. Especially in the case of an angular junction, the stream of material should be guided so that only the velocity component in the direction of the receiving belt is kept intact. This component helps to minimize abrasion as the belt accelerates the material to the same speed at which it travels. On the other hand, the transverse component of the stream should be minimized, if not completely eliminated, to avoid its tendency to push the belt out of line and to reduce the wear of the belt cover.

Occasionally, a fixed grizzly is provided to screen out the fines so that they fall on the belt first. They then form a cushion for the lumps to fall on and thereby minimize possible belt damage. Fishplates may be provided at the bottom of the chute to help guide slabby or lumpy material in the direction of the belt travel (see Fig. B9-8). Adjustable curtains may be used to guide the material and to reduce its impact and the resultant wear.[2] Skirtboards, in lengths of 10 ft or more, guide and contain the material after it has fallen on the belt.

A wide variety of wear plates are available. These are usually bolted inside the chutes with countersunk bolts for ease of replacement. Special rubber liners work well where wear is not accompanied by gouging and sharp cuts. For areas of extreme impact and abrasion, used manganese crusher jaw plates, if available, can outwear any other abrasion-resistant material.

BELT CONVEYORS

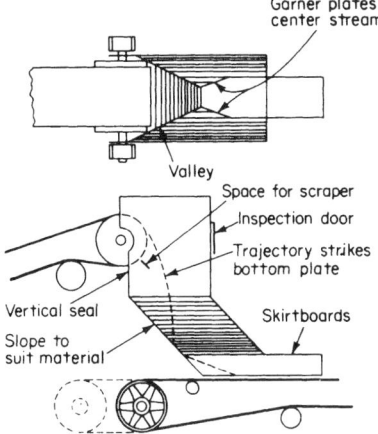

FIGURE B9-8 Typical transfer chute and fishplate (or Garner plate).

Supporting Structures

Supporting structures for a belt-conveyor system consist essentially of the terminal supports and the intermediate sections. At the terminals, structural framework supports the pulleys, belt tension takeups, drives, chutes, screens, feeders, and other associated equipment. The intermediate structures consist of stringers made up of steel channels or angles to support the troughing and return idlers (see Fig. B9-9). Occasionally, deck plates are provided to protect the return belt and, at the same time, provide lateral strength to the supporting structure. Short posts, in turn, support these stringers and anchor them either directly on the foundations or on elevated trusses. These stringers may serve the dual purpose of supporting the idlers and functioning as the top chord of a deck truss.

For belt widths under 42 in, a service walkway on one side of the belt may be sufficient. To adequately service wider belts, walkways on both sides of the conveyor may be necessary.

For conveyor widths up to 48 in, preengineered and prefabricated trusses of standard construction with a wide choice of spans are available. Their use can result in considerable savings, both in fabrication and erection costs.

Wire ropes, which replace steel stringers, have gained popularity for their ease of construction and lower investment cost. They are used with specially designed idlers and supporting stands (Fig. B9-10).

FIGURE B9-9 Typical intermediate sections—troughing idlers supported on steel channels and angle posts.

FIGURE B9-10 Belt conveyor supported by wire-rope stringer.

Accessories

Takeups. These are used to ensure belt tension. The simplest and most positive is the gravity type, which utilizes a counterweight to maintain belt tension. For shorter conveyors, screw-type takeups may be adequate.

Electrically controlled takeups may be used where exceptionally high belt tensions necessitate a sophisticated monitoring and control. Where belt vibrations become a problem, a hydraulic takeup may be the solution.

Scales. Various standard belt scales for continuous weighing are available. These are usually precalibrated and are capable of registering capacity rates in terms of tons per hour. They are set within a limited percentage range of the conveyor's rated capacity. Certain belt scales can offer an accuracy of 0.5 percent over a range from one-third to full load. The sum total of tonnage handled is registered on a totalizer. Electronic remote-controlled devices enable monitoring of tonnage readings in a control center.

Since catenary action of the belt between idlers can materially affect the accuracy of the scales, it is mandatory to maintain the tension in the conveyor belt as constant as possible. Hence, counterweighted takeups are preferred and the scale should be located at the low tension sector along the conveyor.

Belt Cleaners. These devices are provided near the discharge end of the conveyor, immediately after the mainstream of material has been unloaded. Counterweighted or spring-loaded belt scrapers with high durometer rubber inserts are the simplest devices. They are usually located under or behind the discharge pulley so that the scrapings can be collected on the conveyor fed by the same transfer chute.

Rubber or bristle brushes are available for special applications. They may be driven independently or through the conveyor drive. Where water is abundant and drainage presents no problem, strong jet spray does a good job in cleaning the belt.

Rocks which accidentally have fallen onto the return belt can become wedged between the belt and the tail pulley and cause severe damage. A simple belt plow, riding on top of the return belt and located in front of the tail pulley, is a simple and effective safeguard.

Safety Devices. For open long-conveyor installations, it may be necessary to protect the belt from unwanted detraining forces caused by side wind. A vertical steel plate along the belt conveyor provides the simplest wind shield. A more elaborate installation may have a hood cover over the belt. When the presence of wind is sporadic, wind hoops spaced at 30- to 40-ft intervals may be sufficient to prevent the belt from being lifted by gusts.

A rattrap may be installed where occasional boulders tend to roll back on steep conveyors. A rattrap is essentially a simple one-way swing gate which permits the boulders to pass through in the direction of material flow. Any boulders which run backward are stopped at the gate where they are accumulated and wedged together until they move forward again en masse.

Low-cost *wire-mesh guards* are versatile and have been used extensively around drives and moving machinery as well as under conveyor trusses. Pull-cord *emergency switches* are often installed along the entire length of a conveyor. The cord is mechanically tied to switch arms, and the conveyor drive will stop instantaneously when the cord is pulled. The electrical circuits are usually designed so that the conveyor cannot restart until the cause for stoppage is corrected and the switch arm is mechanically reset.

Other safety devices include overload cutout switches, belt-penetration switches (to avoid extensive belt damage), etc. These provide added protection to the system.

ENGINEERING DESIGN

Material Characteristics

A thorough study of the characteristics of the materials to be handled is a prerequisite for a well-engineered conveyor system with properly selected components. Heading the list of pertinent items is size—not only the maximum lump size but also the percentage breakdowns of the various size ranges. A reliable and representative screen analysis will provide this information.

High percentages of fines may create a dust problem, especially at transfer junctions. When moisture is present, cohesive articles tend to cling to belting and machinery. In addition to being a nuisance this buildup can cause detraining of the belt. On the other hand, if the material contains many large lumps, wear becomes the major concern. Rounded boulders have a tendency to roll back on a steeply inclined belt, causing damage to machinery and injury to personnel. Table B9-6 indicates maximum safe inclinations of troughed belt conveyors for handling various bulk materials.

The origin of the material can often provide a hint about its ultimate behavior on a belt conveyor. For example, crushed aggregates are usually angular and will result in more belt-cover wear than normally rounded alluvial material. Well-screened and well-sized lumps, without the benefit of the fines to cushion the impact, cause substantially more wear on the belt and wear plates.

The minimum width of a conveyor belt is usually governed by the maximum lump size of the material carried. A general rule is to have the belt width two to three times the size of a maximum lump. Minimum belt width is more critical for well-sized material than for unsized. In analyzing the maximum lump size, one must pay particular attention to the fracture pattern of the rock. For instance, some rocks tend to be slabby, with one dimension many times smaller than the others. When slabs occur frequently, a conveyor designer must provide sufficient belt width and transfer space to handle them.

TABLE B9-6 Maximum Safe Inclinations of Troughed Belt Conveyors for Handling Various Bulk Materials

Material	Inclination, degrees	Rise in ft per 100 ft
Cement—loose	22	40.4
Clay—fine dry	23	42.4
Clay—wet lump	18	32.5
Concrete—wet	15	26.8
Earth—loose	20	36.4
Gravel—bank run	18	32.5
Gravel—screened	15	26.8
Gypsum—powdered	23	42.4
Limestone	18	32.5
Rock—fine and crushed	22	40.4
Rock—mixed	20	36.4
Rock—sized	18	32.5
Sand—dry	15	26.8
Sand—damp	20	36.4
Sand—tempered and foundry	24	44.5

Capacity, Speed, and Sizing of Equipment

The carrying capacity of a belt conveyor is expressed in terms of tons per hour (tph). This capacity is a function of belt width, speed, and material density. Thus, a conveyor of a given width, running at a fixed speed, has its theoretical capacity cut in half when loaded with a material whose bulk density is one-half that of the material for which the conveyor was designed.

Table B9-7 lists belt-conveyor carrying capacities in short tons per hour for material weighing 100 pcf, belt widths up to 120 in, speeds varying from 100 to 1000 fpm,

TABLE B9-7 Belt-Conveyor Capacities

[for 20 and 35° troughing idlers—capacities in tons (2000 lb) per h]

Belt width, in	Cross-sectional area, ft² 20°	Cross-sectional area, ft² 35°	100 20°	100 35°	200 20°	200 35°	400 20°	400 35°	500 20°	500 35°
24	0.369	0.451	111	135	221	271	443	542	553	677
30	0.599	0.733	180	220	360	440	719	880	899	1,100
36	0.886	1.083	266	325	531	650	1,063	1,300	1,328	1,625
42	1.228	1.502	368	450	737	901	1,473	1,802	1,841	2,252
48	1.625	1.988	488	596	975	1,193	1,950	2,385	2,438	2,982
54	2.078	2.542	624	763	1,247	1,525	2,494	3,050	3,118	3,813
60	2.587	3.164	776	949	1,552	1,899	3,105	3,797	3,881	4,746
72	3.772	4.613	1,132	1,384	2,263	2,768	4,527	5,536	5,659	6,919
84	5.180	6.334	1,554	1,900	3,108	3,800	6,216	7,601	7,770	9,501
96	6.811	8.327	2,043	2,498	4,086	4,996	8,173	9,992	10,216	12,491
108	8.664	10.592	2,599	3,178	5,198	6,355	10,396	12,711	12,996	15,889
120	10.740	13.130	3,222	3,939	6,444	7,878	12,888	15,756	16,109	19,695

Note: Table follows essentially CEMA formulation based on 25° surcharge angle and 100-pcf material density (maximum edge distance being 5.5 percent belt width plus 0.9 in); for other density, tph (listed above for 100 pcf) × (known weight/ft³)/100.

and idler troughing angles at 20 and 35°. As can be seen, the use of 35° deep trough idlers can increase the carrying capacity of a belt conveyor by as much as 25 percent.

A belt conveyor can theoretically attain as high a capacity as desired by merely increasing its speed, and belt conveyors carrying overburden have been run as fast as 1200 fpm. In normal practice, however, belt speeds are limited by many factors. Among these are the character of the materials conveyed, expected operating and maintenance conditions, and vibrations of the supporting structures.

Extremely abrasive materials, sharp-edged rocks, and very large lumps require lower speeds to reduce belt cover wear. Fine granular materials will spread out at high speeds as the belt passes over idlers, particularly at low tension. Lumps then tend to be forced toward the edges, causing spillage. If dust is a problem, high speeds usually aggravate it.

Conditions at the feed point are important for a high-speed conveyor. When material is loaded on a slope, the tendency to roll back is more pronounced on a faster-running belt. This condition is aggravated by the presence of large numbers of round boulders (see Table B9-8 for lump size limitations).

Generally, higher speeds may be allowed if belt-to-belt transfers are in line. Angular transfers introduce a transverse velocity component which causes chute or belt wear. The wear is intensified as the belt speed increases. At extremely high speeds, vibrations of the belt and structures should be carefully examined.

To avoid possible pileup at the junctions, the conveyor which receives the material should never have lower speed than the preceding feed conveyor. Table B9-9 lists recommended maximum belt speeds in feet per minute for various widths of conveyors and types of material.

Design Calculations

A belt conveyor involves a wide variety of components which must function together as a mechanical complex. The belt itself is made up of materials of various

\multicolumn{10}{c}{Speed, fpm (material weight 100 pcf)}										
\multicolumn{2}{c}{600}	\multicolumn{2}{c}{700}	\multicolumn{2}{c}{800}	\multicolumn{2}{c}{900}	\multicolumn{2}{c}{1,000}	
20°	35°	20°	35°	20°	35°	20°	35°	20°	35°
664	812	774	948	885	1,083	996	1,219	1,106	1,354
1,079	1,320	1,259	1,540	1,438	1,760	1,618	1,980	1,798	2,200
1,594	1,950	1,860	2,275	2,125	2,600	2,391	2,925	2,657	3,250
2,210	2,703	2,578	3,153	2,946	3,604	3,314	4,054	3,683	4,505
2,925	3,578	3,413	4,174	3,900	4,771	4,388	5,367	4,875	5,963
3,741	4,576	4,365	5,338	4,988	6,101	5,612	6,863	6,235	7,626
4,657	5,696	5,434	6,645	6,210	7,594	6,986	8,543	7,762	9,493
6,790	8,303	7,922	9,687	9,054	11,071	10,185	12,455	11,317	13,839
9,324	11,401	10,878	13,301	12,432	15,201	13,986	17,102	15,540	19,002
12,259	14,989	14,302	17,487	16,345	19,985	18,388	22,483	20,432	24,981
15,595	19,066	18,194	22,244	20,793	25,422	23,392	28,599	25,991	31,777
19,331	23,634	22,553	27,573	25,775	31,512	28,997	35,451	32,219	39,390

TABLE B9-8 Suggested Minimum Belt Width in Relation to Lump Size

	Ratio of belt width to lump size	Maximum lump size, in						
		Belt width, in						
Material		24	30	36	42	48	54	60	72
Quarry	2¼	11	13	16	19	21	24	27	32
Occasional lump	3	8	10	12	14	16	18	20	24
Crusher run (no slabs)	3½	7	9	10	12	14	15	17	21
Crusher run, fines removed	4	6	8	9	11	12	13	15	18
Sized material from screening operation	4½	5	7	8	9	11	12	13	16

degrees of elasticity and strength. With the many variables involved, an exact mathematical analysis of horsepower and tension is virtually impossible. Empirical design formulas based on limited field and laboratory tests are available and have been satisfactory for most applications.

These formulas are derived on the assumption that the power required to drive a conveyor may be subdivided into two principal parts: (1) power to overcome friction, and (2) power to lift the load.* For certain transfers, additional power is required to accelerate the load.

The power required for lifting the load is directly proportional to the lift and capacity rate in tons per hour.† Power to overcome frictional resistance concerns all of the moving parts such as belts, idlers, pulleys, takeups, drives, and other accessories. The interaction of all these components, in turn, will vary, depending on the configuration of the system. Furthermore, their friction coefficients vary as a function of the temperature, load, tension in the belt, and operating and climatic factors.

In addition to horsepower requirements, a complete belt-conveyor analysis must take into consideration the belt tension. This information is required for the selection of the belt and for the design of the mechanical components, such as pulleys, shafting, reducers, couplings, structures, and other accessories.

Computers have substantially aided the engineering design of complex belt-conveyor systems. Various horsepower, tension, and component design programs have been written and are used extensively for general applications. The solution of a complex belt-conveyor problem may now be derived in a matter of minutes on a computer, as compared to hours or days of hard manual labor. Figure B9-11 shows a typical computer printout.

Other Parameters

Before any capacity requirements and component selections are finalized, the designer must always raise the question of adequacy of the system to accommodate

---

* For a declined conveyor which delivers material below the elevation at which it is received, the load on a belt actually tends to generate power. When the decline is steep and the power derived from lowering the load exceeds the frictional horsepower, the system becomes *regenerative*. The motor becomes a generator and has been frequently used to feed power back to the system.

† A simple rule of thumb for estimating lift horsepower is lift in feet multiplied by capacity in 1000 tph. Thus, 60 hp would be required for a conveyor with a net rise of 20 ft handling 3000 tph.

TABLE B9-9 Maximum Recommended Belt Speeds (fpm) for Standard Service

Material	Maximum speed*	\multicolumn{8}{c}{Belt width, in}
		24	30	36	42	48	54	60	72
		\multicolumn{8}{c}{Granular fines}
Minus 1-in. lump	900	600	700	800	900	900	900	900	900
Occasional lump 10% belt width	800	550	600	700	750	800	800	800	800
		\multicolumn{8}{c}{Half maximum-sized lump}
Rounded pieces	700	500	600	650	700	750	750	750	750
Abrasive and sharp	650	450	550	600	650	650	650	650	650
		\multicolumn{8}{c}{Maximum-sized lump}
Rounded pieces	650	400	500	550	600	650	650	650	650
Abrasive not sharp	600	400	450	500	550	600	600	600	600
Abrasive and sharp	550	350	400	450	500	550	550	550	550

* Higher speeds may be used under special conditions and impact forces must be duly considered.

```
PROGRAM E0140
BELT TENSIONS                                                    CID 1156

OUTPUT DATA           INQUIRY NO

CONTR SYS. NO.              CUSTOMER NAME              ENGR. NO.

CONVEYOR P1
                  SUMMARY OF BELT TENSIONS 1

PROFILE           MAXIMUM TENSIONS FOR ANY LOADING CONDITION
 POINT    ACCELERATE    DECELERATE    RUNNING NF    RUNNING RF    BREAKAWAY
   1        10117.         6343.         6648.         6436.         6996.
   2        10215.         6342.         6689.         6437.         7080.
   3        11576.         6439.         7946.         6591.         8585.
   4        20454.         8796.        12659.        10842.        14496.
   5        22608.        10161.        15245.        13357.        17242.
   6        23590.         9965.        15513.        13535.        17757.
   7        30629.        14450.        27565.        25274.        30529.
   7         7765.         8687.         8472.         8667.         8255.
   8         7842.         8618.         8513.         8600.         8404.
   9         7843.         7995.         7956.         7992.         7920.
  10         7700.         7700.         7700.         7700.         7700.
  11         8076.         7931.         7975.         7934.         8016.
  12         7972.         7045.         7163.         7066.         7288.
  13         8448.         7052.         7196.         7086.         7355.
  14         8453.         6724.         6886.         6766.         7069.
  15         9792.         6289.         6551.         6377.         6861.
  16         9920.         6287.         6553.         6377.         6865.

TMAX        30629.        14450.        27565.        25274.        30529.
TMIN         7700.         6287.         6551.         6377.         6861.
TE          22863.         5763.        19092.        16606.        22274.
ACMAT       98213.        87062.        98213.        98213.        98213.
WK2           508.          450.          508.          508.          508.
```

FIGURE B9-11 Typical computer printout for belt-conveyor design program.

future expansion. More often than not, little or no additional investment is required to make provisions for future expansion if the system is properly planned in advance. Layouts can then be arranged so that minimal alteration to machinery or structures will be required when the future stages of expansion are implemented. Since fieldwork usually is costly and plant downtime is equally hard to come by, early recognition of these requirements reflects sound engineering judgment. In the same connection, standardization of components rarely costs more at the initial stage. Even if some of the components are slightly oversized at a small added cost, overall savings may still result from a reduced spare parts inventory and lower maintenance costs.

Rubber belting is usually the most vulnerable and, at the same time, the most costly component in an entire belt-conveyor system. A well-designed belt-conveyor system is one which pampers the rubber belt to the limit. All the remaining components in a well-engineered system must be selected on the basis of maximum possible belt life.

Once the material is settled on the belt, very little can happen unless it is again disturbed. It follows from this that transfer points are where troubles usually develop. Each transfer merits careful study by an experienced designer. It should be so engineered that it will accommodate the anticipated range of material characteristics without major field alterations.

If the terrain and right-of-way conditions permit, long conveyors with the least number of transfers will minimize drives, chutes, and other junction accessories as well as their inherent maintenance. Minimizing the number of transfer points will also reduce belt wear and degradation of the materials hauled.

Longer conveyors may be subjected to high belt tensions. These may dictate structural requirements for the belt and its associated machinery which are different from those suitable for shorter conveyors elsewhere in the system. In such a situation, in the interest of standardization and minimizing spares inventory, a designer may choose to deliberately limit the length of some conveyors in the system. In so doing, the designer will have to cope with the undesirable consequence of a corresponding increase in the number of transfers.

To minimize cut and fill as well as bridge work, a long overland belt conveyor will approximately follow the existing land contours (Fig. B9-12). The consequent savings in cost should be weighed against any increases in complexity of the design.

One or a number of the components in a belt-conveyor system may become critical as various combinations of design conditions are investigated. For example, when the majority or all of the declined sections of a roller-coaster conveyor are loaded, the conveyor may tend to run away. Conversely, if all the inclined sections are loaded, the system would require the maximum driving force to restart after an accident or emergency stop. A long conveyor system possesses high inertia, especially when loaded, and its tendency to coast or drift could result in serious pileups at junctions. In addition to a properly designed braking system, flywheels at succeeding drive stations may be needed to ensure the continued flow of material in the proper sequence.

OPERATION AND MAINTENANCE

A belt-conveyor system, if properly designed, will give the user years of low-maintenance, easy operation. If automation is carried out to the highest degree, an entire conveyor system representing millions of dollars of investment may be safely operated by one operator in an air-conditioned control center. This operator merely pushes the proper buttons on an illuminated panel, using indicator lights to check on

FIGURE B9-12 A long overland conveyor follows the relatively undisturbed land contour.

the flow sequencing. The operator of any belt-conveyor system must be keenly aware of the capabilities for which the plant was designed. Most important of all is to police the plant feed to ensure that foreign material or materials in excess of design specifications do not enter the system. Violation of this simple rule may result in plugging the key transfer chutes or cutting long conveyor belts, with costly damages as well as the resultant downtime.

Overloading is another common cause of unsatisfactory plant performance. Any component has its design limit and safety margin, and repeated overstress will lead to premature fatigue failures and consequent downtime. Overloading at transfer points causes plugging of chutes and possible damage to the belt, as well as spillage on the conveyors.

It is the operator's prime responsibility to obtain high production. This can best be achieved by making sure that the downtime is minimized, rather than by resorting to periodic overloads which nearly always defeat the purpose. A belt-conveyor system is a continuous bulk-handling device; as such, it must be kept in continuous operation to give the maximum output. If the simple rules of operation are not abused and if the maintenance instructions are followed, a belt-conveyor system will be a faithful, trouble-free workhorse.

In maintenance, the center of attention should again be the rubber belt itself. An acceptable preventive maintenance program must include at least the following.

Belt Alignment. Alignment of the belt's supporting structures must be periodically inspected to ensure that there is no ground settlement or damage from accidents or other causes. A belt conveyor on a canted support cannot be expected to run centered. Alignment of head and tail pulleys, drives, and idlers should be periodically checked for the same purpose.

Transfers. Transfer junctions and chutes should be inspected to see that excessive wear has not changed the flow pattern. The wear plates may have worked loose, causing them to fall on the belt. This also applies to the skirtboards or other guiding devices. Loose bolts which have escaped inspection have proved to be one of the most frequent starters of belt damage.

Cleanup. Accumulation of dripping on rolling parts or under the belt can result in a pileup that makes the training of the belt impossible. It may also cause direct abrasion and wear of machinery or the belt. Plows and scrapers should be inspected periodically to see that excessive wear has not neutralized their effectiveness. If trolley-mounted magnetic devices are used for tramp iron removal, they must be discharged prior to being used again. Oil and grease are also detrimental to rubber belting and should be removed if inspection shows them to be present.

Climatic Considerations. In extremely dry climates, the periodic removal of dust from machinery and electrical components is a necessary part of the preventive maintenance program. Ozone coupled with intensive heat is injurious to rubber; if a long, open conveyor is to be exposed to intensive sunlight for any length of time, it should be run periodically to avoid possible damage to the section of the belt that is continually exposed to intensive heat and sun. The use of the wrong grease for the idler bearings has been a common cause for many conveyors' inability to start in cold weather. This situation may be further aggravated when the equipment is new and the idlers have not yet been run in.

REFERENCES

1. A. T. Yu, "Recent Trends and Advances in Shiploading," *Skillings' Mining Review,* April 27, 1968.
2. A. T. Yu, "After 1¾ Million Tons of Iron Ore, Long Belt Shows Hardly Any Wear," *Engineering and Mining Journal,* March 1964.

CHAPTER 10
CRANES*

James J. O'Brien, P.E.
O'Brien-Kreitzberg, Inc., Pennsauken, New Jersey

INTRODUCTION

Historically, human muscle, along with domesticated animal power, has been the means of raising loads large and small. With the advent of winching machinery, larger loads have been handled faster and more easily. This has permitted construction projects of magnitudes and speeds not otherwise possible. Lifting capacities for hoisting units normally range up to 1443 tons, but there appears to be no maximum limit to future loads, to the height they can be raised, or to the distance through which they can be moved. As the requirement arises, the appropriate equipment will be developed.

Each style of hoisting device has evolved to satisfy a special need within the construction industry. The stiffleg derrick in current use is a unit which, apart from its modern prime mover, is not too different from the derricks used in ancient Egypt. The mobile crane is the same basic structure with the added feature of mobility, so that it can be readily moved from one work site to another. The development of mobile cranes closely parallels that of the excavator industry, as the machinery used is similar and often common. The hydraulic crane was developed to provide superior mobility and reduced setup time in comparison with the mechanical crane, although it sacrifices maximum length of boom and some load-picking capacity in exchange for these advantages. The tower crane was developed to facilitate the rebuilding of large areas of apartments and office buildings which were destroyed during World War II, and it has subsequently been used on a variety of urban projects. The original tower crane designs were European and, mobile tower cranes excluded, most of the units used in this country are still imported.

---

*Some of the material in this chapter is from R. L. Bauer, "Cranes and Hoists," in Stubbs and Havers, *Standard Handbook of Heavy Construction*, 2d ed., McGraw-Hill, New York, 1971.

MOBILE MECHANICAL CRANES

This type of crane has a mechanical, rigid drive from its power plant through transmissions, chains, gears, shafts, friction clutches, and cables. The basic crane motions are raising and lowering the load, swinging the load by rotating the crane body on its carrier mounting, and raising and lowering the crane boom. Other auxiliary functions may be provided as needed.

The power plant in the crane body is usually gasoline or diesel engine, but occasionally an electric motor is employed. Power from the prime mover is transmitted through a clutch to a single- or multiple-speed transmission or to a torque converter. The clutch provides a means of disconnecting the drive train for ease of starting and for maintenance work. A multiple-speed transmission allows the selective use of high line speeds for rapid hoisting with reduced loads and lower line speeds for heavier loads or precision control of hoisting.

Torque converters can be substituted for multiple transmissions to provide comparable line-speed control without the requirement of shifting (see Fig. B10-1). Several variations of torque converters and converter controls are available, and an appropriate selection can be made to match the specific operating requirements. A standard single- or three-stage converter will normally have two governors, one at the engine to control the throttle setting and one at the output shaft to control the winching-machinery speed. The engine governor can be a variable control which is used to regulate engine speed, or it can be set to act only as an emergency control shutoff and thus protect against overspeeding.

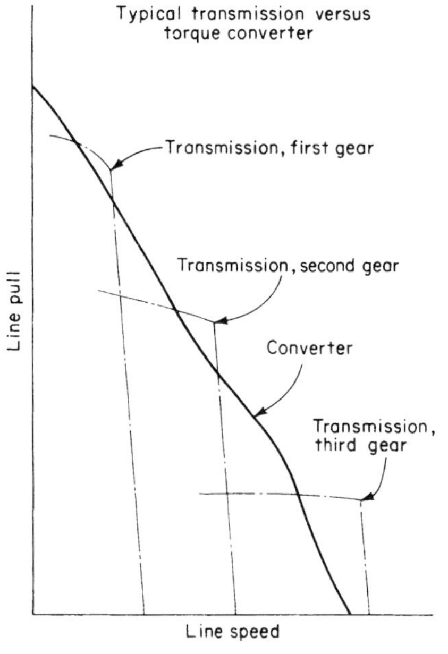

FIGURE B10-1 Line pull versus line speed, multiple-speed transmission, and torque converter.

If the engine governor is a variable one, the machinery or output-shaft governor could be either of the fixed-setting type or eliminated entirely. However, it is general practice to provide a variable-control governor for the output shaft, regardless of the type of engine governor. The variable-control output governor makes it possible to limit machinery speed for precision slow-speed hoist control of loads while still permitting swing or second load lines to be engaged without loss of line pull up to the capacity of the engine. When dual variable-governor controls are installed on a convertor, the one calling for the least speed at the engine will govern.

Split drives, at least one of which will be equipped with a converter, are used to transmit engine power at separate speeds to the different crane functions. Speed and torque output from the converter can remain essentially constant at any one setting, despite some variation in engine speed. This allows one speed control, say on the load hoist line, and a separate speed control on the boom swing motion. The direction of one drive can even be reversed through the converter so it is possible to lower the hoist line without affecting the speed or direction of the swinging motion.

The load hoist motion is controlled with a friction clutch and brake. These components can be actuated mechanically or with power air, power hydraulics, static hydraulics, or a combination such as air over hydraulics. Although each system has its specific merits, the mechanical one is usually less expensive. It also permits the operator to "feel" the reaction of the load more directly than with a power system. The static hydraulic system has similar characteristics. Power air requires rather larger components, since it operates at 80 to 120 lb/in^2, but leakage is not critical. Excessive atmospheric moisture can be troublesome, although it can normally be controlled with alcohol evaporators and moisture ejectors. Power hydraulics can be operated with pressures to 3000 lb/in^2, so its components are smaller than for power air. Any leakage in a power hydraulic system is critical, but this is not a problem when proper components are selected.

Manitowoc's patented VICON® (Variable Independent CONtrol) system provides precision control and independent operation of major functions (see Fig. B10-2). Engine power is divided by transmission case to two controlled torque converters. The front converter powers hoisting drums and boom hoist; the rear converter powers swing and travel machinery. With VICON, clutches engage when no torque is transmitted from the power source, eliminating clutch slippage and wear. After the clutch is fully engaged, converter output is increased to provide infinitely variable speed and torque for smooth, precise control.

VICON provides controlled power load lowering for line pulls exceeding 6000 lb (2724 kg). Load can be held or lowered using the hoist converter's stepless variable output.

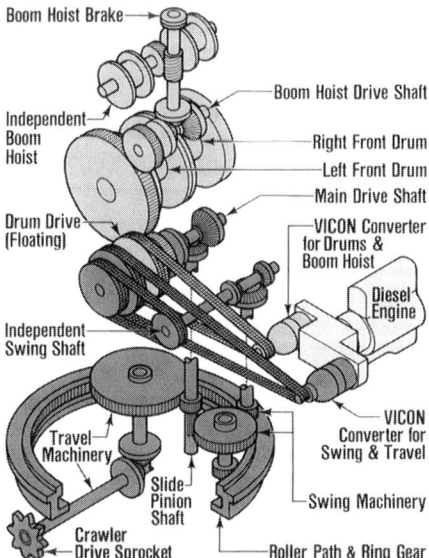

FIGURE B10-2 Mechanical power transmission, VICON (Variable Independent CONtrol). (*Manitowoc.*)

An operational hydraulic motor drives the output shaft of the VICON hoist converter in reverse to provide power lowering for line pulls less than 6000 lb (2724 kg). It permits a full range of lowering speeds with any load from empty hook through maximum capacity.

The crane boom is a latticed structure consisting of three or four main chords which are connected with lacings to form a beam. The chords and lacings are either rolled angle or tubular sections and are usually made of alloy or high-strength steels to minimize their dead weight. The boom foot pivots on, and is supported by, a revolving frame. The boom hoist system is connected to the boom point and thereby supports the boom. It is also used to vary the boom angle and, as a consequence, the load radius. Maximum boom angle is 80° above the horizontal. The maximum rated crane radius is usually equal either to the boom length or to its projected length when raised 10 to 15° above the horizontal. There are several variations of boom tips available: a heavy-duty tip which accommodates the maximum number of parts of hoist-rope reeving and thus allows the handling of maximum crane-rated loads; a light-duty tip which permits a longer overall boom, although this will not have enough sheaves to allow the maximum reeving for the maximum load; a hammerhead or offset tip to permit reduced overhead height and facilitate the handling of bulky loads; and container-handling tips which are designed primarily for ship loading of large containers.

A lighter structural section, known as a *jib,* can be mounted on top of the boom if so desired. Although the maximum lift capability of the crane is thereby reduced, the combined length of the boom and jib will be greater than can be obtained by the boom alone. An offset jib for the boom will also permit larger load radii than an equivalent length of boom and can be advantageous when working close to a high structure.

Figure B10-3 shows four boom end configurations. One (open throat) is for usual lift crane work. Each of the other three offers its own special advantage.

Boom hoisting and lowering are controlled in much the same way as the load-hoist motion. The boom hoist brake is spring set, and normally, but not always, it remains set for the raising motion and is released only for lowering. Lowering must be with power, as a *live boom* is never permitted. A safety ratchet or pawl is engaged to the boom hoist drum at all times except when lowering the boom.

The swing motion of the crane can be controlled with friction clutches, electromagnetic clutches, hydraulic pump and motor, or an electric motor.

A mobile crane can be mounted on a crawler-type unit, a conventional truck, or a self-propelled wheel-type carrier. The crawler unit is propelled by power which is transmitted from the crane body through a shaft at the center of crane body rotation. Positive locks or friction brakes are used to steer the crawler unit and also serve to hold the crane in position while it is in use. The friction brake can be manually operated, or it can be automatically spring set and released when torque is applied from the engine.

The ability of the crawler-type crane to lift loads without tipping is almost always greater over the end than over the side, and caution is required when swinging a load to a less stable position or when luffing it out to a greater radius. Various widths and lengths of tracks are available for most crawler-type carriers to allow for variations in the maximum crane load. A larger track also permits use of larger counterweights, so the gain in stability is twofold. Transporting a large crawler unit can be a problem, and some models have the capability of increasing or decreasing the width between their tracks. These features (crawler retraction and counterweights) are seen in Fig. B10-4. This track separation is fully extended for maximum stability when the crane is handling maximum loads. Crane load ratings are reduced when the tracks are

CRANES

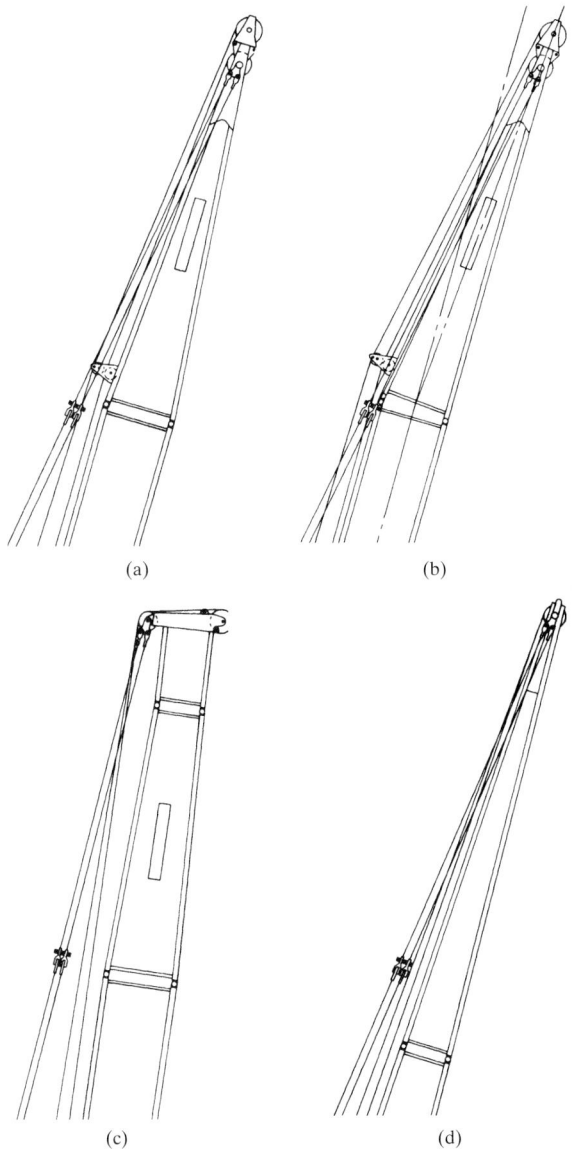

FIGURE B10-3 Different boom end configurations: (*a*) open throat, for normal liftcrane work; (*b*) 4½° offset, for higher load clearance; (*c*) hammerhead, for heavy lifts and superior load clearance; and (*d*) light tapered, for longer reach with lighter loads. (*Manitowoc.*)

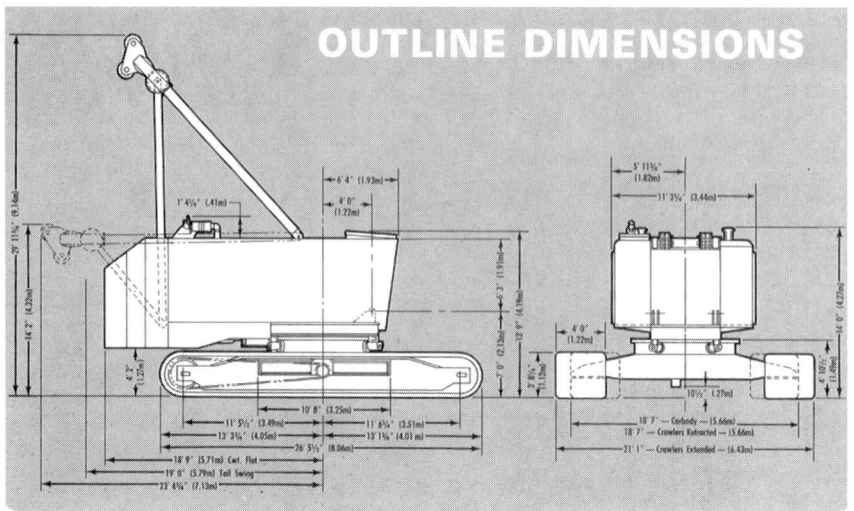

WEIGHTS

	POUNDS*
LIFTCRANE, complete with 70' No. 22A Boom, gantry and backhitch, boom hoist rigging and pendants, hoist wire rope, 15-ton swivel-type hook and weight ball, basic upperworks package, counterweights, 26' 6" long crawlers (48" wide treads), extendible width from 18' 7" to 21' 1", and outside crawler drive chains	356,660
CRAWLERS, with crawler side frames, crawler treads, and outside crawler chains (each 36,400)	72,800
CARBODY, with center pin, roller path, and travel mechanism, without crawlers	49,500
UPPERWORKS, complete with basic machinery, including drums, but not including gantry and backhitch, front end attachments or counterweights	80,500
GANTRY AND BACKHITCH	7,800

	POUNDS*
SELF-REMOVING COUNTERWEIGHT (3-PC)	
Inner	41,900
Middle	41,500
Outer	39,000
BOOM NO. 22A	
BOOM BUTT: (less wire rope and pendants)	5,980
BOOM TOP: (equipped with lower boom point, sheaves, and pendants)	9,760
Add for upper boom point and sheave	1,460
Total	11,220
BOOM INSERTS:	
Insert — 10' (with pendants)	2,000
Insert — 20' (with pendants)	3,100
Insert — 40' (with pendants)	5,340
DRAGLINE FAIRLEAD — REVOLVING TYPE	1,910
DRAGLINE FAIRLEAD — HINGED TYPE	7,420

*Weights are approximate and may vary between machines as a result of design changes and component variations.

POWER PLANTS

	Model	Cylinder	Bore	Stroke	Cubic Inch Displacement	Net HP @ RPM (at flywheel)
BASIC	Cummins NTA-855 Diesel	6	5.500"	6.000"	855	333 @ 2000
OPTIONAL	G.M. 12V-71N Diesel	12	4.250"	5.000"	852	360 @ 2000
	Caterpillar D343-TA Diesel	6	5.400"	6.500"	893	364 @ 2000
Air Compressor: 37.5 CFM.					Fuel Tank Capacity: 315 gallons.	

FIGURE B10-4 Specifications for Manitowoc mechanical crane 4100W. (*Manitowoc.*)

moved closer together, but the task of transporting the crane or moving it on the job site is made easier.

Figures B10-5 and B10-6 illustrate the components and nomenclature for crawler-mounted cranes and accessory equipment as established by the Society of Automotive Engineers (SAE).

Trucks used for crane mounting are 6 by 4 or 6 by 6 for the smaller crane sizes, 8 by 4 for intermediate sizes, and 12 by 6 for the larger sizes. (See Fig. B10-7.) Other combinations are also available for special requirements. Power plants and transmissions are generally matched to give a creeping speed of 1 mi/h and travel speeds of up to 40 or 50 mi/h. The carrier power must be adequate for the required travel speed and gradeability. Outriggers which are manually extended and retracted will normally be supplied as standard equipment, with hydraulic power actuation commonly available as an option on most models. Outriggers, outrigger housings, coun-

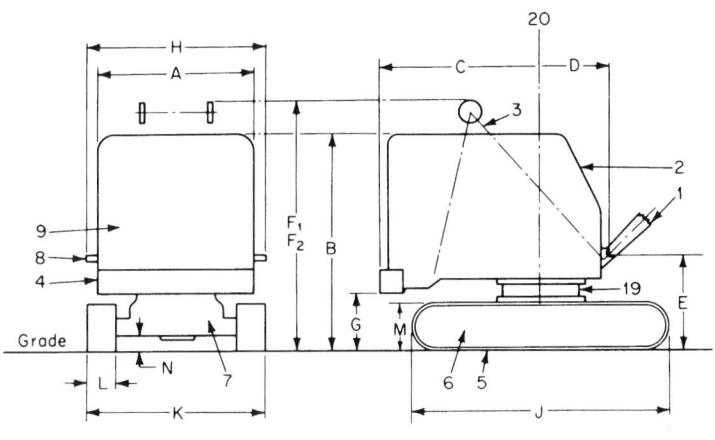

DIMENSIONS

A. Width of cab
B. Maximum height of cab above grade
C. Swing clearance (radius of rear end from axis of rotation)
D. Distance of boom foot pin to axis of rotation
E. Height of boom foot pin above grade
F_1. Gantry height above grade when in lowered position
F_2. Gantry height above grade when in raised operating position
G. Distance under counterweight to grade
H. Overall width when running boards are used
J. Overall length of crawler
K. Overall width of crawler
L. Width of crawler tread shoes
M. Height of crawler tread belt at center of end tumblers
N. Minimum clearance under crawler base to grade

DEFINITIONS

1. Front end attachment
2. Cab
3. Gantry or A frame
4. Counterweight
5. Crawler tread belt
6. Crawler side frame
7. Carbody or crawler base
8. Running board
9. Revolving superstructure
19. Swing circle or roller path
20. Axis of rotation

FIGURE B10-5 Crawler-type crane mounting. (*Society of Automotive Engineers.*)

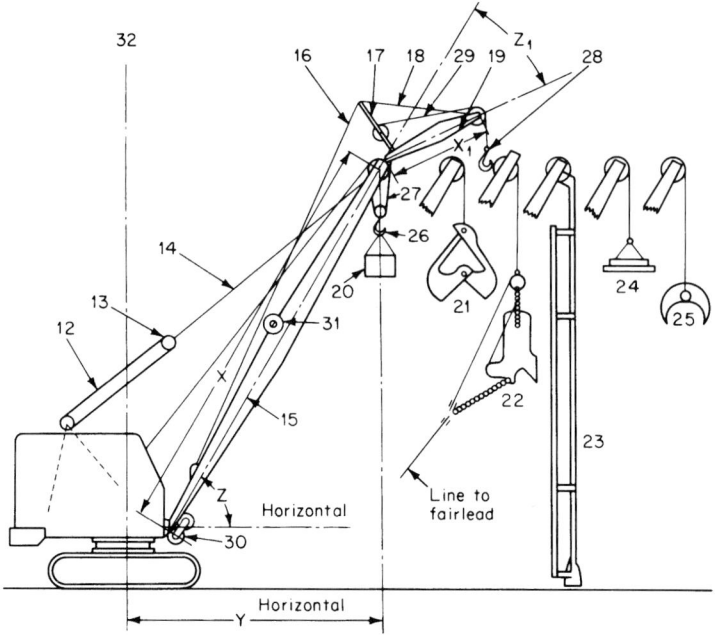

DIMENSIONS

X. Boom length from boom foot pin to boom head sheave pin
X_1. Jib length from jib foot pin to jib head sheave pin
Y. Radius of load (also applies to jib hook load)
Z. Boom angle
Z_1. Offset angle of jib (also can be given as an offset dimension)

DEFINITIONS

12. Derricking or live boom hoist rope
13. Floating harness or bridle
14. Pendants, guys or boom backstays
15. Crane boom
16. Jib backstay lines
17. Jib mast
18. Jib front stay lines
19. Jib
20. Concrete bucket
21. Clamshell bucket
22. Dragline bucket
23. Pile driver leads
24. Magnet
25. Grapple
26. Main lift-hook block
27. Main hoist line
28. Jib or whipline hook
29. Jib or auxiliary hoist line
30. Dragline fairlead
31. Tagline winder or magnet take-up reel
32. Axis of rotation

FIGURE B10-6 Boom equipment for mechanical crane. (*Society of Automotive Engineers.*)

CRANES

B10-9

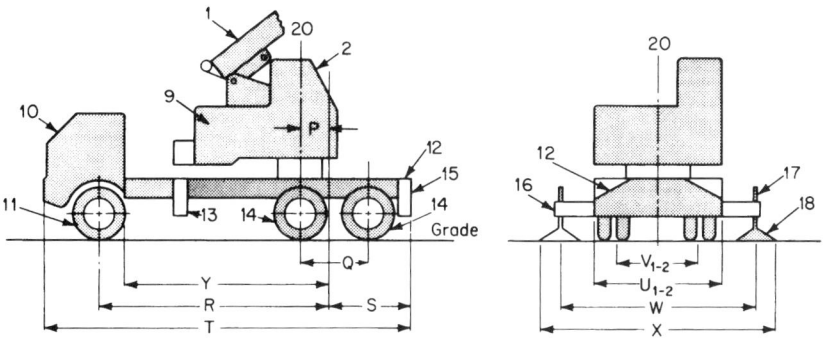

DIMENSIONS

P. Distance from center of rear axle or bogie to axis of rotation
Q. Distance between centers of axles of tandem axle bogie
R. Wheelbase (wheelbase for tandem front axle is measured to tandem center pivot point)
S. Distance from center of rear axle or bogie to rear end of frame
T. Overall length of carrier
U_1. Maximum overall width with retracted outriggers (floats removed)
U_2. Max. overall width with retracted outriggers (floats attached)
V_1. Track or tread width, rear axle
V_2. Track or tread width, front axle
W. Effective length of extended outriggers
X. Overall width over floats with outriggers extended
Y. Distance from back of carrier cab to center of rear axle or bogie (known as CA distance in trucking industry)

DEFINITIONS

1. Boom
2. Cab
9. Revolving superstructure
10. Carrier cab
11. Front axle
12. Carrier frame
13. Front outrigger box
14. Rear axle
15. Rear outrigger box
16. Outrigger beam
18. Outrigger float
19. Swing circle or roller path
20. Axis of rotation

FIGURE B10-7 Truck-type crane mounting (6 by 4 or 6 by 6 shown). (*The Power Crane and Shovel Assoc.*)

terweights, and front-end equipment can be removed prior to traveling, to the extent necessary to satisfy local highway limitations as to total load, axle loads, and load per inch of tire width. The weights of these components are available from the manufacturer. In general, the published lifting capabilities for truck-mounted cranes will be based on outriggers set, over side and over rear; and without outriggers set, over side and over rear. The over-side ratings are usually the lowest.

Figure B10-8 is a 300-ton capacity Manitowoc model M250T truck crane.

Special auxiliary equipment for mobile mechanical cranes is available when desired. This includes foot throttle control, modified hand throttles, and combination throttle controls in addition to the dual throttle controls or torque converters which were discussed earlier. Hoist-drum rotation feelers or indicators can be applied so the crane operator can detect slight drum motions without looking at the hoist drum or hoist line. Power load lowering is available on the hoist drum for easing loads into place. Its use will reduce the reliance otherwise placed on the hoist brake. Dual hoist brakes will reduce brake heating by increasing brake area and are an asset for long-

FIGURE B10-8 300-ton-capacity Manitowoc truck crane Model M250T. (*Manitowoc.*)

distance lowering or fast-cycle operation. Partially setting or dragging swing brakes can help control swing motion when handling loads in winds or on slopes. Inching devices have been applied to hoist and swing motions for fine control. Special brake linings can also modify and improve load control. Boom-angle indicators, mechanical or electrical, are an aid to the operator in spotting loads and in using the crane-capacity chart. Electronic or hydraulic load indicators can also be installed. Although they are not accurate load-weighing devices, they do aid the operator in keeping loads within the crane's capacity when either rated load or rated load radius is approached (see Fig. B10-9).

Overload indicating systems are available. These indicate an approach to an unstable condition or to any other predetermined load limitation. These devices utilize mechanical, hydraulic, or electronic means to sense the load and load radius or boom angle and thereby determine the amount of overturning moment applied by the live load. Visual and audio signals aid in preventing accidental overload of a machine when this equipment is applied.

Cold-weather starting aids can be supplied, and heaters can be provided in either or both upper and lower operator stations. Lights can be powered through individual lighting plants or through increased capacity of the electric system on the engine. Accessory equipment which is normally intended for highway trucks can be applied, such as special mirrors, clearance lights, etc. Counterweight removal devices, magnets, and magnet generators are available. The crane body can also be equipped with various types of digging attachments. Concrete-handling buckets in various configurations and sizes are available, as are load hooks, sheave blocks, etc.

FIGURE B10-9 Operator's cab equipped with electronic controls (EPIC®) positioned for easy visibility. (*Manitowoc.*)

MOBILE HYDRAULIC CRANES

The working motions of the hydraulic crane are powered and controlled through a pressurized hydraulic system. Pumps are used to supply the desired pressures, and the versatility of the crane may be enhanced as the number of pumps is increased. If a single pump is used to control more than one operating motion, then one motion must necessarily have priority over the others. True independence of motion is not obtained unless each motion is controlled by a separate pressure system.

The pumps supply hydraulic oil at pressures up to 6000 lb/in^2 to actuate cylinders or hydraulic motors and thus power the required crane motions. Each pump, motor, and cylinder is sized to give the end force and speed desired. The maximum power which can be supplied to each crane component is limited to that which can be provided by the pump which services it. The structural design of each hydraulic crane component can thus be based on the limiting load which its particular hydraulic subsystem can support. In mechanical cranes, on the other hand, it is possible to direct full crane power to each operating function, and component design must be based on this possibility.

Use of variable-volume pumps and/or motors can provide the hydraulic crane with operating characteristics which are similar to those obtained with mechanical cranes equipped with torque converters. Hoist motion is accomplished through hydraulic-motor power winch with a built-in gear reducer or a special low-speed control hydraulic motor. The winch is mounted on the boom or in the vicinity of the boom pivot point.

The crane boom can be essentially the same as traditional lift-cranes or it can be hydraulic. The hydraulic crane boom consists of two or more telescoping boxes which are fabricated from steel plate or special latticed designs (see Fig. B10-10). These boxes are extended or retracted by the action of one or more hydraulic cylinders. Some boom designs combine mechanical control with hydraulic control; for example, the upper part of the boom is extended mechanically and locked in position, then further extension is accomplished through a hydraulic system. The boom will work in a range from the horizontal to 80° above horizontal. Jibs can be applied to the tip of the boom to gain reach at the expense of a reduced maximum load.

Boom hoist is powered through hydraulic cylinders, which are locked through check valves and act as brakes to hold the boom in the raised position. Swing motion is accomplished with a hydraulic pump and motor with a gear reduction. Several variations of fixed and variable displacement pump and motor are used to obtain the desired characteristics.

The carrier for the mobile hydraulic crane can be fitted with crawlers in much the same way as described for the mobile mechanical crane. Forward motion is then accomplished through a hydraulic motor and gear reduction, located in the crane carrier. A swivel at the center of the crane body rotation carries the hydraulic fluid to this motor and returns it to the tank after use. The hydraulic motor can be used as a partial brake but, due to internal leakage, an external friction brake is used in conjunction with it.

Trucks and other wheel-type carriers can also be utilized, just as they are on mobile mechanical cranes. One variation has a single engine in the carrier, with pump drives to supply hydraulic power to the upper crane body. This eliminates the cost of a separate power plant in the upper crane body but requires a complicated swivel and a more elaborate controls system.

Where applicable, the auxiliary equipment which has been described for the mechanical cranes can also be attached to hydraulic machines. This includes boom-angle indicators, load-weighing devices, and load-moment systems. One system peculiar to hydraulic machines is the crossover of pump flow to allow two-speed application on hoist winches and propel motors.

Crane Sizes

Table B10-1 lists 18 crane models offered by Manitowoc. Six of these are the newer M-Series (hydraulic) and the other 12 are traditional (mechanical). Capacities range from 60 to 1000 tons.

Table B10-2 lists 11 hydraulic rough-terrain cranes by Link-Belt with capacities from 18 to 60 tons. Table B10-3 lists 15 hydraulic truck cranes by Link-Belt with capacities from 14 to 100 tons. Table B10-4 lists 5 lattice-boom truck cranes by Link-Belt with capacities from 35 to 250 tons. Table B10-5 lists 7 lattice-boom crawler cranes by Link-Belt. Three are mechanical models (40 to 50 ton), and five are hydraulic (75 to 165 ton). Table B10-6 lists large models by Link-Belt with capacities from 250 to 500 ton.

The crane capacities given are maximum under specific short-boom and limited-boom angle. Capacities are reduced as boom length and/or angle from the vertical

CRANES   B10-13

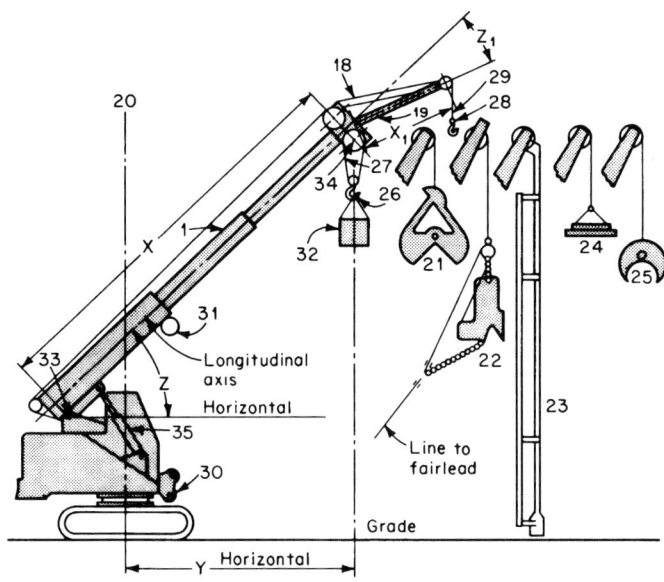

DIMENSIONS

X. Boom length pin
X_1. Jib length from jib foot pin to jib head sheave pin
Y. Radius of load (also applies to jib hook load)
Z. Boom angle
Z_1. Offset angle of jib (also can be given as an offset dimension)

DEFINITIONS

1. Crane boom
18. Jib front stay lines or brace
19. Jib
20. Axis of rotation
21. Clamshell bucket
22. Dragline bucket
23. Pile driver leads
24. Magnet
25. Grapple
26. Main lift-hook block
27. Main hoist line
28. Jib or whipline hook
29. Jib or auxiliary hoist line
30. Dragline fairlead
31. Tagline
32. Concrete bucket
33. Boom pivot pin
34. Load hoist sheave pin
35. Boom hoist cylinder

FIGURE B10-10 Boom equipment for hydraulic crane. (*The Power Crane and Shovel Assoc.*)

TABLE B10-1 Crane Models Offered by the Manitowoc Company

Liftcranes			Lifting Capacities lb[a]		
Model	Boom lengths, ft	Maximum moment[a], kip, ft	Maximum capacity on basic boom	80-ft Bm at 20-ft R	150-ft Bm at 40-ft R
M-series					
M-50W	40–180	1,449.5 at 13-ft R	120,000 at 12-ft R	53,300	19,100
M-65W	40–190	2,149.5 at 15-ft R	150,000 at 14-ft R	86,400	30,600
M-80W	40–200	2,116.8 at 13-ft R	176,400 at 12-ft R	96,200	34,400
M-85W	50–210	2,293.2 at 13-ft R	190,000 at 12-ft R	100,600	35,000
M-250	70–330[d]	11,152.8 at 24-ft R	551,200 at 18-ft R	526,400	200,300
M-250[f] S-2	70–330[d]	11,724.0 at 30-ft R	551,200 at 18-ft R	526,400	249,900
Traditional					
3900	60–210	3,200.0 at 16-ft R	200,000 at 16-ft R	137,000	48,600
3900W	60–250	4,417.5 at 19-ft R	280,000 at 15-ft R	216,600	75,300
3950W	70–260	5,064.4 at 22-ft R	300,000 at 16-ft R	249,500	92,400
4000W	70–220	5,324.8 at 16-ft R	350,000 at 15-ft R	252,600	83,100
4100W S-1	70–260	6,629.1 at 19-ft R	400,000 at 16-ft R	319,400	106,700
4100W S-2	70–260	7,041.6 at 24-ft R	460,000 at 16.5-ft R	346,100	139,200
4600 S-3	80–260	8,787.5 at 19-ft R	480,000 at 18-ft R	421,400	142,700
4600 S-4	80–310	14,000.0 at 19-ft R	700,000 at 20-ft R	700,000	215,200
4600 S-5	80–310	14,050.0 at 26-ft R	700,000 at 20-ft R	700,000	263,700
6000 S-2	75–375	23,340.8 at 32-ft R	1,200,000 at 20-ft R	75-ft Bm 1,200,000	
7000	150–400	33,352.5 at 75-ft R	700,000 at 40-ft R	150-ft Bm at 50-ft R 584,200	250-ft Bm at 70-ft R 413,000
7200[f]	155–505	90,000.0 at 45-ft R	2,000,000 at 45-ft R	155-ft Bm at 55-ft R 1,635,200	305-ft Bm at 80-ft R 800,900

[a] Capacities for W models are with crawlers extended.
[b] Maximum combination raised without assist.
[c] Weight for typically equipped liftcrane with basic boom.
[d] Lengths over 300-ft require long-reach boom top.
[e] With long-reach boom top.
[f] Specifications and capacities are preliminary, subject to test.
[g] 17-ft 6-in with optional 48-in pads.
[h] 48-in pads optional.
Source: Manitowoc.

CRANES

B10-15

200-ft Bm at 60-ft R	Maximum boom + jib[b] Combinations boom plus jib	Jib capacities, lb[a]	Crawlers Length overall	Width overall	Pad width, in	Weight[c] liftcrane, lb
			M-series			
180-ft Bm	160-ft + 30-ft	15,000		14-ft 6-in ext.		
10,000	150-ft + 60-ft	11,100	19-ft 6-in	9-ft 10-in ret.	30	102,055
190-ft Bm	180-ft + 30-ft	15,000		16-ft 8½-in ext.		
16,400	170-ft + 60-ft	10,700	22-ft 3⅜-in	11-ft 5¾-in ret.	36	135,630
18,500	190-ft + 30-ft	20,000		16-ft 3-in ext.		
	180-ft + 60-ft	12,100	22-ft 4⅛-in	12-ft 0-in ret.	30	148,680
18,100	180-ft + 30-ft	20,000		16-ft 8½-in ext.		
	170-ft + 60-ft	18,700	22-ft 3⅜-in	11-ft 5¾-in ret.	36	156,260
112,100[e]	130-ft + 40-ft	100,000				
	250-ft + 120-ft[e]	32,800[e]	30-ft 9-in	26-ft 0-in	48	471,030
137,000	130-ft + 40-ft	100,000				
	250-ft + 120-ft[e]	32,800[e]	30-ft 9-in	26-ft 0-in	48	571,030
			Traditional			
25,900	190-ft + 30-ft	40,000				
	180-ft + 50-ft	20,000	20-ft 4-in	16-ft 8-in[g]	38[h]	229,485
41,800	240-ft + 30-ft	40,000		19-ft 8-in ext.		
	230-ft + 50-ft	20,000	24-ft 0-in	17-ft 2-in ret.	48	261,100
50,600	240-ft + 30-ft	40,000		21-ft 1-in ext.		
	230-ft + 50-ft	20,000	24-ft 0-in	18-ft 7-in ret.	48	301,670
44,200	200-ft + 30-ft	40,000		21-ft 1-in ext.		
	190-ft + 60-ft	10,000	24-ft 0-in	18-ft 7-in ret.	48	312,135
58,600	240-ft + 30-ft	40,000		21-ft 1-in ext.		
	230-ft + 60-ft	10,000	26-ft 5½-in	18-ft 7-in ret.	48	366,575
77,400	240-ft + 30-ft	40,000		21-ft 1-in ext.		
	230-ft + 60-ft	10,000	26-ft 5½-in	18-ft 7-in ret.	48	450,575
78,500	230-ft + 40-ft	80,000				
	230-ft + 50-ft	78,100	26-ft 1-in	21-ft 0-in	60	482,015
118,400	280-ft + 40-ft	88,000				
	260-ft + 80-ft	80,000	30-ft 5-in	25-ft 0-in	60	590,845
148,100	280-ft + 40-ft	88,000				
	260-ft + 80-ft	80,000	30-ft 5-in	25-ft 0-in	60	742,245
	350-ft + 40-ft	160,000				
	325-ft + 100-ft	96,200	39-ft 10-in	32-ft 11-in	83	1,015,945
400-ft Bm at 100-ft R	400-ft + 40-ft	147,800				
188,400	400-ft + 100-ft	79,100	47-ft 5-in	38-ft 10-in	83	1,754,190
505-ft Bm at 110-ft R 330,200	430-ft + 140-ft	225,200	61-ft 5-in	53-ft 0-in	108	2,553,200

Key: R—Radius
Bm—Boom

TABLE B10-2 Hydraulic Rough-Terrain Cranes

Model	Capacity, ton (mt)	Height/ width/ length	Approximate weight/ axle loads	Travel speed, mi/h (km/h)	Wheelbase	Tires	Engine/type
HSP-8018C	18 (16)	10-ft 8 3/16-in (3.26 m) 8-ft (2.44 m) 32-ft 5 15/16-in (9.90 m)	18,018 F (8173 kg) 19,310 R (8759 kg) 37,328 lb (16932 kg)	24.3 (39.1)	10-ft 2-in (3.10 m)	16.0 × 24 17.5 × 25	4BT3.9 Diesel
HSP-8020C	20 (18)	10-ft 8 3/16-in (3.26 m) 8-ft (2.44 m) 32-ft 5 15/16-in (9.90 m)	18,018 F (8173 kg) 19,310 R (8759 kg) 37,328 lb (16932 kg)	24.3 (39.1)	10-ft 2-in (3.10 m)	16.0 × 24 17.5 × 25	Cummins 4BT3.9 Diesel
HSP-8022C	22 (20)	10-ft 8 3/16-in (3.26 m) 8-ft (2.44 m) 32-ft 5 15/16-in (9.90 m)	18,018 F (8173 kg) 19,310 R (8759 kg) 37,328 lb (16932 kg)	24.3 (39.1)	10-ft 2-in (3.10 m)	16.0 × 24 17.5 × 25	Cummins 4BT3.9 Diesel
RTC-8025	25 (23)	11-ft 5-in (3.47 m) 9-ft 5-in (2.87 m) 39-ft 11-in (12.16 m)	22,341 F (10134 kg) 29,725 R (13483 kg) 52,066 lb (23617 kg)	22.9 (36.85)	12-ft 6-in (3.81 m)	20.5 × 25 23.5 × 25	Cummins 6BT5.9 Diesel
RTC-8028	28 (25)	11-ft 5-in (3.47 m) 9-ft 5-in (2.87 m) 39-ft 11-in (12.16 m)	22,341 F (10,134 kg) 29,725 R (13,483 kg) 52,066 lb (23,617 kg)	22.9 (36.85)	12-ft 6-in (3.81 m)	20.5 × 25 23.5 × 25	Cummins 6BT5.9 Diesel
RTC-8030	30 (27)	11-ft 5-in (3.47 m) 9-ft 5-in (2.87 m) 39-ft 11-in (12.16 m)	22,341 F (10,134 kg) 29,725 R (13,483 kg) 52,066 lb (23,617 kg)	22.9 (36.85)	12-ft 6-in (3.81 m)	20.5 × 25 23.5 × 25	Cummins 6BT5.9 Diesel

CRANES

B10-17

Transmission	O.R. spread/base	Boom length	Boom attachments	Winches
Cummins Power shift 6 speeds	18-ft 6¾-in (5.66 m) 18-ft 1-in (5.51 m)	3-section full power 27–70-ft (8.23 m–21.34 m)	25-ft (7.62 m) fixed lattice fly 25-ft (7.62 m) offsettable lattice fly 25-ft–43-ft (7.62 m–13.11 m) offsettable telescoping lattice fly	1 speed 9000 lb (4082 kg) 282 fpm (86 m/min)
(5.66 m) Power shift 6 speeds	18-ft 6¾-in 3-section 18-ft 1-in (5.51 m)	25-ft (7.62 m) full power 27–70-ft (8.23 m–21.34 m)	25-ft (7.62 m) fixed lattice fly 1 speed offsettable lattice fly 25-ft–43-ft (7.62–13.11 m) offsettable telescoping lattice fly	9000 lb (4082 kg) 282 fpm (86 m/min)
Power shift 6 speeds	18-ft 6¾-in (5.66 m) 18-ft 1-in (5.51 m)	3-section full power 27–70-ft (8.23 m–21.34 m)	25-ft (7.62 m) fixed lattice fly 25-ft (7.62 m) offsettable lattice fly 25–43-ft (7.62–13.11 m) offsettable telescoping lattice fly	1 speed 9000 lb (4082 kg) 282 fpm (86 m/min)
Power shift 6 speeds	20-ft (6.10 m) 22-ft 3-in (6.78 m)	3-section full power 28-ft 9-in–70-ft 3-in (8.76 m–21.41 m) 4-section power pinned 28-ft 9-in–91-ft (8.76–27.74 m)	25-ft (7.62 m) fixed lattice fly 25-ft (7.62 m) offsettable lattice fly 25–43-ft (7.62–13.10 m) offsettable telescoping lattice fly	2 speed 10,360 lb (4699 kg) 390 fpm (119 m/min)
Power shift 6 speeds	20-ft (6.10 m) 22-ft 3-in (6.78 m)	3-section full power 28-ft 9-in–70-ft 3-in (8.76–21.41 m) 4-section power pinned 28-ft 9-in–91-ft (8.76–27.74 m)	25-ft (7.62 m) fixed lattice fly 25-ft (7.62 m) offsettable lattice fly 25-ft–43-ft (7.62–13.10 m) offsettable telescoping lattice fly	2 speed 10,360 lb (4699 kg) 390 fpm (119 m/min)
(6.10 m) Power shift 6 speeds	20-ft 28-ft 9-in–70-ft 3-in 22-ft 3-in (6.78 m)	3-section full power offsettable lattice fly (8.76–21.41 m) 4-section power pinned 28-ft 9-in–91-ft (8.76–27.74 m)	25-ft (7.62 m) fixed lattice fly 25-ft (7.62 m) 10,360 lb 25-ft–43-ft (7.62–13.10 m) offsettable telescoping lattice fly	2 speed (4699 kg) 390 fpm (119 m/min)

TABLE B10-2 Hydraulic Rough-Terrain Cranes (*Continued*)

Model	Capacity, ton (mt)	Height/ width/ length	Approximate weight/ axle loads	Travel speed, mi/h (km/h)	Wheelbase	Tires
HSP-8035S		11-ft 6-in (3.51 m)	23,744 F (10,770 kg)			
	35 (32)	9-ft 3-in (2.82 m)	28,527 R (12,940 kg)	22.8 (36.70)	12-ft 6-in (3.81 m)	20.5 × 25 23.5 × 25
		41-ft 11¼-in (12.78 m)	52,271 lb (23,710 kg)			
HSP-8035		12-ft 7¾-in (3.85 m)	34,690 F (15,735 kg)			
	35 (32)	9-ft 8-in (2.95 m)	40,378 R (18,315 kg)	21 (34)	12-ft 7-in (3.84 m)	21.0 × 25
		46-ft 1⅜-in (14.06 m)	75,068 lb (34,051 kg)			
HSP-8040		12-ft 7¾-in (3.85 m)	34,690 F (15,735 kg)			
	40 (36)	9-ft 8-in (2.95 m)	40,378 R (18,315 kg)	21 (34)	12-ft 7-in (3.84 m)	21.0 × 25
		46-ft 1⅜-in (14.06 m)	75,068 lb (34,051 kg)			
HSP-8050		12-ft 6¾-in (3.83 m)	34,200 F (15,513 kg)			
	50 (45)	10-ft 10-in (3.30 m)	43,788 R (18,315 kg)	21 (34)	12-ft 7-in (3.84 m)	26.5 × 25 29.5 × 25
		46-ft 1⅜-in (14.06 m)	77,988 lb (33,829 kg)			
HSP-8060		12-ft 8¼-in (3.87 m)	43,406 F (19,689 kg)			
	60 (54)	10-ft 9½-in (3.29 m) 46-ft 1⅜-in (14.06 m)	43,212 R (19,601 kg) 86,618 lb (39,290 kg)	21 (34)	12-ft 7-in (3.84 m)	29.5 × 25

Source: Link-Belt Company.

Engine/type	Transmission	O.R. spread/base	Boom length	Boom attachments	Winches
Cummins 6BT5.9 Diesel	Power shift 6 speeds	20-ft (6.10 m) 22-ft 3-in (6.78 m)	3-section full power 32-ft–80-ft (9.75–24.38 m) 4-section power pinned 32-ft–101-ft (9.75–30.78 m)	29-ft (8.84 m) lattice fly 21-ft (6.40 m) jib	2 speed 11,700 lb (5307 kg) 413 fpm (126 m/min)
GM 6V-53N Diesel	Power shift 6 speeds	22-ft (6.71 m) 23-ft (7.01 m)	3-section full power 35-ft–85-ft (10.67–25.91 m) 4-section power pinned 35-ft–110-ft (10.67–33.53 m)	33-in (10.06 m) lattice fly 25-ft (7.62 m) jib	2 speed 15,870 lb (7199 kg) 548 fpm (167 m/min)
GM 6V-53N Diesel	Power shift 6 speeds	22-ft (6.71 m) 23-ft (7.01 m)	3-section full power 35-ft–85-ft (10.67 m–25.91 m) 4-section power pinned 35-ft–110-ft (10.67–33.53 m)	33-ft (10.06 m) lattice fly 25-ft (7.62 m) jib	2 speed 15,870 lb (7199 kg) 548 fpm (167 m/min)
GM 6V-53N Diesel	Power shift 6 speeds	22-ft (6.71 m) 23-ft (7.01 m)	3-section full power 35-ft–85-ft (10.67–25.91 m) 4-section power pinned 35-ft–110-ft (10.67–33.53 m)	33-ft (10.06 m) lattice fly 25-ft (7.62 m) jib	2 speed 15,870 lb (7199 kg) 548 fpm (167 m/min)
GM 6V-53N Diesel	Power shift 6 speeds	23-ft (7.01 m) 23-ft (7.01 m)	4-section power pinned 35-ft–110-ft (10.67–33.53 m)	33-ft (10.06 m) lattice fly 25-ft (7.62 m) jib	2 speed 15,870 lb (7199 kg) 548 fpm (167 m/min)

TABLE B10-3 Hydraulic Truck Cranes (Link-Belt)

Model	Capacity ton (mt)	Height/ width/ length	Approximate weight/ axle loads	Travel speed, mi/h (km/h)	Wheelbase	Tires
HTC-814		11-ft 2½-in (3.42 m)	16,174 F (7337 kg)			
	14 (13)	8-ft (2.44 m)	32,369 R (14,683 kg)	47.3 (76.12)	18-ft (5.49 m)	16.5 × 22.5 F 10.0 × 20 R
		40-ft (12.19 m)	48,543 lb (22,019 kg)			
HTC-814XL		11-ft 2½-in (3.42 m)	17,076 F (7746 kg)			
	14 (13)	8-ft (2.44 m)	34,093 R (15,465 kg)	55.1 (88.9)	18-ft (5.49 m)	16.5 × 22.5 F 10.0 × 20 R
		40-ft (12.19 m)	51,169 lb (23,210 kg)			
HTC-830		11-ft 2½-in (3.42 m)	16,174 F (7337 kg)			
	30 (27)	8-ft (2.44 m)	32,369 R (14,683 kg)	47.3 (76.12)	18-ft (5.49 m)	16.5 × 22.5 F 10.0 × 20 R
		40-ft (12.19 m)	48,543 lb (22,019 kg)			
HTC-835		11-ft 2½-in (3.42 m)	16,634 F (7545 kg)			
	35 (32)	8-ft (2.44 m)	34,709 R (15,744 kg)	47.3 (76.12)	18-ft (5.49 m)	16.5 × 22.5 F 10.0 × 20 R
		40-ft (12.19 m)	51,343 lb (23,289 kg)			
HTC-835XL		10-ft 11⅞-in (3.35 m)	27,596 F (12,518 kg)			
	35 (38)	8-ft (2.44 m)	42,573 R (19,311 kg)	50.4 (81.11)	19-ft (5.79 m)	15.0 × 22.5 F 11.0 × 20 R
		45-ft 9¼-in (13.95 m)	70,169 lb (31,829 kg)			
HTC-840		10-ft 11⅞-in (3.35 m)	27,135 F (12,308 kg)			
	40 (36)	8-ft (2.44 m)	43,977 R (19,948 kg)	50.4 (81.11)	19-ft (5.79 m)	15.0 × 22.5 F 11.0 × 20 R
		45-ft 9¼-in (13.95 m)	71,112 lb (32,256 kg)			
HTC-850		10-ft 11⅞-in (3.35 m)	27,217 F (12,346 kg)			
	50 (45)	8-ft (2.44 m)	44,031 R (19,972 kg)	50.4 (81.11)	19-ft (5.79 m)	16.5 × 22.5 F 11.0 × 20 R
		45-ft 9¼-in (13.95 m)	71,248 lb (32,318 kg)			

CRANES B10-21

Engine/type	Transmission	O.R. spread/base	Boom length	Boom attachments	Winches
Cummins 6CTA8.3 Diesel	Fuller RT 6613 13 speeds	18-ft (5.49 m) 18-ft (5.49 m)	3-section full power 32-ft–80-ft (9.75–24.38 m) 4-section power pinned 32-ft–101-ft (9.75–30.78 m)	29-ft 0-in (8.84 m) lattice fly 21-ft (6.40 m) jib	2 speed 11,700 lb (5307 kg) 414 fpm (126 m/min)
Cummins 6CTA8.3 Diesel	Fuller RT 6613 13 speeds	18-ft (5.49 m) 18-ft (5.49 m)	3-section full power 32-ft–80-ft (9.75–24.38 m) 4-section power pinned 32-ft–101-ft (7.21–30.78 m)	29-ft (8.84 m) lattice fly 21-ft (6.40 m) jib	2 speed 11,700 lb (5307 kg) 414 fpm (126 m/min)
Cummins 6CTA8.3 Diesel	Fuller RT 6613 13 speeds	18-ft (5.49 m) 18-ft (5.49 m)	3-section full power 32-ft–80-ft (9.75–24.38 m) 4-section power pinned 35-ft–101-ft (10.67–30.78 m)	29-ft (8.84 m) lattice fly 21-ft (6.40 m) jib	2 speed 11,700 lb (5307 kg) 414 fpm (126 m/min)
Cummins 6CTA8.3 Diesel	Fuller RT 6613 13 speeds	18-ft (5.49 m) 18-ft (5.49 m)	3-section full power 32-ft–80-ft (9.75–24.38 m) 4-section power pinned 32-ft–101-ft (9.75–30.78 m)	29-ft (8.84 m) lattice fly 21-ft (6.40 m) jib	2 speed 11,700 lb (5307 kg) 414 fpm (126 m/min)
Cummins 6CTA8.3 Diesel	Fuller RTO 6613 13 speeds	20-ft (6.10 m) 19-ft 11-in (6.07 m)	3-section full power 35-ft–85-ft (10.67–25.91 m) 4-section power pinned 35-ft–110-ft (10.67–33.53 m)	33-ft (10.06 m) lattice fly 25-ft (7.62 m) jib	2 speed 15,870 lb (7199 kg) 548 fpm (167 m/min)
Cummins 6CTA8.3 Diesel	Fuller RTO 6613 13 speeds	20-ft (6.10 m) 19-ft 11-in (6.07 m)	3-section full power 35-ft–85-ft (10.67–25.91 m) 4-section power pinned 35-ft–110-ft (10.67–33.53 m)	33-ft (10.06 m) lattice fly 25-ft (7.62 m) jib	2 speed 15,870 lb (7199 kg) 548 fpm (167 m/min)
Cummins 6CTA8.3 Diesel	Fuller RTO 6613 13 speeds	20-ft (6.10 m) 19-ft 11-in (6.07 m)	3-section full power 35-ft–85-ft (10.67–25.91 m) 4-section power pinned 35-ft–110-ft (10.67–33.53 m)	33-ft (10.06 m) lattice fly 25-ft (7.62 m) jib	2 speed 15,870 lb (7199 kg) 548 fpm (167 m/min)

TABLE B10-3 Hydraulic Truck Cranes (Link-Belt) *(Continued)*

Model	Capacity ton (mt)	Height/ width/ length	Approximate weight/ axle loads	Travel speed, mi/h (km/h)	Wheelbase	Tires
HTC-860		10-ft 11⅞-in (3.35 m)	30,524 F (13,846 kg)			
	60 (54)	8-ft (2.44 m)	45,198 R (20,502 kg)	50.4 (81.11)	19-ft (5.79 m)	18.0 × 22.5 F 11.0 × 20 R
		45-ft 9¼-in (13.95 m)	75,722 lb (34,348 kg)			
HTC-8650			29,523 F (13,392 kg)			
	50 (45)	8-ft 6-in (2.59 m)	51,444 R (23,335 kg)	50 (80)	19-ft (5.79 m)	18.0 × 22.5 F 12.0 × 20 R
			80,967 lb (36,727 kg)			
HTC-8660			31,263 F (14,181 kg)			
	60 (54)	8-ft 6-in (2.59 m)	54,264 R (24,614 kg)	50 (80)	19-ft (5.79 m)	18.0 × 22.5 F 12.0 × 20 R
			85,527 lb (38,795 kg)			
HTC-1040		11-ft 2⁷⁄₁₆-in (3.41 m)	25,250 F (11,453 kg)			
	40 (36)	10-ft 1½-in (3.09 m)	45,183 R (20,495 kg)	50.4 (81.3)	19-ft 2½-in (5.85 m)	15.0 × 22.5 F 11.0 × 20 R
		45-ft 9½-in (13.96 m)	70,433 lb (31,948 kg)			
HTC-1050		11-ft 2⁷⁄₁₆-in (3.41 m)	25,295 F (11,474 kg)			
	50 (45)	10-ft 1½-in (3.09 m)	45,238 R (20,520 kg)	50.4 (81.3)	19-ft 2½-in (5.85 m)	16.5 × 22.5 F 11.0 × 20 R
		45-ft 9½-in (13.96 m)	70,533 lb (31,994 kg)			
HTC-1060		10-ft 11¹³⁄₁₆-in (3.35 m)	32,636 F (14,804 kg)			
	60 (54)	10-ft 1½-in (3.09 m)	43,639 R (19,795 kg)	50.4 (81.3)	19-ft 2½-in (5.85 m)	18.0 × 22.5 F 11.0 × 20 R
		48-ft 6⅜-in (14.80 m)	76,285 lb (34,603 kg)			

Engine/type	Transmission	O.R. spread/base	Boom length	Boom attachments	Winches
Cummins 6CTA8.3 Diesel	Fuller RTO 6613 13 speeds	20-ft (6.10 m) 19-ft 11-in (6.07 m)	4-section power pinned 35-ft–110-ft (10.67–33.53 m)	33-ft (10.06 m) lattice fly 25-ft (7.62 m) jib	2 speed 15,870 lb (7199 kg) 548 fpm (167 m/min)
Cummins LTA-10-C Diesel	20-ft 6-in Eaton RTO-11609B 9 speeds	(6.25 m) 19-ft 11-in (6.07 m)	4-section power pinned 35-ft 6-in–110-ft (10.82–33.53 m)	34-ft (10.36 m) offsettable lattice fly 34-ft–56-ft (10.36–17.07 m) offsettable 2-piece lattice fly	2 speed 16,800 lb (7620 kg) 498 fpm (152 m/min)
Cummins LTA-10-C Diesel	Eaton RTO-11609B 9 speeds	24-ft 0-in (7.32 m) 19-ft 11-in (6.07 m)	4-section power pinned 35-ft 6-in–110-ft (10.82–33.53 m)	34-ft (10.36 m) offsettable lattice fly 34-ft–56-ft (10.36–17.07 m) offsettable 2-piece lattice fly	2 speed 16,800 lb (7620 kg) 498 fpm (152 m/min)
Cummins 6CTA8.3 Diesel	Fuller RTO 6613 13 speeds	22-ft 4-in (6.81 m) 19-ft 2-in (5.84 m)	3-section full power 35-ft–85-ft (10.67–25.91 m) 4-section power pinned 35-ft–110-ft (10.67–33.53 m)	33-ft (10.06 m) lattice fly 25-ft (7.62 m) jib	2 speed 15,870 lb (7199 kg) 548 fpm (167 m/min)
Cummins 6CTA8.3 Diesel	Fuller RTO 6613 13 speeds	22-ft 4-in (6.81 m) 19-ft 2-in (5.84 m)	3-section full power 35-ft–85-ft (10.67–25.91 m) 4-section power pinned 35-ft–110-ft (10.67–33.53 m)	33-ft (10.06 m) lattice fly 25-ft (7.62 m) jib	2 speed 15,870 lb (7199 kg) 548 fpm (167 m/min)
Cummins 6CTA8.3 Diesel	Fuller RTO 6613 13 speeds	22-ft 4-in (6.81 m) 19-ft 2-in (5.85 m)	4-section power pinned 38-ft–120-ft 6-in (11.58–36.73 m) 4-section power pinned 35-ft–110-ft (10.67–33.53 m)	34-ft 6-in (10.52 m) lattice fly 33-ft (10.06 m) lattice fly 25-ft (7.62 m) jib 30-ft (9.14 m) jib	2 speed 15,870 lb (7199 kg) 548 fpm (167 m/min)

TABLE B10-3 Hydraulic Truck Cranes (Link-Belt) *(Continued)*

Model	Capacity ton (mt)	Height/ width/ length	Approximate weight/ axle loads	Travel speed, mi/h (km/h)	Wheelbase	Tires
HTC-1170						
	70 (64)	11-ft 3¾-in (3.45 m)	34,724 F (15,751 kg)			
		11-ft (3.35 m)	62,156 R (28,194 kg)	48.1 (77.4)	19-ft 5-in (5.92 m)	14.0 × 20 F 14.0 × 20 R
		48-ft 8¾-in (14.85 m)	96,880 lb (43,945 kg)			
HTC-11100						
	100 (91)	11-ft 10-in (3.61 m)	40,555 F (18,396 kg)			
		11-ft 0-in (3.35 m)	72,479 R (32,876 kg)	48.1 (77.4)	19-ft 5-in (5.92 m)	14.0 × 20 F 14.0 × 20 R
		47-ft 1¼-in (14.36 m)	113,034 lb (51,272 kg)			

are increased. Figure B10-11 shows the boom and jib combinations for the 300-ton lift crane Manitowoc M250.

Capacity Enhancements

The use of counterweights was mentioned previously. Other enhancements can be used to increase lift crane capacity. Figure B10-12 shows the 300-ton lift-crane M250 (Manitowoc) with counterweights. These can be varied to suit the task at hand.

Figure B10-13 shows the M250 using the patented X-SPANDER™ which moves the counterweights to increase the load-resisting moment.

Figure B10-14 shows the M250 with the patented MAX-ER™ attachment, hanging counterweight, and 140-ft boom lifting 250 tons at a 26-ft radius during predelivery testing. Requiring just 36 ft of tailswing, the MAX-ER boosts the M250's capacity by up to 250 percent at most-used radii and up to 1000 percent at longer reaches. It permits the M250 to raise boom-and-jib combinations of up to 340 ft plus 120 ft. This is 90 ft more than the basic reach (Fig. B10-11).

Figure B10-15 shows the M250 with eight different configurations, of which four are enhancements: X-SPANDER, MAX-SPANDER, MAX-ER with hanging counterweight, and X-TENDER.

Ringers. The Manitowoc Engineering Company registered the RINGER® concept in 1974. Figure B10-16 shows the M1200 RINGER attachments. The diameter

Engine/type	Transmission	O.R. spread/base	Boom length	Boom attachments	Winches
Cummins NTA-C310 Diesel	Fuller RT 12609A 9 speeds	35-ft 23-ft 6-in (7.16 m) 20-ft 3-in (6.17 m)	38-ft–120-ft 6-in (11.58–36.73 m)	(10.67 m) offsettable lattice fly 35-ft–61-ft (10.67–18.59 m) offsettable telescoping fly 75-ft offsettable lattice jib (22.86 m)	2 speed 15,870 lb (7199 kg) 548 fpm (167 m/min)
Cummins NTA-C400 Diesel	Fuller RT 12609A 9 speeds	27-ft 6-in (8.38 m) 20-ft 3-in (6.17 m)	37-ft–115-ft (11.28–35.05 m)	33-ft (10.06 m) lattice fly 27-ft (8.23 m) jib 88-ft (26.82 m) offsettable lattice jib 103-ft (31.39 m) offsettable lattice jib	2 speed 16,800 lb (7620 kg) 509 fpm (155 m/min)

is 60 ft and the counterweight over 1000 tons. Forty-eight support pedestals provide stability while minimizing ground-bearing pressure.

The RINGER obtains its increased lifting capacity for the following:

- The basic crane's standard boom is utilized as a fixed mast which forms a gantry.
- The basic machine becomes a counterweight in combination with the RINGER counterweight.
- One end of the new carrier is pin-connected to the basic-crane's standard boom hinge (Fig. B10-17) and the other end houses new larger rollers (Fig. B10-18) which rest on a separate, larger ring and roller path (Fig. B10-19).

In Figure B10-20, a 250M crane with an M1200 RINGER assembly, 253-ft boom, and 100-ft jib is lifting 882 tons at a 75-ft radius.

Table B10-7 shows capacities for Manitowoc 4100W and 4600 S-4 cranes at various configurations.

Shipping

The capability to easily disassemble, ship, and reassemble a crane is an important consideration. Figure B10-21 shows the low-clearance folding gantry on a 4100W crane with telescoping backhitch legs when gantry is lowered. Figure B10-22 shows

TABLE B10-4 Lattice-Boom Truck Cranes (Link-Belt)

Truck crane models	Capacity, ton (mt)	PCSA rating	Basic boom, ft (m)	Boom type and size, in (m)	Maximum boom, ft (m)	Maximum boom and jib, ft (m)	Maximum luffing attachment capacity, ton (mt)
HC-78B	35 (32)	10-183	40 (12.19)	0.34 × 34 (0.86 × 0.86) angle 48 × 39 (1.22 × .99) tubular	100 (30.48) 170 (51.82)	100 + 40 (30.5 + 12.2) 150 + 20 (45.7 + 6.1)	NA
HC-108D	50 (45)	10-254	40 (12.19)	0.34 × 34 (0.86 × 0.86) angle	100 (30.48)	100 + 40 (30.5 + 12.2)	NA
HC-228H	125 (114)	12-639	50 (15.24)	62 × 70 (1.6 × 1.8) tubular	240 (73.15)	210 + 70 (64.0 + 21.3)	NA
HC-248H	165 (150)	12-921	50 (15.24)	80 × 68 (2.0 × 1.7) tubular	280 (85.34)	240 + 100 (73.15 + 30.48)	42 (38)
HC-268	250 (227)	12-1471	60 (18.29)	80 × 68 (2.0 × 1.7) tubular	330 (100.58)	300 + 90 (91.4 + 27.4)	42.5 (38.58)

Maximum luffing attachment length, ft (m)	Number of axles	Wheelbase, in (m)	Outrigger spread/base	Travel width	Travel height	Travel weight, lb (kg)
NA	4	183 (4.65)	17-ft 0-in (5.18 m) 15-ft 4-in (4.67 m)	9-ft 0-in (2.74 m)	12-ft 9-in (3.87 m)	60,068 (27,247)
NA	4	224 (5.69)	17-ft 2-in (5.23 m) 18-ft 4-in (5.59 m)	10-ft 4-in (3.15 m)	13-ft 2-in (4.01 m)	70,850 lb (32,138)
NA	4	260 (6.60)	21-ft 0-in (6.40 m) 20-ft 6-in (6.25 m)	11-ft 0-in (3.35 m)	12-ft 0-in (3.66 m)	76,770 (34,823)
190 (57.91) luffing boom 150 (45.72) luffing jib 30 (9.14) fixed jib	4	282 (7.16)	22-ft 8-in (6.91 m) 19-ft 10-in (6.05 m)	11-ft 0-in (3.35 m)	12-ft 6¾-in (3.83 m)	89,710 (40692)
250 (76.20) luffing boom 200 (60.96) luffing jib 30 (9.14) fixed jib	6	288 (7.32)	24-ft 6-in (7.47 m) 24-ft 6-in (7.47)	11-ft 10-in (3.61 m)	12-ft 11-in (3.94 m)	Upper 81,580 (37005) Carrier 87,540 (39708) self-undecking available

TABLE B10-5 Lattice-Boom Crawler Cranes (Link-Belt)
Mechanical models

Mechanical models	Capacity, ton (mt)	PCSA rating	Maximum clamshell capacity (PCSA), lb (kg)	Maximum dragline capacity (PCSA), lb (kg)	Maximum duty cycle boom, ft (m)	Basic boom, ft (m)
LS-98D	40 (36)	10-105	13,600 (6169)	13,600 (6169)	60 (18.29)	40 (12.19)
LS-108D	50 (45)	10-137	13,600 lb (6169 kg)	13,600 (6169)	70 (21.34)	40 (12.19)
LS-110C	50 (45)	10-197	16,800 lb (7620 kg)	16,800 (7620)	70 (21.34)	40 (12.19)

Boom size and type, in (m)	Maximum crane boom, ft (m)	Maximum crane boom and jib, ft (m)	Track gauge	Track length	Crane operating weight, lb (kg)
34 × 34 (0.86 × 0.86) Angle	100 (30.48)	90 + 40 (27.43 + 12.19)	9-ft 6-in (2.90 m)	15-ft 1-in (4.60 m)	73,595 (33,383)
42 × 42 (1.07 × 1.07) Angle	130 (39.62)	100 + 50 (30.48 + 15.24)	Extended 10-ft 8-in (3.25 m) Retracted 8-ft 11-in (2.72 m)	15-ft 1-in (4.60 m)	83,860 (38,039)
42 × 42 (1.07 × 1.07) Angle	140 (42.67)	100 + 50 (30.48 + 15.24)	Extended 11-ft 9-in (3.58 m) Retracted 8-ft 4-in (2.54 m)	18-ft 4-in (5.59 m)	90,670 (41,128)

TABLE B10-5 Lattice-Boom Crawler Cranes (Link-Belt) (*Continued*)

Hydraulic models

Hydraulic models	Capacity, ton (m)	PCSA rating	Maximum clamshell capacity (PCSA), lb (kg)	Maximum dragline capacity (PCSA) lb (kg)	Maximum duty cycle boom, ft (m)	Basic boom, ft (m)	Boom size and type, in (m)
LS-138H							
	75 (68)	12-268	15,800 (7167)	15,800 (7167)	70 (21.34)	40 (12.19)	48 × 48 (1.22 × 1.22) angle
					70 (21.34)	40 (12.19)	44 × 54 (1.12 × 1.37) tubular
LS-208H							
					80 (24.38)	40 (12.19)	48 × 48 (1.22 × 1.22) angle
	75 (68)	12-268	15,800 (7167)	15,800 (7167)	70 (21.34)	40 (12.19)	44 × 54 (1.12 × 1.37) tubular
LS-218H							
	100 (91)	12-413	18,000 (8165)	18,000 (8165)	80 (24.38)	40 (12.19)	50 × 60 tubular (1.27 × 1.52)
LS-248H							
	165 (150)	12-1080	NA	NA	NA	50 (15.24)	68 × 80 tubular (1.73 × 2.03)

Maximum crane boom, ft (m)	Maximum crane boom and jib, ft (m)	Maximum luffing attachment capacity, ton (m)	Maximum luffing attachment length, ft (m)	Track gauge	Track length	Crane operating weight, lb (kg)
150 (45.72)	150 + 40 (45.72 + 12.19)	NA	NA	Extended 13-ft 0-in (3.96 m)	19-ft 5-in (5.92 m)	123,280 (55,920)
180 (54.86)	160 + 50 (48.77 + 15.24)			Retracted 9-ft 3-in (2.82 m)		
150 (45.72)	150 + 40 (45.72 + 12.19)	NA	NA	Extended 13-ft 0-in (3.96 m)	19-ft 5-in (5.92 m)	130,000 (58,968)
180 (54.86)	160 + 50 (48.77 + 15.24)			Retracted 9-ft 3-in (2.82 m)		
200 (60.96)	180 + 60 (4.57 + 18.29)	NA	NA	Extended 13-ft 5-in (4.09 m)	21-ft 0-in (6.40 m)	176,400 (80,015)
				Retracted 9-ft 0-in (2.74 m)		
280 (85.34)	240 + 100 (73.15 + 30.48)	42 (38)	200 (60.96) luffing boom 150 (45.72) luffing jib 30 (9.14) fixed jib	18-ft 10-in (5.74 m)	28-ft 6-in (8.69 m)	275,100 (124,785)

TABLE B10-6 Large Cranes (Link-Belt)

Large models	Capacity, ton (mt)	PCSA rating	Maximum clamshell capacity, lb (kg)	Maximum duty cycle boom, ft (m)	Maximum crane boom, ft (m)	Maximum crane boom and jib, ft (m)
LS-718	250 (227)	17-1494	31,500 (14,288)	130 (39.62)	290 (88.39)	240 + 120 (73.15 + 36.58)
LS-818	300 (272)	17-2188	31,500 (14,288)	130 (39.62)	310 (94.49)	260 + 120 (79.25 + 36.58)
LS-918	400 (363)	18-3048	42,000 (19,051)	160 (48.77)	340 (103.63)	300 + 100 (91.44 + 30.48)
LS-1018	500 (454)	18-4298	42,000 (19,051)	160 (48.77)	340 (103.63)	300 + 100 (91.44 + 30.48)

Maximum tower capacity, ton (mt)	Maximum tower boom length	Track length and gauge	Maximum heavy lift capacity, ton (mt)	Maximum heavy lift boom and jib, ft (m)	Crane operating weight, lb (kg)
50 (45)	250-ft tower (76.2 m) 200-ft boom (60.96 m) 80-ft jib (9.14 m)	29-ft 11-in (9.12 m) 17-ft 6-in (5.33 m)	360 (327)	370 + 120 (112.78 + 36.58)	410,700 (186,294)
50 (45)	250-ft tower (76.2 m) 200-ft boom (60.96 m) 30-ft jib (9.14 m)	31-ft 4-in (9.55 m) 22-ft 6-in (6.86 m)	360 (327)	370 + 120 (112.78 + 36.58)	443,910 (201,358)
100 (91)	254-ft tower (77.42 m) 240-ft boom (73.15 m)	34-ft 9-in (10.59 m) 22-ft 6-in (6.86 m)	NA	NA	629,560 (285,568)
100 (91)	264-ft tower (80.46 m) 240-ft boom (73.15 m)	38-ft 6-in (11.73 m) 26-ft 0-in (7.92 m)	NA	NA	730,100 (331,173)

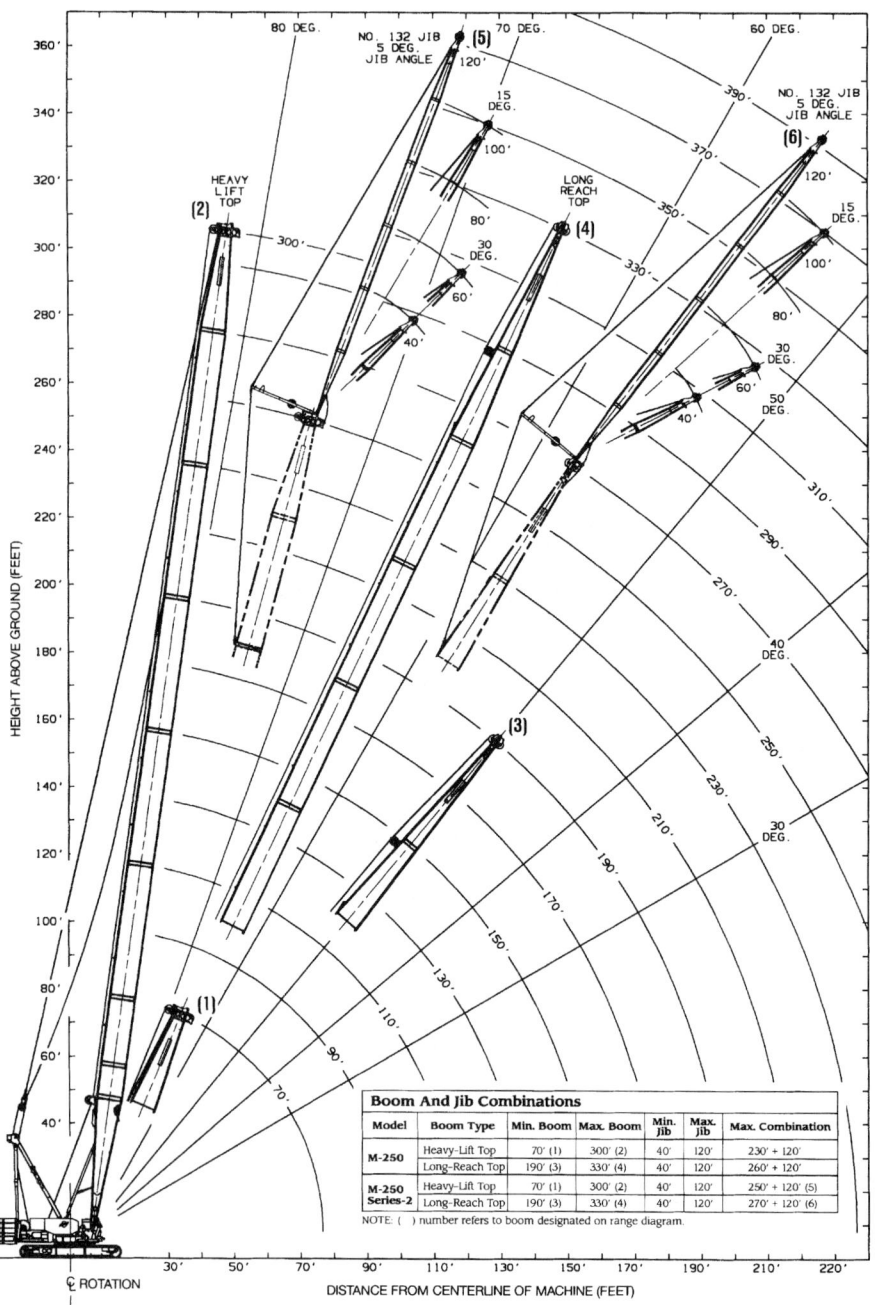

FIGURE B10-11 Boom and jib combinations for the 300-ton lift crane M250. (*Manitowoc.*)

CRANES B10-35

FIGURE B10-12 M250 crane with counterweights. (*Manitowoc.*)

FIGURE B10-13 M250 crane using X-SPANDER™. (*Manitowoc.*)

FIGURE B10-14 M250 crane with MAX-ER attachment and hanging counterweight. (*Manitowoc.*)

an M-Series crane using its own gantry to remove counterweights prior to shipping. Figure B10-23 shows an M-Series crane loaded on a low body trailer. Note that the car body and tracks are fully assembled.

One of the Manitowoc's objectives in designing cranes is to use high-strength, low-weight structures to facilitate shipping. Most models ship with car body, crawlers, upper works, gantry, equalizer, and boom butt intact (see Fig. B10-24). Table B10-8 lists shipability for four M-series crane models.

TOWER CRANES

There are three basic types of tower cranes: (1) mobile, or a modified crane attachment of tower and jib boom on a mobile crane; (2) rail mounted; and (3) climbing or self-erecting. Each of these will do basically the same job, but in a different fashion. The first and second types stay on the perimeter of the working area and reach inside it, while the third type works from a fixed location within the facility which is being constructed. The mobile tower crane has the versatility of quick setup on a job and easy conversion to other modes of operation. The rail type is less versatile, but for large, long-range projects in a confined area it has a lower initial cost. However,

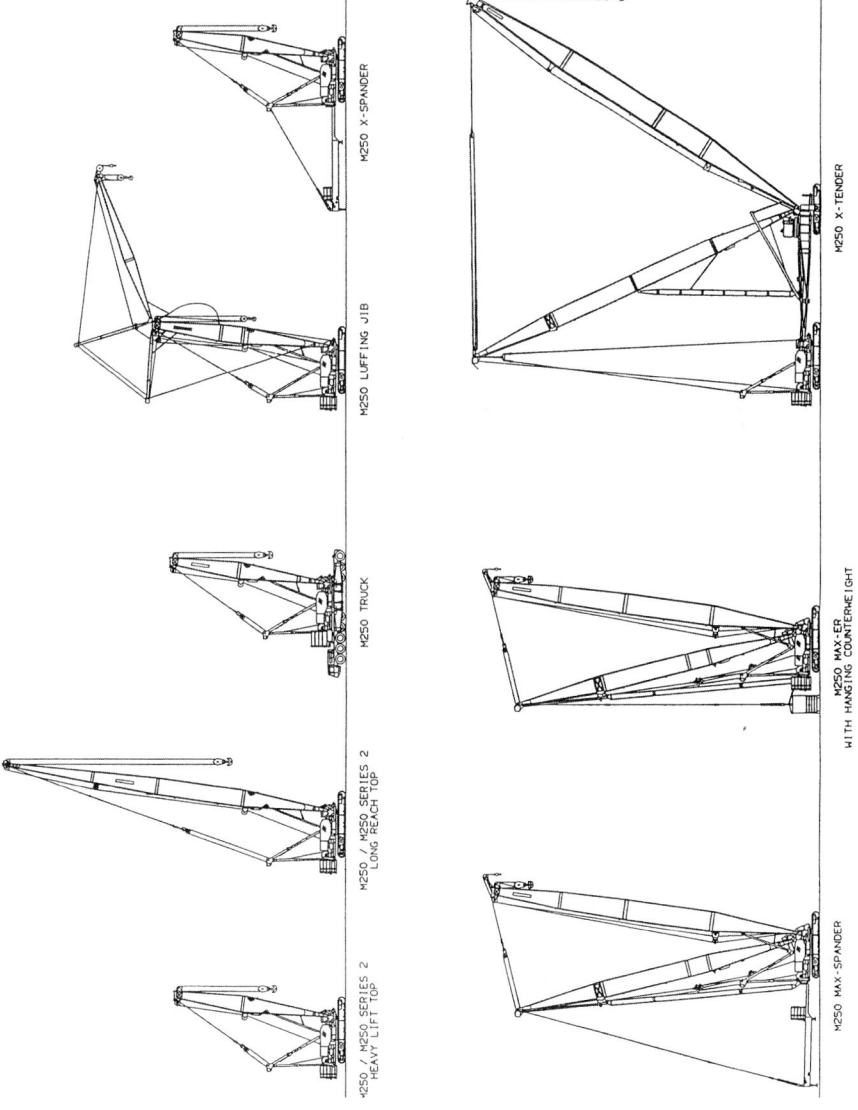

FIGURE B10-15 The M250 in eight configurations, of which four are enhancements. (*Manitowoc.*)

TABLE B10-7 Capacities at Various Configurations Using Ringers

Model	Boom length, ft	Maximum moment, kip ft	Lifting capacities Max. capacity[a]	260-ft boom		Max. boom and jib Combinations boom plus jib, ft	Max. cap.	Aux. cwt.
4100W[b]	140–340	27,355 at 50-ft R	600,000 at 45-ft R	282,500 at 70-ft R	140,500 at 130-ft R	340 + 40 340 + 140[c]	56,600 65,000	275,000
36-ft platform	140–340	27,000 at 50-ft R	600,000 at 45-ft R	271,600 at 70-ft R	140,500 at 130-ft R	340 + 40[d] 340 + 140[d]	65,000[d] 65,000[d]	550,000
4600 S-4[e]	140–400	105,000 at 70-ft R	1,500,000 at 70-ft R	834,500 at 80-ft R	508,900 at 140-ft R	380 + 80 380 + 120	268,300 220,700	978,700
60-ft platform	140–400	105,000 at 70-ft R	1,500,000 at 70-ft R	834,500 at 80-ft R	509,200 at 140-ft R	380 + 80 380 + 120	268,300 220,700	1,296,300

[a] On minimum boom.
[b] 36-ft diameter ring.
[c] 25-ft strut required.
[d] Preliminary, subject to test.
[e] 60-ft diameter ring.
Source: Manitowoc.

FIGURE B10-16 M1200 RINGER®. (*Manitowoc.*)

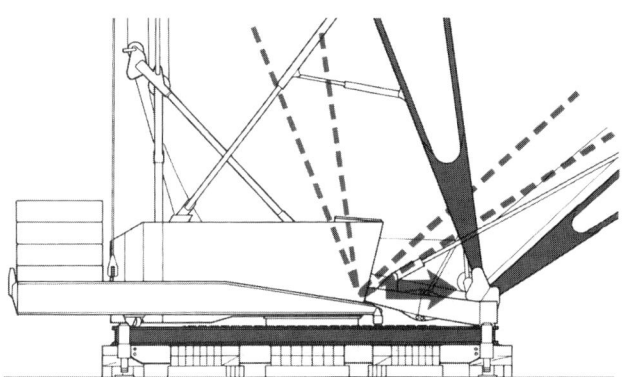

FIGURE B10-17 Crane boom shifted to RINGER. (*Manitowoc.*)

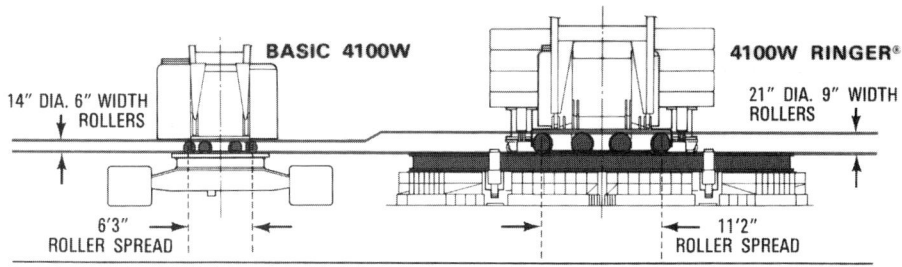

FIGURE B10-18 RINGER rollers are larger than basic crane. (*Manitowoc.*)

B10-39

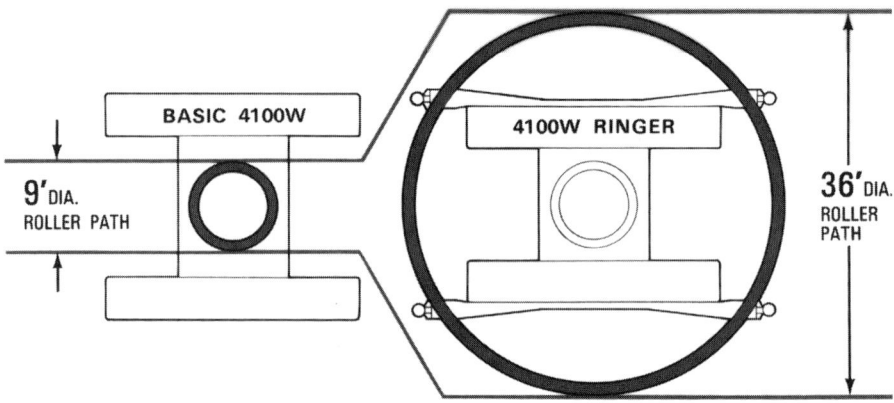

FIGURE B10-19 Ringer roller path four times diameter of basic crane. (*Manitowoc.*)

FIGURE B10-20 M250 crane with M1200 ringer lifting 882 tons. (*Manitowoc.*)

FIGURE B10-21 Low-clearance gantry with telescoping back hitch legs. (*Manitowoc.*)

FIGURE B10-22 4100W crane using its own gantry to remove counterweights. (*Manitowoc.*)

FIGURE B10-23 4100W crane loaded on flat car. (*Manitowoc.*)

FIGURE B10-24 M-series crane self-loading. (*Manitowoc.*)

it is not adaptable to short-term projects. The climbing tower crane is initially the most economical but can service only a relatively small area.

The mobile tower crane has a jib boom at or near the top of a latticed tower structure which is fixed to a rotating crane body. Height is limited by the structural strength of the tower and by the feasibility of raising the structure off the ground.

The rail-mounted tower crane is similar in function to the mobile tower crane just described except for its propulsion mechanism and a simplified machinery arrangement. It is not convertible to other functions such as crane or dragline.

The climbing tower crane has a fixed tower which is intermittently supported to some other structure. Rotation of the jib boom is accomplished at the top of the tower, as illustrated in Fig. B10-25. Load radius is varied by luffing the boom or with a trolley traversing the boom used as a horizontal beam.

Figure B10-26 is a Manitowoc 4600 Series-4 crawler crane with a 254-ft tower, 170-ft boom, and 50-ft jib. The crane is lifting a precast panel.

Table B10-9 lists tower crane capacities for seven crane models.

TABLE B10-8 Shipability for Four M Series Cranes

Model	Capacity, US tons	Weight of main load, lb w/crawlers	Weight of main load, lb w/o crawlers	No. trucks*
M50W	65	72,300	45,380	3
M65W	75	91,200	49,700	3
M80W	88	91,600	57,960	3
M85W	95	97,600	57,695	5

* To ship complete crane with full complement of boom and jib.
Source: Manitowoc.

CRANES

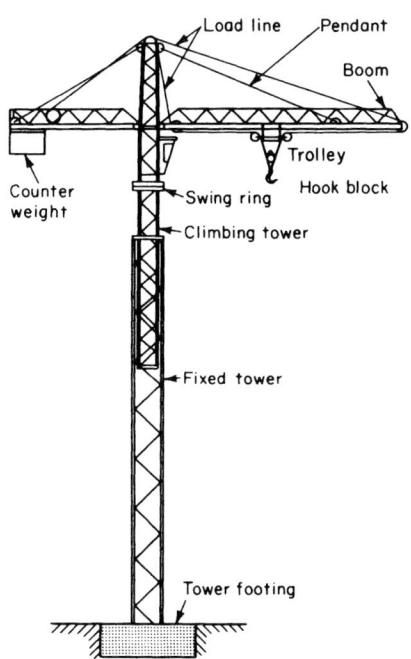

FIGURE B10-25 Tower crane.

EQUIPMENT SELECTION

The selection of equipment to do a job or a series of expected jobs is influenced by several factors. For a variety of small lifts, a hydraulic or mechanical crane with adequate capacity when operating at the maximum expected range might be desirable. A machine rated at 15 to 20 tons would be able to handle 6 tons at a 20-ft radius and 50-ft height, but a larger machine would be required if the same 6-ton load is to be handled at a 200-ft radius and 250-ft height. If tight quarters are expected, such as a high building and narrow streets, a mobile tower crane may be more adaptable than a standard crane. If most jobs are to be relatively light lifts, a small machine can be purchased and larger machines rented for the fewer expected heavy or long-range lifts. Adaptability to other equipment such as dragline, shovel, or hoe can be a factor if there will not be enough crane work to utilize the full time of the equipment. Roadability at the job or enroute to the job can influence the choice of equipment. Economics of initial costs, maintenance cost, expected downtime, taxes, and amortization versus rental charges must be considered. The ability to sell, trade, or rent the equipment to others will be

TABLE B10-9 Tower Crane Capacities (Manitowoc)

	Maximum tower and boom			Capacities (lb) max. tower and boom raised unassisted		
Model	Raised unassisted, ft	Raised with assist, ft	Maximum tip height, ft*	50-ft R	100-ft R	150-ft R
3900T	164 + 150	194 + 150	345.9	21,800	13,800	8,500
3900W	164 + 150	194 + 150	353.6	21,800	13,800	8,500
4000W	163 + 150	213 + 160	371.4	30,000	21,100	14,300
4100W	183 + 170	253 + 170	421.6	32,200[†]	20,400	11,000
4600 S-4	194 + 180	254 + 200	454.6	77,600[‡]	54,200	34,000
6000 S-2	235 + 220	310 + 250	561.8	131,500[§]	113,600	74,900
7200[¶]	235 + 255	285 + 255	554.2	562,600[‡]	398,000	258,300

* On maximum tower and boom raised with assist.
[†] 55-ft radius.
[‡] 60-ft radius.
[§] 70-ft radius.
[¶] Capacities are preliminary, subject to test.

FIGURE B10-26 Tower crane. (*Manitowoc.*)

an additional factor to consider. Speed of load handling and power ability can be factors, and ease of operation and smoothness of load control will influence the selection. Safety features such as operator visibility, adequate guarding of machinery, and clutch and brake size should be investigated.

The information for comparisons should be compiled at the time of an expected purchase or rental in order to have current data. All in all, each job will require some compromises in the choice of lifting equipment. Where alternatives exist, the selection should favor the larger units for improved capabilities and prolonged equipment life. Specialty equipment, except for long-term jobs, should be sought on a rental basis so as to preserve capital for more universally used tools.

CHAPTER 11
MARINE EQUIPMENT

A. D. Pistilli, Ph.D., P.E.
*Retired President, American Dredging Company,
Camden, New Jersey*

E. R. Santhin, M.E.
*Retired Vice President and Chief Engineer, American
Dredging Company, Camden, New Jersey*

L. A. Caccese, C.E., P.E.
*Retired Chief of Operations U.S. Corps of Engineers and
Dredging Consultant, Palmyra, New Jersey*

INTRODUCTION

When one begins to design or plan for the purchase of marine equipment, it is wise to start with consideration of the hull—that essential and costly element of all floating dredges, drill barges, derricks, dumpscows, hopper barges, deck barges, pile drivers, tugs, and other miscellaneous auxiliary vessels. All equipment that floats must be mounted on or in a hull.

The hull must conform to the job which the vessel is to perform. Except in specialized cases, a hull that is too large for the equipment will be costly to maintain and awkward to use. Too small, a hull probably will render poor service and may even be dangerous. The hull must be of steel, welded construction and its shape, method of construction, tonnage, size, and special equipment will govern its cost.

With each year of its life, floating equipment becomes more expensive to maintain because of the effects of use and nature. Natural deterioration by oxidation and corrosion, especially in a salt or brackish water environment, and erosion or damage to protective coatings causes an accelerating heavy toll on steel structures and equipment. Using marine equipment in construction exposes it to damage and wear because of collisions, groundings, excessive loading, strikings, and abuse which sometimes defies imagination.

Marine equipment has a high initial cost and is expensive to maintain. Marine contractors do not always own all the floating plant they need to do a project. Under other circumstances, when a contractor completes a project there may be no further need for a specialized piece of equipment bought for the project. Such fluctuating factors encourage a considerable transfer of marine plant among contractors by purchase or lease.

BARGES AND SCOWS

Selection

Many problems confront a prospective lessee or purchaser of a barge or scow. Is it large enough to remain stable when loaded? In addition to its having an acceptable physical shape and value, how long, how deep, and how wide is it? These questions must be posed and answered as parts of the seek, find, and inspection procedure.

The nomenclature of barges and scows should be explained. Names tend to be interchangeable, and the choice of the name used normally depends on the use and location of the vessel. Generally, a vessel that carries deleterious material such as garbage or trash, dredging material, or debris is called a *scow*—i.e., a garbage scow, a trash scow, or a dump scow. A vessel that carries equipment, farm products (bulk corn, wheat, oats, etc.) or miscellaneous cargo is called a *barge*—i.e., a deck barge, an oil barge, or a hopper barge. However, because the names are interchangeable, certain barges may be called scows and certain scows called barges in different sections of the country, especially on the inland waterways, as distinct from the Great Lakes, gulf coast, or coastal waters.

Most barges or scows today, whether ocean-going or for use on the inland waterways and the Great Lakes, have raked ends (bow and stern), unless they are used in a matched or integrated tow where the bow and/or the stern may be square. Once many ocean going barges used in coastal waters were of the keel type or model bow construction. Today this type vessel normally has a square barge form with skegs at the stern and a raked bow modified to a longer curved rake with an elliptical-shaped bow deck to make for more efficient towing or pushing in a seaway.

Often, in addition to having skegs, the stern is notched and reinforced to permit an integrated tugboat connection for better control of the tow, easier handling, and more efficient propulsion. There was a time when barges were sized according to the use intended and state of the art. Now, all vessels, including dredges, barges, and tugboats, have sizes governed by usage and economics, lock sizes, bridge widths, elevations or openings, and channel depths. Today, size is more an economic determination than a design problem.

Many specialized forms of floating transportation equipment are available to the contractor. Of these, side dump, bottom dump, or pump-out hopper barges and scows merit consideration. Because these vessel types are extremely rugged, built of heavier materials, more heavily framed, and reinforced to withstand the rigorous service of loading, transporting, and unloading material such as sand, clay, gravel, boulders, and rock, they are unusually expensive to construct and maintain. Consequently, dump scows are in short supply on the inland waters, where dredging is normally performed by hydraulic or pump dredges, and material is transported by pipeline, and by dragline dredges that side-cast the dredged material. These dump-type scows or barges are more easily leased for service on the coastal waters or on the Great Lakes, where clamshell or dipper dredging is often done and dredgings must be transported to distant locations, or where upland disposal areas are not available to receive dredgings transported by hydraulic means through a pipeline.

Side-dump barges are used in shallow water and pump-out hopper barges are used where hydraulic dredging can be performed but where disposal areas are too remote for efficient transportation of dredged materials by pipeline. Side-dump barges and pump-out hopper barges, because of their specialized nature, are not easily located or leased.

Inspection

To determine the seaworthiness of a vessel, one should remove it from the water onto a drydock or haul it out on marine ways. Once out of the water, the hull below the waterline may be inspected, as well as that above the waterline. Gaugings should be taken by sonic or other means (holes may be drilled if sonic gauges are not available) to determine the plate thickness and amount of wastage of hull plating. These gaugings should be made in a uniform and orderly manner with special attention to the light (unloaded) waterline, the bilge corners, and the bottom of the rake where the rake end meets the bottom. These areas experience the greatest amount of wastage, the light waterline caused by the interchange of air and water, and the other areas caused by abrasion from groundings, strikings, etc. A few old barges of riveted construction are still in service and some recent barges have riveted straps installed as a design feature for reinforcement and to stop the propagation of any developing cracks. These vessels present special problems and require a close inspection for any loose rivets or rivets with wasted heads which have rusted away or worn off. One must take extreme care when repairing rivets by welding and padding. The heat from welding one rivet can loosen adjacent rivets and cause additional and more extensive leakage to the hull. Because of the special problems associated with riveted hulls, it is best to avoid leasing, purchasing, or using such vessels.

On all barges used in construction work, the bilge area where sides and bottom meet and rake bottoms are subject to considerable abuse from hitting the bottom and working off rock banks, etc. Most barges have some degree of corrugated sides. On older barges and those that have had excessively hard service, the corrugations appear where the side plates are indented between the frame and can run the full length of the vessel. This condition seldom impairs the strength of the vessel unless the framing or scantlings are badly bent or broken. Any damaged hull plate, framing, or scantlings found holed, worn thin, cracked, or broken should be properly repaired before the vessel is returned service.

Barges and other vessels that have worked in such brackish, salt, or acid waters as those in the Gulf, the oceans, the Houston ship channel and other inland waters containing one or a combination of these harsh environmental features may show evidence of excessive deterioration in the form of bad pitting around the light waterline and heavy scaling in other areas. They may also be more rusted and pitted and more heavily scaled than a vessel used in fresh water. Salt or mud deposits on the inside of the hull bottom will betray old or previous leaks, as will wooden plugs or concrete boxes or patches. In salt, brackish, or acid waters, protective coatings will increase the life of the floating equipment.

Money spent for expert inspection before one purchases or leases a barge or other vessel may save considerable and needless later expense. The American Bureau of Shipping, which has an office in almost every major city, can usually furnish inspectors at a reasonable fee. Normally, the marine insurance companies will recommend or perhaps assist in finding an inspector if they are to carry a policy on the hull or cargo.

TUGS AND TOWBOATS

Selection

One of the most useful pieces of floating equipment on a marine construction project is a towboat or tug. The units are similar; the tug has a model bow, a keel running from stem to stern, and usually an elliptical stern, although some have modified square sterns. A towboat has a square raked bow similar to that of a barge, often with pushing knees, and a square stern. Modified towboat bows are those in which the designer has attempted to add streamlining.

For work on the inland waterways, the raked-bow towboat is superior because it can work close to the banks and in shallow water. A towboat requires less draft than a tug of similar size; a tug can take heavier seas and weather in open waters because of its additional freeboard and fairer shape. Stern design affects the ease of shallow-water operation. Because a towboat normally is equipped with a semitunnel for shafts and propellers, the underside of the afterbottom can be reduced in depth to accommodate the propellers. With twin engines, the propeller shafts emerge from the hull on each side of the centerline, and are held steady by struts and bearings at the propellers and at other locations, depending on the length of the shaft. In tugs, the propeller shaft comes through the stern frame above the skeg or keel. Multiple-screw tugs are also constructed with propeller shafts emerging from the hull on each side of the centerline, complete with struts as required. Stainless steel propellers, although more expensive than bronze, are a desirable improvement for long life and reduced wear damage, especially in kort nozzles and where shallow work and expected groundings will be encountered.

The steering system for contractors' boats is normally pneumatic, electrohydraulic, or full hydraulic, depending on the size of the boat and the owners' preferences. Some older boats still have manual steering systems; i.e., the steering wheel is connected to the rudder by shafts, gears, and universal joints or by cables, drums, and sheaves. Manual systems require more effort to operate and, in close work requiring considerable maneuvering, can tire the operator and substantially reduce efficiency. This can produce poorer operation and an increased likelihood of accidents.

A somewhat recent innovation and improvement in tugboat propulsion is the Z-drive unit. The Z-drive encloses the propeller in a Kort nozzle that connects to rotating vertical drives to turn the propeller and rotate the nozzle assembly. This propulsion system permits the nozzle/propeller assembly to rotate a full 360°. It eliminates the need for a rudder and provides better steerage and vessel control than conventional systems. Dual units provide even greater vessel control because each can be rotated independently of the other. Z-drive units are manufactured in small and medium power ranges so that propulsion power of up to 6000 hp can be obtained when dual units are employed. Z-drive units can be installed with direct electric motor drive from the hull or diesel engine drive. Each works through shafts and gears to accommodate the change in direction of supply power required by these units.

The typical crew on a contractors' boat consists of a captain or operator, an engineer if required, and either one or two deckhands, depending on the type of towing or fleet shifting to be done. All modern tugs are constructed with pilothouse controls to eliminate the need for an engineer. Boats whose engines are not pilothouse controlled will require a full-time engineer. Before a boat is placed into operation it may be well to remember that the boat may have to pass a U.S. Coast Guard and other regulatory body inspection and the operator may have to possess a valid U.S. Coast Guard license.

Inspection

When inspecting a boat for possible lease or purchase, one should consider several features, and the services of a surveyor from a reputable shipyard will be valuable. One of the first things to check is the engine, its foundation, and other engine room and deck machinery. Engine room logs should be examined if available. The foundation of the main engine and supports should be heavy and sound. Notice the type, spacing, and size of the framing; this is a key to general condition and design. All structures, plating, and scantlings should be well coated, clean, and free of pitting or scaling. If the beams are small, the girders or channels light, and the welding poor, beware. The main engine and gear should be clean, well maintained, and free of excessive oil leaks. Look for signs of bad oil leaks such as rags stuffed around the engine and oil streaks from heads, bolts, etc. The generators, air compressors, and other auxiliary equipment should be adequate for the purpose intended. Engine room floor plates should be in place and secured. Unsecured or missing floor plates are evidence of trouble areas. Piping should be well run, free of leaks, and adequately supported. Tools and spare parts should be in their places and well stowed. The engine room should be clean and well kept.

Next, the soundness of the hull should be examined. The examination and inspection should follow the guidelines listed under the preceding section covering barges and scows. The deckhouse, quarters, pilothouse, etc., should also be clean and well kept.

The engines are key items and their satisfactory operation requires good service facilities at every port. A shutdown for lack of parts is always expensive and need not be tolerated. Engines manufactured by well-established companies such as either Caterpillar or General Motors are satisfactory because repair parts and service facilities are always available.

DREDGING

Hydraulic Dredging

Hydraulic dredging is the use of power to excavate and transport suspended solid material through pipelines. The main parts of a hydraulic dredge are the hull; the ladder gantry; the dredging ladder with suction pipe and intake or mouthpiece, and cutter with shafting if the dredge is not plain suction; the spud gantry, spuds, and spudwells; the dredge pump and its drive; the hauling and hoisting equipment (usually five drums but often seven or more drums) to handle anchors and spuds; and such auxiliary equipment as generators, compressors, service pumps, etc. Figure B11-1 illustrates typical hydraulic dredge components. The dredge is constructed on a barge or a hull designed according to the service in which it is to be employed, whether on oceans, bays, or rivers.

Hydraulic dredges, rated by the inside diameter of their discharge lines, vary in size from midgets to mammoths. A small dredge with a 6-in dredge pump and drive can be built on a hull only 30 ft long and 14 ft wide with a draft of 2 ft or less. It is ideally suited for dredging in lakes or ponds where small deposits of mud, sand, or gravel are to be removed in limited quantities and transported short distances by pipeline. In contrast, a dredge with a 30- or 36-in dredge pump driven by a 5000 or 10,000 hp or larger pump drive, requires a hull up to 300 ft long and 65 ft wide, with a draft of 10 to 15 ft. Such a dredge can excavate gravel, clay, shell, and boulders and

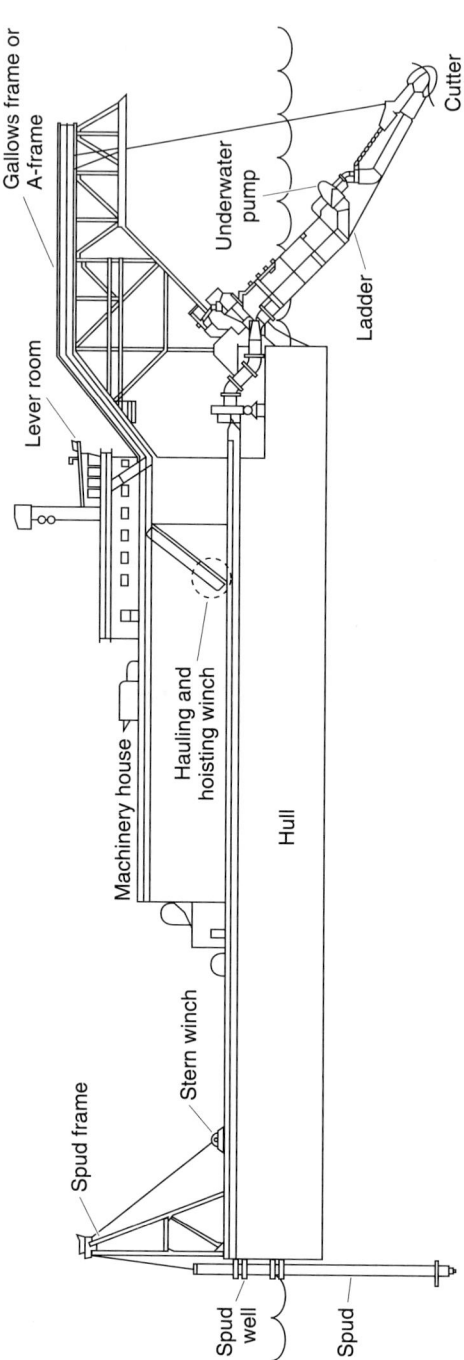

FIGURE B11-1 Typical hydraulic dredge components.

transport the material through pipelines for distances of up to several miles. Table B11-1 lists typical specifications for dredges.

For free-flowing material such as mud, sand, and gravel, a plain suction without cutter or agitator can be used in some applications. The dredge pump suction picks up the bottom material in suspension with the water. Where the material is firm or compacted or does not disintegrate and flow readily, the suction must be augmented with a revolving or rotating cutter. The cutter will break up the material in the immediate vicinity of the suction end so that it can flow with the water. There are several types of cutters, each designed expressly to cut a different type of material—basket cutters for mud, sand, and gravel; a general-purpose cutter; clay cutters for clay; and rock cutters for rock and boulders. The clay and rock cutters have replaceable edges and teeth. The motor and drive shaft for the cutter are mounted on a structural steel framework called the *dredge ladder*. The ladder length, normally operating at a 45° inclination from the horizontal, determines the maximum digging depth. Although ladder lengths of 225 ft have been employed, lengths usually range between 20 and 140 ft. The minimum digging depth is a function of the ladder length, its geometry, and the draft of the dredge. A dredge must be able to dig itself free. If the ladder is too long for the minimum depth required, it will drag on the bottom, cause excessive loading on the hauling or swing winch, and may even restrict or actually stop the swing movement of the dredge. The solutions to this are to dig deeper or shorten the ladder.

The minimum digging width is the width of cut required for the cutter at the end of the dredging ladder to cut a channel bank as wide as the location of the corner of the bow swinging through an arc about the radius from the digging spud. This width, determined by the hull and ladder length, normally is the length of the chord of the arc formed at the point the cutter and hull corner at the bow are tangent to a line parallel to the centerline of the cut. Inboard of that point, the corner of the dredge bow is ahead of the cutter as the dredge swings. Consequently, the dredge hull will hit the bank before the cutter and be locked in the channel, unable to swing. Outboard of that point, the cutter hits the bank first and the dredge can dig itself clear. See Fig. B11-2. The longer the ladder, the less the minimum cut will be.

TABLE B11-1 Typical Specifications for Hydraulic Cutter Dredges

	\multicolumn{6}{c}{Size of dredge discharge, in}
	12	16	20	24	27	36
Length, ft	75	100	120	160	220	280
Beam, ft	30	35	40	50	60	65
Depth, ft	4	6	9	12	15	17.5
Displ, tons	600	1000	1600	2000	2500	3000
Pump, hp	600	1600	2400	5000	7200	10,000
Pump, r/min	500	360	360	360	360	360
Cutter, hp	250	500	750	1000	2000	3000
Cutter, r/min	0–30	0–30	0–30	0–30	0–30	0–30
Spud length, ft	55	65	75	90	110	110
Ladder length, ft	50	55	65	75	100	125
Max. pipeline, ft	5000	7500	10,000	15,000	20,000	30,000
Max. width cut, ft	150	200	250	300	350	400
Min. width cut, ft	50	60	70	90	100	110
Max. dig depth, ft	35	45	50	55	75	95
Min. dig depth, ft	4	6	8	10	12	14

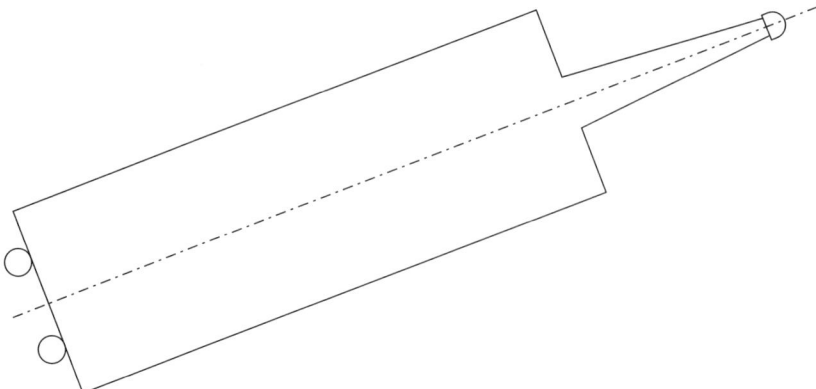

FIGURE B11-2 Minimum digging width.

The hull, the ladder, and the location of the swing anchors determine the maximum digging width. With the anchors in the normal digging position, the dredge can swing to a point off the centerline of the cut where the angle of the swing anchor wire from the opposite side anchor to the dredge ladder will be such that the resultant effective pull of the swing winch from that side will be so low as to effectively limit or stop the return swing of the dredge. The wider the cut, the smaller that angle will be. That point and its distance from the centerline of the cut to each side is the maximum digging width. The longer the ladder, the wider the maximum digging width will be.

The suction pipe carried within the ladder below the cutter shafting is connected to the suction pipe and dredge pump aboard the dredge by a length of special reinforced suction hose. This connection permits adjusting the dredging ladder to the desired dredging depth. The ladder rotates vertically at its aft end about trunnions on each side of the hull, usually in a ladder well, and is supported at its forward end from the ladder gantry or A frame by a block and tackle called a *ladder fall*. The longitudinal centerline and midpoint of the suction hose must pass through the vertical and horizontal centerline of the ladder trunnions when the ladder is at its design digging depth or the hose will crimp, restricting the dredge suction and reducing the output of the dredge. Such crimping also causes the hose to wear excessively and shortens its useful life.

The cutter ladder must be strong enough to withstand the sudden shocks and strains incurred when digging in boulders, compact layers, etc. Its deflection must be less than the deflection of the cutter shaft so that bending will not cause cutter shaft failures. Cutter speeds on most dredges vary between 0 and 40 r/min, and the power applied to the cutter may be as high as 3000 hp or more. A properly designed cutter motor will operate at constant torque from 0 to midrange r/min and at constant horsepower from midrange speed to maximum r/min. Maximum torque, which will occur at 0 r/min will provide the digging power needed for hard, stiff compacted materials. Maximum horsepower will occur at base speed or midrange and higher speeds for other materials. Direct current electric motors and hydraulic motors with proper controls are best suited for this application. Dredging ladders with underwater dredge pumps and/or underwater cutter motors, when used, present special problems for the designer and user.

Any motor, electric or hydraulic, when operated in the environment of the underwater dredge cutter, digging material, shaking and vibrating the end of the ladder it is mounted on, and stirring up all types of material, cannot possibly last as long as a motor out of water. The advantages of underwater pump and cutter drives are such that almost everything has been tried to extend their life and to keep motors from failing prematurely. Motors have been encapsulated in sealed cylinders vented up the ladder through pipes and ducts to the atmosphere, sealed in a housing filled with oil under pressure so any leak will release oil instead of admitting water, and even built as special open motors designed to operate submerged in water. Although all of these motor designs operate for varying lengths of time, they are expensive to purchase, install, and maintain and they still fail prematurely. No design has been able to prevent water and material contamination from entering the motor and causing its premature failure. The special oil used in oil-filled motors even has too much suspended moisture or water when it is delivered from the refinery. The oil must be treated and filtered with more special equipment on the dredge before it can be used. Even with these precautions, the motors will still fail prematurely. The hydraulic motor is probably the best motor for these underwater applications, but size and power limitations require multiple motor assemblies to deliver high horsepower. Special hydraulic systems are designed into these installations, and still the motors fail prematurely because of water and material contamination. Designers should study the cutter or underwater pump drive arrangements and, where feasible, lean toward the advantages of above-water drive for underwater units.

One serious problem with hydraulic motor system installations is the need for flexible connections at the motors and between the piping on the hull and the dredging ladder. Because these systems operate with pressures of 2000 lb/in^2 or higher, numerous leaks and hose failures can occur. When they do, oil spills produce messes and their attendant environmental problems. Despite the trouble and expense associated with these systems, they are in use today because they contribute to markedly increased production.

Certainly no dredge designer should build a new hydraulic dredge without an underwater dredge pump. This pump has brought the single most important advance made in the dredging industry in the past 50 years.

The ladder gantry at the bow of the dredge, sometimes called the *A frame* because of its shape, is either pin-connected to the forward end of the main deck, or built integral with the hull and machinery house or superstructure. The pin connection is best for load distribution and reduced stress problems as well as for ease in erecting and dismantling the gantry. Therefore, it is the arrangement used on all portable dredges, most small- to medium-size dredges, and many large dredges. Some large unitized dredges and small- to medium-size dredges use the gallows frame system, built integral with the hull and house to carry the heavier loads encountered by those dredges and to eliminate the maintenance required for pin connections.

The stern gantry, or spud frame, must be high enough to permit raising the spud so its point is level with the bottom of the hull when top-lift spuds are used. If side-lift spuds are used (a system utilizing a sleeve or collar fitted around the spud with its inside diameter slightly larger than the outside diameter of the spud so it moves freely up and down the spud), the spud frame can be considerably lower. The aft gantry or spud frame can also be either pin-connected to the hull or integral with the hull and house for the same reasons outlined in the description of the ladder gantry or A frame. Some of the larger dredges built in recent years employ this integral structural fabrication which forms a truss on each side of the dredge that runs the full length of the hull and forms a strong and almost indestructible unit.

The side-lift system requires holes spaced along the length of the spuds to accept pins that the collar fetches up against as it is raised by the spud hoist and thereby picks up and raises the spud. Pipes must be welded through the spud at the hole locations to keep the spud free of water, sand, and mud which can increase its weight beyond the hoisting capacity of the spud winch. Some dredges use hydraulic cylinders to raise the spuds and some use a special carriage that serves the dual purpose of raising the spud and also setting the dredge ahead. All of these operate hydraulically. Although these two systems are available to designers and purchasers, they are seldom used. Spuds are cylindrical because the dredge must rotate about them as it swings from side to side. The diameter and wall thickness of the spud are determined by the size of the dredge hull, the distance between the upper and lower spud wells, the depth of the excavation, and the loads imposed by the reactive force of the swing line pull pushing the hull back against the spud. It is considered a cantilevered beam, anchored at the point end.

The spuds allow the dredge to be anchored or maneuvered to "walk" along the length of the project. Walking is accomplished by swinging to one side a set distance and dropping the opposite or setting spud. The swing or digging spud is then raised and the dredge swung in the opposite direction until the required *set* or distance is moved ahead where the swing spud is dropped and the setting spud raised. Power for raising and lowering the spuds usually comes from the main hauling and hoisting winch or from a separate two- or three-drum spud and anchor winch at the stern.

Swing wires are also used to swing or walk the dredge. These wires lead from the hauling and hoisting winch, down the dredging ladder to swing sheaves near the suction end of the ladder, and then to anchors on either side of the dredge set at an angle to its centerline. Swing anchors set too far forward allow the cutter to cross and cut the wire as the dredge swings from side to side; set too far back, the reactive force will pull the dredge back, forcing the suction end and the cutter away from the face of the cut being excavated, with a resulting loss of production. If the anchors are too far back, there will also be a reactive resulting loss of effective line pull in the corners and the operator may not be able to swing the dredge back from the side of the cut. This reduces production. Or the operator may have to move so slowly out of the corner that production is reduced.

The proper positioning and the timely moving of the swing anchors is so important that an inexperienced or lazy operator can affect production more adversely than a worn-out dredge pump or impeller. Such an operator will pump more water than material.

For the swing line to pull the cutter across the face of the cut while it is cutting material, the line pull of the hauling winch must be greater than the tangential force produced by the torque of the cutter motor at the outside diameter of the cutter.

The cutter digging rotation is designed to be in the opposite direction from the swing or digging spud; i.e., if the starboard spud is the digging spud, the cutter should rotate to the port or in the left-hand direction, and vice versa. This means that as the dredge is completing its set or movement ahead, the cutter is rotating into the bank of material or cut, in the proper cutting direction. The design rotation of the cutter must also be such that if the dredge is digging to port, the cutter cuts into the bank, rotating counterclockwise when viewed from the front of the dredge. If the dredge is designed to dig to starboard, the rotation of the cutter must be clockwise. Proper rotational direction is imperative to successful dredge operation. Cutters designed to rotate opposite of the directions described will grab into the top of all classes of such stiff material as hard mud, clay, and hard-packed sands, gravels, etc., and actually pull the dredge across the cut faster than the swing winch can swing the dredge.

In that eventuality, the cutter will not cut the material but merely rotate on top of it with the blades each making a purchase into the material as the cutter turns. Even when the cutter rotation is correct, this condition can occur in stiff material on the return or cleanup swing. The operator must adjust the swing line pull, often retarding the cleanup swing with the brake of opposite swing wire so the pulling wire does not go slack and foul the cutter. See Fig. B11-3.

The discharge line on the deck of the dredge should be on the same side as the swinging or digging spud, and the stern discharge swivel connection should be located close to that spud to eliminate undue movement between the dredge discharge and the floating discharge line as the dredge swings from side to side. The location of the discharge line can be chosen as desired, even though the dredge pump impeller drive normally operates in only one direction, by locating the pump discharge either on the top or bottom of the dredge pump. A discharge located on top of the pump will discharge to the opposite side of the dredge, as one located on the bottom of the pump casing. If the designer prefers one type of discharge over the other, then this choice will govern all other aspects of the dredge design, such as digging direction and cutter rotation.

The location of the spudwells on each side of the centerline of the dredge determines the amount the dredge can set ahead effectively; too close together and the dredge can make only small sets or advancements. The farther apart the spudwells are on each side of the centerline, the longer, or greater, the set or movement ahead the dredge can make. This is an important consideration when the dredge is excavating mud of a liquid nature or a material of a specific gravity approaching 1.0.

The output required, the type of material to be excavated, and the total length of the discharge line dictate the size and power of the dredging pump. The pump and drive assembly must develop enough head and power to lift the material and discharge it at a velocity that will keep the dredged material in a suspension that will carry 15 percent or more of solids. The higher the velocity, the greater the percent of solids that will be carried and the higher the production. On short discharge lines, velocities of 25 or 30 ft/s or higher can be obtained, if the power is available. On long discharge lines, velocities and percent solids will be lower. Material will also affect velocity. Heavier material such as shell, gravel, clay, and boulders have a higher friction factor than mud or sand. The higher friction factor requires more head or pressure to move the same volume of material the same distance. Discharge velocities lower than 12 ft/s cannot be tolerated at any time. Production at those velocities will be so low that mostly water will be pumped, production will be seriously curtailed, and lines will plug and cause stoppages. Where lines are so long that velocity and production drop below an acceptable level, booster pumps must be used to maintain maximum efficiency of the dredge.

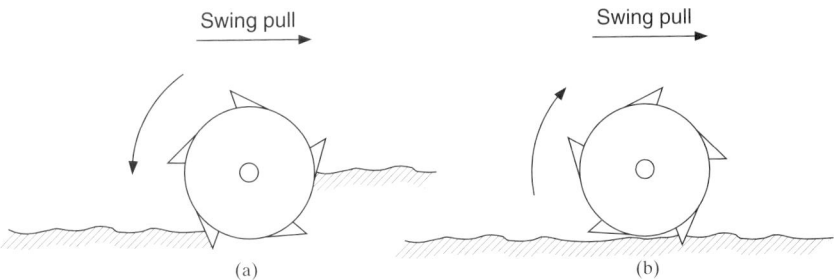

FIGURE B11-3 (*a*) Correct cutter rotation; (*b*) incorrect cutter rotation.

Contrary to popular belief, the dredge pump drive can easily be overloaded on short lines when velocity is very high by the resulting movement of extremely high quantities of water. On long lines with reduced quantities of water, the pump drive will be underloaded. Care must be taken on short discharge line work not to operate the dredge pump drive in sustained overload condition or serious damage can be done to the pump drive. The majority of dredging work uses discharge lines within the range of the design impeller sizes of the dredge pump and the speed and horsepower of the pump drive. Consequently, the system can be operated without damage to the drive.

When a submerged discharge line with medium to long total discharge line length is required on a dredging project, a hole or break in the submerged line has the effect of a sudden overload caused by a short discharge line. The pump must be slowed or stopped as soon as possible to eliminate the overload on the main pump drive. If load indicators are not available to the operator or leverperson, sudden higher-than-normal vacuum or suction pressure readings will indicate this overload.

The brake horsepower for the required pump drive can be computed as follows:

$$\text{bhp} = \frac{OHS}{3960E}$$

where Q = quantity of discharge, gpm
H = total head, ft
S = specific gravity of material
E = pump efficiency expressed as a decimal

The total head is the sum of the suction and discharge heads, including static and dynamic components. Specific gravity of the suspension is approximately: 10 percent solids = specific gravity of 1.1; 20 percent solids = specific gravity of 1.2, etc. For good production, the hydraulic dredged material should run between 15 to 20 percent solids by weight. The efficiency of a dredge pump is approximately 0.55. Some designers and manufacturers claim efficiencies as high as 0.70, but 0.55 is the more realistic number.

Dredge pump and impeller design is a special science involving empirical data normally unavailable to the inexperienced designer. Compromises in impeller vane curvature radii and other impeller and pump volute considerations are based on experience as well as hydraulics. Entrance angles and openings of the impeller vanes must be large enough to accommodate the maximum size of solids, stones, boulders, timber, and trash being dredged, or intolerable frequency of pump stops will be required to clear the pump. Exit angles will be affected by the vane radii employed. Clearances between the pump assembly components must be large enough to facilitate pump assembly and disassembly but not so large as to cause unacceptable efficiency reductions. These designs are best left to experts in the dredge pump design field.

The selection of a drive for a dredge pump involves a number of considerations, not the least of which are first and operating costs. If cost were not a consideration, the choice for pump drive would be simple and straightforward. Years ago, before the oil embargo, when heavy fuel costs were $3.00 per barrel or less, the pump drive of choice was the steam turbine. The steam turbine is the best choice for a dredge pump drive because it delivers full horsepower over a broad range of speeds. Under overloads, the turbine will run away, but overspeed trips will shut it down before any damage occurs.

Unfortunately, steam turbines require high-pressure steam, together with boilers and all of the associated equipment attendant thereto. In addition, more crew members and engineers are required, and, finally, the turbine and reduction gear themselves are too costly.

The next-best choice for a dredge pump drive is the wound rotor electric motor. This type pump drive has characteristics of the steam turbine, i.e., a flattened power curve or almost full horsepower over the normal speed range required for a dredge pump impeller. Speed control is a problem that makes this type drive difficult to use, especially when working with a submerged line. Normally, the electric motor will come to base speed very quickly, which may float the submerged line. The infinite speed range of the steam turbine is impossible to achieve. Still, this type drive is superior to the last but almost universally accepted choice of the diesel engine for dredge pump drives. Because electric motor drives need generators to provide electrical power to the motor, which in turn requires a prime mover—either a steam turbine or a diesel engine of the same or larger size than the ultimate pump drive motor—it follows that the least expensive installation is the diesel engine, directly connected, through a reduction gear, to the dredge pump. The diesel engine drive is not only the lowest first-cost pump drive, it is also less costly to operate, maintain, and repair.

The diesel engine, however, is the least efficient power source for a dredge pump drive because its power curve drops sharply as speed is reduced. Speed can be controlled over a broad range, but because of its power characteristics the diesel engine must run at or near rated speed to utilize the maximum or rated power available at the engine. Consequently, impeller size selection is extremely important. If the impeller is too large, the diesel engine speed must be reduced to eliminate overloads; if too small, the diesel engine will be underloaded at its rated speed. Either case would constitute an inefficient way to operate a costly piece of capital equipment. Because diesel engines and electric motors are like dumb animals—they will continue to operate even under overloads—they can damage themselves severely and reduce their productive life. Hence, the designer should design these drive installations to operate within the limitations imposed by the characteristics of that equipment.

The discharge pipe of the dredge is made of high carbon or alloy steel to withstand the abrasion of the dredged material being transported through it. High velocities cause accelerated wear. The discharge pipe on the dredge usually has a $\frac{3}{4}$-in wall thickness; the floating pontoon discharge pipe, $\frac{3}{8}$- to $\frac{1}{2}$-in wall thickness; and shore discharge pipe, $\frac{3}{16}$- to $\frac{1}{4}$-in wall thickness. At one time, spiral weld pipe was popular in the dredging industry because it cost less than straight welded pipe. The flow of abrasive material at high velocity wears through the spiral welds, however, causing the weld seams to open and the spiral to unravel, resulting in premature pipe failure. An acceptable alternative to high-carbon steel dredge pipe for pontoon and shore pipe is high-tensile, used gasline pipe. The natural gas industry removes this pipe from service in accordance to a schedule based on time in service so as to eliminate any possible pipeline failures. Because of this, and the fact that the pipe has carried natural gas only, the pipe has very little wear. It also has excellent resistance to wear from abrasion. Unfortunately, such pipe is usually in limited supply and is not available in smaller and medium sizes.

Each section or length of floating pipe is mounted on pontoons. The simplest pontoons are steel drums, one on each side of the pipe, spaced as required for the length of pipe used. Usually, rubber hose joints are placed between pontoon assemblies. Larger dredges use fabricated pontoons, sized to support the load of the pipe when it is filled with dredged material, and provided with saddles for the pipe at the center of the top of the pontoon. Flexible ball joints connect pontoon assemblies. Each pontoon is placed under and at a right angle, or normal, to the pipe. The number used depends on the length of the pipe supported. Individual floating discharge pipe lengths of 100 to 150 ft are not unusual. Shore pipe, fitted with a tapered nipple at one end, is rammed into the pipe next to it. The tapered ends face downstream to the flow. Floating pontoon pipe must be added to the discharge line as the dredge advances, and removed from the line as it shifts cuts. Shore pipe, on the other hand, usually remains in place and is added to only at the end of the line as the fill or disposal area requires. Manually operated Y-valves are located in the shore line with two or more branches of pipe extending from the valve to the point of disposal. This arrangement permits one line to be shut off and lengthened while the other branch continues to discharge material. When required, the valve can be shifted and pipe can be added to the opposite line. All dredge pipe should be rotated ⅓ to ¼ turn periodically to distribute wear over the inner circumference and increase pipe life.

Setting up the dredge follows this procedure. The dredge is towed to the desired starting location. The spud and ladder are lowered and the swing anchors are run. The pontoon line is set into position at the stern of the dredge and connected to the dredge discharge. Each piece of floating discharge line is added individually until the desired length is obtained. A landing pipe is connected between the pontoon and shore line which has previously been put in place by a shore crew. The shore line runs to the point of discharge or disposal.

If a hydraulic fill is to be impounded within an enclosed disposal area, rather than pumped overboard or contained in a beach replenishment project, dikes or levees must be constructed, maintained, and raised and provision made to control water disposal or runoff. A spillway is the usual provision so the dredged material will settle in the fill area. A spillway can be simple pipe sections through the dike with vertical pipe nipples to serve as weirs, or more elaborate fabricated steel structures with wood weir boards that can be added or removed to control the amount of water permitted over the weir. The latter structures also have pipe sections in an enclosed bottom area extending through the dike for water runoff. Fabricated steel spillways must be firmly anchored or they may float as the water level rises in the disposal area. The amount of water pumped and the size of the disposal area governs the size of the spillway. Draglines digging inside the disposal area, working from the top of the dike and using the dredged material, raise the dike or levee as required. Bulldozers are sometimes also used to raise dikes or assist in that work. For beach replenishment, bulldozers and/or end loaders control the fill area, pushing up small dikes as needed. The shore crew also use these machines to move, break back, and add pipe as needed.

The number of employees needed to work a hydraulic dredging project depends largely on the size of the dredge. The responsibility for the dredge and its operation rests with the project or dredge superintendent and/or the dredge captain. It is their job to know the dredge, its equipment, and capacity and to be familiar with the project. Dredge operators or leveroperators, one for each shift, serve under the captain. They run the hauling and hoisting machinery, the cutter, and the dredge pump, and ultimately control the amount of material being excavated and pumped. One mate per shift usually works under the operator. If the project requires, an additional chief mate works on the day shift. The chief mate takes direction from the dredge captain and spends time with a bull gang of two to four deckhands, assembling pontoons, repairing pipe, repairing pontoons, and doing other work, usually away from the immediate vicinity of the dredge. The mate who works under the operator on each shift has one or two deckhands working under supervision. This mate operates the floating derrick and, together with the deckhands, shifts the dredge anchors, adds and removes floating pontoon pipe as required, and performs other repair and maintenance work on the dredge and floating derrick. One or more welders and helpers as needed are also included in the crew to do welding and burning as required. The mates usually are also proficient in the use of burning equipment, rigging, and the handling of line and wire rope, including splicing and rolling eyes.

The engine-room crew consists of a chief engineer who works with and takes direction from the dredge superintendent and the captain. The chief engineer must know the engine room and deck equipment, including the hauling and hoisting machinery; main pump drive; the bearings and pump thrust, and/or impeller shafts (but not the dredge pump or the cutter which are the mate's responsibility); the cutter drive, shafting, and bearings; the electrical, pneumatic, piping, and control systems; and all of the attendant auxiliary pumps, compressors, etc. The chief engineer must also know their maintenance and repair. Working under the chief engineer are one shift engineer for each shift, and usually one oiler for each shift. Smaller dredges may only have three engineers, including the chief engineer and one oiler who works on the chief engineer's shift. A shore crew is required at the point of disposal and consists of the dump foreperson and one or more shoreworkers, the number determined by the size of the dredge and the complexity of the job. They operate the end loaders or dozers and maintain and break back and rerun the shore line as needed. The crews normally rotate each week, one shift working a double shift, usually the second and third shift on the last day of the week, so each crew member works the day shift once every three weeks.

The cutter, ladder hoist, swing, spud hoist, dredge pump, and marine signals are operated from the lever room or control house. This centralized control station lies at the forward end of the dredge on top of the machinery house or superstructure so it has a full 360° view of the entire operation. The operator or leverperson, stationed in this room, operates all of the equipment associated with the proper performance of the dredge. Natural and visual ranges placed by the survey crew are utilized to assure that the dredge operates within the limits of the area to be dredged as well as electronic positioning devices. Natural ranges are those topographical features occurring naturally, such as trees, hills, and bushes, as well as structures in the vicinity of the job. The operator is also responsible for displaying and giving the correct marine signals to other vessels.

The newer, better-designed dredges will have a ladder hoist separate from the spud and swing hoists and a separate spud and anchor hoist at the stern. This arrangement permits the operator or leverperson to raise the ladder, swing the dredge, and raise and lower spuds at different speeds all at the same time. This increases the efficiency of the dredge by reducing the time required to set or move the dredge ahead and increasing the dredge operating production time.

Some projects require excavations too deep or in waters too rough for spuds to be used. In those cases, a three-wire anchor fairlead assembly or Christmas tree is used. It consists of a fabricated box girder structure suspended from the main deck on the centerline at the after-end of the hull. This structure has three swivel sheave fairleads located just below the hull bottom. Each can rotate 360°. Cables run from a three-drum spud and anchor winch through these sheaves and thence to anchors, one directly aft of the dredge on the centerline of the dredge cut, the other two running forward under the hull and outside the swing anchor wires. This system serves the purpose of spuds in allowing the dredge to swing or pivot about the fairlead or Christmas tree. Hauling in on the two forward anchor lines and slacking off on the aft anchor line moves the dredge ahead. Although not as easy to operate as a dredge on spuds, a dredge with this system can operate where spuds cannot be used, and the experienced dredge operator or leverperson will quickly learn its use where needed. Some swivel anchor fairleads, especially older designs, are fabricated in short spud sections and inserted and secured in the digging spudwell. This type of fairlead requires more complicated arrangements and causes problems for the three anchor wires leading to it, but its operation is similar to the box girder type.

Recent advances in electrical power and control technology have made direct current electrical supply for winch motors and cutter motors much simpler and more economical to use than was the case previously. DC power can now be rectified from ac power through static devices, thereby eliminating the need for individual dc generators rotating in tandem for each dc motor. The static, solid-state rectifier replaces the Ward Leonard or variable voltage control systems and their attendant hardware. AC generators can be purchased and installed, often a single unit, which supplies ac power to a common buss from which individual rectifiers—one for each dc motor—can be installed for the proper operation and control of the dc motor-driven winches and the cutter. This system can also power propulsion systems on tugs and small ocean-going vessels, but the economic advantage is not available when compared to direct diesel engine drives unless large electrical power systems are required for other reasons. A typical hydraulic dredge is shown in Fig. B11-4.

Ladder or Bucket Dredge

This is a specialized piece of dredging equipment which is adapted for excavating submarine trenches, for use in mining tin or precious metal, and for producing sand and gravel. It consists of a ladder-mounted endless chain of buckets that discharge onto a conveyor belt. Because of its nature the ladder dredge cannot transport its material beyond its point of bucket discharge unless provided with supplementary conveyers. The ladder is also vulnerable to damage by shifting of the barge caused by currents, passing vessels, or rough weather.

FIGURE B11-4 Great Lakes Dredge & Dock Co. hydraulic dredge, Alaska.

Dipper Dredge

These dredges have distinct advantages for excavating dense materials and are especially useful in breaking up ledge rock or excavating blasted rock. The dipper dredge operates with three square spuds, two forward and one at the stern, a crowding, and walking spud. Before dredging starts, a large part of the dredge's weight is raised or hoisted up on the two forward pin-up spuds. This provides a solid foundation from which the dipper can operate. The stern spud prevents the dipper stick from pushing the dredge backward when the dipper stick is crowded into the material being excavated. The dipper bucket is mounted at the forward end of the dipper stick. Its operation is similar to that of a power shovel. The bucket is hoisted by means of hoist cables passing over sheaves at the end of the boom. At the same time, the bucket is crowded into the bank by means of a crowding winch, located where the dipper stick passes through the boom. Crowding effort is provided by wire rope or by rack-and-pinion drive. The boom is mounted on a bull wheel and is supported at its forward end by an A frame. The boom and dipper stick assembly swings around a 180° arc by means of wire rope running from the bull wheel to the swing winch, and the bucket is discharged into a waiting dump scow alongside the dredge. Because the dipper dumping radius is extremely limited, the dump scow must be moved or shifted frequently while the loading operation is taking place. All operations of this type dredge are performed by winches: a main hoist to raise and lower the dipper stick assembly, a swing winch to turn the boom, spud winches to raise and lower the spuds, and scow hauling winches to move the dump scows for loading. These dredges are usually diesel electric with the hoists being driven by dc electric motors. Because hoist speeds are low with high line pull, diesel engine power requirements are reasonable, and first cost, operating, and maintenance costs are not out of line. The reason so few of this type of dredge are available is simply because of the limited amount of work to which they are suited. Their requirement falls into a narrow range between that of the hydraulic dredge and the clamshell dredge.

The crew for a dipper dredge consists of a dredge superintendent and/or captain. Serving under the captain are the operators or leverpersons, one for each shift. The operator operates the spuds, main hoist, crowd, swing, and dump mechanism to load the scows. The operator also moves the dredge by placing the bucket on the bottom and crowding the stick back to pull the dredge ahead. The aft or walking spud is left down while the dredge moves ahead. It pivots in a saddle and helps keep the dredge on range as the dredge moves ahead. When the move is completed, the forward spuds are lowered and the after-spud is raised and reset for the next cut. Some dredges have an auxiliary pullback with a cable leading to the bucket. This pullback helps pull the dredge ahead while the stick is being crowded back. Working under the leverperson are one mate and one or two deckhands per shift. A chief engineer and two shift engineers work in the engine room, each working one shift. The crew will normally change shifts each week as the hydraulic dredge crew does.

Until recently, dipper dredge capacity was in the range of 10 to 15 yd^3 bucket size. Today there is one combination dipper/clamshell dredge in operation with 25 yd^3 dipper bucket capacity. A typical dipper dredge is shown in Fig. B11-5.

Clamshell Dredge

This dredge is exceptionally useful in soft to medium-hard dredging and for deep digging. When spuds are not used, anchors and anchor lines must be placed to keep the dredge in position while the dredging operation is being performed. Gin poles

FIGURE B11-5 Great Lakes Dredge & Dock Co. dipper dredge, Chicago.

or vertical columns with swivel fairleads on top must be used for the anchor lines on the side of the dredge where scows are loaded. They must be high enough for the anchor wires to pass over the dump scow when it is in the light or unloaded condition. For dredging in shallow to medium depths of up to 60 or 70 ft, three square spuds are also used by this type of dredge, two forward and one aft. These spuds are similar in operation to those of the dipper dredge except the dredge is not hoisted or pinned up on the two forward spuds. Clamshell dredges have a live boom that can be raised or lowered by a boom hoist. This permits several advances in the cut before the dredge needs moving. The dredge usually has a large-capacity whirley-type crane or derrick modified for dredging service mounted to the deck of a barge on its tub, pin-connected through the deck to reinforced anchor points. Such dredges are normally diesel electric powered and as with the dipper dredge, all operations are performed by hoists located in the whirley or, in the case of spud and scow hauling winches, in or on the barge hull. The operator in the cab of the whirley can perform all operations including raising and lowering spuds and shifting scows. Shifting scows and raising and lowering spuds can also be performed by the mate on deck.

Standard buckets are power-arm-operated clamshell buckets, some with multiple reevings for harder digging. Bucket sizes range from 10 to 20 yd^3. One large combination clamshell/dipper dredge is in operation today with clamshell capacity of 50 yd^3. Whirley-type clamshell dredges can rotate a full 360°. They can also be outfitted to operate as dragline dredges by extending the boom and modifying the drag and hoist drums. The weight of the bucket itself determines the digging effort. In soft digging, in mud and loose sand, this is not a problem. In harder digging, however, the hoist must take in or pull up on the bucket closing line to close the bucket. Because the upward pull subtracts from the bucket weight and, thus, the digging effort of the clamshell bucket, there are limits to the type of material that can be excavated by a clamshell dredge.

It is not uncommon to secure a mobile tracked crawler crane on the deck of a flat-deck barge and use it as a clamshell dredge. This is perhaps the most economical method of excavating relatively small volumes of material, such as cleaning out powerhouse intakes or removing silt from the face of a dock.

The clamshell dredge crew is similar in size to that of the dipper dredge. The clamshell has advantages over the dipper dredge. It is faster, it can work in closer areas, and its longer boom allows it to load dump scows with less shifting or moving. A typical clamshell dredge is shown in Fig. B11-6.

Hopper Dredge

Self-propelled hopper dredges are required in excavating, improving, and maintaining harbors on coastal waters and their access channels. They also maintain and improve lower reaches of large rivers and bays near the sea. When their hoppers are full, they transport the dredged material to remote disposal areas at sea or other locations and bottom dump or pump off the material to upland disposal areas before returning to the dredging area to repeat the process. A recent application of this type dredge to beach nourishment and improvement has increased its use and purpose.

The erosion of coastal beaches, caused by tides, littoral flow, winds, and storm, has always been a severe problem. Until recently, nourishing or replacing sand and improving these areas was the work of hydraulic cutterhead dredges working in par-

FIGURE B11-6 Great Lakes Dredge & Dock Co. clamshell dredge, Chicago.

tially protected waters such as inlets, bays, and rivers. The class of material—fine to medium sand—found in these areas, although satisfactory for use, is not the best for beach improvement or nourishment. A coarse sand is better suited for the purpose and that type of sand occurs offshore in the ocean itself. Although some recently built hydraulic dredges can operate in an ocean environment, other problems make the seagoing hopper dredge better suited for this work. Because acceptable material usually is not found close to the beach area, sand must be transported from borrow areas more than a mile out to sea. A typical beach nourishment is shown in Fig. B11-7. The seagoing hopper dredge consists essentially of a hopper barge midbody with the hopper divided by transverse bulkheads, evenly spaced, and the hoppers outfitted with hydraulically operated bottom dump doors. The bow and stern are a ship's bow and stern which contain the rudders, propellers, machinery, and pumps necessary to operate as a dredge as well as a seagoing vessel. A superstructure house, located either at the bow or the stern, contains crews quarters and a pilothouse. Specialized equipment provided on a seagoing hopper dredge are a dragarm on each side of the midbody hopper area connected to the dredge pumping system by a sliding trunnion, and winches and davits to raise, lower, and stow the dragarm. The sliding trunnion, at the forward end of the dragarm, permits the suction pipe on the dragarm to be connected to the dredge pump, or pumps, which are normally located in the forward section of the vessel. The dragarm has, at its aft end, a suction or draghead which is lowered to the bottom dredging area by a winch and davit system. About midway down the dragarm is a flexible section or gimbal area which is also lowered by a winch and davit system.

When the draghead contacts with the bottom, the dragtender, or dredging operator, starts the dredging pump or pumps. The operator controls the velocity and per-

FIGURE B11-7 Beach nourishment area; work in progress.

cent of solids of material being excavated and loaded in the hoppers by adjusting the vertical location of the draghead suction in relation to the bottom. The draghead davit systems of the more recently constructed seagoing hopper dredges have automatic swell compensators at the davit winches which maintain the selected relation of the draghead position to the bottom regardless of seas encountered. These compensators can adjust to swells of 10 or 12 ft if necessary. The dragtender works closely with the mate on watch. The mate's job is to stand the helm and maneuver the hopper dredge from the helmsperson's console.

The mate controls the direction and speed of the hopper dredge over the dredging area. A number of navigation and positioning devices are used to do this, including an integrated, automatic ships positioning system, radars, fathometers, a gyrocompass, a speedlog (a ships speedometer), Loran, and dragarm depth indicators. The integrated, automated dredge positioning system is also a plotting system which records the dredge's location and progress over the bottom as well as the after-dredging bottom profile. Variations of this system are also used on hydraulic dredges.

The dredge positioning system is a computerized system that interfaces with many of the navigation and positioning devices, including remote shore-located trisponders. The system is programmed and provided with software that displays on a color monitor the dredging area, the channel or cut outlines, navigation aids, buoys, etc., and any obstructions or other unusual aspects of the project. The monitor, located in front of the helmsperson or mate, usually suspended from the overhead, permits observation of the dredging area without distracting from the proper control of the vessel and its navigation. The monitor also displays a small ship outline which moves over the dredging area shown, in slave fashion to the actual hopper dredge movements. Lines from each side of the vessel trace the travel of the dragheads on each side in colors that contrast to the colors of other aspects shown. The dredging area shoals, or areas of excavation required, appear in still another different color, usually brown. A loading indicator and recorder shows the helmsperson or mate the weight of the material being loaded into the hoppers and the rate of loading.

The dragarm normally operates at a 45° angle from the horizontal, aft end on the bottom. In this it is similar to the ladder of the hydraulic dredge, although it often employs steeper angles. The helmsperson or mate must be especially watchful when operating in cross currents to keep the dredge from crabbing or moving sideways over the dragarm, which would damage or penetrate the hull bottom or break the dragarm.

Many recent seagoing hopper dredges are designed and constructed in a similar manner to the one just described except that these vessels consist of a split hull constructed of two longitudinal sections hinged on the deck at the bow and stern, complete with large hydraulic cylinders located athwart ship at the bottom, below the hinges. The shape of the hull, other than being split, is, essentially, the same as the conventional hopper dredge hull.

Because this type vessel has two separate hulls, the propulsion and the dredge pumping systems must be duplicated in each side, unless a special crossover pump connection is provided. Auxiliary and ships service systems must also be duplicated as necessary. A typical split-hull hopper dredge is shown in Fig. B11-8.

Hopper dredge propulsion and pumping systems can be diesel or diesel electric. Both types are in service today. The hopper dredge superstructure or deckhouse is usually located at the stern, but forward deckhouses have been installed. The deckhouse must be hinged at the deck on split-hull dredges so it stays level when the hull splits for dumping. The hydraulic cylinders at each end of the hull separate the two

FIGURE B11-8 North American Trailing Co. and Great Lakes Dredge & Dock Co. split-hull hopper dredges: Manhattan Island, Sugar Island, and Dodge Island.

hull sections at the bottom, about the deck hinges, opening the entire hull longitudinally like a clamshell. As the hull opens, the material drops out of the bottom. Once the load is dumped, the hull can be closed hydraulically or allowed to close by geometric forces imposed by the sea on the hull sections.

Hopper dredges also have a pump-out system that permits the dredge to unload hydraulically so the material can be transported to remote areas through pipelines. Most hopper dredges pump out with the same pump that is used to load the hopper. Because this pump is designed for low-head, low-horsepower operation (the discharge line from the pump to load the hopper is only several hundred feet long), separate booster pumps must be added ashore even for short discharge lines. Some recent dredges have a separate pump-out system with a high-head, high-horsepower uploading pump so they can discharge over longer distances, as a hydraulic dredge does without booster installations.

The pump-out system consists of a tunnel running full length below the hopper or hoppers on one side, complete with hydraulically operated dump doors spaced along the hopper bottom. One end of the tunnel connects through dredge pipe to the dredge pump or unloading pump; the opposite end to a sea chest at the stern or opposite end, sized to permit unrestricted water flow to the pump. The dragtender opens a hydraulic valve at the sea chest and at the pump, starts the pump, and pumps water until water arrives at the end of the discharge pipeline. The dump doors are then opened as necessary to unload the hoppers properly and allow the water to transport the material to the disposal area in a concentration permitted by the trim of the vessel as it is being unloaded and by the dredge pump characteristics.

Most hopper dredges are ocean-going vessels with dredge pump sizes of 24- or 27-in or larger. Some smaller-size hopper dredges have also been built. The high capital costs involved in their construction normally precludes small vessels from being constructed when cost and return-on-investment comparisons are made. Hopper dredge size is classed according to the hopper size, or capacity:

Large hopper dredge	over 5000 yd^3
Medium hopper dredge	3000 to 5000 yd^3
Small hopper dredge	under 3000 yd^3

All hopper dredges have a large number of hydraulically operated gate valves throughout the vessel in the suction and discharge pipelines to facilitate the direction of flow of the water and dredged material. They also must have spillway structures and weirs of some type to control water runoff and contain the dredged material in the hopper. Several types have been used. Older bottom-dump hopper dredges used systems as simple as openings in the hopper combing along each side to allow water to run onto the deck and over the sides. Later hopper dredges used variations of that system, gradually improving them. Recent dredges have sophisticated weirs in the hopper with a rectangular or similar funnel, telescoped from a vertical pipe extending down through the bottom of the vessel. The funnel can be raised or lowered hydraulically, as required, to adjust the weir to the amount of water runoff desired. Usually there are two weirs, one on each side of the hopper at the forward end. Some newer seagoing hopper dredges also have additional underwater dredge pumps installed on the lower ends of the dragarms. These vessels may or may not have dredge loading and unloading pumps located in the hull itself.

Since the hopper dredge suction draghead does not have a rotating cutter, teeth of various types and shapes are used. These teeth, located on the bottom of the draghead, help break up compacted materials ahead of or in front of the suction inlet. Water jets with water supplied under pressure through pipes from auxiliary pumps on the dredge are provided at the suction inlet to assist in dislodging compacted materials.

Seagoing hopper dredges are as much ocean-going vessels as cargo or passenger ships and are constructed, manned, and operated in a similar manner. They feature all of the advances provided in other seagoing vessels such as bow thrusters, water purification systems, sewage treatment systems, oily water separators, deck cranes, modern navigation and communication systems including satellite dishes, automatic pilots, climate-controlled crew quarters, and automatic monitoring and control systems for main and auxiliary machinery in unattended machinery spaces.

The size of the crew on a seagoing hopper dredge depends on the extent of the automation available and the requirements of the U.S. Coast Guard for vessels of this type service and size. All crew members must have valid U.S. Coast Guard licenses for the position they serve, and the vessel itself must be certified by the American Bureau of Shipping and the U.S. Coast Guard for the route and service intended.

Hopper dredges normally work 7 days a week, 24 hours a day. The person in charge is the master who has complete responsibility for the ship and its operation. The crew normally operates a standard marine watch system, and two crews are on board the vessel at all times. Two crews are always ashore and the normal rotation is to work three weeks on and three weeks off. The time each crew is off the vessel is nonpay time to them, but while working aboard they are paid 12 hours a day, 7 days a week. The time on and off the vessel may vary for different operating contractors. The length of time on and off the dredge has increased over the years out of deference to the desires of the crew members. A captain or master must be on board at all times so two captains are required, one of whom may be the master. They also work three weeks on and three weeks off. Two chief engineers are required who work three weeks on and three weeks off also.

This arrangement provides that in a 52-week year, the equivalent of two crews are paid for 52 weeks, even though four crews are normally on the payroll and available for work. One mate, one engineer, one oiler, and one or more able-bodied or ordinary sailors are required for each crew, making a total of two or more persons for each position on board at all times. A dragtender to operate the dredging system is also provided for each crew. Usually an able-bodied sailor (ABS) trains for this position. The ABS works on deck if the vessel is not dredging during the ABS watch. Cooks, day workers, and boat operators are also included as needed. These persons live on board the vessel during their three-week work period. Survey crews familiar with the computerized automatic positioning and plotting system in addition to normal survey operations are also necessary, but usually do not live on board the dredge. Normally they work only during daylight hours. They operate from the ship's launch which is normally a high-speed vessel used as the ship's tender, crew boat, and survey boat. The launch is equipped with standard survey equipment as well as a duplicate computerized automatic positioning and plotting system to that installed on the dredge. U.S. Coast Guard staffing requirements usually provide for additional crew members to be aboard the vessel when it is on an ocean voyage as opposed to normal dredging operations. In that event, the additional crew members are selected from the crew that would normally be ashore at that time. When engaged in beach nourishment work or upland disposal, a shore crew similar to that used by hydraulic dredges is required.

All dredging operations as well as ship navigation and control are performed from the pilothouse. The dragtender is normally positioned in a dredging pulpit at the forward part of the pilothouse at the centerline of the vessel several feet lower than the pilothouse deck. From the pulpit, there is an unobstructed view of the dredge hopper and the loading and unloading systems, as well as the dragarms and their machinery. The dragtender works seated at a console which contains all the necessary instruments, gauges, and controls to operate the dredging machinery components. The helmsperson's console is located directly behind the dragtender and the mate navigates the vessel from this position. Bridge wing consoles are located on the bridge wings at each side of the ship and all necessary navigation and propulsion controls, instruments, and gauges can be transferred to these locations so the mate can perform duties from the bridge wings if necessary.

The seagoing hopper dredge must have propulsion power and navigation control adequate to operate the vessel while it maintains a two-knot headway with both dragarms on the bottom, a condition similar to an ocean-going vessel operating safely with both anchors dragging.

Dustpan Dredge

The dustpan dredge is a specialized dredging vessel constructed for inland rivers where the material to be excavated is soft mud and light, free-flowing sand. Its purpose is to agitate and pump the shoal material and disperse it into the currents for deposition elsewhere in the waterway. The dustpan dredge is essentially a plain suction dredge with an exceptional form of mouth on its suction tube. This suction mouthpiece may be 30 ft or more wide. Because of its shape or form, the suction tube is called a *dustpan*, which gives the dredge itself the name *dustpan dredge*.

Few dustpan dredges are in use today because of the limited service to which they can be applied. Similar to a hydraulic suction dredge, they are constructed with an A-frame at the forward end that supports the dustpan suction at the bow in a well by means of a block and tackle arrangement. A main, or dustpan hoist, raises and lowers the dustpan the way a dredging ladder is moved on a hydraulic dredge. There is a dredge pump drive complete with auxiliary machinery similar to that on the hydraulic dredge. This dredge pump is normally a low-head, low-power pump, however—more similar to those used on most hopper dredges. The low-head pump is provided because this type dredge discharges the dredged material overboard, directly back into the river several hundred feet from the dredge. Current flow in the river is expected to carry away the dredged solids. Hydraulic dredges operating on the Mississippi and other inland rivers sometimes use this type discharge.

The similarity to a hydraulic suction dredge ends at the dredge pump because the dustpan dredge does not swing through an arc around a spud like the hydraulic dredge. It has spuds to hold the dredge stationary when it is relocated between cuts. Heavy anchor winches, provided with anchors ahead and aft, are used when the vessel is dredging. The anchors and winches pull the dredge straight ahead into the bank with the dustpan lowered in order to excavate the material in front of the dustpan mouthpiece. This movement continues until the dredge moves straight ahead the length of the cut, or the distance permitted by the length of the anchor wires and/or the contours of the river bank. When the dredge has moved forward the limit of its cut, the spuds are lowered and the dredge or anchors are shifted or reset for the next cut. The spuds are then raised and the dredging procedure is repeated.

The short discharge line normally has a curved deflecting blade, which can be turned about a pivot point at the center of the curve. It can be operated manually with a small hand-winch or by other means. The end of the discharge pipe discharges the dredged material into the center of this curved blade, and the blade helps distribute the material over a wide area by means of the position of the curve to the flow of material and the reactive forces of the flow against the blade.

The floating discharge line is similar to that used by a hydraulic dredge utilizing pipe on pontoons with ball joints to permit the flexibility required to distribute the dredged material behind the dredge.

The dustpan suction mouthpiece has water jets spaced along its width to assist in dislodging or loosening the soil being dredged. Pressure supplied to the jets must not be too high because higher pressures may carry the particles of material being dredged away from the suction mouthpiece. Pressures of approximately 1.5 atm are usually sufficient for this type work.

Staffing of the dustpan dredge is similar to that of the hydraulic cutterhead dredge.

Drill Boat

The drill boat is a specialized type of marine equipment used to drill and blast subaqueous rock and other dense material too hard to be dredged by conventional methods. Generally it is a barge hull with several, usually three or four, rail-mounted drill towers located along one side. The drill towers have pneumatic or hydraulic drill motors mounted on a slab back which raises or lowers in a track system that traverses the length of the tower from top to bottom. Drill towers should be tall enough to raise the drill bit out of the hole and allow the bit to be examined or changed without need to remove drill steel sections. The bit should be raised above the tower base or platform level to accomplish its removal, inspection, or storage. This arrangement permits the operation without the time that would be lost if one had to add or remove drill steels to raise the bit out of the hole. The usual drill tower height is 80 to 100 ft. This permits drilling 75 to 95 ft below the water line.

A sand pipe, the purpose of which is to keep the drilled hole open, is lowered into the overburden and/or until the end of it rests on top of the material to be drilled. The sand pipe has a funnel attached to its top to guide the drill bit into the pipe when the bit is lowered under water. The sand pipe assembly is attached to a sand pipe guide consisting of pipes on each side of the sand pipe assembly that run the full height of the tower on each side of the slab back drill motor tracks. These pipe guides travel through short pipe retainers located approximately 10 ft apart along the tower length.

Flanged wheels, chain driven by air tuggers, move the towers on their rails. The towers are positioned to maintain drill hole centers at 8, 10, or 12 ft apart, depending on material density and job conditions. Once the boat is located in the proper position, the driller lowers the sand pipe to the top of the drilling area and then lowers the drill motor slab back assembly so the drill bit enters the sand pipe and continues down to touch the top of the rock to be drilled. The hole is then collared by slowly feeding the bit with the drill motor until the bit forms a spot in the rock.

Delicate feed control is required to keep the spot in line with the sand pipe until the spot is about 6 in deep. Once the hole is properly collared, the driller drills it as fast as the drill can operate and be controlled. The depth of the hole needed is governed by the combined aspects of type of material being drilled, distance between holes in all directions, and the amount and type of blasting material being used. The blasted holes should be far enough apart and deep enough so the blasted material can be removed to the desired grade by conventional dredging methods. If the holes are too close together or too deep, they will require expensive overdrilling; if they are too far apart or not deep enough, the material will not break up enough to be dredged properly.

After each hole is drilled, it is loaded with explosive and primed with blasting caps and electric priming cords. The cords run to an electrical blasting switch located on the side of the superstructure house next to the drill tower rails. Each tower normally drills two, three, or more holes at each setup, depending on the length of the hull between spudwells. The cords are temporarily tied off to a line that runs outboard of the towers and the full length of the drill boat until all holes on each range are drilled, loaded, and primed. Should a drill steel break or a bit jam in the hole, that hole must be abandoned and a new one drilled next to it. The blaster is in charge of loading and priming the holes and setting off the charges.

The drill boat is a raked end deck barge with four pneumatic spud legs, two forward and two aft. Two pneumatically operated high-speed anchor winches are located at each end of the hull on the centerline with anchor fairleads that rotate 180° to accommodate the direction the anchors are run out. In operation, the drill boat sets up on range by lowering the spuds and running out two anchors diagonally opposite and away from each end of the hull. All holes for that range are drilled, loaded, and primed. The spuds are raised and the priming cords are connected to the blasting wire. The drill boat pulls off range a safe distance with its anchors. The blasting wire is energized to the blasting switch and the explosive is detonated. Warning signals, both visual and audible, are sounded and displayed before the charges are detonated. To minimize potential seismic problems, the primer cords are designed so that each hole is blasted a fraction of a second after the one ahead. This eliminates any possibility of a single large explosion. After the boil of the water over the blasted area subsides, the drill boat moves on to the next range using its anchor winches, lowers its spuds, and resumes drilling.

Modern drill motors need enormous amounts of air: 1200 cfm at 100 lb/in^2 or more. This requirement makes the air compressors too large to be placed on the drill tower platform as they were in the past with drills that required less air. Today the air compressors can consist of small units on the drill tower with larger units in a machinery house that usually runs full length of the drill boat on the opposite side of the hull from the tower rails. Air from the compressors in the machinery house accumulates in a piping header. Hose connections, located strategically along this header, connect it to the tower air piping system through hoses that have enough slack or loops to permit adequate tower travel.

Good design will provide at least one compressor in excess of those required for each tower. This will permit continued uninterrupted operation of all towers if one compressor is down for maintenance or repair. Because so much air is produced to operate the drill motors, all large auxiliary equipment, such as anchor winches and spud winches, are also pneumatically operated. Sufficient electrical generating capacity must be provided to operate such small auxiliaries as pumps, bit sharpeners, lights, etc.

The crew of a drill boat consists of the superintendent and the drill foreperson who are charged with the responsibility for the drill boat and its proper operation. They must be familiar with the drill boat, its equipment and operation, and the project. Serving under the foreperson are the drillers, one for each tower, the blaster and a helper, an engineer, a mate, and one or two deckhands on each eight-hour shift. The shifts rotate weekly as on other types of marine equipment.

Several types of drill motors are in use today. One is the independent rotation drill that consists of a drill motor where rotation, percussion, and feed are accomplished independent of each other. This permits the best combination of individual control of these functions for the most efficient hole production in a particular formation. The independent rotation drill is an improvement over the older drill motor that was not independently operated. It is best suited for hard formation drilling in dense material under certain conditions and is faster than other types of drill motors.

A second type of drill motor is the Rotary Kelly Bar-Type drill, identical to the drill system used for oil well drilling. This drill uses drill pipe to connect the drill motor to the drill bit and is not well suited for drilling in such dense material as rock. It is normally used in southern states where less dense materials, such as marl and coral, are encountered during subaqueous excavations. The drill motors, towers, and drill boats operate in a manner similar to that previously described, but the Kelly Bar system cannot drill in dense materials efficiently.

One other type of drill recently introduced in the subaqueous drilling industry is the hydraulically operated drill. This drill can provide high-speed rotation and percussion but is seriously handicapped because a separate pneumatic hole cleanout system is needed, and when operating on rock drilling projects with even a small amount of overburden, the bit frequently jams in the hole. This causes lost time in clearing the drill bit or abandoning and redrilling the hole.

The choice of drill types is extremely important in the efficient drilling and blasting of subaqueous material, and the contractor must weigh the pros and cons of each type for all conditions to be encountered before selecting the system to use. Once a drill type is selected and installed, any change is almost impossible due to the high capital cost involved.

Advances have also been made in the type of material used for subaqueous blasting: from the original dynamite sticks, to gel packets that had to be tamped or rammed, to the newest type of liquid dynamite which is heavier than water and eliminates the need for tamping. These improvements have added to the increased efficiency of the subaqueous drilling and blasting process. For a typical drill boat see Fig. B11-9.

FIGURE B11-9 Great Lakes Dredge & Dock Co. drill boats: Algonquin and Number 8.

FLOATING CRANES, DERRICKS, AND PILE DRIVERS

Cranes and Derricks

Until a successful sky hook is developed, the floating crane or stiffleg derrick is one of the indispensable tools available to the marine contractor. There are several ways of approximating the proverbial hook on a floating hull. The simplest and most economical way is by installing a fixed boom, or A frame, on the bow of a barge. If heavy structural members and multiple reevings of large cable are used, the arrangement can lift extremely heavy loads. These fixed booms are not versatile in handling loads, since the entire barge must shift to position the hook over the load and lift it, then shift again to lower it at the desired location.

Steel stiffleg floating derricks are available in rated capacities of 5 to 800 tons and they provide an economical lifting unit. Such a derrick with its diesel engine–driven three-drum hoist will be less expensive than a whirley-type crane of comparable capacity. The stiff-leg derrick, through the use of a bull wheel and swing drum on the three-drum hoist with swinger, can accomplish a 180° swing. The other drums are used for the main hook, with multiple reevings on the boom hoist and the whip line, which is a single line for light loads and auxiliary work.

Regardless of the marine contracting work involved—whether pile driving, material handling, or steel erection—whirley-type cranes that can revolve a full 360° are a great asset. Barge-mounted whirley-type revolving cranes are expensive, as noted earlier, a contractor may place a crawler-type crane aboard a barge when clamshell operations or light lifting are needed. The crawler crane should be securely lashed to the deck to prevent it from sliding overboard. The operator must pay careful attention to the load and the radius at which it is to be lifted. Many boom failures occur because the capacity of the boom has been exceeded, especially in large-radius operations, and because attempts to pull a load sideways have placed a heavy side load on the boom. Booms are compression members and are not able to withstand loads in bending, as occur in side loading. Whirley-type crawler cranes with ratings of 100 tons or higher are available and are routinely mounted on barges for marine contractor operations.

Even larger capacity whirley-type revolving cranes are available for lease to the marine contractor. These cranes have the tub for the rotating machinery built integral with the hull and capacities of 13,200 tons, utilizing two 6600-ton capacity cranes on the same hull, are in use today. Vessels of this type are semi-submersible. Full floating whirley-type revolving cranes of 3000 tons also are in use and are readily available.

Before attempting to mount any type of lifting equipment aboard a barge, or hull, whether A-frame, whirley-type revolving crane, or stiff-leg type, one must determine whether the deck and framing will support the load which is to be imposed. Planking or timber mats are frequently employed under crawler-type whirley cranes to distribute the loads over the deck area and the hull framing system so as not to impose undue stress on any part of the hull framing. Assuming that the barge has enough capacity to carry the equipment to be mounted aboard, a qualified naval architect or marine engineer should make a stability study to determine what loads can be handled at various radii without diminishing the stability of the vessel and its seaworthiness.

One of the insatiable requirements of progress seems to be that all aspects of business, and especially the marine construction and contracting business, must grow. Larger amounts of capital are required, more staff is needed, and equipment must increase in capacity, size, and complexity. The search for and recovery of oil in the tidelands and offshore provides an excellent example. In contrast to land drilling, this operation requires the construction, transportation, and on-site erection of enormous marine structures, and the use of mammoth, heavy floating equipment. Projects are routinely performed today that would not have been considered just a few years ago. Sunken barges and other vessels with cargo intact, which would have been abandoned in the past, are salvaged successfully. Derrick and crane barges of colossal size are being constructed to perform these services. The largest of these derrick barges is 661 ft long, 323 ft wide, and 162 ft deep. It can work in the open sea in any part of the world and can lift a 13,200-ton load with two 6600-ton revolving cranes. McDermott International, Inc., owns this gigantic crane. The largest single revolving derrick barge is 497 ft long, 150 ft wide, and 41 ft deep, has a lifting capacity of 4400 tons (3500 tons revolving 360°) at 80-ft radius, a boom length of 262 ft, and a lifting height of 290 ft above its operating draft. In addition, McDermott International, Inc., owns a shear-leg crane 450 ft long, 150 ft wide, and 30 ft deep. This crane can lift 5000 tons at 100-ft reach.

Auxiliary equipment for these mammoth hooks includes pile-driving apparatus that can deliver up to 1,647,000 ft-lb to drive piles up to 84 in or 7 ft in diameter. They are also equipped to erect and lay large-diameter pile frames under water and driving piles in water depths of 1500 ft. They have climate-controlled quarters provided for the crew and enough fuel, potable water, and storage space to operate for an extended period without refueling or reprovisioning. Examples of these cranes and derricks are shown in Figs. B11-10 and B11-11.

FIGURE B11-10 Floating cranes and derrick.

FIGURE B11-11 Floating derrick.

By using slings, spreaders, and strongbacks, these giants of the construction industry can team up to lift heavy loads beyond the capacity of any one unit. Extreme care must be taken when using two or more cranes to prevent damage to the equipment and to reduce the possibility of endangering operating personnel. Strain gauges should be used whenever such lifts are undertaken.

Pile Drivers

Pile driving is often required on marine projects. Piles must be driven to create support for all types of structures in the marine environment: piers, quay walls, buildings, marginal wharves, bridge piers, ferry slips, drilling platforms, etc. This support is provided by bearing piles and by sheet piling to restrain fills or to contain concrete pours.

MISCELLANEOUS EQUIPMENT

Floating equipment of a specialized nature, such as compressor barges, pump barges, office barges, work barges, tool storage barges, quarters boats, or equipment for submarine divers, is assembled in accordance with the needs of the project. These requirements should be established by competent engineers or construction superintendents.

AUXILIARY DIVING EQUIPMENT

When a compressor barge is needed to supply air for a pneumatic caisson, someone thoroughly familiar with this type of project should establish the size of the compressors. The compressor capacity needed will depend on the size of the caisson and the frequency of use of the air locks.

Submarine divers will insist that compressors, lines, and tenders have assigned to them an area which will be theirs. To accomplish this, one must reserve for the divers a small barge or pontoon which may be shifted easily. This arrangement has the added advantage of being readily removed when the diver is not needed on the project. Comfortable shelter should also be provided for the diver to change gear during cold or inclement weather. The shelter should be heated and be large enough to house the diver's communication system, which will be monitored by a tender. A decompression chamber should be available for the diver who must work in deep water for protection if an emergency rise is necessary or if stricken with the bends. Only qualified persons should be permitted to operate the decompression chamber.

Transportation

The transportation needed for a project depends on the number of crew to be employed and the distance they must be transported. If the project is near a landing, the job tug or towboat should suffice. If it is remotely located, a motor launch or crew boat capable of moderate to high speeds can be used. Some offshore and inaccessible remote locations may require helicopter transportation for crew changes. Boats used to transport crews should not be overloaded and, depending on conditions, may require a U.S. Coast Guard certification and licensed operators. Except in the most unusual cases of inaccessible and distant projects, quarters boats are rarely used because workers prefer to live ashore. They travel to the site and assemble at a landing to await transportation to the project.

Communications

In today's fast-paced economy, good communications are imperative to the successful operation of any marine project. Supervisory personnel must be able to coordinate the work with subcontractors, the home office, and other construction equipment and gangs on the project. Various types of equipment can be used to facilitate good communications, including radio, mobile telephones, walkie talkies, and transceivers. On a large project, the cost of a base station and several portable units is small compared to the benefits realized. Radio or communication units should be installed in the field office, the superintendent's automobile, all dredges, and every tug, towboat, and crewboat used on the project. Many communications companies, including the mobile service departments of the Bell Telephone Company and Motorola Corp., can provide excellent advice and service about communication systems and can also supply the systems.

Marine Risk Insurance

The contractor whose work requires floating equipment is subject to hazards peculiar to marine operations and construction. Financial loss from damage to or destruction of such equipment as may be owned or leased is at risk; the contractor is also responsible for injury to the person or property of others. Environmental pollution accidents can be especially devastating in the costs involving containment and cleanup of any oil spills encountered. A qualified insurance broker can negotiate the necessary insurance coverages with an underwriter to eliminate all or the greater share of these possibilities of loss.

Because of its special nature, marine insurance is a field of insurance unto itself. Before moving any marine hull or equipment, the contractor should consult a broker well versed and thoroughly qualified in this field and obtain proper insurance.

Marine insurance, in general, pertains to water-borne traffic on oceans, lakes, bays, sounds, rivers and canals, etc. This type insurance protects against loss or damage to hulls and machinery, liability for injury or damage to persons or property, and loss of cargo.

Hull Insurance

This protects the insured for specified damage to marine equipment and is frequently written with a deductible clause. It covers the hull, fittings, machinery, and fixtures of a vessel against such perils of navigation, as collision, stranding, sinking, heavy weather, fire, and explosion. This insurance also protects against loss from salvage charges. The hull policy developed for river marine hazards is a modified form of the one used for ocean service.

Most policies contain a clause that covers latent defects and negligence of the master, officers, and hands. This latent-defect clause does not cover loss due to the negligence of the owners or managers, nor does it cover the cost of the part in which the latent defect existed. Collision liability insures against damage to another vessel, its freight, and cargo. Usually a percentage of the insurance can apply to property separately stored ashore. Port-risk hull coverage is a variation of hull insurance intended to cover vessels laid up for a long time.

On larger and more costly vessels, a premium savings can be made by covering part under a hull policy and the remainder under a disbursements insurance policy. The latter policy then comes into effect in the event of a total loss.

Protection and Indemnity Insurance

This coverage is usually an endorsement to a hull policy. Its terms, similar to the protection afforded by an automobile liability and property damage policy, cover the insured's liability as owner of the insured vessel for loss of life and injury to employees, including the master and the crew. It also embraces those persons who are not employees and the insured's liability for damage to property if those risks are not covered by the collision clause in the hull policy. Pollution insurance protects the owner of the insured vessel for risks of oil spills and environmental accidents. The cleanup cost coverage can be obtained as an endorsement or by a separate policy.

Worker's Compensation Insurance

This is designed and written as prescribed by the worker's compensation statute of the state in which the insured is domiciled or operating. Usually it excludes coverage for masters of vessels operating in navigable waters and crew members, although these excluded employees may be brought under worker's compensation policy by a voluntary compensation endorsement. The insurance company will then pay such amounts as an injured employee would have received had the act been automatically applicable.

Any employee otherwise entitled to compensation under the Jones Act or the Longshore and Harbor Worker's Compensation Act is not obligated to accept the compensation payment set forth in worker's compensation insurance but may instead bring an action in Maritime Law under the Longshore and Harbor Worker's Compensation Act or the Jones Act. In either the Jones Act or the Longshore and Harbor Worker's Act, the injury must occur on navigable waters or on an adjoining pier, wharf, drydock, terminal, building way, marine railway, or other adjoining area customarily used by an employer in loading, unloading, repairing, dismantling, or building a vessel. The Longshore and Harbor Worker's Act provides coverage for a broad range of maritime workers, but expressly excludes a master or member of a crew of any vessel.

Ordinarily, employees covered under the Longshore and Harbor Worker's Act receive compensation benefits but are barred from suing their employers. Under certain circumstances, however, employees may still sue their employers for negligence if the employers are the owners or operators of vessels, and negligent operation of the vessels has caused the injuries to the employees.

Under the Jones Act, any sailor who suffers personal injury in the course of employment may bring an action at law. The injured employee must show that he or she is a sailor. To qualify as a sailor, the employee must independently satisfy all of the following requirements:

1. Be associated with a ship in navigation
2. Have a more or less permanent connection with the ship
3. Show that his or her duties contributed to the function of the vessel or the accomplishment of its mission

If an employee brings an action in Maritime Law under the Jones Act or the Longshore and Harbor Worker's Act, the insurance company will protect and defend the employer against loss under the employer's liability section of its policy. It is important that all owners, leasers, and contractor operators carry these policies properly endorsed.

Dredging and the Environment

The contractor is faced with many problems in the marine construction industry. Marine construction embraces numerous activities, of which dredging is the largest. To realize the scope of dredging in the United States today one need only understand that the Army Corps of Engineers maintains more than 25,000 mi of waterways and spends about $550,000,000 per year to do it. Their responsibility is to maintain and improve our waterways. In the exercise of this responsibility they supervise work involving dredging of approximately 250,000,000 yd^3 each year.

In addition to maintaining and improving the nation's inland waterway transportation system and related work, an ever-increasing amount of beach nourishment is performed annually. Beach nourishment is the process of mining sand from the ocean or an inlet and placing that dredged sand on the beachfront. This work attempts to overcome the relentless erosion which takes place along many of the recreational shorelines of the United States.

In addition to the enormous amount of dredging contracted for by the Army Corps of Engineers each year, private dredging is also performed for marinas, ship berths, private channels, trenches for submerged pipelines, and the cleaning of landlocked lakes and lagoons, power intakes, etc.

Many consider dredging an operation that is harmful to the environment. They perceive it as creating vast areas of turbidity, of releasing toxins to adjacent waters, of consuming oxygen in the water column, destroying of biota, and being detrimental to marine life. These concerns, and others, are associated with both dredging at the site of the excavation and at the disposal areas, whether the disposal occurs in open waters or on upland locations. The dredging contractor believes these concerns, though well intended, are exaggerated in almost every circumstance. The scientific research and literature demonstrates that the effects of dredging and disposal of excavated materials, when properly and reasonable performed, has environmental effects that are local in nature and of short duration. It is evident and easily shown that storms and storm runoff increase turbidity and silt movements substantially more than dredging operations without any permanent harm to the environment. However, today's social culture and environmental concerns have resulted in many restraints being often imposed on dredging contractor operations. Examples are:

1. *Seasonal restrictions:* These restrictions prohibit dredging during certain seasons, normally in the spring, summer, and fall.
2. *Overflow restrictions:* This restricts the ability to dispose of the liquid on top of the load in the hopper at the dredging site. It requires the disposal of the liquid at the disposal site.
3. *Turbidity restrictions:* These restrictions require that silt curtains or other restraining devices be maintained around the dredging and disposal outfall areas.
4. *Capping restrictions:* This requires that fine dredged material, excavated underwater, when disposed of in other underwater areas, must be covered by sand.
5. *Upland restrictions:* This requires that upland enclosed disposal areas be surrounded by wells so that underground water can be examined to assure that underground aquifers are not harmed by the disposal of dredged material.
6. *Aerial surveillance restrictions:* This requires surveillance by air to insure that no turbidity plumes are induced by the dredging process.
7. *Inspection restrictions:* The requirement that an inspector and/or spotter be placed on the dredge to verify compliance with restrictions and to detect if any endangered species enter the dredging area.

8. *Miscellaneous restrictions:* This requires the measuring and recording of salinity, oxygen, turbidity, PH levels, etc., during dredging in the vicinity of the dredging area.

These restrictions illustrate the restraints frequently placed on dredging contractors and their operations. They are imposed by federal and state regulatory agencies by conditions included in the dredging permits and the water quality certificates which are required for the performance of the work. The River and Harbor Act of 1899, the principal legislation that requires and controls the issuance of dredging permits, requires anyone making modifications to a waterway of the United States to have a dredging permit. A dredging project is considered a modification of the waterway within the meaning of this act. The waterways of the United States are defined as any body of water which supports commerce, has supported commerce in the past, or is capable of supporting commerce in its natural state or with improvements. Any body of water within the high water lines of the waterway and any wetland contiguous thereto are also defined as part of the waterbody for federal purposes.

Other legislation directly pertaining to dredging and marine construction is the Federal Clean Waters Act of 1977, as amended. Sections 401 and 404 of this act are pertinent to dredging specifically. Section 401 mandates that the state in which the dredging occurs must certify that the work meets their water quality standards. Obtaining this certification usually requires the contractor to perform one or more of the following tests on the material proposed to be dredged:

1. Bulk sediment analysis of the material to be dredged
2. Testing of the elutriate derived from an EPA-prescribed mixture of the material to be dredged with adjacent waters
3. Bio-assay testing and the performance of prescribed toxicity characteristic leaching procedures on the material to be dredged

These tests may examine more than 100 chemical parameters. Section 404 requires the Army Corps of Engineers to regulate any discharges of dredged material into the waters of the United States under guidelines of the Environmental Protection Agency.

CHAPTER 12
TUNNEL-BORING MACHINES

Sverker Hartwig
President, The Robbins Company, Kent, Washington

INTRODUCTION

As the name implies, *tunnel-boring machines* (TBMs) are machines for boring tunnels. This takes place in a virtually continuous process, as opposed to the cyclic method of conventional drilling and blasting (D & B).

While a rather primitive, compressed air–driven TBM was developed by Beaumont and English and used to—among others—drive two pilot tunnels under the English Channel as early as 1881, the era of the modern TBM did not start until the 1950s, when a machine was put to work excavating a diversion tunnel for the Oahe Dam Project in South Dakota.

Both the Beaumont/English machine (diameter 2.15 m) and the 7.85-m-diameter Oahe Dam machine were suitable for work only in very soft rock formations (chalk and easily bored shale). Since then, development has progressed in two directions:

- Enabling the machines to tackle ever harder and more abrasive types of rock
- Making the machines capable of operating in both stable, competent rock and unstable ground formations which require the tunnel to be lined concurrently with the excavation process

As ground conditions within a single tunneling project may vary widely with regard to rock hardness and ground stability, it is sometimes found necessary to combine the characteristics of typical hard-rock and soft-ground TBMs into hybrid-type machines. For similar reasons—i.e., to make the machines capable of successfully traversing *any* type of ground from start to finish—it often is found essential to engineer the TBM and its backup equipment together with the lining system used behind it.

TBM SELECTION

The type and design of a TBM to be used on a particular tunneling project is invariably governed by, and therefore tightly interwoven with, two main project characteristics:

- The geological conditions expected to be met along the tunnel route
- The overall design of the tunnel and any specific peculiarities of the project as a whole

Geological Considerations

The main geological parameters always to be taken into account in the selection of a TBM for a particular project are the *borability* and the *stability* of the ground. The first will determine what type of boring tools will be used and how much power should be available to rotate and thrust the cutterhead forward in relation to the diameter of the tunnel. The second controls the type of ground support to be used and, thus, whether the TBM can operate on its own or must take an active part in supporting the tunnel and, ultimately, needs to be tied in with the lining, if it is necessary to use that type of support to keep the tunnel from collapsing.

Borability

The types of tools used for TBMs are either drag bits—sometimes also called *picks*—or roller cutters. In both cases, the designation indicates the mode of operation:

- Drag bits loosen the material to be excavated by being dragged through it in circular paths and dislodging the material in between the kerfs they cut.
- Roller cutters roll along the surface of the rock—i.e., the face of the tunnel—under a high thrust load, establishing kerfs by crushing the rock underneath their cutting edges and spalling out larger chips of rock from the area between two neighboring kerfs.

The dragging action of the first type of tools leads to friction between the bit and the material worked and thus to heat development which combines with the abrasivity of the material to cause tool wear. Drag-bit-equipped TBMs therefore cannot be used with economically acceptable results in any but the softest type of grounds. TBMs for all other type of grounds are fitted with roller cutters, which see much less friction and therefore are not as wear-sensitive.

Stability

In stable, competent rock, there will be no need to support the tunnel by anything more than the occasional rock bolt to secure larger blocks of rock delineated by local cracks. As irregularities in the rock—in the form of bedding planes, cracks and fissures, faults, etc.—increase, rock support measures will shift to a regular rock bolting pattern, netting, shotcreting, the installation of ring beams, or combinations of these, and ultimately to a partial or a full lining.

Ground suspected of being in need of more elaborate support or of containing local fault zones can be investigated in advance by probe drilling ahead of the TBM and reinforcing it by grout injection, when the probe holes indicate the need thereof.

Other Geological Influences

Apart from insufficient stability, the ground to be traversed by the TBM may contain such hazards as zones of running ground, abnormal amounts of water under high pressure, methane, etc. Also, these occurrences can be detected by probe drilling, and water may be sealed off or running ground stabilized by injecting grout.

The more elaborate the ground support and reinforcement or sealing measures become, the more they will influence the selection of the type of TBM for a particular project and the detailed design of the machine. While the simpler types of support can be handled in conjunction with the use of a so-called *open-type hard-rock TBM*, support in the form of a lining installed during drivage invariably calls for some type of *shielded TBM*. No matter which type is selected, the occurrence of methane will always influence the design of the machine, either by the installation of a methane warning system or, additionally, the explosion-proof execution of the electrical system.

Tunnel Design and Project Specifics

It goes without saying that both these groups of parameters will influence the design of the TBM, once the selection of the type to be used has been made, based on the available information on the geology.

Tunnel Design. The shape and the dimensions of the tunnel cross section are the first parameters which characterize tunnel design. To lend itself to excavation by boring, a tunnel must in general have a circular cross section, though special types of boring machines have been developed to accommodate noncircular, most often horseshoe-shaped, cross sections. It will be obvious that a TBM's mechanical design and its installed power are directly connected to the diameter of the cross section.

In the case of a very large diameter, it may be better to excavate the tunnel by first boring a pilot tunnel by means of a smaller-diameter, standard TBM and to thereafter ream the pilot bore with a special *reamer TBM*.

If the tunnel is not to be straight from start to finish, the radius of the curves to be bored en route will influence the design of the TBM—and perhaps even more so of the backup equipment which houses all the auxiliary gear a TBM needs for its operation. Most TBMs will cope with curve radii down to some 250 m without any problems. For tighter curves, special layouts may become necessary.

The dip or the incline of the tunnel will influence the design of the boring machine when they exceed approximately 1:10. If the choice exists whether to drive a tunnel going down or up, one will normally select the latter, as this provides a natural drainage. If the tunnel has to be driven downwards, a sump pump must be installed at the front end of the machine to cope with the inflow of water. At declines steeper than 1:4, normal belt conveyors become unsuitable for muck removal and need to be equipped with special belting or exchanged for a different type of conveyor.

Boring upwards poses its problems, too, when the incline becomes so steep that the machine develops a tendency to slide backwards down the slope when it is not gripped in. At that point (usually at inclines steeper than some 1:5, depending on how "slippery" the tunnel invert is) the muck may also present problems, in that it starts to roll on a moving belt. From there onwards, TBMs operating on an incline will have to be provided with a safety holding mechanism to stop it—and the backup—from sliding down, and special muck removal systems must be installed. Inclines of up to 1:1 have been bored successfully by thus modified TBMs on a number of occasions.

Project Specifics. The following items will influence TBM design to a greater or lesser degree:

- *The location of the site.* Here one meets such aspects as access, altitude above sea level, climatic conditions (ambient temperature range, rain/snow falls, humidity), the availability of water for cooling purposes, the availability of skilled personnel to operate and maintain the machine, the availability of maintenance facilities and a reliable spare parts delivery, etc. The absence of electricians with experience in electronics, for instance, makes it less advisable to base the TBM's control system on a programmable logic controller.
- *Access to the starting point of the bore.* Complicated access may necessitate a special machine configuration—for instance, if the components have to be taken up a winding mountain road or lowered down a shaft before they can be assembled close behind the starting point of the tunnel-to-be.
- *Muck transport and dumping possibilities.* These factors may influence perhaps not so much the machine itself as its backup.
- *Environmental aspects.* To quote but one instance: If the avoidance of oil spills is of special importance, the use of a hydraulic cutterhead drive system should be ruled out, due to the risk of hose ruptures.

These examples clearly indicate that varying geological conditions and project-dictated circumstances combined influence the choice of TBM type to be used in general and its individual layout and detailed design to a much larger degree than what is the case for many other types of heavy construction equipment. This, and the large capital expenditure for a TBM and its backup, more often than not lead to a high degree of tailoring of each machine for the specific project it is to be used on. The more comprehensive and the more accurate information the end user can supply when ordering the equipment, the better the chance of receiving a machine that will meet expectations of a profitable tunneling project.

DESCRIPTION OF TBM TYPES

As various types of ground overlap with regard to some of their characteristics, the TBMs used to excavate them can also show an overlap in their applications. If the two main types of ground are called *soft ground* and *hard rock*—however imprecise such names and such a division may be—a coarse classification of the use of various types of TBM can be illustrated diagrammatically as shown in Figure B12-1, which indicates that soft ground will always be tackled by some type of shield machine, while both shield and open (unshielded) type TBMs find application for boring tunnels in hard rock.

Soft-Ground Shields

Per definition, the main characteristic of soft ground is that it requires immediate support to keep it from moving into the excavated space (the tunnel). The only practical way of doing this is by enclosing the boring machinery in a shield which can take the deformative forces acting on the ground.

Sometimes the ground has so little cohesion that it will "flow" freely into open spaces. It is then termed *running ground*. Such a condition is often met in silt, sand, and gravel. Due to the frequently high porosity of those types of formation, an additional problem is that they often are waterlogged and have a groundwater table

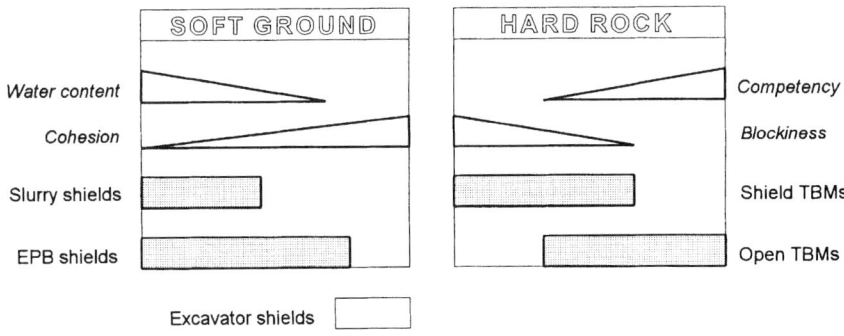

FIGURE B12-1 Comparison of soft ground versus hard rock.

close to surface. Such is often the case in ancient river mouths and estuaries, where a large number of the world's major cities are located.

Unless the inflow of water into the tunnel can be controlled already during the boring process, the tunnel may be flooded and the groundwater table will be lowered along the tunnel route. This may lead to severe damage to surface structures—roads, bridges, and buildings—as well as to earlier tunnels and other underground structures in the vicinity. To control both the movement of the ground and the inflow of water, an old U.K. patent, granted to Lord Cochrane as early as 1830, provided all the major features needed to introduce *compressed air* into subterranean excavations and "... allow workmen to carry out their ordinary operations of excavating, sinking and mining ... and also allow them to pass to and from the enclosed space into the open air...." It was not until 1879, however, that the compressed air method found a first successful practical application in a small tunnel driven in Antwerp, Belgium.

Early shield TBMs also used the, by then, standardized compressed air method for ground and water inflow control, usually with a bulkhead, located at some distance behind the cutterhead, incorporated in the shield itself. However, one inherent problem with using air as a balancing medium is that it has no effective vertical pressure gradient, whereas there is a significant natural pressure gradient in groundwater. On a large-diameter shield machine it is therefore impossible to simultaneously balance the pressures at the invert and at the crown of the tunnel. If the pressure at the invert is balanced, there is the very real risk of blowing out the crown; if the crown is balanced, water will enter at the lower part of the face and through the invert.

For this reason, a liquid or earth pressure balance is preferred. Hence, the development of slurry shields and—somewhat later—earth pressure balance (EPB) shields, which took place since the 1960s, mainly in Japan, where these difficult ground conditions abound.

Slurry Shields

Slurry shields rely on a mud slurry circulating through the cutterhead chamber under pressure to support the face, to prevent the inflow of water from the surrounding strata, and to transport the excavated soil. As such, the process may be regarded as a horizontal, upscaled version of the technique used in oil well drilling, where drilling muds are used for the same purposes. As in the case of oil well drilling, the slurry often contains bentonite to provide a measure of lubrication and better sealing properties, preventing the mud from leaking out of the excavated opening into the surrounding ground.

A slurry shield consists of a cylindrical steel-plate shell—the actual shield part of the machine—with a bulkhead installed at its forward end to maintain the slurry pressure ahead of it. The cutterhead is mounted inside the shield, in front of the bulkhead, and is rotated by either electrically or hydraulically powered drive units, mounted inside the shield, behind the bulkhead. The cutterhead is equipped with drag-bit-type tools for dislodging the ground and has a number of openings in its structure to permit the loosened material to enter into the cutterhead chamber—the pressurized open space delineated by the front of the cutterhead, the bulkhead, and the inner surface of the front part of the shield.

Apart from the cutterhead drive units, the shield contains the thrust cylinders which are placed towards the inside periphery of the shield, pumps for circulating the slurry, and an erector mechanism for the prefabricated segments of which the tunnel lining is constructed in the tail end of the shield. When a ring of segments has been completed, the thrust cylinders are extended to find purchase against the front face of the segments and thus can propel the shield when the cutterhead is rotating. When they have come to the end of their stroke—which is somewhat in excess of the width of a lining ring—the thrust cylinders are collapsed to provide space for setting another ring of segments.

To counteract the tendency of the machine to twist around its longitudinal axis under influence of the torque reaction of the cutterhead drive units, the head and the drive unit are built for bidirectional rotation, and the head will be rotated in opposite directions for subsequent strokes of the machine.

The rear edge of the tail end of the shield always maintains an overlap of the most recently set lining ring, and the ingress of soil and fluids is prevented by special seals between the inside of the shield tail and the outside of the lining.

The shield is steered in the desired direction by the following means, each of which may be used by itself or in combination.

- The thrust cylinders are arranged in groups and control of the relative displacement—i.e., of the oil volume supplied to the cylinders in each group—biases the shield in a certain direction. This is the primary means of directional control.

- Each group can also be operated in the so-called *hold-back* mode. The cylinders in such a group will then be in tension rather than compression, even though they are being extended. This is usually practiced for the cylinders at the top and has the effect of preventing the shield from drooping in very soft ground.

Though successful under favorable conditions, tunneling by means of slurry shields has a number of disadvantages:

- Because the loosened soil is *pumped* out by the mud circulation system, the machine cannot handle boulders which may be found embedded in gravels.

- There are the additional capital outlay for and operating costs of the slurry transport and separation systems and the complexity of the slurry control system.

- A large surface area is required especially for the settling plant, where the soil is removed from the mud. The necessary space often is not available in crowded downtown areas, and there may be soil disposal restrictions, too.

These drawbacks were part of the reasons for developing the next type of shield TBM: the EPB shield.

EPB Shields

Earth pressure balance shields provide ground and water inflow control in a less complicated way than slurry shields and are therefore gaining increased popularity.

As the name indicates, EPB shields balance the pressure of the undisturbed ground and the water which may be contained therein by maintaining a pressure on the excavated soil in the cutterhead chamber. The muck removal system, which is the quintessence of this type of machine, can be described as follows. (Figure B12-2.)

The muck loosened from the tunnel face migrates through openings in the cutterhead into the cutterhead chamber. If it there is a tendency of sticking, it can be dislodged by means of an extendable scraper protruding from the bulkhead structure.

The muck is discharged from the cutterhead chamber by a screw conveyor which has its entrance close to the tunnel invert, i.e., at the point to which the soil will migrate naturally under influence of gravity. The screw conveyor consists of a tubular casing, a screw, a hydraulically powered drive unit, and a discharge gate. The casing enters through and is bolted to the shield bulkhead. The discharge gate can be opened and closed by means of a hydraulic cylinder.

The material inside the screw conveyor acts as a plug, and pressure forces acting on the muck are dissipated by the viscous drag between the soil and the screw and its casing. The pressure in the cutterhead chamber is maintained by controlling the screw rotation speed and the gate opening. An automatic feedback system controls the screw speed in response to the earth pressure measured by sensors on the bulkhead.

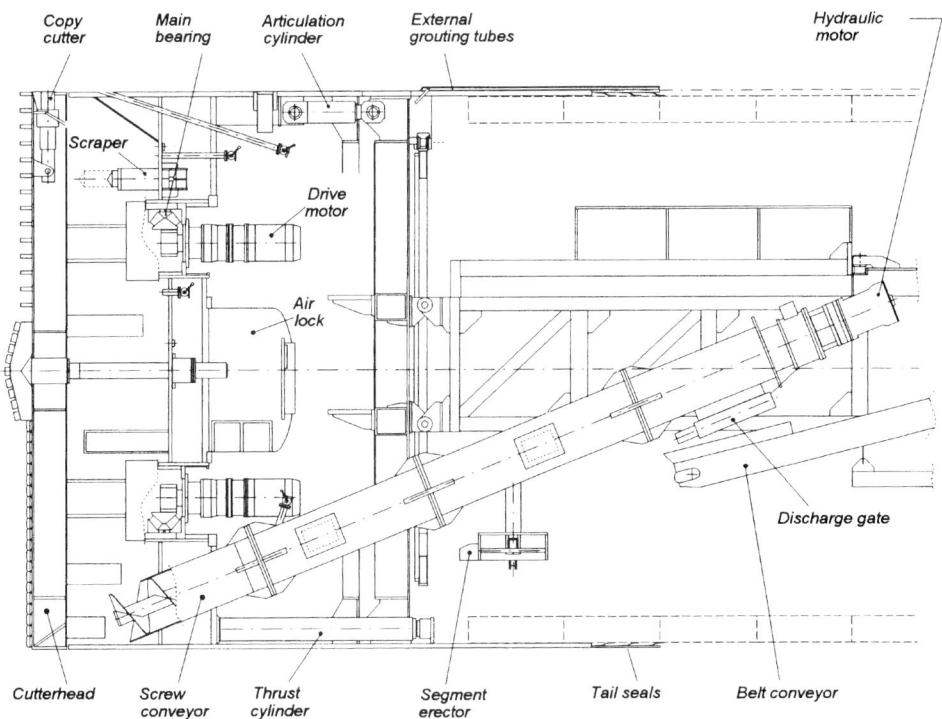

FIGURE B12-2 Diagrammatic layout of an earth pressure balance shield.

Good control of the muck discharge can be maintained by varying the screw speed and the discharge gate opening for pressures up to 2.5 to 3.0 bar (30 m water head). Above that value, an auxiliary mechanical discharging system must be incorporated to avoid the blowing through of large quantities of muck and water. Systems available include reciprocating pumps and rotary pressure dischargers.

Soil conditions may vary widely. The EPB system relies on fines in the soil for fluidity, for blocking the flow of water, and for a controlled flow through the screw conveyor. Clean sand or gravel may contain too few fines and water will then percolate too easily. In that case, fine material in the form of a mud slurry—bentonite or polymer—must be injected into the cutterhead chamber to alter the condition of the excavated material. The mud fills the voids in the sand or gravel, reduces friction, and resists the free flow of water, facilitating the earth pressure balancing function of the screw conveyor.

In the case of clay or dry soils, water percolation may not be a problem, but the material may be sticky or tend to plug. Water or mud can then be injected to fluidize the soil under such conditions. Injection ports are located on the cutterhead, the bulkhead, and the screw conveyor. The fluid is injected by means of mud pumps installed on the backup.

Mixing of the fluid pumped in should preferably be done on surface. The options for delivery of the mix to the pumps are either transportation by an agitator car or pumping through a pipeline. The latter alternative avoids the logistical problems of mud delivery by train.

If boulders are expected, a shaftless ribbon-type screw conveyor is used for muck removal, which provides greater clearance within the conveyor casing. The maximum size of the boulders which can enter the mucking system is controlled by the size of the openings in the cutterhead.

Belt conveyors which carry the muck from the discharge gate to muck cars stationed to the rear are mounted on the decks or the gantries of the backup behind the shield.

Also on EPB machines, the cutterhead and the drive units are laid out for bi-directional rotation to counteract shield roll under influence of the torque reaction. In soft ground, the cutterhead is normally equipped with drag-bit-type tools only. A programmable *copy cutter* can be used to overbore the tunnel over a particular part of the periphery to improve steering capability: As the shield moves forward, it will deviate in the direction of the overbore.

When just a few boulders are expected to occur embedded in the softer material, auxiliary disc cutters may be added, placed in the center line of the spokes' front face. In ground known in advance to contain a large proportion of boulders, it is preferable to use an EPB machine with a domed head equipped with disc cutters as the primary cutting tool.

The shield itself is of welded steel construction and can be either fixed or articulated, i.e., two short shield rings placed one behind the other, as illustrated in Fig. B12-2. Articulation enhances the steering performance of the machine and reduces the bore overcut, which otherwise is required to minimize the tunnel curve radius. The articulation joint is fitted with an elastomeric seal to prevent ingress of soil and water.

The tail of the shield contains rows of greased wire-brush tail seals to prevent the ingress of pressurized groundwater and soil into the tunnel. The number of rows depends on the external pressure.

As in the case of a slurry shield, the propelling system incorporates a number of thrust cylinders placed along the internal periphery of the shield, each of which has its own spreader shoe for pushing against the forward edge of the lining. Also in this case, steering is accomplished by means of the thrust cylinders, either by volume control or in the hold-back mode, but can also be done by using the articulation cylinders, if the shield is of the articulated type, to slightly tilt the front part of the shield (and, thus, the cutterhead) and bias it in the desired direction of travel.

A segment erector in the tail shield is used to place a new lining ring when the thrust cylinders have been retracted on completion of a boring stroke.

Excavator Shields

Excavator shields are not true boring machines in that they are not equipped with rotating cutterheads which bore out the ground. Still, they are sometimes used for driving tunnels in fairly cohesive, nonwaterlogged soft ground which does require immediate support to keep the tunnel open but needs no balancing medium to support the face. Soft marls and stiff clays are typical examples of that type of ground. Boulders embedded here and there can be handled without problems.

Instead of a rotating cutterhead, these shields are equipped with a boom-mounted bucket of the type found on excavators used on surface. Various mounting geometries can be practiced to give the boom and the bucket the necessary flexibility to work the complete face. Due to the absence of a rotating head, the tunnel face will not automatically become circular—nor need it be. In other words, the tunnel can be excavated with a noncircular cross section, if that brings design or construction advantages. The shape of the shield, of course, needs to be adapted to the tunnel cross section.

The absence of a rotating head also makes it possible to let the front edge of the shield protrude in the crown area where it then can act as a cutting edge, if the edge is made sharp and/or provided with cutting teeth, and provide early protection from ground falls from above. Most excavator shields, moreover, provide the possibility of securing the crown in front of the face by forepoling with steel bars of a suitable profile.

The material excavated by the bucket is dumped into a hopper, from where it runs onto a belt conveyor for transport to the rear. Most excavator shields are used for comparatively short drives only, and transport out of the tunnel can then be by further belt conveyors, installed as the machine advances, or by trucks.

To support the tunnel and to provide a purchase for the shield's thrust cylinders, a lining is installed at the rear end of the shield by a segment erector, i.e., in the same manner as practiced for slurry and EPB shields. In this case, there is no need, though, for elaborate sealing arrangements between the shield and the lining to keep fluids and running ground out of the tunnel.

Steering an excavator shield becomes somewhat simpler, too. With the excavator boom, the ground can be *overcut* anywhere along the periphery of the shield, which then automatically will divert into the direction of the overcut as it is advanced. To aid in steering, individual or grouped control of the thrust cylinders will still be practiced, though.

Due to the fact that the excavator bucket works only a small part of the face at any one time, this type of shield cannot advance as fast as a true boring machine, the head of which works the full face all the time. On the other hand, capital costs for this type of shield are low; it can be installed in a short period of time and requires very little in the way of backup equipment. Therefore, it becomes economical to also use this type of equipment for very short drives.

Hard-Rock TBMs

The term *hard rock* can be defined in different ways, and many controversies have existed over the years (and still do) regarding the most appropriate definition of and the best way to measure hardness and express it in a figure. Agreement nevertheless seems to exist, by and large, on using the uniaxial compressive strength (UCS) of a properly prepared rock sample and expressed in MPa, bar, or lb/in^2, as a suitable measure of rock hardness.

Because such samples for UCS determination are small in comparison even to the distance between neighboring cutters on a hard-rock TBM, the UCS figure by itself will not determine how hard a rock may be to bore. Fractures, bedding planes, and other irregularities that do not fall within the sample volume will influence borability. In most cases, they have a weakening effect on the rock—i.e., borability increases.

Finally, there is another factor to be considered: cost. Breaking the rock really is no problem. All it takes is subjecting it to a larger stress than what it can take. Generating that stress then means building a machine with enough built-in strength and power. But, as tunnels, in the end, are to serve an economical purpose, and the cost of driving them therefore has to be minimized, it all boils down to what it costs to bore a particular kind of rock.

Because cutters, through their contact with the rock, do wear, all of a sudden the whole question of borability is no longer simple. There is the rock's UCS to be taken into account, and the structure of the rock mass on a larger scale, as well as the abrasivity of the rock.

For the purpose of dividing TBMs into machines for soft ground and for hard rock, matters need not to be made more complicated than agreeing to accept that hard-rock TBMs are all those tunnel-boring machines working in ground which cannot be bored with economically acceptable results with drag bits. That immediately separates the hard-rock TBMs from the shields previously discussed. However, the latter derive their designation not from the fact that they primarily use drag-bit tools, but from the fact that they are equipped with shields to support the tunnel around them. As will be seen, under certain conditions hard rock may also require this form of support.

That is the reason why hard-rock TBMs are divided into *open* (not shielded) and *shield* machines, even if they use the same type of rolling cutter tools. Each of these will be described here.

Open-Type Hard-Rock TBMs. All hard-rock TBMs loosen the rock by means of roller cutters. Today, these almost universally consist of a hub, rotating on bearings on a shaft mounted in a *saddle* on the front face of the cutterhead. The hub carries a shrunk-on cutter ring, which, in the old days, resembled the shape of a discus in its outer section. Thus, this type of cutters came to be named *disc cutters,* even though the ring profile over the years developed into other shapes.

The cutters are mounted on the cutterhead at a steadily increasing distance from the center of rotation of the head and roll against the face in circular paths with an ever-larger radius. The last cutter—the one placed farthest out—determines the radius of the tunnel cross section. To evenly spread the load and distribute the weight over the cutterhead, the cutters are spread out all over the head in a certain pattern, but they all roll at their predetermined radius and two neighboring cutters are separated by the *spacing*. If all the cutters were to be mounted in a straight line—which, for reasons of balancing and because of space requirements, is a physical impossibility—one would see the *cutter profile,* as illustrated in Fig. B12-3.

Close to the center of the cutterhead, the center cutters (usually four) are mounted in a common saddle for space reasons. The individually mounted face cutters farther out run into the curved gauge area which forms the transition from the tunnel face to the full bore diameter. The gauge cutters are in principle of the same design as the face cutters, but their saddles are different, due to cramped space conditions close to the bore.

In Fig. B12-3, the tunnel face is flat. This is the most common configuration, though a dished face, produced by placing the face cutters on a domed head, is sometimes practiced as well. It is claimed that the domed head has self-centering characteristics. While that is undoubtedly true, it also subjects the cutters to side loads and requires a special steering geometry to direct the machine when boring a curve.

As the cutterhead is rotated and thrust against the tunnel face, the cutters crush the rock in the bottom of the kerf, under the edge of the cutter ring. They penetrate a depth of from 5 to maybe 12 mm per pass, depending on the rock harness and the thrust, or cutterload, applied.

Under influence of the stress induced by the thrust on the cutter, cracks will start from the crush zone and propagate in a radial pattern.

Those cracks which are closest to the free surface will run farthest from their point of origin. When cracks from two adjacent kerfs meet, larger rock chips will spall off the face. Thus, the surface in between the kerfs is lowered, without being really touched by the cutters. This is part of the secret of reaching boring economy in hard rock (Fig. B12-4).

The distance over which the cracks will travel depends on the thrust exerted on the cutters and the resistance the rock offers to the process. Apart from selecting a suitable cutter load, which on modern TBMs can go up to some 250 to 350 kN, depending on the size of cutter used, one can also select various spacings between the kerfs when the machine is being designed. Depending on how hard the rock is to bore, spacings of 70 to 120 mm are commonly used

To react the torque required to rotate the cutterhead and the thrust to make the cutters penetrate the face, the machine must be anchored in the bore. As the weight of the machine alone is by no means large enough to accomplish this, it is done by means of a pair of opposed grippers actuated by hydraulic cylinders. While a single

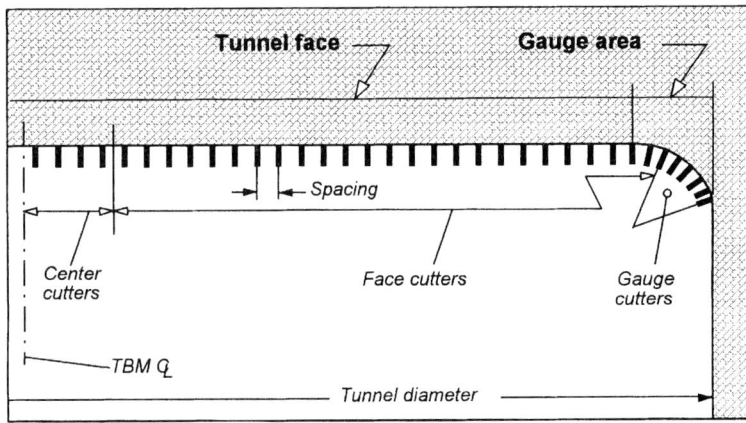

FIGURE B12-3 A typical hard-rock TBM cutter profile.

pair of grippers will suffice to take the force reactions, they will not lend the machine the necessary stability, and one more point of suspension is needed. In practice, this has been solved in two principally different ways (see Fig. B12-5).

One solution is illustrated in Fig. B12-5*a:* This group of machines has been equipped with two pairs of grippers, one at the front and one at the rear of the machine's main body. These are termed *double gripper TBMs*.

The other approach (Fig. B12-5*b*), termed *single gripper TBMs*, provides the machine with a support shoe under the cutterhead which slides on the tunnel invert and further stabilizes the front part of the machine by means of side supports and a roof which are in pressurized, sliding contact with the rock in the tunnel walls and crown when the machine is in operation.

The way the machines are gripped in also dictates the steering method used to make the machine follow the intended line of the tunnel. Due to the fact that the anchoring part of double gripper TBMs is gripped in solidly, while the working part slides forward through the main body, these machines *cannot* be steered during boring. Steering movements are made in between boring strokes, when the main body is reset for the next stroke. Thus, curves have to be bored as a series of chords which, because the stroke is very small in relation to the radius of most curves, is in practice no real drawback.

Single gripper machines are so designed that the tail of the machine can be moved in the vertical and the horizontal plane in relation to the gripper unit. Using the support shoe as a fulcrum, these machines can therefore be steered during boring and bore true, continuous curves.

The disadvantage of the double gripper type—i.e., that a reset will take somewhat longer, as the reset procedure must also include steering manoeuvres—is perhaps balanced by the fact that the single gripper type may have more difficulties if soft or unstable material is encountered in the invert and the front end of the machine will tend to sink down. Both types of machines having been produced in large numbers and in general used to their owners' satisfaction, the obvious conclusion is that both layouts represent viable approaches.

Double Gripper Open TBMs. The anchoring section (Fig. B12-6) consists of the main body assembly, four grippers, and the front lift leg. This section supports the weight of the machine and transfers the thrust and torque reactions to the rock

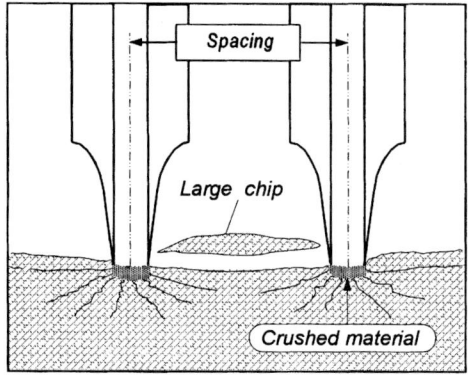

FIGURE B12-4 Spalling action.

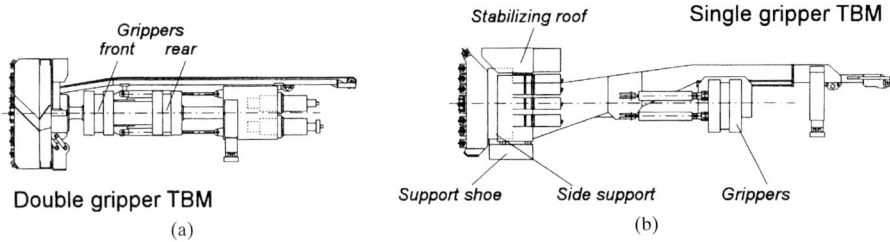

FIGURE B12-5 (*a*) Double gripper TBM; (*b*) single gripper TBM.

during boring. The grippers are mounted horizontally to the sides of the body in a front and a rear pair. The front lift leg is placed underneath the main body, in the same plane as the front grippers and, with them, forms a T-pattern when seen in cross section.

The grippers and the front lift leg telescope out and are pressed hydraulically against the tunnel bore, thus anchoring the TBM in the bore. Individual control of the grippers allows the operator to position the machine correctly on line at the start of a boring cycle, so as to maintain a given direction or to bore a curve.

The front lift leg, in combination with the rear lift legs, is used to position the machine correctly on grade.

The working section consists of the cutterhead, the torque tube, with the bearing housing and the drive train attached thereto, and the muck conveyor. The drive train comprises a number of motor/planetary reducer assemblies which, via a common ring gear in the gear case, rotate the drive shaft which is located along the longitudinal axis of the machine, inside the torque tube, and transmits the rotational power to the cutterhead.

The gear case, with the rear lift legs mounted thereon, is bolted to the rear end of the torque tube. The main bearing, mounted in the bearing housing which is bolted to the front end of the torque tube, carries the cutterhead and transmits the thrust to it. The torque tube has a square cross section and is carried in the main body in slide bearings. Thus, while it is possible to move the working section of the machine axi-

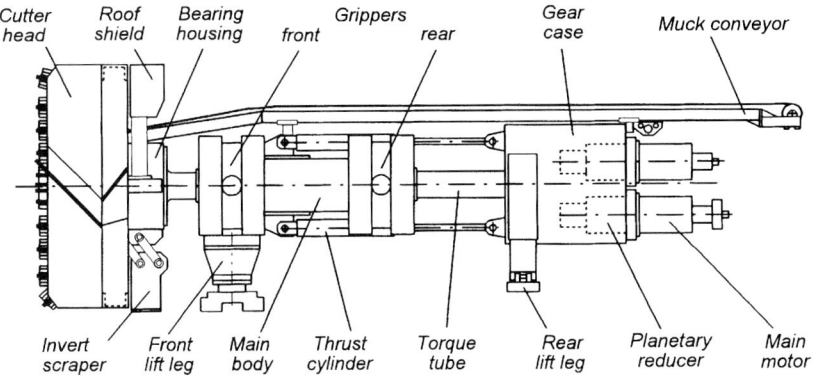

FIGURE B12-6 Main components of a double gripper open-type hard-rock TBM.

ally in relation to the anchoring section, the torque tube at the same time can transfer the drive train torque reaction, via the body and the grippers, to the tunnel walls.

Hydraulic thrust cylinders interconnect the anchoring and the working sections and power the forward movement of the latter, to force the cutterhead against the rock surface of the tunnel face. The rock chips and fine particles formed by the cutting action of the head—the so-called *muck*—fall to the invert where they are moved rearwards through the cutterhead and into buckets by scrapers. The buckets are attached to and rotate with the cutterhead. They deposit their load into the hopper mounted on top of the bearing housing, from where the muck is taken to the rear for transport out by a belt conveyor running over the top for the full length of the machine.

An invert scraper, mounted vertically adjustable underneath the bearing housing just behind the cutterhead buckets, cleans the invert from any muck left behind by the buckets, scraping the muck forward and into the path of the buckets again. The invert scraper also serves to support the weight of the front end of the TBM during resetting, when the anchoring section is moved forward on the torque tube.

Two rear support legs are attached to the gear case to support the weight of the rear end of the machine when resetting the anchoring section. During boring, these legs do not make contact with the rock in the tunnel invert.

The location of heavy components at both ends of the torque tube ensures an even balance of the working section over the main body, and the fact that the main motors with their auxiliaries are located at the rear end of the machine makes for easy access.

The machine is operated from a control panel which can be located either on the one side of the main body, in between the grippers, or in an operator's cabin on the backup. If the latter alternative is selected, a closed-circuit TV system is usually installed to make it possible for the operator to monitor various points on the machine and the backup. Modern TBMs most often have a control system based on a *programmable logic controller* (PLC), which greatly reduces the wiring between the operator's panel and the TBM itself and makes it simple to partly automate the operation and provide the necessary interlocks for safety.

Single Gripper Open TBMs. On these machines, the anchoring section (Fig. B12-7) is formed by the gripper and thrust assembly, which consists of a structural carrier, a gripper assembly, and the thrust cylinders. The carrier mounts to the bottom of the rear end of the main beam, where it slides on machined guideways. The gripper assembly is connected to the carrier by means of a trunnion. That part of the cutterhead torque reaction which is not taken by the cutterhead support shoe is transmitted via the main beam to the carrier and then out to the gripper pads, where it is transferred to the tunnel walls.

The gripper assembly also supports the dead weight of the rear of the machine when the support legs are raised during the boring cycle.

When being extended, the thrust cylinders, which are attached to the front edge of the gripper shoes, direct the thrust force into clevises on the main beam, which transmits it to the cutterhead support, from where it is fed into the cutterhead via the main bearing. Thrust is thus reacted in a direct path through the gripper pads into the tunnel wall, and the gripper cylinders are not subject to the thrust load's bending or shearing forces.

The working section consists of the cutterhead, the cutterhead support with its sliding support shoe and stabilization components, the main bearing, the drive train, the main beam, the support legs, and the muck conveyor.

Cutterhead rotational power is developed by electric motors located at the front end of the machine. They deliver their torque to planetary gear reducers via a

TUNNEL-BORING MACHINES**B12-15**

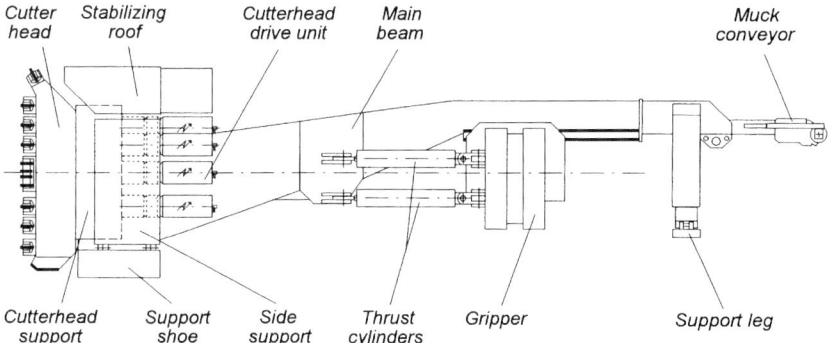

FIGURE B12-7 Single gripper open hard-rock TBM.

hydraulically operated clutch to which they are flanged at the rear face of the cutterhead support. The outgoing pinions of the gear reducers engage a common ring connected to the cutterhead. Both the main bearing and the ring gear are enclosed in the cutterhead support.

The muck that is generated in cutting falls to the invert where it is scooped up by low-profile buckets built into the head. The muck stays in the buckets until they reach their overhead position, when it is dumped out onto the machine conveyor. Any muck missed by the cutterhead bucket scrapers is dozed ahead by the support shoe, and special back scoops on the cutterhead buckets are designed to pick this pile up and deposit it in the muck-conveying system.

The machine is provided with a troughed belt conveyor which transfers the muck from inside the cutting head back through the main beam to the rear of the machine. Here the muck is transferred to the backup muck-handling system for removal from the tunnel.

With most of the bulky components located at the front of the machine, the area around the rear end is uncluttered and access there is better than on the double-gripper-type hard-rock TBMs. Because of the concentration of components at the front, access through the cutterhead to the tunnel face is more cumbersome, on the other hand. Single gripper TBMs also tend to be more nose-heavy which, as already mentioned, may lead to the front end sinking when there is unstable material in the tunnel invert. As single gripper TBMs can be steered during boring, such a tendency may be compensated for by continuously steering the machine somewhat upwards.

Shielded Hard-Rock TBMs.   Rock may be hard to cut, but it need not necessarily be competent or stable enough to use an open-type TBM which offers limited possibilities to provide immediate tunnel support. When the rock is so fractured, blocky, or otherwise unstable that a lining must be used to keep the tunnel open, the machine needs to be shielded to provide that support at a short distance behind the face. This is required not only to keep the tunnel open, but also to protect the machine itself and the personnel operating and servicing it from rock falling out of the crown. A further advantage of a shielded machine is that the weight of the TBM is spread over a much larger area than when it is only carried in gripper pads or in pads and on a support shoe under the cutterhead.

The shielded hard-rock TBM was obviously born out of necessity and its concept was based on putting hard-rock boring and muck-removal components inside a soft-ground shield, from which also the propulsion method—thrust cylinders pushing off the lining—was taken over. In its purest form, this is the so-called *single-shield hard-rock TBM*. A further hybrid development saw the light at a later date, when the single shield's thrust system was combined with a system pushing off a shield section equipped with grippers. This version came to be known as the *double-shield hard-rock TBM*. It combines the best of two worlds and, though it obviously costs more to acquire than a single-shield TBM, it offers the possibility to do without and save the expense of a lining for those stretches of the tunnel which are more competent and do not really require that type of support. In a tunnel excavated by a single-shield hard-rock TBM, lining *must* be put in all the way; no lining, no thrust; thus, no tunnel!

Single-Shield Hard-Rock TBMs. This type of machine (Fig. B12-8) consists of four main structural assemblies:

- Cutterhead
- Cutterhead support and shield
- Conveyor assembly
- Segment erector assembly

The only parts which are *stationary* during the operation of the machine are the thrust cylinder piston rods and their spreader shoes which transmit the thrust to the lining segments. When activated during boring, the thrust cylinders themselves move all other parts forward.

These parts constitute the *working section* of the machine. It consists of the cutterhead, the cutterhead support with its articulation components, the main bearing and ring gear enclosed in the cutterhead support, the drive units, the shield itself, the thrust cylinders, the muck conveyor, and the segment erector.

As was the case for the EPB soft-ground shield discussed earlier, single-shield hard-rock TBMs exist in two basic versions. In one, the shield consists of a one-piece cylindrical structure and the cutterhead support is mounted inside the front end of the shield in such a manner that it, and the cutterhead with it, can be tilted to a slight angle in all directions for steering purposes. This is done by means of the articulation cylinders which act between the shield and the cutterhead support. In the other version, the shield is split into two sections, the front shield and the tail shield, which can be off-angled slightly in relation to each other by means of the articulation cylinders, which in this case act between the two shield sections, while the cutterhead support is mounted in a fixed position inside the front shield.

In both versions, the feature of articulation is an essential part of the steering system of the machine which primarily relies on the use of the thrust cylinders, used with grouped flow control or, in the hold-back mode, in true shield manner.

To counteract the tendency of the shield to roll under influence of the cutterhead torque reaction, the thrust cylinders can be off-angled somewhat by means of an adjustable *skewing ring*.

The outside diameter of the shield is slightly smaller than the excavation diameter of the cutterhead to allow for steering corrections, prevent binding between shield and bore, and allow for gauge cutter wear. At the bottom, the shield will rest on the tunnel invert through gravity. Thus, there will be a slight clearance between the outside of the shield and the rock at the top. To stabilize the machine during boring, the two upper quadrants of the front end of the shield each contain a stabilizer shoe, located in a window in the shield outer skin.

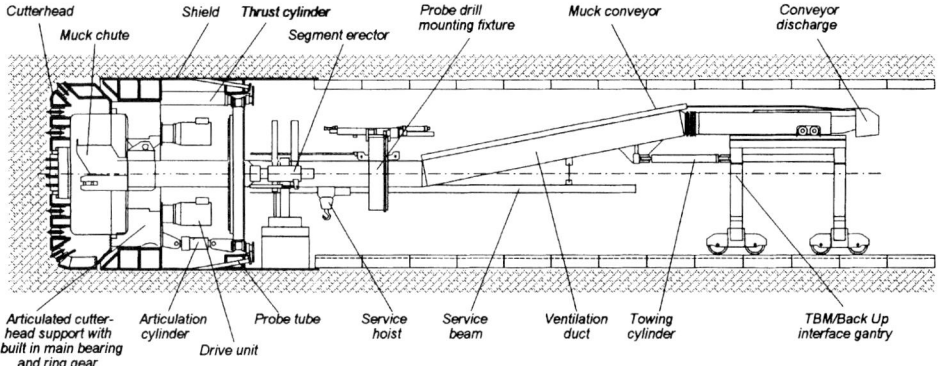

FIGURE B12-8 Main components of a typical single-shield hard-rock TBM.

Stabilization is important to eliminate *washing* of the head, excessive side loading of the cutters, and vibrations. The stabilizers are energized by means of hydraulic cylinders controlled from the operator's panel and slide forward in pressurized contact with the rock during boring.

The rear end of the shield is a cylindrical shell which forms the *tail shield*. It has enough inside space for setting a full ring of lining segments by means of a rotating segment erector located there. Sufficient overlap exists to prevent tunnel wall exposure when the thrust cylinders are fully stroked out.

The machine is provided with a troughed belt conveyor which transfers the muck from the muck chute inside the cutterhead back through the tubular conveyor frame to the rear of the machine. Here, the muck is transferred to the backup muck-handling system for removal from the tunnel.

As a shielded hard-rock TBM lacks the main beam of the open-type hard-rock machines, the rear end of the muck conveyor frame is supported on rollers on top of the TBM/backup interface gantry. As the machine moves forward, the frame rolls forward over the gantry. During resetting, the gantry is pulled up by the towing cylinder which connects it to the conveyor frame.

Double-Shield Hard-Rock TBMs. In Fig. B12-9, the TBM/backup interface gantry, which in principle is of similar design to the one used behind a single-shield hard-rock TBM (see Fig. B12-8), has not been entered on the right-hand side of the figure.

A double-shield hard-rock TBM consists of four main structural assemblies:

- Cutterhead
- Front shield with cutterhead support
- Rear shield with gripper and thrust assembly
- Conveyor assembly

The rear shield forms the *stationary section* of the machine. When activated, the thrust assembly transmits the thrust, which is reacted either via the grippers to the tunnel walls or, if a lining *has* to be installed, against the latter, to the moving section during boring.

The remaining part of the machine is the working or *moving section*. It consists of the cutterhead, the front shield with the cutterhead support (with the main bear-

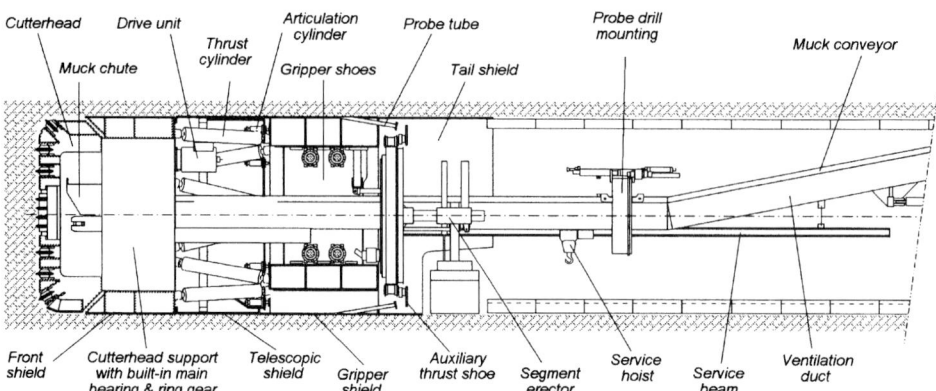

FIGURE B12-9 Main components of a typical double-shield hard-rock TBM.

ing and the ring gear enclosed), the drive units, the shield, and the muck conveyor. The front shield forms an outer box-type structure surrounding the cutterhead support. It consists of a cylindrical shell of welded and bolted segmental construction. On the inside, each segment contains mountings for the piston rods of the thrust cylinders.

As on a single-shield hard-rock TBM, the outside diameter of the front shield is slightly smaller than the excavation diameter of the cutterhead to allow for steering corrections, prevent binding between the shield and the bore, and allow for gauge cutter wear. At the bottom, the shield rests on the tunnel invert through gravity. Thus, there will be a slight clearance between the outside of the shield and the rock in the crown of the tunnel. To stabilize the machine during boring and to help in anchoring the front part during a reset of the thrust system, the two upper quadrants of the front end of the shield each contain a stabilizer shoe, located in a window in the shield outer skin.

Stabilization of the head is important to make the cutters track properly by eliminating washing of the head. It also helps to reduce vibrations and thus enhances main bearing life. The stabilizers are energized by means of hydraulic cylinders controlled from the operator's panel and slide forward in pressurized contact with the rock during boring.

The rear end of the front shield is a cylindrical shell which surrounds the *telescopic shield,* which forms the *transition seal* between the front shield and the rear, or *gripper* shield. The rear or gripper shield is a box-type weldment with a forward shell section—the telescopic shield previously mentioned which slides in the rear end of the front shield—and a shell type extension at the rear, the *tail shield.*

The telescopic shield is held in position by the articulation cylinders, in order to allow it to float within the rear shell of the front shield during steering to maintain sliding clearances. The tail shield provides protection and the necessary space for erecting a ring of lining segments when rock conditions dictate this type of tunnel support.

The gripper assembly consists of two shoes which move radially in windows in the central part of the rear shield when activated by gripper cylinders above and below the spring-line. The large area of the shoes ensures a low specific pressure against the rock.

Though Fig. 12.9 indicates horizontally opposed grippers, they may also be located towards the top of the shield at an angle of about 45° from the vertical centerline. This has the benefit of pushing the shield onto the invert, which makes it more stable and able to resist any vertical forces coming from the front shield.

To leave as much space as possible in the center of the machine for other equipment, the thrust cylinders are arranged around the inside periphery of the shield. Their piston rods deliver the thrust to the rear face of the front shield box structure, when they are activated to react against the front face of the gripper shield box structure and thus, via the gripper shoes, to the tunnel walls.

For operation in broken or otherwise incompetent ground, where gripper shoes are not effective or should not be used so as to avoid damage to the tunnel walls, auxiliary thrust cylinders are provided to react against the segmented lining and provide thrust and torque reaction. Their primary purpose, however, is to hold the lining segments in place as the gripper shield is advanced during reset. In competent ground, where the grippers can be used, the auxiliary thrust system is used only to hold the lining segments in place and assist in pushing the gripper shield forward during reset.

The thrust cylinders are arranged in groups, and control of the relative displacement in each group biases the front shield in a certain direction. Each group can also be operated in the hold-back mode. This is usually practiced for the cylinders at the top and has the effect of preventing the cutterhead from drooping.

Also, side thrust bias produced by the stabilizer shoes in the front part of the shield can be used to assist in steering, by forcing the front shield either to the left or to the right.

The torque of the cutterhead drive system is taken up by the thrust cylinders, which are mounted in lattice arrangement and thus will take the twisting action of the torque reaction and transmit it to the gripper shoes or, in poor ground, via the auxiliary thrust cylinders to the lining. The lattice arrangement also makes it possible to counteract, and if necessary correct, the tendency of the front shield to roll around its longitudinal axis under influence of the drive system torque.

The tail shield provides the necessary space for setting a full ring of lining segments, should ground conditions demand this, by means of a rotating segment erector mounted there. At the bottom, the tail shield can have a cutout (as illustrated in Fig. 12.9), which permits setting the invert segment directly on the rock.

Normally, the design of the cutterhead buckets is of an open configuration to adequately handle coarser muck in fractured formations. However, grill bars can be installed when so required in very broken ground to prevent the entry of large blocks of rock into the muck-removal system.

The machine is provided with a troughed belt conveyor which transfers the muck from the muck chute inside the cutterhead cavity back through the tubular conveyor frame to the rear of the machine. Here the muck is transferred to the backup muck-handling system for removal from the tunnel.

Hybrid TBMs

As was mentioned before, shielded hard-rock TBMs already are a form of hybrid machines. The advantages of boring tunnels instead of excavating them by D & B will continue to lead to an increased use of TBMs and foster the development of more hybrids, so that ever more tunneling projects which have to deal with varied and difficult ground conditions can be tackled by TBMs. A recent example of such development saw use on the French side of the tunnel under the English Channel.

Here it was likely that the TBMs to be used would encounter water under high pressure. The type of machine developed to operate under those conditions was a hybrid between a hard-rock TBM and a slurry shield. It is capable of operating as a soft-ground shield in situations of potential high water inflows at pressures up to 1 MPa—i.e., a head of water of 100 m—and as a high-speed hard-rock TBM in competent rock formations.

Very often the exact geological conditions along the tunnel to be excavated are poorly known in advance. What is good, competent hard rock at the starting point of the bore may turn into a tunnelers' nightmare farther in, which brings an open-type hard-rock TBM to an abrupt stop. Machines have even been lost completely under such conditions and are still buried underground.

The expanding appreciation of the fact that so many geological surprises may lie ahead has seen the increasing installation of probing, pregrouting, and rock-reinforcing and supporting equipment on standard hard-rock TBMs. The wish to really be prepared for the worst will inevitably lead to continued integration of the best characteristics of all types of TBMs into a single machine—i.e., to further hybrids. Such machines will not be the cheapest around, but price difference will more than likely offset the extra costs of a project bogged down for months by ground problems an ordinary TBM cannot cope with.

AUXILIARY EQUIPMENT

The availability of a suitable TBM alone will not get a tunnel through. The machine needs to be served by many supporting services, some of which are delivered by equipment housed on the backup, while others stretch all the way back to the starting point of the tunnel.

Logistics

To operate a TBM, the following are needed.

Electric Power. Though some machines are equipped with hydraulic motors for cutterhead rotation and all machines have a hydraulic system for thrust, gripping, and many subordinate functions, primary power is always supplied in the form of electricity. Even small TBMs of, say, 3.5-m diameter have as much as 2000 kVA capacity installed. Electricity is therefore supplied at high tension—up to 10 or 12 kV—through a cable installed in sections against the wall or hung from the crown of the tunnel as the tunnel progresses.

On the TBM backup, a length of flexible cable is stored, connected to the tunnel cable, which can be paid out automatically from a cable reel or a looped hanger system as the TBM advances. Thus, the machine need only be stopped when a new length of tunnel cable has to be installed. Transformers on the backup lower the high tension to 660 V and lower for the power consumers on the TBM and the backup itself.

Water. Water is needed for cooling purposes and dust suppression. It is piped in from outside the tunnel via a line installed on the tunnel wall and connected to the water system on the backup via a hose reel or other suitable arrangement to allow continuous TBM operation.

Compressed Air. With the increasing use of hydraulically powered rock drills for probing or drilling rock bolt holes, compressed air requirements nowadays are limited to powering maybe a few hand tools or for additional flushing of probing holes. Sometimes, too, compressed air is used to pressurize main bearing cavities to keep dirt and water out. These small amounts are more and more often covered by installing an air compressor on the backup, instead of installing a line from an outside compressor over the full length of the tunnel.

Ventilation. Fresh air must be supplied to the tunnel face for the crew working there and to dilute and remove exhaust fumes from diesel engines. This can be done by a variety of arrangements—Fresh air blown in from outside is commonly used in European countries; "used air" sucked or blown out from the tunnel is often prescribed elsewhere—but all require hanging a vent line throughout the length of the tunnel and regularly extending it where it connects to the ventilation system on the backup.

Tracks. In the majority of tunnels, tracks are installed for railbound transport of people and materials to and from the face and, in most cases, of muck out of the tunnel. This will usually be a single track, as a double track would require partly refilling the invert to obtain a wide enough road bed.

If railbound muck transport is used, stretches of double track must be installed in longer tunnels, either permanently at regular intervals or in the form of a *California switch* which is regularly hauled forward to a point close behind the backup, so that trains of full and empty muck cars can pass each other. Without such an arrangement, TBM performance per shift will suffer as waiting periods of increasing length will occur before new empties stand ready to receive the muck produced by the machine.

Muck Transport. The muck can be taken out of the tunnel by a variety of means or systems, the most common of which was already briefly touched on in the previous paragraph.

Trucks. Trucks will only be used in large-diameter tunnels of short lengths. The large diameter is required to permit two trucks to pass each other in the tunnel and to be able to turn around behind the TBM.

To provide some continuity to the boring operation, a bin should be provided at the rear end of the backup, or the muck can be dumped onto the invert and the trucks loaded by a front-end loader. As the muck, once loaded, will usually be run out by the same truck all the way to the tip—which, especially in urban areas, may be quite some distance away—the system may require quite a large fleet of trucks.

Muck Trains. Trains of muck cars moved by diesel engine or battery-powered locomotives are loaded underneath the off-loading point of the backup conveyor and run out of the tunnel to there be emptied by means which usually are tailored to the local conditions. If the tunnel has a mountainside portal, the train is normally run out to a tip arranged at some short distance outside the portal and the cars can be tipped either sideways or in a rotary overhead tip.

If the tunnel entrance is at the bottom of a shaft, the arrangement causing the least delays is tipping the cars into a holding pocket, from where the muck is removed by skips run in the shaft. A simpler arrangement is hoisting the complete car body up the shaft and tipping the load on surface. This causes longer delays in returning the empty train to the TBM, but it requires virtually no shaft installations other than a mobile crane on surface and is therefore often used for small, short tunneling projects at shallow depths below surface.

If the tunnel entrance is underground and the connection to surface is formed by an adit or a ramp, the cars usually are again tipped into a holding pocket and the muck is moved from there by trucks or a conveyor belt running out of the adit or up the ramp. Another version, which has recently found application on a number of Scandinavian projects, is tipping the cars directly into a truck by lifting the car body off the boggies by means of a lift-up-and-turn-over forklift arrangement operated by the truck driver. This ties the train down at the tip, though, and it is usually necessary to invest in one more train of cars to keep up with the production of the TBM.

It is a fairly simple matter to arrange the continuous filling of the muck cars in one train on the backup. The cars can be moved as they get loaded, either by the loco or by a car spotter (chain driven or equipped with a winching system) to free the loco for other duties. The aim is always to have sufficient cars in the train to take the muck produced during a full boring stroke of the machine. That way, the dead time of the TBM reset can be used to switch the loaded train for an empty one.

When the diameter of the tunnel so permits, a double track is usually installed on the backup, so that a train of empty cars can be held in readiness while the previous train is being loaded. This virtually eliminates all waiting time and ensures optimal conditions for high TBM utilization.

The importance of a smoothly operating muck transport system cannot be stressed enough. With railbound systems it is not unusual to find statistical evidence for delays in muck transport taking some 10 to 15 percent of the total time available per shift, especially when the tunnels get a little longer.

Conveyor Systems. Conveyors are sometimes used for muck transport in very short tunnels, where there is no real need to install a track, as the machine remains within walking distance of the tunnel entrance. In such case, portable belt conveyors may be brought into the tunnel as the machine advances and the one conveyor off-loads onto the next one. It is rather seldom, though, that such short tunnels are driven by TBM, as the time and costs of installing the machine become exorbitant in relation to the length of the tunnel.

Belt conveyor systems are, therefore, more frequently used in long tunnels of such a small cross section, that a muck train system would be hard put to keep up with the machine. In those cases it is normal to use a conveyor design with long lengths of belt being gradually pulled out by the tail pulley which is attached to the TBM backup and thus brought forward when the backup is pulled up to the machine from a belt-tensioning and storage device located outside the tunnel entrance.

Such systems were originally developed in coal mining in the United Kingdom and have found application on a number of civil engineering tunneling projects. The tensioning device can work either by gravity—i.e., with weights—or hydraulically and must hold the full length of belting sections as delivered to the site. Such systems can even be laid out to follow curves in the tunnel rout and offer virtually continuous muck transport, with only short interruptions for adding a new belt section. This down period is then normally used for carrying out TBM maintenance at the same time.

The load-carrying rollers and return idlers for the belt can rest either on prefabricated stands on the invert or on hangers rock-bolted to the tunnel wall. The latter alternative is the more common, to save space on the rather narrow invert for other services. Drilling the rock bolt holes for the hangers can be mechanized by means of a rockdrill mounted at the rear end of the backup.

In 1994, a 3.2-m-diameter tunnel in Australia was driven more than 2300 m in one 30-day period, working six days a week, counted from a point more than 6 km into the drive. It is hard to imagine that a muck train transport system would have allowed that performance under those conditions.

Pumping Systems. The method of moving solids suspended in water or a slurry by pumping them through a pipeline is common in ore treatment plants and lends itself for muck transport under certain conditions:

- *The availability of a sufficient quantity of water.* If source water is not available in sufficient quantities, used water must be reclaimed at the slurry discharge end and recirculated through the system.

- *The excavated material may not be too abrasive.* If it is, high pump wear will occur when the slurry is run through the pump. This can be avoided by letting only water circulate through the pump and injecting the material to be transported in a *locked hopper. Through-the-pump* systems have the advantage of relative simplicity and high capacity but achieve a low head and are consequently limited to short transport distances. *Locked hopper* systems can use conventional high-pressure water pumps and therefore lend themselves to longer distances. Pump maintenance costs will be comparatively low.

- *A high percentage of fines in the excavated material and a specified maximum particle size.* If the required percentage of fines is not present in the excavated material, a crushing operation may have to be installed ahead of the slurry mixing equipment to produce the fines and reduce the muck to maximum allowable size. If there is no concern over the maximum size, but fines are not available in the required quantity, then two other alternatives are open:

1. Fine material can be recirculated in the system, thus building up the required quantity.
2. Drilling muds as used in the oil well drilling industry can be circulated to maintain a sufficiently dense media.

From this it is clear that such a pumping, or hydraulic, system lends itself admirably for use with a slurry shield, which indeed is the normal case for such machines, but would be hard to visualize for use together with hard-rock TBMs. The addition of a crushing and a slurry mixing and pumping system to all the other auxiliary equipment such a machine requires to have installed on the backup, plus the installation of two pipelines in the tunnel—one for slurry out and one for water (or return slurry) in—and of a system for the continuous and uninterrupted lengthening of those lines, can only be warranted under very special, project-dictated circumstances. Other disadvantages of slurry systems have already been mentioned. Those drawbacks formed one of the major reasons for the development of the EPB shields.

Pneumatic Systems. This type of system also derives from the mining industry, where they have been used for many years to fill worked-out stopes in coal mines in Germany to obtain a high fill density with dry material in order to support the ground above and prevent subsidence in the higher workings or even at surface.

Such systems comprise several basic elements: an air source, such as a blower, discharges into an air lock injector for the material to be transported and forces the material into a pipeline, which in turn directs it to its destination. Sufficient air pressure is required to maintain particle transport velocities throughout the system.

As in the case of hydraulic systems, limiting particle size is necessary. Few systems have ever worked satisfactorily with larger than 50- to 75-mm particles, somewhat depending on the specific density of the material. They are also limited in the transport length and require the installation of booster blowers at comparatively short distances along the line. If the material is abrasive, the cost of pipeline maintenance becomes high. In the coal-mining application previously mentioned, ceramic coating was therefore used on the inside of the pipes, particularly at points where the direction of the line changes.

One drawback of pneumatic systems is the rather large amount of power required. The results from numerous tests under various conditions of transport length, transported quantity, particle size, etc., indicate that pneumatic systems consume approximately 500 W per hour per ton per km. Taking a medium-sized TBM (5.0-m diameter), boring at a rate of 5 m/h at a distance of 5 km into the tunnel in rock with a specific gravity of 2.75 (a fair average of most types of rock met in normal tunnel-boring operations), power consumption thus would be some 675 kVA.

The argument offered by some proponents of pneumatic transport—that the high power consumption can be justified by the fact that it does away with the requirement of a ventilation system through the suction of air through the tunnel into the blower behind the TBM—does not appear to be valid. The power requirement for standard ventilation systems is many times smaller than that of a pneumatic transport system for comparable tunnel cross sections and lengths.

As in the case of hydraulic systems, pneumatic systems will be applicable only for muck transport under very specific, project-dictated circumstances.

Conclusion. For normal, run-of-the-mill tunnel-boring projects, muck trains will be the primary choice for a long time to come and can be improved by the addition of smart, automated car movers and of train-switching schemes to take care of the logistics behind most TBMs. Conveyor systems may be warranted for special high-speed operations in confined space, as the example illustrates.

Linings

Lining systems, no matter how simple they are, always serve to support the tunnel and sometimes have the additional purpose of sealing it against water contained in the surrounding ground. The strength of support needed and the urgency in placing it are the main factors in selecting the type of lining to be used. (See Fig. B12-10.)

Full concrete linings offer the greatest strength. Due to the joints between segments, a lining built out of precast segments is in this respect somewhat inferior to a monolithic, cast in situ lining of comparable thickness. On the other hand, a segmented lining can be installed much earlier and, above all, faster and, therefore, is

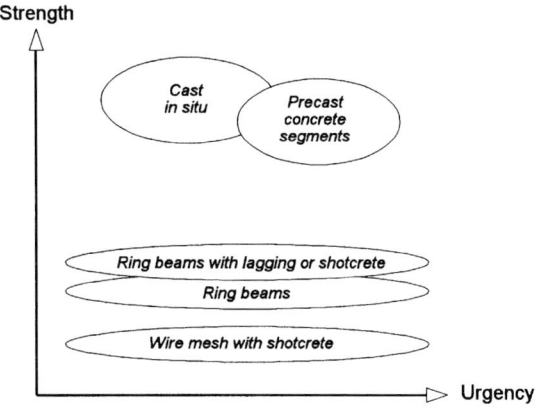

FIGURE B12.10 Tunnel lining schematic.

the only one thinkable for use in connection with shield TBMs, which depend on a lining for their propulsion. As indicated, the remaining types of lining provide weaker support, but can be installed very early, if need be.

In principle, all types of lining can be installed over the full length of the tunnel or only for those stretches which actually require support. Here it must be kept in mind that all shield machines, with the exception of the gripper-equipped double-shield hard-rock TBM, require the *immediate* installation of a full lining directly behind the machine and thus need the lining to be installed for the *full tunnel length*. In those cases, only a lining consisting of precast concrete segments will do.

Some aspects of the lining systems' influence on the TBM and the backup need to be mentioned.

Cast In Situ Linings. If this type of lining is installed concurrently with the excavation of the tunnel, it is normally placed at some distance behind the tunnel face and will therefore not influence TBM operation in other than pure logistical aspects. The materials for construction of the lining must share the same transport facilities in the tunnel as the TBM utilizes.

The shuttering for the lining will cause a narrow stretch for some length in the tunnel which may impede traffic to and from the TBM, and the lining materials stored near the erection point—plus possibly also mixing and pumping equipment—will likewise take up space. Therefore, at the point where the lining is being constructed, there should preferably be enough space for a section of double track *inside* the shuttering.

To save the trouble of erecting shuttering for a section of lining, collapsing it when the concrete has set, bringing it forward, and erecting it again for the next section, the idea of slip-forming a cast in situ lining in a continuous process has been discussed over the years.

In 1985, a laboratory prototype, with a free space diameter inside the lining of approximately 3.5 m, was built in the United States to demonstrate the viability of the method and gather data for the design of a full-scale model. It is presently not known whether slip-forming a lining was ever used on an actual tunneling project.

If that type of lining is adopted for installation concurrently with tunnel excavation by a boring machine, casting the lining should preferably keep pace with the TBM and, in that case, the boring and the lining operations will undoubtedly become more or less integrated and influence each other to a high degree. Advance rates for excavation and lining need not be *exactly* the same, however. The slip-form may, for instance, be towed by the backup at a varying distance through the use of wires and winches. It seems logical to house the mixing and pumping equipment for a slip-formed lining on the backup as well, to avoid creating a cluttered spot at the form itself. Should the need for a lining not have been foreseen before the start of the project, the backup may be extended by a few additional decks, housing the necessary gear and offering materials storage space.

Segmented Linings. This type of lining must be used behind soft-ground shields and single-shield hard-rock TBMs, and will also come into play behind double-shield hard-rock TBMs when the ground becomes incompetent and the machine cannot make use of its gripper-based propulsion system. It therefore influences the layout and operation of those types of machines to a high degree. This has already been illustrated in the sketches earlier in this chapter: Segment erectors are invariably incorporated in the tail shields of those machines.

As the lining installation must keep pace with the boring operation, the logistics of both are highly interwoven. The stroke of the machine is normally selected the same as the width of a lining ring in the direction of the tunnel. A train should therefore contain the right number of cars for the muck volume produced in one stroke and include, at the *inward* end of the train, enough segment cars to hold all the segments for one ring.

The backup must incorporate a system for bringing the segments forward from the cars they arrive on in the muck train and depositing them in the right sequence in or near the tail shield. The sequence is important because the invert segment invariably has another shape than the segments higher up in a ring.

In many cases, segment rings—sometimes also individual segments in a single ring—are bolted together and, when the lining has to withstand water pressure from the surrounding ground, seals must be installed between segments. Also, these materials must be brought forward at the right time and in the proper quantities.

Because segmented linings are erected *inside* the tail shield, the outer diameter of the lining will invariably be a little smaller than the excavated diameter of the tunnel. So as not to disturb the surrounding ground by letting it settle onto the lining, that space is normally backfilled by blowing in *pea gravel,* by grout injection, or by a combination of these two methods. The backup must therefore also accommodate equipment and material storage space for those operations.

Ring Beams. Ring beams, often supplemented by lagging in between the beams or by shotcreting, form a type of lining far less strong than concrete linings and are mostly used for supporting a tunnel bored by an open-type hard-rock TBM only in the weaker sections. While they can be set behind the machine, they are most often installed as soon as possible after excavation, i.e., immediately behind the cutterhead.

At that point, the remainder of the TBM encroaches upon the free working space. Furthermore, the ring beam sections will, in most cases, have a greater weight than what can be handled manually. Under those conditions, to set a ring beam in an efficient and safe manner, the TBM must be equipped with a ring beam erector and a transport system for bringing the ring beam sections—and possibly lagging—forward. The erector should preferably incorporate a possibility for travel in the longitudinal direction of the tunnel, so as to allow the machine to go on boring while a ring is being erected. This allows the two operations to get slightly out of phase and avoids downtime because the one need not wait for the other.

As previously mentioned, ring beams are traditionally set behind the cutterhead. This has one serious disadvantage: as the ring outer diameter will be the same as the internal diameter of the tunnel, the beams will encroach upon the free space in the tunnel and make it *impossible for the cutterhead to pass through them.* In other words, the machine becomes locked in and cannot be pulled back.

In the Alpine regions in Europe, geological conditions often are such that a sudden need arises to quickly secure the ground at or immediately behind the tunnel face. This gave birth to the *New Austrian Tunneling Method* (NATM) for tunneling by D & B which basically comes down to taking hold of the situation, by any suitable means, before it turns into a disaster. The suitable means can be anything from simple rock bolting to draining watery patches, grout injection, forepoling, shotcreting, the placing of ring beams, etc.

Some users of TBMs and proponents of NATM have combined the two and had considerable success in preventing their machines from getting into serious trouble on a number of occasions. The various measures often need to be taken as soon as possible, i.e., *in front of the machine* which is withdrawn to safety and to create working space at the face for a number of strokes. It soon became obvious that shotcreting, netting, and especially ring beams placed in the traditional manner formed an obstruction to the cutterhead when the machine had to be returned to the face again to restart boring operations.

Initially, this problem was solved by manually excavating extra space for those reinforcing elements *outside of the nominal bore diameter*. It did not take long, however, for an open-type hard-rock TBM to appear equipped with *overbore cutters* to temporarily bore the tunnel at a sufficiently larger-than-normal diameter to house such elements *beyond* the nominal diameter (see Fig. B12-11).

The overbore cutters might be regarded as an extra set of gauge cutters—which may be up to six in number—mounted on individual arms which can be swung out hydraulically through hatches in the peripheral plate work of the cutterhead when needed. Depending on how many cutters are swung out, the bore can be overcut by up to some 200 mm on the radius, enough for even very large profile ring beams.

With the overbore cutters active, the machine is advanced a certain distance: one or two strokes, depending on ground conditions. Then the cutters are swung back and the machine is pulled back to create the necessary space for whatever ground treatment measures are necessary at that point in time. Thereafter, the machine is advanced, the cutters are activated again, and the whole process is repeated in subsequent steps until the bad zone is passed. When the machine once again bores at the nominal diameter, the cutters are swung back definitely and the hatches are closed again.

The same system can also be used to place ring beams *behind* the head, if the ground permits waiting that long, in which case the machine does not have to go forward and then step back again, but can advance continuously, boring at the overbore diameter, and the ring beam erector mounted on the machine is then used in the normal manner, but now placing the beams outside of the nominal diameter, where they will not interfere.

Wire Mesh. Wire mesh by itself cannot be regarded as forming a real lining. Secured to the bore by rock bolts, it is only strong enough to support moderately sized loose blocks of rock. When shotcreted over, it acts as reinforcement netting of the concrete and the result then gains in strength and does indeed fulfill the requirements of a moderately strong lining, depending on the size of the bars in the mesh and the thickness of the shotcrete.

To prevent sensitive components of the machine from being covered in a layer of rebound material, any shotcreting should preferably be done either in front of the

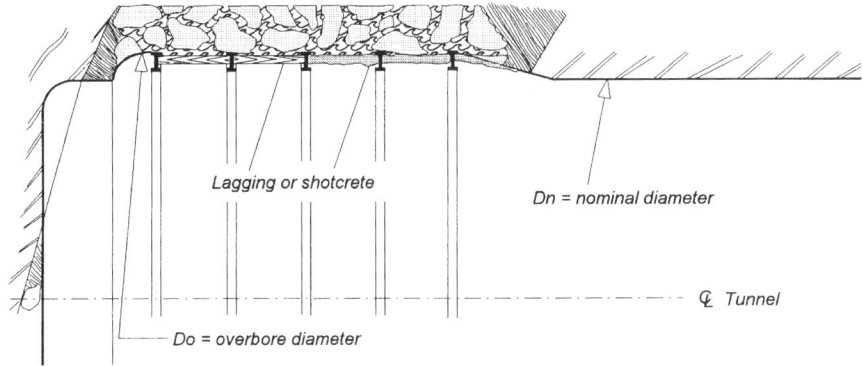

FIGURE B12-11 Overbore diameter.

TBM or behind it, in the usually open area between the machine and the backup. The latter application will not influence the operation of the TBM, nor is there any need of synchronizing the two.

If shotcreting in the area taken up by the TBM cannot be avoided, sensitive parts, such as hydraulic cylinder piston rods or the torque tube of a double gripper open-type hard-rock TBM, should be covered by plastic tarpaulins for protection. In that case, boring will have to be stopped during the shotcreting operation.

The TBM backup must provide the necessary space for storing the shotcreting materials and for the mixing and placement equipment.

P·A·R·T C

HEAVY CONSTRUCTION MATERIALS

CHAPTER 1
CONVENTIONAL CONCRETE*

James J. O'Brien
O'Brien-Kreitzberg, Inc., Pennsauken, New Jersey

---

* Some of the material in this chapter is from Herbert K. Cook, "Materials Selection and Mixture Proportioning," John K. Hunt, "Batching, Mixing, Placing, and Curing," George F. Bowden, "Stationary Forms," J. F. Camellerie, "Slipforms," and John C. King, "Special Concretes and Mortars" in Section 21, "Conventional Concrete," Stubbs and Havers (eds.), *Standard Handbook of Heavy Construction*, 2d. ed., McGraw-Hill, New York, 1971.

SECTION 1
MATERIALS SELECTION AND MIXTURE PROPORTIONING

The major ingredients in concrete are water, cement, and aggregates. In an increasing number of cases, admixtures are also used in concrete. The most widespread use of admixtures is for entraining air in concrete to increase its durability and workability. Admixtures for reducing water content and for the control of set are also in common use.

The type of cement, the source of aggregates, the type or types of admixtures, and the mix proportions for concrete for use in heavy construction have usually been thoroughly explored and final selection has been made before the construction contract has been awarded. It is customary for the job specifications to include the type of cement, the designation of acceptable sources of aggregate, and the requirement that the concrete will be proportioned in accordance with mixture designs provided to the contractor. The contractor may occasionally find such information not supplied, and it is the contractor's responsibility to produce and place concrete of the specified quality.

CEMENT

The greatest proportion of the cement used in the United States is manufactured to meet specifications prepared by the American Society for Testing and Materials. The American Society for Testing and Materials' *Standard Specification for Portland Cement* (ASTM Designation C150) provides for five types. Type I is intended for use in general concrete construction when the special properties specified for the other types are not required. It is the type most widely used and will automatically be provided when no other type is stated in the purchasing agreement. Type II is for use in general concrete construction exposed to moderate sulfate action or where a moderate heat of hydration is required. Type III is intended for use when high early strength is required. Type IV is for use when a low heat of hydration is required. Type V is for use when high sulfate resistance is required. It should be noted that Types IV and V are not usually carried in stock and, in advance of specifying their use, purchasers should determine whether they are or can be made available.

ASTM's *Standard Specification for Portland Cement* also provides for three types of cement which contain an air-entraining agent.[1] These cements are identified as Types IA, IIA, and IIIA. They are similar to the corresponding three types without air except that their strength requirements are somewhat lower and they are required to entrain air in a standard mortar within specified limits. There are no ASTM specifications for air-entraining Types IV and V.

The ASTM has specifications for a number of special cements such as natural cement, portland-pozzolan cement, portland blast-furnace slag cement, and slag cement. Some of these cements are used as replacements for or additions to portland cement and some are used in place of portland cement. Natural cement and slag cement are normally viewed as partial replacements for portland cement. Portland blast-furnace slag cement and portland-pozzolan cement may be considered as combinations of replacement materials and portland cement, preblended at a cement plant. The reasons for using these blends, either at the concrete batching plant or preblended prior to delivery, may include economy, reduction of heat generation in mass concrete, improvement of impermeability or watertightness, and improvement of workability. The manner in which they are to be used will be detailed in the job specifications.

Cements for which these options are applicable usually are made to order and, in advance of specifying their use, the purchaser should determine the manufacturer's ability to produce them. The individual test procedures used to determine compliance with the requirements of the ASTM and Federal Specifications[1,2] are normally of interest primarily to the testing agency, but they become of interest to the purchaser or contractor if there is a question relative to compliance.

AGGREGATES

Aggregate is often regarded as graduated garments of any hard, inert material which can be mixed with water and a cementing material to form concrete. Many of the problems that occur in the production of concrete stem from the fact that some of the materials proposed for use as aggregate turn out in practice not to be adequately hard, adequately inert, or suitably graduated.

That portion of the aggregate which will pass through a No. 4 sieve is called *fine aggregate* (or sand), and that which is retained on this sieve is called *coarse aggregate*. It is also customary and frequently mandatory that the coarse aggregate be divided into several sizes, depending on the maximum size to be used in the concrete. The primary reason for this is to prevent segregation in grading during stockpiling and handling. Standard practice is to separate the coarse aggregate into No. 4 to ¾ in; ¾ to 1½ in; 1½ to 3 in; and 3 to 6 in. On some large projects, particularly for very lean concrete such as is used in gravity dams, it is sometimes required that the fine aggregate also be separated into two sizes and that these be batched separately. The workability of such concrete is very critical and it is mandatory that the grading of the fine aggregate be very closely controlled to ensure that the concrete can be transported and consolidated in the forms in an efficient and uniform manner.

Contract requirements for aggregates, and whether or not acceptable sources are designated, vary more widely than do the requirements for cement. This may be attributed, in part, to the geological differences in deposits from one region to another, to differences in the degree of exposure of the concrete, and to differences in strength and other requirements for the particular structure.

Most specifications for aggregate are based on ASTM[2] Designation C 33. The requirements of these specifications are sometimes modified, either because local materials cannot meet some of the requirements or because special properties dictated by unusual construction or exposure conditions are called for. These standard specifications are suitable for structural concrete and paving. For such structures as large gravity dams, it is customary to prepare more detailed specifications designed to fit more closely the requirements of the individual job, its location, and the suitability of the local aggregates.

The increasing use of both insulating grades and structural grades of lightweight aggregate warrants some discussion of these materials. Lightweight aggregates for concrete for insulation are intended for use in concrete not exposed to the weather. The prime consideration is the thermal insulating value of the resulting concrete. Such aggregates are generally of two types. They may be prepared (1) by expanding materials such as perlite or vermiculite or (2) by expanding, calcining, or sintering products such as blast-furnace slag, diatomite, fly ash, shale, or slate. Requirements for acceptability are contained in ASTM Designation C 332, *Standard Specifications for Lightweight Aggregates for Insulating Concrete.*[2] On the other hand, for lightweight aggregates intended for use in structural concrete, the prime considerations are lightness in weight and sufficient compressive strength in the concrete for structural purposes. They, also, are of two general types which are prepared (1) by expanding, calcining, or sintering materials such as blast-furnace slag, clay, diatomite, fly ash, shale, or slate or (2) by processing natural materials such as pumice, scoria, or tuff. Requirements for acceptability are contained in ASTM Designation C 330, *Standard Specifications for Lightweight Aggregate for Structural Concrete.*[2]

Both of these types of lightweight aggregate are essentially manufactured materials. The selection of suitable and economic sources of supply is therefore a matter of finding appropriate manufacturing plants close to the job site or of otherwise arranging for the supply of materials of suitable quality at a reasonable shipping cost. The number of aggregate manufacturing plants and their geographic locations are such that these materials can generally be made available for most jobs.

Water

Water is a most important material on any heavy-construction job. It is used in the concrete as mixing water and on the surface of the concrete for curing. Considering the attention that is devoted to the testing, inspection, and control of the other ingredients of the concrete, it may seem unusual that so little attention is given to this very important material. Fortunately, the general criterion applies that water which is good enough to drink is satisfactory for use in concrete. The converse is not necessarily true, since water that is not potable may often be quite satisfactory for use as mixing water. In the event of doubt about a source of water, there are more reliable criteria of its acceptability than its potability. If the pH of the water is between 6.0 and 8.0 and if it is free from organic matter, it may be considered to be safe for use. Also, if the 7- and 28-day strengths of mortar cubes made with the questionable water are within 90 percent of similar mortar cubes made with distilled water, the water may be considered acceptable.

Water is essential to hydration, the chemical interaction between the water and the cement, and it is also required to make the plastic concrete sufficiently workable so that it can be handled and consolidated in the forms. Approximately 2.5 to 3 gal of water per sack of cement are required for the chemical hydration process; and this proportion, in theory, will produce maximum strength. However, as a practical consideration, such concrete will be completely unworkable for normal consolidation procedures. It is therefore necessary to add extra water to produce the required workability. The key words are *required workability,* and the addition of water beyond this point will rapidly reduce the strength of the concrete as well as all its other desirable properties.

Water is also used for curing the concrete, particularly in the case of large structures such as dams and bridges where ample supplies of water are almost always readily available. The primary consideration for the acceptability of water for use in curing is that it be nonstaining, especially for structures exposed to public view and where aesthetics or architectural appearance are important. Rather minor amounts of staining materials in the water may cause objectionable surface staining over the extended curing period. The principal objectionable materials in such water are iron and organic matter. The potential staining characteristics of these impurities cannot necessarily be evaluated on the basis of a chemical analysis. They are best evaluated by a performance type of test where a small concrete slab is actually subjected to curing with the water in question.

Admixtures

The use of admixtures has helped to improve the quality and economy of concrete as a construction material. The acceptance of proved materials has grown steadily, and admixtures of one type or another are used in approximately 80 percent of the concrete placed in the United States and Canada. They are selected for specific purposes and have provided qualities which in most cases could not have been obtained as effectively or as economically by any other means.

In the concrete industry, an admixture is defined as any material other than water, aggregates, and hydraulic cement used as an ingredient of a concrete or mortar mixture and added to the batch immediately before or during mixing. When such materials are interground or blended with the hydraulic cement, they are described as additions to cement. For example, an air-entraining agent which is added to a cement is properly described as an *addition*. When the same air-entraining agent is batched directly into the concrete, it is properly described as an air-entraining *admixture*.

In addition to the admixtures which entrain air, there are many others which may be used to procure certain benefits. The most widely used are chemical admixtures to control setting time and/or to reduce water. Others are accelerators, retarders, plasticizers, cementitious and pozzolanic materials, waterproofers, and materials used to inhibit alkali-aggregate expansion.

Air-Entraining Admixtures

The most commonly used admixtures are those employed to entrain air.

Occasionally concrete may be placed where weight is an important factor, but the engineer who specifies density is usually endeavoring to obtain concrete which will be resistant to the passage of water and to the effects of corrosive solutions and which will withstand severe weathering. It is almost universally true that these objectives can be met most readily by concrete which contains entrained air.

Consider, first of all, the action that takes place in a non-air-entrained concrete mix. When the concrete is first placed, the aggregates are uniformly dispersed within a more or less fluid medium. Since they are heavier than the fluid, they tend to settle, even in fairly stiff mixes, and this settling action will force the light ingredients in the concrete to the surface. These materials are the water and weak fines, such as silt. Thus, in the zone where maximum resistance to wear and weathering is needed, there is a layer of the weakest materials in the concrete. In addition, as the water moves to the surface it creates connected channels and voids. These provide easy access for moisture after the concrete has cured. If the aggregate particles continue to settle until they are in intimate contact with one another, they have little opportunity to move with the cement paste as it sets. The paste shrinks as it dries out after curing; and if movement of the aggregate is restricted or prevented, some bond may be lost. When subjected to weathering, such as freezing moisture, this condition may result in pop-outs.

In air-entrained concrete, on the other hand, the minute, disconnected bubbles of air which are dispersed uniformly throughout the mortar act much like water wings, buoying up the heavier pieces of aggregate. Since settlement of the heavier ingredients is reduced, bleeding of water and the light fines to the surface is held to a minimum. Some bleeding does occur, but the channels are fewer and, because of the interference of the air bubbles, they are disconnected.

Because the connected channels and voids are less frequent, permeability and absorption of the hardened concrete are reduced. The penetration of moisture into the concrete in cold weather is retarded, and the force exerted by the moisture when it freezes is absorbed and cushioned by the small bubbles of air which are a permanent part of the concrete. In the plastic state, the air also acts as a cushion that prevents the pieces of aggregate from making close contact. As a result, they are more free to follow the slight movement of the cement paste as it shrinks. Bond between the paste and the aggregates, especially at the surface, is improved. So long as the air is entrained properly and the amount is controlled within optimum limits, the concrete surface no longer consists of high water content and weak fines, and it will have greater resistance to both weathering and moderate wear. Resistance to freezing and thawing, particularly scaling caused by the heavy application of de-icing salts, is very markedly reduced.

One disadvantage of the normal type of air-entraining admixture is the possible loss of strength in medium and rich mixes. The disadvantage can frequently be overcome by the use of a water-reducing admixture in conjunction with air-entraining admixture. Some water-reducing admixtures will concurrently entrain air, others will not but will greatly reduce the amount of air-entraining admixture required to produce a specified amount of air in the concrete. Water-reducing admixtures in each of these categories will reduce or completely overcome the strength loss that is common in medium or rich mixes when only the normal types of air-entraining admixtures are used. In lean mixes, the water reduction produced by normal air-entraining admixtures can often be sufficient to counteract the effect of air on strength.

Accelerators

Calcium chloride, the least expensive of the known accelerators, was widely favored because of simplicity of use and ease of control. However, the use of calcium chloride is not recommended where stray dc electric currents might exist, where embedded items of different metals are in contact or in close proximity, or where the concrete is in contact with stressed wires or cables. These prohibitions apply to the use of calcium chloride in the normally used amounts of 1 or 2 percent by weight of the cement. They also apply to equivalent amounts of any other chloride that may be added to or be present in the materials used in the concrete. Chlorine ions can contribute to corrosion of embedded steel when the chlorine concentration exceeds 0.1 percent. This is only 5 to 10 percent of the standard calcium chloride mix used as an accelerator.

Other accelerating admixtures are available, but in almost every case it would not be practical to use them alone. This may be because of their physical form, because the small amounts required would make dispensing hazardous, because of their expense, or because of their unpredictable nature when used in too small or too large amounts. Those that are used in concrete are generally combined with other materials into a multicomponent admixture. Many of these have the added advantage of water reduction, and some have none of the disadvantages of calcium chloride. They should meet the requirements of ASTM C494, *Chemical Admixtures for Concrete,* for Type C if water reduction is not required and for Type E if water reduction is required or desired—which it normally is.

Retarders

The demand for retarding agents has been stimulated by the fact that concrete without retardation may set too fast to permit achievement of the desired design characteristics of the structure. Examples are bridge decks or other concrete structures placed over cambered steel where the dead load of the concrete must be in place before the concrete sets or where it is necessary to prevent cold joints in mass concrete or other types of construction.

A number of firms specialize in admixtures, including retarders. These patented products are provided in premeasured dosages for site use. They are also available in liquid state for computerized plant batching equipment.

Some retarding admixtures which contain carbohydrate derivatives have been developed. These have been found quite satisfactory in controlling or slowing down the set of concrete as long as they are used in accordance with the manufacturer's instructions. They are capable of moderate water reductions, an important benefit in itself. Some may tend to increase bleeding of the concrete, which is often considered objectionable.

One form of salts of sulfonated lignin has established an excellent record as a retarder as well as a means for controlling the evolution of heat in mass concrete. Under most circumstances, the quantity required for almost exact control of setting time can be determined with minimum preliminary tests. The substance also entrains air in the concrete, thus providing a high degree of durability and reducing water requirements with consequent gain in ultimate strength. These properties assure the contractor that it will have no detrimental effect when properly used.

Water-Reducing, Set-Controlling Admixtures

Admixtures in this category are widely used. ASTM C 494 includes requirements for five types:

Type A	Water-reducing admixtures
Type B	Retarding admixtures
Type C	Accelerating admixtures
Type D	Water-reducing and retarding admixtures
Type E	Water-reducing and accelerating admixtures

Type A admixtures are primarily designed to reduce the water content of concrete without significantly affecting setting time. The other types, as their names imply, are for controlling the setting time with or without water reduction.

Plasticizers

Air-entraining agents, some accelerators, and the retarders have a plasticizing action in concrete. With slump or workability held constant, these materials permit reductions in the water content as compared to a plain mix. By the same token, they permit placement of concrete in heavily reinforced or complicated formwork without increasing the water content of the mix.

Finely divided materials, inert fillers, and cementitious and pozzolanic products might also serve as plasticizing agents in mixes which are deficient in fines. Many of these plasticizers, however, will have no special benefit in mixes containing properly graded aggregates. Adding plasticizers to such mixes may even increase the water demand.

Cementitious and Pozzolanic Materials

This group includes natural cements, slag cements, hydraulic lime, fly ash, volcanic ash, diatomaceous earth, and other materials. Their presence in concrete otherwise lacking fines may be quite beneficial. On jobs where prolonged wet curing is possible, some of them may be used to distinct advantage as a replacement for part of the cement, thus reducing the heat of hydration.

Waterproofers

Agents sold for waterproofing purposes are advertised as helpful in "reducing permeability" or "reducing absorption." Any admixture which reduces water, entrains air, or reduces the voids in the hardened concrete is more or less effective in reducing both absorption and permeability. The extent of its influence as a waterproofer can probably be related directly to its ability to reduce water and bleeding. Those with the lowest water requirements and with the lowest bleeding will produce concrete that is more nearly watertight.

Water repellents, such as fatty acids, are often used to coat the pores and capillaries of the concrete. They help to resist the ingress of water in walls above grade or in slabs laid on the ground, but their value in concrete below grade and subject to water pressure is questionable. If sizeable pressures are involved, they may actually be detrimental. Below-grade concrete is waterproofed better and more economically by the use of water-reducing and air-entraining agents, and the best results above grade or on the ground are procured by combining these with one of the water repellents.

Inhibitors of Alkali-Aggregate Reaction

One of the greatest headaches to construction in some parts of the United States has been the reaction between the alkalis in cement and certain aggregates, causing an expansion which results in cracking of the concrete and occasionally in what might be described as a blowout rather than a pop-out. The use of low-alkali cement is indicated, and some pozzolanic materials are helpful in counteracting the reaction.

For instance, the range of admixtures offered by Fritz Concrete Admixture of Dallas, Texas, includes (but is not limited to) the following:

Supercizer 1—A slump enhancer and high-range water reducer which produces more flowable concrete with up to a 6-in slump increase.

Supercizer 2—An extended-life slump enhancer and high-range water reducer which produces more flowable concrete. Ideal for all hot-weather concrete applications.

Supercizer 3—A high-range water reducer and slump enhancer which produces more durable concrete with higher early strengths.

Supercizer 5—A high-range water reducer which will produce more flowable concrete or reduce water requirements up to 25 percent.

Supercizer 6—A new generation superplasticizer formulated to provide maximum water reduction, up to 40 percent, with ultrahigh early strengths within 6 to 24 h.

Supercizer 7—A premium superplasticizer formulated to provide maximum water reduction, up to 40 percent, and increase compressive strengths at all ages.

Delayed Set—Decreases disposal of waste concrete. Allows concrete to be kept in a plastic state for reuse, extended delivery requirements, and emergency situations.

Mini Delayed Set—Decreases environmental problems associated with concrete wash-water disposal. Reduces the amount of water needed for wash down and cuts labor costs associated with keeping mixer drums and fins clean.

Air Plus and Super Air Plus—Recommended for all types of concrete where an increase in entrained air content at the job site is necessary.

FR-1, FR-2, FR-3—Water-reducing admixtures designed for all concrete applications. These admixtures produce more uniform and workable mixes with less water, yielding stronger, more durable concrete.

Compacted Silicafume, Uncompacted Silicafume—Designed to produce very dense concrete with higher compressive strengths and extremely low permeability.

Fritz Flow #1—Reduces material cost and provides extended working life and improved workability for Class C fly ash slurries and flowable fill.

Mortarcizer—A hydrated lime replacement which plasticizes and improves board life for mortar mixes containing portland cement and mason sand.

Control Finish—Designed to improve concrete surface finishability. This product will facilitate the production of dense, high-quality concrete surfaces without retempering the surface with water.

Hydrocizer 1—An anti-washout admixture formulated to allow direct placement of concrete in underwater applications without segregation.

NS-7—A nonshrink grout fluidifier which produces a more uniform and workable grout with less water. Allows for slow controlled in-place expansion and extended workability.

Slick Pak—A concrete pump primer and pumping aid designed to provide a cost-effective replacement for premium-priced grout slurries and bagged cement primers.

PROPORTIONING CONCRETE MIXTURES

The proportioning of a concrete mixture is the process of determining the most economical and practical combination of materials that will produce a mixture which in the plastic state can be readily handled and consolidated in the forms and which in the hardened state will develop the required strength, durability, volume stability, and watertightness for the job under consideration. ACI 613, *Recommended Practice for Selecting Proportions for Concrete,* published by the American Concrete Institute provides guidelines.[3]

Contract specifications for concrete are generally based on one of three basic approaches:

1. Mixture proportions designed and controlled by the owner
2. Minimum cement content, maximum water-cement ratio, and range of slump specified
3. Concrete strength specified, usually at an age of 28 days

For contracts under condition 1, the contractor has only to comply with the instructions of the owner with respect to the mixture proportions provided. Under condition 2, the contractor may prepare mixture designs, but the cement, water, and slump requirements must be within specification limits. Under condition 3, the contractor has complete freedom in the selection of mixture proportions, subject only to the requirement that the concrete attain the specified strength at the age or ages indicated. The responsibility of the contractor is obviously considerably greater under condition 3 than under conditions 1 and 2. However, in this situation, the contractor has greater control over the cost of concrete production.

Basic Physical Properties of Materials

The detailed specification requirements for cement and the general considerations relative to the selection of aggregate sources from a quality standpoint have been discussed earlier. The specific properties which must be known for computation purposes or which have a bearing on the proportioning of concrete mixtures are reviewed in the following sections. Accepted methods for determining the properties of the materials are contained in references 1 and 2 at the end of this section.

Cement. The physical and chemical characteristics of the cement influence the properties of the hardened concrete. For this reason, cement from the same source and of the same type as that which will be used in actual construction should be used in the laboratory concrete tests. However, the only property of the cement directly concerned in the computation of concrete-mix proportions is specific gravity. For blends of portland cement with other hydraulic materials, such as natural cement or slag cement, the specific gravity of the individual cement should be determined and the specific gravity of the blend to be used on the job should be calculated by direct proportion. For normal portland cements, both plain and air-entraining, the specific gravity may be assumed to be 3.15 without introducing appreciable error in the mix computations.

Aggregates. Gradation, specific gravity, absorption, and moisture content of both fine and coarse aggregate are physical properties essential for mix computations. The dry-rodded weight of the coarse aggregate is also required. Other tests which may be desirable for large or special types of work include petrographic examination and test for chemical reactivity, soundness, durability, and resistance to abrasion, plus test for various deleterious substances. All such tests provide information of value in judging the ultimate quality of the concrete and in selecting appropriate proportions.

Aggregate gradation on particle-size distribution is a major factor in controlling unit water requirements, the proportion of coarse aggregate to sand, and the cement content of concrete mixes for a given degree of workability. While numerous "ideal" aggregate grading curves have been proposed, experience and individual judgment must continue to play important roles in determining acceptable aggregate gradings. Additional workability realized by use of air-entrainment permits, to some extent, the use of less restrictive aggregate gradations.

Undesirable sand grading may be corrected to desired particle-size distribution by separating the sand into two or more size fractions and recombining in suitable proportions, increasing or decreasing the quantity of certain sizes to balance the grading, or reducing excess coarse material by grinding. Undesirable coarse aggregate grading may be corrected by crushing excess coarser fractions, wasting excess material in other fractions, supplementing deficient sizes from other sources, or a combination of these methods to the extent that grading limitation and economy in use of cement permit the proportions of various coarse aggregate should be held closely to the grading of available materials. Whatever processing is done in the laboratory should be practical from the standpoint of field economy and job operation. Samples of aggregates for concrete-mix tests should be representative of aggregate selected for use in the work. For laboratory tests, the coarse aggregate should be cleanly separated into the required size fractions to provide for uniform control of mix proportions.

Elements of the Trial Mix

The first step toward the selection of the proportions of materials to be used in the concrete is to estimate, on the basis of the best information available on the materials and job conditions, the trial mix which will require the least final adjustment. The following basic properties of the materials must be determined by the actual laboratory tests before the trial mixes can be made:

1. Sieve analysis, specific gravity, absorption, and moisture content when batched for both the fine and the coarse aggregate.

2. Dry-rodded unit weight of the coarse aggregate.

3. Whether or not the cement is air-entraining, and its specific gravity. The specific gravity can generally be taken as 3.15 without the introduction of appreciable error in the computations.

After determination of these properties of the materials, the initial trial proportions are established by proceeding through the following seven steps, which are explained in detail immediately thereafter.

1. Select the water-cement ratio from established relationships to meet the specified requirements for durability and strength.

2. Select the limits of slump that will permit proper handling and consolidation of the concrete under job conditions.

3. Determine the maximum available size of coarse aggregate which is suitable for use under the job conditions.

4. Estimate the minimum percentage of sand that will provide adequate workability.

5. Estimate the amount of water per cubic yard of concrete necessary to meet the requirements for slump, maximum aggregate size, and percent sand, as established in steps 2, 3, and 4.

6. If an admixture is used, select the kind and amount required to comply with the conditions established in steps 1 to 5.

7. Calculate the trial-mix proportions that will meet these conditions.

Water–Cement Ratio. This should be selected on the basis of the durability and strength required in the structure. The durability requirements will be governed by the field-exposure conditions, and the strength will be governed by the design loads. Where the contract specifications state a minimum cement factor and a maximum permissible water–cement ratio, these requirements must, of course, be met.

Entrained air is of great benefit in ensuring durable concrete, and should be used when exposure to weathering is expected to be severe. It is also of great benefit in increasing the workability of the concrete, particularly for lean mixes for massive sections. Table C1-1 indicates the maximum permissible water–cement ratios in terms of gallons of water per bag of cement for various types of structures and degrees of exposure. This table guides the user in selecting the water–cement ratio required for adequate durability.

The maximum water-cement ratio or minimum cement factor to produce the required strength can best be determined by making and testing specimens with the concrete proportioned in accordance with the trial mix or mixes and using the same materials, including cement, that will be used in the job. Table C1-2 indicates the approximate strength that can be expected for a range of water–cement ratios using materials of good quality in a well-designed mix. A corresponding table for flexural strength is not given because the flexural strength varies over an excessively wide range with given proportions of different materials. Where flexural strength is specified, the required proportions should be determined by laboratory tests using job materials to ensure that the required strengths will be attained. The compressive strength should also be verified by laboratory tests, but the water–cement ratios shown in Table C1-2 should normally approximate the required strengths.

Strengths for air-entrained concrete for a given water-cement ratio (Table C1-2) are indicated as being 20 percent lower than for non-air-entrained concrete. This reduction applies only when the water–cement ratio is the same for both the plain and the air-entrained concrete. If the cement content and consistency are kept constant, less mixing water is required for the air-entrained concrete for the plain; the resulting decrease in water–cement ratio increases the strength, which partially or entirely offsets the reduction in strength caused by air entrainment.

Slump and maximum aggregate size should now be selected. Concrete proportions should be such that the concrete has the stiffest consistency (lowest slump) that will allow it to be placed efficiently and thoroughly. In the selection of the slump for concrete for heavy construction, full advantage should be taken of the reductions in slump that can be attained by the consolidation of concrete with large heavy-duty vibrators, which are a requirement for such construction. Table C1-3 shows the recommended slumps for various construction conditions.

Within the limits of economy and form and reinforcing spacing, the largest permissible maximum size of aggregate should be used. The use of large aggregate permits a reduction in water and cement requirements because the paste content of the concrete is decreased. The maximum sizes of aggregate recommended for various types of construction are shown in Table C1-4. Aggregate larger than 6 in in size is rarely used because the requirements for processing and handling equipment rapidly approach uneconomical proportions.

Unit water content can be estimated from Table C1-5, which indicates the approximate water requirements for different slumps and maximum sizes of aggregates. It is well to keep in mind the importance of keeping the unit water content to the absolute minimum. The quantity of water per unit volume of concrete required to produce a mix of the desired consistency is influenced by the maximum size, particle shape, and grading of the aggregate and by the amount of entrained air. Within the normal range of mixes, it is relatively unaffected by the quantity of cement.

The quantities of water given in Table C1-5 are sufficiently accurate for estimating initial proportions. They are the maxima which should be expected for the indicated maximum sizes of fairly well-shaped but angular aggregates, graded within the limits of conventional specifications. If otherwise suitable aggregates have higher water requirements than indicated in Table C1-5, it is probable that they are less favorably shaped or are not so well graded as normally may be expected. Unless otherwise indicated by laboratory tests, the proportions of such aggregates should be adjusted or the cement increased, or both, to maintain the desired water–cement ratio.

Cement requirements for a given concrete can be computed using the maximum permissible water–cement ratio selected from Table C1-1 or C1-2 and the water

TABLE C1-1 Maximum Permissible Water-Cement Ratios for Different Types of Structures and Degrees of Exposure

(in gallons per bag)

	Exposure conditions*					
	Severe wide range in temperature or frequent alternations of freezing and thawing (air-entrained concrete only)			Mild temperature rarely below freezing, or rainy, or arid		
		At the water line or within the range of fluctuating water level or spray			At the water line or within the range of fluctuating water level or spray	
Type of structure	In air	In fresh water	In sea water or in contact with sulfates[†]	In air	In fresh water	In sea water or in contact with sulfates[†]
Thin sections, such as railings, curbs, sills, ledges, ornamental or architectural concrete, reinforced piles, pipe, and all sections with less than 1 in concrete cover over reinforcing	5.5	5.0	4.5[‡]	6	5.5	4.5[‡]
Moderate sections, such as retaining walls, abutments, piers, girders, beams	6.0	5.5	5.0[‡]	§	6.0	5.0[‡]
Exterior portions of heavy (mass) sections	6.5	5.5	5.0[‡]	§	6.0	5.0[‡]
Concrete deposited by tremie under water	—	5.0	5.0	—	5.0	5.0
Concrete slabs laid on the ground	6.0	—	—	§		
Concrete protected from the weather, interiors of buildings, concrete below ground	§	—	—	§		
Concrete which will later be protected by enclosure or backfill but which may be exposed to freezing and thawing for several years before such protection is offered	6.0	—	—	§		

* Air-entrained concrete should be used under all conditions involving exposure and may be used under mild exposure conditions to improve workability of the mixture.
† Soil or groundwater containing sulfate concentrations of more than 0.2 percent.
‡ When sulfate-resisting cement is used, the maximum water-cement ratio may be increased by 0.5 gal per bag.
§ Water–cement ratio should be selected on basis of strength and workability requirements.

TABLE C1-2 Compressive Strengths of Concrete for Various Water–Cement Ratios*

Water–cement ratio, gal/bag of cement	Probable compressive strength at 28 days, lb/in^2	
	Non-air-entrained concrete	Air-entrained concrete
4	6000	4800
5	5000	4000
6	4000	3200
7	3200	2600
8	2500	2000
9	2000	1600

* These average strengths are for concretes containing not more than the percentages of entrained and/or entrapped air shown in Table C2-5. For a constant water–cement ratio, the strength of the concrete is reduced as the air content is increased. For air contents higher than those listed in Table C2-5, the strengths will be proportionally less than those listed in this table.

Strengths are based on 6- by 12-in cylinders moist-cured under standard conditions for 28 days. See Method of Making and Curing Concrete Compression and Flexure Test Specimens in the Field (ASTM Designation C 31).

TABLE C1-3 Recommended Slumps for Various Types of Construction*

Types of construction	Slump, in†	
	Maximum	Minimum
Reinforced foundation walls and footings	5	2
Plain footings, caissons, and substructure walls	4	1
Slabs, beams, and reinforced walls	6	3
Building columns	6	3
Pavements	3	2
Heavy mass construction	3	1

* Adapted from Table 4 of the Joint Committee Report on Recommended Practice and Standard Specifications for Concrete and Reinforced Concrete, ACI.

† When high-frequency vibrators are used, the values given should be reduced by about one-third.

TABLE C1-4 Maximum Sizes of Aggregate Recommended for Various Types of Construction

Minimum dimension of section in	Maximum size of aggregate,* in			
	Reinforced walls, beams, and columns	Unreinforced walls	Heavily reinforced slabs	Lightly reinforced or unreinforced slabs
2½–5	½–¾	¾	¾–1	¾–1½
6–11	¾–1½	1½	1½	1½–3
12–29	1½–3	3	1½–3	3
30 or more	1½–3	6	1½–3	3–6

* Based on square openings.

TABLE C1-5 Approximate Mixing Water Requirements for Different Slumps and Maximum Sizes of Aggregates*

Slump, in	Water, gal/yd^3 of concrete for indicated maximum sizes of aggregate							
	⅜ in	½ in	¾ in	1 in	1½ in	2 in	3 in	6 in
	Non-air-entrained concrete							
1–2	42	40	37	36	33	31	29	25
3–4	46	44	41	39	36	34	32	28
6–7	49	46	43	41	38	36	34	30
Approx. amount of entrapped air in non-air-entrained concrete, %	3	2.5	2	1.5	1	0.5	0.3	0.2
	Air-entrained concrete							
1–2	37	36	33	31	29	27	25	22
3–4	41	39	36	34	32	30	28	24
6–7	43	41	38	36	34	32	30	26
Recommended avg. total air content, %	8	7	6	5	4.5	4	3.5	3

* These quantities of mixing water are for use in computing cement factors for trial batches. They are maxima for reasonably well-shaped, angular, coarse aggregates graded within limits of accepted specifications.

If *more* water is required than shown, the cement factor, estimated from these quantities, *should* be increased to maintain the desired water–cement ratio except as otherwise indicated by laboratory tests for strength.

If *less* water is required than shown, the cement factor, estimated from these quantities, *should not* be decreased except as indicated by laboratory tests for strength.

requirements from Table C1-5. The cement factor is obtained by dividing the gallons of mixing water required per cubic yard by the water–cement ratio in gallons per bag of cement. If a minimum cement factor is specified, the corresponding water–cement ratio for estimating strength can be computed by dividing the gallons of water per cubic yard by the cement factor in bags per cubic yard. Selection of proportions for concrete should be based on whichever of the limitations specified (durability, strength, or cement factor) requires the lowest water–cement ratio.

Aggregate requirements can now be examined. The minimum amount of mixing water and the maximum strength will result for given aggregates and cement when the quantity of coarse aggregate used is the largest consistent with adequate placability and workability. The quantity of coarse aggregate that can be used increases with maximum size of aggregate. The quantity of coarse aggregate can be determined most effectively by laboratory investigations of the materials, subject to later adjustment as necessitated by field conditions. An estimate of the best proportions can be made on the basis of established relationships for aggregates graded within conventional limits. Table C1-6 shows these relationships, with the quantities of coarse aggregate expressed as dry-rodded bulk volumes per unit volume (cubic yard) of concrete.

Concrete of comparable workability can be expected with aggregates of comparable size, shape, and grading. This will be true if a given percentage of dry-rodded coarse aggregate is used per unit volume of concrete and if the volume of mortar remains constant, even though the solid volumes of cement, water, air, and sand may be interchanged. In the case of different types of aggregates, particularly those with different particle shapes, the use of a fixed, dry-rodded, solid volume of coarse aggregate automatically makes allowance for differences in mortar requirements as reflected by the

TABLE C1-6 Volume of Coarse Aggregate per Unit of Volume of Concrete*

Maximum size of aggregate, in	Volume of dry-rodded coarse aggregate per unit volume of concrete for different fineness moduli of sand			
	2.40	2.60	2.80	3.00
3/8	0.46	0.44	0.42	0.40
1/2	0.55	0.53	0.51	0.49
3/4	0.65	0.63	0.61	0.59
1	0.70	0.68	0.66	0.64
1 1/2	0.76	0.74	0.72	0.70
2	0.79	0.77	0.75	0.73
3	0.84	0.82	0.80	0.78
6	0.90	0.88	0.86	0.84

* Volumes are based on aggregates in dry-rodded condition as described in Method of Test for Unit Weight of Aggregate (ASTM Designation C 29). These volumes are selected from empirical relationships to produce concrete with a degree of workability suitable for the usual reinforced construction. For less workable concrete such as required for concrete-pavement construction, they may be increased about 10 percent.

void content of coarse aggregate. For example, angular aggregates have a higher void content and therefore require more mortar than rounded aggregates.

Another method of proportioning fine and coarse aggregate is to compute the total solid volume of aggregate in the concrete mix and multiply this volume by the recommended percentage of sand. The percentage of sand in the concrete mix has been used extensively as a means of expressing the proportions of sand and coarse aggregate; the sand is expressed as a percentage of the total solid volume of the entire aggregate. Table C1-7 indicates the approximate air and water contents per cubic yard of concrete and the proportions of fine and coarse aggregate. It duplicates much of the information supplied in Table C1-6, but contains additional data such as the recommended percentages of sand for each maximum size of coarse aggregate.

Computation of Mix Proportions

In the computations that are given here, examples are shown for computing mix proportions both on the basis of estimating the quantity of coarse aggregate from its dry-rodded volume (Example 1) or on the basis of the computed total solid volume of aggregate multiplied by the recommended percentage of sand (Example 2). Either method is satisfactory and under normal conditions will produce approximately the same proportions. Maintaining the solid volume of coarse aggregate constant in concrete has certain advantages: (1) Mix adjustments for changes in air content and cement content are made automatically, and (2) in mixing-plant operation, batch-setting changes for coarse aggregate become unnecessary for changes in other proportions.

The computation of mix proportions and the application of the foregoing discussion can best be explained by showing the calculations for specific examples. For these examples, the following design criteria and mix materials will be assumed:

1. Type I non-air-entraining cement will be used in Example 1 and Type II non-air-entraining cement in Example 2. The specific gravity of each cement will be taken as 3.15 and the weight of a bag as 94 lb.

CONVENTIONAL CONCRETE C1-19

2. The coarse aggregate has a specific gravity, bulk dry, of 2.68; and absorption of 0.5 percent.

3. The fine aggregate has a specific gravity, bulk dry, of 2.64; and absorption of 0.7 percent; and a fineness modulous of 2.8.

4. Both the coarse and fine aggregates are of satisfactory quality. The coarse aggregate is stockpiled in four size groups: No. 4 to ¾ in, ¾ to 1½ in, 1½ to 3 in, and 3 to 6 in.

Example 1. Concrete is required for an architectural reinforced-concrete wall having a minimum thickness of 5 in. The wall will be subjected to wide ranges in air temperature in a northern climate. A 28-day compressive strength of 3500 lb/in^2 is specified for at least 80 percent of the strength tests. The concrete will be consolidated with high-frequency vibrators.

TABLE C1-7 Approximate Air and Water Contents per Cubic Yard of Concrete and the Proportions of Fine and Coarse Aggregate

(For concrete containing natural sand with a fineness modulus of 2.75 and average coarse aggregate and having a slump of 3 to 4 in at the mixer)

Max. size of coarse aggregate, in	Actual weight of coarse aggregate per unit volume of concrete as a percentage of dry-rodded unit weight	Air-entrained concrete			Non-air-entrained concrete		
		Recommended air content, %	Avg. water content, lb/yd^3	Sand, % of total aggregate by solid volume	Approx. percent entrapped air	Avg. water content, lb/yd^3	Sand, % of total aggregate by solid volume
⅜	41	8	322	59	3.0	352	61
½	52	7	306	50	2.5	336	53
¾	62	6	283	42	2.0	316	45
1	67	5	267	37	1.5	300	41
1½	73	4.5	245	33	1.0	280	36
2	76	4	229	30	0.5	266	33
3	81	3.5	204	28	0.3	242	31
6	87	3	164	24	0.2	210	28

Adjustment of values for other conditions

	Effect on values		
Changes in conditions stipulated	Unit water content, %	Sand, %	Dry-rodded coarse aggregate, %
Each 0.1 increase or decrease in fineness modulus of sand	—	±0.5	±1
Each 1-in increase or decrease in slump	±3		
Each 1% increase or decrease in air content	±3	±0.5–1.0	
Each 0.05 increase or decrease in water-cement ratio	—	±1	
Each 1% increase or decrease in sand content	±1	—	±2
For angular coarse aggregate	+7–10	+3–5	
For lower slump concrete as in pavements	–3	–3	+6

Note: If aggregates are proportioned by percent sand method, use the first and second columns; if by dry-rodded coarse aggregate method, use the first and third columns.

The trial-batch proportions are computed as follows:

1. Because of the severity of the exposure, air-entrained concrete will be used. Reference to Table C1-1 indicates that the water–cement ratio should not exceed 5.5 gal/bag.
2. Table C1-2 indicates that a water–cement ratio of 5 gal/bag, with air-entrained concrete, will produce a compressive strength at 28 days of 4000 lb/in^2. Previous experience with job control[4] indicates that approximately this average strength is required to ensure that at least 80 percent of the strength tests will be above the 3500 lb/in^2 specified. For this reason, the lower value of 5 gal/bag will be used.
3. Tables C1-3 and C1-4 show that a maximum slump of 4 in and a maximum aggregate size of ¾ in should be used.
4. The approximate quantity of mixing water for a 4-in slump in air-entrained concrete with ¾-in aggregate is found in Table C1-5 to be 36 gal/yd^3. This table also indicates that the air content should be 6 percent. The entrained air will be obtained by use of an admixture added at the mixer as a solution. The volume of solution added should therefore be included in the computations as part of the volume of mixing water. The quantity of solution added should be sufficient to entrain the required amount of air. The amount recommended by the manufacturer will, in most cases, produce the desired air content. In this example, it is assumed that 1 qt of air-entraining solution per bag of cement is required to provide 6 percent air.
5. From items 2 and 4, the required cement content is computed as follows:

$$\frac{36 \text{ gal water/yd}^3}{5 \text{ gal water/bag of cement}} = 7.2 \text{ bags/yd}^3$$

6. From Table C1-6 it is found that, with a fine aggregate having a fineness modulus of 2.8 and a ¾-in coarse aggregate, 0.61 ft^3 of coarse aggregate on a dry-rodded basis will be used in each cubic foot of concrete. The quantity in a cubic yard will then be 27 × 0.61 = 16.47 (16.5 ft^3). The dry-rodded weight was determined to be 100 lb/ft^3, so the weight of coarse aggregate per cubic yard of concrete will be 100 × 16.5, or 1650 lb.
7. Having established the quantities of cement, water, coarse aggregate, and air, the sand content is calculated as follows:

$$\text{Solid volume of cement} = \frac{7.2 \text{ bags/yd}^3 \times 94 \text{ lb/bag}}{3.15 \text{ (sp gr)} \times 62.4 \text{ lb water/ft}^3} = 3.44 \text{ ft}^3$$

$$\text{Volume of water} = \frac{36 \text{ gal/yd}^3}{7.5 \text{ gal/ft}^3} = 9.87 \text{ ft}^3$$

$$\text{Solid volume of aggregate} = \frac{1650 \text{ lb/yd}^3}{2.68 \text{ (sp gr)} \times 62.4} = 9.87 \text{ ft}^3$$

Volume of air = 0.06 × 27 = 1.62 ft^3
Total solid volume of ingredients except sand = 19.73 ft^3
Solid volume of sand required = 27 − 19.73 = 7.27 ft^3
Required weight of dry sand = 7.27 × 2.64 (sp gr) × 62.4 = 1198 lb

8. The estimated batch weights per cubic yard of concrete are:

Cement = 7.2 bags × 94 lb/bag	=	677 lb
Water = 36 gal × 8.33 lb/gal	=	300 lb
Sand (dry basis)	=	1198 lb
Coarse aggregate (dry basis)	=	1650 lb

9. It should be noted that these batch weights are for 1 yd^3 of concrete on a dry basis, no correction having been made for moisture content. The fine and coarse aggregate used in the field will almost always carry some free moisture. The moisture will also vary from day to day or even from hour to hour. The determination of moisture contents and the correction of batch weights based on the determinations are therefore among the most important operations in the daily production concrete. To illustrate the manner in which such corrections are made, let us assume that the moisture content of the sand has been found by test to be 5.0 percent and the coarse aggregate 1.0 percent. The weight of dry sand was found to be 1198 lb. The amount of moist sand to be weighed out must therefore be 1198 × 1.05 = 1258 lb. On the same basis, the weight of moist coarse aggregate will be 1650 × 1.01 = 1667 lb.

10. The water in the aggregates in excess of the amount which they will absorb must be considered as part of the mixing water and the weight of the mixing water corrected accordingly. The absorption of the sand was previously stated as being 0.7 percent. The amount of free water which the sand contains is therefore 5.0 − 0.7 = 4.3 percent. Correspondingly, the free water on the coarse aggregate will be 1.0 − 0.5 = 0.5 percent.

    The weight of mixing water contributed by the sand will be 0.043 × 1198 = 52 lb, and for the coarse aggregate it will be 0.005 × 1650 = 8 lb. These two weights (52 + 8 = 60 lb) must be subtracted from the indicated weight of mixing water (300 − 60). The weight of water to be batched will then be 240 lb, assuming that the air-entraining admixture is weighed with the mixing water. If the air-entraining admixture is added separately, the weight of mixing water should be reduced correspondingly. In this instance, 1 qt or 0.25 gal of solution is required per bag of cement and the total weight of the solution is 0.25 gal/bag × 7.2 bags × 8.33 lb/gal = 15 lb. The net weight of mixing water per cubic yard of concrete is then 240 − 15 = 225 lb.

11. If a 5-yd batch is required, each of the batch weights will be multiplied by 5, and for batches of other sizes, the individual batch weights will be multiplied by the number of yards required. Occasionally, particularly for mixers of fractional-yard capacities, the batch size is specified in terms of the number of bags of cement required for the batch; for example, if it is desired to use a three-bag batch in a 16-ft^3 mixer, the total volume of concrete will be 27 × 3/7.2 = 13.3 ft^3, and each batch weight of the materials will be 3/7.2 of that given above.

Example 2. In the first example, since coarse aggregate of ¾-in maximum size was used, it was in accordance with good practice to batch the coarse aggregate as a single ingredient. Now assume that concrete is required for a massive footing which will be below ground level and will not be exposed to weathering or other attack. Structural-design requirements call for a 28-day compressive strength of 2000 lb/in^2. No reinforcing is required.

The trial batch proportions are calculated as follows:

1. While the structure will not be exposed to weathering, it is decided to use an air-entraining admixture to provide added workability and to reduce the water content. The water–cement ratio will be established solely on the basis of the required strength. Table C1-2 indicates that a water–cement ratio of 8 gal/bag will produce a strength of 2000 lb/in^2. Since the specifications require only a nominal strength of 2000 lb/in^2 with no minimum requirement with respect to tests, this water–cement ratio will be acceptable.
2. Table C1-3 shows that a slump of 1 to 3 in should be used. Consolidation will be accomplished with heavy-duty vibrators or normal frequency. The mix will be proportioned for an average 2-in slump.
3. Table C1-4 recommends the use of 6-in maximum-size coarse aggregate, which is available.
4. Table C1-5 indicates that, for 6-in aggregate and a 1- to 2-in slump, the approximate mixing water requirement will be 22 gal/yd^3 at a recommended air content of 3 percent.
5. From items 1 and 4, the cement content is found to be 22 gal per yd^3/8 gal per bag = 2.8 bags per yd^3.
6. Because it is difficult to determine accurately the dry-rodded unit weight of 6-in aggregate, the aggregate proportions will be established on the basis of percent sand. Previous experience (see Table C1-7) indicates that approximately 24 percent sand should be satisfactory.
7. On the basis of these established values, the aggregate volumes for 1 yd^3 are calculated as follows:

    Solid volume of cement = $(2.8 \times 94)/(3.15 \times 62.4)$ = 1.34 ft^3
    Volume of water = 22/7.5 = 2.93 ft^3
    Volume of air = 0.03×27 = 0.81 ft^3
    Solid volume of all ingredients except aggregate = 5.08 ft^3
    Total solid volume of aggregates = 27 − 5.08 = 21.92 ft^3
    Solid volume of sand = 21.92×0.24 = 5.26 ft^3
    Solid volume of coarse aggregate = 21.92 − 5.26 = 16.66 ft^3

8. The estimated batch weights per cubic yard of concrete are:

    Cement = 2.8×94 = 263 lb
    Water = 22×8.33 = 183 lb
    Sand (dry basis) = $5.26 \times 2.64 \times 62.4$ = 867 lb
    Coarse aggregate (dry basis) = $16.66 \times 2.68 \times 62.4$ = 2786 lb

9. The coarse aggregate is stockpiled in four size groups and will be weighed in four batches. It is therefore necessary to calculate a combined grading and to determine the percent of each size group and, from this, the batch weight for each group.

    A number of approaches to the optimum combine grading of aggregates have been developed and discussed in the literature. However, it is believed that combining the various sizes on the basis of maximum density is the most widely used and accepted method. The relative proportions of the size groups are determined by taking a complete grading combination from the coarse-aggregate grading curves shown in Fig. C1-1. These curves start at the No. 4 sieve and are derived from the following formula:

$$p = \frac{\sqrt{d} - \sqrt{0.1875}}{\sqrt{d_{max}} - \sqrt{0.1875}} \, 100$$

where d = sieve opening, in
d_{max} = maximum size of aggregate, in
p = cumulative percent passing

$$\sqrt{0.1875} = 0.433013$$

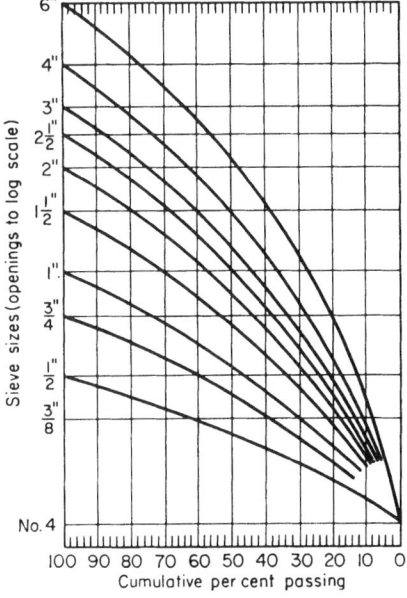

FIGURE C1-1 Grading curves for coarse aggregate. *(Corps of Engineers.)*

Values derived from this equation are also given in Table C1-8. The objective in this example is to arrive at the combination of the four sizes of the coarse aggregate that will approximate the maximum density curve for 6-in maximum-size aggregate shown in Fig. C1-1 or listed in the second column in Table C1-8. Combined gradings using varying percentages of each of the four sizes of coarse aggregate are computed, and the proportions resulting in the combined grading most closely approximating the gradings taken from the curve are used. Table C1-9 illustrates the method of calculation. Columns 1 to 4 represent the grading of the four sizes of aggregate available at the job. From the procedure described above it is found that the proportions which best approximate the maximum density curve are 36 percent of the 3- to 6-in aggregate, 26 percent of the 1½- to 3-in, 16 percent of the ¾- to 1½-in, and 22 percent of the No. 4 to ¾-in. The percentages of material retained on each sieve for each of the four size groups when used in these proportions are shown in columns 5 to 8. Columns 9 and 10 show the combined gradings, and Columns 11 and 12 show the gradings taken from the curve.

It was determined in item 8 that a total weight of 2786 lb of dry coarse aggregate was required for 1 yd³ of concrete. The dry weights of each of the stockpiled sizes are determined as follows:

2786 × 0.36 = 1,003 lb of 3- to 6-in size
2786 × 0.26 = 724 lb of 1½- to 3-in size
2786 × 0.16 = 446 lb of ¾- to 1½-in size
2786 × 0.22 = 613 lb. of No. 4 to ¾-in size

10. As in Example 1, the batch weights for Example 2 are on a dry basis. This is not the condition in which they will be in the field. Further, the moisture contents of

the four sizes of coarse aggregate may not be identical. To illustrate the calculations of corrected batch weights, let us assume the following conditions, all other properties of the materials being as previously stated:

Moisture content of sand = 5.0 percent

Moisture content of 3- to 6-in aggregate = 0.5 percent

Moisture content of 1½- to 3-in aggregate = 0.5 percent

Moisture content of ¾- to 1½-in aggregate = 1.0 percent

Moisture content of No. 4 to ¾-in aggregate = 1.5 percent

Air-entraining admixture at the rate of 0.15 gal/bag of cement is required to entrain 3 percent air; it is batched separately from the mixing water

Batch weights of aggregates in the field:

TABLE C1-8 Proportions of Aggregate of Various Sizes

Sieve size, in	Maximum size, in								
	6	5	4	3	2½	2	1½	1	¾	½
	Cumulative percent passing									
6	100									
5	89.4	100								
4	77.7	86.9	100							
3	64.4	72.0	82.9	100						
2½	56.9	63.7	73.3	88.4	100					
2	48.7	54.4	62.6	75.5	85.5	100				
1½	39.3	43.9	50.5	60.9	69.0	80.7	100			
1	28.1	31.4	36.2	43.6	49.4	57.8	71.6	100		
¾	21.5	24.0	27.6	33.3	37.7	44.1	54.7	76.4	100	
½	13.6	15.2	17.5	21.1	23.9	27.9	34.6	48.3	63.3	100
¾	8.9	9.9	11.4	13.8	15.6	18.3	22.6	31.6	41.4	65.4
No. 4										

TABLE C1-9 Combined Gradings

Sieve retained on	Grading of individual size groups, % retained				36%			22%	Recombined cumulative, %		Grading curve 6 in max, %	
	3–6 in (1)	1½–3 in (2)	¾–1½ in (3)	No. 4–¾ in (4)	3–6 in (5)	1½–3 in (6)	¾–1½ in (7)	No. 4–¾ in (8)	Ret. (9)	Pass. (10)	Ret. (11)	Pass. (12)
6 in	1.0	0.0	—	—	0.4	0.0	—	—	0.4	99.6	0.0	100.0
3 in	91.0	5.0	—	—	32.8	1.3	—	—	34.5	65.5	—	64.4
2 in	8.0	56.0	—	—	2.8	14.6	—	—	51.9	48.1	51.3	48.7
1½ in	0.0	33.0	0.0	—	—	8.5	0.0	—	60.4	39.6	—	39.3
1 in	—	5.0	67.0	0.0	—	1.3	10.7	0.0	72.4	27.6	71.9	28.1
¾ in	—	1.0	28.0	4.0	—	0.3	4.5	0.9	78.1	21.9	—	21.5
½ in	—	0.0	3.0	28.0	—	—	0.5	6.2	84.8	15.2	86.4	13.6
¾ in	—	—	2.0	25.0	—	—	0.3	5.5	90.6	9.4	91.1	8.9
No. 4	—	—	0.0	39.0	—	—	—	8.6	99.2	0.8	100.0	0.0
Smaller than No. 4	—	—	—	4.0	—	—	—	0.8	100.0	0.0		
Total	100.0	100.0	100.0	100.0	36.0	26.0	16.0	22.0				

Sand = 867 × 1.05 = 910 lb
3- to 6-in aggregate = 1003 × 1.005 = 1008 lb
1½- to 3-in aggregate = 724 × 1.005 = 728 lb
¾- to 1½-in aggregate = 446 × 1.01 = 450 lb
No. 4 to ¾-in aggregate = 613 × 1.015 = 622 lb

Corrections in amount of mixing water:
Free water in aggregates:

Sand = (0.05 − 0.007) × 867 lb	= 37 lb
3- to 6-in aggregate = (0.005 − 0.005) × 1003	= 0 lb
1½- to 3-in aggregate = (0.005 − 0.005) × 724	= 0 lb
¾- to 1½-in aggregate = (0.01 − 0.005) × 446	= 2 lb
No. 4 to ¾-in aggregate = (0.015 − 0.005) × 613	= 6 lb
	45 lb

(*Note:* The absorption of 0.5 percent is the same for each of the four sizes of coarse aggregate)
Water in admixture:

0.15 gal/bag × 2.8 bag × 8.33 lb/gal = 3 lb

The total amount of water to be subtracted from the original amount of mixing water is therefore 45 + 3 = 48 lb. The amount of mixing water to be batched in the field is 183 − 48 = 135 lb.

The following table compares the computed batch quantities for Example 2 with those which would actually be used in the field.

	Material quantities, lb		
	Computed	Used in field	
Ingredient	For 1 yd³	For 1 yd³	For 4 yd³
Cement	263 (2.8 bags)	263 (2.8 bags)	1052
Water	183 (22 gal)	135 (20 gal)	540
Admixture	3 (0.4 gal)	3 (0.4 gal)	12 (1.6 gal)
Sand	867 (dry)	910 (moist)	3640 (moist)
3- to 6-in coarse aggregate	1003 (dry)	1008 (moist)	4032 (moist)
1½- to 3-in coarse aggregate	724 (dry)	728 (moist)	2912 (moist)
¾- to 1½-in coarse aggregate	446 (dry)	450 (moist)	1800 (moist)
No. 4 to ¾-in coarse aggregate	613 (dry)	622 (moist)	2488 (moist)

11. Adjustments of batch weights for batches greater or smaller than 1 yd³ are made on a proportional basis in the same manner as discussed in item 11 for Example 1.

Concrete-Mix Tests

The values listed in the tables of this section may be used for establishing preliminary trial mixes. However, they are based on averages obtained from a large number of tests and do not necessarily apply exactly to the materials being used on a particular job. It is therefore excellent policy to make a series of concrete tests to establish the relationships needed for selection of the proper proportions of the materials actually to be used on the project. An example of a series of such tests is shown in Table C1-10. The first mix of the series was a computed trial mix, arrived at in the manner previously discussed and shown in the examples. The second mix was adjusted to increase slump, but it appeared to be oversanded and to contain insufficient coarse aggregate. In the third mix, the amount of coarse aggregate was increased to an amount which was estimated to be the maximum which would still produce a mix of satisfactory workability. The sand content was correspondingly reduced. Three additional mixes were then made, with the water–cement ratio varying over a range of 5 to 7 gal per sack. From these mixes, the relationship between water–cement ratio, cement content, and strength was established for the job materials. The field mixes could then be selected directly from Table C1-10.

In laboratory tests it seldom will be found, even by experienced operators, that the desired adjustments will develop as smoothly as indicated in Table C1-10. Also, it should not be expected that field results will check exactly with laboratory results. An adjustment of the selected trial mix on the job is usually necessary and will be discussed later. Closer agreement between laboratory mixes and field mixes will be assured if machine mixing is employed for making the laboratory mixes. This is especially desirable if air-entraining admixtures are used, since the type of mixing influences the amount of air entrained. Before mixing the first batch, the laboratory mixer should be "buttered" or the mix overmortared as described in ASTM Designation C 192, because a clean mixer will retain a percentage of the mortar. Similarly, any processing of materials in the laboratory should simulate as closely as practical the corresponding treatment in the field.

The minimum series of tests illustrated in Table C1-10 may be expanded as the size and special requirements of the work require. Alternate aggregate sources, aggregate gradings, types and brands of cements, admixtures, maximum sizes of aggregate, and considerations of concrete durability, volume change, temperature rise, and thermal properties are some of the variables that may require a more extensive program.

Adjustments for Field Use

Having established by laboratory test the mix proportions that appear to produce the results desired for job conditions, the next step is to translate these proportions to the batch weights to be used in the job mixer. Most large jobs will use mixers of 4-yd^3 capacity. Items 10 and 11 of Example 1 and items 10 and 11 of Example 2 explained how the batch weights per cubic yard were corrected for moisture content and adjusted to the batch weights for the size batch desired. In this case, for a 4-yd^3 mixer, the corrected batch weights for a 1-yd^3 batch are simply multiplied by 4. Let us assume that the batch material quantities tabulated at the end of item 10, Example 2, were found by laboratory test to be satisfactory for job requirements. The batch weights for a 4-yd^3 batch will then be as shown in the last column of the tabulation.

Note that, in the figures for the 4-yd^3 batch, the number of bags of cement and the number of gallons of water have not been indicated in parentheses, whereas the number of gallons of admixture have been shown. This is because, while it is customary to indicate the number of bags of cement and the gallons of water in a 1-yd^3 batch, in large batching operations it is almost universally required that the cement

TABLE C1-10 Typical Minimum Mix Series to Establish the Properties of Concrete Made with Field Materials

Mix no.	Net water–cement ratio, gal/bag	Water content, gal/yd³	Cement content, bags/yd³	Aggregate content, lb/yd³ Sand	Aggregate content, lb/yd³ Coarse aggregate	Slump, in	Percent air	28-day strength, lb/in² Compression	28-day strength, lb/in² Flexure	Segregation	Workability Rodability	Workability Finish
1	5.6	32.0	5.70	1,186	1,940	1	4.5	3,500	555	None	Good	Good
2	5.6	34.0	6.07	1,112	1,940	3¼	4.5	3,500	555	None	Excellent	Excellent
3	5.6	33.5	5.98	1,064	2,008	2½	4.5	3,500	555	None	Excellent	Very good
4	5.0	33.5	6.70	1,008	2,008	3	4.5	4,000	600	None	Excellent	Excellent
5	6.0	33.5	5.58	1,096	2,008	2¾	4.5	3,200	525	None	Excellent	Very good
6	7.0	33.5	4.79	1,158	2,008	3	4.5	2,600	450	None	Very good	Good

Mix No. 1—Low on slump.
Mix No. 2—Oversanded, increased coarse aggregate and lowered water for Mix No. 3.
Mix No. 3—Workability satisfactory.

and water be batched by weight. On the other hand, because of the small amount used, the admixture may be batched either by weight or by volume. It should also be noted that while the weights for 4-yd^3 have been indicated to the nearest pound, the quantities will actually be weighed to the nearest scale division. The 4032 lb of 3- to 6-in coarse aggregate will probably be weighed to the nearest 50 lb.

After trying a few full-size field batches on the job, it is found that the water content per cubic yard must be increased from 22 to 23 gal to provide adequate workability for consolidation. Consequently, the cement factor should be increased to 23 gal/yd^3 ÷ 8 gal/bag = 2.9 bags/yd^3. The batch quantities should be recomputed accordingly, following the procedures demonstrated in Example 2.

If the field use indicated that less water could be used, it should be reduced. However, as previously recommended, no consequent reduction in cement factor should be made unless laboratory tests indicate that such reduction can be made. It is necessary, however, that some adjustment be made in the batch quantities to compensate for the loss in volume due to the reduced volume of water. This may be done by increasing the solid volume of sand by an amount equal to the volume of the reduction in water. For example, assume that 21 gal of water are required instead of 22 gal for the concrete in Example 2. Then 21/7.5 is substituted for 22/7.5 in computing the volume of water in the batch, or the solid volume of water per cubic yard becomes 2.80 instead of 2.93 ft^3. The difference in solid volume of 0.13 ft^3 is added to the sand, making it 5.39 instead of 5.26 ft^3/yd^3, and this is converted to a weight basis as indicated in the example.

The percentage of air in concrete can be measured directly with an air meter, or it can be computed from theoretical and measured unit weights in accordance with ASTM methods.[2] For any given set of conditions and materials, the amount of air entrained is roughly proportional to the quantity of air-entraining agent used. Increasing the cement content for fines, decreasing the slump, or raising the temperature of the fresh concrete usually decreases the amount of air entrained for a given quantity of agent. The grading and particle shape of the aggregates, particularly the sand, also have an effect on the amount of air entrained. The job mix should not be adjusted for minor fluctuations in water–cement ratio or air content. A variation in water–cement ratio of ±0.25 gal per sack of cement, resulting from maintenance of constant workability, is considered normal. A variation of ±1 percent in air content is also considered normal.

SPECIAL PROPORTIONAL REQUIREMENTS

Six-In Aggregate Concrete

All the procedures used here in arriving at the estimated mix proportions, in making laboratory trial mixes and tests, and in the adjustment of the final mix in the field are as applicable to concrete containing 6-in aggregate as to concrete made with aggregate of smaller size. However, limitations in the size of laboratory testing equipment introduce special problems in the selection of mix proportions and the making of laboratory and routine control tests when 6-in aggregate is used.

Note that the aggregate proportions in Example 2 were selected from Table C1-7 instead of Table C1-6 because of the difficulty of obtaining an accurate dry-rodded weight for 6-in aggregate. Actually, ASTM Method C 29 for determining the unit weight of aggregate is recommended for use only with aggregate of up to 4-in size.[2] The size of the unit weight measure, the capacity of the scales required, and the physical difficulties of moving a full unit weight measure, to say nothing of the technical difficulties of adequately consolidating the aggregate in the measure, make it impractical to determine the unit weight of 6-in aggregate or concrete containing such aggregate.

Similar difficulties are encountered in the determination of slump, in the making of compressive- or flexural-strength specimens, and in the determination of air content of the concrete by either the unit-weight or pressure method. It is generally accepted that any test specimen should be at least five times the diameter of the largest aggregate used. A concrete cylinder for compressive-strength test of concrete containing 6-in aggregate should therefore be 30 in in diameter and 60 in tall. Such a cylinder obviously could not be hand-rodded. It would weigh approximately 3675 lb and, if it developed the relatively low strength of 2000 lb/in^2 would require a testing machine with a minimum capacity of 1,500,000 lb. Several methods of circumventing these difficulties are in common use, although all of them are compromise solutions to the problem. They are satisfactory for job-control purposes if their limitations are recognized and compensated for accordingly.

The most common method for determining slump, strength, and air content is to wet-screen the concrete over a 1½-in sieve and to determine these properties on the concrete which passes the sieve. It is obvious that the slump, for example, of the concrete containing the 1½-in aggregate is not the same as for the concrete with the full-size aggregate, but the relationship of this slump to adequate workability in the forms is usually sufficient for routine control purposes. Many organizations depend on good control of aggregate grading, moisture content, and batching operations, together with close observation of the workability of the concrete in the forms, and do not make slump tests. The Kelly ball test described in Ref. 2 is another means of determining concrete workability. Strength tests are customarily made on 6- by 12-in cylinders containing the 1½-in aggregate concrete. The strength of the job concrete is sometimes verified by testing cores drilled from the structure.

Testing for the air content of the job concrete is one of the more important routine control procedures. Determination of air content on the minus 1½-in portion of the concrete has been used almost universally on large jobs. Extensive investigations have been made of the reliability and the relationship between the air content of concrete which has been wet-screened over a smaller sieve and the air content of the original concrete.

Lightweight Aggregate Concrete

The basic considerations for the proportioning of concrete using normal weight or natural aggregates also apply to the proportioning of concrete using lightweight aggregate. However, excluding those aggregates which have sealed or coated surfaces, most lightweight aggregates have capacities for absorption which may require several days to satisfy. This makes determination of their absorption and specific gravity values impractical, and lightweight concrete mixes are usually established by a series of trial mixes on a cement-content basis at the required consistency.[5]

The pertinent properties of lightweight aggregates will vary widely from one manufacturing source to another. For example, because some lightweight aggregates are rounded with smooth surfaces, the water requirements to produce a 2-in slump in non-air-entrained concrete can vary from 300 to 450 lb/yd^3 for different aggregates. This wide range in water requirements is reflected in a corresponding range in cement content to produce a given strength. The structural strength of different lightweight aggregates also has an important effect on the cement requirement, particularly for higher-strength concretes. Because of the specialized nature of the design of mixture proportions for structural-grade lightweight-aggregate concrete and the special techniques involved in mixing and placing, it is suggested that the contractor and the owner either place their reliance on the recommendations of the lightweight-aggregate manufacturer or consult with others who have a sound knowledge of this kind of concrete. This does not imply that there is anything particularly difficult about the use of lightweight concrete, but it is a specialized field of concrete technology that the average contractor or owner will not have the time or the funds to explore adequately.

SECTION 2

BATCHING, MIXING, PLACING, AND CURING

MATERIAL HANDLING AND STORAGE

One of the more difficult goals in the production of concrete is to have each batch as nearly like each other batch as possible. This means that not only must the same number of pounds of each material be used in each batch but also that the characteristics of the materials used should not vary from batch to batch or from day to day.

Material Characteristics

If the gradation changes within a given aggregate size range, then the concrete will be different. One way to reduce segregation and maintain proper gradation is to have more size ranges, with each range covering a narrower span of sizes. Finish-screening immediately before the aggregate goes into the batching storage bin reduces the effects of segregation and degradation, but this can only screen the materials available and cannot overcome poor gradation.

If all aggregate comes from a single source, the variation in intrinsic strength will probably be small; but if several sources are used, the variations may become significant and must be considered in the concrete design.

A uniform moisture content in all aggregate materials as batched is very desirable in reducing variations in consistency or slump. Stockpiles of sand may need to drain for 24 to 48 h to achieve a stable moisture content. Moisture meters are very useful for detecting a change in moisture content, but they should be checked and calibrated fairly often and particularly any time a different type of sand from a different source is to be used.

The strength of concrete varies considerably depending on the actual cement used to produce the concrete. It has been established that the compressive strength of concrete produced using cement from the same source fluctuates considerably over a period of one year. If cement from more than one source is used, the variations in the compressive strength become much higher. Concrete must be designed based on the cement that is actually to be used in the concrete.

Admixtures must be handled carefully, because a very small change in the amount of a given admixture may have adverse effects on the concrete to an extent far exceeding the actual percentage of error. Since many admixtures are being used in a highly concentrated form, it is extremely difficult to batch them within the necessary tolerances.

Plant Types

Concrete batching plants may be divided into types based on several criteria. They may be either *low-profile* on the one hand or *stack-up, gravity,* or *tower* plants on the other, depending on the flow of material through the plants or their physical configuration. They may be classified as *central-mix, transit-mix,* or *dry-batch* plants, depending on the function. They may be *mass-concrete, paving, ready-mix* or *concrete-products* plants, depending on the ultimate use for which the plant is intended. They may also be classified on the basis of the plant's mobility as *permanent, portable,* or *mobile.*

A stack-up, gravity, or tower plant is one in which the materials are elevated to the storage bin and flow by gravity into the weigh hoppers and into a central plant mixer or directly into a hauling unit. There is no generally accepted definition of a low-profile plant, but if the materials are elevated in the flow from the storage bin to the ultimate receiving unit, then the plant is normally considered to be low-profile. In the most common of these low-profile plants, the material is elevated into the storage bin, discharged by gravity into the weigh hoppers, then discharged onto a belt conveyor and elevated to a central plant mixer or into a transit-mix or dry-batch truck. Another type of low-profile plant elevates the materials to the overhead storage bins, discharges from the storage bin to a belt conveyor, and elevates the materials to the weigh hoppers. From the weigh hoppers the materials may be discharged by gravity into a central plant mixer or into a transit-mix or dry-batch truck.

A plant which has a central plant mixer is called a *central-mix plant.* A transit-mix plant is one where the aggregate, cement, and usually water are batched and then discharged into the transit-mix truck for mixing en route to the job site. In some cases only the aggregate and cement are batched at the plant, and water is added either from the tank of the transit-mix truck or from a supply at the job site. A plant which is used for batching only the aggregate and cement, placing these materials into compartments of trucks for hauling to the job, is normally referred to as a *dry-batch plant* (Fig. C1-2).

Mass-concrete plants are normally large, permanent, central-mix stack-up plants (Figs. C1-3 and C1-4). They are designed to produce a large amount of concrete at a given installation, so ease of transporting the plant is not an important criterion. There are usually between two and six mixers in a mass concrete plant. The mixers may be either 2 or 4 yd^3 each. There are many types of paving plants available today. The most recent ones are utilizing central-mix plant features and eliminating the paver mixer at the job site. The paving plant may be either stack-up or low-profile, portable or mobile, and either central-mix, dry-batch, or transit-mix. The most common paving plant today is a low-profile, mobile, central-mix plant. With the large quantities of concrete that can be produced and paved, there is a definite need for a high-capacity, central-mix, low-profile, mobile plant for the paving industry (Figs. C1-5 and C1-6). Ready-mix plants may be almost any combination of the various types. Many ready-mix producers are utilizing small low-profile mobile plants (Fig. C1-7) to move into a job which is large enough to support a separate small plant. This reduces the travel time and, therefore, also the number of trucks necessary to handle the job. In some cases, a ready-mix producer will use a low-profile mobile plant (Fig. C1-7) to move into an area to sample the market. If the market demands are such that it would appear profitable, then the ready-mix producer may install a large permanent plant (Figs. C1-8 and C1-9) and move the mobile plant to another location. Concrete-product plants are normally designed for one particular location and are normally permanent installations. They may be either low-profile or stack-up plants. They are almost invariably central-mix plants because the material must be mixed in the plant and is usually processed within the same plant area.

Permanent plants quite often take a considerable amount of time for erection. This time may involve several weeks because pieces have to be bolted together.

FIGURE C1-2 Mobile one-stop dry-batch paving plant. *(C. S. Johnson Operations, Road Division, Koehring Company.)*

FIGURE C1-3 Mass batching and mixing plant. *(Manitowoc.)*

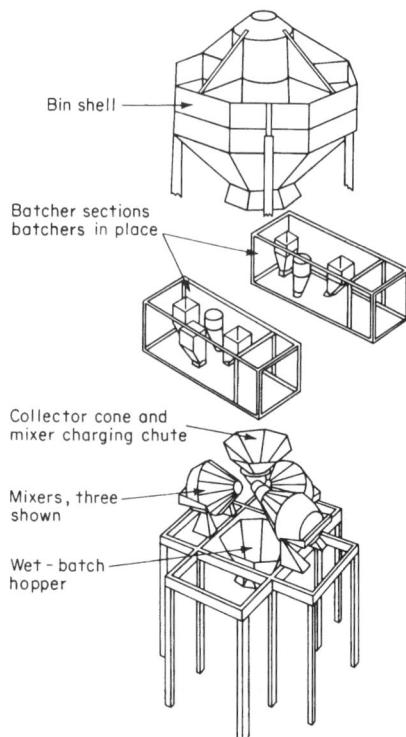

FIGURE C1-4 Schematic diagram of a large batching and mixing plant for a dam construction project. *(Concrete Construction Handbook.)*

There may be preassembled sections which can be erected in large bulk so that there are relatively few pieces to put together and relatively few connections to make, thus reducing erection time and cost. Portable plants can be easily transported or hauled, but several days may still be required to prepare for a move and to set up after being moved. Mobile plants are furnished with their own wheels, axles, and a fifth wheel so that a tractor may be connected to the plant to tow it along the road from one site to another. These are usually used where the plant must be moved frequently and moving time must be cut to an absolute minimum. Sometimes only a few hours are necessary to get a plant ready to move and another few hours may be necessary before the plant is ready to be put into operation after it has been moved.

Material Handling

Concrete batching, mixing, and placing may be considered essentially a problem in material handling. The raw materials—aggregate, cement, water, and admixture—must be delivered to the plant site; either stored and then reclaimed to the batching bins or delivered directly to the batching bins; measured; delivered to a plant mixer, transit mixer, or to a dry-batch truck and then to a paving mixer; and the mixed concrete delivered to its final destination by a truck, transit mixer, concrete bucket, buggy, conveyor belt, chute, or some other means. Many different material-handling methods must be used in the course of getting these different materials through the different phases involved in the production of concrete. All aspects of the material-handling problem, from the arrival of the raw material at the batch-plant site to the final placement of the finished concrete, are important, and any one of them may become the bottleneck for the entire job. Many problems of quality control are inherent in some handling methods and steps must be taken to ensure that proper safeguards are taken, proper methods are used, and proper equipment is on hand to satisfactorily perform each step in this process.

It is essential that the plant be arranged and the delivery of the various materials be scheduled in such a way that the different functions of the material handling do not interfere with each other. The trucks or railroad cars delivering the raw materials must not interfere with the equipment used to transport the finished product from the batch-plant site. This requires a considerable amount of preplanning, forethought, and constant supervision to ensure that there are no interfering processes and that all phases of the work proceed according to plan. Close cooperation and

FIGURE C1-5 Large, mobile, low-profile central-mix paving plant with two tilting mixers. *(C. S. Johnson Operations, Road Division, Koehring Company.)*

FIGURE C1-6 Large, unitized, mobile, low-profile central-mix paving plant for one mixer shown in travel position. *(C. S. Johnson Operations, Road Division, Koehring Company.)*

good coordination between the material suppliers and the batch-plant superintendent are essential.

Since the materials used in the production of concrete have widely different characteristics, several different methods of transporting the raw materials to the batch-plant site must be used. The aggregate may be delivered to the batch-plant site in barges, railroad cars, or trucks. Cement may be delivered to the batch-plant site in railroad cars or trucks. Water is either available at the site or pumped to the site. Admixtures are normally delivered either premixed in containers or in dry bulk form.

FIGURE C1-7 Small, unitized, mobile, low-profile ready-mix plant. *(C. S. Johnson Operations, Road Division, Koehring Company.)*

FIGURE C1-8 Large stack-up central-mix ready-mix plant. *(C. S. Johnson Operations, Road Division, Koehring Company.)*

FIGURE C1-9 Large low-profile central-mix ready-mix plant. *(C. S. Johnson Operations, Road Division, Koehring Company.)*

Each of these materials requires different handling methods to place it into the proper storage facilities. The aggregate may be put into a stock pile or directly into a storage bin by the use of belt conveyors, bucket elevators, or clamshell buckets. Cement is usually placed into a watertight storage silo or bin by pneumatic pumping or by a screw conveyor and bucket elevator. Water and the liquid admixtures are pumped to a storage vessel.

The different materials also require different methods of feeding into the measuring devices. Either a single- or double-clam gate is used for aggregate; a rotary plug valve, a vane feeder, or a screw conveyor is used to feed cement; water may be fed by gravity into a weigh hopper to be measured by weight or may be pumped through a meter for volume measurement; and admixtures may be pumped through a meter, blown into volumetric measuring sight glass, fed by gravity flow through a calibrated orifice based on time, or gravity-fed into a weigh hopper.

After all the ingredients are measured, the means used depend on whether the plant is a central-mix, shrink-mix, transit-mix, dry-batch, or concrete-products plant. In a central-mix plant the material is discharged into a central plant mixer, completely mixed, then discharged into some vehicle for transportation to the final placing site. This vehicle might be an open-top agitating unit, an open-top nonagitating unit, a dump truck, a concrete bucket, or a belt conveyor.

If a shrink-mix plant is used, the materials are fed into a central plant mixer directly after weighing, partially mixed or shrink mixed, and then fed into a transit-mix truck to be mixed for some additional time while en route to the job site.

In a transit-mix plant the materials are put into a transit-mix truck directly after being weighed and are mixed entirely in the truck. Some specifications require that all mixing be done at the plant site and that the truck be operated at agitating speed while en route to the job. Other specifications require that all mixing take place on the job site after the transit-mix truck has arrived there. Still other specifications allow any combination of these two methods, or they may allow mixing while en route to the job site. When the transit mixer has arrived at the job site, some further material handling is necessary. The concrete must be unloaded from the transit-mix truck and directed to the final resting place. Chutes are normally provided with the truck. The concrete may be directed immediately through the chutes into forms or transported farther by means of belt conveyors or concrete pumps.

If a dry-batch plant is used, the material is normally discharged from the weigh hoppers directly into dry-batch hauling units. These may have from two to seven separate compartments, each of which holds one complete batch of aggregate and cement. The dry-batch trucks then transport the material to a paver, where it is mixed with water and discharged into forms or in front of a slip-form paver.

In a concrete-products plant, the material is usually discharged into a mixer and then discharged into forms, a concrete-block machine, or a concrete-pipe machine.

The type of plant and the type of material-handling equipment vary widely depending on the final end use of the concrete being produced, the quantity of concrete being produced in a given time, the specifications governing the production of concrete, the geographic location of the plant, the actual physical location of the plant with respect to land surrounding it, and the amount of room available at the plant site.

Aggregate may be delivered to the plant site by rail, truck, or barge. The method of rehandling aggregate depends on the method of delivery to the plant site. Aggregate may be delivered by rail in bottom-dump cars which may be discharged into a hopper, fed onto a belt conveyor, and then delivered directly into the bin or onto a stockpile. Clamshell buckets may be used to unload the railroad cars and place the material either into the bin or onto stockpiles.

If the aggregate is delivered by truck, it may be placed directly onto stockpiles or into a loading hopper which feeds either a belt conveyor or a bucket elevator. If the batch plant is located at the side of a hill, arrangements may be made so that the truck can drive on an open grill directly out over the storage bin and dump into the compartments of the bin. Barges are usually unloaded with the clamshell buckets.

When a stockpile is built up using a clamshell bucket, the material should be placed, not thrown, in layers in order to reduce segregation and degradation. The stockpile may also be made by using an inclined belt conveyor. In order to build a longer stockpile, either a radial stacker or a parallel stacker is quite commonly used (Figs. C1-10 and C1-11). A reversing shuttle conveyor may also be used in conjunction with an inclined conveyor to build long stockpiles (Figs. C1-12 and C1-13). The parallel stacker builds a stockpile in a straight line and the radial stacker builds a stockpile on a radius. The radial stacker produces somewhat more dead storage if the stockpile is over a tunnel conveyor. Either rock ladders or some provision of lowering or raising the end of the conveyor should be used if the stockpile is to be very high. This is to reduce immediate degradation.

Aggregate may be delivered to the batching plant storage bin by an inclined conveyor fed directly from a hopper (Fig. C1-14) or by a tunnel conveyor (Fig. C1-15) under the stockpile. If a double-clam gate is used to reclaim onto a tunnel conveyor, the material should be discharged into the center of the belt. If a single-clam gate is used, the material should be discharged in the direction of the belt travel. This eliminates the lateral force on the belt. The number of gates to be used under each stockpile will be determined by the height of the stockpile, the length of the stockpile, and the amount of dead storage that can be tolerated.

Aggregate bucket elevators may be used to charge the aggregate storage bin. These are normally used only when space is not available for a belt conveyor. This is due to the additional maintenance and high-wear characteristics of the bucket elevator when used for aggregate.

FIGURE C1-10 Radial stacker with raising and lowering boom. *(Atlas Conveyor Company.)*

FIGURE C1-11 Parallel stacker with raising and lowering boom. *(Barber-Greene Company.)*

The aggregate bin may also be charged with a clamshell bucket, either directly from barges, railroad cars, or reclaimed from a stockpile. When a single inclined conveyor or a bucket elevator is used to feed the storage bin, some means must be provided to direct the material to the proper compartment. When the compartments are in tandem (in a straight line), a shuttle conveyor mounted at the top of the bin may be used. If the charging height of the storage bin is low, either a radial stacker or a parallel stacker may be used to charge the bin directly. A pivoted distributor (turnhead) may be used to direct the material to the proper compartment when the compartments are arranged so they can be fed from a single point. In many cases, more than

FIGURE C1-12 Inclined belt conveyor with reversing shuttle belt conveyor. *(Barber-Greene Company.)*

FIGURE C1-13 Parallel stacker, shuttle belt conveyor, and tunnel belt conveyor. *(Atlas Conveyor Company.)*

FIGURE C1-14 Inclined belt conveyors to plant storage bin. *(Atlas Conveyor Company.)*

FIGURE C1-15 Tunnel belt conveyor. *(Barber-Greene Company.)*

one belt conveyor is used to charge the bin. For high-production plants this is very feasible, particularly if the height of the bin is such that the conveyors are not excessively long. Some low-profile plants can be charged directly with front-end loaders.

In many cases, automatic bin filling is utilized. In a low-profile plant, for example, one belt conveyor may be used to feed each compartment. The belt conveyor is normally fed from a hopper, which in turn is fed by a front-end loader or clamshell bucket. The belt conveyors are kept fully loaded and start and stop at a signal given by a bin-level indicator in the aggregate compartment.

When a tunnel conveyor, inclined conveyor, and pivoted distributor are used, the automatic bin filling becomes somewhat more complicated. The belt conveyors usually run constantly whether any material is being fed to the bin or not. When the material in one bin compartment reaches a predetermined low level, the pivoted distributor rotates to that compartment and the tunnel gate for the correct material opens and begins filling. When the material level reaches a point near the top of the bin, the system is told to stop and the tunnel gate closes. After enough time has elapsed to clear the tunnel conveyor and the inclined conveyor, the pivoted distributor is free to move to another compartment to begin filling it. The system should be interlocked so the tunnel gate cannot open unless the pivoted distributor is at a compartment which indicates a need for material and the tunnel conveyor and the inclined conveyor are running the pivoted distributor should not rotate until the entire system has had sufficient time to clear itself.

A bucket elevator could obviously be used instead of the inclined conveyor. If the bin compartments are in tandem, a shuttle conveyor could be used instead of the pivoted distributor. If there are only two compartments, a flop chute or two-way gate could replace the pivoted distributor.

Plant Storage

Concrete-batching plant storage bins are available in almost every conceivable size and shape. They may be furnished with aggregate only or cement only, with aggregate storage compartments around the outside and a cement storage middle, with aggregate on one end and cement on the other, or with aggregate in the center and cement on both ends. The storage bins may be square, rectangular, hexagonal, octagonal, or round. The compartments may be arranged in a tandem fashion so that the materials are side by side or they may be arranged so that the compartments are in a square. There may be as few as one or as many as 8, 12, or even 16 or more aggregate compartments. There may be only one or as many as two, three, four, six, or more compartments for holding different types of cement. Capacities of the storage bins vary from a few cubic yards of aggregate up to several hundred cubic yards and from a few up to one or two thousand barrels of cement.

Large central-mix paving plants have three or four aggregate storage compartments and one or two cement storage compartments. The aggregate bin is usually arranged for 5 to 20 min of storage. Normally, some system of automatic bin charging is utilized. The cement is normally delivered in pneumatic unloading trucks. Auxiliary storage is commonly furnished with 1000-barrel mobile units. As many as two to six trucks may be unloading at one time. The production capacity varies from about 200 yd^3/h up to 1000 yd^3/h, depending on the number of mixers, the size of the mixer, the mixing time, and whether the plant is equipped with individual single-material aggregate batchers or a cumulative aggregate batcher.

Dry-batch paving plants may have one or two separate plants for cement. Sometimes several sets of batchers are used under these plants so that more than one compartment of the hauling truck may be filled in one cycle. There is usually a very small amount of aggregate storage in the bin. Cement may be delivered by pneumatic unloading trucks or by railroad cars with a screw conveyor and elevator to charge the cement storage facilities. Up to 630 yd^3/h is possible from a dry-batch paving plant. This would require from five to seven 34E dual-drum pavers, depending on the required mixing time. The 630 yd^3/h is based on having four complete sets of single-material batchers charging dry-batch trucks with seven compartments.

Mass-concrete plants are commonly furnished with five aggregate compartments and with facilities for storing cement and pozzolan. The aggregate bin should have sufficient overhead storage for one to two hours' operation at the rated capacity using the maximum-size aggregate. This will ensure continuity of the pour in case there is a breakdown of the material-handling equipment. Normally from three days' to one week's supply of aggregate is kept on hand in stockpiles and from three days' to one week's supply of cement is kept on hand in large steel silos. The production capacity depends on the number of mixers, the mixing time, and the size of the mixers. Up to 480 yd^3/h should be produced from six 4-yd^3 mixers.

Ready-mix plants may have from 3 to 12 or even 16 aggregate compartments and from 1 to 6 or even more cement compartments. Since ready-mix plants are commonly called on to furnish concrete for many different purposes, they require a number of different sizes and types of aggregate and cements. The storage capacity of the ready-mix bin normally depends to a large extent on the peak concrete production in cubic yards per hour (see Table C1-11). The production capacity of ready-mix plants varies up to 300 yd^3/h, with a few plants exceeding this. Very few ready-mix concrete plants produce at the peak capacity for more than a short time in any one day. The peak capacity is normally needed only during the start-up time in the morning when all the trucks must be loaded in a short time. After this, the trucks are usually staggered, so that the production rate drops considerably.

Concrete-products plants normally have provisions of four aggregate compartments and sometimes more if specialty items or ornamental blocks are being produced. Usually one and sometimes two cement compartments are used. The production capacity is normally very small compared to other types of plants. Aggregate and cement storage depends largely on the availability of the materials.

Special Processing

After screening to produce the proper gradation of aggregate, it is sometimes necessary to rescreen the materials to reduce the effects of degradation and segregation. This rescreening procedure can only reject undersize material and cannot correct for poor gradation in the original material supplied. The rescreen may be located on the ground and the material elevated to the storage bin after rescreening. It is usually better to screen on the top of the plant to get as close to the actual batching process as possible. This eliminates one step in the material-handling process after the rescreening is finished. Most specifications require checking at the batcher discharge for a determination of correct gradation. If the screens are located on top of the bin, it is necessary to have the system carefully designed to reduce vibration. Excessive vibration will be transmitted to the lower part of the bin and may have an adverse effect on the controls and scales. Normally the material is sorted into the different sizes and then blended together before it is run through the screen.

On large installations, two double-deck screens, one above the other, are usually adequate. Some cases may require two sets of screens, operating side by side. One triple-deck screen with the bottom deck used for two sizes may be adequate for small installations. Either a sand bypass chute or a separate sand conveyor should be provided. The sand takes about one-third of the total loading time, so a separate sand conveyor increases the effective screen capacity by about 50 percent.

Heating of some of the materials in cold weather is necessary in order to be able to place and protect concrete before it freezes.

Hot water is the most common and simplest method of raising the temperature of concrete. If the water is too hot, there is a danger of flash set when it contacts the

TABLE C1-11 Typical Storage Capacity for Commercial Ready-Mix Plants

Peak concrete production, yd³/h	Aggregate storage, yd³					Cement storage, bbl*			
	Mobile plant, bin	Fixed plant				Mobile plant total storage	Fixed plant†	
		Bin with 3 to 5 compartments	Bin with 6 or more compartments	Total storage			Overhead storage	Total storage
10	0–15	—	—	110	165‡		
25	0–30	30	40	275	165‡	70	350‡
50	35	50	65	550	165‡	130	650
100	40	100	125	1100	350‡	260	1300
150	45	225	270	1650	525	390	1950
200	50	400	480	2200	700	520	2600
300	60	650	780	3300	1050	780	3900

* Cement-storage capacity is shown at 4 ft³/bbl; cubic feet of storage space required is four times the figure shown. Because of fluffing in handling, actual capacity may be 70 to 85 percent of the storage capacity shown.

† Overhead and total cement storage should be increased about 20 percent if there are three or more types of cement.

‡ 165 bbl is the minimum storage for cement delivered in 125-bbl pneumatic trucks. 450 bbl is the minimum storage for cement delivered in 400-bbl rail hopper cars.

cement. Hotter water can be used if the water and aggregate are mixed before the cement is introduced into the mixer. Sparger nozzles in a water reservoir may be used with a steam valve controlled by a thermostat to regulate the water temperature. If boiler feed water does not require treatment, this is the most economical method of heating water. Steam coils may be used in the water reservoir to heat water without loss of boiler water except that lost through leaks in the system. In-line water heaters have steam coils built into a section of the water line. This method heats the water as it passes into the reservoir and requires less area of heating coils than the coils in a tank because both the steam and water are moving.

Frozen aggregate must be heated above 32°F to make warm concrete, but it should not be heated above 100°F because of the danger of flash set. If the temperature of the aggregate can be kept above 32°F, additional heating will be reduced considerably. Live steam jets that run directly into aggregate storage bins can heat the aggregate rapidly. This also increases the moisture content in the aggregate. The material near the jets may be overheated and not enough heat may be transferred to some of the material which is away from the jets. Steam coils, hot-water coils, and hot-oil coils tend to heat the aggregate slowly because the aggregate is a poor heat conductor. One advantage of these methods is that they stabilize the moisture content of the aggregate. These coils should be located near the center of the bin in the line of the aggregate flow, which will also reduce heat loss. These coils can be used to heat stockpiles.

Electric heating units attached to the bottom of the bin can prevent freezing of the aggregates. Heaters to heat the area below the bin bottom also prevent freezing. This does not normally produce enough heat in the storage bin to thaw out the material for any great distance into the bin.

When placing concrete in hot weather, it is often necessary to cool some or all of the materials that are used in the concrete. Using the coldest water available, shading and sprinkling the aggregate stockpiles, and painting storage and handling equipment white to reduce heat absorption from the sun all help to reduce the additional cooling.

If the aggregate stockpiles are sprinkled, enough water must be used to keep the aggregate wet, and there must be more than enough water to replace the moisture evaporated. Usually from ⅛ to ½ in/h is required to keep the piles wet. A very coarse spray should be used since a fine spray increases relative humidity and decreases the water temperature, both of which reduce evaporation and cooling.

The use of ice in cooling concrete is widespread and very effective, but the amount that can be used is limited by the amount of moisture in the aggregates. About 80 percent of the added mix water may be ice, but some water is needed to dilute the admixtures.

Ice may be purchased in blocks and ground before use in the plant, or it may be produced in flake, cube, chunk, or crushed form. Flake ice melts faster in the mixer. Ice is usually elevated to the batching floor by a bucket elevator, stored in an insulated bin and delivered to the weigh hopper by a screw conveyor. A weigh hopper very similar to an aggregate weigh hopper may be used for ice, but the discharge gate must be very large.

Coarse aggregate may be cooled by a vacuum process which removes the air and vapor to maintain a pressure corresponding to the desired temperature, usually about 34°F. This may be done in the overhead storage bin or on the ground before the material is elevated to the storage bin.

Circulating cold air at about 38°F through coarse-aggregate bins can cool the aggregate to about 40°F. Aggregate smaller than about ¾ in has such high air resistance that it is not feasible to use this method.

Inundation of coarse aggregates in about 35°F water has been used to cool the aggregate to about 40°F. Some cold water (about one-third of the tank) is pumped into the tank, which is then filled with coarse aggregate. Cold water is pumped in the bottom of the tank and flows out over weirs near the top. The water is then rechilled and reused.

Sand can be cooled to 34°F by a vacuum cooling process similar to that described for coarse aggregate. Since the sand has a high resistance to the flow of vapor, it must either be cooled quickly in small amounts or circulated to expose the material surface to the vacuum.

Screw-conveyor coolers are available for cooling sand. In these, the screw conveyor is surrounded by a jacket with 35 to 40°F water or 20 to 30°F brine circulating through it. Sand may be cooled to about 50°F with water and to about 40 to 50°F with brine.

Cement and pozzolan cooling is difficult and expensive because both of these materials lose heat slowly. Coolers similar to the screw-conveyor cooler for sand are sometimes used.

BATCHING

After a trial mix has been established, it is important that the same proportions be maintained in every batch of concrete. The volumes of aggregate, cement, and water are the most important factors in producing consistent batches. For several reasons, it is not convenient to measure the amount of materials on a volumetric basis. The aggregates tend to bulk or change volume depending on the moisture content, method of handling, and the size and shape of the particles; the cement volume varies greatly with the amount of aeration in the cement; and the volume of water varies depending on the temperature. It is also not possible to measure as accurately and rapidly using volumetric methods as by weight. Since the specific gravity of most materials is relatively constant, proportioning by weight is the most accurate method of producing consistent concrete. Notable exceptions are cinders, expanded slag, and other lightweight materials where the specific gravity changes drastically.

Most specifications require that cement be weighed on a scale separate from the other materials for greater accuracy. Water is usually weighed on a scale separate from other materials, although often, for economy, it is measured volumetrically through a meter. The different sizes of aggregate may each be measured on a separate scale, or they may be weighed "cumulatively" on a single scale. For cumulative batching, the fill gate for one aggregate material is opened and the desired amount of that material is put into the weigh hopper. The gate is closed and then a second gate is opened and the desired amount of the second material is put into the weigh hopper. This procedure is followed until all the aggregate materials have been weighed.

Typical ranges of material quantities useful for establishing batching-equipment capacities are shown in Tables C1-12 through C1-15. These are based on a large number of batches from many locations. Two formulas out of 100 taken at random can be expected to be less than the second centile value and two can be expected to be more than the 98 centile value. Half of a large number of formulas can be expected to be above and half below the median. Slag aggregates have been excluded but would be about 90 percent of the aggregate weights shown.

TABLE C1-12 Typical Composition of Highway and Airport Paving Concrete
(*Values in lb/yd³ of resultant concrete based on saturated surface-dry aggregates*)

	Lowest observed value	Second centile	Median	Ninety-eighth centile	Highest observed value
Paving mixes, all mixes:					
Cement	385	442	557	674	677
Water	157	150	231	312	316
Sand	975	955	1185	1415	1580
Total coarse aggregate	1660	1790	2030	2300	2380
Total aggregate	3020	3040	3230	3500	3560
Wet concrete	3906	3896	4018	3140	4142
Paving mixes, 2 aggregates:					
Sand	1066	1043	1180	1359	1342
Coarse	1660	1775	2013	2193	2163
Paving mixes, 3 aggregates:					
Sand	975	950	1190	1430	1580
Coarse, smaller quantity	360	520	930	1220	1190
Coarse, larger quantity	956	910	1160	1410	1740
Coarse, smaller aggregate	360	520	1000	1300	1274
Coarse, larger aggregate	683	750	1100	1410	1740
Paving mixes, 4 aggregates:					
Sand	986	—	—	—	1311
Coarse, smallest quantity	143	—	—	—	528
Coarse, middle quantity	733	—	—	—	992
Coarse, largest quantity	850	—	—	—	1203

Aggregate

Aggregates are weighed in a container which is usually suspended from a scale system and fed by a fill gate or fill valve. This fill gate may be either a single- or double-clam gate. The weigh hopper must be of sufficient capacity to hold the material to be batched so that it does not touch the fill gate. The volume of the hopper should be at least 38 ft³/yd³ of rated capacity. A typical fill-gate size is 18 by 18 in, with about 10 by 17 in net opening. This allows about 2.2 cfs for coarse gravel and about 3.6 cfs for fine sand. Considerably larger fill gates are often used for larger aggregate material up to a maximum of 3, 4, or even 6 or 8 in. Larger fill gates are also commonly used on high-speed plants where a batcher of more than 6-yd³ capacity is used. If the gate opening is too large and the material is flowing too fast, the gate opening must be restricted by welding a bar across the gate or by an adjustment on the air cylinder. For special high-speed production, two gates may be used to feed one aggregate material. Unless special care is taken when this is done, the accuracy of weighing may suffer.

Cement

Cement and cementitious materials are particularly difficult to control. When they are well aerated they flow very rapidly and will spurt out through a small pinhole through which water will barely pass. If they are dead (have low air content), it is almost necessary to pull the material from the storage bin. This creates many problems in the accurate measurement of these materials. Obviously, the storage bin and silo must have no holes for the cement to leak from and must be watertight to prevent water from going into the bin or silo.

The hoppers may be either conical or rectangular and should have 50 to 60° net bottom slopes. The volume of the weigh hopper should be 9 ft^3/yd^3 of rated batch capacity plus 3 ft^3 to allow for fluffing and variations in the material. The dust seal between the fill gate and the weigh hopper should be flexible or arranged in some manner so that it will not affect the accuracy of weighing. There must also be an air vent in the weigh hopper to prevent the buildup of air pressure which would also affect the scale system. Normally, a vibrator is attached to the weigh hopper to ensure that all cement is discharged from the weigh hopper. If cement is held in the weigh hopper for more than a minute or so it will probably lose air and will be extremely difficult to discharge.

The cement fill gate may be a rotary vane feeder, a rotary plug valve, or a screw conveyor. The discharge from the weigh hopper may be any of the above types or a conical valve, which is an inverted cone which pulls up into the bottom of the weigh hopper to close the gate opening. Another discharge valve in common use is a flexible tube which is squeezed together to close the discharge opening. An opening is normally provided in the cement weigh hopper to allow an overload to be corrected, but these overload ports are very rarely used in practice. The flow of cement may be controlled by adjusting the pressure of aeration into the storage bin or silo or by adjusting the amount of gate opening.

Water and Admixtures

Water and liquid admixtures are weighed in hoppers of light-gauge corrosion-resisting alloy or galvanized steel. A quick-acting valve should be used to feed the weigh hopper. Many admixtures are viscous and sticky, and steeper sides on the weigh hopper are necessary for better discharge of the material. The admixture hopper often has some provisions for running water through the valves to clean them if one end of the valve is open to the air. No provision is normally made to remove overload from these weight hoppers since the hopper top is open. Water and admixtures are often measured volumetrically in hoppers, tanks, and sight glasses. Since the volume varies with temperature, it is necessary to correct for this difference in volume when there is a considerable difference in temperature, such as when hot water is measured volumetrically. Neglecting this temperature difference in volume results in a 1.6 percent error in the amount of water if 140°F water is used and if the volume was calculated based on 60°F. If water is weighed, this correction does not need to be made.

Water meters are commonly used and within certain ranges are generally accurate within ±1 percent. The repeatability is usually better than the accuracy of these meters. If hot water is to be used in a meter, then a hot-water disk must be furnished. Cold water can be used through a hot-water meter with somewhat less accuracy, but hot water should never be used through a cold-water meter.

A wide variety of admixtures are commonly used with concrete. The quantities of these materials may vary from as little as ¼ oz to as much as 2 qt per sack of cement. The extremely small dosage rates provide problems in accurate dispensing. Some of the admixtures are not compatible with each other, so it may be necessary to add one to the water batcher, one to the sand, and one or more directly into the mixer or mixer truck. It may be possible to dispense them in sequence, so that one goes into the waterline and when it is finished another may be put into the waterline.

TABLE C1-13 Typical Composition of Commercial Ready-Mix Concrete
(Values in lb/yd³ of resultant concrete based on saturated surface-dry aggregates)

	Lowest observed value	Second centile	Median	Ninety-eighth centile	Highest observed value
All mixes except pavement and lightweight mixes:					
Cement	188*	330	552	785	877
Pozzolan	0	—	—	—	75
Cement + pozzolan	230	330	522	785	877
Water	210	209	269	372	383
Sand	985	1019	1342	1596	1670
Total coarse aggregate	1454	1524	1888	2150	2267
Total aggregate	2755	2990	3233	3410	3442
Wet-concrete weight	3809	3845	4011	4148	4180
1½ in max, 2 aggregates:					
Cement	370	352	490	655	658
Water	210	208	246	302	292
Sand	1018	950	1310	1550	1550
Coarse aggregate	1780	1750	1950	2150	2143
Total aggregate	3017	3000	3265	3450	3442
Wet-concrete weight	3844	3820	3990	4165	4165
1½ in max, 3 aggregates:					
Cement	380	426	533	640	750
Sand	985	960	1200	1440	1410
Fine aggregate	750	660	1010	1400	1362
Coarse aggregate	607	650	1010	1250	1155
Total coarse aggregate	1736	1690	2010	2300	2267
Total aggregate	3040	3000	3220	3440	3410
1½ in max, 4 aggregates:					
Cement	432	—	—	—	658
Water	321	—	—	—	339
Sand	1122	—	—	—	1311
Fine aggregate	201	—	—	—	202
Medium aggregate	1167	—	—	—	1170
Coarse aggregate	644	—	—	—	646
Total coarse aggregate	2012	—	—	—	2018
Total aggregate	3134	—	—	—	3329
Wet-concrete weight	4086	—	—	—	4113
¾ to 1 in max, 2 aggregates:					
Cement	188*	305	505	705	750
Water	225	225	270	345	347
Sand	1040	1080	1360	1630	1670
Coarse aggregate	1660	1670	1860	2140	2170
Total aggregate	3021	3040	3235	3430	3440
Wet-concrete weight	3809	3860	4005	4150	4131
⅜ to ⅝ in max, 2 aggregates:					
Cement	393	360	585	910	877
Water	267	250	325	400	383
Sand	1000	1210	1430	1650	1625
Coarse aggregate	1454	1400	1680	1920	1885
Total aggregate	2755	2710	3160	3320	3300
Wet-concrete weight	3960	3970	4020	4065	4180

TABLE C1-13 Typical Composition of Commercial Ready-Mix Concrete *(Continued)*

	Lowest observed value	Second centile	Median	Ninety-eighth centile	Highest observed value
Lightweight insulating or fill concrete:					
Cement	318	300	530	770	700
Water	401	360	460	560	509
Lightweight aggregate	214	180	460	2350	1900
Wet-concrete weight	989	850	1600	3050	2265
Lightweight structural concrete:					
Cement	461	415	590	900	852
Water	250	220	365	510	473
Sand	250	100	1200	1410	1380
Lightweight aggregate	710	640	850	1680	1530
Total aggregate	1592	1450	2030	2270	2260
Wet-concrete weight	2410	2310	3000	3340	3274

* Plus 45-lb pozzolan.

TABLE C1-14 Typical Concrete Mixes for Specific Uses (lb/yd^3)

Use	Cement or cement plus pozzolan	Water	Sand	Total coarse aggregate	Total aggregate	Wet-concrete weight
Grout	590–2360	318–730	316–2700	—	316–2700	3406–3645
Pumped concrete	500–580	280	1555	1485	3040	3820–3900
Nuclear shielding	564	300	900	3860	4760	5624
Prestressed concrete	658–752	267–287	1125–1200	1920	3045–3120	4045–4064
Lightweight prestressed concrete	846	459	180	1293	1473	2780
Regular concrete block	455	177	1425–1950	1425–1950	3375	4007
Lightweight concrete block	340	280–560	—	—	1750–2200	2490–2940
Concrete pipe	640–830	250–420	1120–2230	985–1860	2960–3330	4115–4315
Porous concrete	596	213	—	2835	2835	3644

One of the simplest, least expensive, but least satisfactory methods of dispensing admixture is by using an adjustable orifice which is set so that a specific amount of admixture flows through the orifice in a given length of time. A timer is then set for the number of seconds required for the proper amount of admixture. This method gives no assurance that any admixture flows through the orifice. For example, if the supply tank is empty or if the orifice is clogged up, then no admixture will flow. The height of admixture in the storage tank over the orifice will vary to some extent and this will cause a difference in the flow rate of the admixture, thus producing an error in the amount put into the concrete.

A closed chamber with positive displacement may be calibrated and used to insert admixtures. With this method, admixture is forced into a section of the chamber by an air cylinder. The valve is closed, the air cylinder is operated, and the admixture is forced out into the waterline. This method is superior to the timer but still

TABLE C1-15 Typical Composition of Mass Concrete

[Weights in lb/yd³ of resultant concrete based on saturated surface-dry aggregates. w/c or w/(c + p) indicates water–cement ratio or water cement + pozzolan ratio by weight]

	Lowest observed value	Second centile	Median	Ninety-eighth centile	Highest observed value
Grout mixes:					
Cement	590	—	810	—	1020
Pozzolan	0	—	—	—	260
Cement + pozzolan	590	—	—	—	1020
Water	318	—	325	—	550
w/(c + p), %	32	—	39	—	78
Sand	2065	—	2350	—	2700
Wet weight	3465	—	3600	—	3645
¾-in max mixes:					
Cement	318	410	610	740	711
Pozzolan	0	—	—	—	105
Cement + pozzolan	423	410	610	740	711
Water	225	270	307	355	342
w/(c + p), %	45	44	51	63	59
Sand	991	970	1255	1445	1450
No. 4 to ¾-in. aggregate	1355	1685	1775	1835	2080
Total aggregate	2801	2765	3025	3155	3280
Wet-concrete weight	3794	3790	3920	4090	4043
1½-in max mixes:					
Cement	282	375	590	805	765
Pozzolan	0	—	—	—	100
Cement + pozzolan	374	375	590	885	852
Water	163	160	270	310	310
w/(c + p), %	25	20	45	64	60
Sand	825	795	990	1180	1180
No. 4 to ¾-in. aggregate	830	810	1050	1210	1576
¾- to 1½-in aggregate	444	990	1110	1330	1372
Total coarse aggregate	1844	1960	2180	2330	2300
Total aggregate	2996	2990	3145	3300	3397
Wet-concrete weight	3823	3820	3990	4190	4277
3-in max interior mixes:					
Cement	188	160	280	550	500
Pozzolan	0	—	—	—	89
Cement + pozzolan	255	240	320	550	500
Water	152	140	185	290	252
w/(c + p), %	50	49	57	79	74
Sand	748	680	875	1050	997
No. 4 to ¾-in aggregate	642	590	800	1010	938
¾- to 1½-in aggregate	745	685	830	980	927
1½- to 3-in aggregate	750	840	1010	1160	1115
Total coarse aggregate	2490	2430	2660	2870	2769
Total aggregate	3330	3250	3540	3850	3766
Wet concrete	3980	3930	4090	4250	4193

TABLE C1-15 Typical Composition of Mass Concrete *(Continued)*

	Lowest observed value	Second centile	Median	Ninety-eighth centile	Highest observed value
3-in max exterior mixes:					
Cement	245	250	400	550	625
Pozzolan	0	—	—	—	82
Cement + pozzolan	286	275	425	575	625
Water	160	145	210	280	266
w/(c + p), %	42	41	49	64	61
Sand	685	650	845	1140	1061
No. 4 to ¾-in aggregate	395	300	760	980	910
¾- to 1½-in aggregate	675	650	840	980	927
1½- to 3-in aggregate	750	700	1030	1450	1370
Total coarse aggregate	2445	2420	2610	2920	2884
Total aggregate	3210	3140	3470	3840	3750
Wet-concrete weight	3980	3940	4100	4260	4235
6- to 8-in max interior mixes:					
Cement	148	150	205	400	400
Pozzolan	0	0	56	110	118
Cement + pozzolan	198	200	255	410	400
Water	94	108	165	305	309
w/(c + p), %	36	40	64	85	86
Sand	714	705	825	1210	1170
No. 4 to ¾-in aggregate	410	415	565	725	721
¾- to 1½-in aggregate	333	380	580	770	913
1½- to 3-in aggregate	675	650	815	1140	1096
3- to 6- or 8-in aggregate	433	700	950	1150	1150
Total coarse aggregate	2560	2540	2900	3260	3211
Total aggregate	3542	3520	3750	3980	3925
Wet-concrete weight	3914	3910	4150	4390	4306
6- to 8-in max exterior mixes:					
Cement	247	247	307	441	441
Pozzolan	0	0	50	105	106
Cement + pozzolan	247	247	368	441	441
Water	130	132	175	232	226
w/(c + p), %	41	41	49	56	56
Sand	710	680	790	1030	1022
No. 4 to ¾-in aggregate	280	260	500	650	679
¾- to 1½-in aggregate	201	460	575	670	675
1½- to 3-in aggregate	635	620	790	1060	1069
3- to 6- or 8-in aggregate	764	760	950	1290	1230
Total coarse aggregate	2500	2420	2885	3090	3055
Total aggregate	3440	3410	3630	3970	3913
Wet-concrete weight	4036	4040	4185	4310	4290

leaves the possibility of an empty tank or a valve which is stuck. No air is allowed into this chamber any time so there is no real danger that air will be mixed with the admixture to dispense an erroneous quantity of admixture.

When meters are used to dispense admixtures, many of these problems are eliminated (Fig. C1-16). These meters usually have an electrical impulse device in the head which is calibrated from ½ oz to 2 qt per pulse. These pulses are counted by a preset counter in the control system. When the proper number of pulses has been counted, the meter is stopped. If no admixture goes through the meter, the control system can refuse to discharge the batch or can give the operator an indication that no admixture was received.

Sight glasses can be equipped with a pulsing device (Fig. C1-17) which may be calibrated from ½ oz to 1 qt per pulse. Interlocks may be provided to indicate zero level, to prevent the bottom from being discharged unless the quantity is within tolerance, and to prevent overfilling the bottle.

Weigh hoppers are sometimes used for measuring admixtures. This is by far the most foolproof, most positive, and most accurate system, but it is also the most expensive. It requires a scale system and a complete set of controls very similar to those used for cement, aggregate, or water. There may be almost any number of admixtures used in a given plant and it is not feasible to weigh all these materials on one set of scales. The quantities vary from 1 oz to 20 qt, and a scale system for admixtures should not be used below about 10 percent of its capacity if it is to give the accurate results needed for properly dispensing admixtures. The number of scales needed depends on the range of quantities of the admixtures used.

Scales

Almost all the scales presently used in concrete batching plants are the lever type which are similar to the old steelyard scale. These use the principle of pivoting a lever near one end so that a relatively small weight can balance a considerably larger weight. By combining several levers together in one system in this fashion, it is possible to have a large-capacity weigh hopper suspended from one end of the scale system balanced by an indicated device at the other end of the scale system which requires a load of only 5 to 25 lb.

The hopper is usually suspended by four hanger rods from two main levers (Fig. C1-18). One main lever consists of two simple levers connected by a torque tube to transmit the load from the short lever in the back to the longer lever in the front. The

FIGURE C1-16 Meter for dispensing admixture. *(Protex Industries, Inc.)*

CONVENTIONAL CONCRETE

FIGURE C1-17 Sight glass with pulsing unit for dispensing admixture. *(Protex Industries, Inc.)*

two main levers are connected together by a gathering shackle, and the load from both main levers is transferred to the first extension lever. Since the load from both of the main levers is combined before it is transferred, the two main levers act as a single simple lever. More levers may be used to reduce the load as required by the final indicating device.

The ratio of a lever is the ratio of the weight on the load knife-edge and the weight on the power knife-edge when the lever is balanced. This is the ratio of the distance from the load-knife edge to the fulcrum knife-edge and the distance from the power knife-edge to the fulcrum knife-edge. The ratio is normally expressed as 2:1. 3:1, or 6⅔:1.

The ratio of a scale lever system is the product of the ratios of the individual levers. If the ratio of the main levers is 5:1, the ratio of the first extension lever is 4:1, and the ratio of the second extension lever is 4:1, then the ration of the lever system is $5 \times 4 \times 4$ or 80:1. If the system is balanced and 800 lb is put in the hopper, 10 lb must be added to the power knife-edge to return the system to balance.

The levers are constructed with the knife-edges positioned as accurately as possible. An error of 0.001 in in the location of a single knife-edge may cause an error that would result in the condemnation of the scale system. For this reason there must be some adjustment built into the lever system. A nose iron at the junction point of the main levers provides for changing the ratio of the main levers to compensate for any errors in the actual location of the knife-edges throughout the scale lever system. Adjustable knife-edges are also used to adjust the ratio of the scale lever system. These are mounted in such a way that they can be rotated in their holes to change the location of one knife-edge with respect to the other knife-edges in that lever.

Either a beam scale or a dial scale may be used as an indicating device with the lever system. A dial scale consists of a pendulum weight, a rack, a pinion, a dial pointer, and a dial chart. The pendulum weight produces a counterreaction to the lever system and drives the dial pointer by means of the rack and pinion. The dial scales provide a continuous indication of weight from zero to full-scale capacity, allow quick changes of material quantities, are easier to adapt to graphic or digital recording, and are easier to use with electronic controls.

Controls

Semiautomatic batching is achieved when the operator can preset a weight on the scale and actuate the charging device or fill valve by pulling a lever or pushing a but-

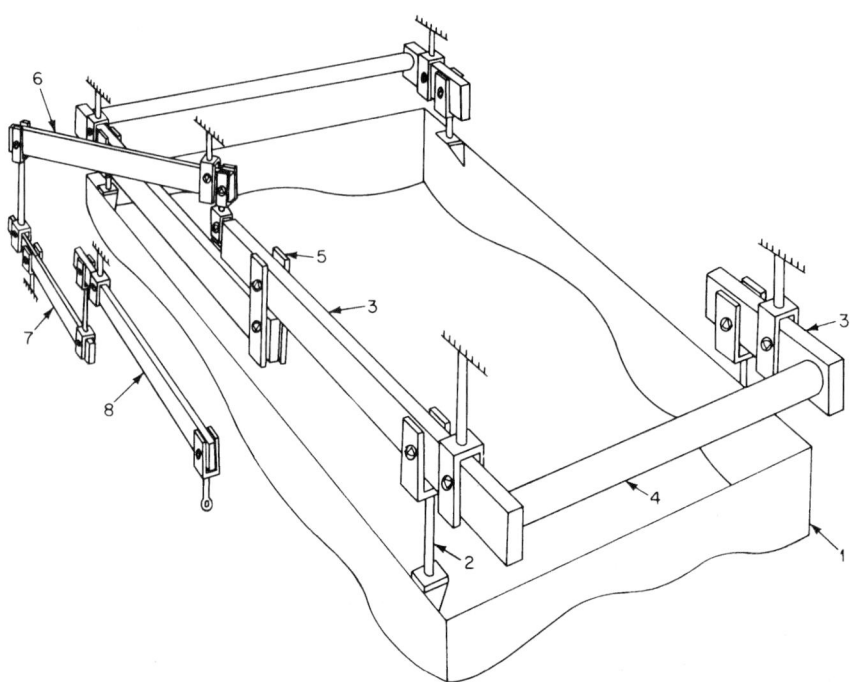

FIGURE C1-18 Typical batcher scale lever system. (1) Weigh hopper, (2) hanger rod, (3) main lever, (4) torque tube, (5) gathering shackle, (6) first extension lever, (7) second extension lever, (8) third extension lever.

ton. When the correct weight is in the weigh hopper, the fill gate will close. This system, of course, frees the operator from the need to watch the batcher closely, but it also introduces a possibility of error when the rate of material flow changes because of a gradation change, a change in moisture content, or any other factor. In these cases, the cutoff points should be adjusted to produce the correct weights with the new set of conditions. A semiautomatic control system may or may not have interlocks to prevent the discharge of the batcher if the materials are not within the desired tolerances.

Many types and styles of automatic control systems are available. Some of the systems may be used with either beam scales or dial scales. Some of the systems may be arranged with mix selection and others are designed to control only one mix design. The basic cutoff element which tells the system that the scale has reached the correct weight may consist of microswitches, reed switches, photoelectric cells, solenoid cutoff units, or electronic devices.

Automatic controls normally produce more consistent and more reliable results than the other types. The operator must select the amount of material desired and push on a button to start the cycle. The control system automatically batches each material, checks to see that it is within the required tolerances, and tells the operator the batch may be discharged.

Since the automatic system is designed to relieve the operator of some of the decisions, it is important that some interlocks be built into the system to be sure an incorrect batch is not discharged. There should be an interlock to prevent the start of a new batch unless the scales are at zero. The fill gate should not be able to open unless the discharge gate is completely closed. The batcher should not be able to discharge unless all materials are within the required tolerance, the fill gates are closed, the mixer is empty, and the mixer is in the correct condition to receive a new batch. The mixer should not be able to discharge until the mixing time has elapsed.

Some provision must be made to take care of the column of material in the air just before the final weight is reached. There are at least two methods of doing this. In one case, the aggregate gate is alternately opened and closed, or jogged, when the weight approaches the final cutoff point so that material comes into the weigh hopper in short spurts. This reduces the column of material in the air so that the error is negligible. With cement, the gate may be jogged or the opening may be cut down so that the cement feeds in slower.

One of the primary advantages of an electronic control system is that a switch may be incorporated which can be set to the number of cubic yards desired. The weights can be programmed into this system as needed for a 1-yd^3 batch, and the batch-size selector switch may be used to select the number of cubic yards desired for a particular batch. This is obviously of special interest and benefit to a ready-mix producer. For a paving contractor who normally uses the same size batch and the same mix design for an entire day or week, it would have relatively little value. Quite frequently, the batch-size selector switch is calibrated in ¼-yd^3 increments. This switch may also be set to give repeat batches by calibrating the switch to call for more cubic yards than can be batched at one time. For example, if a 4-yd^3 batcher is used, the switch can have settings up to 12 yd^3. If the switch is set for 6 yd^3, two batches of 3 yd^3 each would automatically be produced. If the switch is set for 12 yd^3, three batches of 4 yd^3 each would automatically be produced.

Since the moisture content usually varies in the fine aggregate, it is convenient to have some fairly simple method of adjusting both the fine aggregate and water weights to compensate for the amount of moisture in the fine aggregate. In most electronic control systems, a switch may be installed to make this adjustment. It is relatively simple to design a circuit to adjust the sand weight, since the percent moisture is a direct percentage of the sand weight. Adjusting the amount of water to be called for is more difficult because the number of pounds to be corrected is not a direct percentage of the water weight. The control system must determine the number of pounds of water to be deducted, convert this number of pounds into some electrical signal, and then subtract this signal from the amount of water to be batched. This problem is complicated by the fact that the actual amount of water to be compensated is usually a considerable portion of the actual water weight.

A control which can be and frequently is added to the electronic type control system is a *slump* control, or more accurately a *water trim* control. This merely gives the operator a quick and easy method of changing the amount of water called for without changing the preset amount. Another control is a *harshness* control which increases the amount of sand and decreases the amount of one aggregate or increases the amount of one aggregate and decreases the amount of sand to control the harshness of the mix while retaining the same total weight of aggregate plus sand.

The electronic control systems are especially adaptable to remote operation of the complete plant. Since it is not necessary to go to the batcher to change the amount of each material required or the number of cubic yards, the batcher may indeed be several hundred feet from the control system, so that the operator is not bothered by the harsh sounds of the gates opening and closing and the aggregate hitting the steel weigh hoppers. This requires that the control systems be at least as fully interlocked as the normal automatic systems, if not more so. The operator also should have some indicating lights or some form of remote dial indicator to tell what is going on at the batcher.

Tolerances

Since it is not reasonable to expect absolutely perfect batching from any plant or any system, some tolerance must be established to determine the maximum allowable errors permitted during the batching operation. The delivery tolerance for cement and water varies from ±0.5 to ±1 percent of the net weight desired. The most common tolerance is ±1 percent. The normal tolerance for admixture is ±3 percent of the net amount desired, but the tolerance is often stated as ±3 percent or ±1 oz, whichever is greater. This creates a very dangerous possibility when extremely small dosage rates and small-size batches are used. If the dosage rate is ¼ oz per sack of cement, if a four-bag mix is used, and if only 1 yd^3 is being batched, then the total amount of admixture desired is 1 oz. With a tolerance of ±1 oz, all the admixture could be omitted from the batch or twice the required amount could be put into the batch and it would still be within the allowable tolerances. One solution to this problem would be to specify a minimum dosage rate per bag of cement based on the accuracy and repeatability of available dispensing equipment. Another solution would be to have the specifications rigidly enforced as ±3 percent of the required amount and force someone to design equipment capable of dispensing admixtures within ±0.03 oz. Note that with a 4-yd^3 batch under the same conditions, the allowable error on admixture is ±1 oz in a total of 4 oz, which is ±25 percent of the desired amount.

The delivery tolerance for aggregate varies for ±0.5 to ±3 percent. The most common tolerance is ±2 percent for fine aggregate and ±3 percent for the larger aggregate from 3 to 8 in. When a separate scale is used for each size aggregate, the tolerance is a percentage of the net weight desired.

There are many types of tolerances which apply to a concrete batching plant. Since the different tolerances, their interpretation, and their enforcement determine to a large extent the original cost of concrete batching-plant control equipment and the scales to be used, a realistic approach to the entire situation should be taken. The most reasonable way to do this is to compare all the tolerances to see the relative effect that each has on the ultimate quality of the concrete. Any other factors which will affect the quality of the concrete should also be considered at the same time.

The ultimate use of the batching plant, whether it is to be used as a paving plant, ready-mix plant, mass-concrete plant, or concrete-products plant, must be taken into consideration in this evaluation of tolerances. For example, a paving plant should normally be expected to produce batches which are near the capacity of the plant. One exception to this would be when the load limit for the hauling unit requires a lower maximum batch than the plant is designed to produce. Most mass-concrete plants usually produce the maximum-size batch possible. With some exceptions, a concrete-products plant normally would use the maximum-size batch. Ready-mix plants would necessarily produce batches ranging from the maximum size possible down to the minimum batch which can be produced with sufficient accuracy to meet the applicable specifications. Thus, the ready-mix plant is the only one which should be very much concerned about producing a very small batch.

Assume an 8-yd^3 batcher with a 6000-lb cement dial scale with 5-lb gradations, a 30,000-lb aggregate dial scale with 25-lb gradations for weighing aggregate cumulatively, and a 2500-lb water dial scale with 2.5-lb gradations. The tolerances for test weights vary considerably depending on which specifications are to be used, but assume that the tolerance is ±0.01 percent of the net weight of each test weight. This is equivalent to 0.005 lb, of 0.08 oz, or about 22.7 g for a standard 50-lb test weight. The static test of the scales using the calibrated test weights is quite commonly required to be within ±0.1 percent of the scale capacity, which is usually ±1 scale gradation. The actual delivery tolerances which are most common are: cement ±1 percent of the batch weight, aggregate, ±2 percent of the batch weight; and water, ±1 percent of the batch weight. Moisture compensation is another factor to be considered. Usually ±0.5 percent of the sand batch weight is used as a tolerance for the amount of moisture present in the sand. These tolerances are listed in Table C1-16.

Figure C1-19 shows the relative effect of these tolerances for cement on an 8-yd^3 batch of concrete assuming 500 lb/yd^3 or 4000 lb for the entire 8-yd^3 batch. One additional item has been added to this graph. Since the strength of concrete produced varies due to the physical and chemical differences in the cement used in the concrete, a factor of 5 percent variation is shown here. Obviously there is not a 200-lb variation as shown in the graph, but the 5 percent variation of strength is what would be expected if the cement were consistent but if the amount of cement varied by ±5 percent in the batch. The 5 percent figure used here is conservative and in many cases will be considerably more than 5 percent.

Figure C1-20 shows the effect of the various tolerances on an 8-yd^3 batch aggregate assuming 1000 lb/yd^3 or 8000 lb for the complete batch. Sand is shown here so that we may see the effect of the error in moisture compensation. The ±0.5 percent error in determination of moisture content is shown as 40 lb, and of course this assumes an exact compensation, so that there is no allowance made for errors in actual compensation. The entire error shown is due to the inaccuracy of moisture-content measurement. Since gradation of the aggregate obviously affects many of

TABLE C1-16 Typical Tolerances for Batch Operations

Item	Tolerance
Test weight	±0.01% of net weight
Scale check	±0.1% of scale capacity (or ±1 scale graduation)
Moisture compensation	±0.5% of sand batch weight
Delivery tolerances:	
Cement	±1.0% of batch weight
Aggregate	±2.0% of batch weight
Water	±1.0% of batch weight

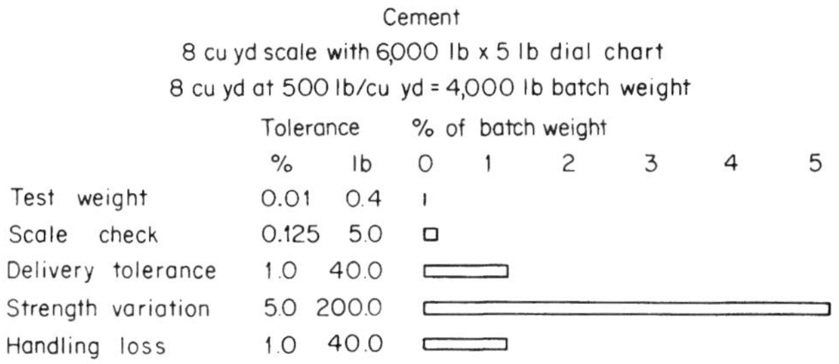

```
                        Cement
         8 cu yd scale with 6,000 lb x 5 lb dial chart
         8 cu yd at 500 lb/cu yd = 4,000 lb batch weight
                    Tolerance    % of batch weight
                    %      lb    0    1    2    3    4    5
Test weight         0.01   0.4   I
Scale check         0.125  5.0   □
Delivery tolerance  1.0    40.0  ⊏══⊐
Strength variation  5.0    200.0 ⊏═══════════════⊐
Handling loss       1.0    40.0  ⊏══⊐
```

FIGURE C1-19 Typical tolerances for an 8-yd^3 batch of cement.

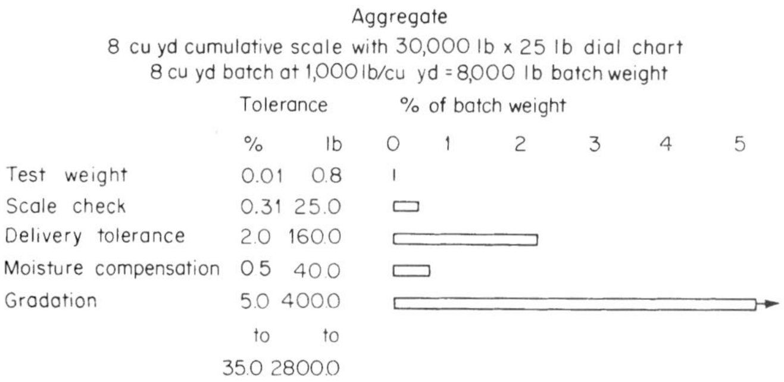

Aggregate
8 cu yd cumulative scale with 30,000 lb x 25 lb dial chart
8 cu yd batch at 1,000 lb/cu yd = 8,000 lb batch weight

	Tolerance		% of batch weight
	%	lb	0 1 2 3 4 5
Test weight	0.01	0.8	
Scale check	0.31	25.0	
Delivery tolerance	2.0	160.0	
Moisture compensation	0.5	40.0	
Gradation	5.0 to 35.0	400.0 to 2800.0	

FIGURE C1-20 Typical tolerances for an 8-yd^3 batch of sand in a cumulative batcher.

the qualities of concrete and since there is and must be some latitude in the actual gradation and in the consistency of gradation, a ±5 to ±35 percent variation due to gradation changes has been indicated. This would correspond to a change of from 400 to 2800 lb of sand in the batch. It would normally be compensated for by 400 to 2800 lb of some different size of material, thus keeping the yield essentially the same.

Figure C1-21 shows the effect of the various tolerances on a batch of water, assuming 250 lb/yd^3 or 2000 lb total amount of water in the batch. The 40-lb error in moisture compensation is due entirely to the error in determining the amount of moisture in the sand shown in the graph for the aggregate. This again assumes a perfect actual compensation for the amount of moisture which is in the sand. If we assume that there is about 5 percent moisture in the sand, then the amount of water to be deducted from the water batch weight would be 5 percent of 1000 lb or 50 lb/yd^3. The actual batch weight of water would be 200 lb/yd^3 for a total of 1600 lb of water for the 8-yd batch. The ±1 percent delivery tolerance is almost always applied to the actual amount of water to be batched rather than to the total amount of water to be included in the batch. Therefore we have ±1 percent of 1600 lb or ±16 lb as a delivery tolerance. As the graph shows, however, the 0.5 percent error in determining the actual moisture content of the sand produces an error of 40 lb in the water batch weight, which is 2.5 percent of the 1600 lb of water to be batched. Another item of concern has been added at the end of this graph. In ready-mix plants it is extremely difficult for the truck driver to remove all wash water from the truck, and it is even more difficult for a driver to determine exactly how much water still is in the truck. A conservative estimate of about 2½ gal of water has been shown here as a variation from the amount of wash water presumed to be in the truck as opposed to the actual amount of wash water in the truck. This would give an error of about 20 lb or 1.25 percent of the 1600 lb of water to be batched. Since the errors introduced by the moisture content in the sand and by the truck wash water both exceed the delivery tolerance, and since these three errors greatly exceed the scale-check tolerance—and even this is considerably more than the test-weight tolerance—it is difficult to understand why so much emphasis is placed many times on these relatively smaller errors while the larger ones are, many times, completely ignored or overlooked.

The situation becomes even worse than when the moisture content increases, and Fig. C1-22 indicates the situation we have when the moisture content of the sand is

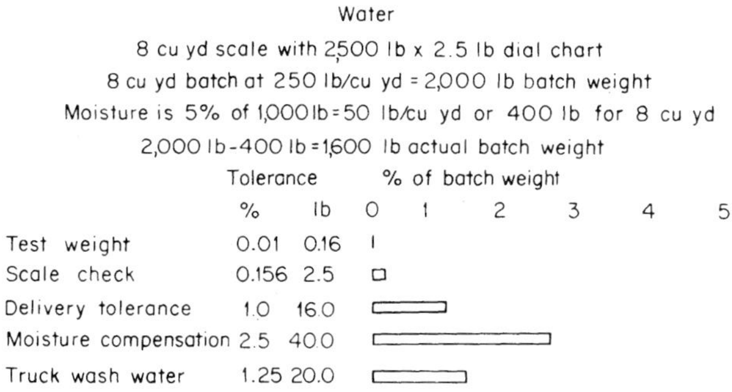

```
                              Water
             8 cu yd scale with 2,500 lb x 2.5 lb dial chart
             8 cu yd batch at 250 lb/cu yd = 2,000 lb batch weight
             Moisture is 5% of 1,000 lb = 50 lb/cu yd or 400 lb for 8 cu yd
                  2,000 lb - 400 lb = 1,600 lb actual batch weight
                            Tolerance      % of batch weight
                             %     lb    0    1    2    3    4    5
        Test weight         0.01  0.16   |
        Scale check        0.156   2.5   □
        Delivery tolerance   1.0  16.0   ▭
        Moisture compensation 2.5 40.0   ▭▭▭
        Truck wash water    1.25  20.0   ▭▭
```

FIGURE C1-21 Typical tolerances for an 8-yd^3 batch of water with 5 percent moisture in the sand.

```
                              Water
             8 cu yd scale with 2,500 lb x 2.5 lb dial chart
             8 cu yd batch at 250 lb/cu yd = 2,000 lb batch weight
            Moisture is 10% of 1,000 lb = 100 lb/cu yd or 800 lb for 8 cu yd
                  2,000 lb - 800 lb = 1,200 lb actual batch weight
                            Tolerance      % of batch weight
                             %     lb    0    1    2    3    4    5
        Test weight          0.01  0.12   |
        Scale check         0.208   2.5   □
        Delivery tolerance    1.0  12.0   ▭
        Moisture compensation 3.33 40.0   ▭▭▭▭
        Truck wash water     1.67  20.0   ▭▭
```

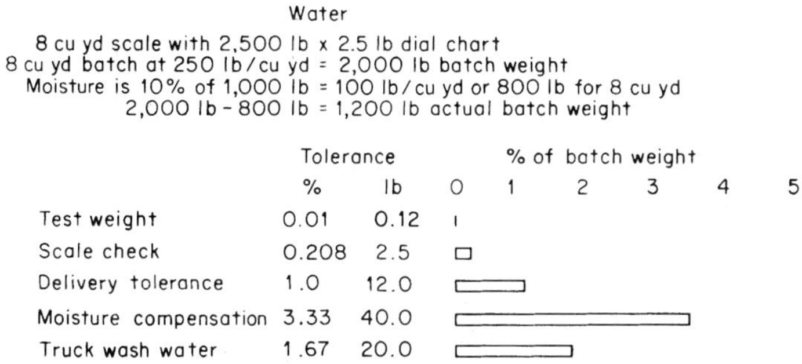

FIGURE C1-22 Typical tolerances for an 8-yd^3 batch of water with 10 percent moisture in the sand.

assumed to be 10 percent. The delivery tolerance is based on the actual amount of water to be batched and not on the total amount of water in the batch (which would include the moisture in the sand). The percentage errors due to the moisture compensation and the truck wash water therefore increase when the amount of water to be batched is decreased.

MIXING

A concrete mixer may be defined as a device or machine used to combine portland cement, water, aggregate, and other ingredients in a homogeneous mixture to produce concrete. The number, size, shape, angle, and arrangement or location of blades in a mixer are largely empirical. Mixers may be divided into three general types. Plant mixers are usually mounted as a stationary part of a concrete batching plant. Paving mixers are mounted on wheels or tracks; they are designed to mix as they move along the roadway so that the concrete can be placed as soon as it is mixed. Transit mixers or truck mixers are designed to mix while they are en route from the batching plant to the site where the concrete will be used.

Tilting Mixers

Tilting mixers have a rotating drum which discharges when the drum is tilted from 50 to 60° downward. The drum may either be horizontal or tilted slightly upward during charging and mixing operations. The inside of the drum is usually weld-coated or supplied with replaceable, abrasion-resistant liner plates.

One-opening tilting mixers with the mixing angle of the drum at 15° with the horizontal usually have a large opening for fast charge and fast discharge (Figs. C1-23 and C1-24). These mixers have been successfully used with almost every type of concrete and for almost every application for ready-mix concrete, mass concrete using large aggregate, cement base course, and pavement concrete.

Nontilting Mixers

Nontilting mixers have either a rotating drum or rotating blades in a stationary drum. These mixers may discharge through gates in the mixing compartment or by chutes that direct the concrete from the rotating drum.

Single-compartment two-opening nontilting mixers have a rotating drum. These mixers are normally charged through the rear opening by means of a chute which projects into the mixer. The mixing blades pick up the concrete and spill it over onto the concrete in the bottom. These blades direct the concrete toward the front of the mixer, where a chute is hinged so that it can be rotated into a position to discharge material as received from the blades.

FIGURE C1-23   One-opening tilting mixers, mixing angle 15° with horizontal. *(C. S. Johnson Operations, Road Division, Koehring Company.)*

FIGURE C1-24 Self-erecting one-opening tilting mixer, mixing angle 15° with horizontal, shown in travel position. *(C. S. Johnson Operations, Road Division, Koehring Company.)*

Vertical-shaft mixers (turbine, pan, or compulsory mixers) have a mixing compartment with rotating blades (Fig. C1-25). There may be one or more rotation vertical shafts with blades attached. In some mixers there are rotating blades and a rotating mixing compartment. The mixer has a high wear factor because of the rapid movement of concrete against the walls and against the blades or paddles. The power requirements are quite high compared to other mixers. These mixers are used for a wide variety of slump requirements, but excessive power is required for low slump and the capacity of the mixer may be reduced for very low slump applications. These mixers are most commonly used in concrete-products plants, but some have been successfully used in paving operations and ready-mix concrete plants. Concrete is discharged through one or more horizontal doors in the bottom of the mixing compartment. A mixer load indicator is normally used with these mixers to give an indication of the actual loading of the mixer motor and gears. This helps to alert if the operator is overloading the mixer. These mixers are sometimes very sensitive to sequence of feeding or charging of material. The aggregate and cement should be blended before reaching the mixer, or they should be fed into the mixer together. The water should be introduced rapidly with the cement and aggregate, and it should be placed well inside the mixer. If water dribbles down the side, the wall may be lubricated and the concrete will slide with the blades around the mixer.

Paving Mixers

Paving mixers are mounted on tracks and move as they are mixing (Fig. C1-26). There are usually two or three drums mounted in one single shell. The cement and aggregate are charged into the first drum from a skip on which the dry-batch trucks dump the material. The water and admixture are put into the mixer from containers mounted on the mixer. The water and admixtures are measured volumetrically at the mixer. The mixing time usually starts when the skip is at its top position. The material is mixed in the first drum and then transferred to the second drum. When all the material has been transferred from the first drum to the second, a new batch may be placed in the first drum. When the material has been mixed, it is discharged into a bucket which is supported by a boom on the mixer. This bucket is then moved out laterally and the concrete is dumped on the ground. A mixer timer with a

FIGURE C1-25 Vertical-shaft mixer. *(C. S. Johnson Operations, Road Division, Koehring Company.)*

mechanical interlock on the bucket is a part of the mixer. The batch size varies from 1.26 to 1.51 yd³.

Truck or Transmit Mixers

See Table C1-17. Truck mixers or transit mixers either mix or agitate concrete in a drum mounted on a moving truck or trailer (Fig. C1-27a). There are three types: inclined-axis revolving drum, horizontal-axis revolving drum, and open-top revolving blade or paddle mixers. The inclined-axis type is the most common. This mixer discharges by reversing the drum rotation so that the blades tend to screw the concrete back to the drum opening. Truck mixers may be used to completely mix the concrete. This may be done in the yard before the mixer goes to the job to place the concrete, it may be done at the job site, or it may be done en route to the job site. Truck mixers are also used to complete mixing when a central plant mixer is used to shrink-mix the concrete. When the truck mixers are used to transport central mixed concrete, they are usually operating as agitating units. Specifications allow the truck mixer to haul about 10 percent more shrink-mix or central-mix concrete than transit-mix concrete.

Mixing speed is not less than 4 and not more than 12 r/min of the drum or blades. Agitating speed is not less than 2 and not more than 6 r/min of the drum or blades. Drums are often operated at higher speeds while charging or discharging to handle the material faster. Concrete should be mixed between 70 and 100 revolutions at mixing speed after all ingredients are in the drum. The mixing water should also be

CONVENTIONAL CONCRETE   C1-63

FIGURE C1-26   Tri-batch paver. *(Koehring Division of Koehring Company.)*

TABLE C1-17   Capacities of Truck Mixers*
(In cubic yards)

Mixing capacity	Agitating capacity[†]
6	7¾
6½	8½
7	9¼
7½	9¾
8	10½
8½	11¼
9	12
10	13¼
11	14¾
12	16
13	17½
14	19
15	20¼
16	21¾

* These sizes are established by standards of the Truck Mixer Manufacturers Bureau, 900 Spring Street, Silver Spring, MD 20910.

[†] Unless otherwise restricted on the manufacturer's data plate on the mixer. Some mixers will agitate this amount only when the drum opening is closed.

FIGURE C1-27a Truck mixer.

in the drum before the mixing begins. If any water is added after the concrete has been mixed, the mixer should be rotated a minimum of 30 additional revolutions at mixing speed. Any additional rotation of the drum should be at agitating speed. The overall total number of revolutions should not exceed 250.

Figure C1-27b shows a new type of transit mix. The Elkin Mobil Mixer meters the dry batch at the job site, adding water through a precision measuring system. The concrete is mixed by an adjustable auger (9 ft 9 in, ½ in thick). The resulting concrete is fresh at the job site and there is no waste.

FIGURE C1-27b Truck mixer, 10 yd^3/60 yd/h. *(Elkin Mfg. Inc.)*

Mixing Time

One of the most important factors in the time necessary for proper mixing is the sequence of charging material into the mixer. If the materials are blended well as they enter the mixer, the mixing time may be substantially reduced. Many low-profile central-mix plants use a conveyor belt to charge the mixer, and the materials are well blended or premixed on the belt before they enter the mixers in most of these plants. The large, mass-concrete plants are arranged so that the material is batched in single material batchers which feed into a collecting hopper and from there to the mixer. The materials are blended in the collecting hopper before they enter the mixer.

Mixing times for tilting mixers are shown in Table C1-18. These mixing times are primarily for mass pour work. Tilting mixers used for highway paving projects normally required mixing times of 40 to 90 s. It is quite common to require 75-s mixing time without mixer performance tests and to reduce this time to 60 s or less if performance tests are made. In a ready-mix plant using a mixer for shrink mixing, 30 s is usually adequate. The turbine-type mixers which are charged adequately will mix in about 30 s, but sometimes 45 s is necessary to produce a consistent mix. The horizontal-shaft mixers used in concrete-products plants have a mixing time from 2 to 5 min. Mixing times for paving mixers range from 50 to 90 s, with 60 s being the most common. The transfer time between drums is usually considered part of the mixing time. Truck or transit mixers require from 8 to 15 min mixing time.

Overmixing should be avoided because it increases fines and thus requires more water to maintain consistency, drives out entrained air, and tends to increase the concrete temperature.

TABLE C1-18 Tilting-Mixer Production Rates—Mass Concrete*[6]

Nominal or rated mixer capacity, yd^3	Typical mixing time, min	Typical cycle time, min	Batches/h	yd^3/h
2	1½	2	**30.0**	**60.0**
3	2	2½	**24.0**	**72.0**
3½	2¼	2¾	21.8	76.4
4	2½	3	**20.0**	**80.0**
4½	2⅝	3⅛	19.2	**86.4**
	2¾	3¼	18.5	**83.1**
5	2¾	3¼	18.5	**92.3**
	3	3½	17.1	**85.7**
5½	2⅞	3⅜	**17.8**	**97.8**
	3¼	3¾	**16.0**	**88.0**
6	3	3½	17.1	**102.9**
	3½	4	15.0	**90.0**
7	3¼	3¾	**16.0**	112.0
	4	4½	13.3	93.3
8	3½	4	**15.0**	**120.0**
	4½	5	12.0	**96.0**

* Above production rates are for mixers with a batch equal to the nominal capacity shown. Mix times shown are typical for specification work in the absence of mixer-performance tests. Mixer sizes in boldface type are rated capacities of Plant Mixer Manufacturers Division standard mixers. They are guaranteed to mix rated capacity when slump is between 1½ and 3 in and aggregate size is not over 3 in.

PLACING

Concrete should always be described as being placed rather than poured. The expression "poured" comes from the early concrete work where it was necessary to have very wet and sloppy concrete in order to get it into the forms. This was prior to the introduction of high-frequency vibration for the proper placing of concrete.

It is very important to prevent separation of the coarse aggregate from the concrete during placing operations. This separation occurs primarily because the materials in concrete differ greatly in particle size and specific gravity. If the concrete is allowed to flow, the coarse aggregate separates from the mortar. If the concrete is confined laterally, the heavier materials tend to settle and the lighter materials tend to rise toward the top. Separation of the aggregate from the concrete occurs most frequently at the ends of chutes and conveyor belts, at hopper gates, at a discharge point from a mixer, and during transportation.

Concrete should be placed as near as possible to its final position and should not be allowed to flow horizontally or on a slope in forms. It should not be placed in separate piles and the piles then leveled and brought together. Concrete should not be placed in big piles then leveled and brought together. Concrete should not be placed in big piles and allowed to run or be worked over a long distance to its final position. Concrete should be placed into the face of previously placed concrete, not away from it. When concrete is placed in layers, each succeeding layer should be placed while the one below it is still plastic, and the top portion of the lower placement should be vibrated to ensure that the concrete is continuous and without separation.[6] It is important to avoid the entrapment of air within partially enclosed spaces which are to be filled with concrete. In wall and column placements, the slump should be reduced as the level of concrete rises in order to offset water gain which will weaken the upper portion of the concrete and make it less durable. The slump should be the minimum that can be vibrated well into the critically exposed portions of the work. When placing concrete in walls, curbs, and slabs, work should proceed from the corners and ends of the forms toward the center, rather than toward the corners and ends, thus avoiding accumulation of mortar and wetter concrete in those parts of structures where exposure is most severe.

The handling equipment and methods used should be selected because they can handle the mix design appropriate to the job at hand; it would be a mistake to design a mix to suit the placing methods to be used rather than the job to be done. The batching, mixing, transporting, and placing should be done at such a rate that the concrete placing may proceed without serious interruptions. Cold joints must be avoided.

Concrete should be placed as soon as possible after mixing. The actual amount of time allowable varies with conditions such as the weather and the ultimate use of the concrete. Forms should be clean, tight, adequately braced, and constructed of materials that will impart the desired texture to the finished concrete. Reinforcing steel should be clean and free of loose rust or mill scale at the time concrete is placed. Any coatings of hardened mortar should be removed from the steel.

Before beginning placement, the contractor should be certain that all transporting, placing, vibrating, finishing, and curing equipment is clean and in proper repair and that it is adequate and properly arranged so that placing may proceed without undue delays. Sufficient personnel should be available to handle all phases of the operation. Concrete placing should not be started when there is a probability of freezing temperatures unless adequate facilities for cold-weather protection have been provided.

Chutes

One of the most important problems in handling or transporting concrete is the segregation or separation of the coarse aggregate from the mortar. Concrete should drop vertically and in the clear regardless of the type of equipment used. If drop chutes or elephant trunks are used, the upper sections may be at an angle to facilitate placing of the concrete, but the lower section should always be vertical. Concrete should not be dropped through reinforcement steel or other objects which tend to separate it, nor should it be directed against the forms.

Unless chutes are constructed and used properly, segregation and loss of slump will result. The chutes must be deep enough to handle concrete in such a way that it will move smoothly, and the chutes' slope should be such that concrete of the required slump will slide, not flow. End control should be provided so that the concrete will drop vertically and without segregation from the end of the chute. Two sections of metal drop-chute elephant trunk will serve to control end segregation. A mere baffle is not adequate. Open chutes which are exposed to wind and the sun tend to accelerate slump loss. Enough chutes should be used, or the chutes should be sufficiently portable, that the concrete may be placed directly into its final position instead of being forced to move horizontally by gravity or through vibration. If the concrete is placed in piles, the coarse aggregate which runs down the side should be shoveled back onto the pile and mixed into the concrete.

Buckets

Buckets are a satisfactory means for handling and placing concrete when they are designed for the job conditions and properly operated. Bucket capacities for handling concrete range from 1 and 2 yd^3 for structural work up to 12 yd^3 for the larger mass-concrete projects. Each bucket should have a capacity of at least one batch of concrete as mixed to avoid splitting of batches in loading buckets. Such splitting results in segregation. Buckets should be capable of prompt discharge of low-slump or lean-mix concrete. The dumping mechanism should permit discharge of a relatively small portion of concrete in one place. Also, the discharge should be controllable so that it will cause no damage to or misalignment of the forms. Buckets should be filled and discharged without noticeable separation of coarse aggregate.

If the discharge is too slow or if the bucket is too low, or if the discharge is not vertical, the low-slump concrete will stack in a cone and the coarse aggregate will separate and roll down the cone. Rock pockets usually are formed as a result.

General-purpose buckets (Fig. C1-28) are usually manually operated and handle from 2- to 6-in slump and about 2½ to 3-in maximum aggregate.

Lay-down buckets (Fig. C1-29) are commonly filled from a transit-mix truck where low filling height is required.

Buggies

Either manual or power-operated buggies (Fig. C1-30) may be used to transfer concrete for limited distances. The hand-operated carts or buggies usually have capacities up to about 6 or 8 ft^3. The power-operated buggies have capacities up to ½ yd^3. The buggies need a relatively level runway or path. Depending on conditions, each hand-operated cart should be able to transport up to about 5 yd^3 concrete per h, and each power-operated buggy should be able to transport up to about 20 yd^3/h.

Pumps

Concrete pumps were originally designed and used to place concrete in tunnels where the space available for placing equipment was very tight and large equipment could not be used. The first models of pumps could not handle lightweight aggregates or harsh mixes and were most effective with high-slump mixes containing large proportions of cement and sand. These limitations have been overcome and concrete pumping is usually just as fast or faster than other types of placing. Normally, fewer people are needed on a pumping job. A concrete-pump system requires one pump operator and two or three workers handling the hose. Concrete pumps may be used in tandem to obtain higher pours, and one pump may feed into another for this purpose. Several pumps may be used side by side for higher production rates.

FIGURE C1-28 General-purpose concrete bucket. *(Gar-Bro Manufacturing Company.)*

The most popular concrete pumps are piston type. Figure C1-31 is an 80-hp concrete pump. The specifications for the pump in Table C1-19 indicate a pumping height of 400 ft (500 ft in high-pressure mode) and 1600 ft in horizontal configuration (2000 ft in high-pressure mode).

Table C1-19 also has specifications for a 115-hp unit: height 300 ft (hp 420 ft); and horizontal 1200 ft (hp 1700 ft).

Table C1-20 has specifications for five truck-mounted pumps and six trailer pumps. Use of booms with truck-mounted pumps has made pumping concrete a highly effective placement method. Figure C1-32 shows three large truck-mounted concrete pumps with booms folded for transport. Figure C1-33 shows a 100-ft reach boom folded and at work. The outriggers are shown in plan.

Figure C1-34 is a 120 ft, four-section concrete placing boom. In elevation, the boom at work is shown: 120-ft maximum height; 110-ft horizontal reach, and 80-ft downward.

Figure C1-35 shows a placing boom mounted for use in high-rise work. The enclosed table shows data in four sizes.

Belt Conveyors

Belt conveyors for placing concrete have been used for many years, but their use has increased considerably in recent years. The belt conveyors are able to move the concrete around or over obstacles which would present serious problems with many other methods of placing. Some units are mobile (Fig. C1-36) so that they may be used in one location for a short placement, then quickly and easily moved to another location for a subsequent placement.

CONVENTIONAL CONCRETE

FIGURE C1-29 Lay-down concrete bucket. *(Gar-Bro Manufacturing Company.)*

Figure C1-37 shows a mobile conveyor belt transferring concrete from the ready-mix truck, over vertical rebar to a chute to a slab pour.

Placing concrete with a conveyor belt costs 20 to 50 percent of pumping cost. It is only 10 percent of crane placement cost. In terms of time, it is twice as fast as a crane up to 100 ft—and even faster beyond that.

Many of these conveyors may be used in series to move concrete over a considerable span. The system of belts may be made a variable length by running each conveyor up over the next one (Fig. C1-38). With provisions for moving each belt conveyor with respect to the next, the contractor may start placing concrete at either end of a job and extend or retract the belt system to continue placing concrete without interruption of the pour. A swinging belt conveyor may be used at the discharge end for versatility in placing as needed.

A side discharge may be used to enable concrete to be placed at any point along the conveyor, which greatly facilitates the placing of concrete over a wider area (Figs. C1-39 and C1-40).

A scraper of rubber or other suitable material should be used to prevent loss of mortar on a return belt. Long belts should be protected from the rain or hot sun by suitable covers.

Belt speeds normally run up to 900 ft/min, and concrete placing at rates over 300 yd^3/h has been done successfully.

FIGURE C1-30 Power-operated concrete buggy. *(Morrison Division of Amida Industries.)*

FIGURE C1-31 80-hp piston-type concrete pump. *(Morgen Mfg. Co.)*

TABLE C1-19 Specifications for 80-hp and 115-hp concrete pumps. *(Morgen Mfg. Co.)*

	Performance specifications		Common specifications	
SMU 115	High-capacity mode	High-pressure mode	Engine power @ 2500 r/min	177 hp (131 kW)
			Maximum aggregate size	2 in (50 mm)
Maximum concrete output	122 yd^3/h (93 m^3/h)	90 yd^3/h (69 m^3/h)	Stroke length	54 in (1370 mm)
Piston-face pressure	912 lb/in^2 (63 bar)	1266 lb/in^2 (87 bar)	Concrete cylinder diameter-SMU115	8 in (200 mm)
Number strokes/min	35	26	Concrete cylinder diameter-SMU80	7 in (180 mm)
Pumping height*	300 ft (91m)	420 ft (128m)	Charging hopper height	49 in (1245 mm)
Pumping distance*	1200 ft (366m)	1700 ft (518m)	Hopper capacity	20 ft^3 (.57 m^3)
SMU 80				
Maximum concrete output	93 yd^3/h (71m^3/h)	69 yd^3/h (53m^3/h)	Hydraulic oil tank capacity	120 US gal (454 L)
Piston face pressure	1200 lb/in^2 (82 bar)	1655 lb/in^2 (115 bar)	Fuel tank capacity	77 US gal (291 L)
Number strokes/min	35	26	Main pump hydraulic flow	104 gpm (394 l/m)
Pumping height*	400 ft (122m)	500 ft (152m)	Main hydraulic pressure	4000 lb/in^2 (276 bar)
Pumping distance*	1600 ft (488m)	2000 ft (610m)	Cast steel swing valve (inlet and outlet)	7 in (180 mm)

* Performance is relative to line size and character of the mix.

TABLE C1-20 Specifications for Truck and Trailer Mounted Concrete Pumps

Truck-mounted pumps	750 HDR	900 HDR	1200-20/120*	1200-23/125*	1200-23/120*
Maximum concrete output (yd^3/h)					
Rod side	69	117	147	171	196
Piston side		67	83	104	111
Maximum pressure on concrete (lb/in^2)					
Rod side	859	839	870	759	658
Piston side		1536	1565	1285	1184
Maximum horizontal pumping distance (ft)[†]					
Rod side	940	950	1000	85	700
Piston side		1700	1700	1400	1300
Maximum vertical pumping distance (ft)[†]					
Rod side	255	260	260	230	200
Piston side		450	480	390	360
Maximum strokes per min					
Rod side	34	30	30	26	30
Piston side		17	17	16	17
Pumping cylinder dia. (in)	7	8	8	9	9
Stroke length (in)	39	63	79	79	79
Maximum aggregate size (in)[‡]	1.5	2.5	2.5	2.5	2.5
Minimum concrete slump (in)[‡]	0	0	0	0	0
Type of valve§	Rock	Rock	Rock	Rock	Rock

* Model 1200 pumps available with optional differential cylinders for specific applications.
[†] Pumping distances shown are to be used as a guide only since they have been considerably exceeded on specific projects. Maximum attainable distances depend on concrete mix design and pipeline diameter. Maximum output and distance cannot be achieved simultaneously.
[‡] Minimum slump and maximum aggregate size are dependent upon concrete mix design and pipeline diameter.
Pump specifications are for standard units. Other units are available.
§ BPL 900 & 1200 pumps available with Gate Valves

Trailer pumps*	750-15R	-18R	2000-20R	3001-18R	5000-18R	8000-18R
Maximum concrete output (yd^3/h)						
Rod side	42	69	117	103	137	138
Piston side			67	62	85	87
Maximum pressure on concrete (lb/in^2)						
Rod side	950	859	870	1238	1238	1870
Piston side			1565	2098	2098	2900
Maximum horizontal pumping distance (ft)						
Rod side	1000	940	900	1250	1250	2500
Piston side			1700	2100	2100	4000
Maximum vertical pumping distance (ft)						
Rod side	300	255	250	350	450	500
Piston side			450	750	750	1200
Maximum strokes per min						
Rod side	30	34	30	30	40	34
Piston side			17	17	25	22
Engine hp (diesel)	51	68	177	213	2 @ 177	2 @ 224
Pumping cylinder dia. (in)	6	7	8	7	7	7
Stroke length (in)	39	39	63	63	63	79
Maximum aggregate size (in)	1.5	1.5	2.5	2.5	2.5	2.5
Minimum concrete slump (in)	0	0	0	0	0	0
Type of valve[†]	Rock	Rock	Rock	Rock	Rock	Rock

* Pump models available in various configurations for specific applications.
Specifications subject to change without notice.
[†] All units except 750 available with gate valves.
Source: Schwing America Inc.

CONVENTIONAL CONCRETE C1-73

KVM 52
- 170-foot, 4 section boom
- available with 1200 HDR pump kit
- 157.5 feet horizontal reach
- 360° slewing range
- Five inch pipe line

The reach of our 4-section 52 meter boom allows versatility for single set-up and efficient placement on highrise and flatwork. This extra long boom is mounted on a three steering axle chassis for maneuverability.

KVM 42
- 138-foot, 4 section boom
- available with 1200 HDR pump kit
- 125 feet horizontal reach
- 400° slewing range
- Five inch pipe line
- Fully articulating roll & fold design

KVM 36
- 118-foot, 4 section boom
- available with 1200 HDR pump kit
- 105 feet horizontal reach
- 400° slewing range
- Five inch pipe line
- Fully articulating roll & fold design

FIGURE C1-32 Three large truck-mounted concrete pumps with booms folded for transport. *(Schwing America Inc.)*

A special application is a belt conveyor which is mounted on a truck mixer (Figs. C1-41 and C1-42). This enables the contractor to reach over walls, ditches, or other obstacles at the job site and eliminates rehandling of concrete in many cases. It also enables the truck to stay a safe distance from forms and excavations. The belt conveyor is adjustable in both horizontal and vertical directions so the concrete may be placed wherever it is needed within the reach of the belt conveyor.

Vibrators (Fig. C1-43)

When low-slump concrete is placed into a form, the concrete is in a honeycombed condition. Depending on the mix, size and shape of the form, the amount of reinforcing steel, and the method of depositing the concrete into the form, from 10 to 30 percent of the volume is irregularly distributed, entrapped air. The problem is to get this heaped-up honeycombed mass to subside into a dense concrete without entrapped air. (Reduce from 10 to 30 percent to 2 percent.)

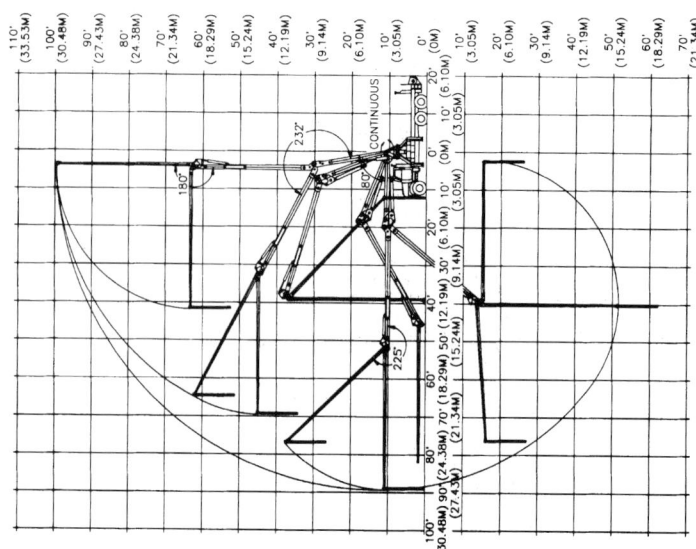

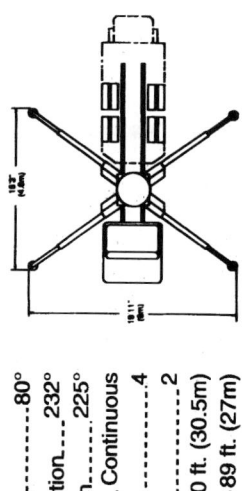

BOOM ARTICULATION:
Maximum angle, first section......80°
Maximum angle, second section....232°
Maximum angle, third section.....225°
ROTATION.....................Continuous
HYDRAULIC OUTRIGGERS..............4
REAR STABILIZERS..................2
VERTICAL REACH........100 ft. (30.5m)
HORIZONTAL REACH......89 ft. (27m)

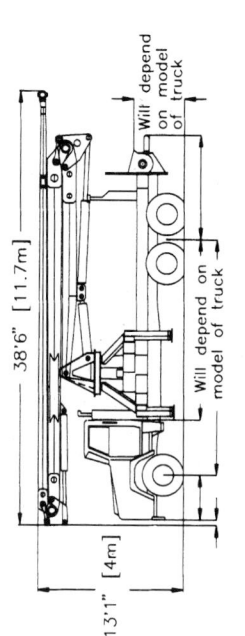

REMOTE CONTROL
Total of 105 ft. (32m) of cable, made up of one 45 ft. section and one 60 ft. section. The control box may be used directly at the control panel on the truck, at 45 ft. or at 105 ft.

FIGURE C1-33 100-ft reach boom. *(Morgen Mfg. Co.)*

C1-74

FIGURE C1-34 120-ft four-section placing boom. *(Morgen Mfg. Co.)*

An internal vibrator causes a violent agitation of the particles in the mix. This reduces the friction between the particles which enabled the concrete to support itself in a honeycombed condition. The mix now becomes unstable and starts to flow under the simultaneous effect of vibration and gravity. The concrete moves tightly against the form and around the reinforcing steel. The mix particles rearrange, the mortar fills the voids between the coarse aggregate particles, and the entrapped air (since it is the lightest ingredient in the mix) rises to the top of the layer.

It appears that even though concrete is a heterogeneous material, the overall combination of coarse and fine aggregate, cement, and water has a considerable amount of resonance. That is, there is a frequency called the resonant or natural frequency, at which a given force is most effective in consolidating the material. This is something like a singer breaking the mirror when the frequency or pitch of the note happens to be at the natural frequency of the mirror. Studies suggest that the resonant frequency of fresh concrete is about 10,000 vibrations per minute. Fortunately, most present-day interval vibrators operate at fairly near this figure. Some vibrators

Concrete placement for floors, walls and columns is easily and quickly accomplished with Schwing Separate Placing Booms mounted on crane towers, fixed pedestals or hydraulic self-climbing pedestals. Booms can be "flown" from one location to another, easily covering the entire area. KVM 28 can be ordered as a truck-mounted unit with pump and removed for pedestal or tower mounting.	SEPARATE PLACING BOOMS	KVM 28	KVM 32	DVM 32	DVM 42
	Dia. Pipeline (in.)	5"	5"	5"	5"
	Maximum Reach (ft.)	79	92	105	138
	Boom Sections	3	4	3	3
	Slewing Range	370°	370°	370°	720°
	Boom Coverage (sq. ft.) at 360° radius	19,606	26,590	34,636	59,828

FIGURE C1-35 Separate placing booms. *(Schwing America Inc.)*

have been built to operate at very high frequencies per minute, but they have not performed as well as hoped.

Why does the effectiveness of vibration drop off at very high frequencies, even though the acceleration is greater? One explanation is that most concrete mixes cannot keep up with such a fast-moving vibrator.

Concrete should not be permitted to flow long distances laterally in the form. There is not much danger of this with low-slump concrete until the worker starts to vibrate it. Vibrating the side of a pile to get concrete to move laterally is very bad practice. It causes segregation. Where there are some mounds in the surface of the concrete as dumped, a vibrator should be stuck into the center of each mound to knock it down.

After a fairly level surface has been attained, the vibrator should be inserted vertically at regular, systematic intervals over the surface. The spacing between insertions can be determined by doubling the radius of vibration for the vibrator being used. If this is not known, then a distance of 14 to 18 in would be recommended. If the spacing is correct, the area visibly affected by the vibrator should overlap with the adjacent just-vibrated area by a few inches.

FIGURE C1-36 Portable conveyor, 30 ft long, discharge to 16 ft high at 50 cy/h. *(Clark Machinery Inc.)*

FIGURE C1-37 Mobile belt conveyor concrete pour. *(Morgen Mfg. Co.)*

The bottom of the vibrator should penetrate 2 to 3 in into the preceding layer, to make sure that the two layers knit together without tell-tale lines between them. The vibrator should be held stationary for about 10 to 25 s (depending on the mix and the force exerted by the particular vibrator used) until the top of the concrete is covered with a thin film of glistening mortar and the entrapped air bubbles are no longer rising to the surface. Then the vibrator should be drawn out slowly with a slight jerking motion so that the concrete will move back into the space vacated by the vibrator. A

FIGURE C1-38 Using belt feeders in series to reach bridge deck. *(Morgen Mfg. Co.)*

well-trained, conscientious vibrator operator (who will consistently maintain the proper spacing of insertions and time of vibration) is absolutely essential to well-vibrated concrete.

In general, vibration should be carried out with the vibrator completely immersed in the concrete. If the vibrator protrudes through the top surface, some of the energy normally transmitted into the concrete will be lost. Additionally, the large force exerted at the surface caused a violent turbulence there. This results in segregation, with a weak mortar layer forming on the surface. Vibrators may also get overheated when they do not have the full benefit of the concrete's cooling effect. To avoid these problems when vibrating thin slabs, the vibrator should be dragged horizontally through the slab.

Some fear that an internal vibrator operated in a horizontal position might not be very effective. However, it is felt that this position is probably more effective than the vertical position because the vibrator waves at right angles to the vibrator are now in the same direction as gravity. Nonetheless, vertical insertions are preferred for most work, because it is simpler and provides more systematic coverage.

For columns, the vibrator or vibrators should be placed in the form inside the reinforcing steel. The form is then filled to required grade with the vibrators remaining idle during this process. The vibrators are turned on and withdrawn at a slow uniform rate until they are free from the concrete. In addition to reducing stratification and honeycombing, this process produces a surface essentially free of air and water voids.

The concrete mix has considerable effect on the vibration requirements. Air entrainment makes the mix a little stickier and a little more vibration is necessary to get the concrete properly compacted. On the other hand, air-entrained mixes

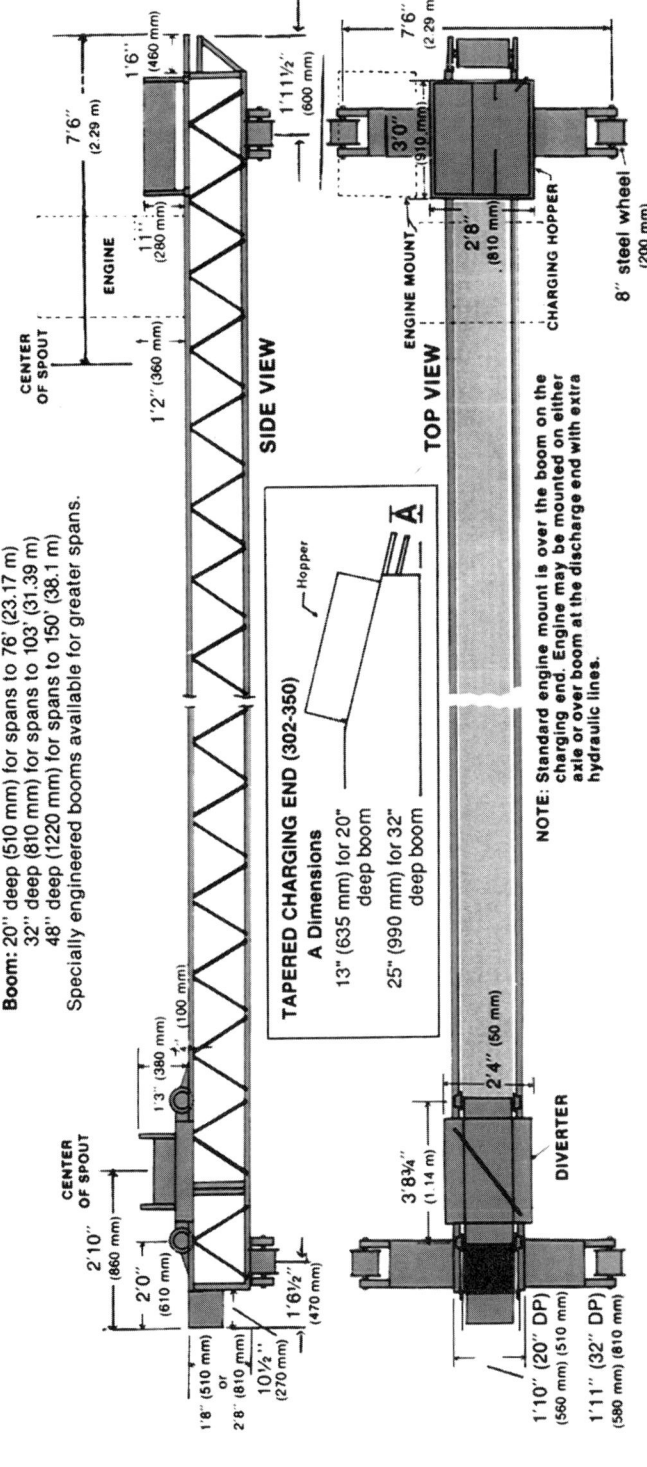

FIGURE C1-39 Side discharge belt conveyor, plan, and elevation. *(Morgen Mfg. Co.)*

C1-80 STANDARD HANDBOOK OF HEAVY CONSTRUCTION

FIGURE C1-40 Side discharge belt conveyor at work. *(Morgen Mfg. Co.)*

FIGURE C1-41 Belt conveyor mounted on ready-mix truck. *(Morgen Mfg. Co.)*

can stand more vibration without producing segregation than can non-air-entrained mixes.

It is not difficult to rescue the entrapped air content of a mix from the original 10 to 30 percent to 1 to 2 percent by vibration. However, to minimize surface pockholes, it may be desirable to drive still more air out of the concrete. Acceptable results can generally be obtained for vertical surfaces if the right vibrator is systematically inserted within a few inches of the form at about 1-ft intervals for an ample time of vibration.

FIGURE C1-42 Truck-mounted belt conveyor in use. *(Morgen Mfg. Co.)*

Some contractors are afraid to use a generous vibration time for fear it would drive air to the form. The vibration process does not drive air to a vertical form. When the concrete becomes temporarily fluid during vibration, the air (being lighter) rises vertically.

It has been reported that under certain combinations of frequency and amplitude, air may be pumped into the surface of a concrete mass through joints and cracks in the form. This could account for some of the cases where field people insist they have done everything humanly possible and still the concrete surfaces are badly pockmarked.

Since vibratory waves cause alternating positive and negative pressures in the concrete, problems sometimes arise when two or more vibrators are operated so closely together that their radii of influence overlap. Under these conditions, it is possible for a point in the concrete to be subjected to positive pressure from one of the vibrators and negative pressure from the other, and they may tend to cancel each other out.

Vibration of reinforcing bars is not likely to impair bond but it can be a cause of bars being displaced and result in high vibrator wear.

Revibrating concrete can improve the quality if it is done at the right time, and there is considerable latitude as to time. Generally, the concrete can be revibrated at any time before the concrete starts to set, i.e., any time when the vibrator can be inserted into the material under its own weight without additional force.

A good general rule to follow in vibration is this: If anyone doubts that the vibration has been adequate, a little more vibration is an excellent investment. Some fear has been expressed that part of the entrained air will be lost by continued vibration. Normal vibration removes only the coarser and less valuable air bubbles. It would be necessary to vibrate for at least four times the proper time to knock out enough entrained air to adversely affect the durability of the concrete.

Screeding

The dictionary defines a screed as a leveling device drawn over fresh concrete. This can be as simple as a 2- × 4-in board. Figure C1-44 is an engine-driven screed that can span a slab and both level and vibrate as it is pulled across (manual or hydraulic).

Figure C1-45 is a laser-guided, ride-on, vibratory screeding machine that established grades by laser, disperses concrete by auger and vibrates/consolidates the concrete. The on-board computer constantly monitors the float profile and adjusts the screed elevation as fast as five times per second.

FIGURE C1-43 Backpack gasoline concrete vibrator. *(WYCO Tool Co.)*

The S-40 has a 12-ft-wide screeding head mounted on a 20-ft telescopic boom. The machine does 240 ft^2 of concrete per pass in just about one minute. Cycle time between setups is about three minutes.

Troweling

After the initial floating, the mix should be allowed to hydrate enough that a worker's weight is supported with only a slight imprint. The surface is then floated or troweled by hand or machine (Figs. C1-46 and C1-47). This process drives aggregate particles located just below the surface to a deeper position, removes slight imperfections, and compacts the concrete at the surface into a more durable, dense working surface.

The best time for floating and troweling is established by experience. Too long a delay results in an unworkable surface. When this occurs, do not permit the spreading of additional mortar or water. Premature floating or troweling is more common and can be countered with patience. Dry cement should not be spread on the surface to permit earlier floating or troweling. This practice will lead to dusting, scaling, or crazing.

Hard steel troweling to a smooth, shiny surface is less popular than it once was, for safety reasons. Certain floors, however, must be troweled for good wear resistance. Surfaces used primarily for walking are often deliberately scored with a broom. Brooming usually follows floating; but after troweling, a surface may accept a fine broom scoring.

FIGURE C1-44 Speed Screed. *(Metal Forms Corp.)*

FIGURE C1-45 The S-240 Laser Screed® sets new standards for slab-on-grade concrete floors.

FINISHING AND CURING

Finishing

The specifications for a job should describe the required finish, since concrete surfaces can be finished in almost limitless variations. The actual finish will be determined by the characteristics of the form or form liners, by any work done on the concrete after the forms are removed, or by the treatment of unformed surfaces. The texture and

FIGURE C1-46 Walk behind trowel. *(Allen Engineering.)*

FIGURE C1-47 Power riding trowel. *(Allen Engineering.)*

appearance of the surface will depend on the rubbing, grinding, or sandblasting of the surface and may range from very rough to satin smooth. The specified finish should be consistent with the purpose of the concrete and the use to which it will be put.

Curing

Concrete hardens because of the chemical reaction, called *hydration,* between portland cement and water. The object of curing is to prevent or replenish the loss of necessary moisture during the early, relatively rapid stage of hydration. Optimum curing is defined as the act of maintaining controlled conditions for freshly placed concrete for some definite period following the placing or finishing operations to assure the proper hydration of cement and the proper hardening of the concrete. Five requirements for proper curing are:[7]

1. Preservation of adequate water content in concrete
2. Maintenance of a fairly constant temperature above freezing
3. Preservation of a reasonably uniform temperature throughout the whole body of concrete
4. Protection from damaging mechanical disturbances
5. Passage of sufficient time for hydration of cement and hardening of concrete

Sufficient water may be maintained by building a small dam around the perimeter and keeping the concrete covered with water. Periodic sprinkling may also be used, but the concrete must not be allowed to dry out between sprinklings. Sand, burlap, canvas, or straw may be placed over the concrete and kept continuously wet.

The concrete may be covered with waterproof paper or liquid, membrane-forming curing compounds to prevent water from escaping. No addition of water is then necessary. The curing compounds should be sprayed on as soon as the concrete has been finished and there is no free water left on the surface.

The time necessary for curing depends on the cement, mix proportions, required strength, size and shape of the concrete mass, weather, method of curing, and future exposure conditions. This time may vary from a few days for very rich mixes in special applications to a month or more for mass concrete.

SECTION 3

STATIONARY FORMS

Formwork, by definition, is the total system of support for freshly poured concrete. It consists of form sheathing plus all supporting members, hardware, and necessary bracing. Objectives of concrete forming include dimensional accuracy, strength, and economy, in the finished product.

In the past, formwork was largely restricted to job-built units. These forms were built in place on the job site, used once, and then scrapped. The trend today is toward increasing prefabrication, reuse of forms, and greater mechanization of assembly and erection. As a general policy, the layout and design of the formwork, as well as its construction, are the responsibility of the contractor.

Forms can be classified on the basis of reuse as *single use* or *multiple use*. In terms of the fabricator, they can be classified as *user-built* (job or shop) or *commercially supplied* (prefabricated forms). Forms in the latter category, when designed for reuse, can ordinarily be either rented or purchased outright.

PLANNING FOR FORMWORK

An initial step in formwork planning is to identify the requirements of the owner or architect with respect to the concrete finish. The contractor can then explore methods which will keep forming costs to a minimum considering job requirements and available materials.

The objective of the contractor, through optimizing all the factors involved in concrete work, is to produce a satisfactory result at the least possible cost. To do otherwise would be contrary to good engineering and design practices. Reuse of a form has a profound effect on the formwork cost which must be allocated against each unit of formed area. Usually it is desirable to obtain the greatest number of form reuses, but this objective must be weighed against labor to reset the form each time, concrete placement methods, and general job scheduling. It is often poor economy to use a form that, although it has a low initial cost, cannot perform with satisfactory results for the planned number of reuses. On the other hand, one could not hope to purchase one form to cast 200 columns on a specific job. The intermittent labor requirement to strip and reset one form and the cost of placing concrete in individual columns as compared to placing several in a set would have adverse effects on overall job costs.

Specification Requirements

It is essential that the owner or architect be explicit as to the finish requirements for the various parts of the building. Requirements should be categorized into specific areas of exposed critical work, areas of exposed semicritical work, and areas that have merely structural requirements. While so doing, the owner/architect should avoid specifying the design of the formwork.

The designer should also avoid specifying work of closer tolerances than those indicated in the current issue of ACI 347, *Recommended Practice for Concrete Formwork*. Otherwise, there is the risk of unnecessarily increasing the cost of the structure. Poor general finish and tolerance specifications both fall in the same category as specifying 4000-lb/in^2 concrete for portions of the building where 3000-lb/in^2 concrete would be adequate or requiring structural-steel members which are much larger than those required to support the anticipated loads. These all are examples of poor design.

Job Requirements for Formwork

Once the designer has specified finish and tolerance limits which are consistent with the intended use of the structure, the contractor can proceed with a detailed analysis of job requirements for formwork. These manifest themselves in terms of formwork timing, number of uses, forming method, quality of performance, and value for salvage. It is important not to overemphasize any one of these factors.

Timing. In general, the time of primary interest to the contractor is that required to complete the entire project. Formwork is only a part of the project, and all construction operations must be analyzed in sequence. It would be inappropriate, for example, to adopt a fast method of forming that prevents subsequent trades from doing their work. In some cases, this means an exchange; the formwork itself may cost more, but subsequent trades are able to accomplish their work more quickly. As a result, the total job can be completed at an earlier date. The selection of a particular formwork material can also be governed by whether or not it can be procured without delaying the job schedule.

Number of Uses. The cost of the formwork is a function of the number of uses. Forming costs are normally compared on the basis of dollars per square foot for the area of concrete which is formed. Therefore, the initial costs and subsequent handling costs for the formwork are divided by the number of uses in order to establish the formwork unit cost. A compromise is usually necessary to arrive at the number of reuses which will result in acceptable formwork costs while avoiding conflict with other major job factors.

Forming Method. The various features of the structure will have bearing on the best forming method to use. Cost comparisons can be made between manual erection and dismantling of small form units versus handling larger sections of forms with cranes or other mechanical devices. The method which is selected must not result in undesirable interference with other construction operations within the specific job environment. For example, the method of placing concrete (chutes, buckets, pumps, tremies) may have a bearing on the requirements for the formwork.

It has sometimes been found that the use of a stronger form, despite its higher initial cost, is justified in terms of job economics by its effect on reducing the labor cost of placing the concrete. For example, substituting a stronger form may permit a crane and a crew of men to double the rate of pour. This can result in a significant reduction in the cost of the in-place concrete. To put it another way, the increased cost of formwork in some cases can be more than offset by reduced labor and equipment costs in placing the concrete. Formwork costs and placing costs should be taken under consideration jointly.

Quality of Performance. In considering all the factors involved in timing, number of uses, and methods of forming and placing concrete, the essential requirement is that of acceptable performance. The finished job must satisfy the specifications. Alignment and deflection must be within permissible limits, and the concrete finish must be as specified.

Salvage Value. This factor injects itself into job requirements in four distinct ways. First, form materials such as steel, aluminum, magnesium, etc., can be sold for their scrap value. Second, forming material may be used as a permanent part of the structure. For example, steel floor forms may be left in place as a composite design. Third, there is the possibility that the forming material can be used on future projects after the completion of the current project. Fourth, the user need not pay the full cost of the material or equipment but can instead pay only a fraction of this in the form of rental. Vertical shoring, horizontal shoring, wall forms, and scaffolding are only a few examples of materials which can be obtained on a rental basis. In this way the contractor can reap the benefits of multiple use for a wide selection of formwork equipment without the requirement of ownership and at a fraction of its total cost.

Material Available for Formwork

Another factor that the contractor must evaluate is the availability of formwork material. Wood, steel, plastic, magnesium, etc., may all be considered if available. In most cases, the form is made of a combination of several of these materials. Where materials have been salvaged from previous formwork, their possible reuse should be studied. In selecting formwork materials, the contractor should keep in mind that the objective is to construct the facility in accordance with specification requirements, within the overall schedule, and at least total cost.

FORM MATERIALS

There are numerous materials available for forming. Several of those which are frequently used will be described in the following paragraphs.

Lumber

Lumber is a commonly available material and has excellent strength, weight, and cost factors. Characteristics of lumber include a pervious surface with varying texture leading to differences in color for the concrete surface which it contacts. The moisture absorption of form lumber will vary depending on its moisture content at the time of concrete casting and also on the presence and type of parting agent.

The useful life of lumber forms, measured in terms of the number of castings that can be made against them with acceptable results, depends on the use of appropriate parting agents or other surface treatments. Lumber can be obtained with smooth, sanded surfaces or it may be left rough-sawn to transfer distinctive textures to the concrete surface. The role of the parting agent will become more critical on the rough surface than on the smooth surface. Without suitable protection, a wooden form surface will deteriorate through alternative wetting and drying and the adhesion of concrete. This deterioration should receive appropriate attention, since its effects can be detected on the finished concrete surface.

Plywood

Plywood is probably the most popular facing material for cast-in-place concrete. It is relatively inexpensive and its performance is good provided that it is an exterior grade. Exterior grade means that the adhesive agents in the plywood will have a satisfactory resistance to the excess moisture in the fresh concrete. An unsuitable type of plywood will have a short life and few uses. Any plywood which is selected by a builder should have a history of good performance in the concrete forming field.

Plywoods can be purchased with various types of surface treatments which are intended to promote better performance over a longer period of time. These treatments include light mill-applied coatings, some of which will provide good parting characteristics through several pours. Other surface treatments provide protection for little more than the first cast, and will have to be renewed prior to each subsequent cast. Plywoods can also be obtained with a plastic coating which will extend the usable life of the material to a considerable degree. In most cases, this higher-cost coating will also need some parting agent to supplement its excellent characteristics. Some of these coatings will provide a nearly impervious surface and will eliminate grain raise and its subsequent transfer to the concrete surface. Therefore, if a grain-raise imprint is desirable on the concrete surface, the impervious coatings should be avoided.

Metals

Steel, aluminum, and magnesium can be used as forming materials. Steel is a commonly used material for formwork because it has a long usable life and excellent strength and cost factors. The impervious steel surface of the form will impart a uniform color to the concrete. Steel-form fabrication is normally done in a remote fabrication plant. Since field modifications of a steel surface are costly and difficult, all use requirements of the form must be anticipated prior to its fabrication.

The steel skin should be thick enough to carry the concrete load between support members without exceeding permissible deflections. It should also be thick enough to resist damage from vibrator contact or through normal field handling of formwork. A parting agent will be necessary for ease in stripping, to eliminate laitance buildup on the form face, and to minimize the possibility of concrete staining due to rust. These last two items can seriously affect the quality of subsequent casts. When white portland cements are to be used in the concrete, the steel skin should first be pickled or sandblasted. This will remove mill scale which may otherwise cause staining, even when the steel forms are well coated with a parting agent.

Aluminum forming material which is in contact with the concrete will be affected to various degrees by alkalis and other salts which are present in the concrete. Parting agents will help to reduce this reaction. The long-run performance of aluminum forms will vary from one location to the next because of the variations in coatings and concrete and the presence or absence of other metals that may cause galvanic action and lead to pitting of the aluminum surface. Because of its excellent strength-to-weight ratio, aluminum has been used to a great extent for structural support portions of the formwork where it is not necessarily in direct contact with the cast concrete.

Magnesium is similar in performance to aluminum. While it is said to be less susceptible to the alkalinity of concrete, its actual performance also varies considerably from job to job. As with aluminum, its behavior is influenced by the nature of the concrete, the parting agents used, and the presence of other materials that can cause a galvanic reaction.

Plastics

They have impervious surfaces that usually impart a smooth finish to the concrete without the discoloration which is characteristic of the absorptive-type materials.

Plastics come in several categories but can be essentially classified as reinforced or unreinforced.

Reinforced plastic consists of a resin matrix with embedded glass fibers in various forms. The presence of these glass fibers greatly increases the strength of the resin material. Reinforced plastic has found considerable acceptance in custom-made forms for specific job requirements, since the possible shapes and sizes of reinforced plastic forms are almost limitless. An appropriate resin must be used on the form surface in order to assure good performance through a reasonable number of uses. The glass fibers should be kept from contact with the concrete.

Unreinforced plastics can be obtained in sheet form with smooth or textured surfaces. The light-textured surfaces transfer an imprint to the concrete, thus toning down the usual characteristics of an extremely smooth surface. Sheet plastic and those plastics that can be heat-formed into rib designs, etc., need appropriate structural backup to enable them to support the concrete pressure loads. Unreinforced plastics are normally used as liners in a form system that provides all the structural requirements of concrete containment. The role of the plastic liner is merely that of changing the characteristics of the cast surface.

Preformed foam plastics can be used in at least two applications. First, the plastic can be used by itself, usually to form recesses in a wall. Preformed foam planks are often used when conditions preclude their salvage, and they can easily be attached to a basic forming system. For example, they can be used to box out walls for slabs. These slab box-outs usually have reinforcing rods and require no draft, thus eliminating the possibility of their withdrawal. Secondly, foam plastic is used in backing up thin, vacuum-formed, plastic form liners where the concrete pressure would otherwise cause deformation.

Absorptive and Impervious Liners

The designer should consider the relative merits of formwork that is impervious and formwork that is absorptive. Each type leaves its characteristic imprint on the concrete surface. The use of an impervious form liner will usually result in a lighter color of concrete and a more uniform appearance. Examples of impervious form liners are steel, plastics, and plywoods which are either of high density or are treated with certain coatings. Absorptive liners, on the other hand, will absorb some water from the fresh concrete as long as the moisture content of the liners is below the saturation point. Such absorption results in a darker color for the hardened concrete and a somewhat softer finish. Variation in concrete color is dependent on the absorptive capacity of the liner. For example, concrete cast against wood will be darker along the spring-growth grains and lighter against the summer-growth grains. Certain parting agents will reduce or nearly eliminate this effect.

FORMWORK REQUIREMENTS

Form Structure

Thickness of Material. The thickness of the various form materials depends on several major factors: the modulus of elasticity of the material, that is, its stretch per unit load; the span that the material must bridge; the estimated loads, including that of the concrete itself; any deterioration of the physical characteristics of the material which will occur during its usable life; and the maximum deflection that will be acceptable on the finished concrete surface.

Bracing. The form face is designed to provide a total envelope to contain the liquid concrete. This form face must maintain the appropriate alignment, whether straight or curved, when the pressure loads of concrete are placed against it. To achieve these requirements, additional members may be placed on the back side of the form face. These members are identified as joists and stringers when they support horizontal sheathing and as studs and wales when they support vertical or near-vertical surfaces. They are usually long, straight members, made of steel or timber, which are capable of holding the form face in proper alignment between its support points. Vertical shores provide the support points for stringers, while wales will receive their support either from external bracing or from concrete ties which act in tension between the wales on opposite sides of a wall. Figures C1-48 and C1-49 illustrate typical bracing systems.

Alignment. The objective is to place the concrete in whatever final position has been prescribed by the designer. In forming horizontal members, alignment is largely a matter of correct initial positioning of the formwork plus control of formwork deflection. The alignment of wall members involves two distinct considerations. First, as with the horizontal formwork, the concrete pressure must be contained so as to keep the deflection of the loaded form within specified limits. Next, this form alignment must be maintained against external forces such as wind and activity by the workers. Bracing to resist external forces is generally placed on one side of the wall only.

FIGURE C1-48 High wall bracing.

FIGURE C1-49 Wood bracing with adjustable turnbuckle for form alignment.

Form Joints

Prevention of Leakage. If water with some cement content is allowed to leak from the form, a surface blemish will result. This blemish is characterized by an aggregate-rich surface which is inconsistent with the normal, dense adjacent surfaces. The area will also have a darker color since a lesser amount of water is available for hydration. This aggregate-rich condition penetrates the concrete mass to a considerable depth and noticeable discoloration may still remain even after relatively deep sandblasting. Grout leakages through fine openings in the formwork must therefore be controlled where appearance is critical. The use of low-slump concrete is helpful in this regard. Form joints can also be made relatively grout-tight by one of several means: (1) placing a lumber batten strip behind plywood joints, (2) wetting wood forms several hours prior to placing the concrete to expand the wood and help in closing joints, (3) lining forms with a separate face, staggering the lining joints with those of the structural form, (4) installing rubber gaskets between sections of steel forms, and (5) caulking the formwork joints.

Fins. Fins are thin projections of set concrete which extend from a poured surface. They occur where grout has tried to escape through a form joint but has been blocked from complete escape. They are usually not considered objectionable, since they can easily be knocked off with little or no effect on the appearance of the finished surface. For some walls, broad fins are specified by the designer to obtain a pleasing appearance. This effect is accomplished by lining a form face with planks which can either be kept some specific distance apart or randomly placed. The fins can remain as they appear after the forms are stripped, or they can be broken back with special hammers.

Textures and Patterns

All forms will have characteristic markings which are normally transferred to the finished concrete surface. Sources of these markings include the following: size of the formwork face (i.e., 4- by 8-ft sheets of plywood); plank widths; various absorptive characteristics of the face; preformed panels; proprietary types of panels with particular perimeter configuration; grain of wooden faces; grain raise due to moisture; number and size of fasteners, etc. Assorted textures and patterns can be produced on the finished concrete surface from these form marks.

A wide variety of textures and patterns for a concrete surface can also be obtained through the use of form liners. This is a very practical approach, since the form facing can be selected and designed entirely apart from structural considerations for the formwork. The contractor then has a great deal of choice as to ways of supplying the necessary backup strength in the forming system.

Wood liners can be used to form feature strips and rustication strips. They can also provide a checkerboard pattern by a 90° change in the direction of grain or planks in adjacent panels.

Plastic liners provide a wide choice of surface textures and designs. Surfaces can be had with smooth or slightly grained (sometimes called *hair-cell*) finishes. These grained finishes impart a soft, pleasing effect to the concrete. Certain plastics can be heat-formed into almost limitless variations in design.

Rubber liners may also be considered for relatively shallow texture concrete surfaces. Rubber can be obtained and used either in sheet form or in solid, extruded shapes. It is suggested that the rubber be checked for its resistance to deterioration from oils which are commonly used as parting agents.

Temperature changes can profoundly affect certain liner materials. Metal and plastic liners, if secured in cool weather, can buckle when exposed to a large increase in temperature. This effect usually is not serious because, as the concrete comes in contact with the material, the liner temperature is immediately reduced to near what it was when installed.

Form Ties

Types. Ties for wall forms will fall generally into one of the following groups: (1) plain coil strapping or wire; (2) preformed strapping, with spacers to ensure specific wall thicknesses and preformed notches to provide breakback in the wall for the desired concrete cover; (3) preformed wire ties for specific wall thicknesses and breakback characteristics; (4) pull ties which are removed completely from the wall after forms are removed; (5) she-bolt ties where an inner male-threaded unit is left in the wall and the outer fastening devices are removed and reused; (6) he-bolt ties where the outer fastening devices are reusable, with an expendable female-threaded unit left in the wall.

Figure C1-50 illustrates various types of commonly used ties. The gang form tie is identical to the standard panel tie except that each of its ends has been extended $2^{13}/_{16}$ in to permit the connection of a gang form bolt. The 6000-lb flat tie is used on single-lift forming and is designed for 3-ft spacing. Its use can reduce labor and material handling, both in erection and in stripping, by as much as one-third. It can also be used on higher walls, where its spacing is depending on job conditions and rate of pour.

Characteristics. Ties are available in almost any desired strength category. Tie strength is usually selected to match the form strength, tie spacing desired, and anticipated pouring rates. The tie spacing must be such that form deflections are kept within acceptable limits in addition to satisfying strength requirements. Each type of tie leaves a characteristic hole in the wall surface. Wire snap ties leave very small holes, usually about ¼ in in diameter and with a normal depth of 1 in. Wood or plas-

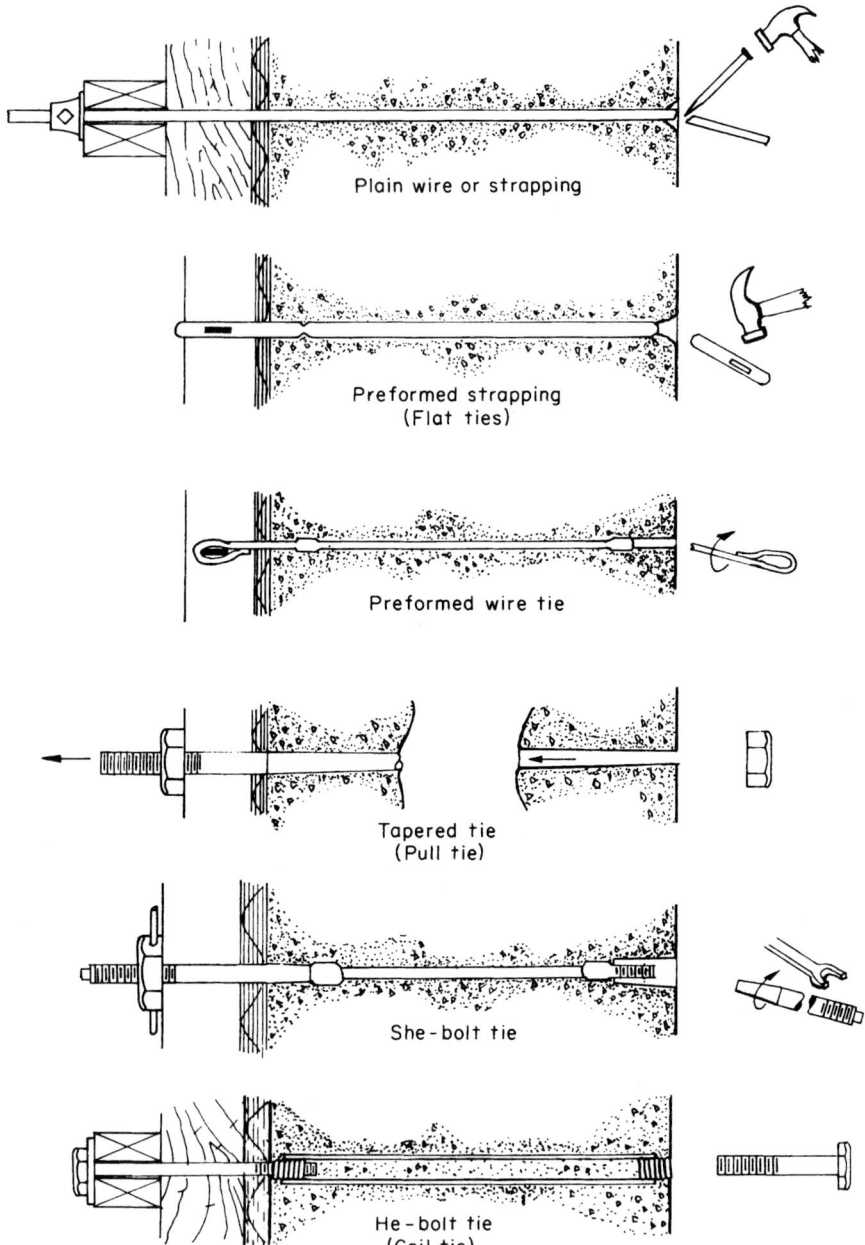

FIGURE C1-50 Common types of form ties.

tic cones can be used with the snap tie when deeper breakbacks (up to 2 in) are required. These will increase the size of the hole, usually to 1-in diameter. Cones are sometimes required to reduce grout leakage at the point where the tie passes through the form.

The characteristic holes of the she-bolts depend on the strength category of the tie. Hole diameter usually falls in the range of ¾ to 1½ in. He-bolts usually require a cone when it is necessary to provide a concrete cover over the portion of the tie which is left in the wall. Diameters of these cones will range from 1 to 2 in. A pull tie will range in diameter from ¼ to 1½ in and will leave a hole of similar size to the rod diameter; this hole passes complete through the wall, leaving no portion of the tie as expendable. All the above-mentioned ties leave round and relatively clean holes for subsequent patching.

There are two types of ties which leave marks other than those just described. The proprietary flat tie is broken off just below the wall surface by a sharp lateral blow. This causes some slight spalling at the concrete surface, the depth of which is usually limited to about ½ in. Next, a wire tie may have a thin washer at the surface of the wall for spreader action. This washer leaves a shallow impression in the face of the wall around the hole, while the wire itself breaks back at a greater depth.

Form Coatings

A form coating is used primarily to prevent the concrete, because of its adhesive characteristics, from bonding to the form face. In using a form coating, the following benefits are achieved: (1) The physical act of pulling the form from the wall is made easier. (2) Spalling of the finished wall is greatly reduced by eliminating any adhesion at vulnerable locations such as corners and edges of feature strips. (3) The form face is protected from the physical deteriorations that could come about by its contact with concrete. For example, a good bond breaker will reduce the tendency of the steel form to rust and will keep moisture from entering wood surfaces with subsequent grain raise, fiber raise, loss of strength, etc. (4) When adhesion of the concrete to the form face is prevented, the form face will remain clean and thus provide a consistent surface for subsequent casts. Conversely, if the concrete is allowed to adhere and build up on the form face, subsequent casts will reflect this buildup. An alternative is to incur additional field labor costs for cleaning the form face prior to each use. It is important to note that parting agents should have no effect on the concrete surface. Assurance of this should be requested from the supplier of the product.

Shores

On all suspended concrete floor and beam systems which are poured in place, various types of shoring are used to support forms and concrete until the concrete has set. Years ago, dimension lumber was used exclusively for shoring purposes. For the majority of work, 4-by-4s were used. They were cut slightly under the length required, and any adjustment in elevation was made by driving wooden wedges under the shore. The wedges were then toenailed.

Because of variations in story heights and the problems involved in adjusting the lengths of shores, many types of patent shoring systems have been developed. These permit adjustments in height, some even with full load. They also have positive load capacities, ease of erection and stripping, and clamping devices which facilitate the attachment of T heads, stringers, etc. These systems are generally rented for a particular job, but they may also be purchased. Patented shoring systems fall into three main categories, each having unique steel clamping and adjusting devices:

1. Dimension lumber combined with structural-steel shapes
2. Steel pipes
3. Tubular-steel scaffolding

Lumber alone is still used in special cases. Shipping costs, materials on hand, and identical story heights on multistory buildings are factors to consider. However, for the main part, proprietary shoring systems are replacing the lumber shores.

JOB-BUILT FORMS

The initial cost of the materials used in single-use job-built forms for simple concrete members is low, and the materials themselves may frequently be reused on other phases of the job. The size of the job may be so small that the introduction of multiple-use forms would be impractical. Most contractors are familiar with single-use job-built forms, and this can reduce delays which might be occasioned if job superintendents had to acquaint themselves with new techniques.

The job location and type introduce other cost factors. The contractor may have only inferior local materials, but rather than pay the costs of freight and rentals for prefabricated forms and incur an undesirable delay while obtaining them, it may be wise to use what is locally available. Much so-called "architectural" concrete lends itself best to job-built forms, since the design is often not repeated. The work may also involve an intricate pattern and appearance which is not easily reproduced with factory-made forms. However, if a contractor has the opportunity for several reuses of forms, it is probable that use of some prefabricated type of forming system will be considered.

Forms must be left in place long enough for the concrete to gain sufficient strength to support its own weight and that of any construction load involved. Materials most commonly used in constructing job-built single-use forms are dimension lumber or plywood. The sheathing is formed from 1-in. boards or from plywood in ⅝- or ¾-in thickness. Studs and wales for simple, straight walls where high concrete pressures are not anticipated can consist of 2-by-4s, used singly or in pairs. Heavier pressures require stronger framing, and steel channels may then be required for the wales.

Form Design

The design of job-built forms is governed by the job condition, the job specifications, and the anticipated conditions of loading.

Wall Forms. The maximum lateral pressures to be anticipated in wall forms are listed in Table C1-21. Fluid concrete exerts a hydrostatic pressure on the forms, the maximum value of which is largely a function of the rate of pour and the length of time required for the concrete to take its set. The values in Table C1-22 apply to structural concrete weighing approximately 150 lb/ft^3 which is placed at controlled rates under prescribed conditions of temperature. The tabulated maximum pressures are valid for internal vibration of normal-density concrete placed at 10 ft/h or less and with a slump no more than 4 in. Depth of internal vibration must be limited to 4 ft below the top of the concrete. Good placement procedures are assumed; i.e., vibration is used for consolidation only and not for lateral movement of the concrete.

Column Forms. The maximum lateral pressures to be anticipated in column forms are listed in Table C1-22.

Slab Forms. In designing forms for slabs, it is essential to consider the following factors: (1) the dead load of concrete and reinforcing steel; (2) a live load, superimposed on the dead load, which may result from the impact of dropping concrete or the weight of temporary heaped concrete, workers, and equipment such as buggies, runways, etc. (40 to 50 lb/ft^2 is often used to cover these live loads for design consideration); (3) shore spacing, shore safe-load capacity, and shore bearing capacity; (4) the time required for concrete to develop sufficient strength for form removal; (5) the ceiling finish and deflection specifications.

Form Hardware

Table C1-23 lists design safety factors and typical uses for various items of form hardware. Snap ties are the common method used to tie and spread user-built wall forms. These ties have guaranteed strengths of from 3000 to 5000 lb and built-in spreaders. For loads greater than 5000 lb, concrete-accessory manufacturers provide tie screws and she-bolts which range in strength from 5000 to 30,000 lb and even higher for special jobs.

Forms are braced externally where interior tying methods are not practical. External bracing, for example, is indicated in cases of inaccessibility—where there is steel piling or an existing structure on the opposite side of the forms. The bracing serves to keep the forms in alignment and to withstand pressures of poured concrete.

TABLE C1-21 Maximum Lateral Pressure for Design of Wall Forms[*,†]

(Based on ACI Committee 347 pressure formulas)

Rate of placement, R, ft/h	p, maximum lateral pressure for temperature indicated, lb/ft^2					
	90°F	80°F	70°F	60°F	50°F	40°F
1	250	262	278	300	330	375
2	350	375	407	450	510	600
3	450	488	536	600	690	825
4	550	600	664	750	870	1050
5	650	712	793	900	1050	1275
6	750	825	921	1050	1230	1500
7	850	938	1050	1200	1410	1725
8	881	973	1090	1246	1466	1795
9	912	1008	1130	1293	1522	1865
10	943	1043	1170	1340	1578	1935

[*] Reprinted from *Formwork for Concrete*, 2d ed., by permission of the American Concrete Institute.
[†] Do not use design pressures in excess of 2000 lb/ft^2 or 150 × height of fresh concrete in forms, whichever is less.

TABLE C1-22 Maximum Lateral Pressure for Design of Column Forms*,†
(Based on ACI Committee 347 pressure formulas)

Rate of placement, R, ft/h	\multicolumn{6}{c}{p, maximum lateral pressure for temperature indicated, psf}
	90°F	80°F	70°F	60°F	50°F	40°F
1	250	262	278	300	330	375
2	350	375	407	450	510	600
3	450	488	536	600	690	825
4	550	600	664	750	870	1050
5	650	712	793	900	1050	1275
6	750	825	921	1050	1230	1500
7	850	938	1050	1200	1410	1725
8	950	1050	1178	1350	1590	1950
9	1050	1163	1307	1500	1770	2175
10	1150	1275	1435	1650	1950	2400
11	1250	1388	1564	1800	2130	2625
12	1350	1500	1693	1950	2310	2850
13	1450	1613	1822	2100	2490	3000
14	1550	1725	1950	2250	2670	
16	1750	1950	2207	2550	3000	
18	1950	2175	2464	2850		
20	2150	2400	2721	3000		
22	2350	2625	2979			
24	2550	2850	3000			
26	2750	3000	\multicolumn{4}{l}{3000 psf maximum governs}
28	2950					
30	3000					

* Reprinted from *Formwork for Concrete*, 2d ed., by permission of the American Concrete Institute.
† Do not use design pressures in excess of 3000 lb/ft^2 or 150 × height of fresh concrete in forms, whichever is less.

PREFABRICATED FORMS

Commercially supplied forms are frequently referred to as *prefabricated forms.* Multiple-use prefabricated forms are normally available either on a purchase or a rental basis, and their merits should be investigated when a number of form reuses is anticipated. These forms are sturdily constructed for a long, usable life, and this factor may reduce their cost per use below that of single-use forms. Additional savings may be realized through lower costs for erecting and stripping, since the prefabricated forms are manufactured in integral units which are still small enough and light enough for one person to handle.

In comparing the costs of user-built forms with those of commercially supplied forms, the contractor should take into consideration the condition of the prefabricated forms, the service provided with them, and the cost of accessories. Form manufacturers will frequently provide engineering layouts at no charge or at a nominal cost, and their field representatives will instruct the contractor's crew in the proper use of the equipment.

Special or extenuating circumstances notwithstanding, prefabricated forming has proved to be consistently beneficial. Because it is preengineered, the risk of on-site or untried design is greatly reduced. The simplicity of its assembly provides uniform results (both in cost and performance) which are little affected by the caliber of the worker. Because of this, both production standards and cost standards become predictable. Interchangeable inventory becomes a reality because similar forming equipment can be brought together from a contractor's various job sites and used without problems of incompatibility.

Maximum salvage is another important benefit. With prefabricated forms, waste, theft, and deterioration of equipment are greatly reduced.

Supplemental equipment for the commercially supplied forming system, under most circumstances, is readily available for rent or purchase. This feature is of considerable value when late decisions are made as to the forming equipment required on a specific project. Prefabricated forming equipment is inventoried in its assembled condition for quick shipment and on-the-job assembly. The user who owns some prefabricated equipment can frequently supplement it with rented units during periods of peak requirements.

Materials and prices for the various types of commercially available forms will vary over wide ranges. Because of this, it is difficult to find a common denominator for comparison. An evaluation should be made by each contractor in terms of total formwork costs, including both material and labor. This total cost will include the job cost of erecting, stripping, and maintenance; form construction, purchase,

TABLE C1-23 Design Capacities of Formwork Accessories[*,†]

Accessory	Safety factor	Type of construction
Form tie	1.5	Light formwork; or ordinary single lifts at grade and 16 ft or less above grade
	2.0	Heavy formwork; all formwork more than 16 ft above grade or unusually hazardous
Form anchor	1.5	Light form panel anchorage only; no hazard to life involved in failure
	2.0	Heavy forms—failure would endanger life—supporting form weight and concrete pressures only
	3.0	Falsework supporting weight of forms, concrete, working loads, and impact
Form hangers	1.5	Light formwork. Design load, including total weight of forms and concrete with 50 lb/ft^2 minimum live load, is less than 150 lb/ft^2
	2.0	Heavy formwork; form plus concrete weight 100 lb/ft^2 or more; unusually hazardous work
Lifting inserts	2.0	Tilt-up panels
	3.0	Precast panels
Expendable strand deflection devices[‡]	2.0	Pretensioned concrete members
Reusable strand deflection devices[‡]	3.0	Pretensioned concrete members

[*] Reprinted from *Recommended Practice for Concrete Formwork* (ACI 347) by permission of the American Concrete Institute.
[†] Design capacities guaranteed by manufacturers may be used in lieu of tests for ultimate strength.
[‡] These safety factors also apply to pieces of prestressing strand which are used as part of the deflection device.

or rental cost divided by the number of uses expected in a reasonable amount of time; cost of ties, hardware, and other tools per job; form adaptability to different jobs which may be bid; and the appearance of the finished surface provided by the forms.

Steal-framed forms are designed for long life and minimum maintenance costs. Aluminum frames are excellent for lightweight handling, but their durability must be evaluated. Steel-faced frames are designed for great durability. Many forms today feature plastic-coated plywood faces which will give a very smooth finish and resist abrasion for a great many pours. The contractor who follows the manufacturer's care and cleaning recommendations may get as many as 200 reuses from these forms. The amount and cost of plastic coating will vary from manufacturer to manufacturer.

For heavy construction, a semiprefabricated form is available which uses 25,000-lb-capacity she-bolts and tapered ties. The system consists of 4- by 4-ft steel frames. The contractor supplies the plywood and bolts it directly to these steel frames. Each tie will support 16 ft^2 of formwork and is generally attached 4 ft on centers at the panel intersection. Due to the system's weight, it is engineered primarily for gang forming on heavy-construction jobs such as dams, bridge piers, and large culverts. Figure C1-51 illustrates the use of semiprefabricated forms on the construction of a 40-ft high retaining wall in New Jersey.

FIGURE C1-51 Use of semiprefabricated forms.

GANG FORMS

Much of gang forming's success can be attributed to the rapid progress of cranes and other mechanical means of transporting forms. Gang forms are most economical where a crane can work to advantage. For example, on a job requiring high walls and columns, after the crane has set the wall forms it can be put to use setting or stripping ganged column forms. The cost of the crane can then be prorated over the large number of square feet formed.

In contrast, either job-built or prefabricated forms would require costly scaffolding to be set up by hand all along the wall. Generally speaking, the cost of this manual labor would be far greater than the cost of mechanical handling for the larger forms. Another saving would be made for the large panels by assembling them on the ground where the carpenters can work to their greatest efficiency. In setting small forms up on a high wall, productivity is reduced because every item must be brought up to the workers. This causes much delay and lost motion. A similar situation exists in stripping forms; the giant forms or gang forms would not require any assembly, while small units would have to be assembled and disassembled every time. In addition to their use in forming retaining walls, giant or gang forms have proved economical on bridges, locks, sewage- and water-treatment plants, utility plants, tunnels, sewers, culverts, multistory buildings, and certain unusual jobs.

The advantage of using a large form lies in the savings in erection. A large form area is erected at one time without the handling of individual parts, lumber, and connecting hardware. Weather, rate of pour, and size of wall will govern the dimensions of lumber used. Normally, plywood facing with 2-by-6 and 2-by-8 studs and/or 2-by-6 or 2-by-8 wales is used. Snap ties, coil ties, and she-bolts with wale extensions are used with large forms. Strengths of individual ties will range from 3000 to 30,000 lb.

Some prefabricated forms lend themselves well to gang forming, especially with advances in manufacturers' gang-form hardware. Then, after the job is completed, the disassembled forms may be put to use on other types of work. With giant forms, a large section of the work is stripped at one time and the form is then moved by crane or rolling scaffolding to its next setup without further handling. Sizes of these large forms range up to 30 by 50 ft and are limited only by the capacity of the crane or by the difficulties of handling.

It is necessary only to move two large forms into place and secure the ties to be ready for the next pour. Speed of erection, therefore, is much greater with large forms than with individual built-up or prefabricated forms. For architectural purposes, such as on highway retaining walls, various feature strips are easily attached to the gang-form face.

FORM APPLICATIONS

Wall Forms

Available prefabricated forming equipment for walls includes the following:

Plastic-coated plywood panels ½ in thick with steel frames

Plywood panels ½ in thick with steel frames

Steel-faced panels welded to steel frames

Aluminum and magnesium framed panels with plywood faces

In prefabricated wall forms, the tie is connected directly to the forms rather than to the wales, thus eliminating the need for a row of wales for every row of ties. The wale is thus used as a means of form alignment rather than as an anchor for the ties. Considerable labor as well as material are saved, and much of the usual hardware is eliminated.

The designs for many types of prefabricated forms are fairly similar, but hardware and ties vary with each manufacturer. Some manufacturer's ties are adaptable to most forming systems, while others can be used only on a particular system. Safe load capacities for ties vary from 1000 to 6000 lb.

Where built-up forms were once common on complicated walls, prefabs are now often used. Complicated walls may include Y and V walls, straight walls with corbels, curved walls that have a batter in them, and other complicating features. Reuse is a key factor. If the wall or variation is repeated, prefabricated forms may be the more economical method of forming.

Column Forms

Columns occur in all sizes and shapes; e.g., round, square, rectangular, octagonal, oval, etc. The forms for such columns can be made in many ways, using wood with steel strapping, wood with proprietary column clamps, all steel, fiberglass, plastic, aluminum staves, etc. Gang forming is often used in reducing the labor involved in erecting and recycling the forms. Certain proprietary column forms are available on a rental basis.

Beam Forms

Beam forms almost invariably are built in place on the job where they are to be used. The soffit or beam bottom is built with 2-in lumber. For the beam sides, ¾-in plywood or 1-in boards reinforced with 1-by-4s are utilized. It is not necessary to use internal tying devices on shallow beams, inasmuch as the beam sides are held in place at the bottom by kickers and at the top by the slab forms. Spandrel beams, on the other hand, may be as much as 4 to 5 ft in height or depth. Internal tying devices such as snap ties would then be used to contain the concrete pressure. The use of prefabricated forms is feasible in many cases where larger beams are being formed, such as the spandrel beams on a multistory reinforced-concrete parking garage.

Table C1-24 is concerned with the spacing of shores under beams of various depths. The deflection of the nominal 2-in beam bottom is the basis for the spacings listed in this table. Modifications must be made if the beam widths run in excess of 18 in. The weight of the concrete and various construction loads will then have to be checked to see that the total weight will not exceed that which the shores can safely support.

Slab and Beam Forms

Flat Slab. The formwork for flat slabs comprises four main parts: the forming brace, joists, stringers or girts, and shores. Three-quarter-inch plywood is usually used for the forming material, with 4-by-4s and 4-by-6s used for joists and stringers, respectively. Other commonly used lumber dimensions are shown in Tables C1-25 and C1-26.

TABLE C1-24 Shore Spacing under Beams*

Depth of beam, in	12	15	18	21	24	27	30	33	36	48	54	60
Shore spacing	4 ft 3 in	4 ft 0 in	3 ft 11 in	3 ft 9 in	3 ft 8 in	3 ft 7 in	3 ft 6 in	3 ft 5 in	3 ft 4 in	3 ft 2 in	3 ft 1 in	3 ft 0 in

* Spacing of shores is limited by ⅜-in deflection of nominal 2-in bottom. Provision is made for 50 lb/ft² construction load.

TABLE C1-25 Spacing of Joists, In—Simple Spans—Two Supports

Slab thick-ness, in	Size of joist, in	4 ft 0 in	4 ft 6 in	5 ft 0 in	5 ft 6 in	6 ft 0 in	6 ft 6 in	7 ft 0 in	7 ft 6 in	8 ft 0 in	8 ft 6 in	9 ft 0 in	9 ft 6 in	10 ft 0 in	10 ft 6 in	11 ft 0 in	11 ft 6 in	12 ft 0 in
4	2 × 4	24	16	11														
	2 × 6				24	20	14	10 / 24	19	15	11 / 24							
	2 × 8											19	15	12 / 22	10 / 19			10
	2 × 10															15	14	
	2 × 12																	
6	2 × 4	20	13	10													
	2 × 6					22	16	11 / 20	15	12 / 24								
	2 × 8										19	15 / 24	12 / 21	10 / 17				
	2 × 10														15	12		
	2 × 12																	
8	2 × 4	17	11															
	2 × 6			24	18	13	10 / 23	17	13	10 / 20	16	12 / 22						
	2 × 8												18	14	12	10		
	2 × 10																	
	2 × 12																	
10	2 × 4	15	10															
	2 × 6			23	16	11 / 24	19	14	11 / 22	17	13 / 24	11 / 19	15	12	10			
	2 × 8																	
	2 × 10																	
	2 × 12																	
12	2 × 4	13	8															
	2 × 6			20	14	10 / 23	17	12 / 24	19	15 / 24	12 / 21	10 / 17	13	11				
	2 × 8																	
	2 × 10																	
	2 × 12																	

Forming surface, ⅝-in plywood.
Deflection limited to ⅛ in.
Construction load, 50 lb/ft².

TABLE C1-26 Spacing of Joists, In—Partially Continuous Spans—Three or More Supports

Slab thickness, in	Size of joist, in	5 ft 0 in	5 ft 6 in	6 ft 0 in	6 ft 6 in	7 ft 0 in	7 ft 6 in	8 ft 0 in	8 ft 6 in	9 ft 0 in	9 ft 6 in	10 ft 0 in
4	4 × 4		24	20	14 / 24	11 / 18	13	10 / 24	19	15	12 / 24	10 / 20
	2 × 6											
	2 × 8											
	2 × 10											
5	4 × 4		24	17	13 / 24	18	13	22	17	14 / 24	11 / 21	17
	2 × 6											
	2 × 8											
	2 × 10											
6	4 × 4		22	16 / 24	11 / 19	14	11 / 24	20	15	12 / 24	19	16
	2 × 6											
	2 × 8											
	2 × 10											
7	4 × 4		21	14 / 24	10 / 17	13	23	18	14	11 / 22	17	14
	2 × 6											
	2 × 8											
	2 × 10											
8	4 × 4	24	19	13 / 22	16	12 / 24	21	16	13 / 24	10 / 20	16	13
	2 × 6											
	2 × 8											
	2 × 10											
9	4 × 4	24	18 / 24	12 / 20	15	11	20	15	12 / 23	18	15	12
	2 × 6											
	2 × 8											
	2 × 10											
10	4 × 4	24	16 / 24	11 / 19	14	10	18	14 / 24	11 / 21	17	14	11
	2 × 6											
	2 × 8											
	2 × 10											
12	4 × 4	21	14 / 23	10 / 16	12	21	16	12 / 24	19	15	12	10
	2 × 6											
	2 × 8											
	2 × 10											

Forming surface, ¾-in plywood.   Maximum deflection, ¼ in.   Construction load, 50 lb/ft².

Joists are commonly spaced on either 16- or 24-in centers, inasmuch as these spacings are suitable for plywood sheets 4 ft wide and 8 ft long. As noted in Tables C1-24, C1-25, and C1-26, any number of combinations may be used in working out the spacing of the stringers and shores. On certain types of flat-slab construction, drop heads, which are additional thicknesses of concrete surrounding a column, may tend to complicate the spacing of stringers and shores.

One manufactured slab-form system permits an average construction speed of a deck a week. The system consists of prefab forms, sliding ledge angles, steel stringers, cross bracing, and adjustable steel shores. The same forms that are used for vertical wall construction are utilized for most decking requirements. Material adaptability, ease of assembly, and speed of erection and stripping are among the many advantages. The deck can readily be adjusted to final grade, and each shore can support up to 60 ft^2 of forms. Reshoring is not necessary, and the usual "forest of shores" is also eliminated.

Forms are set on sliding ledge angles on both sides of the steel stringers without connecting hardware or clamps. In stripping, the sliding ledge angles are lowered to permit form removal with the stringers and shores remaining in place. The normal reshoring operation is thus eliminated.

Almost any deck-height requirement is possible, and forming equipment is quickly released for reuse. Complete forming of 100 ft^2 of deck per worker-hour is easily obtained with the system, resulting in a low cost for carpenters and laborers. Figure C1-52 shows typical slab work with a prefabricated decking system. Shores from previous pours are left in place, and stripped panels are used for the next deck erection.

FIGURE C1-52 Typical slab work.

Pan construction is the forming of slabs with metal or plastic pans. These are poured monolithically with a thin floor-slab section in a series of light concrete joists. The light joists are supported at the ends by concrete beams which are poured at the same time as the floor and joists. Pans can usually be rented or purchased, and in some localities, companies specializing in pan forming will subcontract the complete forming operation. Table C1-27 is concerned with the dead loads for various pan sizes. The soffits are usually supported by 4-by-6 girts or stringers and shores.

Void-in-slab construction is similar to pan construction in that paper tubes or tiles serve to form the voids between the small concrete joists as well as the underside of the thin slab. The tubes or tiles, however, become permanent parts of the floor structure. It is common practice to set the tile in position on falsework which is similar in construction to flat-slab forming. Large paper or plastic tubes are also often used in this same manner. Shoring can consist of 4-by-4s which are wedged to the proper elevation, but these members are slow to install and are awkward to adjust for height. Patented steel and combination steel and wood shores are usually preferred where they are available (see Table C1-28).

CARE OF FORMS

Some suggestions for the proper care of multiple use forms are as follows:

Forms should be thoroughly cleaned and oiled after every use. The first time forms are erected, a light coat of oil should be sprayed on the reverse side. This procedure should be repeated every five to eight pours. Various form oils are available, and the type should be chosen carefully. The intent is to keep the forms clean, leaving no residue of concrete to affect the surface formed in subsequent casts. Some oils have a tendency to discolor concrete and otherwise affect the finished surface. Forms should be piled face to face and back to back in trucking and must not be unloaded by dumping. When not in use, the forms should be piled carefully to minimize weathering and to prevent rotting. The pile should have a slight pitch to shed water, and strips should be used between forms to permit evaporation.

TABLE C1-27 Vertical Load for Design of Slab Forms, lb/ft²*

(Includes weight of concrete and reinforcing steel plus construction live load of 50 lb/ft²; weight of formwork not included)

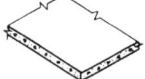

Solid-slab construction

	\multicolumn{10}{c}{Slab thickness, in}
	3	4	5	6	7	8	9	10	11	12
100-lb concrete	75	83	92	100	108	117	125	133	142	150
125-lb concrete	81	92	102	113	123	134	144	154	165	175
150-lb concrete	88	100	113	125	138	150	163	175	188	200

Typical joist-slab construction

Actual weights and dimensions vary slightly from one manufacturer of forming systems to another.

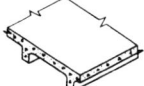

Depth of steel form, in	\multicolumn{4}{c}{20-in-wide forms}	\multicolumn{4}{c}{30-in-wide forms}
	Joist width, in	2-in slab	2½-in slab	3-in slab	Joist width, in	2½-in slab	3-in slab	3½-in slab
6	4	89	95	102	4	91	98	
	5	92	98	105	5	93	100	106
	6	94	100	107	6	95	102	108
8	4	95	101	108	4	95	101	
	5	98	104	111	5	98	104	111
	6	101	107	114	6	100	106	113
10	4	100	106	113	4	100	106	
	5	104	110	117	5	102	108	116
	6	108	114	121	6	105	111	118
12	4	107	113	119	4	104	110	
	5	111	117	124	5	107	113	120
	6	116	122	128	6	111	117	124
14	4	113	120	126	5	112	118	124
	5	118	125	131	6	116	122	130
	6	123	130	136	7	120	126	133

* Reprinted from *Formwork for Concrete,* 2d ed, by permission of the American Concrete Institute.

CONVENTIONAL CONCRETE   C1-109

TABLE C1-27 Vertical Load for Design of Slab Forms, lb/ft² * *(Continued)*
(Includes weight of concrete and reinforcing steel plus construction live load of 50 lb/ft²; weight of formwork not included)

Typical "waffle"-type two-way joist systems
Actual weights and dimensions vary slightly from one manufacturer of forming systems to another.

	Size of pan form, in		Slab thickness above form, in				
Depth	Outside plan	Inside plan	2	2½	3	3½	4½
4	24 × 24	19 × 19	96	103	109	115	
6	24 × 24	19 × 19	109	116	122	128	
8	24 × 24	19 × 19	123	129	135	141	
	36 × 36	30 × 30	111	117	123	130	143
10	24 × 24	19 × 19	129	136	142	148	
	36 × 36	30 × 30	120	128	133	139	152
12	24 × 24	19 × 19	145	152	158	164	
	36 × 36	30 × 30	133	139	145	152	165
14	36 × 36	30 × 30	141	148	154	160	173

TABLE C1-28 Spacing of Shores for Flat Slab Construction*

Slab thickness, in	Shore spacing									
	5 ft 0 in	5 ft 6 in	6 ft 0 in	6 ft 6 in	7 ft 0 in	7 ft 6 in	8 ft 0 in	8 ft 6 in	9 ft 0 in	9 ft 6 in	10 ft 6 in
6	4-6	4-0	4-0	3-6	3-0	3-0	3-0	2-6	2-6	2-6	2-0
7	4-0	4-0	3-6	3-0	3-0	2-6	2-6	2-6	2-6	2-0	
8	4-0	3-6	3-0	3-0	2-6	2-6	2-6	2-0			
9	3-6	3-0	3-0	2-6	2-6	2-6	2-0				
10	3-6	3-0	2-6	2-6	2-6	2-0					
11	3-0	2-6	2-6	2-6	2-0						
12	3-0	2-6	2-6	2-0							

* Spacings based on stock 4-by-6 lengths, with maximum load (including 50-psf construction load) of 3000 lb per shore. Maximum fiber stress in shore is 1500 lb/in². Deflection of 2-by-10 soffits does not exceed ⅛ in.

SECTION 4

SLIP FORMS

The slip-forming method lends itself to the construction of towerlike concrete structures which can be generated by translating a horizontal area vertically upward. In this method, the area of the building is laid out on the foundation and the formwork is constructed anywhere from 3½ to 6 ft high. The forms are filled with concrete and slowly moved upward by jacking. As a set of forms moves upward, additional concrete is placed in the top and hardened concrete is exposed at the bottom. This is a continuous process as compared to the conventional method of placing the concrete in lifts. There is no requirement for horizontal joints and slip forming can be carried on around the clock, although it is usually economical to stop the process over weekends. Stops may also be planned on a daily basis or at predetermined elevations.

The plan shape or extent of the structure is not a limitation. It can be circular, rectangular, cruciform, curved, irregular, solid, hollow, or cellular. The important thing is that this area must be projected straight upward for some appreciable height, usually at least 30 to 40 ft, depending on economic and other considerations. Since the forms must slide past the face of the concrete, no projections beyond this face are possible. Floors, for instance, are placed later using keys, beam pockets, weld plates, dowels, or other devices which can be kept flush to the concrete surface. Sections of the concrete may be deleted during the slide, however. Openings may be formed, inserts cast in place, and even columns and girders may be produced by placing formwork and inserts which fit within the confines of the concrete surfaces.

Slip-formed concrete surfaces are almost always vertical, since any variation therefrom results in complications and increased costs. It is possible to slip-form stepped, tapered, or even vertically curved surfaces; but such construction must be justified in terms of overall job economy, time requirements, or aesthetics. Tapered chimneys have been slip-formed to heights of 1200 ft.

Some of the major construction uses of slip forms, in addition to bins and silos, are in the high-rise commercial buildings. These include central cores which act as backbones and take lateral loading from the structural-steel or precast-concrete framing. The cores usually include elevator shafts, stairwells, toilet facilities, and mechanical runs. Another use of slip forms is in the construction of bearing walls for apartment houses, in combination with cast-in-place or precast concrete slabs. Other uses are water towers, dam intake towers, missile silos, monumental towers (often with restaurants on top), cooling towers, and air traffic-control towers.

The CN Tower in Toronto is the world's tallest building/free-standing structure at 1815 ft (553 m) high. The concrete structure was slip formed. It was completed in June 1976 after a 40-month construction period.

DESIGN OF SLIDING FORMS

The forms may be divided into three basic components: sheathing, wales, and yokes. These forms are subjected to both vertical and lateral loading. Figures C1-53 and C1-54 illustrate a typical form, and Fig. C1-55 shows the working deck with a placing deck above and a finisher's scaffold below. The entire weight of all decks and of the finishing scaffolds is carried on the jack rods. The jack rods will be discussed later, but now it can be said that the vertical loads are transmitted from the sheathing and decks to the wales, through the yokes to the jacks, and into the jack rods. The only function of the concrete is to support its own weight and to prevent the jack rods

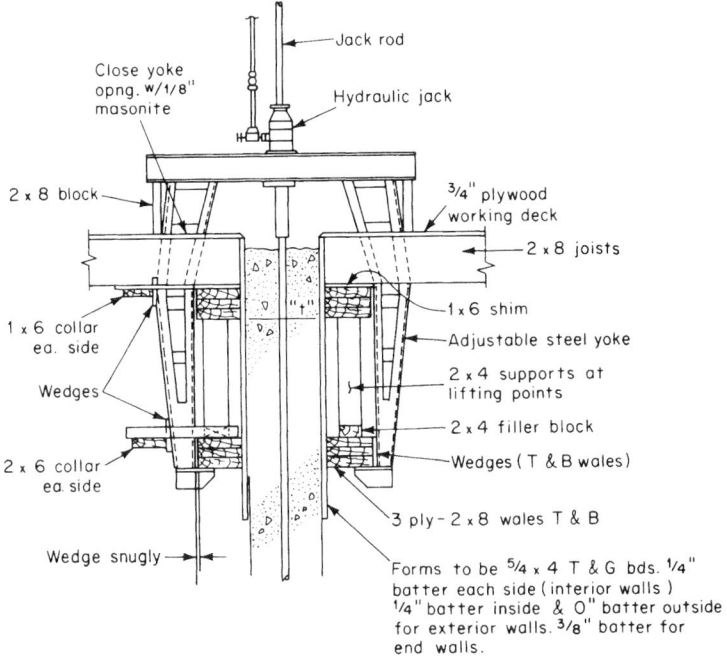

FIGURE C1-53 Section through slip form.

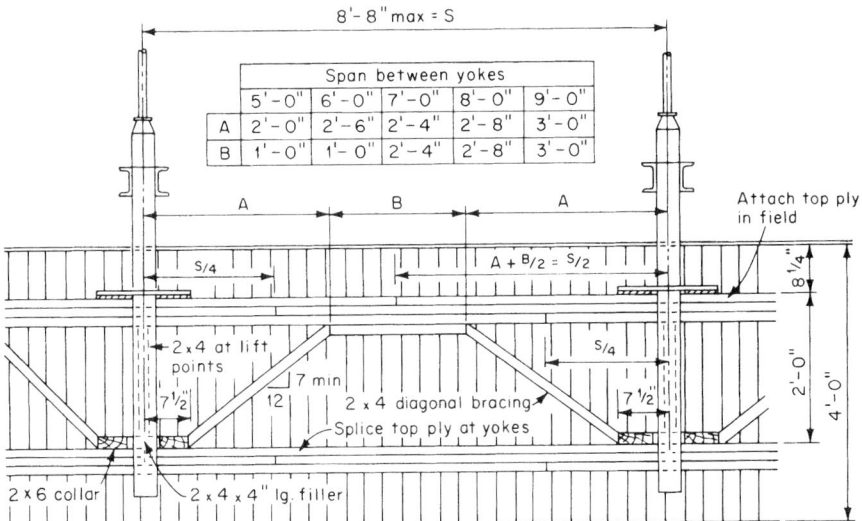

FIGURE C1-54 Side view of slip form.

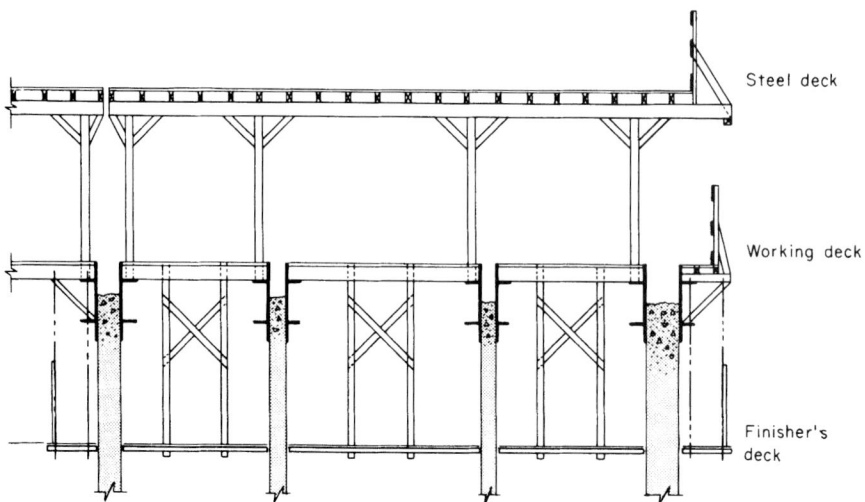

FIGURE C1-55 Slip-form decks.

from buckling. In addition to the dead loads, live loads of at least 40 psf for the deck and 50 lb per lineal foot of scaffold must be included in the vertical design loads. Reinforcing steel, forming boxes, and other materials to be stored on the decks must also be considered.

Another important vertical loading on the forms is the *drag force* or the friction of the concrete against the forms as they are raised. The magnitude of this loading is very difficult to determine since it is affected by inherent properties of the concrete (some concretes are more adhesive than others), by the moisture content, workability, rate of set, temperature, form surface, and the condition of the form surface. In order to reduce this force, the forms are given a slight inward batter of about $\frac{1}{16}$ in/ft of form height. Oiling or plastic treatment of the sheathing is desirable to prevent water absorption by the forms, as excess surface moisture has a beneficial lubricating effect. A drag load of 100 lb per lineal foot of forms is suggested as a good design criterion.

The drag forces are picked up directly by the sheathing and transmitted to the wales as a more or less uniform load. The deck loads are applied directly to the top wales by joists and beams, and the scaffold loadings are also applied to the top wales by the scaffold brackets. These loads must be carried by the forms to the supporting yokes. For short spans, the wales act as vertical beams, with the sheathing distributing the loads between the upper and lower wales. If the sheathing can take shear, as in the case of steel or plywood, the form will act as a girder. For long spans or heavy loads, the form is trussed to carry the loads.

In addition to the vertical loads, the forms must carry the hydrostatic lateral pressure of the plastic concrete. The sheathing must be designed to support this hydrostatic pressure between the wales and as a cantilever at the ends. The wales must, in turn, carry the pressure as lateral beams between yokes. The only time that the lateral pressure on the lower wales is large is when the empty forms are first filled. Under normal conditions the concrete in this area is not plastic, and the upper wales take the lion's share of the lateral pressure.

In figuring the hydrostatic pressure of the concrete, the formula recommended by ACI Committee 347 may be used. Although this formula gives lower values than that used for designing fixed forms, many years of slip-form design have indicated that its use will give adequate safety against lateral failure. In any case, the hydrostatic head of concrete should never be more than three-quarters of the form height. Table C1-29 shows allowable yoke spans for various sizes of doubled wales. Note that for curved walls allowances must be made for the material lost due to cutting.

The sheathing for slip forms may be fabricated from any of several materials. The most commonly used material is wood because of its economy and its ability to stand racking, distortion, and abuse without permanent damage. It also lends itself to easy repair and alteration. The wood section most frequently employed is a 1-in stave in 3- to 6-in widths. The staves may be tongue-and-groove or square ended, and of either soft wood or hard wood, depending on the requirements of the job. Three-quarter-inch plywood sheathing is also used, since it has the advantage of reducing the labor required to build the forms. However, plywood forms need much heavier wale systems and bracing than do staves because of a serious tendency to distort in use. Plywood is also limited to straight surfaces or mild curves because of bending difficulties.

Steel forms are sometimes used for slip-forming. These are several times as expensive as wood but are justified if sufficient reuse is anticipated. They have been used successfully in the construction of underground missile silos which were poured with one-sided forms, the braced sides of the excavation acting as the second form face. Still other types of form material may be used for sheathing as long as they are smooth, strong, somewhat ductile, fairly impervious to water, and not subject to serious changes in properties due to temperature variations.

Wales, except those for steel forms, are usually made of wood and are built up in two or three piles with the joints staggered. Two-ply wales are always built of 2-in-thick material. Three-ply wales may be built of 2-in material or a combination of 2- and 1-in material. Most forms have two wales; some are designed for three wales. For very high forms, perhaps 6 ft in height, additional wales will be required.

The yokes are designed in the shape of an inverted U whose legs are attached to the wales. These legs carry the vertical loads in tension and the lateral loads as cantilever beams. The cross arm of the yoke must be designed as a beam, supported at the center by the jack, and subjected to moments from both vertical and lateral loads. Yokes are usually built of either wood or steel and will have a certain capability for adjustment to permit their use on more than one job. Steel yokes must be adjustable over a considerable range as they are used over and over on many jobs.

JACKING SYSTEM

Propulsion and support of the forms is an important aspect of the sliding form operation. This propulsion and support is affected by means of jacks and the rods on which these jacks are supported. Fifty years ago practically all slip-form jacking in this country was accomplished using manually operated screw jacks, but today they are rare. Manual operation has been largely replaced by electric, hydraulic, and pneumatic jacking. The jacks are more or less cylindrical in shape, with a hole in the center through which the jack rod passes. They usually have two jaw clutches which alternately grip and raise, so that the jack climbs the jack rod very much like a monkey climbs a palm tree.

The extent of each climb is governed by the stroke of the jack, which is commonly 1 in. The speed of the jack is a function of its stroke length times the number of

TABLE C1-29 Allowable Yoke Spans for Various Sizes of Doubled Wales

Rate of slide*					Allowable span, ft							
	9 in/h				12 in/h				15 in/h			
Temperature, °F	40–50	50–60	60–70	70–80	40–50	50–60	60–70	70–80	40–50	50–60	60–70	70–80
					Top wales							
Double 2 × 4	5.88	6.29	6.86	7.60	5.00	5.53	6.09	6.86	4.47	4.89	5.53	6.09
Double 2 × 6	9.12	9.76	10.7	11.8	7.81	8.61	9.43	10.7	6.92	7.61	8.61	9.43
Double 2 × 8	12.2	13.0	14.2	15.7	10.4	11.5	12.7	14.2	9.27	10.1	11.5	12.7
Double 2 × 10	15.4	15.5	18.0	20.0	13.2	14.5	16.0	18.0	11.7	12.8	14.5	16.0
Double 2 × 12	18.6	20.0	21.8	24.1	15.9	17.6	19.3	21.8	14.3	15.5	17.6	19.3
					Bottom wales							
Double 2 × 4	5.15	5.58	6.14	6.91	4.24	4.79	5.37	6.14	3.60	4.12	4.79	5.37
Double 2 × 6	8.01	8.66	9.56	10.7	6.70	7.45	8.33	9.56	5.56	6.40	7.45	8.33
Double 2 × 8	10.6	11.5	12.8	14.4	8.94	9.18	11.1	12.8	7.41	8.60	9.18	11.1
Double 2 × 10	13.6	14.6	16.2	18.2	11.5	12.6	14.1	16.2	10.1	11.2	12.6	14.1
Double 2 × 12	16.4	17.7	19.6	22.0	13.9	15.2	17.0	19.6	11.3	13.1	15.2	17.0

* Rate of slide is maximum expected at any time.

strokes per hour. Jacks in use at this time are all capable of speeds in excess of 20 in/h. Actually, the speed of the sliding form operation is not controlled by the capabilities of the jack but rather by the rate of set of the concrete. If the jacking operation is too rapid, plastic concrete will fall out from the bottom of the forms. If the jacking rate is too slow, the concrete will adhere to the forms and either cause the forms to bind or, what is most probable, the concrete will tear horizontally and lift with the form.

The optimum jacking rate is greatly influenced by concrete properties and temperature. Higher rates of slide are required in hot, dry weather and lower rates in cold, wet weather. Slip-form operations are normally not carried out at slide rates less than 2 in/h; 30 in/h is the highest speed known to the writer. The normal range of operation is between 6 and 18 in/h, resulting in average rates of 8 to 14 in/h on a 24-h basis.

Jack rods are usually either solid steel rods or circular pipe of a fairly high carbon steel. The most common rod diameter is 1 in, although other diameters are used with certain jacks.

Jack rods are normally in axial compression and must be designed to carry this loading without buckling. Where concrete is not present, 4- by 6-in wood posts are used to give the necessary lateral bracing. The jack rods are fastened to these posts with blocks and jay bolts. One end of the bolt is shaped like the letter J and is wrapped around the jack rod, while its other end passes through a hole in the post. A block of wood is placed between the jack rod and the post, and the bolt is then tightened. Under some conditions, jack rods are suspended from a structure above the top of the slide. This places the jack rods in tension and tends to increase the accuracy of the structure as to vertical plum.

The normal spacing of jack rods is between 4 and 9 ft, but this may be varied upward or downward as required. The considerations that affect the jack spacing are the following: maximum allowable span of the wales, curvature of the wall, capacity of the jacks, capacity of the jack rods, distribution of loads to obtain uniform loading on all jacks, and placing of jacks to support all corners. It is usually necessary to concentrate jacks at points of heavy loading such as deck beams, concrete hoppers, and bridge landings. The proper layout of the jacking system is of major importance to the success of the slip-form operation. Jacks are almost always set up to operate simultaneously from a central pressure or power source. When a button is pushed, all jacks on the job climb a certain distance up the jack rods. Most of the jacks have excellent accuracy as to the amount of climb, but field conditions and mechanical imperfections make continuous checking necessary. Leveling adjustments must be made as required, although many of the jacks have devices which make them almost completely self-leveling. In any case, the jacks can be individually raised or lowered by hand or by manipulating valves.

DESIGN OF WORKING DECK AND BRACING

Working decks are supported directly on the forms and rise with them. When the span between forms is too great for the deck joists, beams or trusses are used. The deck sheathing and joists should be designed for dead load plus a local live load of 75 lb/ft^2 or concentrated live loads from concrete buggies or other construction equipment, whichever is greater. Power buggies have not been used on slip-form decks because of the very high lateral loadings they impart. Beams and trusses may be designed for a uniform live load of 40 lb/ft^2.

If the deck is to be used as a slab form at the end of the slide, it must be designed to take the weight of this slab without excessive deflection.

Table C1-30 gives maximum spans for various types of deck sheathing and Table C1-31 gives deck-joist spacings. These tables are based on live loading from nonpowered concrete buggies or 75 lb/ft², whichever is the more critical. The spacings must be modified to take care of any greater loading, if present.

The deck on a slip-form project has the important function of tying all the formwork together so that the structure goes up as a unit. The deck must be designed to maintain the plan dimensions throughout the height of the structure. Distances between walls must remain constant; square corners must stay square; circular arcs must maintain radius. The well-designed and -constructed deck tends to keep itself level. It prevents corners and projections of the structure from moving horizontally.

TABLE C1-30 Maximum Spans for Deck Sheathing

(Loading from nonpowered concrete buggies or 75 lb/ft², whichever is the more critical)

Type of sheathing	Maximum span, in, center to center of joists
1- by 6-in square edge	12
1- by 8-in square edge	16
1- by 6-in tongue and groove	22
1- by 8-in tongue and groove	24
½-in plywood	12
⅝-in plywood	16
¾-in plywood	22
⅞-in plywood	24

TABLE C1-31 Floor-Joist Spacing for Sliding Deck

(Loading from nonpowered concrete buggies or 75 lb/ft², whichever is the more critical)

Span, ft	Spacing, in, for joists of indicated sizes				
	2 × 4	2 × 6	2 × 8	2 × 10	2 × 12
4	38	91	130	164	200
5	20	59	104	131	160
6	12	41	73	110	132
7		27	53	85	114
8		18	41	65	96
9		13	32	52	76
10		9	26	42	61
11			21	34	50
12			18	29	42
13			16	25	36
14				21	31
15				19	27
16				16	24
17					21
18					19
19					17
20					15

It tends to keep straight lines straight. In order to accomplish these functions and, in addition, resist wind loads, the decks must be properly braced in the horizontal plane. This bracing may be wood, steel rods, steel plating, trusses, or combinations of any or all of these.

REINFORCING STEEL

Reinforcing steel for slip-form work must be detailed, placed, and inspected in a somewhat different manner than would be the case for conventional forming. In slip forming, the vertical steel is set up in the forms and held in place by templates attached to the deck and moving with it. The steel is lapped and tied to the rod below; it is held at the top by the templates at heights of from 4 to 10 ft above the deck. The higher the template, the longer the rebar that can be used. Laps are usually staggered; this is better structurally and distributes the workload more evenly for the ironworkers. Vertical bars must be limited in their lengths, as considerable whip may otherwise develop on a windy day. The length of vertical steel is usually kept between 14 and 20 ft, depending on the size of the bars. Bars heavier than No. 6 may be made as much as 5 ft longer than this, but the templates must be high. Bars longer than 20 ft become difficult to handle in the field unless special arrangements are made.

Horizontal steel is set a layer at a time as the work progresses. The bars must fit below the cross arm of the yoke and must be threaded in through the vertical steel and jack rods. Rebar detailing must verify that it will be physically possible to place the bars, but there is seldom any reason why horizontal bars cannot be detailed so that their placing is very easy. In this system, the ironworkers are always working safely and comfortably right at floor level. The actual lengths of bars may be limited by the design of a particular structure and will seldom exceed 20 ft. Bent bars or bars with hooks must be shorter than straight or hoop steel.

Whenever it is possible for the splices in horizontal bars to be inaccurately placed, such as in a continuous hoop of large diameter, it is advisable to add a few inches to the length of each bar to give the ironworkers a reasonable tolerance. The spacing of horizontal steel should be carefully studied to give the easiest possible placing in the field. Spacings of less than 6 in tend to keep the ironworkers too busy and make inspection somewhat difficult. Spacings of 10 to 12 in are normally ideal if this can be arranged. The larger-diameter bars are harder to handle and place, so that it is desirable to use bars of ¾-in diameter or smaller. Of course, it will not always be possible to keep within the suggested size and spacing limitations and yet furnish the required steel area. The designer must use judgment and come up with the best possible solution for the particular structure. Spacings as small as 3 in have been used, and bar sizes up to No. 11 have been employed.

Ties and stirrups must be detailed with particular care so as to ensure easy placement. Hooks bent at 90° are preferred to standard hooks as they can be rotated about the axis of the bar and positioned around a vertical. Care must be taken in placing ties so as not to foul the yokes if a basket-type arrangement is used or if the ties are dropped in from the top. If they cannot be dropped in from the top, the ties must be detailed in pieces that can be easily placed from the sides of the column, pilaster, or wall and yet fully meet the structural requirements.

Inspection of reinforcing steel in slip-form work requires a technique of its own. Large areas of reinforcing cannot be inspected at one time, nor can concreting be delayed until inspection of steel has been completed. The reinforcing is constantly disappearing into the concrete as the forms rise at a rate of 12 in an hour or more. In order to facilitate placing and inspection, all horizontal steel should be designed in horizontal layers; that is to say, all horizontal steel should be at the same vertical spacing or at least in multiples of the same spacing. Using this system, a reinforcing steel crew knows that a full set of steel as shown on the drawings must be installed at certain elevations. The inspector, too, can check layer by layer to make sure that no bar is left out of any set. Particular care must be taken to ensure that lighter steel such as hairpins, ties, corner bars, etc., is not omitted.

Positive means must be established to identify the location of each layer of horizontal steel. One method is to mark the layer spacings on vertical bars at several locations by means of saw cuts or tightly wound tying wire. Keel marks have a tendency to get lost if used alone, but they are very good location markers for saw cuts or wire markings. Some firms embed light angle irons in the concrete at one or two locations. These angles are bolted together in sections as the slide progresses and are drilled with holes at the proper location of each horizontal layer. When possible, short bars which do not interfere with the yokes can be tied several courses in advance. They will then furnish excellent guides for the rest of the bars.

Sometimes the steel and the concrete are placed alternately in layers of equal thickness. This sounds like an excellent system, sometimes it does actually work in the field. However, most of the time it is extremely difficult to maintain, and it tends to slow down both crews and make inspection difficult. A great help to both placing and inspection is to mark the ends of the boards on the decks so that individual measurements are not necessary each time a layer is placed. Since concrete is constantly being placed in the forms, these markers must be of a raised type such as double-headed nails, V notches, steel plates, or wood blocks. At the same time, they must not be a tripping hazard to people working on the deck.

Tying is often greatly reduced on slip-form work, since the steel is placed directly on the concrete. Some variation in spacing is usually not of structural significance unless the effective steel area is reduced over lengths of 3 or 4 ft, or unless maximum allowable spacings for the steel are exceeded. On the other hand, it is very important to keep proper clearance between the steel and the forms. The inspector must see that this clearance is properly maintained and that sufficient wire ties are used to assure proper cover for the steel. Vertical steel presents no problems in placing or in inspection as long as the lengths and ties, if any, are properly detailed. Bars larger than No. 11 must have welded or mechanical connections, and will seriously delay the slip-form operation.

CONCRETE PLACING

In general, the concrete mixes used for slip-forming are the same as those employed for other methods of concrete construction. Concrete to be slip-formed must usually have a higher slump than concrete that is to be formed by more conventional methods, and slumps between 3 and 5 in are most commonly specified. Placing concrete in slip forms with a slump below 3 in becomes very difficult if not impossible. On the other hand, in tropical climates and using certain types of aggregates and cement, slumps up to 7 in have been required. Vibration, retarders, and workability agents can be used under the proper conditions to retain adequate workability while reducing the slump. The slump actually required is difficult to predict and will very continuously with changes in the weather. Since improper slump will result in poor concrete, adequate provisions should be made for its careful control. Several batch mixes should be prepared, covering a range of slumps and all meeting the strength requirements. The batching plant should then be alert and ready to switch from one batch mix to another. The superintendent in charge should be authorized to change slumps, as required, by using one of the previously approved batches.

Another point that requires special consideration is the coarse aggregate to be used in the concrete mix. For walls less than 8 in thick, the maximum size of aggregate must be limited to ¾ in. For most other slip-form work, maximum aggregate size should be limited to 1 in. Crushed stone or gravel is usually specified, but uncrushed gravel has been used successfully. Lightweight and slag concrete present no serious problems.

In slip-form work, concrete is placed in the forms in layers of 6 to 8 in. Each layer is vibrated as it is placed, and care should be taken that the forms are filled evenly and kept as nearly full as possible. Although systematic filling of the forms is advisable, the system should be set up to avoid inducing torsion in the structure as a whole. For instance, a hollow rectangular tower cannot be filled constantly in a clockwise direction as this will tend to make the forms rotate about the centerline of the structure. Placing in alternately clockwise and counterclockwise directions will solve this problem. If the rate of raising the forms is seriously slowed down for any reason, care must be taken to avoid cold joints. Placing the concrete in thinner layers, say 2 to 3 in, will help reduce the time factor between successive placements.

Vibration of concrete was at one time considered dangerous in slip-form work. Thanks to the work of ACI Committee 609 and to favorable experience with vibration, it is now recommended rather than prohibited. The vibrator may be allowed to penetrate as deeply as it will under its own weight, but it should not be forced deeper into the concrete. Further, it should not be used too long in one place.

The slip-form decks must be kept as nearly broom clean as possible in order to prevent hardened concrete spillings from finding their way into the forms. When cleaning the decks, the sweepings should be directed away from the forms. Clean-out openings must be provided in the deck or some other arrangement made for disposal of debris. When daily stops are made, exposed portions of the forms must be thoroughly cleaned before resuming work.

Admixtures have been successfully used with slip-form work. The use of air-entraining agents and workability agents is quite common. Under normal slip-forming conditions, no ill effects result from the use of these admixtures; in fact, their effects are usually quite beneficial. Naturally, the usual rules of careful use and control apply here as elsewhere. Care must be taken to see that any admixtures are properly dispersed throughout the concrete. If, for instance, a retarder gets concentrated on one side of a structure, the concrete in this area is too plastic to allow the forms to move. The concrete on the other side has not been retarded to the same extent and is setting rather rapidly. This results in a very awkward situation: unless the forms are moved, one side will bind; and if they are moved, the concrete will fall out of the bottom of the other side.

Control of set, workability, moisture content, and temperature are of particular importance in sliding-form operations. In hot weather, or when slide rates must be reduced because of many inserts being placed or for any other reason, admixtures are recommended. Under those conditions, retarding workability agents are usually used to delay the set and to increase the plasticity of the concrete without adding excessive amounts of water. In some instances at least, it appears that the use of these workability agents does not completely remove the necessity of increasing the water content of the mix.

A very effective way of controlling the set of concrete during hot-weather placing is by replacing a part of the mixing water with crushed ice. When the necessary ice is available, this method is economical, effective, and entirely safe. The normal precautions of cooling aggregates, painting equipment white, and keeping the equipment in the shade are also in order. In hot weather, in traffic-congested areas, and most especially when these two conditions are combined, the use of plant-mixed concrete becomes dangerous. If this situation exists and a site plant is uneconomical or undesirable, then the trucks should be dry-batched at the plant and the mixing water added at the site. In this way, the concrete does not undergo excessive set in the trucks.

At all times, but especially when the temperatures go over 70°F, the number of ready-mix trucks in use must be such that no delay will result in the concrete placing. Even icing and placing the concrete in 2-in layers will not prevent a cold joint if the concrete supply is interrupted. Auxiliary equipment and parts replacements are musts for the concrete-handling system.

In cold weather, accelerators may be required to provide heat at a greater rate and to reduce the setting time. Careful consideration must be given before high-early cement is used on slip-form work. Unless this is really required, the setting time may be excessively reduced and lead to binding of the forms. It may be sufficient to increase the cement factor and thus accelerate the chemical reaction and obtain early strength.

Insulating the forms is very beneficial, at times absolutely necessary. Lightweight enclosures can also be suspended from and attached to the slip forms. In cold weather, the minimum enclosure should be a tarpaulin enclosure for the deck railing and the finisher's scaffold. Heating is often required for these enclosures. Salamanders can be placed at intervals on the finisher's scaffold, but they constitute a serious fire hazard. If steampipes can be suspended from the forms and supplied from steam generators, the fire hazard will be eliminated and the concrete curing will be greatly improved.

Inspectors may check the rate of slide by plunging steel rods into the concrete and measuring the depth of hard concrete in the forms. This depth should be 12 to 30 in and, if it is less than 12 in, the forms must be slowed down until the concrete sets further. If the depth of hard concrete exceeds 30 in, the rate of slide must be increased to prevent binding of the forms. Experienced personnel can judge the rate of slide by standing on the scaffolds below the forms and scratching the green concrete to see how hard it is as it comes out of the forms.

CONTROL AND TOLERANCE

As a slip form rises there is a tendency for it to translate, or possibly even to rotate, or to combine translation with rotation. The smaller a structure is in plan, the greater this danger becomes. Fortunately, correction is entirely possible, and those structures which most easily go out of plumb are the ones which can most readily be corrected. Any movements are actually of a small order of magnitude, but they could result in a tall building being several inches out of plumb. Adequate provisions must be made to limit and correct any deviations as slip forming proceeds.

If the deck is not level, the building will naturally "grow" in the direction toward which the deck is tilted. Therefore, the first requirement is that the deck be level at all times unless a tilt is deliberately introduced to make a correction. The forms must be carefully built and leveled before the slide starts. Once the slide starts, levels must be checked continuously and all jacks kept within ½ or ¾ in of correct elevation. If a jack gets too far ahead of the others, it is made to miss one climb. If it lags behind, it must be brought up to the proper level manually. Even with automatically leveled jacks, human inspection is still necessary to ensure proper operation of the leveling system.

FINISHING AND CURING

Slip-forming provides an excellent opportunity to finish wall surfaces while the concrete is still quite green. When the concrete is finished below the form it is 4 or 5 h old, rather than 24 h or more. It is therefore possible to do excellent patching and to obtain a very dense surface. The most popular and probably the best finishing technique is the float-and-brush finish, which results in a sandy-textured, attractive surface. When required, as on the inside of flour bins, a steel-trowel finish may be applied.

Because of the combined effect of the deflection of the forms due to the hydrostatic pressure of the concrete and the taper built into the form, a phenomenon called *shingling* often occurs at levels where the form has become empty to a considerable extent and is then filled. This happens, for instance, at planned stops. The forms must be kept moving once the concreting is completed in order to prevent binding and to allow the slide to start the next day or as planned. This movement usually is continued for 2 to 3 h after the last concrete is in place, leaving 1 to 2 ft of empty form. This represents $\frac{1}{16}$ in of draft or a $\frac{1}{8}$-in gap between the forms and the top of the concrete wall. When concreting is started again, the hydrostatic pressure will increase this gap and an unsightly overpour will occur. This can be minimized to almost nothing if the amount of empty form is kept down to 12 or 15 in.

If aesthetic effects are desired, as for architectural concrete, special care must be taken to prevent shingling and cold joints. It is also desirable to keep the depth of plastic concrete in the form as uniform as possible. Special effects can be obtained by setting masonry or plastic form inserts into the forms on the exposed faces.

The main method of curing slip-formed walls is membrane curing. Membrane curing can be applied from a garden-type pesticide sprayer while walking around the structure on the finisher's scaffold. Using a dye will help prevent "missed" areas. Once the membrane is on, no further maintenance is required.

CONNECTIONS FOR BEAMS AND SLABS

Adequate connections must be devised for connecting the floor systems to the previously cast concrete walls. These connecting systems should have maximum tolerance and flexibility.

Slabs are usually attached to walls by means of 1⅝- to 4-in-deep keys. Dowels, if required, can be placed inside the key forms and bent out into the slab after the slide is over. Threaded inserts to take threaded rods, through-holes, or hooked connections may also be used.

Concrete beams can be connected in much the same manner as the slabs except that a pocket or through opening is utilized instead of a key. Beam dowels will usually be of larger diameter and greater length than slab dowels. Stubs for welding or mechanical coupling such as Cadwelding will be required.

Attachment of structural-steel beams to concrete walls can be accomplished by means of pockets and vertical anchor bolts or by weld plates cast into the concrete and welding. Precast beams are attached in the same manner.

SECTION 5
GROUTING

Grouting is a combination of engineering and art wherein fissures, voids, or cavities in rock or soil masses are filled with a fluid that will harden in place to increase the overall strength and/or impermeability of the mass. Engineering judgment is required to determine the feasibility of the work, the type of grout to be used, and the general conditions of execution, including the establishment of limitations. Successful results are also highly dependent on the experience and skill of the workers, their immediate supervisors, and the field engineer.

Grouts may be considered under two broad categories. Those which contain finely divided solids are called *solid-suspension* grouts, and those which are pure solutions are known as *chemical* grouts. The two types are occasionally combined, as in the case of chemical grouts which contain inert fillers or, rarely, portland cement.

Solid-suspension grouts containing reactive solids such as portland cement and pozzolans, swelling materials such as bentonite, and/or inert fines such as clays are generally preferred when the flow channels in the void system are large enough to permit the solids to flow through them. In broad terms, cracks narrower than 0.01 in or granular materials composed of particles finer than ⅙ in will not readily accept solid-suspension grouts.

Chemical grouts, on the other hand, will flow anywhere that water can be made to move. The rate of movement is then dependent on the viscosity of the grout as well as on the pressure gradient. Chemical grouts usually consist of two chemicals which, when combined, will stiffen to a low- or moderate-strength gel. The hardening time and eventual strength of this gel are controlled by the proportioning of the reactive chemicals and the amount of water (if any), and often by a *modifier* which may be considered as a negative catalyst; i.e., the modifier controls the rate of reaction without being otherwise necessary for the basic reaction. Most chemical grouts are proprietary; the user is advised to study the manufacturer's literature for properties, capabilities, proportioning, and methods of placing. For this reason, the major emphasis in the following paragraphs will be on solid-suspension grouts.

MATERIALS

The basic materials for solid-suspension grouts are cement and water. Other ingredients frequently included are sand, pozzolans, clays, and admixtures. Proprietary materials are also available for special purposes.

Cement

If it meets the usual requirements for concrete, the cement will be suitable for most grouting. For pressure grouting fine seams in hard rock, finely ground or air-separated cements are sometimes required.

Water

This should also be suitable for concrete. A minimum specification should call for potability. Sand, except for its grading, should meet the usual requirements (such as ASTM Designation C-33) for any good conventional concrete. Where the minimum size of coarse aggregate is 1 in or less, all the sand should pass a No. 8 screen, 95 percent should pass No. 16, and 10 to 25 percent should pass No. 100. The grading should be as uniform as practicable between these limits to minimum bleeding; the fineness modulus should fall between 1.20 and 2.10. Fly ash and natural pozzolans are frequently used to replace up to 30 percent of the weight of portland cement. Fly ash and natural pozzolans should meet the requirements of ASTM Designation C-618 except that loss on ignition for fly ash should not exceed 5 percent. Either fly ash or natural pozzolan, particularly the former, is favored as a mortar ingredient to improve pumpability, extend grout handling time, and reduce bleeding. Since both materials produce something less than half the heat of hydration which is liberated by an equal weight of portland cement, and since they produce the heat more slowly, their use will also minimize temperature rise.

Bentonite and Other Clays

These materials vary so much from one location to another that each should be investigated individually for its suitability in the particular work. Bentonites are given a *barrel rating* in the oil drilling industry; the higher this rating, the more *mud* of a given consistency it will produce. For example, one ton of *90-barrel minimum* bentonite, a high classification, will produce 90 barrels (45 gal each) of 15-centipoise mud. Bentonites are used for their swelling and gelling action; clays and silts are used as fillers.

Admixtures

These can be highly beneficial in many instances. Water reducers, Types A (normal setting), D (retarding), and E (accelerating), ASTM Designation C-494, will greatly improve the strength and will help with pumpability where concrete strengths are desired. Grout fluidifiers will provide water reduction, retardation, and expansion. Certain proprietary lubricating-suspending admixtures are also available; these will suspend high volumes of sand in relation to cement to provide strengths ranging from over 2000 lb/in^2 to as low as 50 lb/in^2 with virtually no bleeding.

Special-Purpose Materials

High strength with volume stability can be obtained by using special materials which are composed of iron aggregate, catalysts, and plasticizers for job-site mixing with cement and sand. Factory-premixed combinations of these materials are also available. Widely accepted examples of the latter are Embeco Grouts and Mortars, premixed by Master Builders, which will provide 8000 to 12,000 lb/in^2 at ages ranging from 4 to 90 days, depending on the formulation. Premixed grouts of the gas-expansion type will provide strengths on the order of 5000 to 7000 lb/in^2 combined with a measure of bleeding correction.

PROPORTIONING

Grouts are proportioned to satisfy as well as possible a number of overlapping and, to some extent, conflicting conditions. As a fluid, the grout must be pumpable and be able to penetrate the material into which it is to be pumped. After it is in place, it must harden to a stable condition and provide strength and/or impermeability, all as economically as possible. The characteristics of the voids to be filled, together with the desired end result, determine which of these conditions are paramount and which are secondary. Proportioning will therefore be discussed in connection with the type of void to be filled.

Fine Seams and Cracks

These are usually encountered, or at least suspected, in rock under dams and associated structures, around penstocks, and in the bases and walls of deep excavations. The objective of grouting in these areas is to prevent water movement and/or add strength to the rock mass. The openings to be grouted vary from tight discontinuities to spaces an inch or more across, with the narrower widths predominating. This makes penetrability the first objective in proportioning the grout. With this in mind, grouts for such work usually employ only cement and water and start with dilute suspensions on the order of 1 sack of cement to 30 to 40 gal of water. The grout is then gradually thickened by reducing the water–cement ratio. When open cracks are encountered, economy dictates that grouting sand be added in proportions which may range from equal parts sand and cement where high strength is desired to as much as 10 or 12 parts of very fine sand to 1 part cement where only a stable seal is needed. When sand is employed, pumpability demands that the water content be adjusted to provide sufficient consistency to maintain the sand in suspension. Pozzolans in amounts up to 30 to 50 percent by weight of the cementing material may be used with sanded grouts for economy and to improve pumpability. When used in the higher proportions, the pozzolans will delay the hardening of the grout for several hours. Lignosulfonate water-reducing retarders may be used, especially with sand grouts, to reduce the water requirement, control stiffening time, and improve pumpability. Grout fluidifiers will do the same and will add expansion. Retarders and fluidifiers are of little use for water reduction in neat cement grouts when the water content exceeds 6 to 7 gal per sack, but it has been reported that their dispersing and lubricating action increases the take. Chemical admixtures, including fluidifiers, should be proportioned in the mix in accordance with their manufacturers' recommendations.

Cavity Filling

The desirability of a combination of economy and pumpability indicates the use of the sanded grouts just described. Pea gravel, ¼ to ⅜ in, may be suspended in these grouts to help fill space and choke off channels through which water may be flowing. Water-reducing retarders will improve pumpability and suspending properties; calcium chloride in amounts up to 2 to 3 percent of the weight of cement can be employed to overcome the retarding effect of the water reducers and speed up setting characteristics. Grout fluidifiers will provide expansion to hold the grout tightly against roof surfaces while the grout hardens.

Filling Vertical or Steeply Sloping Spaces

In general, all the preceding remarks concerning sanded grouts apply. When the space is such that the upper surface of the grout is not restrained, the expansion contributed by grout fluidifiers is of little value since gravity will hold the grout tightly to the cavity sides.

Mixes for Precision Grouting

This term includes the placing of grout under machinery, where precise alignment requires both volume stability and strength, and under column bases and the like where strength is the prime factor but volume stability is also desirable. Equal weights of activated ferrous aggregate, cement, and sand mixed with water to a pumpable or pourable consistency are generally specified for this work. If the blending is done at the job site, the ferrous aggregate manufacturer's instructions with respect to the selection of components and their proportioning and mixing must be very carefully followed to attain the desired results. The use of factory premixed grouts requiring only mixing with water at the job greatly simplifies this problem. For items requiring less critical support, cement-sand grouts proportioned 1:1 by weight and including a grout fluidifier are occasionally used. Units that are bolted to shims and are to receive their entire support therefrom may be grouted with a plain cement-sand mixture to keep the dirt out and prevent tools and trash from disappearing underneath.

Post-Tensioned Cable Mixes

Where good bond between the grout and cables is desired and the detailer provides fittings (½-in minimum) adequate to permit the ingress of a good grout, cement and water mixed to a 20- to 25-s flow cone consistency can be pumped. Water-reducing retarders will greatly improve both the pumpability and strength. Grout fluidifiers will do the same and will also add expansion, which further improves pumpability and provides protection against cable bursting from freezing. When smaller fittings are provided, as is often the case, a thinner (±15 s) grout must be prepared; this grout will provide corrosion protection but not much strength.

Grouting in Granular Materials

The problem with granular materials is generally one of getting grout to travel as far as possible, so that an area will be consolidated or a grout curtain established from the fewest possible grouting points. This puts a premium on penetrability. To achieve this, dilute grouts are used more frequently than thicker ones; since strength is not generally a controlling factor, clays and silts may be used in amounts several times that of the cement. Where only the arresting of water movement is desired, the cement is sometimes eliminated and clay or bentonite used alone. Water-reducing retarders are occasionally used as lubricants. Proportioning with respect to water-solids ratios can be determined only by field trial and modified as the work progresses. An experienced engineer can estimate the probabilities from a detailed study of grain-size distributions of soil samples and grouting materials.

MIXING AND PUMPING

Ratio of portland cement to other cementing materials may vary from 100 percent portland cement to 70 percent cement, 30 percent fly ash or natural pozzolan by weight. Within these limits, the amorphous silica in the pozzolanic material will react with the lime liberated by the hydrating portland cement in a secondary reaction to produce insoluble cementitious productions. These are similar in strength to the hydrated cement, provided suitable curing conditions are maintained. With portland cement alone, a plot of strength gain versus time will be similar to that for conventional concrete. As the pozzolanic proportion is increased, strength matching the 28-day strengths for the 100 percent portland-cement design will occur at progressively later periods. The time for equal strength occurs at about 90 days for 70-30 blends, and the strength curve is then rising at a steeper slope than for the cement-only curve.

For estimating purposes, the quantity of cement or cementing material will be approximately 6 sacks/yd^3 for structural concrete and less for mass concrete.

Ratio of water to cementing materials by weight should be in the range of 0.40 to 0.45 (4½ to 5 gal per sack) for structural-grade mixes and may run as high as 0.62 (7 gal per sack) for lean, mass concrete.

Grout fluidifier should be proportioned in accordance with the manufacturer's requirements, modified as necessary to provide:

1. A water reduction of at least 5 percent from the same mix without fluidifier.

2. A minimum expansion in a period of 2 to 3 h of (*a*) 2 percent for concrete not requiring freeze-thaw durability or (*b*) 5 to 8 percent for concrete exposed to freezing or a combination of expansion exceeding 2 percent and air-entrainment totaling 5 to 8 percent.

3. Bleeding not exceeding one-half the expansion of 1½ percent, whichever is less. The grout proportioning of sand grading may have to be modified to meet this requirement.

Grout consistency for preplaced coarse aggregate having a minimum size of ½ or ¾ in should be 20 ± 2 s when measured by the flow cone shown in Fig. C1-56. Using the same sand, i.e., fineness modulus (FM) not over 2.1, the flow-cone consistency may be increased to 30 s of 1½-in minimum-sized coarse aggregate and decreased to barely flowable for 6-in stone. The consistency of thick mortars (over 50 s) and those containing coarser sands must be controlled by some other means, such as the flowmeter (Fig. C1-57) developed in Europe for coarse-sanded mortars. In this meter, consistency is measured by the distance a standard volume of mortar flows along a rectangular channel. Flows of from 10 to 22 in are normal.

Having determined the proportions of solids and water to provide the required strength at the desired consistency, the cone or meter is used during the work to maintain the selected water content.

Batching and Mixing Mortar

Weight batching with conventional equipment is used for volume work. Since the coarse aggregate fraction of the concrete is already in place and usually accounts for over 60 percent of the total volume, the mortar batching and mixing equipment will be much smaller than that required for ordinary concrete. Hence, any batching procedure, including hand handling of whole sacks of cement and cubic-foot boxes of sand, is acceptable if suitable controls are set up to maintain accuracy of proportioning.

Figure C1-58 is a self-contained mobile grouter that simultaneously mixes and pumps by air-powered piston pump. Capacity is up to 10 gpm or 2 bags/min.

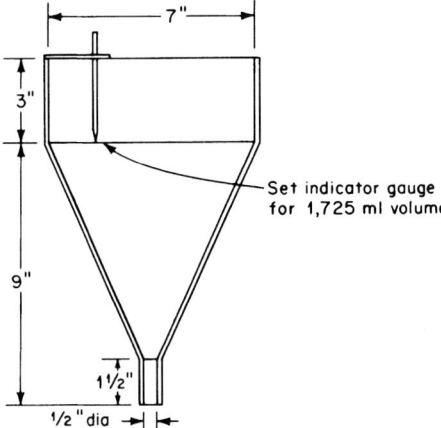

FIGURE C1-56 Flow cone. Instructions for field use: (1) Be sure discharge tube is clean and undamaged. A dirty, nicked, or out-of-round tube will not give an accurate reading. (2) Wet cone with water before using. Shake out excess water. (3) Hold cone firmly and vertically. (4) Place finger over discharge tube and fill to proper level with grout. (5) Remove finger and look directly down into the flow cone. Note time required for cone to empty (when you first see light through discharge tube). (6) Wash flow cone thoroughly after use.

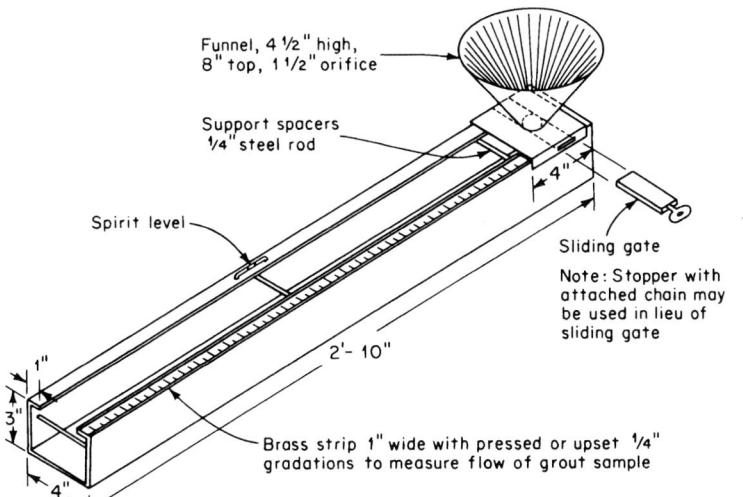

FIGURE C1-57 Flowmeter. First fill with 1.20 U.S. quarts of grout. Then open gate and read flow of grout on scale.

Figure C1-59 is a single-skid plant which can fit in a pickup truck. It pumps at 225 or 2000 lb/in^2. It mixes and pumps neat cement or sand/cement up to 20 gpm. Drive power can be air, electric, hydraulic, gasoline, or diesel.

Mixing for large-volume work is preferably accomplished in horizontal drum mixers with rubber-tipped paddles arranged to sweep the walls of the drum. Capac-

FIGURE C1-58 Mobile self-contained grouter. *(Chem Grout.)*

FIGURE C1-59 Skid-mounted grout plant. *(Chem Grout.)*

ities ranging from 1 to 6 yd^3 have been used. Vertical-shaft paddle mixers have also been used for smaller work. Generally, two mixers or *tubs* are set side by side so that while one is mixing, the other is agitating as it delivers grout by gravity to the funnel of a common pump. In the United States, the mortar has also been batched and mixed at ready-mix plants, or batched there and transit-mixed in conventional concrete trucks en route to the job.

Materials are added in the order of water, grout fluidifier, and other solids for large mixers; water, fluidifier, cementing materials, and sand for small mixers. The mixing cycle is 3 to 5 min, but grout containing fly ash and fluidifier may be held at agitating speeds at temperatures below 70°F for as long as 2 h without harm. Without fly ash, 1 h of agitating time is suitable limit for 3000-lb/in^2 concrete. The controlling factors are maintenance of pumpability and an excess of expansion over bleeding at the time the grout is intruded. For 4000 lb/in^2 concrete and up, the holding limits should either be set at 30 and 15 min, respectively, or determined for the materials actually being used.

A high-speed shear mixer developed in Europe provides an excellent suspension. Water, admixture, if any, and cement are first circulated through a specially designed high-speed centrifugal device which produces a smooth grout having some of the characteristics of a colloidal suspension. For small work (one or two sacks of cement per batch), sand is added to the same drum; for a greater production rate, the grout is passed to a second drum for the addition of sand. The larger equipment is reported to have an output rate of some 240 ft^3/h. Because high-speed mixing requires a lot of power, it is essential that the mixing cycle be limited to 20 to 30 s per batch. If the grout is not quickly discharged to a holding agitator or tank, it will heat up and thus increase the mixing water requirement.

Transporting the Mortar

The mortar may be moved from the mixing plant to the point of use by pumping, trucks, or gravity. Of these alternatives, pumps are most frequently used. In addition to their convenience, they are able to meter the grout, provide controllable pressure and delivery volume, furnish the mortar to otherwise inaccessible locations hundreds of feet above or below the mixers, and provide a good measure of control against the inclusion of air bubbles.

For long hauls, ready-mix trucks or tanks with agitators may also be used. When this method is employed, the mortar should be discharged into a holding tank, also agitated, from which pumps or gravity can deliver to the inserts.

In some instances, grout may be delivered advantageously by gravity, such as down a mine shaft or from a mixing plant located well above the use point. Gravity lines should be free flowing (i.e., without valves) and vented at the upper end to ensure against the creation of a strong vacuum which might dry the mortar and cause a plug or collapse a hose. They should discharge into a receiver tank with an agitator to allow air bubbles to escape before the mortar goes to the pumps.

Grout pumps may be of either the piston displacement or screw types. Air-driven piston pumps equipped with rubber pistons and rubber valves or valve seats are preferred over those driven by engine and crankshaft because the worker at the insert can shut off the flow without damaging the pump. Air-driven pumps can be selected that deliver from less than 1 to over 5 cfm, depending on fluid and piston sizes. Their delivery pressures range from 50 to 1000 lb/in^2 or more depending on the diameter ratios of the fluid and air pistons. Engine-driven pumps can deliver from very small quantities to as much as 25 cfm at pressures to over 5000 lb/in^2. The use of dual or triple pistons will minimize pulsation in the grout lines.

Screw pumps also find wide use, especially for the smaller jobs. The screw pump has the advantage of delivering without throb, but rotor and stator replacement becomes a problem with hard, sharp sand.

For high lifts or long distances between base pump and point of use, relay pumping setups may be necessary. The frequency of relays will depend on available pump capacities. For commonly available equipment, relays at 500 ft vertically to 1000 to 2000 ft horizontally are reasonable starting points for planning. A relay station should include a receiving agitator to feed the next set of pumps and thereby eliminate pulsation interference.

Grout lines should be sized to provide for a mortar velocity in the vicinity of 2 to 4 fps. Although properly proportioned grouts will not segregate, this velocity will keep the mortar "alive" over the whole cross section of the conduit and thus minimize grout buildup around its perimeter. Oversized lines encourage slow choking; undersized lines waste pumping effort. For short lines, 100 ft or less, and for small quantities, 1-in hoses are adequate; for larger volumes and longer lines, pipes or hoses up to 3 or 4 in should be considered. Mortar has been dropped as far as 2500 ft through 2-in vertical pipe in mining applications and pumped it uphill in large volumes through 4-in lines at 30° for 1900 ft (900-ft rise) without difficulty.

All grout should be passed through a ⅜- to ½-in screen at some point between the mixer and pump to remove lumps, stones, tramp iron, pieces of wood, etc., that always seem to get into the mix sooner or later.

Valves of the plug type, providing unrestricted flow, should be installed on the insert ends of all grout supply hoses. Wyes with a valve on each arm may be used to connect a grout supply line to two inserts. Although mortar will be pumped through only one insert at a time, the wye will permit changing connections without pause in the grout flow. Unions or full-section quick couplings are necessary for making connections to inserts. All grout-line joints and valves, whether at the insert or elsewhere, must be drip tight or plugging will occur.

GROUTING PROCEDURES

Grouting procedures vary widely, depending on the type of grout and the particular grouting requirement. Because of this, the coverage supplied herein can only be very general.

Hard-Rock Grouting

Grouting holes are drilled in the rock in a predetermined pattern, usually making use of rotary or percussion drills. Where the split-spacing grouting method is used, a first series of holes will be drilled, washed, and grouted. Successive series of holes are drilled and treated in the same fashion until the curtain or area will accept no more grout. In the progressive grouting method, holes are drilled at a predetermined minimum spacing, then washed and grouted in order from one end to the other. The second permits better washing from one hole to the next of the seams to be grouted, but it requires somewhat more skill. It may also cost more if grout flows easily from one hole to another because many of the holes may have to be redrilled or several adjacent ones may have to be grouted simultaneously.

Washing and grouting may be done in stages as the holes are drilled, a procedure which strengthens the rock from the top down to permit better control of pressure and grout "take" as the holes get deeper. For the successive depth stages, a grout pipe and packer are inserted in the hole to a point in the zone just completed. These are then used to wash and grout the new stage.

The procedure is sometimes reversed: i.e., the hole is drilled to full depth, with grouting proceeding in stages from the bottom up. Some engineers feel this gives them a better knowledge of the rock and its seams before grouting starts; others are concerned that grout may bypass the packer and prevent it from being withdrawn, or that grout will escape all the way to the surface and limit grouting pressures.

In addition to washing the cuttings from a hole, most grouters also apply water pressure in an attempt to wash fine materials from the cracks, either to the next hole or to some distance from the hole. If circulation from hole to hole is obtained, compressed air may be alternated with water to obtain better cleaning.

Mixing may be accomplished at any convenient location. The grout is pumped to the point of use and back to the mixer in a return line, with desired maximum grouting pressure maintained by throttling at the return end. Grout is tapped off this line to the hole; a valve and pressure gauge at the top of the hole are used to control the actual grouting pressure. Closer grout pressure control and a better knowledge of grouting conditions, including take, are effected by drawing from the grouting line to an agitator and pump at the grouting point. A mixing and pumping setup on skids close to the hole gives best overall control, since mixes can be thickened or thinned as the occasion demands without the waste and delay that a long circulating line provides.

Pressures are usually limited to 1 lb/in^2 per foot of depth to minimize the possibility of rupturing the rock. When grouting badly fractured rock, benchmarks may be set and observed for signs of heave during grouting operations.

Grouting Cavities and Open Seams

When these are to be filled with sanded grouts, it is only necessary to gain access at the lowest point above which positive filling is required. Pumping then proceeds from this location. Recirculating lines are generally neither desirable nor necessary, although the crew should make sure that the line is kept alive. This can be checked by wasting a bit of grout every 5 to 10 min when there are delays due to changing connections or for other reasons.

Loss of grout in channels through which water is flowing can sometimes be controlled by loading the grout with the largest-size gravel that the channel will accept along with smaller sizes, using larger grout holes for access when possible. Although at first most of the grout may be lost, the gravel will hopefully hang up on irregularities to build up a reverse filter.

Precision Grouting

Machine and column bases may be grouted with precision by pouring if they are small and flat bottomed (narrowest dimension not over 4 to 6 ft); otherwise, pumping is required. For pouring, the vertical space between the concrete and the plate should be at least 1 in for a width of 2 ft, increasing by 1 in for each fraction of 2 ft wider; for pumping, the space should be 2 in or more. Pea gravel, which contributes to economy and will absorb some of the heat hydration, may be added to grout placed in thicknesses of over 2 in. Any space over 6 in (8 in with gravel in the grout) should be filled in two or more stages, with 24 to 48 h between stages to allow for escape of the heat of hydration.

Prior to setting the bed plate, the concrete should be roughened and the underside of the plate cleaned of grease, dirt, and loose rust. For pouring, the plate should be formed and the grout placed as shown in Fig. C1-60 so that the material flows under the short dimension of the plate. A steel strap should be inserted from the far side and worked back and forth the minimum amount necessary to encourage the grout to flow through the rise at least an inch over the underside of the plate.

If the space is to be pumped, a hose or pipe may be inserted along the centerline in the long direction and slowly withdrawn as the area is filled, using straps from both sides to bring the grout to both edges. The side forms should be vertical, with the top edges even with the top of the plate so that short boards about 2 ft long can be nailed over the edge spaces to retain the grout and force it to flow in the desired direction.

Vibrators should never be used to move grout for precision grouting because they encourage settlement of the heavier particles and bleeding of the mixing water. The latter will leave more space between the grout and plate than the shrinkage-correcting properties of the grout can overcome.

Before grouting is started, the concrete should be flooded with water for at least 6 h so that water will not be drawn from the grout when it is placed. Most of the free water should be blown out before pouring, but minor remaining amounts will be displaced by the grout without harm to the grout.

Procedures for curing, trimming, and protecting the exposed edges of ferrous aggregate grouts vary with the formulation of the grout. The manufacturer's instructions should therefore be carefully studied and followed. Some manufacturers provide free job service with their materials; the user is advised to call for it when available.

Post-Tensioned Cable Grouting

The pump, hoses, and cable duct are flushed with water; the duct is blown clear; then, with the hose disconnected from the duct, grout is pumped until all water is displaced from the pump and hose. Then the hose is connected to the duct and pumping starts.

Key features for successful work and a smooth operation are:

1. Use of a screened grout so that lumps do not plug the small cable grout inlet
2. A slow, steady flow of grout, free from pulsations if possible, continuing without interruption from start to finish for each cable duct.

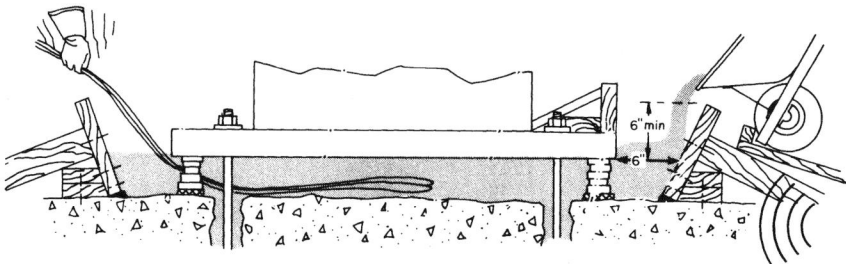

FIGURE C1-60 Procedure for forming and grouting under machine bases. *(By permission of Master Builders.)*

Grout should be pumped until all air and water have been displaced from the duct, as evidenced by an outflow of grout which has the same consistency as that in the mixer and is free of bubbles. The outlet is then closed, followed by closure of the inlet valve with some pressure on the line. If the grout does not contain a gas-expanding agent, such as aluminum powder, some engineers prefer a closing pressure of up to 100 lb/in^2 to compress stray bubbles left in the duct. If the grout does contain an expansion agent, closure pressures over 10 to 20 lb/in^2 will prevent the desired expansion from occurring.

Grouting in Granular Materials

Procedures for grouting granular materials will vary widely with the nature of the material to be grouted and with the end result desired. In most cases, a piping or casing will be driven to the desired depth. Drilling will be necessary to get through boulders, in which event the hole will reduce in size as each drilled area is passed. An alternative method is to wash the fines from a zone 2 to 10 ft deep under the rock and grout the gravel remaining in the pocket thus created with a grout that will provide sufficient strength to permit continued drilling. Chopping bits and churn drills are used to get through ravels and cobbles; if the gravel contains sufficient fines to retain drilling muds, the muds will both temporarily stabilize the walls of the hole and assist in floating out the cuttings. For limited depths in silts, sands, and fine gravels, one can mix grout with the in situ material by pumping grout through a rotating shaft having a mixing head on its lower end.

REFERENCES

1. *ASTM Book of Standards,* part 9, "Cement, Lime, Gypsum," latest issue, American Society for Testing and Materials, Philadelphia, Pa.
2. *ASTM Book of Standards,* part 10, "Concrete and Mineral Aggregates," latest issue, American Society for Testing and Materials, Philadelphia, Pa.
3. "Recommended Practice for Selecting Proportions for Concrete," ACI 613, *ACI Journal,* Proc. vol. 51, latest issue.
4. *Recommended Practice for Evaluation of Compression Test Results of Concrete,* ACI 214, latest issue, American Concrete Institute, Detroit, Mich.
5. *Recommended Practice for Selecting Proportions for Structural Lightweight Concrete,* ACI 613A, latest issue, American Concrete Institute, Detroit, Mich.
6. *Recommended Practice for Measuring, Mixing, and Placing Concrete,* ACI 614, American Concrete Institute, Committee 304.
7. *Curing Concrete,* ACI 308, American Concrete Institute, Committee 612.

CHAPTER 2
REINFORCING STEEL

David P. Gustafson
*Vice President of Engineering, Concrete Reinforcing
Steel Institute, Schaumburg, Illinois*

The main types of reinforcing steel are deformed bars; welded wire fabric; and strands, bars, and wires for prestressing. Descriptions of these types of reinforcement and construction aspects related to their use are presented in this chapter. Information is drawn mainly from several referenced standards and technical reports, documents which are updated quite often. Although the reported information is based on latest available editions of the referenced documents, the editions of standards applicable for a specific construction project should be those cited in the project specifications for the project.

REINFORCING BARS

Standard deformed reinforcing bars are produced in 11 sizes. Deformed bars have protruding lugs on their surface. The ASTM specifications for reinforcing bars include requirements for the deformations. Figure C2-1 shows several of the deformation patterns in use; the patterns are not restricted to the ones shown. Deformed bars are designated on design drawings and in project specifications by a size number. Common practice, especially on drawings, is to include a prefix "#" before the bar number.

Plain reinforcing bars or *plain rounds* are smooth bars without deformations. The ACI 318 Building Code[1] permits the use of plain bars for spirals. Plain bars can be used for spirals and as ties under the AASHTO Bridge Specifications.[2] Plain bars are also used for special purposes such as dowels at expansion joints.

The four ASTM specifications for reinforcing bars are: billet-steel (A615), low-alloy steel (A706), rail-steel (A616) and axle-steel (A617). Table C2-1 shows the standard bar sizes and grades (minimum yield strengths) covered by the four ASTM specifications. The grade number indicates minimum yield strength, ksi, of the steel.

Billet-steel reinforcing bars in Grade 60 are the most widely used type of steel and grade in the United States. The ASTM A615 specification prescribes essentially a set of physical requirements for tensile properties and bending properties. Except for phosphorus content, the ASTM A615 specification does not impose restrictions

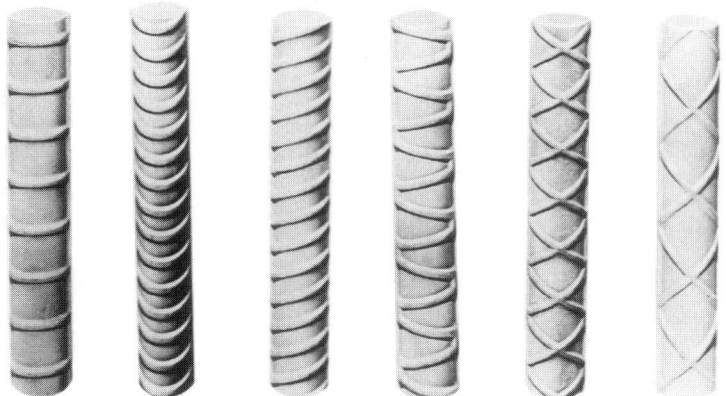

FIGURE C2-1 Deformed reinforcing bars—various types of deformation patterns.

on the chemical composition of the steel nor include any specific provisions to enhance weldability.

Most mills roll billet-steel bars to a standard stock length of 60 ft for all except the smallest and largest sizes. Longer lengths require special rolling arrangements. The absolute maximum length possible varies greatly from mill to mill. Further information on the availability of reinforcing bars and the other types of reinforcing steel is presented in "Steel Reinforcement—Physical Properties and U.S. Availability (ACI 439.4R-89)."[3]

Rail-steel reinforcing bars (ASTM A616) are manufactured from tee rails. Axle-steel bars (ASTM A617) are rolled from carbon steel axles for railway cars and locomotives. Similar to the A615 specification for billet-steel bars, Specifications A616 and A617 prescribe requirements for physical properties. There are no restrictions on chemical composition in the specifications for rail-steel and axle-steel bars.

The ASTM A706 specification for low-alloy steel bars includes requirements for controlled tensile properties and enhanced ductility, because the bars are intended

TABLE C2-1 ASTM Standard Reinforcing Bars

Bar size designation	Nominal weight, lb/ft	Nominal diameter, in	Nominal cross-sectional area, in^2
#3	0.376	0.375	0.11
#4	0.668	0.500	0.20
#5	1.043	0.625	0.31
#6	1.502	0.750	0.44
#7	2.044	0.875	0.60
#8	2.670	1.000	0.79
#9	3.400	1.128	1.00
#10	4.303	1.270	1.27
#11	5.313	1.410	1.56
#14	7.65	1.693	2.25
#18	13.60	2.257	4.00

for use in such applications as earthquake-resistant structures. Since low-alloy steel bars are also intended for welding, the A706 specification includes restrictions on chemical composition. Weldability is accomplished by provisions in the specification which limit the more critical chemical elements: Carbon, maximum = 0.30 percent; and Manganese, maximum = 1.50 percent. The specification also limits *carbon equivalent* or CE to 0.55 percent.

Carbon equivalent is calculated by a formula which is given in the A706 specification. The same CE formula is also used for A706 bars in the ANSI/AWS D1.4 Welding Code for Reinforcing Steel. Further discussion of weldability is given in this chapter under Welded Splices.*

Identification

The ASTM specifications for reinforcing bars require identification marks to be rolled onto the surface of one side of the bar to denote the producer's mill designation, bar size, type of steel, and minimum yield designation. Figure C2-2 shows the standard marking system for deformed bars. Identification marks may also be oriented to read horizontally (at 90° to those shown in Fig. C2-2). Grade mark numbers may be placed within separate consecutive deformation spaces to read vertically or horizontally. If a single line is used for Grade 60, or a double line for Grade 75, the lines must be continued through at least five deformation spaces. A grade line may be placed on the side of the bar opposite the bar marks. Table C2-2 shows bar sizes and grades conforming to ASTM specifications.

Bar Mats

Bar mats are made up of two layers of deformed reinforcing bars. The bars are assembled at right angles to each other in a mat or sheet form. The mat is held together by

---

* Additional information on low-alloy steel bars including availability is presented in "Questions and Answers on ASTM A706 Reinforcing Bars" by D. P. Gustafson and A. L. Felder, *ACI Concrete International*, v. 13, no. 7, July 1991, pp. 54–57.

TABLE C2-2 Reinforcing Bar Sizes and Grades Conforming to ASTM Specifications

Type of steel bars and ASTM designation	Bar sizes	Grade*
Billet steel	#3–#6	40
A615	#3–#11, #14, #18	60
	#6–#11, #14, #18	75
Rail steel	#3–#11	50
A616	#3–#11	60
Axle steel	#3–#11	40
A616	#3–#11	60
Low-alloy steel A706	#3–#11, #14, #18	60

* Minimum yield designation.

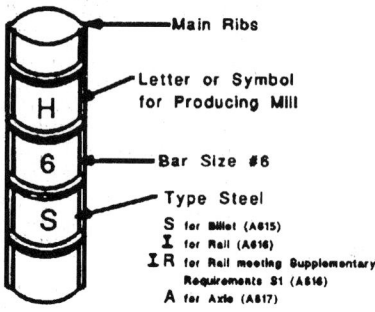

GRADE 40 AND 50

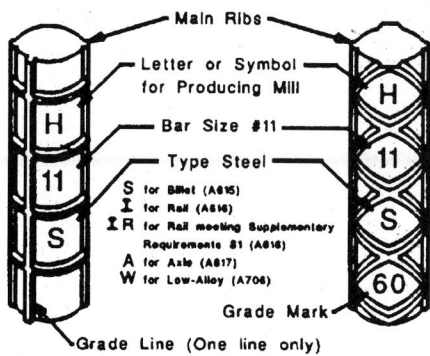

GRADE 60

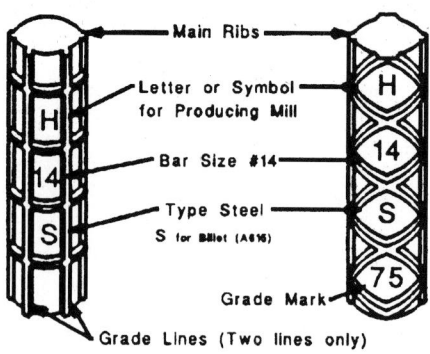

GRADE 75

FIGURE C2-2 Identification marks—ASTM standard reinforcing bars.

clipping or welding the intersections. ASTM Specification A184 covers deformed bar mats. Clipped mats can be fabricated from any of the four types of reinforcing bars. Only bars conforming to ASTM Specifications A615 or A706 can be used for welded mats. ASTM A184 includes requirements for the clipping or welding of the intersections and strength of the connections. Bar mats are most advantageously used when there is considerable repetition in the construction project.

Coiled Stock

Billet-steel and low-alloy steel bars in the smaller sizes can also be furnished in coil form. Coiled stock is commonly used by fabricators with automatic bending equipment for the fabrication of ties and stirrups. Fabricators who specialize in fabricating spirals also use coils of reinforcing bars.

WIRE AND WELDED WIRE FABRIC

Wire for reinforcement of concrete is manufactured to comply with ASTM Specifications A82 (plain wire) or A496 (deformed wire). Plain wire can be used as spiral reinforcement. The AASHTO Bridge Specifications permit the use of plain wire for ties. Deformed wire manufactured in the United States typically has indentations in its surface. Ribbed wire is also commercially available. The ASTM A496 Specification is currently being revised to explicitly recognize ribbed wire. Deformed wire can be used solely like reinforcing bars.

Wire sizes are designated by the letter W or D followed by a number. W denotes a plain wire; D means a deformed wire. The number following the letter W or D indicates the cross-sectional area of the wire in hundredths of a square inch. For example, the designation W2.9 is a plain wire with an area of 0.029 in^2. A wire size D16 is a deformed wire having an area of 0.160 in^2.

Welded wire fabric is manufactured from plain wire or deformed wire. The applicable ASTM specifications are A185 (plain fabric) and A497 (deformed fabric). Deformed welded wire fabric, conforming to the A497 specification, can be made from all A496 deformed wire, or from deformed wire in combination with plain wire.

Welded wire fabric consists of longitudinal and transverse wires arranged to form a square or rectangular mesh. The wires are then welded together at every intersection by a controlled electrical-resistance welding process. Welded wire fabric is available in sheets or in rolled form. Fabric made from larger-size wires is in the form of sheets. Rolls are generally stocked in W1.4 to W4 wire sizes only. Roll widths vary from 5 to 8 ft. Lengths of rolls vary depending on the application and convenience for handling and shipping. The use of sheets rather than rolls should result in more accurate positioning of the reinforcement within the formwork. Thus, rolls should be straightened and cut into sheets before placement. Fabric in sheet form can have widths varying between 7 and 9 ft for building construction. Widths of sheets up to 13 ft are used for pavement construction.

Welded wire fabric is usually denoted on design drawings or in project specifications as WWF, followed by the spacing of the longitudinal wires and then by the spacing of the transverse wires, and last by the sizes of the longitudinal and transverse wires. Common styles of welded wire fabric including their designations are shown in Table C2-3.

Smaller pieces or segments of welded wire fabric can be used as lateral reinforcement in beams and columns.

Additional information on welded wire fabric and industry practices are given in the WRI *Manual of Standard Practice for Structural Welded Wire Fabric,* 4th Ed., 1992.*

PRESTRESSING REINFORCEMENT

In prestressed concrete construction, the prestressing force is introduced into a structural member by tendons. A tendon is defined as a steel element such as a strand, bar, or wire, or a bundle of such elements. Prestressing tendons have very high strength as compared to non-prestressed reinforcement.

---

*Available from Wire Reinforcement Institute, 1101 Connecticut Ave., N.W., Washington, D.C. 20036.

TABLE C2-3 Common Styles of Welded Wire Fabric*

Minimum specified yield strength f_y	Style designation W = plain wire, D = deformed wire	Area of steel, in^2/ft, longitudinal or transverse
65,000 lb/in^2, plain wire only	4 × 4—W1.4 × W1.4	0.042
	4 × 4—W2.0 × W2.0	0.060
	6 × 6—W1.4 × W1.4	0.028
	6 × 6—W2.0 × W2.0	0.040
	4 × 4—W2.9 × W2.9	0.087
	6 × 6—W2.9 × W2.9	0.058
70,000 lb/in^2, plain or deformed wire	4 × 4—W4 × W4 or 4 × 4—D4 × D4	0.120
	6 × 6—W4 × W4 or 6 × 6—D4 × D4	0.080
	6 × 6—W4.7 × W4.7 or 6 × 6—D4.7 × D4.7	0.094
	12 × 12—W9.4 × W9.4 or 12 × 12—D9.4 × D9.4	0.094
72,500 lb/in^2, plain or deformed wire	6 × 6—W8.1 × W8.1 or 6 × 6—D8.1 × D8.1	0.162
	6 × 6—W8.3 × W8.3 or 6 × 6—D8.3 × D8.3	0.166
	12 × 12—W9.1 × W9.1 or 12 × 12—D9.1 × D9.1	0.091
	12 × 12—W16.6 × W16.6 or 12 × 12—D16.6 × D16.6	0.166
75,000 lb/in^2, plain or deformed wire	6 × 6—W7.8 × W7.8 or 6 × 6—D7.8 × D7.8	0.156
	6 × 6—W8 × W8 or 6 × 6—D8 × D8	0.160
	12 × 12—W8.8 × W8.8 or 12 × 12—D8.8 × D8.8	0.088
	12 × 12—W16 × W16 or 12 × 12—D16 × D16	0.160
80,000 lb/in^2, plain or deformed wire	6 × 6—W7.4 × W7.4 or 6 × 6—D7.4 × D7.4	0.148
	6 × 6—W7.5 × W7.5 or 6 × 6—D7.5 × D7.5	0.150
	12 × 12—W8.3 × W8.3 or 12 × 12—D8.3 × D8.3	0.083

*From *Manual of Standard Practice for Structural Welded Wire Fabric,* 4th ed., 1992, Wire Reinforcement Institute.

Strand

Seven-wire strand, conforming to ASTM A416, is widely used in pretensioned precast members, and in post-tensioned construction. Low-relaxation strand is regarded as the standard type. The ASTM A416 Specification covers two strength levels: Grade 250 and Grade 270 (250,000 and 270,000 lb/in^2 minimum breaking strengths). Grade 270 strand is the most widely used grade and is generally available in nominal diameters of ⅜, ⁷⁄₁₆, ½, and 0.6 in.

Indented, stress-relieved seven-wire strand is defined by ASTM Specification A886. The outer wires are indented to enhance bond strength of the strand in concrete. Two grades are covered by A886, designated as Grade 250I and 270I. The grade numbers are the minimum ultimate tensile strengths or breaking strengths in ksi, and the suffix I denotes *indented*. Low-relaxation requirements are delineated in a supplement of the specification. Indented strand is intended for use in pretensioned, prestressed concrete members which require shorter transfer and development lengths.

Other types of seven-wire strands for special applications are also available. These include zinc-coated (galvanized) strand, higher-strength strands, strands with larger nominal diameters than those listed in the ASTM A416 specification, and stainless steel strand. Epoxy-coated strand is discussed under "Coated Reinforcement."

Bars

High-strength alloy steel bars meeting ASTM A722 are used mainly in post-tensioning applications. The specification requires the bars to have a minimum ultimate tensile strength of 150,000 lb/in^2. Both plain and deformed bars are covered by A722. Bars are generally available as part of prestressing systems that include complete tendons and anchorage devices. Deformed bars can be produced with the deformations arranged to permit coupling of the bars with a screw-on type of coupler.

Wire

Prestressing wire is covered by ASTM A421. The use of wire tendons has diminished in recent years, because of the decline in post-tensioned nuclear power plant construction.

Compacted Strand

ASTM Specification A779 covers compacted, seven-wire, stress-relieved strand. Compacted strand is seven-wire strand that is compacted by drawing through a die or a similar compacting process. It is then stress-relieved prior to winding into coils or reelless packs. Compacted strand has a larger area of steel in any nominal diameter as compared to regular seven-wire strand. The larger area results in a larger breaking strength than a regular strand.

For post-tensioned construction, various prestressing systems are commercially available. No attempt will be made here to describe the systems. The reader is referred to the Post-Tensioning Institute's Manual[4] for such information.

COATED REINFORCEMENT

Coated Reinforcing Bars

To protect reinforcing bars from corrosion, they can be coated with epoxy or zinc (galvanizing). Applicable ASTM specifications are A775 and A934 (fusion-bonded epoxy), and A767 (zinc-coated or galvanized).

Epoxy-coated bars are widely used in highway bridge decks where de-icing salts can attack the bars. Other uses include parking garages, wastewater treatment plants, marine structures, and continuously reinforced concrete pavement. The majority of the production of epoxy-coated bars to date has been done by applying the epoxy coating to straight bars. The coated straight bars are then fabricated—cut to the required length, or cut to the required length and bent to the required configuration for the construction project.

An alternate procedure to produce epoxy-coated bars is to fabricate the uncoated bars first, and then apply the coating to the fabricated bars. Specification A934 prescribes the requirements for the coating of bars after fabrication.

Zinc is a metallic coating. Application of the coating, according to the A767 specification, is accomplished by dipping the bars into a molten bath of zinc, hence the term *hot dip galvanizing*. Bars can be galvanized before or after fabrication. The usual practice is to galvanize the bars after fabrication. The length or configuration of bars which can be galvanized depends on the size of the galvanizer's kettle.

Epoxy-Coated Wire and Fabric

ASTM specification A884 covers fusion-bonded, epoxy-coated wire and welded wire fabric. Two classes of coated material are addressed in A884: coated reinforcement for use as a corrosion-protection system in concrete, and coated material for use as reinforcement in earth-reinforced construction, such as mechanically stabilized embankments.

Epoxy-Coated Strand

Epoxy-coated seven-wire prestressing strand is commercially available. Coated strand is intended for use in pretensioned precast members, such as highway bridge girders. It is also intended for post-tensioning applications. ASTM specification A882 prescribes the requirements for epoxy-coated strand.

According to the ASTM A882 specification, the epoxy coating on a strand may be smooth or impregnated with grit. A smooth coating is intended for corrosion protection of the strand. Grit-impregnated coating serves two purposes. The coating protects the strand from corrosion. The grit in the coating enhances the bond of the strand with the surrounding concrete in pretensioned construction, or with the grout in the case of bonded post-tensioned construction.

A supplement to ASTM A882 provides requirements for epoxy-coated and filled strand. In this type of coated strand, the void spaces between the individual wires are filled with epoxy to provide increased corrosion protection.

Since ASTM A882 is a product specification, it does not delineate requirements regarding the use of epoxy-coated strand. Proper use of epoxy-coated strand includes design considerations, handling, installing and stressing of strands, permissible concrete curing temperatures, and procedures for repairing damaged coating and protection of the ends of strands. Information and procedures for such items are presented in a report, "Guidelines for the Use of Epoxy-Coated Strand."[5]

SPLICES OF REINFORCEMENT

Adequate splices are crucial to assure proper performance and integrity of reinforced concrete structures. Complete requirements for splicing reinforcement should be presented on the design drawings or in the project specifications. For example, the ACI Building Code requires the design drawings and project specifications to indicate:

- Anchorage length of reinforcement and location and length of lap splices
- Type and location of welded splices and mechanical connections of reinforcement

To use splices or mechanical connections not indicated or shown on the design drawings or in the project specifications, the contractor should seek authorization from the architect/engineer to do so. For example, the ACI 301 Specifications for Structural Concrete[6] requires the contractor to submit a list and request to use splices or mechanical connections not indicated in the contract documents.

Methods available for splicing reinforcing bars are lap splicing, mechanical connections, and welding.

Lap Splices

The traditional lap splice, when it will satisfy all requirements, is generally the most economical. Bars may be in contact or spaced in a lap splice. Contact lap splices are preferred because, when wired together, they are more easily secured against displacement during concrete casting. Spaced (noncontact) lap splices can be used within certain limitations. The ACI Building Code, for example, limits the clear spacing of bars in noncontact lap splices to one-fifth the lap length but the spacing should not exceed 6 in.

Epoxy-coated bars have less bond strength than uncoated bars. Thus, the required lengths for tension lap splices of epoxy-coated bars are usually longer than those required for comparable size uncoated bars and conditions. Since there are no special requirements in the ACI Building Code for lap splices of zinc-coated (galvanized) bars, the lap lengths will be the same as those for uncoated bars.

Mechanical Connections

Building code requirements or conditions may require or make the use of mechanical connections more feasible or practical than lap splices. Some situations are:

- Where #14 and #18 bars are used; design codes and design specifications do not permit them to be lap spliced, except in compression only with #11 and smaller bars.
- Where spacing of the bars is insufficient to permit lapping of the bars.

- Where the required lengths of tension lap splices are excessively long, especially for bar sizes #9, #10, and #11. It might be more cost-effective to use mechanical connections. Constructability may also be improved.
- To provide for future construction. Mechanical connections would eliminate long bar lengths extending from existing concrete construction.

There are essentially three basic types of mechanical connections: tension-compression; compression-only, which is also called an end-bearing mechanical connection; and tension-only. The tension-compression mechanical connection can resist both tensile and compressive forces. Dowel bar mechanical connections would be included in this category. A variety of proprietary mechanical connection devices are commercially available. Descriptions of available types and their physical features are given in "Mechanical Connections of Reinforcing Bars" (ACI 439.3R-91).[7] The ACI Committee 439 report also includes installation procedures and information on the equipment and tools required to make the connections.

Welded Splices

The use of welded splices might be considered because of code requirements or similar conditions as those which were described previously for mechanical connections. Welded splices might also be feasible in retrofitting and rehabilitation work. The most commonly used manual welding process in the field is electric arc welding.

Welding of reinforcing bars is covered by "Structural Welding Code—Reinforcing Steel," (ANSI/AWS D1.4) of the American Welding Society.[8] Project specifications usually prescribe requirements for welding reinforcing bars by reference to the ANSI/AWS D1.4 Code. The Welding Code's main criterion for welding is *carbon equivalent,* abbreviated as CE Minimum preheat and interpass temperatures are established by the Welding Code for ranges of CE and bar size. See Table C2-4.

To calculate carbon equivalent, it is necessary to have available the mill test report showing the chemical analysis of the bars to be welded, or the CE should be included on the mill test report for A706 bars. CE is then calculated from a formula. The ANSI/AWS D1.4 Code has two formulas for calculating CE.

For all steel bars except A706 low-alloy steel bars, the formula is:

$$CE = \% \ C + \% \ \frac{Mn}{6}$$

TABLE C2-4 Minimum Preheat and Interpass Temperatures, °F

Carbon equivalent range, %	Bar sizes		
	#3–#6	#7–#11	#14–#18
0.40 maximum	0	0	50
0.41 to 0.45	0	0	100
0.46 to 0.55	0	50	200
0.56 to 0.65	100	200	300
0.66 to 0.75	300	400	400
Above 0.75	500	500	500

Source: Structural Welding Code—Reinforcing Steel, ANSI/AWS D1.4-92.

Where C is the abbreviation for the chemical element carbon and Mn is manganese.
For A706 bars:

$$CE = \% \ C + \% \ \frac{Mn}{6} + \frac{\% \ Cu}{40} + \frac{\% \ Ni}{20} + \frac{\% \ Cr}{10} - \frac{\% \ Mo}{50} - \frac{\% \ V}{10}$$

Where Cu is copper, Ni is nickel, Cr is chromium, Mo is molybdenum, and V is vanadium. The elements molybdenum and vanadium enhance weldability. The CE formula for A706 bars in the Welding Code is identical to the formula in the ASTM A706 Specification.

For A706 bars with a carbon equivalent up to 0.55 percent, little or no preheat is required under normal working temperatures for bar sizes #11 and smaller (Table C2-4).

The ANSI/AWS D1.4 Welding Code provides guidance for situations when the chemical composition of bars to be welded is unknown. For billet-steel bars, the Welding Code recommends a preheat of 300°F for bar sizes #6 and smaller, and 400°F for bar sizes #7 and larger. For A706 bars, the Welding Code recommends the use of preheat values for the CE range of 0.46 to 0.55% in Table C2.4.

For compliance with the ANSI/AWS D1.4 Welding Code, written welding procedure specifications are required for all welding that is performed under the Code. This provision requires the Contractor to prepare a welding procedure for each joint as a procedure specification.

Bundled Bars

Heavy reinforced concrete construction might require bundling of the longitudinal bars in columns or beams. Bundled bars are a group of not more than four parallel bars in contact with each other, tied together, to act as a unit. Bundling can improve constructability by reducing congestion of reinforcement. The need for several layers of single, parallel bars in girders can be reduced by bundling. In columns, bundling of longitudinal bars can eliminate many interior ties.

The ACI Building Code limits bundling of bars in beams to bar size #11. The AASHTO Bridge Specifications limits bundling of #14 and #18 bars in beams to two bars in any one bundle. Both documents provide special rules for splices of bundled bars. Bar cutoffs and lap splices of individual bars within a bundle must be staggered. If a mechanical connection or a butt-welded splice is used, a staggered location of the splice point is recommended for practical purposes for erection convenience. Staggering the splice points avoids bunching all mechanical connections or welded splices at one point. The specific requirements for lap splices, mechanical connections, or welded splices should be presented on the design drawings or in the project specifications.

DETAILING AND FABRICATION

Reinforcing steel detailing is the practice of preparing field placing drawings, bar lists, and bending details from the design drawings and project specifications. The contractor should furnish any necessary additional information to the detailer concerning field conditions, field measurements, construction joints, and sequence of placing concrete. Placing drawings are working drawings that show the number, size, length, and bending dimensions of each piece of reinforcing steel and its location in the structure.

Placing drawings are usually prepared by the fabricator (supplier) of the reinforcing steel. Ironworkers use the placing drawings to place (install) the reinforcing steel within the formwork. The bar lists and bending details are used for fabricating the reinforcing steel. Placing drawings are usually submitted by the fabricator to the contractor for approval by the architect/engineer prior to shop fabrication.

The standard for detailing reinforcement is "Details and Detailing of Concrete Reinforcement" (ACI 315-92).* Many state departments of transportation prepare combination design and placing drawings. The combination drawings include schedules of reinforcing materials from which the fabricator prepares shop bar lists. The placer (erector) uses the combination drawings to place the reinforcement. State DOTs that do not use combination drawings follow the procedures described in the ACI 315 standard.

Fabrication of reinforcing bars involves cutting a straight bar to a specified length or cutting and bending a straight bar to a specified length and configuration. Typically, bars are shop fabricated and transported to the job site. In some areas, labor union regulations prohibit shop fabrication, requiring that this be done on the job site.

There are many types of bent bars. Typical bar bends are shown in Fig. C2.3. Standard industry fabricating tolerances for the typical bar bends are shown in Fig. C2.4*a* and *b*. The dimensions and geometry of standard hooks are established by the design codes and design specifications. Dimensions of standard hooks are given in Tables C2.5 and C2.6.

After shop fabrication, the bars are bundled and tagged for delivery to the job site. Practices vary regionally and among suppliers. Common practice is to group bars of one size, length, or mark number (bent bars) into bundles securely tied by wires or bands. Exceptions to the common practice include small quantities that may be bundled together for convenience, or groups of varying bar lengths or marks (bent bars) that will be placed adjacent and may be bundled together.

Units of reinforcing bars tied together for shipment from the fabricating shop to the job site are called *shop lifts*. *Field lifts* are units of reinforcing bars as required for field handling by the contractor. A field lift may consist of single bundles or two or more smaller bundles tied together. A shop lift may consist of one or more bundles, the same as field lifts, or consist of two or more field lifts. Common practice is not to combine straight and bent bars in the same lift. Maximum weight of a bundle or lift is dependent on regional practices and job-site conditions.

Fabricating industry practice limits the shipping width or load limit to 7 ft 4 in for a single bent bar or an L-shaped bar (Fig. C2.5). Since bundles of bars occupy greater space, the limit of 7 ft 4 in is intended to limit the bundle size to an 8-ft maximum load width. When shipping widths exceed 8 ft, permission of authorities is usually required or the material must be shipped under special freight rates.

HANDLING AND PLACING REINFORCEMENT[†]

Handling and Storage

Delivery of reinforcement to the job site should follow an orderly process to meet the construction schedule. Where deliveries of material are scheduled to coincide

---

* The detailing standard, a report entitled "Manual of Engineering and Placing Drawings for Reinforced Concrete Structures," and supporting reference data are included in *ACI Detailing Manual—1994.*
† For further information, see *Placing Reinforcing Bars,* 6th Ed., 1992, Concrete Reinforcing Steel Institute, 933 N. Plum Grove Road, Schaumburg, IL 60173.

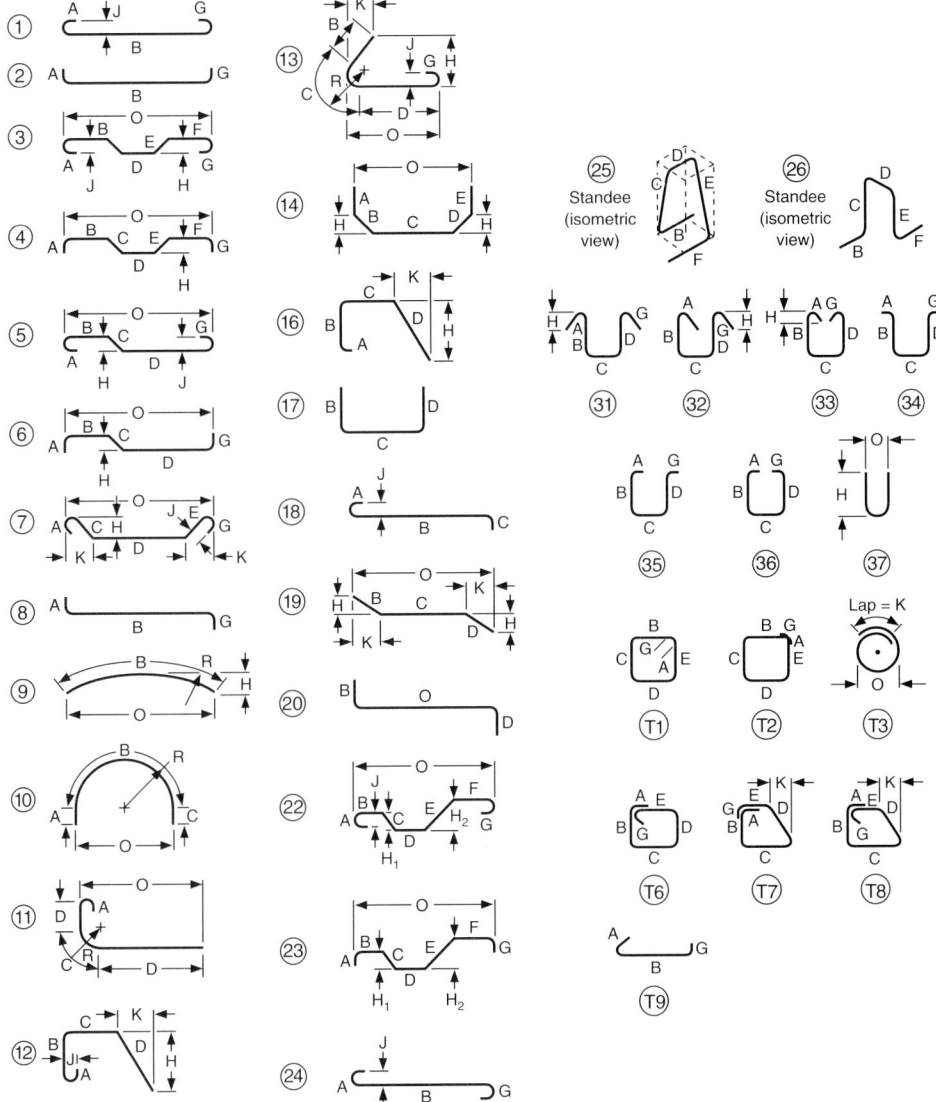

Notes:
1. All dimensions are out-to-out of bar except "A" and "G" on standard 180° and 135° hooks.
2. "J" dimensions on 180° hooks to be shown only where necessary to restrict hook size, otherwise ACI standard hooks are to be used.
3. Where "J" is not shown, "J" will be kept equal to or less than "H" on Types 3, 5, and 22. Where "J" can exceed "H", it should be shown.
4. "H" dimension stirrups to be shown where necessary to fit within concrete.
5. Where bars are to be bent more accurately than standard fabricating tolerances, bending dimensions which require closer fabrication should have limits indicated.
6. Figures is circles show types.
7. For recommended diameter "D", of bends, hooks, etc., see Tables C4.5 and C4.6.
8. Type S1-S6, S11, T1-T3, T5-T9 apply to bar sizes #3 through #8.
9. Unless otherwise noted, diameter D is the same for all bends and hooks on a bar (except for bend types 11 and 13).

FIGURE C2-3 Typical bar bends. *(Courtesy of Concrete Reinforcing Steel Institute.)*

C2-13

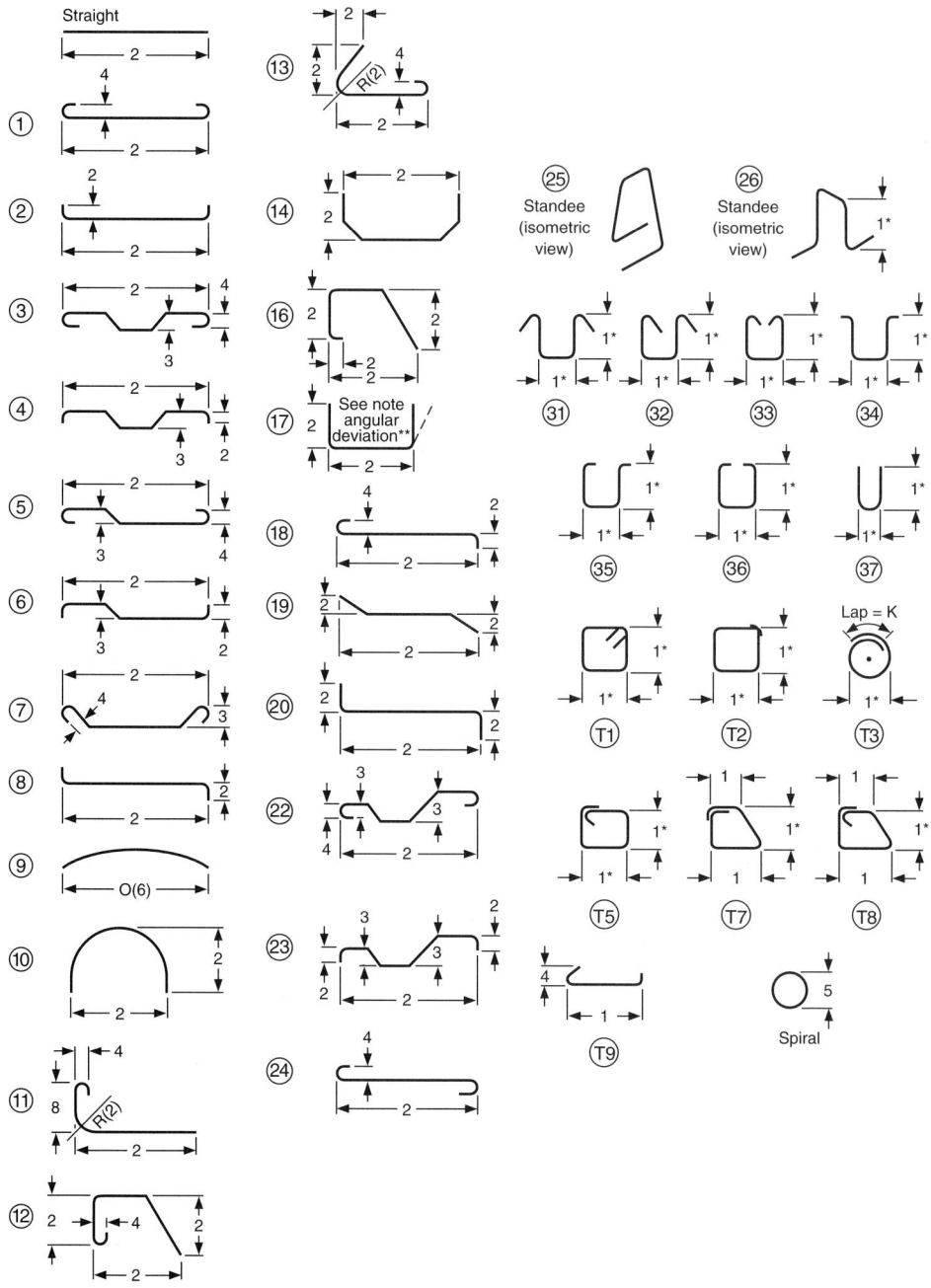

All tolerances single plane and as shown.
*Dimensions on this line are to be within tolerance shown but are not to differ from the opposite parallel dimension more than 1/2".
** Angular Deviation—maximum ± 2-1/2° or ± 1/2"/ft., but not less than 1/2", on all 90° hooks and bends. Tolerances for Types S1–S6, S11, T1–T3, T5–T9 apply to bar sizes #3–#8 inclusive only.

FIGURE C2-4*a* Standard fabricating tolerances for bar sizes #3 through #11. *(Courtesy of Concrete Reinforcing Steel Institute.)*

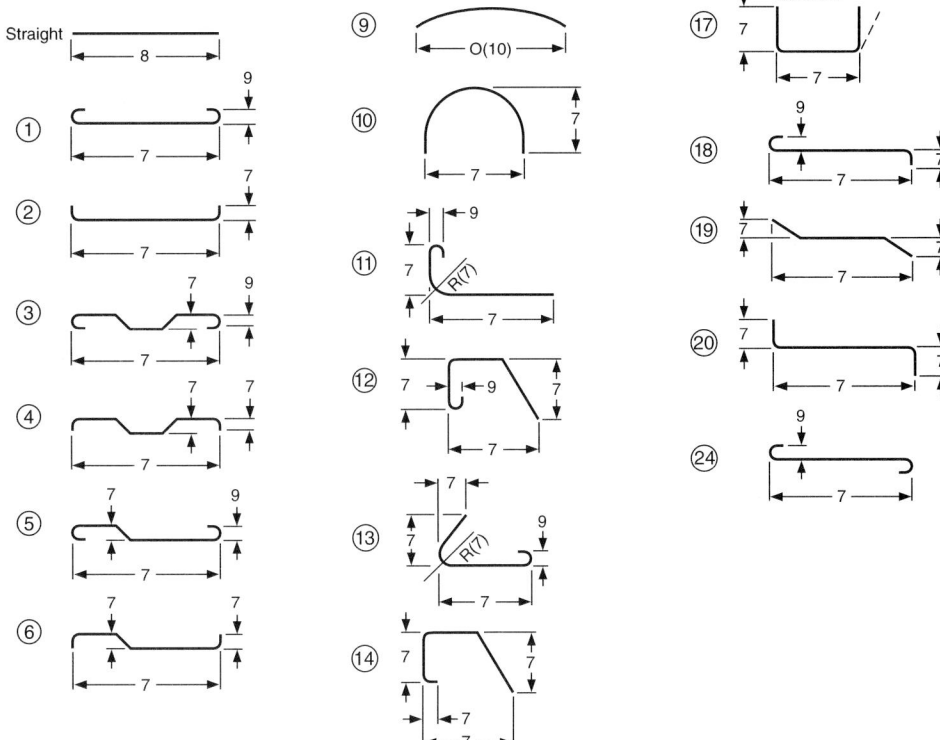

Tolerance symbols for bar sizes #3 through #11

1 = Plus or minus 1/2″ for bar sizes #3, #4, and #5 (gross length < 12′–0″)
1 = Plus or minus 1″ for bar sizes #3, #4, and #5 (gross length ≥ 12′–0″)
1 = Plus or minus 1″ for bar sizes #6, #7 and #8
2 = Plus or minus 1″
3 = Plus 0″, minus 1/2″
4 = Plus or minus 1/2″
5 = Plus or minus 1/2″ for diameter ≤ 30″
5 = Plus or minus 1″ for diameter > 30″
6 = Plus or minus 1.5% × "0" dimension, ≥ ± 2″ mini.***

Tolerance symbols for bar sizes #14 and #18

	#14	#18
7 = Plus or minus	2-1/2″	3-1/2″
8 = Plus or minus	2″	2″
9 = Plus or minus	1-1/2″	2″
10 = Plus or minus 2% × "0" dimension, ≥	± 2-1/2″ min.***	± 3-1/2″ min.***

All tolerances single plane and as shown.
* Saw-cut both ends—overall length ± 1/2″.
** Angular Deviation—maximum ± 2-1/2° or ± 1/2″/ft on all 90° hooks and bends.
*** If application of positive tolerance to Type 9 results in a chord length ≥ the arc or bar length, the bar may be shipped straight.

FIGURE C2-4*b*   Standard fabricating tolerances for bar sizes #14 and #18. *(Courtesy of Concrete Reinforcing Steel Institute.)*

C2-15

TABLE C2-5 Standard Hooks

	Recommended end hooks—all grades, in or ft-in			
		180° hooks		90° hooks
Bar size	D*	A or G	J	A or G
#3	2¼	5	3	6
#4	3	6	4	8
#5	3¾	7	5	10
#6	4½	8	6	1–0
#7	5¼	10	7	1–2
#8	6	11	8	1–4
#9	9½	1–3	11¾	1–7
#10	10¾	1–5	1–1¼	1–10
#11	12	1–7	1–2¾	2–0
#14	18¼	2–3	1–9¾	2–7
#18	24	3–0	2–4½	3–5

* D = Finished inside bend diameter, in
Source: Courtesy of the Concrete Reinforcing Steel Institute.

TABLE C2-6 Standard Hooks—Stirrups and Ties

	Stirrup and tie hook dimensions, in—Grade 40-50-60 ksi				135° seismic stirrup/tie hook dimensions, in—grades 40-50-60 ksi			
		90° hook	135° hook				135° hook	
Bar size	D, in	Hook A or G	Hook A or G	H, approx.	Bar size	D, in	Hook A or G	H, approx.
#3	1½	4	4	2½	#3	1½	4¼	3
#4	2	4½	4½	3	#4	2	4½	3
#5	2½	6	5½	3¾	#5	2½	5½	3¾
#6	4½	1–0	8	4½	#6	4½	8	4½
#7	5¼	1–2	9	5¼	#7	5¼	9	5¼
#8	6	1–4	10½	6	#8	6	10½	6

Source: Courtesy of the Concrete Reinforcing Steel Institute.

with the daily placing requirements, truck loads can be delivered to the points of placement or positioned under a crane for hoisting to the area of placement. When delivered material must be stored on the job site, the reinforcement should be stored above the ground on timbers to keep the bars free from mud and to provide for easy rehandling. It is necessary to maintain identification of the stored reinforcement so tags should remain on bundles.

Epoxy-coated reinforcing bars and welded wire fabric must be handled and stored more carefully than uncoated reinforcement. In unloading epoxy-coated reinforcement from a truck, care must be exercised to minimize scraping of the bundles or bar-to-bar abrasion from sags in the bundles. Equipment for handling the coated reinforcement should have protected contact areas. Nylon slings or padded wire rope slings should be used. Bundles of coated reinforcement should be lifted at multiple pickup points.

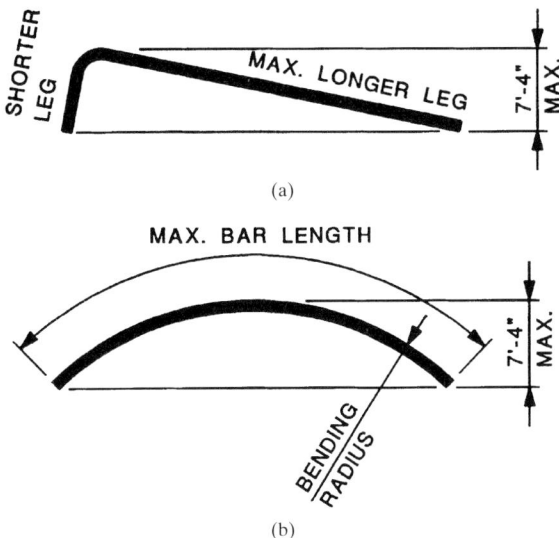

FIGURE C2-5 Shipping limitations: (*a*) height limit; (*b*) length and height limits.

Coated reinforcement should be stored on protective cribbing, as close as possible to the area of placement to keep handling operations to a minimum. Coated reinforcement should not be dropped or dragged. Long-term storage at the job site is not recommended. If it is absolutely necessary to store coated reinforcement for an extended period, the reinforcement should be protected from sunlight, salt spray, and weather exposure. Coated reinforcement should be covered with opaque polyethylene sheeting or other suitable protective material. For stacked bundles, the protective covering should be draped around the perimeter of the stack, secured adequately, and allow for air circulation around the reinforcement to prevent condensation under the covering.

Bond strength of reinforcing bars in concrete can be affected by substances on bar surfaces such as scale, rust, oil, and mud. Scale results from the steel-making process. Loose scale is removed through normal handling of the bars in the fabricating shop, and during loading, unloading, and on the job. Tight scale remaining on bar surfaces should not reduce bond strength.

Open storage of uncoated reinforcing steel will in most cases result in rusting. Loose rust should fall off the surfaces of bars during normal handling. Tight rust usually has a rough surface so it would improve bond strength. When bars exhibit rust on their surfaces, a question arises whether the rust is detrimental: Has the rust reduced the height of deformations or cross section of the bar? The ASTM Specifications for reinforcing bars and the ACI 318 Building Code provide criteria for evaluating the condition. Bars with rust or mill scale or a combination of both are considered acceptable when the minimum dimensions, height of deformations, and weight of a cleaned sample are not less than the ASTM Specifications require.

Guidance for evaluating the rusting on prestressing strand is given in "Evaluation of Degree of Rusting on Prestressed Concrete Strand," by A. S. Sason.[9]

Placing

Reinforcement must be accurately placed within the formwork, securely held in its proper position, adequately tied and supported before concrete is placed, and secured against displacement within placement tolerances during concreting operations. Tolerances for placing should be stated in the project specifications. The project specifications might reference the ACI 117 Standard[10] for the tolerances.

Wire used for tying reinforcing bars is usually no. 16½ or no. 16 gauge black, soft-annealed wire. In some cases, a heavier no. 15 or no. 14 gauge tie wire (or two wires, doubled, of either no. 16½ or no. 16 gauge) may be used when tying bars in heavily reinforced caissons or walls to maintain proper position of the reinforcement. Sufficient intersections of the reinforcement should be tied to prevent shifting of reinforcing steel. It is not necessary to tie every intersection. The tying adds nothing to the strength of the completed structure other than holding the reinforcement in its proper position until the concrete has been placed. Plastic-coated or epoxy-coated tie wire should be used for epoxy-coated reinforcement.

Tack Welding

Connection of crossing bars by small arc welds is called tack welding. Such welding for assembly of reinforcement is not recommended. Building codes and project specifications usually do not permit tack welding unless authorized by the engineer. The ANSI/AWS D1.4 Welding Code also prohibits tack welding except if authorized by the engineer. Unless tack welds are made in conformance with all requirements of ANSI/AWS D1.4, they tend to cause a metallurgical *notch* effect and may affect the strength of the bars. Tack welding is not a substitute for tie wire.

Field Bending and Rebending

At a construction site, invariably it is necessary to bend or rebend reinforcing bars that are partially embedded in hardened concrete. Bars may be accidentally bent and have to be rebent to their intended position. Bars may have to be field bent to provide for openings or embedments. Building codes and project specifications usually require field bending and corrective measures to be authorized by the engineer. Care should be exercised when such bending and rebending is required and the conditions under which it is authorized should be followed.

Bar Supports

Adequate supports are necessary to hold reinforcing steel firmly in its intended position before and during placing of concrete. The bar supports should be heavy enough and numerous enough to support the reinforcement properly. If too few supports are used, the reinforcing steel will sag between supports and will not be correctly positioned.

Bar supports can be made of steel wire, precast concrete, fiber-reinforced concrete, plastic, or other materials. CRSI's *Manual of Standard Practice*[11] presents industry practices for all types of bar supports and examples of their usage.

Bar supports for supporting epoxy-coated or galvanized reinforcing bars should be compatible with the coated reinforcement. When reinforcing bars are used as support bars, the support bars should be coated with coating material that is compatible with the bars being supported. The project specifications should include requirements for the bar supports and support bars.

Coated Reinforcement

When coated reinforcement is used in a project, construction operations should be performed carefully to minimize damage to the coating. Damaged coating will have to be repaired. It is more productive and cost-effective to exercise care in handling and placing coated reinforcement rather than having to repair extensive coating damage. If the coating damage exceeds the limits set by the project specifications, the coated reinforcement will be subject to rejection and will have to be replaced. Replacement or additional coated reinforcement may not be readily available, causing job delay. The ACI 301 Specifications[6] include requirements for coated reinforcement.

METRICATION

At press time, the federal government's mandate on metrication in the construction industry was being implemented. Several standards for reinforced concrete design and construction had been metricated, including the ACI 318M Building Code. Several ASTM metric specifications had been issued: A416M for seven-wire prestressing strand; A615M, A616M, A617M, and A706M for reinforcing bars; A184M for bar mats; A767M for zinc-coated (galvanized) bars; A775M and A934M for epoxy-coated bars; A882M for epoxy-coated prestressing strand; A884M for epoxy-coated wire and welded wire fabric; and A886M for indented, seven-wire prestressing strand. ASTM was in the process of metricating the A82, A185, A496, and A497 specifications for wire and welded wire fabric.

The metric units for length are millimeters (mm) and meters (m). For stresses, the metric unit is megapascals, abbreviated as MPa. In the metric system, mass is used rather than weight. Metric temperatures are expressed in degrees Celsius, abbreviated °C.

In this chapter, the metric equivalents of lengths and areas would be based on the conversion of:

1 in = 25.4 mm
1 ft = 304.8 mm or 0.3048 m
1 in^2 = 645.16 mm^2 or 0.000645 m^2

Converting stresses expressed in lb/in^2 or ksi units to metric units would be based on the conversion of:

1000 lb/in^2 = 1 ksi = 6.895 MPa

Converting weight to mass would require converting pounds to kilograms:

1 lb = 0.4536 kg

Temperatures in degrees Celsius, $t°C$, would be computed from $t°C = (t°F - 32)/1.8$; where $t°F$ is the temperature in degrees Fahrenheit.

REFERENCES

1. "Building Code Requirements for Reinforced Concrete (ACI 318-95)," American Concrete Institute, P.O. Box 19150, Detroit, Michigan 48219.
2. "Standard Specifications for Highway Bridges," 15th Edition, 1992, American Association of State Highway and Transportation Officials, 444 N. Capitol St., N.W., Washington, D.C.
3. ACI *Manual of Concrete Practice,* Part 3.
4. *Post-Tensioning Manual,* 5th ed., 1990, Post-Tensioning Institute, 1717 W. Northern Ave., Phoenix, AZ 85021.
5. "Guidelines for the Use of Epoxy-Coated Strand," *Journal,* Precast/Prestressed Concrete Institute, vol. 38, no. 4, July-August 1993, pp. 26–32. Reprints of the report are available from the Precast/Prestressed Concrete Institute, 175 W. Jackson Blvd., Suite 1859, Chicago, IL 60604.
6. "Specifications for Structural Concrete (ACI 301-96)," ACI *Manual of Concrete Practice,* Part 3.
7. "Mechanical Connections of Reinforcing Bars," ACI 439.3R-91, ACI *Manual of Concrete Practice,* Part 3.
8. "Structural Welding Code—Reinforcing Steel (ANSI/AWS D1.4)," American Welding Society, P.O. Box 351040, 2501 N.W. 7th St., Miami, FL 33125.
9. A. S. Sason, "Evaluation of Degree of Rusting on Prestressed Concrete Strand," *Journal,* Precast/Prestressed Concrete Institute, vol. 37, no. 3, May-June 1992, pp. 25–30. Reprints of the paper are available from the Precast/Prestressed Concrete Institute, 175 W. Jackson Blvd., Suite 1859, Chicago, IL 60604.
10. "Standard Specifications for Tolerances for Concrete Construction and Materials (ACI 117-90)," ACI *Manual of Concrete Practice,* Part 2.
11. *Manual of Standard Practice,* 25th ed., Concrete Reinforcing Steel Institute, 1992.

P·A·R·T D

HEAVY CONSTRUCTION TYPES

CHAPTER 1
ROCK EXCAVATION/BLASTING

Victor A. Sterner
Manager, Technical Services Support,
ICI Explosives USA Inc., Dallas, Texas

INTRODUCTION

The cost of labor and supplies is forever increasing. Efforts to offset these costs by higher productivity from improved equipment, better materials, and the utilization of the latest methods in technology in the planning of a project is paramount for a contractor to be profitable.

Rock excavation—drilling and blasting in particular—has improved significantly in recent years and will continue to get better. Explosives manufacturers offer more powerful products that are less hazardous to use and easier to handle, at relatively lower costs, with computerized analysis methods to optimize economic concerns as well as the performance of alternate blast designs.

Essentially, every construction rock excavation is a custom project, because no two projects are ever exactly the same. We must adapt our knowledge and experience to the problems of working in materials not of our choosing in many cases, and to a lesser extent with preselected material from quarry rock excavations.

PLANNING

Planning a rock excavation project has two broad limitations: one is imposed by specifications and local laws, while the other concerns the nature and location of the rock.

Specifications and Local Laws

Laws exist to protect populated areas and existing installations from vibration, excessive noise, and the dust generated by blasting. Specifications define the grading of the blasted product (fragmentation), the height and slope of the excavated cuts (number of benches), and the duration of the operation. Some specifications elaborate on the local laws as they pertain to vibration and noise levels and go so far as to stipulate formulas for computing the weight of explosive for each delay in a shot. In some cases, they also specify the actual blasthole diameter to be drilled. Some quarry locations can be chosen to minimize the effects of these controls, but construction projects are required where they exist to satisfy a need and, therefore, can be subject to any or all of these restrictions.

Geologic and Engineering Evaluation

A complete geologic and engineering evaluation is essential for the proper planning of a construction project. Geologic exploration of the potential excavation site to evaluate rock structure and overburden features is vital to the design engineer and to the contractor who must perform the work of removing and utilizing that rock.

The overall terrain must be physically examined and a surface map should be prepared which indicates the principal geologic features. Aerial photographs are a valuable guide in the interpretation and evaluation of the observed physical features. The body of the formation may be examined by core drilling to obtain samples or by other tests to determine soundness. These explorations yield geologic information that suggest methods of efficient use of the terrain for development purposes, indicate favorable access routes, and confirm the existence of other natural features such as groundwater, site clearing, and weather conditions which would affect excavation operations.

Site and Equipment Factors

Topography limits the overall approach to the excavation, including the location and the grades of access and haul roads, the face development, and the drainage requirements. System optimization requires that the selection of equipment be based on an evaluation of all features and excavation requirements. In many cases, a contractor may already own or lease certain pieces of rock-handling equipment. In these circumstances it is frequently to economic advantage to use this equipment rather than to buy new, ideally suited units. Compromises are thus made and, as a result, most construction excavations are rarely accomplished using ideal methods or equipment.

When new equipment is being selected for a rock job, the emphasis should be placed on the loading and hauling units and the primary crusher, if one is to be used. The investment for these items will greatly exceed that for drilling equipment. The two operations, drill-shoot and load-haul, meet on the common ground of fragmentation. While it may be possible to drill and shoot rock quite economically when producing a certain maximum size, it may not be economical to purchase and operate loading and hauling equipment which can handle that size. An accommodation is made when the drilling and shooting method is revised, usually at a higher cost, to produce the fragmentation required for the load-haul units.

Operations Planning

Usually, the first step in the development of the site is the removal of the overburden and other weathered materials located over the rock. In quarries, this is usually a continuous, concurrent operation above and ahead of the rock excavation. Overburden removal for quarries is normally minimal, since the ratio of stripping to usable rock is taken into consideration when selecting the site. For construction projects, this operation may be completed for the entire excavation area before rock removal is started.

After the overburden and other soft materials have been removed, the drilling and blasting requirements are more clearly defined. Contour maps should be used to determine the optimum method of working the production face(s). Opening up an area at least 20 ft high, should confirm and supplement the geologic report and enable the final planning of the excavation procedure. Once the overall method of operation is selected, the details of the rock removal can be planned.

Factors to be considered at this stage should include:

1. *Length of face (quarries).* The shorter the face length, the easier it may be to obtain the required volume of rock from a location.
2. *Fragmentation required.* This is dictated by specifications and should be the optimum for the loading and hauling equipment that is to be used.
3. *Drill size.* This selection is based on the fragmentation required, and the type and amount of explosive to be used.
4. *Number of benches.* The selection of equipment is the primary factor in determining the number of benches to be worked. An economic comparison should be made between using a large drill to work one bench versus using smaller drills on two or more benches, with additional haul roads.
5. *Production required.* The overall selection of equipment dictates the excavation method which, in return, corresponds to a certain production. This production must be examined to verify that the work will be performed within the time constraints and at reasonable unit costs, including adequate allowances for equipment depreciation. High-volume, long-term projects, such as commercial quarries, can economically justify the purchase of newer, more efficient equipment; whereas short-term, low-volume construction excavations may not.
6. *Site condition upon termination.* The selected method of operation must also provide for the ultimate treatment of the excavated site. Emphasis is now being placed on increased amounts of site restoration for purposes of safety and aesthetics. The excavation method should facilitate the subsequent performance of such nonproductive work expenses as backfilling, grading, and trimming slopes.

DRILLING

There are various kinds of drilling equipment and methods available to economically overcome the problems encountered with the endless combinations of rock types and structures that must be drilled in construction work. Drilling equipment is grouped into three broad categories: (1) percussion, (2) rotary, and (3) abrasion drills. The percussion and rotary drills are the production tools of the industry. Abrasion drills such as diamond and shot-coring drills are special-application tools and have limited use in construction projects.

Percussion Drills

The percussion drills are the most versatile of all rock drills and can operate economically over a wide range of hardness or abrasiveness. Percussion drills employ a reciprocating hammer (piston) which strikes a rotating bit or a string of rotating drill rods with a bit on its bottom. An air or water system is used to remove rock cuttings from the drill hole. Surface percussion drills are the most widely used type. The percussion hammer operates outside the hole it is drilling and is connected to the bit by drill rods. Surface drills have the advantage of being light in weight and highly maneuverable. Jackhammers and feed leg drills are examples of smaller-type surface drills that can be used in close quarters by one person; whereas percussion drifters require additional mounting equipment, but even these are small in comparison to rotary drills, which must support as much as 90,000 lb of down-pressure thrust. Surface percussion drills can operate at any angle through a complete circle and thus can be used for drilling horizontal, up, down, and angle holes. Figure D1-1 illustrates a top hammer track drill. In operation, the striking bar transmits the blow from the hammer to the drill steel. As holes are drilled deeper, some of the energy is dissipated in the drill steel and couplings, and the air pressure used to remove drill cuttings also drops, with a consequent slowdown in the operation. These factors limit the economical hole depth and have led to the development of the down-the-hole drill. Down-the-hole drills operate at the bottom of a rotating drill string directly above the bit. The outside diameter of the drill is smaller than the hole and thus enables the drill to follow the bit down the hole. Since the drill string is above the drill, the footage life of the drill rods is very good and the energy loss between drill and bit is held at a constant minimum. Figure D1-2 illustrates a DTH drill.

Rotary Drills

Rotary drills provide their drilling action by the application of heavy down-pressure on a roller-cone bit. Separate variable-speed motors, hydraulic and electric, create

FIGURE D1-1 Ingersoll-Rand EMC 490 top hammer drill.

FIGURE D1-2 Ingersoll-Rand T4BH truck-mounted drill for quarrying and construction, equipped with 350 lb/in^2 air compressor for down-hole drilling, 5⅛- to 7⅞-in hole.

the pull-down pressure and the drill rotation to accommodate the material being drilled. Rotary drills may be either diesel or electrically powered. Figure D1-3 illustrates a rotary rig.

The derrick frame of a rotary drill provides storage space for additional drill pipes, each of which can vary in length from 20 ft in the smaller drills to 35 ft in the larger drills. It also serves as the headframe and guide for the pull-down motor and the hoisting drum. The drill pipes are handled mechanically, and adding or removing a section can be done in a few minutes.

Abrasive Drills

Diamond and shot-coring drills are used principally for exploratory purposes when samples are required for physical examination and testing of the rock. These drills are uneconomical for blasthole drilling because of the cost of the abrasive bit.

FIGURE D1-3 Ingersoll-Rand DM-L rotary blasthole drill. Medium-weight rotary drill used in construction, mining, and quarrying, 6- to 9⅞-in hole diameter.

BLAST DESIGN

The variables that influence blast results are to such an extent that blast designs have been usually dependent on trial-and-error methods selected by persons experienced in the art of blasting after giving appropriate consideration to the rock type and its structure, fragmentation required, blasthole geometry, delay pattern, number of free faces, and kind of explosives to be used. This method has worked reasonably well over the years; however, there are more scientific methods now available to analyze the explosive performance on the rock: (1) determination of the mechanics of rock breakage, (2) improvement of the empirical powder-factor method of design by correlating it with blast geometry, and (3) development of scientific methods of blast design. Large economies can be realized when the trial-and-error system is supplemented by computer simulation design such as the SABREX model. Figure D1-4 illustrates a typical computer-generated profile of a blast design.

Characterizing the Rock Mass

Examination of the surface geology will indicate jointing, its direction, spacing, and frequency. Also any faults, openings, cavities, or weaknesses can sometimes be noted. The driller is another source of information of what the rock is like below surface. The hardness and cavities or such that exist can often be noted by the driller. Drill logs are a good record of the drilling conditions and can be important to the blast engineer in designing the blast.

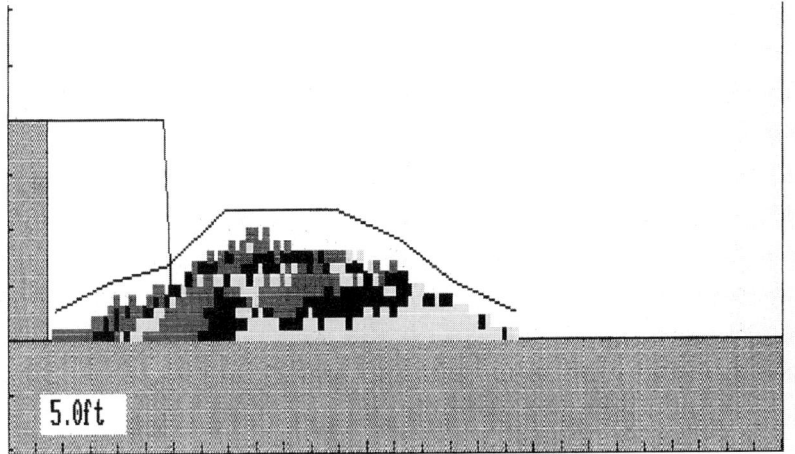

FIGURE D1-4   ICI explosives, SABREX computer model.

Rock Hardness and Density

The design of a blast in hard rock is more critical than with softer material. The harder and denser the material, the more energy is needed to break and displace it. However, if the material is overshot, it can result in flyrock and airblast, and if undershot, it can be very difficult to dig.

Void and Incompetent Zones

Voids and soft layers can allow explosive energy to escape through a path of least resistance. This escape can be violent and dangerous. Care must be taken to load accordingly through such zones and place inert stemming through these soft areas where it is expected that the explosive energy might easily escape. A drill log can be very important in these situations.

Joint Spacing

Close jointing will yield better fragmentation than widely spaced joints, which often result in blocky-type breakage. Small-diameter holes drilled on small pattern dimensions allow a better distribution of the explosive and will usually give improved results when blocky-type breakage occurs.

Bedding Orientation and Location

More energy is needed in the area where a hard layer is encountered. If hard layers exist near the upper portion of a bench, short, small-diameter, satellite holes drilled between the main blastholes may assist in breaking the material.

Where a predominant bedding formation exists, try to make the bench height correspond to that level. This will usually give a smoother floor and reduce or eliminate the need for subdrilling.

The stability of the bench face or final wall must be carefully considered when dealing with dipping formations. Bed thickness and angle of dip will greatly influence toe burden, floor grade, back break, and overhang potential. A change in the blast design to improve one area may adversely affect other areas.

Standard Blast Patterns

A typical blast consists of a number of blastholes arranged in one or more rows in such a way to meet specific requirements. Surface blasts will be one of three types: sinking cut for opening a new area or level; box-type cut for extending an excavation lengthwise; and a corner cut. Regardless of the type of blast, the same pattern design principles apply.

The standard type of drill pattern setups are square and staggered hole placement. The difference between the two is how the rows of holes are defined. The square pattern setup is when holes are positioned directly behind one another. The staggered pattern setup is when holes in successive rows are offset one-half the spacing dimension of adjacent rows. Although the square setup pattern is easier to lay out and drill, the first blasthole on the corner, either in a square or rectangle dimen-

ROCK EXCAVATION/BLASTING

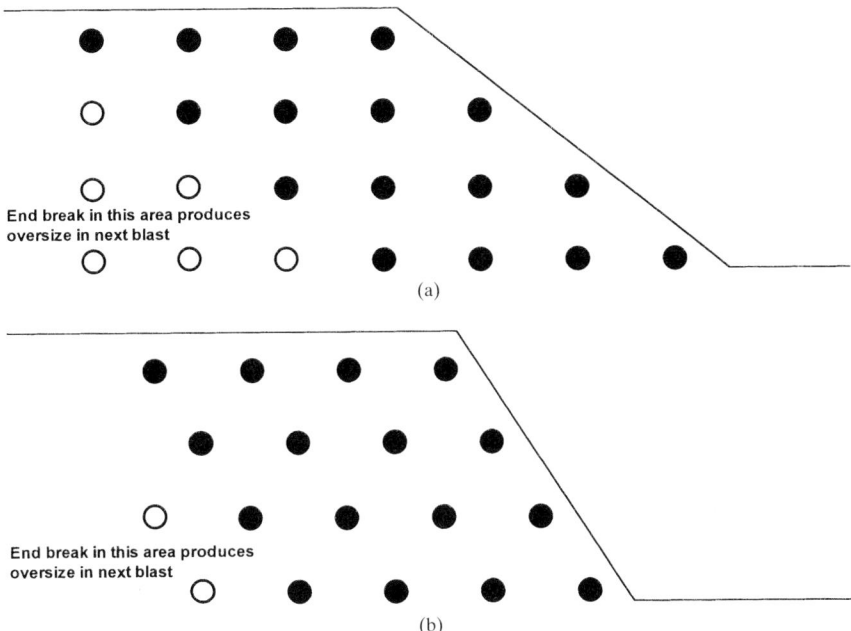

FIGURE D1-5 ICI explosives, reduced end break blast patterns.

sion, becomes an area that is difficult to break satisfactorily. This is because of the square block configuration that the last blast tried to break, but often overbreaks. To avoid this problem, it is better to develop corners with an obtuse angle of the face intersections by planning the proper square or staggered setup drill patterns. Figure D1-5 illustrates this principle of blast design.

The true burden is dependent both on the drill pattern dimensions and the millisecond delay sequence selected. This is illustrated in Fig. D1-6, where the pattern was drilled on a square setup with a 12-ft burden and 12-ft spacing and delayed in an echelon sequence. The true burden in relation to blast direction now becomes 8.5 ft and has a spacing of 16.9 ft. This same principle applies for a rectangle or staggered pattern setups.

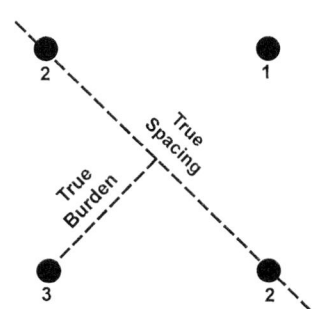

FIGURE D1-6 ICI explosives, true burden and spacing.

Care must be taken in selecting the drill pattern setup and the type of millisecond delay sequence to obtain the desired fragmentation, displacement, and profile of the muck. Generally, square pattern setups are delayed in an echelon sequence which stacks the muck closer to the bench for mucking by a shovel. Staggered pattern setups are usually delayed in a sequence that moves the blasted material in a more

D1-12 STANDARD HANDBOOK OF HEAVY CONSTRUCTION

forward direction that produces a low, well-displaced muckpile conducive for digging by wheel loaders.

There are numerous combinations of drill patterns and millisecond delay sequences. Figures D1-7 and D1-8 illustrate a square pattern and staggered pattern delay timing sequence. Whatever pattern design is chosen, the following rules for locating drill holes still apply:

1. Tight conditions should be avoided whenever possible with adequate space provided for each blasthole to account for swell.
2. Proper burden should exist for each blasthole at the time it fires.
3. Holes should be aligned to take advantage of the rock's geologic structure, especially the perimeter to ensure clean break lines at the edges.

Criteria for Design

Fragmentation must be compatible with the type and kind of loading equipment available. Shovels can handle coarser fragmentation in a tighter muckpile than wheel loaders, and scrapers assisted by dozers require very good fragmentation as well as proper placement of the muck to be economical. Rock to be used in embankments may have maximum specified sizes, and material that is to be crushed should be small enough to minimize crusher delays and maintenance. Experience indicates

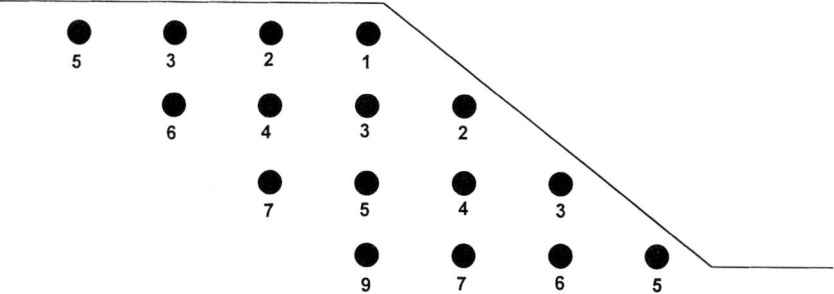

FIGURE D1-7 ICI explosives, typical square pattern design.

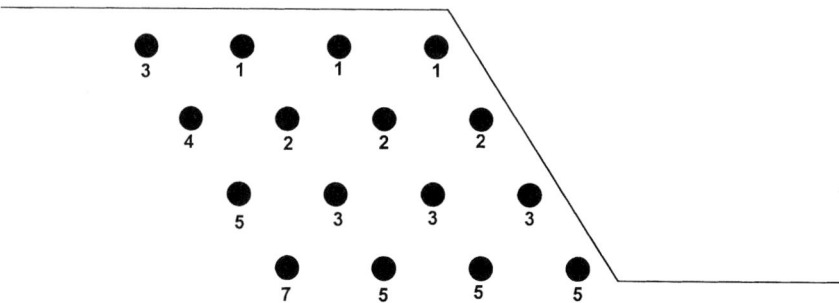

FIGURE D1-8 ICI explosives, typical staggered pattern design.

that it is usually cheaper to shoot all the rock to less than maximum size rather than risk the added costs of secondary shooting or sorting of material on the fill site.

Bench height is principally determined by the drill equipment capabilities. In general, benches should be as high as practical to decrease the number required and thereby minimize equipment movement and bench cleanup for drilling. The height, however, should not exceed the economic drilling depth nor should it result in a hazardous face or muckpile. The optimum bench height for top hammer track drills is generally 20 to 30 ft and 30 to 60 ft for the larger-type drills. The hole depth should be a multiple of the drill-steel length whenever possible. The required amount of subdrilling must be added to the hole depth when designing the bench height; i.e., two 12-ft drill rods and 2 ft of subdrilling will result in a face of 22 ft, not 24 ft. Some larger drills can often be equipped with extended masts so that the bench can be drilled in a single pass without adding drill steel. Drilling high benches in some rock formations may cause the bit to wander, thus producing an inaccurate drill pattern at the bottom of the holes, resulting in poor fragmentation and grade control.

Bench heights are also dictated by the loading equipment to be employed. There is no rule of thumb to determine the best height. This is because different types of rock have varying propensities to "run" to the mucking equipment. A wheel loader cannot safely work as high a face as a shovel; therefore, the muckpile for a loader operation needs to be spread out and lowered by using a different shot design than for a shovel.

Final blast design criteria, including the selection of hole size, should be made only after careful evaluation of all costs involved in drilling, shooting, loading, hauling, fill treatment, and crushing.

RULES-OF-THUMB GUIDELINES

Burden

The burden is the distance from the blasthole to the nearest free face at the time of detonation. The explosive charge diameter, type of material to be blasted, and type of explosive determine the amount of burden. The charge diameter for bulk explosive loaded holes is the borehole diameter; for cartridge explosives, it is the diameter of the cartridge. If the burden is insufficient, airblast and flyrock will occur. If it is too great, fragmentation and digging will be poor, excessive toe is likely, and ground vibration will be higher. A buffer or material against the face will also increase the burden, as it provides resistance for the explosive to overcome. The rule of thumb is the drill pattern burden B should be 20 to 40 times the diameter of the explosive charge D_e:

$$B = 20 \text{ to } 40 \ D_e$$

Every operation and application will have special requirements and the burden will have to be adjusted accordingly. Also note that the delay pattern must be considered when determining the true or effective burden, as internal free faces can occur due to the delay pattern.

Spacing

Spacing S is the distance between adjacent holes and is measured perpendicular to the burden. It is a function of the burden and should be 1 to 1.8 times the burden B:

$$S = 1 \text{ to } 1.8 \, B$$

If the spacing is too great, the fragmented area around each hole will not overlap and may contribute to boulders, tight digging, excessive toe burden, and a scalloped face. If the spacing is too tight, collar stemming ejection and wasted energy may occur.

Subdrill

Subdrill is the distance drilled below floor level to assure that the rock is fractured to a grade line for easy mucking to that elevation. If there is a soft layer of material or geologic parting at the floor, subdrill may not be required.

Too much subdrill is a waste of drill and explosives dollars and increases drill collaring problems on the next bench below because of excessive fracturing. It is also a major cause of vibration complaints since the explosive in that area of the blasthole has the greatest confinement. The rule of thumb is for the subdrill J to be 0.2 to 0.5 times the burden B:

$$J = 0.2 \text{ to } 0.5 \, B$$

Bench Height

Since most operations have little flexibility with drill diameter, it is important that the bench height be in proportion to the burden. The rule of thumb is for the minimum bench height H to be at least twice the burden B dimension to achieve a balance of applied explosive energy:

$$H = 2 \, B$$

There is not a rule for maximum bench height as for blasting. However, if the bench height is too great for the diameter of the drill hole, then accuracy can be a problem at the bottom of the hole, resulting in poor explosives distribution.

Collar Stemming Height

The stemming is the inert material placed on top of the explosive charge in the borehole. It acts to contain the explosive gases during detonation and allow them to do work. Drill cuttings do not make as good a stemming material as crushed stone. If there is too little stemming, the explosive gases could escape into the atmosphere, causing a loss of energy, airblast, and possibly flyrock. Too much can result in poor fragmentation in the top of the blast and increase the potential for excessive ground vibration. Stemming height is very critical and a change of only a foot or more can make for very different results. The height of the top of the explosive should be carefully controlled so the stemming height is exact. The rule of thumb for the stemming height T is 0.7 to 1.3 times the burden B:

$$T = 0.7 \text{ to } 1.3 \, B$$

SPECIAL BLASTING TECHNIQUES

Buffer or Choke Blasting

Buffer blasting involves firing a blast into previously blasted material which lies directly against the face. The buffer of broken material does not hinder the fragmentation mechanisms which rely on explosion-generated strain waves but does restrict rock displacement and the breakage mechanisms that are dependent on rock movement. The overall fragmentation in buffer blasting is usually less than that for blasts which are fired to a free face (with the same design blast parameters) and digging is generally tighter.

The buffer of broken rock in the blast enables the driller to collar the front-row blastholes closer to the crest without endangering people or equipment. The effect of this is to reduce or eliminate a large toe burden that cannot otherwise be drilled and shot. Where there is limited drilling capacity, buffer blasting enables the operator to minimize the movement of drilling equipment from one site to another. The disadvantage of buffer blasts are:

1. Smaller blasthole patterns and higher effective energy factors are required for the same degree of fragmentation and muckpile looseness.
2. More overbreak generally occurs.
3. Higher ground vibrations frequently result.
4. Initiation system cutoffs and misfired explosives potential increases.

Sinking Cuts

A sinking cut differs from normal blasts in that it does not have a vertical free face to break to. The purpose of a sinking cut is to ramp down to a new (lower) bench, and to create a vertical free face to which later blasts can break. Rock movement must be upwards and only to a limited extent sideways. To achieve satisfactory digging usually demands that blastholes are drilled on a comparatively close grid pattern with a high powder factor.

Moderate cut depths of 20 ft are usually taken out in a single lift. A good rule is to space the blastholes about half as far apart as the depth to grade with subdrill about half the spacing dimension. All blastholes should be initiated before the first charge detonates to avoid initiation system cutoffs, and the primers should be placed near the bottom of the blastholes to avoid misfired explosive charges. The firing sequence should be arranged so that the centerline of blastholes open up the shot. Blastholes on later delays break to this free face.

This type of blast generally requires 1.5 to 2 times as much explosive as for a free-face bench blast. All blastholes should be well stemmed to avoid flyrock.

Overbreak Control

Controlled blasting attempts to limit the amount of overbreak damage to the remaining rock so that it will be competent and stand for long periods of time. When the energy of an explosive is generated in a normal production blast, it can destroy, or at least reduce, the structural strength of the rock behind and to the side(s) of the blasted volume. New fractures and planes of weakness are created, and joints and bedding planes, which may have been "tight" before the blast, are opened up. As a result, there is an overall reduction in the rock mass's ability to hold together. This shows up as overbreak and a fractured face (or wall) with a higher likelihood of rock falls.

The successful applications of overbreak control blasting techniques lessen the hazard of rock falls and minimize the need for scaling, rock bolting, or cementing. They also contribute to steeper walls which improve recoverable rock reserves in quarries, and reduce the necessary volume of rock to be removed from other excavations.

Special overbreak control blasting techniques such as presplitting can improve rock wall smoothness and stability. Such techniques are important and, in the appropriate situations, can be very useful. However, efforts should be made to find out to what extent production blasts of modified design can reduce overbreak. Contrary to what one might expect, the redesign of a production blast to reduce overbreak does not necessarily mean poor fragmentation and a tight muckpile. Overbreak can be reduced by observing the following guidelines in the design of production blasts.

1. If a continuous column of bulk explosive is too powerful for back-row blastholes, deck load the explosive and initiate the explosive decks within each blasthole on the same delay sequence. The stemming material between the explosive charge decks should be dry crushed rock.
2. Use explosive cartridges to build up as a continuous column but which are decoupled in the blasthole.
3. The effective burden should not be greater than 25 times the explosive charge diameter, preferably 20.
4. The best spacing between back-row blastholes is the largest spacing that gives a straight face.
5. Drill angled rather than vertical blastholes for the last three to four rows in front of the final wall.
6. In the back row, the length of the stemming column should be increased to prevent surface overbreak, but not so much as to create an overhang.
7. Efforts should be made to minimize subdrilling into a crest or berm of the next bench where it is to remain standing for considerable time.
8. The initiation sequence should be selected such that
   a. there are minimum numbers of blastholes firing on the same delay.
   b. the blastholes along the back row and side(s) of the shot are detonating in a regular delayed sequence with longer intervals than the body of the blast.

Controlled Blasting Techniques

Controlled blasting techniques are important methods of achieving rock walls that are smoother and structurally stronger than those that are left by redesigned production blasts. These techniques are most successful in tough massive rocks and in tight, horizontally bedded formations which are relatively undisturbed by faults, joints, etc. Obtaining a consistently competent wall in unconsolidated, weathered, or highly fissured ground is often difficult to achieve. The technique that produces the desired effect in a tough massive rock may be unsuitable in weak, highly fissured ground. Blasthole patterns and charge loads, therefore, are very dependent on the rock properties encountered.

Due to the increased amount of drilling required, all of the controlled blasting techniques are more costly than regular or redesigned producing blasting. It is essential that drilling be carefully supervised so that blastholes are at the design spacing, in the intended direction, and parallel to one another. Drilling accuracy becomes more important as the bench height increases.

Line Drilling

Line drilling, a largely outdated blasting technique, involves a row of parallel uncharged holes along the design limit. Hole diameter is usually 1.5 to 3 in and the spacing between holes is commonly 2 to 4 times the hole diameter. These holes provide a plane of weakness to which the production blast can break. The burden distance between the last row of blastholes and the line-drill holes usually is 50 and 75 percent of the normal burden distance. Line drilling is used only where other controlled blasting techniques are likely to cause too much overbreak, because the amount of drilling is a very costly operation (as a result of the close hole spacings required).

Presplitting

Presplitting involves drilling a row of closely spaced, 1.5- to 6-in diameter blastholes along the design limit. These blastholes are very lightly charged and then detonated simultaneously before the first production blastholes shoot. Firing of the decoupled presplit charges splits the rock just along the design limit, creating an internal surface to which the production blast can then break. If the presplit holes are spaced too close together or charged too heavily, the presplit charges will themselves create overbreak.

The spacing between presplit blastholes normally increases with the blasthole diameter as shown in Table D1-1. But because rock properties have an overriding effect on blasthole spacing and explosive load, these figures should be considered only as recommended starting values.

Presplit holes are generally charged to within about one burden dimension of the collar. They are stemmed by pushing down a wad of paper or hole plug to the top of the charge and then backfilling above with drill cuttings.

The best possible presplitting action is obtained when presplit charges are fired simultaneously. Where ground vibrations are likely to disturb residents, millisecond

TABLE D1-1 Presplit Blasting Guidelines

Borehole diameter, in	Explosive load, lb/ft	Borehole spacing, in
1.5	0.08	12–18
1.75	0.11	12–18
2.0	0.17	18–24
2.5	0.23	24–30
3.0	0.34	24–36
3.5	0.50	24–36
4.0	0.60	36–48
5.0	0.90	36–60
6.0	1.30	48–72

delays should be utilized so as to obtain the consecutive firing of instantaneous groups of holes and any detonating trunklines covered to minimize noise. Presplit holes should be initiated sufficiently in advance of the first hole in the production blast to enable the presplit fracture to develop to its fullest extent.

The presplit face will be damaged or even destroyed if production blastholes are drilled too close to it. If the distance between the presplit and the last row of production blastholes is too great, rock is left in front of the presplit face which could create a hazardous mucking condition or make it necessary to reblast this area. The best distance between the presplit and last-row production blastholes is usually 30 to 50 percent of the normal burden for production blastholes.

Smoothwall Blasting

Sometimes referred to as slashing, slabbing, or trim blasting, smoothwall blasting consists of drilling a row of closely spaced blastholes with a suitable burden to spacing ratio that trends to split the rock web between holes, giving a smoothwall with minimum overbreak along the final excavation limit. Similar to presplitting, all blastholes are charged with an evenly distributed reduced load of explosives and fired simultaneously or nearly so, to remove the narrow berm left in place after the final production blast in that area has been dug out.

The burden and spacing increase with the blasthole diameter (see Table D1-2). The burden should be greater than the spacing to facilitate reinforcement of explosives energy along the breakage plane. Collar stemming length is typically equal to the burden dimension. Smoothwall blasting does not usually give the spectacular type of result as presplitting in tough massive rocks, but it does produce a reduction in overbreak. Smoothwall blasting tends to give better results in unconsolidated ground than presplitting.

TABLE D1-2 Smoothwall Blasting Guidelines

Diameter, in	Borehole spacing, ft	Burden, ft	Explosive load, lb/ft
1.5	2.0	3.0	0.10
2.0	2.5	3.5	0.15
2.5	3.0	4.0	0.20
3.0	3.5	5.0	0.30
3.5	4.0	6.5	0.45
4.0	5.0	7.0	0.65
5.0	6.0	8.0	0.90
6.0	7.0	9.0	1.25

EXPLOSIVES SELECTION

An important parameter in blasting design is the nature of the explosives used to effect the blast. Low cost, good fragmentation, and displacement of the muckpile are the desired results of the explosives selection process. The principal considerations are:

- Groundwater conditions
- Rock properties—strength, structure, etc.

- Diameter and depth of blastholes
- Drilling costs and drilling capacity
- Relative explosive cost per unit of energy
- Shock and heavy energy partition of the explosive
- Shelf life

Properties of Explosives

An explosive is a chemical compound or mixture of compounds that undergo a very rapid decomposition (detonation) when initiated by energy in the form of heat, impact, friction, or another detonation. The explosive detonation produces shock energy and a large volume of hot gases which generate high pressures in the borehole, both of which cause the rock to be fragmented and displaced.

The principle ingredients in explosives are fuels and oxidizers. Common fuels are fuel oil, mineral oil, wax, carbon, and aluminum. Common oxidizers are ammonium nitrate, calcium nitrate, and sodium nitrate. Additional ingredients are water, gums, thickeners, and emulsifiers. Nitroglycerin, TNT, and PETN are molecular explosives and combine the fuel and oxidizer in the same compound, and are used as sensitizers and energy enhancement.

Strength

A rating system has been developed for most commercial explosives which compares the strength of a mixture of the explosive against ammonium nitrate and fuel oil (anfo). These are the *relative bulk strength* (RBS) and the *relative weight strength* (RWS) values quoted by manufacturers. The RBS is the measure of energy available per volume of explosive as compared to an equal volume of anfo. If the RBS of an explosive is 145 then it would have 45 percent more energy than the same volume of anfo. The RWS is the measure of the energy available per weight of explosive as compared to an equal weight of anfo. If an explosive has an RWS of 105 then it would have 5 percent more energy than the same weight of anfo. Since a borehole represents a volume, the RBS is the more widely used value.

Detonation Velocity

The velocity of detonation (VOD) is the speed at which the detonation front moves through a column of explosive. It varies from about 5000 to 25,000 fps, depending on the explosive. The higher the velocity, the greater the shock energy and the greater the shattering effect from the explosive. Slower-velocity explosives tend to produce their gas energy over a longer period which results in greater heaving action. The VOD is affected by the composition, density, particle size, degree of mixing, confinement, and the primer.

Density

The specific gravity of an explosive is expressed as the weight per unit volume, normally grams per cubic centimeter. It ranges from 0.5 to 1.7 gm/cm^3 for most commercial explosives. Generally, the higher the density the explosive, the greater will be the RBS value of that explosive.

Water Resistance

The ability of an explosive to detonate after exposure to water is very important. Some explosives are very water resistant as a result of their compositions (emulsions), whereas other products are very dependent on their packaging (extra dynamite) for water-resistance protection. The manufacturer should be consulted to determine the water resistance of a product. Both the length of time a product can be exposed to water and the depth of water are important factors in evaluating water-resistance requirements.

Fume Class

Explosives are rated by fume class, with fume class 1 producing the least amount of toxic gases, followed by classes 2 and 3. This is generally of more importance in underground applications or where ventilation is poor. Insufficient priming, insufficient charge diameter, water deterioration, wood spacers, paper hole plugs, and plastic borehole liners can also contribute to increased toxic gases. Adequate time for the fumes to disperse is very important on the surface as well as underground.

Detonation Pressure

The pressure produced in the reaction zone of an explosive is an indicator of the shock energy that an explosive produces. It is rated in kilobars and varies from 5 to about 200. The higher the detonation pressure, the more useful the explosive is in functioning as a primer for another explosive and also at fragmenting hard rock.

Borehole Pressure

The pressure exerted on the borehole walls by the expanding gases after detonation is dependent on the quantity and temperature of the gases and the degree of confinement. It can vary from 30 to 70 percent of the detonation pressure amount, depending on the type of explosive, and it does contribute to some fragmentation but is responsible for most of the heavy movement of the rock. Low-velocity explosives tend to have higher borehole pressures than high-velocity explosives. Borehole pressures range from 10 to 60 kilobars.

Sensitivity and Sensitiveness

Sensitivity is a measure of the explosives' ease to initiation. It is affected by the ingredients, particle size, density, charge diameter, confinement, presence of water, and the product temperature with some products, such as watergels. Manufacturers list the minimum primer to be used with each product.

Sensitiveness is the ability of a product to propagate detonation along its column length once it has been initiated. If it is very sensitive, it may propagate several feet from hole to hole.

Explosive Products

There are three basic categories: nitroglycerin-based explosives, booster-sensitive dry mixtures, and solution-based mixtures, which may be emulsions or watergels.

Nitroglycerin explosives are dynamites and are generally the most sensitive of the commercial explosives available today. Due to this sensitivity, they offer an extra margin of dependability in certain applications. Types of dynamites available are straight dynamite (straight means no ammonium nitrate), ammonia dynamite, blasting gelatin, straight gelatin, ammonia gelatin, and semigelatin. They differ in ingredients, sensitivity, water resistance, performance characteristics, and cost.

Booster-sensitive dry mixtures are defined as any material used in blasting which is not detonator-sensitive and which has no water in the formulation. Anfo is the most common type. The ammonium nitrate (an) is in the form of pills and is not an explosive until the correct amount of fuel is added.

The fuel content is very important; if there is too much, sensitivity and energy output deteriorate, and if there is too little, toxic gases as well as poor energy output occur. Generally 6 percent no. 2 diesel fuel content (by weight) is considered to be the appropriate ratio for consistent performance by manufacturers. Performance characteristics vary with the particle size, density, degree of confinement, charge diameter, and water contamination.

Anfo is hygroscopic and will absorb moisture, which contributes to caking and handling problems. Anfo has no water resistance, therefore a large amount of water can cause the product to fail to detonate; lesser amounts of water will reduce the energy output, and produce more toxic fumes.

The minimum primer for reliable initiation of anfo increases as the charge diameter increases. Large-diameter primers generally produce better, more consistent results than smaller primers.

Emulsions and watergel explosives are basically a mixture of nitrates, a fuel, and water. They can be manufactured as detonator- or booster-sensitive depending on the formulation sensitivity and have excellent water resistance with a wide range of energies. Solution-based explosives can be bulk or packaged.

PRIMER SELECTION

The most fundamental requirement for safe and efficient blasting is to ensure that explosives are initiated effectively. This facilitates release of the maximum amount of effective energy to achieve the desired blasting result. Many poor blasts have occurred when design, drilling, and explosives selection have been correct, but the explosives did not release their full energy potential because of inadequate priming.

Until recently, understanding the exact mechanisms by which primers and explosives are initiated was limited by an inability to take suitable measurements in blastholes in the immediate vicinity of the primer. Improved measurement techniques, supported by recent computer simulation by the explosives manufacturers' technical centers, has greatly improved the understanding of these processes. From this work, the influence of factors such as primer shape, charge diameter, and primer composition is now better understood regarding the initiation process of bulk explosives.

The effectiveness of any explosive in priming or initiating other explosives is a function of the amount of energy that is transmitted into the second explosive, the rate at which this energy is transmitted into the second explosive, and the rate at which this energy is delivered. Explosives that initiate readily from a detonator and have a high density and detonations velocity, such as cast boosters (Fig. D1-9), are the most effective primers.

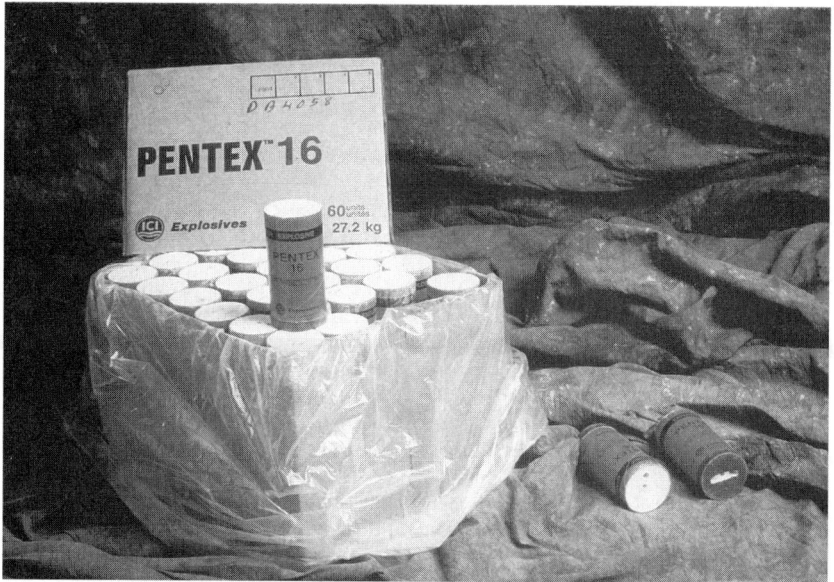

FIGURE D1-9 ICI explosives, PENTEX booster.

For any type of primer, there will be a minimum mass of that explosive that is capable of imparting sufficient energy to start a self-generating detonation front in another explosive charge. Where priming is marginal, there is a risk of producing low transient detonation velocities and reduced blasthole pressures in the run-up zone. In general, manufacturers' recommendations on the size of a primer it takes to initiate another explosive accounts for the nonideal nature of the production blasting, with a factor of safety included to reduce the risk of poor performance and misfires. Some of the factors that contribute to reducing priming effectiveness are as follows:

- Explosives contaminated or diluted
- Length of time loaded in the blasthole
- Presence of ground water
- Primer buried by drill cuttings or mud
- Hydrostatic pressures of deep blastholes
- Dynamic desensitization by pressure transmitted from adjacent blastholes

INITIATION SYSTEMS

Initiation systems fall into two broad categories: electric and nonelectric. In the first category, electric detonators include instantaneous, millisecond delay, and long-period delay intervals. The nonelectric category includes detonating cord with millisecond delay connectors and shock tube–style detonators in millisecond and long-period delay intervals. An initiation system contains three basic parts.

1. An initial energy source. This may be an electric blasting machine, a nonelectric spark generator, or a shotshell starter device.
2. A distributing network to distribute the energy to the blastholes, such as electricity through wires, detonation of detonating cord, or a low-energy dust explosion in a tube.
3. An in-hole component that uses energy from the distribution network to initiate the detonator-sensitive explosive, such as the bridgewire in the electric detonator or a shock spit from the nonelectric tube.

An electric initiation system utilizes an electrical power source with an associated circuit to convey the impulse to the electric detonators. Inside the electric detonator, the electrical energy is converted into heat energy by passing the firing current through a high-resistance bridgewire. The heat energy ignites the pyrotechnic that surrounds the bridgewire of the match assembly. The resulting flash or flame ignites the delay cement. The delay composition then burns for a designated time and ignites the primer charge, which then detonates the base charge. Consult the manufacturer for current requirements of specific products of application and circuit design.

The shock tube used in nonelectric initiation is a self-contained, flexible, abrasion-resistant, plastic tube that contains reactive materials that transmit a firing signal through the tube to the detonator when initiated. It will reliably propagate this detonation around sharp bends and through kinks or knots in the tubing. Because such a small quantity of reactive material is used to sustain the detonation wave in the shock tube, it is therefore compatible with all grades of commercial explosives, including the most sensitive dynamites. Inside the detonator, the resulting firing signal from the shock tube ignites the delay element. The delay composition burns for a designated time and ignites the primer charge, which detonates the base charge.

Detonating cord initiation consists of a core of PETN in a waterproof plastic sheath and reinforced with a wrapping of textiles and plastics. The coreload can vary from 4.5 up to 400 grains per foot. Trunklines (surface lines) and downlines are usually 25 to 60 grain. Millisecond delay connectors are used in surface trunklines to provide the delay intervals. There are various in-hole delay shock tube components that can be used with detonating cord to expand both systems' versatility. It is important to note that detonating cord downlines may cause axial initiation of some explosives or damage to others to the point of failure. Consult the manufacturer of the explosive for compatibility recommendations. Detonating cord also loosens the stemming, making it less effective, and can contribute to premature stemming ejection and airblast. Detonating cord is rugged and relatively insensitive; however, it does cause airblast and should be covered with 1 to 2 ft of material if airblast is a problem.

BLASTHOLE LOADING

Blastholes should be checked prior to loading, using a weighted measuring tape. If the hole is too deep, it should be backfilled to the correct depth. Loading holes that are too deep waste explosives and often contribute to ground vibrations. If the blasthole is not deep enough, it should be cleaned out or redrilled prior to loading any holes on the shot. If a redrill hole is adjacent to a blocked or short hole, the unused hole should be backfilled with stemming to prevent the explosive energy in the redrill hole from prematurely venting to the open hole. Water should be noted on the report, along with hole depth. If there is an opening or void in the hole, it should also be noted so that it can be plugged or have stemming placed in it to prevent excessive explosive being placed in that region of the hole.

General Loading Procedures

The charging of blastholes should be entrusted to those who are trained in this work and who are completely familiar with the safety precautions which must be followed. The fact that certain explosives can be detonated by heat, friction, or impact must be fully appreciated, and particular attention should be paid to the hazards that may exist with extraneous electricity.

Explosives should not be taken to the blast site until all blastholes are ready for loading. Before beginning loading operations, any tools and equipment other than those required for loading the blastholes should be removed. As a general guideline, no work other than that associated with loading should be carried out within 50 ft of all explosives loading and blastholes that have been loaded. Figure D1-10 illustrates an explosives loading warning sign at the blast area.

If a thunderstorm approaches during loading operations, all work in the blast area should be stopped, and the blast crew (and any other personnel in the close vicinity) should move to a place of safety. Even where a nonelectric system is used to initiate the blast, there is always the possibility that lightning could cause a detonation.

Cartridge Explosives

Blastholes are usually loaded with the largest-diameter cartridge that can be safely introduced into them. A clearance of about 1 in is usually sufficient in most conditions. Where blastholes are uniform and smooth, about 0.5 in is sometimes sufficient (Fig. D1-11). Operators should ensure that cartridges are seated properly on top of one another by frequently checking the column rise. It is important to avoid the presence of layers or drill cuttings, sizable chunks of rock, or air or water gaps between adjacent cartridges in the blasthole. If the continuity of the charge is interrupted, a new primer should be loaded to reprime the charge. If the blastholes are very ragged, uneven, or partially blocked, it may be impossible to get full cartridges to their intended positions. Where the blastholes are dry, anfo can be used. In wet holes, NITROPEL, a form of pelletized TNT can be substituted for anfo.

Bulk Explosives

Anfo. Anfo may be poured into blastholes or loaded by means of bulk equipment. Due to its lack of water resistance, anfo should not be loaded directly into any blasthole containing water, but is acceptable after the holes have been dewatered and there is no possibility of further seepage. If water seepage is such that blastholes cannot be dewatered completely, packaged explosives should be used to build out of water, then anfo should be loaded only in the portions of the blastholes that are dry.

ROCK EXCAVATION/BLASTING   D1-25

FIGURE D1-10 ICI explosives, explosive loading warning sign at blast area.

FIGURE D1-11 ICI explosives, cartridge explosive loading.

Where anfo is used above cartridge charge in water, a slit cartridge of emulsion explosive should be used above the water level to seal the hole and prevent the anfo from falling into the water annulus around cartridges and from becoming desensitized up the column by the rising water level. Figure D1-12 illustrates sealing off a hole above the water level to load anfo.

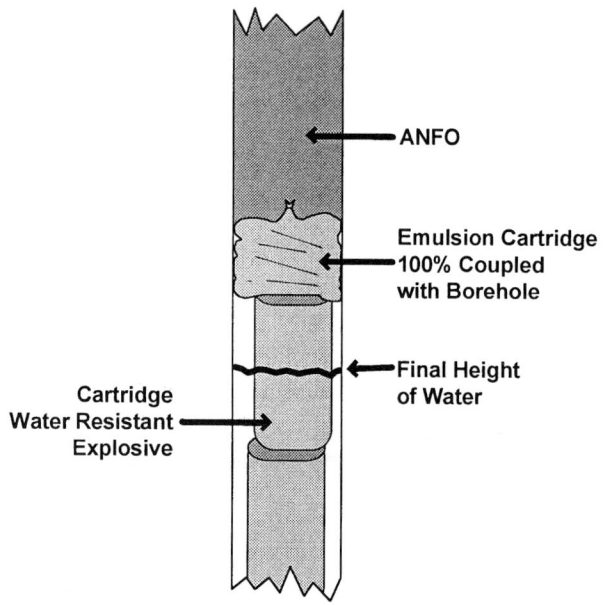

FIGURE D1-12   ICI explosives, borehole sealing concept.

Emulsion/Anfo Blend.   There are a number of systems available for delivering bulk blend explosives to the blasthole. Systems are based on a mix truck which blends anfo and emulsion explosives at the collar of the blasthole immediately prior to loading (Fig. D1-13). The advantage of emulsion/anfo blend explosives is the ability to select the grade of explosive for each blast at the time it is loaded, wet or dry, in hard or soft rock.

When loading through water, care must be taken to ensure that the water is displaced as the explosives column rises. Entrapment of water in the explosive matrix will cause poor blasting performance and possible incomplete detonation.

PRIMING

The overriding factor in making up primers is that the detonator or detonating cord should be securely located in contact with the primer, and should remain there until the moment of firing. Primers should be assembled at the time of blasthole loading in a way which minimizes the chance of damage to signal tube, legwires, or detonating cord.

FIGURE D1-13 ICI explosives, emulsion/anfo blend truck.

Location

The blasting geometry and hardness of rock material may dictate a particular priming horizon for best results. In general, the primer should be at the point of most confinement. In most cases, this will be the bottom of the hole. Where a particularly hard zone of rock exists, there may be a benefit from placing the primer within the hard zone to minimize energy loss into soft material.

In benching operations where subdrill is required, some operators choose to locate the primer below grade-level to ensure that excavating equipment will not strike it in the event of a misfire. Others deliberately raise the primer above grade level, so that the primer can be readily located and dealt with. Fundamental rock-breaking principles indicate that there will be little difference whether the primer is at grade level or a short distance above or below. The overriding consideration must be that the primer is securely located in good, uncontaminated explosives. This generally means at a point somewhat above the bottom of the blasthole, to avoid burial in mud or drill cuttings.

Multiple Priming

If the explosives are loaded according to the manufacturer's specifications and they remain in good condition in a continuous column, a single primer is all that is necessary for complete detonation, regardless of charge length. However, all blastholes are not perfect nor are they loaded perfectly. Therefore, *double-priming* (near the top and near the bottom of the charge) is common practice to provide insurance against misfires. Multiple-priming is most desirable where there is perceived risk of ground movement, damage to downlines, groundwater dilution of explosives, or contamination or separation of the charge. Where multiple primers are initiated in a blasthole using in-hole delay detonators, the top primer in each hole should contain a detonator delay which is one number higher than that in the bottom primer. This makes the top primer an insurance unit only, and in all cases the entire explosives column should detonate before the top detonator functions. Multiple-priming is also common when the investment in each blasthole and the work expected of each explosives charge is seen to clearly outweigh the cost of a second primer unit.

For maximum efficiency, the diameter of the primer should be as near to the borehole diameter as possible, develop at least 100 kilobars of detonation pressure, and have sufficient length to ensure adequate lateral shock pressure development. In large-diameter blastholes, it may be uneconomical to attempt to satisfy these guidelines for priming with a single unit. Therefore, the option of creating a combination primer has proven adequate. This is the practice of using a standard-size primer and surrounding it with an emulsion explosive in the blasthole (Fig. D1-14). The combination then creates a very effective primer for less-sensitive anfo and emulsion/anfo-blend explosive charges.

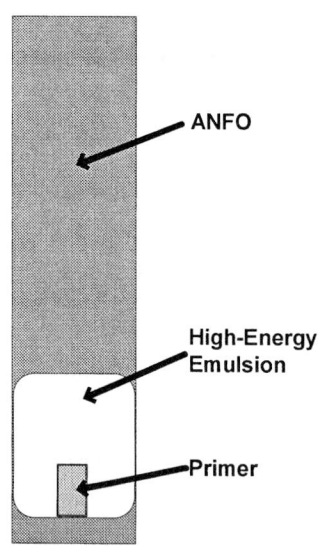

FIGURE D1-14 ICI explosives, combination primer concept.

Side-Initiation

Detonating cord downlines affect the explosive columns through which they detonate. For anfo, this can cause either detonation by side-initiation, or compressions and desensitization of the explosives around the cord. In general, emulsion and watergel explosives are less sensitive to side-initiation than anfo. However, these explosives are more easily compressed and desensitized than anfo, because of the action of the shock wave of the detonating cord on the gas bubbles within these explosives. Until the bubbles re-form after compression, high-order detonation is prevented from propagating beyond a certain distance from the primer. Because of the large number of variables involved, quantifying exact conditions under which bulk explosives can be side-initiated, or the exact degree of desensitization, can be very difficult. Consult the explosives manufacturer for product compatibility recommendations with detonating cord.

INITIATION SEQUENCE

The outcome of any multiple-hole production blast is critically dependent on interactions between blastholes. The sequence in which blastholes are initiated and the time interval between successive detonations has a major influence on overall blast performance. A poor blast design cannot be rectified by good initiation design. A good blast design can, however, be enhanced with appropriate initiation design. The performance of production blasts can only be optimized when blasthole charges are detonated in a controlled sequence at suitable millisecond separation time intervals. The result of a well-designed multihole blast cannot be duplicated by firing the same number of blastholes individually or at random.

The optimum initiation sequence and delay timing for any blast depends on many factors, including the following:

1. Rock properties (strength, density, porosity, structure, etc.)
2. Blast geometry (burden, spacing, bench height, free faces, etc.)
3. Explosive characteristics (type, grade, degree of coupling, decking, etc.)
4. Initiation system (surface or in-hole delays, type of downline, etc.)
5. Primer type and location
6. Environmental constraints (airblast and ground vibration regulations)
7. Desired result (fragmentation, muckpile displacement and profile, etc.)

Delay Time Interaction

The performance of any conventional blast can be significantly influenced by altering delay timing to vary the degree of interaction between adjacent blastholes. This can lead to improved fragmentation and muckpile looseness, reduced overbreak, lower ground vibrations, and better control over the final muckpile position and profile. While absolute values of interrow and intrarow delays are important, the ratio of these times also has a major impact on the final result. This can be explained by the following concepts (which are oversimplifications of a complex subject).

1. The interrow delay controls interaction between adjacent blastholes and determines whether blastholes act independently or together.
2. The intrarow delay controls interaction between dependent blastholes, as it affects the progressive creation of new effective free faces for later-firing blastholes.
3. The ratio of intrarow to interrow delay controls the geometry and orientation of new free faces created as the blast progresses. For a later-firing blasthole, the location, shape, and extent of any effective free face will depend on this ratio of delay times. This influences the direction and extent of displacement of each blasthole's burden and thus the final muckpile shape and position. This is sometimes referred to as the apparent direction of movement of a blasthole or the overall blast.

General rules for delay timing are:

1. Time between rows should be 3 to 5 ms per ft of burden.
2. Time between adjacent boreholes should be 1 to 3 ms per ft of spacing.

3. More time should be added to the last row of burden to allow the blast to pull away from the bench and create less backbreak.

If the delay times are too short, cratering, backbreak, airblast, and flyrock can occur.

The manipulation of delay timing to control blast performance is best illustrated by considering several alternatives of initiating the same blast. Where blastholes are initiated row by row, forward displacement will be enhanced and the general direction of movement will be perpendicular to the rows (Fig. D1-15).

If the same blast is initiated in an echelon pattern, the direction of movement will be perpendicular to the effective rows, towards the free corner (Fig. D1-16).

These sequences can be used for blasting in a wide range of applications, although the best results occur when a free face and/or free end is available. Delay times can be altered to suit different conditions, in conjunction with in-hole delays if necessary.

ENVIRONMENTAL EFFECT OF BLASTING

There are four major environmental effects of blasting: flyrock, ground vibration, airblast, and dust. There is little we can do to control dust; however, flyrock, ground vibration, and airblast are caused by explosives and can be minimized with careful blast design. They are dangerous and annoying and represent wasted energy which did not go into fragmentation or movement of the blast.

A preblast survey is recommended before blasting near dwellings, if the dwelling is within ½ mile of the blast area. The survey helps increase communication between the contractor and community and, more important, it assesses the condition of a structure prior to blasting so that after blasting an equitable resolution of the damage complaints can be resolved if they occur. It protects both the homeowner and the contractor.

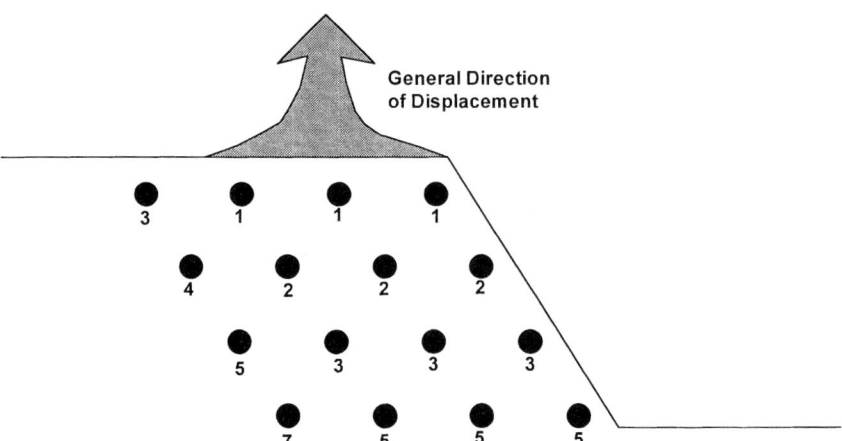

FIGURE D1-15 ICI explosives, typical row delay sequence.

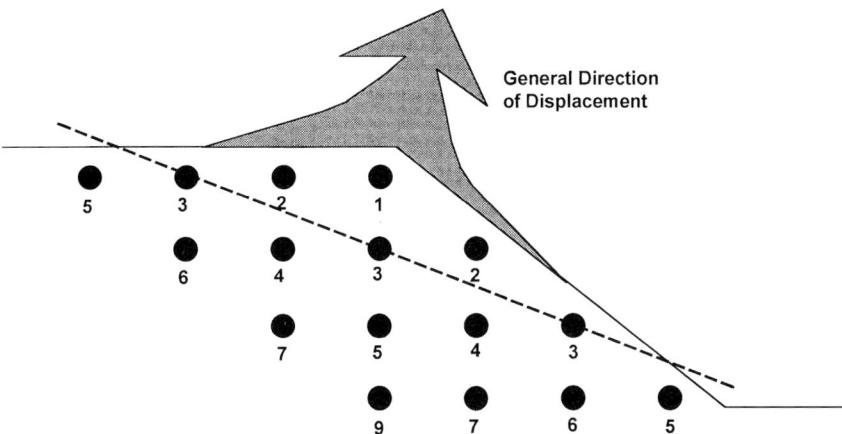

FIGURE D1-16 ICI explosives, typical echelon delay sequence.

Flyrock

Flyrock is the leading cause of injury and property damage in blasting accidents. It is usually attributable to careless loading or poor blast design. Care in the selection of the burden, stemming height, and loading of explosives can reduce the chance of flyrock. To compensate for unknown geologic factors, an adequate blast area safety zone should be cleared of personnel and equipment prior to blasting.

Ground Vibrations

All blasts produce some ground vibration—some more than others. They can range from annoying to damaging, and they are controllable.

Ground vibrations are caused by several factors, which are, generally, too much energy released at insufficient delay interval times between detonation and/or a poor blast design such as overconfinement by too great a burden or excessive subdrill.

Some regulations require a seismograph to record the blast vibration trace, with the responsibility left to the blasting contractor to comply with regulatory limits of peak particle velocity and/or vibration frequencies. If vibrations are not monitored, a scaled distance formula can be used to calculate the maximum pounds per delay based on distance to the dwelling. This value is very conservative and will generally ensure that the ground vibrations are well below the allowable limit.

The most common form of the scaled distance equation is

$$W = \left(\frac{D}{SD}\right)^2$$

where D = distance from blast to structure
       W = maximum weight of explosives per 8-ms time interval
       SD = scaled distance factor

For example, if the building is 2000 ft away, the scaled distance factor allowed would be 55. To find the maximum allowable weight of explosives per delay interval, the equation is used to solve for W.

$$W = \left(\frac{2000}{55}\right)^2 = 1332 \text{ lb}$$

Airblasts

Airblast is a shock wave generated by a blast that travels through the air. Concussion is airblast which has a frequency of less than 20 Hz and cannot be heard. Frequencies above 20 Hz can be heard and are called *sound waves*. It is caused by unconfined explosives such as uncovered surface detonating cord or venting of explosive charges by inadequate stemming, inadequate burden, or gas escaping through cracks or soft seams. Also, the movement of the bench face during a blast can act like a large piston and create concussion airblast in front of the shot.

Generally, if the level is kept below 128 dB at the structure, the airblast will be acceptable. This is not to say that it would not create complaints, but damage is very unlikely to occur. Temperature and wind can enhance airblast and can focus it.

Airblast is often the cause of complaints but is less apt to cause damage than ground vibrations.

BIBLIOGRAPHY

Kennedy, Bruce A., *Surface Mining,* 2d ed., chaps. 2, 5, 6, AIME, 1990.

Morhard, Robert C., *Explosives and Rock Blasting,* Atlas Powder Co., Dallas, Tex., 1987.

Naapuri, Jukka, *Surface Drilling and Blasting,* Tamrock, 1988.

CHAPTER 2
PAVING*

James J. O'Brien, P.E.
O'Brien-Kreitzberg, Inc., Pennsauken, New Jersey

---

*Some of the material in this chapter is from R. W. Beaty, "Bituminous Pavements" and William G. Westall, "Portland Cement Concrete Pavements" in Section 25, "Paving," Stubbs and Haver (eds.), *Standard Handbook of Heavy Construction*, 2d ed., McGraw-Hill, New York, 1971.

SECTION 1
BITUMINOUS PAVEMENTS

This section deals with the construction of bituminous pavements and is specifically directed toward the production and placement of the bituminous-treated materials which are incorporated in the total flexible pavement structure.

PAVEMENT ELEMENTS, MATERIALS, AND TYPES

Pavement Elements

The flexible pavement structure consists of several courses of treated and/or untreated mineral materials placed on the subgrade or prepared earthwork foundation. The quality (and therefore the cost) of the materials involved will usually increase with each successive course, from the subgrade upward.

Subbase Course. This course consists of an imported material of higher quality than the subgrade, but the term *subbase* usually indicates that it does not meet the quality requirements for a base-course material. It is used in the pavement structure for reasons of economy, as in the case of a locally available granular material. Subbase materials may or may not be treated with additives such as hydrated lime, portland cement, or one of several forms of bituminous materials.

Base Course. This course is placed directly on the prepared subgrade or on the completed subbase if such a course is included in the pavement structure. Base-course materials comprise naturally occurring mineral materials or crushed mineral aggregates of a quality superior to those used in subbase construction. The materials may or may not be treated with additives such as portland cement or bituminous materials. The thickness of the base course will usually range from 4 to 10 in, depending on the types of materials used and the engineering determination of the required structural thickness for the particular pavement design involved. The treatment of the base-course materials with portland cement or bituminous additives results in a reduction in required thickness, thereby reducing the total quantity of materials required for a given design strength. The use of bituminous-treated base-course material has been in favor because of its economy and also because its superior binding and waterproofing qualities improve its durability.

Leveling Course. In most cases, the term *leveling course* designates a course of bituminous paving material placed on the base course or on an existing surface for resurfacing projects to eliminate surface irregularities. It follows that the leveling course must be of variable depth in order to remove sags and high spots in the underlying surface. In some states, however, the pavement course directly underlying the surface course is called the leveling course. It is then comparable to the *binder course,* as next described.

Binder Course. For the high-type bituminous pavements (asphalt-concrete pavements), the binder-course material differs from the surface-course material in that the quality may be somewhat lower. For example, since the binder course is not exposed to the action of traffic, requirements for cohesion and/or skid resistance are not prime considerations. The binder course must have stability at least equal to that of the surface course. The thickness of the binder course is usually on the order of 2 to 3 in, depending on the maximum aggregate particle size used and the overall design thickness requirement.

Surface Course. The functions of the surface course of the flexible pavement are several: (1) to provide a smooth, quiet-riding surface; (2) to provide a surface which is cohesive and durable to resist the wearing stresses imposed by traffic; (3) to provide a surface with a high coefficient of friction between the surface and vehicle tires so that adequate skid resistance is provided; and (4) to provide a layer of material which is relatively impervious to the infiltration of water and air, thus preventing accelerated aging of the bituminous binder material. The thickness of the surface course will vary, but it is usually on the order of 1 to 2 in. However, where durable, nonpolishing aggregates are in short supply and therefore expensive, it is not unusual to find surface courses placed as thin as ½ to ¾ in.

For roads subjected to low volumes of traffic, such as secondary roads, the surface course may consist of a surface treatment of about ¾-in thickness. This type of surface course, which will be described subsequently, is placed directly on a base course of treated or untreated crushed stone or gravel.

Pavement Materials

The materials for a flexible pavement consist of aggregates plus one or more types of bituminous material.[1]

Bituminous Materials. The principal bituminous binder material is petroleum asphalt, although a minor amount of flexible pavement construction utilizes the road tars, usually in the subbase and base courses. Asphalt is used in three different forms: (1) asphalt cements of the several penetration grades, which require heating to relatively high temperatures to achieve the fluidity required for application and/or mixing; (2) cutback asphalts, which are diluted with naphtha or kerosene to render them more fluid for handling (usually, heating to a relatively low temperature is also required to achieve the proper viscosity for their application); and (3) emulsions of paving-grade asphalts and water with emulsifying and stabilizing agents. Emulsified asphalts usually require little or no elevation of temperature for application and/or mixing.

Table D2-1 indicates the uses of the various asphalt materials in accordance with Asphalt Institute suggested criteria.[1]

Aggregates. For bituminous pavement construction, a great variety of aggregate types and gradations are utilized. Important characteristics of aggregates for flexible pavement uses are resistance to wear, polishing, and weathering (soundness). Table D2-2 shows recommended aggregate gradations for various bituminous mixtures.[1] It can be seen from this that a wide range of aggregate gradations may be used, varying from coarse, open-graded mixes to dense, fine-graded mixes.

In the case of asphalt concrete, the aggregates are heated and dried before mixing with the heated asphalt cement. For mixtures involving use of the cutback asphalts, the aggregates may or may not be heated and dried. For hot-mix cutback asphalt mixtures, the temperature of the aggregates is substantially lower than for asphalt concrete. The same is true for emulsified asphalt-aggregate mixtures.

TABLE D2-1 Principal Uses of Asphalt

	Paving asphalts					Liquid asphalts Rapid-curing (RC)			
Type of construction	40–50	60–70	85–100	120–150	200–300	70	250	800	3000
Asphalt concrete:									
Highway		x	x	x	x				
Airports		x	x	x					
Parking areas	x	x	x						
Curbs		x	x*						
Industrial floors	x								
Blocks	x								
Plant mix, cold laid:									
Graded aggregate									
Mixed-in-place:									
Open-graded aggregate								x	
Dense-graded aggregate									
Clean sand							x		
Sandy soil									
Penetration macadam:									
Large voids			x					x	
Small voids				x			x		
Surface treatments:									
Aggregate seal				x	x			x	
Sand seal							x		
Slurry seal									
Fog seal									
Penetration treatment							x		
Prime coat, open surfaces						x	x		
Prime coat, tight surfaces									
Tack coat						x			
Dust laying									
Patching mix:									
Immediate use							x	x	
Stock pile									
Hydraulic structures:									
Membrane linings, canals, and reservoirs	x‡								
Hot-laid, graded aggregate mix for groins, dam facings, canal and reservoir linings		x							
Crack filling						x			
Soil treatment, membrane envelope, mixtures	x‡								
Expansion joints	Blown asphalts, mineral-filled asphalt cements, and preformed joint compositions								
Undersealing portland-cement concrete	Blown asphalts								
Roofing	Blown asphalts								
Miscellaneous	Specially prepared asphalts for pipe coatings, battery boxes, automobile undersealing, electrical wire coating, insulation, tires, paints, asphalt tile, wall board, paper sizing, waterproofing, floor mats, ice cream sacks, adhesives, phonograph records, tree grating compounds, grouting mixtures, etc.								

    In northern areas where rate of curing is slower, a shift from MC to RC or from SC to MC may be desired. For very warm climates, a shift to next heavier grade may be warranted.

    * In combination with powdered asphalt.
    † Diluted with water.
    ‡ Also 50-60 penetration blown asphalt and prefabricated panels.
    § Slurry mix.
    ¶ Cationic quick-setting grades are comparable in use to RS-1 and RS-2 Cationic mixing grades are comparable in use to MS-2 and SS-1.
    Source: The Asphalt Institute.

\multicolumn{13}{c}{Liquid asphalts}												
\multicolumn{4}{c}{Medium-curing (MC)}	\multicolumn{4}{c}{Slow-curing (SC)}	\multicolumn{5}{c}{Emulsified¶}	Inverted
70	250	800	3000	70	250	800	3000	RS-1	RS-2	MS-2	SS-1	SS-1 h	emulsion
		x	x			x					x		x
		x	x			x				x			
	x	x			x	x					x		x
	x	x			x							x	
	x				x						x		x
									x				
								x					
		x	x					x	x				x
	x							x					
											x	x	
											x†	x†	
		x			x	x							
	x												
x				x									x
				x							x†	x†	
											x†		
	x	x				x							x
	x				x	x							x
						x						x§	

TABLE D2-2 Compilation of Suggested Mix Compositions*

Mix type	2½ in	1½ in	1 in	¾ in	½ in	⅜ in	No. 4	No. 8	No. 16	No. 30	No. 50	No. 100	No. 200	Percent asphalt
							Mix seal							
II a						100	40–85	5–20					0–4	4.0–5.0
II b					100	70–100	20–40	5–20					0–4	4.0–5.0
							Surface							
II b					100	70–100	20–40	5–20					0–4	4.0–5.0
II c				100	70–100	45–75	20–40	5–20					0–4	3.0–6.0
III a					100	75–100	35–55	20–35		10–22	6–16	4–12	2–8	3.0–6.0
III b				100	75–100	60–85	35–55	20–35		10–22	6–16	4–12	2–8	3.0–6.0
IV a					100	80–100	55–75	35–50		18–29	13–23	8–16	4–10	3.5–7.0
IV b				100	80–100	70–90	50–70	35–50		18–29	13–23	8–16	4–10	3.5–7.0
IV c			100	80–100		60–80	48–65	35–50		19–30	13–23	7–15	0–8	3.5–7.0
V a					100	85–100	65–80	50–65	37–52	25–40	18–30	10–20	3–10	4.0–7.5
V b*				100	85–100		65–80	50–65	37–52	25–40	18–30	10–20	3–10	4.0–7.5
VI a					100	85–100		65–78	50–70	35–60	25–48	15–30	6–12	4.5–8.5
VI b*				100		85–100		65–80	47–68	30–55	20–40	10–25	3–8	4.5–8.5
VII a*						100	85–100	80–95	70–89	55–80	30–60	10–35	4–14	6.0–11.0
VIIIa							100	95–100	85–98	70–95	40–75	20–40	8–16	6.5–12.0
							Binder							
II c				100	70–100	45–75	20–40	5–20					0–4	3.0–6.0
II d			100	70–100		35–60	15–35	5–20					0–4	3.0–6.0
III b				100	75–100	60–85	35–55	20–35		10–22	6–16	4–12	2–8	3.0–6.0
III c				100	75–100		30–50	20–35		5–20	3–12	2–8	0–4	3.0–6.0
III d			100	75–100		45–70	30–50	20–35		5–20	3–12	2–8	0–4	3.0–6.0
IV c			100	80–100		60–80	48–65	35–50		19–30	13–23	7–15	0–8	3.5–7.0
							Leveling							
III b				100	75–100	60–85	35–55	20–35		10–22	6–16	4–12	2–8	3.0–6.0
V b*				100	85–100		65–80	50–65	37–52	25–40	18–30	10–20	3–10	4.0–7.5
VI b*				100		85–100		65–80	47–68	30–55	20–40	10–25	3–8	4.5–8.5
							Base							
I a	100	35–70		0–15				0–5					0–3	3.0–4.5
II d			100	70–100		35–60	15–35	5–20					0–4	3.0–6.0
II e			100	70–100	50–80		25–60	10–30	5–20				0–4	3.0–6.0
III d			100	75–100		45–70	30–50	20–35		5–20	3–12	2–8	0–4	3.0–6.0
III e		100	75–100	60–85		40–65	30–50	20–35		5–20	3–12	2–8	0–4	3.0–6.0
IV d		100	80–100	70–90		55–75	45–62	35–50		19–30	13–23	7–15	0–8	3.5–7.0
						Skip gradations which have been successfully used								
A				100	95–100		50–70	30–50			5–25		2–10	4.0–9.5
B		100	95–100	60–80			30–50	20–40			5–25		1–10	4.0–9.5
C	100	95–100		60–80			25–45	15–35			3–20		0–5	4.0–9.5

* May be used for base where coarse aggregate is not economically available.
Source: The Asphalt Institute.

Pavement Types

Major types of bituminous pavements will be identified briefly in the following paragraphs. Discussions of the materials, construction methods, and equipment used in each of these types of pavements will be found in subsequent portions of this chapter.

Bituminous concrete is a composition of asphalt cement and high-quality aggregates, carefully controlled for asphalt content, aggregate gradation, and void content. These parameters are carefully controlled to produce a pavement mixture which meets the specified criteria for stability, cohesion, and impermeability.

Bituminous plant mix is generally similar to asphalt concrete except that the levels of control for production and placement, quality of materials, and design criteria are lower than for bituminous concrete. Methods of production and placement are similar to those for bituminous concrete.

Bituminous penetration macadam is constructed in multiple lifts of coarse, open-graded crushed stone, gravel, or slag. Each lift receives an application of asphalt, after which a succeeding, smaller-sized aggregate is used to produce a well-keyed, stable layer. If the base lift of the macadam pavement is coated with bituminous material in a central mixing plant prior to placement, the pavement is called a *bituminous macadam* type.

Bituminous road mix, as the name implies, is constructed by mixing aggregates and bituminous materials on the roadway, usually from a windrow. It is generally considered to be the lower-quality type of mixed paving material because of the low level of control of aggregates and bitumen proportioning. This type of pavement is generally regarded as an intermediate-type bituminous pavement.

Surface treatments and seal coats are characterized by an application of liquid asphalt or emulsified asphalt, which then may or may not be covered with a spread of sand, screened gravel, or crushed stone. The term includes such applications as prime coat, tack coat, single or multiple surface treatments of a base course, single seal-coat treatments of an existing pavement surface, fog seal coats, plant-mix seal coats, and special applications for sealing and/or rejuvenating an existing bituminous surface. Single seal treatments applied for the purposes of restoring the integrity and/or skid resistance of an existing surface are sometimes called *armor* coats.

CONSTRUCTION OF SURFACE TREATMENTS, BITUMINOUS PENETRATION MACADAM, AND ROAD-MIXED PAVEMENTS

A discussion of surface treatments, seal coats, macadam-type construction, and road-mix methods is appropriate at this point.[2] The base must be protected from weathering and the effects of traffic. It is entirely possible that engineers may prefer to allow traffic on the completed base for a period of months in order to prove the construction before proceeding with any of the high-type bituminous surfacing methods. Such proving would allow correction for settling and would contribute to a better-riding road.

The construction procedures which will be discussed herein may result in the final pavement surface. They may also comprise one or more of the earlier steps in "stage" construction. In the latter case, with the exception of some types of seal coats and surface treatments, the paving will seldom be the final riding surface for high-type highway construction.

Some types of liquid asphalts which are used in regular seal-coat work have also been employed as corrective treatments on old paving surfaces. These light application treatments are often regarded as regular maintenance operations. There can be several purposes for *seal coating* using bituminous materials, the most important ones being:

To seal the pavement against oxidation and the entrance of moisture

To rejuvenate or enliven a dry, hardened, and weathered surface

To reinforce or build up the pavement and at the same time stop damaging erosion and wear

Light liquid asphalt applications may also be employed as nonskid treatments for a slippery surface, to improve reflective properties, and to mark special traffic lanes. The amount of liquid material applied usually determines the use of a light application of sand, commonly referred to as a *blotter*. In fact, this type of maintenance work is sometimes referred to by road engineers as *blotter work*. The technique may be used as spot spray patching, depending on necessity, or it may be used for a complete coverage.

Tack Coat

A tack coat is a thin application of bituminous material to a nonabsorptive old road surface—usually asphalt, brick, stone block, or concrete—to provide a bond between the old pavement and the new surface.

In applying the tack coat, the surface is first swept or flushed to free it from dust and foreign material. Cutbacks, tars, or emulsions are then applied with a distributor in the thinnest application that will give complete coverage, seldom over 0.1 gal/yd^2.

Low viscosity or light bitumens should be used for thin coverage, but the residual must produce a hard, tacky surface. If a tack coat is not cured to a tacky condition, it acts as a lubricant and defeats its purpose.

The following bituminous materials are usually used:

RC-7	RS-1	RT-7
RC-250	RS-2	RT-8

Prime Coat

A prime coat is a light bituminous material applied to a stone, gravel, stabilized, or similar absorptive base that is to be given a bituminous surface. The purpose of the prime coat is to act as a bonding agent as well as to seal the joint between the base and the new pavement. The capillary rise of moisture into the surface and surface moisture working through to the base are thus retarded.

Sweeping does not remove all the loose dust and, without a prime coat to provide adhesion, the dust acts as a lubricant between the base and surface. Prime-coat materials are more fluid than tack-coat materials, so they penetrate into the base a slight amount to help in holding down the dust. The heat of hot-laid bituminous surfaces or the cutback materials in cold-laid surfaces will soften the hardened prime and give the desired adhesion.

With thicker surface courses, priming is not so necessary because of the inherent stability present in any thick mat. However, during the construction of a surface, a prime coat aids in waterproofing the base so that, in the event of rain, it dries off more quickly, and holds time to a minimum.

The primer is applied with a distributor at an approximate rate of 0.25 to 0.50 gal/yd^2 (see Fig. D2-1). The quantity applied should be absorbed in 24 h, and the normal drying or curing period is approximately 48 h. It is advisable to underprime rather than overprime, as any unabsorbed material is often taken up by the new pavement and surface bleeding may result.

PAVING

D2-9

FIGURE D2-1 Base priming with bituminous distributor.

As with tack coats, traffic is excluded from the primed surface to prevent loss of any of the primer and to prevent dust from collecting. When traffic cannot be excluded, medium-fine sand is spread over the surface to protect it by blotting up any excess bituminous material. Before the bituminous surface is placed, any loose sand remaining is swept off the base. Figure D2-2 illustrates the sequence of activities.

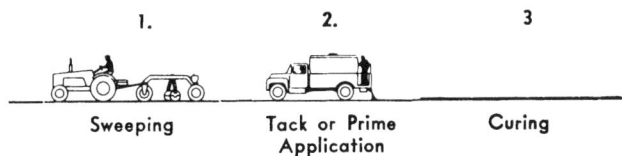

FIGURE D2-2 Sequence of operations for priming or tack coating.

Bituminous materials used for priming are listed as follows. The more viscous materials are used with the coarser bases.

RT-1	SC-70	MC-70
RT-2	SC-250	MC-250
RT-3		MC-800

Rapid-curing cutbacks are not desirable because the distillate has a tendency to separate from the asphalt cement and penetrate into the road base, leaving an excess of asphalt cement on the surface.

Fog Seal Coat

The preceding terms, *tack* and *prime* coat, are normally associated with resurfacing or new paving projects. A fog seal coat can be considered as a remedial or maintenance treatment for deteriorating surfaces, sealing and rejuvenating the mat for continued service. It is also used as a sealing coat in some forms of hot plant-mix work or as an intermediate treatment in new construction. A slow-breaking emulsion is diluted with water in ratios of 1:1 to 1:3 and applied at the rate of 0.1 to 0.2 gal/yd^2. Usually no cover aggregate is required.

Slurry Seal Coat

When the surface to be treated is unusually rough, weathered, and checked, showing cracks sometimes up to ⅛ in wide, applications of liquid treatments are inadequate even when used in conjunction with blotter materials. A *slurry seal treatment*, consisting of an emulsified asphalt combined with a mixture of fine aggregates, provides a much thicker crack-filling material and is applied with a mechanical squeegee. The use of a slow-setting asphalt emulsion eliminates the necessity for heating the material prior to its application. It also provides a workability period prior to breaking of the emulsion and consequent balling of the mixture, which would prevent adequate squeegee application. The mix should be relatively free flowing and of a creamy consistency, permitting it to be squeegeed into all the cracks so as to promote a smooth surface. Such a treatment will contribute little or no structural strength but is designed to reduce maintenance and patching work and retard further deterioration of the mat. Slurry treatment work often precedes a complete coverage by regular seal-coating methods. The slurry-seal method can be used for special deslicking coatings designed for nonskid performance. In such cases, special types of fines aggregates providing hard, sharp edges are employed.

Single-Pass Seal Coats

Single-pass seal coats are thin, bituminous-aggregate courses applied to existing bases or surfaces of any type. The sequence of operations, which is illustrated in Fig. D2-3, is essentially as follows:

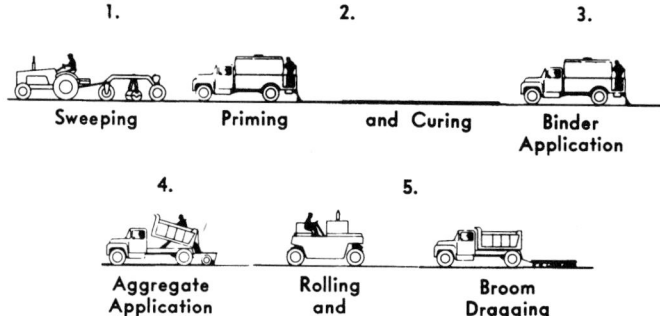

FIGURE D2-3 Sequence of operations for single-pass surface treatment road construction.

1. *Sweeping:* To remove dirt and other foreign matter.
2. *Priming and curing:* Optional, depending on condition of base.
3. *Binder application:* Bituminous material is applied by distributor. See Table D2-3.
4. *Aggregate application:* Applied by mechanical spreaders towed by dump trucks, by spreaders attached to tailgate, or by self-propelled spreaders with receiving hoppers. Aggregate occasionally is cast by hand from a moving truck or from a windrow.
5. *Rolling and broom dragging:* Alternate use of roller (pneumatic or steel wheel) and broom drag to smooth and compact the surface.

Multiple-Pass Surface Treatments

This method of road construction is very similar to the single-pass method except that there are two or more applications of aggregate and bitumen to give a surface having a more substantial load-carrying ability. Larger aggregate particles in the first course are held in place by successive layers of smaller aggregate particles. This method is sometimes designated as *inverted penetration* because the sequence of material application is opposite to that used in penetration macadam construction.

The required steps or operations, as illustrated in Fig. D2-4, are as follows:

1. *Sweeping:* To remove dirt and other foreign matter.
2. *Priming and curing:* Optional, depending on condition of base.
3. *First binder application:* Bituminous material is applied by distributor.

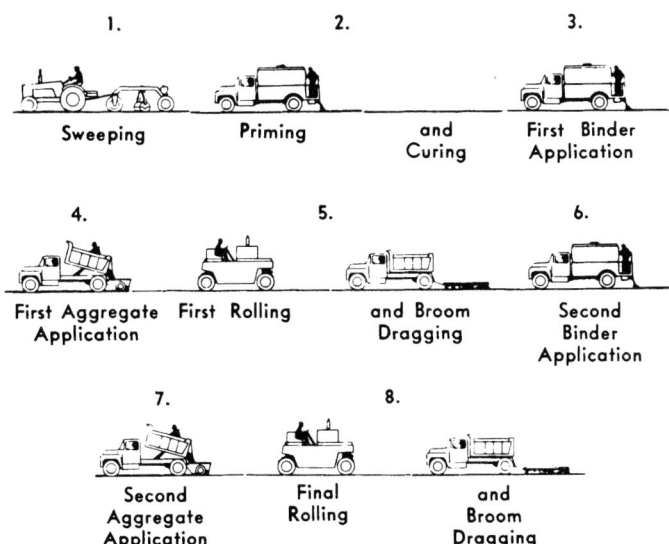

FIGURE D2-4 Sequence of operation for inverted penetration or multiple-pass surface treatment road construction.

TABLE D2-3 Quantities of Asphalt and Aggregate for Single Surface Treatments

Size of aggregate	Pounds of aggregate per yd²*,†,§	Gallons of asphalt per yd²*,†,¶	Hot weather Hard aggregate*	Hot weather Absorbent aggregate*	Cool weather Hard aggregate*	Cool weather Absorbent aggregate*
¾–⅜	40–55	0.28–0.35	120–150 RC3000, RS2	RC3000, RS2	RC800, RS2‡	RC800, RS2‡
¾–no. 8	30–45	0.23–0.30	200–300 RC800, RS2	RC800, RS2	RC800 RS2‡	RC800 RS2‡
½–no. 4	25–35	0.20–0.25	200–300 RC250, 800 RS1, 2	RC250, 800 RS1, 2	RC250, 800 RS1‡ RS2‡	RC250, 800 RS1‡ RS2‡
½–no. 8	25–35	0.20–0.25	RC250, 800 RS1, 2	RC250, 800 RS1, 2 RS2‡	RC250, 800 RS1‡ RS2‡	RC250, 800 RS1‡
⅜–no. 4	20–25	0.20–0.25	RC250, 800 RS1, 2	RC250, 800 RS1, 2 RS2‡	RC250, 800 RS1‡ RS2‡	RC250, 800 RS1‡
⅜–no. 8	20–25	0.20–0.25	RC250, 800 RS1, 2	RC250, 800 RS1, 2 RS2‡	RC250, 800 RS1‡ RS2‡	RC250, 800 RS1‡
¼–no. 8	15–20	0.15–0.20	RC250, 800 RS1, 2	RC250, 800 RS1, 2 RS2‡	RC250, 800 RS1‡ RS2‡	RC250, 800 RS1‡
Sand	10–15	0.10–0.15	RC250, 800 RS1, 2	RC250, 800 RS1, 2	RC250, 800 RS1‡ SS1‡	RC250, 800 RS1‡

* These quantities and types of materials may be varied according to local conditions and experience.
† The lower application rates of asphalt shown in the table should be used for aggregate having gradings on the fine side of the limits specified. The higher application rates should be used for aggregate having gradings on the coarse side of the limits specified.
‡ Caution should be exercised when using this material under poor drying conditions.
§ The weight of aggregate shown in the table is based on aggregate with a specific gravity of 2.65. In case the specific gravity of the aggregate used is less than 2.55 or more than 2.75 the amount shown in the table should be multiplied by the ratio which the bulk specific gravity of the aggregate used bears to 2.65.
¶ Under certain conditions, the heavier grades of MC liquid asphalts may be used in cool weather.

4. *First aggregate application:* Larger-size aggregate is applied by mechanical spreaders towed by dump trucks, by spreaders attached to tailgate, or by self-propelled spreaders.
5. *First rolling and broom dragging:* These two steps are alternated to seat aggregate.
6. *Second binder application:* Same type of bituminous material, again applied by distributor.
7. *Second aggregate application:* Smaller size and quantity of aggregate is spread evenly to fill surface voids in first course.

8. *Final rolling and broom dragging:* Alternate use of roller (pneumatic or steel wheel) and broom drag to smooth and compact surface. Broom dragging fills voids for uniform texture.

The number of applications is governed by the desired thickness; there could be as many as three or four.

Plant-Mixed Seal Coat

An open-graded seal coat which is premixed in a regular mechanical plant and laid with a finishing machine to an approximate depth of ½ in is frequently used in lieu of the seal coats just described. By mixing in a hot-mix asphalt plant, regular 80-100 or 120-150 penetration asphalts can be used. A tack coat is desirable on a dry, hardened mat.

Penetration Macadam

The penetration-macadam method of road construction is older than any other bituminous construction method. The name stems from the use of macadam-type aggregate and the penetration method of applying the binder to the aggregate.

Inverted penetration mats are seldom thicker than 1½ in, but penetration macadam can be 4 in or more thick. For this reason, and because of the nature of their construction and the size of aggregate used, penetration-macadam roads will generally handle heavier traffic loads and may be base courses rather than surface courses. See Fig. D2-5.

The sequence of operations, as shown in Fig. D2-6, is essentially as follows:

1. *Sweeping:* To remove dirt and other foreign matter.

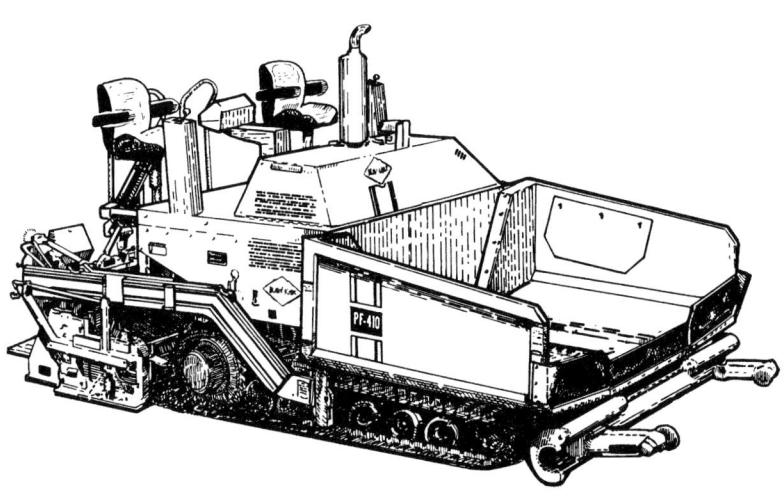

FIGURE D2-5 Finisher equipped with automatic screed controls can place plant-mixed macadam base course. (*Blaw-Knox.*)

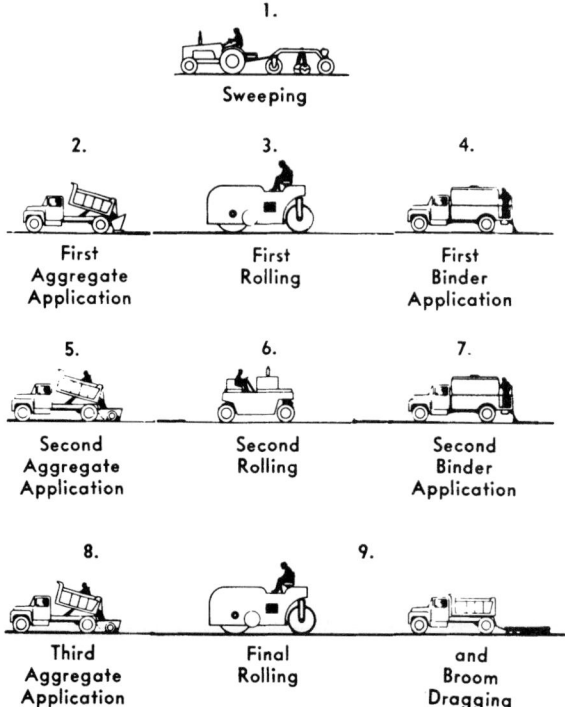

FIGURE D2-6 Sequence of operations for penetration-macadam road construction.

2. *First aggregate application:* Larger-size aggregate applied, usually with spreader box attached to dump truck or with bituminous finisher.
3. *First rolling:* Compacts coarse aggregate and locks it into place (pneumatic or steel wheel roller).
4. *First binder application:* Bituminous material is applied by distributor.
5. *Second aggregate application:* Smaller size and quantity of aggregate, usually called *key* aggregate, is spread evenly to fill voids in first course. Broom drags may be used to aid in distributing keystone.
6. *Second rolling:* Starts immediately following keystone application, while binder is still warm, to bond and compact aggregate.
7. *Second binder application:* Same type of bituminous material is applied in smaller quantity.
8. *Third aggregate application:* A still smaller size and quantity of chips or key aggregate may be immediately applied. Acts as blotter treatment and gives non-skid surface that resists traffic abrasion.
9. *Final rolling and broom dragging:* Alternate use of roller (pneumatic or steel wheel) and broom drag smooths and compacts surface. Broom dragging fills voids for uniform texture.

The three passes which have been outlined are the usual number. However, many roads are constructed with four, and some with only two.

Road Mix

This is a common method of constructing a bituminous road. As the name implies, the aggregate and binder are mixed on the road. The sun and wind remove the moisture from the aggregate.

The three methods of blending the aggregate and binder will be discussed later. The other steps or operations, as shown in Fig. D2-7, are essentially as follows:

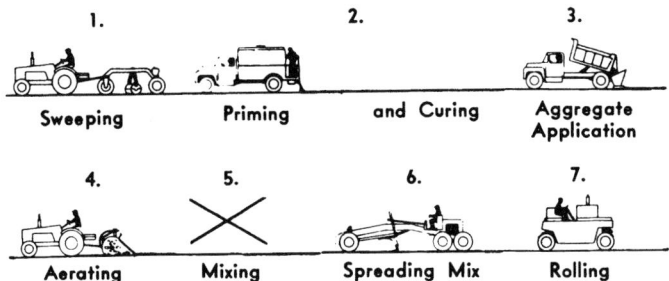

FIGURE D2-7 Sequence of operations for road-mix construction.

1. *Sweeping:* To remove dirt and other foreign matter.

2. *Priming and curing:* Optional, depending on condition of base.

3. *Aggregate application:* New aggregate is dumped into truck-towed windrow shaper or dumped directly onto road for blade shaping. Old aggregate (from existing road), or combination of old and new aggregate, is blade windrowed. An evener may be dragged over windrow to remove any major irregularities.

4. *Aerating:* If windrowed aggregate is too wet (1½ to 2 percent surface moisture by weight is usual limit), it must be aerated with blade, harrow, or some type of tiller to expose individual particles to sun and wind. Following drying, aggregate is usually windrowed again. Windrow may then be spread out by blade prior to mixing operation.

5. *Mixing:* The three methods of mixing aggregate and binder are blade, mechanical, and travel-plant mixing.

6. *Spreading mix:* Blades distribute mix to required loose depth and contour.

7. *Rolling:* Rolling starts at outside edges and progresses toward center with succeeding passes. Roller may be either smooth steel or pneumatic tired or both, depending on type of mixture. The period of time between spreading and rolling is a matter of experience and judgment; it varies from job to job, depending on materials used in mix. On some mixes the roller operates directly behind the blade, while on other mixes it may be necessary to hold the roller back as much as 24 h. In extreme cases, the roller is kept off the mat for two days.

The three methods of mixing will now be discussed.

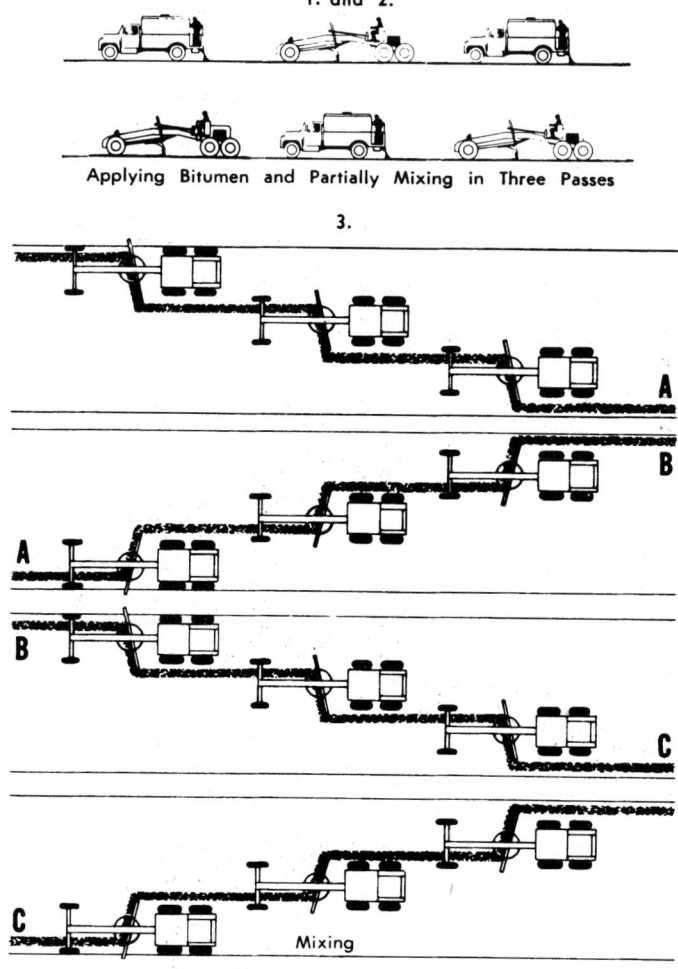

FIGURE D2-8 Blade mixing.

Blade mixing makes use of blade graders or some types of multiple-blade drag to blend the aggregate and binder together. (See Fig. D2-8.) After the aggregate has been dried and spread uniformly over one lane of the road, the following steps will complete the mixing:

1. *Partial application of bitumen:* Ordinarily applied in three separate applications by distributor. Alternates with step 2.
2. *Partial mixing:* Motor grader partially mixes bitumen and aggregate, leaving as little free bitumen as possible on the surface and preventing pools of it from collecting.
3. *Mixing:* Partially mixed material is formed into a windrow and bladed back and forth across the road until aggregate particles are coated and there are as few rich

and lean spots as possible. To reduce the required number of passes, multiple-blade drags are sometimes used instead of graders on open-graded aggregate road mixes.

PRODUCTION OF HOT-MIXED BITUMINOUS CONCRETE

Asphalt Plants[3]

Figure D2-9 illustrates a modern asphalt mixing plant. The flow of material through all asphalt plants is similar up to the point of measuring and mixing (see Fig. D2-10). Aggregate is fed from storage to the dryer, which removes the moisture and heats the aggregate.

The heated, dried aggregate is continuously fed to vibrating screens which divide it into the specified separations. Each size then flows into its respective hot bin. The next steps are measuring and mixing, and the two different systems used for this phase are referred to as *batch* and *continuous* types.

Aggregate Feeding

The first step in achieving maximum plant output is a properly designed cold-feed system. Inadequate feeding methods result in costly delays. Hot-bin starvation, whether caused by bridging of sand or irregular feeding rate of raw aggregate, stops mix production (see Figs. D2-11 and D2-12).

Drying costs, the most expensive item in mix production, are reduced by maintaining a balanced flow of aggregate through the dryer. Only those aggregate sizes which are needed to meet mix requirement should be dried. Every effort should be

FIGURE D2-9 Parallel-flow, portable Ventuir-Mixer asphalt production plant. (*CMI Corporation.*)

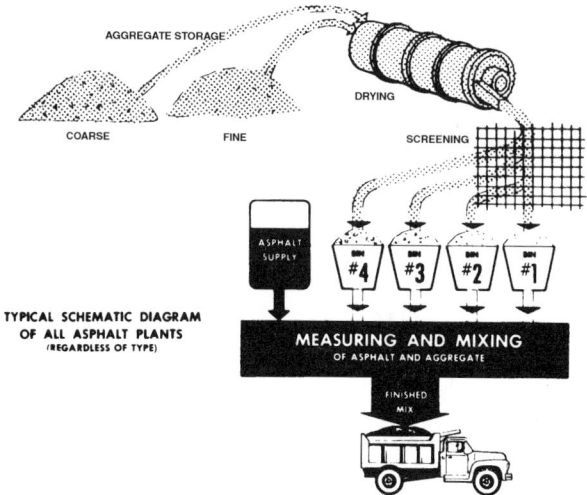

FIGURE D2-10 Typical schematic diagram of all asphalt plants.

made to eliminate costly, wasteful drying of surplus materials that will eventually overflow the hot bins and create rehandling costs as well as littering the plant site.

Any blending of minus no. 8 material necessary to achieve the required density and stability in the mix must be done at the cold feed. All sand, fine and coarse, passes through the bottom-deck screen into the no. 1 hot bin. Therefore proportioning of the various sand fractions must be done at the cold feed.

The increasing use of total and proportional controls, many of which are operated from remote panels, has lowered the cost of maintaining a feed which is properly

FIGURE D2-11 Four-bin cold feed system over gathering conveyor. (*CMI Corporation.*)

FIGURE D2-12 A tunnel conveyor system for reclaiming aggregates from stockpiles offers lowest handling costs, gives quick mix-change flexibility, and is ideally suited for total and proportional control feeding from a remote station.

matched to the needs of the final mix. The following points should be considered in selecting a feed system (see Fig. D2-13).

Tons per Hour Required. The maximum operating capacity of the plant, as well as the relative proportions of each aggregate size, will set the capacity requirements for the cold-feed systems.

Number of Separations of Materials. Mix design and gradation of available aggregates will determine the number of materials to be handled. This will dictate the number of stockpiles or bins required as well as the method of reclaiming.

Types of Materials to Be Handled. Consideration should be given to the characteristics of the aggregates handled. For example, extremely abrasive aggregates, such as trap rock, or sticky materials that may tend to bridge over bin openings or adhere to the feeder will affect the selection of system components.

Aggregate Source. The source of aggregate plus the frequency of delivery will also affect the choice of system components. Frequent delivery will reduce the amount of live storage required at the plant site; less frequent delivery will mean providing sufficient storage and the necessary equipment to stockpile and reclaim this material.

Flexibility Required. The necessary degree of mix flexibility should be considered in the planning. A drive-in operation where a variety of mixes will be needed over a short period of time requires a feed system that will permit quick and economic switching of mixes.

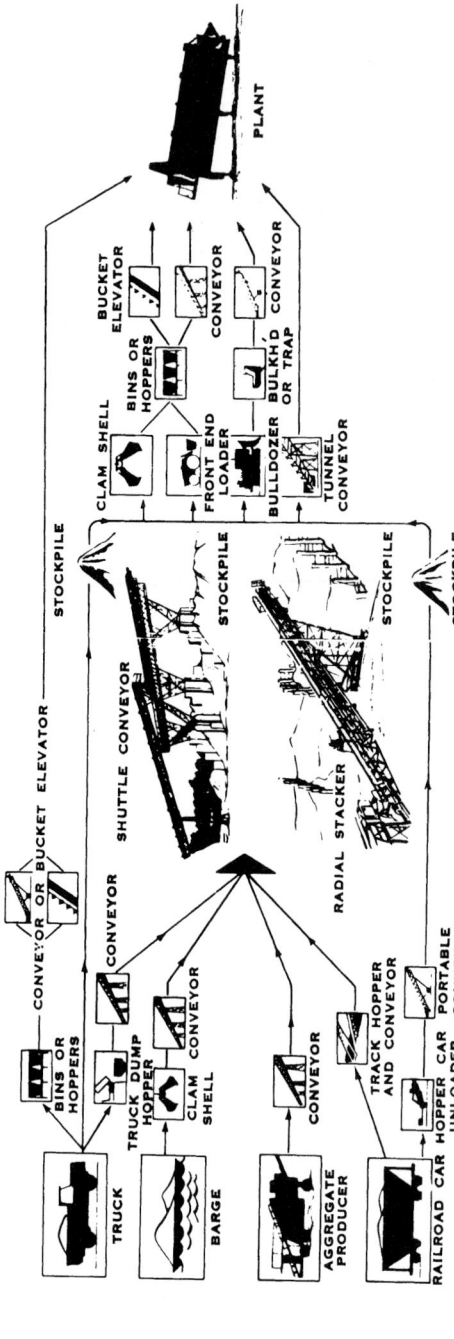

FIGURE D2-13 Flow diagram of most of the basic elements employed in typical aggregate feed systems.

Portability. Whether the plant operation is to be portable, semiportable, or stationary will help determine its configuration.

Aggregate Drying

The most frequently encountered bottleneck in asphalt-plant operation is the dryer. The optimum dryer is the one that will meet the desired performance level at the lowest investment and operating cost. However, the optimum design changes in different geographic areas and under different operating conditions. There are a large number of elements that affect drying, and changing one variable usually causes other factors to vary. The key to optimum dryer performance is proper balance of all these factors to satisfy a particular operating objective.

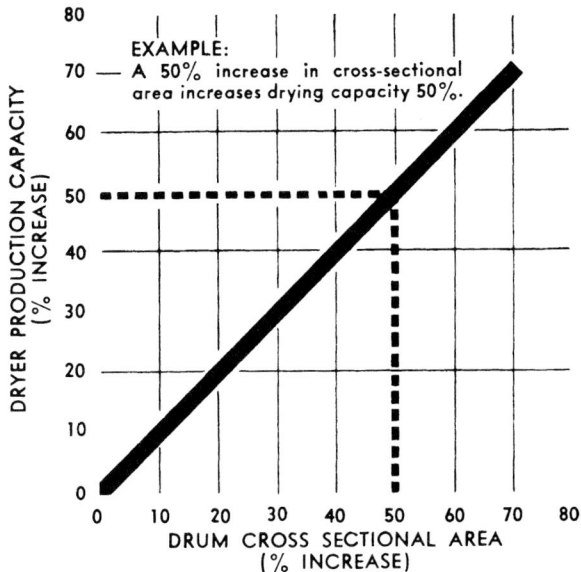

FIGURE D2-14 Dryer production capacity versus drum cross-sectional area.

With the same drum gas velocity* and the same length of dryer, the productive capacity varies in direct proportion to the increase in drum cross-sectional area. This area, of course, is in proportion to the square of the diameter (Fig. D2-14).

With the same drum gas velocity and the same drum diameter, capacity increases with more length, but not at a proportional rate. For instance, Fig. D2-15 shows that a 50 percent increase in the length of a 20-ft dryer results in an increase of only 20.5 percent in production capacity.

---

* Cubic feet of exhaust air is common terminology, but its effect on overall performance is more easily related to the velocity of the gas as it passes through the drum. The *drum gas velocity index* (DGV) is obtained by dividing the total discharge of exhaust (cfm) by the cross-sectional area of the drum (ft^2).

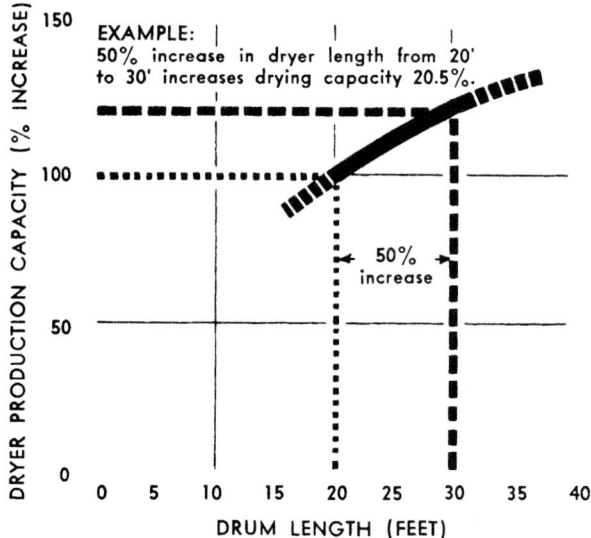

FIGURE D2-15 Dryer production capacity versus drum length.

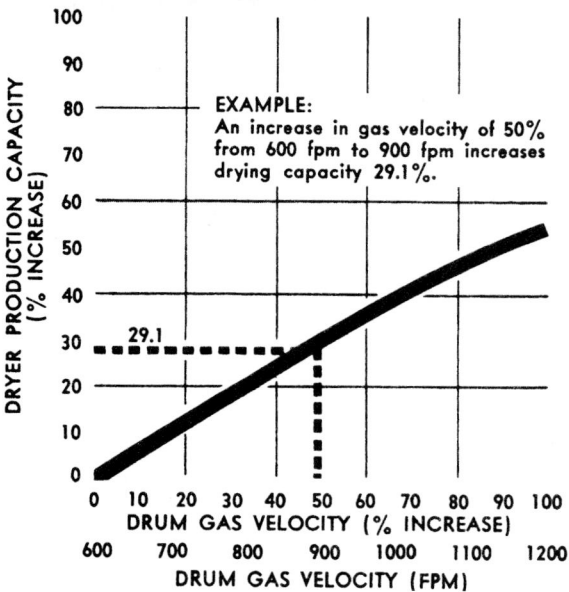

FIGURE D2-16 Dryer production capacity versus drum gas velocity.

For specific size of drum, greater productivity capacity can be obtained by increasing the air flow and burner system, although each successive increase in burner and air is of reduced effectiveness. The result, of course, is a consequent increase in the drum gas velocity in that specific drum. Figure D2-16 shows a 29.1 percent capacity increase due to a 50 percent increase in drum gas velocity.

There are practical limits to such an increase. The dust carry-out under higher gas velocities may be one such limitation, depending on the operating conditions of the job. (See Fig. D2-17). Combustion-chamber and drum-shell temperatures are not critical, because of the cooling effect of the high-velocity air which passes over them. However, the equipment could be severely damaged in the event of an emergency such as a stoppage of the cold feed.

The percentage of moisture to be removed has a major effect on dryer capacity, as indicated in Fig. D2-18. Here dryer capacity is correlated with the moisture to be removed, rather than with total moisture in the aggregate, since the residual amount for different mixes can be variable. To achieve the desired retention time, which can be a key factor in the internal moisture removal, loading of the drum can be controlled through flight design, r/min of drum, slope of drum, etc. Because any length and diameter drum can be balanced out to have the same percent loading, retention

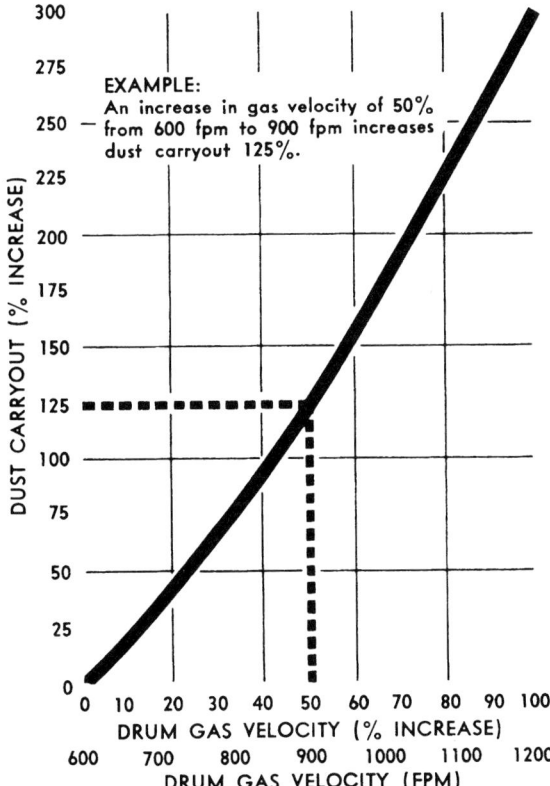

FIGURE D2-17 Dust carry-out versus drum gas velocity. Dust carry-out of aggregate increases by the square of the increase in drum gas velocity, as illustrated.

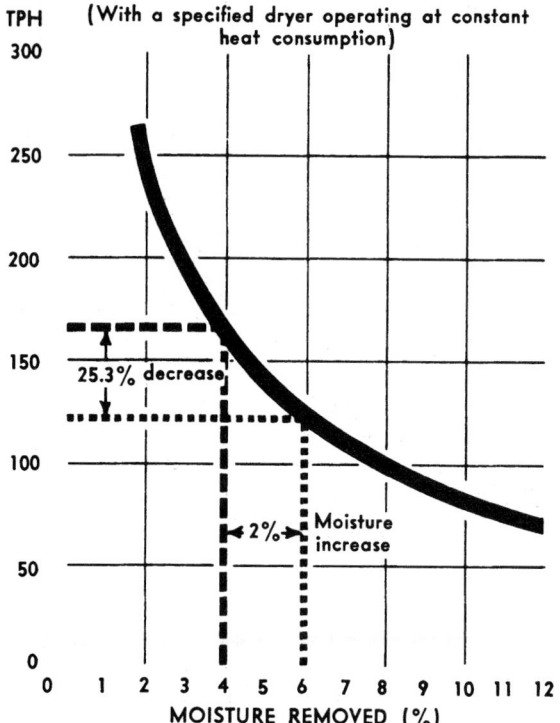

FIGURE D2-18 Tons-per-hour capacity at various moisture contents. Moisture content in the incoming aggregate is often guessed, for example, to be between 4 and 6 percent. And yet, there is a 25.3 percent difference in capacity obtainable in this small 2 percent moisture range.

time becomes directly proportional to the volume of the drum. However, a dryer should not be loaded up with material beyond 16 percent of the drum volume or there is apt to be erratic flow with material surges.

The most effective heat transfer in drying aggregate is through radiation. On the other hand, material spends only 3 to 5 percent of its retention time in actual veil suspension. The rest of the time it is cascading and soaking at the bottom and in the flights of the drum, during which period there is heat transfer from the smaller particles to the larger particles.

Specifications will normally state the amount of residual moisture permitted in the dried aggregate. They also will specify the temperature of the aggregate for mixing. These limits will vary with the type of mix and asphalt being used. Generally, the moisture requirement will be met* whenever a steady temperature is maintained for the aggregate as it is discharged from the dryer.

---

* Exceptions to this include (1) some cutback mixes where it is necessary to cool the aggregate after drying, and (2) aggregate causing problems of stripping or foaming because of internal moisture sweating.

Aggregate temperature is measured by either a thermometer or a thermocouple attached to an indicating pyrometer. The latter has a much faster response to changes in temperature and hence is more desirable. Recording devices are available for either type of sensor and can be supplied with extra pens in the event the temperature of a hot bin also must be recorded.

Automatic controls are available to change the rate of fuel to the burner whenever the aggregate discharge temperature varies beyond a preset range. The controls include safeguards that shut off the fuel in the event of flame failure or excessive air temperatures.

Dust Collection

The dust collection equipment must be considered an integral part of the drying system. It plays a major part in providing an even flow of adequate air volume and in collecting and returning a portion of the fine aggregate to the mix. It is important that a proper balance be achieved between the air flow necessary for optimum drying performance and that which will permit proper dust control. The increasing use of aggregates having borderline fines content is encouraging greater concentration on the efficient collection of these valuable fines which otherwise would be lost, requiring expensive replacement with mineral fines.

The dust collection equipment serves a secondary purpose by reducing nuisance dust in the plant vicinity and minimizing air pollution in the general area.

Dry System. Virtually all primary dust collectors in asphalt plants today are of the cyclone type, in which centrifugal force determines the amount of fines precipitated. Dust-laden air from the dryer has a certain velocity which is induced by the dust-collector fan. When this dust-laden air attempts to follow the sharp curve in the cyclone, centrifugal forces throw the dust out of the airstream and against the wall of the cyclone. As the particles of dust strike the wall, their speed is decreased and they gradually slide down the walls of the cyclone to the collecting screw or discharge chute. (See Fig. D2-19.)

The weight of the particle also affects dust collection. Heavy particles, because of their greater momentum, tend to be deposited more rapidly than light particles. Thus, light particles require more time to leave the airstream than is necessary for heavy particles. To give light particles this additional time, the airstream follows a continuously spiraling, downward path rather than taking a straight course through the dust-collector cyclone. The air is finally exhausted through a cylindrical duct located in the center of the cyclone, thus throwing additional dust out of the airstream as it reverses direction.

To prevent clogging in a small cyclone, the dust or fines must be extremely dry. The material for bituminous plant applications is usually dried to under 1 percent moisture content. This is suitable for mixing and application purposes, but not for dust collection in very small cyclones (approximately 6-in diameter or less). In cyclones of this size, even a small moisture content will cause dust to build up and plug the cyclone. Also, aggregates used for bituminous mixtures often contain twigs, straws, etc., which can quickly cause plugging in a small cyclone. Therefore, although very small cyclones are desirable for dust collection because of their high efficiencies, they are too refined for primary collection in bituminous plant applications. They can be used as secondary units to reclaim additional fines that otherwise would be lost.

To gain proper airflow capacity to handle larger dryers, more cones are added to the collectors instead of increasing diameter. Thus efficiency is maintained.

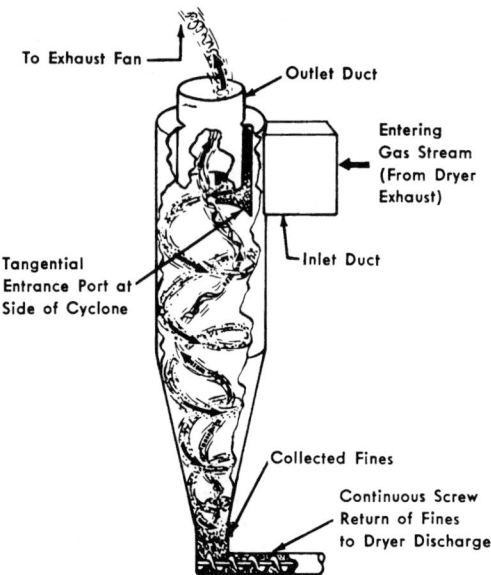

FIGURE D2-19 Cross section of cyclone dust collector.

Wet System. In metropolitan areas, more stringent pollution codes may force the contractor to reevaluate and improve the secondary dust collection system. A wet collector system is frequently used because of its overall economy and lower first cost.

Industrial wet collectors are not new by any means, but most attempts to modify or adapt them to asphalt-plant work have not been successful. Many of the problems of asphalt-plant operation, such as high volume of airflow, high percentage of fines, high temperatures, etc., are not common to industrial collection.

Most wet collectors work on the principle of wetting the dust particles that remain in the exhaust air after it has been put through the dry cyclone dust collector. The method of wetting the particles and separating the resulting sludge varies appreciably between designs. There is a very broad range of wet collector efficiencies as a result of these differences.

Screening

Aggregate gradation variations in an asphalt mix may be traceable to:

1. Stockpile variations
2. Method of reclaiming from stockpile
3. Handling of the material between the cold feeder and the mixer

In high-type mixes, the screening unit is used to reduce the variation and to signal when correction in the aggregate feeding is required. For this reason, a good cold-feed unit is still the most important single contributor to a well-proportioned mix.

Screening Efficiency. While the dryer is the primary potential bottleneck in an asphalt plant, a second major limiting factor of plant capacity is the screen. A screen can pass material at only a limited rate and, when loaded beyond its capacity, material which would normally pass through the screen into the desired hot bin will overrun into the next bin. To control this, standard specifications have established controls on the permissible amount of overrun.

Some specifications simply require a minimum screen efficiency, while others specify a limit on variation within a given bin. One state controls this by specifying that the bin containing the fine material shall not contain more than 10 percent of material retained on the no. 8 sieve or that sieve size used on the bottom deck. Conversely, the material in all other bins shall not contain more than 10 percent of material passing a no. 8 sieve.

Overrun is difficult to control, since it will vary from test to test under supposedly identical conditions. This, of course, is due to minor changes in material gradation and type. The best test of the aggregate is a composite test of all bins. In this way, one can determine whether, despite the overrun, the aggregate is meeting specifications.

Selecting Screen Sizes. The hot-bin splits can be determined by a plot of the total aggregate specification. The split should be made to obtain approximately equal percentages in each bin. For the three-bin split in Fig. D2-20, the smallest particle screen size is selected (no. 10), and the remaining 64 percent is split 33 to 31 by a 3/16-in screen. For a two-bin split in the same example, a 3/16-in screen is used.

To separate aggregate into specified sieve sizes used in testing, screen cloths having slightly larger openings are used on the asphalt-plant vibrating screen. (See Table D2-4.)

TABLE D2-4 Plant Screen Selection

U.S. Standard Sieve size of bin aggregate, in	Actual size of square openings of vibrating screens, in
No. 10 (0.079)	1/12
No. 8 (0.094)	1/10
No. 7 (0.111)	1/8
No. 6 (0.132)	5/32
No. 5 (0.157)	3/16
No. 4 (0.187)	1/4
1/4	5/16
5/16	3/8
3/8	7/16
7/16	1/2
1/2	9/16
9/16	5/8
5/8	3/4
7/8	1
1	1 1/8
1 1/8	1 1/4
1 1/4	1 3/8
1 3/8	1 1/2
1 1/2	1 5/8
1 5/8	1 3/4
1 3/4	2
2	2 1/4
2 1/4	2 1/2

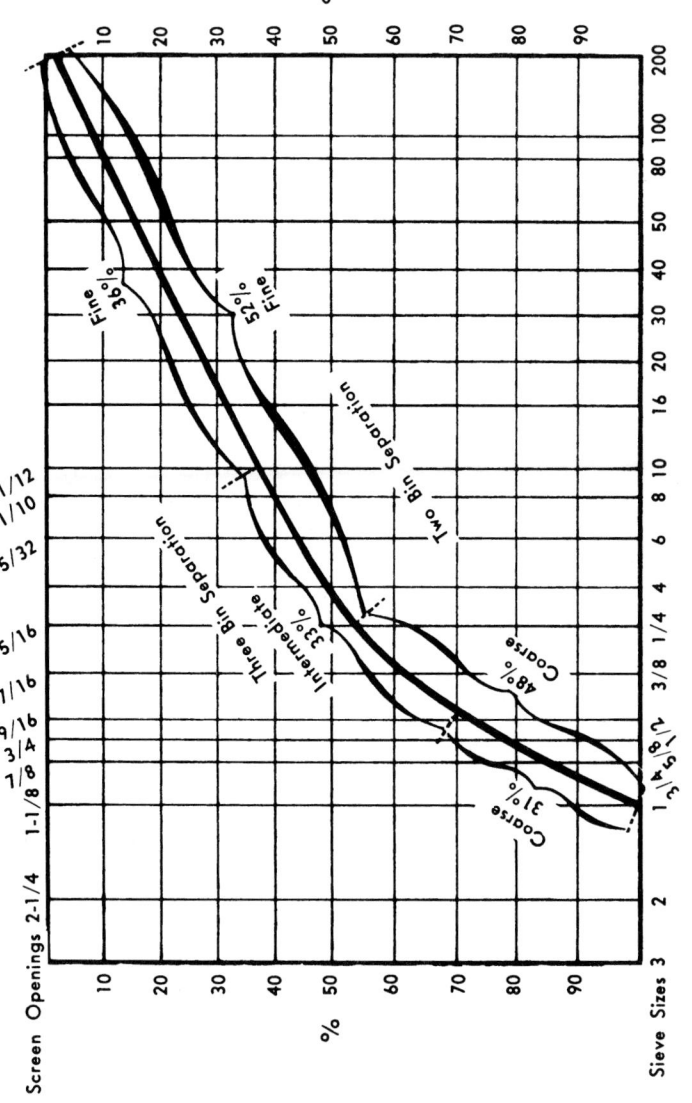

FIGURE D2-20 Cumulative gradation curve.

Screen Deck Arrangements. In Fig. D2-21, a 3⅓-deck screen is shown. A full deck screen area is provided for the two smaller screen cloths, but the top deck is split for scalping and no. 3 bin aggregate openings. In the example, the 9/16-in area is greater than the 1⅛-in area to further balance screening capacity.

A typical two-deck screen arrangement is shown in Fig. D2-22. The bottom deck is split, dividing its area between the no. 1 and no. 2 bins. The fine screen is subjected to heavier loading, since it must carry no. 3 as well as no. 2 bin aggregate.

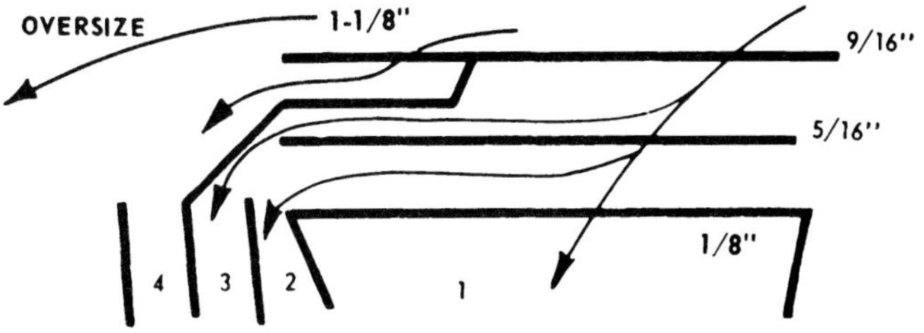

FIGURE D2-21 3⅓-deck screen arrangement.

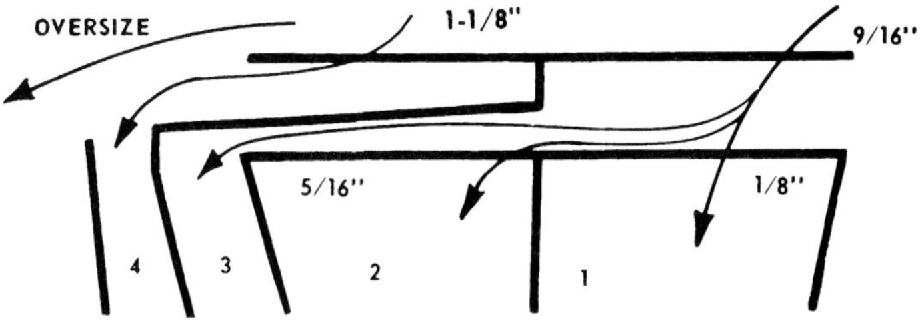

FIGURE D2-22 Double-deck screen arrangement.

Vibrating Screen Capacity. The screen capacities listed in Table D2-5 are representative of a broad range of experience and are neither the maximum

TABLE D2-5 Plant Screen Capacity

Clear square opening of plant screen cloth, in	tph/ft²	tph for deck size indicated			
		3 × 8 ft	4 × 10 ft	4 × 14 ft	6 × 14 ft
1/10	1¼	30	50	70	105
⅛	1½	36	60	84	126
3/16	2	48	80	112	168
¼	2½	60	100	140	210
⅜	3¼	78	130	180	275
½	3¾	90	150	210	315
⅝	4¼	100	170	240	355
¾	4¾	115	190	265	400
⅞	5¼	125	210	295	440
1	5¾	140	230	320	485

nor the minimum that may be expected. The capacity of a screen for a particular operation will be influenced by many factors, including:

1. Aggregate moisture content
2. Gradation of the material
3. Screen load versus amount passing
4. Size of screen wire

Where experience indicates that blinding, moisture buildup, or other screening problems will not be encountered, the data in Table D2-5 may be used for dried aggregates. In preparing this table, it has been assumed that approximately 25 percent of the screen load is retained. For example, an indicated 60-tph capacity corresponds to a total load on the screen of approximately 80 tph, with 20 tph being retained and carried to the next bin.

Measuring and Mixing

The batch-type and continuous-type plants will be discussed separately. Figure D2-23 illustrates a typical batch plant, and Fig. D2-24 shows a continuous-mix plant.

Batch Type. Total batch weight must be made up of the proper amount of aggregate from each hot bin and the correct percentage of asphalt. In a manually operated batch plant, the operator opens the first aggregate gate while watching a dial scale, then closes the gate as the pointer approaches the required weight for that aggregate size. He progressively opens each succeeding bin gate to accumulate the proper

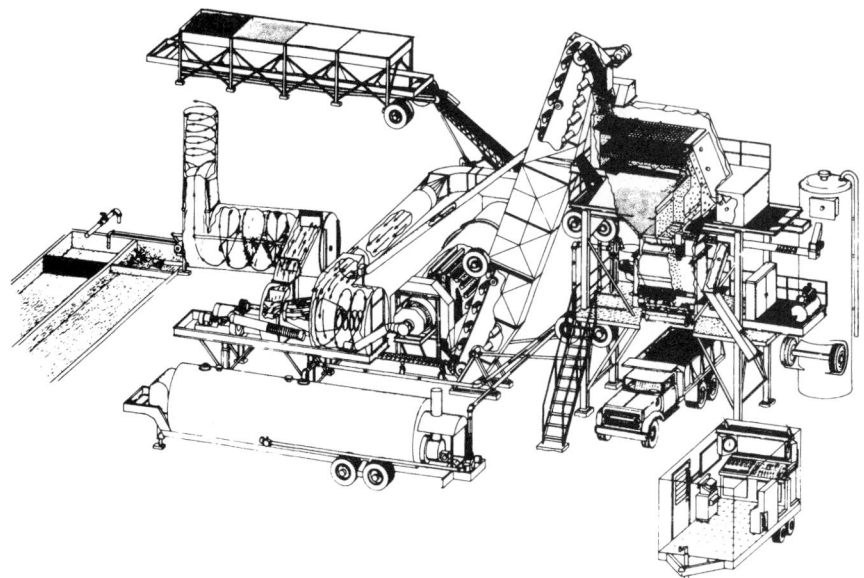

FIGURE D2-23 Cutaway view of a batch-type asphalt plant. (*Iowa Manufacturing Company.*)

FIGURE D2-24 Cutaway view of a continuous-mix asphalt plant.

aggregate batch in the weigh hopper. The total aggregate batch is discharged into the pug mill and dry mixed for a prescribed time. The operator then proceeds to weigh another batch. At the completion of a dry-mix cycle, the proper amount of asphalt is dumped or sprayed onto the aggregate. The asphalt may be measured by a meter or by weighing on its own scale.

The operator's controls may be completely manual or may be power operated. Cycling controls may be incorporated to handle the various operations after the aggregate has been weighed and discharged into the mixer. This is often referred to as *semiautomatic* operation.

Fully automatic operation of batch plants handles the entire measuring and mixing phase automatically and includes safeguards to ensure accuracy, batch after batch. This frees the operator to coordinate other plant operations, such as proper bin balancing through remote cold-feed control and regulation of aggregate temperature through remote burner control.

Continuous Type. The screen and hot bins in a continuous plant are parts of its gradation control unit. Feeders under the bins deliver measured quantities of the hot aggregate to a bucket elevator, where it is elevated to the pug mill and sprayed with asphalt. The paddles in the pug mill are set to propel the mix toward the discharge end as the mixing action takes place.

Calibrated gates located over apron-type feeders control the amount of aggregate which is drawn from each hot bin. The feeders are interlocked with the asphalt metering pump, and the output of both aggregate and asphalt is controlled through a reference shaft and a revolution counter. The aggregate and asphalt fed to the pug mill during each revolution of the counter may thus be thought of as a small batch. Each batch will contain just the right proportion of each aggregate size and the correct amount of asphalt. In actual operation, these batches come in a continuous stream.

Once the gates and pump are set, the operation becomes automatic and accuracy is assured. Continuous plants are equipped with warning devices or automatic cutoffs which are actuated if the supplies of aggregate or asphalt run short. Samples can be taken at any time to check gradation or rate of feed.

A *single-aggregate* type of plant can be used to produce intermediate types of mixes which do not require a split or screening of the aggregate after drying. Any required aggregate blending is accomplished prior to drying, and the heated aggregate is deposited in one bin. Under this bin, a feeder with an adjustable gate is interlocked with the asphalt pump, just as in the highest-type plants previously described.

Intermediate or single-aggregate plants offer the advantages of central-plant drying, precise control of asphalt content, and thorough twin-shaft pug-mill mixing. They are economical units for asphalt mixes intended for secondary roads and other surfaces which do not carry the heavy traffic of primary roads.

Mixing time is the number of seconds necessary to properly combine the aggregate and bitumen and obtain a homogeneous mixture of thoroughly coated particles. The optimum mixing time involves both a rate and an overall cost factor. This optimum time is necessarily influenced by many variables.

1. Environmental
   a. Type of aggregate (chemical and structural variables)
   b. Design of the mixture
   c. Aggregate moisture
   d. Humidity
   e. Viscosity of the bitumen

2. Pug-mill design
   f. Shape of pug mill
   g. Peripheral speed of paddle tips
   h. Number of paddles
   i. Size of paddles
   j. Shape of paddles
   k. Position of paddles or action produced as a result of paddle arrangement

3. Operational and system
   l. Method of introducing bitumen
   m. Method and sequencing of adding aggregate
   n. Condition of pug mill
   o. Size of pug mill versus batch size
   p. Condition of related equipment
   q. Personnel and operating variables

Gradual recognition of improved designs and techniques is evidenced by a trend toward a lowering of mixing times required by specification. A test used to determine the time required for complete mixing has been developed in research carried out by the Barber-Greene Company. This Standard Method of Test for Degree of Particle Coating of Bituminous-Aggregate Mixtures, ASTM Designation D2489, expresses completeness of mixing in terms of the ratio of completely coated coarse particles of aggregate to the total number of coarse particles in the sample.[4]

The total mixing time for batch mixing, as typically specified, is the interval of time between the closing of the aggregate weigh-hopper gate and the opening of the pug-mill gate. This total mixing time is further broken down into a dry-mix period and a wet-mix period. Several studies have indicated that, at most, only a short dry-mix time is needed. In some instances, the dry-mix time contributes nothing to achieving a complete mix. Pug-mill design, aggregate gradation, and the relative segregation of the aggregates as they are introduced into the pug mill are factors to be considered.

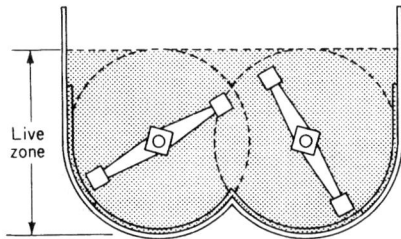

FIGURE D2-25 Batch sizes expressed in percent of live zone. Live zone, which offers a uniform method of indicating pug-mill cubics is the net volume below a line extending across the top arc of the inside body-shell radius. Shafts, liners, paddle arms, and tips are all deducted to determine the net cubic feet. Batch sizes may be expressed as a percent of live zone. For example, if a pug mill has a live zone of 150 ft^3, mixing a batch of 7500 lb (based on 100 pcf aggregate) will utilize 50 percent of the zone cubics.

The required mixing time on a batch-type asphalt plant is one factor in the determination of plant production. The output of the plant and the daily production of the entire operation may be dependent on this mixing time. Assuming sufficient drying and screening capacity, an increase in mixing time means a decrease in production. (See Fig. D2-25).

The continuous-mix asphalt plant differs from the batch type of intermittent asphalt plant in that it is possible to change the mixing cycle or mixing time without affecting the output of the plant. The mixing process in a continuous plant begins by blending the graded aggregates as they come from each feeder gate of the gradation unit to the bucket elevator and thence to the mixer. This preblending of the aggregates achieves a thorough *dry mixing*. The bitumen is sprayed into the curtain of aggregate as it enters the pug mill, and thus a precoating occurs even before the measured mixing cycle starts. The mixing cycle is completed when the material is discharged from the pug mill.

Most specifications require that mixing time by measured by the following formula:

$$\text{Mixing time, s} = \frac{\text{pug-mill dead capacity, lb}}{\text{pug-mill output, lb/s}}$$

Using this formula, the mixing time can be computed by measuring the level of the material below the top of the pug-mill in order to determine the dead load.

Figure D2-26*a* illustrates an "easy" mix where a short mixing cycle gives adequate coating and mixing. The dam is set in a low position, which causes the material to be propelled by the paddles through the pug mill at a relatively shallow depth. This means that any given particle will pass at a relatively rapid rate through the pug mill and over the dam into the discharge hopper.

Figure D2-26*b* shows a mix requiring a longer mixing cycle, which is accomplished by raising the dam. The mass of material is now relatively deep, and this considerably increases the length of time that any given particle remains in the pug mill.

Note that, in both examples, the materials enter the pug mill at 100 tph and the mix is discharged at 100 tph. The operator has additional control mix time since any number of paddles can be set to propel or retard. In any case, the rate of discharge per hour is the same as the rate of charging the pug mill, regardless of the means used to lengthen or shorten the mixing cycle.

Plant Production

Frequently, a plant's capacity is judged on the basis of its pug-mill size alone. An hourly capacity rate is then calculated from the mixer size and required mixing time.

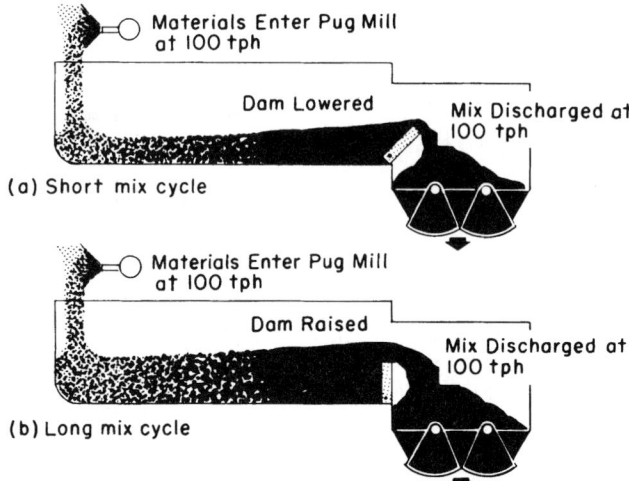

FIGURE D2-26 The mixing cycles of the continuous plant. Each cycle can be increased or decreased without affecting the daily tonnage output. In (a), note that the right end wall of the pug mill proper is an adjustable dam. The height of the dam is controlled from the operator's platform.

However, it is important to recognize the interdependence of all plant components. Under a given set of environmental conditions, aggregate characteristics, and mix requirements, the limiting factor of plant output may be the dryer, the screens, the mixer, and other items as well. Table D2-6 shows a representative selection of components to achieve the indicated tonnages under varying conditions.

For best results, plant components should be selected on the basis of the anticipated conditions under which the plant will be operated.

Fines Feeding and Handling

High-type asphalt plants and stabilization plants often require separate feeding systems for introducing fines materials into the mix. Typical of the materials for these applications are ground limestone, hydrated lime, portland cement, and wet or dry fly ash.

In plant operations where fines usage is high, a system using a storage silo for maintaining several days' supply of fines is often advisable. The ultimate choice of this system is usually dependent on the availability of bulk fines and their price in relation to bagged fines. The bulk-handling system, consisting of receiving hopper, screw conveyor, and dust-tight elevator, is used to charge the storage silo as well as the vane feeder which meters the fines into the plant.

In operations where the volume of fines used does not justify a bulk silo, the plant owner will want to consider a big-feeding system. The system consists of a ground-mounted feeder, dust-tight elevator, surge hopper, vane feeder or screw conveyor, and an overflow chute. A typical batch-plant bag system is illustrated in Fig. D2-27.

Bulk or bag fines systems are equally adapted for continuous-mix plants. Final metering of the fines to the mix is accomplished through a variable-speed vane,

TABLE D2-6 Plant Capacity in Relation to Components

Plant capacity in tph (with 5% asphalt cement)		Cycle times in seconds for batch sizes, lb, of				Minutes of hot bin storage for bin sizes (tons) of				% of total aggregate through ⅜-in sand screen (capacity from Fig. 25-5)					Dryer capability in % moisture removal (example only)	
										4×12 ft double deck with 4×7 ft sand deck	3½ deck					
Total mix	Aggregate only	5000	6000	7000		25	30	50			$4 \times$ 10 ft	$4 \times$ 14 ft			$7 \times$ 30 ft	$9 \times$ 20 ft
125	119	75	90	97		12	15	25		35	50	70			9.1	13.0
150	143	60	72	78		10	12	21		29	42	59			7.2	10.6
175	166	51	62	67		9	11	18		25	36	50			6.0	9.0
200	190	45	54	58		8	9	16		22	31	44			5.0	7.6
225	214	40	48	52		7	8	14		19	28	39			4.2*	6.6
250	238	36	43	47		6	7	12		17	25	35			3.6*	5.7

* Provided maximum dryer conveying capacity is also sufficient for these tonnages. Actually, top tonnage may be only 200–210 tph for example, regardless of how dry the aggregate is.

FIGURE D2-27 A typical batch-plant bagged-fines feeding system.

screw, or belt feeder, depending on the material to be handled and the capacity required. In each case, the fines feed is interlocked with the aggregate and asphalt feed to ensure constant accuracy.

Where excess fines are encountered in the raw aggregate feed, a bypass stem can be employed to receive the fines collected by the dust collectors. The required amount of fines is then fed back to the mix and any surplus amounts are diverted to a storage bin for disposal or other use.

Plant Accessories

Asphalt storage should normally be equal to one day's needs. Specific circumstances may render this unnecessary, such as where a plant is very close to a refinery and transport trucks can deliver the asphalt at or near the mixing temperature.

Two smaller storage tanks are more desirable than one big one. Should there be only one tank, it will not always be full and at proper temperature in the morning. The dumping of a cold load of asphalt on top of a small quantity of hot binder can make it necessary to shut down until the tank has been brought back to the correct temperature. With two smaller tanks, there is always enough flexibility and extra capacity in the heater to take care of one of the tanks during the daytime.

Sizes of tanks, pipes, and heaters can be selected on a preliminary basis from the data supplied in Table D2-7, but they must be influenced by individual plans and circumstances. Distance, method of transportation, heating facilities, asphalt type, and fuel to be used are all factors.

Plant Arrangement.   Many factors are important in selecting a satisfactory plant site. They include sufficient work space, proximity to job or market, traffic considerations, availability of aggregates, good footing and water-table height, and noise abatement and dust nuisance regulations.

Once the site has been selected, the plant should be located with prime consideration given to truck traffic patterns and prevailing winds. If practical, the operator's platform should be located upwind so that any dust is normally blown away from operator and engine locations.

TABLE D2-7 Guide for Selecting Asphalt Tanks, Pipes, and Heaters

Plant capacity, estimated 10-h output	30 250	55 500	80 750	105 1,000	130 1,250	155 1,500	180 1,750	205 2,000	250 2,450	tph tons
Asphalt requirements—storage and pipe size										
Asphalt required per day at 6%	15 3,800	30 7,600	45 11,400	60 15,200	75 19,000	90 22,800	105 26,600	120 30,400	147 37,200	Tons Gal
Storage tanks	1 4,000	2 4,000	2 6,000	2 8,000	2 10,000	3 8,000	3 10,000	3 10,000	4 10,000	No. gal
Asphalt pipe	2	2	2	3	3	3	3	3	3	Diam, in
Fuel requirements										
Estimated fuel oil per day	500	1,000	1,500	2,000	2,500	3,000	3,500	4,000	4,900	Gal
Recommended fuel tank size	2,000	2,000	3,000	4,000	5,000	6,000	8,000	8,000	10,000	Gal
Recommended supply and return fuel line size	2	2	3	3	3	3	3	3	4	Diam,* in
Hot oil heater data										
Hot oil pipe size	2	2	2	2	2	2	2	2	2	Diam, in
Output per h	450	450	750	800	800	1,200	1,600	1,600	2,000	Units of 1,000 Btu
Estimated fuel per day	70	70	70	90	90	90	125	125	175	Gal

* If supply tank is located over 40 ft from mixer unit, it would be advisable to increase recommended diameters of supply and return pipes by 1 in.

Tanks and heaters should be located to minimize piping and to facilitate the delivery of asphalt and fuel without interfering with plant operations. Representative arrangements are shown in Figs. D2-28 and D2-29.

Heaters, tanks, and piping are selected on the basis of the daily asphalt requirements and the amount needed in storage. In turn, the system directly affects the efficiency of the overall operation of the plant. The required heating capacity is dependent on the losses encountered in the plant, piping, and tanks and the desired rate of heating for asphalt which is received at less than application temperature.

In Table D2-8 are the formula and tables to assist in arriving at total Btu requirements for a particular application. However, it is always advisable to contact the heater manufacturer for a firm recommendation.

The heater, storage tanks, and plant equipment should be grouped as compactly as possible to keep the length of supply and delivery lines to a minimum. Hot transfer oil should be circulated through the system with a velocity of about 7 ft/s to promote adequate heat transfer. Unusually long supply lines may require the use of an additional circulating pump. It is usually considered good practice to install hot oil delivery and return manifold lines of at least 2 in in diameter.

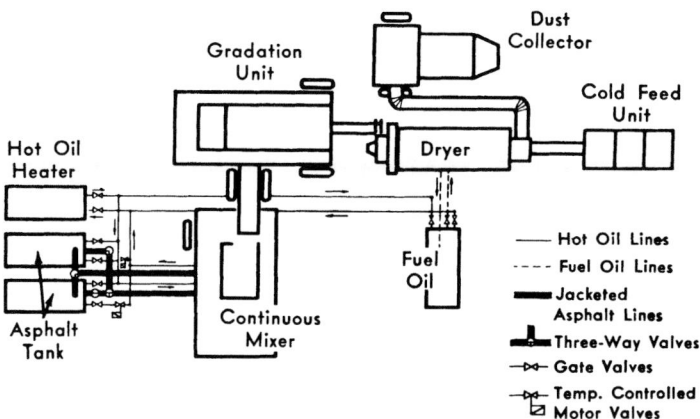

FIGURE D2-28 Typical continuous plant and piping arrangement.

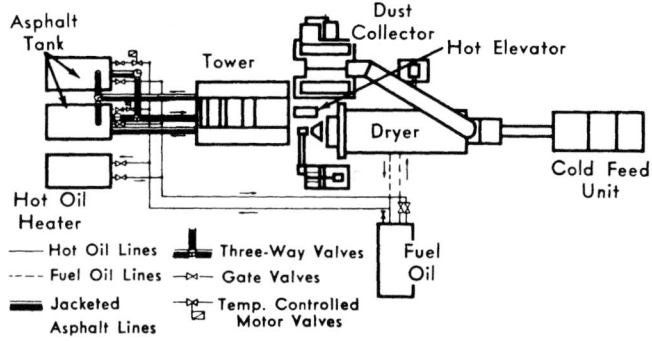

FIGURE D2-29 Typical batch plant and piping arrangement.

TABLE D2-8 Formula/Tables to Determine Btu Requirements

I. Heat losses from mixer in Btu/h
 50-ton capacity   100,000
 100-ton capacity   180,000
 150-ton capacity   265,000
 200-ton capacity   350,000

II. Heat losses from pipe
 Heat loss per 100 ft of 3-in diameter pipe with 4-in jacketed line = 170,000 Btu/h

III. Heat losses from tanks in Btu/h for asphalt temperature of 300°F and air temperature of 50°F*

Tank size, ft	Tank capacity, gal	Heat losses from tanks, Btu/h Uninsulated tank	1-in insulation	2-in insulation
8 × 14	5,000	350,000	31,000	18,000
10 × 15	8,000	500,000	43,000	25,000
10 × 17	10,000	540,000	46,000	28,000
11 × 21	15,000	720,000	62,000	37,000
12 × 24	20,000	900,000	76,000	46,000
12 × 32	25,000	1,150,000	96,000	58,000

* Corrections for heat loss for tank temperature other than 300°F:

Asphalt temperature	Multiply values from table by
350°F	1.25
325°F	1.15
250°F	0.75
200°F	0.50

IV. Heat required to raise asphalt temperature
 Btu/h = specific heat × temp. rise/h × asphalt weight
 Example: For a required heat rise of 15°F/h
 Btu/hr = 0.5 Btu/lb–°F × 15°F/hr × 8 lb/gal × no. gallons

Gal in tank	5,000	8,000	10,000	15,000	20,000	25,000
Btu/h (1,000)	300	480	600	900	1,200	1,500

Note: Asphalt supplier can give exact specific heat and weight. Contractor's need or desire may indicate different temperature rise per hour.

V. Safety factor
 In calculating total Btu heat loss, unknown factors must be taken into consideration. The heat loss which takes place from fuel tanks, transfer pumps, valves and related equipment, wind velocity, etc., must be allowed for. Since it is next to impossible to pinpoint each of these factors, it is recommended that an additional 10 to 15 percent of the above Btu requirements be added. If the unit is to be located in a very windy area, additional allowances may be in order.

Lateral lines from these manifolds to each individual tank and the mixing plant are reduced to about 1 in.

By providing gate valves at the entrance and exit points in each of the various components, hot oil may be stored during transport in these units. It is advisable to provide this for the manifold lines. In addition, these gate valves may be used to balance the oil flow through the entire system, helping to regulate the desired pressure at approximately 30 lb/in^2. Hot oil will always seek the path of least resistance. Uniform system resistance may be balanced by the use of valves. It is also advisable to install small air-venting valves at the high points to assist in completely filling the system.

Piping for an average asphalt plant layout can be quite costly. Flexible-type couplings have become popular in reducing setup time. Plant components, manifold lines, asphalt-jacketed lines, etc., can be positioned and the entire system connected with a minimum loss of time. One or two plant moves will generally pay for the additional cost of these couplings.

The jumper lines from one jacketed supply line to another can also be of the flexible design and should be at least ¾ in and preferably 1 in in size.

CONSTRUCTION OF HOT-MIXED, HOT-LAID BITUMINOUS CONCRETE PAVEMENTS

Bituminous surfaces have been in use in the United States since before the turn of the century. The major drawback in their application in the early days of road construction was the method of spreading and leveling the material on the roadway. Many methods were used, most of them highly inefficient. Perhaps the oldest method used was that of hand spreading. In this operation the mix was dumped from the trucks onto dump boards and then shoveled onto the road, following which the asphalt was raked smooth to grade and contoured by hand. With the increasing cost of hand labor, this method become uneconomical. In addition, a smooth, level, and even-textured surface was very difficult to obtain by this procedure.

Motor graders have been used to spread mix after it has been windrowed or dumped on the road, and this method is still used today for small spreading jobs. However, its disadvantages far outnumber its advantages. Since poor joints frequently accompany blade spreading operations, two graders are usually required to handle the entire width of the road in one pass. There is no compaction gained from blade spreading, and so all compaction of the mat must be obtained from rolling. Segregation of the mix is also a problem in road-grader spreading.

In 1937, after a seven-year period of development, the Barber-Greene Company introduced the tamping-leveling asphalt-finishing machine. This machine lowered the cost and increased the rate of laying asphalt mix.

The Finisher

A modern asphalt finishing machine must be able to handle all types of asphalt mixes. It must evenly spread and compact these materials, leaving a finished surface ready for final compaction by the steel and pneumatic-tire rollers. It must compensate for minor grade irregularities and must be able to form proper crowns and superelevations for modern-day highways. It must be able to push the large trucks that carry the bituminous mix from the hot-mix plant, and it must be able to operate on surfaces which range from old cracked pavements to newly prepared subbases. Therefore, the finisher must be a highly refined yet ruggedly built piece of construction machinery.

A typical asphalt finisher consists of two main units: the tractor unit and the floating-screed unit (see Fig. D2-30). The tractor unit provides the motive power

PAVING

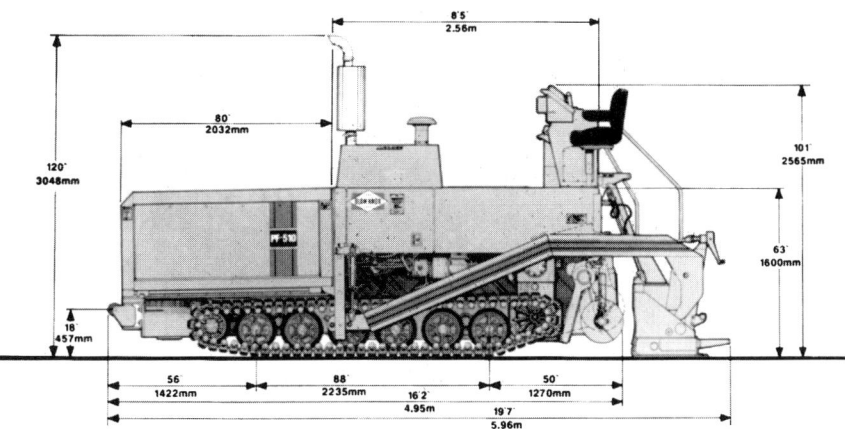

FIGURE D2-30 Asphalt finishing machine. (*Blaw-Knox.*)

through crawlers or rubber tires which travel on the road base. The tractor unit includes the receiving hopper, feeders, distributing augers or spreading screws, power plant, transmission, dual controls, and operator's seat.

The screed unit is towed by the tractor unit and rides on the finished surface. The tamper or vibrator units, thickness controls, crown controls, screed heater, and screed plate are all parts of it.

Finisher Operation

The plan and side views shown in Fig. D2-31 trace the flow of bituminous mix from the receiving hopper at the front of the finishing machine to the finished pavement behind the screed unit at the rear of the machine. The mix is dumped into the receiving hopper from a truck, which is pushed ahead by the finisher. Rollers mounted on the front of the finisher contact the rear tires of the truck and allow the finisher to push the truck while it is dumping into the hopper.

After receiving the material in the hopper, two independently controlled bar feeders carry the mix back through the control gates to the spreading screws. Each spreading screw is synchronized to its respective feeder, permitting the operator to distribute the mix accurately in front of the screed unit.

The screed unit is attached to the tractor by two long screed arms that pivot well forward on the track casing of the tractor unit. These arms provide no support for the screed when it is in operating position. Thus the floating action of the screed, as it travels along the road, compensates for surface irregularities which raise or lower the tractor unit. As the tractor pulls the screed into the material, the screed will seek the level where the path of its bottom surface is parallel to the direction of the pull. The screed can also be adjusted for a new thickness of mat by changing the depth-adjustment controls.

The screed unit consists of the tamping bar, which is the compacting medium as well as the strike-off; the screed plate, which gives the ironing action and provides a stable floating action in the mat laid; the thickness controls by which the tilt of the screed plate is changed in order to increase or decrease the thickness of the mat being placed; and the curved deflector plate.

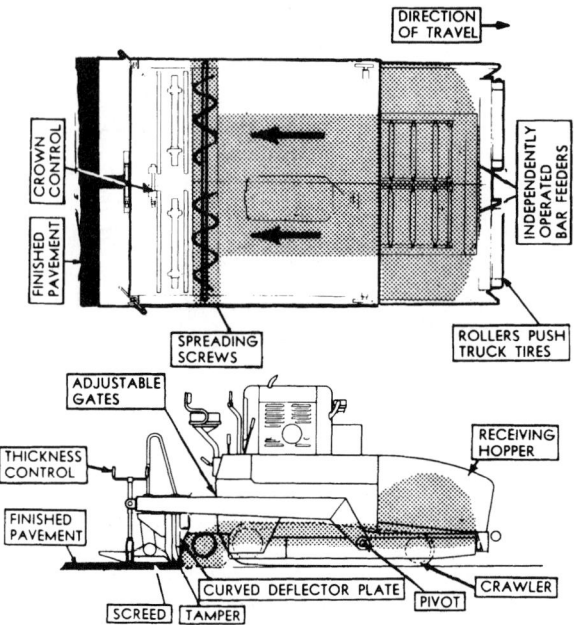

FIGURE D2-31 Material flow diagram.

The spreading screws are mounted on the tractor unit, each screw responding automatically to the amount of mix in the screw chamber along with the bar feeder on the same side. The automatic feeder controls can be moved outward as extensions are added, providing an adequate and constant supply of mix at the ends of the screws at all times. Flow-control gates are located at the rear of the tractor unit and have a crank adjustment within easy reach of the screed person. These gates, properly set, ensure a uniform flow of mix to the screw chamber. This minimizes feeder-clutch actuations and helps to maintain a uniform level of mix ahead of the screed unit.

The screed is also equipped with a heater to prevent the mix from sticking to the screed plate. It is used to heat the screed plate when the air temperature is cold or when starting out first thing in the morning.

The horizontal face on the bottom of the tamper bar strikes off the material so the screed plate can ride smoothly over it. The bevel face of the tamper bar effects the primary compaction of the mix as the paver moves along. The density of the loose asphaltic mix is increased as the material is forced down by the beveled face of the tamper. The horizontal face of the tamper imparts some compaction, but its primary function is in striking off the material.

The horizontal bottom face of the tamper bar will gradually wear to a knife edge with use. There is very little change in its compactive or strike-off action as it wears to this condition.

Vibrating screeds are also used on asphalt pavers requiring compaction and strike-off in the same operation. Strike-off is done by the leading edge of the screed and the compactive effort occurs under the screed plate.

Crown and Superelevation. The crown-control adjustment at the center of the screed provides for the desired contour of the finished pavement. Both crown screws are connected by a chain so that a single lever adjusts the leading and trailing edges of the screed uniformly to the specified crown. The leading edge can also be adjusted independently to achieve the slight additional crown normally required for best surface results. This adjustment is made while paving. It is possible to adjust the crown on the screed from ½ in negative to 2 in positive. (See Fig. D2-32.)

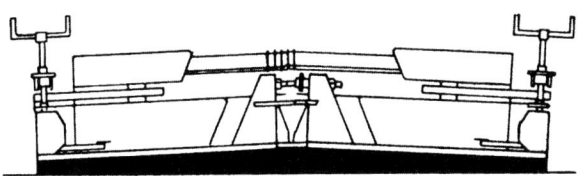

FIGURE D2-32   Positive crown.

The screed can also be adjusted to produce a superelevation by increasing the thickness on one side of the screed through the thickness control as shown in Fig. D2-33.

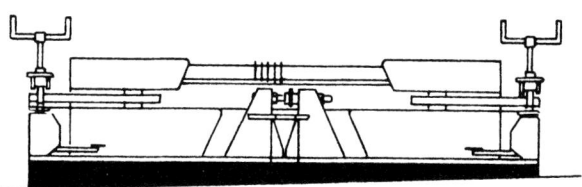

FIGURE D2-33   Superelevation.

Leveling Action.   Inherent in the basic design of most asphalt finishers is the principle of self-leveling. This feature, properly utilized, can produce a smooth and level mat over a rough subbase or a cracked and uneven roadway. The screed, when pulled into the material, will automatically seek a level where its bottom becomes parallel to the direction of pull asserted by the pivot joint of the finisher. This inherent self-leveling principle depends on time, distance, and mix stability rather than on a fixed mechanical ratio. Minor base irregularities are virtually eliminated because of it. Coupled with compaction before strike-off, it assures a smooth-riding mat.

Thickness Controls.   The screed-control operator can also change the thickness of the mat through the use of the thickness controls on the screed. These controls allow the screed to be tilted either up or down. The screed will then either rise or fall until it has reached a new level, with its bottom again parallel with the direction of pull. Figure D2-34 shows the asphalt finisher laying a level mat of uniform thickness over a level base. It will continue to lay such a mat indefinitely without any manual adjustment until (1) the operator changes the thickness control, which causes the machine to automatically lay an ascending or descending ramp until the new thickness has been reached and the screed is again parallel to the direction of the pull (see Fig. D2-35); or (2) an irregularity in the base causes the machine to automatically increase or decrease the thickness of the mat, compensating for the irregularity and producing a level surface over an irregular base. The mat is automatically made thicker to fill in depressions in the base and thinner to smooth out bumps. (See Fig. D2-36).

FIGURE D2-34 Finisher maintains a level mat over a level surface, indefinitely.

FIGURE D2-35 When adjustments for a thicker mat are made, the finisher makes a gradual transition to the new mat thickness. Abrupt changes cannot be made.

FIGURE D2-36 Finisher automatically compensates for base irregularities, laying level mat without operator attention.

The leveling feature of the finisher is further accentuated through the use of the tamper bar, which automatically compacts more material into the depressions in the base. This results in a smooth and level mat after compaction. If paved by a screed not having compaction before strike-off, a depression will reappear after the mat has been rolled and opened for traffic.

Due to the relatively long reaction time of the screed to any change in level of the tractor unit, the level of crawler or tires of the tractor unit may change but the screed will take much longer to come to the new level.

Automatic Screed Controls

Control of screed level to established grade lines and matching of adjacent mats or gutters can be accomplished through manual, semiautomatic, or fully automatic devices. These systems control the angle of attack of the screed. Some systems will automatically control transverse slope and provide the manual adjustment to construction superelevations or gradually varying cross slopes. Thus, side-to-side rocking motion of the automobile can be alleviated even though cost considerations may dictate the need for compromise in longitudinal leveling.

Primary methods used for establishing a longitudinal grade reference include string line to grade, joint-matching shoe, and long ski. Following is a brief review of these and other methods.

String Line to Grade. With this system the grade is established by a survey crew and marked by a string line. The string-line system is used on jobs where leveling to a desired profile is the requirement. For other objectives (i.e., leveling for a smooth-riding surface, joint matching, laying to an existing curb or gutter), there are other reference systems that may be more workable or economical.

When working with a string-line system, special care should be taken in making the survey and translating it into a string line. A poor survey defeats the purpose of a grade-controlled job before paving starts. After the string line is established, care should be exercised to see that a proper amount of tension is maintained on the line to keep it true. A sagging line will mislead the control system and result in a wavy surface.

Joint-Matching Shoe. Where an adjacent line of pavement serves as a grade line, a small joint-matching shoe can be used to sense the surface grade information and pass it on to the control mechanism. The method also finds considerable application in matching curbs or gutters, or in controlling uniform thickness over a surface previously established to a desired contour (see Fig. D2-37).

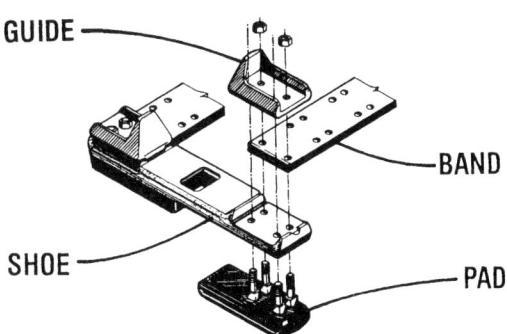

FIGURE D2-37 Joint-matching shoe. (*Blaw-Knox.*)

Long Ski. A semirigid boom or ski is another device that can be used as a grade reference with automatic screed controls. With this system, the ski glides over the high points of the base or old surface, signaling the control mechanism as its midpoint proceeds ahead of the screed. On jobs with serious surface irregularities, a long ski is used. This long ski spans from peak to peak, sending impulses back to the finisher. A mat is thus laid in the low spots between peaks, with a minimum cover over the high spots to eliminate surface irregularities. Where the base surface is relatively level, a shorter ski can be used.

Blaw-Knox uses a floating beam (Fig. D2-38) using 10-ft sections of aluminum beam with spring-loaded road shoes spaced at 30-in intervals; used as a sensing reference for either the Blaw-Kontrol II or automatic level control, the floating beam accurately averages longitudinal errors or irregularities in the grade over which paving is being performed.

FIGURE D2-38 Floating reference beam. (*Blaw-Knox.*)

This eliminates the expense of installing and maintaining a string-line reference while providing the accuracy and versatility required to meet most spec paving requirements. Available in 20-, 30-, and 40-ft lengths to match operational requirements, and in sets (two) for dual reference operations.

Blaw-Knox also has a material reference approach (Fig. D2-39). This consists of 20-ft aluminum beam (two 10-ft sections) with spring-loaded, dual road wheel spaced at 30-in intervals, towed behind the screed and used in conjunction with the floating beam to permit taking 50 percent of the longitudinal grade reference from the freshly laid mat and combining it with the average taken from the existing grade to produce a smooth-riding finished surface. Combing the mat reference with a floating beam mobile reference produces a 55-ft mobile reference system (Figs. D2-40 and D2-41).

Finisher Applications

The asphalt finishing machine has been widely used on all types of paving and resurfacing projects. It is chiefly through its introduction that low, intermediate, and high-type asphalt surfaces are economically feasible.

New Construction. As used here, this term refers to bituminous pavements laid on newly constructed and graded roadbeds—usually a well-drained, stabilized gravel or stone base. This type of construction is a simple task for the asphalt finishing machine since subgrades are improved, steep grades are at a minimum, bases are of a prepared material, and obstructions are seldom encountered. However, the

FIGURE D2-39 Mat reference. (*Blaw-Knox.*)

FIGURE D2-40 55-ft mobile reference. (*Blaw-Knox.*)

asphalt finishing machine may be working on a relatively soft base and will tend to disturb it if proper flotation is not provided.

The *compaction before strike-off* principle of the tamping and leveling finisher enables it to lay a smooth mat over an irregular base which may vary as much as 1 in from established grade. Irregularities and depressions in the base surface are filled in and properly compacted before the screed unit levels off the surface of the mat. Otherwise, these depressions will again show up after rolling. It is always desirable to lay at least two courses on any varying base in new construction. Irregularities remaining after the first course can be minimized in the second course.

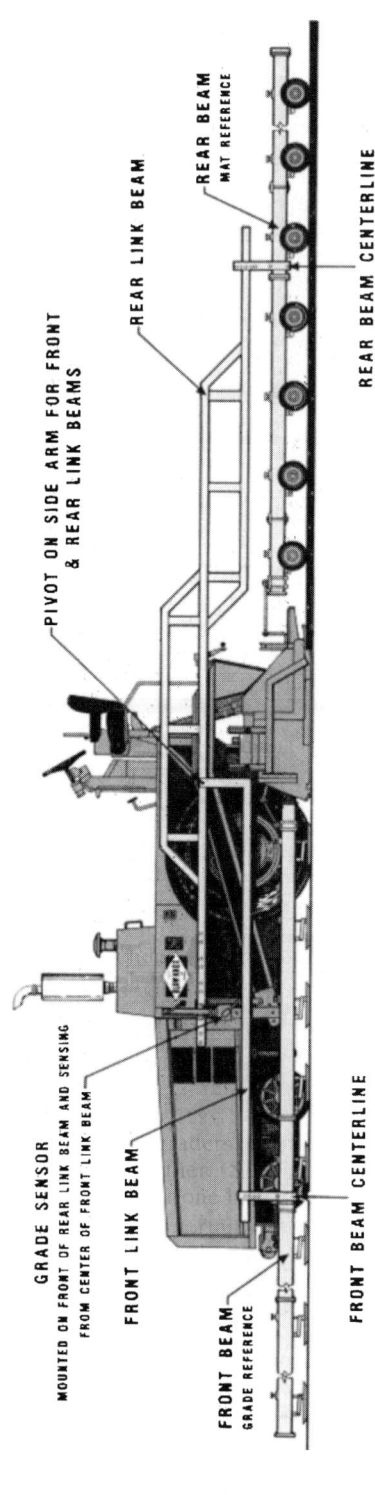

FIGURE D2-41 55-ft mobile reference system. (*Blaw-Knox.*)

Resurfacing. Old pavements may be very irregular and full of potholes and other surface failures. By resurfacing this pavement with bituminous material, the life of this roadway can be extended far beyond its original design life.

Resurfacing means the placing of a smooth bituminous mat (single or double course) on an old concrete, brick, or bituminous road. On some resurfacing projects, especially in city work, the irregularities or potholes require patching ahead of the finisher. Skip patching or scratch courses are used only in extreme pavement failure cases where holes or depressions are large enough to allow the entire side of the machine to drop into them. Also, if deep potholes are allowed to fill as the machine passes over them, later compaction of this material will in turn cause depressions in the final surface.

On resurfacing where an extreme crown exists, it is desirable to first reduce it. This may be done by laying a leveling course consisting of two separate wedge courses on the old road, feathering each course near the center of the crown, as illustrated in Fig. D2-42. The most common material for this type of work is fine-graded aggregates. This is an important point, as fine material should be used on all leveling courses to take care of the thin section laid over bumps in the old pavement.

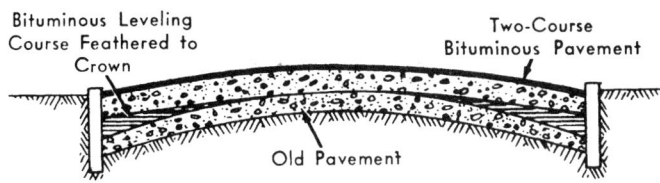

FIGURE D2-42 One method of reducing a high crown.

Bituminous layers are often placed on old pavement surfaces to increase the life of the pavement or to produce a nonskid surface. This is usually a simple operation and is covered in the application just discussed. In the resurfacing operation, the finisher is operating on a solid base provided by the old pavement surface. Therefore, traction becomes the important feature of the tractor unit.

Laying Variable Widths. The first step before starting any laying operation is to decide what width of mat to lay. The machine must then be fitted with proper extensions, cutoff shoes, etc. With the use of cutoff shoes, a standard finisher can be set up to lay pavement widths which can vary over a wide range of dimensions. By adding extensions singly or in pairs, the laying width of the machine can be further increased.

In multiple-lane work, the cutoff shoe should always be used opposite the joint-matching side, and the final lane should be at least the width of the machine. Using these simple rules and knowing the total width of the pavements, the proper width and sequence of lanes can be established. Extensions which are added to the finishing machine should perform the same function as the main part of the screed. If the screed is a tamping one, the extension section should also incorporate a tamper bar. If it is a vibratory screed, it is equally important that the extensions also contain the vibration feature. Otherwise, the mat laid under the extension section will be different from the main portion. There are also screeds which can be extended by power.

Matching Pavements. When matching a joint to a mat previously laid, a small overlap is helpful in preventing any decrease in thickness. If the machine is started out with an overlap of 2 in, a qualified operator should be able to guide the machine with an overlap of not less than 1 in nor more than 3 in, provided care was exercised in laying a straight first lane.

When matching a joint that has been previously laid and rolled, the depth of the overlapping mat must be sufficient so that additional compaction from the roller will bring the new mat down only to the level of the old mat. Excessive overlap prevents sufficient density of material in the new mat ahead of rolling, and it will also be likely to cause bridging and tearing under the screed.

A propane-fueled infrared heater attached on the side of the finisher next to the previously laid mat can be used to preheat the existing mat edge just ahead of the screed to achieve a hot joint between the two mats.

Milling Machines

Several types of milling machines are available to reprofile the paved surface (bituminous or concrete) to be repaved. The advantages are twofold: first, material is removed so that the original cross section is maintained after repaving. Also, embedded elements such as manholes and catch basins do not have to be altered. Second, the material milled can be reclaimed.

CMI started with the PR-500FL ROTO-MILL, a half-lane width pavement profiler. In 1992, CMI added three new half-lane (86-in cut) width pavement profilers with capabilities of a 525-, 650-, or 800-hp machine, each with the capability to cut up to 12 in deep in a single pass and up to 14 ft wide in two adjoining passes.

The machines all employ a three-track, front-loading design; the discharge conveyer of each machine swings a full 75° either side of center, permitting loading directly in front of the machine or ahead in adjacent lanes, and the conveyor is of sufficient length to permit complete loading of both end-dump and bottom-dump type trailers. (Fig. D2-43).

The ROTO-MILLs feature 86-in-wide, heavy-duty, deep-flighted cutters, capable of removing up to 12 in of pavement in a single pass. The deep flighting permits cutting and collecting a high volume of material with minimal flighting and cutting bit wear. Cutter flighting is triple-wrapped which permits maximum operating speeds, up to 150 ft/min, without the chevron or scallop surface patterns which occur with other milling equipment.

In 1993, CMI added the PR-1050 (Fig. D2-44); the three-track, front-loading PR-1050 features two power units totaling 1030 hp. An 800-hp Caterpillar V-12, double-turbocharged and aftercooled diesel powers mechanically driven, 12-ft 6-in cutting drum with the capability to remove up to 12 in of pavement in a single pass.

The second engine, a 230-hp Caterpillar in-line six-cylinder, turbocharged and aftercooled diesel, powers the track drives, conveyors elevation cylinders, and all other auxiliary functions.

The PR-1050's low-profile design provides excellent operator job-site visibility and ease of transport from job to job, especially in urban areas. The cutter housing is modular in design and incorporates quick disconnect features which speed installation and removal.

All of the CMI ROTO-MILLs are highly maneuverable. The three-track design features full-time coordinated steering, front or rear steering, and even crab steering, capabilities which permit fast positioning in a given cut, quick lane changes, the ability to crab up to or away from a curb, cut around a tight radius (like a cul de sac) or do a 180° turn in a limited amount of space. (Fig. D2-45).

PAVING   D2-51

SPECIFICATIONS

Operating Dimensions
Width .... 3057 mm/**10' 3/8"**
Height .... 3000 mm/**9' 10"**
Weight .... 37 671 kg/**83,050 lbs.**
       (with 4920 liters/**1300 U.S. gallons** of water)

Transport Dimensions
Height .... 3000 mm/**9' 10 1/8"**
Length .... 157 mm/**51' 8"**
Width .... 2499 mm/**8' 2 3/8"**
Weight .... 34 473 kg/**76,000 lbs.**

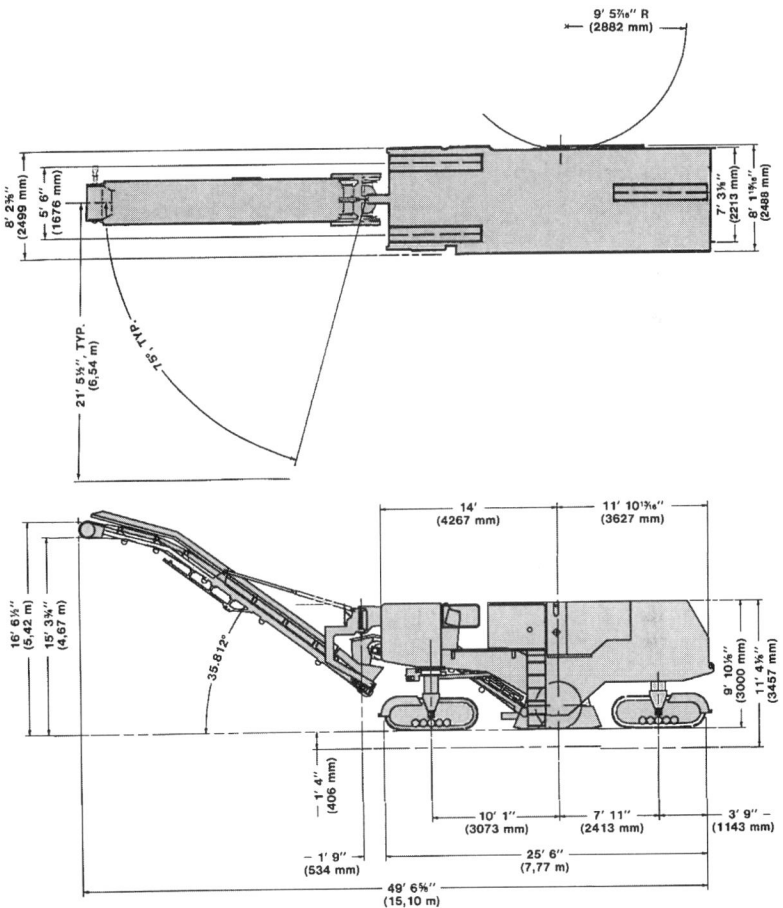

FIGURE D2-43 PR-800-7 pavement profiler. (*CMI Corporation.*)

D2-52 STANDARD HANDBOOK OF HEAVY CONSTRUCTION

FIGURE D2-44 PR-1050 pavement profiler. (*CMI Corporation.*)

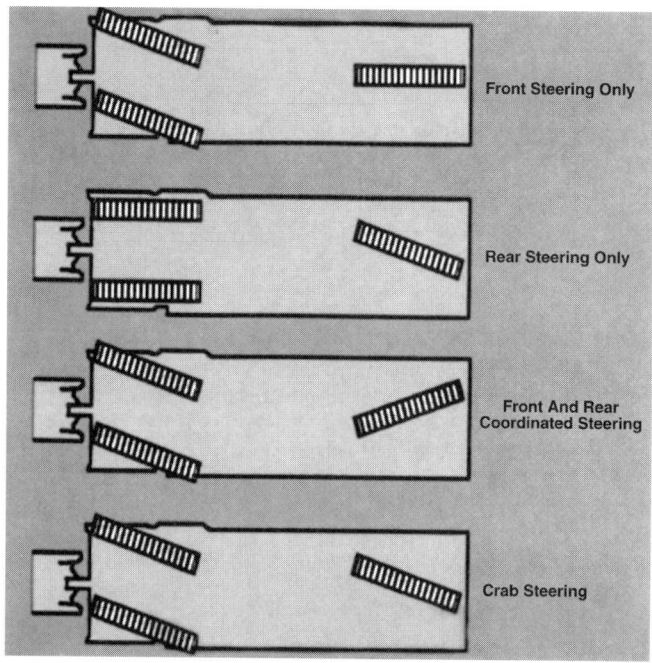

FIGURE D2-45 Profiler three-track traction modes. (*CMI Corporation.*)

The 1050 features a 48-in-wide collecting conveyor and 25-ft-long, 42-in-wide trough-style discharge conveyor, which can swing 50° either side of center and has a vertical reach in excess of 15 ft. The discharge conveyor is covered and equipped with a water spray system for duct control. Conveyor belts are seamless for longer life.

The CMI HYDRAMATION automatic grade control system offers reliability and accuracy to within ⅛-in tolerances. The 1050 features an electric x-slope control with hydraulic x-slope available as an option. Ground-level controls are provided for adjusting both grade and x-slope, a valuable feature in urban environments where there are lots of utility covers and other obstructions.

Hydraulically powered end-slides can be quickly raised and lowered to accommodate maneuvering around utility points, flush cutting next to curbs, or producing clean butt cuts.

A 1500-gal water system with variable pressure from 0 to 200 lb provides more than enough water to keep cutting teeth cool, extending their life and controlling dust. Water is the least expensive daily maintenance item; cutting teeth are the most expensive.

Reclaimers/Stabilizer

In 1994, CMI Corporation introduced the RS-500B Road Reclaimer/Soil Stabilizer, based on the design of the venerable RS-500 (Fig. D2-46) introduced in 1990. Due to its size, power, versatility and advanced technology, highway contractors began putting the RS-500 to work on more diversified and demanding jobs than were imagined at the time of its introduction. The tough new cutter drive assembly, electronic components, and heavier mainframe members of the new RS-500B make it more durable and productive.

FIGURE D2-46 RS-500B Road Reclaimer/Stabilizer. (*CMI Corporation.*)

The RS-500B is a self-propelled, pneumatic-tired machine that is diesel powered and contains a single, heavy-duty multipurpose cutter. The machine finds use in these basic areas:

1. *Cold-mix reclamation* of deteriorated asphalt roadways, parking lots, runways, taxiways, and access roadways.
2. *Soil stabilization*

    In-place modification of soils for their use as the base material for roadways, parking lots, etc.

    Clay soils—require the addition of lime or lime slurry

    Sandy soils—require addition of portland cement and water, asphalt, emulsions, fly ash, chlorides, etc.
3. *Mining*—reducing overlay and burden into materials that can be handled and removed as required

Its microprocessor and monitors provide control of all machine components and functions; the RS-500B makes extensive use of computer-controlled variables such as cutting depth, travel speed, and cross slope. The feedback data and control input sent to and from the RS-500B's computer system significantly enhances machine performance and operating economy.

Powered by a 525-hp, eight-cylinder, turbocharged and aftercooled Caterpillar diesel engine, the RS-500B offers all the power and performance of its predecessor. Four-wheel drive, with high-torque, integral drive that eliminates axles, transmissions, and drive shafts, combines with large, pneumatic tires for superior tractive power on the toughest jobs. Working speeds range from 0 to 150 ft/min; travel speeds to 9 mi/h with a unique modular operator station swinging 180° to enable working forward or backward, without the need to turn the machine.

The RS-500B's 50-in-diameter, 8-ft-wide cutter is fitted with conical teeth for reclamation and stabilization and offers variable cutting depths to 16 in. This eliminates the need for two cutters and conversion time. For interstate shoulder work, an optional cutter can be quickly reconfigured to a 48-in width. In sandy and other loose soils, an optional 58-in-diameter stabilization rotor cuts as deep as 20 in.

Finishers Working in Echelon. This type of operation, illustrated in Fig. D2-47, is frequently used in airport and highway construction. The first finisher usually operates from 50 to 100 ft ahead of the second finisher, which is matching the joint of the first machine. In addition to the high capacity thus obtained, the joint quality is improved because both strips are hot when the joint is made. This assures union of material and a smoother joint with less rolling.

In tandem operations, the roller must keep 6 to 12 in away from the inside edge on the first strip. The second strip can then be laid the same depth as the unrolled part of the first strip, and the roller can compact the joint while the material is hot.

Compacting and Rolling

The several layers of a flexible pavement, including the granular or stabilized base and subbase and the asphalt surface itself, must all be adequately compacted and finally rolled to an established contour. Compaction and rolling equipment for this purpose includes steel and pneumatic-tire rollers as well as more specialized types

FIGURE D2-47 Finishers working in Echelon, New Jersey. (*Blaw-Knox.*)

which are designed to produce the desired effect in different materials under varying conditions.

Rolling of the various bituminous layers in the road requires skill and care. A well-engineered and -executed construction job which employs the most advanced grading and leveling techniques and the best of materials and workmanship can be canceled entirely by slipshod methods in rolling and finishing. Additional considerations include atmospheric conditions, temperature, and working characteristics of the mix.

It is generally agreed that initial rolling should follow the spreading operation immediately or as soon as possible thereafter. On hot summer days it may be necessary to delay the initial pass for a short time to allow the mixture to cool. A common sequence of rolling begins with the steel three-wheel breakdown roller, followed by a pneumatic-tire intermediate rolling operation and a final or finish rolling using a steel-wheel tandem roller. The value of experienced supervision cannot be disputed. Being familiar with materials and equipment and cognizant of atmospheric conditions, the supervisor can readily determine the timing and sequence of rolling patterns. This can be verified on the first pass of the roller and additional changes can be made if required.

Practice has indicated that the large wheels of the three-wheel breakdown roller exert a great compressive force with little displacement of the mix. This is particularly desirable on longitudinal joint work. A definite rolling patter should be observed in order to assure a uniformly and correctly compacted mat. (See Fig. D2-48.)

It is universal practice to start rolling at the outside edges of a newly laid mat and roll inward toward the center, or crown. In the case of superelevated curves, rolling is started at the low side and carried to the high side in the same manner. A careful roller operator will observe these requirements, regulating speed to produce a smooth, finished surface which is free from tears and depressions. Some mixes may be prone to tenderness and will show failure from excessive compaction. A good operator must be aware of these signs and know when to stop the operation.

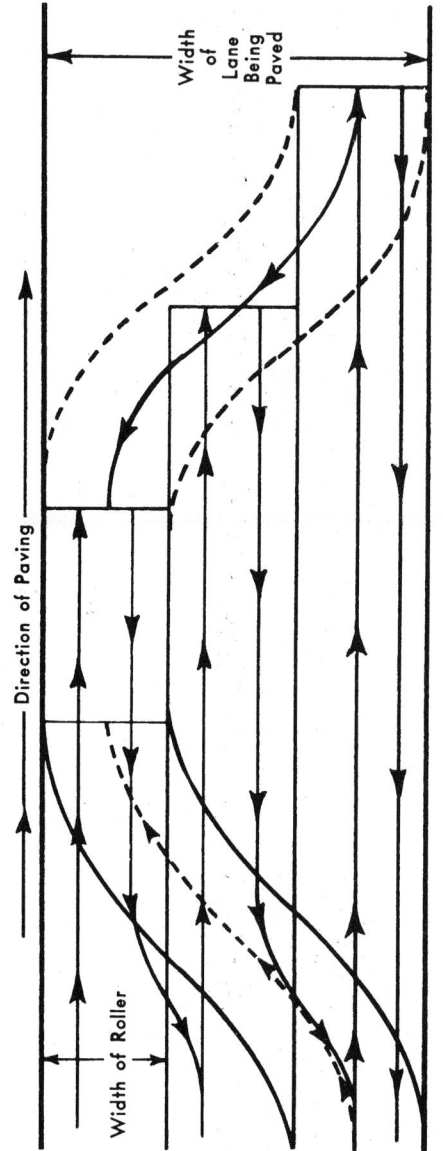

FIGURE D2-48 Correct rolling pattern.

D2-56

Immediately following the initial rolling operation, the surface of the mat should be checked with a straightedge for compliance with specifications. The mat, still being warm and relatively soft, can be corrected for variations at this time.

After the initial rolling and any necessary corrections are completed, the pneumatic-tire roller follows. Having noted that traffic will frequently continue to compact a new pavement, engineers now believe that much of this consolidation should be accomplished during the pavement construction in order to eliminate roughness which will otherwise appear after the first few months of traffic. Pneumatic-tired rollers of the type illustrated in Fig. D2-49 have gone a long way toward accomplishing improved compaction.

FIGURE D2-49 Pneumatic-tire roller. (*Galion Iron Works & Mfg. Co.*)

The pneumatic-tire roller follows the same pattern as the breakdown roller, starting at the outside edge and working toward the center or crown of the road. As soon as uniform coverage has been accomplished and no further densification is apparently occurring, pneumatic-tire rolling should stop. Further rolling will probably produce little or no further compaction and in some cases may encourage pumping of the asphalt cement to the surface. The flexibility of the pneumatic-tire roller is especially good since its compactive effort is a combination of tire pressures, tire size, and the ballast weight. Engineers feel that pneumatic-tire rollers can be used to reproduce, artificially, the same conditions that would occur from heavy truck traffic.

At the conclusion of the pneumatic-tire rolling operation, the surface of the bituminous mat, although almost totally densified and compacted to the specifications, may still present an unfinished appearance. The application of a tandem steel-wheel finishing roller will eliminate any marks left by the pneumatic-tire rolling and will promote a smooth, uniform surface texture. This final rolling must be done before the temperature of the mix has cooled beyond the point permitting such corrections (see Fig. D2-50).

The number and size of the various rolling units are usually specified by the engineer in charge. Pertinent factors include the speed of the paving operation, the nature of the mix being placed, general requirements of the job, and daily atmospheric conditions. The spacing and speed of the three rolling operations will therefore vary, depending on these circumstances. It is not unusual to find the finished

FIGURE D2-50 Tandem steel-wheel roller. (*Galion Iron Works & Mfg. Co.*)

rolling operation a considerable distance behind the paving machine during hot summer days. On the other hand, the three operations should be quite closely grouped in cool weather.

Figure D2-51 illustrates the basic types of compactors. The three rolling units just described are most commonly used, but there are other units which offer special advantages. A three-axle tandem steel-wheel roller is frequently used to assist in leveling operations. Some designs of this equipment will permit the automatic shifting of weight or ground-contact pressures to the center wheel when it goes over high spots, thus assisting in flattening these spots for additional leveling. As soon as the high spots flatten out, the roller weight is redistributed uniformly.

FIGURE D2-51 Types of compactors used in bituminous pavement construction.

A variation of the plain steel-wheel roller employs vibration by eccentric weighted shafts for a combination of static and dynamic compactive efforts. The vibrating force of the eccentric weighted shaft can be varied to suit a variety of situations. In principle, the vibratory forces are transmitted in all directions and to greater depths than static forces. They assist in orienting or relocating particles for greater compaction.

The size and weight of the three rollers most generally used in bituminous work will vary considerably. In general, the three-wheel breakdown roller will vary in weight from 10 to 20 tons on average-sized construction work. The tandem-axle rollers will vary from 8 to 12 tons. Self-propelled pneumatic-tire rollers have an even wider weight range, from about 5 to 35 tons or more. For an individual roller size, the contact pressure may be varied considerably by changing the ballast weight and tire pressures. Some very large pneumatic-tire compactors of the towed type range upward to 50 tons in weight. They are often used for proofrolling, wherein test strips are subjected to high stress for a specified number of loading repetitions.

Paving Tips

Common causes of mat deficiencies are summarized in Table D2-8. Recommendations for paving practice are summarized in the following paragraphs:

TABLE D2-9 Causes of Mat Deficiencies

Location	Cause	Cracking	Tearing	Wavy mats	Segregation
Mix	Excess no. 200 mesh material	x			
	Too hot or too cold	x			
	Too dry or too rich		x		
	Lack of fines		x		
	Too cold		x		
	Improper ratio of mat thickness to aggregate size		x		
Plant	Mixing temperature fluctuations			x	
	Segregated aggregate stockpile				x
	Poor cold feed				x
	No. 1 hot bin segregation				x
	Insufficient dry mix time				x
Trucks	Truck brake set too hard			x	
	Improper loading of truck				x
Roller	Improper rolling			x	
	Overrolling where base deflects	x			
	Turning too abruptly	x			
	Reversing too abruptly	x		x	
Finisher	Buildup in hopper sides				x
	Flushing of fines				x
	Screed overcontrol			x	
	Overloading spreading screws			x	
	Screed rams holding			x	
	Condition of tamper or screed		x		
	Adjustment of tamper or screed		x		

Laying

1. Set truck brake slightly before dumping.
2. Operate finisher at a rate only slightly greater than the capacity of the plant which supplies the mix, thus avoiding frequent stops.
3. End strips with square, vertical edge.
4. Vary screed crown to suit contour at intersections.
5. Clean areas over manholes prior to rolling.
6. Allow at least 4-in clearance between outside of screed and straight curb.
7. On binder, treat gutter flange same as straight curb (4-in clearance). Screed can overhang gutter on top course.
8. Leave uncompacted, hand-raked mix at a higher level than mat placed under screed.
9. Give careful thought to the width of each strip to be laid so as to minimize longitudinal joints and to permit maximum capacity.
10. Lay cutoff strips prior to final strip.
11. Cutoff shoe should be on opposite side of matching joints.
12. Small overlap (1 to 3 in) is helpful when matching joints.
13. Brooming in most cases robs a joint of needed material and causes voids or depressions at the joint.

Rolling

1. Reverse roller slowly and smoothly.
2. Hold roller speed to point where shoving and separation of mat do not occur.
3. On short strips where hot, unrolled joints are to be matched, keep roller 6 in to 1 ft from edge.
4. Avoid stopping roller in same transverse location after each pass.
5. Do not allow roller to stand on hot mat.

SECTION 2

PORTLAND CEMENT CONCRETE PAVEMENTS

This section will discuss the employment of machines and methods for producing durable concrete pavement with good riding qualities. It will not attempt an exhaustive survey of all the methods that may be used to build an acceptable pavement, nor will it try to provide an inspector's guide covering all the minute details of pavement construction. Rather, it will focus primarily on the high-production capabilities of mechanized devices and the advantages of their use on large paving projects.

The procedures and equipment for construction of concrete pavement have undergone revolutionary changes. The old methods, involving an assortment of manually adjusted machines and a great deal of hand labor, have been almost entirely displaced. A few versatile and sophisticated machines have now eliminated most of the hand labor and have greatly increased the rate of production.

The development and adoption of mechanized and automated devices, together with greatly improved control procedures, have gone a long way toward eliminating human error. They have made it possible for contractors to do a better job, with greater production and reduced labor costs.

Evidence of the improved construction methods is the rate at which concrete pavement is now being built. Construction of more than a mile a day of 24-ft pavement is commonplace. Some contractors are achieving more than 2 mi a day with a single paving spread.

PREPARATION FOR PAVING

Preparation for paving will include stabilization of the material that is to provide the pavement foundation, setting forms (when they are to be used), and finishing of the bearing surface to grade within the specified tolerances. The foundation material supporting the concrete may, in rare instances, consist of the on-site soil, but it will more generally consist of selected materials used to improve stability and provide a uniform bearing capacity. Stabilized subbases, using portland cement, lime, or bituminous materials, gained acceptance for several reasons: (1) they not only give additional pavement support and prevent *mud pumping,* but they also furnish the contractor with an all-weather working platform; (2) they provide more precise and stable subbase elevations, which facilitate form setting and are absolute requirements for slip-form paving; and (3) they result in low-yield loss of the concrete.

Compaction Equipment

Certain items of equipment have long been successfully used for compacting subgrade and subbase materials to specified density. Sheepsfoot rollers work well in heavy clay soils, while pneumatic-tired wobble-wheeled rollers work better on gravelly material. Vibratory compactors work well on sandy materials. Tandem-axle steel rollers are well adapted for final compaction.

Automated Earthwork Equipment

Final preparation for paving has been greatly facilitated by the use of multipurpose combination machines that can trim the subgrade, spread the subbase materials, and trim the compacted subbase. These machines offer considerable advantage over older methods that employed a variety of equipment and sizable crews and that required laborious adjustments to produce a compacted surface and bring it within specified tolerances.

A typical machine of this new type has a 30-ft wheelbase and is 28 ft wide, traveling on four 16-in crawler tracks. The 28-ft working width may be extended by the addition of wings that are available for the purpose. The basic components are a single, two-piece, 30-in-diameter rotary cutter; a 30-in-diameter helicoid screw spreader with hydrostatic transmission for rotation in either direction; and tandem two-piece moldboard templates. The machine is automatically controlled for elevation by four control systems, one at each corner, which operate from a neoprene string line. Positive ⅛-in tolerance control is provided by an autolevel control system. Tracers operating from the string lines instantly sense any irregularities from a true plane, and the master control unit immediately makes a hydraulic correction. The machine can trim subgrade and spread subbase material, each at a rate of 75 fpm, or it can trim compacted subbase (fine grading) at a rate of 0.5 mi/h.

The machine is generally put into operation after the bulk earthwork (cutting, filling, etc.) has been completed and the contractor is ready to start cutting the final profile. Recommended procedure is to set the machine to cut 0.3 ft above grade elevation and make an initial pass. If there are low spots, fill material is brought in and another pass is made with the setting 0.1 ft above grade. Any remaining lows are then filled and, if schedule calls for immediate application of subbase material, a final cut to neat line tolerance is then made.

The machine is particularly adapted to spreading stabilized or natural subbase materials that have been windrowed on the subgrade (see Fig. D2-52). With the previously described controls, the machine spreads the select materials uniformly and discharges excess material from either side or from both sides simultaneously. After final compaction is completed on the subbase, the machine trims or fine grades the completed subbase to within ⅛-in tolerance so that forms can be set or the slip-form paver operated without any further grading preparations.

FIGURE D2-52 Combination trimmer-spreader preparing subbase in multiple-lane construction.

Forms and Form Setting

Mechanization of paving operations has resulted in heavier and more complex equipment which, in some instances, will operate directly on the forms. This makes it more important than ever that, when forms are used, they be suitable for the loading and properly set for carrying this equipment as well as for providing a pavement that is built to proper line and grade.

Before any forms are set, they should be individually inspected and approved for use on the project. The top of the rail should be straightedged to check compliance with the requirements of the specifications. The braces, locks, and wedges should be in good order, and the welds or bolts fastening the braces should be tight. Twisted, bent, or otherwise defective forms should be removed from the project.

The use of the trimming and fine-grading equipment previously described should greatly simplify the setting of forms to the proper line and grade. A minor amount of work may be required to ensure the firm seating of forms at the proper elevation. The forms should be set accurately to line and grade, form pins driven, and the wedges and form locks made tight. The forms should then be tamped by hand or machine methods. If the tamping is to be effective, it will be necessary to place a small ridge of material to be tamped under the forms along both the inside and outside of the base. Very little tamping will be required if the trimming and fine-grading equipment was properly adjusted.

PLACING CONCRETE AND STEEL

Concrete paving operations have advanced to the point where spreading, consolidating, and finishing are almost wholly mechanized operations. Machine methods of placing and finishing have several advantages. First, there are the reduced labor requirements and lower construction costs. Second, when the equipment is in good mechanical condition, is kept in proper adjustment, and is staffed by trained and capable operators, the resulting pavement surface will be smoother.

Slip-Form Paving

While the several manufactured models will vary in methods of obtaining results, all acceptable slip-form pavers perform the functions of spreading, vibrating, striking off, consolidating, and finishing the pavement to the prescribed cross section and profile with a minimum of handwork required for an acceptable riding surface. The name *slip-form* is derived from the fact that the side forms of the machine, which vary from 16 to 48 ft in length, slide forward with the paver and leave the slab edges unsupported.

The basic operating process on one widely used machine is illustrated in Fig. D2-53 and consists of molding the plastic concrete to the desired cross section and profile under a single, relatively large conforming screed. This is accomplished with a full-width dead screed, 3 to 6 ft in length, which is maintained at a predetermined elevation and cross slope with hydraulic jacks that are actuated through an automatic control system. The control is referenced in offset grade lines preerected parallel to the planned profile for each pavement edge. Concrete is delivered into the forward open-bottom receiving hopper. It is then, in effect, liquefied by intense high-

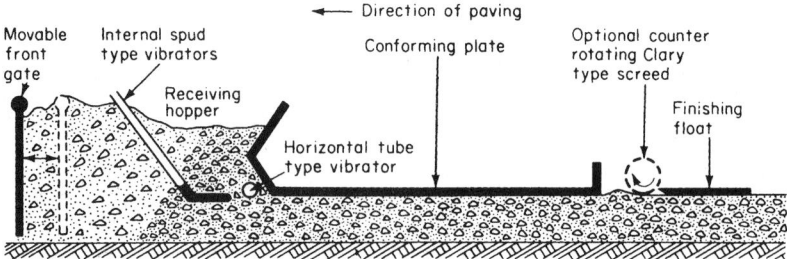

FIGURE D2-53 Schematic diagram of slip-form paver.

frequency internal vibration so that it flows into and completely fills the space between the sliding side forms, conforming plate, and the underlying subbase as the paver moves forward.

The rotating screed located behind the conforming screed is maintained at the proper elevation and cross slope by the same controls as the conforming screed. It cuts away any excess concrete and is made to carry a slight excess of grout for filling tears or other imperfections in the surface. Variations in the amount of grout carried by the rotating screed are a signal to the paver operator that mix consistency or pressure has changed enough to case variations in the surge behind the conforming screed. The operator can then adjust the frequency of vibration to compensate. Vibrating frequency is decreased to compensate for increasing mortar in front of the rotating screed and is increased when the amount of mortar ahead of the rotating screed is decreasing.

All available models of slip-form pavers have front vibration and some type of strike-off screed as means of obtaining initial consolidation. The machines may have a vibrating screed and oscillating screeds following the initial process, and at least one model provides a belting action preceding the final float component. Some

FIGURE D2-54 Slip-form paver with auger spreader. (*CMI Corporation.*)

pavers do not have a receiving hopper, operating instead with concrete depositing directly on the subbase and spread transversely by a paddle or an auger (see Fig. D2-54). All models are equipped with a final float-finisher.

While most slip-form pavers are used to pave the 24-ft paving lanes that are common to two- and four-lane divided highways, machines are available that will place 36- and 48-ft-wide pavements in one pass. These exceptionally wide machines are adaptable to multilane urban expressways and to airport pavements.

Paving Train

For high-production slip-form paving, a train of machines can be used. One such train was used by J. H. Reid, General Contractor, to pave 34 miles of mostly three-lane roadway on Interstate 287 in northern New Jersey.

I-287 was specified by the New Jersey DOT to require two passes on the three-lane portions: one 24-ft pass and one 12-ft. Though such a specification call for a conventional approach, Reid's president, Ken Lindstrom, believed it was twice the effort the job really required. "After studying the route and specs carefully, Jim Reid (the company's founder) and I concluded that one 36 foot pass would produce the right result. We felt that quality would be enhanced because a joint would be eliminated, and everyone involved would save both time and money. We presented our idea to the state people, and they approved," said Lindstrom. "All we needed was the right machine to do it."

Given the width and volume of paving that lay ahead, Reid's officials decided to go with a new machine for the project—one that was durable, reliable, and offered high production capability. "We looked at several manufacturers, but found they could not provide what we wanted, or ship it when we needed it," said Lindstrom. "CMI, on the other hand, had the machine best suited for our work, and they put forth a lot of effort to ensure we had the paver ready to go when the job started." In choosing a CMI paver, two of the SF-450B's most basic features proved most important to the company. "It was the SF-450B's sheer weight, along with its power, that sold us," recalled Ken, "and as expected, the machine holds itself down and doesn't ever hesitate. That's the key to this kind of job."

The SF-450B is an updated version of the SF-450 slipform paver, with design improvements that enhance the contractor's work. Features include a deeper main-frame for superior structural integrity out to a maximum paving width of 44 ft, end frames that are split for dowel bar insertion, and multiposition legs that are lengthened 18 in for improved versatility. Additional improvements include an updated and streamlined hydraulic system that is more efficient and cost productive, a lower engine cowling for improved visibility, and the addition of Series 90 pumps which can be stacked to provide a better arrangement and pump performance.

Reid's SF-450B was delivered in June, equipped with 12-ft extensions to allow for the wide paving width. On the interstate project job site, the paver is preceded by two CMI PST 28-300 placer spreader trimmers. The first of these machines places 3750 lb/in^2 concrete at a 7-in depth and also pulls a CMI mesh cart from which mesh is placed by hand. The second PST 28-300 places an additional 4 to 4½ in. These machines speed placement and spreading in a controlled manner, allowing for better paver performance. The SF-450B follows, equipped with an oscillating surface finisher option to finish the concrete (Fig. D2-55).

Reid's SF-450B came equipped with a Series II paving assembly, one of two available to suit the particular needs of the individual contractor. The Series II features a flanged sectional subframe that allows for a varied arrangement of paving compo-

FIGURE D2-55 Uphill paving train 36 ft wide in single pass. (*CMI Corporation.*)

nents and multiple position crowning. The side-form system features adjustment capability in paving width, vertical angle, and taper. An adjustable float pan provides secondary finishing for a tight surface and precise edge control. Reid utilized the paver's optional two-point crowning capability to produce two-percent crowns on most sections of the job. "We had a 1.5 to the left, a 1.5 to the right, and a 2.0 to the right, thereby presenting the need for a machine that offered two-point capability," said Lindstrom.

After the SF-450B has paved and finished the surface, a CMI TC-250 texturing/curing machine equipped with an 8-ft comb then places the tine marks on the surface at full width. A second TC-250, rigged with a 500-gal tote tank applies white pigment curecrete.

A third PST-28-300 follows, performing a job not common to this piece of equipment. "We basically gutted it, and made it into a traveling saw," said Lindstrom. Equipped with three saws, the machine cuts both transverse and longitudinal joints off a stringline at a rate of about 4 ft/min. It and the other two PST 28-300s were extended to 40½ ft wide, point to point.

An important part of the job, and something unique to New Jersey, is the Type A transverse doweled expansion joint, which in the past has required a great deal of hand work and made slip-form paving difficult. By straddling the pavement with their traveling saw and sawing down to the expansion material 4 in below the surface of the green concrete, Reid eliminated the hand-forming that usually resulted. These joints are placed every 78 ft on the mainline and somewhat closer on bridge approaches. Reid manufactures the joints in their shop and modifies them for easier installation.

At the halfway point in the project, 341,000 yd² had been paved. Average daily production with the SF-450B is 2500 linear feet, though production has reached 2950 linear feet on several days.

Concrete Delivery

Concrete can be dropped directly in front of the paving train by ready-mix truck, end-dump truck, or side-dump truck.

There are a number of special pieces of equipment that can be employed as an intermediary between truck and paver. CMI has a pneumatic-tired material transfer/placing machine. (See Fig. D2-56.) The MTP-400 is capable of continuously processing a wide variety of construction materials at rates of up to 400 yd³/h. The MTP-400 transfers and places material rearward along a 170°, 25-ft radius to a paving or material-spreading operation. The MTP-400 tractor also accommodates conversion from rear placing to front-mounted placing and spreading operations such as road widening. The front-mounted configuration permits placement of material to either side, up to 14 ft, at controlled production rates while maintaining control of grade and slope. Powered by a 180-hp turbocharged diesel engine, the MTP-400 features a posi-track, four-wheel-drive providing traction and drawbar under a wide range of construction site conditions. Unlike track-propelled units, the wide flotation pneumatic tires of the MTP-400 permit use on existing pavement without cracking or scarring the surface. The combination of wide-base pneumatic tires, positive four-wheel-drive and two and four (optional)-wheel automotive steering, provides work-site performance and maneuverability, including travel speeds of up to 10 mi/h.

FIGURE D2-56   MTP-400 material transfer/placing machine. (*CMI Corporation.*)

CMI offers an attachment which converts the TR-225B finegrade trimmer to a high-production concrete placer system. Working from an adjacent lane or shoulder area, the TR-225B can receive concrete from end dumps and conveys it to the grade in front of the paver.

Installation is quick and easy, requiring less than an hour. No disassembly of the TR-225B is required. The trimmer is simply driven up to the hopper and conveyor attachment where it engages the mounting system. Hoses with quick couplings are then connected to provide hydraulic power to the hopper's auger and the side-mounted conveyor.

The attachment's diverter chute system directs concrete from the 36-in-wide side conveyor to the TR-225B's reclaiming conveyor for delivery to the grade. The receiving hopper is equipped with a segmented material screw constructed of all-cast, wear-resistant Ni-hard alloy for extended life. An optional hydraulic-powered truck hitch locks the delivery truck to the TR-225B, which pushes the truck as it discharges material into the hopper.

Utilization and productivity of the TR-225B is increased by combining two machine functions: first as a high-production finegrader in the preparation phase, then as a high-volume concrete placer in the paving phase. The attachment makes the TR-225B ideal when paving over reinforcing dowel baskets, and when specifications or job conditions prohibit driving trucks on the finished grade.

The Blaw-Knox MC-30 (Fig. D2-57) is a self-propelled, wheel-mounted, bulk handling/delivery system with a built-in surge storage capacity of about 30 tons.

FIGURE D2-57 MC-30 mobile conveyor. (*Blaw-Knox.*)

Steel Reinforcement and Dowels

The installation of steel mesh, bar mats, and dowels has long been the source of major problems for the paving contractor. In the past, it was the general practice to construct a formed pavement in two lifts with the reinforcing steel sandwiched between the two layers. Dowels were assembled in a supporting framework and placed at proper elevation on the subbase preceding the first course of concrete. Handling, processing, and installation of the steel components were tedious and expensive procedures and were subject to errors that contributed to pavement defects.

In the quest for more economical and efficient methods of installing dowels and reinforcement, two devices have been developed. These are showing good results, particularly in the construction of fixed-form pavement. Dowel-placing machines, one of the improved devices, were developed several years ago but have been widely used only recently. The dowel placer, one version of which is illustrated in Fig. D2-58, is relatively simple in principle. Dowels are set in a slot device that, when vibrated into the fresh concrete, places the bars in exact position with minimum disturbance of the concrete (see Fig. D2-59). This procedure eliminates the requirement for elaborate supporting framework for the dowels and also ensures that the dowel bars will be uniformly surrounded with concrete. The dowel bars are frequently given a plastic coating so that they will not require painting or greasing prior to insertion into the concrete.

FIGURE D2-58   Dowel-placing machine.

The mechanical mesh placer is a very recent development. It had its beginning when the mesh was taken off the shoulder or back slope where it was spread and placed on a mesh carrier towed by the spreader. The next step was to develop a following device which would force the steel into the concrete. For placing mesh or reinforcing mat between forms, the placing machine is usually mounted on the front end of a long-wheelbase finishing machine. The mesh or mat is depressed in 15-ft-long sections from the slab surface, with the machine standing at rest on the side forms. The placing machine uses vibration in placing the reinforcement, which should eliminate any necessity for other surface vibration prior to finishing the slab surface.

FIGURE D2-59 Dowel placer with bars ready for lowering into concrete.

When permitted by the specifications, the dowel placer and the mesh placer may be used in combination for placing dowels and distributed steel on a formed concrete paving project. Since dowel bars are placed at the midpoint of the slab and the distributed steel is usually placed in the top third of the pavement, the mesh placer can logically follow the dowel placing machine in the construction of formed concrete. Recognizing that the purpose of distributed steel is to prevent surface cracking at long joint intervals, it should be permissible to omit reinforcement immediately adjacent to joints where the internal stresses are minimal. With such discontinuity, the mesh placer will not interfere with the previously installed dowels.

The installation of dowels and distributed steel in pavement built by slip form has usually been done by the two-lift method. Generally this is accomplished by having the mechanical spreader place a 6- or 7-in layer of concrete at a reduced width. A reduction of about 2 in on each side allows passage of the slip-form paver. This machine will form the completed pavement edge as it places the top layer of concrete above the reinforcing steel. In this operation it is necessary to eliminate internal vibration which would interfere with the steel. The mesh or mats are placed by hand, generally from a mesh cart towed by the spreader. With this method of installation, the dowels will usually be placed on the subbase and supported by the proper framework.

Mechanical dowel and mesh placers have been used with the slip-form paver (see Fig. D2-60). Such use is largely dependent on the resourcefulness and ingenuity of the contractor and the flexibility of job specifications.

Machine Pavers

Bid-Well makes a family of machine pavers for bridge deck paving and/or street paving. The backbone is a traveling truss that spans the area to be paved (Fig. D2-61). The truss travels on roller-mounted legs supported on rails. The paving unit is hung from the truss and travels transversely.

The Bid-Well Dual Vibration System is mounted from the paving carriage and consolidates the concrete just ahead of the paving augers. With this system, the vibra-

FIGURE D2-60 Slip-form paver with mesh placer mounted in front.

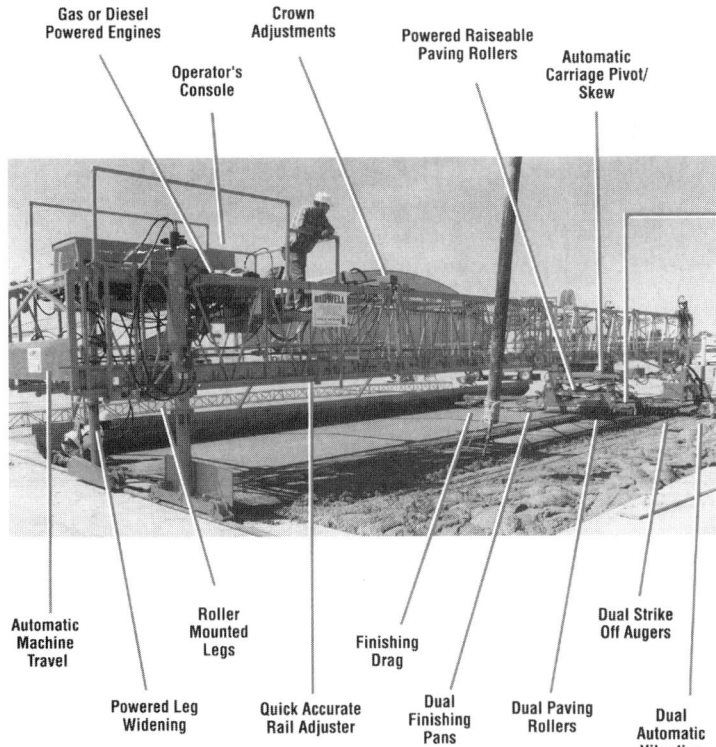

FIGURE D2-61 Bridge deck paving machine. (*Bid-Well.*)

tors are inserted and travel laterally in the concrete at spaces ranging from 6 to 12 in between each vibrating pass, giving uniform consolidation in slabs up to 24 in deep.

The two spud-type hydraulic paving vibrators are mounted on a mechanism that alternately inserts the spuds into the concrete as the carriage approaches the paving form. The tip (which is the most active part of the vibrator) vibrates the concrete at the paving form as the paving carriage stops, advances forward, and starts traveling in the other direction. This imparts vibration directly to the edge of the concrete slab, to maximize consolidation of the concrete near the paving forms.

The Bid-Well Rol-A-Tamp finishing attachment (Fig. D2-62) provides a means of achieving a more uniform concrete surface with desired density. This facilitates sealing difficult-to-finish concrete, because of harsh mix designs, low slump specifications, wind exposure causing abnormal surface drying, and unforeseen and unpredictable delays.

The Rol-A-Tamp which mounts from the finishing carriage, utilizes two freewheeling finned rollers. The finned rollers are positions between the leveling augers and finishing rollers. The finned rollers are vertically adjustable from ½ in above concrete grade to ⅛ in of fin penetration below concrete grade.

FIGURE D2-62 Rol-A-Tamp attachment. (*Bid-Well.*)

JOINTS

The types of joints to be provided in the concrete pavement, their intervals, and dimensions (depth and width), will be covered by the job plans and specifications. The four types of joints used in concrete pavement are (1) construction joints, both longitudinal and transverse; (2) longitudinal center joints, sometimes called *hinged joints,* to relieve curling and warping stresses; (3) contraction joints in the transverse direction across paving lanes; and (4) expansion joints which are full pavement depth, with a compressible filler to permit the joint to close as the pavement expands. Regularly spaced expansion joints are seldom provided because tests and experience have shown that, under normal conditions, contraction joints provide adequate space for pavement expansion. Expansion joints are still used adjacent to fixed structures and at asymmetrical intersections.

Jointing arrangement may vary widely, depending on the local construction practices. Dowels and distributed steel are omitted in some localities where cement-stabilized subbases are used. In these instances, contraction joints will be at short intervals, on the order of 15 to 20 ft. The joints may be skewed diagonally across the pavement, or they may be placed at irregular intervals such as 15, 18, and 21 ft, with repetitions of this randomized grouping. Joints placed at such short intervals may be of rather narrow width, such as ⅛ in, since each joint will have little movement in opening and closing. Where distributed steel reinforcement is used, joints will be spaced at much wider intervals, such as 40 to 100 ft, since wider joint spacing is the primary reason for the use of distributed steel. At these longer intervals, contraction joints will be much wider, from ½ in to as much as 1 in, in order to accommodate a greater amount of movement.

Joint Forming

While there are several acceptable ways of forming joints, sawing is by far the most common and doubtless the most economical method. Saw manufacturers have been very active and resourceful and are able to provide power saws that are adaptable to any jointing situation. Since contraction stresses develop as soon as the concrete hardens, contraction joints must be sawed very early. The exact time of sawing depends on the type of aggregate, curing method, cement factor, and the weather. Generally, all contraction joints should be sawed as soon as this can be done without damage to the surface. A slight amount of raveling is permissible and desirable because it gives the operator a good gauge of timing. If there is no raveling at all, the concrete is too hard and cracks may develop ahead of the saw. All joints should be sawed in succession to provide the plane of weakness at the time of maximum contraction and before the slab gains too much strength. This will ensure cracking and uniform opening of all joints. Sawed joints should be thoroughly flushed or blown clean immediately after sawing to remove all residue, which otherwise may cake on the sides of the joint and hinder proper sealing.

When longitudinal center joints are sawed, the timing is not so critical. These joints may be formed at any convenient time before traffic is allowed on the pavement. A common method of forming longitudinal center joints in slip-form paving is by the use of a continuous polyethylene strip. This is automatically placed into the concrete from a roller carried at the rear of the slip-form paver. Center joints in formed pavement may be formed in a similar manner by devices that are available. Expansion joints, when required, are generally installed by hand methods.

Wherever construction joints occur, they should be grooved to a depth sufficient to retain joint sealer. Sawing is generally the most practical method of providing a joint groove. Some hand edging may be required during construction in order to prevent the plastic concrete from overlapping the hardened concrete already in place.

Joint Sealing

Materials and methods for sealing joints are covered in detail by the specifications. It is important that the joints be sealed as soon as practicable to minimize the intrusion of foreign materials. Joint openings may be protected by a temporary filler such as a section of rope or a strip of tape over the opening. In any case, the joint opening should be thoroughly cleaned just before sealing.

CURING

The hardening and strength gain of concrete are dependent on the availability of moisture for the hydration processes. The durability and long-term utility of the pavement are greatly dependent on the timely use of adequate curing methods.

The specifications will cover the approved curing and protection methods, and it is essential that the curing be accomplished as required. All equipment and materials required for the application of curing must be available on the job before concrete placement is started each day.

The two most common methods of curing are by the use of waterproof membrane and polyethylene sheets. Equipment is available that is adaptable to the application of white-pigmented curing compound on either formed or slip-form pavement. Polyethylene sheeting may be placed on the pavement in long strips, anchored by loose dirt, and rolled up for further use after the curing period has expired. One advantage of the use of polyethylene sheeting is that it protects the pavement from damage by sudden rainstorms.

The Bid-Well TC-3000A and TC-3000M texturing-curing machines (Fig. D2-63) span up to 50 ft. It uses rakes to a textured pattern and sprays on curing compound.

FIGURE D2-63 Texturing-curing machine. (*Bid-Well.*)

REFERENCES

1. *The Asphalt Handbook,* Manual Series No. 4 (MS-4), The Asphalt Institute, College Park, Md.
2. *Bituminous Construction Handbook,* Barber-Greene Company, Aurora, Ill.
3. *Asphalt Plant Manual,* Manual Series No. 3 (MS-3), The Asphalt Institute, College Park, Md.
4. *ASTM Book of Standards,* part II, American Society of Testing and Materials, Philadelphia.

CHAPTER 3
PILES AND PILE DRIVING

Peter K. Taylor
Consultant, Geotechnical Division
Stone & Webster Engineering Corporation

INTRODUCTION

History

Piles are structural elements inserted through soft soils to transmit loads to underlying firmer soils. The use of piling to support structures dates from over 6000 years ago, and the first piles were inverted tree trunks. Among the first recorded uses of piling were the structures built out over lakes by the Swiss Lake dwellers of Central Europe and pile-supported structures built by the Aztec Indians of Mexico. During the Middle Ages many large structures were supported on piling, notably the Royal Palace in Amsterdam, the Berlin Castle and Opera House, London Bridge, the Campanile Tower in Venice, and the Cathedral of the Notre Dame in Paris.

Determining the Need for Piling

Geotechnical engineers determine the need for piling based on a study of the subsurface conditions and a settlement analysis. If estimated building deformations exceed the tolerable deformation, deep foundations or some form of ground improvement is required. If an economic analysis indicates a pile foundation is the most cost-effective foundation, a pile foundation is designed.

Types of Piles

There are two generic types of structural piles: end bearing and friction. An end-bearing pile is one which is supported almost entirely by soils near or at the pile tip. A pile driven to refusal on bedrock is an example of an end-bearing pile. Friction piles derive their support from the friction on the periphery of the pile. A typical example of this type of pile is a long pile driven into a clay stratum. The pile derives its support almost entirely from its surface area. The tip resistance is negligible and can normally be ignored.

Installation

Typically, piles are inserted into the ground by a pile driver, either by driving the top of the pile or by inserting a heavy wall steel mandrel into a thin wall steel shell and dragging the pile into the ground. However, there are certain pile types that are installed by augering or drilling a hole, which is subsequently filled with concrete and in which reinforcing steel can be placed.

Unique Relationship Among Engineer, Contractor, and Owner in Pile Foundations

There is always an element of uncertainty in the installation of pile foundations. This is due to the inherent heterogeneity of the subsurface conditions of most sites where piling is to be installed. In addition, there is the competitive nature of the design and construction industry, which pressures both designers and contractors to minimize total construction costs.

Consequently, the selection of the most cost-effective piling type may entail comparing unknown risks against known cost savings. To provide the owner with the most cost-effective foundation system, all three interested parties—the owner, the designer and the contractor—need to understand the risks as well as the savings and work together during the design and installation to avoid costly delays and litigation expenses.

Summary and Scope

There are too many pile types to discuss each in detail. This chapter will introduce the more popular types of piles, their material properties, and installation equipment and methods, and will provide a summary of items of concern during pile-driving inspection. For a more detailed study of pile driving, refer to Engineering of Pile Installations, by F. M. Fuller.[1]

PILE TYPES

General

Typical pile types are timber, prestressed concrete, cylinder, cast-in-place shell, steel H, pipe or monotube sections, augered or drilled cast in situ piles, and pressure-injected footings. Selection of the appropriate pile type is a function of load and strength requirements, cost of construction, and cost of materials. Illustrations and physical characteristics of these pile types are shown in Fig. D3-1a–d.

Timber Piles

Timber piles are typically pressure-treated southern pine or douglas fir. Timber piles date back to the earliest uses of piling. Douglas fir piling comes from the Northwest in single pieces up to 120 ft on special order, but 60 ft and shorter lengths of Southern pine are readily available. Timber piles normally do not exceed 40 tons in capac-

	Concrete Filled Steel Pipe Piles	Augercast Concrete Piles
Consider for Length of	40 to 120 ft. or more	30 to 60 ft.
Applicable Material Spec.	ASTM A252 for pipe ACI Code 318 for concrete	ACI Code 318 for concrete
Maximum Stresses	9000 psi for pipe shell 33% of 28 day strength for concrete.	33% of 28 day strength for concrete.
Disadvantages	Relatively high initial cost. Displacement pile.	Very dependent on quality of workmanship. Not suitable in very soft organic silts.
Advantages	Best control during installation. High load capacity Easy to splice Easy to inspect	Very economical Non displacement No vibration High skin friction No splicing required
Remarks	Provides high bending resistance in all directions Loading in the range of 40 to 120 tons.	Best suited as a friction pile Loading in the range of 40 to 100 tons
Typical Illustration	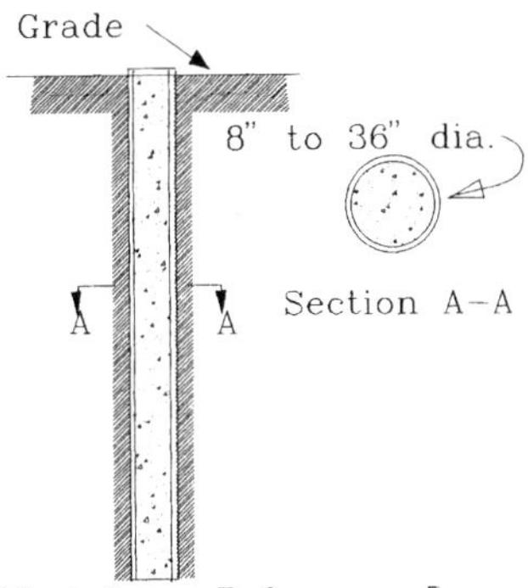 Grade — 8" to 36" dia. Section A-A	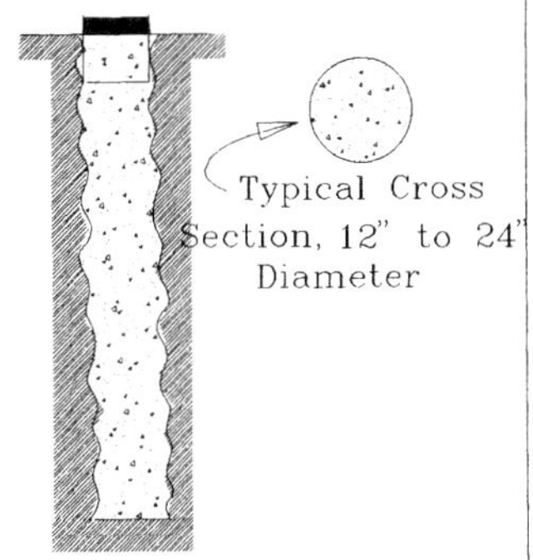 Typical Cross Section, 12" to 24" Diameter

Source: Modified from Reference 5

FIGURE D3-1a Typical pile types.

	Precast Prestressed Concrete	Cast-in-Place Concrete Mandrel Driven Shell Pile
Consider for Length of	60–100 ft.	10–120 ft. but typically in the 50 to 80 ft. range.
Applicable Material Spec.	PCI Standard 112 – Prestressed Concrete MNL 116 – Prestressed Concrete Quality ASTM A615 – Steel Reinforcing	ACI Code 318 for concrete
Maximum Stresses	Fc = 0.33F'c – 0.27Fpe where Fpe = the effective prestress stress on the gross section.	33% of 28 day strength of concrete
Disadvantages	Large Displacement of soil Difficult to splice Difficult to know when it breaks	Redriving is not recommended. Thin shell is vulnerable to collapse due to earth pressure. There is considerable displacement.
Advantages	High load capacity. Corrosion Resistant	The use of the mandrel allows an ecconomy of steel. Piles may be internally inspected after driving..
Remarks	General loading range from 40 to 400 tons	Generally loaded in the 40 – 100 ton range. Suitable as both friction or end bearing piles.
Typical Illustration	(10" TO 24" dia.; 36" to 66" DIA.)	Grade — Cross Section; 8" to 18" DIA. Corrugated Shell Thickness 12 ga. to 20 ga. Straight Sided / Step Tapered

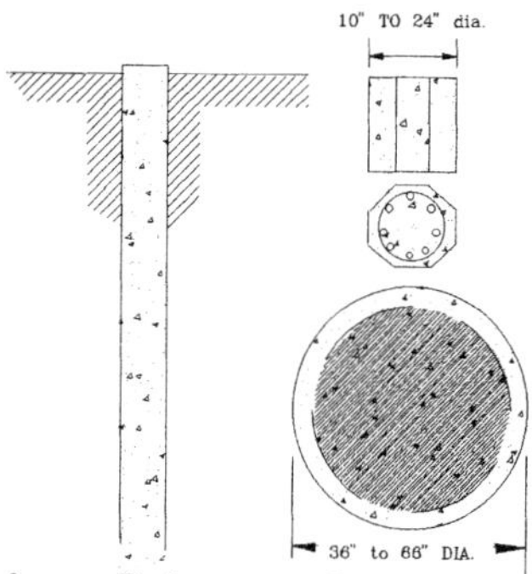

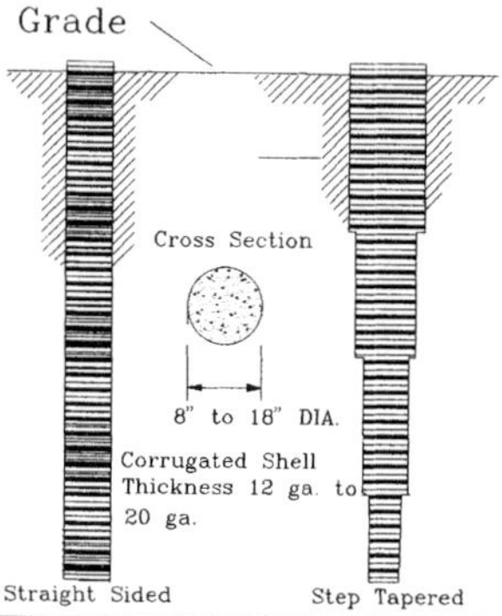

Source: Modified from Reference 2.

FIGURE D3-1b Typical pile types.

	Timber	Steel H Sections
Consider for Length of	30 – 60 ft.	40 – 100 ft.
Applicable Material Spec.	ASTM – D25	ASTM A36
Consider for Design loads	10 to 40 tons	40 to 120 tons
Maximum Stresses	1200 psi in compression at the most critical section for Southern Pine and Douglas Fir	12,000 psi
Disadvantages	Difficult to splice. Vulnerable to damage by hard driving. Tip may have to be protected. Vulnerable to decay if not treated.	Vulnerable to Corrosion. HP Section may be damaged or deflected by major obstructions
Advantages	Low initial cost. Easy to handle.	Easy to splice. Available in various lengths and sizes. Low displacement. Able to penetrate light obstructions. Harder obstructions or end bearing on rock may require point protection.
Remarks	Best suited for friction pile in granular material. Maximum load is 40 tons.	Best suited for end bearing on rock. Load range is 50 to 120 tons.
Typical Illustration	Butt Diameter 12" to 22"; Tip Diameter 5" to 9"	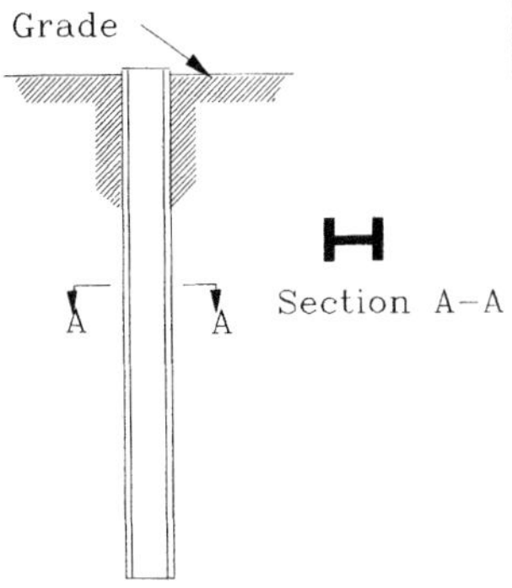 Grade; Section A–A

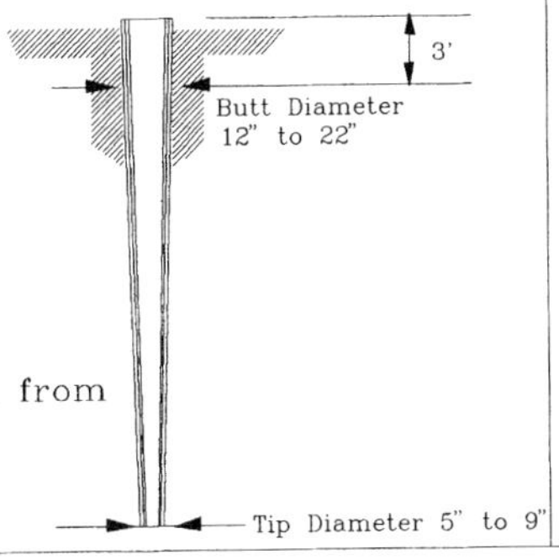

Source: Modified from Reference 2.

FIGURE D3-1c Typical pile types.

	Pressure Injected Footings
Consider for Length of	20 to 50 ft.
Applicable Material Spec.	ACI Code 318 for concrete
Maximum Stresses	33% of 28 day strength of concrete
Disadvantages	Large Displacement of soil Cannot be redriven Very dependent on operator skill
Advantages	Very cost effective for shallow bearing stratum Corrosion Resistant High load capacity. Densifies granular soil during creation of footing
Remarks	General loading range from 60 to 300 tons End bearing
Typical Illustration	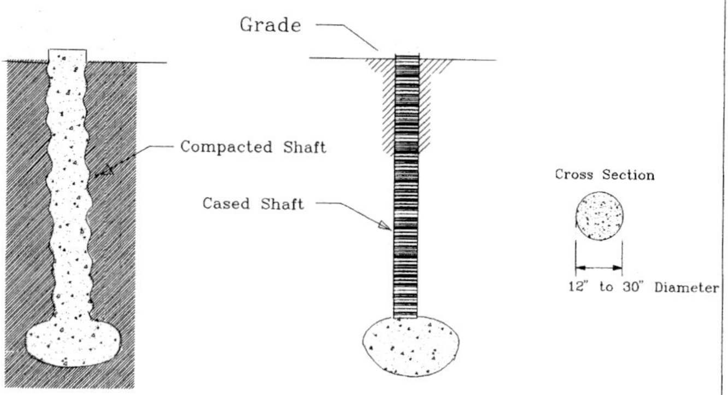 Grade — Compacted Shaft — Cased Shaft — Cross Section — 12" to 30" Diameter

FIGURE D3-1d Typical pile types.

ity. They are very difficult to splice, so it is necessary to select appropriate pile lengths in advance of pile driving. Timber piles tend to break when overdriven; consequently, the contractor must be cautious in selecting driving equipment and the driving criteria must be mutually agreeable to the contractor and the engineer. Most timber piles are pressure-treated to preserve the wood. Untreated timber piling deteriorates quickly above the groundwater table. Timber piles lose strength if subjected to prolonged high temperatures, and should not be used under structures such as blast furnaces, or chemical reaction units.

The quality and dimensions of timber piles are normally specified by referring to the Standard Specification for Round Timber Piles, ASTM D-25, and the pressure-treatment by referring to the American Wood Preservers Association (AWPA) Standard C1, C3, or C18 (for saltwater).

Prestressed Concrete Piles

Prestressed concrete piles are formed, poured, and cured in a casting yard. A compressive stress is locked into the pile during manufacture to enable the pile to withstand tensile stresses. Piles are usually square, round, or octagonal in section, with or without taper, and reinforced to permit handling. They may be cast with a hole in the center to enable them to be advanced by jetting.

Prestressed concrete piles preferably are driven in one continuous piece, if the driven lengths can be predicted with accuracy, and the maximum length does not exceed about 80 to 100 ft. If not, the piles can be made in sections, each section joined together in the field by spliced connections. There are many different spliced connections; the most common are some form of fabricated steel end pieces which are cast into the pile. The two precast sections are joined by either pins or wedges being driven into the corners of the castings, locking the two pieces together. Some typical splice connections are shown in Fig. D3-2.

Concrete Cylinder Pile

A special type of prestressed concrete pile is the cylinder pile. It ranges in diameter from 30 to 66 in and is cast in up to 60-ft lengths that may be spliced together to make any required length. The cylinder pile has a vertical load capacity of up to several hundred tons and can absorb and transmit horizontal forces of considerable magnitude.

Cast-in-Place Concrete Piles

Cast-in-place piles are constructed by pouring concrete into a preformed hole in the ground. There are a variety of ways to create the holes, either by drilling or by driving, and each has unique advantages and disadvantages. One advantage of the cast-in-place pile is that no driving stresses are inflicted on the permanent structural elements of the pile. Another advantage is that the length of the pile is not as critical as for most top-driven piles, as long as the installation equipment is able to install the pile to the design depth. The following is a partial listing of several of the more common types.

Shell Piles. Cast-in-place, concrete-filled shell piles are formed by driving a light-gauge corrugated steel shell with a heavy steel mandrel. The mandrel is similar to a pipe pile which is reusable. It may be either a single piece or split in half, with the two

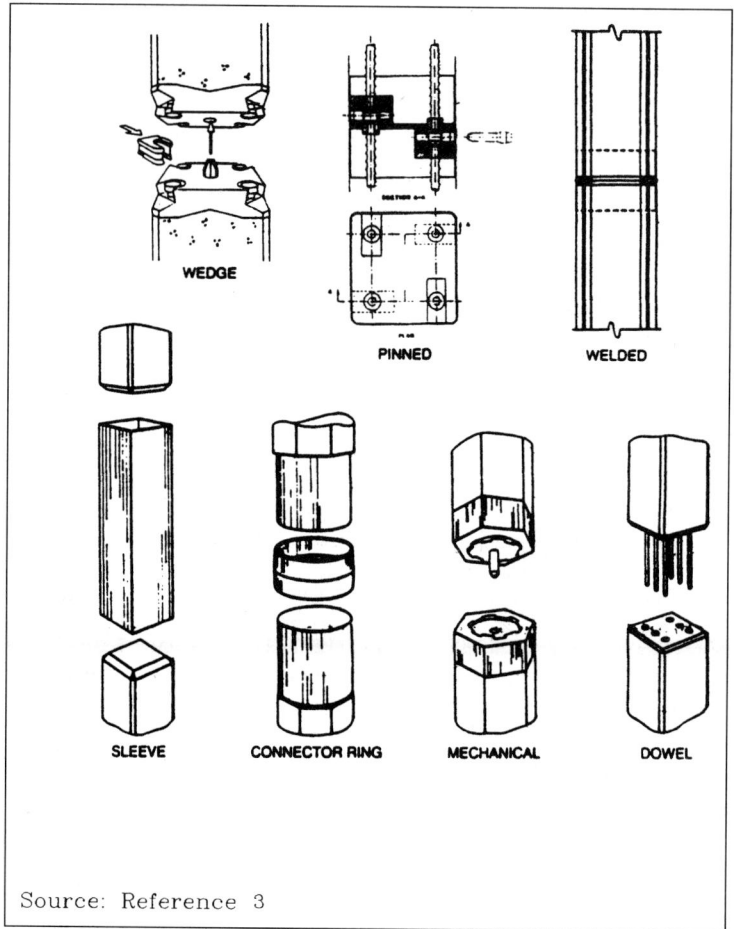

FIGURE D3-2 Typical splices for prestressed concrete piles.

halves capable of being expanded to grip the corrugated shell during driving of the mandrel. The mandrel is contracted and extracted and the shell is filled with concrete. A reinforcing cage may be installed if required. The shells are easily spliced, they may be inspected after driving, and the excess shell that is cut off may be reused.

Bored Piles. This type of pile is installed by drilling a hole into the ground to the required depth and filling the hole with concrete and reinforcing steel. The hole may be either cased or drilled with drilling mud to maintain the stability of the hole. The shaft may be socketed into rock or have an enlarged base for increased capacity. Lengths exceeding 100 ft have been installed in favorable soil conditions, i.e., medium to stiff clays. The principal advantages of this type of pile are the relatively low cost and the lack of vibration and displacement of soil during the installation.

Bored piles are typically reinforced with a cage of longitudinal and spiral steel extending below any zone of significant bending stresses. If piles are to be installed with the hole full of bentonite, care must be taken to avoid trapping the bentonite clay balls in the concrete while filling the pile. If temporary casing is used which is removed as the concrete is tremied into the pile, it is advisable to tack-weld the reinforcing cage to avoid unraveling as the casing is removed.

Augered Cast-in-Place Grout Piles. Augered cast-in-place grout piles are installed by screwing a continuous hollow stem auger into the ground, and inserting grout under pressure into the void created as the auger is slowly withdrawn. Pile sizes typically range from 12 to 24 in in diameter, and pile capacities range up to about 100 tons. This type of pile makes an excellent friction pile; however, they have been used on occasion for end-bearing on rock. The length of this pile is limited to the length of the auger, which normally is less than 100 ft. A single reinforcing bar may be installed through the hollow stem to the bottom of the pile before the grout is injected, and a reinforcing cage can be inserted from the top of the pile immediately after completing the pile. This type of pile may be difficult to install through very soft soils, such as organic silt. Great care must be taken during the installation to ensure that the shaft does not contain voids due to lack of soil support of the grout column or too rapid withdrawal of the auger. The success of this type of pile is very dependent on the skill of the operator.

Pressure-Injected Footings (PIFs). PIFs are installed by driving a steel casing or drive tube into the ground using a drop weight operating inside the drive tube and driving on gravel or a zero-slump concrete plug at the bottom of the tube. The required depth of penetration is identified by the driving resistance as well as the soil stratigraphy. When this depth is reached, the plug is driven out and an enlarged base is formed by adding and driving out small batches of zero-slump concrete.

Reinforcing steel is placed as a welded cage and joined to the base by placing and compacting a small quantity of concrete. The shaft may be formed either by compacting zero-slump concrete uncased or by casing it and filling with a higher slump concrete and either pulling or leaving the casing. The compacted shaft is formed by placing small lifts of concrete and withdrawing the steel tube in small increments while driving out the concrete with the hammer.

The PIF is very economical for transmitting high loads to a relatively shallow, granular bearing stratum. The maximum practical depth for a PIF is about 50 ft. The success of the pile is very dependent on the skill of the operator.

Steel H Piles

Steel H piles are particularly well suited for hard driving into a dense bearing stratum and have a large section modulus about one axis to resist bending moments. These piles may be reinforced with prefabricated pile points to protect them during hard driving. Pile lengths are unlimited since they are easily spliced and the cutoff ends are reusable. Steel H piles are susceptible to corrosion. Each site should be evaluated for corrosion potential and the required protection methods when considering the use of this pile type.

Steel Pipe Piles

Steel pipe piles are similar to H piles in that they are easy to splice, have reusable cut ends, and are subject to corrosion. Pipe piles are better suited to resist bending moments from any direction because of the uniform section modulus. They may be driven either closed-ended or open-ended (and later cleaned out by drilling or jetting) to minimize soil displacement or facilitate penetrating dense strata. Pipe piles may be visually inspected before concreting. Pipe piles may be designed to carry larger compression loads than H piles with comparable steel areas, due to the increased strength of the combined concrete and steel section. This applies only to situations where the piles are driven to a very dense bearing stratum or bed rock.

Monotube Piles

Monotube pile walls are fluted and of varying cross-sectional area. They are a proprietary pile type manufactured by the Union Metal Manufacturing Company. The advantage of this pile type is that it derives an increase in strength over straight-walled piles by the fluting, allowing a lighter steel cross section for the same design load. Monotubes are produced in gauges of 3, 5, 7, 9, and 11.

PILE STRESSES

The following common sources of stress in a pile should be considered when selecting the pile type, material, and size.

Design Loads

Live, wind, and dead loads will cause compression, tension, shear, or bending stresses in a pile. Both tension and compression stresses in piles will be diminished along the length of the pile, depending on the distribution and magnitude of the shearing resistance between the soil and periphery of the pile.

Handling Stress

Piles that are lifted, stored, and transported may be subject to substantial handling stresses. Bending and buckling stresses should be investigated for all conditions, including lifting, storing, transporting, and impact.

Driving Stress

Driving stresses are complex functions of pile and soil properties, influenced by the required driving resistance, size and type of pile-driving equipment, and method of installation. Both tension and compression stresses occur and could exceed the yield strength of the pile material. Dynamic compressive stresses during driving are considerably greater than the stresses incurred by the maximum design load. Analysis of driving stresses has been made possible by development of the wave equation. A thorough discussion of the wave equation theory and application is provided in Ref. 1.

Tension Stresses Due to Swelling Soils

Piles are sometimes subjected to temporary axial tensile stresses due to swelling of certain types of clays when the moisture content increases. Swelling clays should be provided for in the design or their effects minimized in the installation procedures.

Compression or Bending Stresses Due to Negative Skin Friction

Negative skin friction can produce additional compressive loads on the pile resulting from the consolidation of compressible soils through which the pile extends. Consolidation is generally caused by an additional load being applied at the ground surface and continues until a state of equilibrium is reached. Under the negative skin friction conditions, the critical section of the pile may be located at the surface of the bearing strata. The magnitude of the load applied to the pile as a result of negative skin friction is limited by certain factors: shearing resistance between the pile surface and the soil, shear strength of the soil, pile shape, and thickness of the compressible stratum.

PILE MATERIALS

Timber

Timber piles are specified in accordance with ASTM D-25, latest revision. This standard covers timber quality, which includes soundness; straightness; twist of the grain; knots, holes and scars; and checks, shakes, and splits. It also covers piling sizes, either by the pile circumference at a point 3 ft from the butt or by the tip circumference for varying lengths. The dimension of the pile end that is not specified is determined by the natural taper of the pile. The piling size should typically be specified using the minimum permissible cross section based on the allowable stress. The minimum permissible cross section is located at the point where the pile encounters the bearing stratum. For a friction pile this would be at the top of the supporting stratum, and for an end-bearing pile this would be at the pile tip.

All timber piling should be pressure-treated in accordance with the American Wood Preservers' Association (AWPA) Standard C1 and C3. In a marine environment, AWPA C18 should be specified.

Pile tips may be reinforced to prevent breakage. Prefabricated steel tips, either pointed or flat, are available from specialized suppliers of pile fittings.

Steel

Pipe or Monotube. New steel pipe or monotube piles are normally specified by grade in accordance with ASTM A252, Specification for Welded and Seamless Steel Pipe Piles. Occasionally the contractor will propose using used pipe that was not produced under ASTM A252. For this to be acceptable, the contractor must demonstrate that the pile is in good condition, is drivable, and can be welded by standard methods. Some steel is brittle and very difficult to weld. Coupons of the proposed pipe should be tested for yield strength and weldability.

Pipe piles can be delivered in uniform lengths which are controlled by transportation limitations. The diameter and wall thickness of pipe piles are given in Table D3-1.

Monotube piles come in both tapered and straight pieces in the dimensions and wall thickness ordered. Table D3-2 includes information on monotube piles.

The welding of closure plates or pile points should be done on a horizontal rack at the site, unless they are ordered to be attached at the mill. Closure plates are normally from ¾ to 2 in thick and are welded to the pipe with a full-penetration butt weld. Splice welding of pipe piles and monotube piles can be done either in a rack or in the leaders. Proprietary prefabricated splice rings are available from specialty manufacturers to facilitate the splicing of pipe piles in the leaders.

Welding of pipe piles should be in accordance with the American Welding Society (AWS) Structural Welding Code—Steel D 1.1.

Steel H Piles. Steel H piles normally are specified to be in accordance with ASTM A-36, Specification for Structural Steel, or ASTM A-572, Specification for High-Strength Low-Alloy Columbian Vanadium Steel of Structural Quality.

H piles are normally shipped in lengths between 40 and 60 ft. Longer lengths may be obtained by special order.

The size and weight of H piles is normally specified and should meet the minimum requirements of the American Institute of Steel Construction (AISC), HP shapes. Data on various H pile shapes is given in Table D3-3.

Pile tip reinforcement is normal for H piles driven to end bearing in rock. Tip reinforcement is available from specialty piling fitting manufacturers.

Welding of H piles is similar to the requirements for pipe piles. Splice welding of H piles can be done either in a rack or in the leaders. Proprietary prefabricated splice rings are available from specialty manufacturers to facilitate the splicing of H piles in the leaders.

Steel Shell. Steel shells are light-gauge corrugated steel which are driven into the ground with an internal mandrel. The shells can be either straight-sided or step-tapered. The shells are a form to hold the concrete and have no load-bearing function. For either step-taper or straight-sided piles, the tip diameter is usually specified. The butt diameter is determined either by the number of steps or, in the case of straight-sided piles, is the same as the tip diameter. Step-taper shells are available in section lengths of 4-, 8-, 12-, and 16-ft lengths. Straight-sided shell piles are available in lengths up to 80 ft and may be spliced by butt welding. Typical gauges for shell piles range from 12 to 18. Normally, the specification requires the contractor to select the gauge of the shell to prohibit collapse of the pile either during or after installation. See Table D3-4 for the diameter of step-taper shell piles, and Table D3-5 for the typical diameter of straight-sided shell piles. Shell piles must have a boot welded to the tip of the pile to prevent the collapse of the pile tip and prevent dirt and water from entering the shell. Driving in boulders or dense soils may require a steel plate, typically 0.75 in thick, to be welded to the bottom of the boot.

Reinforcing Steel. Reinforcing steel is normally specified to be in accordance with ASTM A615, Specification for Deformed and Plain Billet-Steel Bars for Concrete Reinforcement. Frequently, high-yield-strength bars conforming to ASTM A722, Specification for Uncoated High-Strength Steel Bars for Prestressing Concrete, such as Dywidag bars, are used for tension reinforcement.

Concrete

Prestressed Concrete Piles. Prestressed concrete piles are normally specified to be in accordance with the Prestressed Concrete Institute (PCI) Standard 112, Standard

TABLE D3-1 Typical Properties of Pipe Piles

Size O.D., in	Wall thickness, in	Weight per ft, lb	Moment of inertia, in^4	Radius of gyration, in	Section modulus, in^3	Area external surface, ft^2/lin. ft	Area of steel in cross section, in^2	Inside cross-sectional area, in^2
11	.115	13.36	58.14	3.848	10.57	2.87	3.932	91.10
	.125	14.51	63.02	3.845	11.45	2.87	4.270	90.76
	.134	15.60	67.64	3.841	12.29	2.87	4.591	90.44
	.141	16.30	70.60	3.839	12.83	2.87	4.797	90.23
	.156	18.06	77.99	3.834	14.18	2.87	5.314	89.71
	.164	19.02	82.00	3.831	14.91	2.87	5.596	89.43
	.172	19.87	85.55	3.828	15.55	2.87	5.846	89.18
	.179	20.72	89.07	3.826	16.19	2.87	6.095	88.93
	.188	21.65	92.93	3.823	16.89	2.87	6.369	88.66
	.203	23.40	100.19	3.817	18.21	2.87	6.885	88.14
	.219	25.18	107.50	3.812	19.54	2.87	7.409	87.62
	.230	26.45	112.68	3.808	20.48	2.87	7.782	87.25
	.250	28.70	121.81	3.801	22.14	2.87	8.443	86.59
	.279	31.94	134.86	3.791	24.52	2.87	9.397	85.63
12	.115	14.59	75.68	4.202	12.61	3.14	4.293	108.80
	.125	15.85	82.06	4.198	13.67	3.14	4.663	108.43
	.134	17.04	88.08	4.195	14.68	3.14	5.013	108.08
	.141	17.81	91.95	4.193	15.32	3.14	5.239	107.85
	.156	19.73	101.62	4.187	16.93	3.14	5.804	107.29
	.164	20.78	106.86	4.184	17.81	3.14	6.112	106.98
	.172	21.71	111.51	4.182	18.58	3.14	6.386	106.71
	.179	22.60	116.11	4.179	19.35	3.14	6.658	106.43
	.188	23.72	121.17	4.176	20.19	3.14	6.958	106.13
	.203	25.57	130.68	4.171	21.78	3.14	7.523	105.57
	.219	27.56	140.26	4.166	23.37	3.14	8.096	105.00
	.230	28.98	147.06	4.162	24.51	3.14	8.504	104.59
	.250	31.37	159.05	4.155	26.50	3.14	9.228	103.86
	.281	35.17	177.53	4.144	29.58	3.14	10.354	102.74
	.312	38.95	195.70	4.133	32.61	3.14	11.474	101.62
	.330	41.12	205.75	4.127	34.29	3.14	12.098	100.99
12¾	.115	15.51	90.93	4.467	14.26	3.33	4.564	123.11
	.125	16.85	98.61	4.463	15.46	3.33	4.957	122.71
	.134	18.12	105.86	4.460	16.60	3.33	5.330	122.34
	.141	18.94	110.53	4.458	17.33	3.33	5.570	122.10
	.156	20.98	122.17	4.452	19.16	3.33	6.172	121.50
	.164	22.10	128.49	4.450	20.15	3.33	6.500	121.17
	.172	23.09	134.09	4.447	21.03	3.33	6.791	120.88
	.179	24.07	139.64	4.444	21.90	3.33	7.080	120.59
	.188	25.16	145.75	4.442	22.86	3.33	7.399	120.27
	.203	27.20	157.22	4.436	24.66	3.33	8.001	119.67
	.219	29.28	168.79	4.431	26.47	3.33	8.611	119.06
	.230	30.75	177.00	4.427	27.76	3.33	9.046	118.63
	.250	33.38	191.48	4.420	30.03	3.33	9.817	117.85
	.281	37.45	213.82	4.409	33.54	3.33	11.017	116.65
	.312	41.51	235.83	4.398	36.99	3.33	12.210	115.46
	.330	43.77	248.01	4.392	38.90	3.33	12.876	114.80

TABLE D3-1 Typical Properties of Pipe Piles *(Continued)*

Size O.D., in	Wall thickness, in	Weight per ft, lb	Moment of inertia, in^4	Radius of gyration, in	Section modulus, in^3	Area external surface, ft^2/lin. ft	Area of steel in cross section, in^2	Inside cross-sectional area, in^2
13	.115	15.82	96.44	4.555	14.83	3.40	4.655	128.07
	.125	17.18	104.58	4.552	16.09	3.40	5.056	127.67
	.134	18.48	112.28	4.548	17.27	3.40	5.436	127.29
	.141	19.31	117.23	4.546	18.03	3.40	5.681	127.05
	.156	21.39	129.59	4.541	19.93	3.40	6.294	126.43
	.164	22.53	136.30	4.538	20.96	3.40	6.629	126.10
	.172	23.54	142.25	4.535	21.88	3.40	6.926	125.80
	.179	24.55	148.14	4.533	22.79	3.40	7.221	125.51
	.188	25.65	154.62	4.530	23.78	3.40	7.547	125.18
	.203	27.74	166.80	4.524	25.66	3.40	8.161	124.57
	.219	29.86	179.09	4.519	27.55	3.40	8.783	123.94
	.230	31.36	187.81	4.515	28.89	3.40	9.227	123.50
	.250	34.04	203.19	4.508	31.26	3.40	10.013	122.71
	.281	38.20	226.94	4.497	34.91	3.40	11.237	121.49
	.312	42.34	250.33	4.487	38.51	3.40	12.455	120.27
	.330	44.65	263.28	4.481	40.50	3.40	13.135	119.59
14	.115	17.05	120.68	4.909	17.24	3.66	5.016	148.92
	.125	18.52	130.89	4.905	18.69	3.66	5.448	148.48
	.134	19.92	140.55	4.902	20.07	3.66	5.858	148.07
	.141	20.87	146.76	4.900	20.96	3.66	6.122	147.81
	.156	23.06	162.27	4.894	23.18	3.66	6.784	147.15
	.164	24.29	170.70	4.891	24.38	3.66	7.145	146.79
	.172	25.38	178.17	4.889	25.45	3.66	7.466	146.47
	.179	26.42	185.57	4.886	26.51	3.66	7.785	146.15
	.188	27.66	193.72	4.883	27.67	3.66	8.136	145.80
	.203	29.91	209.04	4.878	29.86	3.66	8.798	145.13
	.219	32.20	224.49	4.873	32.07	3.66	9.470	144.46
	.230	33.83	235.47	4.869	33.63	3.66	9.949	143.98
	.250	36.71	254.84	4.862	36.40	3.66	10.799	143.13
	.281	41.21	284.77	4.851	40.68	3.66	12.121	141.81
	.312	45.68	314.29	4.840	44.89	3.66	13.437	140.50
	.344	50.17	343.39	4.829	49.05	3.66	14.747	139.19
15	.115	18.28	148.68	5.262	19.82	3.92	5.377	171.33
	.125	19.85	161.28	5.259	21.50	3.92	5.841	170.87
	.134	21.35	173.21	5.255	23.09	3.92	6.281	170.43
	.141	22.31	180.87	5.253	24.11	3.92	6.564	170.15
	.156	24.73	200.03	5.248	26.67	3.92	7.274	169.44
	.164	26.04	210.45	5.245	28.06	3.92	7.662	169.05
	.172	27.21	219.69	5.242	29.29	3.92	8.006	168.70
	.179	28.38	228.84	5.240	30.51	3.92	8.348	168.36
	.188	29.66	238.91	5.237	31.85	3.92	8.725	167.98
	.203	32.08	257.86	5.232	34.38	3.92	9.436	167.27
	.219	34.53	276.98	5.226	36.93	3.92	10.158	166.55
	.230	36.28	290.57	5.222	38.74	3.92	10.672	166.04
	.250	39.38	314.57	5.215	41.94	3.92	11.584	165.13
	.281	44.21	351.68	5.204	46.89	3.92	13.005	163.70
	.312	49.01	388.30	5.193	51.77	3.92	14.419	162.29
	.344	53.80	424.45	5.183	56.59	3.92	15.827	160.88

TABLE D3-1 Typical Properties of Pipe Piles *(Continued)*

Size O.D., in	Wall thickness, in	Weight per ft, lb	Moment of inertia, in⁴	Radius of gyration, in	Section modulus, in³	Area external surface, ft²/lin. ft	Area of steel in cross section, in²	Inside cross-sectional area, in²
16	.115	19.50	180.70	5.616	22.58	4.18	5.738	195.32
	.125	21.19	196.04	5.612	24.50	4.18	6.234	194.82
	.134	22.79	210.57	5.609	26.32	4.18	6.703	194.35
	.141	23.82	219.90	5.607	27.48	4.18	7.006	194.05
	.156	26.40	243.24	5.601	30.40	4.18	7.764	193.29
	.164	27.74	255.94	5.599	31.99	4.18	8.178	192.88
	.172	29.06	267.20	5.596	33.40	4.18	8.546	192.51
	.179	30.30	278.35	5.593	34.79	4.18	8.911	192.15
	.188	31.66	290.63	5.590	36.32	4.18	9.314	191.74
	.203	34.25	313.74	5.585	39.21	4.18	10.074	190.98
	.219	36.87	337.08	5.580	42.13	4.18	10.845	190.21
	.230	38.70	353.67	5.576	44.20	4.18	11.394	189.66
	.250	42.05	382.98	5.569	47.87	4.18	12.370	188.69
	.281	47.22	428.32	5.558	53.54	4.18	13.888	187.17
	.312	52.36	473.11	5.547	59.13	4.18	15.401	185.66
	.344	57.48	517.37	5.536	64.67	4.18	16.907	184.15
	.375	62.48	562.08	5.526	70.26	4.18	18.408	182.65
18	.125	23.86	280.38	6.320	31.15	4.71	7.019	247.39
	.188	35.67	416.16	6.298	46.24	4.71	10.520	243.94
	.250	47.39	549.09	6.276	61.01	4.71	13.941	240.48
	.281	53.22	614.52	6.266	68.28	4.71	15.642	238.75
	.312	59.03	679.23	6.255	75.47	4.71	17.337	237.17
	.375	70.59	806.58	6.233	89.62	4.71	20.764	233.71
20	.125	26.53	385.40	7.027	38.54	5.24	7.805	306.29
	.188	39.67	572.70	7.005	57.27	5.24	11.701	302.54
	.250	52.73	756.50	6.983	75.65	5.24	15.512	298.66
	.281	59.23	847.10	6.972	84.71	5.24	17.408	296.78
	.312	65.71	936.70	6.962	93.67	5.24	19.298	294.77
	.375	78.60	1113.50	6.940	111.35	5.24	23.120	291.02
22	.141	32.83	576.95	7.728	52.45	5.76	9.683	370.51
	.188	43.68	764.28	7.712	69.48	5.76	12.883	367.34
	.250	58.07	1010.35	7.690	91.85	5.76	17.082	363.02
	.281	65.24	1131.79	7.680	102.89	5.76	19.173	361.01
	.312	72.38	1252.24	7.669	113.84	5.76	21.258	358.85
	.375	86.61	1489.84	7.647	135.44	5.76	25.476	354.67
24	.156	39.72	830.52	8.430	69.21	6.28	11.686	440.64
	.188	47.68	994.32	8.419	82.86	6.28	14.064	438.34
	.250	63.41	1315.44	8.397	109.62	6.28	18.653	433.73
	.281	71.25	1474.20	8.386	122.85	6.28	20.939	431.42
	.312	79.06	1631.52	8.376	135.96	6.28	23.218	429.12
	.375	94.62	1942.44	8.354	161.87	6.28	27.833	424.51

TABLE D3-2 Monotube Specifications

Tapered sections

Type	Size tip point dia. × butt dia. × lgth.	Weight per ft 9 ga. 0.1495 in	7 ga. 0.1793 in	5 ga. 0.2092 in	3 ga. 0.2391 in	Est. conc. vol. yd³
F taper 0.14 in per ft	8½ in × 12 in × 25 ft 8 in × 12 in × 30 ft 8½ in × 14 in × 40 ft 8 in × 16 in × 60 ft 8 in × 18 in × 75 ft	17 16 19 20 —	20 20 22 24 26	24 23 26 28 31	28 27 31 33 35	0.43 0.55 0.95 1.68 2.59
J taper 0.25 in per ft	8 in × 12 in × 17 ft 8 in × 14 in × 25 ft 8 in × 16 in × 33 ft 8 in × 18 in × 40 ft	17 18 20 —	20 22 24 26	23 26 28 30	27 30 32 35	0.32 0.58 0.95 1.37
Y taper 0.40 in per ft	8 in × 12 in × 10 ft 8 in × 14 in × 15 ft 8 in × 16 in × 20 ft 8 in × 18 in × 25 ft	17 19 20 —	20 22 24 26	24 26 28 31	28 30 33 35	0.18 0.34 0.56 0.86

Extension sections

Type	Dia. × length	9 ga. 0.1495 in	7 ga. 0.1793 in	5 ga. 0.2092 in	3 ga. 0.2391 in	Yd³/Ft
N 12	12 in × 12 in × 20 ft/40 ft	20	24	28	33	0.026
N 14	14 in × 14 in × 20 ft/40 ft	24	29	33	38	0.035
N 16	16 in × 16 in × 20 ft/40 ft	27	32	38	44	0.045
N 18	18 in × 18 in × 20 ft/40 ft	—	37	43	49	0.058

Physical Properties

Steel thickness	Tips 8 in A in²	8½ in A in²	12 in A in²	I in⁴	S in³	r in	14 in A in²	I in⁴	S in³	r in	16 in A in²	I in⁴	S in³	r in	18 in A in²	I in⁴	S in³	r in
9 gauge (0.1495 in)	3.63	3.93	5.81	102	16.3	4.18	6.75	159	22.0	4.86	7.64	232	28.3	5.50	—	—	—	—
7 gauge (0.1793 in)	4.40	4.77	6.97	122	19.5	4.18	8.14	194	26.7	4.89	9.18	278	33.9	5.51	10.4	404	43.5	6.23
5 gauge (0.2092 in)	5.19	5.61	8.18	145	23.0	4.21	9.50	227	31.0	4.88	10.8	329	39.9	5.53	12.2	478	51.2	6.26
3 gauge (0.2391 in)	5.87	6.58	8.96	148	24.2	4.07	10.6	239	33.6	4.77	12.0	348	43.1	5.40	13.6	504	55.4	6.10
Concrete area, in²	42.3	47.3		101				136				176				224		

Notes:
1. Choice of tapered section is usually determined by estimated pile length.
2. Longer lengths are obtained by splicing extension sections N into tapered sections F, J, or Y.
3. Extension sections are essentially constant in diameter, and are furnished 1 ft longer than nominal length to provide for telescopic splice.
4. All tapered sections include forged steel conical nose factory-attached.

Source: Reference 4.

TABLE D3-3 Typical Properties of HP Sections

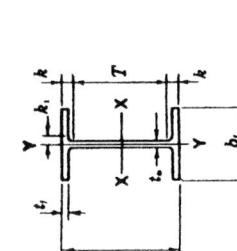

Designation	Area A in²	Depth d in	Web Thickness t_w in	$t_w/2$ in	Flange Width b_f in	Flange Thickness t_f in	Distance T in	Distance k in	Distance k_1 in
HP 14 × 117	34.4	14.21	0.805 (13/16)	7/16	14.885 (14 7/8)	0.805 (13/16)	11 1/4	1 1/2	1 1/16
× 102	30.0	14.01	0.705 (11/16)	3/8	14.785 (14 3/4)	0.705 (11/16)	11 1/4	1 3/8	1
× 89	26.1	13.83	0.615 (5/8)	5/16	14.695 (14 3/4)	0.615 (5/8)	11 1/4	1 5/16	15/16
× 73	21.4	13.61	0.505 (1/2)	1/4	14.585 (14 5/8)	0.505 (1/2)	11 1/4	1 3/16	7/8
HP 13 × 100	29.4	13.15	0.765 (3/4)	3/8	13.205 (13 1/4)	0.765 (3/4)	10 1/4	1 7/16	1
× 87	25.5	12.95	0.665 (11/16)	3/8	13.105 (13 1/8)	0.665 (11/16)	10 1/4	1 3/8	15/16
× 73	21.6	12.75	0.565 (9/16)	5/16	13.005 (13)	0.565 (9/16)	10 1/4	1 1/4	15/16
× 60	17.5	12.54	0.460 (7/16)	1/4	12.900 (12 7/8)	0.460 (7/16)	10 1/4	1 1/8	7/8
HP 12 × 84	24.6	12.28	0.685 (11/16)	3/8	12.295 (12 1/4)	0.685 (11/16)	9 1/2	1 3/8	1
× 74	21.8	12.13	0.605 (5/8)	5/16	12.215 (12 1/4)	0.610 (5/8)	9 1/2	1 5/16	15/16
× 63	18.4	11.94	0.515 (1/2)	1/4	12.125 (12 1/8)	0.515 (1/2)	9 1/2	1 1/4	7/8
× 53	15.5	11.78	0.435 (7/16)	1/4	12.045 (12)	0.435 (7/16)	9 1/2	1 1/8	7/8
HP 10 × 57	16.8	9.99	0.565 (9/16)	5/16	10.225 (10 1/4)	0.565 (9/16)	7 7/8	1 7/16	15/16
× 42	12.4	9.70	0.415 (7/16)	1/4	10.075 (10 1/8)	0.420 (7/16)	7 7/8	1 1/16	3/4
HP 8 × 36	10.6	8.02	0.445 (7/16)	1/4	8.155 (8 1/8)	0.445 (7/16)	6 3/8	15/16	7/8

Source: Courtesy of the American Institute of Steel Construction.

D3-17

TABLE D3-4 Typical Sizes of Step-Tapered Shells

NOMINAL DIMENSIONS

DESIGNATION #	SIZE
8	18 3/8"
7	17 3/8"
6	16 3/8"
5	15 3/8"
4	14 3/8"
3	13 3/8"
2	12 3/8"
1	11 3/8"
0	10 3/8"
00	9 1/2"
000	8 5/8"

Shell lengths vary from 8 ft. to 16 ft

Source: Reference 5.

for Prestressed Concrete Piles. Quality is normally specified to be in accordance with MNL-116, The Manual for Quality Control for Plants and Production of Precast Prestressed Concrete Products.

Prestressed concrete piles are constructed using steel rods, strands, or wires under tension. The pretensioning steel is tensioned either before (pretensioned) or after (posttensioned) the concrete is cast.

The concrete is put into compression by the tension steel, which increases the ability of the concrete to withstand stresses due to handling and driving. Tensile forces that develop during installation are reduced by the prestressing, but compressive stresses are increased.

The most common sizes of prestressed concrete piles range from 10 to 24 in and may be either square or octagonal. Prestressed hollow cylinder piles have been produced by both centrifugal and bed casting in diameters up to 66 in. See Table D3-6 for typical precast concrete pile properties and service loads.

TABLE D3-5 Typical Sizes of Straight-Sided Shell Piles

Size OD & gage in	Inside area X-section, in²	Concrete volume, yd³/ft	$f_c = 1250$ lb/in²	$f_c = 1320$ lb/in²	$f_c = 1650$ lb/in²	$f_c = 2000$ lb/in²	$f_c = 2400$ lb/in²
*8⅝ × 18 16	54.69	0.0141	34.2	36.1	45.1	54.7	65.6
*10⅝ × 18 16 14	84.96	0.0218	53.1	56.1	70.1	85.0	102.0
11⅞ × 18 16 14 12	86.36	0.0222	54.0	57.0	71.3	86.4	103.7
12¾ × 18 16 14 12	106.74	0.0274	66.7	70.4	88.0	106.7	128.0
14 × 18 16 14 12	141.64	0.0364	88.5	93.5	116.8	141.6	170.0
16⅛ × 18 16 14 12	188.90	0.0486	118.1	125.6	155.8	188.9	226.7

Maximum column design loads, tons

Notes: Inside cross-sectional area is based on 16-gauge steel. Outside diameter is constant, regardless of gauge. Corrugations on 8⅝- and 10⅝-in OD shell are ¼-in deep and 1⅛-in pitch. Corrugations on 11⅞-in OD and larger shell are ½-in deep and 2-in pitch.
* These diameters are available on special order only. Larger diameters than those shown are available. f_c is 40 percent of the 28-day strength of concrete.
Source: Reference 6.

TABLE D3-6 Properties/Service Loads for Prestressed Concrete Piles

Size, in	Core diameter, in	Area, in	Weight, plf	Moment of inertia, in	Section modulus, in	Radius of gyration, in	Perimeter, ft	Allowable concentric service load, tons f'_c			
								5,000	6,000	7,000	8,000
				Square piles							
10	Solid	100	104	833	167	2.89	3.33	73	89	106	122
12	Solid	144	150	1,728	288	3.46	4.00	105	129	152	176
14	Solid	196	204	3,201	457	4.04	4.67	143	175	208	240
16	Solid	256	267	5,461	683	4.62	5.33	187	229	271	314
18	Solid	324	338	8,748	972	5.20	6.00	236	290	344	397
20	Solid	400	417	13,333	1,333	5.77	6.67	292	358	424	490
20	11 in.	305	318	12,615	1,262	6.43	6.67	222	273	323	373
24	Solid	576	600	27,648	2,304	6.93	8.00	420	515	610	705
24	12 in.	463	482	26,630	2,219	7.58	8.00	338	414	491	567
24	14 in.	422	439	25,762	2,147	7.81	8.00	308	377	447	517
24	15 in.	399	415	25,163	2,097	7.94	8.00	291	357	423	488
30	18 in.	646	672	62,347	4,157	9.82	10.00	471	578	685	791
36	18 in.	1,042	1,085	134,815	7,490	11.38	12.00	761	933	1,105	1,276

					Octagonal piles						
10	Solid	83	85	555	111	2.59	2.76	60	74	88	101
12	Solid	119	125	1,134	189	3.09	3.31	86	106	126	145
14	Solid	162	169	2,105	301	3.60	3.87	118	145	172	198
16	Solid	212	220	3,592	449	4.12	4.42	154	189	224	259
18	Solid	268	280	5,705	639	4.61	4.97	195	240	284	328
20	Solid	331	345	8,770	877	5.15	5.52	241	296	351	405
20	11 in.	236	245	8,050	805	5.84	5.52	172	211	250	289
22	Solid	401	420	12,837	1,167	5.66	6.08	292	359	425	491
22	13 in.	268	280	11,440	1,040	6.53	6.08	195	240	283	328
24	Solid	477	495	18,180	1,515	6.17	6.63	348	427	506	584
24	15 in.	300	315	15,696	1,308	7.23	6.63	219	268	318	368
					Round piles						
36		487	507	60,007	3,334	11.10	9.43	355	436	516	596
42		581	605	101,273	4,823	13.20	11.00	424	520	616	712
48		675	703	158,222	6,592	15.31	12.57	493	604	715	827
54		770	802	233,373	8,643	17.41	14.14	562	689	816	943
66		1,131	1,178	514,027	15,577	21.32	17.28	826	1,013	1,199	1,386

Source: Reference 7

There is no theoretical limitation on the manufactured lengths of precast concrete piling. Typically, the maximum length of individual piles will be dictated by the mode of transport, maximum manageable weight, and handling stresses. If single piles are planned for the project, additional length should be provided to account for the probability that some piles will drive deeper than planned.

For centrifugally cast hollow-core piles, there is a tendency for the coarse aggregate to migrate toward the outer surface during spinning, resulting in a concrete of nonuniform texture. For bedcast hollow-core piles, it is difficult to hold the void form in place, and inspection is necessary to check for a uniform wall thickness.

All corners and edges of square piles should be chamfered. The width of the chamfer face should be limited to about 1.5 in so that the reduction in the side dimension due to chamfering is not more than about 2 in.

If the use of sectional piles is permitted, a splice joint is normally specified. This will require that special end fittings be cast into the pile. Special end protection such as a steel plate or a 4-ft H section may also be cast into the pile.

Cast-in-Place Concrete. Typically, the specification will require that concrete be supplied in accordance with the American Concrete Institute (ACI) 301, Specification for Structural Concrete for Buildings. The specified minimum 28-day compressive strength normally ranges between 3000 and 7000 lb/in^2.

Cement is specified in accordance with ASTM C 150, Specification for Portland Cement, and the aggregates are normally specified to be in accordance with ASTM C33, Specification for Concrete Aggregates. Type III or high-early-strength cement may be specified for test piles. Types II or V may be specified for areas where sulfate attack is a concern. Either gravel or crushed rock may be permitted. Crushed rock requires more cement and sand to achieve the same workability as gravel. Lightweight aggregates are not normally permitted, and alkali-reactive aggregates or friable rocks are not recommended. Aggregates should be well graded, with a maximum size of ¾ in. Water should be drinking-water quality.

The contractor normally must submit to the engineer for approval the proposed source of materials and the concrete mix proposed to be used. The submittal must include sufficient strength test data from an accredited test laboratory to establish the mix strength. Consistency of the concrete is measured in accordance with ASTM C 143, Method of Slump Test for Portland Cement Concrete, and should be 6 in (±1 in). Higher slumps may be required if concrete is placed after setting rebar cages.

Grout

Grout for augercast piles consists of a mixture of portland cement, fluidifier, sand, and water so proportioned and mixed as to provide a mortar capable of maintaining the solids in suspension without difficulty, and which will laterally penetrate and fill any voids in the foundation soil. The materials should be so proportioned as to provide a hardened mortar having a minimum ultimate compressive strength of 4000 lb/in^2. Consistency limits as determined by ASTM C939, Flow Cone Test, using a ¾-in opening in the flow cone, are normally established by the contractor in the mix design.

PILE-DRIVING EQUIPMENT

The several essential pieces of pile-driving equipment include the hammer, hammer cushion, drive head, pile cushion, the leaders, the spotter, and, sometimes, the mandrel. An illustration of a typical pile driver is shown in Fig. D3-3. The following is a discussion of this equipment.

Hammers

The installation equipment common to all driven piles is the pile-driving hammer. The hammer imparts dynamic energy to the pile by the impact of a falling ram. Most of the energy is kinetic energy due to gravity acting on a falling ram. This can be increased in some hammer types by pressure on the top of the ram which increases the rate of fall.

All hammers have frictional energy losses within the system, so that the actual energy transmitted to the pile may vary from 35 to 90 percent of its rated energy. The amount of the energy loss depends on the hammer type, condition, and operation. When driving piles on a batter, the energy losses increase because of the reduction in the force of gravity as well as increased friction. Double-acting or differential-acting hammers have less energy loss than single-acting hammers.

Hammers can be categorized into types, which include the drop hammer, the single-acting hammer, the double-acting hammer, the differential-acting hammer, the diesel hammer, and hydraulic hammers. A special hammer type is the vibratory driver.

A summary of pile hammer specifications is given in Table D3-7.

The Drop Hammer. The drop hammer is the most fundamental type of pile-driving hammer. It is nothing more than a heavy metal weight which is repeatedly raised by a line and allowed to fall by gravity. The hammer is controlled by guides so that the pile is struck axially and concentrically.

The actual hammer energy transmitted may be reduced by the friction of the line, which must follow the hammer in its fall. Also, if the hammer must follow the pile, there may be friction in the guides. The only pile type which consistently uses a drop hammer as a part of the installation procedure is the PIF.

Single-Acting Hammer. A single-acting hammer is activated by steam, compressed air, or hydraulic fluid which lifts the ram until a port opens and allows the pressure to be released. The ram continues upward a short distance until gravity overtakes its ascent, and then the ram reverses direction and falls under gravity. This type of hammer is characterized by a relatively short stroke, i.e., about 3 ft, is relatively slow at about 60 blows per min, and imparts its energy with a heavy ram and a low-impact velocity. This type of hammer is well suited for certain pile types and soil conditions where a lighter ram and high impact velocity are not efficient. The operating cycle for a single-acting hammer is shown in Fig. D3-4.

Double-Acting Hammer. The double-acting hammer uses either steam or air pressure both to raise and drop the ram. The energy transmitted is the sum of both the kinetic energy and the pressure acting on the top of the ram. The delivered energy is very sensitive to the air, steam, or fluid pressure, and it must be monitored carefully to understand the delivered energy. The double-acting hammer typically has a short stroke and a lightweight ram, and operates at a relatively high speed compared to single-acting hammers. As the weight of the ram increases, the speed of the hammer is reduced.

Differential-Acting Hammer. The differential hammer is powered by air, steam, or hydraulic fluid pressure on both the rise and fall of the ram. The difference between the double-acting hammer and the differential-acting hammer is that the

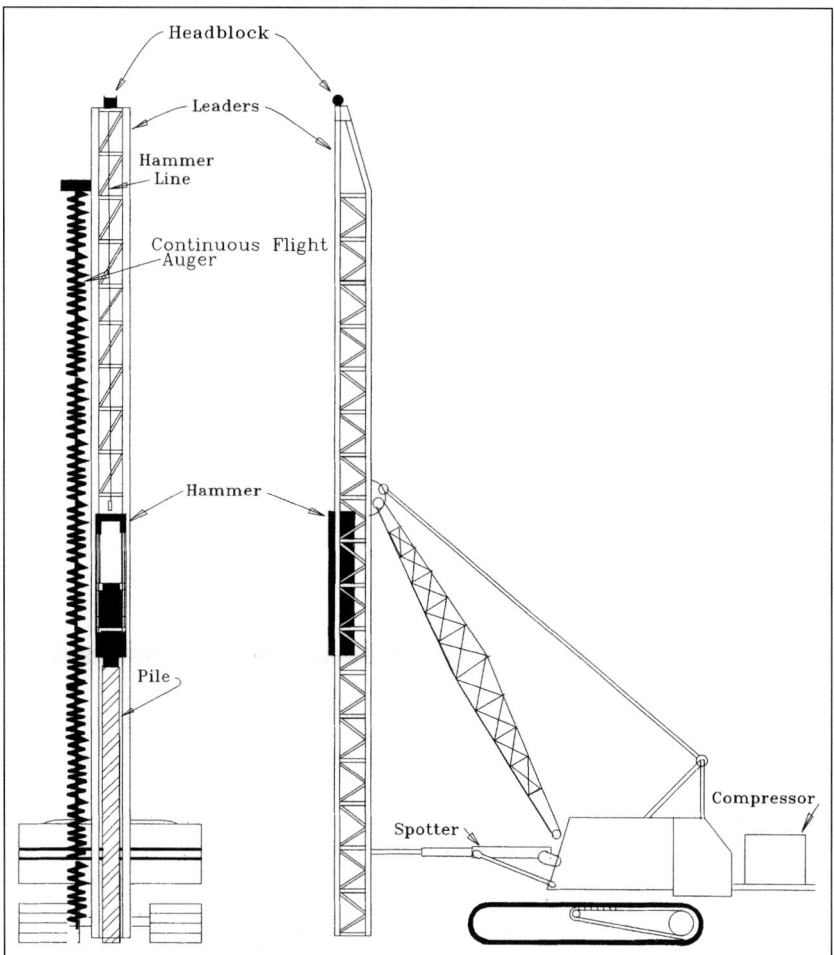

FIGURE D3-3 Typical pile-driving rig.

latter has two different size cylinders. The smaller cylinder under pressure lifts the ram, and the larger cylinder under pressure accelerates the ram as it falls. The difference in the two cylinder areas results in a differential pressure, hence the name. Because of the extremely high operating pressure of this hammer (5000 lb/in^2) the hammer is able to raise relatively large rams while not sacrificing running speed.

The maximum kinetic energy which can be delivered is the product of the hammer weight times the stroke of the ram. Neither the double-acting hammer nor the differential-acting hammer permits the observer to measure the stroke of the ram, so monitoring the operating pressure is necessary in order to understand the hammer energy.

Diesel Hammers. The diesel hammer is an internal combustion hammer which utilizes the explosive force of the diesel fuel in the chamber below the bottom of the ram and the anvil block to raise the ram for the next stroke. As the ram nears the bottom of its fall, it closes off the exhaust port and diesel fuel is injected into the combustion chamber. The fuel is compressed by the falling ram, and, on impact of the ram on the anvil, the hot compressed mixture of fuel and air is ignited.

Diesel hammers are either single-acting (open top) or double-acting (closed top). The single-acting diesel hammer permits the ram to rise through the open top of the cylinder to whatever height results from the combination of the rebound and the explosive force at impact. The rated energy of the single-acting diesel hammer is the product of the weight of the ram and the stroke. The stroke varies with the pile resistance. At low resistance, the pile movement dampens the explosive energy, and the stroke is much smaller than it will be when the pile resistance is high (hard driving). The explosive force is only intended to raise the ram, and at final drive is a negligible portion of the driving energy. Driving in very soft ground may not provide enough resistance to fire the fuel, requiring that the hammer be manually started for each stroke until the resistance is sufficient for the hammer to continue to fire. The stroke can be monitored by observing marks on the hammer piston or by a stroke indicator.

The double-acting diesel hammer compresses air in the top of the cylinder on the upstroke, which provides an acceleration to the ram on the downstroke. The stroke of the double-acting hammer cannot be monitored, so the pressure in the chamber above the ram must be monitored using a bounce chamber pressure gauge, and the pressure is converted to an equivalent stroke.

The energy of the diesel hammer can be affected by abnormal preignition of the fuel. While some hammers are designed to have preignition, for those that are not, it may reduce or, in extreme cases, prevent impact of the ram. This can be caused by an overheated hammer or by using fuel with too low a flashpoint.

A schematic diagram of a double-acting diesel hammer is shown in Fig. D3-5.

Vibratory Driver. Vibratory drivers cause penetration of the pile into the ground by applying an oscillatory force in both compression and tension, to break down the pile-soil friction. This allows the weight of the pile and the hammer to push the pile into the ground. The drivers operate at either low (20 Hz) or high frequency (40 Hz). The dynamic force is produced by pairs of eccentric weights mounted on a rotating shaft, which are positioned to cancel out horizontal forces but amplify vertical forces. Vibratory drivers are powered either by electrical motors, hydraulic forces, or gasoline or diesel engines. Some types of vibratory drivers are modular, and may be operated in tandem to provide additional vibratory energy. The vibratory driver must clamp the pile firmly in order to not dampen the dynamic force transmitted to the pile. A schematic diagram of a vibratory hammer is shown in Fig. D3-6.

Equipment Placed Between Hammer and Pile

The hammer cushion, the drive head, the pile cushion, and the follower are between the hammer and the pile. A discussion of their purpose follows.

Hammer Cushion. The hammer cushion (capblock) is used to cushion the peak force of the ram, and protect both the hammer and the pile from direct impact. Typically, the cushion is made up of alternating discs of aluminum and micarta or some other hard, durable material. These commercially manufactured cushions have fairly uniform elastic and heat transfer properties which hold up relatively well during driving. They have good energy transmission characteristics and a relatively high

TABLE D3-7 Specifications for Pile Hammers

			Diesel hammers							
Energy range, ft-lb	Model	Manufacturer	Single/ Double- acting	Blows per min	Piston weight, lb	Total weight, lb	Total length, ft-in	Maximum stroke, ft-in	Width between jaws, in	Fuel used, gal/h
750,000–384,000	D350	Delmag	Single	36–50	77,161	165,345	34'10"	9'8"	Cage	22.46
300,000–157,740	D100-13	Delmag	Single	34–45	23,612	45,357	20'4"	12'8"	36	7.93
225,000–126,190	D80-23	Delmag	Single	36–45	19,500	37,739	20'4"	11'6"	36	6.60
165,000–78,960	D62-22	Delmag	Single	36–50	14,600	27,055	19'6"	11'4"	32	5.28
149,600–88,000	MH80B	Mitsubishi	Single	42–60	17,600	43,600	19'6"	14'10"	42	8–12
141,000–63,360	MB70	Mitsubishi	Single	38–60	15,840	46,000	19'6"	8'11"	42	7–10
135,100–79,500	MH72B	Mitsubishi	Single	38–60	15,900	44,000	19'6"	14'10"	42	7–10
127,500–90,000	DE150/110	Mkt	Single	40–50	15,000	29,500	19'10"	10'9"	32	7
117,000–62,500	D55	Delmag	Single	36–47	12,100	26,300	17'9"	9'8"	32	5.50
107,170–52,260	D46-32	Delmag	Single	37–53	10,143	19,602	17'4"	10'7"	30	4.23
105,800–36,800	B-5505	Berminghammer	Single	36–60	9,200	24,000	22'0"	11'6"	32	2.2
100,000–40,000	200S	Ice	Single	53–70	20,000	33,600	17'0"	5'0"	32	4.00
93,500–66,000	DE-150/110	Mkt	Single	40–50	11,000	24,550	17'10"	10'9"	32	5.7
91,100	K45	Kobe	Single	39–60	9,900	25,600	18'6"	9'2"	36	4.5–5.5
90,000–36,000	90S	Ice	Single	38–55	9,000	16,800	17'3"	10'0"	29	4.0
87,400–30,400	B-5005	Berminghammer	Single	36–60	7,600	22,400	22'0"	11'6"	32	2.2
87,000–43,500	D44	Delmag	Single	37–56	9,500	22,300	15'10"	9'2"	32	5.50
85,400–50,200	MH45	Mitsubishi	Single	42–60	10,500	24,600	17'11"	11'5"	36	4–6
84,000–37,840	M43	Mitsubishi	Single	40–60	9,460	22,660	16'3"	8'10"	37	4–6
83,880–40,900	D36-32	Delmag	Single	36–53	7,938	17,397	17'4"	10'7"	30	3.04
80,000–32,000	80S	Ice	Single	38–55	8,000	15,400	18'11"	10'	26	2.9
79,500	J44	Ihi	Single	42–70	9,720	21,500	14'10"	8'2"	37	6.86
79,000	K42	Kobe	Single	40–60	9,260	24,000	17'8"	8'6"	36	4.5–5.5
75,900–26,500	B-4505	Berminghammer	Single	36–60	6,600	16,000	18'7"	11'6"	24	1.5
73,000–40,150	3400	F.E.C.	Single	40–60	7,500	14,600	16'0"	—	26	3.3–5.0
70,800	K35	Kobe	Single	39–60	7,700	18,700	17'8"	9'2"	30	3.0–4.0
70,000–28,000	70S	Ice	Single	38–55	7,000	14,100	16'8"	10'0"	26	2.10
70,000–36,100	1070	Ice	Double	64–68	10,000	21,500	17'10"	7'	30	3.5
70,000–42,000	DE70/50C	Mkt	Single	40–50	7,000	14,700	17'11"	11'6"	26	3.3
69,900–35,380	D30-32	Delmag	Single	36–52	6,615	13,252	17'3"	10'7"	26	2.64
65,600–38,600	MH35	Mitsubishi	Single	42–60	7,720	18,500	17'3"	11'	32	3.4–5
64,600–29,040	M33	Mitsubishi	Single	40–60	7,260	16,940	13'2"	8'	32	3.4–5
63,500	J35	Ihi	Single	72–70	7,730	16,900	14'6"	8'3"	32	4.76
63,000–34,650	3000	F.E.C.	Single	40–60	6,600	13,200	15'6"	10'6"	26	2.8–4.2
60,100	K32	Kobe	Single	40–60	7,050	17,750	17'8"	8'6"	30	2.75–3.5
58,250–29,480	D25-32	Delmag	Single	37–52	5,513	12,149	17'3"	10'7"	26	2.11
58,000–26,000	60S	Ice	Single	42–58	7,000	13,750	16'8"	8'4"	26	2.1
57,876	V25-1	Vulcan	Single	—	5,512	12,125	16'4"	10'6"	—	2.4
57,500–20,000	B-4005	Berminghammer	Single	36–60	5,000	14,400	18'7"	11'6"	24	1.5
54,250–23,800	D30	Delmag	Single	39–60	6,600	12,300	14'3"	8'2"	26	2.90
53,750–20,100	B-400	Berminghammer	Single	37–60	5,000	15,000	14'10"	10'9"	—	1.5
50,700	K25	Kobe	Single	39–60	5,510	13,100	17'6"	9'3"	26	2.5–3.0
50,000–30,000	DE70/50C	Mkt	Single	40–50	5,000	12,700	17'11"	11'6"	26	3.3
50,000–27,500	2500	F.E.C.	Single	40–60	5,500	12,100	15'6"	10'6"	20	2.8–4.2
50,000–25,100	660	Ice	Double	84–88	7,564	24,480	17'4"	6'7"	30	3.25
48,500–24,500	D22-23	Delmag	Single	38–52	4,850	11,400	17'2"	10'	26	1.60
46,900–27,550	MH25	Mitsubishi	Single	42–60	5,510	13,200	16'8"	10'9"	26	2.4–4
46,000–16,000	B-3505	Berminghammer	Single	36–60	4,000	12,000	18'1"	11'6"	26	2.1

TABLE D3-7 Specifications for Pile Hammers *(Continued)*

Diesel hammers

Energy range, ft-lb	Model	Manufacturer	Single/ Double-acting	Blows per min	Piston weight, lb	Total weight, lb	Total length, ft-in	Maximum stroke, ft-in	Width between jaws, in	Fuel used, gal/h
45,000–20,240	M23	Mitsubishi	Single	42–60	5,060	11,220	14′1″	8′10″	26	2.4–3.7
44,800	DE50C	Bsp	Single	42–54	4,980	10,300	14′4″	9′	26	2.68
42,800–20,540	D19-32	Delmag	Single	37–53	4,190	7,800	15′6″	10′3″	20	1.45
42,500–30,000	DA-55C	Mkt	Single	40–50	5,000	17,000	17′4″	10′6″	26	2.7
42,000–16,000	42S	Ice	Single	37–55	4,088	7,610	16′1″	10′3″	20	1.3
41,300	K22	Kobe	Single	40–60	4,850	12,350	17′6″	9′2″	26	2.0–2.75
40,300–15,000	B-300	Berminghammer	Single	37–60	3,750	9,520	14′0″	10′10″	—	1.0
40,200–18,870	D16-32	Delmag	Single	36–52	3,528	7,386	15′6″	11′5″	20	1.45
40,000–25,400	640	Ice	Double	74–77	6,000	14,460	15′7″	6′8″	26	3.0
40,000–24,000	DE-33/30/20C	Mkt	Single	40–50	4,000	9,400	15′11″	10′6″	26	3.0
40,000–16,000	40S	Ice	Single	38–55	4,000	7,500	15′9″	10′0″	20	1.20
39,700	D22	Delmag	Single	42–60	4,850	11,200	14′2″	8′2″	26	2.90
39,100–12,000	J22	Ihi	Single	42–70	4,850	10,800	14′0″	10′0″	26	3.2
38,200–31,200	DA-55C	Mkt	Double	78–82	5,000	17,000	17′4″	—	26	3.0
34,500	B-2505	Berminghammer	Single	36–60	3,000	11,000	18′1″	11′6″	26	2.1
34,000–24,000	DA-45C	Mkt	Single	40–50	4,000	14,200	15′1″	10′6″	26	2.5
33,000–19,800	DE-33/30/20C	Mkt	Single	40–50	3,300	8,700	15′11″	10′6″	20	2.0
31,320–15,660	D12-32	Delmag	Single	36–52	2,820	6,260	15′6″	11′2″	20	0.95
30,700–18,500	DA-45C	Mkt	Double	78–82	4,000	14,200	15′1″	—	26	2.8
30,000–17,000	520-30	Ice	Double	80–84	5,070	13,400	13′7″	5′11″	26	1.35
29,250–12,000	B-225	Berminghammer	Single	39–60	3,000	8,730	14′0″	9′10″	—	1.0
28,100–16,550	MH15	Mitsubishi	Single	42–60	3,310	8,400	16′1″	10′3″	26	1.3–2
28,000–16,800	DE33/30/20C	Mkt	Single	40–50	2,800	8,200	15′11″	10′6″	20	2.0
27,100–14,900	1500	F.E.C.	Single	40–60	3,300	7,225	14′2″	10′11¾″	20	1.5–2.3
27,100	D15	Delmag	Single	42–60	3,300	6,615	13′11″	8′2″	20	1.75
27,000	DE30C	Bsp	Single	42–54	3,000	7,600	14′2″	9′	26	1.7
26,300–17,700	520-26	Ice	Double	80–84	5,070	12,545	13′6″	5′3″	26	1.35
26,000–11,800	M14S	Mitsubishi	Single	42–60	2,970	7,260	13′6″	8′9″	26	1.3–2.2
24,400	K13	Kobe	Single	40–60	2,860	7,300	16′8″	8′6″	26	0.75–2.0
23,800–16,800	DA-35C	Mkt	Single	40–50	2,800	10,800	17′	10′6″	20	1.7
23,000	B-23	Berminghammer	Double	80	2,800	9,940	16′0″	4′6″	—	1.9
22,500–12,375	1200	F.E.C.	Single	40–60	2,750	6,540	14′0″	10′9¼″	20	1.5–2.3
22,500	D12	Delmag	Single	42–60	2,750	6,050	13′11″	8′2″	20	1.75
22,500–9,000	422	Ice	Double	76–82	4,000	9,750	13′11″	5′8″	22	1.0
22,500–9,000	30S	Ice	Single	44–67	3,000	6,250	12′4″	7′6″	20	0.80
21,000–15,600	DA-35C	Mkt	Double	78–82	2,800	10,800	17′	—	20	2.7
20,000–12,000	DE33/30/20C	Mkt	Single	40–50	2,000	7,400	15′11″	10′6″	20	2.0
18,100–7,700	440	Ice	Double	88–92	4,000	9,840	13′6″	4′8″	20	1.16
18,000–9,435	D8-22	Delmag	Single	38–52	1,762	4,220	15′5″	10′2″	20	1.00
18,000–7,500	312	Ice	Double	100–105	3,857	10,375	10′9″	4′8″	26	1.1
18,000–8,600	B-200	Berminghammer	Single	39–58	2,000	6,940	13′9″	10′2″	—	1.9
10,500–6,300	D6-32	Delmag	Single	39–52	1,322	3,810	12′6″	7′11″	20	0.70
9,350–6,600	DA-15C	Mkt	Single	40–50	1,100	4,825	13′11″	10′6″	20	1.0
9,100	D5	Delmag	Single	42–60	1,100	2,730	12′6″	8′4″	20	1.32
8,800	DE10	Mkt	Single	40–50	1,100	3,100	12′2″	9′	20	0.90
8,200–6,600	DA-15C	Mkt	Double	86–92	1,100	4,825	13′11″	—	20	1.8
8,100–4,060	180	Ice	Double	90–95	1,725	4,645	11′3″	4′9″	20	0.65
3,630–1,625	D4	Delmag	Single	50–60	836	1,360	7′9″	4′4″	—	0.21
1,815–868	D2	Delmag	Single	60–70	484	792	6′9″	3′8″	—	0.13

TABLE D3-7 Specifications for Pile Hammers *(Continued)*

Air/steam hammers											
Rated energy, ft-lb	Model	Manufacturer	Type	Style	Blows per min	Wt. of striking parts, lbs	Total weight, lbs	Hammer length, ft-in	Jaw dimensions	Inlet pressure, lb/in²	Inlet size, NPT
1,800,000	6300	Vulcan	Sgl.-Act	Open	42	300,000	575,000	30′0″	22″ × 144″ (M)	235	2@6″
1,582,220	MRBS 12500	Menck	Sgl.-Act	Open	36	275,580	540,130	35′9″	Cage	171	2@6″
1,200,000	6200	Vulcan	Sgl.-Act	Open	36	200,000	438,218	30′½″	22″ × 120″ (M)	110	4@4″
1,200,000	2000E6	Conmaco	Sgl.-Act	Open	40	200,000	490,000	35′6″	Cage	155	2@6″
1,050,000	1750E6	Conmaco	Sgl.-Act	Open	40	175,000	465,000	35′6″	14½″ × 120″	135	2@6″
900,000	6150	Vulcan	Sgl.-Act	Open	41	150,000	275,000	35′3″	18¾″ × 88″ (M)	175	4@4″
867,960	MRBS 8000	Menck	Sgl.-Act	Open	38	176,370	330,690	30′10″	Cage	171	8″
750,000	1500E5	Conmaco	Sgl.-Act	Open	42	150,000	283,000	30′6″	14½″ × 120″	135	2@6″
750,000	5150	Vulcan	Sgl.-Act	Open	46	150,000	275,000	26′3½″	22″ × 120″ (M)	175	2@6″
510,000	850E6	Conmaco	Sgl.-Act	Open	40	85,000	173,600	25′2″	18¾″ × 100″	160	2@4″
500,000	5100	Vulcan	Sgl.-Act	Open	48	100,000	197,000	27′4″	22″ × 120″ (M)	150	2@5″
499,070	MRBS 4600	Menck	Sgl.-Act	Open	42	101,410	176,370	27′5″	Cage	142	6″
350,000	700E5	Conmaco	Sgl.-Act	Open	43	70,000	152,000	23′2″	18¾″ × 100	130	2@4″
325,480	MRBS 3000	Menck	Sgl.-Act	Open	42	66,135	108,025	25′0″	Cage	142	5″
300,000	3100	Vulcan	Sgl.-Act	Open	60	100,000	195,500	23′3″	18¾″ × 88″ (M)	130	3@4″
300,000	560	Vulcan	Sgl.-Act	Open	47	62,500	134,060	23′0″	18¾″ × 88″ (M)	150	2@5″
225,000	450E5	Conmaco	Sgl.-Act	Open	45	45,000	103,000	23′3″	14″ × 80″	130	2@4″
200,000	540	Vulcan	Sgl.-Act	Open	48	40,900	102,980	22′7″	14″ × 80″ (M)	130	2@5″
189,850	MRBS 1800	Menck	Sgl.-Act	Open	44	38,580	64,590	22′5″	Cage	142	4″
180,000	360	Vulcan	Sgl.-Act	Open	62	60,000	124,830	19′0″	18¾″ × 88″ (M)	130	2@4″
175,000	535	Vulcan	Sgl.-Act	Open	37	35,000	61,000	22′4″	10¼″ × 54″ (M)	150	3″
150,000	300E5	Conmaco	Sgl.-Act	Open	40	30,000	58,400	20′10″	11¼″ × 56″	135	4″
150,000	530	Vulcan	Sgl.-Act	Open	42	30,000	57,680	20′5″	10¼″ × 54″ (M)	125	3″
150,000	60X	Raymond	Sgl.-Act	Open	60	60,000	85,000	22′7″	10½″ × 43½″	165	3″
120,000	340	Vulcan	Sgl.-Act	Open	60	40,000	98,180	18′7″	14″ × 80″ (M)	120	2@3″
100,000	520	Vulcan	Sgl.-Act	Open	42	20,000	47,680	20′5″	11¼″ × 37″	102	3″
100,000	200E5	Conmaco	Sgl.-Act	Open	46	20,000	48,000	19′1″	11¼″ × 56″	110	4″
100,000	40X	Raymond	Sgl.-Act	Open	64	40,000	62,000	19′1″	10½″ × 43½″	135	3″
93,340	MRBS 850	Menck	Sgl.-Act	Open	45	18,960	27,890	19′8″	Cage	142	3″
90,000	030	Vulcan	Sgl.-Act	Open	54	30,000	53,470	16′4″	11¼″ × 37″	150	3″
90,000	300	Conmaco	Sgl.-Act	Open	52	30,000	55,390	16′10″	11¼″ × 56″	150	3″
81,250	8/0	Raymond	Sgl.-Act	Open	40	25,000	34,000	19′4″	10¼″ × 25″	135	3″
75,000	30X	Raymond	Sgl.-Act	Open	70	30,000	52,000	19′1″	10½″ × 43½″	150	3″
62,500	125E5	Conmaco	Sgl.-Act	Open	41	12,500	22,000	18′0″	9¼″ × 26″	100	2½″
60,000	200	Conmaco	Sgl.-Act	Open	55	20,000	44,560	15′0″	11¼″ × 56″	110	3″
60,000	512	Vulcan	Sgl.-Act	Open	41	12,000	23,480	18′5″	9¼″ × 26″	100	2½″
60,000	S-20	Mkt	Sgl.-Act	Closed	60	20,000	38,650	15′5″	— × 36″	150	3″
60,000	020	Vulcan	Sgl.-Act	Open	59	20,000	41,670	14′8″	11¼″ × 37″	120	3″
57,500	115E5	Conmaco	Sgl.-Act	Open	42	11,500	21,000	17′9″	9¼″ × 26″	100	2½″
56,875	5/0	Raymond	Sgl.-Act	Open	44	17,500	26,450	16′9″	10¼″ × 25″	150	3″
50,000	510	Vulcan	Sgl.-Act	Open	41	10,000	21,480	18′5″	9¼″ × 26″	83	2½″
50,000	100E5	Conmaco	Sgl.-Act	Open	47	10,000	19,500	17′9″	9¼″ × 26″	100	2½″
48,750	160	Conmaco	Sgl.-Act	Open	50	16,250	33,200	13′10″	11¼″ × 42″	100	3″
50,000	200-C	Vulcan	Differ.	Open	95	20,000	39,000	13′11″	11¼″ × 37″	142	4″
48,750	016	Vulcan	Sgl.-Act	Open	58	16,250	30,250	13′11″	11¼″ × 32″	120	3″
48,750	4/0	Raymond	Sgl.-Act	Open	46	15,000	23,800	16′1″	—	120	2½″
48,750	150-C	Raymond	Differ.	Open	95–105	15,000	32,500	15′9″	10″ × 27½″	120	3″
45,200	MRBS 500	Menck	Sgl.-Act	Open	48	11,020	15,210	16′8″	— × 26″	142	2½″
44,000	MS-500	Mkt	Sgl.-Act	Open	40–50	11,000	15,500	15′1″	8½″ × 26″	115	3″
42,000	140	Conmaco	Sgl.-Act	Open	55	14,000	30,750	13′10″	11¼″ × 42″	100	3″
42,000	014	Vulcan	Sgl.-Act	Open	59	14,000	27,500	13′8″	11¼″ × 32″	110	3″
40,600	3/0	Raymond	Sgl.-Act	Open	50	12,500	21,000	15′7″	10¼″ × 25″	120	2½″
40,000	80E5	Conmaco	Sgl.-Act	Open	47	8,000	17,500	17′9″	9¼″ × 26″	80	2½″
40,000	508	Vulcan	Sgl.-Act	Open	41	8,000	19,480	18′5″	9¼″ × 26″	65	2½″

TABLE D3-7 Specifications for Pile Hammers *(Continued)*

Air/steam hammers											
Rated energy, ft-lb	Model	Manufacturer	Type	Style	Blows per min	Wt. of striking parts, lbs	Total weight, lbs	Hammer length, ft-in	Jaw dimensions	Inlet pressure, lb/in²	Inlet size, NPT
37,500	S-14	Mkt	Sgl.-Act	Closed	60	14,000	31,700	13′7″	— × 36″	100	3″
37,375	115	Conmaco	Sgl.-Act	Open	52	11,500	20,830	14′2″	9¼″ × 32″	100	2½″
36,000	140-C	Vulcan	Differ.	Open	101	14,000	27,984	12′3″	11¼″ × 32″	140	3″
32,885	100-C	Vulcan	Differ.	Open	103	10,000	22,200	14′0″	9¼″ × 26″	140	2½″
32,500	100	Conmaco	Sgl.-Act	Open	55	10,000	19,280	14′2″	9¼″ × 32″	100	2½″
32,500	65E5	Conmaco	Sgl.-Act	Open	50	6,500	12,500	16′10″	8¼″ × 20″	95	2½″
32,500	506	Vulcan	Sgl.-Act	Open	46	6,500	13,025	17′5″	8¼″ × 20″	100	2″
32,500	2/0	Raymond	Sgl.-Act	Open	50	10,000	18,550	15′0″	10¼″ × 25″	110	2″
32,500	010	Vulcan	Sgl.-Act	Open	50	10,000	18,780	15′0″	9¼″ × 26″	105	2½″
32,500	S-10	Mkt	Sgl.-Act	Closed	55	10,000	22,380	14′1″	— × 30″	80	2½″
30,800	MS-350	Mkt	Sgl.-Act	Open	40–50	7,716	10,500	15′1″	8½″″ × 26″	105	2½″
26,000	80	Conmaco	Sgl.-Act	Open	56	8,000	17,280	14′2″	9¼″ × 32″	80	2½″
26,000	85-C	Vulcan	Differ.	Open	111	8,525	19,020	12′7″	9¼″ × 26″	128	2½″
26,000	08	Vulcan	Sgl.-Act	Open	50	8,000	16,750	14′10″	9¼″ × 26″	83	2½″
26,000	S-8	Mkt	Sgl.-Act	Closed	55	8,000	18,300	14′4″	— × 26″	80	2½″
25,000	50E5	Conmaco	Sgl.-Act	Open	48	5,000	11,000	16′10″	8¼″ × 20″	70	2″
25,000	505	Vulcan	Sgl.-Act	Open	46	5,000	11,800	17′5″	8¼″ × 20″	77	2″
24,450	80-C	Vulcan	Differ.	Open	109	8,000	17,885	12′7″	9¼″ × 26″	120	2½″
24,450	80-C	Raymond	Differ.	Open	95–105	8,000	17,885	12′2″	10¼″ × 25″	120	2½″
24,375	0	Raymond	Sgl.-Act	Open	50	7,500	16,000	15′0″	10¼″ × 25″	110	2″
24,375	0	Vulcan	Sgl.-Act	Open	50	7,500	16,250	15′0″	9¼″ × 26″	80	2½″
24,000	C-826	Mkt	Compound	Closed	85–95	8,000	17,750	12′2″	— × 26″	125	2½″
19,500	65 E3	Conmaco	Sgl.-Act	Open	61	6,500	12,100	12′10″	8¼″ × 20″	100	2″
19,500	65-C	Raymond	Differ.	Open	110	6,500	14,675	11′8″	9¼″ × 19″	120	2″
19,500	1-S	Raymond	Sgl.-Act	Open	58	6,500	12,500	12′9″	7½″ × 28¼″	100	1½″
19,500	06(106)	Vulcan	Sgl.-Act	Open	60	6,500	11,200	13′0″	8¼″ × 20″	100	2″
19,200	65-C	Vulcan	Differ.	Open	117	6,500	14,886	12′1″	8¼″ × 20″	150	2″
19,150	11B3	Mkt	Dbl.-Act	Closed	95	5,000	14,000	11′2″	8½″ × 26″	100	2½″
16,250	S-5	Mkt	Sgl.-Act	Closed	60	5,000	12,460	13′3″	— × 24″	80	2″
16,000	C-5(STM)	Mkt	Dbl.-Act	Closed	100–110	5,000	11,880	—	— × 26″	100	2½″
15,100	50-C	Vulcan	Differ.	Open	117	5,000	11,782	11′0″	8¼″ × 20″	120	2″
15,000	50 E3	Conmaco	Sgl.-Act	Open	64	5,000	10,600	12′10″	8¼″ × 20″	80	2″
15,000	1(106)	Vulcan	Sgl.-Act	Open	60	5,000	9,700	—	8¼″ × 20″	80	2″
15,000	1	Raymond	Sgl.-Act	Open	60	5,000	11,000	12′9″	7½″ × 28¼″	80	1½″
14,200	C-5(AIR)	Mkt	Compound	Closed	100–110	5,000	11,880	8′9″	— × 26:	100	2½″
13,100	10B3	Mkt	Dbl.-Act	Closed	105	3,000	10,850	9′2″	8½″ × 24″	100	2½″
8,750	900	Bsp	Dbl.-Act	Closed	145	1,600	7,100	7′10″	— × 20″	90	2″
8,750	9B3	Mkt	Dbl.-Act	Closed	145	1,600	7,000	8′4″	8½″ × 20″	100	2″
7,260	30-C	Vulcan	Differ.	Open	133	3,000	7,036	8′11″	7¼″ × 19″	120	1½″
7,260	2	Vulcan	Sgl.-Act	Open	70	3,000	6,700	11′7″	7¼″ × 19″	80	1½″
4,700	700N	Bsp	Dbl.-Act	Closed	225	850	6,630	7′5″	— × 18″	90	2″
4,150	7	Mkt	Dbl.-Act	Closed	225	800	5,000	6′1″	6½″ × 21″	100	1½″
4,000	DGH-900	Vulcan	Differ.	Closed	328	900	5,000	6′9″	Varies	78	1½″
3,000	600N	Bsp	Dbl.-Act	Closed	250	500	4,800	7′2″	— × 15″	90	1½″
2,500	6	Mkt	Dbl.-Act	Closed	275	400	2,900	5′3″	6½″ × 15″	100	1¼″
1,200	500N	Bsp	Dbl.-Act	Closed	330	200	2,520	5′11″	— × 14″	90	1¼″
1,000	5	Mkt	Dbl.-Act	Closed	300	200	1,500	4′7″	6″ × 11″	100	1¼″
386	DGH-100D	Vulcan	Differ.	Closed	303	100	786	4′2″	4¼″ × 8¾″	60	1″
—	3	Mkt	Dbl.-Act	Closed	400	68	675	4′5″	—	100	1″
350	300	Bsp	Dbl.-Act	Closed	400	68	675	4′10″	—	90	1″
160	200	Bsp	Dbl.-Act	Closed	500	48	343	2′9″	—	90	¾″
—	2	Mkt	Dbl.-Act	Closed	500	48	343	2′5″	3¼″ × 8¼″	100	¾″
—	1	Mkt	Dbl.-Act	Closed	500	21	145	3′3″	—	100	¾″

TABLE D3-7 Specifications for Pile Hammers *(Continued)*

Hydraulic vibratory drivers/extractors															
Driving force, tons	Model	Manufacturer	Frequency, VPM	Eccentric moment	Amplitude, in	Max. hydraulic hp	Engine hp	Max. Pull extraction, tons	Pile clamp force, tons	Suspended weight, lb	Shipping weight, lb	Height (w/clamp), ft & in	Width (thickness), ft & in	Length, ft & in	Throat width, in
390	V-140	Mkt	1,400	14,000	1.0	1,400	1,800	150	300	42,000	96,000	14′6″	4′0″	11′10″	48″
390	400B	Ape	400–1,400	14,000	1.5	865	1,000	250	500	20–34,000	45–54,000	8′ (varies)	2′9″	11′11″	32″
310	400A	Ape	400–1,350	12,000	1.125	760	800	250	500	20–34,000	43–48,000	8′ (varies)	2′9″	11′11″	32″
287	HBV260.01	Delmag/Tunkers	1,600	7,938	.86	1,073	1,153	110	287	22,000	44,000	14′9″	—	7′	18.1″
244	110H2	Ptc	1,350	9,550	1 3/16	584	760	88	2 × 165	32,900	57,250	12′9″	2′11″	7′7″	—
244	110H1	Ptc	1,350	9,550	1 1/8	565	760	132	4 × 120	39,050	63,400	10′1″	2′6″	12′4″	—
230	200	Ape	1,800	5,000	1.125	460	503	200	—	13,000	28,000	7′6″	1′7″	9′6″	14″
222	1412-B	Ice	1,250	10,000	1–1 1/2	760	800	150	196	28,500	52,500	14′6″	3′5″	7′11″	32″
218	600	Hpsi	1600	6000	1 1/4	611	800	75	200	17,500	36,000	9′8″	1′2″	7′11″	14″
200	520	Hpsi	1,600	5,200	1 1/2	554	670	75	200	16,500	33,000	9′7″	1′2″	7′11″	14″
200	5600	Vulcan	2,400	—	.50	408	475	51	177	16,000	27,000	12′3″	1′2″	12′	14″
199	60H1	Ptc	1,650	5,210	7/8	523	660	88	2 × 165	29,200	53,550	13′0″	2′3″	7′7″	—
182	V-36	Mkt	1,600	5,000	3/4	550	650	80	100	18,800	36,300	13′1″	1′2″	12′0″	14″
180	180A	Ape	1,800	4,500	1.125	460	503	100	200	6,500–11,000	22–28,000	7′6″	1′7″	7′4″	14″
167	4600A	Vulcan	1,200–1,600	4,600	1–1 3/8	—	560	80	177	14,000	24,600	10′11″	2′6″	7′2″	14″
167	4600	Vulcan	1,600	4,600	1.25	474	600	66	176	16,000	32,500	8′0″	1′6″	6′3″	14″
166	50H3	Ptc	1,650	4,340	7/8	394	510	44	220	23,300	38,750	14′	1′1″	7′10″	—
165	423	Ice	2,300	2,200	.75	452	503	40	125	6,500–11,000	22–28,000	3′–7′6″	1′7″	7′4″	14″
164	1412A	Ice	1,200	8,000	1.50	554	650	100	196	26,900	47,500	12′0″	3′5″	7′11″	32″
164	450	Hpsi	1,600	4,500	1	555	600	75	150	14,100	30,100	9′3″	1′2″	7′11″	14″
162	23HF1	Ptc	2,400	2,000	7/8	302	465	44	220	12,850	30,500	9′10″	2′6″	7′2″	13″
160	V-30	Mkt	1,600	4,400	1.00	510	600	80	100	15,000	32,500	11′1″	1′1″	8′6″	13″
160	815	Ice	1,600	4,400	1	452	503	50	125	15,600	30,500	10′2″	2′4″	8′0″	12″
148	150B	Ape	1,800	2,500	0.875	460	503	100	150	9,500	24,500	7′6″	1′7″	7′0″	14″
148	23HF1	Ptc	2,300	2,000	7/8	248	330	44	220	12,850	25,600	9′10″	2′6″	7′2″	13″
145	400	Hpsi	1,600	4,000	1 1/8	407	503	75	150	13,600	29,750	9′3″	1′2″	7′11″	14″
143	HBV130.01	Delmag/Tunkers	1,600	3,991	.79	547	625	55/110	154	12,600	31,300	13′3″	—	7′0″	17.32″
134	4150	Foster	900–1,500	4,166	5/16–1 1/4	—	570	55	145/200	16,495	31,995	6′6″	2′3″	8′0″	12″
116	V-20	Mkt	1,650	3,000	.66	310	350	60	75	12,500	31,500	12′2″	1′2″	8′0″	14″
115	150B	Ape	1,800	2,500	.875	460	503	100	100	6,500–11,000	22–28,000	3′–7′6″	1′7″	7′4″	14″
115	150A	Ape	1,800	2,500	.875	300	325	100	100	6,500–11,000	22–28,000	3′–7′6″	1′7″	7′4″	14″
110	HVB100.01	Delmag/Tunkers	1,600	2,950	.70	359	503	55	154	10,580	23,600	13′	—	7′0″	15.74″
100	2800	Vulcan	2,400	—	.50	204	250	25	87	8,300	18,800	6′6″	1′	9′8″	12″
100	260	Hpsi	1,600	2,600	.80	277	335	60	150	10,750	22,250	8′4″	1′2″	7′0″	14″

90	612	Ice	1,200	4,400	1.0	254	300	40	125	12,700	23,500	9′4″	2′4″	7′11″	14″
84	2300	Vulcan	1,600	2,300	1	237	335	33	87	8,200	19,700	7′9″	1′6″	3′9″	14″
84	13HF1	Ptc	2,300	1,130	7/8	177	255	33	120	7,150	18,600	6′10″	2′6″	5′8″	13″
84	2300A	Vulcan	1,200–1,600	2,300	3/4–1 1/8	—	360	48	87	8,550	16,550	9′7″	2′4″	5′	14″
83	223	Ice	2,300	1,100	.75	254	300	40	125	5,350	16,150	5′7″	16″	3′11″	12″
83	25H2	Ptc	1,650	2,170	15/16	253	325	44	120	9,350	16,900	8′11″	1′1″	7′6″	—
80	V-17	Mkt	1,600	2,200	.75	254	325	60	70	12,000	31,500	10′0″	1′1″	8′6″	13″
80	416L	Ice	1,600	2,200	75	254	300	40	125	9,900	20,700	8′0″	1′5″	7′11″	14″
78	V-16	Mkt	1,750	1,800	.47	161	210	40	75	9,250	24,000	11′2″	1′2″	6′6″	14″
78	HVB70.01	Delmag/Tunkers	1,630	2,065	74	290	312	28/55	120	7,050	17,850	10′2″	—	5′7″	11.81″
69	25H1	Ptc	1,500	2,170	15/16	185	255	44	120	9,350	16,300	8′11″	1′1″	7′6″	—
66/55	HVB60.05 Dual	Delmag/Tunkers	1,500/1,000	2,065/3,879	.67/1.18	251/191	312	27.5	91	7,050	18,850	5′3″	—	5′7″	11.81″
65	1800	Foster	1,000–1,600	1,800	1 5/16–3/4	—	220	30	90/120	11,000	24,375	6′8″	1′9″	8′0″	12″
57.5	416	Ice	400–1,500	1,800	1/4–1	—	250	40	100	13,100	22,000	8′9″	1′10″	8′0″	12″
—	H-1700	H & M	1,250	—	1/2–3/4	175	210	30	75	7,000	13,000	8′0″	12″	5′0″	12″
50	130	Hpsi	1,600	1,300	.80	148	175	30	48	7,050	14,000	6′6″	1′2″	7′3″	14″
50	V-5B	Mkt	1,800	1,100	.75	118	175	30	62	7,200	11,200	7′8″	1′1″	6′4″	13″
49	1400	Vulcan	2,400	600	.38	123	175	25	50	4,350	10,350	8′8″	1′11″	3′8″	12″
46	V-14	Mkt	1,500	1,500	.32	140	210	40	75	10,000	29,500	11′1″	1′2″	5′3″	14″
44	HVB40.01	Delmag/Tunkers	1,500	1,346	.67	178	253	28	77	5,700	16,100	9′8″	—	5′3″	11.81″
44	13H1	Ptc	1,700	1,085	11/16	112	155	22	60	5,400	12,250	6′5″	1′1″	4′7″	13″
42	1150	Vulcan	1,600	1,150	.75	131	175	33	50	6,500	14,000	6′6″	1′6″	3′9″	14″
42	7H F1	Ptc	2,300	565	5/8	102	150	17	60	3,800	10,650	5′9″	1′7″	3′7″	13″
42	1150A	Vulcan	1,200–1,600	1,150	1/2–3/4	—	155	32	48	6,300	13,800	7′11″	2′1/2″	5′	14″
40	216	Ice	1,600	1,100	.75	146	175	40	50	5,350	12,975	6′6″	1′4″	3′11″	12″
37	1000	Foster	1,000–1,600	1,000	.25–.75	—	185	20	90	6,094	10,500	5′11″	1′8″	5′7″	12″
33	HVB30	Delmag/Tunkers	1,800	712	.79	112	158	7/13	33	2,100	7,935	3′4″	—	3′6″	14.56″
30	V-5	Mkt	1,450	1,000	.50	59	175	30	62	5,200	9,200	9′4″	1′2″	6′10″	14″
—	H-75B	H & M	1,900	—	25–.50	105	210	15	62	4,000	9,000	5′7″	—	3′1″	14″
27	HVB24	Delmag/Tunkers	2,100	434	.55	86	131	7/13	27	2,100	7,935	3′3″	—	3′6″	14.56″
24	65	Hpsi	1,600	650	3/4	76	105	25	48	3,100	5,800	4′11″	1′0″	5′4″	12″
23	6H1	Ptc	1,800	—	.75	47	—	22	—	3,180	6,580	4′1″	1′1″	3′8″	
18	V-2A	Mkt	1,800	400	.75	44	50	8	16	2,400	4,400	3′8″	1′0″	3′8″	12″
17	400A	Vulcan	2,400	200	1/4′–3/8″	—	58	10	6	1,100	4,950	4′0″	2′6″	3′8″	—
17	400	Vulcan	2,400	200	.50	42	75	10	6	1,100	4,100	3′2″	10″	2′2″	10″
17	HVB16	Delmag/Tunkers	2,100	278	.39	64	131	7/13	27	2,000	7835	3′3″	—	3′4″	14.56″
12	HVB10	Delmag/Tunkers	2,100	191	.31	43	131	7/13	27	2,000	7835	3′3″	—	3′4″	14.56
4	30	Ape	3,000	30	0.375	20	(varies)	5	20	225	225	2′	5′2″	2′	3′5″

Source: Reference 7. © Copyright 1994, Pile Buck®, Inc.

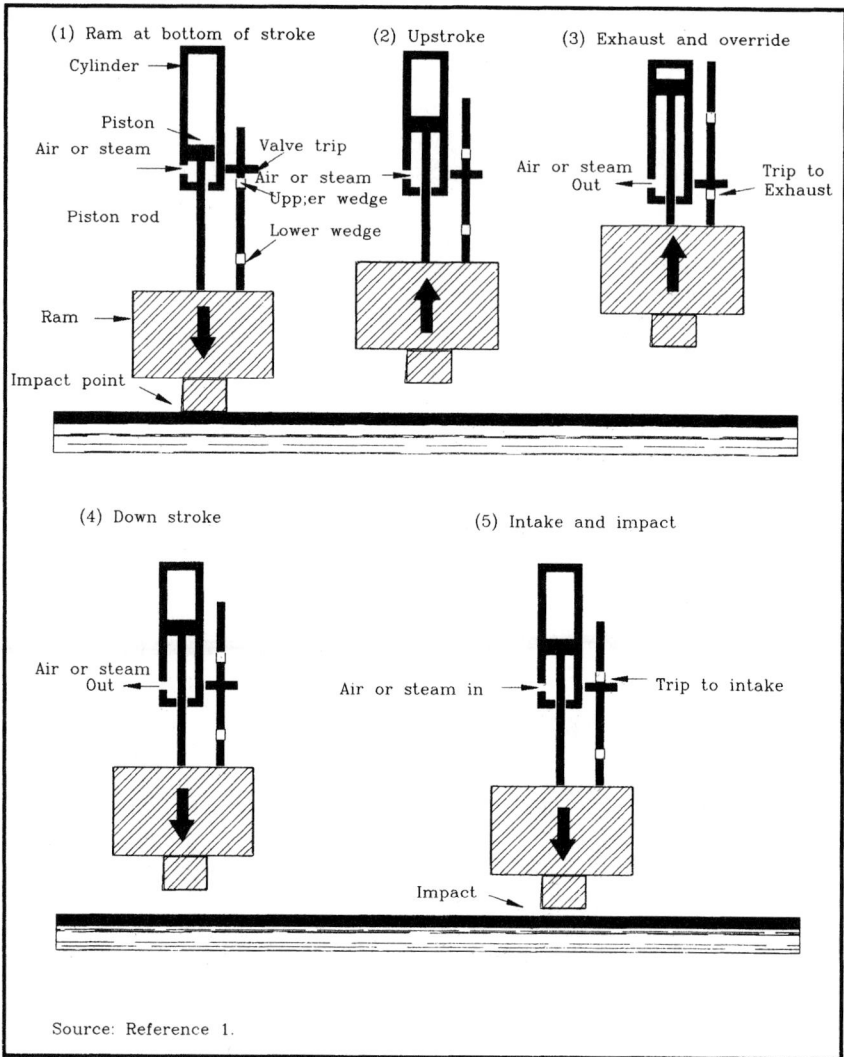

FIGURE D3-4 Operating cycle for a single-acting air or steam hammer.

efficiency. Because of their consistent elastic properties, they are easy to model in performing a pile-driving analysis using the wave equation. (See "Pile Installation.") Prior to the introduction of the manufactured cushion, it was common practice to use either wood chips or steel cable as a cushion. These materials do not have consistent elastic properties and should not be used. Hard wood block cushions, typically 6 in thick, with the grain parallel to the pile axis, and confined within a steel band, are still quite common. They have the disadvantage of not having consistent elastic properties; they frequently burn and must be changed often.

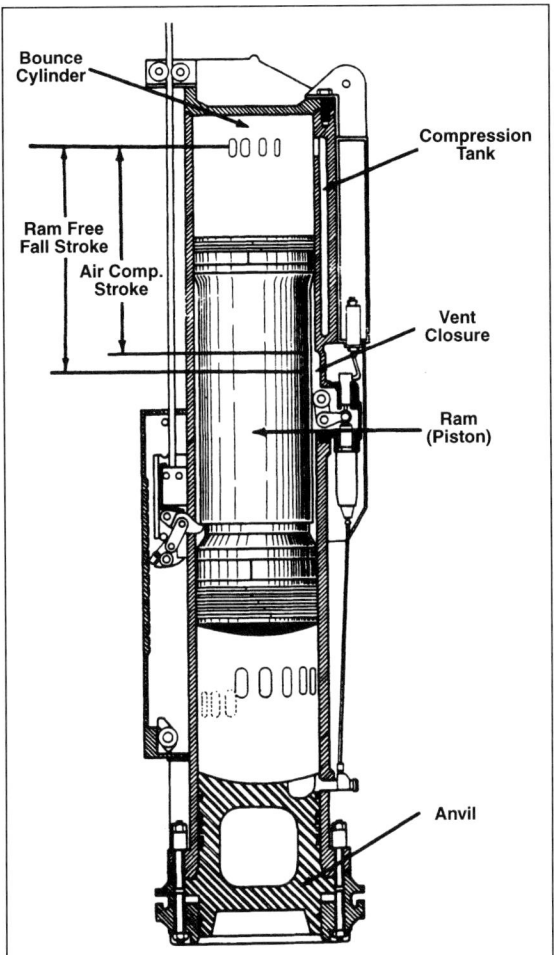

FIGURE D3-5 Double-acting diesel hammer.

The hammer cushion must be inspected regularly during driving and replaced after significant nonelastic distortion has occurred.

Drive Head. The drive head is used to hold the head of the pile in place and to distribute the force of the hammer blow uniformly to the pile head during driving. Pipe pile heads usually are stepped to fit several different pipe sizes, and should be machined to have a uniform and relatively tight fit to prevent crimping or buckling of the pile top during hard driving. They may be either internal or external to the pile. H pile heads should be in the shape and size of the pile they are driving, and heads for precast concrete piles are relatively loose fitting.

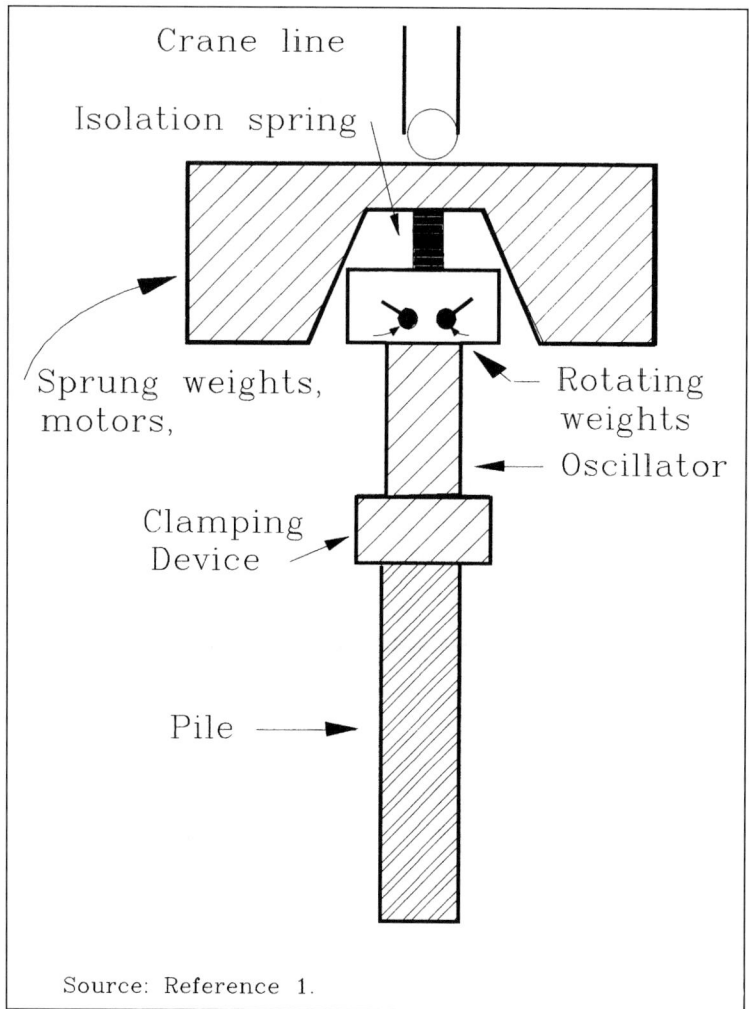

FIGURE D3-6 Schematic diagram of a vibratory hammer.

Pile Cushion.   Typically, precast concrete piles are protected with a cushion which is in addition to the hammer cushion. The purpose is to further cushion the pile from the force of the hammer blow and prevent spalling of the concrete. The cushion usually consists of plywood sheets and either softwood or hardwood boards, and must cover the entire top of the pile. The cushion must be designed to protect the pile head and prevent excessive tensile forces from damaging the pile, yet not reduce the energy such that the pile cannot be driven to the design capacity.

Follower.   A follower is a method of extending the pile so that the pile top may be driven below the ground surface or the surface of the water. The follower must be stiff enough so that the hammer energy may be transmitted to the pile without significant losses. It must also be held in position so that it is in correct alignment with the pile and the hammer. The end of the follower must fit the top of the pile in the same way as the drive cap.

Mandrel. A mandrel is a special piece of equipment which is used to install light-gauge shell piles or thin wall pipe piles. The pipe or shell is pulled over the mandrel, or, conversely, the mandrel is lowered into the shell. The mandrel must then either grip the shell by expanding, if the shell is straight sided, or it may be used to bottom drive or, in the case of step-tapered piles, drive on the shoulder of short segments of casing. During pile driving, the mandrel becomes the pile. It must be stiff enough to transmit the hammer energy efficiently to the pile tip, and in the case of an expandable mandrel, strong enough to grip and drag the shell down without damaging it while under the stresses imposed by the hammer blows. The mandrel permits the contractor to install light-gauge steel shells or thin wall pipe to relatively high capacity loads—i.e., greater than 100 tons—while saving on material costs.

Leaders

Leaders are intended to hold the hammer and the pile in alignment during driving. Leaders have the appearance of a light truss, and when fixed at the bottom with a spotter, may be aligned in three dimensions so that the hammer may be aligned with the pile even if on a batter. Leaders must be held firmly and be rigid enough to prevent the pile from drifting out of plumb during driving. Light leads which are not fixed at the top or bottom (swinging leads) may allow this to happen.

Spotter

The spotter is a horizontal beam or platform attached from the bottom of the leaders to the pile-driving rig, which provides fixity to the leaders and allows the pile to be spotted over the pile location stakes. If the spotter is rigid, all piles must be driven from the same radial distance from the pile-driving rig. If the spotter is telescopic, the spotter may change in length, allowing more flexibility of operation for the pile driver.

Preexcavation Equipment

Preexcavation equipment is typically mounted on the side of the leaders. Generally, this is an auger, which may be either a single section attached to a Kelly bar, or a continuous-flight auger. Alternatively, a drill bit and hollow drill rods may be used with drilling water.

If permitted by the specifications, a jet pipe and water under pressure may be used to preexcavate for the pile. The disadvantage of this method is that it may overexcavate the pile hole, thereby weakening the surrounding soil and lowering the capacity of the pile in friction.

PILE INSTALLATION

General

Piles are installed by either driving, drilling, vibration, or jacking. The proper installation of piles involves an understanding of the subsurface conditions and the selection of the appropriate pile installation system compatible with those conditions. The pile-driving contractor must have sufficient experience to install the piles with a minimum of damage to the pile, in the correct location and alignment. The following is a discussion of the major components and potential problems associated with pile installation.

Subsurface Conditions

A knowledge of the subsurface conditions is imperative for the successful installation of deep foundation elements. A typical subsurface investigation will address the potential for encountering buried obstructions, the depth to the bearing stratum, the location of the groundwater table, and a description of the stratigraphy. The contractor must have reservations about the subsurface conditions as reported by the engineer for the following reasons:

1. Test borings are notorious for not detecting obstructions.
2. Test borings are only 2 to 4 in in diameter. The potential always exists for the conditions between borings to be different from those discovered.

Despite these limitations, the contractor usually has no alternative but to plan the work on the information provided by the engineer, or on previous personal experience or that of others working in the area. On this basis, the contractor must estimate the pile length and installation time required to achieve the pile design capacity.

Handling and Storage

Piles may be damaged during unloading, handling, and storage. Timber, precast concrete, and very long H piles are particularly vulnerable to damage. Timber piles should not be handled with tools which puncture the wood and which will diminish the effectiveness of the preservative treatment. Precast concrete piles have specified lifting points provided by the manufacturer which must be used to prevent overstressing and cracking the pile.

Very long H piles should be lifted at a sufficient number of points to prevent permanent distortion. Steel piles and shells should be stored on blocking off the ground to minimize contamination with standing water.

Planning, Site Preparation, Pile Numbering, and Location

The pile-driving contractor, usually a subcontractor to the general contractor or owner, must plan the work to minimize delays after mobilization. The sequencing of the work may be dictated by the owner or general contractor, but normally the contractor is able to plan the work to complete the installation in a logical and expeditious manner. The sequencing must accommodate the work of the other subcontractors. The excavation, access ramps, dewatering, sheeting, and tieback activities must be coordinated with the pile-driving operation. The expeditious completion of the project is dependent on the satisfactory maintenance of the laydown areas, access ramps, and roads, and on the dewatering of the excavation during the course of the work.

It is important that the pile contractor be given a dry, level site to work on. If the subgrade is soft, the contractor must plan to use timber mats to prevent the pile-driving rig from becoming mired. Movement of the mats must be considered in planning the work.

Frequently, the pile cutoff elevations vary in elevation by a few feet to over 10 ft, but the planned excavation during pile driving is typically to a common grade. The piles may be driven, cut off, and capped 6 in above the excavation grade, and the final excavation and cutoff may be by others. Conversely, the final cutoff may be above the excavation grade. In that case, the placement of backfill to establish the final grade will be by others.

All piles planned for a project should be given a unique designation, and the pile designation plan should be submitted to the engineer or owner at the completion of the job. Pile-driving records always reference the pile designation. The pile designation may be sequential numbers, or each pile cap may be designated by a letter or column lines and each pile designation would start with the cap number, followed by the sequential number in the pile group.

Hammer Selection, Wave Equation, and Pile Load Test

The hammer selected is the one owned by the contractor which most nearly meets the engineer's requirements regarding minimum hammer energy. This is frequently a diesel hammer. The engineer must determine if the proposed hammer/pile system is adequate for the pile design capacity and the soil conditions. The engineer (and sometimes the contractor) performs a wave equation analysis to determine the pile-driving criteria. The wave equation analysis is based on using the stress wave generated from the hammer impact to determine the displacements and stresses in the pile during driving. Such information is useful to ensure that the pile is not overstressed during installation. Solutions to the wave equation for a given hammer and pile are also used to evaluate the pile capacity and equipment compatibility. Since the wave equation analysis is only an analytical tool based on certain limiting assumptions, it is important that the pile-driving criteria be confirmed by a pile load test. The compression pile load test procedure is given in ASTM D1143, Method of Testing Piles Under Static and Compressive Loads. The normal exception to running a load test is if there is previous pile-driving experience with similar piling at the same site.

Preexcavation

Preexcavation is performed in advance of pile driving if there is concern about soil displacement or excessive vibration during pile driving. There may be concern about damaging adjacent structures or buried utilities, or collapse of thin-walled piles, or heave of adjacent piles. Preexcavation may be required by the specifications, or may be determined to be in the best interests of the contractor.

Jetting. Jetting consists of creating a hole using a pipe with water jetting from the bottom under high pressure. The water is directed either downward or to the sides and down. Prejetting is normally used only in granular soils, and care must be taken not to create too large a hole. Also, the larger gravel and cobble-size particles tend to fall to the bottom of the hole and can cause a problem when attempting to drive a pile through this zone. Jetting should stop after the predetermined depth has been reached. Piles should always be driven below the bottom of the jetted hole.

Jetting can also be used while driving, using either internal jets built into a concrete pile or an external jet pipe.

Predrilling. Predrilling is normally accomplished by augering or using a wet rotary drill. It is much more of a controlled drilling technique than jetting. Augering a hole can be accomplished in the dry. The auger is either a continuous flight or a short section at the end of a drill stem or Kelly bar. Wet predrilling requires water or drilling mud circulated through the hollow drill steel to either a spade bit or a tricone bit. If there is concern that the hole will collapse before the pile is inserted in the hole, then the hole is stabilized by the use of drilling mud. The fluid is directed out through side ports in the bit. Drill cuttings are carried to the surface with the drilling fluid, where the cuttings are screened out and the drilling mud is recirculated. Auger drills can excavate either soil or soft rock. If obstructions such as boulders or old foundations are anticipated near the surface, they must be removed before predrilling commences. Removal may be by spudding (driving a heavy wall pipe) or by excavation and removal with a backhoe.

Pile Driving

General. Most pile types can be damaged during the installation process. It is very important that the pile-driving equipment be matched with the pile properties and the subsurface conditions to minimize this potential problem. Piles must have sufficient stiffness to be driven to the ultimate load required and sufficient strength to prevent being damaged during installation.

Timber Piles. Timber piles should not be used in subsurface conditions where obstructions are likely to be encountered. If timber piles are to be driven as point-bearing piles, the tips should be reinforced with steel boots to prevent brooming. If there is a sudden loss in the driving resistance during hard driving which is not explainable by the soil profile, or if there is a sudden movement of the butt, pile damage should be suspected. If pressure-treated timber piles are to be driven at temperatures below freezing, care should be taken not to drive the piles when frozen, which could cause shattering of the pile.

Steel H Piles. Steel H piles are subject to twisting and drifting off center when driven through obstructions or into a hard, sloping rock surface. Tip damage may also occur when driving through obstructions or if the pile is overdriven. This may be avoided to some degree by the use of pile tip reinforcement. Steel H piles should be sized to avoid overstressing during the installation, and overdriving should be avoided. Damage to H piles usually occurs at or near the pile tip and results in either ripping or tearing or folding over of the pile.

Prestressed Concrete Piles. Prestressed concrete piles are vulnerable to spalling, cracking, and breaking. Spalling may occur at the pile butt due to overstressing caused by insufficient cushion or eccentric hammer blows. Spalling may occur along the length of very long piles which may tend to buckle during driving. Spalling may also be caused by lack of spiral reinforcement, the head not being perpendicular to the axis of the pile, or reinforcement protruding through the end of the pile. If piles are properly constructed, spalling may be avoided by the use of proper cushion material and not overdriving the pile.

When piles break through a dense layer into a soft layer, tensile forces are generated in the pile which may crack or break the pile. This can be avoided by immediately reducing the hammer force as breakthrough occurs.

When driving spliced precast concrete piles, care must be taken to ensure proper alignment to prevent bending and buckling at the joints.

Pipe Piles. Pipe piles are usually top-driven with a steel bottom plate and filled with concrete for increased strength after driving. Consequently, the driven pile is subjected to greater stresses and has less ability to resist these stresses than after being filled with concrete. It is important that pipe piles be sized to avoid overstressing during driving. One advantage of pipe piles is that they may be inspected after driving. Typical damage to a pipe pile is buckling, usually at the tip, or at the pile top. If buckling or curling of the pile top occurs, it will dampen the energy transmitted to the pile tip and the pile capacity cannot be predicted based on the driving resistance. Another potential problem associated with pipe piles is pile heave, caused by the movement of soil displaced when driving adjacent piles. Pile movements should always be monitored, and piles which heave must be retapped to the original driving resistance.

Mandrel-Driven Shell Piles. Mandrel-driven shell piles may be damaged during the initial drive or while driving adjacent piles. Damage is caused by overstressing the casing. This damage may be caused by obstructions, driving into very dense or stiff soils, or displacement caused by driving adjacent piles. This damage may be found by inspection of the shell before filling with concrete. Typically, the damage results either in a partial collapse of the casing or a tear which may allow soil and water to enter the casing. These problems may be overcome by preexcavation or by increasing the gauge of the shell. An advantage of shell piles is that the casings are corrugated, which are flexible and can stretch. Consequently, heave due to soil lifting the casing is not usually a problem.

Augercast Piles

The successful installation of augercast piles is predicated on good construction technique. One problem which may occur due to poor technique is discontinuities in the pile shaft. These are caused either by too rapid withdrawal of the auger or displacement while installing adjacent piles near piles which have not yet set up. Other related problems include the inclusion of soil pockets in the grout and a reduction of the cross-sectional area of the pile (necking). Installation of augercast piles in very soft soil, where there is danger of loss of side support, should be avoided.

During the drilling of the augercast pile, the auger should be advanced continuously while the flights rotate. When the required tip elevation is reached, rotation should stop so as to avoid removing excess soil.

The hole is filled with grout by pumping as the auger is withdrawn while rotating continuously. Initially, at the start of grout injection, the auger is lifted about 1 ft and grout is injected until, theoretically, the quantity of grout is equivalent to about 5 ft of pile, and the grout pressure is fully developed (normally not less than 350 lb/in at ground level). The auger is then advanced to the original tip elevation and the formation of the pile begins by the deliberate lifting of the auger in either a continuous operation or in small, controlled increments not exceeding 6 in. During this process, grout is continuously pumped at a rate such that the quantity of grout is approximately equal to the volume of the shaft being constructed. If the auger is lifted too fast, or if the operation is interrupted for any reason, or if there is a drop in grout pressure, the pile construction should stop and the pile should be redrilled.

The quantity of grout injected into the pile should be at least 15 percent greater than the theoretical volume of the pile. If there is a significant increase in the volume of grout injected over the theoretical volume, the cause should be investigated.

After withdrawing the auger, the top of the pile can be formed by the use of a sleeve of the same diameter as the pile.

Pressure-Injected Footings (PIFs)

PIFs, like all cast-in-place piles, require good construction technique for successful installation. Forming the base of the PIF is a very important operation, as the compaction of the surrounding soil creates its high bearing capacity. It is important that the fall of each hammer stroke be carefully monitored to ensure that the full energy is being delivered.

If PIF shafts are constructed without a casing, they are subject to loss of shaft continuity if improperly formed. It is important that the concrete in the drive tube not be driven completely out of the tube. This is monitored by marking the hammer cable so that, as the base of the hammer approaches the bottom of the drive tube, the mark approaches the top of the drive tube. Forming uncased shafts in soft soils should be done with caution.

If heave of the top of the shaft is observed, it may mean that the shaft has separated from the base. In this case the pile should be abandoned and rejected.

If a casing is used, there is a potential for a void between the casing and the outer wall of the drive tube after it is withdrawn. This void should be filled if lateral support of the pile is required.

INSPECTION

General

Inspection of the pile installation is an extremely important function. It must be continuous. It should be performed by engineers aware of the design objectives, the subsurface conditions, and the details of the specification. Most important, inspectors should be experienced with the inspection of similar installations. It is typically performed by the owner's engineer. The inspection program has several objectives:

1. Enforcement of the specification
2. Documentation of important data for each foundation element
3. Documentation of changed conditions
4. Rapid communication to the appropriate persons of actual or potential problems

The inspector must maintain clear, concise, and accurate records which should be submitted in a timely manner to the appropriate persons.

Equipment

Typically, the contractor submits a description of equipment that is intended to be used after bids have been taken and the contractor is being considered for the work. The engineers review the equipment and determine its suitability. When construction commences, the inspector must document that the equipment is as described in the original proposal. Typically, this includes the crane size, length, and fixity of the leads; hammer type and size; cushion material; and type and size of drill or prejetting equipment if required.

If augercast piling is employed, the concrete pump, stroke counter, and pressure meter; auger diameter; and type of auger bit must be documented. It is very important that the concrete pump be calibrated so that the quantity of grout placed with each stroke is known.

A typical check-off list of items to be documented follows:

1. Hammer and associated equipment
   a. Hammer type, make, model, weight, rated energy
   b. If closed-end diesel hammer, bounce chamber pressure at full energy
   c. Stroke of single-acting or drop hammer
   d. Capblock material, dimensions, properties
   e. Pile cushion material, dimension, properties
2. Drilling equipment
   a. Torque, down pressure
   b. Length of augers, diameter, type of bit
3. Crane
   a. Type, size, boom length, and lifting chart
4. Leaders
   a. Length, fixity
5. Mandrel
   a. Length, diameter, condition, number on site
6. Augercast piles
   a. Pump calibration
   b. Stroke counter
   c. Augers—diameter, length, torque, down pressure
7. PIFs
   a. Drive tube length, inner and outer diameter
   b. Drop hammer diameter, length, weight

Materials

Concrete or Grout. The inspector must be aware of the concrete or the grout mix design, must know the results of the strength tests, and must receive a copy of each concrete or grout delivery ticket. The inspector must confirm, if not personally perform, slump tests, flow cone tests, and the making of concrete cylinders or grout cubes. The inspector must document the results of these tests and document any additives which are added to the batch after delivery to the site. The inspector must record the time of arrival of each batch and, if required, the wait time prior to use.

A typical check-off list of items is as follows:

1. Design mix—test breaks
2. Admixtures, type, quantity, where added
3. Delivery tickets
4. Results of flow cone and slump tests

5. Preparation of test cylinders
6. Documentation of strength test results

Steel Piles (Pipe, H Beams, Monotubes). The inspector should confirm that pile dimensions, lengths, taper, and thicknesses are in accordance with the specification. The inspector must be furnished mill certificates which confirm that the material properties meet the requirements of the specification. If there is any doubt about the material, the inspector should require that coupons be cut and submitted for chemical and strength tests. The inspector should document the condition of the piles. If the material is coated for corrosion protection, a certificate of compliance with the specification is required. Also, the condition of the coating should be documented and any damage corrected in accordance with the specification. Corrosion or damage is grounds for rejection of the piles. If special pile tips or cover plates are required, they should be noted. All welding should be in compliance with the specifications. Piles should be stockpiled on blocks above the ground to avoid damage.

A typical check-off list of items is as follows:

1. Mill certificates
2. Lengths, size, thickness
3. Certificate of compliance for special coatings
4. Condition of piles, coatings
5. Welding
6. Special accessories, tips, bottom plates, splice connections

Prestressed Concrete Piles. The inspector must be furnished with certificates stating that the piles have been manufactured in accordance with the specification. In particular, for newly cast piles, the date of manufacture and the expected strength at date of driving must be in accordance with the specification. The length, size, and condition of the piles should be documented. Variations in sizes should meet the minimum specified tolerances. Specifically, piles should be inspected for defects such as honeycombed surfaces, spalling, cracks, exposed steel, and lack of chamfer. Splice connections should be inspected for conformity to the specification and general condition.

Proper handling and storage of precast concrete piles is very important. Pickup points and pickup slings should be placed at the correct locations along the pile to prevent excessive bending stresses. Piles should be stored by blocking at specified locations to prevent excessive bending stresses. In the event that the proper support points are not clearly marked on the pile, the engineer or the manufacturer should be contacted to establish those points.

A typical check-off list of items is as follows:

1. Certificates, date of manufacture, design strength
2. Lengths, size, in accordance with specified tolerances
3. Condition of piles
4. Lift points and blocking points clearly marked
5. Splice connections

Shell Piles. The inspector shall document that the shells have the specified dimensions, gauge thickness, and boot or bottom-plate thickness.

Timber Piles. The pile inspector should receive certificates from the supplier stating the type of wood and that the piles are in accordance with ASTM D-25, latest revision. The pressure-treatment should be in accordance with AWPA Standard C1, C3, or C18. Piles should be inspected for conformance to the specified dimensions, either the circumference of the pile butt (measured 3 ft below the butt) or the circumference of the pile tip. If tip reinforcement is specified, the condition, size, and shape of the tip reinforcement should be documented.

The inspector should also inspect for evidence of decay or insect attack, damage, shakes, cracks, splits, grain twist, and general soundness. Pile straightness and number of knots should also be noted. Any holes or cuts should be treated with the appropriate preservative in accordance with the specifications. Timber piles should be stored off the ground in a manner to prevent damage or decay.

A typical check-off list of items is as follows:

1. Certificates indicating pressure treatment, pile quality
2. Lengths, size, meets specified tolerances
3. Quality of piles—decay, damage, shakes, cracks, splits, grain twist, general soundness, straightness
4. Tip reinforcement
5. Storage

Steel Reinforcement. The inspector should document that reinforcing steel is in accordance with the specification with respect to the ASTM designation, length, and spiral diameter. If reinforcing cages are required, cages should be welded if they are to be used in a drive tube with compacted concrete, or if the cages are to be installed after the concrete is placed. The number of bars in the cage and their length should be compared with the design drawings. The inspector should check for adequate centralizers to provide the minimum concrete cover. Cages should be stored above the ground, and all mud, rust, and scale should be removed before the cages are placed in the concrete.

A typical check-off list of items is as follows:

1. Longitudinal bars, length, size, diameter, spacing
2. Spiral steel, type, size, diameter, spacing
3. Hoops, type, size, diameter, straightness
4. Centralizer

Driven Pile Installation

General. Basic data for each pile should be recorded by the inspector as the piles are installed. Each pile should have a unique pile designation. Piles should be marked before being taken into the leaders. If continuous driving records are required, the pile should be marked at every foot mark, and the distance above the pile tip should be marked on the pile in 5-ft increments. The length of pile taken into the leaders should be recorded, together with any additional pieces which are subsequently added to the original length. The inspector should require 1-in marks on the pile for final driving.

The pile-driving inspector should confirm that the piles are located correctly and that they are plumb or have the correct batter. The location of the pile may be inferred by checking the distance from adjacent piles or pile stakes. The location of pile groups and pile spacing may be compared roughly with the drawings. Major discrepancies should be apparent, small deviations will not be known until the as-built survey is performed. Plumbness may be determined by placing a long carpenter's level against the pile in two vertical planes 90° to one another. If steel H piles are being installed, the pile orientation should be checked against the drawings.

If preexcavation is required, the inspector should record the type of predrilling, i.e., augering, wet rotary drilling, prejetting, etc., and the depth to which it was performed.

The driving resistance should be recorded, and the final acceptance should be signaled to the contractor. The acceptance criterion may be a final driving resistance, a final tip elevation, or a final driving resistance below a minimum tip elevation.

Any indication of pile damage or any delays in completing the drive should be noted on the record. The time of start and finish driving should also be noted.

It is particularly important that test piles have a carefully documented record for the complete driving, including the time and date. If retapping of piles is performed, all retap data should be recorded.

Typically, pile butt elevations are surveyed immediately after driving and again after all piles in the immediate vicinity have been installed. If any pile heave is noted, piles are retapped. The butt elevations and the retap data must be included with the record.

Top-Driven Steel Piles. The inspector should ensure that all welding of pile attachments or splicing is in accordance with the specification. Pipe piles must be inspected prior to concreting for contamination or excessive water. Excessive water is considered to be more than 3 to 4 in. If pile crippling is observed at the pile top, the pile should be carefully trimmed to remove the buckled section and the pile redriven.

Prestressed Concrete Piles. The inspector should inspect the depth of the pile cushion for each pile, and replace the cushion when excessive nonelastic deformation occurs. Any sudden change in driving resistance or movement of the pile butt should be carefully documented. Overdriving the pile to bring the pile butt to the design grade should be avoided.

Shell Piles. The inspector must inspect the shell prior to filling to ensure that the pile is clean and dry. On sunny days, a mirror is an excellent means of illuminating the bottom of the pile. Alternatively, a light on a cord may be used. If the shell is leaking and water is entering the casing faster than it can be removed prior to concreting, it may be necessary to tremie or pump concrete from the bottom up under a static head of water.

Cast-in-Situ Piles. The top of the pile should be formed with a steel sleeve to prevent contamination with soil, and to establish the size and location of the pile center. If rebar must be inserted into the pile after concrete placement, the condition of the rebar must be acceptable to the inspector and the method of placement must be such that the cage remains intact and has sufficient concrete cover. When forming adjacent piles, the concrete level of completed piles not yet set should be monitored to ensure there is no drop in the concrete.

Timber Piles. If pile shoes or butt bands are required, the inspector should document that they have been properly attached. If any holes or pile damage below the butt are observed, the pile should be protected with the specified pile preservative.

PILES AND PILE DRIVING

Timber pile driving must be carefully monitored to document any evidence of breaking. In particular, a sudden decrease in driving resistance or sudden movement of the pile butt should be noted.

Prolonged hard driving should be avoided. If pile breakage is suspected, the engineer and the contractor should be notified immediately. Any damaged or broomed section at the pile butt should be cutoff.

Augercast Pile Installation. Proper construction of the augercast pile is very dependent on the skill of the operator. It is very important that a knowledgeable inspector observe and document the construction of each pile. Leads must be properly marked so that the inspector knows the length of each pile. If auger flights are added as the pile is advanced, the contractor must cooperate with the inspector to confirm the length of the pile. During construction of the pile, the inspector must observe that the auger is withdrawn at a constant rate equal to the inflow of grout. Typically, the inspector will count and record the pump strokes for each 5 ft of pile. If there is a significant reduction from the theoretical volume of the pile, the inspector should require that the pile be redrilled at least 5 ft beyond the point where the reduction in grout volume was first observed.

Each pile should have a grout take at least 15 percent greater than the theoretical volume. If the take is significantly less than this, the pile should be redrilled. If the take is much more than 15 percent of the theoretical volume, it is likely that grout is going into underground openings. The cause for the grout loss should be investigated, and the pile may have to be rejected and replaced.

The inspector should document the placement of the rebar steel. Cages should be tack-welded and centralizers used to keep them in location.

The top of the pile should be formed with a collar. The grout at the top of the pile should be screened to remove any soil lumps and dipped to the proper elevation. If the temperature is below freezing, the pile tops must be protected. If subsidence of the concrete occurs, the pile should be rejected.

A typical check-off list of items is as follows:

1. Required pile lengths
2. Actual pile lengths
3. Volume of hole versus grout pumped, in 5-ft increments
4. Grout pressure drops
5. Placement of reinforcing steel
6. Forming of pile butts
7. Grout subsidence

Pressure-Injected Footing Installation

PIFs depend on creating an expanded base in granular soils. Consequently, it is very important that sufficient subsurface data is available so that the depth of the bearing stratum can be determined in advance at the location of each PIF. If the PIF is being founded in a relatively thin stratum overlying a fine-grained cohesive deposit, the bottom elevation of the bearing stratum is also important.

The concrete in the base should be zero-slump concrete, and placement of concrete in the drive tube should be in measured batches, to allow the establishment of the impact energy per unit volume of concrete. Creation of the expanded base cannot commence until the driving resistance, combined with the subsurface data, confirm that the tip of the drive tube is within the bearing stratum. The specification must provide the required energy per blow, the standard batch volume, and the average number of blows required to inject 1 ft^3 of concrete, during the injection of the last batch. The height of concrete within the drive tube during the injection of the last batch of concrete should not be more than one third of the drive tube inside diameter.

During construction of the base and of uncased shafts, the inspector should be sure that the top of the concrete does not fall below the tip of the drive tube. If this were to occur, it would imply that soil had contaminated the concrete, and the PIF must be rejected. The inspector should observe a mark on the drop cable which would indicate the location of the bottom of the hammer.

Uncased compacted concrete shafts should be constructed with zero-slump concrete and should be compacted in a controlled manner as the drive tube is withdrawn. PIFs placed through fine-grained soils within heave range of one another should be predrilled through such soil. Uncased compacted concrete shafts should not be formed in very soft, fine-grained soils.

The inspector should document the quantity of concrete installed in each base and in each shaft.

Uncased concrete shafts may also be constructed by filling the drive tube with high slump (8-in minimum) concrete. During withdrawal of the drive tube, the level of the concrete should have a positive head with respect to the external soil and water pressures at all times. As a minimum, the shaft should be provided with four full-length number-five reinforcing bars evenly spaced near the shaft perimeter.

PIFs located less than 9 ft from a completed uncased high-slump shaft should not be installed until at least 12 h after pouring the shaft.

If permanent casing is used, the casing should be fastened to the enlarged base so that the two will not separate. The void between the casing and the surrounding soil created by the withdrawal of the drive tube should be backfilled with sand or grout to develop the lateral support of the PIF.

A typical check-off list of items is as follows:

1. Elevation of base, elevation of bearing stratum
2. Quantity of concrete used to form base
3. Number of hammer blows/unit volume for last batch of concrete
4. Placement of reinforcing steel
5. Compacted uncased shaft—number of batches—blows per batch
6. Permanent steel casing—size, length, gauge
7. Void between casing and soil filled with grout
8. High-slump concrete—volume placed

REFERENCES

1. F. M. Fuller, *Engineering of Pile Installations,* McGraw-Hill, New York, 1983.
2. U.S. Department of the Navy, Naval Facilities Engineering Command, NAVFAC DM 7.2, *Foundations and Earth Structures,* Government Printing Office, Washington, D.C., 1986.
3. PCI Committee on Prestressed Concrete Piling, "Recommended Practice for Design, Manufacture & Installation of Prestressed Concrete Piling," Precast/Prestressed Concrete Institute, Chicago, Ill., March-April 1993.

4. Monotube Pile Corporation, P.O. Box 7339, Canton, OH 44705-7339, a subsidiary of the Davidson Pipe Supply Company, Canton, Ohio.
5. Ray Step Pile Company, 61 Chaucer Drive, Berkley Heights, N.J., a subsidiary of the Davidson Pile Supply Company, Canton, Ohio.
6. CONTECH Construction Products, Inc., P.O. Box 800, Middletown, Ohio 45042.
7. PILE BUCK, Inc., "Pile Hammer Specifications Chart," P.O. Box 1056, Jupiter, Fla., 1994.

CHAPTER 4
COFFERDAMS AND CAISSONS

Ben C. Gerwick, Jr.
*Professor Emeritus, University of California at Berkeley and
Chairman, Ben C. Gerwick, Inc. of San Francisco*

INTRODUCTION

Cofferdams and caissons are devices used for the construction of bridge piers and other structures which extend into water or unstable soils. Although they are primarily means to an end and often serve only a temporary purpose, some of these caissons and cofferdams rank as major construction achievements. The design of the larger caissons and cofferdams requires detailed engineering analysis and the utmost judgment and experience as to the loads from water, soils, ice, current, and other external sources, plus evaluation of such problems as erosive scour and barge impact. The designer must be primarily a constructor, for the adverse conditions under which cofferdams and caissons are often built require that a thoroughly practicable method of construction be adopted. While some of the larger caissons have been designed by the bridge engineer, the usual practice is to make the contractor responsible for the design and construction of an adequate temporary structure.

The specified requirements as to allowable tolerances, the methods of pouring concrete and the location of construction joints, etc., have frequently been the controlling factors in determining the type of cofferdam or caisson which can be used. Unrealistic requirements and arbitrary specifications have often made the construction far more costly and hazardous than it need have been. On the other hand, some specifications have been entirely too lax and indefinite. Where serious troubles have been encountered because of changed field conditions, improper selection of method and procedure by the contractor, accidents in construction, or unrealistic requirements, it has frequently been found necessary and expedient to reduce and modify the requirements of the specifications as a field change. Therefore, a great deal of thought should be given to setting forth the criteria on which the contractor is to base methods and designs. To rely entirely on the phrase "subject to the approval of the engineer" is neither sufficient nor equitable. The engineer should specify the limitations or prohibitions needed to safeguard the design and adjoining structures and then allow the maximum flexibility to the contractor. To protect against an inadequate or unsafe design, it may be specified that the design and procedure for the cofferdam or caisson be prepared and signed by a registered engineer.

Perhaps in no other branch of construction have there been so many failures. Some of these have been spectacular and disastrous, while others have been minor and either partially corrected or else compensated for by a design change. It is striking to note that serious troubles have sometimes been encountered by experienced contractors and engineers. This emphasizes that each bridge pier cofferdam or caisson must be treated individually to meet the particular conditions under which it is to be built and that care, skill, and alertness are required.

In retrospect, the cause of every failure has been determinable. Unanticipated characteristics, conditions, and behavior of the soil have caused the majority of the troubles. This indicates that a detailed study of the soil should be undertaken at every important caisson or cofferdam site. The bridge engineer may have taken borings but is primarily interested in the supporting strata, whereas the contractor is interested in the overlying soil profile.

A great number of factors must be considered by the designer or constructor of a large cofferdam or caisson. Most major cofferdams and caissons involve three-dimensional frames; continuity considerations, deflection analyses; column stability; and determination of current pressures, hydrostatic pressures, pile uplift values, soil pressures, soil permeability, and temporary concrete pressures. Superimposed and accidental loads must be considered. The structure must be practicable to build. Spatial constraints of the various structural elements during construction, the method and quality of underwater connections and the stages and sequence of constructing the pier itself will all introduce problems which must be considered.

Many of the serious failures have occurred in the course of attempts to correct minor deviations or to facilitate pier construction. While it seems hard to believe, there are numerous instances where a strut has been left out to expedite excavation or has been notched to accommodate prefabricated forms for concrete. To correct minor listing of caissons, extensive dredging and other radical steps have been taken in haste and without full engineering consideration. Symmetry of delicate caisson-sinking operations has been totally neglected in order that one crew might dredge more yardage than the previous shift. Cofferdams have been pumped down "just a few feet" without realization that a large percentage of the total head is acting, since the pressures vary as a function of the square of the depth. Sheet-pile setting and driving records have been established, only to find one sheet pile driven out of the interlock. So cofferdams have collapsed, and the bottoms have blown, and the caissons have tilted or shifted otherwise, but such results are always traceable to some error in design, procedure, or construction.

A cofferdam or caisson must not only lend itself to rapid construction but be economical as well. Critical path method (CPM) scheduling is particularly well adapted to cofferdam and caisson construction planning. The confined area of operations and the numerous engineering considerations require a step-by-step planning procedure whereby temporary and permanent construction operations are integrated. The proper use of a critical path schedule may show, for example, that selected operations such as sheet-pile driving or excavation should be performed on an overtime or shift basis, or that all forms and reinforcing steel should be prefabricated in order to minimize their installation time. The removal of a cofferdam brings new problems of economic gains versus costs and the interference with and possible effect on the pier itself.

Finally, the human factor must be considered in all dimensions. Safety is paramount, as there have been far too many accidents and deaths in cofferdam and caisson construction. The desired design and construction procedures have to be transmitted to the actual builders, superintendents, and workers in such a way as to ensure compliance. During construction, the actual results must be continually checked with the design and appropriate steps taken to correct any new factors or conditions which arise. Good design and construction procedures must be made effective through constant supervision.

As opposed to major caissons and cofferdams, many hundreds of small cofferdams have been built in shallow water and in soil; these have been used as building foundations and in construction of small bridge piers. Although reduced in size, the same principles, risks, and precautions still apply.

COFFERDAMS

General Considerations

Functions. A cofferdam is a temporary structure designed to keep water and soil out of the excavation in which a bridge pier or other structure is to be built. Usually cofferdams are dewatered, in whole or part, so that the structures may be built substantially in the dry. However, there are many instances where the structure has been partially or wholly built underwater, without dewatering the cofferdam. In the first case, the cofferdam must exclude both water and soil; in the second case, it has to exclude only the soil.

Cofferdams are much more expensive, in cost per cubic yard of material removed, than open-cut excavation. But where free-flowing water, or unstable soils, or heavy surcharge loads are encountered, open-cut excavation is generally impracticable and the use of cofferdams may be the best approach. It is often necessary to seal the bottom of the cofferdam so that it can be dewatered. To accomplish this, underwater concrete is placed in a sufficient thickness so that, by the concrete weight alone or in combination with piles, it can resist the uplift pressure. Other types of seal have been used in special cases, including clay blankets and concrete on the outside; however, the normal and usually the most economical method is to place a seal of underwater concrete inside the cofferdam.

Types. Many successful shallow cofferdams for bridge piers have been built using earth dikes. In some rivers, sand dikes have been used, with wellpoints to dewater both the dike and the excavation.

Steel sheet piling combines moderate watertightness, high strength in bending and shear, high interlock strength, ease of driving and removal, and high salvage or reuse value. While some shallow cofferdams resist the pressure of water and soil entirely by cantilever action of the sheet piles, most cofferdams require additional bracing or similar support. The usual bridge pier requires a rectangular cofferdam, and for this an interior bracing system is normally required (Fig. D4-1). Bracing systems usually consist of horizontal wales and struts, with as many levels as necessary to resist the external forces (Fig. D4-2). Vertical soldier beams, with horizontal wales and struts, have occasionally been used to reduce the number of levels of cross bracing or to enable them to be placed at more favorable elevations. In other cases, timber cribs have been sunk by filling them with rock, then sheet piling is driven on the outside face. This method requires a much larger cofferdam in plan, but it is a good solution where there is bare rock bottom with little or no possibility of obtaining a toehold with the sheet piles and where there may be considerable current or swell from ocean waves.

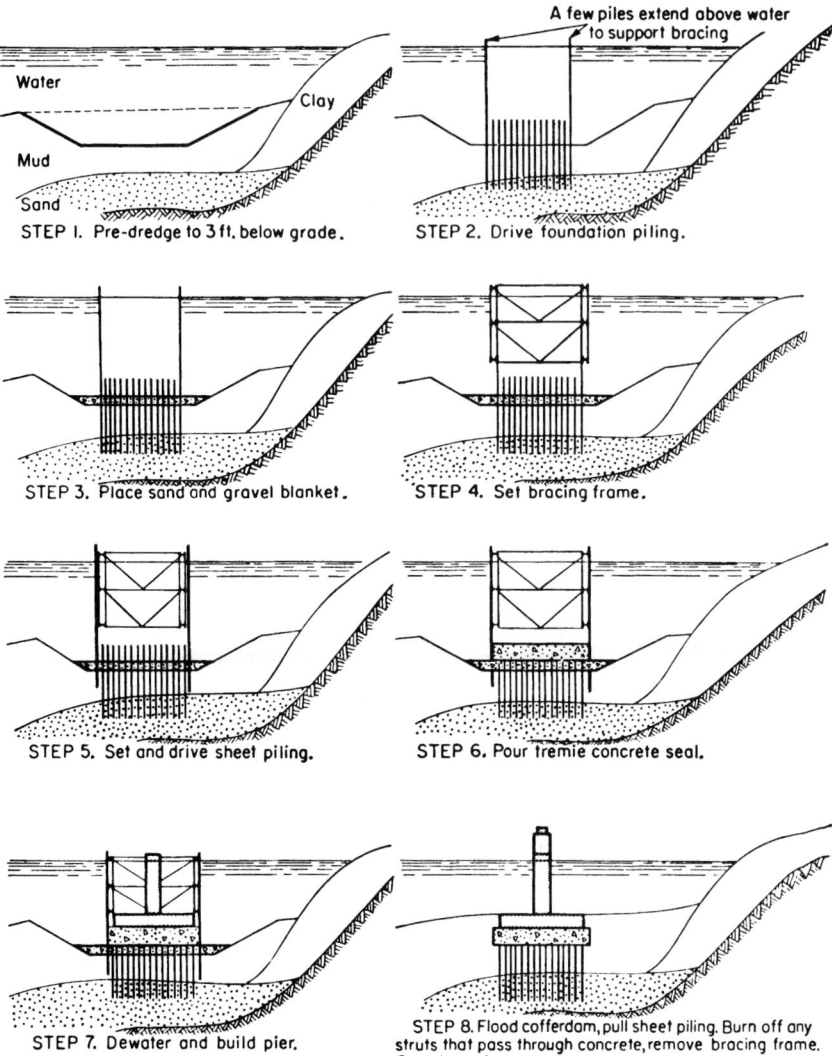

FIGURE D4-1 Typical cofferdam construction sequence.

Some of the largest and deepest cofferdams have been built in a circular or elliptical plan. In some instances, a double row of piling has been driven and the space between the walls filled with concrete. Thus, the cofferdam acts as a ring, and no bracing is required. In other instances, a single wall of sheet piling is driven, with rings of cross bracing. A recent development is to construct a circular cofferdam of reinforced concrete, employing the slurry trench diaphragm wall methods. In this case, the resistance to the lateral loads of water and soil are taken in ring compression.

FIGURE D4-2 Large sheet-pile cofferdam. Note corner and intersection details and electric pump in foreground to maintain dry cofferdam. Steel strut was damaged by digging bucket, a possibility for which allowance must always be made in design.

Another type of cofferdam which is valuable in special cases is a composite box cofferdam-caisson. In this case, the entire bracing system, together with the walls and sometimes with the bottom, is floated or set in place on a previously prepared foundation. This type of cofferdam can be used only where no overburden is present or where it can be removed in advance. If the cofferdam is to be set on rock, its bottom is usually tailored to fit the rock contours closely. The prepared foundation may consist of underwater timber piles cut off to grade, or a dredged hole backfilled and leveled with crushed rock or sand or hardpan carefully dredged to grade and leveled with sand backfill.

Cellular cofferdams of sheet-pile cells can be used for very large and deep bridge piers; however, their use is rare because they do not fit the plan shape of the usual pier and because they are relatively expensive. They may be a practicable solution in very swift water or in exposed locations where progressive construction is necessary. In such cases, they would have the same general function as the circular cofferdams.

Some bridge piers for major structures have been constructed entirely under water, a method which has many advantages in time of construction, economy, and reduction of risk. High-quality underwater concrete can be obtained provided that the mixes and techniques which have been developed in recent years are followed carefully and properly. To realize the benefits of this method, design must be integrated with construction methods.

External Loads. The external loads on cofferdams are principally those of water (hydrostatic) pressure and soil pressure. In addition, the cofferdam may have to resist current pressure, ice loads, log jams, and wind and wave loads. The soils may be higher on one side than the other, as on a steep bank, throwing an unbalanced load on the cofferdam. There can also be scour around and under the cofferdam due to currents.

Hydrostatic pressure is the fluid pressure due to the weight of water. In fresh water, the pressure is 9.8 $KN/m^2/m$ depth (62.4 lb/ft^2 per foot of depth); in ocean salt water, it is 10 $KN/m^2/m$ depth (64.4 lb/ft^3 per foot of depth); and in the brackish water commonly encountered in tidal bays and estuaries, it runs somewhere in between. The variation in unit weight will assume real importance only in an inland salt sea, where the pressure of water may run as high as 12.5 $KN/m^2/m$ (80 lb/ft^2 per foot).

Soil pressure against a structure is a highly technical and complex subject about which many good books and technical papers have been written. The characteristics and properties of the several strata surrounding and underlying the cofferdam must be determined first. The effective pressures can then be computed.

A rough approximation of the soil pressure can be obtained by the method of equivalent fluid pressure. The method assumes that the soil will act as a fluid, which is, of course, seldom true. For piers in water or saturated ground, full hydrostatic pressure is assumed to act from the water level down to the bottom of the excavation within the cofferdam. In addition, the submerged-soil pressure is assumed to be distributed and to act like a fluid. The equivalent fluid pressure of the submerged soil, including the water, will generally be in the range of 11 to 12 $KN/m^2/m$ of depth (70 to 80 lb/ft^3 per foot of depth). This pressure is assumed to act from the top of the soil down to the bottom of the excavation. If the water table is below the top, then soil pressure should be separated from water pressure and the combined values added to determine the total external pressure at any level.

Having determined the total load from the soil on the basis of equivalent fluid pressure, experienced judgment may then be applied in predicting its variation with depth. Instead of a triangular (hydrostatic) pressure distribution, a rectangular, parabolic, or trapezoidal distribution may be postulated. This increases the design loads in the upper levels of the cofferdam and reduces them in the lower levels. In very soft soils, muds, and weak clays, the soil pressures may continue to have a net inward force for 2 to 10 m (6 to 30 ft) below the excavation, thus increasing the unsupported length for the sheet piles.

The great advances of soil mechanics, and particularly the dissemination of knowledge of this subject through the engineering profession, now make it practicable to make a more careful investigation of soil pressures on major cofferdams. The cost of a thorough investigation by a competent engineer will usually be repaid many times in economy of proper design and in safety.

In any method of determining soil pressure, consideration must be given to the effect that driving the sheet piles and building the cofferdam will have on the soils. Driving, and especially jetting, may open a passage through normally impervious strata. Driving will remold and disturb clays. If the bottom of the cofferdam blows, even without complete failure, the condition of the soils and the pressure exerted by them may be suddenly and radically changed.

A cofferdam is a flexible structure, and the deflection of the sheet piles serves to equalize pressures and distribute them. It can also cause the soils to yield and thus exert greater pressures. With a cofferdam in deep water (or in soils which behave like a fluid), considerable pressure is exerted on the full height of the sheet-pile walls just as soon as dewatering is under way. The total pressure varies as the square of the depth, and with an outside depth of, say, 15 m (50 ft), one-half of the total pressure is on the sheet-pile walls when the inside has been pumped down only 5 m (15 ft). While this is obvious, it is frequently proposed to pump down in stages and install each bracing set as its level is reached, without recognizing the pressures acting just prior to installation of the new level of bracing. Again, when the structure inside is partially completed and it is necessary to remove a set of bracing, a proposal is often erroneously made to flood the cofferdam to that level and then take out that set of bracing.

Where deep cofferdams must be constructed through a great depth of soil (as opposed to water), it may be necessary to install each level of bracing as its elevation is reached in excavation. In this case, complete calculations have to be made for each stage of excavation. Additional sets of bracing may be required to take care of the pressures at the several stages as compared to those necessary to take care of the pressures at the final stage.

One technique that has proved economical in such circumstances, is to purposefully fill the excavation with water, pumped in if necessary. Then any remaining bracing sets are lowered through the water as the excavation proceeds and blocked to the sheet piles by a diver. By filling the sheeted excavation with water to as high or even higher than the groundwater elevation outside, considerable control of pressures can be effected during the construction stage.

For small shafts or cofferdams, a bentonite slurry may be used to offset and balance the external pressures. Slurry densities of 70 to 80 pcf are used (higher ones are possible). Tremie concrete can be poured through this bentonite slurry just as it is through water; the major difference is a reduction of bond on the sheet piles. Brine densities of up to 75 pcf can be accomplished by saturated solutions of rock salt (NaCl) and even higher densities can be achieved, although at very high expense, with calcium chloride and sodium silicate. Obviously, subsequent disposal of these solutions must comply with environmental regulations.

Ice pressures, both static and impact, may impose extremely severe loads on a cofferdam. Solid sheet ice exerts pressures of 100 to 300 lb/in^2. This is so great that it is usually uneconomical to design any cofferdam to resist more than a few inches of ice thickness, so the most practicable approach is to create open water around the cofferdam. Moored logs, through their ability to absorb and retain heat and their cushioning effect, have been found very effective in protecting cofferdams against crushing. Similarly, coal dust spread on the ice may absorb radiant heat from the sun, while salt and other chemicals may destroy the structure of the ice. Explosives are generally ineffective in breaking up ice while at the same time endangering the cofferdam. Their use is not recommended.

Moving ice is also dangerous to a cofferdam because of its tendency to raft and pile up. A log-boom shearwater can be effective in deflecting ice cakes round the cofferdam.

Moving *logs and debris* during runoff periods may endanger the cofferdam, both from local impact of a single large log and from rafting and piling up. A log boom, well anchored, can be utilized to keep floating logs and debris away from the cofferdam. During severe flood runoff stages, the cofferdam should be filled with water.

Current pressure produces relatively small loads except when the current is very swift. For example, in a river of 3-m/s (10-fps) velocity (about 6 knots), the average unit pressure on the upstream side of a rectangular cofferdam would be about 3KN/m^2 (130 lb/ft²). However, the total unbalanced current load may distort the cofferdam or even overturn it. The total current load may be computed by the formula

$$P = CAW \cdot \frac{v^2}{2g}$$

where P = total current load
 C = shape factor × roughness factor
 A = area of cofferdam wall exposed to current
 v = velocity of current
 g = acceleration due to gravity
 w = unit weight of water

Units must be consistent, either SI or English.

For rectangular cofferdams constructed of steel sheet piles, use $C = 1.7$. For circular cofferdams of steel sheet piles, use $C = 1.0$. These increased factors over the shape factor are due to the drag caused by the sheet piles, with their castellated sides. The pile-up of debris causes both increased area and increased drag. A number of bridge pier cofferdams have been destroyed by floods in this manner, even though they have been preflooded in an attempt to stabilize them.

Guide structures employing vertical and batter piles may be used to hold the cofferdam in place during the construction operation. A barge moored upstream will break the current on the surface: This will help in setting sheet piles, although it provides little relief for total pressure against the cofferdam. In a swift river, special protection may have to be provided for the cofferdam to reduce the current pressure and to still the water during setting and driving operations—for example, deflectors consisting of independent jacket structures and sheet piles, pinned to the riverbed by piling.

Scour is a frightening phenomenon which may destroy a cofferdam or caisson in short order. The very act of building the cofferdam places a block or dam in the river, and the resultant increased current may be further aggravated by floating barges and moored derricks. Such scour may occur where pier construction tends to partially block even relatively wide rivers, such as the Mississippi. It can also occur in tidal bays and estuaries where the tidal current is locally high. The combination of river flood and runout on ebb tide can be especially destructive at the mouths of river estuaries.

Scour is usually deepest in the eddy at the corners of a rectangular cofferdam. Rounding of the upstream corners and the addition of a few sheet piles tailing back from the two downstream corners like streamlining fins can be of considerable help. This design was successfully used on the sheet-pile cofferdams for dams on the Mississippi. Providing a deflecting nose of sheet piles helps to prevent scour immediately behind the upstream corners (Fig. D4-3). In many rivers or estuaries where currents are swift and scour is anticipated, the prior placing of willow mattresses or of 2 or 3 ft of small rock (3 in minus) over the entire area, before setting the sheet piles, is effective in preventing the erosion from ever starting.

Mattresses may also be fabricated by stitching two layers of filter fabric to each other, one layer of the randomly oriented fibers, with fine mesh, and the other of oriented fibers, selected for strength. These fabrics should be negatively buoyant, i.e., having a specific gravity greater than one, and held in position by dumped stone.

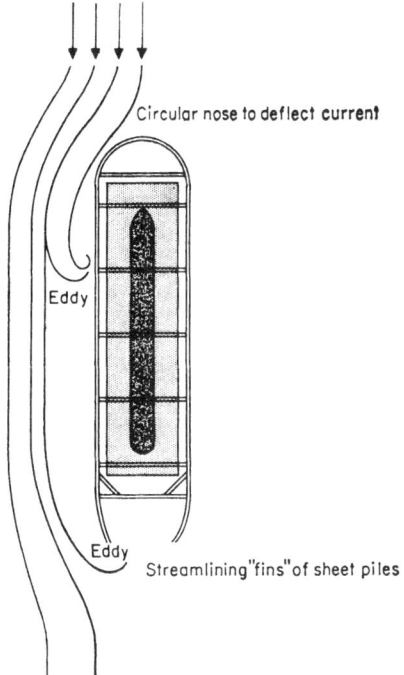

FIGURE D4-3 Schematic drawing of cofferdam designed to reduce scour due to fast river current over sandy bottom.

Sheet Piling

Types and Sections. Most cofferdams today are built using one of the standard rolled sections of steel sheet piling. Timber sheet piling has been widely used in the past, and fabricated-steel sheet-pile sections are occasionally needed and used on very deep or unusual cofferdams.

Standard rolled-steel sheet-pile sections are manufactured by the larger steel companies in the United States. Lighter sections for shallow cofferdams and trenches are manufactured by several other steel companies. There are also German, Japanese, Canadian, Belgian, and English standard sheet-pile sections, some of which are significantly heavier and stronger, so as to suit the need of deep cofferdams (Fig. D4-4). When ordering foreign sheet piles, a few points of difference in practice must be remembered. Most foreign sheet piles do not have as high interlock strength, nor are the interlock tolerances as close as in domestic sheets. Also, at least some of the foreign sheet piles have excessive sweep in the longer lengths. Both these matters can be taken care of by specifying exactly what tolerances are needed and acceptable for the particular job. In listing the section modulus for a sheet pile, the U.S. practice is to list the section modulus for an individual pile. Foreign practice is usually to list the section modulus for a linear foot of wall, assuming adjacent piles turned in opposite directions and full interlock friction. For the same deep-arch steel sheet-pile sections, therefore, the foreign catalog would show approximately twice the section modulus that is shown in the U.S. catalog.

Similar sections of the several U.S. manufacturers will also interlock with each other, although there may be an increased tendency to bind in the common interlock.

The Z type of steel sheet piling is being used more and more for deep cofferdams because of its very high strength in bending. The ratio of section modulus to weight (and cost) is more favorable for Z types than for the deep-arch types. However, in setting and driving there is a tendency for Z piles to roll and twist, and they may bind in the interlocks in both driving and pulling. There also seems to be a greater tendency for Z piles to drive out of the interlocks under very hard driving. Z piles are made with vertical webs and also in sections having inclined webs. The inclined-web piles are much more likely to twist. While a sheet pile of greater strength decreases the number of sets or levels of bracing, the total load to be carried across remains the same. Thus, there may be a saving in the number of bracing levels, but not necessarily a significant reduction in total weight of bracing.

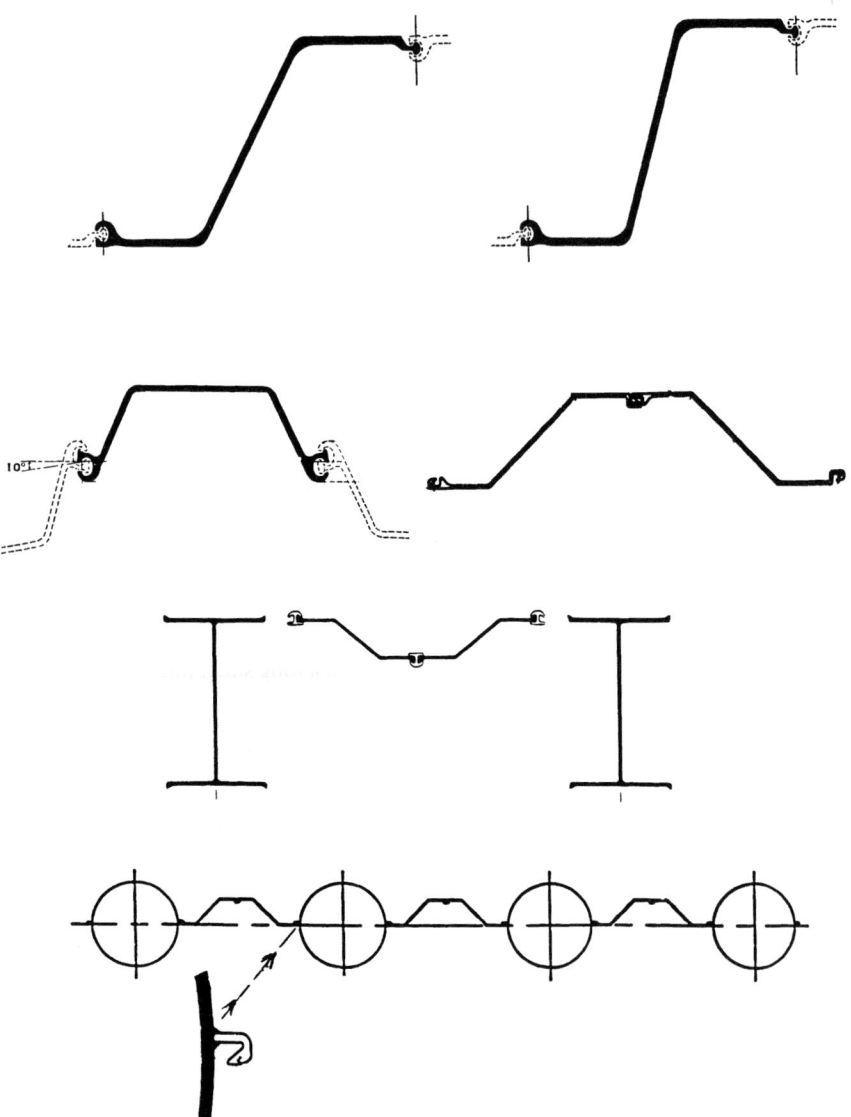

FIGURE D4-4 Typical configurations of steel sheet piles used for braced cofferdams. Consult manufacturer's catalogues for details and properties.

In recent years, even heavier and stronger sheet piles are coming into use; sections fabricated from wide-flanged beams and pipe piles. These enable even greater distances between bracing sets and often permit the cofferdam to be braced with a single above-water set. Although thinner-plate sheet piles show an apparent benefit in weight-to-strength ratio, and, hence, economy in purchase, they are more easily deformed during installation, especially when the tips encounter rocks and hard driving. For the usual bridge pier cofferdam, a thicker and more rugged sheet pile will usually prove most constructible and economical.

The manufacturers' catalogs set forth the properties of the several sections of steel sheet piles and also give a great deal of valuable information on their use. Some manufacturers also publish design procedures and sample configurations.

Fabricated-steel sheet piling is sometimes used for very deep cofferdams or where it is necessary to have a very high-strength sheet pile to span between very wide spread bracing levels. A very strong sheet pile can be made up by welding or bolting an I or WF beam to the back of a flat or shallow-arch sheet pile. Cofferdam walls have been built up using these *soldier beam* sheet piles spaced several feet apart horizontally, with connecting arcs of flat or shallow-arch sheet piles. These arcs act in tension through the interlocks. This scheme was used on the East Bay cofferdam piers for the San Francisco–Oakland Bay Bridge in order to eliminate all underwater bracing. The 36-in wide-flange beams were riveted to a shallow-arch sheet pile on its inner face. Three shallow-arch sheet piles formed the arc between adjoining master piles.

Corner sections, Ys, and Ts are available from the manufacturers of the standard rolled-steel sheet-pile sections. For the arch-type sheet piles, corner sections are rolled. For Z-type sheet piles, corner sections are fabricated, using pieces cut from standard sections and riveted together. Corner sections should be selected with care from the manufacturer's catalog to be sure that the fingers of the interlock are turned in the correct direction. Where there is an odd number of sheet piles in a wall, a different corner piece is required than where there is an even number. Corner sections for arch-type sheet piles can also easily be bent from a section of shallow-arch sheet pile. Most shipyards are equipped to bend even a long sheet pile to any desired angle. Deep-arch sections may be rotated through approximately 10° per pile, so as to form an arc and a rounded corner.

Alloy steel sheet piles have recently been produced by domestic and foreign manufacturers. These give significantly higher yield-point stresses, thus permitting higher design stresses in bending and longer spacing between bracing frames (Table D4-1). This may be of major economic significance when the sheet piles are driven first and the bracing is placed as excavation proceeds. Again, it may permit better spacing of the frames to miss the concrete structure inside, and fewer penetrations. On one major bridge pier cofferdam, the use of alloy sheet piles allowed the bracing frame to be placed as a space frame instead of being lowered piecemeal and bolted in place by divers after excavation.

With the use of higher-yield steel and consequently greater spacing of bracing sets, deflections will be greater, which is not too important in water but which may cause greater movements or higher pressures in soils. The higher yield point reduces damage in driving, both at the tip and in the interlock. Since, presumably, salvage

TABLE D4-1 Properties of Alloy Steel Sheet Piling

Grade	Minimum yield point, MPa (lb/in^2)	Minimum tensile strength, MPa (lb/in^2)	Recommended working stress, MPa (lb/in^2)
ASTM A328 (Standard)	265 (38,500)	500 (70,000)	160 (23,000)
Grade 45	300 (45,000)	450 (65–70,000)	180 (27,000)
Grade 50	350 (50,000)	500 (70,000)	205 (30,000)
Grade 55	400 (55,000)	500 (70,000)	230 (33,000)

value is also proportionally higher, alloy sheet piles are being increasingly used, even for cofferdams at moderate depth.

Table D4-1 is a composite of current U.S. manufacturers' data. For more detailed information, the manufacturers' catalogs should be consulted.

Alloy steel sheet piles are weldable, but E-70xx low-hydrogen electrodes are recommended to eliminate the need for preheating. The increased cost of alloy steels over standard grade is from 10 to 15 percent of mill price. Thus, whenever their increased design stress can be utilized, alloy steels will show both direct and indirect economies. In addition to their higher yield point, some alloy steel sheet piles have a greater resistance to corrosion in the sea-water splash zone.

Setting and Driving. Sheet piling can be ordered with one or two handling holes in one end for ease in lifting and in pulling. These handling holes must subsequently be welded or plugged if high tide or floods can reach them. If sheet piles are spliced, particular care should be taken to make sure that all handling holes are welded tight.

Sheet piling must be handled with care to prevent buckling or damage to the interlocks. A minor crimp of the flange on a steel H pile is of little concern; on a steel sheet pile it could cause the pile to drive out of the interlock. Flat-arch and shallow-arch sheet piles have little strength in bending when lifted flat and must be picked up with equalized slings at several pickup points or with a strongback. Another method is to turn them and pick them up on edge instead of flat. Wind gusts can also cause buckling of long flat-arch piles while they are being picked up.

When the sheet piles will be driven only through loose or moderately dense sands or through soft clays, then it may prove economical to preweld the sheet piles in pairs, so as to be able to set them, drive them, and later remove them as a double pile. This is especially useful when a number of similar cofferdams are to be constructed in sequence.

Guide structures are required for the proper setting and alignment of sheet piles. Accurate, well-built guides will more than repay their cost in ease of setting, driving, and pulling and in the accuracy and safety of the cofferdam. The principal guide may be the top level of the bracing, which is usually located at or near the high-water mark. If the water is deep or the soil soft, the sheet piles, when first set, will run well down, leaving only short lengths sticking above the top bracing. When the top bracing is to be placed after the sheet piles have been set and driven, then a temporary inside or outside guide frame may also be used. When the soils are stronger so that the sheet piles stand high after setting, then upper levels of guides must be built against which to set the sheet piles and to secure them against wind.

Wind loads on a high-standing wall of sheet piling can be very serious. One such wall of high-section-modulus sheet piles was bent over flat by a gust of wind, although it had been set against what was thought to be an adequate guide structure. If the guides are on the inside, which is the usual practice, then it may be necessary to provide temporary wire lines as ties. These can be fastened to the top of the sheet piles through the handling hole or to a clip welded onto the arch of an occasional sheet pile in such a way that the clip will pass the wale without hitting it. Ties and upper sets of guides are progressively removed as the piles are driven to grade.

The entire cofferdam should be set before any appreciable driving is begun. Setting should start at one corner. Every tenth pile or so may be driven 1 or 2 m. Particular care should be taken to get the corners exactly vertical and in the correct location. Minor adjustments can then be made in the runs of standard sheet piles since there is about 6 mm (¼ in) play per sheet pile in the interlock. Care must be taken not to gain too much, particularly with Z piles, where a slight twisting will increase the space occupied by a single pile. Some experienced contractors allow about 3 mm (⅛ in) gain per pile in laying out the cofferdam design.

The piles must be set vertically and, should any tendency to lean be noted, setting must be stopped and the leaning piles reset. Some cofferdam builders set the four corners first and then set each way, making connections in the straight wall where there is considerable flexibility for adjustment. Steel sheet piles can often be set very quickly, and many times inexperienced crews will try to make a record in their setting. By so doing, they may cause far more time to be required in subsequent driving and pulling due to friction and binding in the interlocks.

Sheet Pile Driving. Driving should be done a few feet at a time, working around the cofferdam and bringing all sheet piles down approximately together. The corner piles are very rigid and will not yield. If the adjacent walls are driven down ahead of the corner piles and if there is a misalignment or if the piles are out of vertical in either direction, the corner will very likely be driven out of the interlock. If the corner is driven so that its toe is slightly ahead of the adjacent standard piles, then the latter piles will adjust to fit the corner. They can accomplish this by taking the tolerance in the interlock or by a slight spreading of the arch or twisting of the Z. Unfortunately, it is difficult to fit a sheet-pile hammer on a corner pile which does not have its head above the adjacent piles. A solution is to order the corner piles 1 m longer than the standard sheet piles, so that its top will stand high even when its toe is advanced beyond the adjacent pile tips.

Friction in the interlocks can contribute considerable resistance to both driving and pulling and may be greater than the resistance of the soils. In many cases, greasing the interlocks greatly aids setting, driving, and pulling. In some sands, however, the grease and sand may ball up. Before driving, pile interlocks should be thoroughly cleaned of wedged sand and concrete from previous uses and checked for defects. A special zinc-and-grease lubricant for sheet-piling has given very satisfactory results on a number of projects. One note of caution: Greasing of interlocks should never be employed where friction in the interlocks is a design feature—e.g., sheet-pile cells.

After the piles are substantially down, and when driving has become very hard, then an impact hammer may often be used to advantage—for example, a diesel hammer.

A sheet-pile hammer must be properly lubricated. It is a fast-acting piece of machinery, working under adverse conditions and taking hard and repeated shocks. Insufficient oil or the use of the wrong kind of oil causes inefficient operation, excessive repairs, and excessive downtime. The manufacturer's instructions must be followed explicitly.

Where obstructions are expected, such as boulders, cemented gravel, buried logs, old foundations, or riprap, the thickest section sheet pile is preferable (Fig. D4-5). The tips of these piles can be hard-faced, and each pile in succession should be carefully driven for not more than a foot or so. Alloy steel sheet piles are particularly desirable for such use; their higher yield strength resists local distortion and tearing. Where obstructions are known to exist, then holes should be predrilled along the line of the sheets. Typically, one 300-mm (12-in) diameter hole can be drilled, spaced one to each sheet. Alternatively, a backhoe can pretrench.

A drop hammer is sometimes used in initial setting and driving, with built-in guides or legs that keep it aligned with the sheet pile. The height of fall must then be kept to a practical minimum to avoid damaging the sheet-pile head or tip or driving the pile out of interlock, etc.

Driving of sheet piles through sandy soils can be aided by jetting, but this may produce undesirable side effects. It may cut a pathway through semi-impervious

FIGURE D4-5 Steel sheet pile driven out of interlock because corner pile had not been driven ahead and because hammer used was too heavy.

strata so that the full hydrostatic head will act throughout the entire cofferdam height. It may also create a pathway for run-in under the tips during excavation and pumping, particularly where there is little penetration of the sheet piles or where it is planned to dewater the cofferdam without a seal. In all cases where jetting is used, vibration or driving should be continued 2 or 3 min after jetting ceases, in order to densify the soil around the pile.

Vibratory hammers are widely used for driving (and pulling) sheet piling. These are extremely effective in cohesionless material—sands, silts, etc. They are relatively ineffective in clay, hardpan, sandy clay, gravel, etc. When driving through heterogeneous strata, it is frequently advantageous to use a vibratory hammer in the cohesionless soils and a diesel or steam hammer in the cohesive soils and gravel layers. Jetting assists the vibratory action; or in hard driving, just allowing water to flow freely at the surface will sometimes help. The water vibrating down along the sheet pile face serves to lubricate it.

Most vibratory hammers operate at frequencies of 12 to 20 cycles per second. For some of them, the impulses are directional, i.e., down for driving, up for removal. Vibratory hammers are relatively quiet and thus are suitable for use in urban zones. They are equipped with hydraulic clamping devices to clamp the sheet pile, thus minimizing damage to the sheet-pile head. Transmission of vibration to the rigging, boom, etc., is a potential source of fatigue failure of the handling equipment. Therefore, use of a damping device is necessary. Some vibratory hammers may be assembled in multiple units, as to deliver increased energy.

In some sands, where impact hammers develop excessive skin friction, the vibratory hammers will dramatically outperform their more conventional counterparts. Vibratory hammers transmit some vibration to the surrounding soil, and driving one pair of piles will sometimes cause adjacent sheet piles in the cofferdam to sink simultaneously. Penetration is very rapid in cohesionless soils, ranging up to 2 m (7 ft) per min, so it is important not to run any pair of piles too far ahead of the adjoining piles. Frequent shifting of the hammer is necessary to keep sheet-pile alignment true.

As indicated earlier, there are some soil conditions for which vibratory hammers are unable to advance the sheets. These include dense clay, compacted gravel or conglomerate, cemented materials, and, surprisingly, layers of volcanic ash and peat. For these, impact hammers are needed. Both diesel and steam hammers are used. They must have proper driving heads so as to properly deliver the energy to the piles without deforming the heads (Fig. D4-6). In most cases, the hammer size will require that sheet piling be driven in pairs.

FIGURE D4-6 Sheet pile driven down center of buried brick wall using light, fast hammer. When heavy hammer was put on pile, tip curled and further driving became impossible.

There is one other type of soil which may prove very difficult to penetrate and that is dense glacial silt, which has been very densely compacted. Fortunately, these silts are readily penetrated by a high-pressure jet, which can be run down along the web or between pairs of the sheet piles, while the impact hammer is working.

Length and Penetration. The determination of the proper length of sheet piling to use and the necessary penetration requires careful analysis. Generally, the top of the sheet piling must be above extreme high water. Spring tides in tidal bays and estuaries may be augmented by wind-driven waters and flooding tributaries. The extreme high tides may be ½ m (2 ft) greater in one part of the tidal bay than at the recording station, and prolonged winds can often raise the water level appreciably. While the sheet piles do not have to keep out the splash of waves under these extreme conditions, they must keep out the bulk of the water. It is common to place the top of the sheet piles ½ to 1 m (2 to 3 ft) above the highest recorded tide at the site.

In seasonally fluctuating rivers, it is usually uneconomical to design a cofferdam to hold out flood levels. The cofferdam top is set at an elevation a few feet above the highest anticipated river level for the season, and work is scheduled for completion during that season. Although a cofferdam can always be flooded to meet an unexpected or unseasonal flood, damage to equipment and forms and the costs of pumping out again and cleaning out the silt and debris do not justify cutting the height too close.

The sheet piles must penetrate far enough to develop sufficient bearing capacity to hold up the weight of the pile, bracing frame, and tremie pipes, etc. They must also penetrate a sufficient distance below the lowest level of excavation to prevent the soil from running in underneath. There is a stage during the construction when the sheet piles have been driven and the excavation has been carried to its deepest level, yet the seal has not yet been poured. At this stage, even when there is equal water pressure inside and out, there is a tendency for the soils to run in underneath due to the unbalanced head of soil. The stability of the soils around and below the tips of the sheet piles should be considered in this regard. This tendency is exacerbated if there is a lower water level inside than out, or if the soils inside and under the sheets are disturbed by jetting.

The sheet piles must resist the unbalanced pressures caused by the greater height of the soils outside the cofferdam than inside it. In firm soils, such as hardpan or firm clays, the required passive lateral support may be developed in a meter or two of pile penetration below the inside excavated elevation. In soft mud or unstable sands, it may take 5 to 6 m (15 to 20 ft). The passive resistance can be destroyed, at least temporarily, by run-in of soils under the tips of the sheet piles. More cofferdam failures have been due to inadequate sheet-pile penetration than to any other single cause. The usual reason given for using short sheet piles is to save costs, yet a few extra feet will provide a safeguard far beyond the small additional cost. The tendency of the soils to run in underneath the piles can also be reduced or eliminated by predredging. On the other hand, any surcharge, such as a temporary roadway or crane equipment working alongside, will intensify the unbalanced pressures.

Substantial penetration into relatively firm materials will develop a degree of fixity at the lower reaction of the sheet piles. In many deep cofferdams, the greatest stress in the sheet piles exists after excavation but before the seal is poured. If a fixed reaction can be developed through adequate pile penetration, the critical bending stresses under this loading will be reduced. Similarly, in a later stage, when the seal has been poured and the cofferdam dewatered, adequate penetration to ensure a fixed end moment will reduce critical bending stresses.

The effective section modulus of deep-arch sheet piles will be greatly increased (1.5 to 1.75 times) if sliding along the interlock can be prevented. Adequate penetration into the bottom will help to develop this joint action. In some critical cofferdams, adjoining sheet piles are welded for a meter length or two at the top and are driven to a deep penetration in firm material, so as to prevent sliding in the interlock.

If the cofferdam is to be dewatered without a seal, then a different and much more critical approach is involved. The penetration here must be adequate to prevent boiling, blowing, and piping due to the imbalance of water head and soil on the outside. In normal sediments, muds, soft and medium clays, and sands, the pile penetration required for the safe dewatering of even a moderately deep cofferdam without a seal is usually too great for economy. If the bottom does become quick and lateral resistance at the bottom is lost, the sheet pile will kick in and the cofferdam will collapse. Each such situation can be analyzed by means of flow nets, but in sands, the penetration below the lowest dewatered excavation must be at least one-half the total outside head.

When the cofferdam is founded on clay hardpan or on soft rock where a fair penetration of sheet piling can be obtained and where there is a fairly impervious stratum on the outside, it may be practicable and even desirable to eliminate the seal. A set of bracing should then be installed as low down in the cofferdam as possible as a safeguard against any softening of the bottom and a resulting tendency of the sheet piles to kick in. For example, serpentine is hard and sound until exposed to air, when it quickly slakes to a slimy mud. Sand may be hard packed and dense until continued pumping sucks the fines out and starts a piping action. Placing of a layer of graded sand and gravel, similar to a concrete mix, before pumping down, will often help to stabilize the bottom.

Splicing. Sheet piles sometimes have to be spliced, although this should be avoided for major cofferdams whenever possible. The best splice is a full-penetration butt weld which has been made in accordance with the prescribed welding procedure. Good workmanship is necessary, because a poor weld will fracture under the repeated hammer shocks. Use of low-hydrogen rods is very beneficial in reducing any tendency of welds to fracture under driving. Welding should be confined to the outside of the finger, so as not to deposit metal inside the interlock where it would impede the free movement of the adjoining pile.

It is very important to line up the interlocks perfectly. One practical method is to take each of the two pieces which are to be spliced and temporarily thread them on the interlock of a single section. While they are being welded, the single sections should extend 1.5 m (5 ft) or more on each side of the joint. Splices should be avoided at the point of maximum moment, as there is obviously a reduction in strength because of the incomplete welding of the interlocks. Splice plates are usually not required, since tests have shown that the full strength of the flange of a deep-arch section can be developed with the full-penetration butt weld. If necessary, plates can be added on the inside of the web just above the interlock to restore the full section of the pile.

On Z piles, if splices have to be made at points of maximum moment, splice plates on the flange can restore the section. These plates should be located on the inside of the flange so as not to hit the wale in driving. Experience has shown that well-made splices will stand up satisfactorily in both driving and pulling. Sheet-pile driving is a severe test of the quality of a weld.

Extraction. Sheet piles are extracted either by pulling the entire pile or by cutting of the sheet pile at or above the mud line and then pulling the top portion. The second alternative, which leaves the lower portion of the sheet pile in place, provides excellent scour protection and prevents undue disturbance to the soils around and below the footing block. When a sheet pile is extracted from clay soils, it is usually found that a clay plug in the arch (or in the web of Z piles) is pulled along with the sheet pile itself. This removes an increment of soil from below the edge of the footing block. In sands, jetting will probably be necessary to accomplish the removal of the entire pile, and this also disturbs the soils below the edge of the footing block. Disturbance of soils in the vicinity of a footing block which rests directly on compact sand or on clay hardpan is undesirable.

Scour, either in a sudden flood or gradually through the years, may lower the bed of a river sufficiently to uncover or undermine the footing block. Several serious bridge collapses have occurred due to this cause. Dredging or channel deepening or for the installation of submarine pipelines, sewer siphons, or power cables or for the construction of future bridge piers or other waterfront installations may also uncover and undermine the footing block, exposing timber piles to the action of marine borers and steel piles to abrasion-corrosion. For these reasons, leaving the lower portion of the sheet piling in place may often have engineering advantages that greatly outweigh the additional cost.

Pulling of sheet piles is preferably performed with vibrators, because of their speed. However, when they cannot be removed by the vibrator, then an extractor may be used. Extractors are particularly effective in hard clay and on bent or distorted piles. Extractors are operated by steam or air. A good strain must be kept on the line from which the extractor is suspended, but an excessively heavy direct pull on the top of the extractor can cause the extractor casing to be distorted or even pulled apart. In other words, the extractor itself should do the work. Common practice is to suspend the extractor from a derrick or crane boom and, while maintaining a steady pull, let the extractor bring the pile clear of the bottom and of the tremie seal. The second line from the derrick boom can be shackled directly to the sheet pile. When the extractor has done its job, it is removed and the pile is lifted clear. Sometimes jetting will be necessary to aid the extractor.

Extractors transmit considerable vibration and shock through the rigging to the boom of the crane or derrick; this may damage the boom or cause it to fail in fatigue. Use of an additional hydraulic damping device may be advisable. Even though extractors are already equipped with a special shock-absorbing suspension unit, this becomes ineffective if the maximum crane pull is exceeded.

The use of an A frame or pulling beam may be quite effective, especially in removing the first sheet pile from a cofferdam. The frame or beam is designed to rest on the next adjacent pile or on a completed portion of the pier, and held vertical by one line from the derrick. Heavy blocks are rigged to develop the necessary pull to raise the pile clear of the firm material in which it was embedded. After the pile or a group of piles have been raised clear, the frame is removed and the piles are lifted out directly. The use of a frame transfers the reaction directly to the sheet piles, so that there is no heavy load on the boom itself. Many serious accidents have occurred in pulling directly off the boom, especially where many parts have been rigged. A good general rule is never to pull directly from the boom.

In some cases of unusually high resistance, it has been necessary to use both an extractor and a direct pull from an A frame to the sheet pile (not through the extractor). In these cases, it is questionable whether the cost of removal justifies the salvage obtained, and it may be cheaper to burn off the pile at the mud line.

Vibratory hammers are extremely effective in extracting sheet piling driven in cohesionless materials and even some clays. They will often extract piling readily from material through which the same vibratory hammer could not drive it. These vibrators require a hydraulic damping unit to prevent vibration fatigue and damage to the crane or derrick boom.

Removing the first pile is usually the hardest job. Often it is necessary to try several piles before finding one which can be pulled free. The center piles of a long wall are likely to be the easiest. When a pulling frame is used, this first pile should be broken loose by driving it down a foot or two and then making an attempt to pull it. This same sequence can also be tried if any subsequent pile gives trouble. Pulling is often a longer and more costly operation than driving, and it is at this stage that care in setting, greasing of the interlocks, and similar foresightedness will reduce costs.

High removal costs can be experienced when sheet piles have been driven hard into firm material. As we noted earlier, it is often economical to abandon the lower portion, burning it off and salvaging only the easily removed upper section. In this case, it is likely that the extractor will have to be used only on the first pile; the rest can often be vibrated clear. The burning must be performed by a diver. The electric-arc-oxygen method is the most efficient, although only a few marine divers are familiar with this equipment. With a good diver, the proper equipment, and an efficient top crew, sheet piles can be burned off under water at a reasonable cost.

One problem encountered in hard pulling, particularly when an extractor is used, is the tendency for the holes in the top of the sheet pile to shear out. Most extractors are equipped to pull with two pins, and both of these should be used in cases of hard removal. The holes in the sheet piles should be burned carefully, and in difficult cases should be burned undersize and reamed to exact size. As an alternative to the use of pins, with its consequent risk of shearing the pin holes during hard pulling, both mechanical and hydraulic clamps or tongs have been developed. Some extractors and most vibrators come equipped with these clamps which, in comparison with pins, considerably reduce pulling damage to the tops of the piles. They are also much faster and safer in their operation, and their use is strongly recommended.

Sealing to Existing Structures. Occasionally, a bridge pier must be built immediately adjacent and connected to an existing pier or structure. On one such occasion, the designer of the first bridge had been foresighted enough to leave the common wall of sheet piling in place, with a sheet-pile T at each corner. Such foresight is rare, and the problem often arises of sealing a new sheet-pile wall to an existing concrete pier (Fig. D4-7).

In one case, a seal was successfully constructed to resist a 40-ft hydrostatic head, by forming a cell against the existing concrete wall, using a Y-section sheet pile. As this cell was excavated, pumps kept an excess head of water inside it to prevent loose sand from flowing in. The cell sides were then jetted clean, and the cell was filled with tremie grout. The three sheet piles which form the cell should be longer than the rest so that the excavation and seal can be carried well below the level of excavation planned for the main cofferdam.

On another occasion, the joint between the sheet piling and the concrete wall was carefully caulked by a diver, using soft wood wedges which were placed as excavation proceeded inside the cofferdam. Although this worked successfully, it is not believed to be as safe a method. It does not seal below the excavation level against the possibility of material flowing in underneath.

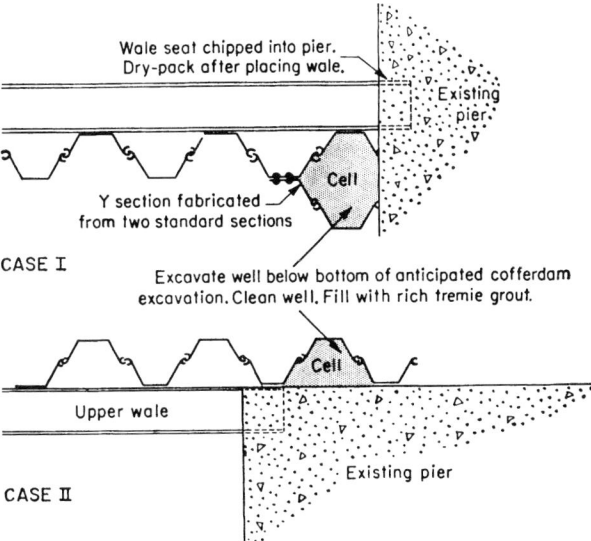

FIGURE D4-7 Method of connecting and sealing sheet-pile wall to existing concrete pier.

D4-20 STANDARD HANDBOOK OF HEAVY CONSTRUCTION

Draping a canvas sheet on the outside of the connecting cell will help to seal the joint above the external mud line. Both cement grout and chemical grouts have also been used effectively below the mud line. Cement grouts have good structural strength but will not penetrate silts and clays thoroughly. Chemical grouts penetrate finer soils and may form a colloidal gel, but they generally have little structural strength. In some instances, the use of cement grout as an immediate structural barrier, followed by chemical grouting behind it, has proved to be the optimum solution.

Bracing Systems

Circular Cofferdams. Several very important cofferdams have been built in a circular or nearly circular plan, so that the external pressures will be transmitted circumferentially around the perimeter (Fig. D4-8). In this way, all cross bracing can be eliminated. The method is an economical and comparatively safe one for constructing very deep cofferdams under difficult conditions, such as swift currents. The cofferdam may be constructed as a relatively rigid continuous wall, such as a heavy

FIGURE D4-8 Setting sheet piling for circular cofferdam. Note two levels of guides to ensure accuracy.

COFFERDAMS AND CAISSONS

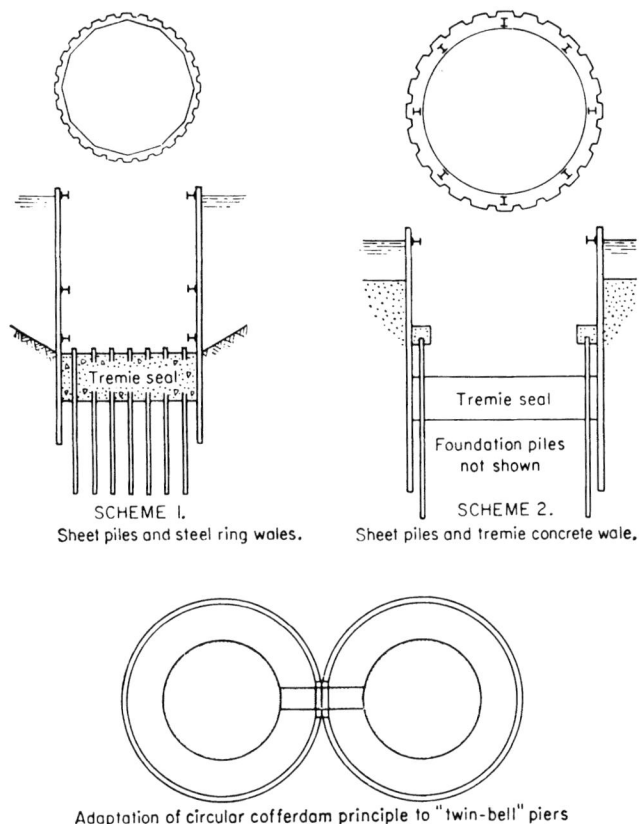

FIGURE D4-9 Circular cofferdams.

concrete cylinder, which is capable of transmitting pressures circumferentially; or it may be constructed as a flexible wall, such as a sheet-pile ring, which is braced by heavy circumferential wales (Fig. D4-9). The condition of equal pressures and true ring action is satisfied only for a cofferdam in water. For a land cofferdam, an imbalance of pressure may result from the soils and from any surcharge, such as construction equipment. Unbalanced pressures will, of course, introduce bending moments into the ring.

The circular cofferdam is probably the only type that is safe to build if the cofferdam is to be unwatered to very great depths. Because of the elimination of cross bracing, construction inside the pier is expedited. Corner sheet piles and corner connections are also eliminated.

One scheme of construction of a circular cofferdam is as follows. Guide piles are driven in the center, and from these is hung a bracing frame of steel rings to act as wales. Sheet piles are set around the frame, and after the last of these is entered, they are driven to grade. Excavation is then completed to grade, and foundation piles are driven by a barge-mounted crane rig, reaching over the sheet piles. There is no cross bracing to interfere with pile driving. Then the tremie seal is poured, the cofferdam unwatered, the piles cut off to grade, and the pier constructed. The cofferdam is flooded, the sheet piles are pulled, and the bracing frame is lifted intact for use on the next pier.

The ring wales must be of a section or configuration that can provide moment capacity to resist unbalanced loads, since there is little redundancy against collapse. Deep wide-flanged beams and trussed rings of steel are suitable. In France, ring wales of reinforced concrete are often utilized. In that case, the concrete wales are separately supported on steel piles.

In extremely exposed conditions and where there are high currents, it may not be practicable to construct the entire cofferdam at one time. In such a case, it may be constructed in units, with each one completed and made stable in itself before the next unit is started. Individual cells of large pipe to which sheet piles have been bolted, or of concrete cylinders with embedded sheet piles, are driven in an intermittent circle, excavated, and filled with concrete. These cells can be connected with double walls of steel sheet piles, which in turn are excavated and filled. It is essential that the primary cells be properly positioned and driven vertically, although there is some play in the connecting arcs of sheet piles. This concept was first used for the cofferdam of the Hetch-Hetchy crossing of San Francisco Bay and was subsequently employed for the South Pier of the Golden Gate Bridge and on a Tidal Power Project in France (Fig. D4-10). Such a cofferdam is, of course, left in place and may serve as protection against ship collision.

Rectangular Cofferdams. The rectangular cofferdam is the type most commonly constructed for bridge piers. Once the loads have been determined, it is necessary to select the section of sheet pile and the configuration and spacing of the bracing system. Usually the selection of the sheet-pile section depends on several independent factors such as availability, driving conditions anticipated, length required and the accompanying handling problems, and strength and stiffness before the seal is poured as well as after dewatering.

The bracing system must be adequate to resist the total external loads, as well as the loadings during the stages of excavation and construction.

The following factors must also be considered in the selection of a bracing system:

1. Elevation of deepest excavation
2. Soil properties throughout soil column and below deepest excavation
3. Top elevation of tremie concrete seal
4. Elevation of ground outside the cofferdam
5. Extreme high water elevation
6. Plan dimensions of cofferdam, with allowances for construction tolerances and deformations of cofferdam walls
7. Configuration of footing block and pier or column shafts
8. Requirements, if any, for bracing system to support vertical loads
9. Procedure for fabrication and installation of bracing
10. Removal procedure and planned reuse

In the past, structural steel bracing systems were usually installed in a series of horizontal panels, prefabricated as practicable, and progressively lowered and fixed as excavation proceeded. More recently, the structural steel bracing system has been prefabricated as a three-dimensional space frame, and placed as a complete unit, using heavy lift equipment or flotation (Fig. D4-11).

COFFERDAMS AND CAISSONS D4-23

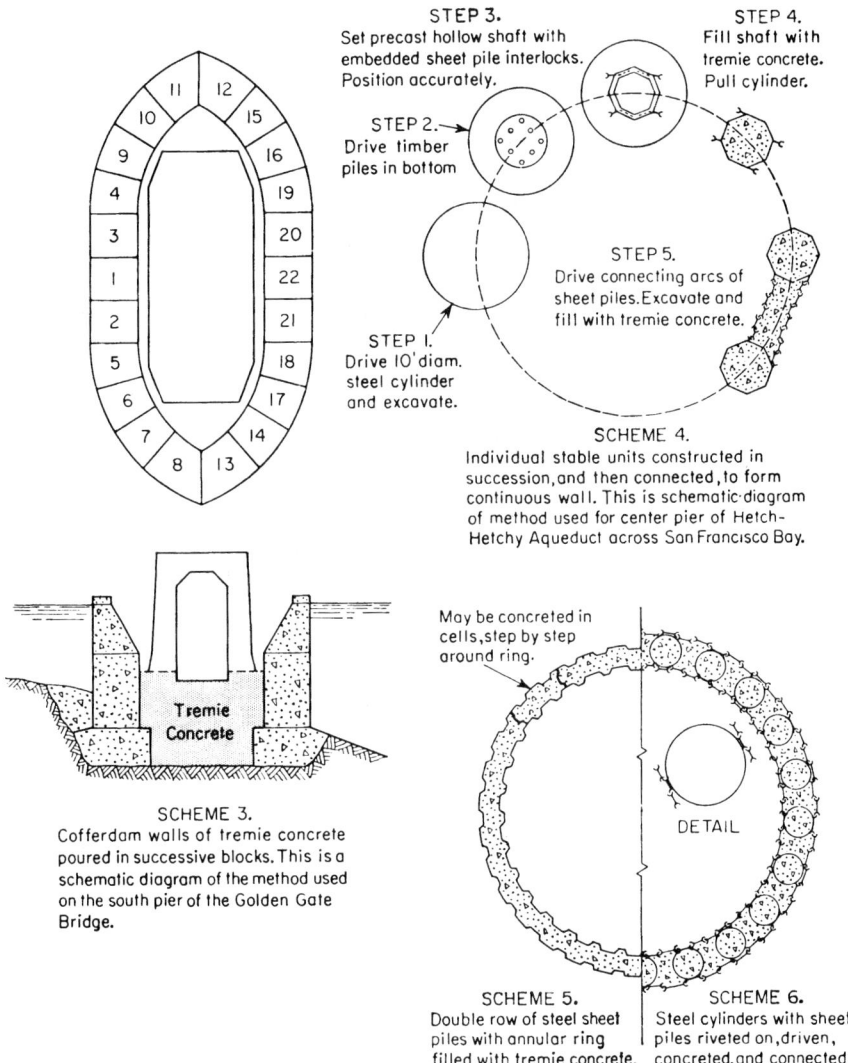

FIGURE D4-10 Circular and elliptical cofferdams.

In one case, two identical piers were to be built in successive seasons. Each pier had two shafts, with a clear space between them above the distribution block. Most of the bracing and most of the trussing of the struts were placed in the open space between the shafts, leaving a few single member struts going through the shafts. Most of the bracing system was salvaged in large, reusable units.

Wales. The number and location of wales can be determined once the loads are known and the sheet-pile section is selected. The theoretical solution is given in many texts and in the publications of the leading manufacturers of steel sheet

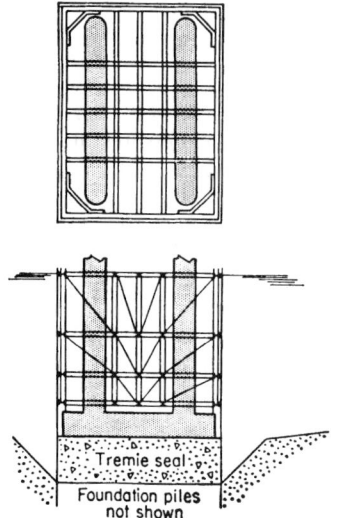

FIGURE D4-11 Cofferdam bracing system.

piling. The author prefers to utilize simple approximations on a trial-and-error basis, since many considerations may influence wale location. A detailed computation is then made as a final check.

One set or level of bracing is desirable at or near the top, even though the sheet piles theoretically might carry the top hydrostatic loads as a cantilever. This top set is useful in aligning the cofferdam and as a guide against which to set the sheet piles. It provides the skeleton for supporting tremie pipes, pumps, compressors, and a pile template. Also, it resists lateral forces applied at the surface, such as the impact of the floating equipment working at the cofferdam. It also prevents excessive inward deflections during the initial stage of excavation. This bracing is therefore much heavier than would be required by the hydrostatic loading alone. One empirical rule is to use the same size of members that is determined for the second level of bracing, although this rule requires judgment in its application. If the struts are to carry any loads, such as equipment, then they must be sufficiently strong to carry both axial and bending loads and will frequently require trussing.

The usual procedure is to sketch the cofferdam, showing the elevations of high tide, outside soils, and bottom and top of tremie seal. A top set of bracing is tentatively established at the high-tide water elevation. Additional levels of bracing are now tentatively set at intermediate elevations, and the approximate bending moments are computed by assuming simple spans between wales. A fairly high bending stress (60 to 70 percent of yield) can be allowed in steel sheet piling provided that all the loads used in the computations represent maximum conditions, including those during the stages of excavation. Dividing the maximum bending moment by the allowable bending stress gives the required section modulus ($S = M/f_b$). By varying the trial levels of bracing, the section modulus for each of the spans can be made approximately equal to the tabulated section modulus for the sheet-pile section in question.

The sheet-pile section may have been previously determined by the availability of one particular section, or by driving conditions, etc. But in most cases it is prudent to lay out tentative bracing levels for a range of sheet-pile sections.

Next, the stresses during construction must be analyzed. Where predredging to grade is performed, the external and internal pressures will be equal until the cofferdam is dewatered. Where predredging is not performed, it is necessary to check the stage of construction when all excavation is completed inside and before the seal is poured. There will then be considerable pressure from the soils outside the cofferdam, acting over the span from the lowest set of bracing to a distance below the bottom where the passive resistance develops zero shear. In typical sands and clays, it may take 1 to 2 m for the soil to provide this reaction. In very soft mud and silt, the required distance may be 3 to 5 m. The effective span is thus longer than the distance from the lower wale to the bottom of the excavation. A short distance below this reaction point (zero shear), the sheet pile may develop fixity in the soil and thus reduce the bending moment in the lower sheet-pile span. Although more refined analyses may be made by geotechnical calculations, in many cases a careful evaluation based on experience is adequate to select the points for use in computations.

Since the bracing frames should preferably be fabricated as complete space frame and set as a structural unit before the sheet piling is installed, predredging will often be indicated. Predredging is low-cost excavation, reduces soil pressures at all stages, makes pile driving easier, reduces or eliminates the need to jet and clean the arches of the sheet piles before pouring tremie concrete, and enables the bracing frame to be set at the best levels to meet other criteria. Where predredging cannot be done or where it is limited by the proximity of structures which might be undermined or the stability of slopes, then complete levels of bracing can be bunched at the top, and lowered, frame by frame, as excavation proceeds, blocking each against the sheet piles.

These operations are difficult and costly to perform satisfactorily. Extreme care must be taken to ensure that the lowered bracing frame is truly level, well blocked to the sheet piles, and properly secured to the vertical supports and to the sheet piles. It is therefore preferable to use a heavier section of sheet piling so as to span from a prefabricated bracing frame which is set at or near the ground line.

The next criterion has to do with the structure inside the cofferdam. To facilitate removal of wales and struts and to obviate extensive shifting of bracing in the cofferdam while building the structure, bracing levels should be located just above construction joints wherever possible. Frequently, the lowest level can be located just above the footing block or distribution block. It is sometimes possible to raise the top level of bracing a little higher than would otherwise be required and thus keep it completely above the concrete which must be poured in the cofferdam.

In some cohesive and impermeable soils, such as marine clays, it may be practicable to eliminate the tremie seal. In this case, the cofferdam may be kept full of water during excavation and a set of bracing may be lowered to the bottom and blocked to the sheet piles at this level, i.e., below the top of the footing block, before dewatering. In this case, of course, the lower set of bracing is not salvageable.

In designing the struts, their action as columns must be considered. Where vertical posts are not practicable, trussing of the several sets can supply the needed strength in the vertical plane. This may require pairs of bracing levels instead of single heavy sets, even though the sheet pile might take the bending with a single set. To use an extra level for this purpose is not too costly since the total load which must be carried is the same. The undesirable features are the extra fabrication cost and the interference during construction; the benefit is the provision for adequate vertical support of the struts.

In computing bending moments in the sheet piles for selection of the proper section of sheet piling and for locating wale levels, an approximation can be made by assuming continuity over the wales and a uniform load equivalency. This assumption is used by many engineering authorities and most cofferdam constructors. Thus, the top span's moments may be approximated by $Wl/10$, the intermediate span's by $Wl/12$, and the lowest span by $Wl/10$ or $Wl/12$, depending on whether l is extended to the point of zero shear or to that of full fixity.

The maxima moments are generally the negative moments over the supports and, in a moment diagram, will show a sharp peak. In actuality, the wale has a finite dimension, usually 300 to 500 mm, and the peak negative moments reduce to those given by the diagram at points opposite the edges of the wales. An exception may be the lowest span, prior to placing the concrete seal or footing, when the positive moment may have the largest value.

More exact calculations can be used to check the final design, but these approximations are adequate for preliminary layout and are usually within 5 to 10 percent of the true values.

The wale section should be a heavy, structural member having torsional strength as well as primary bending strength. The work must be adequate to resist the bending and shear between struts and the crushing and localized concentration at the struts. A section with adequate flange width is desirable as contrasted with a more efficient I-beam section, since, as the sheet piles are driven past, there is a tendency to roll or twist a narrow wale section. A reasonably rugged wale section is also desirable in order to resist localized crimping and damage, such as a blow from the clamshell bucket in excavating. The bending stress in wale design should be kept low (e.g., 50 percent of yield) since the wale will probably carry some axial compressive stress also. If the analysis includes axial compression and the effect of combined stresses, then higher stresses, up to 60 percent or even 70 percent of yield may be permitted. The allowable shear stress in steel wales should be 35 percent of yield. The wale must be adequately stiffened opposite the strut to prevent local buckling. Stiffeners must be welded between the flanges, both top and bottom. These should be installed during prefabrication as it is very difficult to install a stiffener on the underside of a wale in place.

In computing the bending stress in wales, approximate methods are again usually satisfactory. The wale may be considered as a partially continuous ($Wl/10$) beam between struts. The best way to compute the load per linear foot is to draw a diagram of external pressures, and divide it into tributary rectangles and triangles, from which the load coming on any particular wale can be quickly approximated.

With steel wales, the intersection problems require careful detailing. The corner must be designed to transfer thrust both ways since the wales will usually be required to carry axial stress, acting as end struts. To make a rigid corner with a full-moment connection, a short, diagonal H beam with stiffeners between the flanges of the wales and the flanges of the diagonals may be used, supplemented by cover plates if necessary. This type of corner helps to reduce the bending stresses in the wale at the location of greatest axial stress.

Struts. These are essentially columns, so the main problem in design is to provide adequate lateral stiffness and support in both the vertical and horizontal planes. Lateral support can usually be provided by intersecting struts in the other direction, which makes the struts act as short columns in this plane. Vertical support is difficult to furnish and is also the most necessary because of the dead-load deflection of the member itself. When a single set of bracing is used and is above water, properly located piles can be used to support the struts. This vertical support cannot be removed until it has been replaced by a support on the structure itself or until the cofferdam is flooded again. Underwater struts can be supported by preconnecting several sets to a vertical pipe post. A steel pile can then be driven through the pipe and bolted or welded at its top.

Where several sets of bracing are involved, common practice is to truss the struts at two or more levels in a vertical plane using heavy angles or channels. Since the trussing must be provided for struts in both directions, quite a lot of truss bracing is required in large cofferdams. This makes forming and pouring more difficult and also increases the risk of accidental damage by a dredging or concrete bucket (Fig. D4-12).

For these reasons, strut members should be selected with as large a radius of gyration as possible. In general, the depth of section should be approximately the same as the width of the wale against which the strut will bear. Steel H beams or wide-flange beams usually work out best.

If the struts in one direction are detailed so as to lie just above the struts in the other direction, then the wide-flanged beams can be installed with their webs vertical to give them maximum strength against dead-weight bending. If, on the other

FIGURE D4-12 Well-braced cofferdam, using steel bracing system which is trussed in the vertical plane. Columbia River Bridge at Portland, Oregon.

hand, it is desired to have struts from both directions lie in the same plane, then the webs should be horizontal so as to simplify strut intersections. Strut intersections must be carefully detailed to prevent crushing or breaking of flanges. Adequate stiffeners must be installed.

Pipe columns have excellent radii of gyration in relation to their weight, making them useful where long struts are required. In narrow cofferdams, pipe struts may be used to span the entire width. Intersection details are difficult, since full steel area must be provided across the intersection. The only practicable solution is to prefabricate short sections into a cross and then to weld the longer sections of pipe to their corresponding stubs. Square tubing is also an excellent section for struts, although adequate stiffener details are required at intersections.

The allowable stresses in struts are kept low because of the mode of failure and the risk of progressive collapse. Alloy steel is not used because of the potential problem of buckling. Thermal stresses must also be considered. The top set of bracing is the most exposed to temperature variation. This may be minimized by insulation or by painting with titanium oxide (white reflective paint) or by water spray on hot afternoons. The sheet pile walls are fortunately not fully rigid so about 50 percent of the temperature strain can be assumed dissipated. If thermal stresses are considered, then a design value of 30 percent of yield is usual practice for struts, but if not separately considered, then 30 percent of yield will generally lead to a safe design.

The design loads are those due to the external pressure which is transmitted through the wales, plus an allowance for possible impact from a dredging or concrete bucket. The most common way to take care of the impact loads is to use conservative stresses in the design of the struts and to provide more than adequate trussing, etc. Vertical sheathing of wood planks is sometimes hung on the struts, to guide buckets and protect the struts. In any event, care in bucket handling, not speed, must be emphasized. Some designers actually compute the stress arising from the impact of a bucket, but this is only a guide and not an exact answer.

Diagonal struts are frequently used at the cofferdam corners to reduce obstructions within the structure and to shorten the required length of strut. These diagonals carry an increased stress because of their inclination (1.4 times the normal load if on a 45° diagonal). Their reaction must be resisted by an adequate shear connection to each wale, and this is usually provided by a kicker block. This also increases the axial stresses in the wales.

Ties. Frequently, a cofferdam must resist outward pressures during some stages of construction. One source of outward pressure is a high tide which hangs up longer inside the cofferdam than outside; this problem can be eliminated by a tide gate in the sheet piles. However, in a location exposed to long wave swells or surges, or to pulls from mooring lines, external ties may be necessary. These ties can be simply provided by bolting every other sheet pile to the wale and making the strut also act as a tie.

Outward pressures on a cofferdam are also encountered when making a pour of underwater concrete that will extend well above the mud line. These pressures can be countered by backfilling against the sheet piles with sand prior to pouring the concrete, but this may then give too great an inward pressure against the sheet piling. Both design criteria have to be satisfied.

Large cofferdams are often tied with cables at the top to help maintain their alignment. Some small bridge piers have been constructed under water by using a sheet-pile cofferdam which is excavated and filled to its top with underwater concrete. The outward pressure of the fresh concrete on the sheet piles is resisted by wales placed on the outside of the sheets and tied across the top or around the sides.

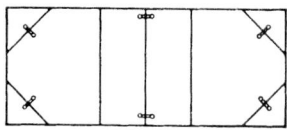

FIGURE D4-13 Temporary support for prefabricated bracing frame.

Setting Bracing. As stated earlier, whenever possible, the bracing frame should be completely prefabricated: Two crane barges can be used when the weight is beyond the capacity of one. The frame may be hung on temporary piles (Fig. D4-13). After the sheet piles have been set and driven, the load of the bracing system can be transferred to the sheet piles (Fig. D4-14).

When the bracing frame becomes too heavy to be handled with crane barges or when heavy-lift derricks are not available, the bracing frame can be erected in units. Each level of wales and struts is bolted or welded together and hung from temporary piles. The uprights and trussing angles which will extend upward to the next lift are fastened on, and the unit is lowered until the uprights are just above water. The next unit is then erected, and again the completed portion is lowered. This procedure is contin-

FIGURE D4-14 Cofferdam for main pier of Dame Point cable-stayed bridge, Jacksonville, Florida.

ued until the complete frame is in place. Alternatively, the bracing can be fabricated into vertical sections (space frames) and bolted together under water by divers. For the anchorage cofferdam on the Second Delaware Memorial Bridge, the bracing was fabricated into three sections, each weighing over 100 tons (Fig. D4-15). However, connecting under water is not only costly but it is difficult to ensure accuracy.

Many ingenious schemes utilizing flotation have been employed. Certain sections of the bracing system can be completely boxed in and made watertight, although care must be taken that they do not unduly impede pile driving or concrete placement. Hollow, watertight box girders can be used for wales, and pipe can be used for struts if proper details are employed at strut intersections. These means and others are frequently employed when lowering the units in stages but, except in inaccessible locations or with extremely large bracing systems, elaborate methods of utilizing flotation are usually more costly than direct handling and setting.

The bracing system can also be erected between two properly spaced and moored barges. It is then lowered in stages from winches on the barges.

Timber and steel bracing cribs have been built on launching ways or in dry dock and then launched and floated to the site. After positioning, they are sunk by weighting until they are hung from temporary supporting piles. The steel sheet piling is then set and driven around the crib.

Removing Bracing. Where many similar piers are to be constructed in sequence, the bracing frame may be so laid out as to permit its being removed wholly or largely intact after the sheet piles have been pulled. In these cases, depending on the bracing weight, it may be lifted up over the structure with the aid of one or more derricks.

Most specifications permit steel struts to be left in the structure below the waterline. Such struts may be boxed by forms for the outside 300 mm (12 in) and then burnt back after flooding the cofferdam. The boxed cavity is subsequently filled with

FIGURE D4-15 Bracing frames for very large cofferdam of Second Delaware Memorial Bridge anchorage pier were set in three sections and connected underwater.

an underwater-setting grout. Alternatively, and much less costly, the exposed steel may be cut flush and thoroughly coated with epoxy. Where specifications require the removal of struts, they may be completely boxed out in the structure, burnt off under water, and pulled out sidewise after all the other bracing has been removed.

A method commonly used in the past, and one which is still necessary occasionally, is to remove the bracing piecemeal before flooding the cofferdam. In this case, after the footing or distribution block is poured, it is carefully blocked to the sheet piles and the next higher set of bracing is taken out. After the next pour, the pier is again blocked to the sheet piles and another level of bracing removed. This method requires a great deal more labor, consumes valuable time in a deep cofferdam, and increases the risk of carelessness or human error. For example, the bracing frame is usually rigid and husky enough to withstand the impact of a dropped beam, whereas the blocking may not be capable of taking a load except in its principal lateral direction. Also, since the full or substantially full pressures remain on the cofferdam until it is completely flooded, the necessity of placing the blocking below the level of the brace to be removed means that the stress distribution and deflections will be different. Complete analyses and calculations are required for each loading condition, and additional bracing may be needed.

Excavation

Predredging. Whenever practicable, it is usually best to predredge the site to or just below the elevation of the bottom of the tremie-concrete seal. When the material at this bottom elevation is still soft, excavation is frequently carried a few feet further down to permit placing a gravel or sand-and-gravel blanket. This blanket provides a level base for the tremie-concrete seal and, if of adequate thickness, will also provide lateral support to the sheet piles in the critical stage before the seal is poured. Predredging is usually performed by a clamshell dredge, although hydraulic and ladder (continuous bucket) dredges have also been used.

Predredging quantities may run three times the *pay quantity* inside the cofferdam, but this is usually offset by a much lower unit cost. Predredging facilitates other operations such as setting the bracing frame and pile driving, although it may raise a problem of outward pressures with deep tremie seals.

One common sequence is to predredge to an elevation 1 m or so below grade. Using a floating driver, the foundation piling is then either driven to or cut off at approximate pile cutoff grade. A sand blanket is then placed, the bracing frame is set, sheet piles are set and driven, the tremie-concrete seal is poured, and the cofferdam is unwatered. Where it is impracticable to predredge all the way to grade, it still is desirable to predredge down to an elevation below the lowest bracing frame.

The selection of type and size of clamshell bucket depends on the material encountered, and thus a change in bucket may frequently be required as different strata are reached. In muds and silts, a bucket without teeth may suffice. In hard clay or sandy clay, teeth may be required and the bucket may have to be weighted. The most difficult material is densely packed sand, which is hard to dig and yet tends to run out of the bucket. Boulders and old riprap and logs are extremely difficult to remove and require patience. One good way is to dig one pocket considerably ahead of the remainder, then work back from this face.

Excavation Inside Cofferdam. After predredging is completed, the bracing frame is put in place and the sheet piles are set and driven. Then further excavation can be performed with clamshell buckets, working through the frame. Care must be exercised to prevent damage to the frame by the bucket. Vertical sheathing or individual sheet piles set alongside the bracing frame, struts, and wales will be of help in this regard.

The clamshell bucket cannot dig close to the sheet piles or under the wales and struts. A long chisel can be run up and down the sheet piling, and jets can be used to break down a wall of clay. Care should be exercised to avoid running chisels and jets below the bottom of the proposed seal. When excavating in the dry, a small crawler-mounted dozer or loader may be lowered into the bottom of the excavation. This equipment can clean beneath struts and work the material to the clamshell buckets.

Dredging should be carried ½ to 1 m below grade and the chiseling and jetting done thoroughly at the sheet piles and especially at the corners. Then a diver should be sent down with a jet to clean the arches of the sheet piles, since many cofferdam blows have occurred where an arch was left full of clay. If not previously installed, the bearing piles are then driven through the bracing, and a sand and gravel blanket is placed to bring the bottom to grade.

Soft, soupy mud which is left in the bottom after the main clamshell dredging should be cleaned up, using an air-lift pump. Success and economy in air-lift pumping can best be achieved by using an excess of air. Often, a diver with a jet can wash material to the pump intake. Also, where piles have to be driven before final excavation, an air-lift pump assisted by a diver with a jet can excavate safely between the piles.

Pile Driving Within the Cofferdam

When the pier site has been predredged, piles may be driven using a floating crane barge or pile driver. Care must be taken in controlling position, in order to prevent the hammer from hitting on a previously driven pile (Fig. D4-16).

Telescopic leads and a hammer capable of working under water were much used in the past. Most recently, the underwater hammer has been largely replaced with an above-water standard hammer and a machined follower, fitted so as to stay centered in the leads and to fit properly over the head of the pile being driven.

The other, more common scenario occurs with the sequence of partial predredging, setting of bracing frame, installation of sheet piles, and final excavation. Next, the piles must be threaded through the bracing. The preferable way, especially when there are multiple piers to be built, is to use extra-long piles, extended so as to be above water even when they have reached their design tip. This enables piles to be spliced or extended as necessary, working in the dry, as it is impracticable to splice piles below water.

Then later, after the concrete seal has been placed and cured, the cofferdam may be unwatered, the piles cut off and the excess length returned to the fabrication yard to make up pile lengths for the subsequent piers.

Particular problems often arise when the peripheral piles of a cofferdam are battered. First, the cofferdam may have to be enlarged in plan to enable the piles to run beneath the tips of the sheet piles. Secondly, pile interference must be checked, especially at the corners, where piles battered in two orthogonal directions may intersect with each other or with the vertical piles. Relocation of some piles may be necessary.

In some recent cases, bored piles (drilled and concreted shafts) have been constructed below the base of cofferdams, in lieu of driven piles. In these cases, the sheet piles have been installed, the cofferdam excavated to grade, and the shafts drilled,

FIGURE D4-16 Steel H piles, 60 m (200 ft) long, being driven for Columbia River Bridge piers at Portland, Oregon.

prior to placing the tremie concrete seal and dewatering. It is, of course, necessary to case through the water column. This then makes it possible to keep a slightly greater water head inside the shafts or, alternatively, to use slurry, so as to prevent loss of soil by sloughing of the bored holes. This sloughing or collapse of the hole must be prevented, as otherwise the cofferdam might collapse due to a loss of passive pressure over the lower portion of the sheet piles. Another way of ensuring against such collapse is to lower a level of bracing to the bottom of the excavation before drilling.

The juncture between the bored pile and the tremie seal must be located and detailed so as to ensure sound concrete through the entire thickness of the seal.

Jetting of piling within a cofferdam can be very dangerous because it may cause sands to run underneath, destroying the lateral support at the bottom of the sheet piling. This is particularly hazardous when jetting batter piles extending underneath the tips of the sheet piles. Where necessary, a set of bracing may be placed on the bottom after excavation just to hold the toe of the sheet piles during jetting. This set should be prefabricated at 10 mm or so smaller in plan dimension than the cofferdam and hung underneath the bottom of the bracing frame. After excavation, it can be lowered down to the bottom and wedged to the sheet piles by a diver.

Bottom Seal

Functions. The functions of the tremie concrete seal are:

1. To seal the bottom of the cofferdam box and to withstand the uplift pressure from hydrostatic forces
2. To provide a rigid support for the sheet piles at the bottom
3. To constitute a solid base on which to form a substructure

In modern design, the seal is frequently used as a part of the permanent structure as a distribution or footing block, for weight in uplift, and for load transfer to the piles. High-quality concrete can be constructed under water by controlled tremie methods. A great deal of the cost and the risk of deep cofferdams can be eliminated by a design that permits underwater tremie concrete construction of the entire footing block. Provision must then be made for placing the reinforcing steel in welded cages hung from the bracing, and piling must be cut off to proper elevation under water. The savings on a deep cofferdam may then run 20 to 50 percent due to use of a lighter section of sheet piling, eliminating the lowest and heaviest set of bracing, and, in many cases, facilitating salvage and reuse of the bracing system as a whole. Time of construction is reduced, and risk is materially lessened.

Resistance to Uplift. The simplest case assumes that the uplift will be equal to the hydrostatic head, so a thickness of concrete seal is specified that will give a weight equal to the water head at the bottom of the seal. With sea water, this thickness of seal becomes $64/140$ or 45 percent of the water depth at the bottom of the seal. The sheet piles below the seal act as a cutoff wall, but, unless effective and reliable drainage or pressure relief is provided, full head will ultimately develop under a tight seal. The placement of a gravel layer and a few bleeder pipes or pump wells under the seal will temporarily relieve much of this pressure.

The sheet piles themselves may have considerable frictional resistance to uplift. If their sides and arches have been cleaned before the seal is poured, they help to resist uplift on the seal. Foundation piles are very effective in resisting uplift. Timber piles taper, so that any tendency for the seal to lift is transferred directly to the pile by wedging as well as through bond. Peeled piles must be used for this purpose in order to develop skin-friction resistance to uplift in soil and to bond with the concrete. In very pervious soils, some leakage may occur through cracks or checks in the timber piles, but these are generally not serious.

With concrete piles or H piles, the uplift can be transferred through bond alone, usually in 1 or 2 m of embedment. 550 KPa (80 lb/in^2) has been sustained in tests. Therefore, 150 KPa (20 lb/in^2) is generally used for design. Greater bond can be developed where shear keys are prewelded on steel pile heads—e.g., values of 400 KPa (50 lb/in^2) can then be used in design. Algae and other marine growths can form quickly under proper conditions of salinity, warmth, and light, even at considerable depths in somewhat muddy water, and these reduce bond unless cleaned off by jetting or wirebrushing. Hence, shear keys are a more positive answer than bond alone.

The uplift value of the piles depends on the bond developed, the dead weight of the pile itself, and the pull-out shear capacity of the "plug" or "cone" of earth around each pile and around the group of piles. Common practice often allows uplift values of 50 percent of the allowable bearing capacities, but this should be more carefully investigated on critical projects. Often the limiting factor is the ability of the unreinforced tremie concrete to span between piles, and this can be empirically set at about twice the thickness of the seal.

Unless special conditions require otherwise, present practice on major bridge-pier cofferdams is to use a combination of dead weight of the seal with uplift on the foundation piles; a minimum thickness of 1 m is considered to be a practicable minimum. Thus, seals with a thickness of 15 to 20 percent of the pumped-out head are common.

Seal Construction by Tremie Method. The tremie method is currently the accepted standard for high-quality structural underwater concrete. Tremie pipes are used so that the fresh concrete is always deposited under the surface of the previously deposited concrete. In this way, only the surface is ever exposed to the water.

The tremie pipe is plugged at its lower end (Fig. D4-17). For depths of water up to 20 m (70 ft) or so, the best plug is a sheet of 25-mm (1-in) plywood with a rubber gasket. This plug is fastened to the end of the pipe by heavy twine. The plugged tremie pipe is lowered to the bottom and charged with tremie concrete to about half the water depth. It is then raised about 300 mm (12 in) off the bottom, breaking the twine and allowing the concrete to flow. For deeper pours, where the buoyancy of the plugged tremie pipe may present problems, it may be necessary to replace the plug by a pipeline *pig,* which is a cylindrical plug with wipers, like squeegees, around its periphery. The pipe is lowered to the bottom and slowly filled with concrete, the weight of which forces the pig down and out the bottom end of the pipe. The first ½ m^3 of concrete should be grout, i.e., with no coarse aggregate. The concrete flows out over the base. If the pig is buoyant, it will float up and can be recovered and reused. From this time on, the lower end of the pipe is kept continually immersed in the concrete. Use of an inflated pig or ball should never be used in any but very shallow cofferdams. Otherwise, when the hydrostatic pressure exceeds the internal pressurization, the ball collapses.

Experience has shown that tremie concrete stiffens and gets initial set in about 8 h in normal water temperatures and with normal cements, unless retarding admixtures are used. The pipe end must be kept in the freshly placed concrete which has not yet taken its set but, on the other hand, it must be kept embedded deeply enough so that the tremie concrete will not all run out as new concrete is placed and the seal

FIGURE D4-17 Placing plug at tip of tremie pipe for underwater concrete seal course for intake structure for Hope Creek nuclear power plant, New Jersey.

be lost. The proper depth of the pipe immersion will depend on the rate of pour and the mix, the water temperature, the time of set of the cement and, of course, the height of the concrete column in the tremie pipe. The concrete inside will essentially balance with the water head outside and the pipe friction when about 60 percent full.

Sometimes a valve at the bottom end of the tremie pipe is specified; however, this seriously disturbs the flow of the concrete and causes segregation and laitance. A smooth and gradual restriction in the bottom of the pipe has been used in Japan and Sweden. Because of the prevalence of problems that have developed with valves due to jamming with aggregate particles, their use is not generally recommended.

Plasticizing admixtures and air entrainment are very beneficial in preventing segregation, reducing laitance, and ensuring a more level surface. Retarding admixtures are desirable, especially on large pours to prevent stuck tremie pipes, and to permit a deeper embedment of the tremie pipe. Retarding admixtures increase the form pressures. Superplasticizers should normally be avoided as they tend to premature slump loss as a result of the heat from the hydrating concrete.

The loss of seal always results in the formation of laitance and should be avoided whenever possible. If the seal is lost, the pipe must be raised, replugged, lowered into the fresh concrete, recharged, and then slightly raised to start the flow. Use of the pig, while very convenient in that the tremie pipe does not have to be raised clear, results in washing the fresh concrete by jet action as the pig is forced down the pipe. Therefore, an end-plug should be used for restart.

Proper mix, proper workability, and substantially continuous placing are essential for obtaining the best results. Tremie concrete has been successfully placed at depths up to 250 m, with practically no laitance whatsoever.

Gravel, not crushed rock, should be preferably employed as the aggregate for tremie concrete, with a maximum size of coarse aggregate of 25 mm (1 in), although 20 mm (¾ in) is best. For small, complex placements, 10 mm (⅜ in) has been employed.

Experience dictates the use of 45 percent sand. A rich mix is best, with a cementitious material content of 400 kg/m^3 (675 lb/yd^3) and a water cement ratio of 0.42 or less. Slump should be 116 mm ± 15 mm (6½ in ± ½ in). Even higher slumps may be used where special thixotropic, nonsegregating admixtures are incorporated in the mix. Replacement of 30 to 50 percent of the cement by PFA will lower the heat of hydration. Coarse-ground blast furnace slag-cement can be used, in the proportions of 65 to 70 percent BFS and 30 to 35 percent portland cement. This will reduce the heat of hydration even more.

Placing should be through good-sized tremie pipes, from 8 to 16 times the size of maximum aggregate. Pipes of 250 to 300 mm (10 to 12 in) in diameter have been very successfully used for cofferdam seals. Since the pipes must be raised gently and uniformly, some method of gradual, continuous hoisting should be provided. Pipes and hoppers can be raised by derrick and blocked up or tied off, but this tends to be jerky. The best way is by the use of air or hydraulic hoists.

For normal cofferdam seals, tremie concrete should be placed by gravity (Fig. D4-18). Pumping as a means of placing all the way to the seal is undesirable because the concrete emerges in pulses which cause disturbance of the surface and produce laitance. However, pumping may properly be used for delivering the hopper at the top of the tremie pipe.

On deep seals, the pipes may have to be raised so far as to present a problem in handling, supporting, and filling. The most common solution is to have removable sections in the tremie pipes. Some means of holding the lower end of the pipe during this operation must be provided. Underwater joints must be gasketed and tight to prevent water leakage, otherwise the joint may suck in water by Venturi action.

Any agitation or disturbance of the fresh concrete produces laitance. Therefore, flow must be kept as uniform as possible. An ideal situation would be to have a continuous flow into the hopper and out the bottom. A diver should never be allowed to walk on the concrete surface until after it has taken its set.

Tremie-pipe spacing depends on depth of pour, practical limitations of size and configuration, obstructions such as piles, slump, and whether or not admixtures are used. Modern practice is to use increasingly larger spacing, up to 6 to 10 m (20 to 30 ft) on placements 2 m (6 ft) deep and more. Some fairly large cofferdams with deep seals have even been poured with only one setting of the tremie; however, this can create a situation of overfalling and thus create pockets and segregation. In a large cofferdam it is usually necessary to shift pipes forward as the pour progresses. In this case, the pipes should be lifted clear, resealed and set gently into the advancing slope, using the end-plug method of sealing. Care must be exercised not to concentrate all the laitance in one corner or end. An airlift pump may be beneficially used to remove it from the far end.

Experience indicates that most of the laitance is formed at the very start of the pour as the concrete flows out over the bottom and that the amount of laitance is

FIGURE D4-18 Placing tremie concrete by gravity flow, Verrazano Narrows Bridge, New York.

dependent more on the area of seal than on its depth. Piles impede and disturb the flow and may increase the amount of laitance.

The slope of the tremie-concrete surface will usually run 1:6 to 1:10. Piles will cause a steeper slope, as will a stiff or unworkable mix. For these reasons, thin tremie seals are not desirable. Even in shallower cofferdams, 1 m (3 ft) should be the practicable minimum, and 2 m (6 ft) is preferable. Because of slope, laitance, and inaccuracies, 150 to 300 mm (6 to 12 in) of concrete thickness should be discounted in determining the weight and strength of seals. Thus, when a very thin seal is specified, good practice would be to start the seal below the specified bottom grade.

In bringing the seal to grade where a footing block or distribution block is to be poured later over the entire surface, and especially if this block is heavily reinforced near the bottom, as is the usual case, the seal should be stopped slightly below the design grade. This cannot be safely done on very thin seals, but it is good practice on reasonably thick or noncritical seals. The reason for this is that it is very expensive and time consuming to chip out tremie concrete above grade. On some major cofferdams, three to eight weeks have been consumed chipping out hard, sound concrete so that the reinforcing steel for the footing block could be placed. When stopped a short distance low, the deficiency can be made up when pouring the distribution block.

After the tremie seal is placed, which sometimes may involve thousands of cubic yards of concrete and many shifts, it is important to schedule one more operation before everyone goes home. This is to send down a diver after the pour is completed and after the concrete has definitely set, to jet off the laitance from the surface. At this stage, it can readily be jetted off and pumped out by an air-lift pump. If left until the cofferdam is dewatered several days later, jackhammers usually may be required to remove it (Fig. D4-19).

Tremie concrete which is properly made, mixed, and placed will develop compressive strengths of 30 to 50 MPa. Tests have shown that it also develops excellent bond to steel or timber piles, or to other concrete, provided that the surfaces are clean. In actual practice, ultimate bond strengths of 0.6 MPa (80 lb/in^2) and more have been developed (Fig. D4-20).

FIGURE D4-19 Excellent results can be obtained when tremie concrete of proper mix is placed by proper procedures. Figure shows West Anchorage of Second Delaware Memorial Bridge at time of dewatering. All laitance was able to be removed by jetting.

FIGURE D4-20 Tremie-concrete seal depends on bond to resist uplift. Deep pier of Columbia River Bridge after dewatering.

Due to its continuous immersion, tremie concrete does not develop drying shrinkage. In good tremie-concrete practice, the yield will run from 100 to 102 percent. Any greater yield is an indication of segregation and formation of laitance.

Tremie concrete, because of its large mass, can develop very high temperatures due to heat of hydration. On subsequent cooling of the surfaces (top and sides), cracks may form. For this reason, many modern tremie concrete mixes employ either a substitution of 30 to 50 percent of the cement by fly ash or the use of blast-furnace slag-cement. Reinforcing steel may be placed as a top or midheight mat in the tremie, in order to constrain thermal expansion and reduce crack width.

Seal Construction by Bucket Methods. Bottom-dump buckets can be used only where no foundation piles project above the bottom. The bucket must be watertight and must be covered. It is lowered gradually to the bottom and the gate is slowly opened, allowing the fresh concrete to flow out smoothly and uniformly.

This method has been widely used in the past, but it is not too satisfactory for cofferdam seals. As each load is deposited and the concrete flows out, some laitance is formed on the surface. The succeeding load may displace this upward, or it may trap it. In any event, experience has shown that an excessive amount of laitance is often formed by this method. In some cases, the depth of laitance above the concrete has actually been equal to the depth of seal itself.

Improved air-operated buckets have been developed by leading manufacturers, and the trend is to large buckets (up to 6 yd^3). The larger bucket and fewer loads result in less exposed surface and less laitance. If this method is used, it is suggested that a slope be kept across the concrete surface and an air-lift pump operated at the low end of the slope to remove as much laitance as possible. An antiwashout admixture should be used.

Seal Construction by Grout-Intrusion Method. This method has been used on a number of cofferdams as well as caissons. A graded gravel, with no fines, is placed in the location of the desired seal, and then a specially compounded grout is pumped through preplaced pipes. The advantages of this method are the much reduced quantity of fluid material to be placed, and the ability to carry out the work at exposed sites with minimal problems due to waves, etc.

However, the disadvantages of this method generally outweigh the other considerations. The principal disadvantage is the unintended blockage of flow of the grout due to inadvertent pockets of chips and fines incurred in placement, resulting in lenses and strata with no grout. Other problems are the trapping of bleed water under the coarse aggregate particles and the contamination of the aggregate by marine growth and siltation.

With this method, as the grout nears the surface of the seal, it finds its easiest route is up to the water above. Thus it is necessary to place 2 to 3 m of gravel above the top of the seal, to confine the grout. This surface layer will then be partially cemented, but must later be removed.

One source of contamination with fines is the practice of cleaning off the transport barge: All the fines due to mechanical abrasion will collect on the deck and, unless care is taken, will be concentrated in one zone when they are placed.

Another problem has been segregation during placing, as the gravel falls through the water. This can be minimized by lowering the bucket of gravel to the bottom before opening it, or by placing through a tremie pipe.

The preplaced grout pipes have to be placed at relatively close spacing, both laterally and vertically, say 2 to 3 m apart horizontally and 4 to 5 m apart vertically.

The admixtures used in the grout are designed to reduce surface tension, promote fluidity while keeping the water cement ratio low, and reduce segregation and bleed.

Although when properly executed, good, strong concrete with no shrinkage can be attained, the many problems associated with this method have led to a current trend to use the standard tremie concrete method instead.

Dewatering

Pumping. Large, temporary pumps are used to pump out a cofferdam initially, and smaller automatic pumps are then used to keep it dry.

It is important to have a large pump capacity in initial dewatering, since leakage through the sheet piles will be excessive until a differential head is established and the sheet-pile interlocks tighten up under pressure. Several large pumps are normally required to give a satisfactory rate of dewatering. While electric pumps are fine if the power supply and connections are available, diesel- or gasoline-powered pumps are also satisfactory since they will be used for a short period only. The exhaust of gasoline- or diesel-powered pumps must be led over the sheet piles and clear of the cofferdam to prevent dangerous concentrations of carbon monoxide in the bottom of the cofferdam.

After the cofferdam has been dewatered, a few small electric float-operated pumps can be used to keep it dry. These pumps are economical in their operation and, being float operated, will automatically control the water level. An excess pump capacity of at least 100 percent should be installed in case a float valve sticks while unattended. The extra pump or pumps should be automatically controlled so as to go into operation if required.

Maintenance pumps are usually installed in small sumps. If a tremie concrete seal is used, a sump can be jack-hammered in a low spot in the tremie-concrete surface after dewatering. Where the distribution block is smaller in plan than the tremie-concrete seal and leakage occurs around its edges, sandbags can be used to channel this water to the sump.

It is often desirable to construct a tide gate in a cofferdam. This is usually done during the driving of the sheet piles by burning off one pile at the low-water line. The top section of this pile can then be raised a few feet to open the gate. By leaving this gate open during construction of the cofferdam, unbalanced pressures are avoided. Otherwise, without the tide gate, the water will fill up inside the cofferdam to the high-tide level through in-leakage, but as the tide starts to fall, the drop in the inside water level will occur at a slower rate. The unbalanced head produces an outwardly directed pressure for which few cofferdams are designed.

When the cofferdam is ready for dewatering, the tide gate is left open until low tide. As the tide starts to rise, the gate is closed and welded or caulked. All available pumping capacity is then thrown into action. This has been found to be the best and sometimes the only way in which a start can be made on dewatering. Most well-built cofferdams should not leak excessively after this start in dewatering has been accomplished. The more rigid and heavily braced cofferdams generally leak more than flexible cofferdams, because in the latter case the interlocks are very tight under stress.

If excessive leakage through the interlocks is encountered, there are several techniques for reducing the flow. Often, some material is dropped through the water just outside the sheet piles so that the inflowing water will suck it into the interlocks. Some of the mixtures which have been used are sand and sawdust; fibers and sand; and cinders, manure, and sand. On several cofferdams which leaked excessively, weighted canvas sheets were draped on the outside of the sheet piles. A diver can place sandbags at the bottom to hold the canvas against the sheet piles.

Serious leaks can occur where sheet piles have been driven out of interlock, and corrective measures are time consuming and expensive. Usually the fact that a sheet pile has been driven out of the interlocks is not known until efforts to pump down have been unsuccessful. A diver can then be sent down to inspect the interlocks but, since any opening is usually at or near the inside bottom, it may be hard to find. Under some conditions, a colored dye in the outside water might help to locate the bad spot. In some cases, cofferdams have been successfully dewatered only to subsequently blow at the location of a gap.

Once a leaking spot has been definitely located, corrective measures can be started. One method is to drive a blister of two or three sheet piles on the outside. A gasket of some soft material is then used to seal the new piles against the old wall, or a diver can drive wedges. Where possible, the soil inside this blister is excavated down to the level of the bottom of the tremie seal, and the blister is then filled with tremie concrete. In other cases, a temporary patch may be made by a diver using an underwater quick-setting hydraulic cement.

Pressure grouting can be used below the mud line in granular, poorly graded materials to help seal areas which cannot otherwise be reached. One technique which has been successfully used in connection with the blister method is to keep the water level inside the blister higher than normal. This excess head tends to keep the soil outside the blister from entering it during excavation. Bentonite slurry can also be used.

Leaks through handling holes have occurred on several major projects, due to carelessness. A routine procedure should be set up to check every pile where spliced or used piles are being driven. If a leak does occur it can be plugged by a diver, assuming that it is above the mud line and can be reached. Otherwise, a blister or similar method must be employed.

Splices will always leak since the interlock cannot be fully welded. Also, the repeated sharp blows in driving will cause poorly made splices to crack and leak. The individual leakage volume at a splice is usually small and is generally important only as a contribution to the total leakage. Occasionally, the added leakage due to splices is enough to make corrective measures (such as canvas) necessary. For this, and for structural reasons, and for efficiency in removal afterward, first-quality, full-penetration welds are essential.

Other Dewatering Methods. When no underwater concrete seal is placed, wellpoints can often be effectively used, particularly in very large cofferdams that are not too deep. Standard wellpoints are most effective in medium and coarse sands, while a vacuum wellpoint system should be used in fine sands.

A clay blanket placed on the outside of a cofferdam has much the same effect as additional penetration of the sheet piles. It increases the flow path of the water, thus decreasing infiltration and reducing the tendency of the bottom to blow. Under proper conditions, the placing of several feet of gravel inside the bottom of the cofferdam before starting the dewatering helps to stabilize the bottom. Gravel lets the water flow through but holds down the sand and other sediments, preventing a real blow. A thick gravel layer will also act as a strut to provide reaction for the sheet piles. Where the excavated bottom is in fine sand, a blended gravel–coarse sand mixture, similar to a concrete mix, will act as a reverse filter and hold the bottom stable while permitting water to flow through.

Cofferdam Difficulties

A few of the more interesting cofferdam difficulties are described in the following paragraphs.

Cofferdam Destroyed by Surge. A cofferdam for a bridge pier was built at the mouth of an estuary, where heavy ocean swells ran in past it almost continuously. During construction, when three walls of the cofferdam had been set, the swell inside would get out of phase with the swell outside. This caused the sheet piles to work back and forth until some particularly heavy swells broke one wall off as the result of fatigue. The cofferdam was rebuilt and each sheet was tied securely with cables and turnbuckles to prevent working. More rigid bracing and guides were used, and the setting and driving of sheet piles were completed as rapidly as possible.

Cofferdam Bottom Blows. The project required precast-concrete piles to be driven as bearing piles for new bridge piers in a tidal lagoon. The soil consisted of medium and fine hard-packed sand. The cofferdam bracing was set, the sheet piles were set and driven, and some 6 m of sand was excavated. The precast concrete piles were then set and driving was started, using swinging leads and an underwater hammer. Because driving was unable to develop adequate penetration of the concrete piles, a jet was employed. This jetting caused the bottom to blow, and fine sand to run in. The situation was aggravated by the lack of a tide gate, so that on a rising tide, the water level was higher outside than inside. The blowing of the bottom caused the sheet piling to kick in so badly that the entire cofferdam had to be removed. It was finally rebuilt, using a larger cofferdam and longer sheet piles so as to get more penetration.

Improper Sheet-Pile Section and Bracing. In this case, the bridge pier was to be built in a river, with 10 m of water overlying 10 m of sand. A 2-m rise of water could be expected after rains, and a river current of as high as 10 knots (5 m/s) was anticipated. The contractor built heavy timber pile guides to resist the current. He selected the lightest deep-arch section in 80-ft lengths which required six sets of bracing, two of which would be below the existing bottom. Deep predredging was impracticable because of the current.

The sheet piles were set and driven, using a jet, and the inside was excavated a few feet and leveled. The bottom set of bracing and frames was lowered to the sand, and the other five sets were erected above it.

As excavation proceeded, the bracing was forced down by weighting and even by driving down, using a spud set on the wales and driven with a pile hammer. When about 3 m of sand had been excavated, the sheet piles deflected inward so far that the bracing could be no longer forced down until additional driving and jacking forces were employed.

Excavation was finally completed, a 6-m-thick seal poured, and pumping started. As the water lowered, the third or middle level of bracing suddenly began to crush and give way. It was necessary to flood the cofferdam and have divers install an additional, heavier set of bracing before again commencing dewatering.

It is probable that the excessive pressures were due to the nature of the sands and their disturbance by jetting. However, the method that was employed is hazardous, whether successful or not. A heavier section of sheet pile such as a Z section would be preferable today, although Z piles were not manufactured when this particular cofferdam was built. The bottom could have been predredged, following which a complete bracing system could have been set as a unit. The sheet piling would then need to be strong enough to take the unbalanced soil pressure during excavation and also the hydrostatic pressure after the seal had been poured and the cofferdam dewatered.

Sheet Piles Stopped by Boulders. In this cofferdam, a pier was to be carried approximately 60 ft deep to glacial till. Because of the length of sheet piles, an auxiliary cofferdam of sheeting was used to allow the main cofferdam to be built with its top at low-tide level. A heavy, deep-arch sheet-pile section was selected for the main cofferdams and 22-m lengths were set and driven, using a heavy steam hammer to overcome the excessive driving resistance. Even so, many of the sheet piles stood high, having encountered refusal on the boulders above the till.

The excavation was started, and dewatering and bracing were carried out as the soil was removed. As the tip elevations of the high piles were reached, sand began to boil in underneath. Wellpoints were employed, both from on top and through holes in the sheet piles.

It was found that it was the boulders which had caused the trouble and that the sheet piles were bent, twisted, and split "beyond description." The boulders were drilled and shot, and those sheet piles which were not damaged too badly were driven farther. Nevertheless, conditions finally got too bad to proceed. The design of the pier was changed to employ foundation piling instead of the prescribed plan for constructing a concrete block on the glacial till.

This problem of boulders floating in silt and sand on top of bedrock or glacial till, is always a serious and costly one. Conglomerate presents similar problems. Excavating, drilling, and shooting under water before driving sheet piles have sometimes been successfully employed. Line jetting or drilling down along a sheet pile wall can be employed. Alloy sheet piles and hard-facing of the tips of the sheet piles also helps to minimize distortion.

Buckling of Long Struts. One large, medium-depth cofferdam employed long, unsupported struts for its single level of bracing. As the cofferdam was dewatered, these struts buckled and the cofferdam collapsed. While the pile struts would have had sufficient column strength if they had been perfectly straight, their dead-load deflection caused them to sag before the loads ever came on; so they obviously had reduced column capacity to resist the water pressure. In this particular case, a few piles driven near the middle of the struts would have been a practicable and economical method of supporting the struts so as to reduce their unsupported lengths.

Improper Procedure in Unstable Soil. Several piers had to be built in relatively shallow water, but excavation was to be carried a considerable distance into the underlying soils. The soil was glacial rock flour (silt), very dense and impervious but unstable in that it would fail easily in shear on curved planes. It became much weaker as soon as it was disturbed. Foundation piles were specified, and a tremie seal was prohibited in order to keep foundation pressures low. The drawings indicated that a long penetration of sheet piles should be obtained.

The contractor used the longest sheet piles which could be obtained in a single length. After setting and driving around a top frame, the contractor proceeded to excavate, dewater, and place bracing progressively. The lower portion on two of the cofferdams collapsed suddenly; fortunately, the failures occurred when no work was going on. Excessive pressures were noted in other cofferdams, and the bracing was doubled up and reinforced. The owner blamed the contractor for using too-short sheet piles, while the contractor made a claim against the owner for his failure to specify a tremie-concrete seal.

The steel bracing frames provided by the contractor were adequate to resist the final pressures in the completed cofferdams. But the procedure adopted, excavating and pumping down level by level and installing bracing as work progressed, permitted excessive deflection of the sheet piles. The remolding effect of driving the sheet piles weakened the soils and reduced their passive resistance, leading to failure. Further, the excessive deflection of the sheet piling caused shear planes to become well established and lubricated, thus multiplying the pressure. In at least one case, there had been slow plastic "squeezing" flow of the soil upward in the bottom (heave). This reduced the toe support for the sheet piles, allowing them to collapse inward. In the other case, the plastic flow of the bottom was apparently simultaneous with the shear failures on the outside.

Based on previous experience in the glacial rock flour which is frequently encountered in Alaskan and Canadian lakes, the author believes that the pier sites could have been predredged some 6 m deep to within 2 or 3 m of grade. Side slopes of about 1:3 to 1:4 would have been stable underwater. Then the bracing frame could have been set as a unit, the sheet piles set and driven, the excavation completed, foundation piles driven under water, a few feet of gravel placed, and the cofferdam dewatered. The sheet piles would require a sufficient penetration below the bottom to cut off flow. In this impervious but weak material and with about a 12-m external water head when dewatered, the penetration should have been 6 to 10 m beyond that. One heavy set of bracing would have been needed near the bottom to provide support. This extra level of bracing would be made up and set as part of the whole bracing unit at the predredged site and placed at as low a level as possible—i.e., at the predredged grade.

An alternative method of construction, if predredging were not possible, would have been to keep the cofferdam pumped full of water during excavation. Divers would then have been employed to lower and wedge the bracing as the excavation progressed.

Tying the sheet piling at the top would help to prevent rotation of the sheet piles about the additional bottom set of bracing. Heavier section sheet piles would have helped in any case, provided that there had been good penetration and tying at the top. A careful study of the soil characteristics would have pointed up the dangerous conditions and probably have led to the selection of a safe procedure at the start.

Scour and Poor Tremie-Concrete Procedure. Pier 169 of the Columbia River Bridge at Astoria, Oregon, was one of four twin cofferdam piers for the crossing of the main channel. The cofferdam construction procedure was to predredge to the bottom of the seals at −20 m, drive the piling with an underwater hammer, set and hang a bracing frame from falsework piles, set and drive sheet piles, then pour the tremie seal. On Pier 169, after the H piles had been driven and the bracing frames set, the sheet piles were set and driven. Suddenly, on the occasion of a high ebb-tide runout which augmented the river's normal flow, the bottom scoured on the channel side to a depth of 6 m. The sheet piles settled vertically and the cofferdam tilted. Attempts were made to deposit sand in the hole by using a hopper dredge, but to little avail. Brush mattresses were made up and sunk with rock, and then riprap, etc., was placed over them. This succeeded in stabilizing the bottom and restoring most of the material on the outside. The inside soil surface had dropped about 1.5 m, so the sheet piles were extended by splicing them at low tide, then driving them down another 1.5 m. Since the steel H piles for the foundation were battered closely under the original tips of the sheet piles, the additional driving forced the tips against the sloping batter piles, so that the sheet pile tips bent outwards, rupturing some interlocks.

The tremie seal was then poured. Only one tremie pipe was used. This pipe was apparently charged and sealed correctly and the start of the pour was properly performed. According to later verbal accounts, after 2 or 3 h with the pour only 2 m thick at the time, the tremie pipe was then raised vertically some 4 to 6 m and the remainder of the concrete was poured through water. A mound eventually built up under the pipe and the tremie concrete was again being properly discharged. It was noted that only 64 to 76 percent of theoretical volume of tremie concrete was being required; that is, the swell was 150 percent. It has since been reported that the water surrounding the entire area was milky white, indicating excessive dispersion of cement.

After dewatering, chipping of high spots revealed generally good concrete on the surface. A meter down, however, uncemented sand and gravel was encountered. The cofferdam began to leak excessively through the seal, and test probes showed that this loose sand and gravel extended to within a few feet of the bottom. Water began to flow freely through channels forming in the seal. The cofferdam was ordered flooded and the contractor was directed to remove and replace the seal. This led to considerable contractual disagreement, termination of the contract, and selection of a new contractor.

Most of the old seal was initially removed from the flooded cofferdam by using high-pressure jets, chisels, and air-lift pumps. At about midheight of the seal, sound concrete was found interbedded between the gravel layers. The decision was made to use explosives to remove it. Small-scale trials proved effective, so a larger charge was tried. Unfortunately, this blew out the side of the cofferdam. A steel blister was subsequently installed around the damaged area, and the work of removing the seal was completed by diver chipping.

The new contractor chose to use grout-intruded aggregate to replace the seals. After he had placed the gravel aggregate in one pier, it was discovered that excessive silt from the river had been deposited in and around this aggregate. It was therefore necessary to remove all the gravel and replace it with clean aggregate. After this had been done, the grout was intruded successfully, the cofferdam dewatered, and the pier finally completed.

Problems with Tremie-Concrete Seal. The East Anchorage of the Second Delaware Memorial Bridge was constructed in a cofferdam 35 m × 80 m in plan, excavated into firm clay at −20 m. The bracing frames were assembled just above high water and were supported on pin piles. Then the sheet piles were set and driven into the hard-clay founding strata. The cofferdam was excavated progressively by clamshell and, as each particular level was reached, a bracing frame was lowered down to position and secured to the pin piles and sheet piles. Sections of bracing were bolted together by divers, which proved to be a very time-consuming task. Further, the hard clay proved difficult to excavate near the sheet piles and required considerable jetting and chiseling.

When the founding elevation was reached, a tremie-concrete footing block 9 m (30 ft) thick was poured in one continuous operation lasting 13 days and nights, totaling 22,000 m^3 (28,000 yd^3). The specifications required that it be poured in 2.5-m-thick layers, starting at one end and proceeding to the other before returning with the next lift. A low cement factor mix was specified in order to reduce the heat of hydration and as a means of economy. However, lean tremie mixes have seldom turned out to be successful, and this was certainly the case here. A better means of reducing the heat of hydration on the basis of today's knowledge would have been to add mineral admixtures such as pulverized fly ash.

In addition to low cement content, the tremie mix had the further flaw of being undersanded (36 percent sand), so that flow could be accomplished only by addition of water. Such a mix was harsh, unworkable, and inherently subject to segregation. The result was excessive zones and pockets of sand and gravel and laitance, sometimes partially covered by good concrete. It required several months of around-the-clock work to cut out the defective material preparatory to continuing work (Fig. D4-21).

Defective Concrete Seal. Unfortunately, a number of such cases have occurred which have proven extremely costly and time-consuming to correct. In one such case, in the northeastern United States, a large tremie-concrete seal was to be placed in a deep cofferdam for a bridge pier. The day of the scheduled operation, a young engineer was sent out to get a "volleyball" to use as a plug for starting the pour (error number 1). He came back with a play ball that was the right diameter but had a large handle attached. When the ball was halfway down the pipe, the handle caused it to stick (error number 2). There was no spare pipe on hand. So a diver was sent down to burn out a hole in the pipe in order to extract the ball (error number 3). Then for four days and nights the pour was continued, all the time sucking in water by Venturi action, and causing washout of cement and segregation even before the mix reached the bottom of the pipe.

FIGURE D4-21 Improper mix design and procedure led to extensive segregation and laitance. East Anchorage of Second Delaware Memorial Bridge. Compare with subsequent performance on West Anchorage, Fig. D4-19.

The cofferdam could not be unwatered. Investigation showed large lenses of sand, large pockets of clean gravel, and, unfortunately, large overlying lenses of hard concrete.

There were other contributors to the disaster: an undersanded harsh mix, excessive preheat of the mix water, and a steeply inclined open conveyor belt for delivery, which allowed the excessively wet mortar to run off the edge.

Scour Around the Cofferdam. A bridge was being built across a narrow stretch of the river, in moderately swift current. The river banks were steeply confined. The contractor built all the five cofferdams at the same time. He had moored crane and supply barges alongside each.

A sudden flood caused the current to double in velocity. The obstruction by the cofferdams and barges caused excessively high local currents, scouring the bottom below the sheet pile tips, so that two cofferdams collapsed.

Flood. Closely related to the preceding were two cofferdams in the Mississippi River. When the summer flood raised the river, the cofferdams were flooded. However, this was an inadequate measure and the cofferdams failed.

First, there were several levels of bracing in each cofferdam but no x-bracing in the vertical plane. Secondly, no thought had been given to the drag of the current on the Z sheet piles, nor on the barges tied alongside, so that the design did not provide adequate global strength. To make matters worse, driftwood piled up against the barges and cofferdam.

Sheet-Piling Deflection. On a deep cofferdam in an estuary in California, the tremie concrete seal was to be 10 m thick. The contractor predredged, drove bearing piles, installed a prefabricated bracing frame, and drove the sheet piles. So far, so good. He then placed his tremie concrete, programming it at a moderate rate so that the concrete below had time to set, so as not to produce excessive pressure on the sheet piles. The internal pressures of the tremie concrete deflected the sheet piles so that they bulged outwards, allowing the newly placed concrete to run down the gap and exert the full fluid head, causing even further bulging. Although the contractor quickly proceeded to backfill outside and managed to save the cofferdam, he placed several thousand yards of additional concrete and was unable to salvage his sheet piles for reuse.

Combined Mistakes.   A pier was to be constructed in water 30 m (90 ft) deep, with only a few feet of overburden into which to seal the sheet piles. The contractor set a well-designed bracing frame, drove the sheet piles, and started placement of a 20-m-thick seal. Sheet piles were spliced piles because of the length and were fabricated of used piling. Problems similar to the preceding case arose, although not to the same degree, since the concrete mix was very harsh (undersanded) and of a low cement content, and the placement rate slow. He placed the concrete by pump all the way to the bottom. A vacuum formed when the concrete plunged down the vertical riser, causing segregation. The concrete pump discharge took place with pulses. The result was the formation of a great deal of laitance in layers between the sound concrete.

The used, spliced sheet piles were not mortar tight, hence mortar was sucked out by the tidal current, leaving large pockets of honeycomb and some actual voids.

Extensive repairs were undertaken by pressure-grouting in closely spaced drilled holes, using a variety of admixtures and procedures to attempt to wash out the laitance and fill the resultant space with sound grout. Eventually, satisfactory results were obtained, but at the price of bankruptcy for the first contractor.

Grout-Intruded Aggregate—Failures.   A bridge pier was being built in a polluted river, where algae growth was very rapid due to the nutrients available from the pollution. The cofferdam was properly constructed, grout pipes were placed, and gravel placed. Then a strike ensued and the cofferdam was left for several months. It was finally grouted, but was unable to be dewatered. The algae growth had contaminated the aggregate, preventing interparticle flow and bond to the coarse aggregate.

Slurry Wall Cofferdams

A number of major cofferdams have been built in recent years using the slurry wall technology in lieu of steel sheet piles. When constructed in an overwater location, a *sand island* has first been created so that the slurry wall can be excavated as on land and subsequently concreted up above water. Rectangular cofferdams have been constructed by this method up to 20 m (70 ft) deep, while circular (ring) cofferdams have been constructed to give a dewatered depth of 80 m (250 ft) in Japan. In the case of the Kawasaki Island ventilation shaft of the Trans-Tokyo Bay Crossing, the slurry wall cofferdam is 100 m in diameter and extends to a depth of 120 m (400 ft). Thus, the slurry wall technique, with its ability to take the high compressive ring loads, is enabling large cofferdams to be built to unprecedented depths, even in very weak soils (Fig. D4-22).

The slurry wall technology is based on the construction of successive small increments, such as primary panels or piles spaced a set distance apart, with secondary panels subsequently constructed to fill the space between them. A panel slot is drilled or excavated by special buckets or multiple drills, using bentonite slurry to prevent caving. Reinforcement consists either of steel shapes such as wide-flanged beams or prefabricated cages of reinforcing steel. The slot is then filled with concrete placed by the tremie method.

FIGURE D4-22 Slurry wall cofferdam for cofferdam 100 m in diameter, dewatered to 70 m depth. Kawasaki Island, Trans-Tokyo Bay Crossing, Japan.

Joints between panels must be carefully cleaned by scraping or brushing. Usually some form of shear key is provided between panels.

For ring cofferdams constructed by this method, it is essential that the wall be a true circle, with very limited tolerances for out-of-round, since it is very difficult to provide a reliable means of circumferential continuity between adjacent panels. Therefore, reliance is placed on ring compression and the use of thick panels, with close control of verticality by inclinometer, taut wire, or electronic instrumentation.

Reinforcing steel cages, if used, must be adequately welded so as to prevent dislocation of bars and distortion of cages when they are lifted and set. Diagonal bracing is required. Spacer pads of concrete are used to keep the cage centered in the excavated slot. The quality of the slurry must be controlled at all stages, as contamination by the soil chemistry and the concrete will cause it to thicken, and thus prevent intimate bond of the concrete with the steel. Contamination may also cause clay balls to form, resulting in voids in the wall.

Cellular Cofferdams

Sheet-pile cellular cofferdams have historically been employed in river diversion and for the construction of major locks and dams. They are also employed for marginal wharves. They are also used as dolphins to provide protection to bridge piers against ship collision. Since their construction presents special problems better dealt with in other chapters, they will be only briefly discussed as they relate to bridge piers.

Sheet-pile cells form a gravity dam, in which the material inside the cell, sand or gravel, is contained by the tension ring formed by the sheet piles. In order to prevent uplift and to prevent shear on a vertical plane through the center of the cell, they are typically dimensioned so that the diameter is approximately equal to the depth of water which they must resist.

Sheet-pile cells, in which the individual sheets are set one by one around a circular template, can be constructed in swiftly flowing water. In such a case, the sheet piles should be tied to the template as they are installed.

Sheet-pile cells, filled with sand or gravel, are strongly resistant to ice, especially moving ice.

For the Verrazano Narrows Bridge in New York, sheet-pile cellular cofferdams were used to form sand islands, through which caissons could be sunk to bedrock.

It is very important that the individual sheets be set vertically and properly spaced. Therefore, the template should preferably have two levels of wales, supported on piling or spuds. When water conditions permit, all the sheets should be set around the circle before major driving operations are undertaken. Where the sheet-pile cells are connected by Ys and closure arcs, the Ys, being stiff, should be set first in exact position and verticality.

Stability is enhanced by densification of the fill material. Therefore, vibrating probes may be used to densify the sand and to drive the excess water up and out. To enhance this escape of water, vertical gravel drains will often be found useful, reducing the energy and time required for densification.

When cellular cofferdams are constructed on bedrock, the most critical zone is the contact area where water inflow may cause loss of stability and/or prevent dewatering. Tremie concrete may be placed, either inside the cofferdam before the sand fill or around the outside toe. In the latter case, sacked concrete may be appropriate.

BOX CAISSONS

General Considerations

A box caisson is a prefabricated box, with sides and a bottom, which is set on a prepared foundation and filled with concrete to form a bridge pier or similar structure. The excavation is completed before the box caisson is set. Box caissons and box cofferdams have many elements in common, the semantic distinction being that the box caisson becomes part of the structure and the box cofferdam is only temporary. Most actual structures contain elements of both, and the question of terminology is unimportant.

Many variations of the box caisson have been developed and used on notable and important bridges. While some of these were not true box caissons, they have possessed many of the same elements: the prefabrication, the completion of excavation before setting, the setting on a prepared foundation, and their incorporation in the completed structure by filling with structural concrete.

Often, the top section of a conventional box caisson is a box cofferdam. After the pier shaft is completed to an elevation above high water, the box cofferdam is flooded and the cofferdam units are removed. A representative box caisson of this type might extend up to 6 m (20 ft) below water level, with the caisson and its fill of underwater concrete remaining in place as the permanent pier base structure. Above this elevation, the box might be a cofferdam and the pier might be formed by shafts poured in the dry. The sides of the box cofferdam would then be removed after completion of the shafts.

Box caissons, box cofferdams, and concrete-filled cribs were much used in the past for comparatively small bridge piers. A box was built, using heavy timbers to form its sides and bottom, and its joints were well caulked and sheathed. This box was towed to the site and sunk to rest on a previously excavated base by filling pockets with rock or concrete. The water was pumped out and the pier built in the dry. Where the bottom was sloping rock, it was usually excavated sufficiently to form a key and the box was tailored to fit the actual bottom.

As the method evolved, many piers were built from box caissons which were supported on foundations of timber piles. A basin was first predredged, following which timber piles were driven. Usually these piles extended above water so that a diver could measure down and accurately cut them off to grade. Guides were then constructed, and the box caisson was built, launched, towed to the site, and sunk to rest on the piles. A concrete seal was placed, the upper portion of the caisson dewatered, and the pier completed in the dry. The excavated hole was then backfilled to protect the timber piles and timber bottom of the box from marine borers, and riprap was placed to protect against scour.

Later box caissons of this type frequently had concrete bases and occasionally had concrete sides with removable timber cofferdam walls. The box could be floated to the site and sunk to grade by filling pockets with concrete. In another development, box caissons with timber bottoms were set on predredged and leveled bases. Timber piles with shoes or steel piles were driven, punching through the timber base. A tremie-concrete seal was placed and the caisson unwatered.

Where bedrock or hardpan can be exposed and cleaned by predredging, a steel or precast-concrete box can be sunk to found directly on the rock. Some structures of this type consist of bottomless sheet-pile boxes which are set in place by derricks. A concrete seal is poured and the pier constructed either by underwater concreting or by dewatering and construction in place. Precast-concrete or structural steel units have also been employed to form box caissons. Such units may be set in sequence and later filled with tremie concrete, thus tying the entire structure together. A number of recent major bridges have been constructed as box caissons in water depths as great as 65 m (200 ft) (Fig. D4-23).

Another variation is a box caisson which uses permanent buoyancy to support a part of the load. The complete underwater pier structure is built of concrete in a dry dock, floated to place, and sunk between guides to seat on previously cutoff piles. The pier is cellular in construction, having many watertight compartments. Buoyancy then supports a large portion of the dead load of the pier, while the foundation piles support the remainder of the dead load and the live load. The compartments near the waterline may be left open to the outside water to eliminate changes in buoyancy due to tide.

Sinking of such a box is accomplished by filling specified cells with water. Extremely accurate control and very sturdy and effective guides are required if this method is to be successful. Stability must be assured at all stages, especially the critical stage when horizontal surfaces first sink below water.

Where bottom conditions are adequate, the piles may be omitted and the box sunk directly on the rock or soil. Any voids under the box may be filled with grout under pressure after the structure is backfilled.

Site and Foundation Preparation

Excavation at the site is performed by clamshell, dipper or ladder (bucket) dredges, or occasionally by hydraulic dredge. If the box caisson is to be set on bedrock, the site must first be stripped of all overburden. Then, if practicable, the rock should be leveled. Underwater drilling and shooting may be required for this purpose, and care

FIGURE D4-23a Steel box caisson afloat for main pier of Akashi Strait Bridge, Honshu-Shikoku Crossing, Japan. *(Courtesy of Kajima Corp.)*

FIGURE D4-23b Steel box caisson (in place) for main pier of Akashi Strait Bridge, Honshu-Shikoku Crossing, Japan. *(Courtesy of Kajima Corp.)*

should be taken to avoid breaking up too much of the rock below grade. If it is not practical to level the rock at the site, it should be stepped and adequate keys should be provided.

Crushed rock may be then placed in layers 1 to 1.5 m thick and consolidated by vibration, then screeded. After the caisson is seated, grout may be injected underneath to fill interstices and fractures in the rock.

Where the box is to be set directly on hardpan or soft rock, predredging should go deep enough to ensure good support, and the site must be leveled accurately. The dredge must be able to cut the final surface to a very small tolerance.

Alternatively, a bottomless box may be set on three or four accurately leveled pads, and tremie concrete placed to fill under the walls and supporting structure.

Where the pier is to be supported by foundation piles, the site should be overexcavated to prevent interference with the setting of the caisson. If the box caisson is to be temporarily supported on the bottom, then a gravel or crushed-rock blanket can be placed and leveled off to grade. This rock should be of an adequate size (25 mm [1 in] is frequently satisfactory) and should cover a sufficient area so that it will not erode or scour as the caisson is lowered. A Venturi effect can cause a substantial increase in bottom current velocity just before the caisson is landed.

Foundation piles are driven by the customary methods, using an underwater hammer riding in telescopic leads or a special follower. In many cases, the piles can be driven to exact grade. In other cases, the piles must be cut off accurately to grade if the box is to be set on them. A diver can cut off timber piles under water, using an air saw, but it is a slow and tedious process. It is necessary to first measure down on each pile and then clamp on girts so that the pile can be cut off accurately and squarely. For this reason, sizable projects will generally employ large underwater mechanical saws.

Steel pipe piles may be cut off by internal casing cutters; steel H piles may be cut by hydraulic shears or by burning.

Fabrication, Launching, and Deployment

The conventional box caisson is designed to be floated to the site and sunk in place. Therefore, it is built much as a barge would be: in a dry dock, on a marine railway, or on a launching ways. A box caisson may also be built on a barge, which is then submerged to launch the caisson.

Since a box caisson is set on a previously excavated and prepared site, there will normally be no soil pressures against its walls. The caisson walls in the zones that are to be filled with underwater concrete must be designed to withstand the net fluid outward pressure of the concrete. The effective pressure of underwater concrete will run about 14 KN/m^2/m depth (80 lb/ft^3 per foot of depth.) The depth of concrete which has not yet taken its initial set must be considered to act as a fluid. Thus, there is a point of maximum lateral pressure which moves progressively up the walls; e.g., with a scheduled rate of pour of ½ m/h and a maximum time of 6 h until initial set, there will be a fluid zone 3 m in depth which moves progressively up the wall. It exerts a maximum pressure at any point of 42 KN/m^2 (800 lb/ft^2). The walls of a circular caisson are designed to resist this load by hoop tension, just like a tank. Steel is well adapted to resist stress in this way, but reinforced concrete can also be used. The design tension in the reinforcing steel should be kept low, say 120 MPa (16,000 lb/in^2) in order to minimize cracks in the concrete. Prestressing can also be employed.

In rectangular caissons, the sides can be designed to take the load in bending horizontally between vertical soldier beams of steel or concrete. Alternatively, the sides may be designed to take the load in bending vertically, spanning between wales. The lateral load is once again caused by the fluid zone of concrete as it moves progressively up the walls. Steel dished plates may be used to take the load in bending.

Deflections during pouring must be watched, particularly with steel, to prevent the fluid concrete from pushing the whole wall outwards and breaking it loose from the already set zone of concrete. Such movement would allow fresh concrete to run down the crack, thus building up a very high fluid head to further deflect the walls outward. Since this could lead to progressive wall failure, anchors or dowels should be provided which will tie the walls to the concrete as it sets. Time of set depends on the cement properties, the temperature of the resultant cement, the temperature of the resultant mix, the rate of heat development due to hydration, and the temperature of the surrounding water. Accordingly, the rates of pour must be intelligently adjusted in the field to prevent excessive fluid heads from developing. A rough check can be made in the field with a rod to determine the depth of fluid concrete.

If any zone of the caisson is to be dewatered, its sides must be strong enough to resist the external hydrostatic pressure. This was a factor previously considered in cofferdam design, but in the case of a box caisson, the load is usually taken in horizontal bending between vertical soldier beams or vertical solid walls. The walls in this zone must be relatively watertight and, if made of steel, should preferably be welded or else bolted and caulked. If the box caisson has a structural bottom slab, this must be designed to distribute the load over the prepared foundation. If the box caisson is to be set on piles, it must resist the punching shear of the piles and support its loads while spanning between supports. Since one pile may be a little high and another a little low, the bottom should be designed to span across at least one missing pile and to resist two or more times the average punching shear. Similarly, if the box is set on compacted and screeded gravel, the bottom must be designed to withstand the concentrated loads from high points, often called *hard points*. The bottom of the box must also resist upward hydrostatic pressure during flotation and all stages of sinking. This force must be taken in bending between exterior and interior walls. Similarly, if the box is to be dewatered, the bottom which is reinforced with tremie-concrete fill must take the upward hydrostatic pressure. In addition to resisting the design stresses, each member and section must be rugged enough to take the stresses and impacts of launching, towing to the site, and sinking.

One type of box caisson consists of several prefabricated units which are set under water. These are usually built in a yard, picked up by crane, set on a barge, towed to the site, and set with a derrick. Later they are joined by tremie concrete infill. To aid in lifting these heavy units, picking eyes or bolts are embedded in the concrete.

An assembly of high-strength alloy steel bars, of the type used in post-tensioning, have been adopted for heavy lifts on some recent bridge construction. It is important to have long lifting bolts and to suit detail mild steel reinforcing so as to prevent the bolt and adjacent concrete from shearing out. A factor of safety suitable to rigging practice, i.e., 5:1 or 6:1, should be used in the design of lifting bolts. Loops of reinforcing steel are not safe to use as picking eyes because deformations cause local stress concentrations.

Caissons up to 7000 tons in weight have been directly lifted from a quay by a large crane barge, transported to the site and set (Fig. D4-24). In other cases, they have been built on a barge, which is satisfactory as long as warping of the barge during pouring does not distort the unit, since when two or more segments are to be constructed on the same barge, the placing of concrete in one segment may warp the barge and thus distort another recently constructed segment. Large caisson segments have sometimes been constructed on girders spanning a basin; a barge has

FIGURE D4-24a For the Great Belt (Denmark) Bridges, very large and heavy caissons of precast and prestressed concrete are prefabricated, then transported to the site and set directly on a prepared foundation. *(Courtesy of COWIconsult.)*

FIGURE D4-24b For the Great Belt (Denmark) Bridges, very large and heavy caissons of precast and prestressed concrete are prefabricated, then transported to the site and set directly on a prepared foundation. *(Courtesy of COWIconsult.)*

then been brought underneath at low tide and cribbed, so as to pick the unit off at high tide. A platform which is just awash at low tide has also been used to support a large caisson segment while it is being constructed. The unit is then lifted at high tide by a large derrick, taking advantage of the buoyancy of the water acting on the submerged lower portion. Portions of the caisson may be made watertight to help the flotation.

Oceangoing ships have been successfully launched sideways off a dock, and very large and long caissons have been launched in a similar manner. The caisson is usually allowed to slide from launching timbers into the waters. A positive means must be used to ensure the simultaneous launching of both ends, since any tendency for one end to get ahead of the other would cause the caisson to rotate.

Several precautions must be taken in towing a floating caisson to its site. Splash boards may be needed to prevent waves from overtopping the sides. Baffles or compartmentation are desirable, particularly in long caissons, to prevent in-leaking water from accumulating at one end and thus causing the caisson to tilt. Stability during towing may be improved by the use of temporary side pontoons bolted to the caisson.

Setting

Guides or moorings are required to hold the caisson against current pressure and other lateral forces during setting. Guides may consist of well-braced dolphins, often joined by heavy floating fenders or heavy trusses. They are built around three sides, the box is floated in, and the fourth side is then closed. Vertical guides are hung from the dolphins; these may be heavy timbers or steel beams which are suspended in exact position and are truly vertical. Some tolerance is necessary to prevent binding of the box in the guides: a space of 300 mm (1 ft) all around is common. The guide must be strong enough to resist the load which results when one edge of the caisson is forced against the guide, either by the current or by tipping during sinking.

If anchors are used, they should be set off at a sufficient distance so that the change in horizontal length of the anchor cable due to tides is well below the tolerance within the guides. The mooring lines can be led down to fairleads at the center of rotation of the caisson, so that pulling forces will not tilt the caisson.

Several types of anchors may be used, including large concrete blocks jetted into place and heavy ship anchors with steel plates welded to the flukes. A steel H pile may be driven underwater, so that its top is at the mud line, and used as an anchor. An anchor cable may be fastened on before driving the pile; this eliminates diver work and permits the cable to be fastened some distance down from the pile top. Another scheme is to use a cast-iron or cast-steel point, say 1 m in diameter, with a cable fastened to it. The point is fitted on the tip of a pile and driven to refusal. The pile is then pulled, leaving the point and attached cable deeply embedded in firm material.

For large caissons in deep water, tugs fanned out in star pattern may be used to provide control. The tug should anchor itself, then pull on the caisson with its towing winch, augmented as necessary by propeller thrust.

Current pressure may be computed by the same formula given in the subsection on cofferdams, namely:

$$P = \frac{CAwv^2}{2g}$$

where P = total load
         A = area of caisson surface normal to current, ft
         v = velocity of current
         g = acceleration of gravity
         w = unit weight of water
         C = shape factor: 0.75 for circular and 1.33 for rectangular for smooth-sided boxes. For boxes constructed of steel sheet piles, add 25 percent to compensate for drag.

Units must be consistent: KN, M/S, M/s^2, KN/m^3, or tons, ft/s, ft/s^2, T/m^3

More sophisticated formulas from Naval Architecture should be applied to odd shapes or where the current makes an angle with the structure.

The guides or anchors are designed to resist the current pressure, acting through the caisson as it bears on the guides. The caisson sinking is normally accomplished by flooding. The placing of tremie concrete into the box adds weight at the bottom and increases stability. It also strengthens the bottom to resist upward pressure during installation and again when the box is dewatered, if such is part of the construction plan. Sinking has also been accomplished by filling pockets with sand or rock. Flooding with water must be controlled by compartmentation such as interior bulkheads to prevent loss of stability. Stability should be calculated for each stage of submergence, especially for the conditions when the water plane suddenly diminishes as it does with a stepped or tapered structure. The concept of metacentric height is used as an approximate check. Metacentric height, $\overline{GM}$ is determined by the formula: $\overline{GM} = \overline{KB} + \overline{BM} - \overline{KG}$ and must have a positive value, typically 1 m, for practical stability. In this formula, the point K is located at the base, on the centerline. The point B is the center of buoyancy (i.e., the center of displaced water at that stage), and G is the center of gravity. $\overline{BM}$ is equal to I/V where I is the moment of inertia of the water plane, about the weak axis and V is the displaced volume at that stage. When water is allowed to flood in or is physically pumped inside, $\overline{BM}$ must be reduced by the summation of I'/V in which I' is the moment of inertia of each internal water plane as confined by bulkheads. Since I' varies as the cube of the short dimension, it is obvious that even one internal bulkhead is quite effective.

After the box caisson is sunk to position, enough additional weight should be added to ensure stability and to resist current pressures. It is safest to support the box at or below its center of rotation as it submerges, either by mooring to anchors or by continuous guides. However, box caissons have been successfully sunk using guides which support the caisson only at the water surface. In the latter case, the box caisson must be inherently stable through all stages of sinking.

As the box caisson nears touchdown, a complex set of phenomena take place. The water temporarily trapped beneath the caisson must escape, and when it does rush out under one side, the thrust forces the box laterally. At the same time, the rush of water plus the normal current may cause scour.

Where the box has been sunk on a previously leveled base, grout is usually pumped under the base to ensure uniform bearing. Care must be taken to keep the grout pressure low enough so as not to raise the whole caisson or overstress any part of its bottom.

The outside perimeter must be protected against scour by riprap or rock fill, placed on an appropriate filter of small rock or filter fabric. Where the caisson is supported on piles, a slurry of cement and sand or sandy loam may be pumped and jetted under the base. This filling of voids had until recently been considered unnecessary in many box caissons—unless there was fear of scour during construction—as long as adequate backfill was provided and the backfill itself was protected from scour by rock. However, a number of recent investigations have revealed that scour beneath caissons has been more serious than previously believed. The best protection is a peripheral steel sheet pile cutoff wall or else an engineered rock and riprap fill plus underbase fill.

In setting a box caisson which is made up of several units that are to be joined under water, it is especially important to get the first or lower units level and correctly positioned. If the base consists of more than one unit, these units should be joined after setting to hold them together properly in alignment during subsequent construction operations. Guides should be provided, such as tapered sockets, or seats, which will tend to position the next higher unit as it is set. Since bearing of one unit on another may be uneven and concentrated, asphalt-impregnated fiber strips, and neoprene pads have been used to distribute the loads more evenly. Sufficient tolerance must be provided to enable minor corrections to be made as the subsequent units are set.

The weight of each unit aids its stability. Strong connections should be used to tie upper and lower units together (Fig. D4-25).

FIGURE D4-25 Setting upper shaft segments of box caisson, San Mateo–Hayward Bridge, California.

If tremie concrete is to be placed inside, it is important that the joints be sealed so as to prevent runout and leaching out of the concrete.

Concreting

Underwater concrete may be placed in the box caisson by any of the three methods described for cofferdams: tremie, bottom-dump bucket, or intrusion grouting. The standard method is tremie concrete. There is little if any shrinkage of underwater concrete. Heat of hydration needs to be considered, and the consequent outward pressures and subsequent inward contraction. Thus the replacement of up to 30 percent of the cement by PFA or the use of approximately 70 percent blast furnace slag should be considered as a means of reducing the heat of hydration.

Good bond can be developed between precast concrete and tremie concrete and between steel and tremie concrete, although the surfaces must be kept clean of algae. Corrugations and mechanical keys have been widely used, but these may partially defeat their purpose unless they are smoothly curved and gradually formed. The rising tremie-concrete surface tends to trap laitance and free water under any abrupt horizontal projection. "Windows" or openings in interior walls may be similarly undesirable, since as the tremie concrete flows through the window, it drops on the far side and forms laitance.

Where precast units are set on top of one another, there must naturally be a seat and a change in the wall section. As far as possible, this seat should be formed on the outside, so that the interior may have an unchanged section at the joint. Then the concrete surface can rise past the joint without undue disturbance and without trapping water or laitance. Where a change in the interior section has to be made either at a joint or elsewhere, there should be a smooth, gradual transition.

Box-caisson piers have been built where the tremie concrete in a lower section must come to a true plane and elevation to support an upper section. In a relatively small unit, it is possible to fill the lower unit and overflow it, screeding off the top surface with a steel beam and cleaning the surface with a high-pressure jet handled by a diver several hours after the concrete has set. This joint will be clean, but somewhat rough. The next precast unit can then be set down and wedged to exact grade. The joint may be sealed with sandbags placed by a diver around the second unit. Weighted canvas hung over the joint will also restrict the flow sufficiently to allow the grout to be injected to seal the joint.

Where the horizontal seat must be absolutely true, a seat of steel or precast concrete can be suspended on the bottom unit. This seat must have sides or bulkheads so that the tremie concrete can be brought up under and around it, and it must be smooth. A half-round pipe of large diameter forms a good seat. It must be rigidly held in position to prevent displacement by flotation or by accidental bumping with a tremie pipe. If a flat surface is required, it should be sloped 7 to 10° from the horizontal, in order to ensure full contact and prevent voids due to trapped bleed water.

Horizontal underwater construction joints can be made effectively with tremie concrete, if carefully constructed. The lower pour is made with workable concrete so as to get as nearly level a surface as possible. Several hours after it has set, the surface should be jetted clean to expose the coarse aggregate, and the laitance and scum are then pumped out. The laitance tends to gather in low spots around the perimeter and behind any foundation piles. Before starting the next pour, the surface should again be jetted clean of any silt. Unless the sediment is extremely heavy, (e.g., sand), removal by pumps may not be necessary as the jetting will carry silt into suspension. As on dry pours, it is best to start the second pour with a rich tremie grout.

If there is a curtain of reinforcing steel around the perimeter, there will be an abrupt drop in the concrete surface behind it. This can be reduced by spacing larger bars farther apart, by reducing the size of the coarse aggregate, and by keeping the steel a little farther away from the side walls. Regions where bars are spliced by lapping are notorious in this regard, so the overlap of reinforcing steel bars should be kept about 1 m above the concrete construction joint.

Special effort and attention should be given to cleaning the pocket outside the reinforcing steel. It is here that laitance from the first pour gathers, along with other sediments. After cleaning, a diver can fill this pocket with tremie grout before commencing the next placement of tremie concrete.

With intrusion grouting, the practice is to stop the mortar at least 1.5 m below the aggregate surface. This protects the surface from any washing action of the water. Many individual pieces of aggregate will be half embedded in the lower pour and will key with the next pour. Any fines washed off the aggregate subsequently placed will be dispersed throughout the 1.5-m or thicker layer so that they will not be concentrated at the construction joint. Particular care to avoid fines at all stages must be taken with this method of underwater concreting, and is absolutely essential in the zone of construction joints.

Problems may occur near the water surface when placing underwater concrete by any method. Tidal and river currents may cause turbulence and leach out cement paste through even very small cracks or seams in the box caisson. Where a near-perfect surface is required, in very shallow placements, and in swift currents or waves, an anti-washout admixture should be added to the mix.

Examples

Richmond–San Rafael Bridge, San Francisco Bay, California. This 4-m bridge required 78 piers, of which 62 were of the so-called *bell-bottom* type. Water depths up to 20 m (60 ft), exposure to storm and waves, very soft mud bottom, tidal currents up to 5 knots (2½ m/s), and extreme distance to bedrock all contributed to the selection of the type of construction (Figs. D4-26, D4-27, D4-28).

Precast-concrete units were assembled under water and filled with tremie concrete to form the piers. During the placement of the tremie concrete, some minor cracks were noted in the precast-concrete shafts, indicating fluid pressures of over 5 MPa (800 lb/ft^2). It is believed that the high stresses were due to extreme variations in the rate of set of cement (up to 7 h) and to localized impact stresses in the diaphragm forms near the tremie-pipe discharge. These impact forces are apparently dissipated when pouring tremie concrete over a medium or large area and become significant only when, as in this case, the pour is made into a narrow wall and especially when discharging directly by pump.

Many mixes of tremie concrete were tried on this project, some with the cement factor as lean as 300 kg/m^3 (500 lb/yd^3) with 37-mm aggregate, but the most workable and satisfactory mix was 400 kg/m^3 (700 lb/yd^3) mix with 20-mm aggregate and 42 percent sand.

Many factors entered into the quality of tremie concrete on this job. The very strong tidal current caused turbulence inside the diaphragm forms, which was solved by making them as watertight as possible and by scheduling the pour for a period of slack or low current. Windows had been built in the shafts to permit the tremie concrete from the shafts and diaphragm to flow together to form a mechanical key. These windows were a cause of considerable laitance because the concrete would flow through from whichever side was higher and spill over.

COFFERDAMS AND CAISSONS

D4-61

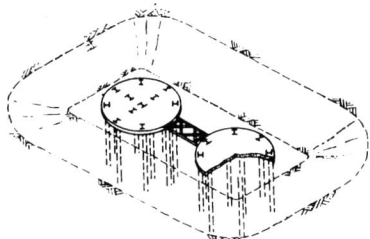

Step 1: Excavation of pier site, illustration of temporary support by timber piles and setting precast base grid.

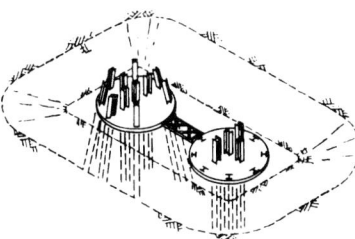

Step 2: Driving of steel H piles through grid. Grouting of H piles to base grid.

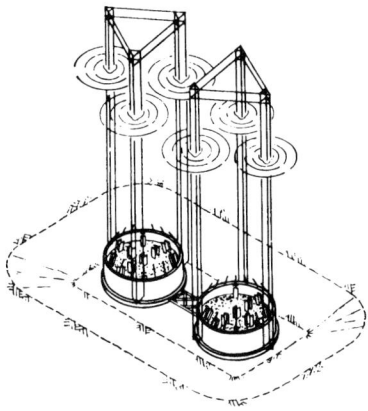

Step 3: Setting precast concrete bottom shells, with temporary steel towers bolted on. Tremie concrete seal poured.

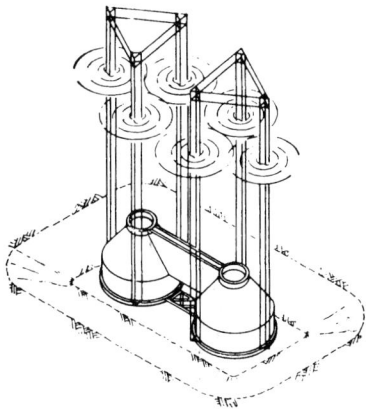

Step 4: Precast concrete cone-diaphragm set.

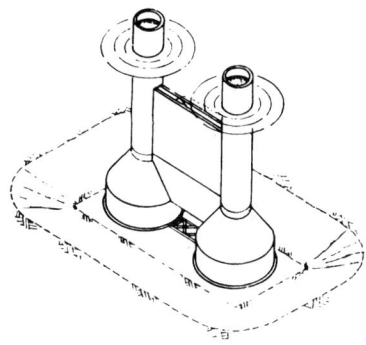

Step 5: Precast concrete shafts set. Prefabricated steel diaphragm set. Entire pier filled with tremie concrete.

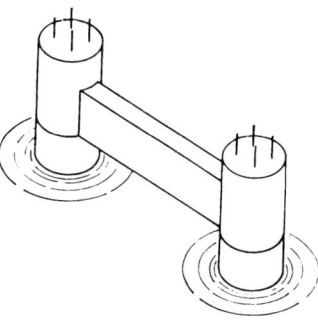

Step 6: Precast concrete spandrel beam set across tops of shafts. Anchor bolt assemblies set. Top of pier shafts concreted.

FIGURE D4-26 Richmond–San Rafael Bridge.

FIGURE D4-27 Richmond–San Rafael Bridge. Setting precast-concrete cone-diaphragm unit.

FIGURE D4-28 Richmond–San Rafael Bridge. Setting precast-concrete shaft. Note temporary steel tower supports and spreader beam used in picking shaft.

FIGURE D4-29 Richmond–San Rafael Bridge. Setting steel shell unit for large four-bell pier. Reinforcing steel cage supported inside shell.

The close spacing of reinforcing steel at splices at underwater construction joints, which were required in certain special piers, caused a drop of 300 mm (12 in) or so in the tremie surface behind the steel. Further cleaning and the filling of this depression with tremie grout before making the next pour was the best solution found, although it was an expensive and time-consuming operation. Increasing the clear spacing of bars to at least twice the maximum aggregate size and locating the splices above the construction joint helped to prevent this depression. The shaft unit had an offset on its interior where it set on the cone. This offset trapped water in a small pocket, and the problem was corrected in later pours by flaring the offset in a smooth transition. Experience on this project showed that the best tremie concrete was obtained when the pour was made at a continuous rate.

Certain of the larger piers had four very large bells forming a base. Dredging, then setting the precast base grid, driving the steel piles, setting the precast bottom shell, and pouring the tremie-concrete seal proceeded as for the more typical piers. Then huge steel forms, designed for 24 KN/m^2 (500 lb/ft^2) fluid concrete pressure and 3 KN/m^2 (60 lb/ft^2) external tidal pressure, were set over the shells (Fig. D4-29). Form plate was 6 mm (¼ in) thick, with angle stiffeners. Truss units were provided on the inside face of the form to resist tidal pressure and to prevent any tendency of the diaphragms to work back and forth because of wave action.

Each prefabricated steel form unit had the reinforcing steel welded to it on the inside. Because of the weights involved, the height of the main channel piers due to the depth of water, and the tidal current forces on the shells while installing them, the forms were made in three sections with horizontal joints. They were set one on top of the other with gasketed joints, and bolted together. All lower sections of these steel units were anchored to the grids by cables and turnbuckles to prevent shifting or overturning due to tidal pressure.

Chesapeake Bay Bridge, Maryland. This bridge required several different types of piers, the most interesting and unusual of which were the bell-bottom piers used at 28 of the pier sites. These piers were similar in many respects to those described for the Richmond–San Rafael Bridge, which was built after the Chesapeake Bay Bridge.

The basic steps of construction were:

1. A hole 3½ m (12 ft) deep was dredged into the bay bottom.
2. Timber falsework piles were driven and cut off underwater, just above the dredged bottom.
3. A large timber platform was floated to the site and sunk by placing precast concrete blocks at the center. These blocks later served to support the diaphragm connecting the two bells. The timber mat had square holes in it, through which the H piles were later driven. Each hole was marked with raised Roman numerals to enable the diver to verify his location in the muddy water.
4. Steel H piles up to 40 m (135 ft) long were driven through the holes in the mat on radial lines and battered outward, using an underwater hammer and telescopic leads.
5. Steel "cans" of 6-mm (¼-in) plate, suitably stiffened, were prefabricated at a shipyard with all reinforcing steel welded in place. They were transported to the site suspended from one or two derricks, depending on the weight, and set on the timber mats. For the very large four-bell piers, the cans for the pier were made in three or four units which were set down successively, one on top of the other, and bolted together by divers.
6. A tremie-concrete seal 1.5 m (5 ft) thick was poured which served to bond the steel H piles to support construction loads. Then the second tremie concrete pour was made, filling the can to just below the waterline.
7. The top 2½ m (8 ft) was then dewatered, the seal cleaned off, and the pier completed in the dry.

Considerable difficulty was reported as a result of the steel H piles damaging the timber mat during driving. Many of the H piles were driven on a batter outward, which gave them a tremendous wedging action against the mat. There were also several cases in which the steel cans were badly damaged. The diaphragm ripped near its junction with the shaft, apparently due to fatigue from its constant working under wave action. Construction progress, other than delays caused by these two problems, was very rapid. This was primarily because the piers were constructed on a production-line basis.

The method just described was similar to that used on the Potomac River Bridge at Dahlgren, Virginia, in 1940. There the largest of the cans were prefabricated complete, with temporary covers welded on. Compressed air was used to provide buoyancy, and a derrick was used to provide stability during the tow and sinking. The can was sunk at the site by releasing the air, meanwhile holding the caisson between guides with derricks since it was quite unstable during sinking.

San Mateo–Hayward Bridge, California. The bell piers for this bridge carried forward the development of the box-caisson concept in some interesting and significant ways. Pier construction was basically the same as for the Richmond–San Rafael Bridge, with these important changes:

1. The bottom precast element consisted of two bells connected with a diaphragm and set as one piece. This ensured that all upper units set subsequently were in proper relative position (Fig. D4-30).

FIGURE D4-30 Precast double-bells (binocular configuration) were seated over previously driven piles and connected by tremie concrete. San Mateo–Hayward Bridge, California.

2. Elimination of cast-in-place underwater diaphragms. These had proved difficult to form and pour on the Richmond–San Rafael Bridge.
3. Reduction of the size of tremie pours to a maximum of about 270 m^3 (350 yd^3). This was in contrast to the large individual pours on the Richmond–San Rafael Bridge of up to 3600 m^3 (4800 yd^3).
4. To ensure the highest possible quality of structural tremie concrete, the following steps were taken:
   a. All tremie-concrete pours were made by the same crew from start to finish, working in daylight hours as far as possible.
   b. Pours were made as soon as practicable after the precast units had been set, to minimize the silting.
   c. Extra cement was provided and a plasticizing admixture was used to ensure a workable, homogeneous mix.
   d. All pours were under the direct supervision of a registered engineer who was qualified as an expert in tremie concrete.
   e. Cores were taken to prove the soundness and homogeneity of each structural tremie pour. Excellent results were obtained. The cores showed good bond between the two tremie concrete lifts and gave compressive strengths well above 4000 lb/in^2.
5. On this project, the long steel H piles were driven with telescopic leads. A large hydraulically operated shears was used to cut off the piles to grade.

Newport Bridge, Rhode Island. The main tower piers for the suspension bridge are located in water depths of 30 to 40 m (100 to 140 ft). The construction was complicated by the high tidal velocities and exposure to weather. A modified form of bell-pier box caisson was selected, with the piers supported on long steel H piles driven to shale and hard sand strata. The first step in construction was the predredging of 6 m (20 ft) of hard-packed silt. Piles were then driven with an underwater hammer (McKiernan-Terry S-14) mounted on special leads, capable of maintaining rigidity and alignment at a depth of 50 m (160 ft). These leads were 1.5 m × 1.5 m, of heavy pipe framing. When they were lowered to the bottom, spuds penetrated about 2 m into the bottom and secured the leads during driving. Although pile driving tolerances permitted piles to extend ½ m (2 ft) above or 3 ft below cutoff, a great many required cutting off. Divers, working in a saturation diving bell, made the cutoff with an electric oxygen arc.

Eight 1-m (36-in) diameter pipes, extending up to above water, were driven to act as guides or spuds for the box-caisson steel forms. These were secured by driving an H pile inside, connecting it to the 1-m pipe by grout at the base, and welding guide cones at the top. Each box caisson, with its internal bracing and reinforcing steel, weighed 400 KN (400 T). It was set in place over the guides by a 500-ton-capacity crane barge. The shaft forms consisted of four cylinders connected by diaphragm walls. They were floated on their sides to the site, upended with the assistance of the heavy-lift derrick, and sunk by flooding.

Reinforcing steel was detailed with particular care to ensure the free flow of tremie concrete. Internal bracing and strutting were incorporated in the steel forms so as to prevent distortion from their own weight, and the pressures of the current and the tremie concrete.

Great Belt Western Bridge, Denmark. Sixty-six piers were required, in depths up to 30 m. Each site was predredged using a ladder dredge, which trimmed the hard glacial clay to a tolerance of 200 mm. Then a jack-up barge moved to the site and placed a layer of gravel 1.5 m thick. A traveling screed leveled the top surface; then a very large vibrating screed was used to compact the gravel to 70 percent relative density (93 percent max density). A thin leveling course of gravel was placed and screeded to ±20 mm.

Meanwhile, the precast concrete caissons were being prefabricated on an assembly-line basis in a fabricating yard. A huge crane barge of 7000-ton capacity picked each caisson from the bulkhead, then traveled with it to its site and, using sophisticated electronic survey systems including GPS, set it in proper position on the screeded base.

These caissons had flat bottoms. The caisson rested directly on the crushed rock, without grout injection.

On the Great Belt Eastern Bridge, some 30 similar caissons were set on crushed rock bases, but grout was injected underneath to provide a full contact. A thixotropic grout was employed, so as to prevent excessive penetration into the crushed rock and loss to the seawater outside.

Second Severn Bridge, England. This bridge is under construction at the time of writing. Precast concrete pier shells, weighing up to 3000 tons, are being set on concrete pads seated in a pit which has been predredged into the glacial till. Then the inside is filled with tremie concrete.

Some difficulty is being experienced in sealing between the lower edges of the pier shell and the dredged bottom. If the gap is not fully sealed, tremie concrete escapes to the sea. As the tidal current sweeps by, it tends to erode the fresh concrete.

A similar problem occurred many years earlier on the Columbia River Bridge near Astoria, Oregon, where the swift ebb currents eroded beneath the shell's lower edge, allowing the tremie concrete to be washed out.

It should be noted that on several of the box caisson projects described earlier, there was a base to the caisson, preventing escape of the concrete, while on the Astoria and Severn Bridges, there is no base. On the Severn Bridge, screens of light metal are being used in an attempt to seal under the edges while the concrete is being placed.

Ma Wan Tower Caisson, Tsing Ma Suspension Bridge, Hong Kong. Two large concrete caissons were constructed on the deck of an oceangoing submersible barge. They were launched by submersion of the barge, then floated to the site.

The site had been prepared by drilling and blasting, then dredging the weak and fragmented surface material to sound rock.

The caissons were seated on concrete pads which had been preset to grade. Then heavy filter fabric was unrolled from the sides, laid out over the gap, and held down by rock, so as to prevent escape of the tremie concrete. After the tremie concrete had hardened, the upper portion was dewatered and mass concrete placed in the dry.

Some voids were found just under overhanging slabs that separated the tremie concrete from the mass concrete, possibly due to premature loss of slump in the concrete, aggravated by bleed. Superplasticizing admixtures always present a potential difficulty with mass pours because of the heat of hydration, which can lead to a sudden loss of slump.

These voids were grouted. After the pier was completed, holes were drilled through tubes set in the concrete and on into the rock below, so that pressure grouting of all fractures in the rock could be carried out.

I-205 Columbia River Bridge, Oregon. This precedent-setting bridge required over 40 bell piers, seated in water depths of 10 to 16 m at low water, 10 m deeper at flood stage. Preexcavation of the surface sands was performed, then the underwater driving of steel H piles into a founding stratum of conglomerate. A template was employed to fix the position of the piles on the sea floor.

Then a prefabricated cage of reinforcing steel was lifted into a large steel removable form and the whole seated over the piles (Fig. D4-31). Tremie concrete was then placed to tie the individual units to the piles. The steel form was then dewatered and hollow pier constructed in the dry. Last, the bell form was loosened and water was injected between the form and the concrete pier. It required 2 or 3 h for the water pressure to gradually separate the form from the shaft so that it could be lifted off for reuse.

The entire process was carried out successfully on schedule. Of particular interest was the high quality of tremie concrete, with minimal laitance, obtained by use of a carefully developed concrete mix that ensured minimal segregation and bleed.

Honshu-Shikoku Bridges, Japan. Japan embarked on a mammoth project in the 1970s which will only be completed by the end of this millennium: the construction of three major crossings between the islands of Honshu and Shikoku. As far as practicable, generic solutions were adopted for the bridge piers. All deep-water piers have been steel box caissons, seated on prepared foundations in rock. These steel box caissons have then been filled with concrete. Grout-intruded aggregate was used for all but the last bridge. For the world record, Akashi Strait Bridge, whose piers were completed in 1992, a special tremie-concrete mix was placed, incorporating an anti-washout, antisegregation admixture, and low-heat cement.

The general process employed at each pier was typified by Anchorage Pier 7A, of the Bisan Seto Bridge. The pier was founded at –50 m on bedrock. The initial operations were the drilling and shooting of the rock, then dredging of overburden sedi-

FIGURE D4-31*a* I-205 Columbia River Bridge, Portland, Oregon. Removable steel bell forms used to construct 40 bell piers. (Rebar cage foreground, steel form in background.)

FIGURE D4-31*b* I-205 Columbia River Bridge, Portland, Oregon. Removable steel bell forms used to construct 40 bell piers. Form being positioned.

ments and fractured rock by a giant 99-ton grab bucket. Then a jack-up barge supported a very powerful grinder which ground the perimeter of the base to a level surface. After video and diver verification of the rock's competence, a steel box caisson was positioned above the site and ballasted down to seat on the rock. Crushable rubber cushions, 2 m high, sealed the periphery.

Grout pipes had been incorporated in the box caisson during manufacture. Coarse aggregate was placed to fill the multiple vertically separated cells of the caisson. An inhibitor was introduced to prevent the growth of algae. Now grout was pumped in to progressively fill the voids in the aggregate. Thermocouples monitored temperature rises due to heat of hydration.

OPEN CAISSONS

General Considerations

Open caissons are a primary option where large, heavily loaded piers are to be founded beneath deep layers of sediments. Open-caisson construction is usually only economical in the range of depths at which cofferdam construction can no longer be practically employed. It is also more costly than pile-supported bell piers and pile-supported box caissons. However, the open caisson may be one of the few practicable solutions where very heavy loadings such as the anchorages for suspension bridges, must be carried to a founding stratum through deep overlying sediments. (Another solution is a slurry wall cofferdam.) The open caisson is also a major contender for the construction of deep silos, e.g., for underground nuclear reactors.

An open caisson is sunk by dredging through its bottom while building up the walls. The lower end of the caisson is the cutting edge; this shears through soils by concentrating the entire weight of the caisson on its perimeter. Sinking is often aided by jets and sometimes by adding weight to the caisson. Since the caisson is to become the basic structure of the pier, it is essential that the pier design and construction procedure be integrated and developed concurrently.

Although the open-caisson method, as exemplified by the "wells" of ancient India, has been practiced over 2000 years as a means of constructing bridge piers in rivers, the sinking of the caisson has until recently been more an art than a carefully planned engineering procedure. However, the lessons of that art have been passed down from the experience of hundreds of examples and should not be set aside, but rather integrated with engineering analysis and programs.

The caisson sinks because the effective weight of the structure exceeds the resisting capacity of the soils on the outside and under the cutting edge. The shear transfer on the outside, often generalized as skin friction, is a function of the friction and the cohesive shear strength of the soil acting on the exterior surface of the caisson. This component of resistance depends on the surface roughness and is influenced by jetting, due to both the lubrication and the increase in effective pore pressure.

The term *sinking* as applied to a caisson means the controlled progressive failure of the soil's capacity. The external shear or skin friction must be overcome and the soil under the cutting edge must fail, which it does as a bearing failure, on shear surfaces curving up to the inside. When failure occurs, the structure moves downward dynamically until a new equilibrium condition is established. Since the skin frictional resistance will change very little during a single stage of sinking, it is the bearing value under the cutting edge which increases with penetration and gradually increases due to its confinement by the inside height of soil above the cutting edge.

Since all operations are purposefully taking place at a condition of imminent soil failure, where the safety factor varies between 0.98 and 1.02, for example, it becomes obvious why art must take over the detailed control of the sinking process. It also becomes apparent why procedures which produce relatively continuous and steady sinking are needed as opposed to large incremental steps.

Lubrication through lubricating jets installed on the periphery is one means of reducing skin friction. Bentonite and compressed air may also be used. There appears to be an advantage in alternating these. Vibration will also reduce skin friction.

As noted, the other major resistance to the sinking of the caisson is the bearing capacity of the soil under the cutting edge. This is a function not only of its shear strength, but of its confinement. The confinement on the inside is determined by how close the excavation has proceeded to the cutting edge. The bearing capacity may be significantly reduced by cutting jets.

The cohesive shear capacity of sediments, especially clay is time dependent. Once the caisson is moving, this resistance is lowered. Then if the caisson is allowed to sit for a period of hours or days, setup occurs, and it takes the application of increased effective weight to break the caisson loose.

The effective weight of the structure is its air weight less buoyancy of water inside. However, buoyancy is really a pressure phenomenon: It exerts an upward pressure on any horizontal surface—in this case, the exterior cutting edge and interior structure such as cross walls—and thus is determined by the height of the water inside. Transient states occur when the inside water level is raised or lowered rapidly relative to the groundwater outside. Lowering the water inside produces an inherently unstable condition since soils may suddenly flow in. This practice, while speeding sinking, has led to tipping and displacement of the caisson.

One method of constructing an open caisson in the water is to build temporary false bottoms in the wells which constitute the subdivisions of the caisson. The entire caisson can then be floated to the site. The false bottoms can continue to support the caisson by buoyancy until it has sunk into the mud a sufficient distance so that its weight is entirely supported on its cutting edges and by skin friction on the outside. Then the temporary or false bottoms can be removed. The temporary bottoms must therefore be designed to resist the upward hydrostatic pressure and the mud pressure as the caisson is sunk (which is equal to the air weight of the caisson at that stage). This method of caisson construction is usually the lowest in cost but also the most hazardous. Guides and anchorages must be provided to hold the caisson while it is being sunk to the bottom, during the removal of the false bottoms, and throughout the initial stages of dredging. While these steps are being performed, the soft material will tend to run in under the cutting edges and cause the caisson to tip. The soil under the cutting edge may also be so overloaded that it shears on one side, moving the caisson sidewise. There is a strong tendency for bottom scour just before a floating caisson is landed, and this can possibly undermine the guides.

The dredging wells should always be flooded prior to removal of the false bottom. If excess mud pressure is suspected, it might be well to overfill the well with water in an attempt to balance the pressure. Other means of relieving excess mud pressure on the false bottom are by jetting or by removing mud with air-lift pumps. The pumps can operate either through small-diameter wells (holes) left in the interior walls or through pipe wells which penetrate the false bottoms. In very soft and unstable soils, steel domes have been placed over the dredging wells and compressed air used to control the stability and rate of sinking of the caissons. This method is slower and much more expensive, but it has been successful in founding very large caissons to great depth in extremely unstable soils.

To protect the false bottoms from accidental tripping during the buoyant phase and to act as a safeguard for men working on forms and concreting the dredging wells high above the false bottoms, heavy safety nets should be hung in the dredging wells. These can be secured to the same inserts used to hold the forms on earlier lifts of concrete.

One basic type of open-caisson construction is used when the existing site is on a land site above water. This can be an artificially created sand island or just the natural site. The cutting edge is laid out in exact location and leveled, and the walls are built up progressively as the soils are excavated through the dredging wells. Difficulties may occur in the early stages when the weight of the caisson is insufficient to overcome the resistance of the soils without supplementary excavation below the cutting edge. Unstable material may then slough in, causing the caisson to tip. As a general rule, excavation should never be carried below the cutting edge. High-pressure jetting and further raising of the caisson walls are preferable methods of causing penetration. A stratum of cobbles or hardpan which is known to be present near the surface should be broken up, or even removed and replaced with sand, prior to starting caisson construction.

The sand-island method envisages the construction of a temporary island on which the cutting edge is built and through which the caisson will ultimately be sunk (Fig. D4-32). The sand island serves as a working platform, enabling forming and concreting operations to be performed at or near ground level with conventional materials and equipment. The sand island may be confined by steel sheet-pile cells and then filled with sand, or it may be a naturally sloped sand island with its slopes protected by riprap. The initial sinking is uniform, since it is through selected sand, and the caisson is firmly held in position and alignment by the time that hard strata are encountered. The principal difficulties with this method are the tendency to scour around and under the sand island and the high cost of the sand island. Also, a

FIGURE D4-32 The caissons for the Verrazano-Narrows Bridge in New York were constructed by the sand-island method.

sand island cannot be built in very deep water or where the underlying soils are incapable of supporting its weight.

The simplest open caisson is circular in plan and consists of a single heavy exterior wall. Larger open caissons are often rectangular in plan and have interior cross walls which divide the caisson into interior dredging wells. Some large open caissons have double walls and are floated to place and sunk by filling the space between the walls with concrete. This method requires more materials because the walls must be strong enough to resist the hydrostatic pressures. However, the caissons are considerably more stable and more easily controlled during sinking. The double-wall method is generally faster in construction and safer, but it does involve the costs of the strong, watertight double walls. This method of double walls also permits the adding and removal of ballast or sand so as to match weight with resistance. Concrete fill and iron ore are other potential materials for controlling the effective driving force.

Many caissons are surmounted by cofferdams to allow the upper portions of the pier to be completed by normal cast-in-place construction, thus permitting a reduction in cross section and weight at these upper levels. The cofferdam also permits minor adjustments in position and attitude and facilitates the installation of granite facing, fenders, and other protective devices which may be specified (Fig. D4-33).

Cutting Edges

The cutting edge functions primarily as the bearing element of the caisson as it shears through the soil, boulders, buried timber, hardpan, and into rock. Thus, the cutting edge must be extremely strong and rugged to resist concentrated local pres-

FIGURE D4-33 Caissons for the Sunshine Bridge, Louisiana, across the Mississippi are surmounted by removable cofferdams.

sures. It must be securely tied to the caisson so as to fully utilize the great stiffness against distortion or warping which is inherent in the caisson walls and box-like structure.

For a caisson which is started while floating in water, the cutting edge must also serve as a rigid floating frame or barge that will not warp or distort during the early stages of building up the caisson. The first construction loads and first concreting cannot be placed exactly symmetrically and simultaneously, so the entire cutting edge must act as the girders or trusses of a deep and heavy barge.

Cutting edges are usually made of steel in combination with concrete. Heavily reinforced concrete has been used for smaller caissons, and it is well suited structurally and economically for large caissons if properly designed and detailed. Reinforced concrete, with steel-plate armor, has been used on some recent large caissons built by the sand-island method. The shoe or bearing element is usually roughly triangular in cross section and must meet the following requirements:

1. It must be strong and rugged to resist extremely high localized pressures, such as might be caused by a boulder.
2. It must be designed to resist twisting, shearing, crushing, and particularly the tendency to spread outward because of its sloping inner surface.
3. Its plates must be adequately stiffened, and outside plates must be heavy.
4. Connection and splice details must be rugged and strong.
5. There must be provision for transferring shear from the concrete seal course to the caisson, such as shear keys.
6. There must be a sufficient number of vertical diaphragms to make the cutting-edge cross section act as a whole. All cutting edges must be tied together in a rigid frame so as to resist distortion.
7. The cutting edge should have provision for penetrations in order to enable access to the soils below. It must also accommodate cutting and lubricating jets. The shoe must be designed to provide for excavation under the cutting edge by special excavating tools such as chisels and drills, or robot excavators.

The shoes for the interior cross walls are usually several feet above the exterior shoes and so do not have to shear through the soils nor resist the high localized pressures. Their tips are often square or slightly beveled steel plate and angle boxes which can resist crushing and damage during excavation.

When a caisson is built on land or on a sand island, the concrete placed in the early stages can be used in conjunction with steel shoes to act as the girders of the cutting edge. Additional reinforcing may be required, and care must be taken to bond the shoes securely so that the entire cutting edge will act as a unit. Adequate support for the shoes must be provided in the early stages to hold the frames true.

Heavy steel girders are used in the exterior and interior walls of cutting edges of floating caissons to hold them rigidly and to act as the bulkheads of a barge. The cutting edge is often made slightly larger in plan than the caisson in order to reduce skin friction. This may be effective in cohesive soils such as hard clay, but it is probably ineffective in granular soils. It may be undesirable even in clays, since the skin friction on the sides is the guiding and supporting force during excavation. The scheme may be suitable only in heavy and hard cohesive soils where skin-friction forces are believed to be excessive in relation to the sinking forces available.

Some cutting-edge shoes are made with flat bottoms, since these are said to give better alignment control when sinking a caisson in very soft materials. This is a matter on which there is a divergence of opinion, and the choice must be based on a thorough analysis of all factors in the particular case.

Setting

A floating caisson must be held against unbalanced external lateral forces. These forces are principally due to current, but consideration must also be given to ice, debris, logs, waves, surge, and the effect of floating equipment tied alongside as well as accidental collisions. Formulas for determining current pressures and force are given in the subsection on box caissons. Current practice is to use large mooring winches mounted on the caisson, with lines led down to the center of rotation of the structure and thence to anchors. In quiet waters, vertical guides may be used, although they can resist only limited lateral forces.

Once a stable caisson is well embedded in the bottom, it develops considerable additional resistance to lateral movement. Eventually it is able to resist external forces without the aid of guides or anchorages. The most critical time is just before the caisson is landed, when its maximum surface is exposed to the current. At this same stage, the current velocity underneath the caisson is usually increased, sometimes radically so. In a narrow channel, the entire body of water may be speeded up by the constriction caused by the caisson.

When anchors are used to hold the caisson and where differences in water elevation will occur, the scope must be long enough to make the change in horizontal length negligible during any one stage of sinking. Thus scopes of 200 to 400 m (600 to 1200 ft) are commonly used. The maximum possible change in water elevation should produce a horizontal difference not to exceed the allowable tolerance, which will usually be of the order of magnitude of 100 to 150 mm (4 to 6 in). Each anchor line is led to a heavy fairlead near the center of rotation of the caisson. From this block, the haul line is led up to a winch on the caisson top. The winches must be shifted upward, one at a time during slack water, as the caisson sides are built up. Enough cables must be provided so that the failure or jamming of one or more will not endanger the caisson. After the caisson is landed, the fairleads may be progressively shifted upward to new pad eyes on the sides of the caisson.

The best features of this method of anchoring are the ability to control the location while landing to within a matter of inches, the freedom from interference with floating equipment, and usually the moderate initial cost. The disadvantages are the necessity of moving the winches up continually.

Guide structures are essentially dolphins which form a cage around the caisson. Usually, three sides are constructed, the caisson is floated in, and the fourth side is then closed. The several guide structures may then be joined at the top so as to develop frame action. Since the guide itself must be truly vertical and in accurate position, the usual practice is to hang the guide on the face of the dolphin and then adjust it accurately. The dolphin often consists of a combination of vertical and batter piles. On major structures, a template or jacket may first be set through which piles are driven. This makes a more rigid dolphin and provides bracing at the bottom. In a strong current, vibration is a source of trouble; therefore, bracing members should be of adequate size and welds of the highest quality to prevent fatigue. The guide beam may extend all the way into the bottom, or it may provide lateral support near the waterline only. In the latter case, the caisson must be designed to remain plumb by reason of its inherent stability. Anchor wires may be attached to the guides at or just below the surface to give additional support in the direction of greatest pressure. Heavy guide beams driven well into the bottom provide increased support against tipping and lateral movement during the most dangerous stages of sinking and dredging. Where more than one guide is used on a side, the guides must be located fairly near the center so that they will all act together to prevent the possibility of progressive failure.

The advantages of guide structures are the elimination of changes during sinking and landing, presumably greater reliability and strength, and the opportunity to utilize them as ancillary support structures for construction equipment or even to become a part of the permanent fender system, etc. The disadvantages are the cost, which is usually greater, and the fact that about 600 mm of tolerance must ordinarily be provided to prevent binding. This latter problem may be easily solved by leading lines from the cutting edge to a sheave fastened to the guide structure near the bottom, then up to a winch on the top of the caisson walls. This allows exact adjustments to be made while landing.

Floating guides which are secured by anchor lines have been used to hold caissons in position. The floating guide must be strong enough in bending and shear to hold the caisson should it bear on one end only. This method has the advantage of low comparative cost and the elimination of constant adjustments. It has the disadvantages of lack of exact control at landing, of interference of anchor lines with floating equipment, and of being inherently less reliable than the other methods. A combination of this and other methods can sometimes be used to advantage; for example, using guide structures at the two ends where the current acts and using anchored floating guides on the sides.

When a floating caisson has been sunk through the water and is just about to land on the bottom, the work must be well programmed so that the landing can be completed as expeditiously as possible. It is at this stage that scour often takes place under the caisson because of the great increase in the velocity of the current. At times this scour has been so rapid and of such magnitude as to undermine the guide structures and endanger the whole caisson. In sand and sandy silt river bottoms, very large willow or filter fabric mattresses are often sunk beforehand by covering with rock, yet even these have been only partially successful. This problem of scour due to increased velocity must be carefully considered and kept to a minimum by positive steps. The number of piers under construction at one time, the amount of floating equipment moored in the river, the stage of the river or tide, and the character of the bottom are all important factors determining scour.

During landing, the exterior walls or cofferdam walls should be extended to a substantial freeboard so that the caisson can be landed promptly even if some scour takes place during landing. At a tidal site, the caisson will probably be landed at low tide, but there must be sufficient freeboard so that water ballast can be added to keep it on the ground during the next high tide. Landing on low tide and then allowing the caisson to float again on high tide is an invitation to scour and serious trouble. With a large, flat-bottomed caisson, the phenomenon of water escaping from under the caisson may cause the caisson to displace laterally. Thus the timing should be controlled long enough to allow the water to escape while short enough to prevent excessive scour.

Sinking

A double-walled caisson has the advantage of discrete spaces which can be used for buoyancy initially and weight during sinking. Concreting consists of filling the space between the double walls in stages.

With single-walled floating caissons, the outside form has usually been brought up in steel, or precast-concrete panels so as to act as a continuous shallow cofferdam as the caisson is sunk. This allows the center of gravity to be kept low and makes more freeboard available. The inside of the dredging wells can be formed with collapsible steel or panel forms which bolt at the bottom to inserts left in the previous pour. These forms preferably span the entire height of the lift to a cross-tie above the top of the lift. They should be designed so that they can be raised directly to their next position without having to be lifted clear. On single-walled caissons built on land or sand islands, or in some cases of floating caissons, the outside walls can be formed with reusable panels. Care must be taken to ensure good construction joints so as to prevent leakage under hydrostatic head.

The removal of material through the dredging wells is usually accomplished by clamshell buckets. Where hard material is anticipated, the heaviest and largest bucket that will operate inside the wells should be used. Weighting of the bucket is preferable and more effective than dropping it. Very soft materials and sands can be removed by air-lift pumps and other suction-type pumps. This method may be applied in the early sinking stages by sucking through pipes which pierce the interior walls or the false bottoms before they are removed.

When the caisson approaches the founding stratum or whenever digging becomes very hard, special rock teeth should be added to the bucket. High-pressure jets can be used through holes in the walls to help loosen the material. The most difficult material to dig is that which lies under interior and exterior walls. Chisels can also be operated through holes in the walls. Free jets can be operated in the dredging wells, held in position by a cage which rides up and down in the well. The direction of the jet nozzle can be controlled from above so as to knock down the clay that stands high under the walls. This jet should be equipped with a reaction nozzle so as to counter the high thrust.

Jet systems for cutting purposes are frequently built into the cutting edges, and sometimes into the interior walls as well. The systems are arranged with headers and valves so that any group of nozzles may be operated individually. While the installation of cutting jets in exterior cutting edges is good practice and is recommended, it has nevertheless been found that these jetting systems are seldom as efficient as planned. They tend to plug as the caisson is sunk, and the direction of the nozzle cannot be adjusted as needed. Also, it is not practicable to use as high a pressure and volume on groups of nozzles as with an individual free nozzle. Thus, a jet through a formed hole in the wall will be found most effective. The latter should also be provided and used to supplement the built-in cutting jet system.

Lubricating jets are used in the exterior walls to reduce skin friction and must be built into the walls. High-velocity nozzles are not desired; the system should be designed to give a high-volume flow of water up the sides. The usual system consists of a row of nozzles around the perimeter of the caisson, several feet above the cutting edge. Additional rows may be added at 3 to 5 m (10 to 15 ft) intervals up the side if excessive skin friction is expected and particularly if sinking through deep sand strata. The nozzles usually are pointed straight out. Turned-up nozzles have been tried but have become plugged by sand falling in when jetting stopped. When a nozzle is directed horizontally, a bead or other projection which partially blocks its lower edge will deflect the jet upward. Compressed air may be injected through these jets, alternating with water. Bentonite slurry may be injected to reduce friction.

The material dredged from the cutting wells should be disposed of in such a way as to prevent any unbalanced load against the caisson. On some caisson projects the material is just dropped into the water alongside. This may be all right for a small amount of fine material in a river or tidal current, but the material must not be allowed to pile up against the caisson side. It will usually be necessary to use barges to haul away the excavated material, especially if it is contaminated.

The sinking of the caisson is a most delicate process and calls for extremely careful control. Excessive dredging in wells along one side, dredging below the exterior cutting edges, and other unsymmetrical operations may cause the caisson to tip. Dredging must be carried on symmetrically about the major axis of the caisson; that is, the axis parallel to the long sides. On relatively long, narrow caissons, first one end and then the other may be dredged if it is not possible to work both simultaneously. Some authorities even recommend this alternating procedure, but the work at each end must be symmetrical with respect to the sides.

Dredging should never be carried below the exterior cutting edges during sinking operations until all other methods have been exhausted. Jetting and chiseling, combined with removal of material from under interior walls, should be used to lower the caisson, rather than digging under the walls. For this reason, the bottom of the interior walls should be built about 2 m (6 ft) above the exterior cutting-edge elevations. Dredging below the exterior cutting edge may initiate sudden and usually unsymmetrical failures of the soil in shear. Such failures may cause tipping of the caisson or a local blowout of material from under the cutting edge.

An extremely critical stage in the sinking process occurs when the cutting edge penetrates from a firm stratum into weaker soils below, e.g., from sand into soft clay. Prior to reaching the soil interface, the caisson's effective weight is primarily resisted by a combination of skin friction on the outside and bearing on the cutting edge. As the cutting edge breaks through, bearing resistance suddenly drops and the caisson will sink rapidly until the necessary support is regained. This will usually be a nonuniform soil failure mode, and since one side fails before the other, the caisson tends to tip out of control.

One sophisticated means of controlling this is to install, prior to reaching the soil interface, a series of tension anchors (ground anchors), evenly spaced around the circumference of the caisson. These ground anchors will consist of post-tensioning tendons, drilled and grouted into the underlying bedrock or founding stratum. Linear jacks may then be installed, reacting against the caisson periphery, providing the additional force necessary to cause the caisson to overcome bearing resistance while the cutting edge is still in the firmer stratum. By limiting the stroke of the jacks, the force will automatically be applied in short increments; thus, erratic plunging is prevented.

The removal of rock and boulders from under the cutting edge can be a very difficult problem. Chiseling and jetting should be used wherever possible. Explosives have been used, both in holes drilled beneath the exterior walls and as shaped charges placed below the cutting edge. Experience has shown that blasting near the cutting edge is exceptionally risky, because, in at least two cases, the caisson walls were seriously damaged. Wherever it is known beforehand that rock will be encountered, the safest procedure, and probably the cheapest in the long run, is to drill and shoot before the caisson is set in position, or at least while the cutting edge is high above the rock elevation. Similarly, when cemented soils will have to be penetrated, breaking them up with a high-pressure jet ahead of time is best.

Some data are available on the actual skin-friction values experienced in open-caisson construction. R. F. Legget, writing in *Civil Engineering* in July 1940, about experience in India on relatively small-diameter caissons across the river Hooghly, quotes values of 14 to 28 KPa (280 to 560 lb/ft^2). At Willingdon, 50 KPa (1000 lb/ft^2) was experienced. George L. Freeman, writing in *Civil Engineering,* says that skin-friction values over 90 KPa (1350 lb/ft^2) were experienced on the Huey P. Long Bridge across the Mississippi. E. S. Blaine, writing in *Engineering News-Record,* Feb. 6, 1947, gives more detailed and explicit values for skin friction experienced on the Baton Rouge Bridge, as follows:

Material	Skin friction, lb/ft^2
Watertight stiff clay	800
Tight clay grading to sandy clay	850
Same, lubricated by jets	647
30% sand and gravel, 25% clay and sand, 40% stiff clay, 5% sand	845
50% sand and clay, 50% sand	736
20% silt, 80% fine sand	1120

Note: 100 lb/ft^2 = 5 KPa

For large caissons to be sunk through medium-dense sand, a value of 60 KPa (1200 lb/ft^2) is a conservative figure for planning. Dense sands and gravels will, of course, give higher values.

Explosives have been used to reduce skin friction while sinking, but this is not recommended. Vibration in the adjoining soil, caused by driving of sheet piling nearby, has also been effective in sinking a caisson through dense sand where jets alone were proving inadequate. Weighting of caissons has been used to help sinking, but this is very time consuming and expensive. Also, it raises the center of gravity excessively if the weights are added above water. It is generally better and more economical to design the caisson to have sufficient net (buoyant) weight to enable it to be sunk without additional weighting or extreme measures. However, note the special use of ground anchors and jacks described earlier for use when passing from firm material to weaker soils below.

One phenomenon which may impede sinking is the formation of mud plugs—which act like a false bottom—in the dredging wells. These can be dangerous, since they may let go suddenly during dredging and cause tipping. Mud plugs can be broken up by high-pressure jets.

Injection of bentonite slurry through the embedded jets is a new development which has proved very successful. The slurry is injected through lubricating pipes into an area just above the enlarged cutting edge. This creates a thin (1- to 8-in) annular ring of bentonite which greatly reduces skin friction. The Lorenz-Fehlmann method, which is extensively employed in Europe, makes use of a rubber membrane, curving upward, to prevent the bentonite from escaping downward. Careful control must be exercised over the mix composition and the injection pressure. After founding, the annular ring must be stabilized by injection of cement grout under high pressure, because, otherwise, the skin friction is permanently reduced by the bentonite. Painting the exterior of a steel shell caisson with a paint having a low value of friction on sand has been tried, apparently without significant results. The use of ground anchors and jacking, previously described as a means of controlling penetration in upper strata, can also be applied to the situation when the final founding is impeded by dense soils.

During the floating stage, stability of the caisson is provided by the righting couple of the buoyancy and gravity, in much the same way as for a ship or a barge. Since the caisson's center of buoyancy is generally quite high in relation to its center of gravity, the unit is usually very stable while it is afloat. The wider the caisson is in relation to its height, the greater is its stability. The several stages during construction and sinking prior to landing must be checked for freeboard and stability, and in some cases it may be necessary to carry the exterior walls up higher than the interior walls or to place a cofferdam on top of the caisson.

As soon as the caisson enters the mud, a definite determination of stability can no longer be made. The material at the cutting edge may act partly as a solid and partly as a fluid, and these states may change suddenly and irregularly as the material is disturbed. The stage at landing and for the next 5 to 8 m (15 to 25 ft) of sinking is the most critical. The buoyant force of the water may be utilized as long as it is possible to keep some dredging wells dewatered.

Tipping and Sliding

The tipping of a caisson while it is being sunk is the major hazard and problem of open-caisson construction. During these early stages of sinking, the forces acting on the caisson which tend to counteract its tipping are the support of the soil under the cutting edge, the skin-friction support of the soil on the sides, the buoyancy of the water, and the lateral support of guide structures or mooring lines. By providing guides capable of resisting lateral forces at the level of the top of the caisson or water surface, a very large moment can be developed because of the long lever arm. In a number of cases, well-constructed caisson guides or anchorages have saved their caissons from catastrophe. Well-braced guide structures, tied together to act as a unit, are therefore very desirable.

Many open caissons tend to slip sidewise in the early stages of sinking, after the caisson has been landed. This is due to overloading of the soil, such as may occur when flooding dredging wells prior to removal of the false bottom. Sidewise movement may also accompany tipping and the subsequent measures for its correction.

To provide external support against lateral slippage, several measures are possible. If anchor lines were originally used to provide lateral support to the caisson, the points of attachment could be shifted up the sides of the caisson during the early stages of sinking and a strain maintained. If guide structures were installed, the lateral support of these will tend to hold the caisson in position. This positioning can be supplemented by installing full-length guide beams, driven well into the bottom and held at the groundline or the top by the guide structure. Spuds driven through spud wells in the caisson have also been used, but there is the danger of their binding if the caisson tips. Therefore, it is recommended that only external guides be used. Guide beams and spuds should be driven to a sufficient penetration to eliminate the chance of their being undermined by the scour which frequently occurs just before the caisson lands.

As stated earlier, the most dangerous time is during the early stages of sinking. The time of removal of false bottoms is particularly hazardous. Once false bottoms have been removed and the caisson has been sunk deeper into firmer material, the danger is decreased. Nevertheless, serious tipping has occasionally occurred after considerable penetration into soils which offered little lateral support.

Caissons tip because support is lost under one side. Rectangular caissons, particularly long, narrow caissons, with a ratio of width to length in the neighborhood of 1 to 2, are very susceptible to tipping about the long axis.

Support is lost under one side because of failure of the soil under the cutting edge and sometimes the loss of skin-friction support on the side. When a dredging well suddenly blows, the soil under the cutting edge becomes momentarily quick. If the false bottoms are removed at too early a stage, the soil may fail in shear under the loading of the cutting edge. If dredging is carried below the cutting edge, sudden failure in shear may take place. If the dredging wells are flooded too early, the mud on which the caisson rests may be overloaded and flow out under the cutting edge.

A study of the history of open caissons leads to the following conclusions:

1. Most cases of initial tipping occur because of attempts to rush the work. For example, a night-shift dredging crew wants to dredge more yards than the day shift and it may therefore excavate too deeply in one pocket.
2. Most cases of serious tipping occur when radical steps are taken to correct the minor initial tipping.
3. False-bottom caissons are the lowest in cost but also the most prone to tip.
4. Dome caissons, such as used on the San Francisco–Oakland Bay Bridge and the Tagus River Bridge, are very expensive. While the domes are of theoretical value in preventing tipping, in practice serious tipping has nevertheless occurred. Domes, therefore do not give positive assurance against its occurrence.
5. Modifications of the false-bottom type which permit substantial penetration in soft soils by sucking out muds and sands underneath before removal of the false bottoms (for example, the caissons used on the Rappahannock River and Second Carquinez Bridges) are of major value in preventing tipping.
6. The double-walled caisson, such as that used on the Mackinac Straits Bridge, is expensive but is very unlikely to tip seriously because of its low center of gravity and high center of buoyancy.
7. Any caisson may, at times, act like a false-bottom type because of the formation of mud plugs in the dredging wells.

General rules to help prevent initial tipping of large, open caissons include the following:

1. Keep the weight of the caisson and its center of gravity as low as possible. This can be done by keeping the interior concrete below the waterline during the dangerous stages.
2. Provide lateral support at as high a level as possible.
3. Maintain symmetry at the plane of the cutting edge. The removal of false bottoms and dredging in the wells should generally be such as to preserve symmetrical conditions, particularly about the long axis.
4. Avoid sudden movements of material. Dredge so as to remove soil from beneath the interior cutting edges while maintaining bearing on the cutting edges.
5. Do not dredge below the exterior cutting edges until the caisson is founded. If the caisson will not settle satisfactorily, use jets on the sides to reduce skin friction as well as cutting jets to loosen the material below the cutting edges.
6. In unstable soils, such as sands, avoid any action which might tend to make the soil under the cutting edge quick, particularly the use of blasting or the pumping down of the dredging wells. Jetting on the sides of the caisson, as well as in the dredge wells if mud plugs tend to form, is usually safer and adequate.

The correction of tipping is, as mentioned earlier, an undertaking which must be handled slowly and carefully. When the caisson first tips, it moves over until the righting forces equal the tipping forces. One of the main righting forces under these circumstances is supplied by the caisson's buoyancy, or more specifically, by the moment couple which acts through the center of gravity and the center of buoyancy.

When steps are taken to move the caisson back out of this tipped position, this buoyancy force gives it considerable rotational momentum which tends to tilt it over in the other direction. If the corrective steps include overdredging and overweighting on the high side, it is obvious that the caisson will tip even more dangerously to the other side as it swings back from the initial tip. Therefore, certain general rules can be suggested for correcting initial tipping:

1. Do not take radical or drastic steps.
2. Provide means of restraining the caisson as it swings back, so as to reduce the momentum of the swing.
3. Keep the sides freed from skin friction by adequate jetting as righting measures are started, so that the caisson will not break free all at once.
4. Do not overdredge on the high side. As the caisson rights, it will rotate about its center of gravity and it may also sink for some distance.
5. Applied righting forces must be capable of rapid removal as the caisson swings back up to vertical. Lateral forces applied high up—e.g., by jacks or pulls to the guide structures or to anchors—are much safer than weights, since their force can be quickly released as the caisson starts to right itself.

In most cases, the practical means of righting the caisson from a minor initial tip will include:

1. Applying a lateral righting pull as high up on the caisson as possible
2. Affixing lateral restraining cables on the low side to keep it from rotating too fast
3. Dredging the high side down to the grade that the cutting edge will come to when it rotates back to vertical

Lateral displacement can usually be corrected or reduced by dredging some material from outside the caisson on one side and dumping mud or gravel on the other side. Dredging slightly more in the wells on the one side than on the other will also help, provided that no tendency to tip is present.

Completing the Installation

In the ideal case, the caisson is sunk until it penetrates a small distance into the bearing stratum. The area enclosed by the cutting edges is excavated a sufficient additional distance into the bearing stratum to secure firm bearing over the entire area and to resist lateral shear.

In actual cases, the conditions may vary considerably from this ideal. If the stratum which stops the cutting edge is decomposed or fractured rock, it may be necessary to excavate a substantial distance below the cutting edge. Often the most practicable way to perform this is to excavate down in sections, pouring the concrete seal in one section before excavating the adjoining ones. Thus, the middle third of a rectangular caisson might be excavated and tremie concrete poured to support the exterior and interior cutting edges in this area. Then the end thirds could be completed in succession.

When the caisson comes to rest on sloping bedrock, serious problems may develop. One means of eliminating these problems, particularly where there is little overburden, is to profile the caisson cutting edge to fit the rock. Another method, applicable only where there is little overburden, is to remove the overburden and level the area by using submarine rock-removal techniques beforehand. However, where there is substantial overburden, lying above sloping rock, the caisson may come to a stop with the cutting edge bearing on only one side or corner. If the overburden is soft, there may be a considerable tendency to tip, and therefore the process of excavation into rock and immediate sealing with underwater concrete must be carried out in very small increments, possibly one dredging well at a time. In some cases, it has been necessary to employ divers to drive short lengths of sheeting to keep out the mud and sand while digging down to rock. Blow-ins may occur under that part of the cutting edge which has not yet reached rock, and it may be necessary to inject grout to stabilize the material sufficiently to permit excavation to be continued.

Excavation in broken bedrock or other firm strata is particularly troublesome immediately under the interior and exterior cutting edges. High-pressure jets and chisels can be used, working through holes formed in the caisson walls or through the dredging wells. It is extremely difficult to control the jets in the dredging wells at great depths, and guides may be affixed to the jet pipe to control its position and direction. Reaction jets handled by divers have been widely used, but in deep water, this is a slow and costly method. Shaped charges placed by divers have been used in a few cases with varying degrees of success and occasional disasters when the charges became misplaced and blew out the cutting edge or the blast wave traveled up the access holes to fracture the walls. As stated earlier, it is probably sounder practice to break up the rock before the caisson is moved to the site or at least while the cutting edge is still well above the bottom. For example, with the caisson well embedded in overlying strata, jet probes through holes in the caisson walls could determine the profile of the rock. Then, through these same wells, holes can be drilled and the rock shot with minimal charges. Holes must be well stemmed between the powder and the cutting edge and the wells filled with sand to prevent the blast effect from traveling up the holes. Underwater robotic excavators similar to headers in tunnel work have recently been developed which can excavate under the cutting edges.

Once excavation is completed, all loose material should be cleaned up. Rehandling buckets which scrape the bottom have been found more effective than standard clamshell buckets. Large air-lift pumps may be effectively employed.

Where the rock is fractured, as when it has been previously drilled and shot, removal of the broken rock from the blast pockets will be quite time consuming and may require the use of divers to guide the buckets. While it is generally necessary that all clay, sand, and silt be removed from the bottom, it may be acceptable to incorporate fractured pieces of rock in the concrete by starting the placement with a fluid tremie grout containing an anti-washout admixture, so that it will flow into the crevices. Alternatively, or as a supplement, after the tremie-concrete seal has hardened, the fractured zone can be pressure-grouted.

In some cases, notably the caissons for the Honshu-Shikoku Bridges in Japan, the fractured rock surface was ground smooth around the perimeter by underwater rock-grinding machines. Large drilling equipment could be similarly employed. When the caisson is to be landed on exposed rock, excavation by mechanical cutting tools such as hydraulic rock cutters, drills, or trenches ahead of time may thus be preferable to drilling and blasting.

Jetting the bottom just before starting the tremie-concrete pour will place light sediments such as sand and silt in suspension and allow the concrete to develop full contact. Algae, other marine growth, and clay should be removed by water jet from the underside and sides of the interior and exterior cutting edges so that bond may be developed with the concrete seal. Nontoxic algae inhibitors, similar to those used in swimming pools, may be used to control algae growth and slime. Covering the top to keep out sunlight will also inhibit algae growth.

The sealing of the caisson usually consists of an underwater-concrete pour which is carried up some distance above the interior cutting edges. Sometimes this seal pour is carried well up the caisson, as in an anchor pier for a suspension bridge. In any event, it must be deep enough to provide full transfer of load from the bedrock to the caisson walls, with an adequate factor of safety. It may be placed by any of the three methods described for cofferdam seals: tremie, bottom-dump bucket, or intrusion grouting; however best results will be attained by properly placed tremie concrete. If there is to be a significant difference in the height of fresh concrete against interior walls, they will have to be adequately reinforced to withstand the unbalanced pressure.

Dewatering and completion of concreting in the dry were common in the past and may still be required on occasion. However, in many recent caissons, the dredging wells have not been dewatered. These wells were left open, filled with water only, to reduce dead weight. Precast-concrete covers may be placed over them to serve as a soffit for the distribution block. The distribution block and upper portions of the caisson are then completed by normal pours. Near the water surface, cofferdam walls are frequently carried up as a continuation of the caisson exterior walls so as to permit concrete placement in the dry. Minor corrections of position can be accomplished in the distribution block and subsequent pours.

Examples

Mackinac Straits Bridge, Michigan. The foundations for this bridge were constructed under extremely adverse conditions of exposure to storm and ice. The construction was highly successful and was completed very rapidly.

The two main tower piers were constructed by means of double-walled steel caissons, circular in plan. They were 35 m (116 ft) in diameter and depths to rock were 60 m (195 ft). Water depth was about 30 m (100 ft), with some 10 m (30 ft) of overburden above the bedrock.

Caissons were prefabricated near shore. The watertight steel double-walled shell provided flotation, thus eliminating the need for false bottoms or domes. 2.5 m (8 ft) of concrete was added as ballast, and the caisson was then towed to position and moored between heavy steel guide towers. The guide towers consisted of a frame of pipe piles and cross bracing, set on the bottom. Through the pipe piles, steel H piles were driven to rock. The H-pile spuds were locked to the pipe frames by filling them with concrete.

To sink a caisson, the space between the double walls was filled with concrete in progressive lifts of 1 to 3 m (3 to 10 ft). As the caisson sank, the steel walls were built up to keep a safe freeboard. The concrete placement was performed by the intrusion-grouting method which permitted large yardage to be placed in a minimum of time, despite rough water.

The caissons were landed in 30 m (100 ft) of water and penetrated about 5 m (16 ft) under the weight added by concreting between the double walls. Sinking through the harder material was aided by dredging through the dredging wells and by jetting. The concrete surface remained well below the water surface during sinking, thus ensuring a very low center of gravity. Absence of false bottoms, a symmetrical layout of the structure, symmetrical dredging, and sturdy lateral support by the guide towers all contributed to the safe and positive manner in which the caissons were installed.

The maximum head to which the double-walled shell was subjected was 20 m (70 ft). This meant that heavy internal bracing had to be built into the walls. Heavy wide-flange soldier beams were placed vertically to support the steel shells, with structural-steel cross struts closely spaced between the soldier beams. After founding, the interior wells were cleaned out and filled with underwater concrete. The top portions of the piers were completed in the dry by dewatering the interior of the caisson, using the steel walls as a cofferdam. This permitted placement of the wrought-iron protection plates in the dry.

The cutting edge was sturdily constructed, using steel plates supported by wide-flanged beams. The cutting edge was 6 m (20 ft) in height. Heavy trussing 4 m (14 ft) in height spanned between the cutting edges.

For the concrete filling operation, coarse aggregate with no fines was placed by means of conveyor from self-unloading barges at rates up to 2500 tons/h. Simultaneously, mortar was pumped from the mixing barge through flexible hoses to vertical 30-mm (1.25-in) pipes spaced 3 to 6 m (10 to 20 ft) on centers each way. These pipes were withdrawn gradually as the mortar height raised. Observation wells were used to keep track of the mortar surface and ensure that it remained between 6 m (20 ft) and 1.5 m (5 ft) below the top of the rock at all times.

Some uncemented layers occurred in the underwater concrete due to thin layers of fines being inadvertently deposited as the aggregate was placed. Apparently, these fines initially accumulated on the decks of the barges due to abrasion during loading and shipment and were then placed in the caisson at the end of each unloading cycle by the self-unloading conveyors. As a consequence, these strata of uncemented fines remained without cementation.

While the double-walled method of caisson construction involves large quantities of fabricated steel and thus high initial cost, it is undoubtedly a safe and positive method of caisson construction. Its use enables completion in a short time under difficult conditions, such as prevailed on this project.

Rappahannock River Bridge, Virginia. This bridge made a notable innovation in the construction of caissons by the false-bottom method. Fifteen caissons were constructed, the largest being 20 m (62 ft) by 14 m (43 ft) in plan and sunk to a maximum depth of 50 m (153 ft). Each caisson was essentially a double-walled one, equipped with a false bottom for full flotation. The external walls were rectangular in plan, and the internal walls were formed by six 5-m (15-ft)-diameter steel dredging wells. As sections were added, they were filled with concrete, progressively sinking the caisson to the bottom of the estuary.

There was a stratum of very soft mud at the site, 5 to 6 m (15 to 20 ft) in thickness. If the conventional false-bottom method had been employed, difficulty might have been encountered due to overloading this mud, leading to sideways slippage. Also, the removal of the false bottoms would have entailed considerable risk because of the low shear strength of the mud. For these reasons, the contractor designed and built removable domes which served as false bottoms for the dredging wells. A 300-mm (12-in)-diameter pipe was connected to the top of each dome and extended up to the top of the caisson walls as constructed at each stage. When the caisson had landed, the soft mud was removed from under the false bottoms by means of jets and air-lift pumps working through the 300-mm pipe. By removing mud uniformly from under all the domes, by adding weight to the caisson by water pumped into the closed dredging wells, and by adding more sections of the steel and concrete walls, the caisson was sunk through the soft mud and into firmer sediments under excellent control and with a minimum of risk.

The false-bottom domes were then removed. This was accomplished without any difficulty, since the caisson was well embedded at this stage, with no mud pressures under any false bottom. The caisson was then sunk to the bearing stratum of firm sand as an open caisson, aided by jets operating through sounding wells.

Fraser River Bridge, British Columbia. This bridge was constructed across the Fraser River at Vancouver, British Columbia. The river at this site had a current velocity of up to 4.5 m/s (8.5 k).

The false-bottom method was used for floating caissons constructed of timber with steel cutting edges. These were towed to the site, moored inside timber-pile enclosures, and sunk by filling the caisson walls with concrete. Considerable scour was experienced as the caissons were landed on the bottom. This phenomenon of underscour has occurred on many caissons in many different types of material and must always be reckoned with. In this case, scour was checked by dumping gravel around the caisson until it was well embedded. The caisson was then sunk to elevation −40 m (−132 ft), foundation piles were driven underwater, and a tremie seal was poured.

During construction of pier 5, after the caisson had been founded and the tremie seal placed, a flood occurred in the river and the bottom scoured out to below the elevation of the bottom of the tremie seal. There was fortunately no settlement of the pier, and remedial steps were based on the following requirements:

1. Restoration of bearing value of foundation piling
2. Permanent protection against deep scour
3. Streamlining of the pier to reduce scour

The scoured hole was first backfilled with sand and gravel to the elevation of the bottom of the tremie seal. Fine sharp sand was then jetted in under the tremie-concrete seal, and steel sheet piling was driven around the outside of the pier to form a streamlined nose on each of its ends. Within the enclosures (the spaces between the sheet-pile noses and the pier itself), bearing piles were driven and a tremie-concrete slab was poured. The underlying gravel was pressure-grouted, and the entire hole was then filled with gravel to the original elevation of the riverbed.

Second Carquinez Bridge, California. This bridge required three open caissons to be sunk in about 26 m (90 ft) of water with currents up to 4 to 5 m/s (8 to 10 k). A fender system of concrete-filled pipe piles was constructed first and used as an aid in anchoring. Sixteen concrete anchors were also used to hold each caisson. These weighed 25 tons each and were sunk into place by jets and air lifts. Anchor cables were then run through sheaves on the cutting edge and up to the top of the walls, thus taking their pull low so as to prevent any tendency to tip.

The caisson walls were extended upward progressively by precast-concrete panels that served as forms and became an integral part of the walls. The concrete caissons were made buoyant by false bottoms in the shape of inverted timber domes with a 750-mm (30-in)-diameter steel-pipe dredging well in each cell. After the caisson reached the bottom, a hydraulic jet monitor and air-lift pump was lowered down each of these pipe wells. The jet was a 150-mm (6-in) pipe fed by two 10-m^3/s (1500-gpm) pumps at 2.5 MPa (350 lb/in^2) pressure. It could be rotated horizontally and angled in the vertical plane, and effectively loosened the sand and gravel for removal by the air lift. This means of excavation was so effective that the false bottoms were left on until near final founding on the bed rock.

After removal of the false bottoms, hard material was removed by clamshell bucket. In the caisson for the north pier, the rock contours sloped across the pier. As rock was reached at the high corner, shaped charges were placed by divers to break up the rock. These, unfortunately, became misplaced and when they fired, they did considerable damage to the lower walls and cutting edge. It is believed that a much better plan would have been to drill and shoot this rock before the caisson was even brought to the site or, at least, while the cutting edge was well above the rock and thus protected by a thickness of sand and gravel.

Sunshine Bridge, Donaldsonville, Louisiana. This bridge required four major open caissons. While they were being constructed in a shipyard, willow mattresses were sunk at the site and weighted with rock. Three sides of the "pen" were built at each pier site, using a tubular braced frame through which pin piles were driven, and the caissons were then towed in and the pen closed. The caisson wells had false bottoms which were bolted on and contained central pipes through which air lifting could be performed. After the caisson had been sunk through the mat and the soft overlying material, the bolts were removed by diver on a few wells at a time, so that bucket excavation could be used. When the caissons were well down, their tops were extended by a sheet-pile cofferdam.

The first three piers went extremely well. The fourth and last was located near the west levee bank. In earlier caissons, some difficulties had been experienced in removing the false bottoms due to excessive upward soil pressure, so it was decided to remove some of the inshore false bottoms after only about 6 m (20 ft) of penetration. As these were being removed, the soil rushed in and the bank suddenly sheared. The caisson tilted some 15° toward the levee.

The corrective action taken was well planned and executed. Heavy steel beams were bolted across the top of the caisson and made into a truss extending out over the high side through the structure of the pen. Weights consisting of steel piles were placed on top of the pen, and jacks were inserted to push downward on the beams with a force of several hundred tons. Thus, a very large leverage was exerted to right the caisson. Dredging was then carried out in two of the central wells on the high side. Work was performed slowly and methodically, and the caisson gradually was righted. Extreme care was taken not to overdredge on the river side as the caisson returned to vertical. The caisson was ultimately sunk to founding elevation, less than ½ m riverward in plan from the originally designed location.

Verrazano-Narrows Bridge, New York. The two main piers for this bridge were constructed as open caissons inside sand islands. These islands were formed by sheet-pile cells filled with sand. They served as work platforms for construction and as docks for mooring sand and gravel barges. The caissons were 70 m (229 ft) by 40 m (129 ft) in plan, contained 66 cells, and were founded at depths of 30 to 50 m (105 to 170 ft). Each cell was 5 m (17 ft) in diameter. The area inside the cells was filled with sand to an elevation of −4 m (−12 ft) and was kept dry by a wellpoint system during the assembly of the cutting edge and the start of the wall concrete. The 4-m (13-ft) cutting edges were made of reinforced concrete, protected by a heavy armor of fabricated steel. To reduce skin friction, the cutting edge was stepped back 100 mm (4 in) at its upper edge.

High-pressure water jets were installed in the outer walls, two tiers in the Staten Island caisson and three tiers in the Brooklyn caisson. Nozzles were discharged pointing upward 60° from the horizontal. Risers were 60 mm (2.5 in) in diameter; headers were 150 mm (6 in) in diameter. Nozzles were spaced 4.5 m (14 ft) on centers and were connected and operated in groups of four or five. Operating water pressure was 2 to 2.5 MPa (300 to 350 lb/in^2). Provisions were made for the injection of bentonite slurry as a lubricant, but the water jets alone proved adequate.

The contractor took advantage of the great inherent stability of these caissons and the control made possible by the sand-island method to build the caisson walls up in lifts of approximately 12 m (40 ft). This procedure made adequate weight available for sinking, with the aid of jetting and clamshell excavation in the wells. Since the soils through which the caisson was to be sunk had high shear strengths, there was no danger in making high lifts; on the contrary, the contractor felt there was less risk, as he never had to excavate beneath the cutting edge. With the greater effective weight afforded by the high lift, he could obtain a satisfactory rate of caisson penetration without excessive disturbance of the surrounding soil by jetting, etc. Clamshell buckets were used to excavate. To minimize stresses in the caisson walls and to ensure a condition of uniform support, jetting was employed during sinking, before the next stage was built up.

Because of the large size of the caisson, it was impracticable to jet all around the periphery at the same time, so jets were manifolded in groups and operated in succession.

Bridge Across the Garonne River at Bordeaux, France. The main pier for this suspension bridge was constructed on a sand island formed by a 55-m (170-ft)-diameter steel sheet-pile ring. The concrete caisson was then constructed in the dry. It was sunk through sand, gravel, and hard shale and sandstone lenses by injecting an annular ring of bentonite just above the cutting edge. During excavation and sinking, the caisson was supported almost entirely on the cutting edge.

San Francisco–Oakland Bay Bridge, California. The caissons for this bridge were both the largest and the deepest which had ever been sunk at the time, and they were responsible for the selection of this bridge as one of the seven wonders of the modern civil engineering world.

On the East Bay Crossing, pier E-4 was a hollow-walled caisson with a steel cutting edge. It was floated and sunk as a false-bottom caisson. It was designed so that during the flotation stage the concrete surface was well below the water surface. The necessary weight for landing might have been furnished by water ballast in the dredging wells, but since this would have greatly increased the pressures on the timber walls, the necessary weight was obtained by pouring the concrete walls higher.

The caisson landed on a low tide, raised on the subsequent high tide, and alternated between landing and floating for several tide changes before finally entering the bottom. Then 5000 tons of concrete was added to the walls, and the dredging wells were filled with 5500 tons of water, giving a unit pressure on the semifluid mud of 20 tons/m^2 (2 tons/ft^2). The caisson sank 5 m (16 ft), pushing up a 1-m (4-ft) mud wave outside, and rocked back and forth while the false bottoms were removed.

At pier E-3, constructed in similar fashion, the caisson landed on nonuniform material, with the mud under one corner being very soft. The caisson listed diagonally and skewed. Its motion was checked and corrected by piling mud outside the corner and by dredging at the opposite corner.

Extensive jetting through built-in jets greatly facilitated the sinking operations. Free jets, including reaction jets, and jets held in position by sliding frames working in the dredging wells were effective in cleaning off the bedrock some 80 m (240 ft) below sea level.

For the West Bay Crossing, the caissons were sunk as open caissons, with control by air pressure in the temporarily domed dredging wells. The air pressure was used to provide buoyancy in the water and semifluid muds, and to control the sinking. Domes were removed from one or more wells at a time for the extension of the well shaft and for dredging operations, and then replaced so that other domes could be removed. There were 55 dredging wells with domes in the West Anchorage caisson.

After pier W-6 had been landed and had penetrated 6 m (20 ft) into the soft bottom, it hung up. The concrete walls were well above water. The caisson started to settle on the west side so that dredging was carried below the cutting edge on the east side in an attempt to correct the slight list. Suddenly it tilted 6° to the east, but fortunately was held at this inclination by the floating fender which was secured between two work platforms and guide towers.

The caisson was straightened up by dredging in the west wells, while using domes and compressed air to make sure it would not swing too far when it rotated back to vertical. Anchor lines were led to the west to help pull it back, and jetting was carried out along the west cutting edge. To provide adequate freeboard while the caisson was tipped without unduly increasing the height of the center of gravity, the concrete in the exterior walls was carried up above the interior walls.

Rocking the caisson down alternately one end and then the other while jetting under the cross walls was successful from then on, although at 13-m (45-ft) penetration, the north end suddenly dropped 1.5 m (5 ft). This was corrected by dredging in the south end.

Pier W-3 was landed at −20 m (−70 ft) and sank to −23 m (−79 ft). The concrete walls were built up to +10 m (+34 ft), adding considerable weight and causing the caisson to sink to −30 m (−94 ft) while forcing mud up 10 m (30 ft) into the cylinders. This mud formed a plug or mud false bottom, and material was forced out under the cutting edge. The air pressure was temporarily lowered in the dredging wells without any results. When the pressure was again lowered, the caisson dropped 2 m (6 ft) and listed. This list was corrected by varying the air pressure in the domed cylinders. Dredging was commenced in the dredging wells to break up these mud plugs. The mud suddenly rose 7 m (24 ft), and the caisson listed to the north.

Because the tipping was endwise instead of sidewise, it was not so serious and was corrected without excessive difficulty. If the concrete walls had been kept lower, it is probable that the mud plugs could have been removed or broken up before bottom conditions became so adverse. Also, the center of gravity would have been lower and the control by air would have been more effective.

At pier W-4, when there was 6-m (20-ft) penetration into the bottom, the mud had risen 8 m (25 ft) in the cylinders and had plugged, forming a false bottom of mud. The caisson then rocked back and forth for 3 weeks, with tilt up to 1 m (4 ft). At 10-m (30-ft) penetration, with the caisson level, and while dredging the center wells, the mud displaced sideways under the cutting edges, causing the pile-supported working platforms to heave. The caisson tipped slowly 2.5 m (8 ft) to the north.

The exterior concrete walls were built up to give more freeboard. Domes were put on the north wells and air pressure was raised on the north and removed on the south, with no effect because of the restraint of the mud plugs. Extra concrete was added to the south walls, and the south wells were dredged. The caisson then slowly swung back and rotated on over to a 9-ft list to the south. The caisson was finally righted by pumping the water down in the north wells and by dredging in the north wells. It appears that the caisson was too heavy at this stage for the soil to support it. The concrete should have been kept lower to reduce weight and to lower the center of gravity.

By way of contrast, pier W-5 was the narrowest pier and in the softest mud. The concrete was kept low after landing. The anchor lines, after the cutting edge was embedded, were moved up above the existing waterline on the caisson and tensioned to 30 tons per line. Symmetry was maintained in all operations, and the mud was kept excavated down to, but not below, the cutting edge to eliminated mud plugs. No difficulty or tipping was encountered with this caisson.

The cleaning and sealing of the bottoms, of all caissons after founding in bedrock, were carried out in three sections, one-third at a time. At pier W-4, the material tended to flow in under the cutting edge even at this great depth; so the caisson was sunk into rock, using *gads* or chisels and high-pressure jets. Under the cutting edge, where broken-rock fragments were running in, grout pipes were jetted down outside the caisson, and the outside material was thoroughly grouted.

Tagus River Bridge, Lisbon, Portugal. The open caisson for one of the main piers of this bridge was founded at elevation –80 m (–260 ft), the deepest caisson yet constructed. Because of the great depth to rock, the swift currents, open water, heavy overburden, and earthquake danger, an air-dome method similar to that used on the San Francisco–Oakland Bay Bridge was selected. Water depth was 28 m (90 ft), followed by 30 m (100 ft) of mud and 20 m (60 ft) of sand before bedrock was reached. Twenty-eight dredging wells were used, each 5 m (16 ft) in diameter. The air domes were alternately removed and replaced so that dredging could be performed. Balance was maintained with compressed air at 140 KPa (20 lb/in^2). A computer was used to control the entire process. Excavation was performed by clamshell bucket working through the wells. Although one caisson tipped 15° during initial sinking, it was successfully righted by use of the air pressure and dredging (Fig. D4-34).

PNEUMATIC CAISSONS

These were much used in the early half of this century but because of labor problems and consequent high costs, they have generally been replaced by open and box caissons and by slurry-wall cofferdams. An exception has been in Japan where pneumatic caissons continue to be used for major bridge piers, the latest being the piers for the Rainbow Suspension Bridge in Tokyo.

Innovative robotic machines were developed for excavating the soil, reacting against the roof, which was a heavy steel and concrete slab located a few meters above

FIGURE D4-34 Tagus River Bridge, Portugal. Air pressure in domes was successfully used to counter tipping of caisson in soft sea floor sediments. *(Courtesy of Tudor Engineers.)*

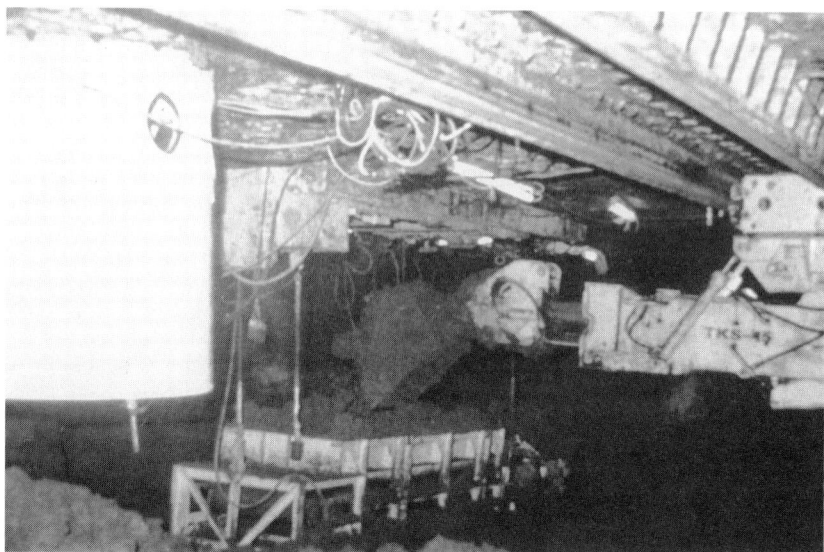

FIGURE D4-35 Robotic excavator excavates under cutting edge of pneumatic caisson. "Rainbow Bridge," Tokyo. *(Courtesy of Shiraishi Engineering Corp.)*

the cutting edges. It was strong enough to resist the air pressure. A robotic header cutting wheel operated beneath the cutting edges. The operation of these machines was controlled by personnel in a diving bell, and by instrumentation (Fig. D4-35).

The use of a pneumatic caisson of this type permits excellent control of the sinking rate, although stability against tipping is reduced by the free surface. Therefore, a very strong and rigid pen was constructed to control the position of the caisson. The heavy roof of the pneumatic chamber, being located only a few meters above the cutting edge, helps the stability by lowering the center of gravity.

The principal reason for adopting a pneumatic caisson instead of an open caisson was the ability to fine-tune the weight on the cutting edge to soft soils and the prevention of sand run-ins under the cutting edge. It also enabled the controlled excavation of the conglomerate and decomposed rock under the cutting edge. Thus, despite its large initial and operating costs, it is a viable method for use in critical soils, and critical locations, for example, where a caisson must be sunk adjacent to and below an existing structure.

CHAPTER 5
CONSTRUCTION DEWATERING*

James J. O'Brien, P.E.
O'Brien-Kreitzberg, Inc., Pennsauken, New Jersey

INTRODUCTION

Construction projects frequently involve deep excavations which extend below the groundwater table. In certain instances, excavation may be made in the wet, such as in the cases of open caissons or bridge-pier cofferdams. Generally, however, it is preferable to remove the water from the excavation by the use of an appropriate dewatering method.

The dewatering method should effect the removal of the water from the excavation without causing instability in the side slopes or bottom of the excavation. The system may also be required to provide pressure relief in pervious strata underlying the bottom of the excavation.

The selection of the method of dewatering and the design of the system should be based on a careful study of all available soil and groundwater data. Frequently, such information can be obtained from governmental agencies, water-supply well logs, previous construction records in the general area, etc. The design engineer will generally provide the contractor with drill logs on projects involving dewatering problems. Exploration holes should be sufficient in number to indicate continuity of strata. Samples should be taken for permeability evaluation. On projects involving major dewatering problems, pump tests may be required for the purpose of obtaining accurate permeability data.

All too frequently, exploration holes are not sufficiently deep to detect the presence of pervious strata requiring pressure relief. *Exploration should extend to a depth below the water table equal to at least twice the depth of the excavation bottom below the water table.* The importance of this rule cannot be overstressed.

Dewatering operations can be divided generally into two main classes: (1) surface pumping methods, and (2) predraining methods. In surface pumping, the water is pumped from open reservoirs or sumps. Predraining methods are charac-

---

*Most of the material in this chapter is from R. Y. Bush, "Construction Dewatering," in Stubbs and Havers (eds.), *Standard Handbook of Heavy Construction*, 2d ed., McGraw-Hill, New York, 1971.

terized by pumping from below the ground surface from wells or wellpoints prior to excavation.

DEWATERING METHODS

Surface Pumping

Sumps and ditches are frequently used for dewatering. The method simply involves pumping from sumps located inside the excavation area. It is applicable to soils whose stability is not significantly reduced by the action of seepage, for example, stiff clay or a soil having a high percentage of gravel or boulders.

The dewatering of an excavation using this method (frequently referred to as *sumping*) usually consists of making an initial excavation to the water level, followed by the excavation of ditches and sumps around the perimeter of the inside of the excavation. Sumps are, if possible, excavated to an elevation below the deepest final grade within the excavation area. Excavations extending to appreciable depths below the water table require the use of a system of pumps progressively excavated to deeper elevations, with pumps located in leap-frog fashion until final excavation grade is reached.

The concurrent removal of water and soil from an excavation results in seepage entering the excavation through the bottom and the side slopes. In the case of fine-grained granular soils such as fine sands and silts, the bottom of the excavation frequently becomes "quick" and the side slopes unstable as a consequence of the seepage forces imparted to the soil grains by the water entering the excavation. The dewatering of such soils generally required the application of predraining methods.

The problem of instability due to seepage is frequently compounded by the contractor's desire to expedite the dewatering operation. Water is pumped from the excavation at a rapid rate, and the water level inside the excavation lowers considerably faster than the groundwater level in the adjacent ground. To minimize the head differential between the inside and outside of the excavation, it is important to pump at a relatively slow, *continuous* rate rather than at a rapid intermittent rate (the interruptions occur when pump capacities exceed the rate of inflow into the sumps).

Horizontal drainage systems consist of a series of ditches leading to sumps. These ditches are frequently filled with gravel (the so-called, *French drain*) to prevent their caving and/or to permit later concreting or backfilling above the excavation level. Due to the coarseness of the gravel generally used, such drains are subject to clogging by removing fines from the adjacent natural material. Further, their hydraulic carrying capacity is quite low. A more trouble-free and efficient drain can be provided by utilizing either perforated or porous drainpipe surrounded by a properly graded sand filter, or by wrapping the stone in geotechnic filter fabric.

To permit continuous pumping at later construction stages during backfilling, horizontal drains can be connected to vertical or sloping pipe *risers*. These risers are extended as the backfilling progresses, thus making it possible to control the water level below the backfill. After the completion of construction, the drainpipe and risers may be grouted if this is considered necessary.

Cofferdam unwatering is utilized for structure, such as dams, powerhouses, and bridges, which involve construction in or immediately adjacent to "open" water. To perform this construction, it is first necessary to build a cofferdam around the work area.

Cofferdams such as those used for stream diversion to permit dam construction may be constructed of soil or rock materials. The design of such cofferdams should be based on the principles of earth-dam design, with particular attention given to stability considerations inside the cofferdam.

Steel sheet piling is frequently used in cofferdam construction. It may be used in connection with an earth-dam-type embankment to reduce seepage, i.e., to act as a cutoff. Cellular-type cofferdams, consisting of circular cells of interlocking sheet piling filled with granular material, are used where it is necessary to enclose large work areas, particularly in swiftly flowing streams. Excavations of small areas, such as those required for bridge piers, are frequently made by using rectangular cofferdams consisting of sheet piling cross-braced with a system of wales and struts.

Excavation inside small, rectangular, sheet-pile cofferdams may be made in the wet by using a clamshell. By maintaining the same water level both inside and outside the cofferdam, the bottom of the excavation remains hydraulically stable; i.e., there is no upward flow of water. Excavation is followed by the underwater tremie placement of a concrete plug. The vertical thickness of this plug must be sufficient to withstand the hydrostatic uplift after pumping out the impounded water (correctly referred to as *unwatering*) inside the cofferdam. In this case, the pumping rate selected will be dependent mainly on the desired duration of the pumping period. Leakage through sheet-pile interlocks and/or the concrete plug will generally be nominal if the cofferdam is constructed carefully. Occasionally, remedial measures are required (e.g., grouting in voids of the tremie plug, sealing interlocks with straw, etc.) to reduce the leakage into the cofferdam.

The unwatering of cofferdams which do not incorporate a concrete plug or a positive cutoff will require the removal of continuous infiltration in addition to the impounded water. In the case of an earth cofferdam, this seepage will occur through, as well as under, the embankment. The depth of sheet piling in a rectangular cofferdam may significantly affect the pumping rate as well as the stability of the bottom. Considering the frequency of cofferdam failure attributable to insufficient embedment of the sheet piling, it is prudent to use a conservatively large embedment.

An extremely important consideration in the design of cofferdams, as well as in the design of dewatering systems in general, is that of bottom pressure relief. Figure D5-1 illustrates a cross section of a rectangular sheet-pile cofferdam which requires bottom pressure relief. To effect this relief, wells or wellpoints should be installed inside the cofferdam. If such relief is not provided in the example, it is apparent that the hydrostatic uplift pressure on the bottom of the silty clay layer will be twice that of the downward pressure due to the saturated soil weight, after excavation has reached subgrade! This obviously dangerous condition exists in spite of the fact that a reasonably large (what would normally be considered conservative) sheet-pile embedment is used—i.e., 10 ft for a 15-ft cofferdam depth.

If the underlying pervious soil in this example were a thin, continuous stratum, the sheet piling should be driven sufficiently deep to cut off this pressure zone and thereby reduce the need for wells or wellpoints inside the cofferdam.

Figure D5-1 also serves to illustrate the recommendation that exploratory borings should extend to a depth below water equal to twice that of the excavation depth below water (for an "average" saturated-soil unit weight of 125 lb/ft^3, which is about twice the unit weight of water).

Well Method

Predraining methods of dewatering generally require the use of wells of some type. These wells are installed around the excavation or adjacent to it, as in the case of trench excavation. The necessity for predraining is dictated by the presence of cohe-

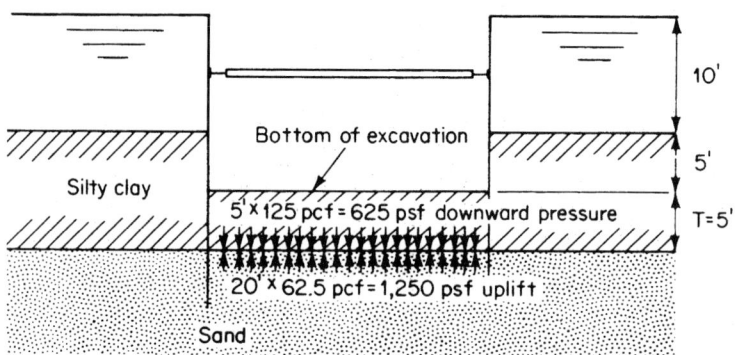

FIGURE D5-1 Cross section of cofferdam requiring bottom pressure relief. Note that when T = 10 ft, downward pressure = 1250 lb/ft². Therefore, if depth to previous stratum is greater than or equal to twice excavation depth (both measured below water level), uplift pressure will be less than or equal to downward pressure.

sionless soils which are subject to becoming unstable and quick if surface-pumping methods are used.

Predraining systems range from large-diameter (12 in or more) deep wells to smaller-diameter wells or wellpoints which extend to relatively shallow depths of less than 30 ft.

Ideal soil conditions for economical dewatering with wells are a high permeability and a reasonable degree of homogeneity below the water table. If these conditions extend to an appreciable depth below excavation level, widely spaced deep wells of relatively high capacity are applicable. The dewatering of stratified or finer-grained soils requires the use of smaller-capacity, more closely spaced wells. Wells are not suitable for dewatering down to the top of an impervious layer due to the reduction in the wells' collection capacity with the lowering of the water level. Either conventional or ejector-type wellpoints should be used in this situation.

Well-drilling methods used in construction dewatering include cable-tool, rotary, and reverse-rotary methods. Support of the sides of the drilled hole may be provided by (1) a steel casing, as in the cable-tool method, (2) using *drilling mud* which is introduced into the drilled hole to obtain a fluid having a specific gravity slightly higher than that of water, or (3) maintaining a higher water level in the drilled hole than that of the groundwater. This excess head provides the necessary support, as in the case of the reverse-rotary method.

The cable-tool method is applicable in relatively coarse-grained materials which do not require gravel packing. The steel casing is installed progressively with the drilling of the well and provides permanent support to the sides of the hole after the completion of drilling. The casing is perforated either prior to installation or in place by using special perforating equipment.

The use of temporary steel casing is generally unsatisfactory for large-diameter gravel-packed wells which extend to appreciable depths (in the range of 100 ft). The necessity for installing and removing the casing in short sections, requiring welding and cutting, makes the method slow and frequently expensive. The pulling of the casing is particularly troublesome, necessitating the use of large jacks to overcome the friction between the casing and the natural ground and gravel pack.

The rotary method is frequently used with drilling mud to support the sides of the hole. Occasionally the *bucket-auger* method is used with drilling mud. By eliminating the necessity for casing support, these methods are generally quite efficient and relatively economical. The main disadvantage of methods utilizing drilling mud is the difficulty in removing the mud from the completed well. During drilling the mud forms a *cake* along the walls of the hole, and it is doubtful that this cake can ever be completely removed even though the well is treated with extensive bailing and surging. The mud that is not removed will reduce, in many cases appreciably, the collection capacity of the well.

The reverse-rotary method is used with an excess hydrostatic pressure of water in the hold to provide support. Clean water is used, and the problem of *mudding the hole* is reduced to that resulting from the suspension created by the drilling. Due to the size and expense of reverse-rotary drilling equipment, there are areas where it is not possible to find local drillers experienced with the method. Where equipment and qualified drillers are available, the method is one of the best to use for dewatering wells. Although the unit cost of a given size of well will generally be higher with this method, the greater productivity of the wells will frequently more than offset the additional drilling expense. In many cases, the yield of reverse-rotary wells will be several times that of wells of comparable size drilled with mud.

Excellent dewatering wells have been installed using jetting methods. A specially designed casing is jetted in the ground with appropriate high-pressure pumping equipment. A crane of adequate size and boom length is used to handle the long casing during jetting and removal after gravel packing. Because of the fairly extensive amount of special equipment required, the jetting method is generally applicable only on projects requiring the installation of a large number of wells.

After the completion of drilling, the perforated casing or well screen is installed. The upper portion, above the final water level, is unperforated. The size of the perforations as well as the gradation of the gravel-pack material must be selected so that the well will perform efficiently and that, at the same time, the pumping of fines is prevented. Criteria for determining gravel-pack gradation and the size of well-casing perforations are supplied in this section under Design of Dewatering Systems.

Deep-well turbine- or submersible-type pumps are generally used for pumping from the wells. Occasionally, horizontal-type pumps are used for where required draw-down in the well does not exceed the suction limit of the pump. Because of their high efficiencies and relatively low maintenance, deep-well turbine-type pumps are most frequently used for prolonged pumping on large-capacity wells.

Small clearances (less than 1 in) between the inside diameter of the perforated casing and the maximum diameter of the pumping equipment installed inside it should be avoided. Larger clearances reduce the danger of sand or gravel particles lodging between the pump (or pump bowls) and the casing, which would make the ultimate removal of the pumping equipment quite difficult.

Wellpoint Method

Wellpoint dewatering may be used for a wide variety of soil types and conditions. Wellpoints are installed on relatively close spacings, usually in the range of 2 to 8 ft, making them well suited to the dewatering of fine-grained or stratified soils. The broad range of application of the method is illustrated by Figs. D5-2 and D5-3, representing, respectively, a sewer-trench excavation dewatered with a single stage of wellpoints and a dam cutoff trench excavation dewatered with a multistage wellpoint system.

A bottom-suction self-jetting wellpoint, similar to the type generally used on construction dewatering operations, is illustrated in Fig. D5-4. Such wellpoints are ordinarily approximately 3 ft long, having an outside screen diameter of about 3 in. The

FIGURE D5-2 Single-stage wellpoints for sewer trench.

FIGURE D5-3 Multistage wellpoints for dam core trench.

wellpoint is connected to a riser pipe, the length of which is governed by the amount of water lowering required. Generally, these riser lengths vary from 10 to 21 ft. The connection between the header manifold and the top of the riser is made by means of a *swing joint* which may be either of the flexible hose type or an assembly of pipe nipples and elbows. Each swing joint is provided with a valve which permits the regulation of the flow of water and air into the wellpoint. The water from the wellpoints is collected in the header manifold and carried to the wellpoint pump.

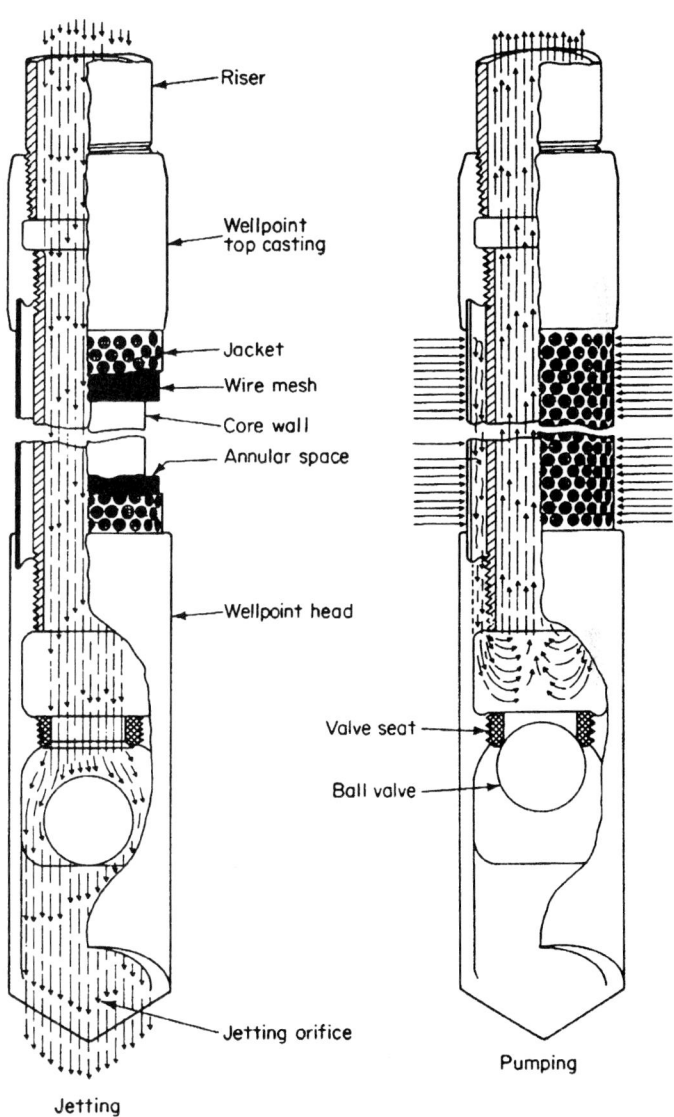

FIGURE D5-4 Self-jetting type wellpoint.

Wellpoint pumps are designed to handle air as well as water. All water entering the pump passes through an air-separation tank from which the air is removed by means of a supplementary vacuum pump. These features keep the system "primed" at all times by maintaining a continuous high vacuum. The water-handling pump of the unit is a horizontal-centrifugal type.

Figure D5-5 illustrates a typical wellpoint installation and shows all the component parts assembled and operating. Installation of the wellpoint system is started by excavating to, or slightly above, groundwater level, and installing the header manifold at this elevation. In some cases, where the final depth of excavation below the ground surface is less than the practical suction lift, the header is installed on the ground surface. In either case, it is necessary to provide a level grade or one resulting in a slight slope of the header up to the pump. By eliminating "high" points in the header line, the accumulation of air at these points is prevented. Having installed the header with all the necessary elbows, gate valves, tees, and other fittings, the swing joints are connected to the threaded outlets at the required spacing. Unused outlets are plugged.

At this point, the wellpoints, attached to the risers, are installed. If the soil below the water table is fairly coarse grained and homogeneous, wellpoints are usually self-jetted into the ground. This operation is illustrated in Fig. D5-6. Where a fine-grained and/or stratified soil profile exists, it is necessary to provide a sand filter around the wellpoint and riser. This filter increases the effective radius of the wellpoint and permits the dewatering of water-bearing strata which are separated by relatively impervious horizontal layers. The installation of filtered wellpoints is usually performed by utilizing a *sand casing*. This casing is jetted into the ground as shown in Fig. D5-7. After jetting, the wellpoint and riser are placed inside the casing and the annular space is filled with the filter sand. The casing is then removed, leaving a filtered wellpoint such as those shown in Fig. D5-5.

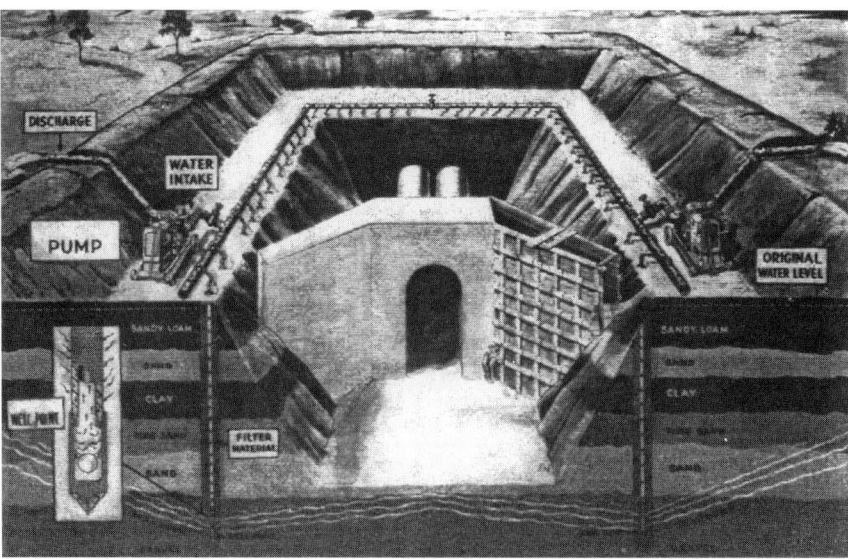

FIGURE D5-5 Typical wellpoint installation. (*John W. Stang Corp.*)

CONSTRUCTION DEWATERING D5-9

FIGURE D5-6 Wellpoint being self-jetted into the ground.

FIGURE D5-7 Jetting sand casing into the ground.

The wellpoint pumps are carefully set on firm soil, or on timbers if it is necessary to place them on soft soil. To effect a maximum drawdown, the vertical distance between the header and the pump suction intake should be kept at a minimum. The discharge line from the pump should extend a sufficient distance to minimize the reentry of water into the excavation.

All pipe connections on the suction side of the pumps must be airtight to maintain a maximum vacuum on the entire system. Couplings used on the header manifold must be provided with positive seal gaskets, and all threaded pipe connections on risers, swing joints, and wellpoints should be treated with pipe compound.

When pumping is started, it is frequently necessary to open the swing-joint valves gradually over a period of several hours to avoid excessively high initial entry velocities. Adjustment of these valves is also necessary during the later stages of drawdown to permit a uniform lowering of the water level and to prevent some of the wellpoints from going dry. The regulation of these valves is extremely important to maintain a high vacuum on the line and also to prevent clogging the wellpoint screens.

With a properly designed wellpoint system, the desired drawdown is usually obtained in several hours when dewatering pervious soils. In less pervious materials, the rate of drawdown is slower, frequently requiring a period of several days or longer. After the water level is lowered below the deepest final elevation, the excavation is completed. In some cases, the excavation is performed as the water level is being lowered, but at no time should it be carried to a depth below the receding water level. Disregarding this rule will result in a wet *silting* condition in the bottom of the excavation which is frequently difficult to dewater satisfactorily. In the case of a multistage wellpoint operation, the excavation is continued to the lower water level, at which point the next stage of wellpoints is installed. The wellpoint system is pumped continuously until excavation and construction including backfilling up to the original water level are completed, at which time the wellpoint system is removed.

On pipeline operations, sufficient wellpoint equipment is required to permit the predraining of the trench prior to excavation and also to allow for the reinstallation of a portion of the wellpoint equipment. Where only a short length of pipeline requires excavation below the groundwater level, it is generally more economical to install wellpoints along the entire length of the "wet" trench. Such an operation is quite efficient since the necessity for simultaneously reinstalling the wellpoint equipment along with the excavation and pipe-laying operation is eliminated.

Ejector Method

Ejectors (also referred to as *eductors*) can be used to lower the water table beyond the limits of a conventional wellpoint system. The ejectors are identical in principle to the so-called *jet pumps* used for domestic water supply where it is necessary to pump from depths beyond the suction limit. When water under pressure passes through the nozzle of the ejector at a high velocity, pressure is reduced below atmospheric. By connecting an ejector to a wellpoint screen, the *vacuum* is applied at this level rather than at the header manifold elevation as in a conventional wellpoint system.

An ejector system requires two manifolds: one, the high-pressure line which supplies the ejectors, and the other, the collection manifold which carries the water discharged by the ejectors. This includes the groundwater collected by the screens. Each ejector is connected to each of the manifolds by separate small-diameter pipes, and water is circulated in the system by a horizontal centrifugal pump. The water picked up by the system is discharged at an overflow tank located on the suction side of the pump.

The design of an ejector system must be given careful consideration. It is particularly important to select properly sized ejectors for maximum efficiency at the anticipated head and discharge conditions. Maximum overall efficiencies of ejector systems are low—generally not more than about 20 percent—and their use should not be considered where large rates of pumping are anticipated.

Other Methods

Less frequently used methods for controlling groundwater in connection with excavations will now be discussed briefly. For the most part, these methods are not, strictly speaking, dewatering ones, since water removal is not a prime requisite. Generally, they would be more accurately described as either (1) stabilization methods, such as electroosmosis; (2) cutoff methods, such as chemical grout injection; or (3) a combination of stabilization and cutoff, such as the freezing method.

Electroosmosis is applicable in extremely fine-grained, low-strength soils. The method is based on the principle that, if a pair of electrodes is driven into water-bearing ground and connected to a source of direct current, the current passing through the ground actuates a flow of water from anode to cathode. Simultaneously, there is a reduction of *pore-water pressure* which changes the stress in the water contained in the interstices of the soil from compression to tension. This phenomenon results in the stabilizing of the soil since the water, being in a state of tension, actually tends to pull the soil particles together.

On a typical electroosmotic installation, the anodes and cathodes are placed alternately around the perimeter so that a wall of stabilized soil is formed by the action of the current flow. Ordinary wellpoints are frequently used as cathodes since they permit the collection of the water flowing to the cathodes. The amount of water collected is quite small as compared with the conventional dewatering methods. The anodes may be pipe or reinforcing steel of suitable length.

The spacing between anodes and cathodes, as well as the capacity and number of the generators, can best be established by preliminary field testing to determine the conductivity of the soil. This conductivity varies over a wide range and appears to depend on water content, mineralogical composition of the soil, and the amount of chemicals in the water.

Particular emphasis should be placed on the fact that this method should be utilized only where excavation below the water table is required in *extremely fine-grained soils*. Soils that are coarser grained than fine silt can generally be more effectively dewatered by utilizing one of the methods of predraining discussed previously.

Cutoff methods should be considered in areas where pervious strata are underlain by a continuous, relatively impervious layer. Such methods are frequently used where permeabilities are too high to permit economical dewatering by pumping. The necessity for providing bottom pressure relief, as previously discussed, must be checked when cutoff methods are used.

Interlocking steel sheet piling is frequently used in cofferdam construction where driving conditions permit. A reasonably tight cutoff is provided by driving the sheet piling into an impervious stratum. Nominal pumping is generally required to remove seepage through the sheet-pile interlocks. This is a more serious problem when the sheet piling is unstressed, as in the case of a sheet-pile cutoff in an earth or rock cofferdam. Installation of the sheet piling through granular, pervious soils can frequently be performed effectively by using supplementary jetting with conventional dynamic pile-driving equipment or by using vibratory pile-driving equipment.

Cement grout cutoffs are used occasionally but are restricted to extremely pervious soils such as gravel. They are frequently required (too often overlooked, unfortunately) in gravel bedding under structures which extend below groundwater level where future extensions of the initial construction are contemplated.

The use of grouted cutoffs has significantly increased since the development of the relatively new chemical grouts. These grouts may be used in less pervious soils due to their relatively low viscosities. The newer chemical grouts, when used with wetting agents, have viscosities approximately equivalent to that of water. Only minor strength increase is generally obtained in the soil mass, but an impervious gel is formed which provides the cutoff.

Chemical grouting may be either *two-shot* or *one-shot*. In two-shot grouting, successive injections of two chemicals are made. An example is the Joosten process, which utilizes sodium silicate followed by calcium chloride. In the one-shot process, a single mixture of the main gel-producing chemical with a catalyst is injected. The amount of the catalyst regulates the polymerization time—the time required for the formation of the gel.

Chemical grouts are generally quite expensive and, to use them effectively, it is important that soil properties (particularly permeability) be carefully investigated. In addition, the grouting should be done by specialists experienced in this field. To illustrate this point, if the polymerization time is estimated on the basis of a permeability that is lower than the correct amount, the injection solution will disperse over too wide an area and the result will be an ineffective cutoff.

Slurry-trench cutoffs have been widely used in recent years. The method basically consists of providing internal support to an excavated trench by introducing a bentonite slurry, similar to the method of supporting a rotary-drilled hole with drilling mud. The cutoff may be solely for temporary construction purposes, in which case the trench may be backfilled with a mixture of bentonite and natural material. This mixture will have a much lower permeability than the natural material. If the cutoff is to serve as a cofferdam wall and ultimately become a part of the permanent structure, the trench is usually excavated by special drilling or clamshell equipment. Excavation is followed by placement of a structural or reinforced-concrete bulkhead.

Freezing methods are also used to stabilize soil, thereby eliminating the necessity for dewatering. This method was originally used in the mining industry and was introduced in construction starting about 1960. Generally, it consists of freezing an ice wall which serves as a cutoff to water and provides support to the walls of the excavation or shaft due to the increase in soil strength resulting from freezing. This method has been used most frequently in connection with the construction of shafts, many extending to considerable depths. Frozen shafts of as much as ±1500 ft have been sunk in Saskatchewan, Canada, for the mining of potash.

Freezing pipes are installed in holes drilled on spacings which usually range from about 3 to 6 ft. A refrigeration unit circulates low-temperature brine solution through the freezing pipe. The process may be considered as analogous to a dewatering problem to the extent that heat, rather than water, is pumped out of the ground. (The analogy between the principles of heat and water flow through soils is extremely helpful in making analyses of such problems.)

The freezing time required to obtain an ice wall of sufficient thickness to satisfy cutoff and structural considerations is usually in the range of from 2 to 4 months. This period may be shortened by reducing the spacing of freezing pipes. The design of the ice wall thickness is performed in a manner similar to the analysis of a comparable unreinforced-concrete structure. Ultimate compressive strengths for frozen soil range from about 500 lb/in^2 for clay to about 2000 lb/in^2 for sand. For a given soil, strength increases with an increase in water content and with a decrease in temperature.

PUMPS FOR DEWATERING

Pumps most frequently used for dewatering are of the centrifugal type. A centrifugal pump is basically an impeller or rotor with an intake at its center. Water entering the impeller is rotated and discharged by centrifugal force into the casing which surrounds the impeller. The head or pressure developed by the pump is a result of the velocity imparted to the water by the rotating impeller.

Centrifugal pumps may be single- or multistage, the stages representing the number of impellers in the pump. Multistage pumps are required where discharge pressures are high. Centrifugal pumps are also classified as horizontal or vertical, depending on the position of the pump shaft. Centrifugal pumps are characterized by their ability to operate over relatively wide ranges of pumping rate and pressure. These ranges may be further increased by varying the pump speed.

Horizontal Centrifugal Pumps

Many sizes and types of horizontal centrifugal pumps are available to the contractor. Some of these are the conventional self-priming *contractors' pumps*, wellpoint pumps, and "trash" pumps. They are usually available either engine- or electric-motor driven. Due to their relatively high efficiencies (±70 percent) they are particularly well suited for large-capacity pumping requirements at moderate heads.

Horizontal centrifugal pumps have a suction lift limitation which is characteristic of suction-type pumps and is due to the restriction of atmospheric pressure. Generally, the maximum practical suction limit for these pumps is about 20 ft at sea level. This limit can be exceeded, in some cases up to a lift of 25 ft, at the expense of reduced pumping capacity and accelerated pump wear due to cavitation.

Horizontal pumps are not capable of running dry for extended periods without mechanical damage. In addition, if a continuous supply of water is not provided, the repriming of the pump becomes a problem. For this reason, suction strainers should always be provided with foot valves. Wellpoint pumps are occasionally used for surface-pumping dewatering because the supplementary vacuum pumps with which they are equipped eliminate the problem of priming.

Trash pumps have special, open-type impellers, which are capable of handling considerable amounts of suspended solids without excessive wear. Because of this feature, the units are frequently used for the bypass pumping of sewer lines which are undergoing repair.

Vertical Pumps

The most frequently used type of vertical pump is the so-called *deep-well turbine*, which is a vertical centrifugal pump similar in performance characteristics to the horizontal centrifugal pump. Propeller and axial-flow vertical pumps are used occasionally for large-capacity pumping requirements at low heads.

A vertical pump installation consists of (1) a vertical electric motor, or gasoline or diesel engine with an *angle-gear drive;* (2) a discharge head which provides a mounting for the electric motor or angle-gear drive, and the connection for discharge piping; (3) water column, tubing, and shafting for connection between the pump bowls and discharge head; (4) pump impellers and bowl assemblies which are located at the bottom end of the column, tubing, and shafting. A strainer is usually provided at the bottom of the bowl assembly.

These features permit the use of vertical pumping equipment for a wide range of head requirements. Multiple impeller and bowl assemblies are used for high head requirements. The column, tubing, and shafting are usually manufactured in 10-ft lengths, which are combined to provide the necessary *pump setting*—the distance between discharge head and strainer.

Vertical pumps are ideally suited to pumping from deep wells and are also used for dewatering cofferdams. Due to their inability to run dry without excessive wear, vertical pumps are frequently installed with float or electrode switches which automatically control the pump motors with the rise and fall of the water at the pump intake.

Submersible Pumps

The term *submersible* is used to describe centrifugal pumps which are driven by electric motors close-coupled to the impellers. The motor and impeller assembly is contained in one housing, which normally operates below the water surface. Submersible pumps used for construction dewatering are turbine type or contractors' type.

Submersible turbine pumps are similar to the previously described deep-well turbine except for the drive mechanism and their smaller size. They are frequently used as multistage pumps. Most manufacturers build the units to a maximum-size electric motor of about 5 hp. Due to the close impeller tolerances, these pumps will not handle suspended solids without excessive wear.

Contractors' submersible pumps are available in a wide range of pumps sizes up to 8 in with electric motors of a maximum size of about 65 hp. Most contractors' submersible pumps can run dry for limited periods without mechanical damage and therefore require little attendance during their operation. Due to larger impeller clearances, they can handle dirty water, but large amounts of suspended particles will result in excessive maintenance. Contractors' submersibles are *not* dredge pumps.

Contractors' submersible pumps have certain disadvantages, among which are relatively low efficiencies (generally less than 60 percent), the need for engine-driven electrical generators on moving jobs such as pipeline construction, and their relatively high initial and maintenance costs.

Diaphragm Pumps

Diaphragm pumps have been for many years the workhorses of pipeline contractors. Their mobility, simple construction, and low maintenance are some of the reasons for their popularity. They are usually engine driven, eliminating the need for generators. Diaphragm pumps are capable of handling dirty water with relatively little maintenance.

Their use is restricted to small pumping rates, due to their relatively low efficiencies and small available sizes (up to about 4 in). Suction lift limits are similar to those of horizontal centrifugal pumps.

Air Lifts

Air-lift pumping requires the use of an air compressor rather than conventional pumping equipment. Although the method is relatively inefficient, it is often used because of its simplicity and its ability to handle large amounts of suspended materials. For the latter reason, an air lift is frequently used for well-developed pumping. It may also be applied for limited dredging, such as the excavation of material inside a small bridge-pier cofferdam in a river.

The method operates by injecting compressed air into the water inside the discharge pipe at a point below the water level. The unit weight of the created mixture of air and water is less than that of water alone and results in a flow of the air, soil, and water mixture to the top of the discharge pipe. This vertical discharge pipe, along with the smaller air line inside it, must be submerged for at least 40 percent of its length. Pipe diameters for discharge and air line, respectively, range from about 2 in and ½ in for 60 gal/min up to 8 in and 2½ in for 700 gal/min.

Pump Selection

An analysis of a dewatering problem will result in estimates of the required pumping rate and head, which are used for selecting an appropriate pumping plant.

The pumping conditions for which a given pump is applicable are graphically represented by its characteristic curves. These curves are compiled by the manufacturer on the basis of accurate tests and provide a picture of the relationships between capacity, head, break horsepower, and efficiency.

Figure D5-8 illustrates a typical set of characteristic curves for a centrifugal pump. The procedure for using these curves can best be illustrated by an example.

Example. Provide a horizontal centrifugal pump capable of handling 2000 gal/min if the pump has 15 ft of suction lift and the discharge point is located 55 ft above the pump. Water is discharged from the pump through a 10-in pipeline 300 ft long.

Item	Calculated head, ft
Suction lift, supply reservoir to inlet	15
Suction inlet to discharge outlet, vertical distance	1
Discharge outlet to discharge point, vertical distance	55
Friction loss, from tables, 3.59×3	11
Velocity head, $v^2/2g = 8.16^2/64.4$	1
Minor losses, entrance, exit, elbows, etc.	2
Total dynamic head	85

The pump selected for this operation should be capable of efficiently pumping 2000 gal/min at a total dynamic head of 85 ft. Figure D5-8 indicates that this 10-in pump fulfills these requirements very well since it will be operating in the range of maximum efficiency of about 72 percent. The brake-horsepower curve indicates that an electric motor or engine capable of providing 60 bhp would be required.

DESIGN OF DEWATERING SYSTEMS

The analysis of any dewatering problem requires a knowledge of (1) soil conditions, in particular, the permeability of the soil or the permeabilities of individual strata if the soil profile is variable; (2) the distance of open water, such as a river, lake, etc.; and (3) the depth to a continuous impervious stratum. On the basis of this informa-

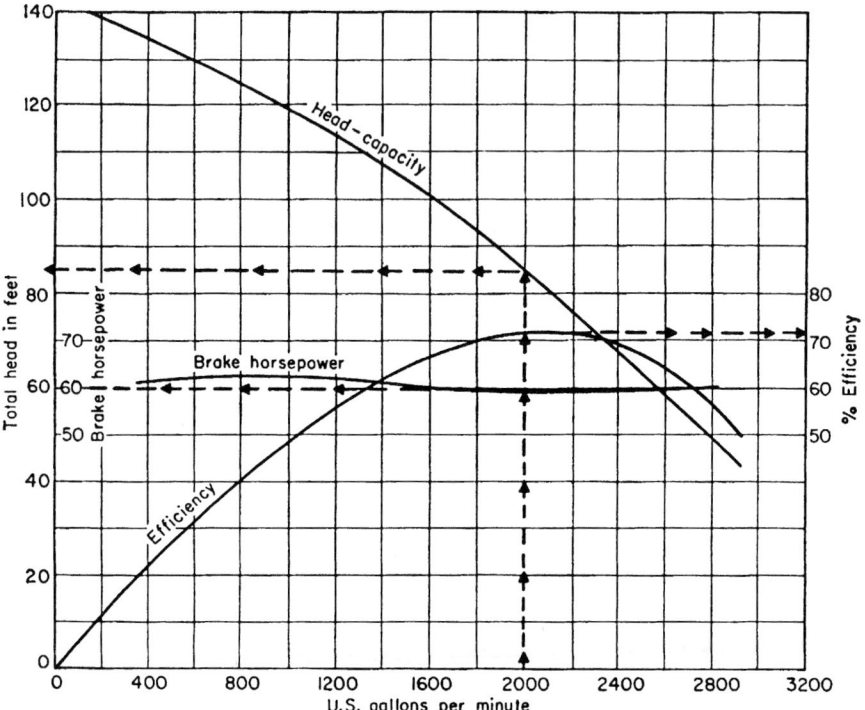

FIGURE D5-8 Centrifugal-pump performance curve. (*John W. Stang Corp.*)

tion, the quantity of water to be pumped is calculated for the amount of water lowering required. The dewatering system selected must provide a collection capacity equal to or greater than the calculated rate of inflow. In other words, any dewatering problem must be considered from the standpoints of yield from the ground and collection capacity.

The nomenclature used in this section is defined as follows, along with the units used:

Q = rate of flow, gal/min

Q_w = rate of flow from a single well, gal/min

q = rate of flow per ft of perimeter, gal/min

k = permeability; velocity of unit hydraulic gradient, ft/min

i = hydraulic gradient (dimensionless)

H = vertical distance from undisturbed groundwater level to an impervious stratum, ft

h = vertical distance from lowered water level to an impervious stratum, ft

h_0 = depth of immersion of well, ft

$H - h$ = drawdown, ft

m = aquifer thickness, artesian condition, ft

R = radius of influence of depressed water table, ft
r_1 = effective radius of a ring of wells or wellpoints, ft
r_w = effective radius of a well or wellpoint, ft
W = spacing between wells, ft
p = horizontal distance from center of ring of wells to shoreline of open water, ft

The flow of water through soils is governed by the basic relationship known as *Darcy's law,* which states that, for a given soil, the velocity of flow is directly proportional to the hydraulic gradient and the permeability of the soil:

$$\text{Velocity} = ki$$

The permeability factor k can be considered as a coefficient representing the *hydraulic conductivity* of a given soil. Having the units of velocity (ft/min), it can be considered as the rate of water movement through the soil at a hydraulic gradient of 1. The physical significance of Darcy's law can best be illustrated by Fig. D5-9, which shows a constant-head permeability test. The hydraulic gradient i is equal to the difference in piezometric head ΔH divided by the length of the sample.

Determination of Permeability

The design of a dewatering system requires a reasonably accurate evaluation of the permeability of the soils below the water table. Laboratory testing may be used, such as the constant-head test illustrated schematically in Fig. D5-9. However, due to the difficulty of obtaining relatively undisturbed samples of the granular pervious soils which are of principal concern in a dewatering analysis, permeability based on laboratory tests is frequently found to be misleading and significantly different from the in situ permeability.

Permeability is frequently estimated on the basis of grain size using a method originally proposed by Hazen.[1] Hazen found that, for uniformly graded filter sands, permeability (ft/min) is roughly equal to $2(D_{10})^2$, where D_{10} is in millimeters. Here D_{10} is the so-called 10 percent size, which is defined as that grain size representing the division between the finer fraction which accounts for 10 percent of the total

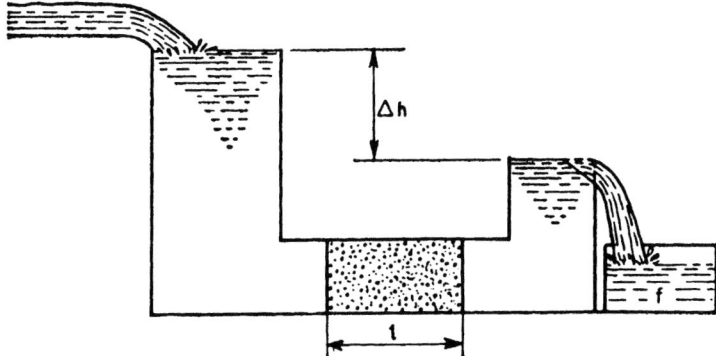

FIGURE D5-9 Constant-head permeability test.

sample weight and the coarser fraction which accounts for 90 percent of the total sample. It must be emphasized that any method of estimating permeability based on grain size alone should be used cautiously, particularly if applied to fine-grained sands and silts.

The approximate relationship between permeability and the D_{10} size indicates that the fine fraction of a soil has a significant effect on its permeability. As an illustration of this important point, a relatively clean concrete sand will have a permeability of around 0.04 ft/min. If it contains the maximum permissible amount of fines (10 percent passing a no. 100 sieve), its permeability can be reduced to the range of from 0.7×10^{-4} to 0.7×10^{-5} ft/min.[2]

Permeability may be more accurately evaluated by performing pumping tests, and these are strongly recommended for projects involving large dewatering costs. Water is pumped from a well, preferably one which fully penetrates the aquifer and is located at a central point in the area to be dewatered. The rate of pumping should be kept constant during the test period since this permits the application of nonequilibrium well formulas relating permeability, pumping rates, and drawdown rate. When the drawdown surrounding observation wells approaches a static condition, pumping can be stopped. Water elevations are measured in the observation wells during and following pumping. The pump test data should be evaluated on the basis of the nonequilibrium methods, both drawdown and *rebound,* and using the classical equilibrium well formulas discussed in this section. A review of the nonequilibrium methods is beyond the scope of this section.

Well Formulas

The basic well formula for a single gravity well, fully penetrating the pervious stratum, is

$$Q_w = \frac{7.5\pi k(H^2 - h^2)}{\ln(R/r_w)}$$

The following expressions for multiple gravity wells can be derived by following Muscat's procedure.[3,4]

For two wells a distance W apart:

$$Q_w = \frac{7.5\pi k(H^2 - h^2)}{\ln(R^2/r_w W)}$$

For three wells in a pattern of an equilateral triangle:

$$Q_w = \frac{7.5\pi k(H^2 - h^2)}{\ln(R^3/r_w W^2)}$$

For three wells equally spaced in a straight line:

$$Q_w = \frac{7.5\pi k(H^2 - h^2)\ln(W/r_w)}{2 \ln R/W \ln W/r_w + \ln W/2r_w \ln R/r_w}$$

$$Q_w = \frac{7.5\pi k(H^2 - h^2)\ln(W/2r_w)}{2\ln(R/W)\ln(W/r_w) + \ln(W/2r_w)\ln(R/r_w)}$$

For four wells in a square pattern:

$$Q_w = \frac{7.5\pi K(H^2 - h^2)}{\ln(R^4/\sqrt{2}r_w W^3)}$$

The basic well formula can be applied for obtaining the flow into a ring consisting of an infinite number of wells:

$$Q = \frac{7.5\pi k(H^2 - h^2)}{\ln(R/r_1)}$$

The simplifying assumption that a continuous ring of wells exists permits the determination of the flow per foot of perimeter for estimating the spacing of wellpoints. For normal wellpoint spacings, the assumption does not introduce any appreciable error.

The preceding well formulas are all based on the assumption that a circular boundary exists having a radius of influence of R. Equations can be derived for various other boundary conditions. For example, the case of a ring of an infinite number of wells bounded by a line source (shore line) is[5]

$$Q = \frac{7.5\pi k(H^2 - h^2)}{\ln(2p/r_1)}$$

A high degree of precision in the estimation of the radius of influence R (or the distance to a line source p), and the effective radius of the ring of wells or wellpoints r_1, is not required because the ratio of these dimensions appears as a logarithmic term in the well formulas. Therefore, the formulas can be applied with reasonable accuracy to typical excavations which are square or rectangular in plan rather than circular. Approximate r_1 values can be readily determined on the basis of the dimensions of the excavation.

For excavations that are extremely long in plan—for example, a pipeline trench with wellpoints on one or both sides of the trench—the selection of appropriate r_1 values is more difficult. The effective radius of a rectangular installation of wells or wellpoints having a length-to-width ratio greater than 10 is equal to approximately one-sixth of the length.

All the preceding well formulas are applicable to fully penetrating gravity wells. It should be noted, however, that for a given soil profile and a required drawdown $(H - h)$, the pumping rate Q from partially penetrating wells will be less than that for wells that are fully penetrating. Therefore, well formulas which are strictly applicable only to fully penetrating wells can be used as a conservative basis for the design of dewatering systems which involve partially penetrating wells. The necessity for applying a correction for partial penetration is difficult to justify in many cases because of uncertainties regarding permeability and the depth to imperviousness.

The well formulas considered here are for *gravity flow* conditions; i.e., the drawdown effected by pumping occurs in an *unconfined* aquifer—the condition most frequently encountered in practical dewatering problems. Pressure relief problems, on the other hand, require pumping from a *confined* aquifer, such as the underlying layer shown in Fig. D5-1. Such conditions require the application of *artesian* well formulas.

Gravity well formulas may easily be adjusted for application to artesian conditions by considering the aquifer thickness through which seepage passes, in the following manner. The term $(H^2 - h^2)$ in the gravity well formulas may be written as $(H - h)(H + h)$, which could be considered as the drawdown $(H - h)$ times *roughly* twice the average wetted thickness of the aquifer $2 \times (H+h)/2$. In a confined aquifer, the wetted thickness of the aquifer is equal to the aquifer thickness except in the rare instance when the water level is lowered below the top of the aquifer. (For this last case, the wetted thickness is reduced slightly.) The basic artesian well formula may be expressed as

$$Q_w = \frac{7.5\pi k(H-h)(2m)}{\ln(R/r_w)}$$

Again, $2m$ is twice the average (which in this case is a constant) wetted thickness of aquifer. From this it is apparent that the only difference between gravity and artesian well formulas is a factor which expresses (approximately) the variable of wetted aquifer thickness. All the previously discussed gravity well formulas may be expressed as corresponding artesian well formulas by replacing the term (H^2-h^2) by $(H-h)(2m)$.

Capacity of Wells

The rate of flow into a pumped well or wellpoint depends on the area and permeability of the ground immediately outside the well and on the hydraulic gradient causing the flow. Applying Darcy's law, the following expression can be written:

$$Q = 15 k i \pi r_w h_o$$

Sichardt found that the entry gradient into the well can be expressed in terms of the permeability of the soil.[6] His results are contained in the empirical formula

$$i = \frac{1}{1.07\sqrt{k}}$$

From this it follows that

$$Q \approx 14\sqrt{k} \pi r_w h^o$$

This equation permits the determination of the capacity of a well or wellpoint for any conditions. The effective radius of a well would be the radius of the perforated casing if not gravel packed and the radius of the filter envelope in the case of gravel-packed wells. For self-jetted wellpoints, the effective radius depends on the soil, since the cavity formed in the jetting process varies in size in different types of soil. In coarse-grained soils, a larger cavity is formed than is the case in fine-grained soils. During the jetting process, the coarser particles settle around the wellpoint screen, filling this cavity and forming a natural filter having a radius ranging up to approximately 6 in. Where wellpoints are used in fine-grained materials, it is generally advisable to use the sand casing discussed under Wellpoint Method. In this case, the effective radius of the wellpoint would be the radius of the sand casing. In estimating the capacity of a given well or wellpoint, it is important to allow for the reduced inflow area which results from the lowering of the water table; in terms of the foregoing equation, this is a reduction in h_o.

Gravel or sand filters are used on wells or wellpoints for the purpose of preventing an excessive amount of fines from entering the screens or perforated casing, increasing the collection capacity, and providing vertical drainage in stratified soils.

The gradation of the filter with regard to the natural soil is generally made on the basis of the Terzaghi criterion,[7] which is expressed in the following relationship:

$$\frac{D_{15} \text{ (of filter)}}{D_{85} \text{ (of natural soils)}} < 4 \text{ to } 5 < \frac{D_{15} \text{ (of filter)}}{D_{15} \text{ (of natural soil)}}$$

The subscripts designate percent, as in the case of the previously defined D_{10} size. The slot width or hole diameter used in the well casing or screen should be equal to or smaller than the D_{70} size of the adjacent material.

COST OF DEWATERING OPERATIONS

The following items must be considered in estimating the costs involved in any dewatering operation:

1. *Cost or rental of pumping equipment, jetting equipment, pipe, valves, fittings, etc.* Except on projects of unusually long duration, it is generally the contractor's advantage to rent such equipment. The manufacturer generally provides free estimates covering equipment rental along with other cost estimates covering the additional items entering into the total cost of dewatering.

2. *Transportation costs.* This item includes freight charges plus an allowance for unloading and loading the equipment at the site.

3. *Installation and removal costs.* In the case of the well method, these include the cost of drilling the well, gravel packing and its installation, perforated casing and its installation, setting of the pump, and installation of all necessary pipe. In the case of a wellpoint installation, these costs include laying the header and discharge lines, installation of the pumps, and jetting the wellpoints. The costs of the jetting operation will generally range from ¼ to 2 worker-hours per wellpoint, depending on the nature of the soil and whether the self-jetting or sand-casing method is utilized. A light crane is generally required during the jetting operation. An allowance must be made for miscellaneous materials, such as filter sand and fuel to operate the jet pump.

4. *Operational costs.* (a) Fuel, grease, and oil. The cost of fuel or power can be estimated on the basis of the horsepower rating of the power unit and/or the performance rating provided by the manufacturer. On continuously operating pumping units with engine drives, oil changes are generally required at least once every 48 h. (b) Pump operators are generally required around the clock on continuous pumping operations.

5. *Miscellaneous.* Allowance must be made for shortages, damages, maintenance, etc. These items are extremely indeterminate but can generally be satisfactorily estimated on the basis of the nature of the project and its anticipated duration.

REFERENCES

1. Allen Hazen, *Some Physical Properties of Sands and Gravels,* Massachusetts State Board of Health, 24th Annual Report for 1892, 1893.
2. H. R. Cedergren, "Seepage Requirements of Filters and Pervious Bases," *Trans. Am. Soc. Civil Engrs.,* vol. 127, part I, pp. 1090–1113.

3. M. Muskat, *The Flow of Homogeneous Fluids through Porous Media,* chap. 9, J. W. Edwards, Publisher, Inc., Ann Arbor, Mich., 1946.
4. Harold E. Babbitt and David H. Caldwell, "The Free Surface around, and Interference between, Gravity Wells," *Univ. Illinois Bull.,* vol. 45, no. 30, Jan. 7, 1948.
5. Stuart B. Avery, Jr., "Analysis of Ground Water Lowering Adjacent to Open Water," *Proc. Am. Soc. Civil Engrs.,* vol. 77, December 1951.
6. W. Sichardt, Springer-Verlag, Berlin, 1928.
7. Karl Terzaghi and Ralph Peck: *Soil Mechanics in Engineering Practice,* 1st ed., chap. 2, John Wiley & Sons, Inc., New York, 1948.

P·A·R·T E

HEAVY CONSTRUCTION PROJECTS

CHAPTER 1
AIRPORTS

SECTION 1

INTRODUCTION

P. Clay Baldwin
*Senior Vice President, O'Brien-Kreitzberg, Inc.,
New York, New York*

Jerome Gold
*Vice President, O'Brien-Kreitzberg, Inc.,
New York, New York*

The purpose of the air transportation system is the safe efficient movement of people and goods by air. An important element in this process is the airport. Airports can be categorized as either commercial service or general aviation airports.

In 1993, North American traffic volume exceeded 900,000,000 passengers and 17,000,000 metric tons of cargo. Worldwide volume exceeded 1,800,000,000 passengers and 35,000,000 metric tons of cargo. Preliminary projections by the International Air Transport Association indicate the worldwide growth pattern to continue for the foreseeable future, with an initial forecast for an average 9.0 percent annual rise in freight through 1998, accompanied by a slightly lower annual increase in international passenger traffic of 6.6 percent.

Seven of the ten busiest airports in the world are located in the United States.

HISTORY

To promote development of a system of airports to meet the nation's needs, the Federal Airport Act of 1946 was enacted. This act established the requirements for the formulation and annual revision of the five-year National Airport Plan. The act of 1946 also created the Federal-Aid Airport Program (FAAP) to bring about the establishment of an adequate national system of public airports. The act drew its funding from the general fund of the treasury.

In 1970, a more comprehensive program was established with the passage of the Airport and Airway Development Act of 1970. This act provided grants for both airport planning and development programs. These programs were funded from a newly established airport and airway trust fund, into which were deposited revenues from several aviation user taxes on such items as airlines fares, air freight, and aviation fuel. This program expired on September 30, 1981.

The current grant program, known as the Airport Improvement Program (AIP), was established by the Airport and Airway Improvement Act of 1982. The AIP has been revised and reenacted every year since 1982.

Two commercial developments have significantly affected airport development: (1) the introduction of passenger jet aircraft in 1958 and (2) the deregulation of airlines in 1978. The former had a major impact on existing facility requirements with regard to airside construction in particular, and runway, taxiway, and apron construction. The latter fostered growth in the aviation industry by opening up air routes to all interested parties. Deregulation resulted in realignment of the point-to-point travel philosophy to that of a hub and spoke. Further, competitive factors resulted in the creation of new airlines and the demise of others.

Passenger jet aircraft significantly increased the payload capacities. For example, the Lockheed Super Constellation had a maximum take-off weight of 160,000 lb while a Boeing 747-400 configured for carrying freight has a take-off weight of 873,000 lb. In addition to payload capacity and take-off weights, airfield pavement design and construction must take into account the load transfers through the landing gears. Presently, the new Boeing 777 aircraft is the critical design aircraft since it has approximately 65 percent of the 747-400 take-off weight with one-half the landing gear.

Other aircraft developments that can and will have an effect on airport construction include the proposed *new large aircraft* (NLA) which will be a double-decked aircraft capable of carrying 600+ passengers, and aircraft support systems such as preconditioned air, fixed ground power, etc.

The NLA, as envisioned, would be approximately 13 m longer than the 747-400, and its wingspan 15 m greater. The NLA would require longer runways for take-off, as well as greater separation of runways and taxiways. In addition, runways and taxiways may have to be widened to accommodate the NLA. Airport terminals would have to be reconfigured to enable two-level hold rooms and loading bridges to be developed.

CASE STUDIES INTRODUCTION

The case study of heavy construction projects for John F. Kennedy International Airport was selected to provide a brief overview of the complexities of performing major construction projects within the environment of a very busy operating airport. The case studies were composed of several elements of the JFK Redevelopment Program which presented challenges of maintaining daily passenger and flight operations during construction.

SECTION 2

CASE STUDY: JOHN F. KENNEDY INTERNATIONAL AIRPORT, JFK REDEVELOPMENT PROGRAM

Richard Smyth
Vice President, O'Brien-Kreitzberg, Inc.,
New York, New York

Joseph Dixon
Project Manager, O'Brien-Kreitzberg, Inc.,
New York, New York

PROGRAM GOAL/DEFINITION

The program was intended to focus primarily on the airport's passenger-related facilities in the *central terminal area* (CTA), where the passengers enter the airport, proceed in and around the core, and to and from the terminals. The goal was to reduce traffic congestion, provide faster and more convenient processing of passengers, and facilitate the transfer of passengers among the CTA terminals.

AIRPORT HISTORY

The airport opened in 1948 as the New York International Airport, also known as Idlewild Airport, to serve the New York and New Jersey metropolitan area. The airport was planned as a major gateway for international travel in the 1950s. Access to the terminal areas was provided by a grand circular boulevard entering from the Van Wyck Expressway. The first major terminal was the International Arrivals and Wing Buildings (IAB), built in 1957. Other major airline terminals were built shortly thereafter and into the early 1960s. The ring of terminals was known as "Terminal City" (Fig. E1-1). The facilities were designed for a capacity of 15 million annual air passengers. The airport was essentially completed as planned by the early 1960s in a configuration that remained without significant change into the 1990s (Fig. E1-2). The airport was rededicated on December 24, 1963, as the John F. Kennedy International Airport.

(a)

(b)

FIGURE E1-1 Aerial views of JFK—1950s.

FIGURE E1-2 Aerial view of JFK—1990s.

The roadway system was designed to provide access to the individual terminals by routing vehicles into the central terminal area, then past other terminals, either inbound or outbound. Individual recirculation roads in front of each terminal provided a level of convenience to allow for vehicles to discharge departing passengers prior to parking or to allow for vehicles to return from parking to the terminal to pick up arriving passengers.

Airport traffic continued to develop throughout the 1960s as air travel expanded, jets were introduced, and passenger volumes continued to grow. By 1962, however, passenger volume had reached 100 percent of design capacity. In the late 1960s, the introduction of wide-bodied aircraft, with significant increases in passenger capacity, meant greater demands on the infrastructure without an increase in the number of aircraft movements.

A number of different planning studies were initiated through the 1970s and early 1980s to address the inadequacies of the CTA facilities. Minor changes were made to the roads to improve local circulation and flow patterns, and other infrastructure systems were improved, but did not address a comprehensive approach to rectify shortcomings of the CTA facilities.

The studies showed that despite the interim improvements provided, the airport was functioning poorly. This was partly due to the flight characteristics of this international airport, which exhibits a significant variation or *peak* in flight schedules and, consequently, passenger processing volumes at certain times of the day. Many European countries imposed strict windows during which flights take off or land, thereby affecting the arrival and departure of flights from JFK. Additionally, there is peaking of travel at different times of the year. Volumes increase substantially in the summer, especially in August, and also during the holiday seasons.

By the 1980s, JFK terminal facilities and roadways had become overwhelmed by the continued growth of wide-bodied jet aircraft traffic, which more than doubled the peak passenger volume for which the airport had been initially designed. Further, many of the airport infrastructure systems were approaching the end of their effective lifespans. On the airside of the CTA, demand for 747 aircraft contact gates exceeded the supply, particularly at the TWA, IAB, and Pan Am (now Delta) terminals.

Studies showed that the volume of passenger traffic was exceeding 30 million passengers annually in the late 1980s. At double the design capacity, the facilities were very crowded, as was quite evident to anyone who used the airport. There were severe restrictions to passenger movement to and from the CTA terminals on the approach roads and also for passenger movement among the terminals for those *interlining* or transferring to different airlines.

To address this situation, the Port Authority commissioned the development of a comprehensive master plan in 1984 that would establish action programs addressing every major landside component of the CTA and some airside components, thus providing JFK with the capacity to handle a projected 45 million annual air passengers. With the existing facilities already strained, and the forecast that the airport would see an additional 50 percent growth, it became apparent that immediate action was needed to keep the airport functioning efficiently.

The redevelopment of JFK was conceived and plans initiated to implement needed changes. The program was intended to symbolize the approaching millennium and a new dawn for air travel convenience at this facility in the 21st century. The inspiration was to transform the airport facilities into a modern, international gateway that would return the airport to a preeminent status in aviation transportation well into the next century.

The primary goal of this plan was to improve the passenger's experience. This could be achieved by relieving congestion on the approach and frontage roads to facilitate vehicle flow into, out of, and around the CTA. Another major component was the processing of passengers into and around the terminals, and improving the efficiency of aircraft movements on the airside. The JFK Redevelopment Program was intended to provide a major transformation of the way passengers would perceive their arrival at the airport and convenience of service. The master plan for revitalization of the CTA was developed into a specific plan of action.

JFK REDEVELOPMENT PROGRAM SCOPE

The scope of the program included several major components to facilitate passenger movement in the CTA and ancillary projects to expand and refurbish the infrastructure. The roadway system was to be reconfigured from the loop roadway around the CTA to a quadrant system. The quadrant system would split the CTA into four major areas, each containing one or more unit terminals. Passenger vehicles could proceed directly to the portion of the CTA where their individual terminal was located. This would alleviate the congestion and backups caused by many vehicles traveling past terminals that were not their destination, either entering or leaving the CTA. Concurrently, the traffic volumes necessitated that the terminal frontage roads also be expanded. A people-mover system would transport people to and from their unit terminals in automated monorail-type vehicles. This system would move people between terminals in the CTA and would also connect with the proposed airport access system connecting JFK to other transportation centers, including La Guardia Airport, the LIRR station in Jamaica, and Manhattan. Other components to be built in conjunction with the reconfiguration of the CTA included a structured parking garage having five levels, built to support a multilevel hotel on top of the garage; a new ATCT to improve visibility over existing and planned buildings; and new and relocated utilities to support and facilitate construction of CTA elements.

Implementation of design and construction of the comprehensive initial plan commenced in 1988. Through the early 1990s, major components of the plan were advanced into construction: a new airside air traffic control tower, a completely revised quadrant roadway system and other major improvements to the CTA infrastructure including new utility systems, and one of the two planned CTA structured parking garages.

During the initial phase of implementation, the effects of changes in the airline industry and a concurrent weakening of the national economy resulted in adjustments in the priority and phasing of select components of the 1986 Master Plan.

The need for a new development focus for the central terminal area of the airport was regarded to be of critical importance to effectively defining a new physical and functional image for JFK. In response to this need, a concurrent reassessment of the desirable program of uses for the CTA led to further analysis and evaluation of alternative, cost-effective approaches to significant development of this highly visible area. Development components regarded as highly desirable for the CTA included a new hotel, a phased construction of the second parking garage as demand requires, the ultimate development of the people mover which would be similar in design and function to the initially proposed system, and a redeveloped International Arrivals Building. Planning for the PDS was resumed and a project for the Redevelopment of the IAB, including related amenities and a proposed hotel site, was initiated as a new component in 1993.

AIRPORT CONSTRUCTION

Airport construction can be separated into four generic categories which define all construction on airports:

Landside	Terminals—access/egress, passenger-related facilities
	Cargo areas—freight receiving, shipping, storage
Airside	AOA—runways, taxiways, passenger-related facilities
	Cargo areas—hangars, storage, loading/unloading

Additionally, the work can fall into either *building* or *civil construction*. Each of these areas has specific requirements and restrictions which must be addressed during design and construction. The operating environment of the airport demands strict compliance with regulatory and code requirements for construction and operations.

OPERATING ENVIRONMENT

Many significant challenges were identified in the planning stages for the redevelopment of the airport. Among those requiring attention were the general operating environment, which included passenger considerations, restricted construction areas, compliance with a wide range of regulations, and special requirements of construction on Aeronautical Operations Areas (AOA).

Additional challenges included the local environmental conditions such as the construction of the airport on a hydraulic sand landfill over a meadow mat of salt-marsh grasses. The shallow water table and presence of a thick (6 to 8 ft), dense organic layer of decaying marsh in the tidal estuary of Jamaica Bay make construction of foundations and other inground structures an unusual and difficult working environment.

A very common restraint encountered during heavy construction projects is the need to perform effectively and efficiently in a controlled operating environment. The need to maintain ongoing airport operations during construction is complicated by the multitude of users and/or regulations which could impact the prosecution of the construction work. The list of factors affecting the construction process is lengthy.

PASSENGER CONSIDERATIONS

JFK airport operates 24 hours a day, 365 days a year. As mentioned earlier, there are significant peaks which result in large volumes of passengers and aircraft movements. The first and smaller peak is in the morning, from about 0800 to 1100 hours. The larger peak starts in the afternoon and continues well into the evening (1400 to 2200 hours). Additionally, the peaks vary considerably throughout the year, the summer being busier, and the winter less so. These characteristics are noted because contractors are sometimes precluded from performing work at certain times of the day or year to prevent disruption of operations or inconvenience to passengers. These no-work periods can last for hours, days, or weeks, especially around holiday periods when passenger traffic increases substantially.

Staging of work is often a critical aspect of construction. Minimizing passenger inconvenience is an important consideration for work in the CTA. Both the Port Authority and the airlines insist that any work not impede or inconvenience passengers. Roadway work necessitates a large amount of traffic control. Temporary roads and routing must be well planned and constructed prior to taking other areas out of service. Signage must be erected and clearly delineate the traffic flows and access/egress routes. One method of communicating upcoming changes to the traffic patterns is through *construction bulletins,* which describe the changes, frequently with graphics, to show the areas affected (Fig. E1-3). These are disseminated to a variety of groups and users on and off the airport who may have business or need access in the area of construction. These bulletins can describe the traffic change or construction work, identify how the area is affected, and whether the work is temporary or permanent.

RESTRICTED CONSTRUCTION

The area available for construction is severely limited for certain work at the airport. The CTA has been extensively developed and utilization approaches capacity at certain times of the day and year. Large parking areas afford the opportunity to establish construction field office sites in areas remote from the greatest activity;

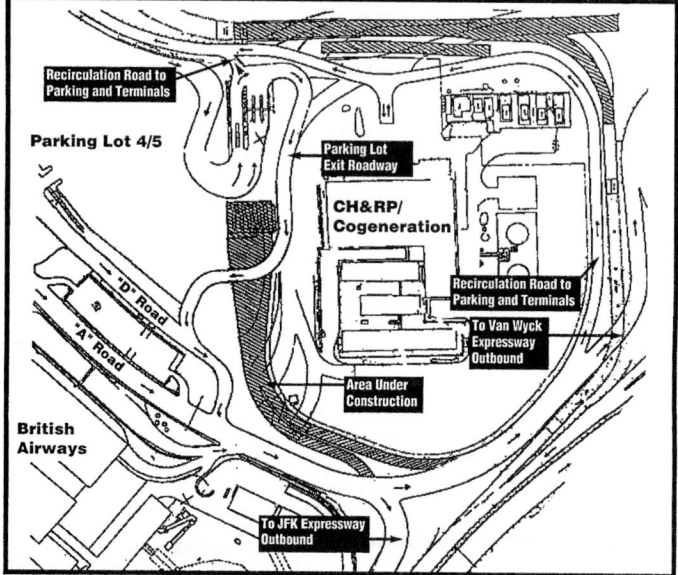

FIGURE E1-3 Construction bulletin.

however, this can impact the availability of space for passenger vehicle parking at peak times. Construction close to the terminals has a greater opportunity to disrupt passengers and their processing. As would be expected, work sites are minimized to the physical requirements for construction. Concurrently, laydown areas in the CTA are kept to the minimum required and returned to their original use as soon as possible. Contractors are frequently offered the use of laydown areas outside the CTA, but still on the airport. The small work sites and remote laydown areas sometimes result in double-handling of materials.

SPECIAL REQUIREMENTS

The Port Authority of New York and New Jersey (PANYNJ) has prepared Division 1 Specifications—General Provisions—to address requirements for working in an airport environment. These specifications are tailored to the individual needs of the work to be performed under contracts issued by the Port Authority.

A sampling of selected passages that might be included follows:

- *Area available for contractor's use*—"Subject to the conditions elsewhere stated herein, those areas to be occupied by the permanent construction will be made

available to the Contractor upon the commencement of his first operations at the construction site, together with an area shown cross-hatched on Contract Drawing No. G-5 and designated 'Area Available for Contractor's Use.' "
- *Daily cleanup of areas*—"The Contractor shall daily clean up the areas made available to him so that they are free at all times of refuse, rubbish, scrap material or debris."
- *Operations of others*—"During the time that the Contractor is performing the contract, other persons will be engaged in other operations on or about the construction site including general airport operations and maintenance, all of which must remain uninterrupted."
- *Definitions*—" 'Air Operations Area (AOA)' means that portion of the airport designated and used for landing, takeoff, parking or surface maneuvering of aircraft. 'Night' means the time between the end of evening civil twilight, and the beginning of morning civil twilight, as published in the American Air Almanac, converted to local time."

AOA RESTRICTIONS

The AOA is a tightly controlled environment necessitated by the special considerations of aircraft operations. Safety and need to maintain continuous operations are the primary considerations for any activity on the AOA. Security is a high priority in support of these considerations. The Federal Aviation Administration (FAA) has promulgated regulations which are enforced by the Port Authority (or other authority at different airports) to control access to the AOA. For longer-term access, personnel must be security-badged and vehicles must have special license plates to be allowed onto the AOA. Special agents from the Operations Department are assigned to escort vehicles (not preauthorized) and personnel onto the AOA for individual, short-term access.

Airport operations mandates that safety requirements be followed explicitly. Aircraft have the right of way and proceed on taxiways which do not correspond to the vehicle roadways on the AOA. Special training (provided by the Authority) is required to obtain an AOA driver license. Further complicating the AOA are a multitude of service vehicles including baggage tugs with carts, auxiliary power units, tanker trucks, catering trucks, etc. Jet blasts can overturn vehicles, so care must be taken when driving airside. Personnel and loose materials can also be blown by aircraft exhaust, causing injury or damage to other aircraft or other structures. Therefore, all materials must be securely protected from inadvertent dislodging, both during and after work hours. If materials come loose, the contractor must secure them immediately or the Authority will do so and backcharge the contractor. Due to the large open area, it is not uncommon for very strong winds to blow at the airport, sometimes in excess of any expected by general weather forecasts.

Aircraft require large clearances for safe movement. Construction equipment must be placed carefully so as not to impede or damage aircraft. Their flight operations also necessitate that very strict height restrictions are placed on any equipment on the AOA. The use of cranes must be carefully coordinated with the resident engineer, operations personnel, and the FAA. Specific clearances as defined by slope diagrams are provided in the contract documents to control the placement of equipment (Fig. E1-4).

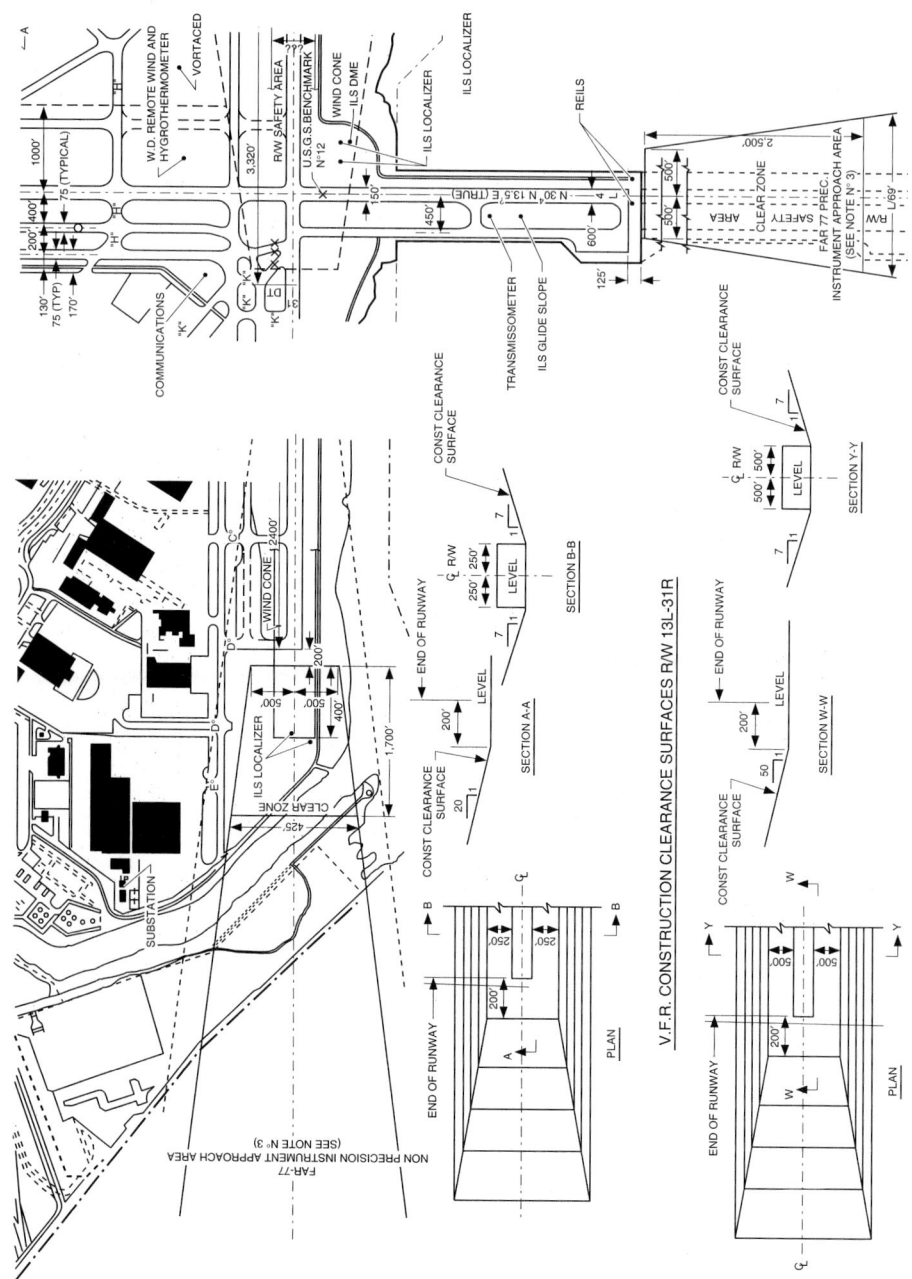

FIGURE E1-4 AOA construction clearance diagram.

ADMINISTRATIVE PROGRAMS

The Port Authority administers two programs which pertain to construction work at JFK. All construction work contracted by the Port Authority is covered by Port Authority–provided insurance. As the airport operator, the Port Authority is also concerned that all work by the tenants complies with appropriate codes and standards. Additional information is provided as follows.

Insurance

Insurance for construction work under Port Authority contract is provided under an *owner-controlled insurance program* (OCIP). This places the cost of insurance with the Port Authority at more favorable and consistent rates than may be possible through individual contractors. As such, the costs are not incurred by the contractor nor included in the contractor's cost. The policy provides for certain limits on primary public liability, excess public liability, worker's compensation and employer's liability, and builder's risk insurance. However, this does not limit the contractor's obligations or relieve the contractor from those obligations.

Tenant Construction or Alterations

The Port Authority provides a *Tenant Construction Review Manual* for all tenant work. It presents the technical criteria to be considered by tenants at the Port Authority facilities in connection with construction work. The manual covers the scope-of-review of design documents submitted by the tenants to the Port Authority in connection with proposed construction or alterations. The review does not address aesthetic or functional aspects of the design. The documents are reviewed solely for compliance with applicable codes and Port Authority engineering standards.

The tenant is required to ensure that all construction or alteration submissions conform to the criteria included in the manual. The responsibility for engineering design remains with the tenant's engineer or architect.

BUILDING CONSTRUCTION

Air Traffic Control Tower Project Description

The Air Traffic Control Tower (ATCT) is a completely new 321-ft-tall concrete reinforced tower, the tallest tower operating in North America in 1994. It allows outstanding sightlines for controllers, contains state-of-the-art technology, and permits significantly improved efficiency in handling the growing aircraft operations at JFK with continued safety.

The responsibility for controlling aircraft flying in the immediate vicinity of the airport (within 5 mi) and on the taxiways resides with the FAA. The Port Authority controls aircraft on the IAB ramp for 50 different international airlines. Both these functions are performed from the ATCT. As the airport operator, the Port Authority took the lead in replacing the 30+-year-old control tower as part of the JFK Redevelopment Program. Primary considerations for the design included height, visibility, and stability.

The 321-ft height is measured to the top of the radar equipment; the tower shaft is 32 ft 8 in square. The wall thickness at the base is 20 in and at the top, 10 in, with the change in thickness accommodated internally at several levels. Operational space for IAB ramp control is located within a structural steel cantilevered truss at midlevel (113 ft). The FAA occupies the cab on top of the tower and has mechanical and electronic equipment and office space within the overhanging, balanced truss systems at the upper levels.

Construction Logistics. The location of the new tower was established very early in the program with the concurrence of the FAA. The selected location was just south of the IAB, in the "U" of the International Arrivals Building. This location was relatively close to the existing tower in order to maintain good lines of sight to the entire AOA. A study was conducted to determine the visibility of runways and all taxiways from the proposed location, and they concluded that the new tower location and height would improve the visibility of these areas (Fig. E1-5). One factor which complicated the proposed site was the blockage of view from the existing tower to runway intersection 4L/22R 13R/31L in the southeast portion of the airfield during construction (Fig. E1-6). This was resolved by placement of a CCTV camera on the new tower shaft as the new tower height rose to block the view. Air traffic controllers could see the intersection on a monitor and had the ability to control the view with pan and zoom features on the camera. This system remained in operation until the new tower was commissioned.

The construction site was airside and required extensive coordination with operations personnel to get construction workers, materials, and equipment to the work site. Contractor access to the site was limited to one security gate with a specific route identified for all vehicles. Escorts were arranged on an as-needed basis, often several times daily to move equipment and materials. Personnel were identified by individualized badges for security purposes and display was required to comply with FAA regulations. The site was enclosed within a plywood fence on all sides for security and protection.

The location of the new tower adjacent to the IAB necessitated that two aircraft gates be taken out of service permanently in order to permit construction and operation. The proximity to the existing building created a concern for settlement and vibration from the foundation pile-driving operation. This was monitored closely throughout the driving portion of the contract (Fig. E1-7).

Wind and weather were major factors in the process of the work. Proper precautions were mandatory and strictly enforced. Work was suspended in cases of high wind and the stability of the crane was affected by high winds. The weather concerns required that a close watch be kept on the weather several days in advance. Operations were affected if the winds exceeded 25 mi/h and the crane's mast had to be lowered if winds exceeded 75 mi/h. Although the boom could be raised relatively quickly, the size of the mast necessitated a long lead time to raise and lower. During construction of the highest levels, a hurricane threatened the East Coast and the crane was lowered as a precautionary measure.

Access to the work on the tower shaft was accomplished by the erection of an exterior hoist. This was erected and secured to the shaft as the shaft went up and permitted the construction workers to get to several different working levels. The hoist remained in service until the internal electric elevators were operational. The elevators were used for access and material transfer before they were finished.

Foundation. The foundation consisted of a monotube steel pile system with a reinforced-concrete pile cap. Prior to excavating and driving piles, a portion of the

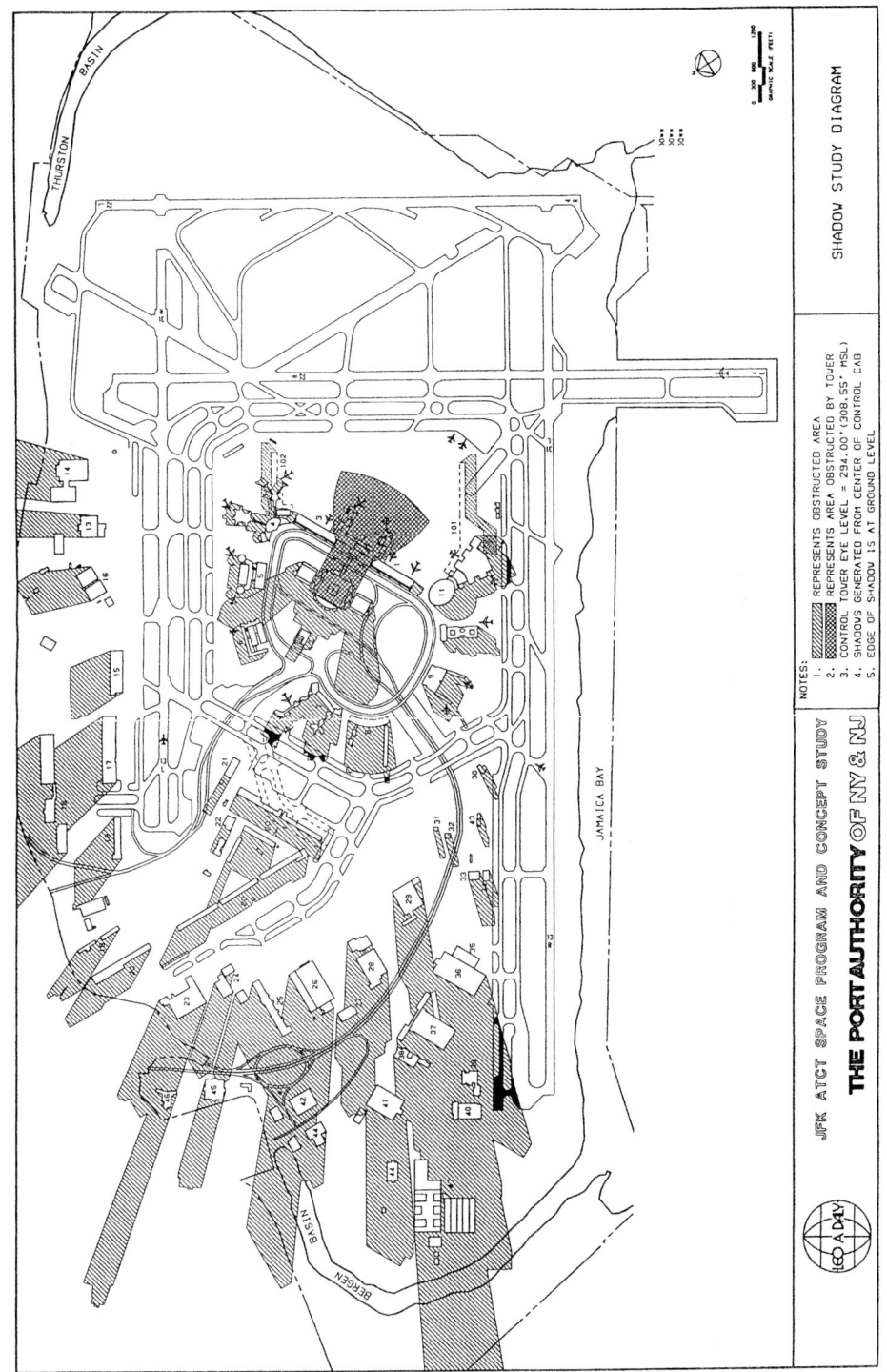

FIGURE E1-5 ATCT shadow study diagram.

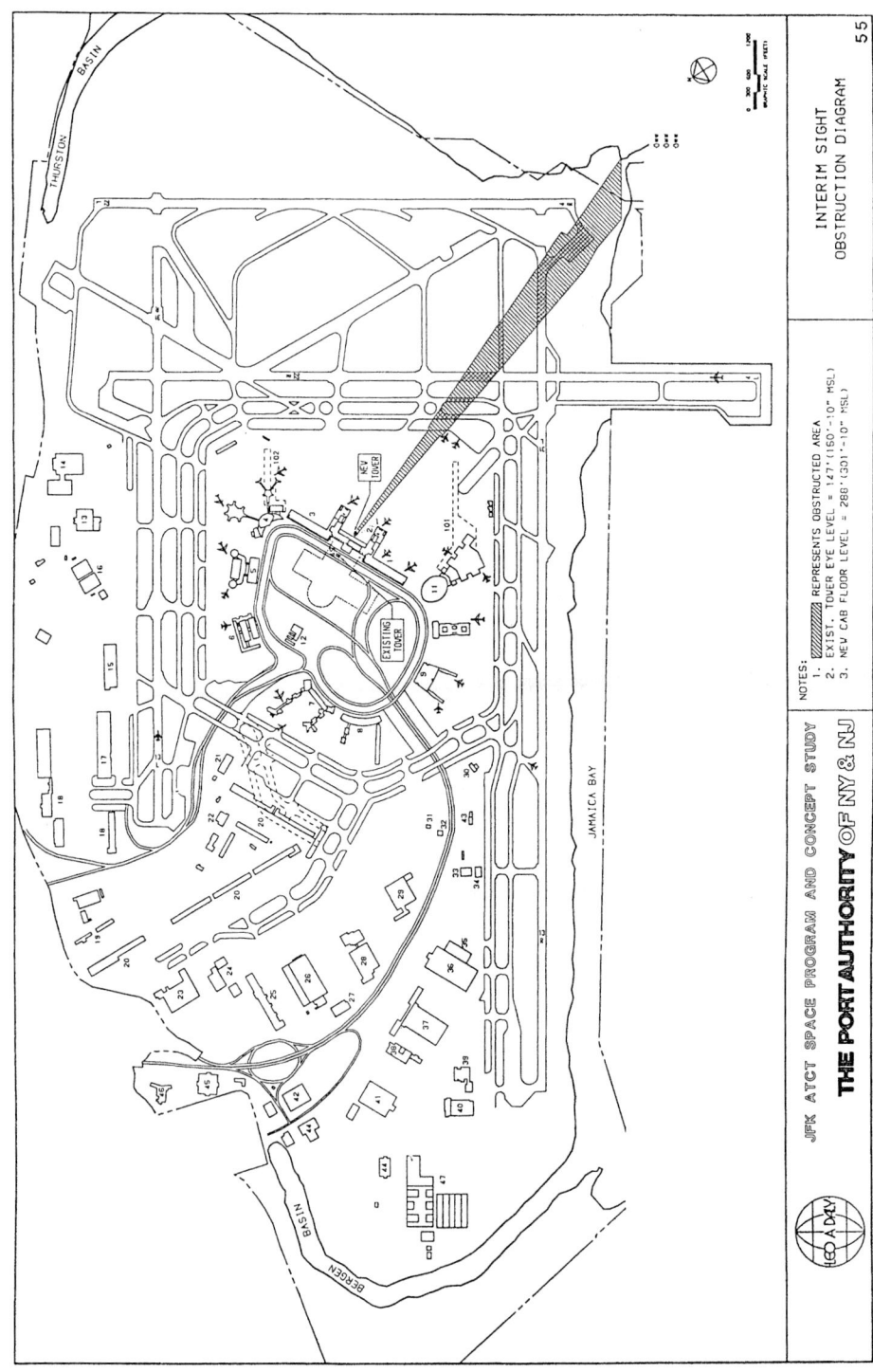

FIGURE E1-6 ATCT interim sight obstruction diagram.

FIGURE E1-7 ATCT pile driving, adjacent to building.

underground hydrant fueling system had to be decommissioned and capped. Steel sheet piling was driven around the perimeter of the foundation area with the intention of leaving it in place, 2 ft below grade with clearances of 6 in around utility penetrations. The area was excavated approximately 15 ft before commencing driving. The 140 piles supporting the tower each have a capacity of 100 tons (Fig. E1-8).

The contract requirements called for either steel or monotube piles to be driven by 32,500 ft-lb of force. Steel piles required 15 blows per inch and a tip elevation of between −45 and −70 ft. Monotube piles required 10 blows per inch at a tip elevation of −30 ft. The contractor elected to use monotube piles. A test load of 250 tons was applied to the piles. A further test was required to demonstrate at least 90 tons uplift, outside the foundation area. The pile cap is reinforced concrete, having a thickness of 7 ft 3 in and 46 ft^2. It was placed as a continuous pour, no joints, with the top 6 ft below grade. Water and electric utilities were installed under the floor slab. The electric supply at 5 kV was brought into the main transformers from two directions for redundancy. An extensive electrical grounding counterpoise system was installed in the ground as part of the foundation contract.

Due to the depths required for the foundation, localized dewatering was required for the pile cap and grade beams. The zone of influence on water levels extended beyond the immediate area so the dewatering was activated gradually over several days to draw down the water levels in the immediate vicinity. Following completion

FIGURE E1-8a ATCT foundation.

of the deep work, the dewatering was deactivated over a similar time frame to allow gradual recharging of the groundwater.

Superstructure. The superstructure is constructed of reinforced concrete with a uniform white cement finish to give the appearance of precast panels, which is the FAA standard. This required that all control joints be located within the reveals of the shaft design. In order to confirm proper construction techniques and ensure appearance, a mock-up of the wall was constructed off-site in the contractor's laydown area. This allowed the designers to confirm the concrete mix, the quality of the finish and techniques utilized to produce the details and reveals called for in the design. The quality of the finish was observed to ensure that the porosity and air pockets were kept to acceptable limits.

Reinforced concrete was chosen to provide a very stiff structure as a stable platform for the FAA electronic equipment. A steel-framed structure was deemed too flexible due to the height of this tower. The materials were suitable for rapid construction. A jump-lift methodology was utilized for the placement of vertical lifts. A Periform gangform system (8 forms tied together) was used with external whalers and trusses, for each 11-ft 10-in lift. Finnform panels were used to provide a smooth appearance, and the forms were lowered, cleaned, and recoated between lifts. The jump-lift was accomplished by crane and erecting scaffolding on the inside and outside of the forms (Fig. E1-9). The schedule was set at three days between lifts, although this stretched to seven days on occasion. Once the shaft got above 60 ft, concrete was lifted to the height of the pour by three yard buckets. The weather was a factor because the concrete was placed during the winter season and into the summer. In order to obtain accurate strength tests, the test cylinders were cast into the

FIGURE E1-8*b*   ATCT foundation piles.

pours and left to cure. The concrete was required to develop 1800 lb/in^2 before the next jump, and having the cylinders cure within the poured shaft wall provided assurance of concrete strength based on actual field conditions for the entire lift. Additives to the concrete included superplasticizer for flow and workability, air entrainment additives, and a high range water reducer for winter conditions. A total of 2000 yd was used in the construction of the tower.

A crane for high work came on site when the tower shaft reached above 60 ft. Several different-sized cranes were used at different times, according to the needs of

FIGURE E1-9*a*   ATCT shaft—lower portion.

FIGURE E1-9*b*   ATCT shaft—lower portion.

the construction. The crane for the highest work was a self-propelled Manitowoc 4100W crawler unit that required precise calculations for stability and load. The mast was 183 ft tall, with a 170-ft boom and 40-ft jib (Fig. E1-10). It was returned to the storage position at night. The crane's capacity varied with reach; however, the uppermost trusses weighed approximately 20 tons. In addition to the steel and concrete, the crane was used to lift equipment such as the air-handling units, condensers, window wall components, and the FAA radar dome (Fig. E1-11).

The support system for the cantilevered steel truss consists of reinforcing steel, studs, and steel plates cast into the walls of the tower shaft, supplemented by post-tensioning devices at the 113-ft height. The rebar is #9 weldable bar, and the studs are ¾-in diameter, 8-in long, all cast in place. The post-tensioning devices consist of 1⅜-in-diameter bars, four on each side in 3-in-diameter conduit to support the upper chord of the overhanging trusses. These were tightened to designed stresses and then grouted in place.

The steel trusses were partially assembled on the ground and lifted to their locations on the shaft. The uppermost trusses were set atop the shaft and balanced by overhangs on both the north and south sides of the shaft. These were secured in place by welding, studs cast in place, and bolting of trusses resting on bearing plates.

FIGURE E1-10 ATCT shaft—nearing full height, with crane and hoist.

FIGURE E1-11 ATCT shaft—at full height, with crane placing radar dome.

The specific construction requirements at the highest levels and crane limitations necessitated that some of the uppermost portions of the concrete shaft were deferred until truss sections at the opposite side of the shaft were set in place.

This contract also required that the contractor construct a mock-up of the window wall system for testing. The test was conducted at a remote facility specifically for the purpose of proving performance. Air pressures were applied to the interior and exterior of the sections with water spray to simulate high winds and rain. Leaks observed in the mock-up generated corrective action and techniques to improve system performance. There were four gaskets installed as a complete system between each section: an outer silicone gasket for weatherproofing, a pair of "loaded" vinyl sound barriers, and an inner neoprene finish gasket. The wall was constructed of an extruded aluminum frame with hinged joint anchorages to allow for thermal variations, anchored to the steel trusses. The shape was a compound curved structure on three sides of the shaft. It curved on a radius of approximately 100 ft from side to side and also from top to bottom, starting at vertical and sloping inward toward the shaft. This shape of the window wall proved to be a difficult manufacturing challenge and made field assembly rigorous with extensive hanging scaffolding and many field adjustments.

Certain special specifications were incorporated into the design. Due to the location near Jamaica Bay and the ocean, seacoast corrosion protection requirements were specified for equipment in contact with outside air. This included coatings on portions of or complete systems for outdoor air-cooled condensing units, evaporator units, and fans moving outside air. Isolation of dissimilar metals was required as well.

Fire protection for the ATCT consisted of two systems: a wet system for the base building and a dry pipe sprinkler system for the tower. The dry pipe was required because the unheated shaft was subject to cold temperatures. The dry pipe is maintained under air pressure. If a sprinkler head released, the pressure would drop and the fire pumps would start and supply water throughout all levels of the tower within one minute. The FAA cab was protected by a preaction dry pipe system which required both a sprinkler head to fuse and activation of an electronic smoke detector prior to the release of water to prevent any damage to FAA electronics.

The Port Authority turned over the FAA spaces of a functional building to the FAA for installation of their electronic equipment. The FAA fit-out required many months of installations and extensive communications work. Comprehensive testing of the many systems confirmed that the tower was ready for commissioning. Extensive precautions were taken to ensure that all equipment was functioning properly before placing the new tower in service.

East Garage

Project Description. The East Garage was constructed as the first of two planned parking structures. The need for additional parking was identified with the plans to utilize a large portion of Parking Lot #2 for new facilities. The symmetrical West Garage will be constructed when growth of passenger volume demands.

The East Garage has a square footprint 300 ft long on each side. The structure is steel columns with precast-concrete decks for parking. The garage has 1400 parking spaces on five levels. Access to and from the upper levels is through a pair of circular helixes, one up and one down.

Construction Logistics. The first order of business for construction of the garage was to clear the footprint of existing utilities. Although this area was relatively free of underground services, the main thermal distribution high-temperature hot water (HTTW) and chilled water (CW) pipelines ran through the site. These supplied the unit terminals with thermally conditioned water in the appropriate season. The system piping was rerouted prior to start of foundation work.

The precast reinforced concrete decking for the parking structure was specified in order to minimize the time for erection. The planks were ordered early by the contractor soon after award of the structure contract as a long-lead procurement item. This ensured that they would be available when the contractor required them. They were delivered and placed as the structure was ready. Staging of the trailer trucks carrying the decking required close coordination due to the large numbers of planks.

Foundation. The foundation construction proceeded in a similar fashion to the ATCT foundation, except that the garage foundation was much larger, and individual pile caps for support columns were utilized. The area was excavated to the depths required for each footing, which was at or near the water table, so only localized dewatering was required. Piles were driven using two pile-driving rigs (Fig. E1-12). The contract documents offered several options for the type of pile to be installed. The drawings were based on 26-in-diameter steel piles. Other piles sug-

FIGURE E1-12 East Garage—foundation, pile driving, two rigs.

gested included 12- or 18-in-diameter steel for 100- or 150-ton loads, 3NJ8x14 monotube piles for 100-ton loads, or 36-in-diameter bored piles for 250-ton loads. The piles selected for this application were steel monotube piles. The foundation contains 74 pile caps in total, with the number of piles in each varying from 13 to 41, depending on the ultimate future loads (Fig. E1-13).

Parking Structure. The supporting columns were concrete-encased steel, constructed in place. The steel framing was coated with spray-on fire protection. The decking planks were constructed with prestressed reinforcement to enhance their structural characteristics. The most common, normal-weight double-Tee planks for this structure were 9 ft by 58 ft, with 5-in decks and 2½-ft-deep webs. The decking was placed in a specific sequence to permit the crane to reach each plank location as the structure rose to its full height (Fig. E1-14). Care in lifting and placing the planks was required due to their span and susceptibility to damage, especially on edges and corners.

The sides of the structure are open to provide adequate ventilation and access for firefighting activities. Metal inserts were cast into the columns to allow the addition of finish panels to the exterior at some time in the future.

The helixes were cast-in-place reinforced concrete, cantilevered out from the center support column. Framing and false work were used to support the ramp roadway during construction (Fig. E1-15). Architectural concrete was used for the helix features in order to conform to the architectural theme for the CTA. Special lighting was designed to highlight the attractive features of the structural helixes.

FIGURE E1-13 East Garage—foundation, pile driving.

CIVIL CONSTRUCTION

Roadways

Project Description. The CTA roadway project consists of a completely reconstructed roadway network enabling travelers to proceed directly to their desired terminal quadrant, thus reducing congestion and providing clear and direct access to each terminal. The division of the CTA into quadrants is accomplished by a bow-tie interchange in the center of the CTA. This feature separates the CTA into four distinct areas or quadrants, leading passengers directly to their terminal and out again without bypassing other terminals unnecessarily. A special roadway for *high-occupancy vehicles* (HOVs) allows certain vehicles (such as interline buses) to proceed around the ring of terminals continuously to permit drop-off and pick-up at each terminal.

The roadway system consists of *on-grade* and *on-structure* roads. Many of the structured roads are long span steel beam bridges supported by single, centerline, reinforced concrete columns. The structured roadway is cast-in-place concrete over galvanized steel decking. The parapets are cast-in-place concrete with painted steel pipe railing.

Construction Logistics. The project was divided into several contract packages to control work areas and keep the scope of work to manageable levels. The complexity of the work approaches that found in congested urban areas. Numerous bridges

FIGURE E1-14 Parking structure—crane lifting decking.

FIGURE E1-15 Helix construction, with false work.

E1-28

and flyovers were incorporated in the plans. The difficult working conditions were compounded by the tight work sites and the necessity to maintain full operations at the terminal frontage roads immediately in front of the buildings. This can be very difficult at certain times of the year due to the high volume of traffic. Additionally, pedestrian traffic, frequently with large amounts of baggage, must be accommodated to and from the parking lots and taxi stands. Compliance with ADA requirements and international passengers who do not understand the language further complicate staging.

The Port Authority chose to prepurchase the long-lead materials, in this case steel, for the first roadway contract to minimize the time impacts for manufacture and delivery of the bridge beam steel. The steel was manufactured, painted, and inspected at the factory by Port Authority personnel to ensure quality. It was delivered to the contractor as work progressed and it was needed (Fig. E1-16).

The management of traffic during construction was paramount to successful performance. The heavy traffic at the airport necessitated extensive staging and phasing plans for each of the contracts. The contract documents required close coordination with the facility to ensure that the passengers would not be unnecessarily inconve-

FIGURE E1-16 Roadways—lifting steel.

nienced by the work. This required that temporary work be reviewed by facility personnel and be placed in service several days prior to major travel days, such as weekend holidays, as specified in the documents.

CTA Roadway Construction. The roadways were designed to comply with current AASHTO Standards to provide safe, efficient highways and roads. Port Authority standards were also incorporated to enhance the final product in conformance with specific airport needs. Among the features included were breakdown lanes, adequate length acceleration and deceleration lanes, controlled weave distances, and a geometry to support conventional airport speed limits.

The on-grade roadways and parking areas were composed of four pavement types. Type I, the most common for roads, consisted of a 6-in dense graded aggregate base over the subbase, a 4-in plant mix Macadam, a tack coat, and a 4-in asphalt concrete top coarse. Excavation took the grade down to the appropriate levels, or fill built up to the proper elevations where required. In all instances, the subbase was placed and compressed to proper densities. To ensure quality and timely response for reports, the Port Authority established a testing laboratory on the airport to conduct ongoing quality assurance tests. Proctor density tests were performed in the field to ensure proper compaction. The lab performed concrete compression tests at the initial 7-day, intermediate, and 28-day time frames. Asphalt core samples were taken for testing.

The roadway bridge abutments were supported by monotube piles and constructed of reinforced concrete. Select fill was placed in this area to provide a proper base for the approach slab. Materials excavated which were suitable for use elsewhere on the project were stockpiled in staging areas on the airport designated by the Port Authority. Unsuitable materials were sent to offsite approved dumpsites. Additional fill was brought in as required, including lightweight fill for certain locations, such as over utilities.

For the structural roadways, the concrete columns were also mounted on vertical and batter piles to accommodate transverse forces on the roads. The columns are 5 ft in diameter and topped with bearing plates for the steel beams. The expansion joints were cast into the concrete decks. The reinforced concrete decking was supported by galvanized steel pans. The contract required epoxy-coated rebar in the bridge decking and all above-grade structures to prevent spalling and prolong the useful life of the surface. The concrete was placed in three lateral sections, all tied by rebar. The main deck was placed, then the cast-in-place parapets were poured on each side. The decking was poured longitudinally (with allowable lane line joints) for large portions of the length, leaving several spaces for shrinkage of the pours. These were provided to allow the concrete to shrink and be filled in later to prevent cracking from shrinkage. The form spacers were made of styrofoam board at a height to allow the bulk placement of concrete pours and continuous leveling by the screeding machine.

Other features of the roadway construction were similar to conventional highway projects.

Utilities

Project Description. The utilities project consists of the installation of new and relocation of existing utilities in the CTA to support the redevelopment program. The utilities included storm and sanitary drainage, water lines, electrical ductbanks and cables, and portions of the thermal distribution system.

Construction Logistics. Early action was required for utility relocations to clear the footprint of initial contracts for the roadway bow-tie and East Garage. Subsequently, the construction of utilities was combined with roadway work to be performed in the same area to attain greater efficiencies of cost and time. The goal was to perform heavy construction work in each area only once. Several utility components installed were not placed into full operation for several years; however, the early effort prevented subsequent disruption to vehicle and passenger flows, and kept new road work intact. In addition to the Port Authority–contracted utility work, there were other utilities operating on the airport with infrastructure services. All work requires proper coordination with the local utility companies to identify and/or relocate gas lines, communication lines, etc. Although utility installation on the airport must address the specific issues of tidal salt marsh, dewatering, proper staging, and coordination as previously noted, most utility work is fairly straightforward. One contract which required special considerations due to its airport location was the installation of high-voltage ductbanks on the AOA.

High-Voltage Ductbanks. The installation of underground electrical ductbanks on the AOA included specific circumstances pertaining exclusively to airport construction: the routing of the ductbank under an active runway. Major concerns included the timing of the work to keep the runway available for use and the possibility that there would be settlement of the ground, causing the runway to be damaged, thus affecting aircraft operations. Runway 13L/31R was equipped for instrument landings using sophisticated electronics that required strict precautions to be taken to prevent interference from construction equipment. Misplaced equipment or materials could cause inaccurate or false information which could affect flight operations.

The restrictions on glideslopes previously noted were also enforced for this work. The contract contained strict dates (April 1 through 30) for the work, based on the best possible weather conditions (historically) and least likely need for the instrument landing system. The unavailability of this runway would affect airport operations, especially at peak times.

The potential settlement of the runway could also seriously impact the airport operations and was unacceptable. The size of the sleeve and 500-ft distance between pits on each side precluded jacking as a possible technique. The relatively new technique of microtunnelling was proposed for this installation. Microtunnelling is the use of boring equipment to bore a hole horizontally with great precision. This entails the digging of two open pits: the sending pit and the receiving pit. These pits were 40 ft by 50 ft and 15 ft deep, protected by steel sheeting and required localized dewatering. The sending pit contained the boring equipment and area for lowering and assembling (through welding) the "following" pipe (Fig. E1-17). Each section of pipe was 20 ft long with the 36-in diameter of the pipe only a fraction of an inch less than the cutting head. Following the placement of the pipe, this annular ring was grouted under pressure to prevent settling of the soil above and around the pipe. The direction of the cutting head was controlled by hydraulic steering mechanisms to maintain alignment.

The ductbank was designed as three parallel conduit runs approximately 20 ft apart. After all the pipe work was completed, the conduit for the cabling was installed. A total of 18 conduits, 6 for each of the three ductbanks, was placed in the pipe. The spacing for the conduit was maintained by high-density polyethylene bore spacers or *spiders* mounted on rollers and placed at 5-ft intervals to keep them aligned. The entire pipe was then filled with a concrete grout to stabilize the ducts. Confirmation of complete grouting was ascertained by the installation of relief con-

FIGURE E1-17 Microtunnelling—starter pit, welding pipe.

duits through the length of the larger pipe at the high point. Assurance that the grout had completely filled the pipe was attained when the grout flowed from the relief conduits.

Conventional concrete-encased ductbank emanated from each side of the pipe ducts, with manhole access at each side. The depth was adjusted at the manhole to typical depths for this work. The rest of the contact work was not remarkably different from other ductbank work.

Cogeneration

Project Description.   The Port Authority entered into a 25-year agreement with KIAC Partners for the purchase, construction, and operation of a new "privatized" cogeneration facility and adjacent central heat and refrigeration plant. The existing plant was rehabilitated and expanded in conjunction with the new construction. The cogeneration plant will burn natural gas (or alternate fuels) to generate electrical and thermal energy for use on the airport. Excess electrical generation is wheeled back into the public utility system.

The thermal energy is distributed as low-temperature hot water for heating in new underground piping and chilled water in existing piping for cooling. KIAC elected to change the operating temperature of the hot water system. The original system operated under pressure at high temperatures, 380 to 400°F. The high temperatures increased the system efficiency; however, the salt water environment at the airport created ideal conditions for corrosion of the system, and numerous leaks were developing in the distribution system prior to replacement. The new system operates at low temperatures, 200 to 250°F which reduces the corrosive action of the salt water. The new system is also equipped with a sophisticated leak detection system to isolate any leaks to within several feet to facilitate the excavation and repair in the future. The piping was welded in place and then subjected to x-ray testing to confirm full penetration and quality of the welds.

Some facts about the project are provided to give an idea of the scope of work. The primary equipment for cogeneration consists of two General Electric LM-6000 gas turbines, coupled to two steam boilers that recover waste heat from the turbines. One additional steam turbine generator was also installed to generate electricity from the gas turbine waste heat. Under the business arrangement, the electrical routing initiates at the plant where the generator output is transformed to 138 kV for transmission to the Con Ed substation in Jamaica through high-voltage lines and back to the airport's four main substations through 27-kV lines. The Cogen plant can be connected to the airport substation in the event that the Con Ed substation is out of service. The capacity is rated at 100 MW. The primary fuel is natural gas, although alternate fuels, such as jet fuel from the airport distribution system, can be used. The heating capacity is 225 million btu/h distributed via the new low-temperature hot water system. The cooling capacity is 28,000 tons (13,000 tons new and 15,000 tons refurbished). A variety of cooling unit sizes and types are utilized to take advantage of the most efficient combination of energy sources: absorption chillers using waste heat and electrically driven compressor units. The sizes are organized to satisfy small to large thermal energy demands with efficiency, minimizing the electrical demand charges for the electric units. Approximately 30,000 ft of new low-temperature hot water piping was installed around the CTA in a loop with branches to the terminals. The piping varied in size from 12 to 16 in and was protected by polyethylene sheathing and sacrificial anodes for cathodic protection.

Construction Logistics. The change in operating temperature necessitated that the nine individual airline terminals change their heat exchangers to maintain adequate transfer rates. Prior to changing the heat exchangers, an asbestos abatement project was undertaken to provide a safe working environment for the changes. Extensive instrumentation was installed to monitor and control the flow of thermal energy to the greatest efficiency. A significant change in the operation of the unit terminals was the change from HTTW heat exchangers to electric boilers for domestic hot water.

The connection of the new system to the unit terminals depended on the installation of the direct buried, bidirectional loop piping work around the entire CTA and the need to complete tie-in of the branch piping to the unit terminals before the heating season. The loop was installed predominantly in the parking lots, but the branch lines to the individual unit terminals required intensive planning and coordination to minimize disruptions to the frontage roads and the airlines. This was a tremendous amount of work and resulted in multiple shifts to make the tie-ins.

SECTION 3

CASE STUDY: MANCHESTER AIRPORT, NEW TERMINAL PROJECT

Richard Fennema
Assistant Vice President, O'Brien-Kreitzberg, Inc.,
New York, New York

The Manchester Airport is located in a rural area of New Hampshire on the outskirts of the city of Manchester. The surroundings are low hills covered with trees, farms, and low-density suburban communities. The airport contains a light industrial commercial park and is host to an Air National Guard unit. The property is flat and well isolated from the neighboring community.

The existing airport facilities consisted of a small commuter airline terminal which was out of date by today's standards for public access, maintainability, landside support, air-side support, and economic self sufficiency. The Airport Authority and the state of New Hampshire embarked on a development program with the goal to transform a small regional airport into a first-class gateway to New Hampshire. The focus of the new terminal was to expand air access to the state and to enhance the economic growth. It was also to provide a competitive alternative to Boston Logan Airport for business and passengers alike.

The project consisted of the new terminal, taxiway improvements, FAA improvements, and airport access. The new terminal consisted of seven jet gates, seven regional airline-grade level gates, waiting lounges, baggage handling, freight handling, ticketing counters, rental car counters, business center, airport management offices, security, public assembly, restaurant, and entertainment spaces. The ancillary improvements included enhanced fuel handling, waste disposal, snow removal, and increased parking capabilities. The project had a stated goal throughout its various developmental stages that the facility would not only meet the various Americans with Disabilities Act requirements for new public facilities but strive for state-of-the-art ADA accommodation wherever possible.

The new terminal is 10 times the size of the original facility and successfully incorporated the *gateway* theme. The facilities exterior is clad with aluminum panels on nonpublic building faces and clad with vast expanses of glass curtain wall on its public views. The structure is of reinforced concrete, steel frame, composite deck, and membrane roofing. The interior spaces consist of 30-ft ceiling in ticketing, baggage claim, public assembly, and central foyer areas. Finishes consist of tile floors with brick bands, acoustical plaster ceilings, floor-to-ceiling glass partitions between land- and air-side areas, tactile wall coverings, and fiberglass-enclosed columns. The economic viability of the new terminal was enhanced by large portions of the wall space being utilized for advertising and interior space utilized for a restaurant/lounge, fast food service, gift shop, video arcade, and hotel books facility. Waiting areas are located after security clearance in two wings which access the jet gates. The regional airline gates are located in a secondary extension of the building which brings the airline access down to grade level.

The project was executed through a managed design/build process. The site development, building structure, building enclosure and finish, airplane gate apron, taxiway, airport access, and landscaping were all developed concurrently in design and then implemented in separate construction contracts. The contract packaging allowed for tailored requirements associated with airport construction such as scheduling of activities on the air side of the terminal versus the public/land side of the terminal. The design and contract packages were tailored to enhance schedule compaction and promote bid competition among specialized contractors.

The scope of the construction consisted of 160,000 ft^2 of terminal building; 14,009 tons of structural steel; 25,000 ft^2 of glass; 30,000 yd^3 of concrete; 67,000 tons of asphalt; 33,000 ft of granite curb; 6000 trees and plants; 2500 light fixtures; 25,000 ft^2 of wall covering; 1500 gal of paint; 90,000 ft of wire; 45,000 ft of piping; 1.2 million worker-hours; and a 13-month construction schedule. Concrete was batched at an on-site transportable batch plant to facilitate the aggressive schedule and minimize community impact from the intrusion of concrete transit mix trucks. The construction effort was accomplished through 11 prime contracts which employed 80 subcontractors; 50 of the contracts were to firms local to New Hampshire. The project was completed on time and within its $65 million budget.

CHAPTER 2
WATER TREATMENT FACILITIES

Wesley F. Mikes
*Senior Vice President, O'Brien-Kreitzberg, Inc.,
Pennsauken, New Jersey*

James J. O'Brien
*Vice Chairman, O'Brien-Kreitzberg, Inc.,
Pennsauken, New Jersey*

INTRODUCTION

There are two basic categories of water treatment facilities: potable or drinking-water treatment facilities and water pollution control facilities (WPCFs). Drinking-water facilities must meet federal and state standards in terms of purity. WPCFs must meet similar standards which are measured in the effluent quality. Effluent quality must meet limits in suspended solids (parts per million), *biological oxygen demand* (BOD), and scum (ppm).

WPCFs are classed as primary, secondary, or tertiary (also known as *advanced water treatment,* or AWT). Primary treatment was the norm for many years. The Federal Clean Water Act of 1972 made secondary treatment (which is really primary plus secondary) the standard.

Tertiary or AWT is used where there is not a sufficient river capacity available to handle the effluent. AWT effluent can be used to recharge groundwater supplies.

Wastewater Chronology

Wastewater is accepted in sanitary sewers which flow by gravity directly to the WPCF, or more commonly to a pumping station where the influent is pumped either directly to the WPCF or into another gravity sewer line that leads to yet another pumping station.

At the WPCF, the influent enters the wetwell of the headworks of the plant. Here preliminary treatment takes place. The influent passes through screens and then raw sewage pumps. Coming out of the raw sewage pumps, the influent goes through cyclones to degrit the water.

The influent enters the primary settlement tanks by gravity through channels. The primary tanks are about 25 ft deep. Mechanical, continuous chain-driven, wooden boards (known as flights) pull the settled materials to a cross-collector trench. The wastewater leaves the primary tanks at a high level, maintaining gravity head.

The water enters the aeration tanks. The water is agitated and mixed with oxygen. The oxygen is generated by on-site air separation plants. The oxygen is stored in tanks. It is drawn off as needed, compressed, and fed into the tanks through diffusers low in the tanks.

The aerated wastewater is led into the secondary tanks through a high weir. The secondary tanks contain activated sludge. The activated sludge reacts with the raw sewage in the wastewater. Settled sludge is collected by chain-driven flights and cross collectors.

Clear effluent leaves the secondary tanks over a weir. It goes to the chlorine contact tanks where it is disinfected, and then flows into an outfall, and then to the river.

The collected sludge is transported to sludge digesters. In the digester, heat is added and the reaction continues until the sludge is digested. The digested sludge is then dewatered by filter press, centrifuge, or similar mechanical process. Final disposition can be spreading on farm/park land, composting, or landfill.

Clean Water Chronology

Clean water flow is similar to that of wastewater. Raw water is pumped from a source (lake, river, or reservoir) to the headworks. The influent is rough-filtered and introduced to settlement tanks. The tanks use chain-driven flights to collect settlement. Settlement is enhanced by the introduction of a flocculating agent such as alum.

From this point, the process is different. The next phase is filtration using either sand filters, carbon filters, or a combination of both.

Water is then collected in the clearwell. Chlorine disinfection can be accomplished there. From the clearwell, water is pumped to elevated storage tanks. From the tanks, the water is fed into the water distribution system.

Sludge treatment is similar to WPCF, but the composition of the sludge is much less concentrated. The quantity is much less.

Construction Characteristics

There are many components in a water treatment project. The major component is concrete tanks. The tanks require foundations to support dead weight (tank plus water), but also buoyancy. Empty tanks are just like a ship.

CASE STUDY: CAMDEN COUNTY, NEW JERSEY

In 1972, Camden County formed the Camden County Municipal Utilities Authority (CCMUA). At that time, the County had 37 municipalities operating 34 sewage treatment plants—all primary treatment type. CCMUA purchased a 19-mgd (million gallons per day) plant from the city of Camden. A plan was developed to regionalize the county treatment and upgrade it to secondary type.

The regional plan was based on an upgrade of the Camden Plant (Delaware #1) to 75 mgd and construction of a 17 mgd plant (Delaware #2) upriver. Two new interceptors totaling 64 mi would bring influent to Delaware #1, and an existing collection system would service Delaware #2.

In 1978, construction began on a 75-mgd preliminary treatment building at Delaware #1. This building was built at the north end of the existing primary tanks without interrupting plant operation. When the preliminary treatment building was completed in 1983, it was placed in service. Then the obsolete preliminary treatment building was demolished.

In 1982 to 1983, O'Brien-Kreitzberg, Inc. was conducting value engineering reviews of the regional designs. Savings of more than $7 million were confirmed. There were even more ramifications. Changes in the first pumping station were replicated in at least 15 more. A VE idea of replacing Delaware #2 with a large pumping station (to pump influent to Delaware #1) was not included in these savings. That idea was implemented later.

Delaware #1 WPTF

The expansion of Delaware #1 was done in two phases. Phase one expanded the treatment capacity to 38 mgd. This started in 1984 and was concurrent with the construction of the Cooper River interceptor system. The system is more than 22 miles of pipeline ranging in size from 12-in ductile iron to 96-in reinforced concrete. There are five pumping stations in this system ranging from 3 to 57 mgd. The cost of this system was $76 million. This system replaced 21 small sewage treatment plants.

Figure E2-1 is an aerial view of the completed phase one. The preliminary treatment building is at the right. The 10 primary settlement tanks are in the foreground. Next, toward the river, are the aeration tanks with 60 ton/day oxygen plant. The final sedimentation tanks are nearest to the river.

Plant Characteristics

Ultimate plant capacity 75 million gal/day

Population to be served 418,300

Design year 2000

Wastewater characteristics in 1 (ppm)

	Influent		Effluent (permit limits)
Suspended solids	200		25.7
BOD 5	290		30
Scum		6–12	
Phase 1 capacity		38 million gal/day	

FIGURE E2-1 Aerial view of CCMUA Delaware #1 after phase one.

Preliminary treatment facility	
Grit removal	
No. of tanks	3
Size of tanks	43 ft^2
Type of tanks	Detritor
Grit particle size removed at average flow	0.1 mm
Grit separators	3
Separator size and type	18-in cyclone
Grit removal	6 ft^3/million gal

Secondary treatment
Channel aeration

	Ultimate	Phase I
Length of channels, ft	1,250	625
Air requirements, scfm	3,750	1,875
Capacity of each blower, scfm	1,250	625
Number of blowers	2	2

Final sedimentation tanks

	Ultimate	Phase I
Wastewater flow, mgd	75	38
Size of each tank		
$l \times w$, ft	270×78	270×78
effective length	240	240
average liquid depth, ft	12	12
Number of tanks	8	4
Total effective settling area, ft^2	150,000	75,000
Total volume, mil. gal	15.1	7.6
Surface loading rate, gal/ft^2/day	500	505
Solids loading, lb/ft^2/day	30	30
Total weir length, ft	5,000	2,500
Weir overflow rate, gal/ft/day	15,000	15,200
Displacement time, h	4.8	4.7
Final sedimentation tank effluent		
suspended solids, mg/l	30	30
BOD 5, mg/l	25.5	25.5

Chlorination
Chlorine storage

	Ultimate	Phase I
Type	Tank cars	Tank cars
Capacity, tons	55	55
Chlorine contact tanks, size		
$l \times w$, ft	328×31	328×31
average liquid depth, ft	14	14
Number of tanks	2	1
Total volume, mil. gal	2.1	1.05
Displacement time, min		
annual average flow	40	39
maximum rate	20	20

Sludge management
Sludge characteristics

	Ultimate	Phase I
Primary sludge		
dry tons/day	19	38
percent solids	5	5
Waste-activated sludge		
dry tons/day	19	38
percent solids	1.5	0.5
Scum, dry tons/day	4	2
Sludge storage (existing)		
no. of tanks	2	2
volume per tank	860,000	860,000

	Sludge dewatering	
No. of filters	6	4
Capacity/unit, gpm	125	125
Capture efficiency, %	90	90

As part of the phase one construction effort, structural work required 130,000 yd^3 of excavation, 4800 piles with a driven total of 149,000 lineal ft, 45,000 yd^3 of concrete, 3400 tons of reinforcing steel, and nearly 400 tons of structural steel. In addition, the mechanical and electrical disciplines on the project required the use of approximately 42 mi of electrical conduit, 250 mi of wire, and more than 14 mi of pipe. The pipe utilized on the project ranged in size from ¼-in pneumatic control tubing to 96-in raw sewage conduit.

Figure E2-2 shows walls in progress, phase one aeration tanks (left) and Final Settling Tanks (right). Figure E2-3 shows walls in progress, phase one final settling tanks.

Figure E2-4 shows pile driving in progress in the phase two final settling tanks (FST) area. Figure E2-5 shows concrete slabs and walls in progress in the phase two FST area.

Figure E2-6 shows the two oxygen plants and oxygen storage tanks. Figure E2-7 shows concrete superstructure work in the final sedimentation tanks. Figure E2-8 shows the completed final settling tanks with the chain-driven flights in place. Figure E2-9 shows the crew pouring the roof on the aeration tanks.

FIGURE E2-2 Walls in progress, phase one aeration tanks (*left*) and final settling tanks (*right*).

FIGURE E2-3 Walls in progress, phase one final settling tanks.

FIGURE E2-4 Pile driving in progress in the phase two final sedimentation tank area.

FIGURE E2-5 Concrete slab and wall work in phase two FST area.

FIGURE E2-6 Oxygen plans and tank farm. Phase two in progress.

FIGURE E2-7 Final sedimentation tank superstructure concrete.

FIGURE E2-8 Completed final sedimentation tank with chain-driven flights in place.

FIGURE E2-9 Pouring roof on the aeration tank.

FIGURE E2-10 Gallery at final sedimentation tank.

FIGURE E2-11 Sludge dewatering building.

FIGURE E2-12 Sludge pumps in basement of sludge dewatering building.

FIGURE E2-13 Dewatering equipment and conveyors at second level of sludge dewatering building.

FIGURE E2-14 96-in influent line interface with existing wetwell.

FIGURE E2-15 Two electrical transformers, slab mounted.

FIGURE E2-16 Three underground electrical manholes.

Figure E2-10 shows the typical gallery at the end of the final sedimentation tanks. The overhead pipe runs include waste sludge, return sludge, and plant water. The conduit runs supply electrical power. Sludge pumps are mounted on the floor at the left.

Figure E2-11 shows brickwork in progress at the sludge dewatering building. Figure E2-12 shows sludge pumps in the basement of the sludge dewatering building. Figure E2-13 shows dewatering equipment and conveyor belts on the second level of the sludge dewatering building.

There is a substantial amount of heavy construction work on the site. Figure E2-14 shows construction at the 96-in influent line interface with the wetwell chamber.

Figure E2-15 shows two slab-mounted transformers. Figure E2-16 shows three underground electrical manholes.

CHAPTER 3
HIGHWAYS

Thomas A. Bryant II, P.E.
O'Brien-Kreitzberg, Inc., Denver, Colorado

Michael Giaramita
O'Brien-Kreitzberg, Inc., New York, New York

INTRODUCTION

Highways are the backbone of the economy in the United States. Our nation's highways now carry over 32 percent of the total revenue ton-miles of freight. In 1988 for the first time highways carried more freight than railroads. The per-capita vehicle miles traveled in the United States now exceeds 8600 per year, 20 percent more than the United Kingdom and almost double any other country. We depend on highways for the movement of goods, for shopping, for social and recreational activities, for access to and from work, for other family and personal business, and for many other functions necessary in our complex society.

The network of roads varies from the controlled-access interstate system connecting major population centers and urban expressways to local rural roads and streets. Surprisingly perhaps, of these 3,880,151 miles of roads, 42 percent are still unpaved. It wasn't until 1977 that paved mileage exceeded the unpaved. There are more than 190 million registered motor vehicles—more than in all the other countries combined—operating on the U.S. highway system. Approximately one in every six workers in the United States is employed in highway transportation and its related fields. These statistics focus attention on the importance of highways in general and heavy highway construction in particular in the U.S. economy.

DEFINITION

Heavy highway construction is construction requiring the use of large machinery such as cranes, excavators, dozers, end loaders, off-road trucks, and, generally, equipment with a GVW in excess of 80,000 lb. Although for wage rate determination in contracts, highway construction of almost any type is rated as heavy and the higher wage rates are applicable.

HISTORY

The interstate highway system we take for granted today had its origin in ancient Mesopotamia, not the Federal-Aid Highway Act of 1956. Great public works projects, including embankments, pavements, bridges and tunnels, were created in the alluvial land between the Euphrates and Tigris rivers. Neither stone nor lumber was readily available for their construction efforts. Generally, the Mesopotamians used the materials at hand: clay, sand, asphalt, and reeds. Even before the invention of the wheel, which is popularly supposed to have occurred some ten thousand years ago, mass movements of people took place. The earliest travel was on foot; later, pack animals were utilized, travois and crude sleds were developed, and simple wheeled vehicles came into being. Many of the migrations of the early historical period involved large numbers of people and covered relatively great distances. Regularly traveled routes developed, extending to the limits of the then-known world.

As various civilizations reached a higher level, many of the ancient peoples came to a realization of the importance of improved roads. They invented the brick, dried for interior use and baked, and sometimes glazed, for exposed use where permanence was desired. With sand and the natural asphalt that oozed from the earth in numerous places, they compounded mastic for mortar and waterproofing. In some structures, they used the asphalt mastic alone as the mortar. In others, a layer of brick was covered with a layer of asphalt, followed by a layer of clay, and topped with another course of brick. This sequence was repeated for four courses, then a reed mat was placed on the asphalt layer in place of the clay, and so on, until the structure was completed.

An example of the use of brick and asphalt in public works was the massive wall that surrounded the mighty city of Babylon. It contained both sun-dried and baked brick set in asphalt mortar. The inner wall, 22.4 ft wide, was separated by a 38-ft space from the middle section, 25 ft thick. The space between was filled to the top with rubble to make a roadway wide enough for four-horse chariots to pass each other. Between the middle wall and the 12-ft-thick outerwall there was a moat-like space, designed to trap any enemy that happened to scale the outer rampart. Construction methods utilized during this period are just recently being discovered. Of course, massive use of human labor was the major ingredient to their efforts. Ingenious use of ramps, levers, and pulleys was a key to their success. The definition of heavy highway construction at the start of this chapter would exclude these early major projects; however, they are included for their historical significance and the size of the completed work.

The excavators of Babylon found the remains of a bridge that once spanned the Euphrates, before the river shifted its course. Originally, the bridge was 370 ft long. Its piers were constructed entirely of brick set in asphalt mortar and the base of each was daubed with asphalt to waterproof it.

Upstream and downstream from the bridge were protective embankments. These Babylonian levees were similar to one at Assur built by King Adad Nirari I about 700 years earlier. The Assur embankment ran 5000 ft along the banks of the Tigris River. After placing a brick cofferdam, the engineers raised a retaining wall of limestone blocks (brought from the mountains to the north) and mortared it with asphalt mastic.

As early as 2500 B.C., Mesopotamian engineers paved processional streets between palaces and temples with colorful limestone, breccia, and glazed brick. And about 810 B.C. they even bored a tunnel over 3000 ft long, under the Euphrates at Babylon, to be used for escape if necessary. History also records the construction of a magnificent road to aid in the building of the Great Pyramid in Egypt in nearly 3000 B.C. Traces of early roads have been found on the island of Crete, and it is known that the early civilizations of the Chinese, Carthaginians, and Incas also led to extensive road building.

By far the most advanced highway system of the ancient world was that of the Romans. The Appian Way was a main element. Highway design engineers in the United States today are beginning to utilize thicker base and pavement designs like the Romans and modern-day Europeans in an effort to gain a longer life for new highways. Looking back on the interstate system of highways construction program it is easy to second guess the design standards which were based on only a 20-year life. In many cases, the addition of only incremental thickness (1 to 2 in) of pavement can add decades to the life of highways. Some of the Roman engineers' designs have lasted over two millennia; however, they were subject to much smaller loads and fewer numbers of trips. The longevity of the ancient system is a testimony to their engineering expertise.

After the decline and fall of the Roman Empire, road building, along with virtually all other forms of scientific activity, practically ceased during the Dark Ages. Even as late as the early portion of the 18th century, the only convenient means of travel between cities was on foot or on horseback. Stagecoaches were introduced in 1659, but travel in them proved exceedingly difficult in most instances because of the extremely poor condition of rural roads.

Interest in the art of road building was revived in Europe in the late 18th century. During this period, Tresaguet, a noted French engineer, advocated a method of road construction utilizing a broken-stone base covered with smaller stones. The regime of Napoleon Bonaparte in France (1800–1814) gave a great impetus to road construction, chiefly for military purposes, and led to the establishment of a national system of highways in France.

At about the same time in England, two Scottish engineers, Thomas Telford and John L. McAdam, developed similar types of construction. Telford urged the use of large pieces of ledge stone to form a base with smaller stones for the wearing surface. McAdam advocated the use of smaller broken stone throughout. This latter type of construction is still in extensive use, being the forerunner of various types of modern macadam bases and pavements.

DEVELOPMENT OF HIGHWAYS IN THE UNITED STATES

During the early history of the American colonies, travel was primarily local in character, and rural roads were generally little more than buffalo trails or cleared paths through the forests. Toward the end of the 18th century, public demand led to the improvement of various roads by private enterprise. These improvements generally took the form of toll roads or *turnpikes,* and were principally located in areas adjacent to the larger cities. The first American turnpike made use of existing roads in Virginia to connect Alexandria with settlements in the Blue Ridge Mountains. Another important toll road of this period was the Philadelphia and Lancaster Turnpike Road in Pennsylvania.

Another famous early road in American history was the Wilderness Road, blazed by Daniel Boone in 1775, which led from the Shenandoah Valley in Virginia through the Cumberland Gap into Kentucky.

In 1806, the federal government entered the field of highway construction for the first time with the authorization of construction of the National Pike, or Cumberland Road. This road generally followed the alignment of present day U.S. 40. Many of the concrete location markers can be found standing as sentinels to a history of westward-bound travelers. A few of the second-phase bridges built in the late 1800s are still in use also. The longevity of these structures is a testimony to the strength and durability of the stone-arch method of construction. The first contract on this route was let in 1811, and by 1816 the road extended from Cumberland, Maryland, to Wheeling, West Virginia, on the Ohio River. When original construction was completed (1841), it had reached Vandalia, Illinois. This route, which was surfaced largely with macadam, was nearly 800 mi in length and was built at a cost of approximately $7 million.

With the discovery of gold in California in 1849, the great western expansion of the country began, and many roads or trails figured prominently in this development. Probably the most famous of the great trails was the Oregon Trail. Other famous western roads, each of which has a fascinating history, included the Santa Fe, Morman, California, and Overland Trails.

In 1830, however, the steam locomotive demonstrated its superiority over horse-drawn vehicles, and interest in road building began to wane. By the time the first transcontinental railroad (the Union Pacific) was completed in 1869, road-building activities outside the cities virtually ceased.

By 1900, a strong popular demand for highways again existed. The principal demand came from farmers, who sought farm-to-market roads, to more readily move their agricultural products to the nearest railhead. The need to go to the county seats of government to pay taxes and vote also created the need for improved routes connecting the larger centers of populations. Our local government centers today are still spaced so the citizens can reach them by traveling a day or two on horseback, even though we now have the automobile as our principal means of travel. During this period, certain states began to recognize the need for financial aid for road construction. The first state-aid law was enacted by New Jersey in 1891, and, by 1900, six other states had enacted similar legislation.

The year 1904 marked the beginning of a new era in highway transportation in America with the advent of motor vehicles in considerable numbers. Almost overnight a tremendous demand was created for improved highways, not only for farm-to-market roads but also for through routes connecting the metropolitan areas. Additional state-aid laws were enacted, and, by 1917, every state participated in highway construction in some fashion. By this time, most states had also established some sort of highway agency and had delegated to these bodies the responsibility for the construction and maintenance of the principal state routes.

Before 1917 and until 1933, in some areas, each able-bodied male citizen was required to work on the road a certain number of days each year if he owned property. This assured at least a minimum level of maintenance but proved unworkable as society became more urbanized.

This period was also marked by radical changes in road-construction methods, particularly with regard to wearing surfaces. The early roads of American history were largely natural earth. Since timber was readily available, plank and corduroy roads were numerous, while later, wood blocks were used. Some gravel was used on early surfaces, while many city streets were paved with cobblestones. The invention of the power stone crusher and the steam roller led to the construction of a considerable mileage of broken-stone surfaces. Development of this type of surface generally paralleled that in Europe.

In the cities, relatively large concentrations of wheeled vehicles and the need for abatement of noise and dust brought various improved surfaces into being. The first brick pavement in this country was supposed to have been built in Charleston, West Virginia, in 1871. This original brick pavement on Virginia Street in Charleston has recently been overlaid with asphaltic concrete. In 1867, asphalt was used for paving Pennsylvania Avenue, in Washington, D.C. Concrete pavements were introduced about 1893, and the first rural road of concrete was built in 1909 in Wayne County, Michigan.

Federal participation in highway affairs on a continuing basis began in 1893, when Congress established the Office of Road Inquiry (ORI), an agency whose work was primarily educational in nature. World War I intensified highway needs in the United States and led the federal government to again actively enter the field of highway construction. The ORI later became the Bureau of Public Roads (BPR) of the Department of Agriculture, and in 1939 it became the Public Roads Administration under the Federal Works Agency. In 1949, this agency was transferred to the Department of Commerce and its name again changed to the Bureau of Public Roads.

In 1967, the BPR became the Federal Highway Administration (FHWA) of the U.S. Department of Commerce. The BPR name was retired in 1970 in the course of a reorganization effort.

The modern era of interstate highway construction began with the passage by Congress of the Federal Aid Highway Act of 1956 and its signing into law by President Dwight D. Eisenhower. As a Lieutenant in the U.S. Army, Eisenhower had recognized the need for a good system of highways when it took him 60 days to travel across the country in a military convoy in 1919. As a general during World War II, he was tremendously impressed by the German autobahns and the ease with which the Germans could move troops around. This knowledge caused President Eisenhower to emphasize road building and secure congressional approval for the interstate system. This act marked the beginning of the largest public works program in the history of the world. The original act called for the completion of a 41,000-mile national system of interstate and defense highways. Young engineers of draft age were given job deferments from the military because their work was considered vital for the national defense.

The system was expected to cost $38.5 billion to complete; however, inflation, increased safety improvements, environmental constraints, the extended period for construction, and increased complexity have seen the costs spiral upward to an estimated $125 billion to complete. The mileage has increased to 42,796 also.

HIGHWAY SYSTEM

There is no federal system of highways in the United States that is built and maintained by the federal government except for the roads in national parks and forests (only portions), Indian and military reservations, etc. The interstate system is owned, built, and maintained by individual states.

The interstate system connects 90 percent of the principal cities and carries 22.3 percent of the traffic. This backbone of the national system is only 1.2 percent of the total of 3,880,151 miles of highways and streets in the United States.

Federal-Aid Systems

The federal system comprises only 4.6 percent of the mileage which includes roads in national forests and parks, and on military and Indian reservations. Surprisingly, to some, the roads under jurisdiction of the federal government (179,220 mi) do not include the interstate system.

State Systems

The individual states have their own systems made up of interstate, primary, and secondary routes of statewide importance. This system includes some 798,532 mi of rural routes and urban extensions or 20.6 percent of the total system.

Local Road Systems

The largest percentage of mileage (74.8 percent) is under control of local government agencies. There are approximately 2,902,579 miles of county, township, town, city, and village roads. Most of this mileage (1,540,000) is under the jurisdiction of counties. In New England and Pennsylvania, the primary unit of control is the borough or township, respectively.

Delaware, North Carolina, Virginia, and West Virginia have responsibility for all local roads. These four states even maintain some streets and alleys in unincorporated small towns.

City Streets

The city street system comprises more than 344,000 mi of streets and alleys.

Toll Roads

Because of the lack of local revenue and the necessity for providing high-type facilities to relieve traffic congestion, many toll roads or turnpikes were built between 1950 and 1958. Most of these toll roads were built by special authorities created by state governments located in the northeast and midwest. Revenue bonds were sold with the future highway improvement and land as collateral, and the income from tolls collected used to pay the interest and principal on the debt.

Some 2200 mi of the 3200-mi system have been included in the interstate system. Even though they are included in the interstate system, many of the turnpike authorities accept no federal funds and choose to go it alone.

Interest in toll-road construction dimmed between 1958 and the late 1980s. With increased congestion and the substantial completion of the interstate system, interest in toll-road construction has recently blossomed. Several states are now actively pursuing the start-up of new toll facilities. Most are using a design/build-operate-transfer concept to avoid the large capital investment required initially and to place the burden of operation on the successful contractor until the debt is paid and profit accumulated from the tolls.

The Transportation Corridor Agency in Southern California has three toll roads in various stages of design or construction. Caltrans has four privatization ventures authorized by AB 680. Programs are in various stages from preenvironmental to construction. Denver, Colorado, has seen the completion of 5.5 mi of E-470 and the E-470 Authority plans to add 25 mi. The Oklahoma Turnpike is planning $1.7 billion worth of improvements and expansion. The Virginia DOT recently sold bonds for the construction of the I-66 corridor in northern Virginia. Efforts are also under way to complete projects in Florida and to a lesser degree in Arizona.

The early 1990s have been marked by the completion of almost all of the interstate system. The system is currently 99.3 percent complete. On October 14, 1993, the last interstate project in California was completed and opened to traffic. I-105, Glen Anderson Freeway, is the most expensive freeway project in U.S. history. This high-tech, urban freeway is the model for future freeway construction. A case study on the construction of I-105 is covered later in this chapter.

Another major closing link in the interstate system occurred on October 14, 1992, with the completion of I-170 near Glenwood Springs, Colorado. I-70 traverses the heartland of the United States from Philadelphia, Pennsylvania, near the East Coast to Cove Fort, Utah, where it joins I-15. The I-70 Glenwood Canyon Final Link is the ultimate example of major highway construction in rough terrain. The completion of I-70 through Glenwood Canyon represents the culmination of over 90 years of efforts to safely get from one side of the Rocky Mountains to the other by wagon or automobile. A separate case study on this award-winning construction is detailed later in this chapter.

Four other case studies are included for the reader. One is on construction in intermediate terrain with frequently spaced cuts and fills and bridges over streams. The Beaver Valley Expressway in Lawrence and Beaver Counties, Pennsylvania, was chosen for this case study. Based on the size of the major cut (2,000,000 yd^3) on the section chosen, this case study could easily have been put in the rough-terrain category.

The case study for flat-terrain construction selected was the building of Superstition Freeway east of Phoenix, Arizona. The geology of this desert terrain combined with the low rainfall and heat presented unique problems for the contractors and their equipment.

The case study selected for swamp reclamation-type construction was through Alligator Alley in South Florida. The completion of I-75 on November 25, 1992, presented surprising challenges.

Every major interstate or highway construction project in the country has geologic or environmental features which present challenges to the design engineers and contractors. It is intended that these case studies selected will represent most of the wide variety of conditions the engineers and contractors will face.

CASE STUDIES INTRODUCTION

Case studies of heavy highway construction projects were selected from throughout the country to give the reader an idea about the complexity of modern projects and possible alternative solutions to unique problems on their respective projects. A variety of projects were selected to represent the range of geology and topography facing the designer and constructors. Projects were also selected based on their location to represent a variety of weather and population factors.

CASE STUDY: URBAN CONSTRUCTION, I-105 GLEN ANDERSON FREEWAY

History

First proposed in 1958, the design began in 1968 and the project was added to the interstate system. The federal government required the formation of a multidiscipline design team including consultants, local agency members, Federal Highway Administration (FHWA), and Caltrans. Consultants added urban planners and architects, sociologists, economists, environmental acoustic experts, urban development specialists, and others not normally part of transportation projects at that time.

Nineteen separate studies by the design team lead to design public hearings with over 5600 participants. A total of 25 freeway agreements were executed. The city of Hawthorne was opposed to the route location at that time.

The National Environmental Policy Act (NEPA) became effective on January 1, 1970. The act required an *environmental impact statement* (EIS), among other things. California adopted a similar act in 1970. FHWA determined that ongoing projects should be reassessed. After substantial community involvement, the design team determined that an EIS was not required.

In 1972, a class action lawsuit was filed and a preliminary injunction issued requiring:

1. A formal environmental impact statement
2. Additional corridor and design public hearings focusing on air and noise pollution impacts
3. Additional housing availability studies
4. Specific assurance that the state could provide relocation assistance and payments to displaced persons

During December, *1974* a draft EIS was circulated for public review. There was broad public support for an eight-lane freeway/transitway project.

1977   The final EIS was completed calling for an eight-lane freeway plus a transitway, initially as a busway, and it included a realignment through Hawthorne.

1978   The federal government conditionally agreed to allow Caltrans to proceed with the project.

1979   The consent decree was issued in October.

1981   Consent decree was amended. Federal government questioned the adequacy of funds to complete the freeway. A proposal to reduce the scale of the project was agreed on by federal, state, and local officials. The Amended Consent Decree was approved by all parties and the court. Main features of the project became:

      1. Six lanes for general traffic, two *high-occupancy vehicle* (HOV) lanes, and two light rail tracks.
      2. Ten transit stations and park-and-ride lots.
      3. Ten local interchanges, associated with the 10 transit stations.
      4. Two interchanges at the east-west ends of the project.
      5. Ramp metering and HOV bypass lanes.
      6. Landscaping and noise attenuation.
      7. Relocation and rehabilitation, or new construction of at least 3700 housing units.

8. Continuation of the Employment Action Plan (1979 consent decree).
9. Continuation of the Office of the Advocate (1979 consent decree).

1982 Construction finally began with the first of 117 projects with ground breaking on May 1, 1982, in the city of Lynwood. Before substantial completion and opening to traffic on October 14, 1993, the I-105 project became the most expensive freeway project in U.S. history. The I-105 cost $2.2 billion to build (excluding the light rail costs). Eleven years of freeway construction cost $1.2 billion, $680 million for right-of-way and $360 million for replacement housing. The light-rail line known as the Metro Green Line will cost another $1 billion. Even this massive I-105 project will be surpassed by the Central Artery I-93 construction in Boston which is expected to cost $7.2 billion to complete.

Description

The 17.3-mi-long Interstate 105 Glen Anderson (Century Freeway) project was designed to incorporate high-occupancy vehicle (HOV) lanes and a light-rail line, transit stations, and park-and-ride lots.

I-105 runs east-west from Norwalk to Los Angeles International Airport, connecting four major freeways: the San Gabriel River Freeway (I-605), the Long Beach Freeway (I-710), the Harbor Freeway (I-110), and the San Diego Freeway (I-405). It provides improved access to the airport and relieves traffic from the Artesia Freeway (SR-91), I-405, and the Santa Monica Freeway (I-10), the nation's busiest, carrying more than 340,000 vehicles a day. By 2010, the I-105 is expected to carry about 230,000 vehicles a day, with projections of 25,000 more people riding the Green Line on opening day and 40,000 by 2000.

Vehicle flows are managed electronically by Caltrans' Traffic Operations Center (TOC). Sensors embedded in the pavement transmit real-time traffic information to the TOC, which synchronizes metered on-ramps, updates changeable freeway-condition message signs, and dispatches incident response teams. In the future, cars equipped with in-vehicle navigation systems, such as those tested in Caltrans' Pathfinder program, could also be tied into the TOC's information bank.

Within the freeway's minimum 230-ft-wide right-of-way are three 12-ft mixed-flow lanes and one HOV lane in each direction, with 10-ft portland cement concrete shoulders on the outside in both directions. The HOV lanes are separated from regular traffic by a 4-ft-wide buffer and have a 10-ft-wide inside shoulder on the left. The shoulders are designed to be integral with the driving lanes so lanes can be added to the freeway or HOV facility by restriping if necessary.

Of the entire freeway length, only 0.5 mi is built at grade, 6.1 mi is below grade, and 10.7 mi is above grade. A typical below-grade pavement section about 2 ft thick, consisting of a 10-in nonreinforced concrete pavement placed on a 4-in asphalt-treated permeable base (ATPB) and a 4.8-in lean concrete base, with 6-in aggregate base. The ATPB was chosen because of its ability to carry water to edge drains, which generally consist of 4-in slotted polyvinyl chloride (PVC) pipe wrapped with filter fabric made of polyester, nylon, or polyphrophlene material to prevent fines from being carried away by water. A unique feature of the drainage system is the inverted basket grates protruding from the light-rail line ballast in the median. Also, all of the ballast is placed on filter fabric to collect the sheet flow and prevent the surface water from penetrating the highway base.

FIGURE E3-1 I-105 Glen Anderson Freeway Interchange construction with Metro Green Line light rail ties being placed. *(Photo Courtesy of Tad Teferi, Caltrans.)*

I-105 is the first freeway to be designed from the start to accommodate mass transit. The 23-mi-long Metro Green Line is part of a planned 300-mi network of subways, trolleys, and commuter rail to be completed by 2021 in Los Angeles County (see Fig. E3-1). It will run for 16.5 mi in the median of I-105, then separate into a 3-mi north-coast extension to Los Angeles International Airport and a 3.5-mi southern branch to El Segundo. Construction on the line began in January 1991 and is scheduled for opening by fall 1994 or spring 1995. All but the north-coast extension is currently under construction.

The line will run in the median of I-105 either at grade or on overhead structures, powered by overhead electrical wires, and will be served by 10 transit stations and park-and-ride lots accessible by stairs, escalators, and elevators. Trains will run every 2 min during rush hour and every 6 min at other times, at speeds of up to 65 mi/h. Stations will be monitored around the clock by closed-circuit video cameras.

The I-105 project includes four freeway-to-freeway interchanges. The most complex connects I-105 and the $450 million I-110 Harbor Freeway, also under construction. This interchange is actually two separate, intertwined freeway-to-freeway interchanges. In addition to connecting ramps in the normal right-side positions, there are also separate ramps connecting I-105's HOV lanes, with similar lanes being constructed in the median of I-110. This will allow buses and car pools on I-105 to make the 6-mi trip north to downtown Los Angeles without changing lanes or merging with regular traffic. A Southern Pacific Railroad line also passes through the interchange diagonally, east to west, just south of I-105, and the Green Line, with upper-level I-105 and lower-level I-110 stations in the middle.

Some 90,000 ft of cast-in-drilled-hole concrete piling support 66,000 yd^3 of structural concrete and 14.8 million lb of reinforcing steel used to construct the cast-in-place concrete box-girder bridge overpasses. Steel pipe falsework was used to a height of 120 ft for the bridges. Most of the other contractors on the project used 12-by-12 wooden posts for falsework. Another interchange between I-105 and I-405 is the largest single construction contract ever awarded in California. The $134 million project takes up 100 acres of real estate near Los Angeles International Airport (LAX).

The interchange consists of 7 mi of direct connectors and ramps. It has five levels, a first for Los Angeles, and stands taller than a seven-story building. The project involved removal of four bridges, construction of 11 new bridges, and two bridge widenings. The interchange also included three 3500-ft-long cut-and-cover tunnels excavated 45 ft below the freeway. Each tunnel has a portal size of approximately 44 ft wide by 18 ft high. Abutment walls tapering from 2 ft 9 in at the top of the footings to 5 ft 9 in at the bottom of the wall support each tunnel side. Depths vary from 6 to 10 ft, since the tops of tunnel support the piers for the Airport Viaduct portion of I-105 that crosses over them. The third wall, called a *vent wall,* was built 4 ft inside the one abutment wall, creating a 4-ft by 18-ft void that serves as the exhaust air vent. Thousands of 14-in^2 precast-concrete piles support each abutment wall. The tunnels form sweeping arcs designed for cars traveling 50 mi/h and require special pumping stations for drainage and ventilating systems to prevent the accumulation of auto exhaust fumes.

While I-105 was built to aid in traffic flow, its construction was not allowed to impede traffic on other freeways and major streets. Extraordinary measures were taken to keep traffic moving. For example, a special 1-mi-long detour of the eight-lane I-405 freeway was built and then traffic was routed around the interchange construction zone. The restriping job was done overnight, and most motorists barely noticed the changes. Once the interchange was completed, the lanes were restriped and returned to the original alignment. The detour cost $20 million.

The urbanization of South Los Angeles forced the designers to route the interstate through the least populated areas. Many of these undeveloped areas were landfills where uncontrolled dumping had occurred. Extensive finds of old dumps and hazardous materials required special contracts to neutralize and haul the material to approved landfills.

In one case, $30 million was spent to remove tainted soil from a parcel of land near Normandie Avenue used as an illegal neighborhood dump before Caltrans acquired the land. Tests found that about 173,000 yd^3 of soil contaminated with heavy metals and petroleum hydrocarbons was among 600,000 yd^3 of material to be excavated. In the cleanup operation, bulldozers and end loaders loaded the tainted soil onto a 20-car train, which was lined and then covered with polyethylene sheeting for transport to a disposal site near rural Kettlemen City, 160 mi north of Los Angeles. While being loaded, the soil was kept wet to minimize dust particles escaping into the air. Monitoring stations upwind and downwind ensured that the air quality remained within guidelines set forth in the cleanup permit. Samples were collected daily. Soil that was not removed was capped. A typical capping process involved a layer of clay, lined with an impervious geomembrane to prevent seeping, which was then covered with soil and planted.

Fortunately, groundwater at the Normandie site was deep enough that there was no penetration. There were many unique situations for the designers, contractors, Caltrans engineers, and other agencies to solve before the I-105 freeway could be built. Following is a partial list of the unique features:

- 10 transitway stations with park-and-ride lots
- 4 freeway-to-freeway interchanges
- The first freeway to employ connector-ramp metering on full interchanges from initial design
- Automatic adjustment of meters by computers to optimize traffic movement
- High-tech traffic management systems with closed-circuit television monitoring
- High-speed light railway
- Noise-abatement sound walls over most of its length
- Connection to Caltrans District 7 Operations Center for traffic-flow monitoring
- 30 percent participation by minority and women-owned business enterprises in construction contracts
- Separate HOV-to-HOV (2 people per vehicle minimum) connections at the I-110 (Harbor Freeway) and I-405 (San Diego Freeway) interchanges
- Transit rider connection with I-110
- 2 local street interchanges at both ends of the freeway
- 10 diamond interchanges with local streets
- Probably the last interstate freeway project in California
- Enforcement zones (designated parking space for California Highway Patrol)
- Design for safety of maintenance
- Changeable message signs controlled at Freeway Operations Center

Some unique features of bridge construction included:

- Transverse post-tensioning at the I-105 and I-110 interchange.
- Retrofit of a post-tensioned bridge constructed early in the program to handle the light-rail loading imposed by the redesign. Two additional girder webs were installed.
- Piers supported by tunnels.
- Tiedowns placed at the temporary ends of post-tensioned decks to prevent the cantilevered sections from rising and to allow the staged removal of falsework.
- Built I-105 viaduct over existing Imperial Highway by placing piers in the median, etc. Major utility relocation was accomplished and traffic maintained with reduced lane widths.

Although currently in use in the Los Angeles Basin freeway system, this is the first time *highway advisory radio* (HAR) was designed and built in for a new freeway.

Other features include the use of a fly-ash mix design for the portland cement concrete (PCC) pavement. Fly ash provides five advantages:

1. It contributes to durable, high-strength concrete.
2. It increases resistance to sulfate attack.
3. It minimizes the effects of alkali-aggregate reaction.
4. It reduces permeability.
5. Its use is both economically and environmentally sound.

Also to aid in preserving the environment and save cost, contractors crushed over 350,000 yd^3 of existing concrete to utilize as part of the 6-in aggregate base course.

Even though it has been placed elsewhere, full width, this was the first use of PCC pavement for the full roadway, including shoulders, in California.

The biggest challenge to the I-105 involved the settlement of 1972 lawsuits and resumption of work after the consent decree of 1979. The project was downsized from 10 lanes to 8 and participation goals for minority and women business enterprises were established. A job-training program, which created and afforded replenishment housing, provided by the Century Freeway Housing Program (CFHP) and funded by Caltrans and the Federal Highway Administration, is the largest program of its kind in the nation. Caltrans acquired the 5029 right-of-way parcels needed for I-105. Starting in the early 1980s, CFHP rehabilitated existing housing and then financed the construction of more than 2000 units of affordable housing in the freeway corridor, resulting in a total investment in the community of $175 million. The CFHP continues to administer a $142 million trust fund being used to pay for land acquisition, construction, and permanent take-out financing for an additional 2000 housing units, to be constructed by 1995. See Fig. E3-2.

I-105 Fact Sheet

Number of I-105 contracts	117
Cubic yards of dirt moved	16 million
Cost of storm drains rebuilt	$15 million
Major storm drains installed	15
Cost of sewers relocated	$2 million
Cubic yards of concrete placed	2.5 million
Tons of steel used	115,000
Cost of hazardous waste cleanup	$80 million
Cost of railroad relocation	$5 million
Miles of sound walls 8- to 14-ft high	25
Length of berms—4 ft high	8 mi
Schools rebuilt	6
Historical site relocated	1
Acres of land used	1000

CASE STUDY: INTERMEDIATE TERRAIN, BEAVER VALLEY EXPRESSWAY SEGMENT OF THE PENNSYLVANIA TURNPIKE

History

The James E. Ross Highway (Toll 60) segment of the Beaver Valley Expressway is the first new extension of the Pennsylvania Turnpike in more than 30 years, and only the second branch route since the highway was opened a half-century ago.

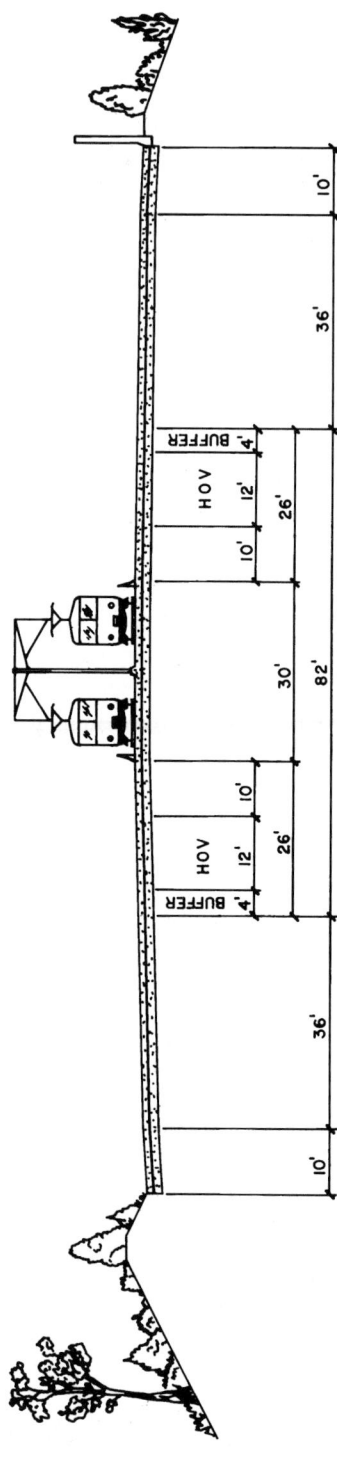

FIGURE E3-2 I-105 Freeway—Transitway.

As the nation's first four-lane, limited-access, long-distance superhighway, the turnpike was a transportation pioneer. Conceived during the Depression of the 1930s as a means of relieving unemployment, its first section—160 mi from Carlisle, near Harrisburg, to Irwin, near Pittsburgh—opened October 1, 1940. After World War II, the toll road was extended east to Valley Forge, then west to Ohio, then east again to the Delaware River with a bridge connection with the New Jersey Turnpike. With the opening of that bridge on May 23, 1956, the east-west turnpike stretched for 360 mi and soon became the center link of a chain of toll roads extending from New York to Chicago. Ultimately, more than a dozen states followed Pennsylvania's example and created toll-road authorities. In 1957 the Northeast Extension was opened, from a junction with the east-west highway near Norristown to Allentown, Wilkes Barre, and Scranton.

In September 1985, the Pennsylvania General Assembly passed Act 61, the Turnpike Organization, Extension and Toll Road Conversion Act. Act 61 authorized and empowers the Turnpike Commission to undertake the construction of new projects and to operate them as part of the Pennsylvania Turnpike system. The language of Act 61 segregates the improvement and extension authorizations into four major groups of projects and specifies when construction for each can commence.

One of the key elements of the Act 61 projects was the Beaver Valley Expressway (Toll 60—The James E. Ross Highway) in Lawrence and Beaver Counties northwest of Pittsburgh. The 16.5-mi highway completes a gap between two completed sections of S.R. 60 and an improved U.S. 422. The highway connects I-80 and nearby I-79 with Greater Pittsburgh International Airport.

Economic Development

The most important reason for constructing the Beaver Valley Expressway was to help Beaver and Lawrence Counties revive economically. Since 1980, traditional steel mills and steel-dependent industries laid off thousands of workers, or went out of business entirely. In 1983, Beaver County's unemployment rate peaked at 30 percent. Community leaders, hoping to replace these lost jobs with new positions in diversified fields, found that companies seeking to locate in the area expressed two primary needs.

With more emphasis on commercial, service, office, and light-manufacturing activities, these firms needed medium-sized buildings, not the huge, vacant factories left over from the heavy-industrial past. And they needed good highway transportation for product distribution as well as for access to the workplace by their labor force.

By setting up economic incentives, creating industrial parks, and analyzing and marketing sites for development, officials in both counties have encouraged new small and medium-sized businesses to locate in the area. By opening a region that has long been bypassed by other major highway corridors, the Beaver Valley Expressway provides the access new area businesses need to succeed. Access is also provided to the new Greater Pittsburgh International Airport which opened in October 1992.

Access to the airport is important for both the traveling public and commercial air freight. When the highway opened, the airport became less than a 40-min drive from New Castle. The expressway has realistically broadened the airport's service area to as far north as Sharon, Pa., and Mercer County, and west to the city of Youngstown, Ohio.

At approximately the midpoint of the expressway, a new interchange connects the Beaver Valley Expressway with the Turnpike's 360-mi east-west route across

Pennsylvania. By giving the local area access to major arteries of commerce, the expressway makes it easier for local industries to compete in markets in Ohio and Canada.

Most development in Beaver County is concentrated in the central and southern parts. With the expressway's completion, the direction of new commercial development and residential growth is expected to shift northward.

As a result of the expressway's completion, officials in Lawrence County expect to see more residential growth than all-new industrial or commercial development.

For any business relying on large trucks, or a trucking-dependent company that is considering locating in the area, the expressway makes life much easier. Truckers traveling north-south highways, such as PA 18 found two-lane roads, stop-and-start traffic, signals and intersections, residential neighborhoods, sharp curves, and some hills. Although the community of Rochester had a bypass, Beaver, Beaver Falls, and New Brighton did not. A trip that took a half-hour because of congestion now takes as little as 10 min with the expressway opening as a limited-access road.

By closing a gap between two completed parts of S.R. 60, the expressway cut travel and commuting time to the major trading and employment centers of Pittsburgh, Sharon, New Castle, and Beaver, Pa. and Youngstown, Ohio. It also opened a previously inaccessible area to development and improved traffic flow internally on local roads, as well as provided a direct route to Greater Pittsburgh International Airport.

Schedule

Preliminary planning and property acquisition have been underway since 1985. Groundbreaking ceremonies for the Beaver Valley Expressway took place on October 20, 1989, for the Mahoning River Bridge, and on June 14, 1990, for the roadway.

Completed in late 1991, the 70-ft-high Mahoning River Bridge carried a price tag of $20.9 million. Its 1700-ft length means that it displaced the turnpike's nearby Beaver River Bridge (1545 ft) as the highway's fourth-longest bridge.

Toll Collection

This was only the second time in the turnpike's 53-year history that the turnpike had used coin-drop toll facilities. Drivers pay as they go rather than carry tickets.

Land Acquisitions

In 90 cases, property acquisition meant that individuals or families had to be displaced from their residences and had to find new homes. The Turnpike Commission staff negotiated fair settlements with each land owner.

In only three cases, property acquisition resulted in the closing of businesses—a garage, a print shop, and a fish hatchery.

Environmental Damage and Corrective Measures

In one way or another, all large construction projects affect the natural environment, and the building of the Beaver Valley Expressway was no exception. Engineering and environmental specialists took steps to avoid or minimize any damage using methods to compensate for changes brought on by the project. These people paid particular attention to wetland replacement, farmland protection, noise abatement, air quality, and landscaping.

Wetlands/Farmlands. The expressway route runs north-south approximately parallel to the Beaver River in an area where all but one of the streams drains from west to east into the Beaver and Mahoning Rivers, meaning that the right-of-way crossed wetlands in many places. In all, construction required the filling-in of nearly 14 acres of wetlands. A total of 33 separate wetlands were affected to some degree.

Environmental agencies require that wetlands be replaced on an acre-for-acre basis. Priority in this replacement process was given, first, to locations adjacent to the impacted wetlands, second, to sites within the same watershed, and, last, to locations outside the affected watershed. The contractors for the Pennsylvania Turnpike met and exceeded this acre-for-acre replacement requirement.

Noise Abatement. Noise barriers were installed at nine locations along the route. Eight are natural earth mounds ranging in height from 8 to 20 ft, and one is a 16-ft-high structural wall. They reduce noise by about 5 to 6 dB.

Air Quality. The project ranks as having minimal impact on air quality with worst-case situations based on highest peak-hour traffic volume resulting in carbon monoxide levels less than half the National Ambient Air Quality Standard level of 35 ppm for a 1-h average.

Historic Sites. The state historic preservation officer determined that road construction would not affect three historic structures in the vicinity or their historic integrity.

Endangered Species. Only four animals considered endangered species have traveling ranges in the project area and none uses the region as a permanent habitat.

Landscaping. The Turnpike Commission has a full-time landscape architect who provided expert planning for the Beaver Valley Expressway.

Landscaping practices were carried out with a good neighbor policy in mind. Among the chief aims of this policy are safety, low maintenance costs, appearance, roadside aesthetics, and environmental sensitivity.

A 30-ft shoulder clearance, for example, allows natural solar ice and snow clearing while providing adequate safe sight distance to see wildlife movement, and at the same time reducing the threat of blockage by fallen tree limbs. The amount of mowing was reduced by the use of ground covers.

A section of PA 60 was chosen for detailed description based on the innovative planning utilized by the contractor, Geupel Construction Company, Inc., of Columbus, Ohio, to successfully win the project with their low bid and completed the project one year ahead of schedule.

Description of Work

Work included construction of a section of S.R. 60 on a new location, including ramps, with cement concrete pavements; rehabilitation and reconstruction of other roads with various bituminous courses or concrete pavements; concrete and bituminous shoulders; structures; storm drainage facilities; guide rail; right-of-way fence; concrete barriers; landscaping; sanitary sewer relocation; water main construction; erosion and sedimentation control; and other miscellaneous items of construction, all within a project length of 10,900 linear ft (2.06 mi).

Bid date: April 4, 1990

Notice to proceed: May 30, 1990

Completion date: November 30, 1992

Actual completion dates:

    Substantial completion: November 1991

    Opened to traffic: June 18, 1992

Cost: Over $19.4 million

Existing S.R. 60 was a four-lane controlled-access highway which terminated at S.R. 51 in a partial interchange. This project started with a pair of new bridges over S.R. 51 (BV No. 101) and the completion of the interchange. Eighty percent of the earthwork was located in one large 2500-ft-long cut which started on the south end of the project, adjacent to the new bridges across the S.R. 51. At the north end of this cut, S.R. 588 crossed the right-of-way. A new bridge (BV No. 102) was proposed to carry S.R. 588 over the new road in its same location. Existing traffic was to be maintained on a temporary roadway adjacent to the new bridge. The next 2000 ft of construction was close to final grade and was the site of a toll plaza to be constructed under another contract. This was followed by the project's only major embankment a 2,000,000 yd^3 fill. A 7-ft-6-in by 6-ft-6-in box culvert (BV No. 103) was to carry storm drainage beneath this embankment. This structure, plus sanitary sewer relocations, had to be completed prior to any fill being placed in this embankment. The last 2000 ft of the project was essentially at grade with a new bridge (BV No. 104) to carry S.R. 251 over the right-of-way. This road could be closed during construction of the new bridge.

Utilizing critical path scheduling, the contractor determined that the critical path of this project went through BV No. 102, the new bridge for S.R. 588. This bridge was located at the north end of the major cut. It was 55 ft above the grade of the new roadway and consisted of two 180-ft spans. The center pier could not be built until the S.R. 60 roadway had been excavated to grade. The entire 2,000,000 yd^3 of excavation had to be hauled through this site, over and across the temporary detour and to the only embankment in the center of the project. It was impossible to simultaneously maintain a haul road through the new bridge site and across the temporary detour and erect the steel for the new bridge. The proposed construction sequence consisted of installing the temporary detour and the box culvert in the summer of 1990. All of the excavation and embankment could then be performed between the fall of 1990 and late summer of 1991. The substructure work on bridge BV No. 102 would be done in the fall of 1991, with the deck completed in the spring of 1992. The detour could be removed in early summer of 1992, the last piece of earthwork could be performed beneath the detour, and the project could then be paved in the summer of 1992.

The contractor's innovative solution was to shorten the schedule so that the earthwork and BV No. 102 could be constructed simultaneously. The temporary detour was to be constructed in the cut at approximately the same elevation of the existing road. Approximately 60,000 yd^3 of excavation could not be removed until the detour was eliminated. There was not enough room to haul underneath the steel of the new bridge and over the detour. The contractor proposed constructing a temporary bridge as part of the detour. Traffic would be maintained on the temporary bridge and the contractor's haul trucks would travel beneath the bridge. This eliminated the need to construct a ramp to cross the detour at grade, which in turn eliminated the obstacles to building BV No. 102 simultaneously with performing the excavation.

A Mabey Universal panel bridge with a 120-ft clear span was chosen for the temporary bridge. See Fig. E3-3. This allowed for two-way haul road traffic beneath the bridge. The bridging system was rented for eight months and erected at the site on concrete abutments on spread footings that were prepared prior to the arrival of the bridge components.

The cost of the temporary bridge was borne by the contractor. Its cost was offset by eliminating the need for flaggers where haul trucks would have had to cross the temporary detour at grade. The bridge also eliminated a hump in the haul road and thus improved haul road speed and shortened truck cycle time.

All excavation beneath the temporary detour was drilled and shot prior to construction of either the detour or the temporary bridge. This eliminated the need to blast adjacent to the new BV No. 102 after it was completed. All excavation to the north of the detour was performed and a slot beneath the temporary bridge was excavated. The temporary bridge was installed along with temporary utilities, water, and gas. The detour was opened to traffic the first week of November 1990. During this time, the sanitary sewer relocations and the box culvert were completed beneath the major embankment. Water was diverted through the box culvert by mid-November. Excavation from the major cut could now proceed to the embankment. The first objective was to excavate in the immediate vicinity of the new bridge so that the center pier could be started. Work on the center pier started in mid-December and was completed in 30 days. Structural steel was delivered and erected in January. The deck was poured in March 1991 and the new bridge was ready to

FIGURE E3-3 Beaver Valley Expressway. Steel in place for structure BV-102 with Mabey Temporary Bridge to right. *(Photo courtesy of Charles Wooster, Geupel Construction Company.)*

open by May 30, 1991. Meanwhile, excavation work continued two shifts per day, six days per week until mid-June 1991, when it was completed.

The last piece was the removal of the excavation below the temporary detour. Concrete paving was performed in September and October 1991 and the project was substantially complete by Thanksgiving, nearly 12 months ahead of schedule. The project was formally opened to traffic on June 18, 1992 when construction of the next project was also completed.

Equipment

The contractor chose a Caterpillar 992C end loader and Caterpillar 777B 85-ton trucks to perform the bulk of the excavation on this project. The material to be excavated was predominantly a medium-hard gray shale which required drilling and blasting prior to removal. Drilling was done with a Reed SK40 rotary drill.

Geologic Features

An unmined coal seam was located near grade throughout most of the cut. Prior to submitting its bid, the contractor drilled four core holes to recover intact samples of the coal seam. These confirmed that the seam averaged, 3½ ft in thickness and that its quality was 12,200 btu's and less than 1 percent sulphur. The volume of the coal amount to 60,000 tons and its value for the job site was $17.00/ton. This allowed the contractor to anticipate receiving $1,000,000 additional compensation for the removal and sale of the coal.

Major work items included the following:

- 2,485,258 yd^3 of Class I excavation (Dirt and rock)
- 201,141 yd^2 of subbase (various depths)
- 13,448 linear ft of drainage pipe (various sizes)
- 62,991 yd^2 of plain cement concrete pavement 12-in depth
- 9,653 linear ft of guide rail
- 6,346 linear ft of sanitary sewer relocations
- 33,522 yd^2 of concrete shoulders
- Structure No. BV-101, Dual three-span composite steel multigirder bridge, S.R. 60 over S.R. 51
- Structure No. BV-102, two-span continuous curved steel multigirder bridge, S.R. 588 over S.R. 60
- Structure No. BV-103, 7½-ft by 6½-ft reinforced concrete box culvert
- Structure No. BV-104, two-span composite steel multigirder bridge, S.R. 251 over S.R. 60

CASE STUDY: FLAT TERRAIN, SUPERSTITION FREEWAY, ARIZONA

Superstition Freeway, Arizona's old S.R. 360, now U.S. 60, was completed and opened to traffic on August, 16, 1991. The opening of the final 11 mi of the 27-mi-long freeway culminates the construction which began in 1969. The long construction period was caused by the Arizona Department of Transportation's (ADOT) pay-as-you-go philosophy. Construction was expedited when the first bonds were approved in 1979; however, limited funding continued to control the pace of construction. The 13 contracts cost approximately $140 million to complete.

Superstition Freeway is named for the Superstition Mountains of Lost Dutchman Gold Mine fame, located just north of the eastern end of the freeway. It is Arizona's premier urban highway which links Apache Junction, Mesa, Chandler, Gilbert, and Tempe with Phoenix via its connection with Interstate 10 and other Maricopa Association of Government freeways under construction.

The Superstition Freeway was depressed to solve visual and noise problems in flat terrain. This necessitated the design of an extensive and complex drainage system with major pumps at 1-mi intervals to prevent flooding.

The freeway has had a positive economic and social impact on the East Valley as documented in a 1987 survey and study prepared for the Transportation Research Center at Arizona State University in Tempe. The survey showed that 76 percent of the homeowners in Tempe living within a half mile said the overall impact of the freeway was good. Only 10 percent complained about the freeway's noise and air pollution. The positive feelings about the freeway can be attributed largely to its depressed design, extensive irrigated landscaping, noise walls, and wide right-of-way.

Since old U.S. 60 parallels the Superstition Freeway throughout its length, ADOT is in the process of turning the old road over to the cities and Maricopa County for maintenance. There have been losses for businesses on the old expressway but many are adjusting by changing their businesses to serve local neighborhoods. Just as businesses adjust to changed conditions, ADOT has changed its specifications to adjust to new information and technological innovation.

Superstition Freeway projects became an experimental ground for pavement types as shown here:

- 9 in of portland cement concrete pavement (PCCP) over a 4-in cement treated base (1981 construction)
- 9 in of PCCP over a 4-in lean concrete base (LCB)
- 8 in prestressed concrete pavement over base
- 13 in of PCCP on 4-in of aggregate base (1991 construction)

The long construction period also saw the evolution of pavement texturing from the original bull float, burlap drag finish, to a swirling broom finish, to transverse brooming, and, finally, to transverse tining. The transverse tining has become the standard for all ADOT concrete pavement surface finishing.

The last 11-mi paving project saw the introduction of a new specification to improve the ride quality. Essentially, the pavement acceptance specification has an incentive/disincentive clause. The paving contractor on the last section received a bonus of $1/yd^2 for the excellent smoothness of 2 in/mi using the old BPR roughness index. The specifications required a smoothness of 7 in/mi and offered contractors a $.20/yd^2 bonus for each inch the ride is better than 7 in.

The last 11-mi-long projects on Superstition Freeway required extensive quantities of construction pay items to complete, as shown here:

- 9 million yd^2 of excavation
- 17 mi of mostly concrete pipe
- 725,000 yd^2 of portland cement concrete paving

Fortunately for ADOT, the excavation was relatively easy, with most of the material being a sandy loam; however, there were some expansive clays. The Pulice Construction Company of Phoenix used some of the largest haulers ever used on a highway project to move the dirt. See Fig. E3-4.

The completed freeway is a modern six- and four-lane highway with PCCP and shoulders throughout, except for a bituminous concrete shoulder area. This bituminous concrete area will be removed and replaced with full-depth PCCP as traffic generates the need to widen the six lanes in the future. All bridges that span the freeway were built wide enough to accommodate future lanes that may be needed later.

Bridge construction in the Arizona area is unique. ADOT specifies mostly post-tensioned longitudinal box girder–type bridges. Contractors since the early 1980s have built a soffit fill to support the bridge construction. Piers and abutments are built first, which is typical for most bridge projects, then protected by urethane foam as the fill is compacted around the substructure units. The grade for the top of the soffit fill closely matches the bottom of bridge elevations with 2½ in minimum clearance. A lean concrete waste slab is then placed on top of the soffit fill. The waste slab extends 2 ft beyond the edge of the deck and is 3 in thick in this area. Slab edges are thickened to 6 in where vertical planes join the abutment and pier faces. The waste slab is covered with two coats of an approved bond breaker prior to concrete placement. Construction then proceeds in normal fashion with the placement of reinforcing steel and the conduit for the tendons. Concrete is placed in four increments with the bottom slab first, the webs and exterior fascia second, the deck third, and lastly the parapets.

After the concrete has cured and the superstructure is post-tensioned, the soffit fill and waste slab is excavated under the bridge. This method has proven to be more economical than conventional cast-in-place bridge construction falsework. It also happens to be much safer.

FIGURE E3-4 Superstition Freeway hauling units. *(Photo courtesy of Arizona Department of Transportation.)*

Construction in the desert area requires special controls and adjustments because of the excessive heat. Phoenix and the Southwest frequently record over 120 days per year where the temperature exceeds 100°F. Contractors are required by the ADOT Standard Specifications for PCCP to:

- Add ice to mixing water.
- Water-cool aggregate piles.
- Fog-application with an atomized mist of water when ambient temperature is above 85°F until sawing is completed.
- Place concrete only between 8 P.M. and 8 A.M. when daytime ambient temperatures are expected to exceed 100°F.

Contractors, when not paving, frequently work from 5:30 or 6:00 A.M. to 1:30 or 2:00 P.M. to avoid the oppressive afternoon heat. This also avoids the afternoon's peak traffic flows.

Work on the Superstition Freeway's original eastbound section at the I-10 interchange is under construction, with completion of the first $24 million widening project expected in mid-1994. The second project for the widening of westbound Superstition and widening of I-10 will begin in 1994. See Figs. E3-5 and E3-6.

The success of the Superstition Freeway project is verified in its first-place award for the Outstanding Concrete Paving Project by the American Concrete Pavement Association in 1991.

CASE STUDY: SWAMP RECLAMATION, I-75 ALLIGATOR ALLEY

Alligator Alley, the final link of Interstate 75, as it bisects the Midwest from Sault Ste. Marie, Michigan, veering south through Detroit, Cincinnati, Ohio, and southeast through Lexington, Kentucky, Knoxville and Chattanooga, Tennessee, Atlanta, Georgia, Tampa, Florida, and east to Fort Lauderdale, was opened to traffic in January 1993. Alligator Alley extends from a point east of Florida S.R. 858 (east of Naples) to a point west of S.R. 25 (west of Fort Lauderdale).

The existing two-lane facility was upgraded to interstate standards with the addition of two lanes north of S.R. 84 (old Alligator Alley).

The old Alligator Alley alignment crossed land formations having sensitive environmental features. The major formations are the Fakahatchee Strand, Big Cypress Swamp, and the Florida Everglades. The concern of environmentalists has been, and continues to be, the preservation of these major wetland areas. The three wetland areas are extremely flat and are generally characterized by surface water movement in a southerly direction. These areas contain unique natural vegetation and biologic systems.

When the I-75 corridor along S.R. 84 (Alligator Alley) was approved, certain environmental commitments were specified to preserve these unique environmental features.

These environmental commitments/improvements were incorporated into the project design. Examples of these improvements are (1) the provision of a unique water distribution system, (2) the blockage of existing canals to prevent overdrainage of the surrounding area, (3) the construction of a right-of-way fence to prohibit indiscriminate access to and from the highway, and (4) the construction of wildlife crossings.

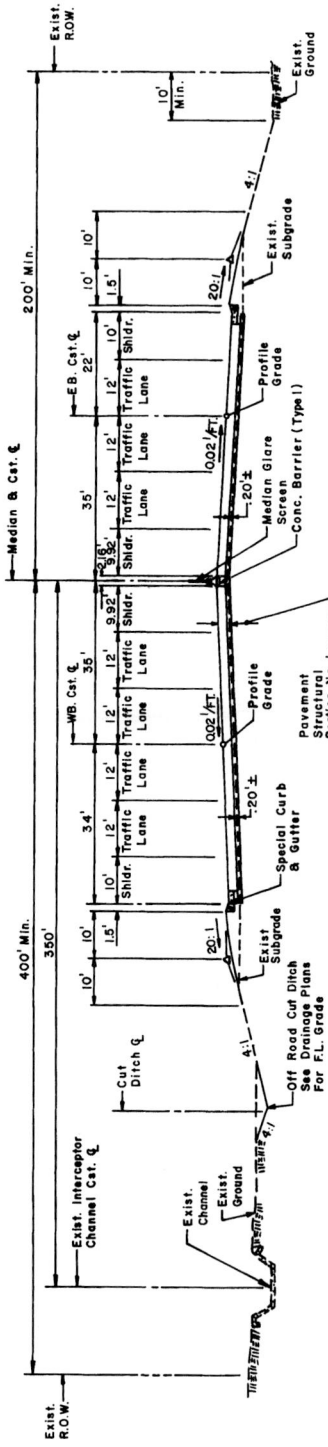

FIGURE E3-5 Superstition Freeway: typical section—eastbound and westbound roadway.

E3-24

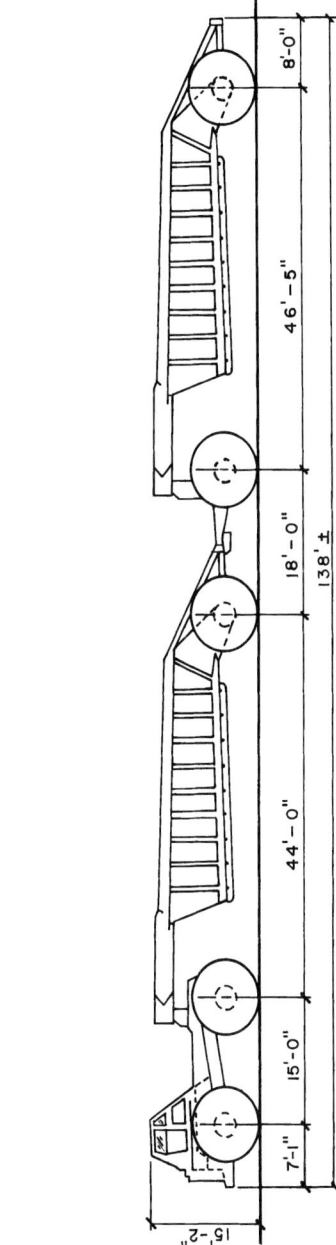

FIGURE E3-6 Superstition Freeway—diagram of the hauling units.

E3-25

The conversion of old Alligator Alley to an interstate highway was divided into 11 construction projects. The major features of I-75, Alligator Alley, are summarized as follows:

- 75 mi long (50 mi in Collier and 26 mi in Broward Counties).
- 4-lane divided interstate with 88-ft median (for future lanes or rail if needed).
- Design speed 70 mi/h.
- Terrain varies from sawgrass wetlands to wetland prairie, including piney woods and cypress domes.
- Interchanges—Snake Road Interchange on the Miccosukee Tribe Indian Reservation in Broward County.
  —S.R. 29 Interchange in Collier County.
- 36 wildlife crossings (Florida panther, etc.).
- 12-ft wildlife fence.
- Parallel canals for water distribution and borrow needs.
- Two rest area/recreational access facilities and four additional recreational access-only facilities.
- Tolls in effect until non-federal-aid commitments paid.

For the contractors, most of the challenges revolved around environmental issues rather than engineering concerns.

Most design and construction issues involved earthwork. A-8 Material (muck) varied from 1 to 15 ft in depth and had to be removed beneath the median and roadway and from over the parallel canals, which were excavated for water distribution and borrow needs. This material was stockpiled and used as a muck blanket for growing grass.

For excavation in wet areas, contractors generally used large drag lines and crawlers instead of wheel loaders. The tracks were usually the widest and longest available. The crawler units were also the smallest that could still do the work.

Naturally, the ability of a machine to stay on top of soft ground is affected by its ground pressure measured in lb/in^2 of ground contact; *shear,* which is the load on the edge of the track or tire; and total weight. Total weight affects deep mud which may creep or flow from under the machine.

Cleats cut and churn up their footing but may be necessary for traction on wet, slippery surfaces. For Alligator Alley construction, the contractors generally used cleatless pads for their crawler excavators because of the flat terrain.

As the muck was excavated along Alligator Alley, it was also used to form a berm between the roadway and the canal to prevent any toxic spills (overturned tankers, etc.) from entering the canal. This was a major concern because of the depth of the canals in relation to the Florida Aquifer. Extremely hard cap rock in much of Collier County required blasting to excavate and crushing to utilize the rock in the embankment. Also, montmorillonite clay was encountered in western Broward and eastern Collier Counties. This material, which retains moisture and is very plastic, was utilized in the embankment but required careful handling and extensive drying.

Embankment construction for the new westbound roadway required the standard interstate highwater clearance to be used. The criterion provided 3 ft of clearance between the highwater elevation (50-year flood) and the bottom of the base. When the original two lanes of S.R. 84 (Alligator Alley) were constructed, the highwater clearance criterion required only 2 ft between the bottom of the base and the highwater elevation. The existing roadway maintained this criterion; however, the pavement was resurfaced to provide the desired structural strength and to develop the proper cross slope. Complete reconstruction of segments of the eastbound roadway were required where changes in vertical alignment were made.

The improvement to the drainage patterns specified in the Environmental Impact Statement (EIS) required that a discontinuous water distribution system be developed during the construction of I-75. This system partially restored the natural drainage patterns in the area.

The construction of the water distribution system was the major drainage consideration along the project. The purpose of the system is to allow the waters collected north of the highway to circulate under the roadways, and be gradually dispersed from the system south of the highway.

In order to develop the water distribution system, the following work was performed with the I-75 project:

1. The existing canals which parallel Alligator Alley were altered.
2. Additional excavation to complete the system was performed south of the highway in Collier County and north of the highway in Broward County.

A second requirement of the EIS specified the blockage of existing canals to prevent overdrainage (and loss of wildlife habitat, etc.) of the surrounding area. To fulfill this requirement, blocks in the existing parallel canals were built at intersections with all north-south canals. Major damming sites were constructed at the Barron River Canal, Turner River Canal, L-28 Interceptor Canal, L-28 Miami Canal, and the L-68 Canal. The blocks at these canals were necessary to prevent the rapid southerly drain-off of the surface water.

In addition to the specific locations previously outlined, supplemental blocks were built to complete the water distribution system. These blocks were spaced in the canals to take advantage of existing water crossings under Alligator Alley.

A 12-ft-high, chain link right-of-way fence was constructed for the entire length of the interstate. At each water and wildlife crossing, the fences tie to the structures to allow unobstructed crossing under the interstate by the wildlife.

The fence and wildlife crossings (36) constructed have improved the survival chances for the endangered Florida panther. Only approximately 30 are known to exist today. Several other rare and endangered species of animals have had their chances of survival improved by the restoration of former wetlands and the restriction of access to the area.

Successful contractors in Florida and elsewhere are becoming more and more conscious of their activities' impact on the environment. They are increasing the scope of their services to provide environmentally safe and socially acceptable intrusions that some contractors have previously callously ignored.

The success of the Alligator Alley project can be measured by the many sightings of wildlife (including the Florida panther) using the new crossings, and the improved east-west transportation corridor across south Florida. (See Figs. E3-7 and E3-8.)

CASE STUDY: ROUGH TERRAIN, GLENWOOD CANYON

Glenwood Canyon, the final link of Interstate 70 as it crosses the mid-section of the nation from Baltimore, Maryland, to Cove Fort, Utah, was opened to traffic on

FIGURE E3-7 Alligator Alley. Bobcat using one of the animal crossings. *(Photo courtesy of Jim Valentine, Florida Department of Transportation.)*

October 14, 1992. The opening marked the end of a 12-year, $490-million construction effort. Glenwood Canyon is probably the most environmentally sensitive highway project completed to date in the United States. The beauty of the 12.5-mi-long gorge was preserved as segments of the four-lane interstate highway were constructed.

Glenwood Canyon, about 150 mi west of Denver in the White River National Forest of Colorado, is a gorge over 2000 ft deep, that was carved by the Colorado River over 70 million years ago. Because of the formidable barrier imposed by the surrounding terrain, the canyon has served as a transportation corridor for over 100 years. A railroad was built at the base of the south cliffs in 1887, followed by a primitive road at the turn of the century along the north bank. This road was improved in the 1930s to a narrow two-lane highway (U.S. 76) which has served ever since as the principal ground transportation link from Denver to the far western regions of the nation.

The years of controversy, confrontation, and frustration that followed the designation of Glenwood Canyon as the route for I-70 developed into a design and construction process which resulted in a segment of highway that has no equal. What environmentalists and other opponents of the project demanded was a complex transportation/environmental/recreational infrastructure that pushed the imaginations and talents of some of the finest professionals in the country to the maximum. Environmental concerns aside, the extremely difficult physical conditions and heavy traffic volumes, combined with potential administrative and funding obstacles, drove a process that can only be fully appreciated by those who were intimately involved.

The project is a prototype for future highways in environmentally sensitive terrain. Serving as a major commuter corridor, this 12.5-mi stretch of winding highway was the most dangerous in Colorado. Accident rates and fatalities were higher than on any comparable facilities. The bottleneck imposed by Glenwood Canyon discouraged the casual traveler and tourist, impeded the efficient flow of goods and services (with adverse economic effects both locally and nationally), and restricted the free enjoyment of recreational activities which are endemic to this special place. The

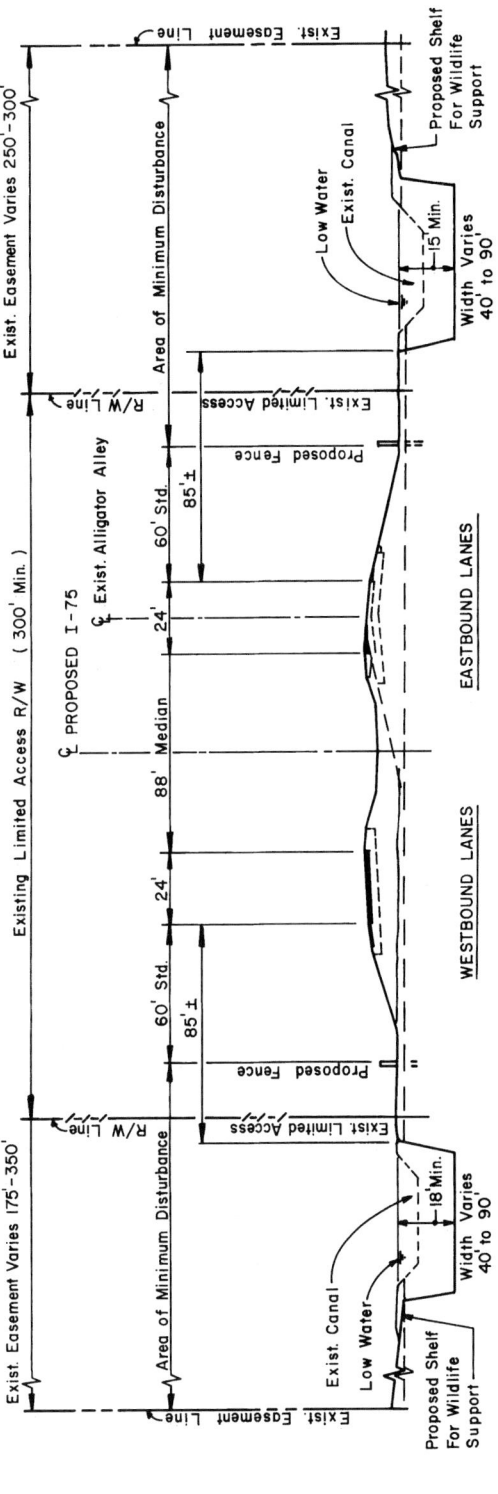

FIGURE E3-8 Alligator Alley. Typical section in Broward County.

E3-29

damage done to the canyon during earlier construction resulted in aesthetic and physical destruction of the environment and caused serious rockfall problems along the entire route. The construction of I-70 through the canyon has corrected all of these deficiencies.

The Colorado Department of Transportation (CDOT) design team used resourcefulness to plan the solutions to the severe limitations placed on the project. These project objectives were agreed to initially:

- Be flexible to accommodate changing conditions.
- Address and resolve constructability issues.
- Accommodate the concurrent needs of contractors and motorists.
- Utilize available funding in a meaningful manner.
- Afford construction continuity through seasonal variations.
- Provide environmental protection and maintain recreational access.

This was accomplished by breaking the project into more than 40 individual construction contracts. These were further broken down into segments which could be manipulated in order to optimize prevailing conditions. For example, construction operations at one location could affect how traffic is detoured through an adjacent or remote contract several miles away. The scheduling ultimately became so complex that no individual contractor could be allowed to operate independently on his contract. The interface between, and overlap of, concurrent and adjacent contracts frequently resulted in one contractor's no-work zone becoming another's designated work site. Because of the severe restrictions pertaining to available operating space imposed by environmental needs, traffic operations, seasonal river hydraulics and geological features, imaginative adaptations were required.

A typical solution involved multiphase construction of many major elements, especially retaining walls and bridge substructures. Care was used to make sure contractors would not build themselves into a corner. Detailed constructability studies analyzed each operation with respect to other operations as well as the delivery of materials and the ability of equipment to operate. Especially important was the capability of cranes to lift and reach when placing heavy steel and concrete components. While these studies did not direct contractors how to construct their projects, they demonstrated that they could be built, which significantly helped in the preparation of meaningful and responsive bids.

To solve the problem of traffic control (18,000 vehicles per day) CDOT took all traffic control away from the individual contractors. This resulted in total coordination between all contractors, traffic control personnel, and the traveling public. The reduced delays, improved construction efficiencies, and significant safety enhancements were benefits. Without the early recognition and planning for traffic impacts, it is doubtful that this project would have succeeded while satisfying the commitment to complete construction under traffic.

Pioneering in the Use of Materials and Methods

Innovations abounded in Glenwood Canyon. Following are examples of unique applications involving the use of materials and methods.

Earthwork operations were scheduled for materials balance. With limited materials resources available in the canyon or nearby and no suitable place to dispose of spoils, the scheduling and packaging of the project had to consider recycling of materials as a major aspect. For example, major sections of retaining walls had to be constructed before excavation of the tunnels could begin. As a result, the tunnel contractor only had to handle the excavated material as backfill behind the walls. Any large boulders that resulted from rock excavations were used for landscaping purposes as well as for placement in the river to create and improve fish habitat.

Injection grouting was used to stabilize talus slopes which had to support bridge foundations. As construction of the first bridges in the canyon got underway, engineers encountered numerous large voids in the talus slopes which created the potential for major settlement problems. In one instance, one abutment of a cast-in-place box girder bridge experienced sudden and potentially catastrophic settlement after the falsework was removed and the weight of the superstructure was transferred to the abutment. To resolve this situation, grout was injected under the abutment, thus creating a stable footing and saving the bridge from major damage.

Rockfall protection structure (fences) that partially deflect under impact were used. Utilizing *flex posts* developed by CDOT for the project, the impact of rolling/falling rocks is absorbed and in some cases the excess energy transferred into the fence is enough to toss the rock back up the slope. This unique technology involves construction posts by bundling flexible steel strands (identical to those used for prestressing concrete) and encasing them in a steel pipe while the bottom is encased in a concrete foundation. The remaining unencased section allows the post to bend under impact without experiencing permanent deformation. The fence itself is made of ordinary wire mesh.

Core-drilling equipment and explosives were used to construct bridge foundations. At one location requiring extremely heavy steel bridges, existing foundation materials consisted of an armored layer of river gravel, intermixed with clay and boulders, that was not stable enough to carry the enormous loads, yet was impenetrable to conventional pile-driving or caisson-drilling operations. The solution called for the use of core-drilling equipment, which could penetrate this material, to drill a deep hole into which explosives were placed and then detonated. This resulted in a column of pulverized material through which steel H piles could be driven with relative ease. This process was successfully repeated hundreds of times until all piles were in place.

This project saw revolutionary use of reinforced earth retaining wall technology. Reinforced earth technology (using steel strips or mesh embedded in the backfill to hold precast retaining wall panels in place) has been used for many years. But until implemented on the Glenwood Canyon project, the use of full-height wall panels had never been considered as an alternative to the traditional tiered arrangement of smaller panels placed in increments as the backfill and metal strips are placed. Furthermore, this is the first application of this type of wall system in conjunction with cantilevered pavement slabs.

First-of-its-kind tunnel technology in North America. A trend-setter for "smart highway" technology of the next century, the Hanging Lake Tunnels have been fitted with an incident detection and management system that is just being introduced in the United States. A complicated computer algorithm allows numerous systems to operate in concert with each other based on information gathered by various types of operational, environmental, and physical condition detectors located in the tunnels and on the approaches. Vehicles are identified by each registering a unique electronic footprint upon entering the tunnel; the vehicle is then tracked by the computer until it has safely emerged from the other end. The computer predicts each vehicle's arrival in each of the tunnels' 16 traffic control zones; failure of a vehicle to arrive in a zone triggers the appropriate incident management response. The tunnel control complex was also designed to accommodate future *intelligent transportation system* (ITS) technology.

Innovations in Construction

The Glenwood Canyon project invited—even demanded—innovation in various areas of construction, since in some cases it would have been impossible to build some of the structural elements using conventional methods. While in most cases traditional equipment was used to accomplish the task, this equipment was frequently called on to carry out activities for which it was not designed. Following are some of the special construction techniques that evolved out of this project.

Terraced cross section involving precast retaining walls and cast-in-place, post-tensioned pavement slabs cantilevered beyond the face of the walls by six ft were used. Tensioning strands were installed parallel and perpendicular to diagonal expansion joints such that two-way post-tensioning could take place along only one side of the slab. This saved substantial amounts of time in that subsequent placement of slab pours could take place before stressing of previous sections.

Special pier caps permit architectural continuity. Designers early in the process standardized many visual elements of the project. This included single-column piers for all bridges and viaducts. In the case of steel girder bridges, this created difficulty in that the outside girders would not be resting on the narrow pier. To support the girder, an integrally framed girder/pier cap detail was developed. The pier cap was then filled with concrete and transversely post-tensioned. This gives the impression that the outside girders are floating unsupported from pier to pier, yet resulted in a structurally sound configuration.

Rock reinforcement was used as permanent support for the Hanging Lake Tunnels. Never before used in the United States for a highway tunnel, this technique requires no active structural support. The favorable geologic conditions, confirmed after construction of an exploratory pilot bore, permitted designers to devise an operation of sequential excavation and installation of rock bolts for permanent support. The bolts, placed around the circumference of the tunnel, reinforce the rock surrounding the opening, thereby creating a self-supporting rock arch. Only a thin concrete lining, finished with ceramic tiles, was required for the interior.

Temporary straddle piers make possible construction of the Hanging Lake Viaduct. While erection gantries have been used extensively for overhead bridge construction, the Hanging Lake Viaduct presented an unusual challenge. In one location, encompassing a length of about four spans (total of 800 ft), the entire superstructure had to be supported by outrigger or *straddle* piers which spanned about 28 ft. This was necessary to carry out the construction directly over the old highway which continued to carry two lanes of traffic, squeezed between the four pairs of pier columns. Since these piers were not compatible with the standard single pier, provisions were made to install the final, permanent piers and remove the temporary ones—after traffic was removed from the old roadway and placed onto the viaduct.

Impact on Physical Environment, Unusual Aspects, and Aesthetic Values

Virtually every element constructed, as well as the construction techniques themselves, in Glenwood Canyon was in some way influenced by environmental and aesthetic factors. Following are brief descriptions of environmental and preservation measures utilized:

1. Terraced system of precast retaining walls (15 mi) and cantilevered pavement slabs including.
   a. Painting of walls to match canyon cliffs.
   b. Natural shadowing to lessen wall heights.

2. Aesthetic use of bridges (39 exceeding 6 mi in total length) to preserve natural landforms.
   a. Crossing natural draws.
   b. Spanning side hill talus slopes.
3. Overhead erection of bridges and viaducts (see Fig. E3-9).
   a. Overhead erection gantry imported from France was the chosen method.
   b. Limited vegetation clearing.
   c. Strategic pier placement saved natural rock formations.
4. Landscape protection.
   a. Landscape appraiser utilized.
   b. Plan to minimize destruction.
   c. Inventory of native vegetation.
   d. Liquidated damages for unnecessary destruction.
   e. Fencing during construction.
5. Preservation and restoration of landform.
   a. Alignment adjustment.
   b. Limited and controlled blasting.
   c. Rock sculpting.
   d. Staining of newly exposed rock for a naturally weathered appearance.
6. Revegetation and reclamation.
   a. Revegetation test plots.
   b. Installation of over 150,000 new plants.
   c. Irrigation systems.
7. Earth-sheltered comfort stations in rest areas.

The success of the project can be attributed to the many men and women who dedicated as many as 12 years of their careers to this monumental undertaking. CDOT won the 1993 Outstanding Civil Engineering Achievement Award from the American Society of Civil Engineers for completing the final line of Interstate 70 through Glenwood Canyon. Numerous other awards testify to the projects' success including:

- The Richardson Medal for the Hanging Lake Viaduct presented at the 1993 International Bridge Conference in Pittsburgh, Pennsylvania
- Engineering Excellence Award, Florida Institute of Consulting Engineers
- Prestressed Concrete Institute Design Award
- Excellence in Highway Design, Merit Award, 1986 FHWA

Supporting Information

The construction of I-70 in Glenwood Canyon began in 1980 and has involved 15 construction contractors for more than 40 separate construction contracts over the 12-year construction periods. Following are pertinent facts about the project. See Fig. E3-10, illustrating a typical section.

FIGURE E3-9 Erection gantry allows construction of French Creek Viaduct without disturbing the landscape—I-70 balanced cantilever bridge construction with gantry crane. *(Photo Courtesy of Ray Schmahl, Flatiron Structures Company.)*

Highway Facts

- Independent eastbound and westbound roadways each 12.5 mi long.
- A total of 39 bridges and viaducts with a combined length of 6.1 mi.
- The Hanging Lake Tunnels, twin bores each 4000 ft long, are the most technologically advanced tunnels currently operational in the United States.

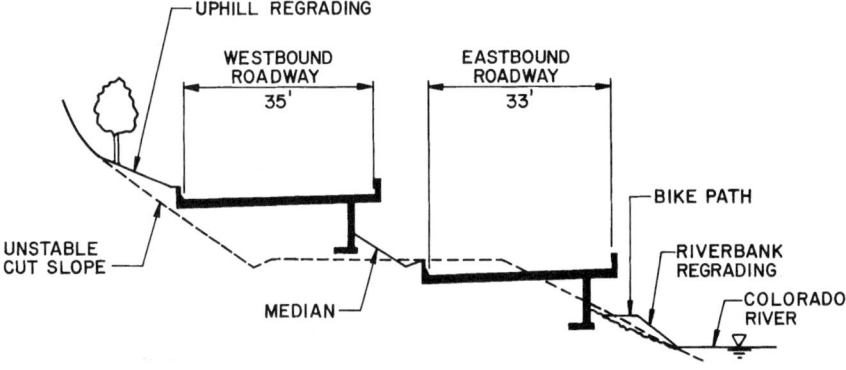

FIGURE E3-10 I-70 typical section—terraced concept.

- More than 30 million lb each of structural steel and reinforcing steel were used on the project.
- Over 400,000 yd^3 of concrete were required.
- Tunnel excavation resulted in 250,000 yd^3 of material that had to be used elsewhere on the project.
- Recreation facilities include a full-length bike path, four full-service rest areas, and special facilities for launching rafts, boats, and kayaks.
- Total project cost: $490 million
- Project owner: Colorado Department of Transportation

FUTURE CONSTRUCTION

The future of new heavy highway construction in the United States is dependent on a number of factors including:

- Natural supply of oil
- Development of alternative fuels
- Available funding
- High cost of maintenance
- Congestion mitigation and air-quality policies
- Availability and economic feasibility of other transportation modes
- Environmental restrictions
- Basic law of supply and demand
- Economic factors

Work opportunities for the members of the highway community should increase if Trust Fund obligations are less severely restricted. There will also be many other work opportunities in the United States as the private sector is tapped as a source for funding highway improvements.

Funding for highways in the United States comes from motor fuel and use taxes which are deposited in the Federal Highway Trust Fund, state and local agency accounts. The Trust Fund is made up of two accounts—highway and mass transit—and is dedicated for funding of federal surface transportation programs. Several states also have the related taxes dedicated for capital improvements and maintenance of the highway system. Some states place these highway taxes in their general funds and the State Departments of Transportation must compete with other agencies for their use.

Federal Highway Trust Fund receipts will net over $19 billion per year in the mid-1990s. Obligations of federal-aid highway dollars have been restricted by Congress as a tool to balance the federal budget for a number of years. Obligations under the 1991 Intermodal Surface Transportation Efficiency Act (ISTEA) of 1991 have been approximately $15 billion per year. The balance in the Trust Fund averages more than $10 billion.

ISTEA was described as the tool to create jobs, reduce congestions, rebuild the infrastructure, maintain mobility, help state and local governments, address environmental issues, and ensure America's ability to compete in the global marketplace of the 21st century. Samuel K. Skinner, Secretary of Transportation under President

George Bush, stated that ISTEA would maintain and expand the nation's transportation system, foster a sound financial base for transportation, keep the industry strong and competitive, promote safety, protect the environment and improve the quality of lief, and advance U.S. technology and expertise. Effective October 1, 1993, a new motor fuel tax of 4.2 cents per gallon became law. All of this extra tax collected will be used to offset the federal deficit so there will be no additional funds available for construction. Federal gasoline taxes totaling 18.2 cents per gallon are extended through September 30, 1999.

The provisions of ISTEA are effective through fiscal year 1997 and will, of course, be revised by Congress at that time. ISTEA's major provisions affecting highways are outlined in the following paragraphs.

Program

Instead of four federal-aid systems, there are two systems:

- The National Highway System (NHS)
- The Interstate System, which is a component of the NHS

Plus, a new block grant type program, the Surface Transportation Program, will be available for all roads not functionally classified as local or rural minor collector. Thus the federal-aid program will encompass about 920,000 mi and will be based on a new framework.

National Highway System. The National Highway System will consist of 155,000 mi (plus or minus 15 percent) of major roads in the United States. Included will be all interstate routes, a large percentage of urban and rural principal materials, the defense strategic highway network, and strategic highway connectors.

Interstate System. Although a part of the NHS, the Interstate System will retain its separate identify and will receive separate funding. Provided are:

- Complete funding of Interstate Construction ($7.2 billion).
- Interstate Substitute highway projects ($960 million). (Interstate Substitute transit projects are funded at $325 million in Title III.)
- An Interstate Maintenance program, at a total of $17 billion, finances projects to rehabilitate, restore, and resurface the Interstate System. Reconstruction is also eligible if it does not add capacity. However, high-occupancy-vehicles (HOV) and auxiliary lanes can be added.

Surface Transportation Program

The Surface Transportation Program (STP) is a new block grant type program that may be used by the states and localities for any roads (including NHS) that are not functionally classified as local or rural minor collectors. These roads are now collectively referred to as federal-aid roads. Bridge projects paid for with STP funds are not restricted to federal-aid roads but may be on any public road. Transit capital projects are also eligible under this program.

The total funding for the STP over the six years is $23.9 billion. This may be augmented by the transfer of funds from other programs. Each state must set aside 10 percent for safety construction activities—i.e., hazard eliminations and rail-highway crossings—and 10 percent for transportation enhancements, which encompass a broad range of environmental-related activities. The state must divide the remaining funds by formula to each area over 200,000 population and the remaining areas of the state.

Congestion Mitigation and Air Quality Improvement Program

The Congestion Mitigation and Air Quality Improvement Program directs funds toward transportation projects in Clean Air Act nonattainment areas of ozone and carbon monoxide. These projects will contribute to meeting the attainment of national ambient area air quality standards. If a state has none of these nonattainment areas, the funds may be used as if they were STP funds.

Total funding for the program is $6 billion. The funds are distributed based on each state's share of the population of air quality nonattainment areas weighted by degree of air pollution. A ½ percent minimum apportionment is guaranteed to each state.

Bridge Replacement and Rehabilitation Program

The Bridge Replacement and Rehabilitation Program is continued at a total authorization level of $16.1 billion or $2.7 billion/year for the six-year program. This will be used to provide assistance for any bridge on a public road. Newly eligible projects under the program are bridge painting, seismic retrofitting, and calcium magnesium applications. A bridge discretionary program is continued, with a new timber bridge component, with $400 million in funding.

Federal Lands

The Federal Lands Program authorization, previously available through four categories, is now provided through three categories.

- Indian reservation roads
- Parkways and park roads
- Public lands highways, which incorporate the previous forest highway category

Total funding for Federal Lands is $2.6 billion. The funds are allocated on the basis of relative needs. The forest highway portion of Public Lands Highways and the indian reservation roads authorizations are allocated by administrative formula.

Special Programs

There are 539 congressionally designated highway projects in six broad groups:

- High-cost bridge
- Congestion relief
- High-priority corridors on NHS
- Rural and urban access

- Priority intermodal
- Innovative projects

There are other special projects and provisions throughout the act that receive separate funding, some with contract authority from the Highway Trust Fund and some requiring annual appropriations. Many of these special projects were added to ISTEA during the joint committee meetings between the House and Senate Transportation Committees to iron out a compromise bill. Since the vote was close in the House of Representatives, most of these 66 projects came from the House side of Congress. Most of the projects will serve the needs of the public in a positive way; however, their priority may not have been as high as ISTEA provides them. Several of the projects would have been completed 10 years or more hence versus the six-year funding period of ISTEA. With funding obligations restricted to 80 percent of dollars collected, these projects could displace higher-priority projects.

National High-Speed Ground Transportation Programs. A magnetic levitation (Maglev) prototype development program is authorized at a sum of $725 million ($500 million from the Trust Fund and $225 million from the general fund).

Scenic Byways Program. Grant funds totalling $50 million are authorized for the planning, design, and development of state scenic byway programs. In addition, an interim Scenic Byways grant program is funded at $30 million to allow states to undertake scenic byways projects.

There are many other features of ISTEA too numerous to list here; however, two others are the new *management systems* category and *toll road* provisions.

Management Systems. The state must develop, establish, and implement management systems for the following categories:

- Highway pavement
- Bridge
- Highway safety
- Traffic congestion
- Public transportation facilities and equipment
- Intermodal transportation facilities and systems

Toll Roads. Tolls are permitted to a much greater degree than in the past on federal-aid facilities—i.e., roads, bridges, and tunnels. Types of work that may be done are:

1. Initial construction of toll facilities (except for interstate)
2. 4R work on all facilities
3. Reconstruction or replacement of free bridges or tunnels and conversion to toll facilities
4. Reconstruction of free highways (except interstate roads) to convert to toll
5. Preliminary studies to determine the feasibility of the above work

For the first time private entities may own the toll facilities.

The future of new heavy highway construction elsewhere in the world is a different story. There is a huge demand for new highways in Mexico, China, Russia, and

India, to name a few key countries. The continent of Africa frequently has famine relief shipments shipped in from points around the world only to have it bogged down in the inadequate highway network.

Future opportunities for new heavy highway construction in third-world developing nations depends on the success of emerging free-trade agreements and the support of the United States and the international community in a mutually beneficial joint venture. Several countries are able to trade mineral rights for the capital necessary to build highways and communities.

The following case study was provided by Michael Giaramita.

CASE STUDY: URBAN FREEWAY RECONSTRUCTION, U.S. 59/SOUTHWEST FREEWAY HOV LANE PROJECT, HOUSTON, TEXAS

History

The Southwest Freeway is a portion of U.S. 59 South located in the southwestern quadrant of the city of Houston, Texas. It begins at Shepherd Drive and ends at Beltway 8/Sam Houston Tollway, a distance of approximately 11 mi (see Fig. E3-11). It is the only freeway from the southwest side of Houston and it serves four major activity/employment centers: the Galleria, Greenway Plaza, the Medical Center, and the Central Business District. This freeway was constructed in the early to mid-1960s and consisted of four lanes in each direction inside IH 610, West Loop, and three lanes southwest, outside of the West Loop. Due to the explosive growth in Houston in the late 1970s and early 1980s, level of service F (bumper-to-bumper traffic, less than 15 mi/h) was commonplace on the Southwest Freeway. During the mid-1980s, vehicles traveling on the Southwest Freeway exceeded 220,000 per day; the original roadway was designed for much less. Besides overusage, the roadway was deteriorating at an ever-compounding rate due to the quantity of heavy truck traffic. In the late 1970s, TxDOT began planning for the reconstruction of the Southwest Freeway to add capacity and bring the freeway up to then-current design standards. Essentially at this same time, the Metropolitan Transit Authority of Harris County implemented the Accelerated Transitway Program. This program consisted of constructing transitway, now referred to as *high-occupancy vehicle* (HOV) lanes, facilities on the Southwest and Northwest Freeways in Harris County, Texas.

In 1983, a joint agency partnership was formed between the TxDOT and METRO. Tasks and responsibility were assigned to each agency and, in 1983, TxDOT completed schematic designs for the reconstruction. In 1984, METRO and TxDOT began the environmental process which resulted in the preparation of a full Final Environmental Impact Statement (FEIS). The FEIS received a Record of Decision, approval, in 1985

For design and construction purposes, the project was divided into four segments. METRO managed the detailed design of Segments I, II, and III; TxDOT designed Segment IV with their own in-house forces. Design began in 1986 and was completed in 1988. Construction began in 1989 and was completed in late 1992.

There were a total of approximately 130 parcels of land to be acquired for the project. All parcels were acquired prior to the start of construction.

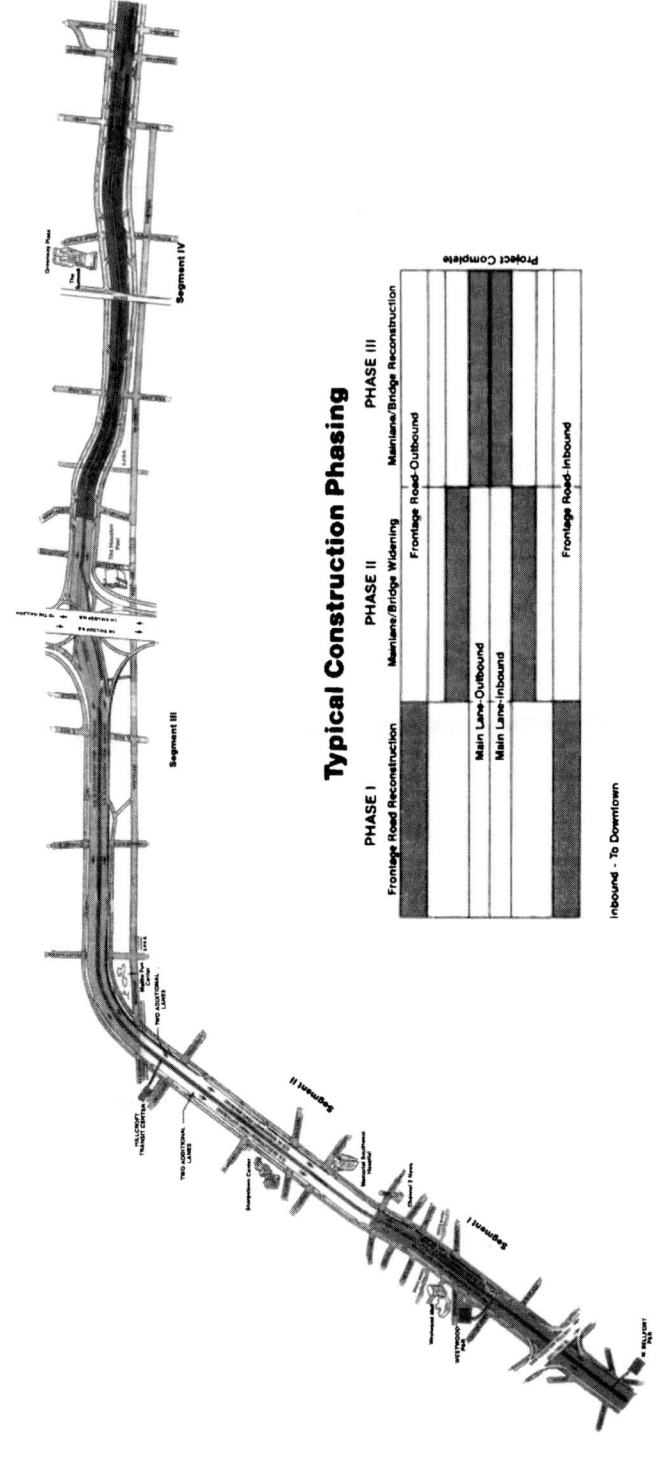

FIGURE E3-11 The U.S. 59 Southwest Freeway Project—typical construction phasing.

Project Description

The four segments of the project are described as follows:

Segment	Limits	Length (mi)	No. of bridges
Segment I	Beltway 8 to Beechnut	2.7	4
Segment II	Beechnut to Westpark	2.6	3
Segment III	Westpark to IH 610	2.6	4
Segment IV	IH 610 to Shepherd Drive	2.7	7
		10.6	18
Transitway lane		11.5	3 T-ramps

Each segment was further divided into three phases, for maintenance and protection of traffic purposes. Phase one consisted of widening and reconstructing the frontage roads. Phase two included construction of additional outside main lanes, including bridges and all egress and ingress ramps. Phase three completed construction of the center of the freeway, including the HOV Lane (see Figs. E3-12 and E3-13).

In areas of the project, the main-lane capacity was doubled, from three to six lanes in each direction. However, on average, two additional lanes were added in each direction for the length of the project. In addition to the added main-lane capacity, a 20-ft, 6-in wide, one-way reversible HOV lane was added throughout the project, and beyond. The HOV lane extends 0.9 mi past the main-lane reconstruction to serve the West Bellfort park-and-ride lot.

Of the 11 main-lane bridges in Segments I, II, and III, nine were demolished and reconstructed. Rehabilitation became cost prohibitive due to the condition of the bridges and amount of work required to bring them up to then-current design standards. Of the remainder, one bridge was widened and overlaid and one bridge was added over a major thoroughfare. All bridges in Segment IV were widened and overlaid with 4 in of concrete. Three of the existing seven bridges were raised 1 ft to provide adequate clearance underneath them. Three T-ramp bridges and one flyover ramp on the HOV facility were constructed to provide access to two park-and-ride lots, one transit center, and a major thoroughfare/activity center.

New frontage roads were constructed, adding at least one lane and as much as two lanes (four lanes total) in some areas with three lanes (five lanes total) added at the major street intersections. In addition, major utilities were relocated, new storm drainage facilities were constructed, high-mast lighting replaced typical street lights, all signing was updated, and sound barrier walls were constructed to abate noise.

Based on high traffic volume, quantity of heavy trucks, and a paving life of 30 years, design recommended a 13-in pavement thickness for the mainlanes, 9-in pavement thickness for the frontage roads, and an average of 7-in pavement thickness for bridges. In total, 1.7 million yd^2 of concrete paving/bridge slabs was placed.

Goals of the Project

The following project goals were established during the FEIS and design stage of the project by TxDOT, FHWA, businesses, and the traveling public.

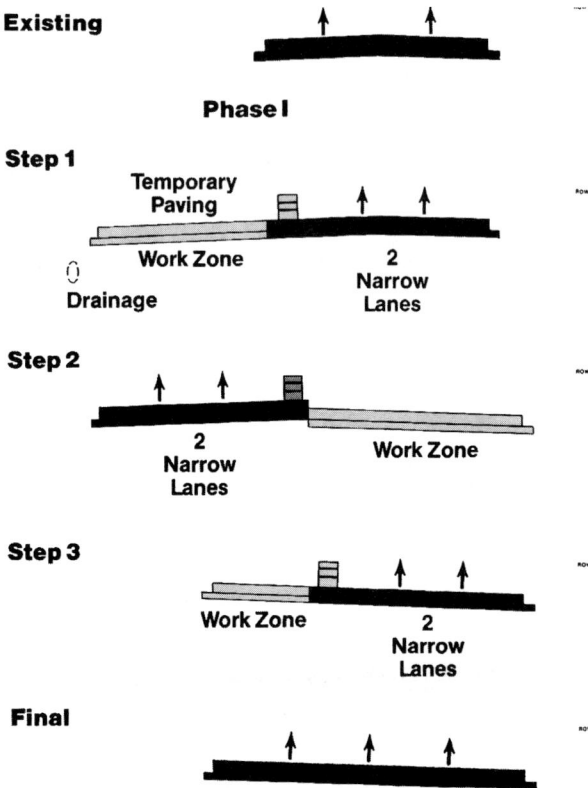

FIGURE E3-12 Southwest Freeway/Transitway typical construction sequencing, southbound frontage roads.

- Complete the project on time.
- Complete construction with no claims.
- Minimize the inconvenience to the traveling public.
- Minimize disruptions to the businesses located along the route (the largest-volume mall in Houston was located directly in the center of the project) and also minimize inconveniences to the shopping public.
- Provide a uniform appearance throughout the project and minimize swerving transitions.
- Take a very proactive management approach to the project.

All of these goals were met.

Project Costs/Funding

The total construction cost was approximately $213,000,000. The portion constructed under contract to TxDOT was $200,000,000, including $4,000,000 in change orders, as follows:

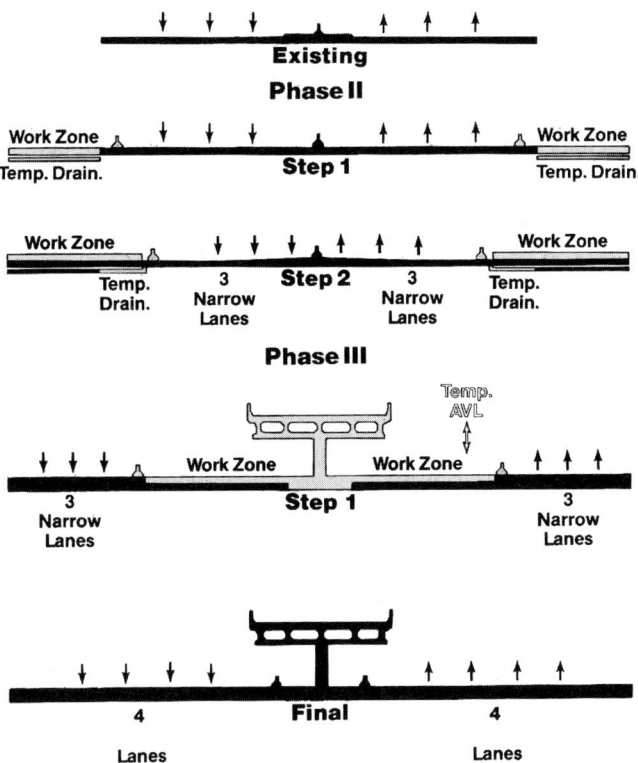

FIGURE E3-13 Southwest Freeway/Transitway Project typical construction sequencing—main lines/AVL.

Segment I	$51,000,000 (including West Bellfort T-ramp)
Segment II	$49,000,000
Segment III	$60,000,000
Segment IV	$40,000,000

This cost included reconstruction of the main lanes, frontage roads, bridges, the HOV lane, and associated T-ramps. The cost for the construction of METRO's off R.O.W. facilities, including one new 1000-space park-and-ride lot, rehabilitation of one park-and-ride lot, and one new transit center which has 800 parking spaces was $13,000,000.

Funding for the rehabilitation of the freeway was provided by the Federal Highway Administration, (75 percent) and the state of Texas (25 percent). Funding for the HOV Lane and ancillary facilities was provided by the Federal Transit Authority (66 percent), TxDOT (18 percent, based on total cost), and METRO (16 percent).

Innovations in Construction and Management

Establishment of a Central Project Management Office. Due to the size of the project and the coordination between all of the project participants, TxDOT/

METRO established a *central project office*. The central project office housed the supervising resident engineer, the office engineering staff, surveying and quality control staffs, schedule control, and community outreach function for the entire project. The office was located in a shopping mall in the center of the project. Project status displays were mounted in the storefront window for public information. This was the first central office ever established by TxDOT.

Maintenance and Protection of Traffic Plan Development. One of the goals on the project was not to lose any of the existing capacity on the main lanes or frontage roads during peak traffic periods. To achieve this end, a Maintenance and Protection of Traffic committee, with members from each of the three design consultants, TxDOT, and METRO, was established. The result of this committee was a reduction from the eight-phase traffic control plan originally established in the FEIS, to three coordinated phases. Over 1,000 drawings were prepared directing the contractors where to work and what work to perform.

Switching Traffic along the Entire Project, Simultaneously. With the aid of the CPM schedules, accurate predictions were made of when all four contractors would finish construction of the new inbound main lanes. A series of meetings was held with the four contractors and *Operation Big Switch* was born. Two months in the future, September 7, 1991, was the date established when all four contractors along the 10.6 mi of freeway would move traffic from the existing three inbound mainlanes located at the center of the highway to the newly constructed three inbound mainlanes located at the freeways outer edge. On September 7, 1991 the switch was made, resulting in a safer facility and cost savings of over $600,000, because contractors didn't need to build costly, weaving transitions between new and existing pavements. In fact, one contractor stated that because of the work concentration on the inbound side, a gain of almost two months was made on the project schedule. Productivity among all the contractors increased, resulting in greater profits for the work accomplished.

Use of Construction Management. Because of the potential to delay the traveling public and high road user costs estimated at over $183,000 per day, the project personnel focused not only on quality and cost issues, but also on the management of time and claims. Special management provisions were written and included in the bid documents. These provisions were strictly enforced. Project personnel from TxDOT, the contractors, and METRO met every two weeks to assess progress against the approved schedule. Action item lists were prepared and everyone on the project was made accountable for his or her actions in order to keep the project on track. The project was completed on time and $30,000,000 under the original budget.

Development of a Customized CPM Scheduling System. An off-the-shelf, high-end, computerized CPM scheduling system was customized in order to accept and provide feedback on resources needed to construct each activity. Items customized included the ability to input quantities of material to complete each activity, production rate to install material, labor/crew size, and hours per day to be worked. All this information was used to calculate the activities duration. From this information, worker-hour curves were also generated which depicted labor-intense periods. This helped in managing the needed resources for the project. The information contained in the system also allowed the management staff to easily assess the status of each activity and was the single most effective source of information in the mitigation of claims.

Use of CPM Scheduling for Determination of Contract Time. CPM techniques were used to develop preconstruction schedules. Detailed schedules were developed for each segment which established the time requirements for each phase and the entire project based on the Maintenance and Protection of Traffic plans and quantities of materials to be installed. These schedules enabled the project's expected duration to be pared back by 20 months, from 60 to 40 months. The schedules also established the intermediate milestone dates and the work sequence for the entire project. These milestone dates were written into the contract documents.

Development of Special Provisions for Management. In order to increase the level of management control on the project, special provisions were developed for items such as work times (hours of work), definition of a day (calendar versus work day), liquidated damage provisions (Phase I—$10,000, Phase II—$15,000, and Phase III—$10,000 per day. Traditionally, liquidated damage assessments were in the $100 to $500 range). Provisions were also made for all the contractors to use the same customized CPM scheduling system which the TxDOT purchased for each contractor, as well as requirements for its use and reporting, project and work sequence descriptions, and a special provision on how to prosecute and progress the project.

Construction of Bridges from the Roadway. Many of the existing bridges were demolished and reconstructed due to inadequate geometry, mainly involved with flattening the vertical curve to improve sight distance. In order to expedite reconstruction of many of the project's bridges, contractors constructed the abutment walls from the top of the existing main lanes. The auger pile rig drilled from the top and placed the drilled shaft to the bottom of the pier cap, thus eliminating the need for forming the superstructure. In fact, many contractors used the existing fill as the form for the bottom of their pier cap.

Use of Retained Earth Technology. In order to construct a new bridge on the project, the embankment leading up to the bridge section had to be stabilized because of the way the traffic control was established—construct the outside lanes first, then reconstruct the middle. The contractor built up the earthen embankment using the Hilfiker method—a fabric/mesh reinforced earth panel. The traveling public watched for several months as the earthen embankment rose up inches from their travel lanes. Both inbound and outbound sides were constructed simultaneously. The walls were several thousand feet long and were approximately 22 feet high.

T-Ramp Bridges. The three HOV T-ramps were constructed as one-way reversible facilities. However, the design allows for conversion of the current one-way reversible facility to a two-way facility within the existing median area, if TxDOT allows the use of the inside shoulders for this purpose. The current bridge width allows for two-way operation; the ramp-up sections will require widening for two-way operation. The conversion of the current one-way operation to two-way is shown in Figs. E3-14 and E3-15.

Concrete Paving Design. When TxDOT began evaluating pavement design, the condition of the existing pavement, current maintenance requirements, traffic volumes (current and projected), and a design life of 30 years were all considered. Based on this criteria, the resultant pavement thickness was 13 in. TxDOT standards require two layers of reinforcing to control pavement cracking in continuously reinforced concrete pavement when pavement thickness is 13 in or greater.

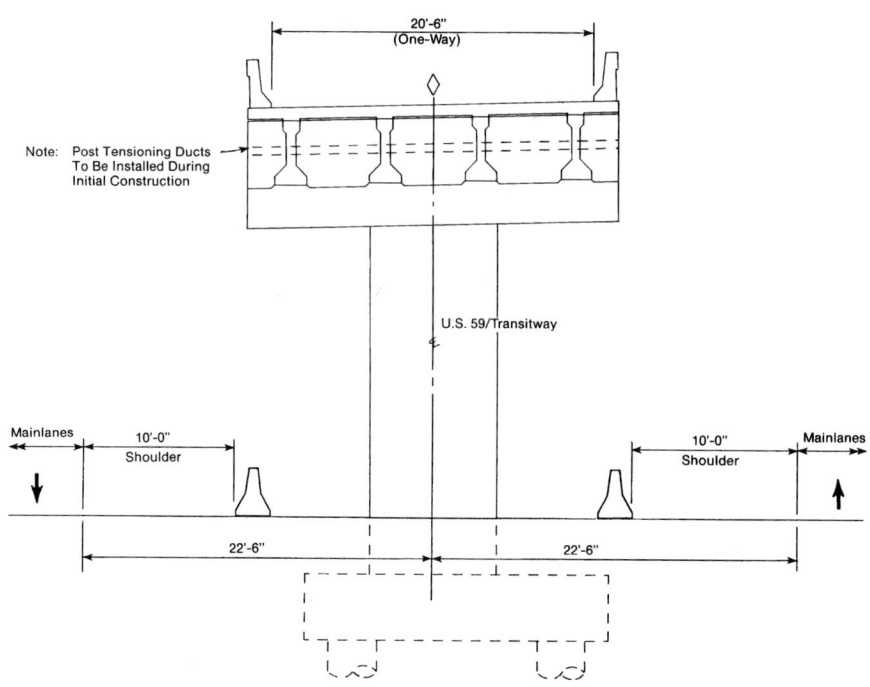

FIGURE E3-14 Initial one-way operating—structure typical section 20-ft, 6-in roadway.

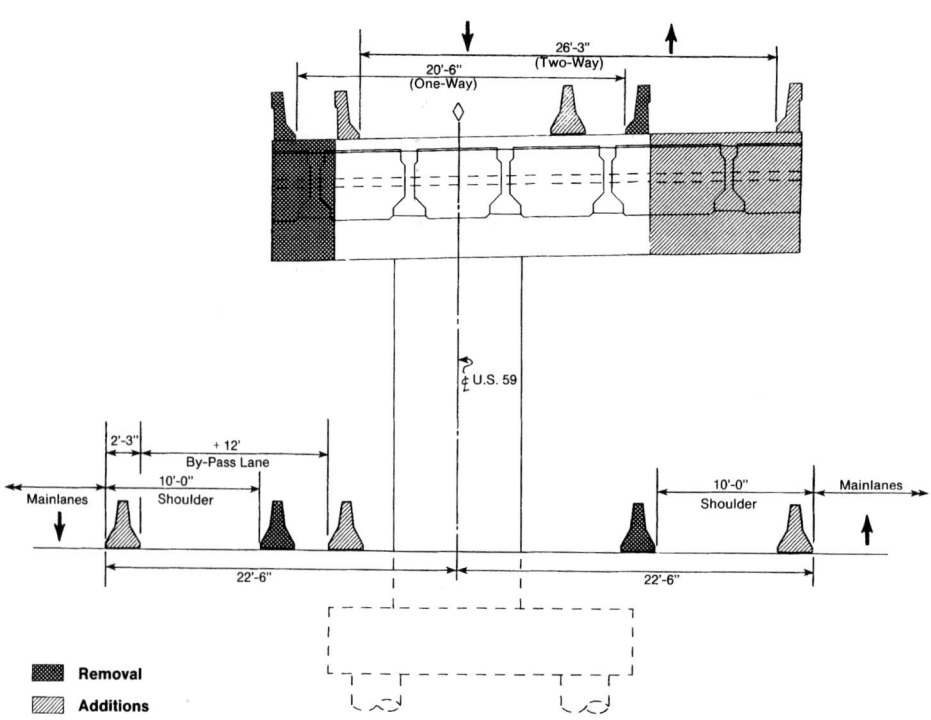

FIGURE E3-15 Typical section 20 ft, 6 in, widening to 26 ft, 3 in.

E3-46

Reversing of All Egress and Ingress Ramps. All of the egress and ingress ramps on the freeway were reversed to aid in the management of traffic. Egress ramps that were originally constructed to egress traffic just prior to the intersection were reversed so that traffic would ingress the freeway. This provided for safer traffic operations, allowed for continuous driving on the frontage roads without having to stop for a traffic light, and eliminated the weaving normally associated with egress ramps out away from the intersection.

Braided Ramps. Due to the limited amount of rights-of-way, TxDOT constructed braided ramps to allow egress and ingress to the freeway at certain locations. Basically, a braided ramp is two ramps which cross each other—one on top of the other—to simultaneously allow traffic to egress and ingress the freeway at the same location.

Special Awards/Significant Notoriety

- Feature article in December 2, 1991, issue of ENR
- Selected by TxDOT, District 12, for FHWA Golden Award
- Received Best Quality Construction award for 1992 by American Concrete Paving Association
- Selected and won the Project Management Institute, Houston Chapter, Project of the Year Award for management excellence, placed 2d in regionals

CHAPTER 4
PIPELINES*

James J. O'Brien, P.E.
O'Brien-Kreitzberg, Inc., Pennsauken, New Jersey

---

* Some of the material in this chapter is from Laurence L. Lyles, "Underground Utility Construction," and Harold C. Price, "Transmission Pipelines" in Section 26, "Pipelines," Stubbs and Havers, *Standard Handbook of Heavy Construction,* 2d ed., McGraw-Hill, New York, 1971.

SECTION 1

UNDERGROUND UTILITY CONSTRUCTION

It is quite likely that utility construction is the second-oldest type of construction that humanity has undertaken. Probably the first thing humans sought was shelter, where they improvised and improved their living conditions. It is also likely that the second convenience humans wanted was ready access to water, a need which could be satisfied by digging a ditch and having the water come closer to the cave. The Greeks and Romans developed sophisticated utilities and had both running water and workable sewer installations in some homes. Lead pipes for carrying water were found in the ruins of Pompeii.

The types of work that will be discussed in this section will include all utilities installed underground in a city to deliver energy or service to a business or home. These utilities specifically include water service; sanitary sewers; gas, electricity, and telephone services; and storm sewers. The construction problems involved with each of these utilities are very similar, although there are variations in the actual installation of the different types of pipes and conduit. Therefore, most of the steps in construction will be discussed in terms that will apply to all utilities. Each phase of utility construction will be discussed in its normal sequence.

PRELIMINARY WORK

Planning

Planning is the first and most important step in the construction of underground work; it is also the point at which the most money can be made. Thorough planning effort is necessary for a profitable job. The actual order in which various operations will be performed is rather simple because it is obvious that a trench must be dug (or a tunnel bored) before pipe can be installed. The timing—when each operation is to occur and how long it will take—is the key to a smooth and profitable operation. Jobs go well because they are planned well and not because of luck.

The proper execution of an underground job requires a balanced crew. This means a crew made up of the right equipment and labor to have all phases of the project progress at the same speed. To obtain a balanced crew, goals or objectives must first be determined: the number of feet of pipe to be installed per day, week, or whatever. Then consideration must be given to the labor and equipment to be used, to see if they will provide the required production. Invariably, there will be one phase of the work that will control the speed of the whole operation. When this is found, the work must be planned in such a way that the controlling phase will progress at the requisite speed. For example, the excavation of the trench with the equipment planned may not give the footage desired. One may then decide to change types of excavators or possibly use more than one machine.

Once the production requirements for the controlling work phase have tentatively been established, other pieces of equipment and some of the labor on the job that do not have full days of production will invariably be found. For example, the pipe-installation crew may need only four hours to install all the pipe that the excavating crew can provide for in eight hours. To maintain a balanced crew, additional work must be found in their crafts, such as making tie-ins or setting fire hydrants, for those workers who would otherwise not have a full, productive day. If this is impossible, the possibilities of increasing speed of the controlling phase of the work must be looked into so that, in the end, full production may be obtained from all the workers and most of the equipment.

Locating Crosslines

Crosslines, existing underground structures that are in the work area, are probably the key item in setting production objectives because of the time required to excavate under them with any type of equipment. The number of crosslines on the job can be determined during the bidding stage; however, their exact location and depth is a job problem. Determining the location of all crosslines should be accomplished before the entire crew is brought on the job because, especially in the old areas of a city, you will uncover surprises.

To do the job properly and safely, owners of all possible crosslines in the area of construction should be contacted. The utility companies, city departments, fire-alarm services, local TV antenna companies, etc., will be just as anxious as you are that no lines be cut and usually will give full cooperation. Many will come to the construction area and mark the location of their lines. Others will have you bring in a map and then indicate their lines on it.

After obtaining all available information, the only sure way to safeguard the existing lines is to have a small crew of workers actually pothole and expose the lines at the points where the trench will cross them. The crosslines should be exposed at least one full day in advance of excavation so that they will not cause delays for an expensive crew. This is particularly important where the utility line is to be set to a given grade, because you must be sure that there is sufficient clearance at the crossline for the utility line to be correctly installed.

PAVEMENT REMOVAL

Pavement Breaking

Pavement breaking is the shattering or cutting of the pavement into pieces that can be economically removed from the area to be excavated. Techniques vary according to the type and thickness of the material to be broken. Several manufacturers have developed fine, self-propelled machines that break pavement very economically. It goes without saying that the tougher the breaking job, the larger the machine should be. The most commonly used types of machines are the ones where a 1000-lb or larger weight with a cutting or shattering tool is lifted about 9 ft in the air and allowed to fall free and strike the pavement. Other machines use pneumatic power to accomplish the same result.

An experienced operator can cut a straight line as marked on the pavement. Generally, the most economical procedure is to cut the pavement about 1 in outside the planned trench on both sides and then break or shatter the material in between the outside cuts. Applying this procedure of three passes, 4-in pavement can be broken at the rate of about 110 ft of trench per hour and 6-in pavement at about 70 ft of trench per hour.

The use of hand pavement breakers (jackhammers) should be limited to jobs where the amount of pavement to be broken is very small. Since it is not possible to determine the efficiency of the operation of pavement breakers by watching or listening, it is advisable to have them tested periodically to determine the impact they are producing as compared to new tools. Such a test will show many tools that are only 20 to 40 percent efficient.

By removing any protruding pavement material of 2-in or larger size after breaking, it is often possible to allow traffic to run over the broken pavement prior to its removal. This permits a better traffic flow in the construction area.

Pavement Removal

Pavement removal can be accomplished by the use of a rubber-tired skip loader of a 1½-yd or larger size with a scoop attached to the front of the bucket. This scoop should be smaller than the width of the trench and slightly deeper than the pavement to be removed. If the pavement has been thoroughly shattered, this is an economical method for removing the broken pavement and loading it into dump trucks. As an example, an operator should be able to remove pavement from a 3-ft-wide trench and load it into a dump truck at the rate of about 7 yd^3 of pavement per hour. Keeping in mind the balanced crew, a full 8-h shift may be considerably more than is needed for this particular task. Other duties which can occupy the same workers and equipment will then have to be planned, such as removal of excess excavated material, placing base rock, etc.

Sawing Pavement

If the pavement is sawed at all, it can be sawed either before or after pavement breaking. The procedure to be followed by the contractor is usually established either by the specifications or by the customary practice in the community where the work is being done. The purpose of sawing pavement is to give an even line to repave against and leave the street looking neat, even though patched.

The advantage of sawing *before* breaking pavement is that the sawing takes the place of cutting the edge with another piece of equipment. As a result, the combined cost of pavement breaking and pavement sawing is reduced.

The disadvantages of sawing ahead are twofold: (1) in old and fractured pavement, pieces of pavement outside the sawed area are likely to become dislodged and break out; (2) the excavating equipment might catch a tooth on the edge of the pavement and break it back outside the sawed area. Both of these can leave you with uneven edges, the very thing that sawing was meant to prevent.

The advantage of sawing *after* pavement breaking and just prior to repaving is that you can always leave a neat, straight line to repave against. The disadvantages again are twofold: (1) the trench to be repaved is usually wider than it would otherwise have been, and (2) after sawing it is necessary to chip off the narrow piece between the sawed line and the trench. Both of these add cost to the repaving process.

Good, diamond-bladed saws are made by several manufacturers. The deeper the cut and the greater the quantity of sawing that has to be done, the larger and more powerful the saw should be for economical operation.

Using an 8-hp saw, one may expect to saw 1 in deep in asphaltic concrete at the rate of approximately 165 linear ft/h, and 2 in deep at the rate of about 120 linear ft/h. For sawing concrete, these rates should be reduced by 40 percent.

EXCAVATING THE TRENCH

Equipment and Methods

Excavation of a trench in its simplest form is merely a question of how to remove the earth from the trench for the lowest cost per cubic yard. When choosing the type of equipment to excavate a trench, the contractor will, of course, give first consideration to the equipment the company owns. However, it is not always cheaper to use a machine that the company owns, especially when it is the wrong machine for the job. The types of excavating machines that will be discussed are wheel-type trenchers, ladder-type trenchers, and backhoes.

Where they can be used, wheel-type trenchers are the most economical means of digging a trench. The limiting factors on wheel trenchers for utility work are the width they will dig (usually a 36-in maximum) and the depth they will dig (usually 7 ft or less). In addition to these limitations, they leave a large quantity of earth underneath each crossline intersected. As a result, before you can say that a wheel trencher is the best machine for a certain job, the cost of the trenching plus the cost of removing the extra dirt at crosslines must be compared to other means of excavating the trench.

The ladder type of trenching machine is the next most economical method of removing earth from a trench. Generally speaking, the cost per cubic yard for excavation with a ladder trencher is approximately 50 percent higher than with a wheel trencher. Ladder machines come with almost any width and depth of cut and, compared to backhoes, leave an even-sided trench that can be shored economically and safely. In general, where there are few crosslines, the ladder trencher will give more feet of trench per day than a backhoe. Ladder trenching machines are relatively slow in lowering and raising the bucket line and, as a result, jumping crosslines and turning corners becomes expensive. They also leave a sizable amount of earth after jumping a crossline, but this can be reduced if the operator will back the machine so that the ladder digs back under the crossline. The remaining amount of earth will then be less than a wheel trencher leaves.

When using either the wheel or the ladder trencher, the largest buckets and the least amount of side-cutter width should be used within practical application. The reason for this is that the larger the buckets on the machine, the more earth will be excavated every time the wheel or ladder makes a complete revolution.

Backhoes serve several useful purposes in excavating for utilities. First, a backhoe will dig over and around a crossline with a minimum of delay. By straddling the trench already dug with a trenching machine, a backhoe is often used to remove the excess earth left at the crossline. The cost per cubic yard of excavating a complete trench with a backhoe is more than the cost with a trenching machine for the part that a trenching machine can do. However, when the cost of the complete backhoe excavation is compared to that of excavating with a trencher plus a backhoe to excavate under numerous crosslines, there are many jobs where the use of a backhoe for all excavation is most economical.

Hydraulically controlled backhoes seem to be the fastest moving and most economical ones on the market. They come in all sizes, from the smallest ones that can straddle an existing trench for the purpose of excavating around crosslines to the large and very large ones. Backhoes can be found to dig almost any depth that a contractor would need for utility work. Where all the excavated material has to be hauled away from the construction site, it is easy to excavate and load the material into dump trucks in the same operation. There are three main disadvantages to a backhoe: (1) it is more expensive in cost per cubic yard to get the earth out of the trench, (2) it is very difficult to shore a trench dug with a backhoe because of the uneven sides of the trench, and (3) the bottom of the trench is left uneven.

Exact production figures for the trench excavation are impossible to give due to the many variables such as soils, working conditions, etc. The following examples are given as guides.

Assuming that the trench to be excavated is 2½ ft wide and 5 ft deep and will not be dug to grade, the production in cubic yards per hour will approximate the figures shown in Table E4-1.

If the trench were to be excavated to grade, as required for sewer lines, etc., the production would be decreased by about 25 percent.

Assuming that the trench to be excavated is 4 ft wide and 12 ft deep and will be dug to grade, the production in cubic yards per hour will approximate the figures shown in Table E4-2.

Up to this point the discussion of trench excavation has been limited to normal, stable soils. A discussion of some of the problem soils follows.

Rock Excavation

Rock excavation refers to the removal of materials encountered in trenching that are too hard to be removed with available excavating equipment. When rock has to be blasted, there are a few procedures that can maximize safety. First, when there are any buildings within 200 ft of the blasting, special precautions should be taken. The dirt overburden in the trench area should be removed with a backhoe so that the trench thus formed can be used to help minimize the lateral force of the blast. Adequate mats made of hemp, wire, or timbers should be placed over the charge to prevent any rocks from being thrown into the air. The proper type of blasting material should be chosen to minimize the possibility of damage to building foundations. Holes for the blasting material should be drilled about 6 in below the bottom of the proposed trench. Only experience will tell how close the holes should be drilled to

TABLE E4-1 Production/Excavation, yd^3/h

Conditions	Wheel trencher	Small backhoe, 44-hp diesel
Good soil—no crosslines	200	50
Hard soil—no crosslines	100	20
Good soil—crossline every 50 ft	95	35
Hard soil—crossline every 50 ft	60	16

TABLE E4-2 Production/Excavation, yd^3/h

Conditions	Ladder trencher	1½ yd^3 hydraulic backhoe
Good soil—no crosslines	180	160
Hard soil—no crosslines	100	90
Good soil—crossline every 50 ft	100	125
Hard soil—crossline every 50 ft	60	60

each other. Alternating the holes from one side of the trench to the other will usually produce a trench that is easier to clean out.

After the charge has been set off, the most economical means of removing the shattered rock is with a backhoe. If the overburden that has been removed before drilling and blasting is fairly deep, it may be both safer and cheaper to completely backfill the ditch after blasting and then excavate it with a backhoe.

An alternate to blasting is the use of a hoeram machine.

Dewatering

The method used for removal of underground water will, of necessity, vary depending on the amount of water entering the trench, the type of soil, and the hardships or hazards caused by the water. Methods of dealing with these problems will be discussed, starting with the most severe water conditions and then going on to the less serious conditions. In almost every case, it is advisable to remove all the underground water possible before starting excavation.

The most common and most successful method for "drying up" an area is by means of wellpoints. Since the use of wellpoints is an accepted practice, details that are applicable under all conditions will not be repeated. The first consideration for wellpoints is to determine the direction from which the underground water is coming. If it is seeping in from just one side of the proposed trench, then one row of points set to intercept the water will be sufficient. However, if the underground water is a pool or is able to seep into the excavation from more than one direction, it will very likely be necessary to place a row of points on each side of the excavation. The positioning of the wellpoints should give adequate consideration to the space required for the excavator, for storage of the excavated material, and for the workers and equipment needed to install the utility lines.

Where the quantity of water to be removed is not great and where the soil is sufficiently permeable so that well pumping can lower the water level for a fairly good distance (25 ft or more), the use of a series of wells with individual pumps has proven economical. This system requires the drilling and graveling or sanding of wells at appropriate intervals along the proposed trench. The wells can be any size from 12-in diameter up and should have a center casing into which the suction hose can be placed. These wells should be at least 6 or 8 ft below the depth to be excavated, because the water level, as it is pumps down, will slope toward the suction inlet in a conical shape. Before the excavation work approaches each well, a pump (usually gasoline powered) should be set up at the well and put into operation with the suction hose near the bottom of the well. The water pumped from the wells should be discharged at a location where it will not flow back into the work area. State environmental regulations may require a plan for collecting and settling out solids before discharge to storm sewers. This system has proved economical where there are short areas to be dewatered and in sandy areas that are easy to dewater.

In some areas, the soil may be compact enough to leave vertical banks that, after shoring, are safe even with water entering the trench. Under the circumstances, the trench can be overexcavated by a depth of 6 to 12 in and the overexcavated section backfilled with coarse gravel. Water will then flow through the gravel to the lowest point of the trench, where it can be pumped out. This system, the *sump method,* will work in areas where a relatively small amount of water is entering the trench at one time due to the firm, tight soil through which the trench was excavated.

Shoring

Trench shoring should first be examined from the viewpoint of its requirements. Alternatives for shoring installation and removal can then be explored.

Requirements. Aluminum shores, using hydraulic jacks to hold them in place, have proved to be the most economical means of shoring trenches. They are light in weight, and the worker-hours saved in moving and installing them more than offset their high initial cost. They have a long life and can be used repeatedly, if properly maintained. In addition, because they can be installed from the surface, they afford greater safety to the installer than wood shoring (see Fig. E4-1).

Installation. As previously mentioned, the workers who are installing aluminum shoring with hydraulic jacks can all stay on the surface. If timber shoring is to be installed, the same principles of safety should be applied. All possible work should be done from the surface; i.e., the vertical shores should be placed from the surface and the first horizontal brace installed just below the surface from above. Then, using a ladder, a worker should go down into the trench just enough to install the next lower brace or trench jack, etc. By this means, the trench is made safe above as the worker descends to install additional horizontal braces. Trench jacks and horizontal braces should never be used as a ladder for getting in or out of a trench, as they are not designed to take a vertical load.

FIGURE E4-1 Aluminum-hydraulic trench shoring. *(Allied Steel & Tractor Products, Inc.)*

Figures E4-2 and E4-3 are examples of wood sheeted trenches.

A trench should never be left open without shoring. Shoring should be installed immediately behind the excavator.

Removal. When planning for the removal of shoring, it is safest to anticipate the possible collapse of the trench sides. The newly installed utility line will then be safeguarded routinely by being covered with loose or compacted fill before the shores are removed. For conditions where the trench will almost certainly cave in on removal of the shores, the trench can also be filled up to the bottom horizontal brace. It is then safe to go down on a ladder and remove this brace, after which the additional trench space can be filled up to the next horizontal brace or screw jack.

If the trench is expected to stand after the removal of the shoring, all the shoring can be removed just ahead of the backfilling. This should not be done until all work within the trench has been completed and the newly installed utility line has been protected or covered. A worker can then use a ladder to descend to the bottom horizontal trench jack and remove it. The remaining horizontal jacks are removed as the worker ascends the ladder. Shoring removal is hazardous work, despite the best safety precautions, and one person should never be permitted to work at this alone.

Sand Excavation

Methods of excavation in sand should be carefully studied. Adequate protection must be provided for crosslines.

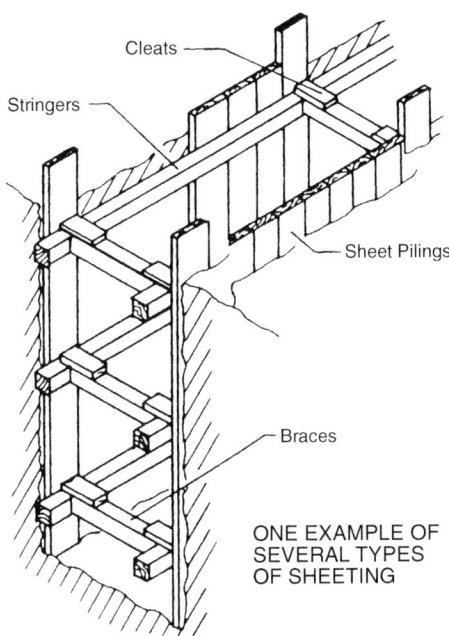

FIGURE E4-2 Example of wood sheeted trench. *(Source: OSHA.)*

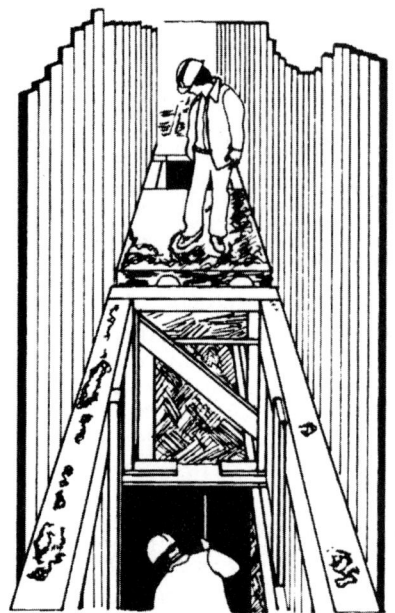

FIGURE E4-3 Example of wood sheeted trench. *(Source: OSHA.)*

Methods. Considering both economy and safety, trench excavation in sand is the greatest hazard that a contractor encounters in utility work. If the trench is to be excavated in a new and undeveloped subdivision or on a beach where the width of the trench is not important, the answer is both simple and relatively economical: a large backhoe or dragline may be used to bail out the sand. However, the trench to be excavated is usually located on a city street or highway where there is a limit to the width of the excavation that can be made. It may be necessary to drive steel sheeting prior to the trench excavation. This method has long been used, and under the most severe conditions, is the only safe means of getting the utility line installed. However, two other methods are more popular when they can be applied:

1. The trench-shield systems consist of a steel shield attached to a ladder trencher. The workers then have a working area which is completely protected on four sides by steel siding, appropriately braced. The length of this shield must be sufficient to allow for the installation of a length of pipe and its joining with the last piece laid within the confines of the shield. The back side of the box, away from the trencher, has an opening large enough to clear the pipe that has been installed. As the trenching and pipe laying proceed, the excavated material must be placed in the trench immediately behind the shield by means of a conveyor or a skip loader. Where this system can be used, it is probably the safest and best means available. It has one major disadvantage: The size of the trenching machine ladder and the shield are such that it is very slow and expensive to remove it from the trench, and thus crosslines cannot be economically jumped. However, in many cases, crosslines can be temporarily cut and removed economically to allow the use of the trencher-shield system for excavating sand (see Fig. E4-4).

2. A more popular system for excavating in sand makes use of a similar shield with a backhoe used as the excavator. In this system the shield (see Fig. E4-5) is not attached to the backhoe but instead rides free on the bottom of the trench as a protective box for the workers. Cutting edges on both of its sides extend forward toward the backhoe, thus minimizing the unbraced space between the shield and the backhoe bucket. To move the shield, the backhoe operator reaches back with the backhoe bucket and hooks a cable onto it. The shield is then pulled toward the backhoe the distance desired. The disadvantage of this system is that the shield is not stable because it is freestanding. The sides of the excavation are rough-trimmed, and sufficient caving can occur to knock the shield out of line. In addition, controlling the grade of the pipe and the depth of the shield is difficult. The backhoe-shield has the advantage that, since it is not attached to the backhoe, it can be dragged under shallow crosslines if the sides of the excavation above the shield can be sloped to a safe angle of repose.

PIPELINES   E4-11

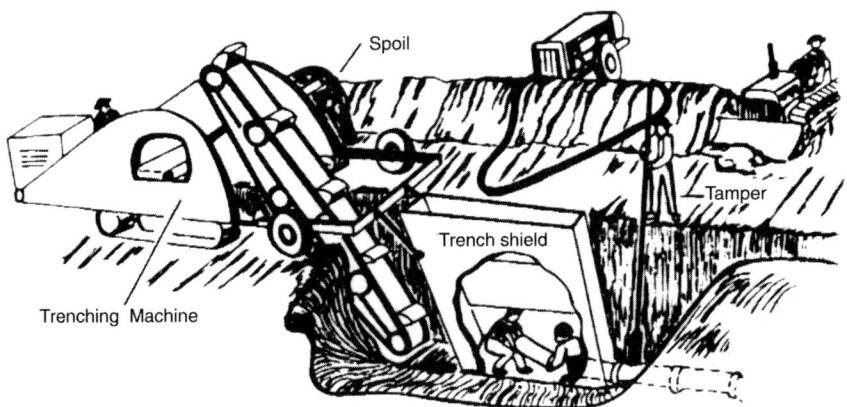

FIGURE E4-4 Trench shield. *(Source: OSHA.)*

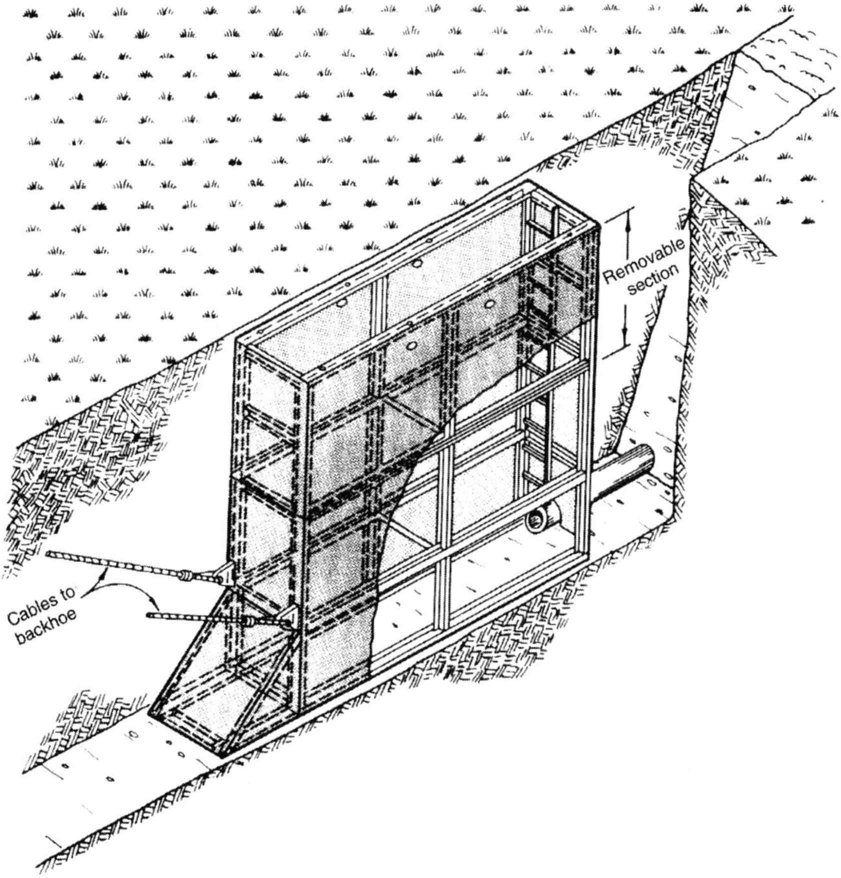

FIGURE E4-5 Trench shield. *(W. M. Lyles Co.)*

INSTALLATION OF PIPE OR CONDUIT

Water Mains

The most commonly used types of pipe for water mains are cast iron and asbestos cement, both of which have many years of successful use behind them. In larger sizes, steel-cylinder concrete pipe is widely used. Plastic pipe is also gaining acceptance.

Due primarily to the ever-increasing cost of labor, almost all types of pipe are now being produced with a rubber or neoprene type of joint. These joints have several advantages in addition to the labor-saving advantage. They allow flexibility at each joint and, in the case of steel pipe, allow the joining of the lengths of pipe without destroying the interior pipe protection by welding.

Generally, the fittings, valves, and fire hydrants for the line can be obtained with the same type joint as the pipe. However, some localities prefer a flanged joint for valves and fire hydrants to make replacement easier.

The use of flexible joints has made the proper installation of thrust blocks even more important. The size of the bearing surface of the thrust block is generally specified. The main items to which attention should be given are (1) that a good bearing surface with the pipe is obtained and (2) that the bearing surface against the trench side is against native soil or well-compacted backfill. In soft, wet, or sandy soils, larger thrust blocks may be needed. If the waterline is to be tested at pressures over 150 lb/in^2, extra attention should be given to thrust blocks.

Installation. Manual installation of water pipe is usually the most economical when the sections are 20 ft in length or less and weigh 150 lb or less. The pipe is first strung close to the trench or is rolled within reach of the trench. The workers who are standing in the ditch then reach over and lower the pipe into position. Using this method, with adequate trench prepared ahead and not too many crosslines, 2000 ft of water main can be installed in an 8-h day. In most cases, the fittings and valves will be too heavy to install manually. When possible, the best method is to prefabricate fittings and valves and have them located adjacent to the trench where they will be installed. When the laying crew reaches the location for a fitting or valve, they can cut the pipe to length and, with the help of a piece of equipment, lower the fitting into place. This will minimize the installation time for valves or fittings.

The installation of heavier lengths of pipe requires equipment to lift the pipe and lower it into place. For pipes up to 1200 lb per joint, a swing crane mounted on a small tracklaying tractor has proved economical. Side-boom tractors have proved economical for larger-sized pipe. These two types of equipment have the economical advantage of requiring only one operator and no helper or oiler.

The production attainable for installing large-diameter or heavy pipe is directly related to the number of obstacles encountered. In wide-open laying without shoring, an average of 2000 ft per day is often maintained. However, the greater the number of crosslines and the greater the amount of shoring, the more slowly the installation will proceed.

Testing. There are several practices that will make the testing of water mains easier and more economical:

1. When installing the pipe, each joint should be gauged with a depth gauge to see that the rubber gasket or ring is positioned properly in the completed joint.
2. When possible, all joints should be left exposed for testing.

3. The line to be tested should be filled to operating pressure and allowed to remain at operating pressure for 1 or 2 days.
4. Testing should not be done against old valves, or, unless it is absolutely necessary, even against new valves. Small amounts of water leaking past valves can give an unsuccessful test when in reality the line is watertight.
5. When testing, the pressure should be raised gradually to the required test pressure.
6. The line should be walked and every exposed joint should be checked visually immediately after operating pressure is put on the line and again after the test pressure is put on the line.

There are several means of bringing a water main up to pressure economically. The simplest and most economical, if it will produce the desired pressure, is the use of a small, conventional water pump. The use of high-pressure water pumps will do the job too; however, these have to be kept in perfect repair and, with infrequent use, they often prove costly. An inexpensive means can be the use of a 200- to 200-gal pressure vessel (such as a butane tank) as a reservoir for the required water and a bottle of nitrogen as the pressure source. Connect the bottom of the pressure tank to the water main by means of tubing of proper strength. Connect the nitrogen bottle to the pressure tank through a pressure regulator. By regulating the pressure of the nitrogen admitted to the pressure tank, an even, controlled pressure can be applied to the water main.

Water services building connections generally are installed with copper, galvanized steel, or plastic pipe. When possible, the use of a very small trenching machine or narrow backhoe has proved the cheapest means of excavating the necessary trench.

In areas with good pavement, curbs, gutters, and crosslines, the installation of service lines by boring may prove more economical than open-cut excavation. This can be done by the use of an air motor that rotates a bore pipe (usually ¾ or 1 in) with a cutting head attached. A short trench is excavated, usually in the location where the service will terminate, about 15 ft long and a little deeper than the planned depth of service. The bore pipe is screwed together and should be at least 20 ft longer than the length of bore desired. It is then aligned in the trench by means of metal guides to point it in the proper direction both horizontally and vertically. The air motor is attached to the rear end of the bore pipe. Two to four workers, depending on the hardness of the soil, then start the motor and push the bore pipe in the direction of the desired bore. Water is usually allowed to flow through the bore pipe, thus helping with the cutting by lubricating the hole bored and flushing out the excess dirt. With a soft soil, two workers can install a 30-ft bore in less than an hour. Harder soils take more workers and a longer time. Where water is not allowed in the boring process, a less satisfactory but still economical result can be obtained by the use of compressed air instead of water.

Sanitary and Storm Sewers

These two types of sewers will be discussed together because the installation methods are identical except for the different types of pipe.

The most important item to look into prior to commencing work on a sewer job is the location of the proposed pipe in relation to existing pipes. Because the exact depth of the lines has been predetermined to match the elevation of existing or future lines, it is usually necessary for them to be installed at the elevations indicated. Known crosslines should be exposed prior to the start of excavation to ensure that existing lines will not interfere with the space allocated for the new line.

The second most important item to watch in sewer construction is that the trench be excavated to grade as closely as possible. This will allow quicker and more economical fine grading of the ditch and also give a more uniform base for installing the pipe. Grades for pipe can vary from 0.1 to 6 percent, depending on the terrain. Needless to say, careful fine grading is necessary when working with minimal grades. When working with close tolerances and checking to see if pipe is laid to grade, it is important to remember that the flow line of the pipe is the critical point to check, not the top or bell of the pipe. The reason for this is that often the allowable variation in the diameter or straightness of the pipe will be as much as the variation allowed in installing the pipe.

The type of pipe most commonly accepted for sanitary sewers is vitrified clay. It has the characteristics necessary to withstand the chemical action from sewer gases and it is made throughout the country. If the clay pipe has plastic joining surfaces such as Speed-seal or Wedge-lock, a watertight line can easily be installed. Epoxy-lined asbestos-cement pipe is also popular and, with its pliable joint, makes a watertight line.

Storm sewers generally are installed with concrete pipe which, depending on the load conditions, is either reinforced or nonreinforced. In many instances, leakage is not a factor and cement mortar joints are all that is necessary. When a leakproof joint is necessary, a rubber gasket pipe is usually required.

One word of caution is in order concerning testing of installed lines for leaks. With both types of sewers, it is a poor practice to bid on lines with mortar joints that are to be pressure-tested. The reason is that the mortar shrinks away from the joint when drying, often cracks if it dries too fast, and will crack if there is any appreciable contraction or expansion of the pipe.

The smaller sizes of pipe weighing less than 150 lb per section can best be laid by hand by two layers. For plastic joints, it may be necessary to use a steel bar to push the pipe home and make the joint completely tight. In trenches over 5 ft deep, it is advisable to have the pipe lowered to the layers with a rope by one or two workers standing on the surface. Under conditions where the trench can be prepared, a 3- or 4-person crew can install an average of about 2000 ft of pipe per day. Quality installation requires that the body of the pipe be laid on a uniform foundation, with holes dug to make room for the bells.

A similar method of installation should be used for heavier pipe, except that equipment should be used to lower the pipe into position. In selecting the type of equipment, the key items to consider are (1) equipment that will require only one worker to operate, and (2) equipment that can move the pipe from the surface to the trench bottom with a minimum of movements.

Gas Distribution

The gas distribution installation that will be discussed is the medium-pressure, 20- to 50-lb/in^2, system generally used for distributing gas within cities. Gas transportation between cities is usually effected by high-pressure lines operating at pressures of several hundred pounds per square inch, and the gas pressure must be reduced to the distribution pressure as it enters the distribution system. The pressure is further reduced at each service connection by means of a regulator to approximately 8 in of water or a little over 4 oz of pressure.

The most commonly used pipe for gas distribution is black steel pipe, which is usually wrapped or plastic coated to protect the steel from corrosion or electrolysis. Plastic pipe is used under some circumstances. Copper pipe has been used successfully but generally is too expensive.

Pipe sizes for distribution mains will vary with the requirements; however, 2-in pipe is the most common size. Joining of the lengths of steel pipe is usually done by welding to make the joints permanent and leakproof. A qualified welder and helper can average about 50 welds of 2-in pipe in an 8-h day. Pipe is usually supplied in 20- or 40-ft lengths. Welding of distribution mains and house services is almost always done with the pipe above ground. This is possible because the pipe is usually light enough to be lowered in the trench and under crosslines easily. The pipe is rested on skids or other objects to facilitate its alignment and allow for its rotation during welding.

Pipe of 2 in or less can be welded most economically with acetylene, while 4-in and larger pipe can be welded most economically with arc welding. Pipe of 3-in diameter can be done either way, depending on the welders and equipment available. Welds which are properly made are actually stronger than the pipe itself. A utility company will generally require that a welder's competence be demonstrated by means of a test.

Most pipe for distribution is already wrapped when it reaches the job, so the contractor is chiefly concerned with the wrapping of the joints after the welds are made. There are, however, several points that should be stressed concerning wrapped pipe. It should be handled carefully to protect the wrap and should not be dragged. Joint wrapping during installation should be equal to or better than the yard wrapping. In areas where electrolysis and corrosion are a problem, high-voltage electrical current will be used to test the pipe wrapping for defects just prior to the lowering of the pipe into the trench. The bottom of the trench should be cleared of all stones, hard objects, and metal that might penetrate the wrapping. The same care should apply to the earth placed adjacent to the pipe. Where suitable material is not available from the excavation, imported sand is often required to give the necessary protection to the pipe wrapping.

Contact between dissimilar metals in underground piping causes the creation of electrical current. When different types of pipe are joined, such as copper services from a steel main line, an insulated connection should always be used. Distribution systems are usually tested for leaks with a pressure of 100 lb. A recording chart is installed after the pressure has been put in the line, and no leaking or drop in the line pressure is allowed during a 24-h period. If the line is to be buried before acceptance testing, the preliminary testing of the welds with soapy water before wrapping and while the line is under 100 lb of air pressure is a good precaution.

Excavating the trench for gas distribution lines can usually be most economically done with a wheel-type trencher with narrow buckets. If the trench is about 3 ft or less in depth, a ditch 6 in wide is sufficient. For trenches more than 3 ft deep, the ditch should be dug 14 in wide to allow a worker to stand in the bottom of the trench to clean out any debris.

Electrical and Telephone

The direct-burial methods for installing underground electrical and telephone lines are almost identical.

Plowing the cable in has proved very economical. There are machines in use that are easily moved from one location to another and which have sufficient traction and power to plow in a cable up to 42 in deep (see Fig. E4-6).

The installation of a duct system for underground electrical distribution has several peculiarities. First of all, there are invariably both primary and secondary circuits to be installed, and these normally are separated by 1 in of concrete. Secondly,

FIGURE E4-6 Cable plow. *(W. M. Lyles Co.)*

the usual structure for electrical distribution is a multiple combination of round duct conduits of the proper size and in the proper quantity, which are subsequently encased with pea-gravel concrete. Each electrical distribution structure is formed by stacking these ducts with shaped separators between them and then tying the ducts together with a soft steel wire. If this is done properly, with boards of adequate width placed between the structures and the trench sides, there is little change of the structure moving during the concrete pour. To ensure that the duct structure does not float, a worker can walk on top of it and work the concrete in place with a shovel. The concrete should be just wet enough to work into the voids between the conduits but not so wet as to encourage floating of the ducts.

A variety of materials can be used for the conduits which are to be encased. Plastic, asbestos-cement, and fiber ducts have all been used extensively, and all are easily installed in accordance with the manufacturer's recommendations.

Pull boxes, manholes, and transformer vaults are now being precast and hauled to the job. These offer several advantages over the old cast-in-place method. They allow the street to be opened and closed faster. They are less hazardous to the workers in caving conditions as there is less labor necessary below the surface. And, lastly, by being built under controlled conditions, they usually provide a more economical finished product. The only disadvantage to precasting is that when the casting yard is too far from the job, the hauling cost sometimes offsets the savings.

The installation of telephone conduits and structures is very similar to that just described for electrical structures. Telephone ducts are usually precast or formed in sections that can be hauled to the site and installed. Vitrified-clay multiple duct was for years the standard for the industry. It is still in use in many locations, although precast-concrete multiple ducts have become more popular. Most multiple-duct sections are too heavy to lay by hand, so equipment is generally used to lower them into the trench.

An ice-tong type of clamp that will go around the middle of the structure will allow the installation of the duct in its final position without the necessity of moving it manually. As a rule, a final duct structure for telephone need not be watertight, as the cables that will be installed in the conduits will be waterproof. However, all joints should be tight enough to prevent silt from washing into the duct.

BACKFILLING THE TRENCH

Shading and Bedding

There are two different reasons for shading and bedding, so they are best discussed separately.

Shading, which is the covering of the pipe with loose or compacted soil, is primarily for the purpose of protecting the pipe and preventing its movement during backfilling. Soil can be placed over the pipe by several means. Shading by hand is the old reliable means and is probably the only sure way in conditions where the soil is lumpy or rocky. Hand shading is usually necessary when the material is placed in thin layers for compaction. Shading pipe where the soil is sandy, fine, and not to be compacted in layers can often be accomplished by the use of an angledozer. Damage to or movement of the pipe will not occur if the dirt does not fall directly on the pipe. The angledozer blade should move the dirt so that backfilled material will roll ahead, around and over the pipe (see Fig. E4-7).

Bedding usually refers to the support conditions under the bottom half of the pipe. The type of bedding will affect the ability of the pipe to support structural loading without adverse effect. The importance of bedding is then a matter of the additional strength needed by the pipe to fulfill its designed purpose. The larger the pipe, naturally, the greater the need for and the greater the expense of proper bedding. The problem of getting firm dirt under the pipe and the small working space available to work alongside of the pipe complicate proper bedding.

For smaller-diameter pipe, where a good foundation is all that is needed or where a low relative compaction is accepted, flooding the area around the pipe is the best and least expensive means of gaining the desired results. Of course, this method is unsatisfactory with soils that swell when saturated. When using the flooding method, sufficient material should be placed over the pipe to ensure that it will not float when the pipe area is flooded.

When a greater degree of compaction is needed to give support to the pipe, two systems are often used. The first employs mechanical compaction of the soil below the spring line (center) of the pipe. In spite of the fact that this system is often

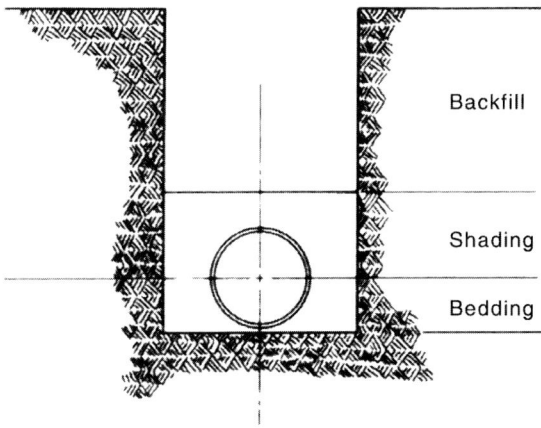

FIGURE E4-7 Trench cross section.

required, it may give only fair results in many soils. No matter how much compactive effort is placed vertically along the side of the pipe, the earth can only be pushed back under the pipe and compacted for a limited distance. The method has gained acceptance because the degree of compaction at the sides of the pipe can be easily tested; however, this is really not the location that is important for giving support to the pipe. A worker using a 33-lb pneumatic tamper should be able to compact about 4 yd^3/h when properly moistened soil is being placed in layers by others.

The second system which is in use is that of filling the trench to the spring line with sand and then compacting the sand. Dry sand can be compacted mechanically very well and will get well under the pipe, but moist sand compacted mechanically will not be forced under the pipe. The most effective means of compacting moist sand is by saturation and internal vibration with a concrete-type vibrator. By this means, relative compaction values of 80 percent are attainable. Care must be taken to be sure that sufficient water is added to saturate the sand but not so much at any point as to cause the pipe to float.

Equipment and Methods

Backfilling methods which utilize machines, where they can be employed, are the most economical. The greater the amount of earth to be moved, the larger the machine should be, etc. Where the excavated material is to be used for backfill, a track-type angledozer is generally the most economical unit. Where there is pavement that can be damaged by a tracklayer, rubber-tired equipment should be used. The angledozer has a definite advantage over the straight dozer. It can roll the dirt into the ditch if it can be angled as much as 38°, and it stays closer to the trench as compared to just shoving in a bladeful at a time. There is a definite relationship between the cost of machine backfilling and machine trenching. As a rule of thumb, it will cost about one-third as much to replace the backfill mechanically as it does to excavate mechanically. This factor seems to hold true even in the complicated or deep trenches where the excavation costs can be quite high.

Once satisfactory bedding and shading have been accomplished, the methods of backfilling are largely determined by the compaction requirements. However, there are two warnings that should be given no matter what type of compaction is necessary, and these concern care of exposed crosslines: (1) Large quantities of dirt must not be pushed in on unsupported crosslines, and (2) large quantities of uncompacted earth must not be placed under exposed crosslines. In general, great care should be exercised with regard to crosslines while backfilling, or the consequences of breaks, damage, etc., may have to be dealt with at a later date.

Compaction

Requirements. The required degree of compaction is usually spelled out in the specifications. If it is not, then the best policy where settlement of the trench is undesirable is to obtain a compaction equal to that of the existing ground.

Probably one of the most misused specifications with which contractors must comply is the one requiring 90 to 95 percent relative compaction in trenches where the existing ground has a relative compaction of anywhere from 65 to 85 percent. The result of compliance with such specifications is an island of compacted earth in a sea of relatively soft earth, and no one benefits. However, comply the contractor must, so the following paragraphs will deal with the means of this compliance.

There are several precautions that may be taken where future settlement of the trench is undesirable. First, attention should be given to compacting the earth below the spring line of the pipe. Otherwise, voids under the pipe will allow later settlement. Secondly, the depth of trench to be backfilled at one time must be controlled. The depth should be the optimum depth which, for the equipment to be used, will produce the specified compaction. A plan must be formulated for obtaining compaction under crosslines, around structures, and throughout other areas peculiar to the job.

Methods. Flooding has long been the most extensively used method for obtaining compaction. It is probably the best means available in pure sand and may prove acceptable in relatively free-draining soil such as sandy loam. Some future settlement will occur in soils other than pure sand, no matter how good a job of flooding or jetting is done, and relative compaction of greater than 80 percent is unusual. Compaction by flooding may also be acceptable for clay-like materials in farm fields, but it is not satisfactory for work in existing or future streets. As a rule of thumb, trenches should be flooded or jetted in lifts of not over 5 or 6 ft.

Proper moisture is the key to all other types of compaction, since methods of placing and compactive effort will have little effect if the proper amount of moisture is not present. Knowledge of the optimum moisture for the type of soil is essential. If this optimum is not known, it may be necessary to experiment with varying moisture contents on the job. Laboratory tests may be helpful in indicating the probable range of required moisture.

Hand compaction with the use of pneumatic tampers has long been a standard method. It is now generally considered the most expensive means because of labor costs and the necessity of keeping the thickness of the layers of material to be compacted to about 6 in. It will ultimately be used only on very small jobs or in areas where other equipment is not suitable. As an example, a worker operating a pneumatic tamper weighing about 33 lb would do well to compact to 90 percent relative compaction more than 5 yd^3/h.

There are several hand-vibratory compactors on the market which can be used in place of the pneumatic tamper when hand compaction is required. These vibrators use a combination of weight and vibration and can obtain the same results as the pneumatic tamper in much thicker layers. Needless to say, the heavier the machine, the greater the impact and usually the greater the vibrating surface and effectiveness.

Compaction of trenches with machines too heavy to be hand held is becoming more popular as a means of obtaining the desired results economically. Use of vibrating rollers where the trench is wide enough and where the crosslines are infrequent is probably the most efficient method. The sheepsfoot roller is also effective but usually requires more passes.

Since most trenches do not lend themselves to a roller type of compaction, the impact type of compactor (the same machines as those described for pavement breaking) has been developed to compact trenches economically. These machines, which basically drop about a 1000-lb weight a distance of 9 ft, have proved to be the most economical means of obtaining relative compactions in the 90 percent range.

The depth of material that such machines can compact and the speed at which the compaction can be obtained are dependent on the type of soil and the degree of compaction desired. In sandy loams, compaction in 24- to 36-in lifts is not uncommon. In clay-like material, lifts of as little as 8 in may be necessary. For compacting a suitable soil in 12-in lifts, an average of 15 yd^3/h with a machine would be reasonable. As much as 40 yd^3/h can be compacted if 3-ft lifts are used.

The advantages of such machines are (1) they are highly mobile, (2) they require only one operator, (3) they can easily compact on both side of a crossline, and (4) good results are obtained.

The disadvantage of these machines is that, if care is not exercised near the pipe or conduit, the installed pipe can be damaged or crushed. Continued tamping in an effort to obtain unrealistic results can actually move the trench sides and even raise the existing pavement on the sides of the trench.

PAVING THE TRENCH

Removing excess material is the first step in trench repaving. When compaction tests are required on the backfill, it is quite common to refill the trench to the road surface so that the street can be used until the results of testing are available. When this is done, it is best to fill the top portion of the trench with the required depth of base material. Temporary paving can be placed over this fill and permanent paving can later be installed with a minimum of effort. On lightly traveled streets, base rock is often left for a week or so without temporary pavement.

The removal of the material necessary for paving the trench can best be done with a loader, usually rubber tired. A scoop that will remove the proper width and depth of material should be installed on the bottom of the bucket edge, and the loader should be used to dig out the material and load it into a dump truck. This same method can be used with a self-loading scraper when the quantity of earth to be removed is sufficient and when the disposal site is near enough to allow for transportation by the scraper.

Cleaning pavement edges is, of course, a must if a good bond is to be secured between the old and new pavement. Either a stiff broom or an air jet will serve the purpose. For a really good job, the edges of the pavement should be given special attention just prior to the placing of the new pavement. For asphalt pavement, a bitumal spray or slurry will help provide a good bond. With concrete, wetting the edge just before the concrete is placed will improve the bond.

The types of material to be used will normally be specified, so the placing of materials will be the next item of importance. On very wide trenches the procedures for handling of materials are quite similar to those used in paving a street, so no discussion of them is necessary.

For narrow trenches, a dump truck with a narrow gate built into the tailgate can economically place materials such as base rock and asphaltic pavement. The gate allows the release of the material in a width less than that of the trench and at the same time allows the control of the quantity of material being released. Trench-paving boxes are manufactured which can be pulled behind any dump truck and give the same controls. Both methods work well, with the latter probably being the most economical on trenches of any length.

Production quantities are impossible to give due to the multitude of varieties of work required. Probably the greatest insurance of good production is the use of an experienced crew. Low costs and good-quality work can best be obtained by keeping the same workers on a crew assigned to do all the trench paving.

SECTION 2
TRENCHLESS PIPELINES

BORING

Webster describes the verb bore as "to pierce with, or as with, a rotary tool." TT Technologies, Inc. developed an underground piercing tool designed to create bore holes for the installation of cable or conduit. This tool, named the *GRUNDOMAT*®, has a patented reciprocating chisel head within a stepped cone. The tool is driven by compressed air and is reversible.

The compressed air is at 85 to 100 lb/in^2. For winter operations, the air is prewarmed to 248°F to prevent internal freezing of the tool. Table E4-3 gives the relative characteristics of the eight standard sizes.

The GRUNDOMAT chisel head (Fig. E4-8) pushes forward from the main casing at up to nine times per second. This creates a small, accurate pilot bore which the main casing follows:

The direction of the box can be pinpointed exactly before starting. The target point is sighted with an aiming frame from the starting pit to a predetermined mark on the surveyor's staff in the target pit.

The tool can also be configured to pull or drive steel or plastic pipe into position. Boring rate is up to 200 ft/h depending on diameter of bore and soil conditions.

The tool can be used vertically in a variation called GRUNDPILE.

TT Technologies, Inc., has developed a family of hydraulic-driven drilling systems called *Bor-mor*. The model numbers (1500, 1000, 500, 100) equate to the typical length of bore in feet. The Bor-mor 1500 is an easy-to-operate directional boring system that is truck-mounted for fast site setup. It features separate dedicated hydraulic circuits to provide for longer bores up to 1500 ft (Fig. E4-9). The Bor-mor sets up within 5 min for along-the-road boring and is totally self-contained. No trailer is required to haul pumps or a fluid tank.

Because the Bor-mor 1500's directional capability, surface, and underground obstacles are no problem, you can bore under highways, streams and ponds, wetlands, and irrigation ditches, as well as around buildings or alongside winding streets, virtually without disturbing the surface. The Bor-mor 1500 can be used for installing telephone, power, water, CATV, or similar utilities, whether for industrial or residential service.

TABLE E4-3 GRUNDOMAT Characteristics

Tool diameter, in	Length, in	Weight, lb	Air cons., cfm	Strokes/min
1¾	36	18	16	570
2	44	31	21	500
2½	51	55	25	480
3	58	75	35	420
3¾	67	148	46	320
4¼	74	212	60	300
5	68	280	95	350
5¾	73	396	158	300
7	74	640	215	290

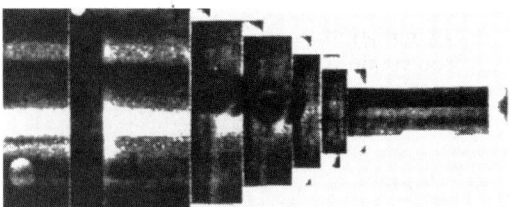

FIGURE E4-8 Reciprocating chisel head assembly.

FIGURE E4-9 Bor-mor setup.

The Bor-mor 1500 operates off the host truck's engine, enabling it to deliver up to 30,000 lb of thrust/pullback power and a two-speed spindle drive providing 1550 and 3100 ft·lb of spindle torque. It can bore up to 1500 ft at a shot and travel at up to 1 ft/s, depending on soil conditions. Boring depth can range up to 30 ft. The bore head rotates at 220 r/min, producing a 3+-in pilot bore. Its maximum backream capability is 18 in, depending on soil conditions.

Setup is practically as easy as parking the truck. This is because the unit is completely self-contained. Within just five minutes, a three-worker crew can have the unit parked, set up, and ready to start work. Hydraulic motors lower the drill frame into position from the turret on the truck. Teardown can be accomplished just as swiftly. In addition, there's virtually no site restoration or cleanup required.

Features include the following.

- *Less labor required:* Bor-mor's ease in setting up is matched by its ease in operating. While conventional trenching methods require a five- or six-person crew to complete a job, a Bor-mor system can be operated efficiently by just three people: the machine operator, a person to manage the locating system while maintaining contact with the operator, and a person to install bore rods.
- *Adjusts for angles:* Different boring angles can be used from a single site by adjusting the turret which sweeps 180° horizontally. Its vertical angle range is 16 to 46°.
- *Stable:* The drill frame requires minimal anchoring since the unit is attached to the truck.
- *Operator overview:* Bor-mor's operator station is located in the truck bed, providing a clear overview of the machine while operating.

- *Simplified controls:* After setup, operation of the unit is reduced to two main hydraulic valves that control thrust and rotation, and a flow control valve that regulates fluid flow and pressure.
- *Safety shut-off:* Self-centering valves provide instant shutdown as a safety precaution.
- *Fuel efficiency:* The Bor-mor system uses less fuel than competitive systems. It can operate for up to two days on less than 50 gal of fuel.
- *Low fluid usage:* Bor-mor provides accurate fluid monitoring. Normal fluid flow is only 1 to 2 gal/min, with a maximum of 9 gal/min, depending on soil conditions. The amount of fluid used is usually insignificant.

MICRO TUNNELING

Three examples of larger trenchless pipelines were key to the Camden County Municipal Authority program.[1]

Construction was well under way on the Cooper River Interceptor Program. As of May 30, 1986, nearly 15 mi of the 22-mi pipeline were installed. Construction was also proceeding concurrently on 5 major pump stations and 10 metering and sampling stations constituting the first phase of the Camden County Municipal Utility Authority's $600 million program. Overall, construction on the Cooper River Interceptor Program was 60 percent complete.

The construction of the gravity lines and force mains composing the Cooper River system had, of necessity, to cross numerous busy highways and rail systems. In an effort to minimize disruption to the motorists, commercial rail interests, and commuters, many of these crossings were designed to be made in tunnels or borings. In order to accommodate the interceptor pipeline, these tunnels and borings range in size from 42 in (3½ ft) in diameter to 144 in (12 ft) in diameter, and up to 130 ft in length. A total of 14 tunnels are required for the Cooper River system.

Two basic methods of tunneling were employed: one method utilizing liner plates and the second method employing a jacked casing in conjunction with an auger. Both methods required the construction of an access shaft and a receiving shaft at opposite ends of the tunnel. These shafts are constructed by driving sheeting, excavating, and bracing the sheeting after excavation. With the liner plate tunnel, a hole is cut in the sheeting at the bottom of the shaft and the liner plate is advanced on a ring-by-ring basis, either employing hand excavation or the use of a hydraulically jacked shield, with the liner plate rings being bolted together behind the shield as it is advanced through the earth. The liner plate rings are constructed of individual gasketed plates, each approximately 1 ft by 2 ft and curved so that, when bolted together, they form a complete circle or ring of the appropriate diameter. The ring sections are advanced, and grout is pumped into the space between the liner plates and the excavated earth to provide solid bearing (Fig. E4-10).

In the case of the jacked casing utilizing an auger, a casing of the appropriate diameter is hydraulically jacked through an opening cut in the sheeting and the earth within the casing is removed by use of the auger. The casing is advanced ahead of the auger to ensure against potential cave-ins. In order to reduce friction between the outside of the casing and the surrounding earth, a bentonite slurry is often used as a lubricant.

On completion of the tunnel or casing, the concrete cradle is poured on the invert (or bottom) of the liner plate or casing, and the carrier pipe is then pushed through the tunnel in sections and connected to portions of the pipeline abutting the tunnel already placed by conventional open-cut methods. The space between the carrier

FIGURE E4-10 Cruz Construction Company forces beginning excavation on 7-ft liner plate tunnel under Route 30 at Dwight Ave., Collingswood. (Note: Shaft construction three liner plate rings in place. All excavation by hand and pneumatic shovel.)

pipe and the liner plate tunnel or casing is then filled with grout or pea gravel, providing a solid structural support for the carrier pipe.

A third approach investigated technology which has been successfully employed in Europe and Japan for direct jacking of the pipe in a manner similar to the hydraulically jacked casings. This method of pipeline construction, called *thrustboring*, was utilized on a section in Cherry Hill Township. This method utilized shafts approximately 500 ft apart, with the pipe being hydraulically jacked between shafts behind a thrustboring machine of a diameter approximately equal to the pipe, preceding the pipe, excavating the earth, and mixing it with water to produce a slurry, which is pumped out of the excavation, dried, and hauled away. This procedure was also used on a project in Bridgeport, Connecticut, where approximately 3500 ft of 60-in sanitary sewer was being placed with the thrustbore method (see Fig. E4-11).

PIPE-PUSHING MACHINES

Tramac's pushing method of installing steel pipes provides an economical alternative to open-cut trenching or augering. It is possible, with the proper adaptor, to accurately punch steel pipes 4 to 60 in in diameter. Tramac's steel-pipe-pushing machines are driven by compressed air (see Fig. E4-12). Steel pipes can be driven into all types of soil (sand, loam, gravel, boulders, and clay mixtures). During the

FIGURE E4-11 Bridgeport, Connecticut—66-in gravity line being installed with a thrustbore machine. Pipe is hydraulically jacked through the earth behind a mole machine (thrustbore).

FIGURE E4-12 Tramac pipe-pushing machines.

process of driving the pipe, the soil inside the pipe keeps the pipe level and on course. When the pushing process is completed, the soil inside the pipe can be removed by means of compressed air or water, according to the soil conditions.

Tramac's steel-pipe-pushing machines are used for the undercrossings of streets, railroads, embankments, rivers, and any other obstacles that can't be trenched. The pipe pusher is ideal for installing water or gas pipes, electric cables, and all other utilities.

With Tramac's steel-pipe-pushing machines, boring costs are kept low because jacking from fixed abutments is not required, setup time is minimal, less labor is required, and augers are not needed (see Table E4-4).

TABLE E4-4 Tramac's Steel-Pipe-Pushing Machines

Technical specifications	PP-120	PP-160	PP-220	PP-280	PP-350	PP-500
Outer diameter, in	5	6	8.5	11	14	20
Length, in	57	65	71	83	98	100
Application up to dia., in	3–6	8–12	8–16	12–24	16–40	24–60
Air consumption, ft^3/min	110	160	250	425	715	1250
Operating pressure, lb/in^2	100	100	100	100	100	100
Strokes/min	380	360	320	300	250	200
Weight, lbs	254	408	992	1654	3087	7000

TT Technologies has a ramming machine called GRUNDORAM® (see Fig. E4-13). The GRUNDORAM machine and first pipe section are lowered into the starting pit and placed on adjustable bearing stands. The pipe is then accurately aligned and the machine firmly secured to the pipe by means of chains and strops to ensure that full impact is transmitted to the pipe. The GRUNDORAM is connected to a conventional air compressor and the pipe can then be driven in. This process is continued until the bore is completed.

FIGURE E4-13 GRUNDORAM pipe-ramming machine.

TABLE E4-5 GRUNDORAM Technical Data

Ramming machine	Titan	Olympus	Hercules	Gigant	Koloss	Goliath	Taurus
Diameter, in	5.70	7.00	8.70	10.50	13.80	17.70	24.00
Length, ft	5.1	5.5	6.3	6.6	7.7	9.3	11.9
Weight, lb	302	496	811	1,356	2,601	5,434	10,582
Air consumption, cfm	141	177	282	424	706	1,236	1,765
Number strokes/min	285	305	340	310	220	180	180
From pipe size, in	4	4	4.75	8	11	15	15

Steel pipes are laid as protective sleeves for carrier mains or as pipes for drainage. The ideal length is 20 ft; however, if the ramming pit is smaller, shorter lengths can be used. By means of a conical ramming head and taper-locked ram cones, the machines can be adapted to current diameters from 8 to 55 in. The cones are locked together by the first few strokes of the machine.

During the ramming operation, the soil in the pipe can be partially removed through the cutaway ports of a ram cone adaptor. When the job is completed, the soil remaining in the pipe is flushed out by pressure jetting or is forced out under pressure by a combination of compressed air and water injected into the pipe through a pressure plate. In average soil conditions, the rate of propulsion is 9 to 50 ft/h. However, speeds of up to 65 ft/h can often be achieved. Table E4-5 lists technical data for seven models of the GRUNDORAM.

SECTION 3
TRANSMISSION PIPELINES

The United States today has over 850,000 mi of pipeline for natural gas alone. These pipelines move over 30 trillion ft^3 of gas into over 40 million homes, factories, and stores. This volume of gas represents about one-third of the nation's annual consumption of fuel energy. Accordingly, this section will deal principally with the construction of the large-diameter pipelines used for natural gas.

The vast network of oil pipelines further illustrates the importance of pipelines. An oil pipeline is present wherever there is, or has been, a productive oil well. Small feeder lines of 2-in pipe or larger bring oil from the wells in a field to a larger crude line for delivery to a refinery. This may be a short distance or a matter of hundreds of miles. After processing in the refinery, the petroleum products are carried by products pipelines to market centers for distribution.

In the delivery of natural gas from the producing field to the consumer, there are three principal and distinct stages. These are production, transmission, and distribution. The manner in which natural gas is gathered follows an almost identical pattern to that for oil. Field gathering lines take the natural gas into a central plant for removal of impurities and then into the big transmission systems. The long-distance natural gas transmission arteries are necessarily of larger diameter than those used for oil, now being primarily 22- to 42-in pipe as compared with the 8- to 36-in pipe used in the major oil products lines. Since the transmission phase carries the greatest amount of gas the longest distance, it is the construction of these pipelines that we will describe.

PLANNING TRANSMISSION LINES

Approximately 85 percent of the proved natural gas reserves in the United States are located in the Southwest, and the transmission lines which serve the highly industrialized markets of the northern and eastern states are necessarily of great length and cross all kinds of terrain. Building such pipelines is a mammoth task and involves huge sums of money. Before a new transmission line is even put on the drafting board, both assured markets and supplies must be determined. An investment of $300 million or more is involved in the construction of a pipeline of about 1000 mi in length, and the financing and normal contract arrangements on both the producing and consuming ends are usually set up on a 20-year basis. The planning of such important pipelines must go beyond the initial capacity to satisfy present demand. Allowance must be made for expansion, too, in order to satisfy a foreseeable increase of demand.

Planning by a transmission company generally includes allowance for increasing capacity in two ways. The first is to use a sufficiently large pipeline so that additional compressor stations may be constructed along its route to push the gas in greater quantity at a faster rate. The second is the advance planning for additional sections, called *loops,* to be built later to run parallel to the main line. Only after months and often years of planning does actual construction begin.

Once a starting date for construction has been set, the aim is toward the earliest possible completion of the pipeline. It is to the advantage of both the consumer and those who have invested their money in the pipeline to start the gas flowing at the earliest possible moment. In view of this and the need for assured permanence through the quality of construction, there is an absolute necessity for specialization in building pipelines.

Just as proper planning is perhaps the most important single part of developing a new transmission system, so it is with the actual construction of the pipeline. The different terrain over which a pipeline passes demands flexibility, and both equipment and personnel must be geared to immediate and frequent change while building a uniform product.

The proper building of a pipeline requires the advance consideration of a great number of factors. A few of these will be considered, and the main problems which might arise from each will be pointed out (see Fig. E4-14).

Length

It is axiomatic that the larger the project, the lower the overhead in connection with moving equipment, labor, and supplies to a new location.

Location

The location of the pipeline right-of-way determines its accessibility. Moving the vast array of equipment is a big job, and it is important to move it by rail as close as possible to the starting point of new work.

The number and adequacy of access roads to the pipeline right-of-way are other important factors determined by location. The more access roads there are, the easier it is to string the fabricated pipe sections along the route as well as to move special equipment onto the right-of-way. Closer supervision, too, is another advantage of a sufficient number of access roads. In mountainous or remote rural areas where the number of such roads is insufficient, additional roads must be built by the pipeline constructor. This, of course, is an important item of cost.

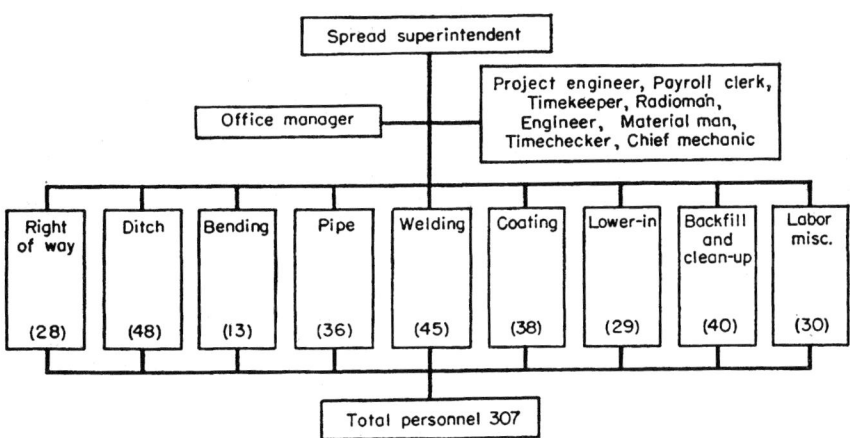

FIGURE E4-14 Organization chart for typical pipeline spread.

Terrain

The type of land which must be crossed in building a pipeline is one of the biggest cost factors. In cutting a ditch across open farmland or treeless hills, a ditching machine can make relatively swift progress. There is also little problem in clearing the right-of-way to give room and footing for the construction crews to operate. On the other hand, both effort and cost increase sharply where heavily wooded and rocky areas must be crossed. Under the latter conditions, the entire right-of-way must be cleared of all brush, trees, and stumps and the irregular surface smoothed out. For this one task alone, upward of a hundred workers may be required.

When rock is encountered, the pipeline ditch must be prepared by blasting and the use of backhoes. Where steep cliffs are encountered, the heavy equipment must often by supported by cables.

Weather

The worker who keeps the records of seasonal rainfall is important in planning the rate of progress. A great deal of aggravation may be avoided if an approximate idea can be obtained of the number of days on which the big equipment will be stalled because of slick underfooting.

Labor Availability

Advance arrangements are necessary to ensure an uninterrupted supply of labor. Building and moving at the same time, a pipeline construction unit is often faced with varying labor conditions with regard to both supply and local union relationships. The simple act of crossing a county line can present a major problem unless advance arrangements have been made.

Right-of-Way

Another point of major difference between pipeline construction and other types of construction is in the ownership of the site of construction. The property actually belongs to a third party, the landowner, and right-of-way privileges are purchased by the gas company building the new pipeline. Special consideration of the property is vitally important. Excessive property damage can be costly to both the constructor and the pipeline owner.

CONSTRUCTING TRANSMISSION LINES

Most current pipeline projects involve the construction of loops of existing facilities. Because these loops are necessarily scattered out from the main line, it is not unusual to move workers and equipment 350 mi while building only 50 mi of pipeline. Thus, while the right-of-way is being prepared on the most distant loop, welding is being done on the middle loop and pipe is lowered into the ditch on the first loop.

This basic pattern of activity remains unbroken from start to finish in pipeline construction. Wet weather is perhaps the major cause of delay, although other difficulties are common. There may be a turnover in labor, a change in supplies, or a move by headquarters. To pipeline construction specialists, these things are all in a day's work.

Preliminary Work

Long before construction actually starts, the exact nature of the terrain has been determined from a personal inspection of the proposed route by company executives. This inspection is essential for the determination of the workers, equipment, and supplies for the job. After meeting with the labor unions of the locality and arranging for housing, the next step is moving in a construction spread. The equipment requirements have already been determined during the construction planning phase. This equipment will be moved, preferably on railroad flatcars, as close as possible to the proposed starting location.

The office manager of the spread pulls the headquarters office trailer into the first headquarters location. Arrangements have been made in advance for warehousing and yard space to accommodate the equipment and the bustling activity that goes with the start of a new job.

The office manager and spread superintendent immediately set about completing arrangements for an adequate supply of fuel and oil to power the machinery. They also determine the location of the closest repair parts terminals and do other tasks necessary to the smooth-running organization.

Clearing

Even before all the workers and equipment have been moved into location, the first construction group has already started clearing the right-of-way. Appropriately dubbed the *right-of-way gang,* its job is to clear a smooth highway over the pipeline route for the pipeline building crews that follow. In thick forest country such as found in the mountains of West Virginia, upward of a hundred workers may be put to work clearing underbrush, cutting trees, tearing out stumps, and filling treacherous pockets or holes. Speed is the essence of pipelining, and there must be room and footing on which to work.

Cooperating with the landowners, this crew salvages all marketable timber and stacks it neatly alongside the right-of-way for recovery and sales. It then collects and chips all the underbrush, treetops, and limbs.

Trenching

The second unit that goes out on the right-of-way is the ditching crew. If a 36-in pipe is to be installed, a huge ditching machine will be used to dig a continuous trench, 52 in wide and 6 ft deep. The excavated material is piled on one side of the ditch. On a typical spread, the ditch crew has a complement of approximately 50 workers, far more than are required simply to operate the ditching machine itself.

Working ahead of the ditching machine, a stake-setting crew sets out markers to show the exact line to be followed, as determined earlier by the engineering department of the gas company. A small bulldozer accompanies the ditching machine to further smooth out any humps or ruts left in the right-of-way, and there is often a tow tractor to pull the ditching machine up steep hills.

In mountainous terrain there is far more to preparing the ditch than the simple job of the trenching machine. Loosening of hard rock often calls for large amounts of explosives. Tandem wagon drills, mounted on steel frames and powered by 900-cfm air compressors, can be used to make the necessary boreholes.

Once the rock is loosened by careful blasting under the direction of the blasting foreperson, backhoes are moved in to put the finishing touches to the tougher sections of ditch. Special auger-boring machines can be used to tunnel under main highways and railroads. In such crossings a section of thick-walled steel pipe of larger diameter than the line itself is put into place for casing as the boring machine cuts through the crossing.

Stringing

After the ditch crews have passed along the right-of-way and prepared the bed for the big transmission artery, the sections of steel pipe make their appearance. The *stringers* haul the pipe from where it has been received and stockpiled along a railroad spur and lay it along the ditch side. Most of the large pipeline construction companies now subcontract the stringing to a specialist in this field.

Bending

With the pipe now in place along the ditch, the next equipment on the route is the bending machine. Accompanied by a crew of 10 to 15 workers with side-boom tractors to lift the pipe into place, the bending machine applies just the right amount of pressure to bend the pipe to the specified degree. Each section of pipe is thus actually form-fitted to the contour of the ditch.

After the bending crew has made a start, the tempo of activity along the line begins to pick up to the fast, continuous pace that is typical of efficient pipeline construction. The pipeline is now starting to take shape.

Welding

Perhaps the key job in accomplishing speed in transmission line construction is up to the next crew. Called the *pipe gang*, it features two or four experienced and fast-working welders who are accompanied by two side booms and a battery of tractor-mounted electric welding generators. It is their job to make the first bead on the welds which link the sections of pipe into a continuous line. In dry weather with good footing along the right-of-way, it is not uncommon to see these welders actually trotting from joint to joint as they make these first beads. Completion of the welds is left to crews coming along behind.

As each section of pipe is lifted into place, the inside lineup claim puts the two ends into exact position and the welders start the first welding bead. Lift, lineup, weld. The routine is the same over the entire route, progressing swiftly toward the distant goal of the last tie-in weld at the far end of the line.

A portable machine for down-the-line double-jointing has been developed by an H. C. Price Co. special-projects engineer. This machine, which is adaptable to any type of welding, makes it possible to join two 40-ft pipe sections on the right-of-way in flat country and thus cut welding and lineup time approximately in half. The use of this sled-mounted machine eliminates the problems encountered in hauling 40-ft lengths of yard-jointed pipe. Such hauling is not only more expensive but is prohibited in some states.

The pipe gang is followed by two or four more welders, with their normal accompaniment of helpers and buffers to clean slag and flakes from the previous welds. This is the second of three welding units that pass along the line, and the completed weld will contain a total of four or five beads. This second bead is called the *hot pass*, and gives additional strength and protection until the welding crews can complete the full weld.

The main welding gang of 10 to 20 additional welders follows the second, or hot-pass, bead. This welding gang, with assisting labor, side booms, and generators, all flows out in a long string and works in an atmosphere of feverish activity.

Welding is perhaps the most important single job in the building of a permanent and strong pipeline. Close checks on each individual weld are made by expert inspectors of the pipeline company and by the contractor's welding foreperson. Each welder is thoroughly tested before being hired, and sample welds are put through rigorous testing. Welded-steel straps are pulled apart by a hydraulic tensile-testing device and, in order for the weld to pass the test, the steel of the pipe itself must pull apart before the weld gives way.

Now, 10 to 15 percent of the joining is done by metal-inert-gas (MIG) welding with an inert-gas machine. This process completes a pipe joint faster than stick welding. Also, there are high-strength wire electrodes for MIG welding that can join pipe of higher tensile strength than can be joined by stick welding. The same testing standards apply to MIG welding.

Cleaning

Next the welders are followed along the pipeline by the pipe-cleaning machine. This large machine moves along the pipe itself, much as an upright unirail train moves along its track, but with its weight supported by another side-boom tractor. On this machine, several steel-wire brushes move in a circular motion around the pipe, scraping it clean of the dirt and rust that have accumulated between the factory and the field.

As soon as the metal is burnished clean, a primer is flowed evenly over the bright steel by the primer machine in preparation for the coating and wrapping gang which follows.

Coating and Wrapping

In well-timed construction, a swift pace is maintained toward completion as the coating and wrapping machine performs its work. This machine applies the final protective coating against corrosion and, in much the same manner as the cleaning and prime-coat machines, it moves along the pipe itself. It is followed closely by tractor-pulled sleds loaded with supplies.

As hot tar is cascaded down over the pipe, an assembly on the back of the machine revolves completely around the pipe. This assembly feeds a wide and continuous glass fiber or felt wrapping, or both, from rolls mounted on spindles on the revolving section. As the coat-and wrap crew moves along, a side boom lifts the pipe ahead of it with the use of a pipe cradle. The side boom moves forward, with the pipe sliding easily over the small wheels in the bottom of the lifting cradle, and the pipe is thus raised to the proper height for the coat-and-wrap machine.

Tapes applied over a primer are being used by some companies. The application is by much the same principle, except that the need for hot materials and outer wrap is eliminated.

Lowering In

After the coat and wrap is completed, the pipe may be lowered back onto carefully stacked wood skids along the ditch where it awaits the next-to-the-last process of lowering in. Alternatively, the pipe may be lowered into the ditch directly behind the coating and wrapping machine. Prior to lowering in by either method, a final inspection is made of the ditch to make sure that it is in proper condition. Where the ditch has been blasted out of solid rock, crews haul soft earth to prepare a soft bed in its bottom and thus avoid any damage to the pipe or its coating.

Once everything is in readiness, side booms in groups of four to eight and spaced approximately 50 ft apart are used to lower the long sections of pipe into the prepared ditch. In a hilly or mountainous area, the contractor will have a greater number of horizontal and vertical bends and the lowering in must be done smoothly and easily. Each section of pipe must fit into the ditch without knocking the pipe sections ahead off their skids or causing unnecessary strain on the pipe.

Normally, the welders have left pipe unjointed about every $\frac{1}{2}$ mi to facilitate lowering in and to ensure an exact fit of the pipe to the ditch contours. Once two sections are in the ditch, end to end, a tie-in weld is made right in the ditch to joint the sections.

Backfilling

As much care is taken to see that a soft direct cover goes directly on top of the pipe as was taken in preparing the ditch bed to receive it. A soft and deep protective pad of earth is first pushed into the ditch and the wide-bladed backfiller then completes the job. In solid-rock areas, this often entails moving the earth from distant locations.

Cleanup

By no means the least important job is the cleanup. Major pipeline contractors operate under the theory that a good job of cleanup of the right-of-way is an excellent job of public relations for both their own company and for the transmission company. Bulldozers and hand labor in generous quantity leave a smooth and neat right-of-way at the finish of each job.

Testing Completed

Sections of the pipelines are pressure-tested to ensure the final integrity of the system. Hydrostatic testing is used most often, and water is used more than any other fluid. Sometimes gas or air is used for testing natural gas lines.

The owner company specifies the pressure the contractor is to use in testing, the length of the section to be tested, and the span of time this pressure is to be held in the section of pipe. The pressure is based on a predetermined percent of the minimum yield strength for which the pipe has been designed.

A scraper or *pig* is first pushed through the pipe to clean the inside, and the section of pipe to be tested is then filled with filtered water. Sometimes chemical additives are also required. Pressure is then built up to the percentage of operating pressure which has been designated for the test. This pressure must be maintained for the period of time specified by the owner company.

If the pressure remains within the specified limits during the test period, the section is usually accepted and approved by the owner company. If there has been a reduction in pressure, the location of suspected leaks must be found, the faulty pipe or weld must be cut out, and repairs made or a section of pipe replaced. After this is done, the test must be repeated before that section of pipe can be accepted by the owner company. This testing is carried out on sections of the pipe until the entire pipeline has been tested and approved as ready for use.

REFERENCE

1. *Pipeline*, Camden County (N.J.) Municipal Utilities Authority, June 1986.

INDEX

AASHTO (*see* American Association of State Highway and Transportation Officials)
Abrasive drills, **D1**-7
Adamiecki, Karol, **A3**-2
Admixtures, **C1**-7–**C1**-11, **C1**-47–**C1**-52, **C1**-123
   accelerators, **C1**-8
   air-entraining, **C1**-7
   cementitious materials, **C1**-10
   inhibitors of alkali-aggregate reaction, **C1**-10–**C1**-11
   plasticizers, **C1**-9–**C1**-10
   pozzolanic materials, **C1**-10
   retarders, **C1**-9
   waterproofers, **C1**-10
   water-reducing, **C1**-9
AEC, **A1**-12
Aggregates, **C1**-5–**C1**-11, **C1**-12–**C1**-13, **C1**-46, **D2**-3
   lightweight, **C1**-29–**C1**-30
   six-in aggregate, **C1**-28–**C1**-29
   water, **C1**-6–**C1**-7
Agreement between owner and contractor, **A2**-6–**A2**-7
   cost-plus, **A2**-7
   lump-sum contract, **A2**-6
   unit price work, **A2**-6–**A2**-7
Air lifts, **D5**-14–**D5**-15
Airports, **E1**-4–**E1**-37
   case studies, **E1**-5–**E1**-37
   history, **E1**-4–**E1**-5
Alternative Disputes Resolution (ADR), **A1**-15, **A11**-6, **A11**-11–**A11**-12
American Association of State Highway and Transportation Officials (AASHTO), **C2**-5, **C2**-20
American Bureau of Shipping, **B11**-3
American Consulting Engineers Council (ACEC), **A2**-2
American Institute of Architects (AIA), **A2**-3
American National Standards Institute (ANSI), **A7**-8
American Society of Civil Engineers (ASCE), **A2**-2, **A7**-2
American Society of Mechanical Engineers (ASME), **A7**-8, **A7**-9–**A7**-17
American Society of Testing and Materials (ASTM), **A7**-8, **C1**-4
ANSI (*see* American National Standards Institute)
Anti-indemnity, **A10**-14–**A10**-15
Army Corps of Engineers, U.S., **B11**-34, **B11**-35
Arrow diagramming method, **A3**-6
ASME (*see* American Society of Mechanical Engineers)
Association of General Contractors (AGC), **A2**-2
ASTM (*see* American Society of Testing and Materials)
At-risk construction management, **A1**-5
Auditing requirements, **B2**-22
Augercast piles, **D3**-39–**D3**-40
Auxiliary diving equipment, **B11**-31–**B11**-35

Bachelors' enlisted quarters (BEQ) VE case study, **A6**-29–**A6**-41
Backfilling, **E4**-17–**E4**-20
Backhoes, **B5**-1–**B5**-31
Bar charts, **A3**-1
Barges and scows, **B11**-2–**B11**-3
Baseline, **A3**-3–**A3**-5
   cost loading, **A3**-6
   durations, **A3**-4
   resource loading, **A3**-4
   work breakdown structure, **A3**-5
   work items/activities, **A3**-3–**A3**-4

Basic accounting forms, **A4**-17– **A4**-19
  cost ledger, **A4**-19
  foreperson's time card, **A4**-19
  labor distribution schedule, **A4**-19
  monthly analysis, **A4**-18
  payroll journal, **A4**-18
  time card, **A4**-18
Batching plants, **C1**-31–**C1**-34
  material handling, **C1**-34–**C1**-41
  plant storage, **C1**-41–**C1**-43
  special processing, **C1**-43–**C1**-45
Beaver Valley Expressway, **E3**-13–**E3**-20
Belt conveyors, **B9**-1–**B9**-27, **C1**-68–**C1**-73
  components, **B9**-11–**B9**-19
  engineering design, **B9**-19–**B9**-25
  operation and maintenance, **B9**-25–**B9**-27
  overland haulage systems, **B9**-8–**B9**-9
  plant feed system, **B9**-3–**B9**-4
  portable systems, **B9**-9–**B9**-11
  ship and barge loaders, **B9**-9
  storage-reclaim system, **B9**-4–**B9**-8
Belts, **B9**-11–**B9**-13
Bidding documents, **A2**-3–**A2**-5
  addenda, **A2**-4
  advertisement, **A2**-4
  award of bid, **A2**-5
  bid security, **A2**-5
  guide to preparation of instructions for bidders, **A2**-4
  invitation to bid, **A2**-4
  qualification of bidders, **A2**-4
  schedule, **A2**-4
Bidding procedures, **A2**-3–**A2**-4, **A2**-5
Bituminous pavements, **D2**-2–**D2**-60
  construction of, **D2**-7–**D2**-17
  elements, **D2**-2–**D2**-3
  hot-mixed bituminous concrete:
    construction of, **D2**-40–**D2**-60
    production of, **D2**-17–**D2**-39
  materials, **D2**-3–**D2**-5
  road-mixed pavements, **D2**-7–**D2**-17
  surface treatments, **D2**-7–**D2**-17
  types, **D2**-6–**D2**-7
Blasting, **D1**-3–**D1**-32
  blasthole loading, **D1**-23–**D1**-26
  design, **D1**-8–**D1**-13
  drilling, **D1**-5–**D1**-7
  environmental effect, **D1**-30–**D1**-32
  explosive selection, **D1**-18–**D1**-21
  guidelines, **D1**-13–**D1**-14
  initiation sequence, **D1**-29–**D1**-30
  initiation systems, **D1**-23

Blasting (*Cont.*):
  planning, **D1**-3–**D1**-5
  primer selection, **D1**-21–**D1**-22
  priming, **D1**-26–**D1**-28
  special techniques, **D1**-15–**D1**-18
Bonds and insurance, **A2**-10–**A2**-11
Box caissons, **D4**-50–**D4**-60
  examples, **D4**-60–**D4**-69
Bridges, **D4**-60–**D4**-69, **D4**-83–**D4**-89
  across the Garonne River, **D4**-87
  Chesapeake Bay Bridge, **D4**-63–**D4**-64
  Fraser River Bridge, **D4**-85
  Great Belt Western Bridge, **D4**-66
  Honshu-Shikoku Bridges, **D4**-67–**D4**-69
  I-205 Columbia River Bridge, **D4**-67
  Ma Wan Tower Caisson, Tsing Ma Suspension Bridge, **D4**-67
  Mackinac Straits Bridge, **D4**-83–**D4**-84
  Newport Bridge, **D4**-65–**D4**-66
  Rappahannock River Bridge, **D4**-84
  Richmond–San Rafael Bridge, **D4**-60–**D4**-63
  San Francisco–Oakland Bay Bridge, **D4**-87–**D4**-89
  San Mateo–Hayward Bridge, **D4**-64–**D4**-65
  Second Carquinez Bridge, **D4**-85–**D4**-86
  Second Severn Bridge, **D4**-66–**D4**-67
  Sunshine Bridge, **D4**-86
  Tagus River Bridge, **D4**-89
  Verrazano-Narrows Bridge, **D4**-86–**D4**-87
Brown, Jed, Brigadier General, **A1**-18
Bucket dredge, **B11**-16
Budget estimate, **A4**-2
Bulldozers, **B3**-9–**B3**-18, **B8**-2
  applications, **B3**-14–**B3**-16
  blade types, **B3**-9–**B3**-12
  production estimating, **B3**-16–**B3**-18
  tip, tilt, and angle, **B3**-12–**B3**-14

Caissons, **D4**-50–**D4**-90
  box caissons, **D4**-50–**D4**-60
    examples, **D4**-60–**D4**-69
  open caissons, **D4**-69–**D4**-83
    examples, **D4**-83–**D4**-89
  pneumatic caissons, **D4**-89–**D4**-90
Camden County Municipal Utilities Authority, **E2**-2–**E2**-14
Cash flow forecasting, **A5**-5
Cast-in-place concrete piles, **D3**-7–**D3**-9
Cellular cofferdams, **D4**-49–**D4**-50

INDEX

Cement, **C1**-4–**C1**-5, **C1**-12, **C1**-46–**C1**-47, **C1**-122
Central-plant stabilization, **B8**-7–**B8**-9
Certified value specialist, **A6**-28
Change orders, **A2**-7
Changes, **A2**-14–**A2**-15
Circular cofferdam, **D4**-20–**D4**-22
Civil Engineering Magazine, **A7**-1
Claim, **A11**-2
Claims avoidance and defense, **A5**-5, **A11**-12–**A11**-13
Clamshell dredge, **B11**-17–**B11**-19
Clamshells, **B5**-24–**B5**-31
Classification of accounts, **A4**-4
　commitment record, **A4**-8
　cost report, **A4**-10
　purchase order register, **A4**-8
　purchase requisitions, **A4**-6
　sample classifications, **A4**-5
Coast Guard, U.S., **B11**-14
Coding structures, **A5**-11–**A5**-16
　cost and schedule controls systems criteria (CSCSC), **A5**-11
　organization breakdown structure (OBS), **A5**-11
　project commitment structure (PCS), **A5**-14–**A5**-16
　work breakdown structure (WBS), **A5**-11, **A5**-12
Cofferdams, **D4**-3–**D4**-50
　bracing systems, **D4**-20–**D4**-30
　dewatering, **D4**-40–**D4**-42
　difficulties, **D4**-42–**D4**-48
　excavation, **D4**-31
　pile driving, **D4**-32–**D4**-33
Commitment, **A4**-6
Commitment record, **A4**-8
Compaction equipment, **B8**-10–**B8**-17
Concrete cylinder piles, **D3**-7
Configuration control, **A5**-5
Construction cost control, **A4**-1–**A4**-36
Construction Industry Institute, **A1**-11
Construction Management Association of America (CMAA), **A10**-5
Construction planning and scheduling, **A3**-1–**A3**-18
Construction project delivery system, **A1**-3–**A1**-7
　ancient construction organization, **A1**-4
　at-risk construction management, **A1**-5
　classic owner-designer-constructor organization, **A1**-4

Construction project delivery system (*Cont.*):
　construction documents, **A1**-4
　design-construct CM, **A1**-6–**A1**-7
　no-risk construction management, **A1**-5–**A1**-6
Construction Specification Institute (CSI), **A4**-5
Contract coordination and interface, **A5**-5
Contract documents, **A2**-1–**A2**-3
　EJCDC construction-related contract documents, **A2**-2–**A2**-3
　national standard forms, **A2**-2
Contractor management organization, **A1**-7–**A1**-9
　field office organization, **A1**-10
　functional contractor organization, **A1**-8
　home office organization, **A1**-9
Contractor's organization, **A1**-3–**A1**-21
Contractor's responsibilities, **A2**-11
Conventional concrete, **C1**-3–**C1**-133
　batching of, **C1**-45–**C1**-59
　batching plants, **C1**-31–**C1**-34
　　material handling, **C1**-34–**C1**-41
　　plant storage, **C1**-41–**C1**-43
　　special processing, **C1**-43–**C1**-45
　curing, **C1**-84–**C1**-85
　finishing, **C1**-83–**C1**-84
　grouting, **C1**-122–**C1**-133
　mixing, **C1**-59–**C1**-65
　placing, **C1**-66–**C1**-82
　proportioning mixtures, **C1**-11–**C1**-28
　　adjustments, **C1**-26–**C1**-28
　　basic physical properties, **C1**-12–**C1**-13
　　computation (of mix proportions), **C1**-18–**C1**-25
　　tests, **C1**-25–**C1**-26
　　trial-mix, **C1**-13–**C1**-18
　slip forms, **C1**-110–**C1**-121
　special requirements, **C1**-28–**C1**-30
　stationary forms, **C1**-86–**C1**-109
Conventional scrapers, **B4**-1–**B4**-11
Conveyor accessories, **B9**-18–**B9**-19
Conveyor drives, **B9**-15
Conveyor pulleys and idlers, **B9**-13–**B9**-15
Cost and schedule control systems criteria (CSCSC), **A5**-11
Cost coding system, **A4**-4–**A4**-5
Cost control forms, **A4**-19–**A4**-24
　cost report—field labor, **A4**-23–**A4**-24
　foreperson's quantity report, **A4**-19–**A4**-20
　worker-hour and quantity report, **A4**-20–**A4**-23

Cost engineering, **A4**-2–**A4**-36
  definition, **A4**-2
  objectives, **A4**-2
  principles, **A4**-3
  responsibilities, **A4**-2
Cost models, **A6**-7, **A6**-9
Cost-plus contract, **A1**-14, **A2**-6
Cost report, **A4**-10
Cranes, **B10**-1–**B10**-44
  equipment selection, **B10**-43–**B10**-44
  mobile hydraulic cranes, **B10**-11–**B10**-35
  mobile mechanical cranes, **B10**-2–**B10**-10
  tower cranes, **B10**-36–**B10**-42
Crawler-type loaders, **B6**-1–**B6**-2
Critical path method (CPM), **A3**-1–**A3**-18
  baseline, **A3**-3–**A3**-5
    cost loading, **A3**-6
    durations, **A3**-4
    resource loading, **A3**-4
    work breakdown structure, **A3**-5
    work items/activities, **A3**-3–**A3**-4
  float, **A3**-2, **A3**-10
  logic, **A3**-5–**A3**-6
  network, **A3**-1–**A3**-12
    construction, **A3**-13–**A3**-18
    postconstruction, **A3**-18
    preconstruction, **A3**-12–**A3**-13
  network diagramming, **A3**-6
    arrow diagramming method, **A3**-6
    calculations, **A3**-8–**A3**-11
    events, **A3**-6
    fragnets, **A3**-8
    nodes, **A3**-6
    precedence diagramming, **A3**-6
    restraints, **A3**-7
  slack, **A3**-2
  updating, **A3**-11–**A3**-12
C-spec (*see* Cost and schedule control systems criteria)

Deming, W. Edwards, **A1**-12, **A1**-21, **A8**-1
Deming management method, **A1**-15–**A1**-20
  14 points, **A1**-17–**A1**-20
Department of Labor, U.S., **A10**-9
Derricks, **B11**-28–**B11**-31
Development/recommendation phase, **A6**-18
Dewatering, **D5**-1–**D5**-22, **E4**-7
  cost of, **D5**-21
  design of, **D5**-15–**D5**-21
  methods, **D5**-2–**D5**-12
  pumps for, **D5**-13–**D5**-15

Diaphragm pumps, **D5**-14
Diesel engines, **B1**-9–**B1**-12
Dipper dredge, **B11**-17
Dispute, **A11**-1
Distributable accounts, **A4**-24–**A4**-32
  cleanup account, **A4**-32
  construction equipment, **A4**-29
  employer's labor expenses and benefits, **A4**-32
  field office, **A4**-26
  general condition account, **A4**-32
  insurance—injuries and damages, **A4**-26–**A4**-28
  premium pay, **A4**-29–**A4**-32
  survey account, **A4**-32
  temporary construction, **A4**-28–**A4**-29
Dozers (*see* Bulldozers)
Draglines, **B5**-24–**B5**-31
Dredging, **B11**-5–**B11**-27
Drill boat, **B11**-25–**B11**-27
Du Pont, **A6**-2, **A6**-10
Dustpan dredge, **B11**-24–**B11**-25

Earth pressure balance (EPB) shields, **B12**-7–**B12**-9
Elevating scrapers, **B4**-11–**B4**-15
Energy model, **A6**-8–**A6**-10
Engineering fundamentals, **B1**-3–**B1**-26
Engineers Joint Contract Documents Committee (EJCDC), **A2**-2–**A2**-15
Engineers' Joint Council, **A8**-4
Environmental Protection Agency, **B11**-35
Equipment:
  acquisition, **B2**-1–**B2**-2
  characteristics, **B1**-9–**B1**-24
  economics, **B2**-1–**B2**-30
  estimating costs, **B2**-26–**B2**-27
  financing, **B2**-12–**B2**-13
  maintenance, **B2**-13–**B2**-17
    downtime, **B2**-13–**B2**-16
    repairs, **B2**-17
    servicing, **B2**-16
  monitoring costs, **B2**-29–**B2**-30
  records, **B2**-27–**B2**-29
  replacement, **B2**-8–**B2**-12
  selection, **B1**-3–**B1**-4, **B2**-2–**B2**-8, **B4**-18–**B4**-22, **B8**-9–**B8**-10, **B8**-16–**B8**-17
Evaluation/analytical phase, **A6**-16–**A6**-18
Excavators, **B5**-1–**B5**-31
  attachments, **B5**-2–**B5**-8
  shields, **B12**-9
Exculpatory clauses, **A7**-27–**A7**-29

Executive information system (EIS), **A5**-16–**A5**-23
　development and implementation, **A5**-22–**A5**-23

FAST diagramming, **A6**-15
Federal Acquisition Regulations (FAR), **A11**-5
Field labor cost reports, **A4**-16–**A4**-24
　cost report—field labor, **A4**-17, **A4**-23–**A4**-24
　foreperson's quantity report, **A4**-17, **A4**-19–**A4**-20
　worker-hour and quantity record, **A4**-17, **A4**-20–**A4**-23
　worker-hour report, **A4**-17
Field office, **A4**-26
Final payment, **A2**-7
Financial audit, **B2**-22–**B2**-26
Fixed-price contract, **A4**-1
Floating cranes, **B11**-28–**B11**-31
Front-end loaders, **B6**-1–**B6**-16
　applications, **B6**-11–**B6**-16
　buckets and attachments, **B6**-10–**B6**-11
　ratings, **B6**-6–**B6**-10
Functional analysis, **A6**-12–**A6**-15
Functional analysis system technique (FAST) (*see* FAST diagramming)

Gas turbines, **B1**-9–**B1**-13
Gasoline engines, **B1**-9–**B1**-12
Glen Anderson Freeway, **E3**-8–**E3**-13
Glenwood Canyon, **E3**-27–**E3**-35
Grouting, **C1**-122–**C1**-133
　materials, **C1**-122–**C1**-123
　mixing, **C1**-125–**C1**-130
　procedures, **C1**-130–**C1**-133
　proportioning, **C1**-123–**C1**-125
　pumping, **C1**-125–**C1**-130
Guaranteed maximum price (GMP), **A1**-5, **A2**-5

Hard-rock TBMs, **B12**-10–**B12**-20
Harmonygraph, **A3**-2
Hauling units, **B7**-1–**B7**-25
　estimating production, **B7**-20–**B7**-25
　performance, **B7**-13–**B7**-19
Hayden, William, **A1**-11
Heavy construction claims, **A11**-1–**A11**-14
　acceleration claim, **A11**-3
　defective design, **A11**-3–**A11**-4
　delay claim, **A11**-2–**A11**-3

Heavy construction claims (*Cont.*):
　force majeure, **A11**-3
　litigation, **A11**-10–**A11**-11
　preparation, **A11**-5
　proving and pricing, **A11**-5–**A11**-9
　refutation, **A11**-6
Heavy construction contracts, **A2**-1–**A2**-15
Highways, **E3**-1–**E3**-47
　case studies, **E3**-7–**E3**-47
　　flat terrain, **E3**-20–**E3**-23
　　intermediate terrain, **E3**-13–**E3**-20
　　rough terrain, **E3**-27–**E3**-35
　　swamp reclamation, **E3**-23–**E3**-27
　　urban construction, **E3**-8–**E3**-13
　　urban reconstruction, **E3**-39–**E3**-47
　definition, **E3**-1–**E3**-2
　development, **E3**-3–**E3**-5
　future construction, **E3**-35–**E3**-39
　history, **E3**-2–**E3**-3
　system, **E3**-5–**E3**-7
Home office, **A4**-2
Hopper dredge, **B11**-19–**B11**-24
Horizontal centrifugal pumps, **D5**-13

I-75 Alligator Alley, **E3**-23–**E3**-27
Indemnity, **A2**-13, **A10**-14
Indirect costs, **A11**-8
Information phase, **A6**-6–**A6**-15
Intermodal Surface Transportation Efficiency Act (ISTEA), **E3**-35

John F. Kennedy International Airport:
　administrative programs, **E1**-15
　AOA restrictions, **E1**-13–**E1**-14
　construction:
　　airport, **E1**-10
　　building, **E1**-15–**E1**-26
　　civil, **E1**-27–**E1**-33
　　restricted, **E1**-11–**E1**-12
　history, **E1**-6–**E1**-9
　program goal, **E1**-6
　program scope, **E1**-9–**E1**-10
　Redevelopment Program, **E1**-6–**E1**-33
　special requirements, **E1**-12–**E1**-13
Juran, **A1**-15, **A8**-1

Life cycle cost model, **A6**-11–**A6**-12
Liquidated damages, **A11**-9
Load factor, **B1**-5, **B1**-13
Local area network (LAN), **A5**-6–**A5**-8
Lump-sum contract, **A2**-6, **A4**-12–**A4**-13

INDEX

Machine pavers, **D2**-70–**D2**-72
Maintenance (equipment):
 downtime, **B2**-13–**B2**-16
 repairs, **B2**-17
 servicing, **B2**-16
Management by exception (MBE), **A1**-13–**A1**-15, **A1**-21, **A5**-19–**A5**-22
Management by objective (MBO), **A1**-13–**A1**-15, **A1**-21
Manchester Airport, **E1**-34–**E1**-35
Marine equipment, **B11**-1–**B11**-35
 auxiliary diving equipment, **B11**-31–**B11**-35
 barges and scows, **B11**-2–**B11**-3
 derricks, **B11**-28–**B11**-31
 dredging, **B11**-5–**B11**-27
 floating cranes, **B11**-28–**B11**-31
 insurances, **B11**-32–**B11**-33
 pile drivers, **B11**-28–**B11**-31
 tugs and towboats, **B11**-4–**B11**-5
Markel, Joseph M., **A1**-14
Masterformat™, **A4**-5
Material investigations, **B1**-9
Mining shovels, **B5**-9–**B5**-10
Mission statement, **A8**-1–**A8**-2
Mobile hydraulic cranes, **B10**-11–**B10**-35
Mobile mechanical cranes, **B10**-2–**B10**-10
Monotube piles, **D3**-10
Motors and controls (conveyors), **B9**-15–**B9**-16
Motor graders, **B8**-3–**B8**-5
Multiple-pass stabilizer, **B8**-7
Multiple scrapers, **B4**-15–**B4**-16

National Electrical Manufacturers Association (NEMA), **B9**-16
National Safety Council, **A10**-9
National Society of Professional Engineers, **A2**-2
Navy, U.S., **A3**-2
Network, **A3**-1–**A3**-12
 construction, **A3**-13–**A3**-18
 postconstruction, **A3**-18
 preconstruction, **A3**-12–**A3**-13
Network diagramming, **A3**-6
 arrow method, **A3**-6
 calculations, **A3**-8–**A3**-11
 events, **A3**-6
 fragnets, **A3**-8
 nodes, **A3**-6
 precedence, **A3**-6
 restraints, **A3**-7

New York Times, **A1**-14
Nontilting mixers, **C1**-60–**C1**-61

O'Brien-Kreitzberg, Inc., **A7**-8, **A7**-18–**A7**-27
Occupational Safety and Health Act (OSHA), **A10**-2, **A10**-10–**A10**-11
Off-highway hauling units (types), **B7**-7–**B7**-13
Off-highway tractors, **B3**-1–**B3**-9
On-highway tractors, **B7**-3–**B7**-7
Open caissons, **D4**-69–**D4**-83
 examples, **D4**-83–**D4**-89
Operating costs (of equipment):
 administrative and overhead cost, **B2**-21–**B2**-22
 loss of value, **B2**-19–**B2**-21
 value accumulation, **B2**-17–**B2**-19
Other work, **A2**-13–**A2**-14
Overland haulage conveyor systems, **B9**-8–**B9**-9
Owner's responsibilities, **A2**-14

Pareto, **A6**-7
Partnering, **A1**-12–**A1**-13, **A2**-1, **A9**-1–**A9**-13
 benefits, **A9**-5–**A9**-6
 defined, **A9**-2
 fundamentals, **A9**-3–**A9**-5
 implementation, **A9**-8–**A9**-12
 key elements, **A9**-6–**A9**-8
Paving, **D2**-1–**D2**-74
 bituminous pavements, **D2**-2–**D2**-60
  construction of, **D2**-7–**D2**-17
  elements, **D2**-2–**D2**-3
  hot-mixed bituminous concrete, construction of, **D2**-40–**D2**-60
  hot-mixed bituminous concrete, production of, **D2**-17–**D2**-39
  materials, **D2**-3–**D2**-5
  road-mixed pavements, **D2**-7–**D2**-17
  surface treatments, **D2**-7–**D2**-17
  types, **D2**-6–**D2**-7
 portland cement concrete pavements, **D2**-61–**D2**-74
  curing, **D2**-73–**D2**-74
  joints, **D2**-72–**D2**-73
  placing concrete and steel, **D2**-63–**D2**-72
  preparation, **D2**-61–**D2**-63
Paving mixers, **C1**-61
Paving train, **D2**-65–**D2**-67
Payload, **B1**-5
Penalties, **A11**-9

Pennsylvania Turnpike, **E3**-13–**E3**-20
Percussion drills, **D1**-5–**D1**-6
Performance evaluation and review technique (PERT), **A3**-2
Performance measurement, **A5**-5
Pile driving, **D3**-22–**D3**-47
  cofferdam (within), **D4**-32–**D4**-33
  equipment, **B11**-38–**B11**-31, **D3**-22–**D3**-35
Piles, **D3**-1–**D3**-47
  inspection, **D3**-40–**D3**-46
  installation, **D3**-35–**D3**-40
  materials, **D3**-11–**D3**-22
  stresses, **D3**-10–**D3**-11
  types, **D3**-2–**D3**-10
Pipelines, **E4**-1–**E4**-35
  transmission, **E4**-28–**E4**-35
    constructing, **E4**-30–**E4**-35
    planning, **E4**-28–**E4**-30
  trenchless, **E4**-21–**E4**-27
    boring, **E4**-21–**E4**-23
    micro tunneling, **E4**-23–**E4**-24
    pipe-pushing machines, **E4**-24–**E4**-27
  underground utility construction, **E4**-2–**E4**-20
    backfilling, **E4**-17–**E4**-20
    excavation, **E4**-5–**E4**-11
    installation of pipe, **E4**-12–**E4**-16
    pavement removal, **E4**-3–**E4**-4
    paving, **E4**-20
    preliminary work, **E4**-2–**E4**-3
Plant feed conveyor system, **B9**-3–**B9**-4
Pneumatic caissons, **D4**-89–**D4**-90
Polaris ballistic missile, **A3**-2
Portable conveyor systems, **B9**-9–**B9**-11
Portland cement concrete pavements, **D2**-61–**D2**-74
  curing, **D2**-73–**D2**-74
  joints, **D2**-72–**D2**-73
  placing concrete and steel, **D2**-63–**D2**-72
  preparation, **D2**-61–**D2**-63
Power:
  ratings, **B1**-5–**B1**-17
  required, **B1**-20–**B1**-24
  transmission and control, **B1**-14–**B1**-15
  units, **B1**-9–**B1**-14
  usable, **B1**-17–**B1**-19
Power shovels, **B5**-1–**B5**-31
Premium pay, **A4**-29–**A4**-32
Prestressed concrete piles, **D3**-7
Progress payments, **A2**-7, **A4**-13

Project management information system (PMIS), **A5**-1–**A5**-31
  benefits, **A5**-5
  choosing, **A5**-9–**A5**-11
  components of, **A5**-2–**A5**-3
  costs of, **A5**-3–**A5**-5
  establishing, **A5**-5–**A5**-9
  executive information system, **A5**-3, **A5**-16–**A5**-23
  features, **A5**-23–**A5**-24
  functions of, **A5**-3
Purchase order register, **A4**-8
Purchase requisitions, **A4**-6

Quality assurance/quality control, **A7**-1–**A7**-31
Quality assurance, **A7**-2, **A8**-3
  design phase, **A7**-3, **A7**-5–**A7**-7
  predesign activities, **A7**-3–**A7**-5
Quality assurance program requirements for nuclear facilities, **A7**-8, **A7**-9–**A7**-17
Quality in the constructed project, **A7**-2–**A7**-3
Quality control, **A7**-3–**A7**-4
  construction phase, **A7**-3, **A7**-7–**A7**-8
  furnish and move in, **A7**-3, **A7**-7–**A7**-8
  inspection team, **A7**-29–**A7**-31

Records of payments, **A4**-6
Rectangular cofferdams, **D4**-22–**D4**-23
Reinforcing steel, **C2**-1–**C2**-20
  bars, **C2**-1–**C2**-5
  coated reinforcement, **C2**-8–**C2**-9
  detailing, **C2**-11–**C2**-12
  fabrication, **C2**-11–**C2**-12
  metrication, **C2**-19–**C2**-20
  placing, **C2**-18–**C2**-19
  prestressing, **C2**-6–**C2**-7
  splices, **C2**-9–**C2**-11
  storage, **C2**-12–**C2**-18
  welded wire fabric, **C2**-5–**C2**-6
  wire, **C2**-5–**C2**-6
Retainage, **A2**-7
Revised estimate, **A4**-32–**A4**-36
Rippers, **B3**-18–**B3**-22
  operation, **B3**-20
  production estimating, **B3**-20–**B3**-22
  types, **B3**-18–**B3**-19
Rock excavation, **D1**-3–**D**-32
  blasthole loading, **D1**-23–**D1**-26
  design, **D1**-8–**D1**-13
  drilling, **D1**-5–**D1**-7

Rock excavation (*Cont.*):
　environmental effect, **D1**-30–**D1**-32
　explosive selection, **D1**-18–**D1**-21
　guidelines, **D1**-13–**D1**-14
　initiation sequence, **D1**-29–**D1**-30
　initiation systems, **D1**-23
　planning, **D1**-3–**D1**-5
　primer selection, **D1**-21–**D1**-22
　priming, **D1**-26–**D1**-28
　special techniques, **D1**-15–**D1**-18
Rockefeller, John D., **A1**-13
Rotary drills, **D1**-6–**D1**-7
Rubber-tired compactor, **B8**-10–**B8**-12

SAE (*see* Society of Automotive Engineers)
Safety, **A2**-12, **A10**-1–**A10**-16
Scope control, **A5**-5
Scrapers, **B4**-1–**B4**-24, **B8**-2
Screeding, **C1**-82
Segmented pad compactor, **B8**-13
Shewart, **A1**-15
Ship and barge loader conveyor systems, **B9**-9
Shop drawings, **A2**-12–**A2**-13
Shrinkage factor, **B1**-5–**B1**-8
Single-purpose excavators, **B5**-9–**B5**-31
Slip forms, **C1**-110–**C1**-121
　bracing, **C1**-115–**C1**-117
　concrete placing, **C1**-118–**C1**-120
　control and tolerance, **C1**-120
　curing, **C1**-121
　design, **C1**-110–**C1**-113
　finishing, **C1**-121
　jacking system, **C1**-113–**C1**-115
　paving, **D2**-63–**D2**-65
　reinforcing steel, **C1**-117–**C1**-118
　working deck, **C1**-115–**C1**-117
Slurry shields, **B12**-5–**B12**-6
Slurry wall cofferdams, **D4**-48–**D4**-49
Society of Automotive Engineers (SAE), **B10**-7
Soft-ground shields, **B12**-4–**B12**-5
Soil compaction, **B1**-8–**B1**-9
Soil density, **B1**-8
Soil surfacing equipment, **B8**-1–**B8**-17
Speculative/creative phase, **A6**-16, **A6**-17
Spreading equipment, **B8**-1–**B8**-6
Stabilization equipment, **B8**-6–**B8**-10
Stationary forms, **C1**-86–**C1**-109
　applications, **C1**-101–**C1**-107
　care of, **C1**-107–**C1**-109
　gang forms, **C1**-101
　job-built forms, **C1**-96–**C1**-98

Stationary forms (*Cont.*):
　materials, **C1**-88–**C1**-90
　planning, **C1**-86–**C1**-88
　prefabricated forms, **C1**-98–**C1**-100
　requirements, **C1**-90–**C1**-96
Steel H piles, **D3**-9
Steel pipe piles, **D3**-9–**D3**-10
Steel-wheel rollers, **B8**-15–**B8**-16
Storage-reclaim conveyor system, **B9**-4–**B9**-8
Stripping shovels, **B5**-22–**B5**-24
Subcontract cost reports, **A4**-13–**A4**-16
　commitment records, **A4**-13–**A4**-14
　cost report—subcontracts, **A4**-14–**A4**-16
　subcontractor register, **A4**-13
Subcontracts, **A4**-12–**A4**-16
　lump-sum price contract, **A4**-12–**A4**-13
　progress payments, **A4**-13
　unit price contracts, **A4**-13
Submersible pumps, **D5**-14
Superstition Freeway, **E3**-20–**E3**-23
Supporting structures (conveyors), **B9**-17

Tamping-foot roller, **B8**-13
Tax Reform Act of 1986, **A1**-20
TBMs (*see* Tunnel-boring machines)
Telescopic grading hoes, **B5**-8–**B5**-9
Temporary construction, **A4**-28–**A4**-29
Threshold review analysis (TRA), **A11**-14
Tilting mixers, **C1**-60
Timber piles, **D3**-2–**D3**-7
Tires, **B1**-24–**B1**-26
Toronto Transit Commission rapid transit expansion program (RTEP), **A5**-17
Total cost management, **A6**-7
Total quality management (TQM), **A1**-10, **A8**-1–**A8**-13
　elements, **A8**-3–**A8**-12
　foundation, **A8**-2–**A8**-3
　mission statement, **A8**-1–**A8**-2
　typical, **A8**-3
Towed wheel scrapers, **B4**-1–**B4**-11
Tower cranes, **B10**-36–**B10**-42
Tractors, **B3**-1–**B3**-22
Transfer chutes (conveyor), **B9**-16
Transmission pipelines, **E4**-28–**E4**-35
　constructing, **E4**-30–**E4**-35
　planning, **E4**-28–**E4**-30
Trenchless pipelines, **E4**-21–**E4**-27
　boring, **E4**-21–**E4**-23
　micro tunneling, **E4**-23–**E4**-24
　pipe-pushing machines, **E4**-24–**E4**-27
Troweling, **C1**-82

Truck/transmit mixers, **C1**-62–**C1**-64
Tugs and towboats, **B11**-4–**B11**-5
Tunnel-boring machines (TBMs), **B12**-1–**B12**-28
 auxiliary equipment, **B12**-20–**B12**-28
 selection, **B12**-1–**B12**-4
 types, **B12**-4–**B12**-20

Underground utility construction, **E4**-2–**E4**-20
 backfilling, **E4**-17–**E4**-20
 excavation, **E4**-5–**E4**-11
 installation of pipe, **E4**-12–**E4**-16
 pavement removal, **E4**-3–**E4**-4
 paving, **E4**-20
 preliminary work, **E4**-2–**E4**-3
Unit prices, **A2**-5, **A2**-6, **A4**-13
UNIVAC, **A3**-2
U.S. 59/Southwest Freeway, **E3**-39–**E3**-47

Value engineering, **A6**-1
 application, **A6**-3–**A6**-6
 case study—bachelors' enlisted quarters (BEQ), **A6**-29–**A6**-41
 for construction managers, **A6**-1–**A6**-41
 development/recommendation phase, **A6**-18
 evaluation/analytical phase, **A6**-16–**A6**-18
 information phase, **A6**-6–**A6**-15
  cost models, **A6**-7, **A6**-9
  energy model, **A6**-8–**A6**-10
  FAST diagramming, **A6**-15

Value engineering, information phase (*Cont.*):
  functional analysis, **A6**-12–**A6**-15
  life cycle cost model, **A6**-11–**A6**-12
 report phase, **A6**-18–**A6**-23
 sequence, **A6**-23–**A6**-27
  postworkshop, **A6**-23, **A6**-27
  preworkshop, **A6**-23, **A6**-26
  workshop, **A6**-23, **A6**-26–**A6**-27
 speculative/creative phase, **A6**-16, **A6**-17
Value work, **A6**-2–**A6**-3
VE team coordinator (VETC), **A6**-28–**A6**-29
Vertical pumps, **D5**-13–**D5**-14
Vibratory compactors, **B8**-13–**B8**-15
Volume measure, **B1**-4–**B1**-5

Walking draglines, **B5**-10–**B5**-22
Warranty, **A2**-13
Washington Metropolitan Area Transit Authority (WMATA), **A10**-14
Water treatment facilities, **E2**-1–**E2**-14
 case study, **E2**-2–**E2**-14
Welded wire fabric, **C2**-5–**C2**-6
Wheel tractor-scrapers, **B4**-1–**B4**-11
Wheel-type loaders, **B6**-2–**B6**-6
Wide area network (WAN), **A5**-6, **A5**-8–**A5**-9
Windows, **A5**-17
Wire, **C2**-5–**C2**-6
Wire Reinforcement Institute (WRI), **C2**-6
Workers' compensation, **A10**-12–**A10**-14
WRI (*see* Wire Reinforcement Institute)

ABOUT THE EDITORS

James J. O'Brien, P.E., is Vice Chairman of the Board of O'Brien-Kreitzberg, Inc., a highly respected construction management company. His firm has handled such high-visibility projects as the renovation of San Francisco's cable car system and the redevelopment of New York's JFK International Airport. The recipient of numerous honors and awards over the years, Mr. O'Brien's previous books include *Contractor's Management Handbook*, Second Edition; *Scheduling Handbook*; *Value Analysis in Design and Construction*; *Construction Management: A Professional Approach*; *CPM in Construction Management*, Fourth Edition; and *Preconstruction Estimating*.

John A. Havers is Professor Emeritus of Construction and Engineering Management at Purdue University. He was editor in chief of the second edition of this volume.

Frank W. Stubbs, Jr., was Professor Emeritus of Civil Engineering at Purdue University, and edited the classic original edition of *Standard Handbook of Heavy Construction*, which first appeared in 1959. He died in 1967.